eleventh **edition**

Human **Biology**

Sylvia S. Mader

With contributions by

Susannah Nelson Longenbaker
Columbus State Community College

Kimberly Lyle-Ippolito
Anderson University

Linda D. Smith-Staton
Pellissippi State Technical Community College

Higher Education

Boston   Burr Ridge, IL   Dubuque, IA   New York   San Francisco   St. Louis
Bangkok   Bogotá   Caracas   Kuala Lumpur   Lisbon   London   Madrid   Mexico City
Milan   Montreal   New Delhi   Santiago   Seoul   Singapore   Sydney   Taipei   Toronto

Higher Education

HUMAN BIOLOGY, ELEVENTH EDITION

Published by McGraw-Hill, a business unit of The McGraw-Hill Companies, Inc., 1221 Avenue of the Americas, New York, NY 10020. Copyright © 2010 by The McGraw-Hill Companies, Inc. All rights reserved. Previous editions © 2008, 2006, and 2004. No part of this publication may be reproduced or distributed in any form or by any means, or stored in a database or retrieval system, without the prior written consent of The McGraw-Hill Companies, Inc., including, but not limited to, in any network or other electronic storage or transmission, or broadcast for distance learning.

Some ancillaries, including electronic and print components, may not be available to customers outside the United States.

This book is printed on acid-free paper.

1 2 3 4 5 6 7 8 9 0 QPD/QPD 0 9

ISBN 978–0–07–337798–8
MHID 0–07–337798–8

Publisher: *Janice Roerig-Blong*
Executive Editor: *Michael S. Hackett*
Director of Development: *Kristine Tibbetts*
Senior Developmental Editor: *Lisa A. Bruflodt*
Marketing Manager: *Tamara Maury*
Senior Project Manager: *April R. Southwood*
Lead Production Supervisor: *Sandy Ludovissy*
Senior Media Project Manager: *Jodi K. Banowetz*
Designer: *Laurie B. Janssen*
Cover/Interior Designer: *Christopher Reese*
(USE) Cover Image: © *Chris Noble, Gettyimages*
Senior Photo Research Coordinator: *Lori Hancock*
Supplement Producer: *Mary Jane Lampe*
Photo Research: *Evelyn Jo Johnson*
Compositor: *Electronic Publishing Services Inc., NYC*
Typeface: *10/12 Palatino*
Printer: *Quebecor World Dubuque, IA*

The credits section for this book begins on page C–1 and is considered an extension of the copyright page.

Library of Congress Cataloging-in-Publication Data

Mader, Sylvia S.
  Human biology / Sylvia Mader ; with contributions by Susannah Nelson Longenbaker, Kimberly Ippolito, Linda Smith Staton. -- 11th ed.
      p. cm.
  Includes index.
  ISBN 978–0–07–337798–8 — ISBN 0–07–337798–8 (hard copy : alk. paper) 1. Human biology. I. Title.
  QP36.M2 2010
  612--dc22
                                                          2008040563

www.mhhe.com

Brief Contents

Contents

Readings

Bioethical **Focus**

Health **Focus**

Science **Focus**

Historical **Focus**

Preface

During my career as an educator, I discovered very early that students' attention was captured quickly when the topic was themselves: how their bodies worked, how to keep them healthy, and how they can occasionally malfunction. In addition, students in all fields of study are becoming increasingly concerned with the state of the environment. *Human Biology* integrates the topics of health, wellness, and concern for the environment in a way that perfectly suits the nonmajors' course.

Regardless of profession, citizens are frequently called upon to make health and environmental decisions. Therefore, it would not be appropriate for a college graduate to lack a basic knowledge of anatomy, physiology, genetics, and biotechnology. Students should also understand how the human population can become more fully integrated into the biosphere. Further, every educated individual should appreciate how scientists think and know how research is properly conducted. Wise choices require adequate knowledge and can help ensure our continued survival as individuals and as a species.

In this edition, as in previous editions, the text presents concepts using simple, concise, and clear descriptions. Detailed, high level scientific data and terminology are excluded, because I believe that all learners should have a working understanding of concepts rather than technical facility. This approach ensures that students will feel confident and capable of achieving an adult level of understanding.

The Eleventh Edition of *Human Biology*

Human Biology continues to grow and evolve to better suit the needs of a changing student population. Compelling new features will engage learners of all disciplines and interests. Clear, concise explanations have been teamed with attractive illustrations and sound pedagogy. Features from previous editions have been refined and supplemented where appropriate. Factual information has been updated to reflect current findings. As always, this new edition seeks to keep its sound basic content, making changes to improve relevancy and student appeal.

Producing this fresh, vibrant update was achieved with the very able assistance of three highly talented professors of nonmajors—Susannah Nelson Longenbaker from Columbus State Community College, Kimberly Lyle-Ippolito from Anderson University, and Linda Smith-Staton from Pellissippi State Technical Community College. Together, they are recognized for their significant contributions on the title page of the book. Many other professors also lent their talents, and their names are listed in the acknowledgment section.

Engaging New Chapter Case Studies

The new case study feature that opens each chapter will immediately encourage student interest in the content of the chapter. Each story unfolds at the chapter's beginning and continues throughout the chapter. Accompanying each introduction are photographs that effectively compliment the story. These case studies present real-life scenarios related to each chapter's content, and each is designed to appeal to every learner. In addition, the case studies will have additional appeal to specific disciplines. For example, students in African American studies and women's studies will find the special health needs of African American women described in the case study of Louise Hairston (Chap. 5). The topic of special education is addressed through the story of Jeremy Callen, a young man with fragile X syndrome (Chap. 20). The work of Andrew Scott and Jamie Barrett (Chap. 22) details the discipline and hard work of field anthropologists. Further, case studies dealing with sports themes and those addressing modern wellness issues (e.g., heart disease, diabetes, obesity, and cancer) will interest both students and their professors.

The "Thinking Critically About the Concepts" feature completes each chapter, and once again chapter case studies are incorporated to continue the learning process. Questions combine case concepts with chapter content. Students are challenged to thoughtfully integrate these ideas. The answers to the questions are given in Appendix B.

Updated and Reorganized Chapters and New Applications

Changes to this eleventh edition of *Human Biology* has been undertaken with several goals in mind. Constantly improving student involvement in the text is a primary aim. Equally important, this revision seeks to provide accurate, timely information. As you enjoy the book, you will notice:

Cutting Edge Data

The factual content for each chapter has been edited to reflect the most current findings available, so that professors can rely on the text to provide up-to-date information. Information about different forms of contraception presents all options—both existing and new—available to couples (Chap. 16). Treatments for Alzheimer disease describe the actions of modern drugs (Chap. 17). Data from the American Cancer Society reports the latest statistics on the types and incidence of the disease in both men and women (Chap. 19). These examples and many others show ongoing dedication to reporting state-of-the-art technologies and information.

Infectious Diseases Supplement

The AIDS supplement has been reorganized and titled "Infectious Diseases Supplement." The goal of this effort was two-fold. Recent findings regarding the AIDS epidemic were necessary to provide students with information critical to their health and safety. In addition, descriptions of new and emerging diseases will enable classroom discussion of present-day health concerns. The return of tuberculosis is explained, along with the symptoms and epidemiology of the disease. Antibiotic resistance will inevitably affect most, if not all, present and future populations. Its evolution, as well as strategies to overcome resistant organisms, is also addressed in this supplement.

Have You Ever Wondered . . .

A new feature has been added. Reading and studying new information, especially in the health sciences, often leads students to wonder about their bodies and how they work. **HAVE YOU EVER WONDERED,** presents the type of impulsive, off-the-cuff questions that might be asked in a typical human biology classroom. Questions can be sober and serious or comical and silly:

HAVE YOU EVER WONDERED . . . How do lungs stay open and keep from collapsing? (Chap. 2)
. . . How do you use an automatic external defibrillator, like the ones you see in the airport? (Chap. 5)
. . . Can you drink through your nose? (Chap. 8)
. . . Why does that annoying song you hear seem to replay in your head all day? (Chap. 14)

Inquiries like these are asked and answered several times in each chapter throughout the text. Each will capture attention—informing, entertaining, and educating at the same time.

New Boxed Readings

All boxed readings have been revised and updated. Many are new to this edition. All topics were chosen for relevancy and interest to students.

- *Science Focus* readings, which pertain to biological topics of interest, remain a popular feature of the text. New *Science Focus* readings include a discussion of the genetics of breast cancer (Chap. 3), recent news of a face transplant (Chap. 4), and the problem of diminishing honeybee populations (Chap. 24), among others.
- *Health Focus* articles discuss topics of disease and wellness that are important to all students. New *Health Focus* articles describe how to determine trans-fat content in food (Chap. 8), and how to obtain help for a disabled child (Chap. 18).
- *Bioethical Focus* issues present modern ethical concerns regarding health, culture, and the environment. For example, a new article, "Male and Female Circumcision: Medical Option, Cultural Practice, or Child Abuse?" (Chap. 16) addresses female circumcision as both a legal and moral issue. "Guaranteeing Access to Safe Drinking Water" (Chap. 23) will help students to think about the moral responsibility to provide potable water to all nations.
- *Historical Focus,* a brand-new feature of this text, will allow students to enjoy human biology in a historical context. This unique highlight will appeal to learners in all disciplines: history, philosophy, sociology, women's studies, African American studies, and many others. Individuals such as Vivien Thomas, who helped to develop modern cardiac surgery ("Heart Surgeon Without a Degree," Chap. 5) and Ignaz Semmelweis, who made safe childbirth possible ("An End to Laudable Pus," Chap. 17) will interest and inspire students. Sports fans will discover the story of Lou Gehrig ("The Iron Horse," Chap. 12). Those interested in European history will enjoy "Hemophilia: The Royal Disease" (Chap. 20).

Excellent Pedagogical Features

"During my career as an educator, I discovered very early that students' attention was captured quickly when the topic was themselves: how their bodies worked, how to keep them healthy, and how they can occasionally malfunction."

Sylvia Mader

Check Your Progress features end each section in every chapter. The questions function as a "mini-quiz," testing student understanding before the student moves on to the next section. *Check Your Progress* questions are answered in Appendix B.

Chapter Summaries An extensive review is organized according to the major sections of the chapter. Brief statements, lists, and tables help students re-examine the important topics and concepts. Artwork is included to provide a visual reminder of the important ideas presented. Key terms give students a working vocabulary for the chapter. Finally, a complete set of objective questions is a self-test that will allow the student to determine where further study might be needed.

Thinking Critically About the Concepts Each chapter's case study provides a framework for critical thinking. Students are first prompted with factual questions, then asked to consider future implications for the individuals described in the case study. For example, Chapter 5 presents the case of Louise Hairston, an African American woman who suffers a heart attack. *Critical Thinking* questions then ask the reader to furnish the not-so-typical symptoms often seen when a woman suffers a heart attack. Additional questions focus on ways to avoid a second heart attack. Answers to this style of question are presented in Appendix B. Subjective inquiries with no right or wrong answer prompt learners to form opinions about a health or wellness issue. Chapter 21 first describes recombinant growth hormone, then asks the reader to reflect on situations when the hormone should be used.

Homeostasis and Working Together Illustrations Because of their popular appeal, we have retained the homeostasis sections that include an illustration demonstrating how systems work together. These five sections make use of real-life situations to show how homeostasis is maintained in the body. As an example, see Section 6.6.

Vivid and Engaging Illustrations

The vivid and engaging illustrations in *Human Biology* bring the study of biology to life! The figures have been rendered to convey realistic detail and close coordination with the text discussions.

Combination Art

Drawings of structures are paired with micrographs to provide students with two perspectives: the explanatory clarity of line drawings and the realism of photos.

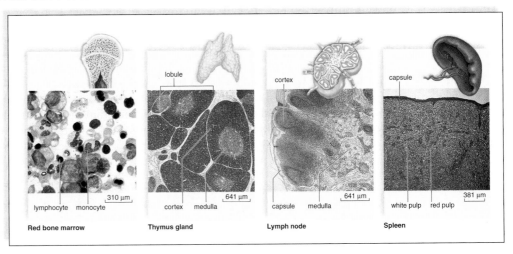

Multilevel Perspective

Such illustrations guide students from the more intuitive macroscopic level of learning to the functional foundations revealed through microscopic images.

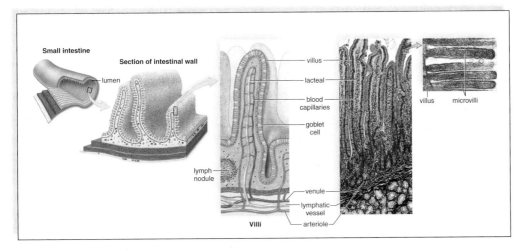

Icons

Icons orient students to the whole structure or process by providing small drawings that help students visualize how a particular structure is part of a larger one.

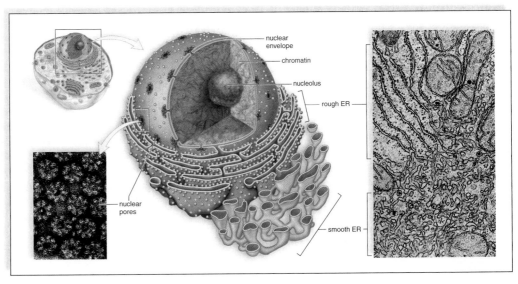

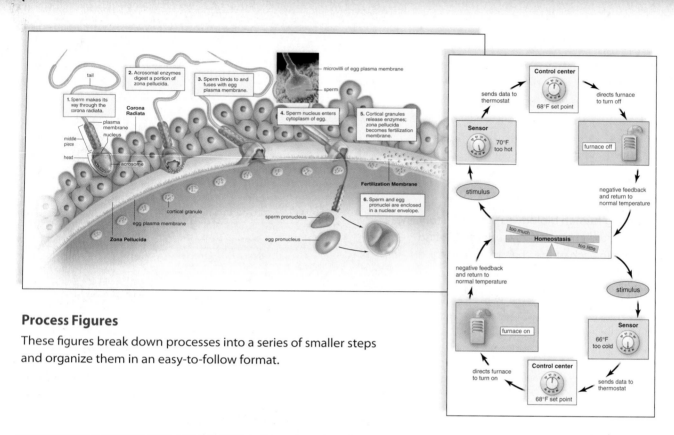

Process Figures

These figures break down processes into a series of smaller steps and organize them in an easy-to-follow format.

Human Systems Work Together

Working together illustrations use brief concise statements to tell students how various other systems help a featured system achieve homeostasis.

In this edition the working together illustrations have been integrated into homeostasis sections making a united whole. The homeostasis sections show how the systems achieve homeostasis despite real-life experiences that could alter the internal environment. For example, see page 269.

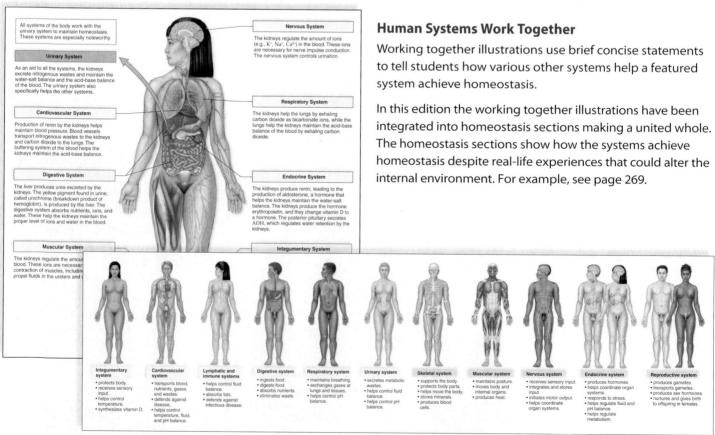

The Learning System

Proven pedagogical features that will facilitate your understanding of biology.

Chapter Concepts

The chapter outline contains a concise preview of the topics covered in each section.

Case Studies

New case studies bring human biology to life. Each story line continues throughout the chapter, and students will find themselves absorbed in each of the characters. *Critical Thinking* questions in the chapter end matter connect the case study with the chapter concepts.

Readings

Human Biology offers four types of boxed readings that put the chapter concepts in the context of modern-day issues:

- **Health Focus** readings review procedures and technology that can contribute to our well being.

- **Science Focus** readings describe how experimentation and observations have contributed to our knowledge about the living world.

- **Bioethical Focus** readings describe modern situations that call for value judgments and challenge students to develop a point of view.

- **Historical Focus** articles will allow the learner to enjoy human biology in a historical context.

Have You Ever Wondered ...

This unique feature presents the types of spontaneous inquiries that students may have as they study the workings of the human body. Questions and answers can be serious or funny, but each will capture the student's attention.

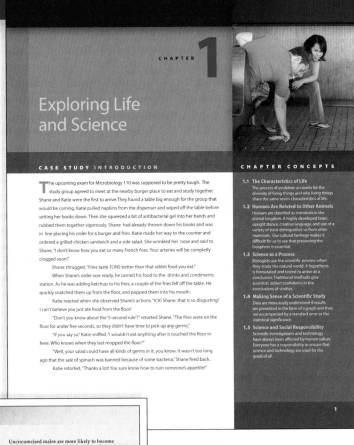

Check Your Progress Boxes

Questions follow main sections of the text and help students assess their understanding of the material presented. Answers to these questions appear in Appendix B.

Summarizing the Concepts

A bulleted summary is organized according to the major sections in the chapter and includes art to helps students review the important topics and concepts.

Understanding Key Terms

The boldface terms in the chapter are page referenced, and a matching exercise allows students to test their knowledge of the terms.

Testing Your Knowledge of the Concepts

Objective and art-based questions allow students to review material and prepare for tests. Answers to these questions appear in Appendix B.

Thinking Critically About the Concepts

This set of questions encourages students to apply what they've just learned to the case study in the chapter.

> **Check Your Progress 17.4**
> 1. What chemical factors are responsible for the many physiological changes in a pregnant woman?
> 2. Maternal blood carbon dioxide levels fall by 20% during pregnancy. How does this benefit the fetus?
> 3. Describe the three stages of labor.

Summarizing the Concepts

17.1 Fertilization
The acrosome of a sperm releases enzymes that digest a pathway for the sperm through the zona pellucida. The sperm nucleus enters the egg and fuses with the egg nucleus.

17.2 Pre-Embryonic and Embryonic Development
- Cleavage, growth, morphogenesis, and differentiation are the processes of development.
- The extraembryonic membranes (chorion, allantois, yolk sac, and amnion) function in internal development.

17.3 Fetal Development
- At the end of the embryonic period, all organ systems are established, and there is a mature and functioning placenta. The umbilical arteries and umbilical vein take blood to and from the placenta, where exchanges take place.
- Exchanges supply the fetus with oxygen and nutrients and rid the fetus of carbon dioxide and wastes.
- The venous duct joins the umbilical vein to the inferior vena cava.
- The oval duct and arterial duct allow the blood to pass throug...

- During the child is born.
- During stage 3, the afterbirth is expelled.

17.5 Development After Birth
Development after birth consists of infancy, childhood, adolescence, and adulthood.
- Aging encompasses progressive changes from about age 20 on that contribute to an increased risk of infirmity, disease, and death.

Hypotheses of Aging
- Aging may have a genetic basis.
- Aging may be due to changes that affect the whole body

Understanding Key Terms

acrosome 356
birth control method 365
birth control pill 366
bulbourethral gland 353
cervix 358
chancre 373
chlamydia 372
circumcision 354
contraceptive 365

luteinizing hormone (LH) 356
male condom 367
menopause 362
menstruation 363
oogenesis 360
ovarian cycle 360
ovary 357
oviduct 357
ovulation 360
Pap test 358
penis 354
placenta 364
progesterone 362
prostate gland 353
scrotum 353
semen 353
seminal vesicle 353
seminiferous tubule 354
Sertoli cell 355
sperm 355
spermatogenesis 354
testes 353
testosterone 356
tubal ligation 367
urethra 353
uterine cycle 362
uterus 358
vagina 358
vas deferens 353
vasectomy 367
vulva 358
zygote 357

Testing Your Knowledge of the Concepts

1. Describe how polyspermy is prevented during fertilization. (page 394)

2. Name the four embryonic membranes and give a human function for each one. (page 396)

3. Justify the division of development into pre-embryonic, embryonic, and fetal development. (pages 395–407)

4. What are the three primary germ layers, and what body... me from each germ layer? (page 399)

...narize the weekly events of embryonic ... (pages 398–400)

...narize the monthly events of fetal development.

...blood circulates to and from the placenta... How is blood shunted away from the ...401, 403)

...ones involved in the development of the male ...ternal and external sex organs and state their ...ges 406-07)

...e of the changes that occur in the mother during ...pages 408)

...marks the end of each stage of birth?

...hypotheses concerning aging. How can you ...major changes that can occur in the body as ...es 411-13)

...rm enters an egg because
...ve an acrosome.
...a radiata gets larger.
...occur in the zona pellucida.
...lasm hardens.
...se are correct.

...se statements is correct?
... organs are formed during embryonic ...ent.
...s and feet begin as paddlelike structures.
...t is at first tubular.
...nta functions until birth occurs.
...se are correct statements.

...ee germ layers are present (ectoderm, endoderm, ...rm), what event has occurred?
...on
...nation
...on
...on

Thinking Critically About the Concepts

Amber and Kent used a home pregnancy test to determine if she was pregnant. These tests detect the level of hCG (human chorionic gonadotropin; see page 398) in the urine. This hormone is released following implantation of the embryo into the uterus, usually around six days after fertilization. Some tests claim that they are sensitive enough to detect hCG on the date that menstruation is expected to begin. However, doctors recommend waiting until menstruation is one week late. If pregnant, a woman's level of hCG rises with each passing day, and testing is more likely to be accurate. However, even with a negative test result, the woman may still be pregnant if hCG levels are too low to be detected at the time of the first test. The test should be repeated later if menstruation doesn't begin. The home pregnancy tests contain a positive control. This is a visual sign (usually a line or a +) that appears if the test is working correctly. If this line does not appear, the test is not valid and must be repeated.

1. At home, pregnancy tests check for the presence of hCG in a female's urine. Where does hCG come from? Why is hCG found in a pregnant woman's urine?

2. A blood test at a doctor's office can also check for the presence of hCG in a female's blood.
   a. Why would you expect to find hCG circulating in a pregnant female's blood?
   b. hCG is a protein, so how does hCG affect its target cells?

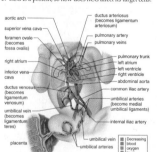

aortic arch
superior vena cava
foramen ovale (becomes fossa ovalis)
right atrium
inferior vena cava
ductus venosus (becomes ligamentum venosum)
umbilical vein (becomes ligamentum teres)
placenta

ductus arteriosus (becomes ligamentum arteriosum)
pulmonary artery
pulmonary veins
pulmonary trunk
left atrium
left ventricle
right ventricle
abdominal aorta
common iliac artery
umbilical arteries (become medial umbilical ligaments)
internal iliac artery
umbilical vein
umbilical arteries

Decreasing blood oxygen level

... matching definitions.

...ease of an oocyte from the ovary.

...male sex hormone that causes the endometrium ...o become secretory during the uterine cycle; ...rogen, it maintains secondary sex ...in females.

...ck, whitish fluid consisting of sperm and secretions ...lands of the male reproductive tract.

...rrow end of the uterus, which projects into ...

...at the anterior end of a sperm that partially ...cleus and contains enzymes that help the sperm ...egg.

Dedicated to providing high-quality and effective supplements for instructors and students, the following supplements were developed for *Human Biology.*

For Instructors

Laboratory Manual

The *Human Biology Laboratory Manual,* eleventh edition, is written by Dr. Sylvia Mader. With few exceptions, each chapter in the text has an accompanying laboratory exercise in the manual. Every laboratory has been written to help students learn the fundamental concepts of biology and the specific content of the chapter to which the lab relates and to gain a better understanding of the scientific method.

ISBN (13) 978-0-07-723513-0
ISBN (10) 0-07-723513-4

Connect

Connect is a complete online tutorial, eletronic homework, and course management system designed for greater ease of use than any other system available. The program enables students to complete their homework online, as assigned by their instructor. Connect allows instructors to automatically grade and report easy-to-assign homework and quizzing, build their own assignments, track student progress, and share course materials with colleagues. Connect also provides instructors with the ability to create or edit questions from the question bank or import their own content. The fully integrated grade book can be downloaded to Excel, WebCT, or Blackboard.

Companion Website

The companion website contains the following resources for instructors:

- *Presentation Tools*   Everything you need for outstanding presentation in one place! This easy-to-use table of assets include
  - **Enhanced image PowerPoints**—including every piece of art htat has been sized and cropped specifically for superior presentations as well as tables that you can edit. Also included are tables, photographs and unlabeled art pieces
  - **Animation PowerPoints**—Numerous full-color animations illustrating important processes are also provided. Harness the visual impact of concepts in motion by importing these files into classroom presentations or online course materials.

- **Lecture PowerPoints** with animations fully embedded.
- **Labeled and unlabeled JPEG images**—Full-color digital files of all illustrations that can be readily incorporated into presentations, exams, or custom-made classroom materials.
- *Presentation Center*   This online digital library contains photos, artwork, animations, and other media types that can be used to create customized lectures, visually enhanced tests and quizzes, compelling course websites, or attractive printed support materials. All assets are copyrighted by McGraw-Hill Higher Education, but can be used by instructors for classroom purposes. The visual resources in this collection include:
  - **Art** Full-color digital files of all illustrations in the book can be readily incorporated into lecture presentations, exams, or custom-made classroom materials. In addition, all files are preinserted into PowerPoint slides for ease of lecture preparation.
  - **Photos** The photos collection contains digital files of photographs from the text, which can be reproduced for multiple classroom uses.
  - **Tables** Every table that appears in the text has been saved in electronic form.
  - **Animations** Numerous full-color animations illustrating important processes are also provided. Harness the visual impact of concepts in motion by importing these files into classroom presentations or online course materials.
- *Instructor's Manual*   The instructor's manual contains learning objectives, extended lecture outlines, student activities, and classroom discussion topics.
- *Computerized Test Bank*   A comprehensive bank of test questions is provided within a computerized test bank powered by McGraw-Hill's flexible electronic testing program "EZ Test Online." "EZ Test Online" allows you to create paper and online tests or quizzes in this easy-to-use program. A new tagging scheme allows you to sort questions by difficulty level, topic, and section. Imagine being able to create and access your test or quiz anywhere, at any time, without installing the testing software. Now, with "EZ Test Online," instructors can select questions from multiple McGraw-Hill test banks or author their own, and then either print the test for paper distribution or give it online.

Test Creation
- Author/edit questions online using the 14 different question templates
- Export your tests for use in WebCT, Blackboard, PageOut, and Apple's iQuiz

Online Test Management

- Set availability dates and time limits for your quiz or test
- Assign points by question or question type with drop-down menu
- Provide immediate feedback to students or delay feedback until all finish the test

Online Scoring and Reporting

- Automated scoring for most of "EZ Test's" numerous question types
- Allows manual scoring for essay and other open-response questions
- Manual rescoring and feedback are also available

Support and Help

- Flash tutorials for getting started on the support site
- Support Website: **www.mhhe.com/eztest**
- Product specialist available at 1-800-331-5094
- Online Training: **http://auth.mhhe.com/mpss/workshops**

Go to www.mhhe.com/maderhumanbiology11e to learn more.

McGraw-Hill: Biology Digitized Video Clips

ISBN (13) 978-0-312155-0
ISBN (10) 0-07-312155-X

McGraw-Hill is pleased to offer an outstanding presentation tool to text adopting instructors—digitized biology video clips on DVD. Licensed from some of the highest-quality science video producers in the world, these brief segments range from about 5 seconds to just under 3 minutes in length and cover all areas of general biology from cells to ecosystems. Engaging and informative, McGraw-Hill's digitized videos will help capture students' interest while illustrating key biological concepts and processes such as mitosis, how cilia and flagella work, and how some plants have evolved into carnivores.

Student Response System

Wireless technology brings interactivity into the classroom or lecture hall. Instructors and students receive immediate feedback through wireless response pads that are easy to use and engage students. This system can be used by instructors to take attendance, administer quizzes and tests, create a lecture with intermittent questions, manage lectures and student comprehension through the use of the grade book, and integrate interactivity into their Power-Point presentations.

For Students

Companion Website

Students can readily access a variety of digital learning objects that include:

- Chapter-level quizzing with pretest and post test
- Bio Tutorial Animations with quizzing
- Vocabulary flashcards
- Virtual Labs
- Biology Prep, also available on the companion site, helps students to prepare for their upcoming coursework in biology. This website enables students to perform self assessments, conduct self study sessions with tutorials, and perform a post-assessment of their knowledge in the following areas: introductory biology skills, basic math, metric system, chemistry, and lab reports.

Electronic Books

If you or your students are ready for an alternative eversion of the traditional textbook, McGraw-Hill eBooks offer a cheapter and eco-friendly alternative to traditional text-books. By purchases ebooks from McGraw-Hill, students can save as much as 50% on selected titles delivered on the most advanced E-Book platform available. Contact your McGraw-Hill sales representative to discuss E-book packaging options.

How to Study Science

ISBN (13) 978–0–07–234693–0
ISBN (10) 0–07–234693–0

This workbook offers students helpful suggestions for meeting the considerable challenges of a science course. It gives practical advice on such topics as how to take notes, how to get the most out of laboratories, and how to overcome science anxiety.

Photo Atlas for General Biology

ISBN (13) 978–0–07–284610–2
ISBN (10) 0–07–284610–0

Atlas was developed to support our numerous general biology titles. It can be used as a supplement for a general biology lecture or laboratory course.

Acknowledgments

A wonderful piece of poetry seems to me to be a fitting opening for my acknowledgments. "No man is an island, entire of itself. . . ." This idea certainly describes the effort required to create a fresh and innovative revision. A project such as this could never be completed without the work of a coordinated group. As always, the McGraw-Hill professionals guided this revision, assisting in all aspects. From beginning brainstorming sessions to completed text, this team supplied creativity, advice, and support whenever it was needed. Developmental Editor Lisa Bruflodt, Senior Project Manager April Southwood, Publisher Janice Roerig-Blong, and Executive Editor Michael Hackett collaborated to steer the book through the production process. Together, they have helped me to bring you a text and ancillaries that will serve your needs in every way.

Fresh, appealing new photos are a feature of this book, which students and professors alike will enjoy. Jo Johnson and Lori Hancock did a superb job of finding just the right photographs and micrographs. Marketing manager Tamara Maury directed the marketing team whose work is second to none.

I am extremely grateful to this edition's team of three talented contributing authors: Susannah Nelson Longenbaker from Columbus State Community College, Kimberly Lyle-Ippolito from Anderson University, and Linda Smith-Staton from Pellissippi State Technical Community College. These writers assisted me with this project from beginning to end. Together, they supplied ideas and content for the many updates, new features, and new illustrations that enrich this eleventh edition of *Human Biology*. Their hard work was vital to this effort. Thanks, too, to Mr. Jacob Coate for his help with the glossary.

Finally, the eleventh edition of *Human Biology* would not have been the same excellent quality without the suggested changes from the many reviewers listed below.

360° Development

McGraw-Hill's 360° Development Process is an ongoing, never-ending, market-oriented approach to building accurate and innovative print and digital products. It is dedicated to continual large-scale and incremental improvement driven by multiple customer feedback loops and checkpoints. This is initiated during the early planning stages of our new products and intensifies during the development and production stages, then begins again upon publication in anticipation of the next edition.

The process is designed to provide a broad, comprehensive spectrum of feedback for refinement and innovation of our learning tools, for both student and instructor. The 360° Development Process includes market research, content reviews, course- and product-specific symposia, accuracy checks, and art reviews. We appreciate the expertise of the many individuals involved in this process.

Ancillary Authors

Lecture Outlines/Image PowerPoints – Rennee Moore, *Solano Community College*
Instructor's Manual – Terri Pope, *Cuyahoga Community College*
Test Bank – Pat Pendarvis, *Southeastern Louisiana University*
Practice Tests – Pat Pendarvis, *Southeastern Louisiana University*
Media Asset Correlations – Donna Potacco, *William Paterson University*
BioInteractive Questions – Alicia Steinhardt, *Hartnell Community College*

11ᵗʰ Edition Reviewers

Tamatha R. Barbeau, *Francis Marion University*
Bill Radley Bassman, *Touro College, Stern College*
Frank J. Conrad, *Metropolitan State College of Denver*
Valentina David, *Bethune-Cookman University*
Maria M. Dell, *Santa Monica College*
Charles J. Dick, *Pasco-Hernando Community College*
Thomas J. Franco, *Erie Community College, North Campus*
Judith E. Goedert, *City College of San Francisco*
Melodye Gold, *Bellevue Community College*

Mary Louise Greeley, *Salve Regina University*
Virginia Gutierrez-Osborne, *Fresno City College*
Martin Hahn, *William Paterson University*
Rebecca J. Heick, *St. Ambrose University*
Jonathan P. Hubbard, *Hartnell College*
Edwin Klibaner, *Touro College*
Robert A. Krebs, *Cleveland State University*
Nicole Okazaki, *Weber State University*
Phillip A. Ortiz, *Empire State College, State University of New York*
Polly K. Phillips, *Florida International University*

Nancy K. Prentiss, *University of Maine at Farmington*
Nicholas Roster, *Northwestern Michigan College*
Megan E. Thomas, *University of Nevada, Las Vegas*
Wendy Vermillion, *Columbus State Community College*
Jagan Valluri, *Marshall University*

Previous Edition Reviewers and Contributors

Rita Alisauskas, *County College of Morris*

Deborah Allen, *Jefferson College*

Elizabeth Balko, *SUNY-Oswego*

Tamatha R. Barbeau, *Francis Marion University*

Marilynn R. Bartels, *Black Hawk College*

Erwin A. Bautista, *University of California, Davis*

Robert D. Bergad, *Metropolitan State University*

Hessel Bouma III, *Calvin College*

Frank J. Conrad, *Metropolitan State College of Denver*

William Cushwa, *Clark College*

Debbie A. Zetts Dalrymple, *Thomas Nelson Community College*

Diane Dembicki, *Dutchess Community College*

Charles J. Dick, *Pasco-Hernando Community College*

Kristiann M. Dougherty, *Valencia Community College*

David A. Dunbar, *Cabrini College*

William E. Dunscombe, *Union County College*

David Foster, *North Idaho College*

David E. Fulford, *Edinboro University of Pennsylvania*

Sandra Grauer, *Limestone College*

Mary Louise Greeley, *Salve Regina University*

Esta Grossman, *Washtenaw Community College*

Gretel M. Guest, *Alamance Community College*

Martin E. Hahn, *William Paterson University*

Rosalind C. Haselbeck, *University of San Diego*

Timothy P. Hayes, *Marshall University*

Mark F. Hoover, *Penn State Altoona*

Anna K. Hull, *Lincoln University*

Laurie A. Johnson, *Bay College*

Mary King Kananen, *Penn State Altoona*

Patricia Klopfenstein, *Edison Community College*

J. Kevin Langford, *Stephen F. Austin State University*

Lee H. Lee, *Montclair State University*

Edwin Lephart, *Brigham Young University*

Martin A. Levin, *Eastern Connecticut State University*

Nardos Lijam, *Columbus State Community College*

William J. Mackay, *Edinboro University of Pennsylvania*

Terry R. Martin, *Kishwaukee College*

Deborah J. McCool, *Penn State Altoona*

V. Christine Minor, *Clemson University*

Nick Nagle, *Metropolitan State College of Denver*

Roger C. Nealeigh, *Central Community College-Hastings*

Polly K. Phillips, *Florida International University*

Shawn G. Phippen, *Valdosta State University*

Mason Posner, *Ashland University*

Donna R. Potacco, *William Paterson University*

Mary Celeste Reese, *Mississippi State University*

Jill D. Reid, *Virginia Commonwealth University*

Kay Rezanka, *Central Lakes College*

April L. Rottman, *Rock Valley College*

Deborah B. Schulman, *Cleveland State University*

Lois Sealy, *Valencia Community College*

Jia Shi, *Skyline College*

Mark Smith, *Chaffey College*

Alicia Steinhardt, *Hartnell Community College West Valley Community College*

Lei Lani Stelle, *Rochester Institute of Technology*

Kenneth Thomas, *Northern Essex Community College*

Chad Thompson, *SUNY-Westchester Community College*

Jamey Thompson, *Hudson Valley Community College*

Doris J. Ward, *Bethune-Cookman College*

Susan Weinstein, *Marshall University*

Dave Cox, *Lincoln Land Community College*

Patrick Galliart, *North Iowa Area Community College*

Sandra Grauer, *Limestone College*

Sharron Jenkins, *Purdue University North Central*

Jill Kolodsick, *Washtenaw Community College*

Edwin Lephart, *Brigham Young University*

Susannah Nelson Longenbaker, *Columbus State Community College*

Debbie J. McCool, *Penn State Altoona*

Jodi Rymer, *Christine Wildsoet Laboratory University of California, Berkeley*

Linda D. Smith-Staton, *Pellissippi State Technical Community College*

Linda Strause, *University of California–San Diego*

Michael Thompson, *Middle Tennessee State University*

CHAPTER 1

Exploring Life and Science

The upcoming exam for Microbiology 110 was supposed to be pretty tough. The study group agreed to meet at the nearby burger place to eat and study together. Shane and Katie were the first to arrive. They found a table big enough for the group that would be coming. Katie pulled napkins from the dispenser and wiped off the table before setting her books down. Then she squeezed a bit of antibacterial gel into her hands and rubbed them together vigorously. Shane had already thrown down his books and was in line placing his order for a burger and fries. Katie made her way to the counter and ordered a grilled chicken sandwich and a side salad. She wrinkled her nose and said to Shane, "I don't know how you eat so many French fries. Your arteries will be completely clogged soon!"

Shane shrugged. "Fries taste TONS better than that rabbit food you eat."

When Shane's order was ready, he carried his food to the drinks and condiments station. As he was adding ketchup to his fries, a couple of the fries fell off the table. He quickly snatched them up from the floor, and popped them into his mouth.

Katie reacted when she observed Shane's actions. "ICK! Shane, that is so disgusting! I can't believe you just ate food from the floor!"

"Don't you know about the '5-second rule'?" retorted Shane. "The fries were on the floor for under five seconds, so they didn't have time to pick up any germs."

"If you say so," Katie sniffed. "I wouldn't eat anything after it touched the floor in here. Who knows when they last mopped the floor?"

"Well, your salad could have all kinds of germs in it, you know. It wasn't too long ago that the sale of spinach was banned because of some bacteria," Shane fired back.

Katie retorted, "Thanks a lot! You sure know how to ruin someone's appetite!"

1.1 The Characteristics of Life

Life Is Organized

Figure 1.2 illustrates that **atoms** join together to form the **molecules** that make up a cell (Fig. 1.2). A cell is the smallest structural and functional unit of an organism. Some organisms are single cells. Human beings are **multicellular** because they are composed of many different types of cells. A nerve cell is one of the types of cells in the human body. It has a structure suitable to conducting a nerve impulse.

A **tissue** is a group of similar cells that perform a particular function. Nervous tissue is composed of millions of nerve cells that transmit signals to all parts of the body.

Several types of tissues make up an **organ,** and each organ belongs to an organ system. The organs of an **organ system** work together to accomplish a common purpose. The brain works with the spinal cord to send commands to body parts by way of nerves. **Organisms,** such as trees and humans, are a collection of organ systems.

The levels of biological organization extend beyond the individual. All the members of one **species** (group of interbreeding organisms) in a particular area belong to a **population.** A tropical grassland may have a population of zebras, acacia trees, and humans, for example. The interacting populations of the grasslands make up a **community.** The community of populations interacts with the physical environment to form an **ecosystem.** Finally, all the Earth's ecosystems make up the **biosphere.**

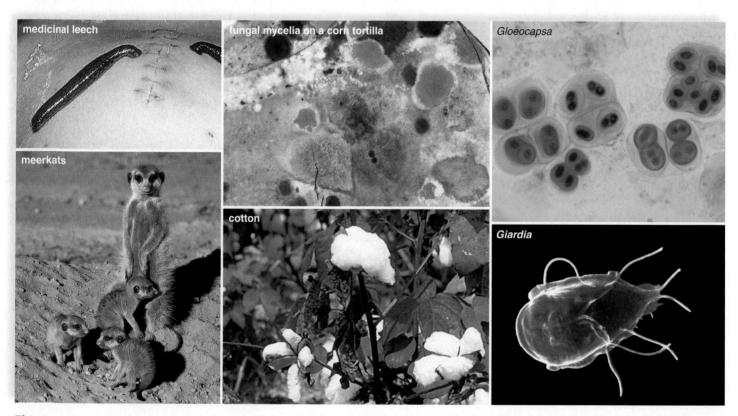

Figure 1.1 What characteristics are shared by the living organisms pictured? According to Figure 1.5, how would you classify each organism?

All living organisms are composed of cells, the smallest unit of life. An organism is classified in part according to the organization of its particular cells.

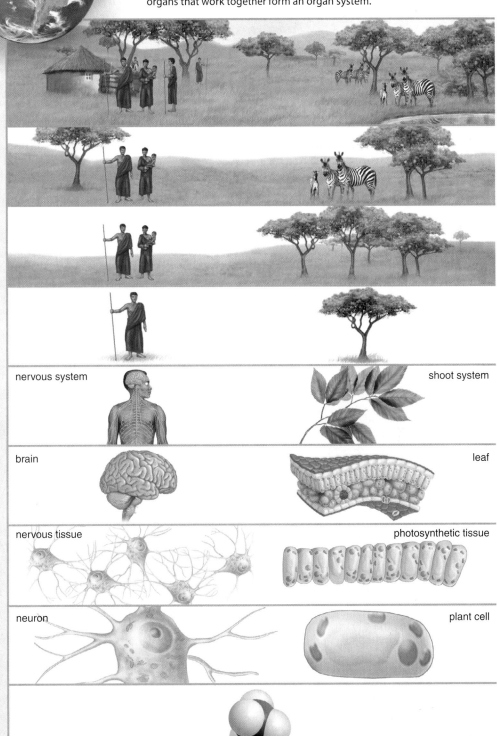

Biosphere
Regions of the Earth's crust, waters, and atmosphere inhabited by living things

Ecosystem
A community plus the physical environment

Community
Interacting populations in a particular area

Population
Organisms of the same species in a particular area

Organism
An individual; complex individuals contain organ systems

Organ System
Composed of several organs working together

Organ
Composed of tissues functioning together for a specific task

Tissue
A group of cells with a common structure and function

Cell
The structural and functional unit of all living things

Molecule
Union of two or more atoms of the same or different elements

Atom
Smallest unit of an element composed of electrons, protons, and neutrons

Figure 1.2 **What other levels of organization must exist for there to be an organ system?**
Living organisms are organized. The smallest unit of living organisms is the cell. Groups of cells form tissues. Different types of tissues form organs and organs that work together form an organ system.

nervous system

shoot system

brain

leaf

nervous tissue

photosynthetic tissue

neuron

plant cell

Acquiring Materials and Energy

Human beings cannot maintain their organization or carry on life's activities without an outside source of materials and energy. Human beings and other animals acquire materials and energy when they eat food (Fig. 1.3).

Food provides nutrient molecules, used as building blocks or for energy. It takes energy (work) to maintain the organization of the cell and of the organism. Some nutrient molecules are broken down completely to provide the necessary energy to convert other nutrient molecules into the parts and products of cells.

Most living things can also convert energy into motion. Self-directed movement is even considered by some to be one of life's characteristics.

Reproducing

Reproduction is a fundamental characteristic of life. Cells come into being only from preexisting cells, and all living things have parents. When living things **reproduce**, they create a copy of themselves and ensure the continuance of their own kind (Fig. 1.3).

Figure 1.3   How do humans and other animals acquire materials and energy?
a. Humans eat plants and animals they raise for food. **b.** A red-tailed hawk has captured a rabbit, which it is feeding to its young.

a.

b.

The presence of genes, in the form of DNA molecules, allows cells and organisms to make more of themselves. DNA contains the hereditary information that directs the structure of each cell and its **metabolism,** all the chemical reactions in the cell. Before reproduction occurs, DNA is replicated so that exact copies of **genes** are passed on to offspring. When humans reproduce, a sperm carries genes contributed by a male into the egg, which contains genes contributed by a female. The genes direct development so that the organism resembles the parents. Red-tail hawks (Fig. 1.3a) only produce red-tail hawks, and humans only produce humans, for example.

Growing and Developing

Growth, recognized by an increase in size and often the number of cells, is a part of development. In humans, **development** includes all the changes that occur from the time the egg is fertilized until death and, therefore, all the changes that occur during childhood, adolescence, and adulthood. Development also includes the repair that takes place following an injury.

All organisms undergo development. Figure 1.4a illustrates that an acorn progresses to a seedling before it becomes an adult oak tree. In humans, growth occurs as the fertilized egg develops into the newborn (Fig. 1.4b).

Being Homeostatic

Together, the organ systems maintain **homeostasis,** an internal environment for cells that usually varies only within certain limits. For example, human body temperature normally fluctuates slightly between 36.5 and 37.5°C (97.7 and 99.5°F) during the day. In general, the lowest temperature usually occurs between 2 A.M. and 4 A.M., and the highest usually occurs between 6 P.M. and 10 P.M. However, activity can cause the body temperature to rise, and inactivity can cause it to decline. The body's ability to maintain a normal temperature is somewhat dependent on the external temperature. Even though we can shiver when we are cold and perspire when we are hot, we will die if the external temperature becomes overly cold or hot.

This text emphasizes how all the systems of the human body help maintain homeostasis. The digestive system takes in nutrients, and the respiratory system exchanges gases with the environment. The cardiovascular system distributes nutrients and oxygen to the cells and picks up their wastes. The metabolic waste products of cells are excreted by the urinary system. The work of the nervous and endocrine systems is critical because they coordinate the functions of the other systems.

Responding to Stimuli

Living things respond to external stimuli, often by moving toward or away from a stimulus, such as the sight of food. Living things use a variety of mechanisms to move, but movement in humans and other animals is dependent upon their nervous and musculoskeletal systems. The leaves of plants track the passage of the sun during the day, and when a houseplant is placed near a window, its stems bend to face the sun. The movement of an animal, whether self-directed or in response to a stimulus, constitutes a large part of its behavior. Some behaviors help us acquire food and reproduce.

Homeostasis would be impossible without the ability of the body to respond to stimuli. Response to external stimuli is more apparent to us because it does involve movement, as when we quickly remove a hand from a hot stove. However, certain sensory receptors detect a change in the internal environment, and then the central nervous system brings about an appropriate response. When you are startled by a loud noise, your heartbeat increases, which causes your blood pressure to increase. If blood pressure rises too high, the brain directs blood vessels to dilate, helping to restore normal blood pressure.

Life Has an Evolutionary History

Evolution is the process by which a species changes through time. When a new variation arises that allows certain members of the species to capture more resources, these members tend to survive and to have more offspring than the other, unchanged members. Therefore, each successive generation will include more members with the new variation that represents an **adaptation** to the environment. Consider, for example, a red-tailed hawk (see Fig. 1.3*a*), which catches and eats rabbits. A hawk can fly, in part, because it has hollow bones to reduce its weight and flight muscles to depress and elevate its wings. When a hawk dives, its strong feet take the first shock of the landing, and its long, sharp claws reach out and hold onto the prey. All these characteristics are a hawk's adaptations to its way of life.

Evolution, which has been going on since the origin of life and will continue as long as life exists, explains both the unity and the diversity of life. All organisms share the same characteristics of life because their ancestry can be traced to the first cell or cells. Organisms are diverse because they are adapted to different ways of life.

> ### Check Your Progress 1.1
> 1. What are the seven characteristics of life?
> 2. Why would you expect all living things on Earth to exhibit these characteristics?

Figure 1.4 How are oak trees and humans similar during their development?
a. A small acorn becomes a tree, and (**b**) a two-celled embryo becomes a human being by the process of growth and development.

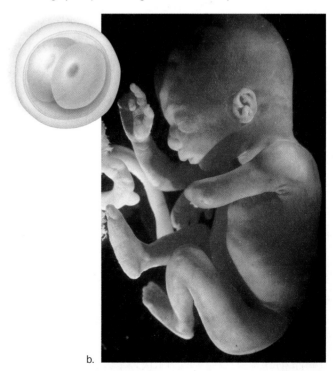

a.

b.

1.2 Humans Are Related to Other Animals

The classification of living things mirrors their evolutionary relationships. Living things are now classified into three domains (Fig. 1.5*a*). The domain Eukarya contains organisms that can be classified into one of four **kingdoms** (see Fig. 1.5*a*). Most organisms in kingdom Animalia are invertebrates like the sea star and earthworm pictured in Figure 1.5*a*. Humans have a nerve cord protected by a vertebral

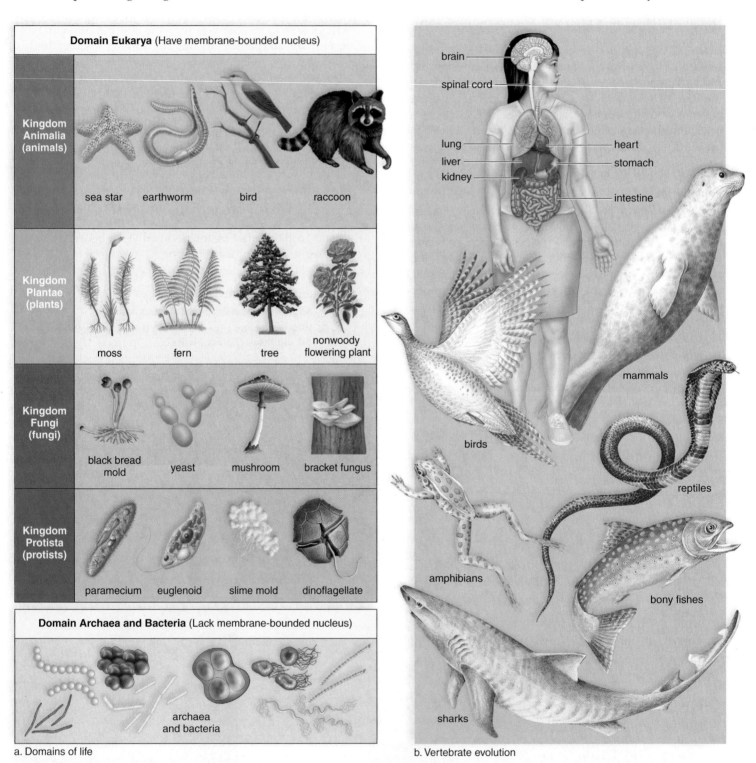

a. Domains of life

b. Vertebrate evolution

Figure 1.5   **How are humans classified with regards to domain and kingdom? Besides the animals pictured, can you name other examples of vertebrates?**

a. Living organisms are classified into three domains: Eukarya, Archaea, and Bacteria. **b.** Humans are in the domain Eukarya and the kingdom Animalia. The presence of a backbone further classifies humans as a type of vertebrate.

column, which makes them vertebrates (Fig. 1.5*b*). **Vertebrates** with hair or fur and mammary glands are classified as mammals. Humans, raccoons, seals, and meerkats are examples of mammals.

Human beings are most closely related to apes. We are distinguished from apes by our (1) highly developed brains, (2) completely upright stance, (3) creative language, and (4) ability to use a wide variety of tools. Humans did not evolve from apes; apes and humans share a common apelike ancestor.

Today's apes are our evolutionary cousins, and we couldn't have evolved from our cousins because we are contemporaries—living on Earth at the same time. Our relationship to apes is analogous to you and your first cousin being descended from your grandparents.

Humans Have a Cultural Heritage

Human beings have a cultural heritage in addition to a biological heritage. **Culture** encompasses human activities and products passed on from one generation to the next outside of direct biological inheritance. Among animals, only humans have a language that allows them to communicate information and experiences symbolically. We are born without knowledge of an accepted way to behave, but we gradually acquire this knowledge by adult instruction and imitation of role models. The previous generation passes on their beliefs, values, and skills to the next generation. Many of the skills involve tool use, which can vary from how to hunt in the wild to how to use a computer. Human skills have also produced a rich heritage in the arts and sciences. However, a society highly dependent on science and technology has its drawbacks as well. Unfortunately, the cultural development may mislead individuals into believing that humans are somehow not part of the natural world surrounding them.

Humans Are Members of the Biosphere

All living things on Earth are part of the biosphere, a living network that spans the surface of the Earth into the atmosphere and down into the soil and seas. Although humans can raise animals and crops for food, they depend on the environment for many services. Without microorganisms that decompose, the waste we create would soon cover the Earth's surface. Some populations of natural ecosystems can clean up pollutants like heavy metals and pesticides.

Freshwater ecosystems, such as rivers and lakes, provide fish to eat, drinking water, and water to irrigate crops. The water-holding capacity of forests prevents flooding, and the ability of forests and other ecosystems to retain soil prevents soil erosion. Many of our crops and prescription drugs were originally derived from plants that grew wild in an ecosystem. Some human populations around the globe still depend on wild animals as a food source. And we must not forget that almost everyone prefers to vacation in the natural beauty of an ecosystem.

a. b.

Figure 1.6 How do humans impact ecosystems?
a. When humans build cities, diversity is lost. Notice the absence of a variety of plants/trees. **b.** The natural ecosystem pictured on the right shows a diverse number of plants that implies a diversity of other organisms.

Humans Threaten the Biosphere

The human population tends to modify existing ecosystems for its own purposes (Fig. 1.6). Humans clear forests and grasslands to grow crops. Later, houses are built on what was once farmland. Clusters of houses become small towns that often grow into cities. Human activities have altered almost all ecosystems and reduced biodiversity (the number of different species present). The present **biodiversity** of our planet has been estimated to be as high as 15 million species. So far, under 2 million have been identified and named. It is estimated that we are now losing as many as 400 species per day due to human activities. Many biologists are alarmed about the present rate of **extinction** (death of a species). They believe it may eventually rival the rates of the five mass extinctions that occurred earlier in our planet's history. The dinosaurs became extinct during the last mass extinction 65 million years ago.

One of the major bioethical issues of our time is preservation of the biosphere and biodiversity. If we adopt a conservation ethic that preserves the biosphere and biodiversity, we will ensure the continued existence of our species.

> **Check Your Progress 1.2**
> 1. What anatomical feature(s) tells us that humans are vertebrates and mammals?
> 2. How do humans differ from the other mammals, including apes?
> 3. Why should humans want to preserve the biosphere, where a variety of organisms live?

1.3 Science as a Process

Science is a way of knowing about the natural world. When scientists study the natural world, they aim to be objective, rather than subjective. Objective observations are supported by factual information, while subjective observations involve personal judgment. For example, the fat content of a particular food would be an objective observation of a nutritional study. Reporting about the good or bad taste of the food would be a subjective observation. It is difficult to make objective observations and conclusions because we are often influenced by our prejudices. Scientists must keep in mind that scientific conclusions can change due to new findings. New findings are often made because of recent advances in techniques or equipment.

Importance of Scientific Theories in Biology

Science is not just a pile of facts. The ultimate goal of science is to understand the natural world in terms of scientific theories. **Scientific theories** are concepts that tell us about the order and the patterns within the natural world; in other words, how the natural world is organized. For example, these are some of the basic theories of biology.

Theory	Concept
Cell	All organisms are composed of cells, and new cells only come from pre-existing cells.
Homeostasis	The internal environment of an organism stays relatively constant.
Genes	Organisms contain coded information that dictates their form, function, and behavior.
Ecosystem	Populations of organisms interact with each other and the physical environment.
Evolution	All organisms have a common ancestor, but each is adapted to a particular way of life.

Evolution is the unifying concept of biology because it makes sense of what we know about living things. For example, the theory of evolution enables scientists to understand the variety of living things and their relationships. It explains common structural features, physiology, patterns of development, and behaviors. The theory of evolution has been supported by so many observations and experiments for over a hundred years, so some biologists refer to the **principle** of evolution. This term is preferred terminology for theories generally accepted as valid by an overwhelming number of scientists.

The Scientific Method Has Steps

Unlike other types of information available to us, scientific information is acquired by a process known as the **scientific method.** The approach of individual scientists to their work is as varied as the scientists. For the sake of discussion, it is possible to speak of the scientific method as consisting of certain steps (Fig. 1.7).

After making initial observations, a scientist will, most likely, study any previous **data,** results and conclusions reported by previous research. Imagination and creative thinking also help a scientist formulate a **hypothesis.** The hypothesis becomes the basis for more observation and/or experimentation. The new data help a scientist come to a **conclusion** that either supports or does not support the hypothesis. Hypotheses are always subject to modification, so they can never be proven true; however, they can be proven untrue. When the hypothesis is not supported by the data, it must be

rejected; therefore, some think of the body of science as what is left after alternative hypotheses have been rejected.

Science is different from other ways of knowing by its use of the scientific method to examine a phenomenon. Any suggestions about the natural world not based on data gathered by employing the scientific method cannot be accepted as within the realm of science. Scientific theories are concepts based on a wide range of observations and experiments.

How the Cause of Ulcers Was Discovered

Let's take a look at how the cause of ulcers was discovered so we can get a better idea of how the scientific method works. In 1974, Barry James Marshall was a young resident physician at Queen Elizabeth II Medical Center in Perth, Australia. There he saw many patients who had bleeding stomach ulcers. A pathologist at the hospital, Dr. J. Robin Warren, told him about finding a particular bacterium, now called *Helicobacter pylori,*

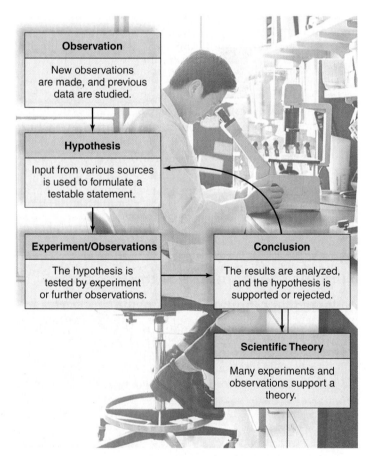

Figure 1.7   What is the process of scientific investigation?
On the basis of new and/or previous observations, a scientist formulates a hypothesis. The hypothesis is tested by further observations and/or experiments, and new data either support or do not support the hypothesis. The return arrow indicates that a scientist often chooses to retest the same hypothesis or to test a related hypothesis. Conclusions from many different but related experiments may lead to the development of a scientific theory. For example, studies pertaining to development, anatomy, and fossil remains all support the theory of evolution.

near the site of peptic ulcers (open sores in the stomach). Using the computer networks available at that time, Marshall compiled much data showing a possible correlation between the presence of *Helicobacter pylori* and the occurrence of both gastritis (inflammation of the stomach) and stomach ulcers. On the basis of these data, Marshall formulated a hypothesis: *Helicobacter pylori* is the cause of gastritis and ulcers.

Marshall decided to make use of Koch's[1] postulates, the standard criteria that must be fulfilled to show that a pathogen (bacteria or virus) causes a disease.

Koch Postulates:

- The suspected pathogen (virus or bacterium) must be present in every case of the disease;
- the pathogen must be isolated from the host and grown in a lab dish;
- the disease must be reproduced when a pure culture of the pathogen is inoculated into a healthy susceptible host; and
- the same pathogen must be recovered again from the experimentally infected host.

The First Two Criteria

By 1983, Marshall had fulfilled the first and second of Koch's criteria. He was able to isolate *Helicobacter pylori* from ulcer patients and grow it in the laboratory. (Success was achieved only after a petri dish was inadvertently left in the incubator for six, instead of two, days.) Further, he had determined that bismuth, the active ingredient in Pepto-Bismol, could destroy the bacteria in a petri dish.

Despite presentation of these findings to the scientific community, most physicians continued to believe that stomach acidity and stress were the cause of stomach ulcers. In those days, patients were usually advised to make drastic changes in their lifestyle or seek psychiatric counseling to "cure" their ulcers. Many scientists believed that no bacterium would be able to survive the normal acidity of the stomach.

The Last Two Criteria

Marshall had a problem in fulfilling the third and fourth of Koch's criteria. He had been unable to infect guinea pigs and rats with the bacteria because the bacteria just didn't flourish in the intestinal tract of those animals. Marshall wasn't able to use human subjects because our society does not condone the use of humans as experimental subjects in dangerous or life-threatening research. Marshall was so determined to support his hypothesis that, in 1985, he decided to perform the experiment on himself! To the disbelief of those in the lab that day, he and another volunteer swallowed a foul-smelling and -tasting solution of *Helicobacter pylori*. Within the week, they felt lousy and were vomiting up their stomach contents.

Examination by endoscopy showed that their stomachs were now inflamed, and biopsies of the stomach lining contained the suspected bacterium (Fig. 1.8). Their symptoms abated without need of medication, and they never developed an ulcer. At Marshall's next talk, he challenged his audience to refute his hypothesis. Many tried, but ultimately the investigators supported his findings.

The Conclusion

In science, many experiments that involve a considerable number of subjects are required before a conclusion can be reached. By the early 1990s, at least three independent studies involving hundreds of patients had been published showing that antibiotic therapy could eliminate *Helicobacter pylori* from the intestinal tract and cure patients of ulcers wherever they occurred in the tract.

Dr. Marshall received all sorts of prizes and awards, but he and Dr. Warren were especially gratified to receive a Nobel Prize in Medicine in 2005. The Nobel committee commented, "Thanks to the pioneering discovery by Marshall and Warren, peptic ulcer disease is no longer a chronic, frequently disabling condition, but a disease that can be cured by a short regiment of antibiotics and acid secretion inhibitors."

How to Do a Controlled Study

The work that Marshall and Warren did was largely observational. Often, scientists perform an **experiment,** a series of procedures to test a hypothesis. As an example, let's say investigators want to determine which of two antibiotics best treats an ulcer. When scientists do an experiment, they try to vary just the **experimental variables,** in this case, the medications being tested. A **control group** is not given the medications, but one or more **test groups** are given the medications. If by chance, the control group shows the same results as a test group, the investigators immediately know the results of their study are invalid because it would mean the medications may have nothing to do with the results.

Figure 1.8 How did Dr. Barry Marshall show that bacteria are the cause of stomach ulcers?
Dr. Barry Marshall, pictured here, fulfilled Koch's postulates to show that *Helicobacter pylori* is the cause of peptic ulcers. The inset shows the presence of the bacterium in the stomach.

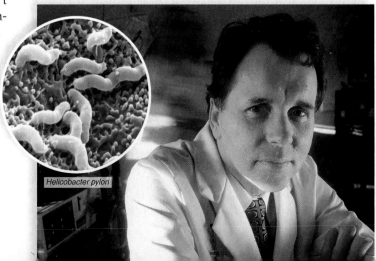
Helicobacter pylori

---

[1]Robert Koch was a German microbiologist who helped verify the germ theory of disease and established the standard as to whether an organism causes a particular disease.

Have You Ever Wondered...

Why some people get diarrhea when taking antibiotics?

Many antibiotics are called broad spectrum antibiotics, because they kill many different types of bacteria. A broad-spectrum antibiotic taken for a bacterial ear infection may also kill some of the beneficial bacteria living in your intestine. When the balance of beneficial and harmful bacteria in your intestine is disturbed, the harmful bacteria tend to flourish and diarrhea often results. The good news is that it usually goes away once you finish the entire course of antibiotics. Eating yogurt can also help to restore the balance of intestinal bacteria.

The study depicted in Figure 1.9 shows how investigators may study this hypothesis:

Hypothesis: Newly discovered antibiotic B is a better treatment for ulcers than antibiotic A, which is in current use.

Investigators who perform clinical research must obtain informed consent from their subjects before proceeding with the research. The informed consent ensures that subjects know details about the research, and that their participation is voluntary. The risks and benefits involved in participating in the research are all outlined. It is important to reduce the number of possible variables (differences) such as sex, weight, other illnesses, and so forth between the groups. Therefore, the investigators *randomly* divide a very large group of volunteers (Fig. 1.9a) equally into the three groups. The hope is any differences will be distributed evenly among the three groups. This is more likely to occur if the investigators have a large number of volunteers.

The three groups will be treated (Fig. 1.9b) as follows:

Control group: Subjects with ulcers are not treated with either antibiotic.
Test group 1: Subjects with ulcers are treated with antibiotic A.
Test group 2: Subjects with ulcers are treated with antibiotic B.

After the investigators have determined that all volunteers do suffer from ulcers, they will want the subjects to think they are all receiving the *same* treatment. This is an additional way to protect the results from any influence other than the medication. To achieve this end, the subjects in the control group can receive a **placebo,** a treatment that appears to be the same as that administered to the other two groups but contains no medication. In this study, the use of a *placebo* would help ensure the same dedication by all subjects to the study.

The Results

After two weeks of administering the same amount of medication (or placebo) in the same way, the intestinal tract of each subject is examined to determine if ulcers are still present. Endoscopy, depicted in the photograph in Figure 1.9c, is one possible way to examine a patient for the presence of ulcers. This procedure is performed under sedation and involves inserting an endoscope—a small, flexible tube with a tiny camera on the end—down the throat and into the stomach. It allows the doctor to see the lining of the stomach and check for possible ulcers. Tests performed during an endoscopy can also determine if *Helicobacter pylori* is present.

CASE STUDY ASKING QUESTIONS

During their microbiology class the next day, Katie and Shane asked Dr. Williams to solve their dispute. Shane inquired, "Do you know about the '5-second rule'? Is it true? Katie thinks I'm a real pig for eating fries that dropped on the floor."

Dr. Williams smiled, "I know about the '5-second rule,' but I'm not familiar with its validity. You two need to go to the library and do a literature search. If the '5-second rule' has been investigated in a scientific fashion, the methods used and results of the investigations should show up in a literature search."

She continued, "You could do an investigation of the '5-second rule' for your class project. What you find during your search of the literature should help you design your experiment."

After class, Shane and Katie headed straight for the library to begin their literature search. They found two studies that investigated the transfer of germs to food dropped on the floor. One study found that *E. coli,* bacteria common to the large intestine, were indeed transferred in 5 seconds or less. The investigators found the bacteria on candy and cookies dropped onto ceramic tiles. Another study quantified the transfer of species of *Salmonella* bacteria from three floor surfaces. Tile, wood flooring, and nylon carpet all managed to transmit *Salmonella* to bread and to bologna if the food was left for 5 seconds. That same study showed that a great deal more *Salmonella* were transferred from the three types of flooring if the food was left for a full minute.

"Shane, those studies are really scary," Katie commented. "Do you remember the *E. coli* food poisoning cases a few years ago? People died from it. And *Salmonella* bacteria comes from water that has raw sewage in it. That is nasty!"

Shane shrugged. "Well, I didn't see any French-fry studies, and I ain't dead yet from eating those fries. I think you're just being paranoid. Let's go show these papers to Dr. Miller."

Endoscopy is somewhat subjective so it is probably best if the examiner is not aware of which group the subject is in. Otherwise, the prejudice of the examiner may influence the examination. When neither the patient nor the examiner is aware of the specific treatment, it is called a *double-blind* study.

In this study, the investigators may decide to determine effectiveness of the medication by the percentage of people who no longer have ulcers. So, if 20 people out of 100 still have ulcers, the medication is 80% effective. The difference in effectiveness is easily read in the graph portion of Figure 1.9*d*.

Conclusion: On the basis of their data, the investigators conclude that their hypothesis has been supported.

Publication of Scientific Studies

Scientific studies are customarily published in a scientific journal, so that all aspects of a study are available to the scientific community. Before information is published in scientific journals, it is typically reviewed by experts. These people ensure the research is credible, accurate, unbiased, and well executed.

Another scientist should be able to read about an experiment in a scientific journal, repeat the experiment in a different location, and get the same (or very similar) results. Each article begins with a short synopsis of the study so scientists can quickly find the articles of greatest interest to them. The materials and methods of performing each study are clearly outlined so that researchers can more easily repeat the work. Some articles are rejected for publication by reviewers when they believe there is something questionable about the design or manner in which an experiment was conducted.

Further Study

As mentioned previously, the conclusion of one experiment often leads to another experiment. Scientists reading the study described in Figure 1.9 may decide that it would be important to test the difference in the ability of antibiotic A and B to kill *Helicobacter pylori* in a petri dish. Or, they may want to test which medication is more effective in women than men, and so forth. The need for scientists to expand on findings explains why science changes and the findings of yesterday may be improved upon tomorrow.

Scientific Journals Versus Other Sources of Information

The information in many scientific journals is highly regarded by scientists because of the review process and because it is "straight from the horse's mouth," so to speak. The investigator

Figure 1.9 **What is the difference between a control group and a test group?**
In this controlled laboratory experiment to test the effectiveness of a medication in humans, subjects were divided into three groups. The control group received a placebo and no medication. Test group 1 received antibiotic A and test group 2 received antibiotic B. The results are depicted in a graph, and it shows that antibiotic B was found to be a more effective treatment than antibiotic A for the treatment of ulcers.

a.

State Hypothesis:
Antibiotic B is a better treatment for ulcers than antibiotic A

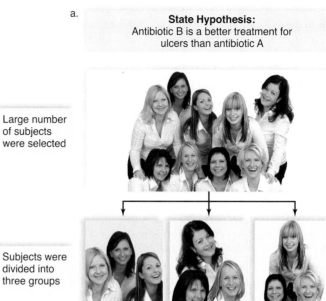

Large number of subjects were selected

Subjects were divided into three groups

b.

Perform Experiment:
Groups were treated the same except as noted

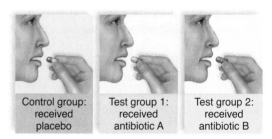

Control group: received placebo

Test group 1: received antibiotic A

Test group 2: received antibiotic B

c.

Collect Data:
Each subject was examined for the presence of ulcers

d.

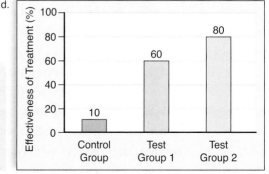

Conclusion: Hypothesis is supported: Antibiotic B is a better treatment for ulcers than antibiotic A

who did the research is generally the primary author of a published study. Reading the results tends to prevent the possibility of misinformation and/or bias. Remember playing "pass the message" when you were young? When someone started a message and it was passed around a circle of people, the last person to hear the message rarely received the original message. As each person passed along the message, information was added or deleted from the original. That same thing happens to scientific information when it is published in magazines, books, or reported by someone other than the original investigator.

Unfortunately, the studies in scientific journals may be technical and difficult for a layperson to read and understand. The general public typically relies on secondary sources of information for their science news. The information may be out of context or misunderstood by the reporter, and the result is transmission of misinformation. Ideally, a reference to the original source (scientific journal article) will be provided so that the second-hand information can be verified. Remember also that it often takes years to do enough experiments for the scientific community to accept findings as well founded. Be wary of claims that have only limited data to support them and any information that may not be supported by repeated experimentation.

People should be especially careful about scientific information available on the Internet, which is not well regulated. Reliable, credible scientific information can often be found at websites with URLs (Web addresses or uniform resource locators) containing .edu (for educational institution), .gov (for government sites such as the National Institutes for Health or Centers for Disease Control), and .org (for nonprofit organizations such as the American Lung Association or the National Multiple Sclerosis Society). Unfortunately, quite a bit of scientific information on the Internet is intended to entice people into purchasing some sort of product for weight loss, prevention of hair loss, or similar maladies. These websites usually have URLs ending with .com or .net. It pays to question and verify the information from these websites with another source (primary, if possible).

Check Your Progress 1.3

1. In general, what steps do investigators use when using the scientific method?
2. What is the standard experimental design for a controlled study?
3. Why might information in scientific journals be more reliable than that found in magazines or on the Internet?

1.4 Making Sense of a Scientific Study

When evaluating scientific information, it is important to consider the type of data given to support it. Anecdotal data, which consists of testimonials by individuals rather than results from a controlled, clinical study, are never considered reliable data. An example of anecdotal data would be someone who claims that a particular diet helped them lose weight. Obviously, this doesn't mean the diet will work for everyone. Testimonial data are suspect because the effect of

---

CASE STUDY DOING RESEARCH

With Dr. Williams' help, Shane and Katie designed a version of a '5-second rule' experiment. The experimental methods used by investigations found during their literature search helped, too. First, Katie and Shane created a survey with a series of questions, asking their classmates if they had ever used the '5-second rule.' Most students reported using the rule on a regular basis, for many types of food and on many different types of flooring.

Shane specifically wanted to investigate the transfer of bacteria to French fries at the burger place, because that was where the '5-second rule' first came into question. The pair obtained the owner's permission to conduct their experiment at the burger place.

Dr. Williams had gathered sterilized equipment for Katie and Shane's experiment. They began by selecting four French fries from one batch. One fry was selected for use as the *control*, which would test whether French fries that hadn't been dropped would grow bacteria. Katie held the fry with sterile tweezers and snipped off a 5 cm piece with sterile scissors. She dropped the piece into a test tube of nutrient broth, used in the microbiology lab to grow bacteria. After replacing the cap on the test tube, Katie prepared four more test tubes in the same way.

Shane dropped the other three fries on to the floor. After 3 seconds, he used sterile tweezers to pick up one fry. He handed it to Katie, who cut it into pieces and put each piece into the broth. Shane picked up another fry after 5 seconds and the last fry after 20 seconds. As each fry was recovered, the pieces were carefully dropped into separate tubes of nutrient broth. As they repeated the process, the test tube racks filled quickly. "This is sure a lot of test tubes," Katie commented. "It seems like a waste to use so many."

Shane grinned, "Great scientists, like us, repeat their experiments over and over. And don't forget, we need a whole bunch of evidence so I can prove that '5-second' floor fries are not only delicious, they're also safe!"

Katie laughed. "You're so gross!"

The pair headed back to Dr. Williams's lab to place their tubes into an incubator that would keep their test tubes at a constant warm temperature.

whatever is under discussion may not have been studied with a large number of subjects or a control group.

We must also keep in mind that just because two events occur at the same time, one factor may not be the cause of the other. Dr. Marshall had this problem when his data largely depended on finding *Helicobacter pylori* at the site of ulcers. More data was needed before the scientific community could conclude that *Helicobacter pylori* was the cause of an ulcer. Similarly, that a human papillomavirus (HPV) infection usually precedes cervical cancer could be viewed only as limited evidence that HPV causes cervical cancer. In this instance, HPV did turn out to be a cause of cervical cancer, but not all correlations (relationships) turn out to be causations. For example, scientific studies do not support the well-entrenched belief that exposure to cold temperatures results in colds. Instead, we now know that viruses cause colds.

What to Look For

Although most everyone who examines a study is tempted to first read the abstract (synopsis) at the beginning and then skip to the conclusion at the end of a study, we should also examine the investigators' methodology and results before going to the conclusion. The methodology tells us how they conducted their study, and the results tell us what facts (data) they discovered. Always keep in mind that the conclusion is not the same as the data. The conclusion is an interpretation of the data. It is up to us to decide if the conclusion is justified by the data.

Graphs

Data are often depicted in the form of a bar graph (see Fig. 1.9*d*) or a line graph (Fig. 1.10). A graph shows the relationship between two quantities, such as the taking of an antibiotic

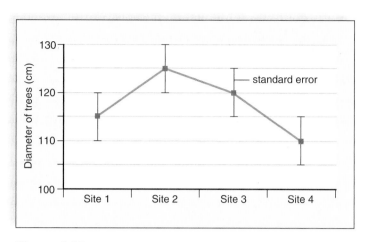

Figure 1.10 What information does this line graph convey? This line graph shows that the diameter of tree trunks varied at four different places.

and the disappearance of ulcer. As in Figure 1.9, the experimental variable (study groups) is plotted on the horizontal (x-axis), and the result (effectiveness) is plotted along the vertical (y-axis). Graphs are useful tools to summarize data in a clear and simplified manner. For example, Figure 1.9*d* immediately shows that antibiotic B produced the best results.

The title and labels can assist you in reading a graph; therefore, when looking at a graph, first check the two axes to determine what the graph pertains to. For example, in Figure 1.10, we can see that the investigators were studying tree trunk diameters at four sites (places). By looking at this graph, we know that trees with the greatest diameter are found at site 2, and we can also see to what degree the tree trunk diameters differed between the sites.

CASE STUDY DRAWING CONCLUSIONS

Two days later, Katie and Shane returned to the microbiology laboratory to check each test tube. Dr. Williams had explained that when transparent broth turned cloudy, bacteria were present. The broth containing the control fries showed no growth, but the remaining tubes of broth were cloudy. Shane carefully removed the top from a clouded tube and grimaced at the awful smell. "Yeah, it's growing bacteria, all right!"

"ICK, ICK, ICK," chanted Katie again and again, as she recorded the results. "You ate fries that picked up bacteria from the floor. What do you think of your '5-second rule' now?"

Katie and Shane met with Dr. Williams the following day to discuss the results of their experiment. "I think the results you recorded after yesterday's observations are accurate," Dr. Williams confirmed. "It looks like the control fries showed no bacterial growth in the broth, while all your experimental fries had some amount of growth."

"So do you think we could get our results published, too?" asked Shane excitedly.

"Let's not get too carried away," replied Dr. Williams. "To publish, you would first have to repeat that same experiment over and over to confirm your results. And you might have other sources of error in your experiment. For example, what if the French-fry oil was hot enough to kill all bacteria on the first fry, and the other fries grew bacteria just because they had cooled off? Maybe hot floor fries wouldn't have grown bacteria. And before you publish, other scientists must read your paper. They must agree that the results are accurate and important enough to publish."

She continued, "Perhaps you should continue to investigate the bacteria that grew in the broth. What kinds of bacteria are they? And most important, are they even harmful bacteria? Remember, the majority of bacteria are beneficial, not harmful. Maybe Shane can keep on eating floor fries, if they just pick up harmless bacteria!"

"Oh, please, don't give him any ideas!" Katie laughed.

Statistical Data

Most authors who publish research articles use statistics to help them evaluate their experimental data. In statistics, the **standard error** tells us how uncertain a particular value is. Suppose you predict how many hurricanes Florida will have next year by calculating the average number during the past ten years. If the number of hurricanes per year varies widely, your standard error would be larger than if the number per year is usually about the same. In other words, the standard error tells you how far off the average could be. If the average number of hurricanes is four and the standard error is ± 2, then your prediction of four hurricanes is between two and six hurricanes.

Statistical Significance   When scientists conduct an experiment, there is always the possibility that the results are due to chance or due to some factor other than the experimental variable. Investigators take into account several factors when they calculate the probability value that their results were due to chance alone. If the probability value is low, researchers describe the results as statistically significant. A probability value of less than 5% (usually written as $p<.05$) is acceptable, but, even so, keep in mind that the lower the p value, the *less* likely that results are due to chance. Therefore, the lower the p value, the *greater* the confidence the investigators and you can have in the results. Before you take a medication for ulcers, you probably want a significance as low as $p<.001$.

> **Check Your Progress 1.4**
>
> 1. What is wrong with (a) anecdotal data (testimonial data) and (b) correlation data?
>
> 2. a. What is the difference between the data and the conclusion of a study? b. Which is more likely to be questionable?
>
> 3. How do scientific studies benefit from the use of graphs and statistics?

1.5 Science and Social Responsibility

As we have learned in this chapter, science is a systematic way of acquiring knowledge about the natural world. Biologists are scientists that study living things, from the tiniest microbes to the tallest trees (Fig. 1.11). Religion, aesthetics, and ethics are other ways in which human beings seek order in the natural world. Science differs from these other ways of knowing and learning by its process, based on the scientific method. Science considers hypotheses that can be tested only by experimentation and observation. Only after an immense amount of data has been gathered do scientists arrive at a scientific theory, a well-found concept about the natural world. Knowing this should help you realize that all scientific theories have merit.

Science is a slightly different endeavor from technology, but science is the driving force behind technology. **Technology** is the application of scientific knowledge to the interests of humans. Western civilization has always believed that science and technology offer us ways to improve our lives. Our ability to build houses, pave roads, grow crops, and cure illnesses all depend on technology. Just think of how many things you use each day made of plastic and you will know how much technology means to your daily life. Even the field of nuclear physics is of direct benefit to human beings. The findings of nuclear physics play a role in cancer therapy, medical imaging, and homeland security. It has also given us nuclear power and the atomic bomb. In other words, science and technology are not risk free; and uncontrolled technology can result in unanticipated side effects.

Science and Technology, Benefits Versus Risks

Investigations into cell structure and genes led to the biotechnology revolution of current times. We now know how to manipulate genes. If you are a diabetic and are using insu-

a.

b.

Figure 1.11   **Which step of a scientific study appears to be going on in these pictures?**
Data collection can be done **(a)** in the laboratory or **(b)** in the field. Biologists discover basic information about the natural world, including the effects of technology on human health and the environment.

Historical Focus

The Syphilis Research Scandal of Tuskegee University

Several sections of this chapter have covered the process of science, the way that legitimate research should be conducted, and the importance of informed consent when using human research subjects. As professionals, scientists have a responsibility to design moral and ethical research. Unfortunately, as with all professionals, not all scientists are ethical. Documented cases of risky, life-threatening, and, in some cases, inhumane research on humans (often without the subject's consent or knowledge) blot scientific history. One of the most extreme examples of such "research" was that done by Dr. Josef Mengele, the handsome Nazi doctor called the "Angel of Death." Mengele tortured concentration camp prisoners in multiple horrible ways. Some were slowly frozen to death, other poisoned, still others bled to death—all to fulfill Mengele's obscene notion of scientific inquiry.

Regrettably, the history of research in the United States is also stained by misconduct. One notorious example of unethical research involving human subjects began in the United States in 1932 and continued until 1972. This research was carried out by the Public Health Service (PHS). Investigators wished to study the progression of syphilis, a sexually transmitted bacterial disease, in African-American males. African-American males were of interest because of diseases associated with African Americans, such as sickle-cell anemia. There was interest in determining if race had an impact on the progression of syphilis.

There are three distinct stages of an untreated syphilis infection. During the first stage, sores or ulcers appear on the genitals of the infected individual. The bacteria associated with the sores can infect individuals who come into contact with the sores. Stage two syphilis typically develops several weeks later. Its symptoms include those associated with the flu: fever, headaches, and joint pain. If the disease shows up in a third stage, severe complications associated with the nervous system occur. Paralysis and insanity are common, and organ failure will eventually kill. Untreated pregnant women can pass syphilis to a fetus. Congenital syphilis in an infant causes physical deformity and mental retardation.

The syphilis study began in 1932 at the Tuskegee Institute. Initially, 600 African-American males were enrolled, with 399 males infected with syphilis and 201 uninfected males. The men were poor, mostly illiterate, sharecropper farmers. None of the men were informed of their participation in a research study,

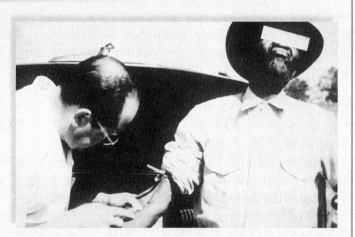

Figure 1A Poorly educated African Americans were recruited for the Tuskegee project with promises of free medical care.

nor about available treatment options. The men were told that investigators were testing for and treating "bad blood." The phrase described a number of common illnesses, including anemia, that were widespread at the time. While they participated in the study, the men were offered medical exams, transportation to and from clinics, treatments for other ailments, food, and money for their burial expenses if necessary.

When the study first began, there were few available treatments for syphilis. Compounds containing mercury and arsenic were used, but all were toxic to the patient.

Originally, the study sought to determine if these toxic compounds helped, or if untreated individuals fared the same as those treated. The study was intended to last 6 months, but persisted for decades. Research continued on these men, even after penicillin became the accepted treatment for syphilis in 1947. Treatment with penicillin was never offered to the study's participants, even though it would have cured them of the disease.

Objections to the Tuskegee study began in the mid-1960s. One critic deemed the project "bad science" because some men had been partially treated during the study. In July 1972, the Associated Press (AP) publicized the story of the syphilis project. The study was finally discontinued. A class-action lawsuit settlement covering medical and burial expenses for all living participants was reached in 1974. President Clinton offered a national apology to survivors and their families in 1997. Even now, there are some family members of the study's participants receiving benefits from that lawsuit.

lin, it was produced by genetically modified (GM) bacteria. Despite its benefits, there are risks associated with biotechnology. Ecologists are concerned that GM crops could endanger the biosphere. For example, many farmers are now planting GM cotton that produces an insect killing toxin. Though insect pests are their intended target, toxins may

also kill the natural predators that feed on insect pests. It may not be wise to kill off friendly predatory insects. Or closer to home, people are now eating GM foods and/or foods that have been manufactured by using GM products. Some people are concerned about the possible effects of GM foods on human health.

Bioethical Focus

QUARANTINED . . . Could it happen to you?

Efforts to quarantine individuals with contagious diseases appear throughout human history. The Old Testament of the Bible tells us that lepers were required to live in isolated colonies to prevent the spread of leprosy. The first formal method of governmental quarantine was created in Venice in the 1300s, to minimize the transmission of the bubonic plague. "Typhoid Mary" is possibly the most infamous case of quarantine that took place in the United States. Mary Mallon was a carrier of typhoid fever, and her story is described in Chapter 7.

The recent case of Andrew Speaker raises the issue of forced quarantine again. Speaker, a carrier of a drug-resistant strain of tuberculosis (TB), traveled back and forth to Italy. Speaker was told by the Centers for Disease Control (CDC) to remain in Italy after he left the United States. He chose to travel back to the United States via Prague and Canada instead. He was taken into custody and quarantined at a Denver hospital. Speaker claimed he was told he was not contagious. However, he was also advised not to travel overseas and did so anyway. He may not have been contagious at the time, but should Speaker have traveled overseas by plane while knowing he carried a drug-resistant strain of TB?

The cases of Mary Mallon and Andrew Speaker raise the question of whether the civil liberties of an individual can be denied if the health of the general public is at risk. The answer is likely to be a balance of educating the carrier, while offering economic security and fair treatment. Mary Mallon may not have returned to cooking after

Figure 1B Quarantined until further notice.

her release from the initial quarantine if she'd been retrained for a job comparable in status and income. Andrew Speaker may have remained in Italy, if the CDC had offered him a way to get back to the United States.

How would you react if you had eaten food prepared by Mary Mallon at the hospital or if you'd been on the plane with Andrew speaker? How would you react if authorities took you away from your family, friends, and home because of risk to the general public?

In medicine, gene technology raises even more difficult ethical issues. These include whether humans should be cloned or whether gene therapy should be used to modify the inheritance of people, even before they are born. A current debate centers around the use of stem cells to cure human illnesses, such as spinal cord injuries or Alzheimer disease. You may not appreciate this debate until you know that early human embryos are composed only of cells called embryonic stem cells. Embryonic stem cells are genetically capable of becoming any type of tissue needed to cure a human illness. Should human embryos be dismantled and used for this purpose? It means they will never have the opportunity to become a human being.

Everyone Is Responsible

Science, technology, and society have interacted throughout human history, and scientific investigation and technology have always been affected by human values. Studying science, such as human biology, can give citizens the background they need to fully participate in ethical debates. All citizens should assume this responsibility because everyone, not just scientists, need to be involved in making value judgments about the proper use of technology.

Should windmills be placed in Nantucket Sound, or other bodies of water, to reduce our use of oil, which causes global warming? To participate in this debate, you have to know what windmills might do to the ecology of Nantucket Sound, what global warming is, and what the evidence for global warming is. In other words, you need to review the data of scientists and only then participate in the debate about how technology should be used in this particular instance.

Should the rise in human population be curtailed to help preserve biodiversity? You might hear on the news that we are in a biodiversity crisis—the number of extinctions expected to occur in the near future is unparalleled in the history of the Earth. To help answer this question, you will want to know how large the human population is, what the benefits of biodiversity are, how it is threatened, and what the threat is to humankind if biodiversity is not preserved.

Scientists can inform and educate us, but they need not bear the burden of making these decisions alone because science does not make value judgments. This is the job of all of us.

Check Your Progress 1.5

1. What are some examples to show that technology has both benefits and risks?
2. Which members of society should be responsible for deciding how technology should be used?
3. Is this a good reason for everyone to become scientifically competent?

Summarizing the Concepts

1.1 The Characteristics of Life

Living things, often called organisms, share common characteristics. Organisms

- have levels of organization—atoms, molecules, cells, tissues, organs, organ systems, organisms, populations, community, ecosystem, and biosphere;
- take materials and energy from the environment;
- reproduce;
- grow and develop;
- are homeostatic;
- respond to stimuli; and
- have an evolutionary history and are adapted to a way of life.

1.2 Humans Are Related to Other Animals

The classification of living things mirrors their evolutionary relationships. Humans are mammals, a type of vertebrate in domain Eukarya. Humans differ from other mammals, including apes, by their

- highly developed brains;
- completely upright stance;
- creative language; and
- ability to use a wide variety of tools.

Humans Have a Culture Heritage

Language, tool use, values, and information are passed on from one generation to the next.

Humans Are Members of the Biosphere

Humans depend on the biosphere for its many services, such as absorption of pollutants, sources of water and food, prevention of soil erosion, and natural beauty.

Humans Threaten the Biosphere

Unfortunately, humans threaten the biodiversity of the biosphere, most likely to their own detriment.

1.3 Science as a Process

The scientific method consists of

- making an observation;
- formulating a hypothesis;
- carrying out experiments and observations;
- coming to a conclusion; and
- developing a scientific theory.

How the Cause of Ulcers Was Discovered

Dr. Marshall followed Koch's postulates to show that *Helicobacter pylori* causes ulcers. When he was unable to infect an animal with the bacterium, he and another volunteer infected themselves. His persistence led to clinical studies that showed antibiotics can cure ulcers.

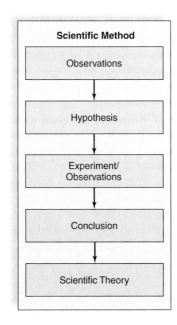

Scientific Method

Observations

↓

Hypothesis

↓

Experiment/ Observations

↓

Conclusion

↓

Scientific Theory

How to Do a Controlled Study

- a large number of subjects are divided randomly into groups;
- the test group(s) is exposed to an experimental variable;
- a control group is not exposed to an experimental variable and is given a placebo;
- all groups are otherwise treated the same, and it is best if the subjects and the technicians do not know what group they are in; and
- the results and conclusion are published in a scientific journal.

Scientific Journals Versus Other Sources of Information

Primary sources of information are best, and if second-hand sources are used, the reader needs to carefully evaluate the source. The Internet is not regulated, although URLs that end in .edu, .gov, and .org most likely provide reliable scientific information.

1.4 Making Sense of a Scientific Study

- Beware of anecdotal and correlation data because they need further study to be substantiated.
- Both bar and line graphs clearly show the relationship between quantities.

  Statistical data. The standard error tells us how uncertain a particular value is. The statistical significance tells us how trustworthy the results are. The lower the statistical significance, the less likely the results are due to chance and the more likely they are due to the experimental variable.

1.5 Science and Social Responsibility

- Scientific information is based on observation and experimentation. Therefore, scientists need not make value judgments for us.
- The use of modern technology has it risks, and all citizens need to be able to make informed decisions regarding how and when technology should be used.

Understanding Key Terms

adaptation 5	evolution 6	organ system 2
atom 2	experiment 9	placebo 10
biodiversity 7	experimental	population 2
biology 2	variable 9	principle 8
biosphere 4	extinction 7	reproduce 4
cell 2	gene 4	science 7
community 2	homeostasis 4	scientific method 8
conclusion 8	hypothesis 8	scientific theory 8
control group 9	kingdom 6	species 2
culture 7	metabolism 4	standard error 14
data 8	molecule 2	technology 14
development 4	multicellular 2	test group 9
domain 6	organ 2	tissue 2
ecosystem 4	organism 2	vertebrate 6

Match the key terms to these definitions.

a. _____ Zone of air, land, and water at the surface of the Earth in which living organisms are found.

b. _____ Smallest unit of a human being and all living things.

c. _____ Concept supported by a broad range of observations, experiments, and conclusions.

d. _____ An internal environment that normally varies within only certain limits.

e. _____ An artificial situation devised to test a hypothesis.

Testing Your Knowledge of the Concepts

1. Name the 11 levels of biological organization from simple to complex, and briefly describe each. (pages 2–3)

2. What is homeostasis, and how is it maintained? Give some examples that show how systems work together to maintain homeostasis. (pages 4–5)

3. How do human activities threaten the biosphere? (page 7)

4. Discuss the importance of a scientific theory, and name several theories basic to understanding biological principles. (page 8)

5. With reference to the steps of the scientific method, explain how scientists arrive at a theory. (page 8)

6. What are Koch's postulates, and what are they used for? (page 9)

7. What is a control group, and what is the importance of a control group in a controlled study? (page 9)

8. How is technology different from scientific knowledge? (pages 14–16)

In questions 9–12, match each description with the correct characteristic of life from the key.

Key:

a. Life is organized.
b. Living things reproduce and grow.
c. Living things respond to stimuli.
d. Living things have an evolutionary history.
e. Living things acquire materials and energy.

9. Human heart rate increases when scared.

10. Humans produce only humans.

11. Humans need to eat for building blocks and energy.

12. Similar cells form tissues in the human body.

13. The level of organization that includes two or more tissues that work together is a(n)
a. organ.   c. organ system.
b. tissue.   d. organism.

14. The level of organization most responsible for the maintenance of homeostasis is the _____ level.
a. cellular   b. organ
c. organ system   d. tissue

15. The level of organization that includes all the populations in a given area along with the physical environment would be a(n)
a. community.   c. biosphere.
b. ecosystem.   d. tribe.

16. Which best describes the evolutionary relationship between humans and apes?
a. Humans evolved from apes.
b. Humans and chimpanzees evolved from apes.
c. Humans and apes evolved from a common apelike ancestor.
d. Chimpanzees evolved from humans.

17. The kingdom that contains humans is the kingdom
a. Animalia.   c. Plantae.
b. Fungi.   d. Protista.

In questions 18–20, match the explanation with a theory in the key.

Key:

a. homeostasis   c. evolution
b. cell   d. gene

18. All living things share a common ancestor.

19. The internal environment remains relatively stable.

20. Organisms contain coded information that dictates form and function.

In questions 21–25, match each description with a step in the scientific method from the key.

Key:

a. conclusion   d. observation
b. experiment   e. theory
c. hypothesis

21. Rejection or acceptance of a hypothesis based on collected data

22. Statement to be tested

23. Procedure designed to test a hypothesis

24. Hypothesis supported by many studies

25. Fact that may lead to the development of a hypothesis

Thinking Critically About the Concepts

1. Viruses are generally lumped into a "germs" grouping with bacteria. Viruses are only some genetic material (DNA or RNA) wrapped in a protein coat. Can something so simple be considered a living organism? Why aren't viruses mentioned in the system of classification covered in Section 1.2? Hint—consider the shared characteristics of all living organisms.

2. Research on human subjects, as described in Section 1.3, divided the subjects into two or more groups. One group is a "control," and receives a placebo instead of the treatment. Imagine that during the research, the treatment is wildly successful and produces spectacular results in the treated subjects. Should all human subjects, including the control group, then receive the treatment? Why or why not?

3. What kind of data would you be collecting if you surveyed your classmates about their knowledge and successful use (failure to become ill) of the 5-second rule?

4. The research Katie and Shane found during their literature search showed that dropped food yielded *E. coli* and *Salmonella* bacteria after being recovered from the floor. Investigate recent news stories to explain why researchers and health care workers are deeply concerned about these bacteria.

CHAPTER

2

Chemistry of Life

After almost two years of working together, Matt and his wife Libby Molyneaux agreed that there was a light at the end of a long, long tunnel. First came the terror of escaping Hurricane Katrina and the boredom and frustration of living in the Houston Astrodome with their two children. Finally, the floodwaters in New Orleans receded. Discovering that Matt's father had drowned in the second floor of his home was a heartbreak. So was the need to throw away brand-new upholstered furniture, Libby's wedding dress, mold-spoiled family pictures, and practically everything else that had been on the first floor.

Wearing facial masks to prevent inhaling mold spores, the couple pulled all the wallboard from their 1900s-era home. The beautiful first-story hardwood floors, warped and curled beyond saving, were next to go. Every other surface on both floors of their home was carefully scrubbed with bleach. A fresh coat of paint had the outside of the classic home looking beautiful once again. Their goal of being able to move back into the second story of their home was finally achieved.

It was a warm spring day and Libby was applying tape and plaster to seams in the wallboard, while 5-year-old Christy and 3-year-old Callie played upstairs. Situating the two little girls in the hallway allowed Libby to duck around the new stairway to check on the girls every few minutes. Suddenly, Libby's work stopped abruptly when Christy screamed, "MOM! Callie is throwing up!"

Libby raced up the stairs to find Callie violently vomiting. Her little girl wobbled toward her on unsteady legs, as though drunk. As Libby watched in horror, Callie fell to the ground, twitching and writhing in a seizure, and Christy dissolved into tears. "Christy, honey, go get Mommy's purse on the kitchen table," Libby gasped as she scooped up the little girl. "We have to take Callie to the hospital."

CHAPTER CONCEPTS

2.1 From Atoms to Molecules
All matter is composed of atoms, which react with one another to form molecules.

2.2 Water and Living Things
The properties of water make life, as we know it, possible. Living things are affected adversely by water that is too acidic or too basic.

2.3 Molecules of Life
Carbohydrates, lipids, proteins, and nucleic acids are macromolecules with specific functions in cells.

2.4 Carbohydrates
Glucose is blood sugar, and humans store glucose as glycogen. Cellulose is plant material that is a source of fiber in the diet.

2.5 Lipids
Fats and oils, which provide long-term energy, differ by consistency and can have a profound effect on our health. Other lipids function differently in the body.

2.6 Proteins
Proteins have numerous and varied functions in cells. Their shape suits their function.

2.7 Nucleic Acids
DNA that makes up our genes and RNA that serves as a helper to DNA are nucleic acids. ATP releases energy used by the cell to do metabolic work.

2.1 From Atoms to Molecules

Matter refers to anything that takes up space and has mass. It is helpful to remember that matter can exist as a solid, a liquid, or a gas. Then we can realize that not only are we humans composed of matter, but so are the water we drink and the air we breathe.

Elements

An **element** is one of the basic building blocks of matter; an element cannot be broken down by chemical means. Considering the variety of living and nonliving things in the world, it's remarkable that there are only 92 naturally occurring elements. It is even more surprising that over 90% of the human body is composed of just four elements: carbon, nitrogen, oxygen, and hydrogen. Even so, other elements, such as iron, are important to our health. Iron deficiency anemia results when the diet doesn't contain enough iron for the making of hemoglobin. Hemoglobin serves an important function in the body because it transports oxygen to our cells.

Every element has a name and a symbol. For example, carbon has been assigned the atomic symbol C, and iron has been assigned the symbol Fe. Some of the symbols we use for elements are derived from Latin. For example, the symbol for sodium is Na because *natrium,* in Latin, means sodium. Likewise, *ferrum* means iron, and Fe is the symbol for iron. Chemists arrange the elements in a *periodic* table, so named because all the elements in a column show *periodicity*. To show periodicity means that each column of elements behaves similarly during chemical reactions. For example, all the elements in column VII (7) undergo the same type of chemical reactions, for reasons we will soon explore. Figure 2.1 shows only some of the first 36 elements in the periodic table, but a complete table is available in an appendix at the end of the book. As you study each box in the table, note that the atom's symbol is in the center. The number above the symbol is the **atomic number.** The elements in each row are arranged in order, according to increasing atomic number. Finally, the number below each symbol is that element's **atomic mass.** Atomic mass is the average mass of all atoms of a particular element.

Atoms

An **atom** is the smallest unit of an element that still retains the chemical and physical properties of the element. The same name is given to the element and the atoms of the element. While it is possible to split an atom, an atom is the smallest unit to enter into chemical reactions. For our purposes, it is satisfactory to think of each atom as having a central **nucleus** and pathways about the nucleus called **orbitals.** Even though an atom is extremely small, it contains even smaller subatomic particles. The subatomic particles, called **protons** and **neutrons,** are located in the nucleus, and **electrons** orbit about the nucleus in the orbitals (Fig. 2.2). Most of an atom is empty space. If we could draw an atom

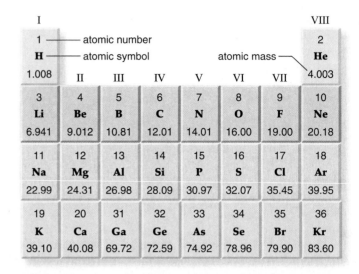

Figure 2.1   **What are the numbers on a periodic table, and what does the letter symbol stand for?**
The number on the top of each square is the atomic number, which increases from left to right. The letter symbols represent each element and are sometimes abbreviations of Greek or Latin names. Below the symbol is the value for atomic mass. A complete periodic table can be found in the Appendix.

the size of a football stadium, the nucleus would be like a gumball in the center of the field, and the electrons would be tiny specks whirling about in the upper stands.

Protons carry a positive (+) charge, and electrons have a negative (−) charge. To be electrically neutral, an atom of any given element must have the same numbers of protons and electrons, so that the positive and negative charges are balanced. The atomic number (the number written *above*

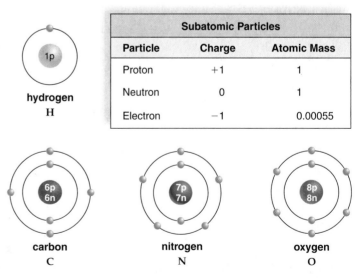

Figure 2.2   **What do scientists think a single atom looks like?**
The models show the arrangement of protons and neutrons in the nucleus. Electrons circle the nucleus in orbitals. The orbitals are grouped into shells. Protons, neutrons, and electrons differ between hydrogen, carbon, nitrogen, and oxygen.

each chemical symbol in Fig. 2.1) tells how many protons—and therefore how many electrons—an atom has when it is electrically neutral. For example, the atomic number of carbon is six; therefore, a neutral atom of carbon has six protons and six electrons.

How many electrons are there in each orbital of an atom? Any given orbital can hold only 2 electrons. Each orbital is filled with electrons according to its energy level, with the lowest energy orbitals filled first (just like the lowest seats in the football stadium fill first, so spectators can get a good look at the team!). The innermost orbital has the lowest energy level and holds the first two electrons. After that, orbitals are arranged into groups called shells. The first shell has only one orbital. After that, each shell for the first 18 atoms illustrated in Figure 2.1 can hold up to eight electrons. Using this information, we can determine how many electrons are in the outer shells of the atoms shown in Figure 2.2. Hydrogen (H) has only one orbital that contains one electron. Helium fills the first orbital—the first shell—with its two electrons. Carbon (atomic number of six) has two shells and the outer shell has four electrons. Nitrogen (atomic number of seven) has two shells and the outer shell has five electrons. Oxygen (atomic number of eight) has two shells and the outer shell has six electrons. Elements with greater than 18 electrons possess additional shells, and thus can accommodate additional electrons.

The **mass** of an atom represents its quantity of matter. The subatomic particles are so light that their mass is indicated by special designations called atomic mass units (abbreviated amu.) Protons and neutrons are each assigned one atomic mass unit, while electrons have an exceedingly small mass unit (0.00055 amu.) The sum of protons and neutrons in the atom's nucleus is called the **mass number.** (Don't confuse an element's *mass number* with its *atomic mass,* the number below each symbol in Fig. 2.1.) For example, the mass number for one form of carbon is 6 protons + 6 neutrons = 12. The mass number is written as a superscript before the atomic symbol, so the symbol for carbon with a mass number of twelve is ^{12}C.

Isotopes

Isotopes of the same type of atom have the same number of protons (and thus, the same atomic number) but different numbers of neutrons. Therefore, their mass numbers are different. For example, the element carbon 12 (^{12}C) has six neutrons, carbon 13 (^{13}C) has seven neutrons, and carbon 14 (^{14}C) has eight neutrons. You can determine the number of neutrons for an isotope by subtracting the atomic number (it's six for carbon, remember?) from the mass number.

Unlike the other two isotopes of carbon, carbon 14 is unstable and breaks down over time. As carbon 14 decays, it releases various types of energy in the form of rays and subatomic particles, and therefore it is a **radioisotope.** The radiation given off by radioisotopes can be detected in various ways. You may be familiar with the use of a Geiger counter to detect radiation.

Low Levels of Radiation

The importance of chemistry to biology and medicine is nowhere more evident than in the many uses of radioisotopes. A radioisotope behaves the same chemically as the stable isotopes of an element. This means that you can put a small amount of radioisotope in a sample and it becomes a **tracer,** by which to detect molecular changes.

Specific tracers are used in imaging the body's organs and tissues. For example, after a patient drinks a solution containing a minute amount of iodine 131, it becomes concentrated in the thyroid—the only organ to take up iodine to make the hormone thyroxine. A subsequent image of the thyroid indicates whether it is healthy in structure and function (Fig. 2.3a). Positron-emission tomography (PET) is a way to determine the comparative activity of tissues. Radioactively labeled glucose, which emits a subatomic particle known as a positron, can be injected into the body. The radiation given off is detected by sensors and analyzed by a computer. The result is a color image that shows which tissues took up glucose and are metabolically active (Fig. 2.3b). A PET scan of the brain can help diagnose a brain tumor, Alzheimer disease, epilepsy, or whether a stroke has occurred.

High Levels of Radiation

Radioactive substances in the environment can harm cells, damage DNA, and cause cancer. The release of radioactive

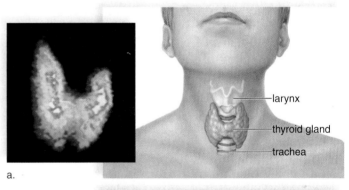

larynx

thyroid gland

trachea

a.

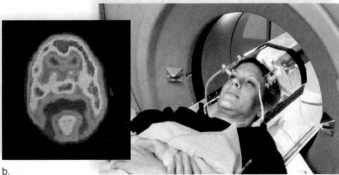

b.

Figure 2.3   Are there medical uses for low-level radiation?
a. The missing area (upper right) in this thyroid scan indicates the presence of a tumor that does not take up radioactive iodine. **b.** A PET (positron-emission tomography) scan reveals which portions of the brain are most active (red surrounded by light green) .

particles following a nuclear power plant accident can have far-reaching and long-lasting effects on human health. However, the effects of radiation can also be put to good use. Radiation from radioisotopes has been used for many years to kill bacteria and viruses, and sterilizing medical and dental products. More recently, the technology has evolved for use in food safety as well. Food can be sterilized, without irradiating the food.

Radiation can be used to ensure public safety against bacterial infection. In the wake of the terrorist attacks of 9/11, radiation was used for a time to sterilize the U.S. mail, because of fear of contamination by possible pathogens such as anthrax bacteria.

The ability of radiation to kill cells is often applied to cancer cells. Radioisotopes can be introduced into the body in a way that allows radiation to destroy only cancer cells, with little risk to the rest of the body (Fig. 2.4). Another form of high-energy radiation, X-rays, can be used for medical diagnosis and cancer therapy, as described in the Historical Focus on page 23.

Molecules and Compounds

Atoms often bond with one another to form a chemical unit called a **molecule.** A molecule can contain atoms of the same type, as when an oxygen atom joins with another oxygen atom to form oxygen gas. Or, the atoms can be different, as when an oxygen atom joins with two hydrogen atoms to form water. When the atoms are different, a **compound** is formed.

Two types of bonds join atoms: the ionic bond and the covalent bond.

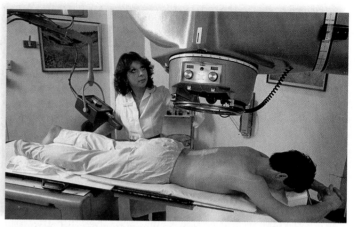

Figure 2.4   Is high-level radiation always dangerous?
Radiation causes cancer, but it can also be used to cure cancer.

Ionic Bonding

Atoms with more than one shell are most stable when the outer shell contains eight electrons. During an ionic reaction, atoms give up or take on an electron(s) to achieve a stable outer shell.

Figure 2.5 depicts a reaction between a sodium (Na) atom and a chlorine (Cl) atom. Sodium, with one electron in the outer shell, reacts with a single chlorine atom. Why? Once the reaction is finished and sodium loses one electron to chlorine, its outer shell will have eight electrons. Similarly, a chlorine atom, which has seven electrons already, needs to acquire only one more electron to have a stable outer shell.

Figure 2.5   What does it mean to form an *ionic bond* between atoms in a compound?
a. During the formation of sodium chloride, an electron is transferred from the sodium atom to the chlorine atom. At the completion of the reaction, each atom has eight electrons in the outer shell, but each also carries a charge as shown. **b.** In a sodium chloride crystal, ionic bonding between Na^+ and Cl^- causes the ions to form a three-dimensional lattice configuration in which each sodium ion is surrounded by six chloride ions and each chloride ion is surrounded by six sodium ions.

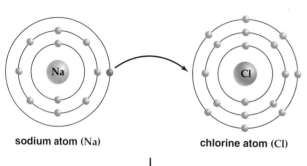

sodium atom (Na)   chlorine atom (Cl)

+   −

sodium ion (Na^+)   chloride ion (Cl^-)

sodium chloride (NaCl)

a.

Na^+ Cl^-

b.

Historical Focus

Serendipity: Wilhelm Roentgen

Serendipity is a word that means "a talent for making desirable discoveries by accident," and it certainly describes the work of Wilhelm Konrad Roentgen (1845–1923). A physicist by training, Roentgen was fascinated by the Crookes tube, a vacuum-sealed glass tube. These gadgets emitted beams of electrons (negatively charged particles) when charged with electricity. Working alone one night in his laboratory, Roentgen noticed that another invisible energy caused a fluorescent cardboard screen to glow brightly. With increasing excitement, Roentgen shone the beam through all sorts of objects—wood, decks of playing cards, and metal sheets. None of these objects could completely block the beam. The metal sheets did the best job of partially blocking the beam. Finally, Roentgen held up a piece of lead into the beam, showing that lead was the only metal to completely block the mystery rays.

Serendipitous, Roentgen's fingers got in the way. It was only then that the real possibilities for X-rays (so-called by Roentgen, because to him they were a mystery) became apparent. His hand was illuminated by the rays. Within the outline of his hand, the bones were visible, because bone has a greater density than surrounding tissue. One last experiment remained. Roentgen asked his wife to place her hand on a piece of photographic film. He directed the rays on the hand. When the film was developed, there was a complete outline of the hand and all its bones, with her wedding ring visible on her ring finger.

Roentgen's discovery earned the Nobel Prize for Physics in 1901. Later researchers discovered that when a patient swallowed a barium solution, the X-rays could be used to illuminate the hollow organs with no harm to the patient. During World War I, X-rays were used extensively to diagnose battlefield injuries, and their many applications in medicine and health care were soon apparent. Yet early researchers

Wilhelm Konrad Roentgen (1845–1923).

also noticed that heavy exposure to X-rays was dangerous. Large doses caused radiation burns and the development of cancer. However, X-rays were used carelessly and frivolously until the 1950's—an X-ray machine was even used in shoe stores to ensure that shoes fit properly!

The discovery of X-rays has lead to ever more complex and useful technologies. A CT scan is a series of X-rays, taken sequentially and then restructured by a computer. In addition, by showing that body tissues have different densities, X-ray technology undoubtedly inspired later scientists. Using different energy sources, magnetic resonance imaging (MRI, which uses radio waves and powerful magnets) and ultrasound (which uses sound waves) were derived from the talent (and luck!) of Wilhelm Roentgen.

Ions are particles that carry either a positive (+) or negative (−) charge. When the reaction between sodium and chlorine is finished, the sodium ion carries a positive charge because it now has one less electron than protons, and the chloride ion carries a negative charge because it now has one more electron than protons:

Sodium Ion (Na^+)	Chloride Ion (Cl^-)
11 protons (+)	17 protons (+)
10 electrons (−)	18 electrons (−)
One (+) charge	One (−) charge

The attraction between oppositely charged sodium ions and chloride ions forms an **ionic bond.** The resulting compound, sodium chloride, is table salt, which we use to enliven the taste of foods.

In contrast to sodium, why would calcium, with two electrons in the outer shell, react with two chlorine atoms? Whereas calcium needs to lose two electrons, each chlorine, with seven electrons already, requires only one more electron to have a stable outer shell. The resulting salt ($CaCl_2$) is called calcium chloride. Calcium chloride is used as a deicer in Northern climates.

The balance of various ions in the body is important to our health. Too much sodium in the blood can contribute to high blood pressure. Calcium deficiency leads to rickets (a bowing of the legs) in children. Too much or too little potassium results in heartbeat irregularities and can be fatal. Bicarbonate, hydrogen, and hydroxide ions are all involved in maintaining the acid-base balance of the body (see pages 28–29).

Covalent Bonding

Atoms share electrons in **covalent bonds.** The overlapping outermost shells in Figure 2.6 indicate that the atoms are sharing electrons. Just as two hands participate in a handshake, each atom contributes one electron to the shared pair. These electrons spend part of their time in the outer shell of each atom; therefore, they are counted as belonging to both bonded atoms.

Double and Triple Bonds Besides a single bond, in which atoms share only a pair of electrons, a double or a triple bond can form. In a double bond, atoms share two pairs of electrons, and in a triple bond, atoms share three pairs of electrons between them. For example, in Figure 2.6b, each oxygen atom (O) requires two more electrons to achieve a total of eight electrons in the outer shell. Four electrons are placed in the outer overlapping shells in the diagram.

Structural and Molecular Formulas Covalent bonds can be represented in a number of ways. In contrast to the diagrams in Figure 2.6, structural formulas use straight lines to show the covalent bonds between the atoms. Each line represents

a pair of shared electrons. Molecular formulas indicate only the number of each type of atom making up a molecule. A comparison follows:

Structural formula: H—O—H, O=O

Molecular formula: H_2O, O_2

What would be the structural and molecular formulas for carbon dioxide? Carbon, with four electrons in the outer shell, requires four more electrons to complete its outer shell. Each oxygen, with six electrons in the outer shell, needs only two electrons to complete its outer shell. Therefore, carbon shares two pairs of electrons with each oxygen atom, and the formulas are as follows:

Structural formula: O=C=O

Molecular formula: CO_2

> ### Check Your Progress 2.1
> 1. **How is an atom organized?**
> 2. **One isotope of calcium is ^{40}Ca, while a second is ^{48}Ca. How many neutrons are found in each isotope? (Hint: The atomic number of calcium is 20.)**
> 3. **What are radioisotopes, and how can they benefit humans?**
> 4. **What are two basic types of bonds formed between atoms to form molecules?**

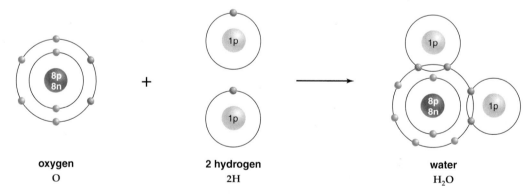

oxygen
O

2 hydrogen
2H

water
H_2O

a. When an oxygen and two hydrogen atoms covalently bond, water results.

Figure 2.6 What does it mean to form a *covalent bond* between atoms in a compound?
An atom can fill its outer shell by sharing electrons. To determine this, it is necessary to count the shared electrons as belonging to both bonded atoms. Hydrogen is most stable with two electrons in the outer shell; oxygen is most stable with eight electrons in the outer shell. Therefore, the molecular formula for water is **(a)** H_2O, and for oxygen gas it is **(b)** O_2.

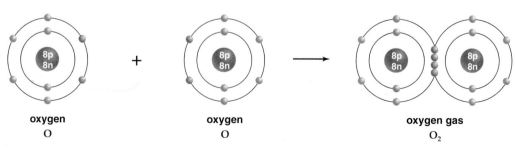

oxygen
O

oxygen
O

oxygen gas
O_2

b. When two oxygen atoms covalently bond, oxygen gas results.

2.2 **Water and Living Things**

Water is the most abundant molecule in living organisms, usually making up about 60–70% of the total body weight. Further, the physical and chemical properties of water make life as we know it possible.

In water, the electrons spend more time circling the oxygen (O) atom than the hydrogens. This is true because oxygen, the larger atom, has a greater ability to attract electrons than do the smaller hydrogen (H) atoms. The negatively charged electrons are closer to the oxygen atom, so the oxygen atom becomes slightly negative. In turn, the hydrogens are slightly positive. Therefore, water is a **polar** molecule; the oxygen end of the molecule has a slight negative charge (δ^-), and the hydrogen end has a slight positive charge (δ^+):

The diagram on the left shows a structural formula of water, and the one on the right is called a space-filling model.

Hydrogen Bonds

A **hydrogen bond** occurs whenever a covalently bonded hydrogen is slightly positive and attracted to a negatively charged atom some distance away. A hydrogen bond is represented by a dotted line because it is relatively weak and can be broken rather easily.

In Figure 2.7, you can see that each hydrogen atom, being slightly positive, bonds to the slightly negative oxygen atom of another water molecule.

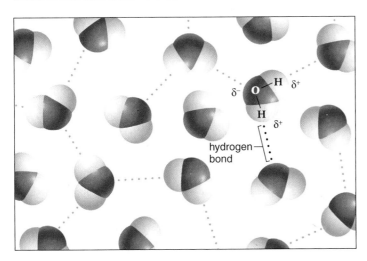

Figure 2.7 **What happens when the electrons in water (H₂0) spend more time circling the oxygen atom?**
Water is a polar molecule because electrons are not equally shared. Electrons move closer to oxygen, creating a partial negative charge, while hydrogen has a partial positive charge. Partial charges allow hydrogen bonds (dotted lines) between water molecules.

Properties of Water

Water molecules are cohesive, meaning that they cling together, because of their polarity and hydrogen bonding. Polarity and hydrogen bonding cause water to have many characteristics beneficial to life.

1. Water is a liquid at room temperature. Therefore, we are able to drink it, cook with it, and bathe in it.

Compounds with a low molecular weight are usually gases at room temperature. For example, oxygen (O_2), with a molecular weight of 32, is a gas; but water, with a molecular weight of 18, is a liquid. The hydrogen bonding between water molecules keeps water a liquid and not a gas at room temperature. Water does not boil and become a gas until 100°C, one of the reference points for the Celsius temperature scale (see inside back cover). Without hydrogen bonding between water molecules, our body fluids—and indeed our bodies—would be gaseous!

2. The temperature of liquid water rises and falls slowly, preventing sudden or drastic changes.

The many hydrogen bonds that link water molecules cause water to absorb a great deal of heat before it boils (Fig. 2.8*a*). A **calorie** of heat energy raises the temperature of 1 g of water 1°C.[1] This is about twice the amount of heat required for other covalently bonded liquids. On the other hand, water holds heat, and its temperature falls slowly. Therefore, water protects us and other organisms from rapid temperature changes and helps us maintain our normal internal temperature. This property also allows great bodies of water, such as oceans, to maintain a relatively constant temperature. Water is a good temperature buffer.

3. Water has a high heat of vaporization, keeping the body from overheating.

It takes a large amount of heat to vaporize water; that is, change water to steam (Fig. 2.8*a*). (Converting 1 g of the hottest water to steam requires an input of 540 calories of heat energy.) This property of water helps moderate the Earth's temperature so that life can continue to exist. Also, in a hot environment, most mammals sweat, and the body cools as body heat is used to evaporate sweat, mostly liquid water (Fig. 2.8*b*).

4. Frozen water is less dense than liquid water, so ice floats on water.

Remarkably, water is more dense at 4°C than at 0°C. Most substances contract when they solidify. By contrast, water expands when it freezes because in ice, water molecules form a lattice in which the hydrogen bonds are farther apart than in liquid water (Fig. 2.8*c*). This is why cans of soda burst when placed in a freezer or why frost heaves make northern roads bumpy in the winter. Also, because ice is

[1]Be aware that a calorie—spelled with a lower case "c"—is not the same as a capital C Calorie, used to measure the heat energy of food. A Calorie is 1,000 calories, or one kilocalorie. A little crazy, isn't it?

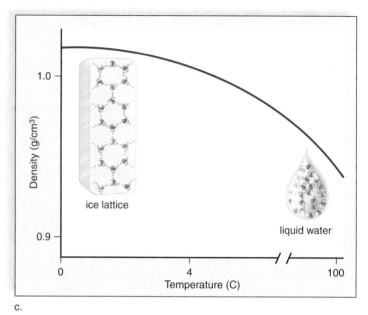

a.   b.   c.

Figure 2.8 **How do water's structure and characteristics affect its behavior in the environment?**
a. Water becomes gaseous at 100°C. If it was a gas at a lower temperature, life could not exist. **b.** Body heat vaporizes sweat. In this way, bodies cool when the temperature rises. **c.** Ice is less dense than water, so it forms a protective layer on top of ponds during the winter.

lighter than cold water, bodies of water freeze from the top down. The ice acts as an insulator to prevent the water below it from freezing. Thus, aquatic organisms are protected and have a better chance of surviving the winter.

5. Water molecules are cohesive, yet still flow freely. Therefore, liquids fill vessels, such as blood vessels.

Water molecules cling together because of hydrogen bonding. You've seen that cohesive property of water in your everday life if you've ever washed drinking glasses, and then stacked them while wet. Big mistake! The glasses will stick together as if glued, because of hydrogen bonding in the film of water on the glass. Yet while its mole-

cules cling to each other, water still is able to flow freely. Water's cohesive property allows dissolved and suspended molecules to be evenly distributed throughout a system. Therefore, water is an excellent transport medium. Within our bodies, blood fills our arteries and veins because it is 92% water. After blood transports oxygen and nutrients to cells, these molecules are used to produce cellular energy. Blood also removes wastes, such as carbon dioxide, from cells.

6. Water is a solvent for polar (charged) molecules, and thereby facilitates chemical reactions both outside and inside our bodies.

CASE STUDY IN THE EMERGENCY ROOM

"We're starting an IV solution for Callie right away," the physician's assistant (P.A.) on duty explained. "We'll put the fluid right into a vein to replace the fluid and blood sugar that Callie lost by vomiting. Little ones often get dehydrated very quickly by vomiting. Just being dehydrated can cause small children to have seizures. And see how Callie is in a stupor—sort of half awake? That can be caused by dehydration, too."

"It's important to remember that adults are 60% water by weight," the P.A continued. "In kids, the percentage is even higher, and

children are more sensitive to fluid loss. That fluid is supposed to fill Callie's blood vessels. Because Callie is dehydrated, her blood pressure is falling. There isn't enough fluid to fill up the vessels and create the pressure. Her little heart is racing, trying to get enough blood to her tissues. This IV I'm starting ought to perk her right back up. I think she probably just has a stomach bug, but the doctor is going to order some blood tests just to make sure." Frightened, Libby and Matt hastily signed the necessary forms.

Have You Ever Wondered...

How do lungs stay open and keep from collapsing?

Our lives depend on water's cohesive property. A thin film of water coats the surface of the lungs and the inner chest wall. This film allows the lungs to stick to the chest wall, keeping the lungs open so we can breathe.

When ions and molecules disperse in water, they move about and collide, allowing reactions to occur. Therefore, water is a solvent that facilitates chemical reactions. For example, when a salt such as sodium chloride (NaCl) is put into water, the negative ends of the water molecules are attracted to the positively charged sodium ions, and the positive ends of the water molecules are attracted to the negatively charged chloride ions. This causes the sodium ions and the chloride ions to separate and to dissolve in water:

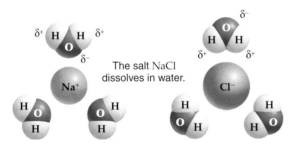

The salt NaCl dissolves in water.

Ions and molecules that interact with water are said to be **hydrophilic** (that is, "water-loving"). Nonionized and nonpolar molecules that do not interact with water are said to be **hydrophobic** (that is, "water-fearing").

Acids and Bases

When water molecules dissociate (break up), they release an equal number of hydrogen ions (H^+) and hydroxide ions (OH^-):

$$H\text{—}O\text{—}H \rightleftharpoons H^+ + OH^-$$
water   hydrogen ion   hydroxide ion

Only a few water molecules at a time dissociate, and the actual number of H^+ or OH^- is 10^{-7} moles/liter. A **mole** is a unit of scientific measurement for atoms, ions, and molecules.[2]

Acidic Solutions (High H⁺ Concentrations)

Lemon juice, vinegar, tomatoes, and coffee are all acidic solutions. What do they have in common? **Acids** are substances that dissociate in water, releasing hydrogen ions (H^+). For example, an important inorganic acid is hydrochloric acid (HCl), which dissociates in this manner:

$$HCl \longrightarrow H^+ + Cl^-$$

Dissociation is almost complete. Therefore, HCl is called a strong acid. If hydrochloric acid is added to a beaker of water, the number of hydrogen ions (H^+) increases greatly. Hydrochloric acid is produced by the stomach, and aids in food digestion.

Basic Solutions (Low H⁺ Concentrations)

Milk of magnesia and ammonia are commonly known basic substances. **Bases** are substances that either take up hydrogen ions (H^+) or release hydroxide ions (OH^-). For example, an important base is sodium hydroxide (NaOH), which dissociates in this manner:

$$NaOH \longrightarrow Na^+ + OH^-$$

Dissociation is almost complete. Thus, sodium hydroxide is called a strong base. If sodium hydroxide is added to a beaker of water, the number of hydroxide ions increases. Sodium hydroxide is also called lye and is contained in many drain-cleaning products.

You should not taste a strong acid or base because they are destructive to cells. Any container of household cleanser, such as ammonia, has a poison symbol and carries a strong warning not to ingest the product.

pH Scale

The **pH scale**[3] is used to indicate the acidity and basicity (alkalinity) of a solution. Pure water with an equal number of hydrogen ions (H^+) and hydroxide ions (OH^-) has a pH of exactly 7.

The pH scale was devised to simplify discussion of the hydrogen ion concentration [H^+] and consequently of the hydroxide ion concentration [OH^-]. It eliminates the use of cumbersome numbers. To understand the relationship between hydrogen ion concentration and pH, consider the following:

[H^+] (moles per liter)		pH
0.000001	= 1×10^{-6}	6
0.0000001	= 1×10^{-7}	7
0.00000001	= 1×10^{-8}	8

You'll notice that to determine pH, each number is first expressed in scientific notation. The negative value of the exponent (remember, the exponent is the superscript number) equals pH.

---

[2] In chemistry, a mole is defined as the amount of matter that contains as many objects (atoms, molecules, ions) as the number of atoms in exactly 12 grams of ^{12}C.

[3] pH is defined as the negative log of the hydrogen ion concentration [H^+]. A log is the power to which 10 must be raised to produce a given number.

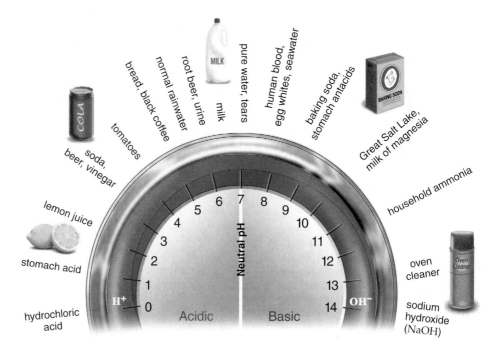

Figure 2.9   What does the abbreviation "pH" mean?
The dial of this pH meter indicates that the pH of solutions ranges from 0 to 14, with 0 being the most acidic and 14 being the most basic. A solution at pH 7 (neutral pH) has equal amounts of hydrogen ions (H^+) and hydroxide ions (OH^-). An acidic pH has more H^+ than OH^- and a basic pH has more OH^- than H^+. The pH of familiar solutions can be seen on the meter.

So for a solution of 0.000001 moles per liter of hydrogen ions, the scientific notation is 1×10^{-6}, and the pH is $-[-6]$ or pH 6.

Of the two values above and below pH 7, which one indicates a higher hydrogen ion concentration than pH 7 and, therefore, refers to an acidic solution? A number with a smaller negative exponent indicates a greater quantity of hydrogen ions (H^+) than one with a larger negative exponent. Therefore, the pH 6 solution is an acidic solution.

Basic solutions have fewer hydrogen ions (H^+) compared with hydroxide ions. Of the three values, pH 8 solution is a basic solution because it indicates a lower hydrogen ion concentration [H^+] (greater hydroxide ion concentration) than the pH 7 solution.

The pH scale (Fig. 2.9) ranges from 0 to 14. As we move toward a higher pH, each unit has 10 times the basicity of the previous unit, and as we move toward a lower pH, each unit has 10 times the acidity of the previous unit.

Buffers

In living things, the pH of body fluids needs to be maintained within a narrow range or else health suffers. Normally, pH stability is possible because the body and the environment have **buffers** to prevent pH changes. A problem arises when precipitation in the form of rain or snow becomes so acidic, the environment runs out of natural buffers in the soil or water. Rain normally has a pH of about 5.7, but rain with a pH of 2.1 is recorded every year in the northeastern Appalachian Mountains. Rain becomes acidic because the burning of gasoline emits sulfur dioxides (SO_X) and nitrogen oxides (NO_X) into the atmosphere. They combine with water to produce sulfuric acid (H_2SO_4) and nitric acid (HNO_3). Acid deposition destroys statues and kills forests (Fig. 2.10). It also leads to fish kills in lakes and streams.

Buffers help keep the pH within normal limits because they are chemicals or combinations of chemicals that take up excess hydrogen ions (H^+) or hydroxide ions (OH^-). For example, carbonic acid (H_2CO_3) is a weak acid that minimally dissociates and then re-forms in the following manner:

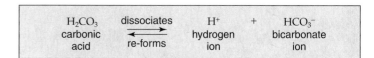

H_2CO_3	dissociates	H^+	+	HCO_3^-
carbonic acid	⇌ re-forms	hydrogen ion		bicarbonate ion

The pH of our blood when we are healthy is always about 7.4—just slightly basic (alkaline). Blood always contains a combination of some carbonic acid and some bicarbonate

a.   b.

Figure 2.10   Why be concerned about so-called "acid rain"?
Acid rain formed during the burning of fossil fuels causes **(a)** statues to deteriorate and **(b)** trees to die.

ions. When hydrogen ions (H^+) are added to blood, the following reaction occurs:

$$H^+ + HCO_3^- \longrightarrow H_2CO_3$$

When hydroxide ions (OH^-) are added to blood, this reaction occurs:

$$OH^- + H_2CO_3 \longrightarrow HCO_3^- + H_2O$$

These reactions prevent any significant change in blood pH.

> ### Check Your Progress 2.2
> 1. What characteristics of water help support life?
> 2. A solution contains 0.001 moles per liter hydrogen ions [H^+]. What is its pH? Is it an acid or a base?
> 3. How does the hydrogen ion concentration and pH change between water, acids, and bases?
> 4. Why does the pH of the environment (soil and water) and the body rarely change?

2.3 Molecules of Life

Four categories of **organic molecules,** called carbohydrates, lipids, proteins, and nucleic acids, are unique to cells. In biology, **organic** doesn't refer to how food is grown; it refers to a molecule that contains carbon (C) and hydrogen (H) and is usually associated with living things.

Each type of organic molecule in cells is composed of subunits. When a cell constructs a **macromolecule,** a molecule that contains many subunits, it uses a **dehydration reaction,** a type of synthesis reaction. During a dehydration reaction, an —OH (hydroxyl group) and an —H (hydrogen atom), the equivalent of a water molecule, are removed as the molecule forms (Fig. 2.11a). The reaction is reminiscent of a train whose length is determined by how many boxcars it has hitched together. To break down macromolecules, the cell uses a **hydrolysis reaction** in which the components of water are added (Fig. 2.11b).

> ### Check Your Progress 2.3
> 1. What are the four classes of molecules unique to cells?
> 2. What type of reaction occurs during synthesis of macromolecules?

2.4 Carbohydrates

Carbohydrate molecules are characterized by the presence of the atomic grouping H—C—OH, in which the ratio of hydrogen atoms (H) to oxygen atoms (O) is approximately 2:1. This ratio is the same as the ratio in water (and *hydros* in Greek means "water"), so the name "hydrates of carbon" seems appropriate. **Carbohydrates,** first and foremost, function for quick and short-term energy storage in all organisms, including humans.

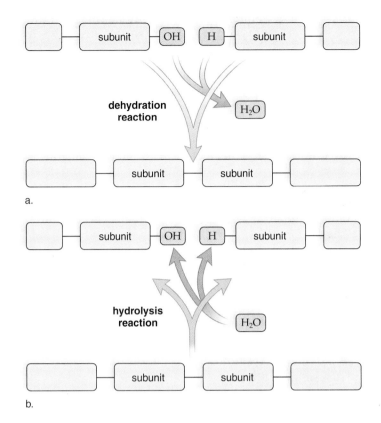

a.

b.

Figure 2.11 **How are small subunits tied together to make large macromolecules? How are macromolecules broken down? a.** Subunits bond in a dehydration reaction. Water is given off and a macromolecule forms. **b.** In the reverse reaction, macromolecules are divided into subunits by hydrolysis.

Simple Carbohydrates

If a carbohydrate is made up of just one ring, and its number of carbon atoms is low (from five to seven), it is called a simple sugar, or **monosaccharide.** The designation **pentose** means a 5-carbon sugar, and the designation **hexose** means a 6-carbon sugar. **Glucose,** the hexose our bodies use as an immediate source of energy, can be written in any one of these ways:

$$C_6H_{12}O_6$$

Other common hexoses are fructose, found in fruits, and galactose, a constituent of milk.

Have You Ever Wondered...

What happened to the paper gum wrapper you ate as a child, when you couldn't get the wrapper completely off the gum?

Paper is made from plant cellulose. Unlike cows or termites, we humans don't have the helpful bacteria that lets these critters digest cellulose. That paper you swallowed was a not-very-tasty piece of roughage that passed through the digestive tract. (By the way, the gum you swallowed back then passed through the digestive tract, too.)

A **disaccharide** (*di*, two; *saccharide*, sugar) is made by joining only two monosaccharides together by a dehydration reaction (see Fig. 2.11*a*). Maltose is a disaccharide that contains two glucose molecules:

maltose $C_{12}H_{22}O_{11}$

When our hydrolytic digestive juices break down maltose, the result is two glucose molecules. When glucose and fructose join, the disaccharide sucrose forms. Sucrose, ordinarily derived from sugarcane and sugar beets, is commonly known as table sugar.

Complex Carbohydrates (Polysaccharides)

Macromolecules such as starch, glycogen, and cellulose are **polysaccharides** that contain many glucose units.

Starch and **glycogen** are readily stored forms of glucose in plants and animals, respectively. Some of the macromolecules in starch are long chains of up to 4,000 glucose units. Starch and glycogen have slightly different structures. Starch has fewer side branches, or chains, than does glycogen. A side chain of glucose branches off from the main chain, as shown in Figures 2.12 and 2.13. Flour, used for baking and typically obtained from grinding wheat, is high in starch. Pasta and potatoes are also high-starch foods.

After we eat starchy foods, such as potatoes, bread, and cake, glucose enters the bloodstream, and normally the liver stores glucose as glycogen. The release of the hormone insulin from the pancreas promotes the storage of glucose as glycogen. In between eating, the liver releases glucose so that, normally, the blood glucose concentration is always about 0.1%.

The polysaccharide **cellulose** is found in plant cell walls. In cellulose, the glucose units are joined by a slightly different type

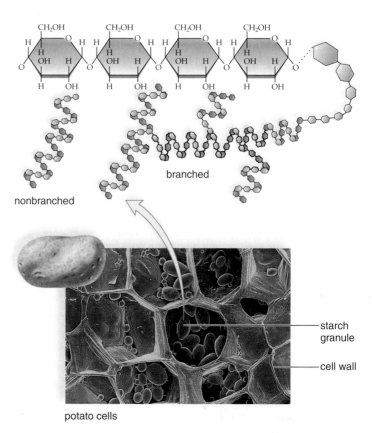

Figure 2.12   Starch is stored food for plants. What is its chemical structure?
Starch has straight chains of glucose molecules. Some chains are also branched, as indicated. The electron micrograph shows starch granules in potato cells.

of linkage than that in starch or glycogen. While this might seem to be a technicality, it is important because we are unable to digest foods containing this type of linkage; therefore, cellulose largely passes through our digestive tract as fiber, or roughage.

Low-Carb Diet

If you, or someone you know, has lost weight by following the Atkins or South Beach diet, you may think "carbs" are unhealthy and should be avoided. However, the vast majority of nutrition experts, including those from the American Heart Association and the Mayo Clinic, consistently recommend that one's diet contain 60-70% high-fiber carbohydrates. Fiber includes various nondigestible carbohydrates derived from plants. Food sources rich in fiber include beans, peas, nuts, fruits, and vegetables. Whole-grain products should be preferred over foods made from refined grains. During refinement, fiber and also vitamins and minerals are removed from grains, so that primarily starch remains. Bleached wheat (used to make white bread) and polished white rice are two examples of refined grains.

Two types of fiber are found in foods. Insoluble fiber promotes water absorption into fecal material and also adds

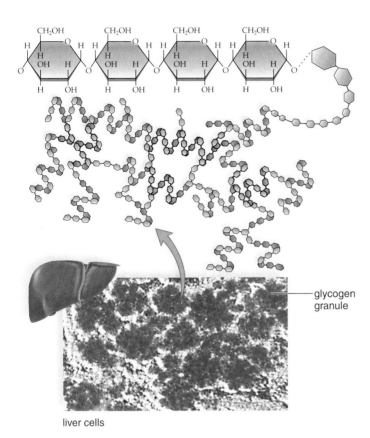

glycogen granule

liver cells

Figure 2.13 **Glycogen is stored food for humans and other mammals. Is its structure different from that of starch?** Glycogen is more branched than starch. The electron micrograph shows glycogen granules in liver cells.

bulk. Constipation is prevented, because additional bulk stimulates movements of the large intestine and additional water makes stool easier to pass. Soluble fiber combines with cholesterol in the small intestine and prevents it from being absorbed. This may be protective against heart disease. In general, Americans should just about double the amount of fiber in their diet.

Check Your Progress 2.4

1. What is the usual function of various carbohydrates in humans?
2. What is the difference in structure between a simple carbohydrate and the various complex carbohydrates?
3. Of what benefit is fiber in the diet?

2.5 Lipids

Lipids are diverse in structure and function, but they have a common characteristic: They do not dissolve in water. Their low solubility in water is due to an absence of polar groups. They contain little oxygen and consist mostly of carbon and hydrogen atoms.

Lipids contain more energy per gram than other biological molecules, and therefore, fats in animals and oils in plants function well as energy storage molecules. Others (phospholipids) form a membrane so that the cell is separated from its environment and has inner compartments as well. Steroids are a large class of lipids that includes, among other molecules, the sex hormones.

Fats and Oils

The most familiar lipids are those found in fats and oils. **Fats,** usually of animal origin (e.g., lard and butter), are solid at room temperature. **Oils,** usually of plant origin (e.g., corn oil and soybean oil), are liquid at room temperature. Fat has several functions in the body. It is used for long-term energy storage, insulates against heat loss, and forms a protective cushion around major organs. Steroids are formed from smaller lipid molecules and function as chemical messengers.

Emulsifiers can cause fats to mix with water. They contain molecules with a nonpolar end and a polar end. The molecules position themselves about an oil droplet so that their polar ends project outward. Now the droplet disperses in water, which means that **emulsification** has occurred. Emulsification takes place when dirty clothes are washed with soaps or detergents. Also, prior to the digestion of fatty foods, fats are emulsified by bile. The liver manufactures bile and the gallbladder stores it.

Fats and oils form when one glycerol molecule reacts with three fatty acid molecules (Fig. 2.14). A fat is sometimes called a **triglyceride** because of its three-part structure, or the term neutral fat can be used because the molecule is nonpolar and carries no charges.

Waxes are molecules made up of one fatty acid combined with another single organic molecule, usually an alcohol (chemists refer to "alcohols" as an entire group of molecules that includes drinking alcohol and rubbing alcohol). Waxes prevent loss of moisture from body surfaces. Cerumen, or ear wax, is a very thick wax produced by glands lining the outer ear canal (see Chap. 14, page 317). It protects the ear canal from irritation and infection by trapping particles, bacteria and viruses. When ear wax is completely washed away by swimming or diving, the result is a painful "swimmer's ear."

Saturated, Unsaturated, and Trans-Fatty Acids

A **fatty acid** is a carbon–hydrogen chain that ends with the acidic group —COOH (Fig. 2.14, *left*). Most of the fatty acids in cells contain 16 or 18 carbon atoms per molecule, although smaller ones with fewer carbons are also known. Fatty acids are either saturated or unsaturated. **Saturated fatty acids** have no double bonds between the carbon atoms. The chain is saturated, so to speak, with all the hydrogens it can hold. **Unsaturated fatty acids** have double bonds in the carbon chain wherever the number of hydrogens is less than two per carbon.

In general, oils, present in cooking oils and bottle margarines, are liquids at room temperature because the presence of a double bond creates a bend in the fatty acid chain. Such kinks prevent close packing between the hydrocarbon chains and account for the fluidity of oils. On the other hand,

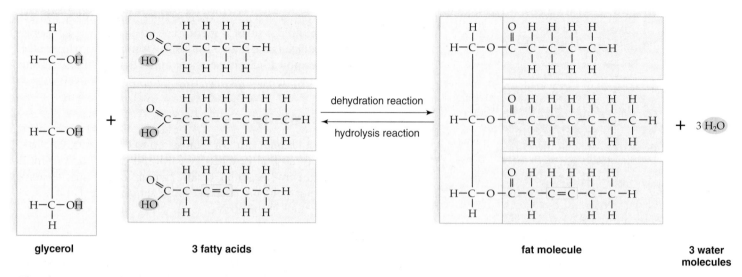

dehydration reaction
hydrolysis reaction

+ 3 H₂O

glycerol   3 fatty acids   fat molecule   3 water molecules

Figure 2.14 **What is the difference between saturated and unsaturated fats? How does the body assemble fats? How are fats digested?**
Saturated fats have no double bonds between carbons in the fatty acid. Unsaturated fats have one or more double bonds (colored yellow) in the fatty acid. Three fatty acids combine with glycerol to assemble a fat and three water molecules are given off. The reverse reaction starts digestion of fat; hydrolysis introduces water, and fatty acid-glycerol bonds are broken.

butter, which contains saturated fatty acids and no double bonds, is a solid at room temperature.

Saturated fats, in particular, contribute to the disease *atherosclerosis*. Atherosclerosis is caused by formation of lesions, or *atherosclerotic plaques,* on the inside of blood vessels. The plaques narrow blood vessel diameter, choking off the blood and oxygen supply to tissues. Atherosclerosis is the primary cause of cardiovascular disease (heart attack and stroke) in the United States. Even more harmful than naturally occurring saturated fats are the so-called **trans fats,** created artificially using vegetable oils. Trans fats are partially hydrogenated to make them semisolid. Complete hydrogenation of oils causes all double bonds to become saturated. Partial hydrogenation does not saturate all bonds. It reconfigures some double bonds, and the hydrogen atoms end up on different sides of the chain. (*Trans* in Latin means across):

Unsaturated
(oils)

Saturated
(butter)

Trans fats
(hydrogenated oils)

Trans fats are found in shortenings and solid margarines. They also occur in processed foods (snack foods, baked goods, and fried foods).

Current dietary guidelines from the American Heart Association (AHA) advise replacing trans fat with unsaturated oils. In particular, monounsaturated oils (like olive oil, with one double bond in the carbon chain) are recommended. Polyunsaturated oils (many double bonds in the carbon chain) such as corn oil, canola oil, and safflower oil also fit in the AHA guidelines.

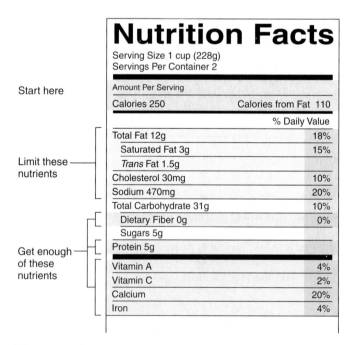

Figure 2.15 **If last night's dinner is macaroni and cheese from a box, how does that meal stack up nutritionally?**
It's pretty tasty, but mac and cheese is almost 50% fat and completely lacks the fiber so essential to a healthy diet. It is pretty high in calcium because of the cheese sauce, but also contains trans fats. There is no % daily value for trans fats, and they should be avoided in the diet. Bottom line: it is ok for an *occasional* treat!

Dietary Fat

For good health, the diet should include some fat but, for the reasons stated, the first thing to do when looking at a nutrition label is to check the total amount of fat per serving. The total recommended amount of fat in a 2,000 calorie diet is 65 g. That information results in the % daily value (DV) given in the sample nutrition label for macaroni and cheese in Figure 2.15. As of January 2006, food manufacturers are required to list the amount of trans fats greater than 0.5 g in the nutrition label for a food.

In the meantime, food manufacturers, fast-food restaurants, and national chain restaurants are starting to get the message about the harmful effects of trans fat. Proctor and Gamble, the manufacturer of Crisco shortening (used in baking), has largely replaced trans fats in its product. Fully hydrogenated cottonseed oil (to keep the product solid at room temperature) is blended with liquid soy or safflower oil (to improve texture) and only a trace of trans fat remains in the product. Likewise, Starbucks®, Denny's®, Applebee's®, McDonald's®, and other chain restaurants have replaced trans fats in the oils used for baking and frying.

Phospholipids

Phospholipids have a phosphate group (Fig. 2.16). They are constructed like fats, except that in place of the third fatty acid, there is a phosphate group or a grouping that contains both phosphate and nitrogen. These molecules are not electrically neutral, as are fats, because the phosphate and nitrogen-containing groups are ionized. They form the so-called polar head of the molecule, while the rest of the molecule becomes the hydrophobic tails. Phospholipids are the primary components of cellular membranes. They spontaneously form a *bilayer* (a sort of molecular "sandwich") in which the hydrophilic heads (the sandwich "bread") face outward toward watery solutions, and the tails (the sandwich "filling") form the hydrophobic interior (Fig. 2.16c). Remember that hydrophilic means "water-loving" and hydrophobic is "water-fearing."

Steroids

Steroids are lipids that have an entirely different structure from those of fats. Steroid molecules have a backbone of four fused carbon rings. Each one differs primarily by the attached molecules, called *functional groups*, attached to the rings. Cholesterol is a component of an animal cell's plasma membrane and is the precursor of several other steroids, such as the sex hormones estrogen and testosterone. The liver usually makes all the cholesterol the body needs. Dietary sources should be restricted because elevated levels of cholesterol, saturated fats, and trans fats are linked to atherosclerosis.

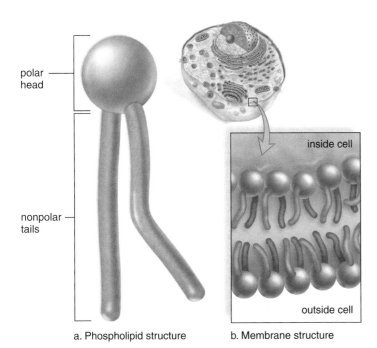

a. Phospholipid structure   b. Membrane structure

Figure 2.16 **Phospholipids make up a sandwich when they form cell membranes. What does the sandwich look like?**
a. Phospholipids are structured like fats with one fatty acid is replaced by a polar phosphate group. Therefore, the head is polar, while the tails are nonpolar. **b.** This causes the molecules to arrange themselves in a "sandwich" arrangement when exposed to water—polar phosphate groups on the outside of the layer, nonpolar lipid tails on the inside of the layer.

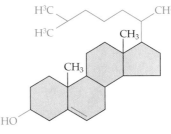

a. Cholesterol

Figure 2.17 **Can you see the difference?**
a. All steroids are made from cholesterol and have four carbon rings. Compare the structure of **(b)** testosterone and **(c)** estrogen, and notice the slight changes in their attached groups (shown in blue). What a profound difference made by a few unique molecules!

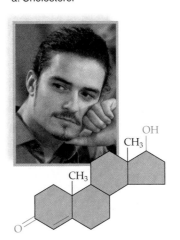

b. Testosterone

c. Estrogen

Recall that atherosclerosis is a blood vessel disease in which fatty plaques accumulate inside blood vessel linings and reduce blood flow (see page 32).

The male sex hormone, testosterone, is formed primarily in the testes, and the female sex hormone, estrogen, is formed primarily in the ovaries. Testosterone and estrogen differ only by the functional groups attached to the same carbon backbone. However, they have a profound effect on the body and the sexuality of humans and other animals (Fig. 2.17). The taking of anabolic steroids, usually to build muscle strength, is illegal because the side effects are harmful to the body (see page 270).

> **Check Your Progress 2.5**
>
> 1. a. What is the main function of fats and oils, and (b) of what are they composed?
> 2. What are the uses of phospholipids and steroids in the body?

2.6 Proteins

Proteins are of primary importance in the structure and function of cells. Here are some of their many functions in humans.

Support Some proteins are structural proteins. Keratin, for example, makes up hair and nails. Collagen lends support to ligaments, tendons, and skin.

Enzymes Enzymes bring reactants together and thereby speed chemical reactions in cells. They are specific for one particular type of reaction and only function at body temperature.

Hair is a protein.

Transport Channel and carrier proteins in the plasma membrane allow substances to enter and exit cells. Some other proteins transport molecules in the blood of animals; **hemoglobin** in red blood cells is a complex protein that transports oxygen.

Defense Antibodies are proteins. They combine with

Hemoglobin is a protein.

foreign substances, called antigens. In this way, they prevent antigens from destroying cells and upsetting homeostasis.

Hormones Hormones are regulatory proteins. They serve as intercellular messengers that influence the metabolism of cells. The hormone insulin regulates the content of glucose in the blood and in cells. The presence of growth hormone determines the height of an individual.

Motion The contractile proteins actin and myosin allow parts of cells to move and cause muscles to contract. Muscle contraction accounts for the movement of animals from place to place.

The structures and functions of vertebrate cells and tissues differ according to the type of proteins they contain. For example, muscle cells contain actin and myosin; red blood cells contain hemoglobin; and support tissues contain collagen.

Muscle contains protein.

Amino Acids: Subunits of Proteins

Proteins are macromolecules with **amino acid** subunits. The central carbon atom in an amino acid bonds to a hydrogen atom and also to three other groups of atoms. The name amino acid is appropriate because one of these groups is an —NH$_2$ (amino group) and another is a

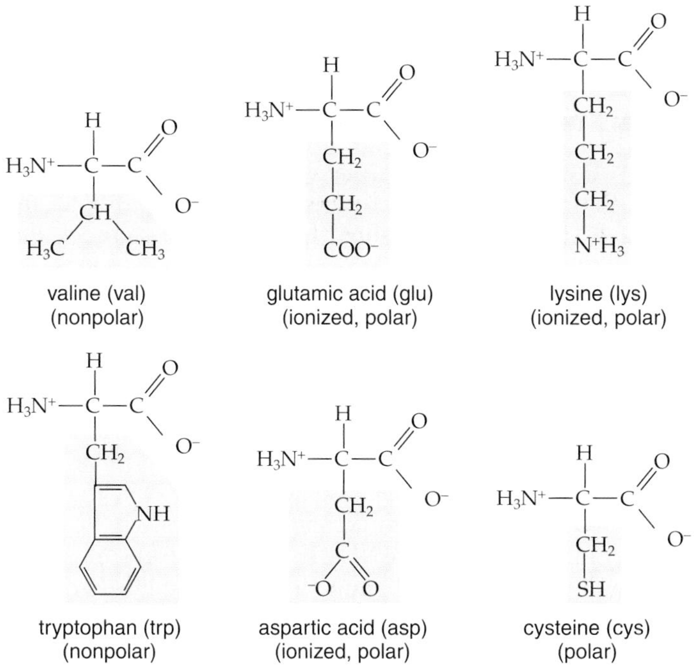

valine (val)
(nonpolar)

glutamic acid (glu)
(ionized, polar)

lysine (lys)
(ionized, polar)

tryptophan (trp)
(nonpolar)

aspartic acid (asp)
(ionized, polar)

cysteine (cys)
(polar)

Figure 2.18 **There are 20 different amino acids. How are they alike/different?**
Amino acids all have an amine group (H_3N^+), an acid group (COO^-), and an R group, all attached to the central carbon atom. The R groups (screened in blue) are all different. Some R groups are nonpolar and hydrophobic; others are polar and hydrophilic. Still others are polar and ionized.

—COOH (carboxyl group, an acid). The third group is the R group for an amino acid:

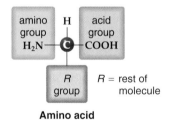

Amino acid

Amino acids differ according to their particular R group. The R groups range in complexity from a single hydrogen atom to a complicated ring compound. Some R groups are polar and some are not. Also, the amino acid cysteine ends with an —SH group, which often serves to connect one chain of amino acids to another by a disulfide bond, —S—S—. Several amino acids commonly found in cells are shown in Figure 2.18.

Peptides

Figure 2.19 shows how two amino acids join by a dehydration reaction between the carboxyl group of one and the amino group of another. The covalent bond between two amino acids is called a **peptide bond.** When three or more amino acids are linked by peptide bonds, the chain that results is called a **polypeptide.** The atoms associated with the peptide bond share the electrons unevenly because oxygen attracts electrons more than nitrogen. Therefore, the hydrogen attached to the nitrogen has a slightly positive charge (δ^+), while the oxygen has a slightly negative charge (δ^-):

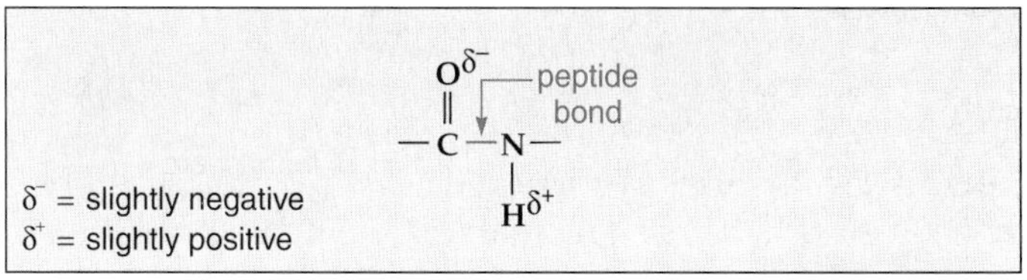

δ^- = slightly negative
δ^+ = slightly positive

Shape of Proteins

Proteins cannot function unless they have their usual shape. When proteins are exposed to extremes in heat and pH, they undergo an irreversible change in shape called **denaturation.** For example, we are all aware that the addition of vinegar (an acid) to milk causes curdling. Similarly, heating causes egg whites, which contain a protein called albumin, to coagulate. Denaturation occurs because the normal bonding between the R groups has been disturbed. Once a protein loses its normal shape, it is no longer able to perform its usual function. Researchers hypothesize that an alteration in protein organization has occurred when Alzheimer disease

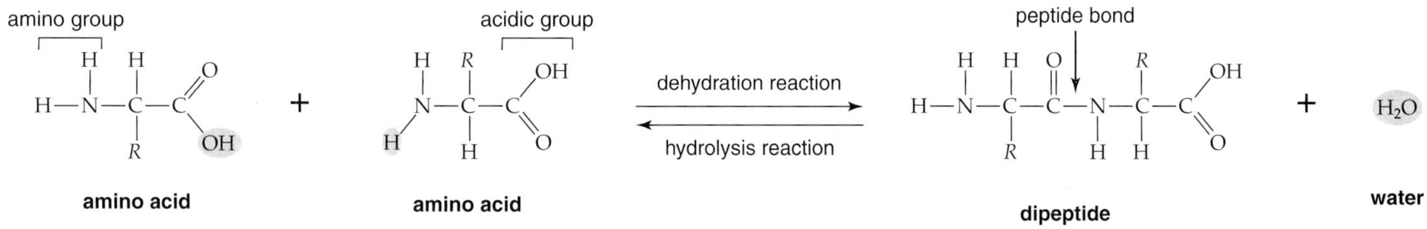

amino group   acidic group

amino acid   amino acid

dehydration reaction

hydrolysis reaction

peptide bond

dipeptide   water

Figure 2.19 **How does the body assemble proteins from amino acids? How are proteins digested?**
Amino acids join by peptide bonds using a dehydration reaction, and a water molecule is given off. In the reverse reaction, peptide bonds are broken by hydrolysis, and a water molecule is introduced.

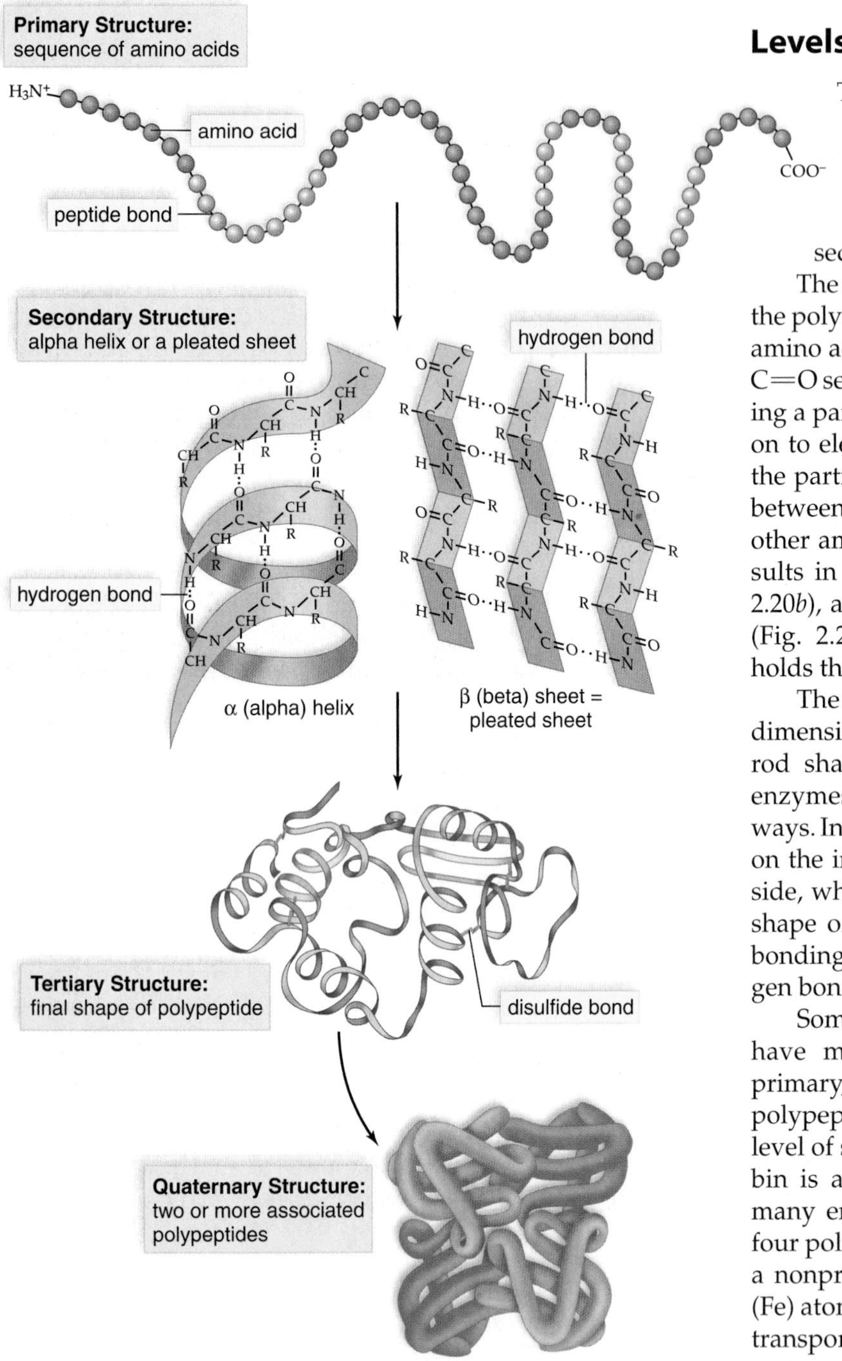

and Creutzfeldt-Jakob disease (the human form of mad cow disease) develop.

Levels of Protein Organization

The structure of a protein has at least three levels of organization and can have four levels (Fig. 2.20). The first level, called the *primary structure,* is the linear sequence of the amino acids joined by peptide bonds. Each particular polypeptide has its own sequence of amino acids.

The *secondary structure* of a protein comes about when the polypeptide takes on a certain orientation in space. Once amino acids are assembled into a polypeptide, the resulting $C=O$ section between amino acids in the chain is polar, having a partial negative charge. (Remember that oxygen holds on to electrons longer than carbon, and that's what causes the partial negative charge.) Hydrogen bonding is possible between the $C=O$ of one amino acid and the N—H of another amino acid in a polypeptide. Coiling of the chain results in an α (alpha) helix, or a right-handed spiral (Fig. 2.20*b*), and a folding of the chain results in a pleated sheet (Fig. 2.20*c*). Hydrogen bonding between peptide bonds holds the shape in place.

The *tertiary structure* of a protein is its final three-dimensional shape. In muscles, myosin molecules have a rod shape ending in globular (globe-shaped) heads. In enzymes, the polypeptide bends and twists in different ways. In most enzymes, the hydrophobic portions are packed on the inside, and the hydrophilic portions are on the outside, where they can make contact with water. The tertiary shape of a polypeptide is maintained by various types of bonding between the *R* groups; covalent, ionic, and hydrogen bonding all occur.

Some proteins have only one polypeptide, and others have more than one polypeptide, each with its own primary, secondary, and tertiary structures. These separate polypeptides are arranged to give these proteins a fourth level of structure, termed the *quaternary structure.* Hemoglobin is a complex protein having a quaternary structure; many enzymes also have a quaternary structure. Each of four polypeptides in hemoglobin are tightly associated with a nonprotein *heme* group. A heme group contains an iron (Fe) atom that binds to oxygen, and in that way, hemoglobin transports O_2 to the tissues.

Primary Structure:
sequence of amino acids

amino acid

peptide bond

Secondary Structure:
alpha helix or a pleated sheet

hydrogen bond

hydrogen bond

α (alpha) helix

β (beta) sheet =
pleated sheet

Tertiary Structure:
final shape of polypeptide

disulfide bond

Quaternary Structure:
two or more associated polypeptides

Figure 2.20 **Are all proteins structurally similar?**
The structure of proteins can differ significantly. Primary structure, the sequence of amino acids, determines secondary and tertiary structure. Quaternary structure is created by assembling smaller proteins into a large structure.

Check Your Progress 2.6

1. What are the major functions of proteins in organisms?
2. How does an amino acid get its name?
3. How does the shape of a protein relate to its function?

2.7 Nucleic Acids

The two types of nucleic acids are **DNA (deoxyribonucleic acid)** and **RNA (ribonucleic acid)** (Fig. 2.21). Early investigators called them nucleic acids because they were first detected in the nucleus of cells. The discovery of the structure of DNA has had an enormous influence on biology and on society in general. DNA stores genetic information in the cell and in the organism. DNA replicates and transmits this information when each cell and organism reproduces. Researchers are beginning to understand how genes function, and are working on ways to manipulate them. The science of biotechnology is largely devoted to altering the genes in living organisms.

Function of DNA and RNA

Each DNA molecule contains many genes, and genes specify the sequence of the amino acids in proteins. RNA is an intermediary that conveys DNA's instructions regarding the amino acid sequence in a protein. If DNA's information is faulty, illness can result. The relationship between a gene, a protein, and an illness is illustrated by sickle-cell disease. (A sickle is a long, C-shaped knife used to cut long grass.) In sickle-cell disease, the individual's red blood cells are sickle shaped. This occurs because in one particular spot in the hemoglobin molecule, an amino acid called valine substitutes for an amino acid called glutamine. Exchanging one amino acid for another—a seemingly small change—makes red blood cells lose their normal round, flexible shape and become weak and easily torn. Profound effects on the person's health result. When these abnormal red blood cells go through small blood vessels, they clog the flow of blood and break apart. Sickle-cell disease is another cause of anemia, and it also results in pain and organ damage.

CASE STUDY DIAGNOSIS

"We're going to admit Callie right away," the chief pediatrician told Matt and Libby. Their daughter was now fully conscious and her blood pressure and heart rate were close to normal. However, their normally happy, inquisitive child was extremely irritable, crying fitfully, and pushing away the toys that her daddy offered. "We'll take some small blood samples and an X-ray and CT scan first. You two can fill out the rest of the paperwork, then you can walk upstairs with Callie." She squatted down, at eye level with the older child. "Christy, we have a little teddy bear store here. You can pick out two, one for you and one for Callie. But can you answer some questions first? Ask Mommy and Daddy if it's okay."

Matt and Libby nodded hesitantly. "I'm thinking that Christy might just have some clues about why Callie is sick," the doctor whispered.

"Christy, did you ever hit Callie?" The little girl denied it vehemently. "Did she ever do anything she wasn't supposed to do?" Looking suddenly ashamed, Christy hung her head and whispered, "She went into the bathroom downstairs. She eated the curly candy on the wall. I tried a little piece too, but it tasted yucky. Callie liked it."

"Callie has lead poisoning," the doctor later confided to her parents. "Blood tests and the X-rays confirm it. See these white spots on the X-ray? The black background on her digestive system is air that lets X-rays through, but lead doesn't allow X-rays to penetrate. And remember your periodic table from college? Lead is called a transition metal, like iron. It interferes with formation of hemoglobin, and that's why she is anemic.* Hemoglobin is the protein in red blood cells." Both parents nodded.

"Lead causes irritation of brain tissue, and that's why she started vomiting. It also causes more calcium to be pushed into Callie's bones. We need to get it out of her blood as soon as we can, or she'll stop growing," she continued. "The good news is that the CT scan showed that her brain isn't swelling up—yet. That's the worst part about lead poisoning."

"How in the world was Callie exposed to lead?" Matt demanded.

"Christy told me that Callie ate 'the curly candy' on the bathroom wall. Unless I miss my guess, that was lead-based paint commonly used in old houses years ago. Unfortunately, lead compounds taste sweet, especially to little kids. Christy told me she tried a small piece, so we're going to test Christy's blood, too." She continued, "Don't beat yourselves up over this . Kids do all kinds of crazy things. Eating paint, swallowing lead bullets or fishing weights—we see those kinds of tricks all the time."

*Anemia: Condition caused by decreased amount of hemoglobin or decreased red blood cell count.

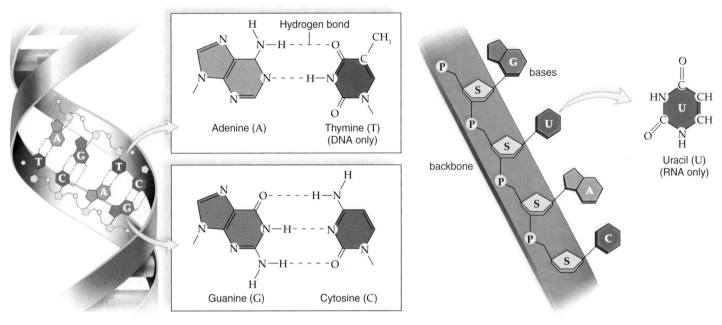

a. DNA structure with base pairs: A with T and G with C   b. RNA structure with bases G, U, A, C

Figure 2.21   **How does DNA form a double helix, like a spiral staircase? Why can't RNA form a double helix?**
In DNA, adenine and thymine are a complementary base pair. Note the hydrogen bonds that join them (like the "steps" in a spiral staircase.) Likewise, guanine and cytosine can pair. RNA has uracil instead of thymine, so complementary base pairing isn't possible.

How the Structure of DNA and RNA Differs

Both DNA and RNA are polymers of nucleotides.

Nucleotide Structure

Every **nucleotide** is a molecular complex of three types of subunit molecules—phosphate (phosphoric acid), a pentose (5-carbon) sugar, and a nitrogen-containing base:

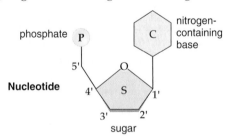

The nucleotides in DNA contain the sugar deoxyribose, and the nucleotides in RNA contain the sugar ribose; this difference accounts for their respective names (Table 2.1). There are four different types of bases in DNA: **adenine (A), thymine (T), guanine (G),** and **cytosine (C).** The base can have two rings (adenine or guanine) or one ring (thymine or cytosine). In RNA, the base **uracil (U)** replaces the base thymine. These structures are called bases because their presence raises the pH of a solution.

Polynucleotide Structure

The nucleotides link to make a polynucleotide called a strand, which has a backbone made up of phosphate-sugar-

Table 2.1	DNA Structure Compared to RNA Structure	
	DNA	**RNA**
Sugar	Deoxyribose	Ribose
Bases	Adenine, guanine, thymine, cytosine	Adenine, guanine, uracil, cytosine
Strands	Double-stranded with base pairing	Single-stranded
Helix	Yes	No

phosphate-sugar. The bases project to one side of the backbone. The nucleotides of a gene occur in a definite order, and so do the bases. After many years of work, researchers now know the sequence of the bases in human DNA—the human genome. This breakthrough is expected to lead to improved genetic counseling, gene therapy, and medicines to treat the causes of many human illnesses.

DNA is double-stranded, with the two strands twisted about each other in the form of a *double helix* (see Fig. 2.21a). In DNA, the two strands are held together by hydrogen bonds between the bases. When coiled, DNA resembles a spiral staircase. When unwound, it resembles a stepladder. The uprights (sides) of the ladder are made entirely of phosphate and sugar molecules, and the rungs of the ladder are made only of **complementary paired bases.** Thymine (T) always pairs with adenine (A), and guanine (G) always pairs with cytosine (C). Complementary bases have shapes that fit together.

Complementary base pairing allows DNA to replicate in a way that ensures the sequence of bases will remain the

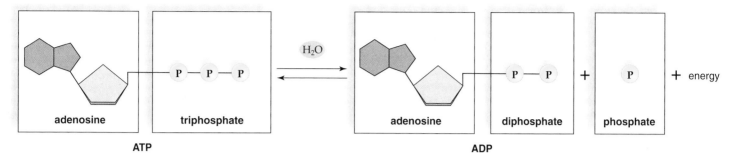

Figure 2.22 **ATP is the universal energy currency of cells. How is energy given off, and how does the cell recycle ATP?**
ATP is composed of the base adenosine and three phosphate groups (called a triphosphate.) When cells need energy, ATP is hydrolyzed (water is added) forming ADP and ℗. Energy is released. To recycle ATP, energy from food is required and the reverse reaction occurs: ADP and ℗ join to form ATP, and water is given off.

same. This is important because it is the sequence of bases that determine the sequence of amino acids in a protein. RNA is single-stranded. When RNA forms, complementary base pairing with one DNA strand passes the correct sequence of bases to RNA (Fig. 2.21*b*). RNA is the nucleic acid directly involved in protein synthesis.

ATP: An Energy Carrier

In addition to being the subunits of nucleic acids, nucleotides have metabolic functions. When adenosine (adenine plus ribose) is modified, by the addition of three phosphate groups instead of one, it becomes **ATP** (adenosine triphosphate). ATP is an energy carrier in cells.

Structure of ATP Suits Its Function

ATP is a high-energy molecule because the last two phosphate bonds are unstable and easily broken. Usually in cells, the last phosphate bond is hydrolyzed, leaving the molecule **ADP (adenosine diphosphate)** and a molecule of inorganic phosphate ℗ (Fig. 2.22). The energy released by ATP breakdown is used by the cell to synthesize macromolecules, such as carbohydrates and proteins. In muscle cells, the energy is used for muscle contraction; and in

nerve cells, it is used for the conduction of nerve impulses. After ATP breaks down, it can be recycled by adding ℗ to ADP. Notice in Figure 2.22 that an input of energy is required to re-form ATP.

Glucose Breakdown Leads to ATP Buildup

A glucose molecule contains too much energy to be used as a direct energy source in cellular reactions. Instead, the energy of glucose is converted to that of ATP molecules. ATP contains an amount of energy that makes it usable to supply energy for chemical reactions in cells. Muscles use ATP energy and produce heat when they contract. This is the heat that warms the body.

As we shall see in Chapter 3, oxygen is involved in the breakdown of glucose. Insufficient oxygen limits glucose breakdown and limits ATP buildup.

> **Check Your Progress 2.7**
>
> 1. How does the (a) function and (b) structure of DNA and RNA differ?
>
> 2. What type of bonding joins the bases within a DNA double helix?
>
> 3. What is the structure and function of ATP?

CASE STUDY TREATMENT

Repeated flooding of Callie's entire digestive tract, using a solution designed to be similar to her blood, finally cleared Callie's system of lead paint chips. Follow-up X-rays confirmed it. The hospital's toxicologist—a physician specializing in treatment of poisoning—was called in next. He carefully monitored the use of a *chelator,* a chemical used to bind lead dissolved in blood. The kidneys could finally eliminate blood lead as part of urine. A week of aggressive treatment finally had Callie out of the intensive care unit

and into a normal pediatric ward, bouncing on her bed and watching DVDs of the *Wiggles* and *Big Comfy Couch.*

"We can't stop now," the pediatrician told Libby and Matt two days later as Callie was discharged from the hospital. "Bring her back weekly so we can test her blood for lead, X-ray her for more signs of lead, and take CTs of her brain. And one more thing—the whole bunch of you move back in with family and get out of that house. Hire professionals to 'get the lead out'!"

Summarizing the Concepts

2.1 From Atoms to Molecules

- Matter is composed of elements; each element is made up of just one type of atom.
- An atom's mass is based on the number of protons and neutrons in the nucleus, as well as the electrons orbiting the nucleus.
- An atom's chemical properties depend on the number of electrons in its orbitals and outer shell.
- Atoms react by forming ionic bonds or covalent bonds.

2.2 Water and Living Things

- Water is a liquid, instead of a gas, at room temperature.
- Water heats and freezes slowly, moderating temperatures and allowing bodies to cool by vaporizing water.
- Frozen water is less dense than liquid water, so ice floats on water.
- Water is cohesive and fills tubular vessels, such as blood vessels. A thin film of water allows the lungs to adhere to the chest wall.
- Water is the universal solvent because of its polarity.
- Water has a neutral pH.
- Acids increase H^+ but decrease the pH of water.
- Bases decrease H^+ but increase the pH of water.

Organic molecules	Examples	Monomers	Functions
Carbohydrates	Monosaccharides, disaccharides, polysaccharides	**Glucose**	Immediate energy and stored energy; structural molecules
Lipids	Fats, oils, phospholipids, steroids	**Glycerol** / **Fatty acid**	Long-term energy storage; membrane components
Proteins	Structural, enzymatic, carrier, hormonal, contractile	**Amino acid**	Support, metabolic, transport, regulation, motion
Nucleic acids	DNA, RNA	**Nucleotide**	Storage of genetic information

2.3 Molecules of Life

Carbohydrates, lipids, proteins, and nucleic acids are macromolecules with specific functions in cells.

2.4 Carbohydrates

- Simple carbohydrates are monosaccharides or disaccharides.
- Glucose is a 6-carbon sugar used by cells for quick energy.
- Complex carbohydrates are polysaccharides. Starch, glycogen, and cellulose are polysaccharides containing many glucose units.
- Plants store glucose as starch.
- Animals store glucose as glycogen.
- Cellulose forms plant cell walls. Cellulose is dietary fiber.

2.5 Lipids

- Fats and oils, which function in long-term energy storage, contain glycerol and three fatty acids.
- Fatty acids can be saturated or unsaturated.
- Plasma membranes contain phospholipids.
- Steroids are complex lipids composed of three interlocking rings. Testosterone and estrogen are steroids.

2.6 Proteins

- Some proteins are structural (keratin, collagen), hormones, or enzymes that speed chemical reactions.
- Some proteins account for cell movement (actin, myosin), enable muscle contraction (actin, myosin), or transport molecules in blood (hemoglobin).
- Proteins are macromolecules with amino acid subunits.
- A peptide is composed of two amino acids.
- A polypeptide contains many amino acids.
  A protein has levels of structure:
- A primary structure is determined by the sequence of amino acids that form a polypeptide.
- A secondary structure is an α (alpha) helix or pleated sheet.
- A tertiary structure occurs when the secondary structure forms a three-dimensional, globular shape.
  A quaternary structure occurs when two or more polypeptides join to form a single protein.

2.7 Nucleic Acids

- Nucleic acids are macromolecules composed of nucleotides. Nucleotides are composed of a sugar, a base, and a phosphate. DNA and RNA are polymers of nucleotides.
- DNA contains the sugar deoxyribose; contains the bases adenine, guanine, thymine, and cytosine; is double-stranded; and forms a helix.
- RNA contains the sugar ribose; contains the bases adenine, guanine, uracil, and cytosine; and does not form a helix.
- ATP is a high-energy molecule because its bonds are unstable.
- ATP undergoes hydrolysis to ADP + $\circledP$, which releases energy used by cells to do metabolic work.

Understanding Key Terms

acid 27
adenine (A) 38
ADP (adenosine
    diphosphate) 34
amino acid 34
atom 20
atomic mass 20
atomic number 20
ATP (adenosine
    triphosphate) 39
base 27
buffer 28
calorie 25
carbohydrate 29
cellulose 30
complementary paired
    bases 38
compound 22
covalent bond 24
cytosine (C) 38
dehydration reaction 29
denaturation 35
disaccharide 30
DNA (deoxyribonucleic
    acid) 37
electron 20
element 20
emulsification 31
fat 31
fatty acid 31
glucose 29
glycogen 30
guanine (G) 38
hemoglobin 34
hexose 29
hydrogen bond 25
hydrolysis reaction 29
hydrophilic 27
hydrophobic 27

ion 23
ionic bond 23
isotope 21
lipid 31
macromolecule 29
mass 21
mass number 21
matter 20
mole 27
molecule 22
monosaccharide 29
nucleus 20
neutron 20
nucleotide 38
oil 31
orbital 20
organic 29
organic molecule 29
pentose 29
peptide bond 35
phospholipid 33
pH scale 27
polar 25
polypeptide 35
polysaccharide 30
protein 34
proton 20
radioisotope 21
RNA (ribonucleic acid) 37
saturated fatty acid 31
starch 30
steroid 33
thymine (T) 38
tracer 21
trans fat 32
triglyceride 31
unsaturated fatty acid 31
uracil (U) 38

Match the key terms to these definitions.

a. _____ Breaking up of fat globules into smaller droplets by the action of bile salts or any other emulsifier.

b. _____ Charged particle that carries a negative or positive charge.

c. _____ Chemical bond in which atoms share one or more pairs of electrons.

d. _____ Type of molecule that interacts with water by dissolving in water and/or forming hydrogen bonds with water molecules.

e. _____ Weak bond that arises between a slightly positive hydrogen atom of one molecule and a slightly negative atom of another molecule or between parts of the same molecule.

Testing Your Knowledge of the Concepts

1. Name the subatomic particles of the atom. Describe their charge, atomic mass, and location in the atom. (pages 20–21)

2. Why can a radioisotope be used as a tracer in the human body? Give an example. (pages 21–22)

3. Explain the difference between an ionic bond and a covalent bond. (pages 22–24)

4. Relate the properties of water to its polarity and hydrogen bonding between water molecules. (pages 25–27)

5. On the pH scale, which numbers indicate a basic solution? An acidic solution? A neutral solution? What makes a solution basic, acidic, or neutral? (pages 27–29)

6. What are buffers, and why are they important to life? (pages 28–29)

7. Nitrogen would be found in which categories of molecules unique to organisms? (page 28)

8. Name some monosaccharides, disaccharides, and polysaccharides, and state some general functions of each. What is the most common subunit for polysaccharides? (pages 34–38)

9. What are the subunits of a triglyceride? What is the difference between a saturated and an unsaturated fatty acid? What are the functions of fats in the body? (pages 31–33)

10. How does the structure of a phospholipid differ from a triglyceride? Describe the arrangement of phospholipids in the plasma membrane. (page 33)

11. What is the subunit of a protein, and how do two subunits join to form a peptide bond? (pages 34–35)

12. Discuss the primary, secondary, and tertiary structures of proteins. Why are these structures so important? (pages 35–36)

13. Describe the double-helix structure of DNA and the single-stranded structure of RNA. (pages 37–39)

14. What type of reaction releases the energy of an ATP molecule? Explain. (page 39)

15. The atomic number gives the
    a. number of neutrons in the nucleus.
    b. number of protons in the nucleus.
    c. weight of the atom.
    d. number of protons in the outer shell.

16. Isotopes differ in their
    a. number of protons.
    b. atomic number.
    c. number of neutrons.
    d. number of electrons.

17. Which type of bond results from the complete transfer of electrons from one atom to another?
    a. covalent
    b. ionic
    c. hydrogen
    d. neutral

18. Which of the following properties of water is not due to hydrogen bonding between water molecules?
    a. Water prevents large temperature changes.
    b. Ice floats on water.
    c. Water is a solvent for polar molecules and ionic compounds.
    d. Water flows and fills spaces because it is cohesive.

19. If a chemical accepted H^+ from the surrounding solution, the chemical would be a(n)
    a. base.
    b. acid.
    c. buffer.
    d. Both a and c are correct.

20. What is true of a solution that goes from pH 5 to pH 8?
    a. The H^+ concentration decreases as the solution becomes more basic.
    b. The H^+ concentration increases as the solution becomes more acidic.
    c. The H^+ concentration decreases as the solution becomes more acidic.

21. An example of a polysaccharide used for energy storage in humans is
    a. cellulose.   c. cholesterol.
    b. glycogen.   d. starch.

22. Saturated and unsaturated fatty acids differ in the
    a. number of carbon-to-carbon double bonds.
    b. consistency at room temperature.
    c. number of hydrogen atoms present.
    d. All of these are correct.

23. The difference between one amino acid and another is found in the _____ group.
    a. amino   c. *R*
    b. carboxyl   d. All of these are correct.

24. An example of a hydrolysis reaction is
    a. amino acid + amino acid $\longrightarrow$ dipeptide + H_2O.
    b. dipeptide + H_2O $\longrightarrow$ amino acid + amino acid.
    c. denaturation of a polypeptide.
    d. Both b and c are correct.

25. The helix and pleated sheet form of a protein is its _____ structure.
    a. secondary   c. primary
    b. tertiary   d. quaternary

26. An RNA nucleotide differs from a DNA molecule in that RNA has
    a. ribose sugar.   c. uracil base.
    b. phosphate molecule. d. Both a and c are correct.

In questions 27–30, match each subunit with the molecule in the key.

Key:
    a. fat   c. polypeptide
    b. polysaccharide   d. DNA, RNA

27. Glucose

28. Nucleotide

29. Glycerol and fatty acids

30. Amino acid

In questions 31–34, match the molecular category with a molecule in the preceding key.

31. Lipids

32. Proteins

33. Nucleic acids

34. Carbohydrates

35. Label this diagram using these terms: dehydration reaction, hydrolysis reaction, subunits, macromolecule.

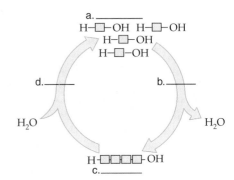

Thinking Critically About the Concepts

Throughout this chapter, you have followed along with the story of the Molyneaux family, New Orleans residents whose small daughter was afflicted with lead poisoning. Lead is one of the transition metals on the periodic table (see Appendix, page A-1). Study these metals carefully and see which ones are familiar to you from everyday life. Many of these metals—including lead—are capable of denaturing body proteins. (Review the term denature and its definition on page 35 if you've forgotten). Some of the metals, when elements, are beneficial to the body. Others are toxic only when in the form of a compound. Still others, like lead and mercury, may be toxic as *both* elements and compounds.

1. One effect of lead in Callie's body was to produce anemia by interfering with hemoglobin production by red blood cells. What transition metal could be replaced by lead in the hemoglobin molecule? If lead becomes part of hemoglobin, how do you think that affects oxygen transport?

2. Explain, in terms of ATP production, why a symptom of Callie's anemia was feeling tired.

3. There are many other causes of anemia. What is a genetic cause for anemia discussed in this chapter?

4. Believe it or not, very expensive wedding cakes are sometimes covered with a thin film of gold. The gold can be safely eaten along with the icing. Do you think the gold is more likely an element or a compound?

5. Some people get a rash when exposed to inexpensive nickel jewelry, watchbands, snaps, and so on. A process called anodizing covers nickel with a different metal, so rashes are less likely to occur. What transition metals can be used for anodizing? (Hint: All will make nickel much more expensive!)

6. Read the label on your jar of multivitamins. What transition metals do you see there? Are large quantities needed for good health?

3

Cell Structure and Function

Josh Citrigno jumped in the car right after dinner and headed to the local YMCA. Tuesday was basketball night, time for a friendly game with some college classmates. Their recreational team was up two games in the league standings, and Josh hoped that he and his teammates could increase their lead tonight.

"Whew, I am tired," he reflected as he stopped at the red light. He looked up and noticed that he was having trouble focusing on the stop light. He squinted, but the light didn't come into focus. He shook his head and blinked a few times. The light was still blurred. He pulled to the side of the road and rubbed at his eyes.

"I must have spent too much time staring at that stupid computer screen!" Josh thought to himself. He liked his job as a copy editor for the student newspaper, and figured his experience was bound to look good on a resume. Still, he did spend a lot of time sitting at his desk. He really wanted to get some exercise to help relax after a long day of work and classes. Besides, Josh was feeling like a couch potato!

He changed out of his street shoes and ran onto the basketball court to warm up. One of his teammates passed him the ball as they began their drills. Josh caught the ball, then twisted to the right to shoot. Suddenly, he couldn't focus on the basket. His teammate stole the ball easily, then good-naturedly slapped him on the back. "Losing it, aren't you Josh? And only 22 years old. Too bad, buddy."

Josh stopped and stared at the basket. He really couldn't see it out of his right eye. Surely he wasn't old enough to be losing his vision! He wasn't having any pain in his eye, but he decided that maybe he needed to see the optometrist for glasses.

3.1 What Is a Cell?
Cells are the basic units of life, and new cells only come from preexisting cells. Cells are small because it gives them a favorable surface-to-volume ratio.

3.2 How Cells Are Organized
Human cells have a plasma membrane, cytoplasm, and a nucleus. The cytoplasm contains several types of organelles.

3.3 The Plasma Membrane and How Substances Cross It
The plasma membrane regulates its permeability. Diffusion, carrier transport, and vesicle formation allow substances to get into and out of cells.

3.4 The Nucleus and the Production of Proteins
DNA within the nucleus specifies the sequence of amino acids in a protein, manufactured at ribosomes, often found attached to the endoplasmic reticulum. The Golgi apparatus serves as the "post office" of the cell.

3.5 The Cytoskeleton and Cell Movement
The cytoskeleton is composed of fibers that maintain the shape of the cell and assist the movement of organelles.

3.6 Mitochondria and Cellular Metabolism
Mitochondria are the site of cellular respiration, the complete enzymatic breakdown of glucose with the buildup of ATP molecules.

3.1 What Is a Cell?

Organisms, including humans, are composed of cells. This isn't apparent until you compare unicellular organisms with the tissues of multicellular ones under the microscope. The cell theory, one of the fundamental principles of modern biology, wasn't formulated until the invention of the microscope in the seventeenth century.

Most cells are small and can be seen only under a microscope. The small size of cells means that they are measured using the smaller units of the metric system. Cells are about 100 micrometers (μm) in diameter, about the width of a human hair. Their contents are smaller yet. The units of the metric system, explained on the inside back cover, are common to people who use microscopes professionally.

The Cell Theory

As stated by the **cell theory,** *a cell is the basic unit of life.* Nothing smaller than a cell is alive. A unicellular organism exhibits the seven characteristics of life we discussed in Chapter 1. There is no smaller unit of life able to reproduce, respond to stimuli, remain homeostatic, grow and develop, take in and use materials from the environment, and become adapted to the environment. In short, life has a cellular nature.

All living things are made up of cells. While it may be apparent that a unicellular organism is necessarily a cell, what about multicellular ones? Humans are multicellular. Is there any tissue in the human body not composed of cells? At first, you might be inclined to say that bone is not composed of cells. What if you were to examine bone tissue under the microscope? You would be able to see that it, too, is composed of cells surrounded by material they have deposited. Cells look different—a blood cell looks different from a nerve cell. They both look different from a cartilage cell (Fig. 3.1). Despite having certain parts in common, cells in a multicellular organism are specialized in structure and function.

New cells arise only from preexisting cells. Until the nineteenth century, most people believed in spontaneous generation; that is, that nonliving objects could give rise to living organisms. For example, maggots were thought to arise from meat hung in the butcher shop. Maggots often did arise from meat to which flies had access. However, people did not realize that the maggots (living) did not arise from the meat (nonliving). In 1864, the French scientist Louis Pasteur conducted a now-classic set of experiments using bacterial cells. His experiments proved conclusively that spontaneous generation of life from nonlife was not possible.

When mice or humans reproduce, a sperm cell joins with an egg cell to form a zygote. This is the first cell of a new multicellular organism. Parents pass a copy of their genes onto their offspring. The genes contain the instructions that allow the zygote to grow and develop into the complete organism.

Figure 3.1 Do all cells look the same?
No. Cells vary in structure and function, but they all exchange substances with their environment.

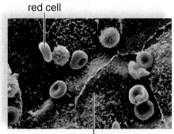
red cell

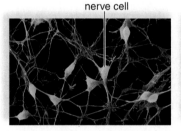
blood vessel cell

nerve cell

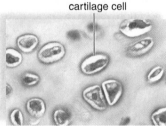
cartilage cell

Cell Size

A few cells, such as a hen's egg or a frog's egg, are large enough to be seen by the naked eye, but most are not. The small size of cells is explained by considering the surface area-to-volume ratio of cells. Nutrients enter a cell, and wastes exit a cell at its surface. Therefore, the greater the amount of surface, the greater the ability to get material in and out of the cell. A large cell requires more nutrients and produces more wastes than a small cell. In other words, the volume represents the needs of the cell. Yet, as cells become larger in volume, the proportionate amount of surface area actually decreases. You can see this by comparing the *two cubes* in Figure 3.2.

We would expect, then, that there would be a limit to how large an actively metabolizing cell can become. Once a hen's

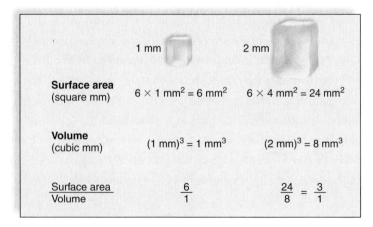

	1 mm	2 mm
Surface area (square mm)	$6 \times 1\ mm^2 = 6\ mm^2$	$6 \times 4\ mm^2 = 24\ mm^2$
Volume (cubic mm)	$(1\ mm)^3 = 1\ mm^3$	$(2\ mm)^3 = 8\ mm^3$
$\dfrac{\text{Surface area}}{\text{Volume}}$	$\dfrac{6}{1}$	$\dfrac{24}{8} = \dfrac{3}{1}$

Figure 3.2 Why are cells so small?
Large cells are unable to efficiently exchange nutrients and waste with their environment because of a decreased surface area-to-volume ratio. As the cell increases in size, the surface area increases by the square of the width. Cell volume is the width cubed (expressed as W^3).

egg is fertilized and starts metabolizing, it divides repeatedly without increasing in size. Cell division restores the amount of surface area needed for adequate exchange of materials.

Microscopy

Micrographs are photographs of objects most often obtained by using the compound light microscope, the transmission electron microscope, or the scanning electron microscope (Fig. 3.3).

A compound light microscope uses a set of glass lenses and light rays passing through the object to magnify objects. The image can be viewed directly by the human eye. The transmission electron microscope makes use of a stream of electrons to produce magnified images. The human eye can't see the image. Therefore, it is projected onto a fluorescent screen or photographic film to produce an image that can be viewed.

The magnification produced by a transmission electron microscope is much higher than that of a light microscope. Also, the ability of this microscope to make out detail in enlarged images is much greater. In other words, the transmission electron microscope has a higher resolving power—the ability to distinguish between two adjacent points. Following is a comparison of the resolving power of the eye, the light microscope, and the transmission electron microscope:

Eye:	0.2 mm	=	200 μm	=	200,000 nm
Light microscope: (1,000×)	0.0002 mm	=	0.200 μm	=	200 nm
Transmission electron microscope: (50,000×)	0.00001 mm	=	0.01 μm	=	10 nm

A scanning electron microscope provides a three-dimensional view of the surface of an object. A narrow beam of electrons is scanned over the surface of the specimen, coated with a thin layer of metal. The metal gives off secondary electrons, collected to produce a television-type picture of the specimen's surface on a screen.

As you no doubt will discover in the laboratory, the light microscope has the ability to view living specimens—this is not true of the electron microscope. Electrons cannot travel very far in air, so a strong vacuum must be maintained along the entire path of the electron beam. Usually for light microscopy and always for electron microscopy, cells are treated so that they do not decompose. They are then embedded into a matrix that allows the specimen to be thinly sliced. Cells are ordinarily transparent; therefore, sections are often stained with colored dyes before they are viewed by a light microscope. Certain components take up the dye more than other components; therefore, contrast is enhanced.

The application of electron-dense metals provides contrast with electron microscopy. The latter provides no color, so electron micrographs can be colored after the micrograph is obtained. The expression "falsely colored" means that the original micrograph has been colored after it was produced.

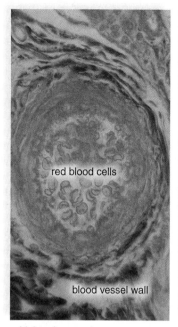

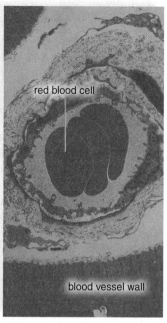

a. Light micrograph   b. Transmission electron micrograph

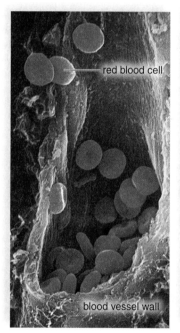

c. Scanning electron micrograph   d. Microscopist

Figure 3.3 **What does a cell look like when magnified by different types of microscopes?**
a. Light micrograph (LM) of many cells in a large vessel (stained). **b.** Transmission electron micrograph (TEM) of just three cells in a small vessel (colored). **c.** Scanning electron micrograph (SEM) gives a three-dimensional view of cells and vessels (colored). **d.** Scientist using an electron microscope.

Check Your Progress 3.1

1. What does the cell theory state?
2. Why are cells so tiny?
3. How do the light microscope and electron microscopes differ from one another?

3.2 How Cells Are Organized

Cells can be classified as either prokaryotic or eukaryotic. Both types of cells have a **plasma membrane,** an outer membrane that regulates what enters and exits a cell. The plasma membrane is a phospholipid bilayer. This bilayer is a "sandwich" made of two layers of phospholipids. Their polar phosphate molecules form the top and bottom surfaces of the bilayer, while the nonpolar tails lie in between. The phospholipid bilayer is selectively permeable. This means it allows certain molecules, but not others, to enter the cell. Proteins scattered throughout the plasma membrane play important roles in allowing substances to enter the cell.

All types of cells also contain **cytoplasm.** The cytoplasm is a semifluid medium that contains water and various types of molecules suspended or dissolved in the medium. The presence of proteins accounts for the semifluid nature of the cytoplasm. The cytoplasm contains **organelles.** Originally the term organelle referred to only membranous structures, but we will use it to include any well-defined subcellular structure. Eukaryotic cells have many different types of organelles.

Internal Structure of Eukaryotic Cells

The most prominent organelle within the **eukaryotic cell** is a nucleus, a membrane-enclosed structure where DNA is found. Each type of eukaryotic organelle has a specific function as noted in Figure 3.4. Many organelles are surrounded by a membrane, which allows compartmentalization of the cell. This keeps the various cellular activities separated from one another. **Prokaryotic cells** (such as bacterial cells) lack a **nucleus.** While the DNA of prokaryotic cells is centrally placed within the cell, it is not surrounded by a membrane.

Evolutionary History of the Eukaryotic Cell

Figure 3.5 shows that the first cells to arise were prokaryotic cells. Prokaryotic cells today are represented by the bacteria and archaea, which differ mainly by their chemistry. Bacteria are well known for causing diseases in humans, but they also have great environmental and commercial importance. The archaea are known for living in extreme environments that may mirror the first environments on Earth. These environments are too hot, too salty, and/or too acidic for the survival of most cells. The eukaryotic cell is believed to have evolved from the archaea.

The internal structure of eukaryotic cells is believed to have evolved, as shown in Figure 3.5. The nucleus could have formed by *invagination* of the plasma membrane, a process whereby a pocket is formed in the plasma membrane. The pocket would have enclosed the DNA of the cell, thus forming its nucleus. Surprisingly, some of the organelles in eukaryotic cells may have arisen by engulfing prokaryotic cells. The engulfed prokaryotic cells were not digested; rather, they then evolved into different organelles. One of these events would have given the eukaryotic cell a mitochondrion. Mitochondria are organelles that carry on cellular respiration. Another such event may have produced the chloroplast. Chloroplasts are found in cells that carry out photosynthesis.

> **Check Your Progress 3.2**
>
> 1. **a.** What are the three main parts of a eukaryotic cell? **b.** Which two of these also occur in prokaryotes? **c.** Which one does not occur in prokaryotes?
>
> 2. How might the nucleus and two of the organelles in eukaryotic cells have arisen?

CASE STUDY AT THE OPTOMETRIST'S OFFICE

Josh called the optometrist, Dr. Yee, and explained that he was having some vision loss problems. The eyesight in his right eye had not improved, even with a good night's rest. "You need to come in today," Dr. Yee stated emphatically.

Dr. Yee did some standard vision tests, including a color blindness test. Then, he added drops of medication to dilate Josh's pupils. "Those drops will make your eyes extra-sensitive to light for a short time, but now I can see internal structures in your eye," Dr. Yee explained. He continued, "The ophthalmoscope allows me to check the retina and vitreous humor of the eye. The retina lines the back of the eyeball and contains the cells that respond to light. The vitreous humor is the gel that fills the eyeball."

Dr. Yee finished his examination and turned off the ophthalmoscope. "I'm pretty concerned here, Josh," he began. "The blood vessels at the back of your eye are abnormal, and you've lost color vision in your right eye. You need to see an ophthalmologist, a medical doctor who specializes in eye diseases. I'm going to see if we can get you an appointment for today or tomorrow."

Josh left the optometrist's office with an appointment card for the next day. He was starting to get really worried. He had hoped that he'd just need glasses or contacts, but this was starting to sound serious.

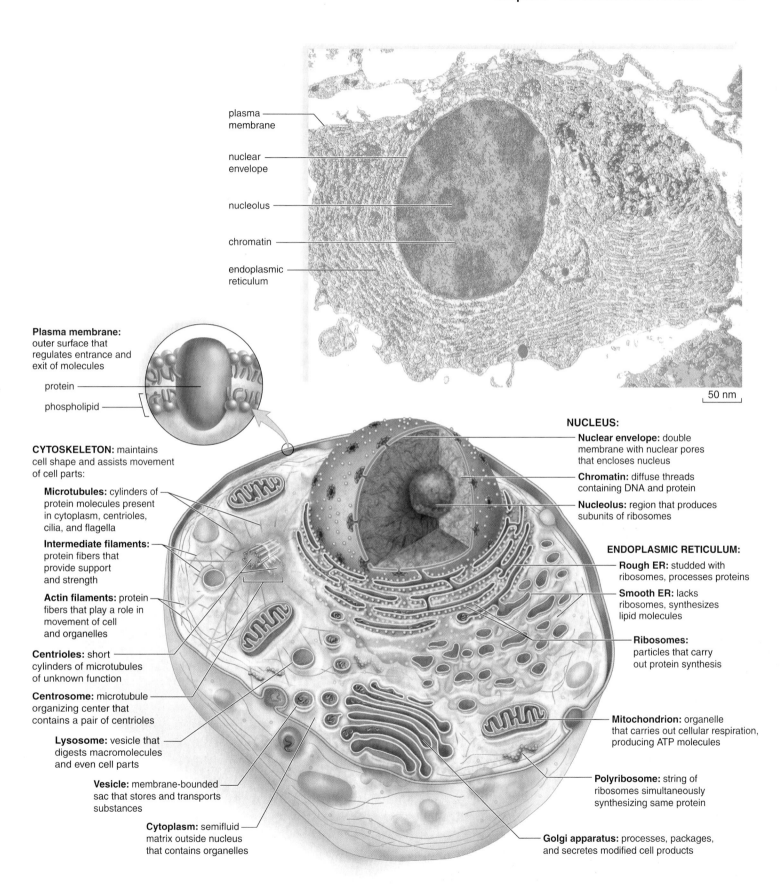

plasma membrane

nuclear envelope

nucleolus

chromatin

endoplasmic reticulum

50 nm

Plasma membrane: outer surface that regulates entrance and exit of molecules

protein

phospholipid

CYTOSKELETON: maintains cell shape and assists movement of cell parts:

Microtubules: cylinders of protein molecules present in cytoplasm, centrioles, cilia, and flagella

Intermediate filaments: protein fibers that provide support and strength

Actin filaments: protein fibers that play a role in movement of cell and organelles

Centrioles: short cylinders of microtubules of unknown function

Centrosome: microtubule organizing center that contains a pair of centrioles

Lysosome: vesicle that digests macromolecules and even cell parts

Vesicle: membrane-bounded sac that stores and transports substances

Cytoplasm: semifluid matrix outside nucleus that contains organelles

NUCLEUS:

Nuclear envelope: double membrane with nuclear pores that encloses nucleus

Chromatin: diffuse threads containing DNA and protein

Nucleolus: region that produces subunits of ribosomes

ENDOPLASMIC RETICULUM:

Rough ER: studded with ribosomes, processes proteins

Smooth ER: lacks ribosomes, synthesizes lipid molecules

Ribosomes: particles that carry out protein synthesis

Mitochondrion: organelle that carries out cellular respiration, producing ATP molecules

Polyribosome: string of ribosomes simultaneously synthesizing same protein

Golgi apparatus: processes, packages, and secretes modified cell products

Figure 3.4 **What structures are found inside a cell?**

The transmission electron micrograph (*above*) shows the interior structures of a cell. The generalized drawing (*below*) labels and describes these structures.

Original prokaryotic cell

DNA

1. Cell gains a nucleus by the plasma membrane invaginating and surrounding the DNA with a double membrane.

2. Cell gains an endomembrane system by proliferation of membrane.

3. Cell gains protomitochondria.

proto-mitochondrion

4. Cell gains protochloroplasts.

mitochondrion

protochloroplast

chloroplast

Animal cell

Plant cell

Figure 3.5   How did eukaryotic cells evolve?
Invagination of the plasma membrane of a prokaryotic cell could have created the nucleus. Later, the cell gained organelles, some of which may have been independent prokaryotes.

3.3 The Plasma Membrane and How Substances Cross It

A human cell, like all cells, is surrounded by an outer plasma membrane (Fig. 3.6). The plasma membrane marks the boundary between the outside and the inside of the cell. Plasma membrane integrity and function are necessary to the life of the cell.

The plasma membrane is a phospholipid bilayer with attached or embedded proteins. A phospholipid molecule has a polar head and nonpolar tails (see Fig. 2.16). When phospholipids are placed in water, they naturally form a spherical bilayer. The polar heads, being charged, are hydrophilic (attracted to water). They position themselves to face toward the watery environment outside and inside the cell. The nonpolar tails are hydrophobic (not attracted to water). They turn inward toward one another, where there is no water.

At body temperature, the phospholipid bilayer is a liquid. It has the consistency of olive oil. The proteins are able to change their position by moving laterally. The **fluid-mosaic model** is a working description of membrane structure. It states that the protein molecules form a shifting pattern within the fluid phospholipid bilayer. Cholesterol lends support to the membrane.

Short chains of sugars are attached to the outer surface of some protein and lipid molecules. These are called glycoproteins and glycolipids, respectively. These carbohydrate chains, specific to each cell, help mark it as belonging to a particular individual. They account for why people have different blood types, for example. Other glycoproteins have a special configuration that allows them to act as a receptor

Figure 3.6

How is a plasma membrane organized?

A plasma membrane is composed of a phospholipid bilayer in which proteins are embedded. The hydrophilic heads of phospholipids are a part of the outside surface and the inside surface of the membrane. The hydrophobic tails make up the interior of the membrane. Note the plasma membrane's asymmetry—carbohydrate chains are attached to the outside surface, and cytoskeleton filaments are attached to the inside surface. Cholesterol lends support to the membrane.

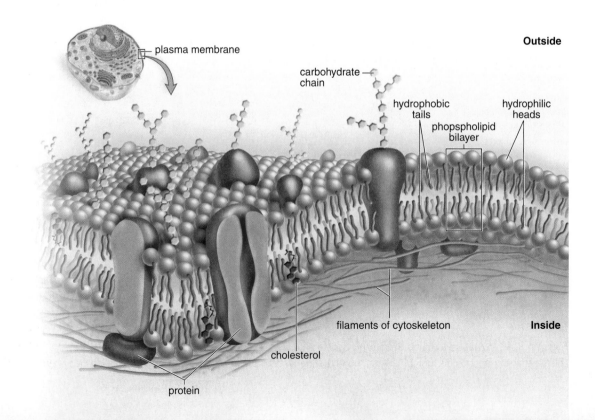

plasma membrane

Outside

carbohydrate chain

hydrophobic tails

hydrophilic heads

phopspholipid bilayer

filaments of cytoskeleton

Inside

cholesterol

protein

for a chemical messenger, such as a hormone. Some plasma membrane proteins form channels through which certain substances can enter cells. Others are enzymes that catalyze reactions or carriers involved in the passage of molecules through the membrane.

Plasma Membrane Functions

The plasma membrane keeps a cell intact. It allows only certain molecules and ions to enter and exit the cytoplasm freely. Therefore, the plasma membrane is said to be **selectively permeable** (Fig. 3.7). Small, lipid-soluble molecules, such as oxygen and carbon dioxide, can pass through the membrane easily. The small size of water molecules allows them to freely cross the membrane by using protein channels called *aquaporins*. Ions and large molecules cannot cross the membrane without more direct assistance, to be discussed later.

Diffusion

Diffusion is the random movement of molecules from the area of higher concentration to the area of lower concentration, until they are equally distributed. Diffusion is a passive way for molecules to enter or exit a cell. No cellular energy is needed to bring it about.

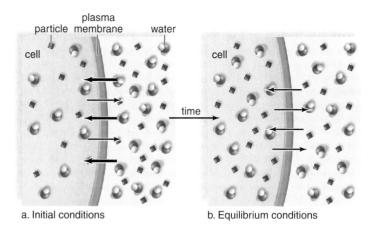

a. Initial conditions   b. Equilibrium conditions

Figure 3.8 **When a molecule diffuses across membrane, which direction does it take?**
a. When a substance can diffuse across the plasma membrane, it will move back and forth across the membrane, but the net movement will be toward the region of lower concentration. **b.** At equilibrium, an equal number of particles and water have crossed in both directions, and there is no net movement.

Certain molecules can freely cross the plasma membrane by diffusion. When molecules can cross a plasma membrane, which way will they go? The molecules will move in both directions. But the *net movement* will be from the region of higher concentration to the region of lower concentration, until equilibrium is achieved. At equilibrium, as many molecules of the substance will be entering as leaving the cell (Fig. 3.8). Oxygen diffuses across the plasma membrane, and the net movement is toward the inside of the cell. This is because a cell uses oxygen when it produces ATP molecules for energy purposes.

Osmosis

Osmosis is the net movement of water across a semipermeable membrane, from an area of higher concentration to lower concentration. The membrane separates the two areas, and solute is unable to pass through the membrane. Water will tend to flow from the area that has less solute (and therefore more water) to the area with more solute (and therefore less water.) **Tonicity** refers to an osmotic property of a solution across a particular membrane, such as a red blood cell membrane.

Normally body fluids are *isotonic* to cells (Fig. 3.9a). There is the same concentration of nondiffusible solutes and water on both sides of the plasma membrane. Therefore, cells maintain their normal size and shape. Intravenous solutions given in medical situations are usually isotonic.

Solutions that cause cells to swell, or even to burst, due to an intake of water are said to be *hypotonic*. A hypotonic solution has a lower concentration of solute

Figure 3.7 **How is the plasma membrane selectively permeable?**
Small, uncharged molecules are able to cross the membrane while large or charged molecules cannot. Water travels freely across membranes through aquaporins.

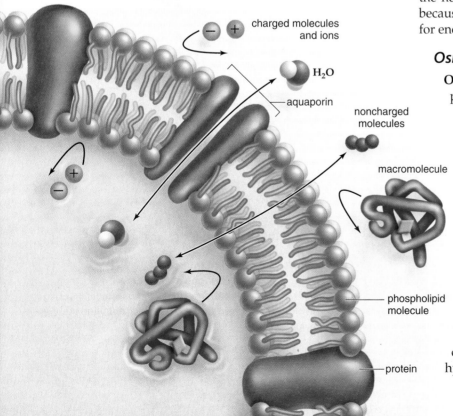

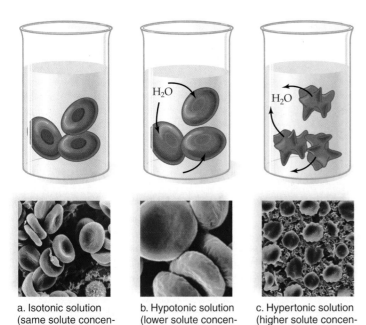

a. Isotonic solution (same solute concentration as in cell)

b. Hypotonic solution (lower solute concentration than in cell)

c. Hypertonic solution (higher solute concentration than in cell)

Figure 3.9   **What happens to red blood cells in solutions of different tonicity?**
a. In an isotonic solution, cells remain the same. **b.** In a hypotonic solution, cells gain water and may burst (lysis). **c.** In a hypertonic solution, cells lose water and shrink (crenation).

and a higher concentration of water than the cells. If red blood cells are placed in a hypotonic solution, water enters the cells. They swell to bursting (Fig. 3.9*b*). *Lysis* is used to refer to the process of bursting cells. Bursting of red blood cells is termed hemolysis.

Solutions that cause cells to shrink or shrivel due to loss of water are said to be *hypertonic.* A hypertonic solution has a higher concentration of solute and a lower concentration of water than do the cells. If red blood cells are placed in a hypertonic solution, water leaves the cells. They shrink (Fig. 3.9*c*). The term *crenation* refers to red blood cells in this condition. These changes have occurred due to osmotic pressure. **Osmotic pressure** controls water movement in our bodies. For example, in the small and large intestines, osmotic pressure allows us to absorb the water in food and drink. In the kidneys, osmotic pressure controls water absorption as well.

Facilitated Transport

Many solutes do not simply diffuse across a plasma membrane. They are transported by means of protein carriers within the membrane. During **facilitated transport,** a molecule is transported across the plasma membrane from the side of higher concentration to the side of lower concentration (Fig. 3.10). This is a passive means of transport because the cell does not need to expend energy to

move a substance down its concentration gradient. Each protein carrier, sometimes called a transporter, binds only to a particular molecule, such as glucose (green). Diabetes type 2 results when cells lack a sufficient number of glucose transporters.

Active Transport

During **active transport,** a molecule is moving from *lower* to *higher* concentration. For example, iodine collects in the cells of the thyroid gland. Sugar is completely absorbed from the

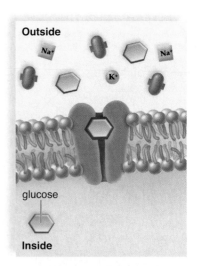

Figure 3.10   **What is facilitated transport?**
This is a passive form of transport in which substances move down their concentration gradient through a protein carrier. Glucose (green) enters by facilitated transport, and the result will be an equal distribution on both sides of the membrane.

gut by cells that line the digestive tract. Sodium (Na⁺) is
sometimes almost completely withdrawn from urine by cells
lining kidney tubules.

Active transport requires a protein carrier and the use
of cellular energy obtained from the breakdown of ATP.
When ATP is broken down, energy is released. In this
case, the energy is used to carry out active transport. Proteins involved in active transport often are called pumps.
Just as a water pump uses energy to move water against
the force of gravity, energy is used to move substances
against their concentration gradients. One type of pump
active in all cells moves sodium ions (Na⁺) to the outside
and potassium ions (K⁺) to inside the cell (Fig. 3.11). This
type of pump is especially associated with nerve and
muscle cells.

The passage of salt (NaCl) across a plasma membrane
is of primary importance in cells. First, sodium ions are
pumped across a membrane. Then, chloride ions diffuse
through channels that allow their passage. Chloride ion
channels malfunction in persons with cystic fibrosis. This
leads to the symptoms of this inherited (genetic) disorder.

Endocytosis and Exocytosis

During endocytosis, a portion of the plasma membrane
invaginates, or forms a pouch, to envelop a substance and
fluid. Then, the membrane pinches off to form an endocytic
vesicle inside the cell (Fig. 3.12a). Some white blood cells are
able to take up pathogens (disease-causing agents) by endocytosis. Here the process is given a special name: **phago-**

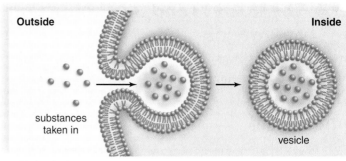

a. Endocytosis

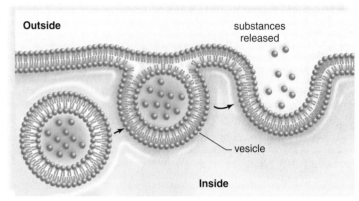

b. Exocytosis

Figure 3.12   How do large substances cross the membrane?
a. Large substances enter a cell by endocytosis, and **(b)** large substances
exit a cell by exocytosis.

cytosis. Usually, cells take up molecules and fluid, and then
the process is called *pinocytosis.* An inherited form of cardiovascular disease occurs when cells fail to take up a combined lipoprotein and cholesterol molecule from the blood
by pinocytosis.

During exocytosis, a vesicle fuses with the plasma membrane as secretion occurs (Fig. 3.12b). Later in the chapter,
we will see that a steady stream of vesicles move between
certain organelles, before finally fusing with the plasma
membrane. This is the way that signaling molecules, called
neurotransmitters, leave one nerve cell to excite the next
nerve cell or a muscle cell.

> **Check Your Progress 3.3**
>
> 1. What is the structure and overall function of the plasma
>   membrane?
>
> 2. **a.** How do isotonic, hypotonic, and hypertonic solutions
>   differ, and **(b)** how do they affect cells?
>
> 3. **a.** What are the various ways substances can enter and
>   exit cells? **b.** Which are passive, and which are active
>   ways of crossing the plasma membrane?

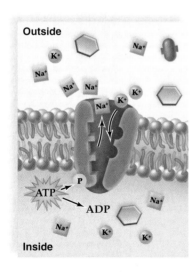

Figure 3.11   What is active transport?
This is a form of transport in which a molecule moves from high
concentration to low concentration. It requires a protein carrier and
energy. Na⁺ exits and K⁺ enters the cell by active transport, so Na⁺ will be
concentrated outside and K⁺ will be concentrated inside the cell.

3.4 The Nucleus and the Production of Proteins

The nucleus and several organelles are involved in the production and processing of proteins.

The Nucleus

The nucleus, a prominent structure in cells, stores genetic information (Fig. 3.13). Every cell in the body contains the same genes, segments of DNA. Each type of cell has certain genes turned on, while others are turned off. DNA, with RNA acting as an intermediary, specifies the proteins in a cell. Proteins have many functions in cells, and they help determine a cell's specificity.

Chromatin is the combination of DNA molecules and proteins that make up the **chromosomes.** Chromatin can coil tightly to form visible chromosomes during meiosis (cell division that forms reproductive cells) and mitosis (cell division that duplicates cells). Most of the time, however, the chromatin is uncoiled. Individual chromosomes cannot be distinguished and the chromatin appears grainy in electron micrographs of the nucleus. Chromatin is immersed in a semifluid medium called the **nucleoplasm.** A difference in pH suggests that nucleoplasm has a different composition from cytoplasm.

Micrographs of a nucleus do show one or more dark regions of the chromatin. These are nucleoli (sing., **nucleolus**), where ribosomal RNA (rRNA) is produced. This is also where rRNA joins with proteins to form the subunits of ribosomes.

The nucleus is separated from the cytoplasm by a double membrane known as the **nuclear envelope.** This is continuous with the **endoplasmic reticulum (ER),** a membranous system of saccules and channels discussed in the next section. The nuclear envelope has **nuclear pores** of sufficient size to permit the passage of ribosomal subunits out of the nucleus and proteins into the nucleus.

Ribosomes

Ribosomes are organelles composed of proteins and rRNA. Protein synthesis occurs at the ribosomes. Ribosomes are often attached to the endoplasmic reticulum; they also occur free within the cytoplasm, either singly or in groups called **polyribosomes.** Proteins synthesized at ribosomes attached to the endoplasmic reticulum have a different destination from that of proteins manufactured at ribosomes free in the cytoplasm.

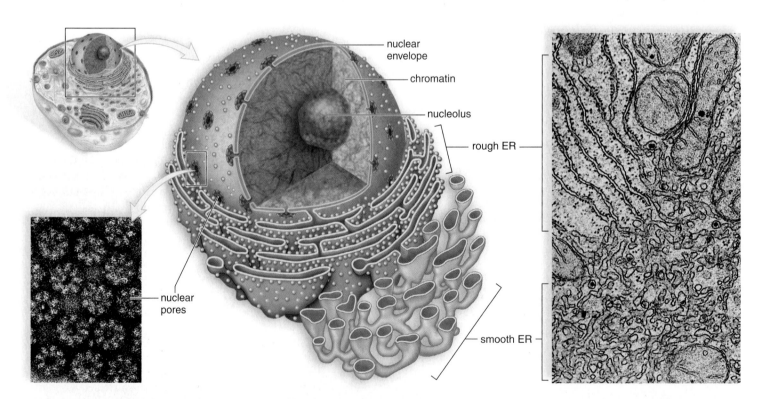

Figure 3.13   **What organelles are involved in the production and processing of proteins?**
The nucleus contains chromatin. Chromatin has a special region called the nucleolus, where rRNA is produced and ribosome subunits are assembled. The nuclear envelope contains pores (TEM, *left*) that allow substances to enter and exit the nucleus to and from the cytoplasm. The nuclear envelope is attached to the endoplasmic reticulum (TEM, *right*), which often has attached ribosomes, where protein synthesis occurs.

The Endomembrane System

The **endomembrane system** consists of the nuclear envelope, the endoplasmic reticulum, the Golgi apparatus, lysosomes, and **vesicles** (tiny membranous sacs) (Fig. 3.14).

The Endoplasmic Reticulum

The endoplasmic reticulum has two portions. Rough ER is studded with ribosomes on the side of the membrane that faces the cytoplasm. Here, proteins are synthesized and enter the ER interior, where processing and modification begin. Some of these proteins are incorporated into membrane, and some are for export. Smooth ER, continuous with rough ER, does not have attached ribosomes. Smooth ER synthesizes the phospholipids that occur in membranes and has various other functions, depending on the particular cell. In the testes, it produces testosterone. In the liver, it helps detoxify drugs.

The ER forms transport vesicles in which large molecules are transported to other parts of the cell. Often, these vesicles are on their way to the plasma membrane or the Golgi apparatus.

The Golgi Apparatus

The **Golgi apparatus** is named for Camillo Golgi, who discovered its presence in cells in 1898. The Golgi apparatus consists of a stack of slightly curved saccules, whose appearance can be compared to a stack of pancakes. Here, proteins and lipids received from the ER are modified. For example, a chain of sugars may be added to them. This makes them glycoproteins and glycolipids, molecules often found in the plasma membrane.

The vesicles that leave the Golgi apparatus move to other parts of the cell. Some vesicles proceed to the plasma membrane, where they discharge their contents. Altogether, the Golgi apparatus is involved in processing, packaging, and secretion.

Lysosomes

Lysosomes, membranous sacs produced by the Golgi apparatus, contain *hydrolytic enzymes*. Lysosomes are found in all cells of the body but are particularly numerous in white blood cells that engulf disease-causing microbes. When a lysosome fuses with such an endocytic vesicle, its contents

Figure 3.14 **What is the function of the endomembrane system?**
The organelles in the endomembrane system work together to produce, modify, secrete, and digest proteins and lipids.

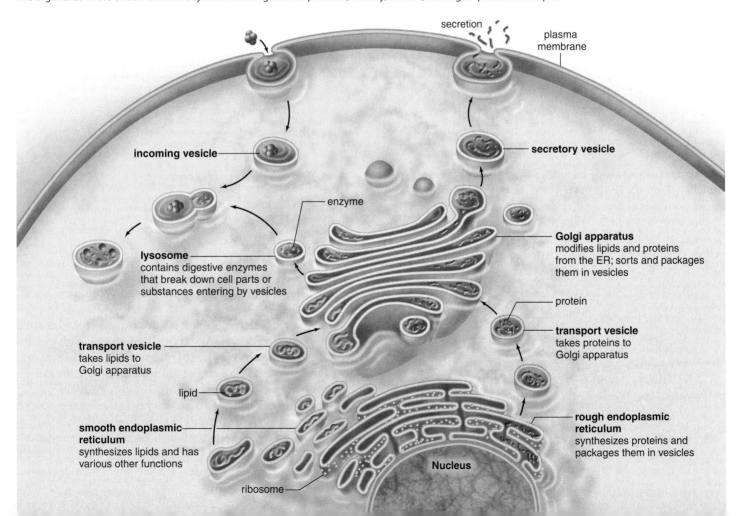

Science Focus

Female Mosaics, Barr Bodies, and Breast Cancer

Most people are familiar with calico cats, whose fur contains patches of orange, black, and white. These cats are genetic *mosaics.* A mosaic is formed by combination of different pieces to form a whole (a stained-glass window is one example.) Likewise, in genetics, a mosaic refers to an individual whose cells have at least two—and sometimes more—different types of genetic expression. In the case of the calico cat, the fur colors are due to expression of different genes. Some of the hair cells of these cats express the paternal copy of the gene. If an orange-haired father's copy of the gene is activated, a patch of orange hair develops. In other cells, the maternal gene is activated. A calico kitten with a black mother will grow black patches of hair scattered among the orange. Were you aware that human females are also mosaics?

The nucleus of human cells contains 46 chromosomes arranged into a set of 23 pairs. One chromosome from each pair is maternal, while the other is paternal. Each of the chromosomes in the first 22 pairs resembles its mate. Further, each member of a pair contains the same genes as the other member. Sex chromosomes that determine a person's gender are the last pair. Females have two X chromosomes, and males have one X and one Y chromosome. The Y chromosome is very small and contains far fewer genes than the X chromosome. Almost all of the genes on the X chromosome lack a corresponding gene on the Y chromosome. Thus, females have two copies of X genes, whereas males have only one.

The body compensates for this extra dose of genetic material by inactivating one of the X chromosomes in each cell of the female embryo. Inactivation occurs early in development (at approximately the 100-cell stage). The inactivated X chromosome is called a Barr body, named after its discoverer. Barr bodies are highly condensed chromatin that appear as dark spots in the nucleus. Which X chromosome is inactivated in a given cell? This appears to be random. But every cell that develops from the original group of 100 cells will have the same inactivated X chromosome as its parent cell. Some of a woman's cells have inactivated the maternal X chromosome and other cells have inactivated the paternal X chromosome: she is a mosaic.

Problems with inactivation of the X chromosome in humans could be linked to the development of cancer. For example, women who contain one defective copy of the breast cancer gene BRCA1 have a greatly increased risk of developing breast and ovarian cancer. The BRCA1 protein produced from the gene is called a *tumor suppressor.* When the protein is functioning normally, it suppresses the development of cancer. This same protein is involved in X chromosome inactivation, although its exact role is uncertain. Presumably, increased cancer risk occurs because abnormal BRCA1 protein can neither inactivate the X chromosome, nor function as a tumor suppressor.

are digested by lysosomal enzymes into simpler subunits that then enter the cytoplasm. Even parts of a cell are digested by its own lysosomes (called autodigestion). Some human diseases are caused by the lack of a particular lysosome enzyme. Tay-Sachs disease occurs when an undigested substance collects in nerve cells, leading to retarded development of a child and early death.

> ### Check Your Progress 3.4
> 1. The nucleus, ribosomes, and rough endoplasmic reticulum make what contribution to protein synthesis?
> 2. The endoplasmic reticulum and the Golgi apparatus make what contribution to the processing of proteins?
> 3. What is the function of smooth endoplasmic reticulum and lysosomes?
> 4. What role do transport vesicles play in the endomembrane system?

3.5 The Cytoskeleton and Cell Movement

It took a high-powered electron microscope to discover that the cytoplasm of the cell is crisscrossed by several types of protein fibers collectively called the **cytoskeleton** (see Fig. 3.4). The cytoskeleton helps maintain a cell's shape and either anchors the organelles or assists their movement, as appropriate.

In the cytoskeleton, **microtubules** are much larger than actin filaments. Each is a cylinder that contains rows of a protein called tubulin. The regulation of microtubule assembly is under the control of a microtubule organizing center called the **centrosome** (see Fig. 3.4). Microtubules help maintain the shape of the cell and act as tracks along which organelles move. During cell division, microtubules form spindle fibers, which assist the movement of chromosomes. **Actin filaments,** made of a protein called actin, are long, extremely thin fibers that usually occur in bundles or other groupings. Actin filaments are involved in movement. Microvilli, which project from certain cells and can shorten and extend, contain actin filaments. **Intermediate filaments,** as their name implies, are intermediate in size between microtubules and actin filaments. Their structure and function differ according to the type of cell.

Cilia and Flagella

Cilia (sing., **cilium**) and flagella (sing., **flagellum**) are involved in movement. The ciliated cells that line our respiratory tract sweep debris trapped within mucus back up the throat. This helps keep the lungs clean. Similarly, ciliated cells move an egg along the oviduct, where it will be fertilized by a flagellated sperm cell (Fig. 3.15).

A cilium is about 20 times shorter than a flagellum but both have the same organization of microtubules within a plasma membrane covering. Motor molecules, powered by ATP, allow the microtubules in cilia and flagella to interact and bend, and thereby move.

The importance of normal cilia and flagella is illustrated by the occurrence of a genetic disorder. Some individuals have an inherited genetic defect that leads to malformed microtubules in cilia and flagella. Not surprisingly, these individuals suffer from recurrent and severe respiratory

Have You Ever Wondered...

How fast does a human sperm swim?

Individual sperm speeds vary considerably and are greatly influenced by environmental conditions. However, in recent studies researchers found that some human sperm could travel at top speeds of approximately 20 cm/hour. This means that these sperm could reach the female ovum in less than an hour. Scientists are interested in sperm speed so that they can design new contraceptive methods.

infections. The ciliated cells lining respiratory passages fail to keep their lungs clean. They are also unable to reproduce naturally due to the lack of ciliary action to move the egg in a female or the lack of flagella action by sperm in a male.

> *Check Your Progress 3.5*
> 1. **The cytoskeleton contains what three types of fibers?**
> 2. **Which of these three is found in cilia and flagella?**
> 3. **Compare the function of the cytoskeleton to that of cilia.**

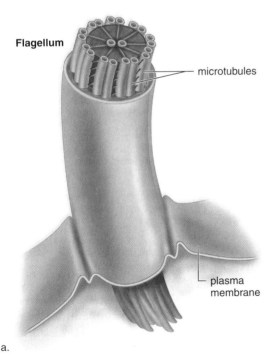

Flagellum

microtubules

plasma membrane

a.

cilia

sperm

flagellum

secretory cell

b.

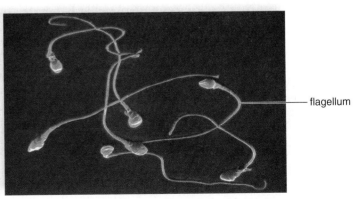

flagellum

c.

Figure 3.15 **How are cilia and flagella involved in reproduction?**
Human reproduction is dependent on the normal activity of cilia and flagella. **a.** Both cilia and flagella have an inner core of microtubules within a covering of plasma membrane. **b.** Cilia within the oviduct move the egg to where it is fertilized by a flagellated sperm. **c.** Sperm have very long flagella.

3.6 Mitochondria and Cellular Metabolism

Mitochondria (sing., **mitochondrion**) are often called the powerhouses of the cell. Just as a powerhouse burns fuel to produce electricity, the mitochondria convert the chemical energy of glucose products into the chemical energy of ATP molecules. In the process, mitochondria use up oxygen and give off carbon dioxide. Therefore, the process of producing ATP is called **cellular respiration.** The structure of mitochondria is appropriate to the task. The inner membrane is folded to form little shelves called cristae. These project into the matrix, an inner space filled with a gel-like fluid (Fig. 3.16). The matrix of a mitochondrion contains enzymes for breaking down glucose products. ATP production then occurs at the cristae. Protein complexes that aid in the conversion of energy are located in an assembly-line fashion on these membranous shelves.

The structure of a mitochondrion supports the hypothesis that they were originally prokaryotes engulfed by a cell. Mitochondria are bounded by a double membrane as a

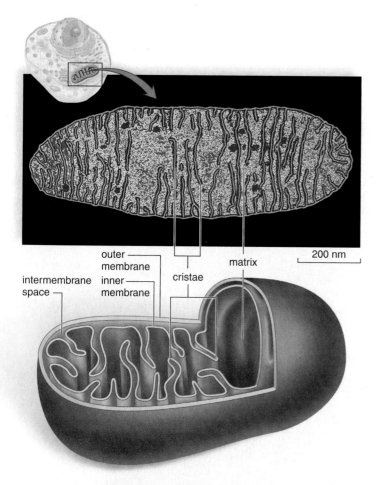

Figure 3.16 **What is the structure of a mitochondrion?**
A mitochondrion (TEM, *above*) is bounded by a double membrane, and the inner membrane folds into projections called cristae. The cristae project into a semifluid matrix that contains many enzymes.

outer membrane
intermembrane space
inner membrane
cristae
matrix
200 nm

prokaryote would be if taken into a cell by endocytosis. Even more interesting is the observation that mitochondria have their own genes, and they reproduce themselves!

Cellular Respiration and Metabolism

Cellular respiration is an important component of **metabolism,** which includes all the chemical reactions that occur in a cell. Often, metabolism requires metabolic pathways, carried out by enzymes sequentially arranged in cells:

$$\overset{1}{A} \rightarrow \overset{2}{B} \rightarrow \overset{3}{C} \rightarrow \overset{4}{D} \rightarrow \overset{5}{E} \rightarrow \overset{6}{F} \rightarrow G$$

The letters, except A and G, are **products** of the previous reaction and the **reactants** for the next reaction. A represents the beginning reactant(s), and G represents the product(s). The numbers in the pathway refer to different enzymes. *Each reaction in a metabolic pathway requires a specific enzyme.* Enzymes are so necessary in cells, therefore their mechanism of action has been studied extensively.

Enzymes

The reactant(s) that participates in the reaction is (are) called the enzyme's **substrate(s).** Enzymes are often named for their substrates. For example, lipids are broken down by lipase, maltose by maltase, and lactose by lactase.

Enzymes have a specific region, called an **active site,** where the substrates are brought together so they can react. An enzyme's specificity is caused by the shape of the active site. Here the enzyme and its substrate(s) fit together in a specific way, much as the pieces of a jigsaw puzzle fit together (Fig. 3.17). After one reaction is complete, the product or products are released. The enzyme is ready to be used again. Therefore, a cell requires only a small amount of a particular enzyme to carry out a reaction. A chemical reaction can be summarized in the following manner:

$$E + S \rightarrow ES \rightarrow E + P$$

(where E = enzyme, S = substrate, ES = enzyme-substrate complex, and P = product). An enzyme can be used over and over again.

Coenzymes **Coenzymes** are nonprotein molecules that assist the activity of an enzyme and may even accept or contribute atoms to the reaction. It is interesting that vitamins are often components of coenzymes. The vitamin niacin is a part of the coenzyme **NAD⁺ (nicotinamide adenine dinucleotide),** which carries hydrogen (H) and electrons.

Cellular Respiration

After blood transports glucose and oxygen to cells, cellular respiration begins. Cellular respiration breaks down glucose to carbon dioxide and water. Three pathways are involved in the breakdown of glucose. They are called glycolysis, the citric acid cycle, and the electron transport chain (Fig. 3.18). These metabolic pathways allow the

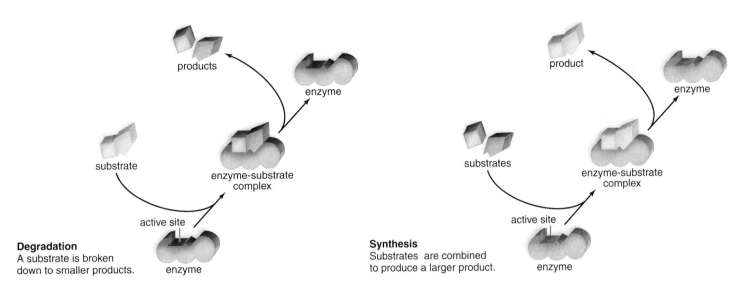

Figure 3.17 How do enzymes work?

An enzyme has an active site, where the substrates and enzyme fit together in such a way that the substrates are oriented to react. Following the reaction, the products are released, and the enzyme is free to act again. Some enzymes carry out degradation; the substrate is broken down to smaller products. Other enzymes carry out synthesis; the substrates are combined to produce a larger product.

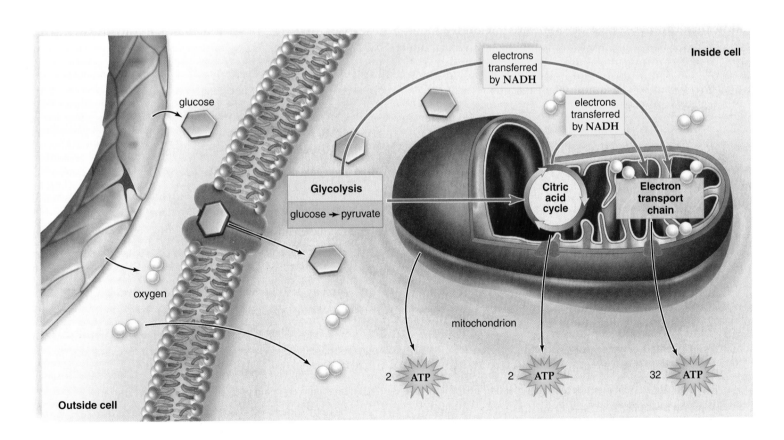

Figure 3.18 How is ATP produced?

Glucose enters a cell from the bloodstream by facilitated transport. The three main pathways of cellular respiration (glycolysis, citric acid cycle, and electron transport chain) all produce ATP, but most is produced by the electron transport chain. NADH carries electrons to the electron transport chain from glycolsis and the citric acid cycle. ATP exists a mitochondrion by facilitated transport.

energy within a glucose molecule to be slowly released, so that ATP can be gradually produced. Cells would lose a tremendous amount of energy if glucose breakdown occurred all at once. Much energy would be lost as heat. When humans burn wood or coal, the energy escapes all at once as heat. But a cell gradually "burns" glucose and energy is captured as ATP.

Glycolysis   The term **glycolysis** means sugar splitting. During glycolysis, glucose, a 6-carbon (C_6) molecule, is split so that the result is two 3-carbon (C_3) molecules of *pyruvate*. Glycolysis, which occurs in the cytoplasm, is found in most every type of cell. Therefore, this pathway is believed to have evolved early in the history of life.

Glycolysis is termed **anaerobic,** because it requires no oxygen. This pathway can occur in microbes that live in bogs or swamps or our intestinal tract, where there is no oxygen. During glycolysis, hydrogens and electrons are removed from glucose, and NADH results. The breaking of bonds releases enough energy for a net yield of two ATP molecules.

Pyruvate is a pivotal molecule in cellular respiration. When oxygen is available, the molecule enters mitochondria and is completely broken down. When oxygen is not available, fermentation occurs (see next page).

Citric acid cycle   The **citric acid cycle** completes the breakdown of glucose. As this cyclical series of enzymatic reactions occurs in the matrix of mitochondria, carbon dioxide is released. Hydrogen and electrons are carried away by NADH. In addition, the citric acid cycle also produces two ATP per glucose molecule.

CASE STUDY DIAGNOSIS

"We'll be doing a lot of tests, Josh," the ophthalmologist, Dr. Licht, told him at their visit. "I want a CT scan of your head, and an electroencephalography or EEG, to measure brain waves. That's the test where we paste electrodes all over your head. Then, we'll do an electrocardiogram to measure the heart's electrical activity."

"Do I have to get stuck for any of these tests?" Josh asked.

"I'm afraid so," Dr. Licht answered. "We have to do blood tests and some genetic tests as well. Do you know of any relatives with vision problems? What about heart problems? Muscle problems?" At each question, Josh shook his head.

"Let's get those tests out of the way as quickly as possible," Dr. Licht concluded.

Two weeks after his original episode of vision loss, Josh returned to Dr. Licht's office and found the ophthalmologist looking very sober and concerned. "I don't have very good news, Josh," he began. "The tests we did show that you have Leber's Hereditary Optic Neuropathy. It's called LHON for short."

"What's LHON?" Josh asked.

"It's an inherited, or genetic, disease, "Dr. Licht replied.

"You mean like sickle-cell anemia? That's a genetic disease, right?"

"Right, but LHON is different because it isn't inherited from the chromosomes in a cell's nucleus," Dr. Licht answered. "Instead, this disease is caused by a defective chromosome in mitochondria. Those are the cell organelles that make energy, and they have one circular chromosome inside," Dr. Licht replied, then asked, "Does your mom have vision problems?"

"I don't think so, but I'm not sure. She died when I was little," Josh answered. "Why is that important?"

"You inherited LHON from your mom," Dr. Licht answered. "I'm sure you remember from biology class that the father's sperm cells have only his DNA inside. They're too small to have other organelles. A mother's egg provides all of the mitochondria for the embryo. Your mom's mitochondria have this defective gene. The genetic studies showed that."

"So now what?" Josh fretted. "Am I going to go blind?"

"The good news is that your brain, heart, and muscles were all fine," the ophthalmologist explained. "The bad news is that the nerves found in the retinas of your eyes are slowly going to die. So yes, you will most likely eventually become blind."

"But there's medicines, right?" Josh cried. "You can fix it can't you?"

"I'm sorry, Josh, but there is no treatment and no cure for LHON," Dr. Licht answered sadly. "If you have brothers or sisters, they will get LHON too, because they all received mitochondria from your mom. You won't pass the disease on to your kids, though, because your sperm won't have mitochondria."

"Isn't there anything you can do, then?" Josh retorted.

The ophthalmologist shook his head. "All I can do is to recommend that you don't drink or smoke. I don't want your mitochondria to have to break down alcohol or tobacco and perhaps become even more damaged."

So far, we have considered only carbohydrates as a possible fuel for cellular respiration. But what about fats and proteins? Fats are digested to glycerol and fatty acids. When our cells run out of glucose, they primarily substitute fatty acids for glucose as an energy source. Fatty acids yield C_2 molecules that can enter the citric acid cycle.

Proteins are digested to amino acids whose carbon chain can easily enter the citric acid cycle. However, first the amino group has to be removed from the carbon chain. This step is primarily carried out in the liver because the liver has the appropriate enzymes to process amino groups. The liver converts the amino groups to urea, excreted by the kidneys. When you eat a lot of protein, your body begins to use amino acids as an energy source. Your kidneys work overtime excreting all that nitrogen.

Electron transport chain   NADH molecules from glycolysis and the citric acid cycle deliver electrons to the electron transport chain. The members of the electron transport chain are carrier proteins grouped into complexes. These complexes are embedded in the cristae of a mitochondrion. Each carrier of the **electron transport chain** accepts two electrons and passes them on to the next carrier. The hydrogens carried by NADH molecules will be used later.

High-energy electrons enter the chain and, as they are passed from carrier to carrier, the electrons lose energy. Low-energy electrons emerge from the chain. Oxygen serves as the final acceptor of the electrons at the end of the chain. After oxygen receives the electrons, it combines with hydrogens and becomes water.

The presence of oxygen makes the citric acid cycle and the electron transport chain **aerobic.** Oxygen does not combine with any substrates during cellular respiration. Breathing is necessary to our existence, and the sole purpose of oxygen is to receive electrons at the end of the electron transport chain.

The energy, released as electrons pass from carrier to carrier, is used for ATP production. It took many years for investigators to determine exactly how this occurs, but the details are beyond the scope of this text. Suffice it to say that the inner mitochondrial membrane contains an ATP synthase complex that combines ADP + Ⓟ to produce ATP. The ATP synthase complex produces about 32 ATP per glucose molecule.

Each cell produces ATP within its mitochondria, and therefore, each cell uses ATP for its own purposes. Figure 3.19 shows the ATP cycle. Glucose breakdown leads to ATP buildup, and then ATP is used for the metabolic work of the cell. Muscle cells use ATP for contraction, and nerve cells use it for conduction of nerve impulses. ATP breakdown releases heat.

Fermentation

Fermentation is an anaerobic process, meaning that it does not require oxygen. When oxygen is not available to cells,

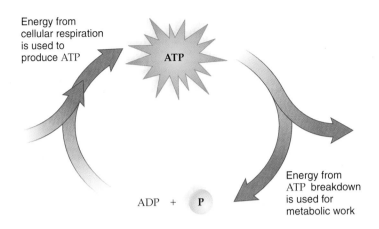

Figure 3.19   What is the ATP cycle?
The breakdown of glucose by cellular respiration transfers energy to form ATP. ATP is used for energy requiring reactions, such as muscle contraction. ATP breakdown also gives off heat. Additional food energy rejoins ADP and P to form ATP again.

the electron transport chain soon becomes inoperative. This is because oxygen is not present to accept electrons. In this case, most cells have a safety valve so that some ATP can still be produced. Glycolysis operates as long as it is supplied with "free" NAD^+—NAD^+ that can pick up hydrogens and electrons. Normally, NADH takes electrons to the electron transport chain and, thereby, becomes "free" of them. However, if the system is not working due to a lack of oxygen, NADH passes its hydrogens and electrons to pyruvate molecules, as shown in the following reaction:

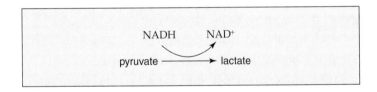

This means that the citric acid cycle and the electron transport chain do not function as part of fermentation. When oxygen is available again, lactate can be converted back to pyruvate, and metabolism can proceed as usual.

Fermentation can give us a burst of energy for a short time, but it produces only two ATP per glucose molecule. Also, fermentation results in the buildup of lactate. Lactate is toxic to cells and causes muscles to cramp and fatigue. If fermentation continues for any length of time, death follows.

Fermentation takes its name from yeast fermentation. Yeast fermentation produces alcohol and carbon dioxide (instead of lactate). When yeast is used to leaven bread, carbon dioxide production makes the bread rise. When yeast is used to produce alcoholic beverages, it is the alcohol that humans make use of.

Lactate and the Athlete

Exercise is a dramatic test of homeostatic mechanisms. During exercise, the mitochondria of our muscle cells require much oxygen. They also produce an increased amount of carbon dioxide. No doubt, if you run as fast as you can, even for a short time, you are out of breath. You are in oxygen deficit—your muscles have run out of oxygen and have started fermenting instead. Aerobic exercise occurs when you can manage to get a steady supply of oxygen to your muscle cells so that oxygen deficit does not occur. Athletes are better at this than nonathletes. Why?

The number of mitochondria is higher in the muscles of persons who train. Therefore, an athlete is more likely to rely on the citric acid cycle and the electron transport chain to generate ATP. The citric acid cycle can be powered by fatty acids, instead of glucose. Therefore, the level of glucose in the blood remains at a normal level, even though exercise is occurring.

Muscle cells with few mitochondria don't start consuming O_2 until they are out of ATP and ADP concentration is high. After endurance training, the large number of mitochondria start consuming O_2 as soon as the ADP concentration starts rising. This is due to muscle contraction and breakdown of ATP. This faster rise in O_2 uptake at the onset of exercise means that the O_2 deficit is less, and the formation of lactate due to fermentation is less.

As mentioned, the body is able to process lactate and change it back to pyruvate. This is the oxygen deficit—the amount of oxygen it takes to rid the body of lactate. Athletes incur less of an oxygen deficit than nonathletes.

Have You Ever Wondered . . .

Why are there holes in Swiss cheese?

The holes in Swiss cheese, known as "eyes" in the cheese-making business, are a product of fermentation. To make cheese, an enzyme called rennet is added to warm milk. Special bacteria are added next, and these bacteria ferment, producing propionic acid and carbon dioxide. The carbon dioxide forms bubbles in the mixture, which result in holes. Cheese makers can vary the size of the holes by aging the cheese longer or by varying temperature and acidity during fermentation.

Check Your Progress 3.6

1. Why does cellular respiration have so many small enzymatic reactions instead of one big step?
2. **a.** Which of the three pathways of cellular respiration occurs in the matrix of a mitochondrion? **b.** Which occurs at the cristae?
3. What's the difference between an enzyme and a coenzyme?
4. **a.** What's the role of oxygen in cellular respiration? **b.** Where does the carbon dioxide we breathe out come from?
5. **a.** What is the benefit of fermentation? **b.** What's the drawback?

CASE STUDY ONE YEAR LATER

"Hey slug, hurry up and get out here!" a teammate yelled as Josh tied on his basketball shoes. Josh smiled wryly. It was great to be able to perform repetitive tasks that didn't require good eyesight, like tying shoes. Sure enough, the sight in Josh's right eye had deteriorated during the six months after diagnosis, just as Dr. Licht had warned. About two months after he first noticed the problem, Josh began to lose sight in his left eye as well.

It had been very difficult for Josh to adjust to LHON. He had needed occupational therapy to adapt to his loss of vision. More important, he had needed his counselor to deal with anger and depression. It was still a struggle. The hardest part of adjusting to losing vision, and being legally blind, was being unable to hop in a car and drive wherever he wanted.

Yet it was a tremendous relief to Josh that LHON had not caused problems with his nervous, cardiovascular, or muscular systems. As Josh studied LHON, he learned that all mitochondrial diseases affect tissues with high energy requirements: brain, heart, muscle, and retinas. Further, one year later after diagnosis, his left eye was still only slightly affected. Josh also discovered that vision improved in one-third of individuals with LHON, and that improvement could occur years after the original diagnosis. Josh still longed for his vision to improve someday.

Josh's studies also showed that research into LHON showed promise for the future. Experimental therapies involved injecting cells with normal copies of the mutated mitochondrial chromosomes that had caused Josh's LHON. He had become a regular visitor to websites tracking research into LHON and other inherited eye diseases. Josh was hopeful and excited that new therapeutic options might become available.

Right now, everyday living was Josh's priority. He found dozens of products tailored for people with low vision. A clever screen magnifier and special keyboard allowed him to keep his job. Playing cards that had giant numbers kept him a part of poker night. Best of all, Josh still played basketball weekly. True, he couldn't catch a long pass or sink a three-point shot. Instead, using the vision remaining in his left eye, Josh caught short passes inside the key, where he could still do a decent layup.

"Shut up, jerk!" Josh laughed as he ran out to the court.

Stem Cell Research

In the human body, stem cells are analogous to immortal "parents." Their "offspring," called daughter cells, can remain as stem cells and potentially divide indefinitely. However, most daughter cells differentiate further, forming mature cells called end cells. Research using stem cells has remained a source of controversy since 1998, when scientists discovered how to isolate and grow human stem cells in the laboratory.

There are primarily two different types of stem cells: embryonic and adult. Advantages and disadvantages exist for each type. Embryonic stem cells are derived from fertilized embryos at various stages of development. Fertilized human ova stored in infertility clinics are often used as the source of embryonic stem cells. The use of these cells for research has sparked tremendous controversy, because many people believe these cells have the potential to become a human being. Adult stem cells are undifferentiated cells found in various body tissues, whose purpose is to repair or replace damaged tissues. The use of adult stem cells is generally accepted. However, adult stem cells lack the flexibility of embryonic stem cells. Adult cells form far fewer types of end cells.

Much research has gone into learning how to control the differentiation of stem cells into desired end cells. Scientists have experimented with adding various chemicals, companion cells, and/or genes to the stem cells to induce the correct type of differentiation. According to the National Institutes of Health, "[I]f scientists can reliably direct the differentiation of embryonic stem cells into specific cell types, they may be able to use the resulting, differentiated cells to treat certain diseases at some point in the future. Diseases that might be treated by transplanting cells generated from human embryonic stem cells include Parkinson's disease, diabetes, traumatic spinal cord injury, Purkinje cell degeneration, Duchenne's muscular dystrophy, heart disease, and vision and hearing loss."*

With all the time and money spent on stem cell research, how close are we to using stem cells for the cure of these diseases? Let's use Parkinson's disease as an example. Parkinson's disease is a progressive motor control disorder, triggered by the death of certain neurons in the brain (Fig. 3A). These neurons are responsible for releasing the neurotransmitter dopamine onto specific brain cells that control movement. (This is why Parkinson's patients are often treated with L-Dopa, which is converted into dopamine.) It is now possible to cause stem cells in the laboratory to differentiate into neurons that produce dopamine. To be used for transplant purposes, however, the stem cells must produce enough end cells for transplant. Further, the cells must survive after the transplant and function cor-

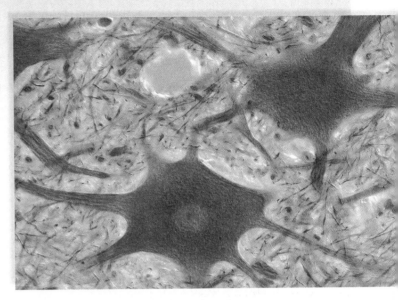

Figure 3A   What type of stem cells are needed to cure Parkinson's disease?

Parkinson's disease results in the loss of neurons such as these. Stem cells that differentiate into dopamine-producing neurons could be transplanted into the brains of Parkinson's patients.

rectly for the remainder of the patient's life. Finally, transplanted cells must not harm the patient. The usual risks of surgery would still exist for the transplant recipient: damage to healthy tissue, bleeding, infection. To date, not one stem cell transplant has even been attempted. Again, according to the National Institutes of Health, "[T]o summarize, the promise of stem cell therapies is an exciting one, but significant technical hurdles remain that will only be overcome through years of intensive research."*

Decide Your Opinion

1. How much time and money should be spent on a therapy that may only work after "years of intensive research"? Would this money better be spent on therapies that have a higher likelihood of success?
2. Should the president remove the ban on certain types of stem cells, so that this research can proceed faster?
3. Which of the listed diseases should we focus on first? What criteria did you use in your decision?

*Stem Cell Basics: What are embryonic stem cells? In *Stem Cell Information* [World Wide Web Site]. Bethesda, MD: National Institutes of Health, U.S. Department of Health and Human Services, 2006 [cited Wednesday, September 12, 2007] Available at <http://stemcells.nih.gov/info/basics/basics3>

Summarizing the Concepts

3.1 What Is a Cell?

- Cells, the basic units of life, come from preexisting cells.
- Microscopes are used to view cells, which must remain small to have a favorable surface area-to-volume ratio.

3.2 How Cells Are Organized

The human cell is surrounded by a plasma membrane and has a central nucleus. Between the plasma membrane and the nucleus is the cytoplasm, which contains various organelles. Organelles in the cytoplasm have specific functions.

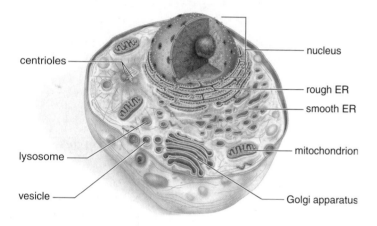

3.3 The Plasma Membrane and How Substances Cross It

The plasma membrane is a phospholipid bilayer that

- selectively regulates the passage of molecules and ions into and out of the cell.
- contains embedded proteins, which allow certain substances to cross the plasma membrane.

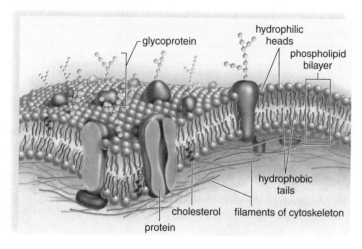

Passage of molecules into or out of cells can be passive or active.

Passive mechanisms (no energy required) are

- diffusion (osmosis) and facilitated transport.

Active mechanisms (energy required) are

- active transport and endocytosis and exocytosis.

3.4 The Nucleus and the Production of Proteins

- The nucleus houses DNA, which specifies the order of amino acids in proteins.
- Chromatin is a combination of DNA molecules and proteins that make up chromosomes.
- The nucleolus produces ribosomal RNA (rRNA).
- Protein synthesis occurs in ribosomes, small organelles composed of proteins and rRNA.

The Endomembrane System

The endomembrane system consists of the nuclear envelope, endoplasmic reticulum (ER), Golgi apparatus, lysosomes, and vesicles.

- Rough ER has ribosomes, where protein synthesis occurs.
- Smooth ER has no ribosomes and has various functions, including lipid synthesis.
- The Golgi apparatus processes and packages proteins and lipids into vesicles for secretion or movement into other parts of the cell.
- Lysosomes are specialized vesicles produced by the Golgi apparatus. They fuse with incoming vesicles to digest enclosed material, and they autodigest old cell parts.

3.5 The Cytoskeleton and Cell Movement

The cytoskeleton consists of microtubules, actin filaments, and intermediate filaments that give cells their shape and allows organelles to move about the cell. Cilia and flagella, which contain microtubules, allow a cell to move.

3.6 Mitochondria and Cellular Metabolism

- Mitochondria have an inner membrane that forms cristae, which project into the matrix.
- Mitochondria are involved in cellular respiration, which uses oxygen and releases carbon dioxide.
- During cellular respiration, mitochondria convert the energy of glucose into the energy of ATP molecules.

Cellular Respiration and Metabolism

- A metabolic pathway is a series of reactions, each of which has its own enzyme.
- Enzymes bind their substrates in the active site.
- Sometimes enzymes require coenzymes (such as NAD^+), nonprotein molecules that participate in the reaction.

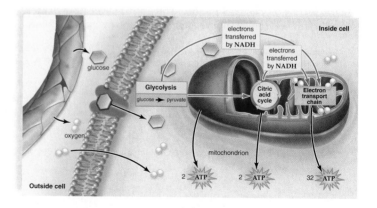

- Cellular respiration is the enzymatic breakdown of glucose to carbon dioxide and water.
- Cellular respiration includes three pathways: glycolysis (occurs in the cytoplasm and is anaerobic), the citric acid cycle (releases carbon dioxide), and the electron transport chain (passes electrons to oxygen).

Fermentation

- If oxygen is not available in cells, the electron transport chain is inoperative, and fermentation (which does not require oxygen) occurs.
- Fermentation produces very little ATP.
- Lactate buildup puts the individual in oxygen deficit.

Understanding Key Terms

actin filament 54	intermediate filament 55
active site 56	lysosome 53
active transport 50	metabolism 56
aerobic 59	microtubule 54
anaerobic 58	mitochondrion 56
cell theory 44	NAD+ (nicotinamide adenine
cellular respiration 56	dinucleotide) 56
centrosome 54	nuclear envelope 52
chromatin 52	nuclear pore 52
chromosome 52	nucleolus 52
cilium 55	nucleoplasm 52
citric acid cycle 58	nucleus 46
coenzyme 56	organelle 46
cytoplasm 46	osmosis 49
cytoskeleton 54	osmotic pressure 50
diffusion 49	phagocytosis 51
electron transport chain 59	plasma membrane 46
endomembrane system 53	polyribosome 52
endoplasmic reticulum (ER) 52	product 56
eukaryotic cell 46	prokaryotic cell 46
facilitated transport 50	reactant 56
fermentation 59	ribosome 52
flagellum 55	selectively permeable 49
fluid-mosaic model 48	substrate 56
glycolysis 58	tonicity 49
Golgi apparatus 53	vesicle 53

Match the key terms to these definitions.

a. _____ Protein molecules form a shifting pattern within the fluid phospholipid bilayer.

b. _____ Diffusion of water through a selectively permeable membrane.

c. _____ The cell will allow some substances to pass through while not permitting others.

d. _____ Anaerobic breakdown of glucose that results in a gain of two ATP and end products, such as alcohol and lactate.

e. _____ Metabolic pathways that use energy from carbohydrate, fatty acid, and protein break down to produce ATP molecules.

Testing Your Knowledge of the Concepts

Choose the best answer for each question.

1. Explain the three tenets of the cell theory. (page 44)
2. Which type of microscope would you use to observe the swimming behavior of a flagellated protozoan? Explain. (page 45)
3. Describe how the eukaryotic cell gained mitochondria and chloroplasts. (page 46)
4. Invagination of plasma membrane produced what structures in eukaryotic cells not present in prokaryotic cells? (page 46)
5. What are glycoproteins, and what functions do proteins, including glycoproteins, have in the plasma membrane? (pages 48–49)
6. A plant wilts if its roots are placed in which type solution (i.e., a hypertonic, a hypotonic, or an isotonic solution)? Explain. (pages 48–49)
7. What is endocytosis and exocytosis, and how do they occur? (page 51)
8. For the following cell organelles, describe the structure and function of each: nucleus, nucleolus, ribosomes, endoplasmic reticulum (rough and smooth), Golgi apparatus, lysosomes, centrioles, and mitochondria. (pages 52–56)
9. Describe the structure and function of the cytoskeleton. (pages 54–55)
10. Describe an enzyme and coenzyme. Explain the mechanism of enzyme function, particularly the relationship of shape to its activity. (pages 56–58)
11. Which stage of cellular respiration produces the most ATP? Explain. (pages 58–59)
12. Running for the bus may produce an oxygen deficit. Explain. (page 60)
13. The cell theory states
    a. cells form as organelles, and molecules become grouped together in an organized manner.
    b. the normal functioning of an organism does not depend on its individual cells.
    c. the cell is the basic unit of life for all living things.
    d. only animals are made of cells.
14. The small size of cells is best correlated with
    a. the fact that they are self-reproducing.
    b. an adequate surface area for exchange of materials.
    c. their vast versatility.
    d. All of these are correct.
15. A phospholipid has a head and two tails. The tails are found
    a. at the surfaces of the membrane.
    b. in the interior of the membrane.
    c. spanning the membrane.
    d. where the environment is hydrophobic.
    e. Both b and d are correct.

16. Facilitated diffusion differs from diffusion in that facilitated diffusion
    a. involves the passive use of a carrier protein.
    b. involves the active use of a carrier protein.
    c. moves a molecule from a low to high concentration.
    d. involves the use of ATP molecules.

17. When a cell is placed in a hypotonic solution,
    a. solute exits the cell to equalize the concentration on both sides of the membrane.
    b. water exits the cell toward the area of lower solute concentration.
    c. water enters the cell toward the area of higher solute concentration.
    d. solute exits and water enters the cell.

In questions 18–21, match each function to the proper organelle in the key.

Key:

a. mitochondrion   c. Golgi apparatus
b. nucleus   d. rough ER

18. Packaging and secretion

19. Powerhouse of cell

20. Protein synthesis

21. Control center for the cell

22. Vesicles carrying proteins for secretion move from the ER to the
    a. smooth ER.   c. Golgi apparatus.
    b. lysosomes.   d. nucleolus.

23. Lysosomes function in
    a. protein synthesis.
    b. processing and packaging.
    c. intracellular digestion.
    d. lipid synthesis.

24. Mitochondria
    a. are involved in cellular respiration.
    b. break down ATP to release energy for cells.
    c. contain hemoglobin and cristae.
    d. have a convoluted outer membrane.
    e. All of these are correct.

25. The active site of an enzyme
    a. is identical to that of any other enzyme.
    b. is the part of the enzyme where the substrate can fit.
    c. can be used over and over.
    d. is where the coenzyme binds.
    e. Both b and c are correct.

26. The metabolic process that produces the most ATP molecules is
    a. glycolysis.
    b. the citric acid cycle.
    c. the electron transport chain.
    d. fermentation.

27. The oxygen required by cellular respiration becomes part of which molecule?
    a. ATP
    b. H_2O
    c. pyruvate
    d. CO_2

28. Use these terms to label the following diagram of the plasma membrane: carbohydrate chain, filaments of the cytoskeleton, hydrophilic heads, hydrophobic tails, protein (used twice), phospholipid bilayer.

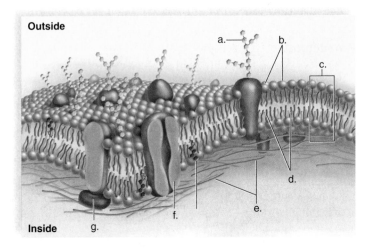

29. Use these terms to label the following diagram: substrates, enzyme (used twice), active site, product, and enzyme-substrate complex.

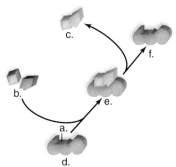

Thinking Critically About the Concepts

In the chapter case study, Josh Citrigno is affected by LHON. This genetic disease occurs because of a defect in the mitochondrial chromosome. This single ring of DNA is passed down from mother to child in the mother's egg. Scientists speculate that LHON and other mitochondrial diseases occur as a result of a *spontaneous mutation*—a change in DNA that occurs for no apparent reason. Spontaneous mutations may affect the reproductive cells (eggs or sperm) and produce a change in the offspring. If body cells (called somatic cells) are affected, the change is seen in the individual instead of the offspring.

1. Josh's condition will affect his mother and any siblings, but not his father or any offspring he might have. Explain why this is true.

2. In some cases, LHON can be fatal. Why?

CHAPTER **4**

Organization and Regulation of Body Systems

J ayme Wesseling was in the bathroom getting ready for school. It was difficult to find bathroom time in the apartment that she shared with three other girls. She and her roommate, Kristen Spracklen, both had 8:00 AM classes and so they had to plan carefully. Right now, Kristen was making some oatmeal for breakfast, so Jayme was trying to finish up in the bathroom. "OWWWW!" Kristen suddenly yelled.

Jayme froze. Kristen's yell had sounded like real pain. Then she heard the crash of a metal pan and the breaking of glass. She ran into the kitchen. "Kristen, what happened?! Are you okay?!"

Kristen had her hand under the faucet with the cold water running full force. There was oatmeal all over the counter and the bowl was broken on the floor. Tears were streaming down Kristen's face and she was moaning. "Kristen, what happened?!" Jayme shouted.

"I was studying my vocab cards for French and wasn't really paying attention," Kristen sobbed. "I tried to pour the oatmeal out of the pan into the bowl. Somehow I missed the bowl and got oatmeal all over the back of my hand. I think I've burnt it pretty badly."

Jayme walked over to the sink, avoiding the broken glass, and took Kristen's hand. It was bright red and already beginning to blister. "I think we'd better go to Student Health Services right now, Kristen. I'll walk with you." They got a plastic bag and filled it with ice. Kristen was obviously in tremendous pain.

It wasn't far to health services, which had already opened by the time they got there. Taking one look at Kristen's face, the student receptionist took them immediately to a room in the back.

4.1 Types of Tissues

Recall the biological levels of organization (Fig. 1.2). Cells are composed of molecules; a tissue has like cells; an organ contains several types of tissues; and several organs are found in an organ system. In this chapter, we consider the tissue, organ, and organ system levels of organization.

A **tissue** is composed of specialized cells of the same type that perform a common function in the body. The tissues of the human body can be categorized into four major types:

Connective tissue binds and supports body parts.
Muscular tissue moves the body and its parts.
Nervous tissue receives stimuli and conducts nerve impulses.
Epithelial tissue covers body surfaces and lines body cavities.

Cancers are classified according to the type of tissue from which they arise. Sarcomas are cancers arising in muscle or connective tissue (especially bone or cartilage). Leukemias are cancers of the blood. Lymphomas are cancers of lymphoid tissue. Carcinomas, the most common type, are cancers of epithelial tissue. The chance of developing cancer in a particular tissue is related to the rate of cell division. Epithelial cells reproduce at a high rate, and 2,500,000 new blood cells appear each second. Thus, carcinomas and leukemias are common types of cancers.

> **Check Your Progress 4.1**
>
>   1. **What are the four major tissue types found in the human body?**

4.2 Connective Tissue Connects and Supports

Connective tissue is diverse in structure and function. Even so, all types have three components: specialized cells, ground substance, and protein fibers. These components are shown in Figure 4.1, a diagrammatic representation of loose fibrous connective tissue. The ground substance is a noncellular material that separates the cells. It varies in consistency from solid to semifluid to fluid.

The fibers are of three possible types. White **collagen fibers** contain collagen, a protein that gives them flexibility and strength. **Reticular fibers** are very thin collagen fibers, highly branched and forming delicate supporting networks. Yellow **elastic fibers** contain elastin, a protein that is not as strong as collagen but is more elastic. Inherited connective tissue disorders arise when people inherit genes that lead to malformed fibers. For example, in Marfan syndrome, there are mutations in the fibrillin gene. Fibrillin is a component of elastic fibers. The mutation results in decreased elasticity in connective tissues normally rich in elastic fibers, such as the aorta. Individuals with this disease often die from aortic rupture, when the aorta cannot expand in response to increased blood pressure.

Fibrous Connective Tissue

Fibrous tissue exists in two forms: loose fibrous tissue and dense fibrous tissue. Both loose fibrous and dense fibrous connective tissues have cells called **fibroblasts** located some distance from one another and separated by a jellylike ground substance containing white collagen fibers and yellow elastic fibers (Fig. 4.2). **Matrix** is a term that includes ground substance and fibers.

Figure 4.1   What does connective tissue look like?
All connective tissues have three components: specialized cells, ground substance, and protein fibers. Loose fibrous connective tissue is shown here.

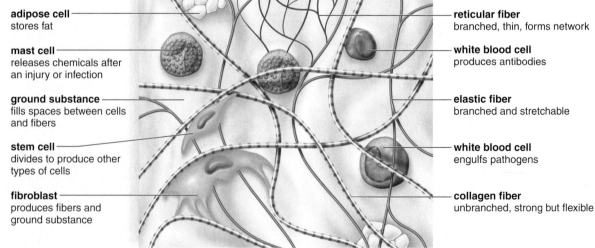

adipose cell
stores fat

mast cell
releases chemicals after an injury or infection

ground substance
fills spaces between cells and fibers

stem cell
divides to produce other types of cells

fibroblast
produces fibers and ground substance

reticular fiber
branched, thin, forms network

white blood cell
produces antibodies

elastic fiber
branched and stretchable

white blood cell
engulfs pathogens

collagen fiber
unbranched, strong but flexible

Loose fibrous connective tissue, also called areolar tissue, supports epithelium and many internal organs. Its presence in lungs, arteries, and the urinary bladder allows these organs to expand. It forms a protective covering enclosing many internal organs, such as muscles, blood vessels, and nerves.

Adipose tissue is a special type of loose connective tissue in which the cells enlarge and store fat. Adipose tissue has little or no extracellular matrix. Its cells are crowded, and each is filled with liquid fat. The body uses this stored fat for energy, insulation, and organ protection. Adipose tissue is found beneath the skin, around the kidneys, and on the surface of the heart.

Dense fibrous connective tissue contains many collagen fibers packed together. This type of tissue has more specific functions than does loose connective tissue. For example, dense fibrous connective tissue is found in tendons, which connect muscles to bones, and in ligaments, which connect bones to other bones at joints.

Supportive Connective Tissue

Cartilage and bone are supportive connective tissues. In both tissues, the extracellular matrix is solid. Chondroblasts and chondrocytes produce the matrix in cartilage, while osteoblasts and osteocytes form bone matrix.

Cartilage

In cartilage, the cells lie in small chambers called lacunae (sing., lacuna), separated by a solid, yet flexible, matrix. Unfortunately, because this tissue lacks a direct blood supply, it heals slowly. There are three types of cartilage, distinguished by the type of fiber.

Hyaline cartilage (Fig. 4.2), the most common type of cartilage, contains only fine collagen fibers. The matrix has a glassy, translucent appearance. Hyaline cartilage is found in the nose and at the ends of the long bones and the ribs, and it forms rings in the walls of respiratory passages. The fetal skeleton also is made of this type of cartilage. Later, the cartilaginous fetal skeleton is replaced by bone.

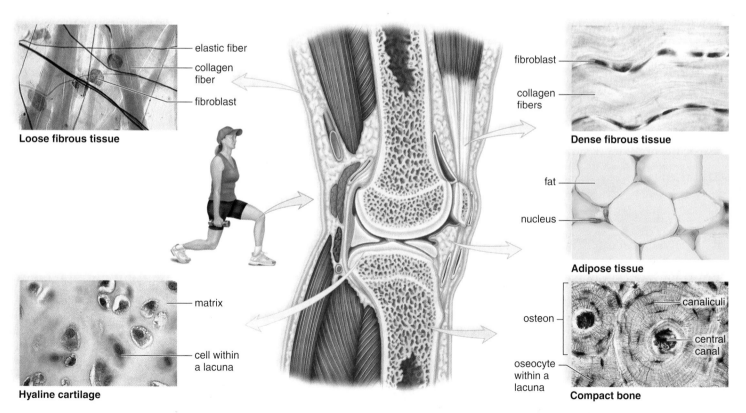

Figure 4.2 What types of connective tissue are associated with the knee?
The human knee provides examples of most types of connective tissue.

Elastic cartilage has more elastic fibers than hyaline cartilage. For this reason, it is more flexible and is found, for example, in the framework of the outer ear.

Fibrocartilage has a matrix containing strong collagen fibers. Fibrocartilage is found in structures that withstand tension and pressure, such as the disks between the vertebrae in the backbone and the cushions in the knee joint.

Bone

Bone is the most rigid connective tissue. It consists of an extremely hard matrix of inorganic salts, notably calcium salts. These salts are deposited around protein fibers, especially collagen fibers. The inorganic salts give bone rigidity. The protein fibers provide elasticity and strength, much as steel rods do in reinforced concrete.

Compact bone makes up the shaft of a long bone (Fig. 4.2). It consists of cylindrical structural units called osteons (Haversian systems). The central canal of each osteon is surrounded by rings of hard matrix. Bone cells are located in spaces called lacunae between the rings of matrix. In the central canal, nerve fibers carry nerve impulses and blood vessels carry nutrients that allow bone to renew itself. Thin extensions of bone cells within canaliculi (minute canals) connect the cells to each other and to the central canal.

The ends of the long bones are composed of spongy bone covered by compact bone. Spongy bone also surrounds the bone marrow cavity. This is, in turn, covered by compact bone forming a "sandwich" structure. **Spongy bone** appears as an open, bony latticework with numerous bony bars and plates. These are separated by irregular spaces. Although lighter than compact bone, spongy bone is still designed for strength. Just as braces are used for support in buildings, the solid portions of spongy bone follow lines of stress.

Fluid Connective Tissues

The body has two fluid connective tissues: blood and lymph.

Blood

Some people do not classify blood as connective tissue; instead, they suggest a separate tissue category called vascular tissue. **Blood,** which consists of formed elements (Fig. 4.3) and plasma, is located in blood vessels. Blood transports nutrients and oxygen to **tissue fluid.** Tissue fluid bathes the body's cells and removes carbon dioxide and other wastes. Blood helps distribute heat and also plays a role in fluid, ion, and pH balance. The systems of the body help keep blood composition and chemistry within normal limits.

The formed elements each have specific functions. The **red blood cells (erythrocytes)** are small, biconcave, disk-shaped cells without nuclei. The presence of the red pigment hemoglobin makes the cells red. In turn, this makes the blood red. Hemoglobin is composed of four units. Each unit is composed of the protein globin and a complex

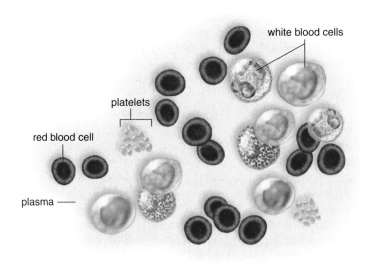

Figure 4.3   What are the formed elements in blood?
Red blood cells, which lack a nucleus, transport oxygen. Each type of white blood cell has a particular way to fight infections. Platelets, fragments of a particular cell, function in helping to seal injured blood vessels.

iron-containing structure called heme. The iron forms a loose association with oxygen and, in this way, red blood cells transport oxygen.

White blood cells (leukocytes) may be distinguished from red blood cells because they have a nucleus. Without staining, leukocytes would be translucent. White blood cells characteristically vary from mostly bluish to pink because they have been stained. White blood cells fight infection, primarily in two ways. Some white blood cells are phagocytic and engulf infectious agents. Other white blood cells produce antibodies, molecules that combine with foreign substances to inactivate them. Certain specialized white blood cells are able to destroy other cells outright.

Platelets (thrombocytes) are not complete cells. Rather, they are fragments of giant cells present only in bone marrow. When a blood vessel is damaged, platelets form a plug that seals the vessel, and injured tissues release molecules that help the clotting process.

Lymph

Lymph is also a fluid connective tissue. Lymph is a clear, watery, sometimes faintly yellowish fluid derived from tissue fluid. It contains white blood cells. Lymphatic vessels absorb excess tissue fluid and various dissolved solutes in the tissues. They transport lymph to particular vessels of the cardiovascular system. Lymphatic vessels absorb fat molecules from the small intestine. Lymph nodes, composed of fibrous connective tissue, occur along the length of lymphatic vessels. Lymph is cleansed as it passes through lymph nodes, in particular, because white blood cells congregate there. Lymph nodes enlarge when you have an infection.

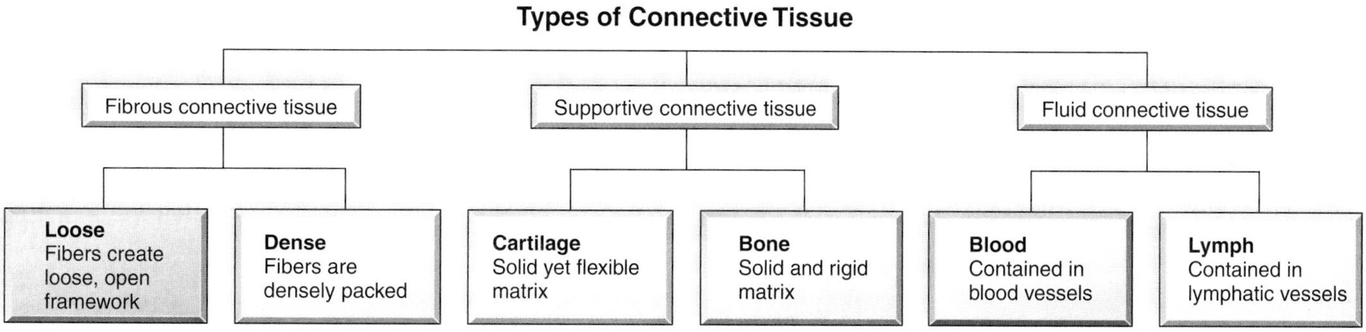

Types of Connective Tissue

Connective tissue is divided into:
- **Fibrous connective tissue**
  - **Loose** Fibers create loose, open framework
  - **Dense** Fibers are densely packed
- **Supportive connective tissue**
  - **Cartilage** Solid yet flexible matrix
  - **Bone** Solid and rigid matrix
- **Fluid connective tissue**
  - **Blood** Contained in blood vessels
  - **Lymph** Contained in lymphatic vessels

Figure 4.4 **What are the types of connective tissue?**
Connective tissue is divided into fibrous, supportive, and fluid types.

> **Check Your Progress 4.2**
> 1. What are the three types of connective tissue, and what are some examples of each type (Fig. 4.4)?
> 2. Compare and contrast loose fibrous connective tissue with dense fibrous connective tissue.
> 3. Compare and contrast spongy bone with compact bone.
> 4. Compare and contrast blood with lymph.

4.3 Muscular Tissue Moves the Body

Muscular (contractile) tissue is composed of cells called muscle fibers. Muscle fibers contain protein filaments, called actin and myosin filaments. The interaction of actin and myosin accounts for movement. The three types of vertebrate muscular tissue are skeletal, smooth, and cardiac.

Skeletal muscle is also called voluntary muscle (Fig. 4.5a). It is attached by tendons to the bones of the skeleton. When it contracts, body parts move. Contraction of skeletal muscle is under voluntary control and occurs faster than in the other muscle types. Skeletal muscle fibers are cylindrical and long—sometimes they run the length of the muscle. They arise during development when several cells fuse, resulting in one fiber with multiple nuclei. The nuclei are located at the periphery of the cell, just inside the plasma membrane. The fibers have alternating light and dark bands that give them a **striated,** or striped, appearance. These bands are due to the placement of actin filaments and myosin filaments in the cell.

Smooth (visceral) muscle is so named because the cells lack striations (Fig. 4.5b). The spindle-shaped cells each have a single nucleus. These cells form layers in which the thick middle portion of one cell is opposite the thin ends of adjacent cells. Consequently, the nuclei form an irregular pattern in the

---

CASE STUDY AT HEALTH SERVICES

It was only a minute before the nurse appeared. She and Jayme helped Kristen sit on the table and remove the ice bag. "What happened?" she asked.

"I spilled hot oatmeal all over the back of my hand," Kristen gasped. "It's really killing me right now."

The nurse inspected the back of Kristen's hand. She was careful not to touch the area or cause Kristen any more pain. "It looks like you have a partial thickness burn," she explained. " It's good that you have pain right now. That means that the nerve endings in your skin weren't completely destroyed. We'll get you some pain medication and treat that burn."

The nurse returned in a few minutes with two white tablets and a glass of water. Kristen swallowed the pills and then lay back on the table. "You hang in there. The doctor will be here in about 15 minutes to start cleaning that burn," she reassured Kristen. "By that time, the pain meds should have kicked in, and it'll hurt a lot less."

"I'm not sure I want anyone to touch my hand," Kristen objected.

"Well, the doctor will need to clean the area very carefully. One of the greatest risks for a burn is infection," the nurse asserted. "He may or may not open up the blisters. He'll make that judgment when he starts to clean the area. He'll cover the burn with antibiotic ointment and wrap it up."

"What about my afternoon classes?" Kristen fretted.

The nurse reassured her, "You'll be able to leave after you're bandaged up, although you probably won't want to go to class today, especially with those pain meds in you. Your thinking won't be very clear. I'll write you a note for your classes."

The nurse looked at Jayme. "I'll also write you a note so that you can stay with her today. With those medications, you probably don't want to leave her alone."

Skeletal muscle
- has striated cells with multiple nuclei.
- occurs in muscles attached to skeleton.
- functions in voluntary movement of body.

Smooth muscle
- has spindle-shaped cells, each with a single nucleus.
- cells have no striations.
- functions in movement of substances in lumens of body.
- is involuntary.
- is found in blood vessel walls and walls of the digestive tract.

Cardiac muscle
- has branching, striated cells, each with a single nucleus.
- occurs in the wall of the heart.
- functions in the pumping of blood.
- is involuntary.

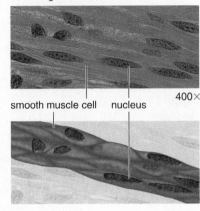

muscle fiber

striation   nucleus   250×

smooth muscle cell   nucleus   400×

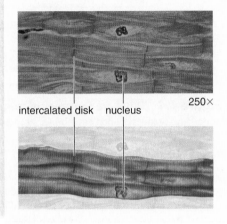

intercalated disk   nucleus   250×

a.   b.   c.

Figure 4.5   **What are the three types of muscular tissue?**
a. Skeletal muscle is voluntary and striated. **b.** Smooth muscle is involuntary and nonstriated. **c.** Cardiac muscle is involuntary and striated. Cardiac muscle cells branch and fit together at intercalated disks.

tissue. Smooth muscle is not under conscious or voluntary control. Therefore it is said to be involuntary. Smooth muscle is found in the walls of viscera (intestine, bladder, and other internal organs) and blood vessels. It contracts more slowly than skeletal muscle but can remain contracted for a longer time. When the smooth muscle of the bladder contracts, urine is sent into a tube called the urethra, which takes it to the outside. When the smooth muscle of the blood vessels contracts, blood vessels constrict, helping to raise blood pressure.

Cardiac muscle (Fig. 4.5c) is found only in the walls of the heart. Its contraction pumps blood and accounts for the heartbeat. Cardiac muscle combines features of both smooth and skeletal muscle. Like skeletal muscle, it has striations, but the contraction of the heart is involuntary for the most part. Cardiac muscle cells also differ from skeletal muscle cells in that they usually have a single, centrally placed nucleus. The cells are branched and seemingly fused one with another. The heart appears to be composed of one large interconnecting mass of muscle cells. Cardiac muscle cells are separate and individual, but they are bound end to end at **intercalated disks**. These are areas where folded plasma membranes between two cells contain adhesion junctions and gap junctions (see pages 74–76).

Check Your Progress 4.3

1. How does the structure and function of skeletal, smooth, and cardiac muscle differ?
2. Where do you find these muscles in the body?

4.4 Nervous Tissue Communicates

Nervous tissue consists of nerve cells, called neurons, and neuroglia, the cells that support and nourish the neurons.

Neurons

A **neuron** is a specialized cell that has three parts: dendrites, a cell body, and an axon (Fig. 4.6). A dendrite is an extension that receives signals from sensory receptors or other neurons. The cell body contains most of the cell's cytoplasm and the nucleus. An axon is an extension that conducts nerve impulses. Long axons are covered by myelin, a white fatty substance. The term *fiber*[1] is used here to refer to an axon along with its myelin sheath, if it has one. Outside the brain and spinal cord, fibers bound by connective tissue form **nerves.**

The nervous system has three functions: sensory input, integration of data, and motor output. Nerves conduct signals from sensory receptors to the spinal cord and the brain, where integration occurs. The phenomenon called sensation occurs only in the brain, however. Nerves also conduct signals from the spinal cord and brain to muscles and glands, and other organs. This triggers a characteristic response from each tissue. For example, muscles contract and glands secrete. In this way, a coordinated response to the original sensory input is achieved.

[1]In connective tissue, a fiber is a component of the matrix; muscle tissue, a fiber is a muscle cell; in nervous tissue, a fiber is an axon.

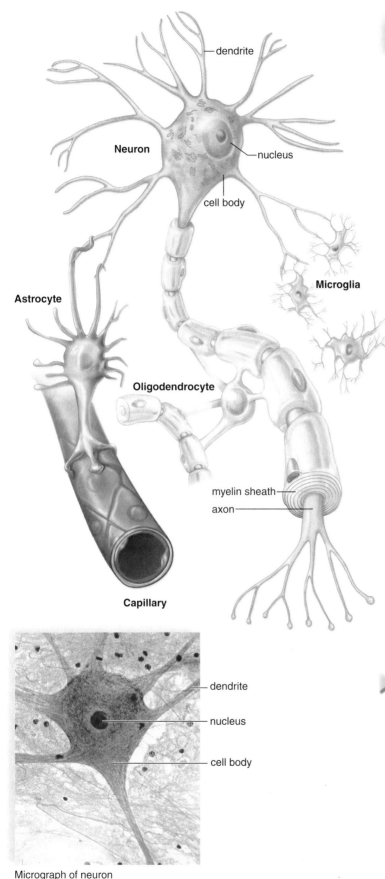

dendrite

Neuron

nucleus

cell body

Astrocyte

Microglia

Oligodendrocyte

myelin sheath

axon

Capillary

dendrite

nucleus

cell body

Micrograph of neuron

Have You Ever Wondered...

How fast is a reflex?

A reflex is a built-in pathway that allows the body to react quickly to a response. One example, the knee jerk, or *patellar* reflex, is tested by tapping just below the knee cap. The lower leg will then involuntarily kick forward. The reaction is designed to protect the thigh muscle from excessive stretch. The knee-jerk reflex is an example of a simple stretch reflex. There is only one pathway required: the stretch sensation (caused by tapping the knee), to the spinal cord, to the leg muscle. The whole circuit is complete within milliseconds—or 1/1000th of a second!

Neuroglia

In addition to neurons, nervous tissue contains neuroglia. **Neuroglia** are cells that outnumber neurons nine to one and take up more than half the volume of the brain. Although the primary function of neuroglia is to support and nourish neurons, research is being conducted to determine how much they directly contribute to brain function. Types of neuroglia found in the brain are, for example, microglia, astrocytes, and oligodendrocytes (Fig. 4.6, *top*). Microglia, in addition to supporting neurons, engulf bacterial and cellular debris. Astrocytes provide nutrients to neurons and produce a hormone known as glial-derived growth factor. This growth factor someday might be used as a cure for Parkinson's disease and other diseases caused by neuron degeneration. Oligodendrocytes form the myelin sheaths around fibers in the brain and spinal cord. Outside the brain, Schwann cells are the type of neuroglia that encircle long nerve fibers and form a myelin sheath. Neuroglia do not have long extensions. Researchers are gathering evidence that neuroglia do communicate among themselves and with neurons, even without these extensions.

> ### Check Your Progress 4.4
>
> 1. a. What are the three parts of a neuron, and (b) what does each part do?
> 2. What are some specific functions of neuroglia?

Figure 4.6 **What types of nerve cells are in the brain?**
Neurons conduct nerve impulses. Neuroglia consists of cells that support and service neurons and have various functions: Microglia are a type of neuroglia that become mobile in response to inflammation and phagocytize debris. Astrocytes lie between neurons and a capillary. Therefore, substances entering neurons from the blood must first pass through astrocytes. Oligodendrocytes form the myelin sheaths around fibers in the brain and spinal cord.

Science Focus

Nerve Regeneration and Stem Cells

In humans, axons outside the brain and spinal cord can regenerate, but not those inside these organs (Fig. 4A). After injury, axons in the human central nervous system (CNS) degenerate, resulting in permanent loss of nervous function. Interestingly, about 90% of the cells in the brain and the spinal cord are not even neurons. They are glial cells. In nerves outside the brain and spinal cord, the glial cells are Schwann cells that help axons regenerate. The glial cells in the CNS are oligodendrocytes and astrocytes, and they inhibit axon regeneration.

The spinal cord does contain its own stem cells. When the spinal cord is injured in experimental animals, these stem cells do proliferate. But instead of becoming functional neurons, they become glial cells. Researchers are trying to understand the process that triggers the stem cells to become glial cells. In the future, this understanding would allow manipulation of stem cells into neurons.

In early experiments with neural stem cells in the laboratory, scientists at Johns Hopkins University caused embryonic stem cells to differentiate into spinal cord motor neurons, the type of nerve cell that causes muscles to contract. The motor neurons then produced axons. When grown in the same dish with muscle cells, the motor neurons formed neuromuscular junctions, and even caused muscle contractions. The cells were then transplanted into the spinal cords of rats with spinal cord injuries. Some of the transplanted cells survived for longer than a month within the spinal cord. However, no improvement in symptoms was seen and no functional neuron connections were made.

4.5 Epithelial Tissue Protects

Epithelial tissue, also called epithelium (pl., epithelia), consists of tightly packed cells that form a continuous layer. Epithelial tissue covers surfaces and lines body cavities. Usually, it has a protective function. It can also be modified to carry out secretion, absorption, excretion, and filtration.

Epithelial cells are exposed to the environment on one side. On the other side, they are bounded by a **basement membrane.** The basement membrane should not be confused with the plasma membrane or the body membranes we will be discussing. It is a thin layer of various types of carbohydrates and proteins that anchors the epithelium to underlying connective tissue.

Figure 4.7   What are the basic types of epithelia?
Basic epithelial tissues found in humans are shown, along with locations of the tissue and the primary function of the tissue at these locations.

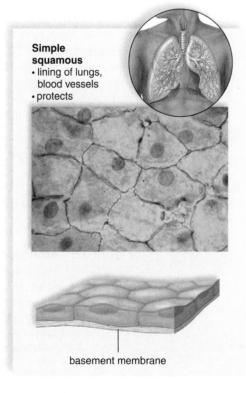

Simple squamous
• lining of lungs, blood vessels
• protects

basement membrane

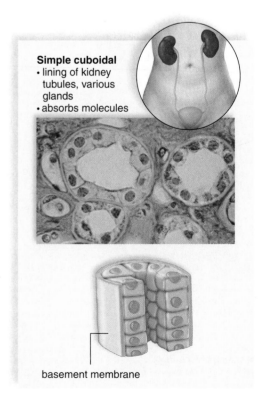

Simple cuboidal
• lining of kidney tubules, various glands
• absorbs molecules

basement membrane

In later experiments, this same research group tried the experiment again. This time, the paralyzed rats were first treated with drugs to overcome inhibition from the central nervous system. A nerve growth factor was added as well. These techniques significantly increased the success of the transplanted neurons. Amazingly, axons of transplanted neurons reached the muscles, formed neuromuscular junctions, and provided partial relief from the paralysis.

Research is being done on the use of the body's own stem cells within the CNS to repair damaged neurons. Studies on the transplantation of laboratory-grown stem cells continue as well. Many questions remain, but the current results are promising.

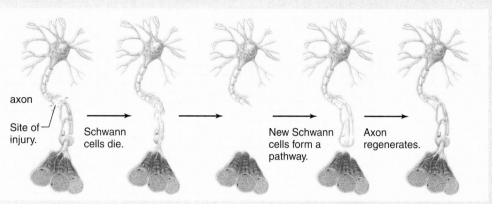

Figure 4A How do nerves regenerate?
Outside the CNS, nerves regenerate because new neuroglia called Schwann cells form a pathway for axons to reach a muscle. In the CNS, comparable neuroglia called oligodencrocytes do not do this.

Simple Epithelia

Epithelial tissue is either simple or stratified. Simple epithelia have only a single layer of cells (Fig. 4.7) and are classified according to cell type. **Squamous epithelium,** composed of flattened cells, is found lining the air sacs of lungs and walls of blood vessels. Its shape and arrangement permit exchanges of substances in these locations. Oxygen and carbon dioxide exchange occurs in the lungs, and nutrient-for-waste exchange occurs across blood vessels in the tissues.

Cuboidal epithelium consists of a single layer of cube-shaped cells. This type of epithelium is frequently found in glands, such as the salivary glands, the thyroid gland, and the pancreas. Simple cuboidal epithelium also covers the ovaries

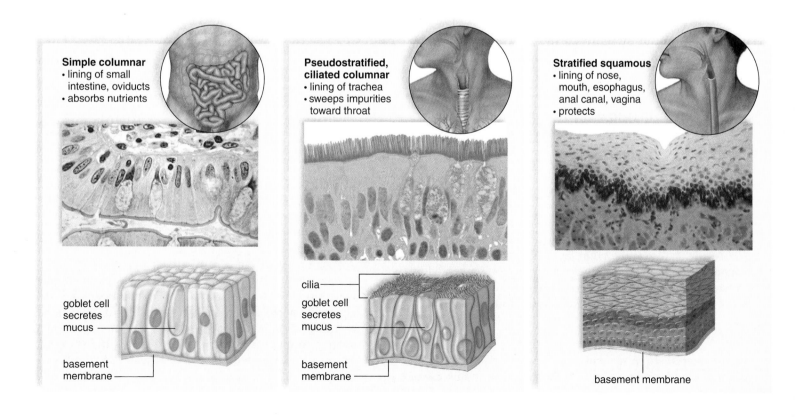

and lines kidney tubules, the portion of the kidney in which urine is formed. When cuboidal cells are involved in absorption, they have microvilli (minute cellular extensions of the plasma membrane). These increase the surface area of the cells. And when cuboidal cells function in active transport, they contain many mitochondria.

Columnar epithelium has cells resembling rectangular pillars or columns, with nuclei usually located near the bottom of each cell. This epithelium is found lining the digestive tract, where microvilli expand the surface area and aid in absorbing the products of digestion. Ciliated columnar epithelium is found lining the oviducts, where it propels the egg toward the uterus, or womb.

Pseudostratified Columnar Epithelium

Pseudostratified columnar epithelium (Fig. 4.7) is so named because it appears to be layered (*pseudo*, false; *stratified*, layers). However, true layers do not exist because each cell touches the basement membrane. In particular, the irregular placement of the nuclei creates the appearance of several layers, where only one exists. The lining of the windpipe, or trachea, is pseudostratified ciliated columnar epithelium. A secreted covering of mucus traps foreign particles. The upward motion of the cilia carries the mucus to the back of the throat, where it may either be swallowed or expectorated (spit out). Smoking can cause a change in mucous secretion and inhibit ciliary action, resulting in a chronic inflammatory condition called bronchitis.

Transitional Epithelium

The term transitional epithelium implies changeability, and this tissue changes in response to tension. It forms the lining of the urinary bladder, the ureters (tubes that carry urine from the kidneys to the bladder), and part of the urethra (the single tube that carries urine to the outside). All are organs that may need to stretch. When the bladder is distended, this epithelium stretches, and the outer cells take on a squamous appearance.

Stratified Epithelia

Stratified epithelia have layers of cells piled one on top of the other (see Fig 4.7). Only the bottom layer touches the basement membrane. The nose, mouth, esophagus, anal canal, the outer portion of the cervix (adjacent to the vagina), and vagina are lined with stratified squamous epithelium. Cancer of the cervix is detectable by doing a *pap smear*. Cells lining the cervix are smeared onto a slide later examined to detect any abnormalities.

As we shall see, the outer layer of skin is also stratified squamous epithelium, but the cells have been reinforced by keratin, a protein that provides strength. Stratified cuboidal and stratified columnar epithelia also are found in the body.

Glandular Epithelia

When an epithelium secretes a product, it is said to be glandular. A **gland** can be a single epithelial cell, as in the case of mucus secreting goblet cells, or a gland can contain many cells. Glands with ducts that secrete their product onto the outer surface (e.g., sweat glands and mammary glands) or into a cavity (e.g., pancreas) are called **exocrine glands.** Ducts can be simple or compound:

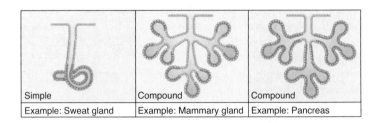

| Simple | Compound | Compound |
| Example: Sweat gland | Example: Mammary gland | Example: Pancreas |

Glands that have no duct are appropriately known as the ductless glands, or endocrine glands. **Endocrine glands** (e.g., pituitary gland and thyroid) secrete hormones internally, so they are transported by the bloodstream.

> **Check Your Progress 4.5**
> 1. Distinguish between the three types of simple epithelium and give a location for each.
> 2. a. How would you recognized pseudostratified epithelium, and (b) what function does this tissue perform in the trachea?
> 3. a. What is stratified epithelium, and (b) where would it be found?

4.6 Cell Junctions

The epithelial cells, and sometimes the muscle and nerve cells, of a tissue are connected by cell junctions. These junctions help a tissue perform its particular function. Just as cement holds bricks together, cell junctions join cells together into a cohesive hold. Cell junctions arise when plasma membranes are joined in these particular ways:

- **Tight junctions** allow epithelial cells to form a layer that covers the surface of organs and lines body cavities. The layer of cells becomes an impermeable barrier because adjacent plasma membrane proteins join, producing a zipperlike fastening (Fig. 4.8a). In the stomach and intestines, digestive secretions do not leak between the cells into the body. In the kidneys, the urine stays within kidney tubules because epithelial cells are joined by tight junctions.
- **Adhesion junctions** firmly attach cytoskeletal fibers of one cell to that of another cell. In a desmosome, featured in Figure 4.8b, the cytoskeletal fibers are anchored to a cytoplasmic plaque adjacent to the plasma membrane. Adhesion junctions are common in tissues subject to mechanical stress. They allow the skin, for example, to stretch and bend in response to mechanical stress.

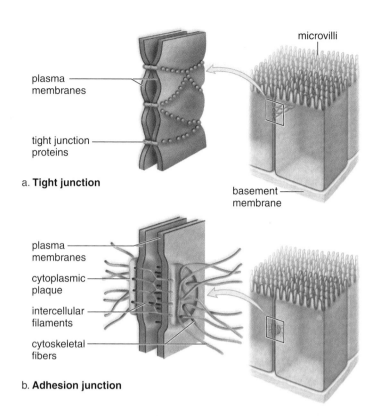

a. **Tight junction**

b. **Adhesion junction**

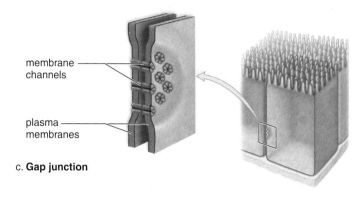

c. **Gap junction**

Figure 4.8 How do the three types of cell junctions work?
a. Tight junctions allow epithelial cells to form a layer that prevents leakage from one side of the sheet to the other. **b.** Adhesion junctions keep cells anchored to one another, but the layer of cells can still bend and stretch. **c.** Gap junctions allow the passage of small molecules and ions from one cell to the other.

CASE STUDY AT HEALTH SERVICES

The doctor came in about twenty minutes later. Kristen was feeling a little woozy but she wasn't completely "out of it." The doctor addressed both Kristen and Jayme, since he wasn't sure Kristen would remember everything he said. "I'm going to give you ladies a lesson in Skin 101. The skin is made up of three layers—the epidermis, the outermost dead layer; the dermis, the underlying layer that is alive; and the subcutaneous tissues underneath the dermis. That means when you look at a person's skin, you're looking at a bunch of dead cells," he smiled.

"That's really gross," Kristen commented weakly.

He explained. "There are three types of burns. We used to call them first-, second-, and third- degree burns, but now we describe them according to the thickness of the burn: superficial, partial, and full."

Both Jayme and Kristen nodded. "I remember that from a middle-school health class." Jayme replied.

The doctor continued, "In a superficial, or first-degree, burn only the epidermis is involved. You've probably had those types of superficial burns before. That's like a sunburn. A full burn is very serious. It involves all of the epidermis and dermis and may include underlying tissues such as muscle. In a full burn, the nerves, sweat glands, and blood vessels are destroyed. Those burns require skin grafting. What you have is a partial burn, what used to be called a second-degree burn. Partial burns can be further divided into superficial or deep. In a superficial partial burn, there is redness and blistering, but it is moist and the hairs are still present. In a deep partial burn, the hair is usually gone. It may not be painful if the nerve endings have been destroyed.

You still have pain and the burn is not very extensive, so I think we are dealing with a superficial partial burn," he reassured. "I'm going to clean this and put some antibiotic ointment on it. Then I'll cover it up and send you home. Tomorrow you can take acetaminophen or ibuprofen if you need to for pain. Just follow the directions on the bottle. I want you to come back tomorrow afternoon. The nurse can check and make sure there are no signs of infection and the skin is starting to heal."

The doctor began to clean the back of Kristen's hand. Kristen didn't think that she'd had enough pain meds. Jayme held her other hand and Kristen gripped it hard. There were tears in Kristen's eyes.

Jayme kept trying to encourage her. "He's almost done. There. He's bandaging it up. You did great."

Kristen did not feel so great. Jayme called a friend, who came and picked them up so Kristen did not have to walk back to the apartment. When they arrived at the apartment, Jayme saw that the other roommates had already cleaned up the mess in the kitchen. Kristen went to bed and Jayme lay down on the couch where she could hear Kristen if she called. Neither one of them slept very well that night.

Face Transplantation

In 2005, a French surgical team, led by Professors Bernard Devauchelle and Jean Michel Dubernard, was able to perform the world's first partial face transplant. The recipient was a woman, Isabelle Dinoire, severely disfigured by a dog mauling. Muscles, veins, arteries, nerves, and skin were transplanted onto the lower half of Isabelle's face (Fig. 4B, *left*). The donor's lips, chin, and nose were transplanted. The donor was a brain-dead patient whose family had agreed to donate all their loved one's organs and tissues. The donor shared Isabelle's blood type and was a good tissue match.

Although the ability to do this type of transplant had previously existed, doctors were concerned with the ethical aspects of the procedure. Organ transplantation has always involved some moral concerns, because the donor must still be alive when the organs are harvested. A face transplant is a "quality-of-life" issue and not a "life-or-death" surgery. In addition, face transplants have concerns regarding the patient's self-image and social acceptability. Ms. Dinoire was badly deformed and had difficulty speaking and eating due

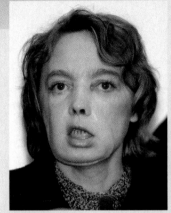

Figure 4B   What does a partial face transplant look like? The recipient of the first partial face transplant poses for photographers.

to her injuries. She underwent extensive counselling before the transplant, to prepare her for the adjustments to a "new" face. She will require immunosuppressive treatment for the rest of her life to prevent rejection of the donor face. However, 18 months after the surgery (Fig. 4B, *right*), Isabelle was able to eat, drink, and smile.

- **Gap junctions** (Fig. 4.8*c*) occur when adjacent plasma membranes converge and leave a tiny channel between them. Small molecules and ions can diffuse through a gap junction from the cytoplasm of one cell to another. During development, signaling molecules pass between embryonic cells at gap junctions and influence the differentiation of cells. Gap junctions between cardiac muscle cells at intercalated disks allow the heart to beat as a coordinated whole.

> **Check Your Progress 4.6**
> 1. a. What are the three types of junctions between cells, and (b) how do they differ in structure and function?
> 2. Why would you expect cell junctions more often in epithelia than in other types of tissues?
> 3. a. Which type of cell junction would you expect to find between muscle cells in the heart wall? b. Why?

4.7 Integumentary System

Specific tissues are associated with particular organs. For example, nervous tissue is associated with the brain. But an **organ** is composed of two or more types of tissues working together to perform particular functions. The skin is an organ comprised of all four tissue types: epithelial, connective, muscle, and nervous tissue. An **organ system** contains many different organs that cooperate to carry out a process, such as the digestion of food. The skin has several accessory organs (hair, nails, sweat glands, and sebaceous glands),

and, therefore, it is sometimes referred to as the **integumentary system**.

Skin is the most conspicuous system in the body because it covers the body. In an adult, the skin has a surface area of about 1.8 square meters (19.51 square feet). It accounts for nearly 15% of the weight of an average human. The skin has numerous functions. It protects underlying tissues from physical trauma, pathogen invasion, and water loss. It also helps regulate body temperature. Therefore, skin plays a significant role in homeostasis, the relative constancy of the internal environment. The skin even synthesizes certain chemicals that affect the rest of the body. Skin contains sensory receptors such as touch and temperature receptors. Thus it helps us to be aware of our surroundings and communicate with others.

Regions of the Skin

The skin has two regions: the epidermis and the dermis (Fig. 4.9). A **subcutaneous layer**, sometimes called the hypodermis, is found between the skin and any underlying structures, such as muscle or bone.

The Epidermis

The **epidermis** is made up of stratified squamous epithelium. New epidermal cells for the renewal of skin are derived from stem (basal) cells. The importance of these stem cells is observed when there is an injury to the skin. If an injury, such as a burn, is deep enough to destroy stem cells, then the skin can no longer replace itself. As soon as possible, the

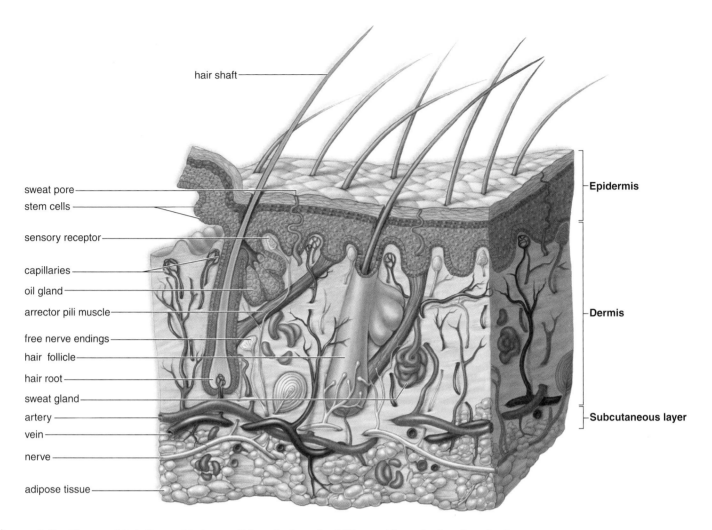

hair shaft

sweat pore
stem cells

sensory receptor

capillaries
oil gland
arrector pili muscle

free nerve endings
hair follicle
hair root
sweat gland
artery
vein
nerve

adipose tissue

Epidermis

Dermis

Subcutaneous layer

Figure 4.9 **Human skin is formed in layers. What do these look like, and how do they function?**
Skin consists of two regions: the epidermis and the dermis. A subcutaneous layer lies below the dermis. Skin has numerous functions. For example, it forms a protective covering over the entire body, safeguarding underlying parts from trauma, pathogen invasion, and water loss. The skin contains sensory receptors that communicate with the central nervous system and make us aware of external conditions. The skin also produces vitamin D, which has important metabolic functions, and the skin helps regulate body temperature.

damaged tissue is removed. Skin grafting is begun. The skin needed for grafting is usually taken from other parts of the patient's body. This is called autografting, as opposed to heterografting. In heterografting the graft is received from another person. Autografting is preferred because rejection rates are low. If the damaged area is extensive, it may be difficult to acquire enough skin for autografting. In that case, small amounts of epidermis are removed and cultured in the laboratory. This produces thin sheets of skin that can be transplanted back to the patient.

Newly generated skin cells become flattened and hardened as they push to the surface (Fig. 4.10a). Hardening takes place because the cells produce keratin, a waterproof protein. Dandruff occurs when the rate of keratinization in the skin of the scalp is two or three times the normal rate. A thick layer of dead keratinized cells, arranged in spiral and

concentric patterns, forms fingerprints and footprints that are genetically unique. Outer skin cells are dead and keratinized, so the skin is waterproof. This prevents water loss. The skin's waterproofing also prevents water from entering the body when the skin is immersed.

Two types of specialized cells are located deep in the epidermis. **Langerhans cells** are macrophages, a type of white blood cell that phagocytize infectious agents and then travel to lymphatic organs. There they stimulate the immune system to react to the pathogen. **Melanocytes,** lying deep in the epidermis, produce melanin. This is the main pigment responsible for skin color. The number of melanocytes is about the same in all individuals, so variation in skin color is due to the amount of melanin produced and its distribution. When skin is exposed to the sun, melanocytes produce more melanin. This protects

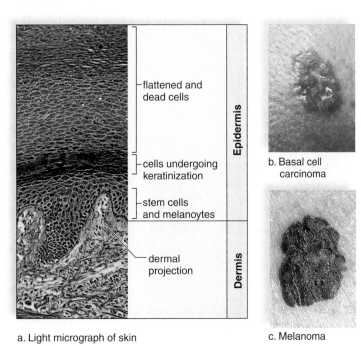

a. Light micrograph of skin

b. Basal cell carcinoma

c. Melanoma

Figure 4.10   **How do normal and cancerous skin differ?**
a. Epidermal ridges following dermal projections are clearly visible. Stem cells and melanocytes are in this region. **b.** Basal cell carcinoma derived from stem cells and **(c)** melanoma derived from melanocytes are types of skin cancer.

the skin from the damaging effects of the ultraviolet (UV) radiation in sunlight. The melanin is passed to other epidermal cells, and the result is tanning. In some people, this results in the formation of patches of melanin called freckles. Another pigment, called carotene, is present in epidermal cells and in the dermis. It gives the skin of certain Asians its yellowish hue. The pinkish color of fair-skinned people is due to the pigment hemoglobin in the red blood cells in the blood vessels of the dermis.

Some ultraviolet radiation does serve a purpose, however. Certain cells in the epidermis convert a steroid related to cholesterol into **vitamin D** with the aid of ultraviolet radiation. Only a small amount of UV radiation is needed. Vitamin D leaves the skin and helps regulate both calcium and phosphorus metabolism in the body. Calcium and phosphorus are important to the proper development and mineralization of the bones.

Skin Cancer   While we tend to associate a tan with health, it signifies that the body is trying to protect itself from the dangerous rays of the sun. Too much ultraviolet radiation is dangerous and can lead to skin cancer. Basal cell carcinoma (Fig. 4.10*b*), derived from stem cells gone awry, is the more common type of skin cancer. It is the most curable. Melanoma (Fig. 4.10*c*), the type of skin

cancer derived from melanocytes, is extremely serious. To prevent skin cancer, you should stay out of the sun between the hours of 10 AM and 3 PM. When you are in the sun

- use a broad-spectrum sunscreen that protects from both UV-A and UV-B radiation and has a sun protection factor (SPF) of at least 15. (This means that if you usually burn, for example, after a 20-minute exposure, it will take 15 times longer, or 5 hours, before you will burn.)
- wear protective clothing. Choose fabrics with a tight weave and wear a wide-brimmed hat.
- wear sunglasses that have been treated to absorb UV-A and UV-B radiation.

Also, avoid tanning machines because, even if they use only high levels of UV-A radiation, the deep layers of the skin will become more vulnerable to UV-B radiation.

The Dermis

The **dermis** is a region of dense fibrous connective tissue beneath the epidermis. (Dermatology is a branch of medicine that specializes in diagnosing and treating skin disorders.) The dermis contains collagen and elastic fibers. The collagen fibers are flexible but offer great resistance to over-stretching. They prevent the skin from being torn. The elastic fibers maintain normal skin tension but also stretch to allow movement of underlying muscles and joints. (The number of collagen and elastic fibers decreases with age and with exposure to the sun. This causes the skin to become less supple and more prone to wrinkling.) The dermis also contains blood vessels that nourish the skin. When blood rushes into these vessels, a person blushes. When blood is minimal in them, a person turns "blue." Blood vessels in the dermis play a role in temperature regulation. If body temperature starts to rise, the blood vessels in the skin will dilate. As a result, more blood is brought to the surface of the skin for cooling. If the outer temperature cools, the blood vessels constrict, so less blood is brought to the skin's surface.

The sensory receptors primarily in the dermis are specialized for touch, pressure, pain, hot, and cold. These receptors supply the central nervous system with information about the external environment. The sensory receptors also account for the use of the skin as a means of communication between people. For example, the touch receptors play a major role in sexual arousal.

The Subcutaneous Layer

Technically speaking, the subcutaneous layer beneath the dermis is not a part of skin. It is a common site for injections. This layer is composed of loose connective tissue and adipose tissue, which stores fat. Fat is a stored source of energy in the body. Adipose tissue helps to thermally insulate the

Have You Ever Wondered...

Which parts of the body have the most touch receptors?

Touch receptors are not distributed equally over the body. A test called a two-point discrimination test determines how numerous touch receptors are in different areas of skin. On a subject with eyes closed, two sharp points are gently touched to the skin surface. As the points are moved closer together, the subject has increasing difficulty recognizing the two points as being separate. The greater the number of touch receptors in an area of skin, the better the person can feel two separate points. In the tongue and fingertips there may be as many as 100 touch receptors per sq. cm. Areas with the fewest touch receptors include the back and arms.

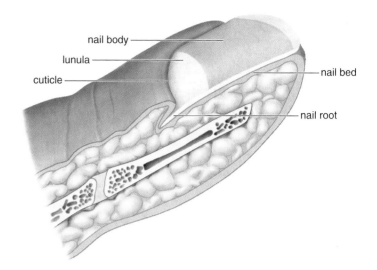

Figure 4.11　How is a nail formed?
Cells produced by the nail root become keratinized, forming the nail body.

body from either gaining heat from the outside or losing heat from the inside. A well-developed subcutaneous layer gives the body a rounded appearance and provides protective padding against external assaults. Excessive development of the subcutaneous layer accompanies obesity.

Accessory Organs of the Skin

Nails, hair, and glands are structures of epidermal origin, even though some parts of hair and glands are largely found in the dermis.

Nails are a protective covering of the distal part of fingers and toes, collectively called digits (Fig. 4.11). Nails grow from special epithelial cells at the base of the nail in the portion called the nail root. The cuticle is a fold of skin that hides the nail root. The whiteish color of the half-moon-shaped base, or *lunula,* results from the thick layer of cells in this area. The cells of a nail become keratinized as they grow out over the nail bed.

Hair follicles begin at a bulb in the dermis and continue through the epidermis where the hair shaft extends beyond the skin (see Fig. 4.9). A dark hair color is largely due to the production of true melanin by melanocytes present in the bulb. If the melanin contains iron and sulfur, hair is blond or red. Graying occurs when melanin cannot be produced, but white hair is due to bubbles in the hair shaft.

Contraction of the *arrector pili muscles* attached to hair follicles causes the hairs to "stand on end" and goosebumps to develop. Epidermal cells form the root of a hair, and their division causes a hair to grow. The cells become keratinized and die as they are pushed farther from the root.

Each hair follicle has one or more **oil glands** (see Fig. 4.9), also called sebaceous glands, which secrete sebum. Sebum is an oily substance that lubricates the hair within the follicle and the skin. The oil secretions from sebaceous glands are acidic and retard the growth of bacteria. If the sebaceous glands fail to discharge, the secretions collect and form "whiteheads" or "blackheads." The color of blackheads is due to oxidized sebum. Acne is an inflammation of the sebaceous glands that most often occurs during adolescence due to hormonal changes.

Sweat glands (see Fig. 4.9), also called sudoriferous glands, are numerous and present in all regions of skin. A sweat gland is a tubule that begins in the dermis and either opens into a hair follicle or, more often, opens onto the surface of the skin. Sweat glands play a role in modifying body temperature. When body temperature starts to rise, sweat glands become active. Sweat absorbs body heat as it evaporates. Once body temperature lowers, sweat glands are no longer active.

> ### Check Your Progress 4.7
> 1. Compare the structure and function of the epidermis and dermis.
> 2. a. What factors are involved in skin color, and (b) how can you protect skin from the ultraviolet rays of the sun?
> 3. a. Contrast the location and function of sweat glands and oil glands. b. How do oil glands protect us from bacteria?
> 4. a. Explain why the subcutaneous layer is not considered a part of the skin. b. What is its function?

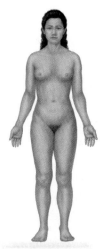
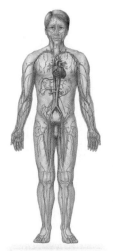
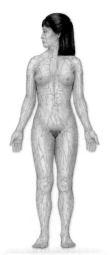
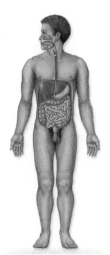
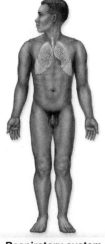
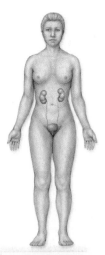

Integumentary system

- protects body.
- receives sensory input.
- helps control temperature.
- synthesizes vitamin D.

Cardiovascular system

- transports blood, nutrients, gases, and wastes.
- defends against disease.
- helps control temperature, fluid, and pH balance.

Lymphatic and immune systems

- helps control fluid balance.
- absorbs fats.
- defends against infectious disease.

Digestive system

- ingests food.
- digests food.
- absorbs nutrients.
- eliminates waste.

Respiratory system

- maintains breathing.
- exchanges gases at lungs and tissues.
- helps control pH balance.

Urinary system

- excretes metabolic wastes.
- helps control fluid balance.
- helps control pH balance.

Figure 4.12   What are the organ systems of the body? What functions does each perform?

4.8 Organ Systems

This text has several illustrations such as the one on page 85 that show how the various systems cooperate to maintain homeostasis. In one sense, it is arbitrary to assign a particular organ to one system when it also assists the functioning of many other systems. The functions of the various systems of the body are listed in Figure 4.12.

Integumentary System

The integumentary system contains skin. It also includes nails, hairs, muscles that move hairs, the oil and sweat glands, blood vessels, and nerves leading to sensory receptors. As discussed in Section 4.7 this system has many homeostatic functions.

Cardiovascular System

In the **cardiovascular system,** the heart pumps blood and sends it out under pressure into the blood vessels. In humans, blood is always contained in blood vessels, never running free unless the body suffers an injury.

While blood is moving throughout the body, it distributes heat produced by the muscles. Blood transports nutrients and oxygen to the cells and removes their waste molecules, including carbon dioxide. Despite the movement of molecules into and out of the blood, it has a fairly constant volume and pH. This is particularly due to exchanges in the lungs, the digestive tract, and the kidneys. The red blood cells in blood transport oxygen, while the white blood cells fight infections. Platelets are involved in blood clotting.

Lymphatic and Immune Systems

The **lymphatic system** consists of lymphatic vessels, lymph nodes, the spleen, and other lymphatic organs. This system collects excess tissue fluid and plays a role in absorbing fats and transporting lymph to cardiovascular veins. It also purifies lymph and stores lymphocytes, the white blood cells that produce antibodies.

The **immune system** consists of all the cells in the body that protect us from disease. The lymphocytes, in particular, belong to this system.

Digestive System

The **digestive system** consists of the mouth, esophagus, stomach, small intestine, and large intestine (colon). It also includes these associated organs: teeth, tongue, salivary glands, liver, gallbladder, and pancreas. This system receives food and digests it into nutrient molecules, which can enter the cells of the body. The nondigested remains are eventually eliminated.

Respiratory System

The **respiratory system** consists of the lungs and the tubes that take air to and from them. The respiratory system brings

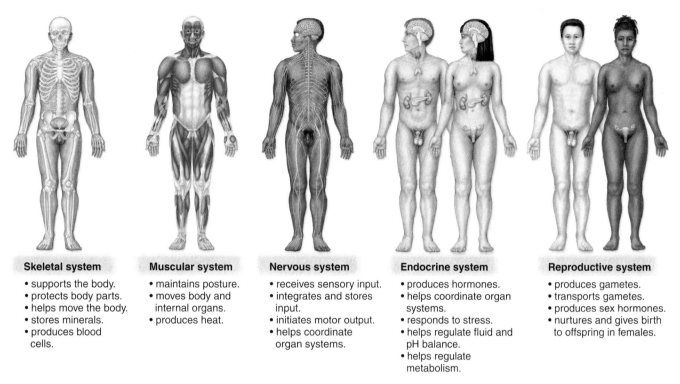

Skeletal system	Muscular system	Nervous system	Endocrine system	Reproductive system
• supports the body. • protects body parts. • helps move the body. • stores minerals. • produces blood cells.	• maintains posture. • moves body and internal organs. • produces heat.	• receives sensory input. • integrates and stores input. • initiates motor output. • helps coordinate organ systems.	• produces hormones. • helps coordinate organ systems. • responds to stress. • helps regulate fluid and pH balance. • helps regulate metabolism.	• produces gametes. • transports gametes. • produces sex hormones. • nurtures and gives birth to offspring in females.

Figure 4.12   Organ systems of the body—continued.

oxygen into the body and removes carbon dioxide from the body at the lungs. The removal of carbon dioxide helps adjust the acid-base balance of the blood.

Urinary System

The **urinary system** contains the kidneys, the urinary bladder, and the tubes that carry urine. The kidneys rid the body of metabolic wastes, particularly nitrogenous wastes. They also help regulate the salt-water balance and acid-base balance of the blood.

Skeletal System

The bones of the **skeletal system** protect body parts. For example, the skull forms a protective encasement for the brain, as does the rib cage for the heart and lungs. The skeleton helps move the body because it serves as a place of attachment for the skeletal muscles.

The skeletal system also stores minerals, notably calcium, and produces blood cells within red bone marrow.

Muscular System

In the **muscular system,** skeletal muscle contraction maintains posture and accounts for the movement of the body and its parts. Cardiac muscle contraction results in the heartbeat. The walls of internal organs, such as the bladder, contract due to the presence of smooth muscle. Muscle contraction releases heat, which helps warm the body.

Nervous System

The **nervous system** consists of the brain, spinal cord, and associated nerves. The nerves conduct nerve impulses from sensory receptors to the brain and spinal cord, where integration occurs. Nerves also conduct nerve impulses from the brain and spinal cord to the muscles and glands, allowing us to respond to both external and internal stimuli.

Endocrine System

The **endocrine system** consists of the hormonal glands, which secrete chemical messengers called hormones into the bloodstream. Hormones have a wide range of effects, including regulation of cellular metabolism, regulation of fluid and pH balance, and helping us respond to stress. Both the nervous and endocrine systems coordinate and regulate the functioning of the body's other systems. The endocrine system also helps maintain the functioning of the male and female reproductive organs.

Reproductive System

The **reproductive system** has different organs in the male and female. The male reproductive system consists of the testes, other glands, and various ducts that conduct semen to and through the penis. The testes produce sex cells called sperm. The female reproductive system consists of the ovaries, oviducts, uterus, vagina, and external genitals. The ovaries produce sex cells called eggs. When a sperm fertilizes an egg, an offspring begins development.

Health Focus

Pursuing Youthful Skin

More and more members of the "baby boomer" generation are willing to spend lavishly for a youthful appearance. Over 30 million Americans have turned to Botox®, laser treatments, and/or tanning to help obtain that vigorous, "healthy" look. But how safe and effective are these treatments?

Botox®

Botox® is a drug used to reduce the appearance of facial wrinkles and lines. Botox® is the trade name for a derivative of botulinum toxin A, a protein toxin produced by the bacterium *Clostridium botulinum.* Botox® stops communication between motor nerves and muscles, causing muscle paralysis. Botox® treatments were approved by the U.S. Food and Drug Administration (FDA) for use as a cosmetic treatment in 2002. Treatments are direct injections under the skin, where the toxin causes facial muscle paralysis. The injections reduce the appearance of wrinkles and lines that appear as a result of normal facial muscle movement. However, Botox® treatment is not without side effects. Excessive drooling or a slight rash around the injection site are among the milder side effects of treatment. Spreading of Botox® from the injection site may also paralyze facial muscles unintended for treatment. In a few cases, muscle pain and weakness have resulted. While rare, more serious side effects, including allergic reactions, may also occur. When performed in a medical facility by a licensed physician, Botox® treatment is generally considered safe and effective.

Laser Treatments

Laser treatments can be used for treatment of acne, wrinkles, and spider veins, as well as for removal of birthmarks, scars, tattoos, and hair. Lasers use a concentrated beam of light energy directed at the tissue. When the light beam hits the tissue, it is absorbed and converted to heat, and the tissue is destroyed. Laser treatments vary as to the energy of the laser and the depth of the skin removed. The beam of light is so precise that it can be used to treat small areas without affecting adjacent areas of the skin. Each treatment is short, usually between 15 and 60 minutes, but several treatments are necessary. New, more youthful appearing skin grows over the treated area. Deeper treatments, such as for wrinkle treatments or scar removal, may require extensive healing times (a month or longer). Redness, blistering, and the possibility of infection are all side effects. Other problems may include lightening or darkening of the skin, but these are usually temporary. The effectiveness of the procedure depends on the type of laser chosen, and the experience and skill of the person operating it. As with Botox® treatment, it is important for the procedure to be performed in a medical facility by a licensed physician.

Figure 4C   What are the dangers of tanning?
In addition to aging of the skin, tanning can cause cancer. Excessive tanning may be addictive.

Tanning

Whether you get that dark, glowing look from a tanning salon or from the beach, there is no such thing as a safe tan. Darker, tanned skin is due to the extra production of the pigment melanin, a sign of skin damage. Instead of leading to a youthful appearance, tanning can lead to premature wrinkling as well as skin cancer. Damage to the skin accumulates over time—the longer the exposure, the greater the damage. Almost 90% of the changes people associate with aging, including leathery skin, dark spots, and wrinkles, are commonly attributed to sun damage. Hardly the youthful complexion originally hoped for! More than 90% of skin cancers are due to sun exposure. Excessive sun exposure is especially dangerous for young people—those in their teens and twenties. The use of tanning beds among young people increases the risk of melanoma by 75%. Getting a dark tan or a burn while young may result in skin cancer in later life.

Sunless tanning products have been around since the 1960s. Bronzers, a type of make-up, tint the skin. Tanning sprays use a colorless sugar known as DHA. The sugar interacts with the dead cells in the epidermis producing a color change. Neither of these products protects a person from damaging UV radiation, so sunscreen is still needed before sun exposure.

Why do people continue to tan, even after they know it could cause premature aging and skin cancer? One research study suggests that people who tan 8 to 15 times a week are addicted to tanning (Fig. 4C). Exposure to UV rays may cause an increase in endorphins, chemicals that make a person feel better.

Jayme and Kristen went back to Health Services the next day. The nurse removed the bandage from Kristen's hand.

"The burn is healing very nicely and there are no signs of infection. It doesn't look like this is going to scar. I'm going to give you a tube of antibiotic ointment to apply twice a day, and I want you to continue to keep the burn covered. If there are any signs of infection—any red streaks, swelling, or pus, then you should come back immediately. If you start to run a fever, then you need to come back. You've had a tetanus shot before you came to school, so you don't need another one. I want you to come back in three days for me to recheck this."

The nurse covered the burn with the ointment and rebandaged it. She gave Kristen the tube and some clean gauze.

"By the way, don't put ice on a burn," she continued. "It is best just to use cool, wet compresses—something like a towel soaked in cold water. Ice could potentially cause more skin damage. It is just an old wives' tale that you should use butter or mayonnaise. Those can increase the risk of infection. It is best to avoid burns by being careful in the kitchen. In other words, don't pour hot oatmeal on your hand!"

Kristen had been holding the oatmeal pan with her right hand, so it was her left hand that was damaged. Thankfully, she wrote with her right hand, so she could still take notes in class. But she was amazed at how many things she had trouble doing. For example, it was hard to button her blouses with only one hand. She continued to keep the area clean and to use the antibiotic ointment. It took about three weeks for the burn to heal completely. There was very little indication that she had ever been burned, but she found that she was a little afraid of fixing oatmeal for breakfast. She has switched to cold cereal—much safer.

Body Cavities

The human body is divided into two main cavities: the ventral cavity and the dorsal cavity (Fig. 4.13a). Called the coelom in early development, the ventral cavity later becomes the thoracic, abdominal, and pelvic cavities. The thoracic cavity contains the lungs and the heart. The thoracic cavity is separated from the abdominal cavity by a horizontal muscle called the **diaphragm.** The stomach, liver, spleen, pancreas, gallbladder, and most of the small and large intestines are in the abdominal cavity. The pelvic cavity contains the rectum, the urinary bladder, the internal reproductive

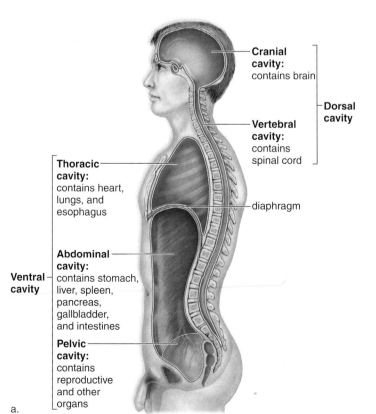

Cranial cavity: contains brain

Vertebral cavity: contains spinal cord

Dorsal cavity

Thoracic cavity: contains heart, lungs, and esophagus

diaphragm

Abdominal cavity: contains stomach, liver, spleen, pancreas, gallbladder, and intestines

Ventral cavity

Pelvic cavity: contains reproductive and other organs

a.

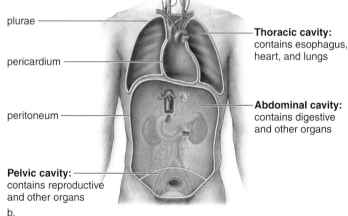

plurae

pericardium

peritoneum

Thoracic cavity: contains esophagus, heart, and lungs

Abdominal cavity: contains digestive and other organs

Pelvic cavity: contains reproductive and other organs

b.

Figure 4.13 How many body cavities are there?
a. Side view. The posterior, or dorsal, (toward the back) cavity contains the cranial cavity and the vertebral canal. The brain is in the cranial cavity, and the spinal cord is in the vertebral canal. In the anterior, or ventral, (toward the front) cavity, the diaphragm separates the thoracic cavity from the abdominal cavity. The heart and lungs are in the thoracic cavity; the other internal organs are either in the abdominal cavity or in the pelvic cavity. **b.** Frontal view of the thoracic cavity, showing serous membranes.

organs, and the rest of the small and large intestine. Males have an external extension of the abdominal wall, called the scrotum, containing the testes.

The dorsal cavity has two parts: The cranial cavity within the skull contains the brain. The vertebral canal, formed by the vertebrae, contains the spinal cord.

Body Membranes

Body membranes line cavities and the internal spaces of organs and tubes that open to the outside. The body membranes are of four types: mucous, serous, synovial, and meninges.

Mucous membranes line the tubes of the digestive, respiratory, urinary, and reproductive systems. They are composed of an epithelium overlying a loose fibrous connective tissue layer. The epithelium contains specialized cells that secrete mucus. This mucus ordinarily protects the body from invasion by bacteria and viruses. Hence, more mucus is secreted and expelled when a person has a cold and has to blow her/his nose. In addition, mucus usually protects the walls of the stomach and small intestine from digestive juices. This protection breaks down when a person develops an ulcer.

Serous membranes line and support the lungs, the heart, and the abdominal cavity and its internal organs (Fig. 4.13b). They secrete a watery fluid that keeps the membranes lubricated. Serous membranes support the internal organs and compartmentalize the large thoracic and abdominal cavities.

Serous membranes have specific names according to their location. The pleurae (sing., **pleura**) line the thoracic cavity and cover the lungs. The pericardium forms the pericardial sac and covers the heart. The peritoneum lines the abdominal cavity and covers its organs. A double layer of peritoneum, called mesentery, supports the abdominal organs and attaches them to the abdominal wall. Peritonitis is a life-threatening infection of the peritoneum.

Synovial membranes composed only of loose connective tissue line the cavities of freely movable joints. They secrete synovial fluid into the joint cavity. This fluid lubricates the ends of the bones so that they can move freely. In rheumatoid arthritis, the synovial membrane becomes inflamed and grows thicker, restricting movement.

The **meninges** are membranes found within the dorsal cavity. They are composed only of connective tissue and serve as a protective covering for the brain and spinal cord. Meningitis is a life-threatening infection of the meninges.

> **Check Your Progress 4.8**
>
> 1. What is the overall function of each of the body systems?
> 2. a. What are the two major body cavities? b. What two cavities are in each of these?
> 3. What are four types of body membranes?

4.9 Homeostasis

Homeostasis is the body's ability to maintain a relative constancy of its internal environment by adjusting its physiological processes. Even though external conditions may change dramatically, we have physiologic mechanisms that respond to disturbances and limit the amount of internal change. Conditions usually stay within a narrow range of normalcy. For example, blood glucose, pH levels, and body temperature typically fluctuate during the day, but not greatly. If internal conditions should change to any great degree, illness results.

The Internal Environment

The internal environment has two parts: blood and tissue fluid. Blood delivers oxygen and nutrients to the tissues and carries carbon dioxide and wastes away. Tissue fluid, not blood, bathes the body's cells. Therefore, tissue fluid is the medium through which substances are exchanged between cells and blood. Oxygen and nutrients pass through tissue fluid on their way to tissue cells from the blood. Then carbon dioxide and wastes are carried away from the tissue cells by the tissue fluid, where they are brought back into the blood. The cooperation of body systems is required to keep these substances within the range of normalcy in blood and tissue fluid.

The Body Systems and Homeostasis

The nervous and endocrine systems are particularly important in coordinating the activities of all the other organ systems as they function to maintain homeostasis (Fig. 4.14). The nervous system is able to bring about rapid responses to any changes in the internal environment. The nervous system issues commands by electrochemical signals rapidly transmitted to effector organs, which can be muscles, such as skeletal muscles, or glands, such as sweat and salivary glands. The endocrine system brings about slower responses but generally have more lasting effects. Glands of the endocrine system, such as the pancreas or the thyroid, release hormones. Hormones, such as insulin from the pancreas, are chemical messengers that must travel through the blood and tissue fluid to reach their targets.

The nervous and endocrine systems together direct numerous activities that maintain homeostasis, but all the organ systems must do their part to keep us alive and healthy. Picture what would happen if the cardiovascular, respiratory, digestive, or urinary system failed (Fig. 4.14). If someone is having a heart attack, the heart is unable to pump the blood to supply cells with oxygen. Or think of a person who is choking. The trachea (or windpipe) is blocked, so no air can reach the lungs for uptake by the blood. Unless the obstruction is removed quickly, cells will begin to die as the blood's supply of oxygen is depleted. When the lining of the digestive tract is damaged, as in a severe bacterial infection, nutrient absorption is impaired and cells face an energy crisis. It is important not only to maintain adequate nutrient levels in the blood, but

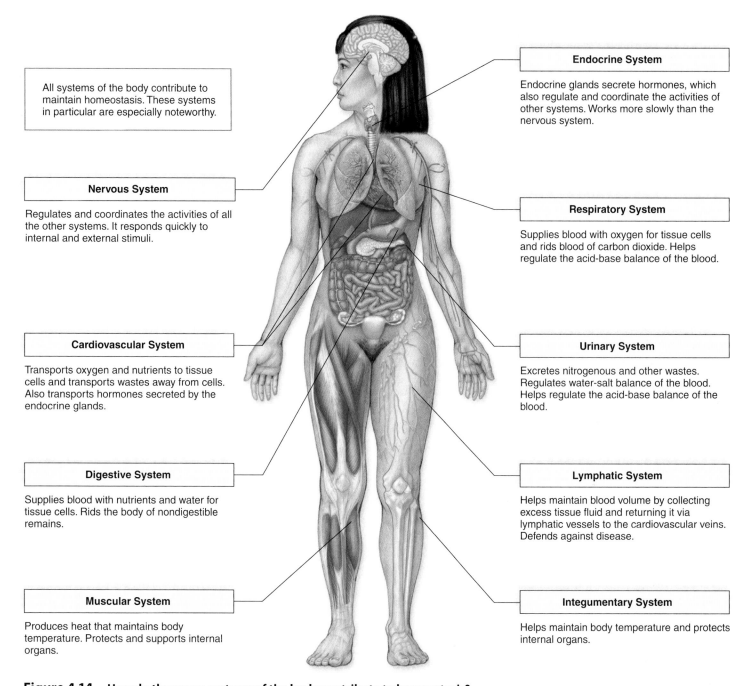

All systems of the body contribute to maintain homeostasis. These systems in particular are especially noteworthy.

Endocrine System

Endocrine glands secrete hormones, which also regulate and coordinate the activities of other systems. Works more slowly than the nervous system.

Nervous System

Regulates and coordinates the activities of all the other systems. It responds quickly to internal and external stimuli.

Respiratory System

Supplies blood with oxygen for tissue cells and rids blood of carbon dioxide. Helps regulate the acid-base balance of the blood.

Cardiovascular System

Transports oxygen and nutrients to tissue cells and transports wastes away from cells. Also transports hormones secreted by the endocrine glands.

Urinary System

Excretes nitrogenous and other wastes. Regulates water-salt balance of the blood. Helps regulate the acid-base balance of the blood.

Digestive System

Supplies blood with nutrients and water for tissue cells. Rids the body of nondigestible remains.

Lymphatic System

Helps maintain blood volume by collecting excess tissue fluid and returning it via lymphatic vessels to the cardiovascular veins. Defends against disease.

Muscular System

Produces heat that maintains body temperature. Protects and supports internal organs.

Integumentary System

Helps maintain body temperature and protects internal organs.

Figure 4.14 **How do the organ systems of the body contribute to homeostasis?**
All the organ systems contribute to homeostasis in many ways. Some of the main contributions of each system are given in this illustration.

also to eliminate wastes and toxins. The liver makes urea, a nitrogenous end product of protein metabolism. Urea and other metabolic wastes are excreted by the kidneys, the urine-producing organs of the body. The kidneys rid the body of nitrogenous wastes and also help to adjust the blood's water-salt and acid-base balances.

A closer examination of how the blood glucose level is maintained helps us understand homeostatic mechanisms. When a healthy person consumes a meal and glucose enters the blood, the pancreas secretes the hormone insulin. Now glucose is re-

moved from the blood as cells take it up. In the liver, glucose is stored in the form of glycogen. This storage is beneficial because later, if blood glucose levels drop, glycogen can be broken down to ensure that the blood level remains constant. Homeostatic mechanisms can fail. In diabetes mellitus, the pancreas cannot produce enough insulin, or the body cells cannot respond appropriately to it. Therefore, glucose does not enter the cells and they must turn to other molecules, such as fats and proteins, to survive. This, along with too much glucose in the blood, leads to the numerous complications of diabetes mellitus.

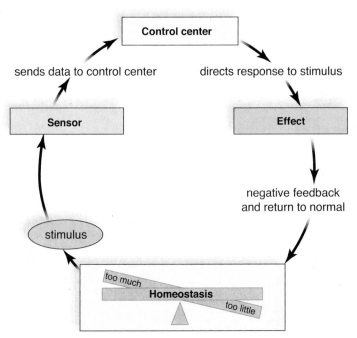

Figure 4.15   **What are the components of a negative feedback mechanism?**
This diagram shows how the basic elements of a feedback mechanism work. A sensor detects the stimulus, and a control center brings about an effect that resolves, or corrects, the stimulus.

Another example of homeostasis is the ability of the body to regulate the acid-base balance of the body. When carbon dioxide enters the blood, it combines with water to form carbonic acid. However, the blood is buffered, and pH stays within normal range as long as the lungs are busy excreting carbon dioxide. These two mechanisms are backed up by the kidneys, which can rid the body of a wide range of acidic and basic substances and, in that way, adjust the pH.

Negative Feedback

Negative feedback is the primary homeostatic mechanism that keeps a variable, such as the blood glucose level, close to a particular value, or set point. A homeostatic mechanism has at least two components: a sensor and a control center (Fig. 4.15). The sensor detects a change in the internal environment. The control center then brings about an effect to bring conditions back to normal. Now, the sensor is no longer activated. In other words, a **negative feedback** mechanism is present when the output of the system resolves or corrects the original stimulus. For example, when blood pressure rises, sensory receptors signal a control center in the brain. The center stops sending nerve signals to muscle in the arterial walls. Now, the arteries can relax. Once the blood pressure drops, signals no longer go to the control center.

Mechanical Example

A home heating system is often used to illustrate how a more complicated negative feedback mechanism works (Fig. 4.16).

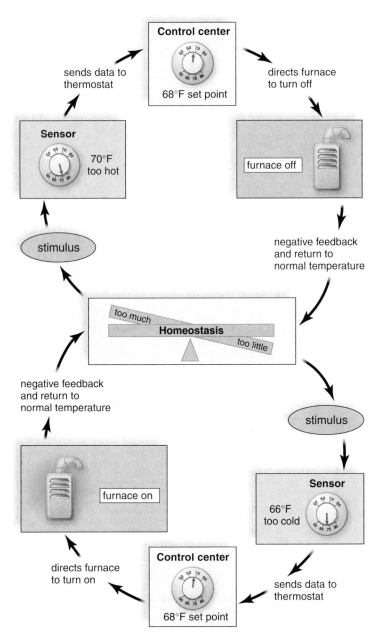

Figure 4.16   **How does a complex negative feedback mechanism work?**
This diagram shows how room temperature is returned to normal when the room becomes too hot (*above*) or too cold (*below*). The thermostat contains both the sensor and the control center. *Above:* The sensor detects that the room is too hot, and the control center turns the furnace off. The stimulus is resolved, or corrected, when the temperature returns to normal. *Below:* The sensor detects that the room is too cold, and the control center turns the furnace on. Once again, the stimulus is resolved, or corrected, when the temperature returns to normal.

You set the thermostat at 68°F. This is the *set point*. The thermostat contains a thermometer, a sensor that detects when the room temperature is above or below the set point. The thermostat also contains a control center. It turns the furnace off when the room is warm and turns it on when the room is cool. When the furnace is off, the room cools a bit. When the furnace is on, the room warms a bit. In other words, typical of negative feedback mechanisms, there is a fluctuation above and below normal.

Human Example: Regulation of Body Temperature

The sensor and control center for body temperature are located in a part of the brain called the hypothalamus. A negative feedback mechanism prevents change in the same direction. Body temperature does not get warmer and warmer because warmth brings about a change toward a lower body temperature. Likewise, body temperature does not get continuously colder. A body temperature below normal brings about a change toward a warmer body temperature.

Above Normal Temperature When the body temperature is above normal, the control center directs the blood vessels of the skin to dilate (Fig. 4.17, *above*). This allows more blood to flow near the surface of the body, where heat can be lost to the environment. In addition, the nervous system activates the sweat glands, and the evaporation of sweat helps lower body temperature. Gradually, body temperature decreases to 98.6°F.

Below Normal Temperature When the body temperature falls below normal, the control center directs (via nerve impulses) the blood vessels of the skin to constrict (Fig. 4.17, *below*). This conserves heat. If body temperature falls even lower, the control center sends nerve impulses to the skeletal muscles and shivering occurs. Shivering generates heat, and gradually body temperature rises to 98.6°F. When the temperature rises to normal, the control center is inactivated.

Positive Feedback

Positive feedback is a mechanism that brings about an increasing change in the same direction. When a woman is giving birth, the head of the baby begins to press against the cervix (entrance to the womb), stimulating sensory receptors there. When nerve signals reach the brain, the brain causes the pituitary gland to secrete the hormone oxytocin. Oxytocin travels in the blood and causes the uterus to contract. As labor continues, the cervix is increasingly stimulated, and uterine contractions become stronger until birth occurs.

A positive feedback mechanism can be harmful, as when a fever causes metabolic changes that push the fever higher. Death occurs at a body temperature of 113°F because cellular proteins denature at this temperature and metabolism stops. However, positive feedback loops such as those involved in childbirth, blood clotting, and the stomach's digestion of protein assist the body in completing a process that has a definite cut-off point.

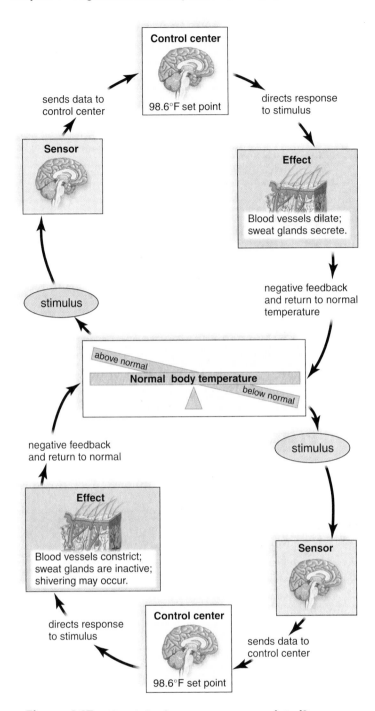

Figure 4.17 How is body temperature regulated?
Above: When body temperature rises above normal, the hypothalamus senses the change and causes blood vessels to dilate and sweat glands to secrete so that temperature returns to normal. *Below:* When body temperature falls below normal, the hypothalamus senses the change and causes blood vessels to constrict. In addition, shivering may occur to bring temperature back to normal. In this way, the original stimulus is resolved, or corrected.

Check Your Progress 4.9

1. **a.** What is homeostasis, and **(b)** why is it important to body function?
2. How do body systems contribute to homeostasis?
3. How do negative feedback and positive feedback contribute to homeostasis?

Summarizing the Concepts

4.1 Types of Tissues

Human tissues are categorized into four groups:

Connective	Muscular	Nervous	Epithelial

4.2 Connective Tissue Connects and Supports

Connective tissues have cells separated by a matrix that contains ground substance and fibers (e.g., collagen fibers). The four types are

- loose fibrous connective tissue, including adipose tissue; and dense fibrous connective tissue (tendons and ligaments);
- cartilage and bone (matrix for cartilage is solid, yet flexible; matrix for bone is solid and rigid); and
- blood (matrix is a liquid called plasma) and lymph.

4.3 Muscular Tissue Moves the Body

Muscular tissue is of three types: skeletal, smooth, and cardiac.

- Skeletal and cardiac muscle are striated.
- Cardiac and smooth muscle are involuntary.
- Skeletal muscle is found in muscles attached to bones.
- Smooth muscle is found in internal organs.
- Cardiac muscle makes up the heart.

4.4 Nervous Tissue Communicates

- Nervous tissue is composed of neurons and several types of neuroglia.
- Each neuron has dendrites, a cell body, and an axon. Axons conduct nerve impulses.

4.5 Epithelial Tissue Protects

Epithelial tissue covers the body and lines its cavities.

- Types of simple epithelia are squamous, cuboidal, and columnar.
- Certain epithelial tissues may have cilia or microvilli.
- Stratified epithelia have many layers of cells, with only the bottom layer touching the basement membrane.
- Glandular epithelia secretes a product either into ducts or into the blood.

4.6 Cell Junctions

Three types of junctions are common between epithelial cells:

- Tight junctions are zipperlike fastenings between cells.
- Adhesion junctions permit cells to stretch and bend.
- Gap junctions allow small molecules and signals to pass between cells.

4.7 Integumentary System

Skin and its accessory organs comprise the integumentary system. Skin has two regions:

- The epidermis contains stem cells, which produce new epithelial cells.
- The dermis contains epidermally derived glands and hair follicles, nerve endings, blood vessels, and sensory receptors.
- A subcutaneous layer lies beneath the skin.

4.8 Organ Systems

Organs make up organ systems, summarized in the table on this page. Some organs are found in particular body cavities.

4.9 Homeostasis

Homeostasis is the relative constancy of the internal environment, tissue fluid and blood. All organ systems contribute to homeostasis.

- The cardiovascular, respiratory, digestive, and urinary systems directly regulate the amount of gases, nutrients, and wastes in the blood, keeping tissue fluid constant.
- The lymphatic system absorbs excess tissue fluid and functions in immunity.
- The nervous system and endocrine system regulate the other systems.

Negative Feedback

Negative feedback mechanisms keep the environment relatively stable. When a sensor detects a change above or below a set point, a control center brings about an effect that reverses the change and brings conditions back to normal again. Examples include

- regulation of blood glucose level by insulin,
- regulation of room temperature by a thermostat and furnace, and
- regulation of body temperature by the brain and sweat glands.

Positive Feedback

In contrast to negative feedback, a positive feedback mechanism brings about rapid change in the same direction as the stimulus and does not achieve relative stability. These mechanisms are useful under certain conditions, such as during birth.

Organ Systems	
Transport Cardiovascular (heart and blood vessels) Lymphatic and Immune (lymphatic vessels)	**Integumentary** Skin and accessory organs
Maintenance Digestive (e.g., stomach, intestines) Respiratory (tubes and lungs) Urinary (tubes and kidneys)	**Motor** Skeletal (bones and cartilage) Muscular (muscles)
Control Nervous (brain, spinal cord, and nerves) Endocrine (glands)	**Reproduction** Reproductive (tubes and testes in males; tubes and ovaries in females)

Understanding Key Terms

adhesion junction 74
adipose tissue 67
basement membrane 72
blood 68
cardiac muscle 70
cardiovascular system 80
cartilage 67
collagen fiber 66
columnar epithelium 74
compact bone 68
connective tissue 66
cuboidal epithelium 73
dense fibrous connective
    tissue 67
dermis 78
diaphragm 83
digestive system 80
elastic cartilage 68
elastic fiber 66
endocrine gland 74
endocrine system 81
epidermis 76
epithelial tissue 72
exocrine gland 74
fibroblast 66
fibrocartilage 68
gap junction 76
gland 74
hair follicle 79
homeostasis 84
hyaline cartilage 67
immune system 80
integumentary system 76
intercalated disks 70
lacuna 67
Langerhans cell 77
ligament 67
loose fibrous connective
    tissue 67
lymph 68
lymphatic system 80
matrix 66
melanocyte 77

meninges 84
mucous membrane 84
muscular system 81
muscular (contractile)
    tissue 69
nail 79
negative feedback 86
nerve 70
nervous system 81
nervous tissue 70
neuroglia 71
neuron 70
oil gland 79
organ 76
organ system 76
platelet (thrombocyte) 68
pleura 84
positive feedback 87
pseudostratified columnar
    epithelium 74
red blood cell
    (erythrocyte) 68
reproductive system 81
respiratory system 80
reticular fiber 66
serous membrane 84
skeletal muscle 69
skeletal system 81
skin 76
smooth (visceral) muscle 69
spongy bone 68
squamous epithelium 73
striated 69
subcutaneous layer 76
sweat gland 79
synovial membrane 84
tendon 67
tight junction 74
tissue 66
tissue fluid 68
urinary system 81
vitamin D 78
white blood cell (leukocyte) 68

Match the key terms to these definitions.

a. _____ Dense fibrous connective tissue that joins bone to
                bone at a joint.

b. _____ Outer region of the skin composed of keratinized
                stratified squamous epithelium.

c. _____ Cancer of epithelial tissue.

d. _____ Relative constancy of the body's internal
                environment.

e. _____ Porous bone found at the ends of long bones where
                blood cells are formed.

Testing Your Knowledge of the Concepts

1. What are the functions of the four major tissue types?
   (page 66)

2. What features do all connective tissues have in common?
   (page 66)

3. Explain why you would have expected skeletal muscle
   to be voluntary, but smooth muscle and cardiac muscle to
   be involuntary. (pages 69–70)

4. What are the two types of cells found in nervous tissue?
   Briefly describe each type. (pages 70–71)

5. How are epithelial tissues classified? Describe each
   major type, and give at least one location for each type.
   (pages 72–74)

6. Which of the three types of cell junctions permits
   communication between cells? Explain. (pages 74–76)

7. Explain why the skin is sometimes referred to as the
   integumentary system. (page 76)

8. Referring to Figure 4.12, list each organ system, the major
   organs, and major functions of each. (pages 80–81)

9. What organs of the body are found in the thoracic cavity?
   The abdominal cavity? (pages 83–84)

10. List the types of membranes found in the body, their
    functions, and their locations. (page 84)

11. Why is homeostasis defined as the "*relative* constancy of
    the internal environment?" Does negative feedback or
    positive feedback tend to promote homeostasis?
    Explain. (pages 88–91)

12. Which of the following is not a type of fibrous
    connective tissue?
    a. hyaline cartilage   c. tendons and ligaments
    b. areolar tissue   d. adipose tissue

13. What type of cartilage is found in the rib cage and walls
    of the respiratory passages?
    a. fibrocartilage   c. ligamentous cartilage
    b. elastic cartilage   d. hyaline cartilage

14. Blood is a(n) _____ tissue because it has a _____.
    a. connective; gap junction
    b. muscular; ground substance
    c. epithelial; gap junction
    d. connective; ground substance

15. Skeletal muscle is
    a. striated.   c. multinucleated.
    b. under voluntary control.   d. All of these are correct.

16. Which is true of both cardiac and smooth muscle?
    a. striated   c. involuntary control
    b. multinucleated   d. spindle-shaped cells

17. Which of the following forms the myelin sheath around
    nerve fibers outside the brain and spinal cord?
    a. microglia   c. Schwann cells
    b. neurons   d. astrocytes

18. Which of these is not a type of epithelial tissue?
    a. simple cuboidal and stratified columnar
    b. bone and cartilage
    c. stratified squamous and simple squamous
    d. pseudostratified and transitional
    e. All of these are epithelial tissues.

19. What type of epithelial tissue is found in the walls of the urinary bladder to provide it with the ability to distend?
    a. simple cuboidal epithelium
    b. transitional epithelium
    c. pseudostratified columnar epithelium
    d. stratified squamous epithelium

20. Tight junctions are associated with
    a. connective tissue.   c. cardiac muscle.
    b. adipose tissue.   d. epithelial tissue.

21. Without melanocytes, skin would
    a. be too thin.   c. not tan.
    b. lack nerves.   d. not be waterproof.

22. Which of the following is a function of skin?
    a. temperature regulation
    b. manufacture of vitamin D
    c. protection from invading pathogens
    d. All of these are correct.

23. Fluid balance is a primary goal of which system?
    a. endocrine   c. digestive
    b. lymphatic   d. integumentary

24. The skeletal system functions in
    a. blood cell production.   c. movement.
    b. mineral storage.   d. All of these are correct.

25. Which system helps control pH balance?
    a. digestive   c. respiratory
    b. urinary   d. Both b and c are correct.

26. Which type of membrane is found lining systems open to the outside environment, such as the respiratory system?
    a. serous   c. synovial
    b. mucous   d. meningeal

27. The correct order for homeostatic processing is
    a. sensory detection, control center, effect brings about change in body.
    b. control center, sensory detection, effect brings about change in environment.
    c. sensory detection, control center, effect causes no change in environment.
    d. None of these is correct.

28. Which allows rapid change in one direction and does not achieve stability?
    a. homeostasis   c. negative feedback
    b. positive feedback   d. All of these are correct.

29. Which of the following is an example of negative feedback?
    a. Uterine contractions increase as labor progresses.
    b. Insulin decreases blood sugar levels after eating a meal.
    c. Sweating increases as body temperature drops.
    d. Platelets continue to plug an opening in a blood vessel until blood flow stops.

Thinking Critically About the Concepts

While studying the chapter, you've learned about the four types of tissues: connective, muscular, nervous, and epithelial. Organs and, by extension, organ systems are structures constructed from these four tissues. The skin is an organ system referred to as the integumentary system, or the integument. Like the skin, many of its accessory organs are conspicuous. Nails, sweat glands, sebaceous glands, and hair follicles assist the skin in its many functions: protecting the rest of the body from pathogen invasion, water loss, and physical trauma. Skin can help express emotion—blushing when you're embarrassed or sweating when you are anxious or upset. It is also vital to temperature regulation. Without the ability to sweat, the individual rapidly dies from *heat stroke* (extremely high body temperature). Likewise, hypothermia (extremely low body temperature) causes death if a person is unable to shiver.

1. The case study details the burn injury to Kristen Spracklen. From the doctor's explanation of "Skin 101," describe:
   a. the three layers that comprise skin.
   b. what a superficial burn looks like.
   c. how a full burn is diagnosed.

2. What accessory structures and tissues are destroyed in a full burn?

3. Describe how the doctor was able to determine that Kristen's burn was a superficial partial burn, instead of a deep partial burn. What accessory organs and tissues remained after the burn?

4. a. How does shivering increase body temperature?
   b. Once body temperature is returned to normal, does shivering continue? Why or why not?

5. a. What two systems are especially important in coordinating the efforts of the various organ systems as they work to maintain homeostasis?
   b. How do these two systems issue their "commands" to the other systems?

6. a. What different tissues enable the trachea to function as an open tube for the passage of inhaled air?
   b. The bladder has a layer of smooth muscle like any other hollow organ, but it also contains transition epithelium found nowhere else in our body. Transitional epithelial cells are able to elongate when stretched. Why would you expect to find transitional epithelium in the bladder?

7. Review the types of proteins in Section 2.6, on page 34.
   a. What specific proteins make our body's reactions possible?
   b. How are these proteins affected by increased body temperature? (Hint: Consider what happens when egg protein is heated on a stove burner.)
   c. What other extreme condition might affect these proteins? What two organs prevent such extremes from occurring?

CHAPTER **5**

Cardiovascular System: Heart and Blood Vessels

Louise Hairston was beginning to wonder if her bad case of the flu would ever go away. She had awakened two days earlier, feeling very nauseated, light-headed, and dizzy. After vomiting several times, she'd been unable to eat or drink much afterward. A nagging ache, like severely stiff neck and shoulder pain, had plagued her too. She'd fallen asleep on the living room couch, unable to drag herself to her bedroom. This particular morning, Louise barely had the energy to let her little dog outside.

She thought about calling the doctor, but knew she couldn't drive herself in for a visit. Besides, although this smart, capable young man was Dr. Cooper's son, in her mind he couldn't take his retired dad's place. He was always after her, too—take your blood pressure and diabetes medicine, get some exercise, prick your finger three times a day to test your blood sugar, stop smoking, lose weight. Louise loved Southern cooking and her cigarettes. Furthermore, a man young enough to be her son shouldn't be disrespecting an elder by suggesting she was too fat. However, she now felt so awful that a doctor visit seemed worth the aggravation.

Her daughter phoned, and Louise summoned the energy to sit up to answer it— barely. "Mom? The kids and I will be just a few minutes late today." That's when Louise remembered. Today was her day to watch little Isaac and Noelle while Antoinette worked.

"Oh honey, I don't think I can do it today. I'm so sick, I can barely get off the couch." Louise whispered.

Alarmed, Antoinette replied, "We're coming right over. Don't argue with me, Mom. Bobby can watch the kids while I take you to the emergency room." Louise didn't have the energy to argue.

CHAPTER CONCEPTS

5.1 Overview of the Cardiovascular System
In the cardiovascular system, the heart pumps blood, and blood vessels transport blood about the body. Exchanges at capillaries refresh the blood and tissue fluid.

5.2 The Types of Blood Vessels
Arteries, which branch into arterioles, take blood away from the heart to capillaries. Capillaries empty into venules. Venules join to form veins that return blood to the heart.

5.3 The Heart Is a Double Pump
The heart's right side pumps blood to the lungs, and the left side pumps blood to the rest of the body.

5.4 Features of the Cardiovascular System
Blood pressure, which moves blood in the arteries, falls in capillaries where exchange takes place. Skeletal muscle contraction largely accounts for movement of blood in the veins.

5.5 Two Cardiovascular Pathways
One cardiovascular pathway takes blood from the heart to the lungs and back to the heart. The second carries blood from the heart to all other organs, and then returns blood to the heart.

5.6 Exchange at the Capillaries
At tissue capillaries, nutrients and oxygen are exchanged for carbon dioxide and other wastes, permitting body cells to remain alive and healthy.

5.7 Cardiovascular Disorders
Hypertension and atherosclerosis are disorders that can lead to stroke, heart attack, and aneurysm. Ways to prevent and treat cardiovascular disease are constantly being improved.

5.1 Overview of the Cardiovascular System

The cardiovascular system consists of (1) the heart, which pumps blood, and (2) the blood vessels, through which the blood flows. The beating of the heart sends blood into the blood vessels. In humans, blood is always contained within blood vessels.

Circulation Performs Exchanges

Even though circulation of blood depends on the beating of the heart, the purpose of circulation is to serve cells. Recall that cells are always bathed by **tissue fluid**—in other words, blood exchanges substances with tissue fluid and not directly with cells. Blood removes waste products from tissue fluid. Blood also brings tissue fluid the oxygen and nutri-

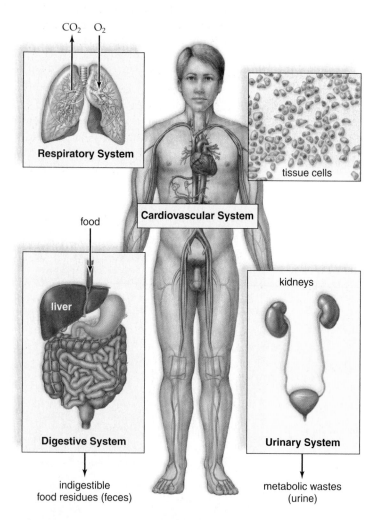

CO₂ O₂

Respiratory System

tissue cells

Cardiovascular System

food

liver

Digestive System

indigestible
food residues (feces)

kidneys

Urinary System

metabolic wastes
(urine)

Figure 5.1 How does the cardiovascular system function to ensure balance throughout the body?
The cardiovascular system transports blood about the body and, with the help of other systems in the body, it maintains favorable conditions for the cells of the body.

ents cells need to continue their existence. Blood would not be able to continue to perform this function if it did not become refreshed in the lungs, at the intestines, and at the kidneys (Fig. 5.1).

We can compare blood vessels to streets that allow cars to move about. Imagine that you are driving around to return birthday gifts you don't want for items you need. Just like blood, at some stops you give up the useless item (wastes) and at these or other stops you get something you need (nutrients and oxygen). Similarly, at the lungs, carbon dioxide leaves blood and oxygen enters it, as indicated by the two arrows in and out of the lungs in Figure 5.1. Blood is purified of its wastes at the kidneys, while water and salts are retained as needed. Nutrients enter blood at the intestines. The liver is important, because it takes up amino acids from the blood and returns proteins to it. Liver proteins transport substances, such as fats, in the blood. The liver removes poisons that may have entered the blood at the intestines, and its colonies of white blood cells destroy bacteria or other pathogens.

Functions of the Cardiovascular System

In this chapter, we will see that

1. contractions of the heart generate blood pressure, which moves blood through blood vessels.
2. blood vessels transport the blood from the heart into arteries, capillaries, and veins. Blood then returns to the heart so the circuit can be completed.
3. exchanges at the capillaries (the smallest of the blood vessels) refresh blood and then tissue fluid, sometimes called interstitial fluid.
4. the heart and blood vessels regulate blood flow, according to the needs of the body.

Lymphatic System

The **lymphatic system** assists the cardiovascular system because lymphatic vessels collect excess tissue fluid and return it to the cardiovascular system. When exchanges occur between blood and tissue fluid, fluid collects in the tissues. This excess fluid enters lymphatic vessels, which start in the tissues and end at cardiovascular veins in the shoulders. As soon as fluid enters lymphatic vessels, it is called lymph. Lymph, you will recall, is a fluid tissue, as is blood (see page 68).

> ### Check Your Progress 5.1
> 1. **What are the two parts of the cardiovascular system?**
> 2. **What are the functions of the cardiovascular system?**
> 3. **How does the lymphatic system help the cardiovascular system?**

Have You Ever Wondered..

If you must stand all day, why do shoes that are loose in the morning feel so tight by evening?

Capillaries are the smallest blood vessels in tissues. Capillary walls are very thin, so they leak. When you stand for long periods, gravity increases the pressure inside the capillaries. Tissue fluid leaks into tissue spaces—and your shoes get too snug. When you lie down at night, your feet are level with your heart. Excess tissue fluid that caused swelling now filters back into lymphatic vessels which empty into shoulder veins, and fluid returns to the cardiovascular system. By morning, your feet have lost their excess water, and those shoes are loose again.

5.2 The Types of Blood Vessels

The structure of the three types of blood vessels (arteries, capillaries, and veins) is appropriate to their function (Fig. 5.2).

The Arteries: From the Heart

The arterial wall has three layers. The innermost layer is a thin layer of cells called endothelium. Endothelium is sur-

rounded by a relatively thick middle layer of smooth muscle and elastic tissue. The artery's outer layer is connective tissue. The strong walls of an artery give it support when blood enters under pressure; the elastic tissue allows an artery to expand to absorb the pressure. **Arterioles** are small arteries just visible to the naked eye. While the middle layer of arterioles has some elastic tissue, it is composed mostly of smooth muscle. These muscle fibers encircle the arteriole. When the fibers contract, the vessel constricts; when these muscle fibers relax, the vessel dilates. The constriction or dilation of arterioles control blood pressure. When arterioles constrict, blood pressure rises. Dilation of arterioles causes blood pressure to fall.

The Capillaries: Exchange

Arterioles branch into capillaries. Each capillary is an extremely narrow, microscopic tube with a wall composed only of endothelium. Capillary endothelium is formed by a single layer of epithelial cells with a basement membrane. Although each capillary is small, their total surface area in humans is about 6,300 square meters. Capillary beds (networks of many capillaries) are present in all regions of the body, so no cell is far from a capillary. In the tissues, only certain capillaries are open at any given time. For example, after eating, the capillaries supplying the digestive system are open, while most serving the muscles are closed. Rings of muscle called **precapillary sphincters** control the blood flow through a capillary bed (see Fig. 5.2). Constriction of

Figure 5.2 How does a capillary bed connect an arteriole to a venule?
A capillary bed is a maze of tiny vessels that lies between an arteriole and a venule. When precapillary sphincter muscles are relaxed, the capillary bed opens and blood flows through the capillaries. Otherwise, blood flows through a shunt directly from the arteriole to the venule. As blood passes through a capillary in the tissues, it gives up its oxygen (O_2). Therefore, blood goes from being O_2-rich in the arteriole to being O_2-poor in the vein. (Note: Blood vessels are colored red if they carry O_2-rich blood and blue if they carry O_2-poor blood. Gas exchange is reversed in the pulmonary circuit, as described in Concept 5.5.)

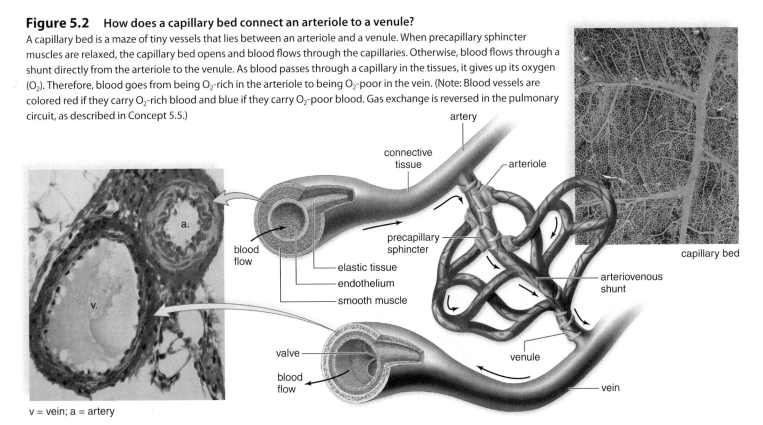

v = vein; a = artery

the sphincter closes the capillary bed. When a capillary bed is closed, the blood moves directly from arteriole to venule through a pathway called an **arteriovenous shunt.**

The Veins: To the Heart

Venules are small veins that drain blood from the capillaries and then join to form a vein. The walls of venules (and veins) have the same three layers as arteries. However, there is less smooth muscle in the middle layer of a vein and less connective tissue in the outer layer. Therefore, the wall of a vein is thinner than that of an artery.

Veins often have **valves,** which allow blood to flow only toward the heart when open and prevent the backward flow of blood when closed. Valves are found in the veins that carry blood against the force of gravity, especially the veins of the lower extremities.

The walls of veins are thinner, so they can expand to a greater extent. At any one time, about 70% of the blood is in the veins. In this way, the veins act as a blood reservoir. If blood is lost due to hemorrhaging, nervous stimulation causes the veins to constrict, providing more blood to the rest of the body.

> **Check Your Progress 5.2**
> 1. What are the types of blood vessels?
> 2. What is the structure and function of these vessels?

5.3 The Heart Is a Double Pump

The **heart** is a cone-shaped, muscular organ located between the lungs, directly behind the sternum (breastbone). The heart is tilted so that the apex (the pointed end) is oriented to the left (Fig. 5.3). To approximate the size of your heart, make a fist, then clasp the fist with your opposite hand. The major portion of the heart, called the **myocardium,** consists largely of cardiac muscle tissue. The muscle fibers of myocardium are branched. Each fiber is tightly joined to neighboring fibers by structures called intercalated disks (Fig. 5.4b). The heart is surrounded by the **pericardium,** a thick, membranous sac that supports and

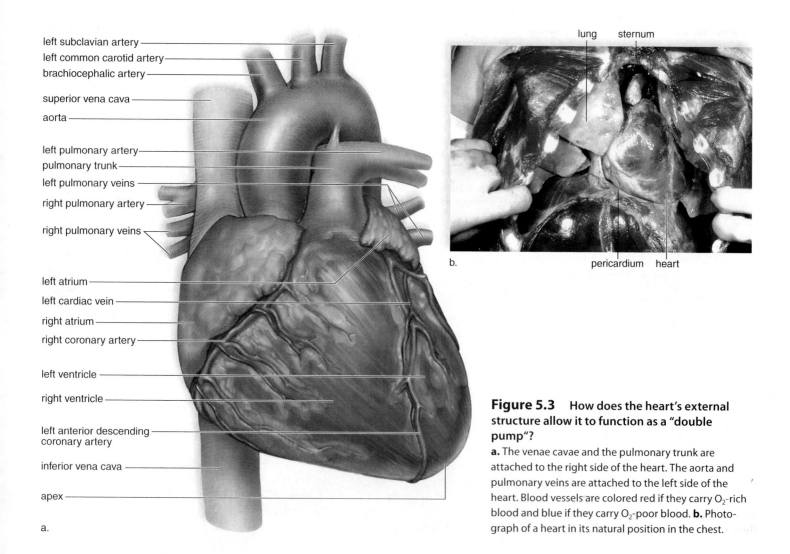

left subclavian artery
left common carotid artery
brachiocephalic artery
superior vena cava
aorta
left pulmonary artery
pulmonary trunk
left pulmonary veins
right pulmonary artery
right pulmonary veins
left atrium
left cardiac vein
right atrium
right coronary artery
left ventricle
right ventricle
left anterior descending coronary artery
inferior vena cava
apex
a.

lung   sternum

b.   pericardium   heart

Figure 5.3   **How does the heart's external structure allow it to function as a "double pump"?**
a. The venae cavae and the pulmonary trunk are attached to the right side of the heart. The aorta and pulmonary veins are attached to the left side of the heart. Blood vessels are colored red if they carry O_2-rich blood and blue if they carry O_2-poor blood. **b.** Photograph of a heart in its natural position in the chest.

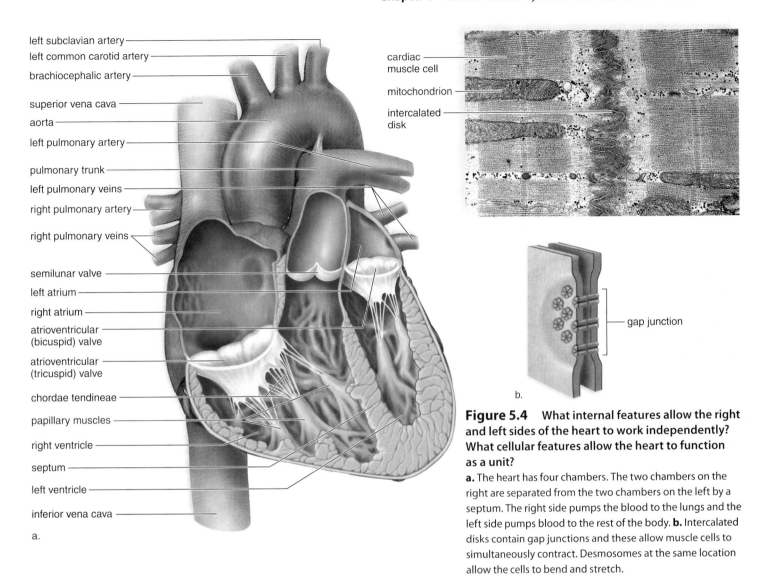

left subclavian artery
left common carotid artery
brachiocephalic artery
superior vena cava
aorta
left pulmonary artery
pulmonary trunk
left pulmonary veins
right pulmonary artery
right pulmonary veins
semilunar valve
left atrium
right atrium
atrioventricular (bicuspid) valve
atrioventricular (tricuspid) valve
chordae tendineae
papillary muscles
right ventricle
septum
left ventricle
inferior vena cava

a.

cardiac muscle cell
mitochondrion
intercalated disk

gap junction

b.

Figure 5.4 **What internal features allow the right and left sides of the heart to work independently? What cellular features allow the heart to function as a unit?**
a. The heart has four chambers. The two chambers on the right are separated from the two chambers on the left by a septum. The right side pumps the blood to the lungs and the left side pumps blood to the rest of the body. **b.** Intercalated disks contain gap junctions and these allow muscle cells to simultaneously contract. Desmosomes at the same location allow the cells to bend and stretch.

protects the heart. The inside of the pericardium secretes a lubrication fluid, and the pericardium slides smoothly over the heart's surface as it pumps the blood.

Internally, a wall called the **septum** separates the heart into a right side and a left side (Fig. 5.4*a*). The heart has four chambers. The two upper, thin-walled atria (sing., **atrium**) are called the right atrium and the left atrium. Each atrium has a wrinkled, ear-like flap called an auricle. The two lower chambers are the thick-walled **ventricles,** called the right ventricle and the left ventricle (see Fig. 5.4*a*).

Heart valves keep blood flowing in the right direction and prevent its backward movement. The valves that lie between the atria and the ventricles are called the **atrioventricular valves (AV valves).** These valves are supported by strong fibrous strings called **chordae tendineae.** The chordae tendinae are attached to papillary muscles that project from the ventricular walls. Chordae anchor the valves, preventing them from inverting when the heart contracts. The AV valve on the right side is called the tricuspid valve because it has three flaps, or cusps. The AV valve on the left side is called the

bicuspid valve because it has two flaps. The bicuspid valve is commonly referred to as the mitral valve, because it has a shape like a bishop's hat, or mitre. The remaining two valves are the **semilunar valves,** with flaps shaped like half-moons. These valves lie between the ventricles and their attached vessels. The semilunar valves are named for their attached vessels: The pulmonary semilunar valve lies between the right ventricle and the pulmonary trunk. The aortic semilunar valve lies between the left ventricle and the aorta.

Coronary Circulation: The Heart's Blood Supply

Myocardium receives oxygen and nutrients from the coronary arteries. Likewise, wastes are removed by the cardiac veins. The blood that flows through the heart contributes little to either nutrient supply or waste removal.

The **coronary arteries** (see Fig. 5.3) serve the heart muscle itself. These arteries are the first branches off the aorta. They originate just above the aortic semilunar valve. They lie on the exterior surface of the heart, where they divide into diverse arterioles. Because they have a very small diameter, coronary arteries may become clogged. The coronary capillary beds join to form venules. The venules converge to form the cardiac veins, which empty into the right atrium.

Passage of Blood Through the Heart

Recall that intercalated disks (see Fig. 5.4b) join fibers of cardiac muscle cells, allowing them to communicate with each other. By sending electrical signals between cells, both atria and then both ventricles contract simultaneously. We can trace the path of blood through the heart and body in the following manner.

- The superior vena cava and the inferior vena cava (see Fig. 5.4a) carry oxygen-poor blood from body veins to the right atrium.

- The right atrium contracts, sending blood through an atrioventricular valve (the tricuspid valve) to the right ventricle. (Remember, the left atrium contracts at the same time.)
- The right ventricle contracts, pumping blood through the pulmonary semilunar valve into the pulmonary trunk. The pulmonary trunk, which carries oxygen-poor blood, divides into two **pulmonary arteries,** which go to the lungs.
- Pulmonary capillaries within the lungs allow gas exchange. Oxygen enters the blood; carbon dioxide waste is excreted from the blood.
- Four **pulmonary veins,** which carry oxygen-rich blood, enter the left atrium.
- The left atrium pumps blood through an atrioventricular valve (the bicuspid [mitral] valve) to the left ventricle.
- The left ventricle contracts, sending blood through the aortic semilunar valve into the aorta. (The right ventricle is contracting at the same time.)

CASE STUDY IN THE EMERGENCY ROOM

Antoinette and Bobby, her son-in-law, had to help Louise to the car. Then, she had to put up with the further indignity of being wheeled on a gurney, just to make it to the emergency room. Though the young nurse on duty was very nice, he needed to stick Louise twice with the needle to start an intravenous solution. Soon after, the emergency room doctor joined them and inspected her IV carefully. "This IV will replace the nutrients and minerals your body needs after being sick for two days," she reassured Louise. "And your blood pressure is a little low, so the IV will help that, too, by putting more fluid into your blood vessels."

"Just as long as I don't have to be stabbed again," she joked weakly.

Her doctor smiled back. "The tube in your nose is to put more oxygen into your blood. We can rub vaseline on it if you're uncomfortable. This finger clip connects you to a pulse oximeter, a gadget to measure blood oxygen." After carefully warming her stethoscope, she listened to Louise's heart and lungs.

"Please turn on your side, Mrs. Hairston, so I can listen to your back, OK?" Louise complied and tried to take several deep breaths when prompted. She ended up gasping and coughing.

"My chest feels so tight, it's hard to breathe," she whispered.

'We're going to take an ECG right away," the doctor asserted. "That's to record the electricity coming from your heart. The technician is a lady, and she'll only uncover your chest for a few minutes. We'll send you for an X-ray right after that. In the meantime, the lab will do blood tests." Then she asked, "Do you or your family have any history of heart disease?"

"Both of my folks died from heart trouble," Louise nodded weakly. "Is that what this is all about? Something wrong with my heart?" Louise gasped.

"Let's just go one step at a time," the doctor reassured. "First, do you have diabetes? What about high blood pressure? Are you taking medicine for your high blood pressure? Do you smoke?" After each of Louise's answers, she made careful notations on her chart.

"I've been worrying about her since my dad died," Antoinette interrupted. "She was under a lot of stress while he was so sick. Now she doesn't take her high blood pressure and diabetes medications. She never smokes around the kids, but they've told me she goes outside to smoke when they watch afternoon cartoons. Don't get mad at me, Mom," she continued as Louise scowled at her. "You know it's true."

"I don't want to scare you, Mrs. Hairston," the doctor responded, "but I think you might have suffered a mild heart attack. Ladies your age sometimes don't have crushing chest pain like men do. Instead, they mistake their symptoms—nausea, vomiting, weakness, muscle aches—for the 'flu.' "

"One thing's for sure," she continued. "I'm going to send you to the cardiac intensive care unit as soon that ECG and X-ray are done. I'll go hurry them along."

- Large arteries, smaller arteries, and arterioles supply tissue capillaries. Tissue capillaries drain into increasingly larger veins. Veins drain into the superior and inferior vena cava, and the cycle starts again.

From this description, you can see that oxygen-poor blood never mixes with oxygen-rich blood and that blood must go through the lungs to pass from the right side to the left side of the heart. The heart is a double pump because the right ventricle of the heart sends blood through the lungs, and the left ventricle sends blood throughout the body. Thus, the left ventricle has the harder job of pumping blood.

The atria have thin walls, and each pumps blood into the ventricle right below it. The ventricles are thicker, and they pump blood into arteries (pulmonary artery and aorta) that travel to other parts of the body. The thinner myocardium of the right ventricle pumps blood to the lungs, nearby in the thoracic cavity. The left ventricle has a thicker wall than the right ventricle, and this enables it to pump blood to all the other parts of the body.

The pumping of the heart sends blood under pressure out into the arteries. The left side of the heart pumps with greater force, so blood pressure is greatest in the aorta. Blood pressure then decreases as the total cross-sectional area of arteries and then arterioles increases (see Fig. 5.8).

A different mechanism, aside from blood pressure, is used to move blood in the veins.

The Heartbeat Is Controlled

Each heartbeat is called a **cardiac cycle** (Fig. 5.5). Recall that when the heart beats, first the two atria contract at the same time. Next, the two ventricles contract at the same time. Then, all chambers relax. **Systole,** the working phase, refers to contraction of the chambers, and the word **diastole,** the resting phase, refers to relaxation of the chambers (Fig. 5.5). The heart contracts, or beats, about 70 times a minute, and each heartbeat lasts about 0.85 second. A normal adult rate at rest can vary from 60 to 80 beats per minute.

To understand the "lub-dup" sounds of a heartbeat, concentrate on the ventricles. The first sound, "lub," occurs when increasing pressure of blood inside a ventricle forces the cusps of the AV valve to slam shut. The force of the blood is just like a strong wind that blows a door (the valve cusps) shut (Fig. 5.5*b*). In contrast, the pressure of blood inside a ventricle causes the semilunar valves (pulmonary and aortic) to open. The "dup" occurs when the ventricles relax, and blood in the arteries flows backward momentarily, causing the semilunar valves to close (Fig. 5.5*c*).

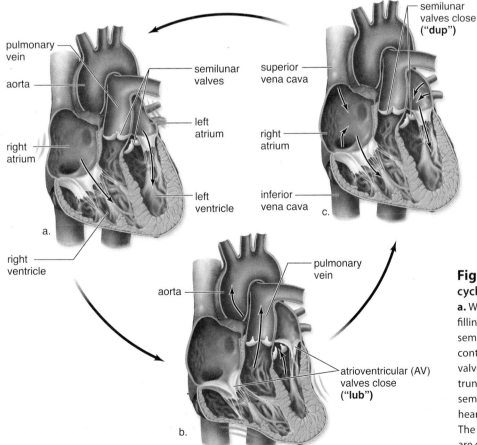

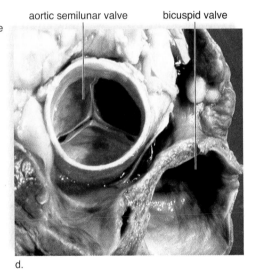

Figure 5.5 **What are the stages of the cardiac cycle?**
a. When the atria contract, the ventricles are relaxed and filling with blood. Atrioventricular valves are open; semilunar valves are closed. **b.** When the ventricles contract, atrioventricular valves are closed; semilunar valves are open. Blood is pumped into the pulmonary trunk and aorta. **c.** Backward flow of blood against the semilunar valves causes the "dup" sound. When the heart is relaxed, both atria and ventricles fill with blood. The atrioventricular valves are open; semilunar valves are closed. **d.** Aortic semilunar valve and bicuspid valve.

A heart murmur, or a slight swishing sound after the "lub," is often due to leaky valves, which allow blood to pass back into the atria after the AV valves have closed. Rheumatic fever (commonly caused by an untreated bacterial infection) is one possible cause of a faulty valve, particularly the bicuspid valve. Faulty valves can be surgically corrected.

Internal Control of Heartbeat

The rhythmic contraction of the atria and ventricles is due to the internal (intrinsic) conduction system of the heart. Nodal tissue is a unique type of cardiac muscle located in two regions of the heart. Nodal tissue has both muscular and nervous characteristics. The **SA (sinoatrial) node** is located in the upper dorsal wall of the right atrium. The **AV (atrioventricular) node** is located in the base of the right atrium very near the septum (Fig. 5.6a). The SA node initiates the heartbeat and automatically sends out an excitation signal every 0.85 second. This causes the atria to contract. When signal impulses reach the AV node, there is a slight delay that allows the atria to finish their contraction before the ventricles begin their contraction. The signal for the ventricles to contract travels from the AV node through the two branches of the **atrioventricular (AV) bundle** before reaching the numerous and smaller **Purkinje fibers.** The AV bundle, its branches, and the Purkinje fibers work efficiently because gap junctions (tiny channels built into intercalated disks) allow electrical current to flow from cell to cell (see Fig. 5.4b).

The SA node is called the **pacemaker** because it usually keeps the heartbeat regular. If the SA node fails to work properly, the heart still beats due to signals generated by the AV node. But the beat is slower (40 to 60 beats per minute). To correct this condition, it is possible to implant an artificial pacemaker, which automatically gives an electrical stimulus to the heart every 0.85 second.

External Control of Heartbeat

The body has an external (extrinsic) way to regulate the heartbeat. A cardiac control center in the medulla oblongata, a portion of the brain that controls internal organs, can alter

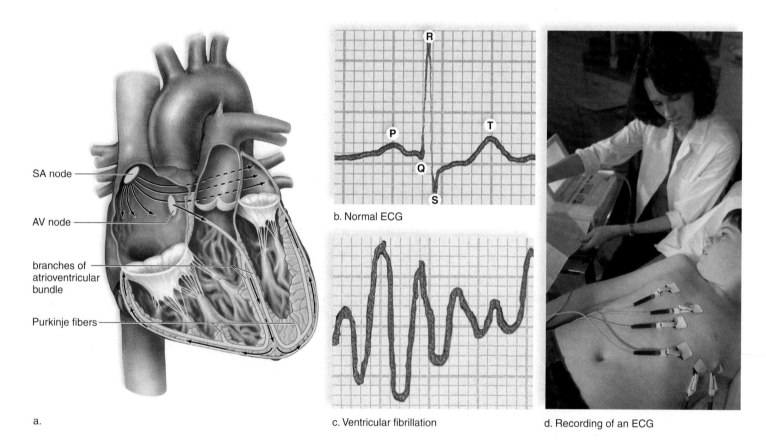

SA node

AV node

branches of atrioventricular bundle

Purkinje fibers

a.

R

P   T

Q

S

b. Normal ECG

c. Ventricular fibrillation

d. Recording of an ECG

Figure 5.6   **What pathway does an electrical signal follow as it travels through the heart?**
a. The SA node sends out a stimulus (black arrows), which causes the atria to contract. When this stimulus reaches the AV node, it signals the ventricles to contract. The electrical signal passes down the two branches of the atrioventricular bundle to the Purkinje fibers. Thereafter, the ventricles contract. **b.** A normal ECG indicates that the heart is functioning properly. The P wave occurs just prior to atrial contraction; the QRS complex occurs just prior to ventricular contraction; and the T wave occurs when the ventricles are recovering from contraction. **c.** Ventricular fibrillation produces an irregular ECG due to irregular stimulation of the ventricles. **d.** The recording of an ECG.

the beat of the heart by way of the parasympathetic and sympathetic portions of the nervous system. As studied in Chapter 1, the parasympathetic division promotes those functions associated with a resting state. The sympathetic division brings about those responses associated with fight or flight. It makes sense that the parasympathetic division decreases SA and AV nodal activity when we are inactive. By contrast, the sympathetic division increases SA and AV nodal activity when we are active or excited.

The hormones epinephrine and norepinephrine, released by the adrenal medulla, also stimulate the heart. During exercise, for example, the heart pumps faster and stronger due to sympathetic stimulation and the release of epinephrine and norepinephrine.

The Electrocardiogram Is a Record of the Heartbeat

An **electrocardiogram (ECG)** is a recording of the electrical changes that occur in myocardium during a cardiac cycle. Body fluids contain ions that conduct electric currents. Therefore, the electrical changes in myocardium can be detected on the skin's surface. When an ECG is being taken, electrodes placed on the skin are connected by wires to an instrument that detects the myocardium's electrical changes. Thereafter, a pen rises or falls on a moving strip of paper. Figure 5.6b depicts the pen's movements during a normal cardiac cycle.

When the SA node triggers an impulse, the atrial fibers produce an electrical change called the P wave. The P wave indicates the atria are about to contract. After that, the QRS complex signals that the ventricles are about to contract. The electrical changes that occur as the ventricular muscle fibers recover produce the T wave.

Various types of abnormalities can by detected by an ECG. One of these, called ventricular fibrillation, is caused by uncoordinated, irregular electrical activity in the ventricles. Compare Figure 5.6b with Figure 5.6c and note the irregular line illustrated in 5.6c. Ventricular fibrillation is of special interest because it can be caused by an injury, heart attack, or drug overdose. It is the most common cause of sudden cardiac death in a seemingly healthy person over age 35. Once the ventricles are fibrillating, coordinated pumping of the heart ceases and body tissues quickly become oxygen starved. Normal electrical conduction must be reestablished as quickly as possible or the person will die. A strong electrical current is applied to the chest for a short time, in a process called defibrillation. In response, all heart cells discharge their electricity at once. Then, the SA node may be able to reestablish a coordinated beat.

Have You Ever Wondered...

Automatic External Defibrillators, or AEDs, can be found in airports and other public places. These are for heart emergencies, such as heart attacks. Should they be used only by a paramedic?

When an emergency happens in a public place like an airport or shopping mall, anyone can use an AED. If a person collapses, perhaps suffering from ventricular fibrillation, the computerized device will make the decisions. It will explain, step-by-step, how to first check for breathing and pulse. Next, it will explain how to apply pads to the chest to deliver the shock, if necessary.

Once the chest pads are attached, the computer analyzes heart activity to determine if a shock is needed. Strong electrical current is applied for a short time to defibrillate the heart. The rescuer moves back, pushing a button when prompted. Voice instructions also explain how to do cardiopulmonary resuscitation (CPR) until paramedics arrive.

However, it's a good idea to first be familiar with CPR and AED use. The Red Cross and many hospitals regularly offer introductory and refresher classes. With training, you might be able to save a person's life!

Check Your Progress 5.3

1. Why is the heart a double pump?
2. What causes the "lub" and the "dup" of the heart sounds?
3. What keeps the heartbeat regular?

5.4 Features of the Cardiovascular System

When the left ventricle contracts, blood is sent out into the aorta under pressure. A progressive decrease in pressure occurs as blood moves through the arteries, arterioles, capillaries, venules, and finally the veins. Blood pressure is highest in the aorta. By contrast, pressure is lowest in the superior and inferior venae cavae, which enter the right atrium (Fig. 5.4a).

Pulse Rate Equals Heart Rate

The surge of blood entering the arteries causes their elastic walls to stretch, but then they almost immediately recoil. This rhythmic expansion and recoil of an arterial wall can be felt as a **pulse** in any artery that runs close to the body's surface. It is customary to feel the pulse by placing several fingers on the radial artery, which lies near the outer border of the palm side of a wrist. A carotid artery, on either side of the trachea in the neck, is another accessible location for feeling the pulse.

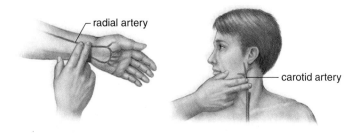

Normally, the pulse rate indicates the heart rate because the arterial walls pulse whenever the left ventricle contracts. The pulse rate is usually 70 beats per minute but can vary between 60 and 80 beats per minute.

Blood Flow Is Regulated

The beating of the heart is necessary to homeostasis because it creates the pressure that propels blood in the arteries and the arterioles. Arterioles lead to the capillaries where exchange with tissue fluid takes place.

Blood Pressure Moves Blood in Arteries

Blood pressure is the pressure of blood against the wall of a blood vessel. A sphygmomanometer (blood pressure instrument) can be used to measure blood pressure, usually in the brachial artery of the arm (Fig. 5.7). The highest arterial pressure, called the **systolic pressure,** is reached during ejection of blood from the heart. The lowest arterial pressure, called the **diastolic pressure,** occurs while the heart ventricles are relaxing. Normal resting blood pressure for a young adult should be slightly lower than 120 mm mercury (Hg) over 80 mm Hg, or 120/80, but these values can vary somewhat and still be within the range of normal blood pressure (Table 5.1). The higher number is the systolic pressure, and the lower number is the diastolic pressure. Blood pressure is higher than 120/80 in the aorta and lower than 120/80 in the venae cavae. High blood pressure is called hypertension, and low blood pressure is called hypotension.

Both systolic and diastolic blood pressure decrease with distance from the left ventricle because the total cross-sectional area of the blood vessels increases—there are more arterioles than arteries. The decrease in blood pressure causes the blood velocity to gradually decrease as it flows toward the capillaries.

Blood Flow Is Slow in the Capillaries

There are many more capillaries than arterioles, and blood moves slowly through the capillaries (Fig. 5.8). This is

Table 5.1	Normal Values for Adult Blood Pressure	
	Top number (systolic)	**Bottom number (diastolic)**
Hypotension	Less than 95	Less than 50
Normal	Below 120	Below 80
Prehypertension	120–139	80–89
Stage 1 Hypertension*	140–159	90–99
Stage 2 Hypertension*	160 or more	100 or more

*Patient will need medication and must maintain a healthy lifestyle.

Figure 5.7   What's a sphygmomanometer, and how is it used to determine blood pressure?
The sphygmomanometer cuff is first inflated until the artery is completely closed by its pressure. Next, the pressure is gradually reduced. The clinician listens with a stethoscope for the first sound, indicating that blood is moving past the cuff in an artery. This is systolic blood pressure. The pressure in the cuff is further reduced until no sound is heard, indicating that blood is flowing freely through the artery. This is diastolic pressure. The procedure can be done manually; computerized sphygmomanometers can also perform the task.

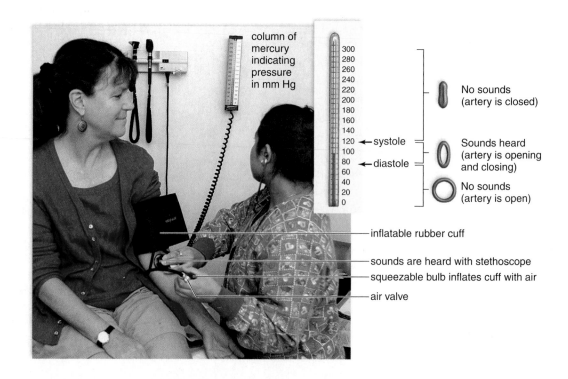

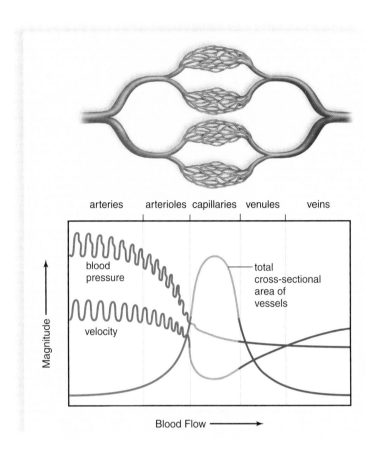

Figure 5.8 **Which blood vessels in the body have the greatest cross-sectional area? If cross-sectional area increases, what happens to blood's pressure and velocity?**
In capillaries, blood is under minimal pressure and has the least velocity. Blood pressure and blood velocity drop off because capillaries have a greater total cross-sectional area than arterioles.

important because the slow progress allows time for the exchange of substances between the blood in the capillaries and the surrounding tissues.

Blood Flow in Veins Returns Blood to Heart

By studying Figure 5.8, it can be seen that the velocity of blood flow increases from capillaries to veins. As an analogy, imagine a single narrow street emptying its cars into a fast multilane highway. Likewise, as capillaries empty into veins, blood can travel faster. However, it's also apparent from Figure 5.8 that blood pressure is minimal in venules and veins. Blood pressure thus plays only a small role in returning venous blood to the heart. Venous return is dependent upon three additional factors:

1. the **skeletal muscle pump,** dependent on skeletal muscle contraction;
2. the **respiratory pump,** dependent on breathing;
3. valves in veins.

The skeletal muscle pump functions every time a muscle contracts. When skeletal muscles contract, they compress

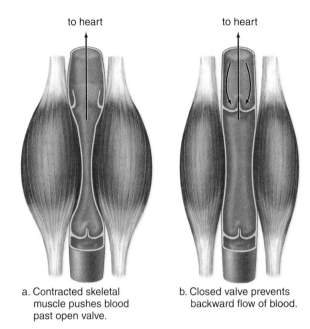

a. Contracted skeletal muscle pushes blood past open valve.

b. Closed valve prevents backward flow of blood.

Figure 5.9 **Veins have a very low pressure. What pushes venous blood, causing it to return to the heart? How do vein valves help?**
a. Pressure on the walls of a vein, exerted by skeletal muscles, increases blood pressure within the vein and forces a valve open. **b.** When external pressure is no longer applied to the vein, blood pressure decreases, and back pressure forces the valve closed. Closure of the valves prevents the blood from flowing backward. Therefore, veins return blood to the heart.

the weak walls of the veins. This causes the blood to move past a valve (Fig. 5.9). Once past the valve, blood cannot flow backward. The importance of the skeletal muscle pump in moving blood in the veins can be demonstrated by forcing a person to stand rigidly still for an hour or so. Fainting may occur because lack of muscle contraction causes blood to collect in the limbs. Poor venous return deprives the brain of needed oxygen. In this case, fainting is beneficial because the resulting horizontal position aids in getting blood to the head.

The respiratory pump works much like an eyedropper. When the dropper's suction bulb is released, the bulb expands. Fluid travels from higher pressure in the bottom of glass tube, moving up the tube toward the lower pressure at the top. When we inhale, the chest expands, and this reduces pressure in the thoracic cavity. Blood will flow from higher pressure (in the abdominal cavity) to lower pressure (in the thoracic cavity). When we exhale, the pressure reverses, but again the valves in the veins prevent backward flow.

> **Check Your Progress 5.4**
> 1. **What does the pulse rate of a person indicate?**
> 2. **What accounts for the flow of blood in the arteries?**
> 3. **What accounts for the flow of blood in the veins?**

Have You Ever Wondered...

What causes varicose veins to develop? Why do some pregnant women develop them, while others don't?

Veins are thin-walled tubes, divided into many separate chambers by vein valves. Excessive stretching occurs if veins are overfilled with blood. For example, if a person stands in one place for a long time, leg veins can't drain properly and blood pools in them. As the vein expands, vein valves become distended and fail to function. These two mechanisms cause the veins to bulge and be visible on the skin's surface. Obesity, sedentary lifestyle, female gender, genetic predisposition, and increasing age are risk factors for varicose veins. Hemorrhoids are varicose veins in the rectum.

Pregnancy in particular can cause leg veins to distended, because the fetus in the abdomen compresses the large abdominal veins. Once again, leg veins can't drain properly and the valves malfunction. Women who don't develop varicose veins during pregnancy may just be genetically "lucky."

Most varicose veins are only a cosmetic problem. Occasionally they will cause pain, muscle cramps, or itching and become a medical problem. There are safe and effective ways to get rid of them, so see your doctor.

5.5 Two Cardiovascular Pathways

The blood flows in two circuits: the **pulmonary circuit,** which circulates blood through the lungs, and the **systemic circuit,** which serves the needs of body tissues (Fig. 5.10). Both circuits are necessary to homeostasis, as we shall see.

The Pulmonary Circuit: Exchange of Gases

The path of blood through the lungs can be traced as follows: Blood from all regions of the body first collects in the right atrium and then passes into the right ventricle, which pumps it into the pulmonary trunk. The pulmonary trunk divides into the right and left pulmonary arteries, which branch as they approach the lungs. The arterioles take blood to the pulmonary capillaries, where carbon dioxide is given off and oxygen is picked up. Blood then passes through the pulmonary venules, which lead to the four pulmonary veins that enter the left atrium. Blood in the pulmonary arteries is oxygen-poor but blood in the pulmonary veins is oxygen-rich, so it is not correct to say that all arteries carry blood high in oxygen and all veins carry blood low in oxygen (as people tend to believe). Just the reverse is true in the pulmonary circuit.

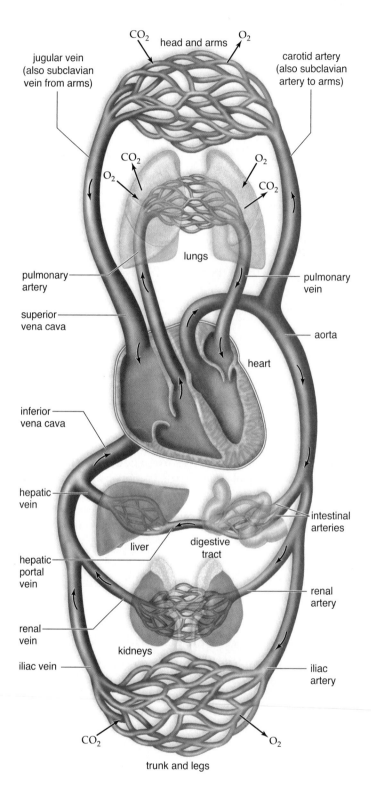

Figure 5.10   How can the cardiovascular system be simplified in a single diagram?
The blue-colored vessels carry blood high in carbon dioxide, and the red-colored vessels carry blood high in oxygen; the arrows indicate the flow of blood. Compare this diagram, useful for learning to trace the path of blood, with Figure 5.11 to realize that arteries and veins go to all parts of the body. Also, there are capillaries in all parts of the body. No cell is located far from a capillary.

The Systemic Circuit: Exchanges with Tissue Fluid

The systemic circuit includes all of the arteries and veins shown in Figure 5.10. (For simplicity, some blood vessels are not shown.) The heart pumps blood through 60,000 miles of blood vessels to deliver nutrients and oxygen and remove wastes from all body cells.

The largest artery in the systemic circuit, the **aorta,** receives blood from the heart, and the largest veins, the **superior** and **inferior venae cavae,** return blood to the heart. The superior vena cava collects blood from the head, the chest, and the arms, and the inferior vena cava collects blood from the lower body regions. Both enter the right atrium.

Tracing the Path of Blood

It's easy to trace the path of blood in the systemic circuit by beginning with the left ventricle, which pumps blood into the aorta. Branches from the aorta go to the organs and major body regions. For example, this is the path of blood to and from the lower legs:

> left ventricle—aorta—common iliac artery—
> femoral artery—lower leg capillaries—
> lower leg veins--femoral vein—common iliac vein—
> inferior vena cava—right atrium

When tracing blood, mention the aorta, the proper branch of the aorta, the region, and the vein returning blood to the vena cava. In many instances, the artery and the vein that serve the same region are given the same name (Fig. 5.11). What happens in between the artery and the vein? Arterioles from the artery branch into capillaries, where exchange takes place, and then venules join into the vein that enters a vena cava.

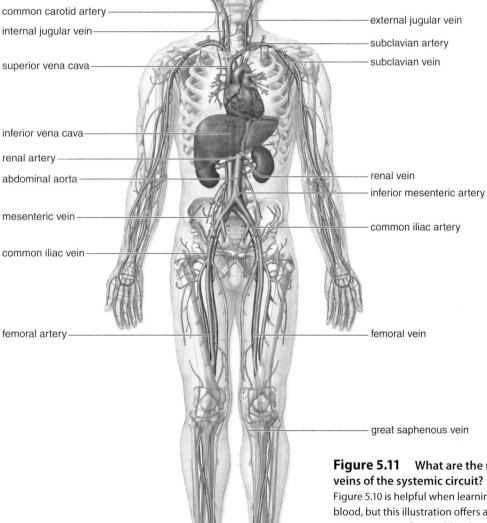

common carotid artery
internal jugular vein
superior vena cava
inferior vena cava
renal artery
abdominal aorta
mesenteric vein
common iliac vein
femoral artery

external jugular vein
subclavian artery
subclavian vein
renal vein
inferior mesenteric artery
common iliac artery
femoral vein
great saphenous vein

Figure 5.11   What are the major arteries and veins of the systemic circuit?
Figure 5.10 is helpful when learning to trace the path of blood, but this illustration offers a more realistic representation of major blood vessels of the systemic circuit. It shows that systemic arteries and veins are present in all parts of the body. The superior and inferior venae cavae take their names from their relationship to which organ?

Hepatic Portal System: Specialized for Blood Filtration

The **hepatic portal vein** drains blood from the capillary beds of the digestive tract to a capillary bed in the liver. (A so-called *portal* system always lies between two capillary beds.) The blood in the hepatic portal vein is oxygen-poor, but rich in glucose, amino acids, and other nutrients absorbed by the small intestine. The liver stores glucose as glycogen, and it either stores amino acids or uses them immediately to manufacture blood proteins. The liver also purifies the blood of toxins and pathogens that have entered the body by way of the intestinal capillaries. After blood has filtered slowly through the liver, it is collected by the **hepatic vein** and returned to the inferior vena cava.

> ### Check Your Progress 5.5
>
> 1. What is the relative oxygen content of the blood flowing in the pulmonary artery compared with that in the pulmonary vein? Explain.
> 2. What is the pathway of blood in the pulmonary circuit?
> 3. Trace the path of blood from the heart to the digestive tract and back to the heart by way of the hepatic portal vein.

5.6 Exchange at the Capillaries

Two forces primarily control movement of fluid through the capillary wall: blood pressure, which tends to cause fluids in the blood to move from capillary to tissue spaces. Osmotic pressure tends to cause water to move in the opposite direction. At the arterial end of a capillary, blood pressure (30 mm Hg) is higher than the osmotic pressure of blood (21 mm Hg) (Fig. 5.12). Osmotic pressure is created by the presence of solutes dissolved in plasma, the liquid fraction of the blood. Dissolved plasma proteins are of particular importance in maintaining the osmotic pressure. Most plasma proteins are manufactured by the liver. Blood pressure is higher than osmotic pressure at the arterial end of a capillary, so water exits a capillary at the arterial end.

Midway along the capillary, where blood pressure is lower, the two forces essentially cancel each other, and there is no net movement of fluid. Solutes now diffuse according to their concentration gradient: Oxygen and nutrients (glucose and amino acids) diffuse out of the capillary; carbon dioxide and wastes diffuse into the capillary. Red blood cells and almost all plasma proteins remain in the capillaries. The substances that leave a capillary contribute to tissue fluid, the fluid between the body's cells. Plasma proteins are too large to readily pass out of the capillary. Thus, tissue fluid tends to contain all components of plasma, except much lower amounts of protein.

At the venule end of a capillary, blood pressure has fallen even more. Osmotic pressure is greater than blood pressure, and fluid tends to move back. Almost the same amount of fluid that left the capillary returns to it, although some excess tissue fluid is always collected by the lymphatic capillaries (Fig. 5.13). Tissue fluid contained within lymphatic vessels

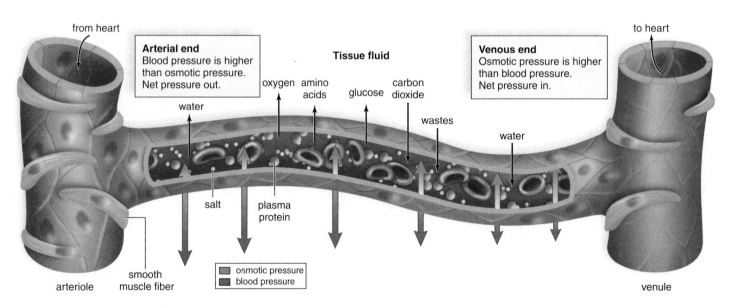

Figure 5.12 **What forces determine the movement of solutes in a tissue capillary? How do these forces form the tissue fluid (interstitial fluid)?**

The capillary shows the exchanges that take place and the forces that aid the process. At the arterial end of a capillary, the blood pressure is higher than the osmotic pressure. Tissue fluid tends to leave the bloodstream. In the midsection, solutes, including oxygen (O_2) and carbon dioxide (CO_2), diffuse from high to low concentration. Carbon dioxide and wastes diffuse into the capillary while nutrients and oxygen enter the tissues. At the venous end of a capillary, the osmotic pressure is higher than the blood pressure. Tissue fluid tends to re-enter the bloodstream. The red blood cells and the plasma proteins are too large to exit a capillary.

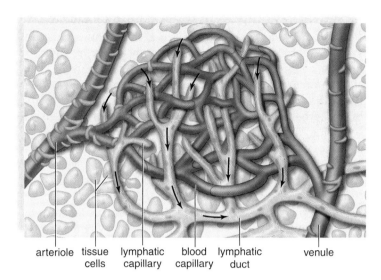

arteriole   tissue   lymphatic   blood   lymphatic   venule
      cells   capillary   capillary   duct

Figure 5.13 **When excess tissue fluid forms, how does it get back into the cardiovascular system?**
Lymphatic capillary beds lie alongside blood capillary beds. When lymphatic capillaries take up excess tissue fluid, it becomes lymph. Lymph returns to the cardiovascular system through cardiovascular veins in the chest. Precapillary sphincters can shut down a blood capillary, and blood then flows through the shunt.

is called **lymph.** Lymph is returned to the systemic venous blood when the major lymphatic vessels enter the subclavian veins in the shoulder region.

> **Check Your Progress 5.6**
> 1. How does the exchange of materials take place across a capillary wall in the tissues?
> 2. What happens to excess tissue fluid created by capillary exchange?

5.7 Cardiovascular Disorders

Cardiovascular disease (CVD) is the leading cause of untimely death in the Western countries. Modern research efforts have resulted in improved diagnosis, treatment, and prevention. This section discusses the range of advances that have been made, first in correcting vascular disorders, and then in correcting heart disorders. The Bioethical Focus on page 111 describes risky behaviors to avoid to prevent CVD from developing in the first place.

Disorders of the Blood Vessels

Hypertension and atherosclerosis often lead to stroke or heart attack, due to an artery blocked by a blood clot or clogged by plaque. Treatment involves removing the blood clot or prying open the affected artery. Another possible outcome is an **aneurysm,** a burst blood vessel. Aneurysm can be prevented by replacing a blood vessel about to rupture with an artificial one.

High Blood Pressure

Hypertension occurs when blood moves through the arteries at a higher pressure than normal. Also called high blood pressure, hypertension is sometimes called a silent killer. It may not be detected until it has caused a heart attack, stroke, or even kidney failure. Hypertension is present when the systolic blood pressure is 140 or greater or the diastolic blood pressure is 90 or greater. While both systolic and diastolic pressures are considered important, diastolic pressure is emphasized when medical treatment is being considered.

The best safeguard against developing hypertension is regular blood pressure checks and a lifestyle that lowers the risk of CVD. If hypertension is present, prescription drugs can help lower blood pressure. Diuretics cause the kidneys to excrete more urine, ridding the body of excess fluid. In addition, hormones (the body's chemical messengers) that raise blood pressure can be inactivated. Drugs called beta-blockers and ACE (angiotensin-converting enzyme) inhibitors help to control hypertension caused by hormones.

Hypertension is often seen in individuals who have **atherosclerosis.** Atherosclerosis is caused by formation of lesions, or *atherosclerotic plaques,* on the inside of blood vessels. The **plaques** narrow blood vessel diameter, choking off blood and oxygen supply to the tissues (Fig. 5.14). In most instances, atherosclerosis begins in early adulthood and develops progressively through middle age, but symptoms may not appear until an individual is 50 or older. To prevent the onset and development of atherosclerosis, the American Heart Association and other organizations recommend a diet low in saturated fat and cholesterol but rich in omega-3 polyunsaturated fatty acids.

Atherosclerotic plaques can cause a clot to form on the irregular, roughened arterial wall. As long as the clot remains stationary, it is called a **thrombus.** If the thrombus dislodges

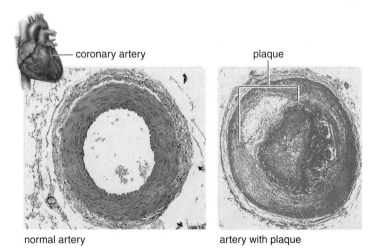

coronary artery   plaque

normal artery   artery with plaque

Figure 5.14 **If an atherosclerotic plaque forms inside a coronary artery, what happens to blood flow in the artery?**
When an atherosclerotic plaque forms in a coronary artery, a heart attack is more apt to occur because of restricted blood flow.

and moves along with the blood, it is called an **embolus.** A **thromboembolism** consists of a clot first carried in the bloodstream, that then becomes completely stationary when it lodges in a small blood vessel. If thromboembolism is not treated, the life-threatening complications described in the next section can result.

Research has suggested several possible causes for atherosclerosis aside from hypertension. Chief among these, as discussed in the Bioethical Focus reading on page 111, are smoking and a diet rich in lipids and cholesterol. Research also indicates that a low-level bacterial or viral infection that spreads to the blood may cause an injury that starts the process of atherosclerosis. Surprisingly, this infection may originate with gum diseases or be due to *Helicobacter pylori* (the bacterium that causes ulcers.) People who have high levels of C-reactive protein, which occurs in the blood following a cold or injury, are more likely to have a heart attack.

Stroke, Heart Attack, and Aneurysm

Stroke, heart attack, and aneurysm are associated with hypertension and atherosclerosis. A cerebrovascular accident (CVA), also called a **stroke,** often results when a small cranial arteriole bursts or is blocked by an embolus. Lack of oxygen causes a portion of the brain to die, and paralysis or death can result. A person is sometimes forewarned of a stroke by a feeling of numbness in the hands or the face, difficulty in speaking, or temporary blindness in one eye.

A myocardial infarction (MI), also called a **heart attack,** occurs when a portion of the heart muscle dies due to lack of oxygen. If a coronary artery becomes partially blocked, the individual may then suffer from **angina pectoris.** Characteristic symptoms of angina pectoris include a feeling of pressure, squeezing, or pain in the chest. Pressure and pain can extend to the left arm, neck, jaw, shoulder, or back. Nausea and vomiting, anxiety, dizziness, and shortness of breath may accompany the chest discomfort. Nitroglycerin or related drugs dilate blood vessels and help relieve the pain. When a coronary artery is completely blocked, perhaps because of thromboembolism, a heart attack occurs.

An aneurysm is a ballooning of a blood vessel, most often the abdominal artery or the arteries leading to the brain. Atherosclerosis and hypertension can weaken the wall of an artery to the point that an aneurysm develops. If a major vessel such as the aorta should burst, death is likely. It is possible to replace a damaged or diseased portion of a vessel, such as an artery, with a plastic tube. Cardiovascular function is preserved because exchange with tissue cells can still take place at the capillaries. In the future, it may be possible to use vessels made in the laboratory by injecting a patient's cells inside an inert mold.

Dissolving Blood Clots

Medical treatment for thromboembolism includes the use of t-PA, a biotechnology drug. This drug converts plasminogen, a protein found in blood, into plasmin, an enzyme that dissolves blood clots. t-PA, which stands for tissue plasminogen activator, is the body's way of converting plasminogen to plasmin. t-PA is also being used for stroke patients, but with less success. Some patients experience life-threatening bleeding in the brain. A better treatment might be new biotechnology drugs that act on the plasma membrane to prevent brain cells from releasing and/or receiving toxic chemicals caused by the stroke.

If a person has symptoms of angina or a stroke, aspirin may be prescribed. Aspirin lowers the probability of clot formation. There is evidence that aspirin protects against first heart attacks, but there is no clear support for taking aspirin every day to prevent strokes in symptom-free people. Physicians warn that long-term use of aspirin might have harmful effects, including bleeding in the brain.

CASE STUDY DIAGNOSIS

Less than an hour later, Louise was visited by the heart specialist on duty in the cardiac intensive care unit. "Mrs. Hairston, the results of your ECG and blood tests show you've had a heart attack. A blood clot in your coronary arteries—those are the blood vessels that bring blood to the heart—has choked off oxygen supply and killed some heart muscle. The blood tests show proteins that belong in your heart muscle cells, and that's how we know the muscle is dead. The heart cells burst when they die, and release their contents into the blood."

She continued, "Your heart attack seems to be pretty mild, but we're going to give you a drug right away that will help to dissolve the blood clot. It's called t-PA, and you may have heard it called the 'clot-busting drug.' We'll also be giving you pain medicine and medicine to help open up the coronary arteries. After we've made you more comfortable and helped your breathing, we'll need to do a test called a coronary angiography. That'll help us figure out where the blocked artery is located and what to do about it."

"Will I need to have surgery?" Louise gasped painfully.

"I hope not, but it's really too soon to tell," the specialist replied. "We've got some new ways to help fix the coronary arteries without major surgery. If we can use those new techniques, you'll just have to have a little incision on your leg."

"Can I see my daughter now?" Louise asked.

"Sure," the doctor replied. "but remember, she can stay only five minutes and no more, OK? You need to rest, and we need to start your drug treatment right away."

Historical **Focus**

Surgeon Without A Degree: Vivien Theodore Thomas (1910–1985)

On a warm summer day in 1941, professors, students, and visitors at Johns Hopkins University Hospital stopped to stare at an unusual sight. Dressed in a laboratory jacket, a young African American crossed the campus. In that era of strict racial segregation and Jim Crow laws, African Americans on a college campus were maids or janitors. Bathrooms, drinking fountains, hotels, even baseball teams were segregated by race. African American patients were treated at the hospital, but had to sneak in by a rear entrance. To see a black man outfitted as a researcher was an extraordinary sight—but Vivien Thomas was an extraordinary man.

Thomas had determined in high school that he would pursue a medical career. After a year in college as a premedical student, the stock market crash and bank failures of 1929 shattered his dream. Thomas' college education and medical school were never completed. Instead, Thomas worked at Vanderbilt University in Nashville, serving as a laboratory assistant to surgeon Alfred Blalock.

After putting in sixteen hour days in the laboratory, Thomas tutored himself in anatomy (the study of body structures) and physiology (the study of bodily function). Next, he taught himself surgical techniques using experimental animals, including how to administer anesthesia and sew blood vessels. Thomas became a brilliant researcher, devising a number of surgical instruments. His respirator, which took over breathing when the chest was cut open, allowed surgeries that had not been possible previously.

When Blalock was offered a position at the prestigious Johns Hopkins University Hospital in Baltimore, he accepted upon one condition—that Vivien Thomas be allowed to accompany him. There, Vivien and Blalock continued their collaboration in medical research and surgical methods. Vivien helped to devise the Blalock-Taussing surgery, named after Blalock and the heart physician Helen Taussig.* This operation helped to save the lives of many babies with severe heart defects. During many complicated surgeries, Viven stood at Blalock's right hand. There, he patiently advised his boss on proper technique. Blalock's contemporary referred to Vivien as "the surgical glove on Blalock's experimental hand."

Though they could enjoy each other's company in the hospital lab, the two men's close relationship ended at the laboratory door. Vivien was paid and classified as a janitor, earning extra money by serving drinks at Blalock's parties. His salary was so low that at one point he threatened to seek work as a carpenter to make more money. Only then did Blalock see to it that his assistant received a raise.

In 1976, Johns Hopkins University awarded Vivien Thomas an honorary degree. Today, his portrait hangs where it should, right next to that of Alfred Blalock in the Blalock Clinical Sciences Building at Johns Hopkins. To the end of his life, Thomas remained a very humble man. Of his life and work, Thomas wrote: "As for me, I just work here … I've thoroughly enjoyed the role I have played and only tried to be me."

*Helen Taussig was also a pioneer. A specialist in the care of children with heart defects, she was also the first female president of the American Heart Association.

Treating Clogged Arteries

Cardiovascular disease used to require open-heart surgery and, therefore, a long recuperation time and a long unsightly scar that could occasionally ache. Now, bypass surgery can be accomplished by using robotic technology (Fig. 5.15a). A video camera and instruments are inserted through small cuts, while the surgeon sits at a console and manipulates interchangeable grippers, cutters, and other tools attached to movable arms above the operating table. Looking through two eyepieces, the surgeon gets a three-dimensional view of the operating field. Robotic surgery is also used in valve repairs and other heart procedures.

A **coronary bypass operation** is one way to treat an artery clogged with plaque. A surgeon takes a blood vessel—usually a vein from the leg—and stitches one end to the aorta and the other end to a coronary artery past the point of obstruction. Figure 5.15b shows a triple bypass in which three blood vessel segments have been used to allow blood to flow freely from the aorta to cardiac muscle by way of the coronary artery. Instead of a

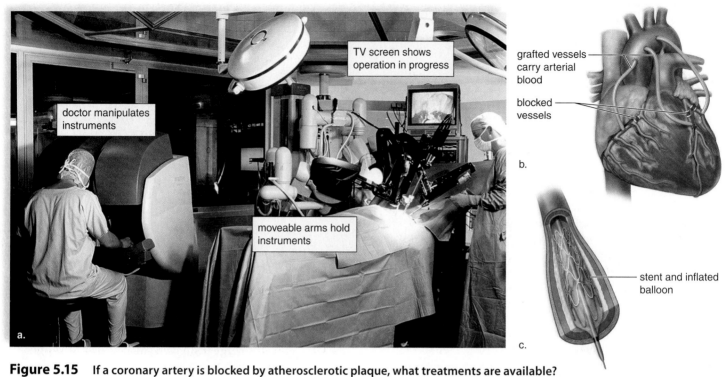

Figure 5.15   If a coronary artery is blocked by atherosclerotic plaque, what treatments are available?
a. Robotic surgery techniques can assist in coronary bypass operations. **b.** In this procedure, blood vessels (leg veins of arteries in the chest) are stitched to coronary arteries, bypassing the obstruction. **c.** In a stent procedure, a balloon-tipped catheter guides the stent (a cylinder of expandable metal mesh) to the obstructed area. When the balloon is inflated, the stent expands and opens the artery.

coronary bypass, gene therapy has been used since 1997 to grow new blood vessels that will carry blood to cardiac muscle. The surgeon need only make a small incision and inject many copies of the gene that codes for vascular endothelial growth factor (VEGF) between the ribs directly into the area of the heart that most needs improved blood flow. VEGF encourages new blood vessels to sprout out of an artery. If collateral blood vessels do form, they transport blood past clogged arteries, making bypass surgery unnecessary. About 60% of patients who undergo the procedure do show signs of vessel growth within two to four weeks.

Another alternative to bypass surgery is available; namely, the stent (Fig. 5.15c). A stent is a small metal mesh cylinder that holds a coronary artery open after a blockage has been cleared. Until a couple of years ago, stenting was a second step following angioplasty. During **angioplasty,** a plastic tube is inserted into an artery of an arm or a leg and then guided through a major blood vessel toward the heart. When the tube reached the region of plaque in an artery, a balloon attached to the end of the tube was inflated, forcing the vessel open. Now, instead, a stent plus an inner balloon is pushed into the blocked area. When the balloon inside the stent is inflated, it expands, locking itself in place. Some patients go home the

same day; at most, after an overnight stay. The stent is more successful when it is coated with a drug that seeps into the artery lining and discourages cell growth. Uncoated stents can close back up in a few months, but not ones coated with a drug that discourages closure. Blood clotting might occur after stent placement, so recipients have to take anticlotting medications.

Disorders of the Heart

When a person has **heart failure,** the heart no longer pumps as it should. Heart failure is a growing problem because people who used to die from heart attacks now survive, but are left with damaged hearts. Often the heart is oversized, not because the cardiac wall is stronger but because it is sagging and swollen. One idea is to wrap the heart in a fabric sheath to prevent it from getting too big. This might allow better pumping, similar to the way a weight lifter's belt restricts and reinforces stomach muscles. But a failing heart can have other problems, such as an abnormal heart rhythm. To counter that condition, it's possible to implant a cardioverter-defibrillator (ICD) just beneath the skin of the chest. These devices can sense both an abnormally slow and an abnormally fast heartbeat. If the former, they generate the missing beat like a pacemaker does. If the latter, they send the heart a sharp jolt of electricity to slow it down. If the heart rhythm should become erratic, it sends an even stronger shock like a de-fibrillator does.

Heart Transplants

Although heart transplants are now generally successful, many more people are waiting for new hearts than there are organs available. Today, only about 2,200 heart transplants are done annually. Many thousands of people could use one. Genetically altered pigs may one day be used as a source of hearts, because of the shortage of human hearts. Also, bone marrow stem cells have been injected into the heart, and researchers report that they apparently become cardiac muscle!

Today, a left ventricular assist device (LVAD), implanted in the abdomen, is an alternative to a heart transplant. A tube passes blood from the left ventricle to the device, which pumps it on to the aorta. A cable passes from the device through the skin to an external battery the patient must tote around. Soon, as many as 20,000 people a year may be outfitted with an LVAD. Instead of

CASE STUDY TREATMENT

Louise was resting comfortably when the specialist returned half an hour later, accompanied by Antoinette and Bobby. "I feel better some, and I'm getting a little hungry," Louise told the three.

"I'm sorry, Mrs. Hairston," the doctor apologized, "but I think we're going to have to hold off on your lunch a little longer. We need to get you into surgery for that coronary angiogram, and you can't have anything to eat or drink until it's over."

"What's a coronary angiogram?" Bobby asked.

The doctor replied, "It's a pretty simple surgery, and your mom won't even be under general anesthesia while it's going on. Instead, we'll give her some medicine to feel relaxed and a little groggy, then numb a small area in her groin. We want to reach her femoral artery—that's the big artery in her leg. We can put a small tube called a catheter into the artery, and then guide the catheter up to her heart. Injecting dye through the catheter makes her coronary arteries show up on X-ray, so we can see where the arteries are blocked. If just one or two arteries are blocked, we can fix it right away."

"Is that when you need to do a major surgery?" Antoinette inquired.

"No," the doctor replied. "Once we locate the blockage, we thread a small guide wire to the blocked area. Then we use a second catheter that has a tiny balloon on the end and slide it over the guide wire. A wire cage called a stent sits over the balloon. We blow up the balloon and smash that blockage." She slapped one palm against another for emphasis, and Louise laughed weakly.

She continued, "When the balloon is inflated, the stent springs open. A stent looks a little like a tiny chicken wire cage. When it's open, it props the artery walls open for good. The whole operation is called a coronary angioplasty. That's what I mean by fixing the problem right away. We'd like to start as soon as possible, after we get some forms completed."

She nodded to Bobby and Antoinette, "You guys can wait just outside. I'm sure we have some old magazines out there for you to enjoy," she smiled.

The young couple sat anxiously in the waiting room outside the coronary intensive care unit. Each took turns pacing the floor, until at last the cardiac specialist reappeared in her surgical scrubs.

"Everything went great," she reassured Antoinette and Bobby. "We found a blockage in one of the branches of a large coronary artery. The coronary angioplasty and stent opened it back up. We're going to monitor your mother in the coronary unit for the next several days. But before she gets discharged, she and I are gonna have some words."

Louise was dressed, more than ready to leave the hospital with Antoinette five days later. The specialist came by. "No driving, no lifting, no heavy exercise, but be sure to walk slowly around the house a few times a day for the next two weeks," she began. "You have to take your blood thinners every day, or you might have a blood clot form."

"And Mrs. Hairston," she continued, "unless you make some changes in your life, you just wasted my time, because you'll wind up dead. You must quit smoking, and I want you to use the quit smoking program here at the hospital to help. The hospital also has a weight loss and exercise program. If you lose weight and get some exercise, you might not even need those diabetes medicines you 'forget' to take. But take them anyway until your family doctor tells you not to, and you'd better not forget those blood pressure pills anymore!"

"I just bought her a pill box with a timer on it," Antoinette said.

"Great," the doctor smiled. "We'll see you again in a week for a checkup."

Have You Ever Wondered...

When a heart becomes available for transplant, how do doctors decide who gets it?

All potential heart transplant patients must be evaluated to decide if the person is a good candidate. If so, the candidate is placed on a national registry. When a heart becomes available, the registry will consider:

- Is the candidate near the transplant center? (A donor heart typically only survives for approximately 4 hours.)

- Does the candidate have the greatest immediate need for transplant, or are others more critically ill?

- Do the donor and recipient's blood types match? Are there other proteins in the recipient's blood that might cause rejection and make the transplant fail?

- Is the heart approximately the right size to fit the recipient?

- How much time has this candidate spent on the waiting list?

The national registry tries to make these very difficult decisions in a fair and impartial manner. Potential recipients must carry a transplant center pager and have a cell phone charged and available at all times. When that very important call comes, the patient must respond and be ready for surgery immediately!

---

an LVAD, a few pioneers have volunteered to receive a Jarvik 2000, a pump inserted inside the left ventricle. The Jarvik is powered by an external battery no larger than a C-size battery.

Fewer patients still have received a so-called **total artificial heart (TAH),** such as that shown in Figure 5.16. An internal battery and controller regulate the pumping speed, and an external battery powers the device by passing electricity through the skin via external and internal coils. A rotating centrifugal pump moves silicon hydraulic fluid between left and right sacs to force blood out of the heart into the pulmonary trunk and the aorta. All recipients, thus far, have been near death, and most have lived for only a short time. It's possible that once healthier patients receive a TAH, survival rate will improve. Different types of TAHs are being investigated in animals.

Check Your Progress

1. a. What cardiovascular disorders are common in humans and (b) what treatments are available?

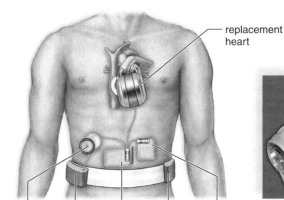
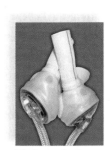

| wireless energy transfer system | external wireless driver | internal controller | external battery pack | rechargable internal battery | Photograph of artificial heart |

Artificial heart inside body

Figure 5.16 **How does an artificial heart work?**
The CardioWest™ temporary Total Artificial Heart moves blood in the same manner as a natural heart. The controller is implanted into the patient's abdomen. The heart is powered by an internal rechargeable battery, with an external battery as a backup system.
Courtesy of SynCardia Systems, Inc.

Cardiovascular Disease Prevention: Who Pays for an Unhealthy Lifestyle?

Cardiovascular disease (CVD) is not only the number one killer in the United States today, it is also one of the most expensive. More than $431 billion is spent annually for health care and lost productivity, according to the American Heart Association. Family history of heart attack under age 55, male gender, and ethnicity (African Americans are at great risk) are unalterable risk factors for CVD. However, most cases of CVD are preventable (Fig. 5A). Preventable risk factors include:

- Use of tobacco: Nicotine constricts arterioles, increasing blood pressure. As a result, the heart must pump harder to propel blood. Carbon monoxide in smoke decreases the blood's oxygen-carrying ability. (Tobacco use also causes many different cancers, as you'll discover in Chapter 19.)
- Drug and alcohol abuse: Stimulants, like cocaine and amphetamines, can cause irregular heartbeat and lead to heart attacks. IV drug use may cause cerebral blood clots and stroke. Alcohol abuse can destroy body organs, including the heart. (However, low to moderate alcohol consumption actually decreases CVD risk.)
- Obesity and a sedentary lifestyle: Extra body tissue requires an additional blood supply, increasing the heart's workload. Hypertension develops as the heart works harder to pump blood. Obesity also increases the risk of type 2 diabetes. Diabetes causes blood vessel damage and atherosclerosis. Lack of exercise contributes to obesity.
- Poor diet: A diet high in saturated fats and cholesterol is a risk for CVD. Cholesterol is ferried in the blood by two proteins: LDL (low-density lipoprotein, the "bad" lipoprotein) and HDL (high-density lipoprotein, the "good" lipoprotein). LDL carries its cholesterol to deposit in tissues, but HDL carries cholesterol to the liver where it can be metabolized. Elevated blood LDL and/or low blood HDL levels can contribute to cardiovascular disease. A diet low in saturated fat and cholesterol can help to restore LDL and HDL to recommended levels. Drugs called statins can further lower LDL level if necessary.
- Stress: A stress-filled lifestyle contributes directly to CVD, and may also cause the person to overeat, avoid exercise, start or increase smoking, or abuse drugs and alcohol.
- Poor dental hygiene: Avoiding the dentist or dental clinic results in gum inflammation. Microbes from infected gums can enter the bloodstream and trigger formation of atherosclerotic plaques.

With few exceptions, these risk factors are personal choices.

Who Pays for Treatment?

People who practice risky behaviors already pay more: for example, paying extra for health or life insurance. Tobacco and alcohol taxes

Figure 5A **What risk factors for CVD are under an individual's control?**
Unhealthy lifestyle choices—smoking, drinking alcohol to excess, obesity, and a sedentary lifestyle—are all factors that a person has the power to change.

also help to defray health care costs. However, insurance companies may not cover pre-existing conditions, including CVD. And what about the uninsured? In these cases, treatment costs are borne by taxpayers. Ultimately, we all pay for CVD treatment—and we all have an incentive to ensure that individuals adopt healthy lifestyles, or pay more for treatment if they refuse.

Is Legislation Needed?

Several organizations, such as the World Health Organization, have advocated legislation to help pay for CVD and its consequences. One proposed example is the so-called "fat tax" on foods with high fat and/or poor nutritional value. The tax would pay for treatment of obesity-related disease, including CVD. Other initiatives stress education of both adults and children regarding healthy lifestyle choices. However, critics oppose this type of legislation, insisting that rewarding healthy behaviors and penalizing risky behaviors is beyond the scope of government.

Decide Your Opinion

1. Would you support charging higher insurance premiums or taxes on people who don't practice a healthy lifestyle?
2. Should public funding be used for prevention programs, if money could be saved in the future?
3. The public does subsidize health care that disproportionately benefits the less healthy, so do you believe that financial interest trumps personal freedom in this matter? Why or why not?

Summarizing the Concepts

5.1   Overview of the Cardiovascular System

The cardiovascular system consists of the heart and blood vessels.

The heart pumps blood and blood vessels take blood to and from capillaries, where exchanges of nutrients for wastes occur with tissue cells. Blood is refreshed at the lungs, where gas exchange occurs; at the digestive tract, where nutrients enter the blood; and the kidney, where wastes are removed from blood.

[handwritten: 1 lungs 2 digestive 3 kidneys liver]

5.2   The Types of Blood Vessels

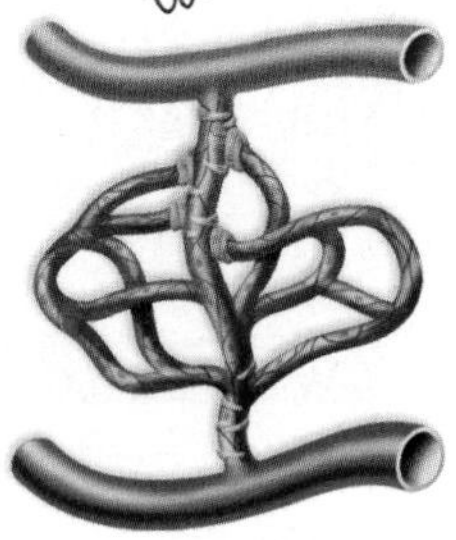

The Arteries   Arteries (and arterioles) take blood away from the heart. Arteries have the thickest walls, which allows them to withstand blood pressure.

The Capillaries   Capillaries are where exchange of substances occurs.

The Veins   Veins (and venules) take blood to the heart. Veins have relatively weak walls with valves that keep the blood flowing in one direction.

capillary bed

5.3   The Heart Is a Double Pump

The heart has a right and left side. Each side has an atrium and a ventricle. Valves keep the blood moving in the correct direction.

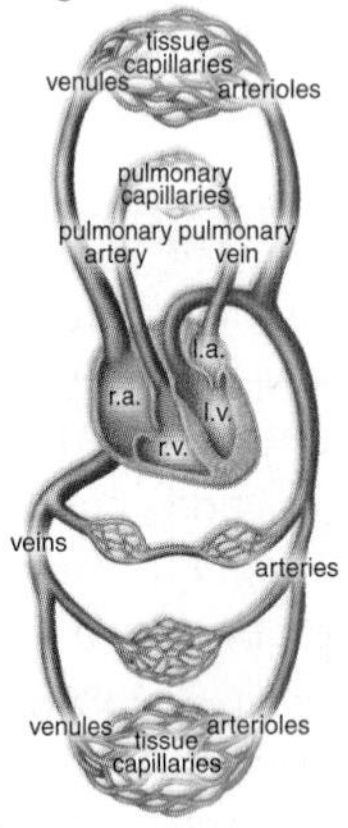

Passage of Blood Through the Heart

- The right atrium receives O_2-poor blood from the body, and the right ventricle pumps it into the pulmonary circuit (to the lungs).
- The left atrium receives O_2-rich blood from the lungs, and the left ventricle pumps it into the systemic circuit.

The Heartbeat Is Controlled

During the cardiac cycle, the SA node (pacemaker) initiates the heartbeat by causing the atria to contract. The AV node conveys the stimulus to the ventricles, causing them to contract. The heart sounds, "lub-dup," are due to the closing of the atrioventricular valves, followed by the closing of the semilunar valves.

5.4   Features of the Cardiovascular System

Pulse   The pulse rate indicates the heartbeat rate.

Blood Pressure Moves Blood in Arteries   Blood pressure caused by the beating of the heart accounts for the flow of blood in the arteries.

Blood Flow Is Slow in the Capillaries   The reduced velocity of blood flow in capillaries facilitates exchange of nutrients and wastes in the tissues.

Blood Flow in Veins Returns Blood to the Heart   Blood flow in veins is caused by skeletal muscle contraction, the presence of valves, and respiratory movements.

5.5   Two Cardiovascular Pathways

The cardiovascular system is divided into the pulmonary circuit and the systemic circuit.

The Pulmonary Circuit: Exchange of Gases

In the pulmonary circuit, blood travels to and from the lungs.

The Systemic Circuit: Exchanges with Tissue Fluid

In the systemic circuit, the aorta divides into blood vessels that serve the body's organs and cells. Venae cavae return O_2-poor blood to the heart.

5.6   Exchange at the Capillaries

This diagram illustrates capillary exchange in tissues of the body—not including the gas-exchanging surfaces of the lungs.

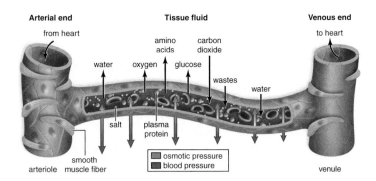

- At the arterial end of a cardiovascular capillary, blood pressure is greater than osmotic pressure; therefore, fluid leaves the capillary.
- In the midsection, oxygen and nutrients diffuse out of the capillary, while carbon dioxide and other wastes diffuse into the capillary.
- At the venous end, osmotic pressure created by the presence of proteins exceeds blood pressure, causing most of the fluid to re-enter the capillary. Some fluid remains as interstitial (tissue) fluid.

  Excess fluid not picked up at the venous end of the cardiovascular capillary enters the lymphatic capillaries.
- Lymph is tissue fluid contained within lymphatic vessels.
- The lymphatic system is a one-way system. Its fluid is returned to blood by way of a cardiovascular vein.

5.7   Cardiovascular Disorders

Cardiovascular disease is the leading cause of death in the Western countries.

- Hypertension and atherosclerosis can lead to stroke, heart attack, or an aneurysm.
- Following a heart-healthy diet, getting regular exercise, maintaining a proper weight, and not smoking reduce cardiovascular disease risk.

Understanding Key Terms

aneurysm 105	lymphatic system 92
angina pectoris 106	myocardium 94
angioplasty 108	pacemaker 98
aorta 103	pericardium 94
arteriole 93	plaques 105
artereovenous shunt 94	precapillary sphincter 93
atherosclerosis 105	pulmonary artery 96
atrioventricular (AV) bundle 98	pulmonary circuit 102
atrioventricular (AV) valve 95	pulmonary vein 96
atrium 95	pulse 99
AV (atrioventricular) node 98	Purkinje fibers 98
blood pressure 100	respiratory pump 101
cardiac cycle 97	SA (sinoatrial) node 98
chordae tendineae 95	semilunar valve 95
coronary artery 95	septum 95
coronary bypass operation 107	skeletal muscle pump 101
diastole 97	stroke 106
diastolic pressure 100	superior vena cava 103
electrocardiogram (ECG) 99	systemic circuit 102
embolus 106	systole 97
heart 94	systolic pressure 100
heart attack 106	thromboembolism 106
heart failure 108	thrombus 105
hepatic portal vein 104	tissue fluid 92
hepatic vein 104	total artificial heart (TAH) 110
hypertension 105	valve 94
inferior vena cava 103	ventricle 95
lymph 105	venule 94

Match the key terms to these definitions.

a. _diastole_ Relaxation of a heart chamber.

b. _IVC_ Large systemic vein that returns blood from body areas below the diaphragm.

c. _pulse_ Rhythmic expansion and recoil of arteries resulting from heart contraction; can be felt from outside the body.

d. _venule_ Vessel that takes blood from capillaries to a vein.

e. _systemic cu_ That part of the cardiovascular system that serves body parts and does not include the gas-exchanging surfaces in the lungs.

Testing Your Knowledge of the Concepts

1. What are the two parts of the cardiovascular system, and what are the functions of each part? (page 92) _heart/blood vessels_

2. Explain where exchanges occur in the body and the importance of those exchanges. (pages 93–94) _lungs capillary beds lenn_

3. Which of the three types of blood vessels are most numerous? Explain. (page 93)

4. Describe the structure of the heart, including the chambers and valves. (pages 94–96)

5. What is the function of the septum in the heart? What would happen if the heart had no septum? (page 95)

6. Trace the path of blood through the heart, including chambers, valves, and vessels the blood travels through. (pages 96–97)

7. Describe the cardiac cycle, using the terms systole and diastole. What is the role of the SA node and the AV node in the cardiac cycle? (pages 97–99)

8. Distinguish between the internal and external controls of the heartbeat. Explain how an ECG relates to the cardiac cycle. (pages 97–99)

9. In what vessel is the blood pressure the highest? The lowest? In what vessel is blood flow rate the fastest? The slowest? Why is the pressure and rate in the capillaries important to capillary function? (pages 100–01)

10. Explain why skeletal muscle contraction has an effect on venous flow but not arterial flow. (pages 100–01)

11. Distinguish between the two cardiovascular pathways. (pages 102–03)

12. Trace the pathway of blood to and from the brain in the systemic circuit. (pages 102–04)

13. Describe the process by which nutrients are exchanged for wastes across a capillary, using glucose and carbon dioxide as examples. (pages 104–05)

14. What is the most probable association between high blood pressure and a heart attack? With this association in mind, what type of diet might help prevent a heart attack? (page 105)

In questions 15–18, match the descriptions to the circuit in the key. Answers may be used more than once.

Key:

  a. pulmonary circuit
  b. systemic circuit
  c. both pulmonary and systemic

15. Arteries carry O_2-rich blood. _d b_

16. Carbon dioxide leaves the capillaries, and oxygen enters the capillaries. _bt c a_

17. Arteries carry blood away from the heart, and veins carry blood toward the heart. _c_

18. This contains the hepatic portal system. _b_

In questions 19–24, match the descriptions to the blood vessel in the key. Answers may be used more than once.

Key:

  a. venules   d. arteries
  b. veins   e. arterioles
  c. capillaries

19. Drain blood from capillaries _a_

20. Empty into capillaries _e_

21. May contain valves _b_

22. Take blood away from the heart _d_

23. Sites for exchange of substances between blood and tissue fluid _c_

24. Rate of blood flow is the slowest _c_

25. During ventricular diastole,
  a. blood flows into the aorta.
  b. the ventricles contract.
  c. the semilunar valves are closed.
  d. Both a and b are correct.

26. When the atria contract, the blood flows
    a. into the attached blood vessels.
    b. into the ventricles.
    c. through the atrioventricular valves.
    d. to the lungs.
    e. Both b and c are correct.

27. Heart valves located at the bases of the pulmonary trunk and aorta are called
    a. atrioventricular valves.
    b. semilunar valves.
    c. mitral valves.
    d. chordae tendineae.

28. Which of these associations is mismatched?
    a. left ventricle—aorta
    b. right ventricle—pulmonary trunk
    c. right atrium—vena cava
    d. left atrium—pulmonary artery
    e. Both b and c are incorrectly matched.

29. Which statement is not correct concerning the heartbeat?
    a. The atria contract at the same time.
    b. The ventricles relax at the same time.
    c. The AV valves open at the same time.
    d. The semilunar valves open at the same time.
    e. First the right side contracts; then the left side contracts.

30. If a person's blood pressure is 115 mm Hg over 75 mm Hg, the 75 represents
    a. systolic pressure.
    b. diastolic pressure.
    c. pressure during ventricular relaxation.
    d. Both b and c are correct.

31. Accumulation of plaque in an artery wall is
    a. an aneurysm.
    b. angina pectoris.
    c. atherosclerosis.
    d. hypertension.
    e. a thromboembolism.

32. Label the following diagram of the cardiovascular system using this alphabetized list:

aorta	hepatic portal vein
carotid artery	hepatic vein

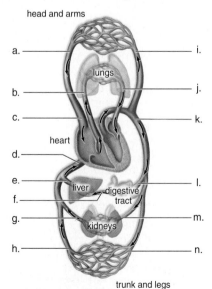

iliac artery	pulmonary artery
iliac vein	pulmonary vein
inferior vena cava	renal artery
jugular vein	renal vein
mesenteric arteries	superior vena cava

33. Label the following diagram showing the forces involved with capillary exchange. Use either blood pressure or osmotic pressure to label arrows a–d.

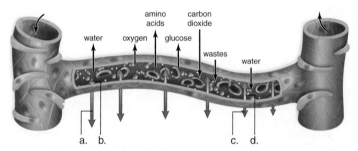

Thinking Critically About the Concepts

The cardiovascular system is an elegant example of the concept that structure supports function. Different types of blood vessels each have a specific job. Each vessel's physical characteristics enable it to do that job. The muscle walls of the right and left ventricle vary in thickness depending on where they pump the blood. When organ structure is damaged or changed (as arteries are in atherosclerosis) the organ's ability to perform its function may be compromised. Furthermore, homeostatic conditions such as blood pressure may be affected as well. Dietary and lifestyle choices can either prevent damage or harm the cardiovascular system.

1. In Case Study: Louise Hairston, Mrs. Hairston displays some not-so-typical symptoms of a heart attack that often occur in women. Can you identify these symptoms?

2. What were Louise Hairston's dietary and lifestyle choices that put her at risk for cardiovascular disease?

3. Why is a blood test important when diagnosing a heart attack?

4. Jonathan Larson, the author and composer of the Broadway musical *Rent,* died in 1996 from an aortic aneurysm.
   a. Why don't aortic aneurysms occur more frequently?
   b. Why do aortic aneurysms typically result in death?

5. a. What happens to the ventricular walls of someone with hypertrophic cardiomyopathy?
   (Hint: Consider the prefix hyper—what does it mean when used to describe a person?)
   b. How would this disease affect the heart's ability to pump blood efficiently?

6. Lymphatic vessels transport tissue fluid back to the cardiovascular system. The pressure in the lymphatic vessels is low. Do you expect lymphatic vessels to have valves? Why?

7. Think of an analogy for capillaries. Remember they are small blood vessels with very thin walls, and this facilitates their ability to perform exchanges with body cells. What are the similarities between capillaries and the object you have chosen to compare them to?

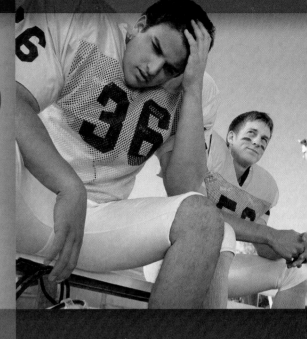

CHAPTER 6

Cardiovascular System: Blood

CASE STUDY BEN MONTEIRO

If he didn't start feeling better soon, Ben feared he would have to quit football. He had played all through high school and loved the game. Fast enough to be the starting wide receiver, he could also drill the ball through the uprights for a field goal or an extra point. Academic standing had gotten him admitted to the state university. But at only five-foot-nine, he was too short to receive a football scholarship. When he walked on to the team as a freshman, his great left-footed kick earned him a spot on the team as a place kicker. In his first year, he had scored two field goals and had never missed the kick for an extra point. As a sophomore, Ben hoped that he could earn his place as the first-string kicker.

The fatigue had seemed to sneak up gradually, along with painful lumps on his neck, armpits, and groin. Each worsened a little more with each passing day. He went to bed completely exhausted, unable to finish assignments. Likewise, his performance in team practice suffered, and he was barely able to complete the drills. After a week of terrible practices, sitting the bench today was a relief—even if it was humiliating. He sat with his head cradled in his hands, wishing for the game to be over.

Without warning, drops of blood fell from his nose onto his white uniform pants. In a matter of moments, the nosebleed worsened until blood ran down his nose and chin, staining his jersey. Ben grabbed the nearest towel he could find and covered his nose, pinching it shut at the same time. It didn't seem to help. Weakly, he signaled to another player. "Hey man, can you get the team doctor over here? I'm not feeling so hot." His teammate barely caught him as Ben slumped to the ground, unconscious.

CHAPTER CONCEPTS

6.1 Blood: An Overview
Blood, a liquid tissue, is a transport medium with various other functions. It fights infections and helps regulate body temperature and the pH of body fluids.

6.2 Red Blood Cells and Transport of Oxygen
Red blood cells contain hemoglobin, which transports oxygen and helps transport carbon dioxide. Red blood cell production is regulated by a hormone sometimes abused by athletes.

6.3 White Blood Cells and Defense Against Disease
White blood cells collectively fight infection. Each of five types has specific functions.

6.4 Platelets and Blood Clotting
Platelets are cell fragments that clump and seal a break in a blood vessel. Blood clotting, which follows, prevents loss of blood.

6.5 Blood Typing and Transfusions
Before a blood transfusion can be given, the blood donation must be typed. Only certain type(s) of blood can be received by each person without ill effects.

6.6 Homeostasis
Various systems assist the cardiovascular system, whose proper functioning is critical to homeostasis.

6.1 Blood: An Overview

In Chapter 5, we learned that the cardiovascular system consists of the heart, which pumps the blood, and the blood vessels that conduct blood around the body. In this chapter, we learn about the functions and composition of blood.

Functions of Blood

The human heart is an amazing muscular pump. With each beat, the human heart pumps approximately 75 mL of blood. On average, the heart beats 70 times per minute. Thus, the heart pumps roughly 5,250 mL per minute—75 mL/beat × 70 beats/minute—circulating the body's entire blood supply once each minute! If needed (when exercising, for example), the heart can cycle blood throughout the body even faster. The functions of blood fall into three categories: transport, defense, and regulation.

Blood is the primary transport medium. Blood acquires oxygen in the lungs and distributes it to tissue cells. Similarly, blood picks up nutrients from the digestive tract for delivery to the tissues. In its return trip to the lungs, blood transports carbon dioxide. Every time a person exhales, the carbon dioxide waste is eliminated. Blood also transports other wastes, such as the excess acid found in soda drinks, to the kidneys for elimination. In this way, capillary exchanges keep the composition of tissue fluid within normal limits (see Fig. 5.12).

Various organs and tissues secrete hormones into the blood. Blood transports these to other organs and tissues, where they serve as signals that influence cellular metabolism.

Blood is well suited for its role in transporting substances. Proteins in the blood help transfer hormones to the tissues. Both HDL and LDL (see Chap. 5, page 49) carry lipids, or fats, throughout the body. Most important, hemoglobin (found in red blood cells) is specialized to combine with oxygen and deliver it to cells. Hemoglobin also assists in transferring waste carbon dioxide back to the lungs.

Carbon monoxide—a colorless, odorless gas whose chemical formula is CO—can also bind to hemoglobin. Hundreds of people die accidentally from CO poisoning, caused by malfunction of a fuel-burning appliance or by improper venting of CO fumes. When carbon monoxide is present, it takes the place of oxygen in hemoglobin. As a result, cells are starved of oxygen. Tragically, treatment may be delayed, because symptoms of poisoning—headache, body ache, nausea, dizziness, drowsiness—can be mistaken for the "flu." Government guidelines recommend that *any* equipment producing CO be checked regularly to ensure proper ventilation. Finally, CO detectors, like smoke detectors, should be installed and used properly.

Blood defends the body against pathogen invasion and blood loss. Certain blood cells are capable of phagocytizing and destroying pathogens (see Chap. 3, page 111, for a review of phagocytosis).

Other white blood cells produce and secrete antibodies into the blood. An **antibody** is a protein that combines with and disables pathogens. Disabled pathogens can then be destroyed by the phagocytic white blood cells.

When an injury occurs, blood clots and defends against blood loss. Blood clotting involves platelets (described in Concept 6.4) and proteins. For example, prothrombin and fibrinogen are two inactive blood proteins. They circulate constantly, ready to form a clot if needed. Without blood clotting, we could bleed to death even from a small cut.

Blood has regulatory functions. Blood helps regulate body temperature by picking up heat, mostly from active muscles, and transporting it about the body. If the body becomes too warm, blood is transported to dilated blood vessels in the skin. Heat disperses to the environment, and the body cools to a normal temperature.

The liquid portion of blood, its plasma, contains dissolved salts and proteins. These solutes create blood's *osmotic pressure*, which keeps the liquid content of the blood high (see Chap. 3, page 49, for a review of the concept of osmotic pressure). In this way, blood plays a role in helping to maintain its own water-salt balance.

Blood buffers, body chemicals that stabilize blood pH, regulate the body's acid-base balance and keep it relatively constant.

Composition of Blood

Blood is a tissue, and, like any tissue, it contains cells and cell fragments (Fig. 6.1). Collectively, the cells and cell fragments are called the **formed elements.** The cell and cell fragments are suspended in a liquid called **plasma.** Therefore, blood is classified as a liquid tissue.

The Formed Elements

The formed elements are red blood cells, white blood cells, and platelets. These are produced in red bone marrow, which can be found in most bones of a child but only in certain bones of an adult. Red bone marrow contains multipotent stem cells, the parent cells that divide and give rise to all of the various types of blood cells (see Fig. 6.1). Stem cells continue to be the focus of a tremendous amount of research. Scientists hope that under the right conditions in the laboratory, stem cells could be "coaxed" into becoming a greater variety of cells—cells that could cure human diseases such as diabetes and Alzheimer disease, among many others (see Chap. 21 for a complete discussion of stem cell research).

Red blood cells are two to three times smaller than white blood cells, but there are many more of them. There are millions of red blood cells and only thousands of white blood cells in a mm^3 of blood, about this size: ■

Plasma

Plasma is the liquid medium for carrying various substances in the blood. It also distributes the heat generated

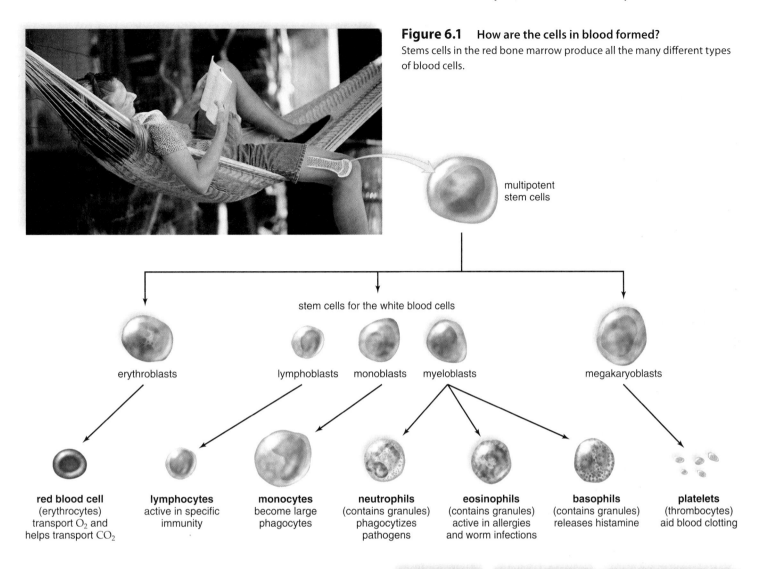

Figure 6.1 **How are the cells in blood formed?**
Stems cells in the red bone marrow produce all the many different types of blood cells.

multipotent stem cells

stem cells for the white blood cells

erythroblasts lymphoblasts monoblasts myeloblasts megakaryoblasts

red blood cell
(erythrocytes)
transport O_2 and
helps transport CO_2

lymphocytes
active in specific
immunity

monocytes
become large
phagocytes

neutrophils
(contains granules)
phagocytizes
pathogens

eosinophils
(contains granules)
active in allergies
and worm infections

basophils
(contains granules)
releases histamine

platelets
(thrombocytes)
aid blood clotting

as a byproduct of metabolism, particularly muscle contraction. About 91% of plasma is water (Fig. 6.2). The remaining 9% of plasma consists of various salts (ions) and organic molecules. Salts are dissolved in plasma. As mentioned previously, salts and plasma proteins maintain the osmotic pressure of blood. Salts also function as buffers that help maintain blood pH. Small organic molecules such as glucose and amino acids are nutrients for cells; urea is a nitrogenous waste product on its way to the kidneys for excretion.

The most abundant organic molecules in blood are called the **plasma proteins.** The liver produces the plasma proteins, with one exception to be mentioned. The plasma proteins have many functions that help maintain homeostasis. Like salts, they are able to take up and release hydrogen ions. Therefore, they help keep blood pH around 7.4. Plasma proteins are too large to pass through capillary walls. They remain in the blood, establishing an osmotic gradient between blood and tissue fluid. This **osmotic pressure** is a force that prevents excessive loss of plasma from the capillaries into tissue fluid.

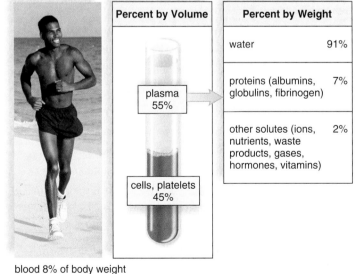

Percent by Volume	Percent by Weight	
plasma 55%	water	91%
	proteins (albumins, globulins, fibrinogen)	7%
cells, platelets 45%	other solutes (ions, nutrients, waste products, gases, hormones, vitamins)	2%

blood 8% of body weight

Figure 6.2 **What is the composition of blood plasma?**
Plasma, the liquid portion of blood, is mainly water and proteins. However, many solutes such as nutrients, vitamins, and hormones are transported in plasma.

Three major types of plasma proteins are the **albumins, globulins,** and **fibrinogen.** Albumins are the most abundant plasma proteins and contribute most to plasma's osmotic pressure. They also combine with and help transport other organic molecules. The globulins are of three types called alpha, beta, and gamma globulins. Alpha and beta globulins also combine with and help transport substances in the blood such as hormones, cholesterol, and iron. Gamma globulins are the antibodies mentioned previously. These plasma proteins are produced by white blood cells called lymphocytes, not by the liver. Gamma globulins are important in fighting disease-causing pathogens (see Concept 6.3). Fibrinogen is an inactive plasma protein. Once activated, fibrinogen forms a blood clot (see Concept 6.4).

> ### Check Your Progress 6.1
>
> 1. a. What are the functions of blood, and (b) what are its two main portions?
> 2. a. What is the composition of plasma, and (b) what are the functions of the plasma proteins?

6.2 Red Blood Cells and Transport of Oxygen

Red blood cells (erythrocytes) are small, biconcave disks that lack a nucleus when mature. They occur in great quantity; there are 4–6 million red blood cells per mm^3 of whole blood.

How Red Blood Cells Carry Oxygen

Red blood cells (RBCs) are highly specialized for oxygen (O_2) transport. They lack a nucleus and, instead, contain many molecules of hemoglobin. **Hemoglobin (Hb)** is a pigment that makes red blood cells and blood red. The globin portion of hemoglobin is a protein that contains four highly folded polypeptide chains. The heme part of hemoglobin is an iron-containing group in the center of each polypeptide chain (Fig. 6.3). The iron combines reversibly with oxygen. This means that heme accepts O_2 in the lungs and then lets go of it in the tissues. By contrast, CO combines with the iron of heme and then will not easily let go.

CASE STUDY IN THE EMERGENCY ROOM

Ben awoke in the emergency room of the university hospital. His nose was packed with cotton and an IV tube hung from his arm. Several football teammates stood around awkwardly, looking anxious. His trainer sat nearby, along with special-teams coach Bill Edwards. Both men jumped up as Ben looked around. "I'm in the hospital, right?" Ben asked weakly.

"Yeah, buddy, you sure are," Coach Edwards answered. "You were out a long time. You didn't even budge when they stuck that needle in your arm." Punching Ben's arm gently, he asked, "Why didn't you tell me you felt bad?"

Ben smiled. "Did we win? Can I leave now? I gotta get back to the dorm and start my calculus homework."

"Yeah, we won, even though it was hard without you there," Edwards smiled. "But there's no way you're leaving until someone can tell me what the heck's wrong with you. Dr. Schadler is talking with the emergency room doctor. We called your folks, and they're on the way." Nodding to the other players, he continued, "Why don't you other guys go on back home? Ben needs to rest. I'll send an e-mail to the whole team as soon as I have some news."

The team physician, Dr. Schadler, entered the cubicle accompanied by a second doctor. Each gently examined Ben's neck, armpits, and groin, pushing on the lumps that Ben had noticed before.

Ben winced in pain each time. When each doctor pushed along his lower rib cage, even gentle touches made Ben cry out. "Well, Ben, we're going to put you in the hospital until we finish your blood tests," the ER doctor said.

Ben struggled to sit up. "Can't I just go back to the dorm? Or maybe I could just go home with my parents? I can't miss any more classes. I have so much work to catch up on!"

Dr. Schadler frowned. "Ben, there is no way you're leaving this hospital until those blood tests are done and we figure out what's wrong with you. When we pushed along your ribs, we could tell your liver and spleen were enlarged. You're a biology major, right? Do you remember what a spleen does?"

"Yeah, it's for storing white blood cells and destroying old used up red blood cells," Ben answered.

"Good job," Dr. Schadler answered. "A spleen is almost like a sponge packed with little blood vessels. Yours is way, way too large and that's why it hurts when I press on it. Right now, if you accidentally fall down and hit your spleen on something hard, you could die from internal bleeding. We can't take that chance. But here's some good news—we can take that stuff out of your nose."

"That would be good," Ben sighed as he laid back and closed his eyes.

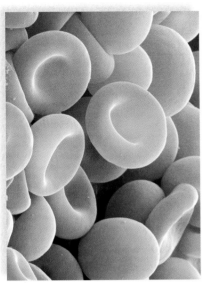

a. Red blood cells

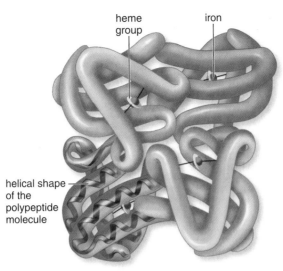

heme group

iron

helical shape of the polypeptide molecule

b. Hemoglobin molecule

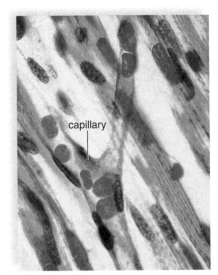

capillary

c. Blood capillary

Figure 6.3 **What do red blood cells look like? How do they work?**
a. Red blood cells are biconcave disks containing many molecules of hemoglobin. **b.** Hemoglobin contains two types of polypeptide chains (blue, purple), forming the molecule's globin portion. An iron-containing heme group is in the center of each chain. Oxygen combines loosely with iron when hemoglobin is oxygenated. **c.** Red blood cells move single file through the capillaries.

How many molecules of O_2 can each red blood cell transport? Each hemoglobin molecule can transport four molecules of O_2, and each RBC contains about 280 million hemoglobin molecules. This means that each red blood cell can carry over a billion molecules of oxygen.

Red blood cells are an excellent example of structure suiting function. They have no nucleus; their biconcave shape comes about because they lose a nucleus during maturation (see Fig. 6.1). The biconcave shape of RBCs gives them a greater surface area for the diffusion of gases into and out of the cell. All of the internal space of RBCs is used for transport of oxygen. Aside from having no nucleus, RBCs also lack most organelles, including mitochondria. RBCs anaerobically produce ATP, and they do not consume any of the oxygen they transport.

When oxygen binds to heme in the lungs, hemoglobin assumes a slightly different shape and is called **oxyhemoglobin**. In the tissues, heme gives up this oxygen, and hemoglobin resumes its former shape, called **deoxyhemoglobin**. The released oxygen diffuses out of the blood into tissue fluid and then into cells.

How Red Blood Cells Help Transport Carbon Dioxide

After blood picks up carbon dioxide (CO_2) in the tissues, about 7% is dissolved in plasma. If plasma CO_2 was higher than 7%, plasma would be carbonated and bubble like a soda. Instead, hemoglobin directly transports about 25% of CO_2, combining it with the globin protein. Hemoglobin carrying CO_2 is termed carbaminohemoglobin.

The remaining CO_2 (about 68%) is transported as the bicarbonate ion (HCO_3^-) in the plasma. Consider this equation:

$$CO_2 + H_2O \rightleftharpoons H_2CO_3 \rightleftharpoons H^+ + HCO_3^-$$

carbon dioxide · water · carbonic acid · hydrogen ion · bicarbonate ion

Carbon dioxide moves into RBCs, combining with cellular water to form carbonic acid (arrows pointing left to right illustrate this part of the reaction). An enzyme found inside RBCs, called carbonic anhydrase, speeds the reaction. Carbonic acid quickly separates, or *dissociates*, to form hydrogen ions (H^+) and bicarbonate ions (HCO_3^-). The bicarbonate ions diffuse out of the RBCs to be carried in the plasma. The H^+ from this equation binds to globin, the protein portion of hemoglobin. Thus, hemoglobin assists plasma proteins and salts in keeping the blood pH constant. When blood reaches the lungs, the reaction is reversed (arrows pointing right to left). Hydrogen ions and bicarbonate ions reunite to re-form carbonic acid. The carbonic anhydrase enzyme speeds this reverse reaction too. CO_2 diffuses out of the blood and into the airways of the lungs, to be exhaled from the body.

Red Blood Cells Are Produced in Bone Marrow

The RBC stem cells in the bone marrow divide and produce new cells that differentiate into mature RBCs (see Fig. 6.1). As red blood cells mature, they lose their nucleus and acquire hemoglobin. Possibly because they lack a nucleus, red blood cells live only about 120 days. When they age, red blood cells split open in the liver and spleen. Their destruction is completed by macrophages, white blood cells derived from monocytes.

It is estimated that about 2 million RBCs are destroyed per second, and therefore, an equal number must be produced to keep the red blood cell count in balance. When red blood cells are broken down, hemoglobin is released. The globin portion of hemoglobin is broken down into its component amino acids, which are recycled by the body. The iron is recovered and returned to the bone marrow for reuse. (Even so, some iron is lost and must be replaced in the diet.) The rest of the heme portion of the molecule undergoes chemical degradation and is excreted by the liver. Should the liver fail to excrete heme, it accumulates in tissues, causing a condition called *jaundice.* In jaundice, the skin and whites of the eyes turn yellow. Likewise, when skin is bruised, the chemical breakdown of heme causes the skin to change color from red/purple to blue to green to yellow. The body has a way to boost the number of RBCs when insufficient oxygen is being delivered to the cells. The kidneys release a hormone called **erythropoietin (EPO),** which stimulates the stem cells in bone marrow to produce more red blood cells (Fig. 6.4). The liver and other tissues also produce EPO for the same purpose.

Blood doping is any method of increasing the normal supply of RBCs for the purpose of delivering oxygen more efficiently, reducing fatigue, and giving athletes a competitive edge. To accomplish blood doping, athletes can inject themselves with EPO some months before the competition. These injections will increase the number of RBCs

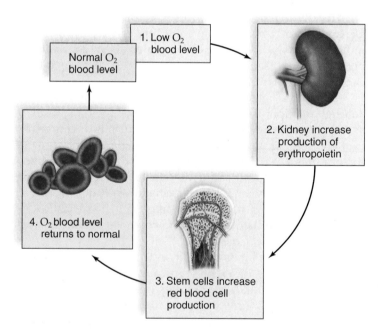

Figure 6.4 **If blood oxygen concentration decreases, how do the kidneys respond?**
The kidneys release increased amounts of erythropoietin whenever the oxygen capacity of the blood is reduced. Erythropoietin stimulates the red bone marrow to speed up its production of red blood cells, which carry oxygen.

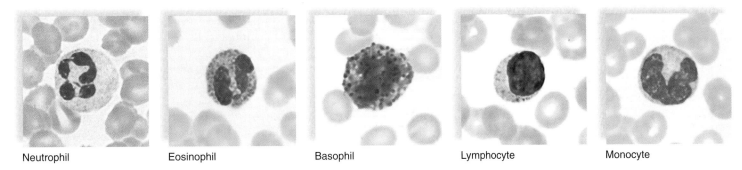

| Neutrophil | Eosinophil | Basophil | Lymphocyte | Monocyte |

Figure 6.5 How do white blood cells differ?
Neutrophils, eosinophils, and basophils have granules. Neutrophil granules stain a light pink, eosinophils granules are orange-red, and basophil granules stain a dark blue-purple color. Lymphocytes and monocytes have few, if any, granules.

in their blood. Several weeks later, four units of their blood are removed and centrifuged to concentrate the RBCs. The concentrated RBCs are reinfused shortly before the athletic event. Blood doping is a dangerous, illegal practice. Several cyclists died in the 1990s from heart failure, probably due to blood that was too thick with cells for the heart to pump.

Disorders Involving Red Blood Cells

When there is an insufficient number of red blood cells or the cells do not have enough hemoglobin, the individual suffers from **anemia** and has a tired, run-down feeling. Iron, vitamin B_{12}, and the B vitamin folic acid are necessary for the production of red blood cells. *Iron-deficiency anemia* is the most common form. It results from inadequate intake of dietary iron, which causes insufficient hemoglobin synthesis. A lack of vitamin B_{12} causes *pernicious anemia*, in which stem cell activity is reduced due to inadequate DNA production. As a consequence, fewer red blood cells are produced. *Folic acid deficiency anemia* also leads to a reduced number of RBCs, particularly during pregnancy. Pregnant women should consult with their health-care provider about the need to increase their intake of folic acid, because a deficiency can lead to birth defects in the newborn.

Hemolysis is the rupturing of red blood cells. In hemolytic anemia, the rate of red blood cell destruction increases. **Sickle-cell disease** is a hereditary condition in which the individual has sickle-shaped red blood

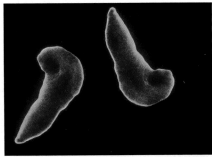

1,600×, colorized SEM

cells that tend to rupture as they pass through the narrow capillaries. The RBCs look like these to the right: The problem arises because the protein in two of the four chains making up hemoglobin is abnormal. The life expectancy of sickle red blood cells is about 90 days instead of 120 days.

Hemolytic disease of the newborn, discussed on pages 128–29, is also a type of hemolytic anemia.

> **Check Your Progress 6.2**
> 1. What substance allows red blood cells (RBCs) to transport oxygen?
> 2. Why do RBCs have a biconcave shape?
> 3. Name three disorders associated with RBCs.

6.3 White Blood Cells and Defense Against Disease

White blood cells (leukocytes) differ from red blood cells in that they are usually larger, have a nucleus, lack hemoglobin, and are translucent unless stained. White blood cells are not as numerous as red blood cells. There are only 5,000–11,000 per mm^3 of blood. White blood cells are derived from stem cells in the red bone marrow, where most types mature. There are several types of white blood cells (Fig. 6.5), and the production of each type is regulated by a protein called a **colony-stimulating factor (CSF).** In a person with normally functioning bone marrow, the numbers of white blood cells can double within hours, if needed. White blood cells are able to squeeze through pores in the capillary wall, and therefore they are also found in tissue fluid and lymph (Fig. 6.6).

White blood cells fight infection, and are an important part of the immune system. The **immune system,** discussed

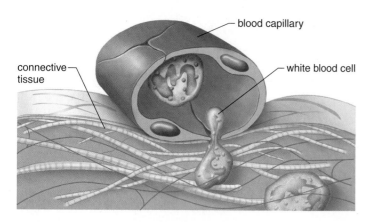

connective tissue

blood capillary

white blood cell

Figure 6.6   **How do white blood cells leave the bloodstream?** White blood cells can squeeze between the cells of a capillary wall and enter body tissues.

in Chapter 7, consists of a variety of cells, tissues, and organs that defend the body against pathogens, cancer cells, and foreign proteins. Many white blood cells live only a few days—they probably die while fighting pathogens. Others live for months or even years.

White blood cells have various ways to fight infection. Certain ones are very good at phagocytosis (see Fig. 3.10a). During phagocytosis, a projection from the cell surrounds a pathogen and literally engulfs it. A vesicle containing the pathogen is formed inside the cell. Lysosomes attach and empty their digestive enzymes into the vesicle. The enzymes digest the pathogen to debris. Other white blood cells produce antibodies, proteins that combine with antigens. An **antigen** is a cell or other substance foreign to the individual. The antibody-antigen pair is then marked for destruction, again by phagocytosis. We will be describing antigens and antibodies involved in blood typing and coagulation in Concept 6.5.

Types of White Blood Cells

White blood cells are classified into the **granular leukocytes** and the **agranular leukocytes.** Granulocytes have noticeable cytoplasmic granules, which can be easily seen when the cells are stained and examined with a microscope. Granules, like lysosomes, contain various enzymes and protein. They help white blood cells defend against diseases. Agranulocytes contain only sparse, fine granules, not easily viewed under a microscope.

Granular Leukocytes

The granular leukocytes include neutrophils, eosinophils, and basophils.

Among granular leukocytes, **neutrophils** account for 50–70% of all white blood cells. Therefore, they are the most abundant of the white blood cells. They have a multilobed nucleus, so they are called *polymorphonuclear leukocytes,* or "polys." The granules of neutrophils are not easily stained with acidic red dye, nor with basic purple dye. (This accounts for their name, neutrophil.) Neutrophils can be recognized by their numerous light pink granules. Neutrophils are usually first responders to bacterial infection, and their intense phagocytic activity is essential to overcoming an invasion by a pathogen.

Eosinophils have a bilobed nucleus. Their large, abundant granules take up eosin and become a red color. (This accounts for their name, eosinophil.) Not much is known specifically about the function of eosinophils, but they increase in number in the event of a parasitic worm infection or an allergic reaction.

Basophils have a U-shaped or lobed nucleus. Their granules take up the basic stain and become a dark blue color. (This accounts for their name, basophil.) In the connective tissues, basophils (and similar cells called **mast cells**) release histamine associated with allergic reactions. Histamine dilates blood vessels, but constricts the air tubes that lead to the lungs. This is what happens during an asthma attack when someone has difficulty breathing.

Agranular Leukocytes

The agranular leukocytes include the lymphocytes and the monocytes. Lymphocytes and monocytes do not have granules and have nonlobular nuclei. They are sometimes called the mononuclear leukocytes.

Lymphocytes account for 25–35% of all white blood cells. Therefore, they are the second most abundant type of white blood cell. Lymphocytes are responsible for specific immunity to particular pathogens and toxins (poisonous substances). The lymphocytes are of two types: B cells and T cells. Mature B cells called plasma cells produce antibodies, the proteins that combine with target pathogens and mark them for destruction. Some T cells (cytotoxic T cells) directly destroy pathogens. The AIDS virus attacks one of several types of T cells. In this way, the virus causes immune deficiency, an inability to defend the body against pathogens. B lymphocytes and T lymphocytes are discussed more fully in Chapter 7.

Monocytes are the largest of the white blood cells. After taking up residence in the tissues, they differentiate into even larger macrophages. In the skin, they become dendritic cells. Like the neutrophils, macrophages and dendritic cells are active phagocytes, destroying pathogens, old cells, and cellular debris. Macrophages and dendritic cells also stimulate other white blood cells, including lymphocytes, to defend the body.

Disorders Involving White Blood Cells

Immune deficiencies are sometimes inherited. Children have **severe combined immunodeficiency disease (SCID)** when the stem cells of white blood cells lack an enzyme called adenosine deaminase. Without this enzyme, B and T lymphocytes do not develop and the body cannot fight infections. About 100 children are born with the disease each year. Injections of the missing enzyme can be given twice weekly, but a bone marrow transplant from a compatible donor is the best way to cure the disease.

Cancer is due to uncontrolled cell growth. **Leukemia,** which means "white blood," refers to a group of cancers that involve uncontrolled white blood cell proliferation. Most of these white blood cells are abnormal or immature. Therefore, they are incapable of performing their normal defense functions. Each type of leukemia is named for the type of cell dividing out of control. For example, lymphocytic leukemia involves abnormal lymphocyte proliferation.

An Epstein-Barr virus (EBV) infection of lymphocytes is the cause of **infectious mononucleosis.** It's called infectious mononucleosis because lymphocytes are mononuclear. EBV, a member of the herpes virus family, is one of the most common human viruses. Symptoms of infectious mononucleosis are fever, sore throat, and swollen lymph glands. Although symptoms usually disappear in one or two months without medication, EBV remains dormant and hidden in a few cells in the throat and blood for the rest of a person's life. Stress can reactivate the virus. Reactivation means that a person's saliva can pass on the infection to someone else, as with intimate kissing. This is why mononucleosis is called the "kissing disease."

> **Check Your Progress 6.3**
> 1. What are the different types of white blood cells?
> 2. What is the structure and function of each type of white blood cell?
> 3. Name and describe three disorders of white blood cells.

6.4 Platelets and Blood Clotting

Platelets (thrombocytes) result from fragmentation of large cells, called **megakaryocytes,** in the red bone marrow. Platelets are produced at a rate of 200 billion a day, and the blood contains 150,000–300,000 per mm^3. These formed elements are involved in the process of blood **clotting,** or coagulation. Also involved are the plasma proteins prothrombin and fibrinogen, manufactured in the liver and deposited in the blood. Vitamin K is necessary to the production of prothrombin.

Blood Clotting

When a blood vessel in the body is damaged, platelets clump at the site of the puncture and seal the break, if it is not too extensive (Fig. 6.7). A large break may also require a blood clot to stop the bleeding. At least 12 clotting factors participate in the formation of a blood clot, but we will mention only a few of these. Calcium ions (Ca^{2+}) are also needed.

Clot formation is initiated when platelets and damaged tissue release **prothrombin activator,** which converts the plasma protein **prothrombin** to thrombin. **Thrombin,** in turn, acts as an enzyme "scissors," to cut two short

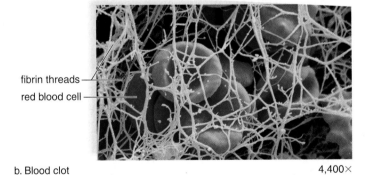

1. Blood vessel is punctured.

2. Platelets congregate and form a plug.

3. Platelets and damaged tissue cells release prothrombin activator, which initiates a cascade of enzymatic reactions.

4. Fibrin threads form and trap red blood cells.

Prothrombin activator

Prothrombin $\xrightarrow{Ca^{2+}}$ Thrombin

Fibrinogen $\xrightarrow{Ca^{2+}}$ Fibrin threads

a. Blood-clotting process

fibrin threads

red blood cell

b. Blood clot   4,400×

Figure 6.7 **What are the steps in the formation of a blood clot?**
a. Platelets and damaged tissue cells release prothrombin activator, which acts on prothrombin in the presence of Ca^{2+} (calcium ions) to produce thrombin. Thrombin acts on fibrinogen in the presence of Ca^{2+} to form fibrin threads. **b.** A scanning electron micrograph of a blood clot shows red blood cells caught in the fibrin threads.

Health Focus

What to Know When Giving Blood

Congratulations! You've decided to help another person—perhaps an accident victim, a newborn infant, someone with sickle-cell anemia, or a leukemia patient. Donating blood is fairly easy and will take approximately one hour of your time.

The Procedure

After you register, an attendant will ask private and confidential questions about your health and lifestyle and answer your questions. Your temperature, blood pressure and pulse will be recorded. A drop of your blood is tested to ensure that you're not anemic.

The supplies used for your donation are sterile and are used only for you. You can't be infected with a disease when donating blood. When the actual donation is started, you may feel a brief "sting." The procedure takes about 10 minutes, and you will have given about a pint of blood (Fig. 6A). Your body replaces the liquid part (plasma) in hours and the cells in a few weeks.

You will have several opportunities prior to giving blood and afterwards to let Red Cross officials know whether your blood is safe. Immediately after you donate, you are given a number to call. After you leave, you can call that number if you decide that your blood may not be safe to give to another person. Donated blood is tested for syphilis bacteria and AIDS antibodies, as well as hepatitis and other viruses. You are notified if tests are positive, and your blood won't be used if it could make someone ill. However, you should *never* use the process of a blood donation to get tested for any medical condition, especially AIDS. It is possible to have a negative result for AIDS antibodies and yet still spread the virus, because forming antibodies takes several weeks after exposure.

The Cautions

Don't give blood **if you have:**

been ill during the week prior to your donation;
taken or are taking, drugs that slow blood clotting, including aspirin;
ever had hepatitis; had malaria; taken drugs for malaria prevention; or traveled to malaria-prone countries;
been treated for syphilis or gonorrhea in the last 12 months.

AIDS or one of its symptoms:

unexplained weight loss (4.5 kg or more in less than two months), fever and night sweats;
blue or purple spots on or under skin, or unusual sores in mouth;

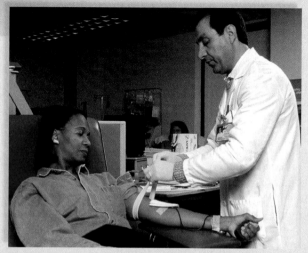

Figure 6A   **Why should you consider being a blood donor?** Donating blood can help save a life—perhaps even that of a family member or friend.

lumps in neck, armpits, or groin for over a month;
diarrhea lasting over a month;
persistent cough and shortness of breath.

If you are at risk for AIDS—that is, if you have

ever injected illegal drugs;
taken clotting factor concentrates for hemophilia;
tested positive for AIDS virus or antibody;
been given money or drugs for sex since 1977;
had a sexual partner within the last year who did any of the above things;
(for men) had sex *even once* with another man since 1977, *or* had sex with a female prostitute within the last year;
(for women) had sex with a male or female prostitute within the last year, *or* had a male sexual partner who had sex with another man *even once* since 1977.

When You're Finished

An area is provided to relax after donation, with drinks and snacks to enjoy. Plan to rest for a few minutes.

Most people feel fine while they give blood and afterward, but a few donors have an upset stomach or feel faint or dizzy after donation. Resting, drinking fluid, and eating a snack usually help. Occasionally, bruising, redness, and pain occur at your donation site, so avoid strenuous exercise and lifting for a day or so. Very rarely, a person may have muscle spasms and/or suffer nerve damage. You can call toll-free for assistance if this should ever happen.

For more information about blood donation, visit the Red Cross website at www.RedCross.org.

amino acid chains from each fibrinogen molecule. The fibrin fragments then join end to end, forming long threads of **fibrin.** Fibrin threads wind around the platelet plug in the damaged area of the blood vessel and provide the framework for the clot. Red blood cells trapped within the fibrin threads make the clot appear red. A fibrin clot is temporary. Once blood vessel repair starts, an enzyme called plasmin destroys the fibrin network so tissue cells can grow.

After blood clots, a yellowish fluid called **serum** escapes from the clot. It contains all the components of plasma except fibrinogen and prothrombin.

Disorders Related to Blood Clotting

An insufficient number of platelets is called **thrombocytopenia.** Thrombocytopenia is either due to low platelet production in bone marrow or increased breakdown of platelets outside the marrow. A number of conditions, including leukemia, can lead to thrombocytopenia. It can also be drug-induced. Symptoms include bruising, rash, and nosebleeds or bleeding in the mouth. Gastrointestinal bleeding or bleeding in the brain are possible complications.

If the lining of a blood vessel becomes roughened, a clot can form spontaneously inside an unbroken blood vessel. Most often, roughening occurs because an atherosclerotic plaque has formed (see page 105). Rarely, the vessel lining is damaged during placement of an intravenous tube. The spontaneous clot is called a **thrombus** (plural—**thrombi**) if it remains stationary inside the blood vessel. Sitting for long periods, as when traveling, can also cause thrombus formation. Should the clot dislodge and travel in the blood, it is called an embolus. If **thromboembolism** is not treated, blood flow to the tissues can stop completely. Heart attack or stroke can result, as discussed in Chapter 5.

Hemophilia is an inherited clotting disorder that causes a deficiency in a clotting factor. There are many forms of the disorder. Hemophilia A, caused by deficiency of clotting factor VIII, is more likely to occur in boys than in girls. Hemophilia A is caused by an abnormal copy of the Factor VIII production gene, found on the X chromosome. This hemophilia arises when a boy has an abnormal gene on his single X chromosome. Girls only need one normal gene between their two X chromosomes to make the normal amounts of clotting Factor VIII (see page 484). In hemophilia, the slightest bump can cause bleeding into the joints. Cartilage degeneration in the joints and absorption of underlying bone can follow. The most frequent cause of death is bleeding into the brain with accompanying neurological damage. Regular injections of Factor VIII can successfully treat the disease.

> ### Check Your Progress 6.4
> 1. Name the major participants in blood clotting and their functions.

CASE STUDY DIAGNOSIS

"Just be patient," Ben's mother told him. "When the doctors have figured out what's wrong with you, then maybe we can take you back to school." His father nodded in agreement. Just then, a soft knock on the hospital room door caused all three Monteiros to look up.

A doctor entered with three students following. "Mr. Benjamin Monteiro? I'm Dr. Garth. I'm an oncologist, a doctor that treats cancer. We've had a look at your test results. I'm afraid I have bad news. You have cancer of the white blood cells, or acute lymphocytic leukemia."

"Right now, your blood is packed with immature, malformed lymphocytes," Dr. Garth continued, "They're crowding out the red blood cells, and you're anemic. That's why you've been so run down and tired. They're also crowding out megakaryoctyes, the cells that make platelets."

"So that's why my nose started bleeding?" Ben asked. "Because the platelets in my blood are too low?"

"Exactly right," Dr. Garth answered, "If platelets in your blood fall too low, you can bleed spontaneously, like what happened with your nosebleed. Also, whole collections of abnormal lymphocytes are causing those swellings in your neck, armpits, and groin. But here's the good news. We caught your case early, and you're a healthy, young guy. Treatment is going to be pretty uncomfortable, but we stand a very good chance of curing you. First, we're going to give you a blood transfusion right away to help with your anemia. Then we're going to put you into the oncology unit and get started on that treatment."

It was all Ben could do to keep from bursting into tears, just as his mother had. "What about my classes? What about football?"

Dr. Garth smiled sympathetically. "We can't let you go back to school. Those abnormal cells are choking out all the other cells and that includes normal white blood cells. You could die from an infection. For right now, let's focus on your cure, and then see what the future brings."

6.5 Blood Typing and Transfusions

A **blood transfusion** is the transfer of blood from one individual to another. For transfusions to be done safely, blood must be typed so that **agglutination** (clumping of red blood cells) does not occur when blood from different people is mixed. Blood typing usually involves determining the ABO blood group and whether the individual is Rh⁻ (negative) or Rh⁺ (positive).

ABO Blood Groups

Only certain types of blood transfusions are safe because the plasma membranes of red blood cells carry glycoproteins that can be antigens to other individuals. An antigen is any substance that is foreign to a person's body. ABO blood typing is based on the presence or absence of two possible antigens, called type A antigen and type B antigen. Whether these antigens are present or not depends on the particular inheritance of the individual.

In Figure 6.8, type A antigen and type B antigen are given different shapes and colors. Study each figure to see the type(s) of antigens found on the red blood cell plasma membrane. As you might expect, type A blood has the A antigen (blue spheres) while type B blood has the B antigen (violet triangles). If you guessed that type AB blood has both A and B antigens, you're correct. Finally, notice that type O blood has neither antigen on its red blood cells.

Each blood type has antibodies (yellow Y-shaped molecules) that correspond to the *opposite* blood type. Thus, an individual with type A blood has anti-B antibodies in the plasma, while a person with type B blood has anti-A antibodies in the plasma. Further, a person with type O blood has both antibodies in the plasma while someone with type AB blood lacks both antibodies (Fig. 6.8). Anti-A and/or B antibodies are not present at birth, but they appear over the course of several months.

Finally, observe that each antibody has a *binding site* that will combine with its corresponding antigen in a tight lock-and-key fit. The anti-B antibodies have a triangular binding

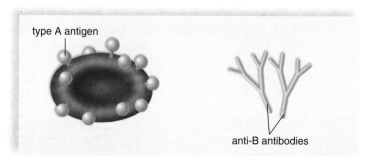

Type A blood. Red blood cells have type A surface antigens. Plasma has anti-B antibodies.

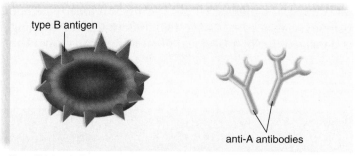

Type B blood. Red blood cells have type B surface antigens. Plasma has anti-A antibodies.

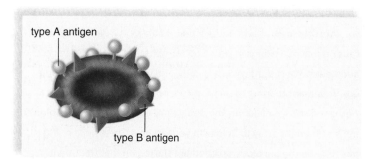

Type AB blood. Red blood cells have type A and type B surface antigens. Plasma has neither anti-A nor anti-B antibodies.

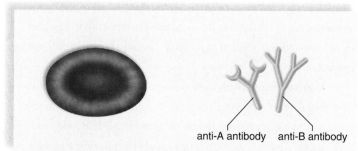

Type O blood. Red blood cells have neither type A nor type B surface antigens. Plasma has both anti-A and anti-B antibodies.

Figure 6.8   How are different blood types determined by using the ABO system?

In the ABO system, blood type depends on the presence or absence of antigens A and B on the surface of red blood cells. In these drawings, A and B antigens are represented by different shapes on the red blood cells. The possible anti-A and anti-B antibodies in the plasma are shown for each blood type. An anti-B antibody cannot bind to an A antigen, and vice versa.

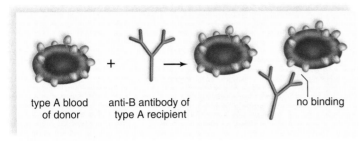

a. No agglutination

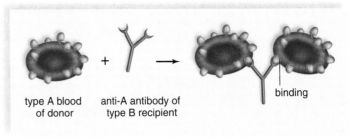

b. Agglutination

Figure 6.9 **When is a blood transfusion unsafe?**
No agglutination occurs in **a.** because anti-B antibodies cannot combine with the A antigen. Agglutination occurs in **b.** when anti-A antibodies in the recipient combine with A antigen on donor red blood cells.

site on the top of each Y-shaped molecule. This binding site fits snugly with the purple, triangular B antigen. Similarly, the anti-A antibodies have a spherical binding site shaped to form a perfect fit with the A antigen. The presence of these antibodies can cause agglutination.

Blood Compatibility

Blood compatibility is very important when transfusions are done. The antibodies in the plasma must not combine with the antigens on the surface of the red blood cells or else agglutination occurs. With agglutination, anti-A antibodies have combined with type A antigens, and anti-B antibodies have combined with type B antigens. Therefore, agglutination is expected if the donor has type A blood and the recipient has type B blood (Fig. 6.9). What about other combinations of blood types? Try out all other possible donors and recipients to see if agglutination will occur.

Type O blood is sometimes called the *universal donor* because the red blood cells of type O blood lack A and B antigens. Type O donor blood should not agglutinate with any other type of recipient blood. Likewise, type AB blood is sometimes called *universal recipient* blood because the plasma lacks A and B antibodies. Type AB recipient blood should not agglutinate with any other type of donor blood. In practice, however, there are other possible blood groups, aside from ABO blood groups. Before blood can be safely transfused from one person to another, it is necessary to physically combine donor blood with recipient blood on a glass slide, then observe whether agglutination occurs. This procedure, called blood-type cross-matching, takes only a few minutes. It is done before every blood transfusion is performed. Type O blood donation without cross matching is performed only in an emergency, when blood loss is severe and the patient's survival is at stake.

Rh Blood Groups

The designation of blood type usually also includes whether the person has or does not have the Rh factor on the red blood cell. Rh⁻ individuals normally do not have antibodies to the Rh factor, but they make them when exposed to the Rh factor.

If a mother is Rh⁻ and the father is Rh⁺, a child can be Rh⁺. During a pregnancy, Rh⁺ can leak across the placenta into the mother's bloodstream. The presence of these Rh⁺ antigens causes the mother to produce anti-Rh antibodies (Fig. 6.10). Usually in a subsequent pregnancy with another Rh⁺ baby, the anti-Rh antibodies may cross the placenta and destroy the unborn child's red blood cells. This is called hemolytic disease of the newborn (HDN) because hemolysis starts in the womb and continues after the baby is born. Due to red blood cell destruction, the baby will be severely

Have You Ever Wondered...

What is the most common ABO blood type? What is the least common ABO blood type?

It depends on your genetic inheritance. In the United States and other countries with a large immigrant population, the most common ABO type is type O. However, type A blood is most common in several European countries, including Portugal, Armenia, Switzerland, and Norway. Worldwide, the least common ABO type is type AB.

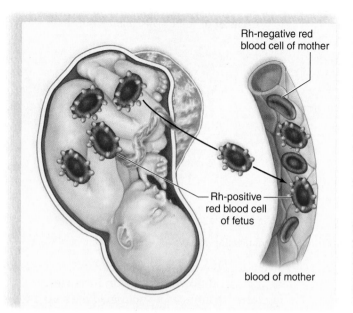

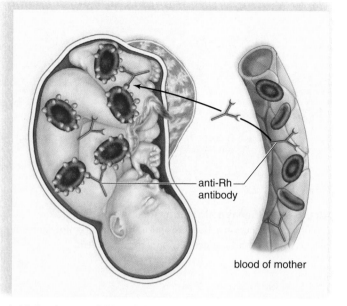

a. Fetal Rh-positive red blood cells leak across placenta into mother's bloodstream.

b. Mother forms anti-Rh antibodies that cross the placenta and attack fetal Rh-positive red blood cells.

Figure 6.10 **Why do some babies develop Rh factor disease (hemolytic disease of the newborn)?**
a. Due to a pregnancy in which the child is Rh-positive, an Rh-negative mother can begin to produce antibodies against Rh-positive red blood cells.
b. Usually in a subsequent pregnancy, these antibodies can cross the placenta and cause hemolysis of an Rh-positive child's red blood cells.

Historical Focus

Making Blood Transfusion Possible: Karl Landsteiner (1868–1943)

"Time is so short and there is so much to do. We must hurry."
Karl Landsteiner

It's hard to imagine, in this age of rapidly developing medical technology, that not too long ago blood transfusion was impossible. Injuries with severe bleeding, such as an injury to the chest or abdomen, were usually fatal. Likewise, most perished after limb amputations (though amazingly enough, historical accounts and primitive prosthetic limbs show that a lucky few survived). Uncontrolled bleeding afer childbirth claimed the lives of many women. Early transfusion experiments involved the blood transfer from animals to humans and were a dismal failure. Similarly, human-to-human transfusion was likely to kill the blood recipient. Microscopic examinations of transfused red blood cells showed that the cells clumped together, in a process called agglutination. Subsequently, the cells broke open and released their hemoglobin, causing shock and death. After a few such incidents, transfusions were abandoned.

The work of Dr. Karl Landsteiner made blood transfusion possible. Using human blood samples, Landsteiner first separated the cells from the liquid (called serum). Mixing the cells from one donor, with serum from a second donor, sometimes caused blood cells to agglutinate and then break. However, at other times, when blood and serum from different donors were mixed, the cells remained intact. Landsteiner concluded that molecules found on the red blood cell membranes divided blood into three groups, which he labeled A, B, and O. Mixing type A with type B always produced agglutination, but mixing bloods of the same type (for example, type A with type A) never did. Blood type O could be safely mixed with all blood samples without producing agglutination. Later researchers described the fourth blood group, AB, as well as the Rh factor and other molecules now used for blood typing.

Landsteiner's work enabled the first successful human transfusion in 1907. Adding anticoagulants (chemicals that prevent blood from clotting) made blood storage and blood banking possible. Transfusions saved many lives during both world wars. Victims of traumatic injury or hemorrhage following childbirth could now survive, even after a tremendous blood loss. Chest and abdominal surgeries, previously impossible because of potential blood loss, became routine.

The use of blood typing to identify criminals and their victims, called forensic serology, also evolved from Dr. Landsteiner's work. Paternity testing based on blood typing was now possible as well (although it has since been largely replaced by much more definitive DNA testing). Landsteiner, called the Father of Modern Serology (the science of blood identification), was awarded the Nobel Prize in Medicine in 1930.

anemic. Excess hemoglobin breakdown products in the blood can lead to brain damage and mental retardation, or even death.

The Rh problem is prevented by giving Rh⁻ women an Rh immunoglobulin injection no later than 72 hours after giving birth to an Rh⁺ child. This injection contains anti-Rh antibodies that attack any of the baby's red blood cells in the mother's blood before these cells can stimulate her immune system to produce her own antibodies. This treatment is termed Rho Gam, and does not harm the newborn's red blood cells. Treatment is not beneficial if the woman has already begun to produce antibodies. Therefore, the timing of the injection is most important.

> ### Check Your Progress 6.5
> 1. a. What are the different blood types, and (b) what determines blood type?
> 2. Among the ABO types of blood, who can give blood to whom? Why?
> 3. When does hemolytic disease of the newborn occur?

6.6 Homeostasis

Think of how important it is to homeostasis for all the human systems to work together. For example, imagine one of a student's worst nightmares. After studying all night for a final exam, you oversleep the next morning.

You wake up just a few minutes before class begins. Immediately, your heart starts beating faster and your blood pressure rises, due to symphathetic *nervous system* stimulation (Chap. 13). Hormones, such as epinephrine (adrenaline) from the adrenal glands, pour into the bloodstream to prolong your body's physical response to stress. The *endocrine system* is at work (Chap. 15).

Frantic, you throw on your clothes and sprint to the bus. Your *muscular* and *respiratory systems* meet the demand (Chaps. 9 and 12). As you run, your muscles require more oxygen. Breathing and heart rate increase, to meet the demand. Carbon dioxide generated by hard-working muscles is carried in the blood to the lungs so it can be expelled. To eliminate the heat produced by your muscles, you sweat and blood flow to your skin increases.

The nightmare continues. In your mad dash up the steps to class, you fall, skinning your knee (Fig. 6.11). At first, blood flows from the wound, but a clot soon seals off the injured area. As a scab forms, healing has already begun. Any injury that breaks the skin can allow harmful bacteria into the body, but the immune system usually deals quickly with invaders. Later, you might notice that your knee is not only skinned, but also black and blue. Bruises form when blood leaks out of damaged vessels underneath the skin, and then clots.

You make it to class just in time. As you settle down to take the exam, your heartbeat and breathing gradually slow, because the muscle need is not as great. As you concentrate, blood flow increases in certain highly active parts of your

CASE STUDY FIVE YEARS LATER

The hot black robe itched, and Ben was glad he'd worn just a T-shirt underneath. His lightest pair of dress pants and those miserable black dress shoes were to please his parents and grandparents. He smiled to himself despite his discomfort. What a long five years it had been.

To say the chemotherapy was "uncomfortable" had been Dr. Garth's biggest understatement. The month he'd spent in the hospital after his initial diagnosis was a nightmare he wished he could forget. Each drug treatment caused severe vomiting and extreme fatigue. He'd lost the entire first semester of his sophomore year at the university, along with all of his hair. But there had been some wonderful memories from that first agonizing month, too. As soon as he was allowed to have visitors, his closest football buddies crowded into his hospital room. Wearing their surgical gowns and masks, they brought videos of each game and spent hours with Ben, analyzing and commenting on each play. Tears came to his eyes as he remembered his return to classes, to find that his teammates had

shaved their heads, so that the entire team was bald. Yearly campus fundraisers, conducted in his name, had raised almost $50,000 for cancer research at the university hospital.

More chemotherapy and spinal taps continued, to make sure all cancer cells had been eliminated. Aggressive treatment had persisted for three years. But very slowly, his strength had returned. Taking classes year round had enabled him to complete his degree. Further, he was finally able to return as a place kicker on the football team. True, he was still second string, but the university had won the conference during his last fall semester, and he'd been there. What a sweet memory that was.

But the sweetest memory had been when Dr. Garth had told him that he could consider himself cured. It had taken four long years of treatment, blood tests, and spinal taps and another year of anxious waiting to see if the cancer returned. He would need checkups throughout his life. But for now, he was healthy, and he was finally graduating. He bowed his head, then laughed as the mortarboard slipped forward and fell in his lap. He was *here*. Life was good.

Figure 6.11   **How many human systems participate to help you in an emergency situation?**
Maintaining homeostasis requires cooperation between all organ systems.

brain. The brain is a very demanding organ in terms of both oxygen and glucose.

You start to feel hungry. There is little blood glucose to provide cellular fuel, so glycogen stored in the liver is broken down to make glucose available to blood and body cells. Finally, you return the exam, and eat breakfast afterward. When most nutrients from your meal are absorbed by the *digestive system,* they enter the bloodstream (Chap. 8). However, fats enter lacteals, vessels of the *lymphatic system,* for transport to the blood (Chap. 7).

Notice in this scenario how many organ systems interacted with the cardiovascular system. The nervous, endocrine, muscular, respiratory, digestive, and lymphatic systems helped to serve cellular needs. Meanwhile, the urinary system (described in Chap. 10) made urine for excretion by the kidneys. No organ system works alone where homeostasis is concerned!

How Body Systems Work Together

Figure 6.12 summarizes how these and other systems cooperate with the cardiovascular system. You previously learned that the body's internal environment contains blood and tissue fluid. Tissue fluid originates from blood plasma, but contains very little plasma protein. Tissue fluid is absorbed by lymphatic capillaries, after which it is referred to as lymph. The lymph courses through lymphatic vessels, eventually returning to the venous system. Thus, the cardiovascular and lymphatic systems are intimately linked.

Homeostasis is possible only if the cardiovascular system delivers oxygen from the lungs, as well as nutrients from the digestive system, to the tissue fluid surrounding cells. Simultaneously, the cardiovascular system also removes metabolic wastes, delivering waste to excretory organs.

The three components of the muscular system make essential contributions to blood movement. Cardiac muscle contractions circulate blood throughout the body. Contraction or relaxation of the smooth muscle in blood vessel walls changes vessel diameter and helps to maintain the correct blood pressure. Further, skeletal muscle contraction compresses both cardiovascular and lymphatic veins. Lymph returns to cardiovascular veins, and blood in the cardiovascular veins drains back to the heart. The circulation of tissue fluid and blood is then complete.

The skeletal and endocrine systems are vital to cardiovascular homeostasis. Red bone marrow produces blood cells. Without the needed blood cells, the person becomes anemic and lacks an immune response. In addition, bones contribute calcium ions (Ca^{2+}) to the process of blood clotting. Without blood clotting, bleeding from an injury (even something as simple as skinning a knee) could be fatal. Both blood cell production and bone calcium release are regulated by hormones. Once again, the endocrine system cooperates with the cardiovascular system.

Finally, we must not forget that the urinary system has functions besides producing and excreting urine. The kidneys help regulate the acid-base and salt-water balance of blood and tissue fluid. Erythropoietin, a hormone produced by the kidneys, stimulates red blood cell production. The urinary system joins the muscular, skeletal, and endocrine systems to maintain the internal environment.

Check Your Progress 6.6

1. Tell how the functions of the cardiovascular system contribute to homeostasis.
2. Give one significant contribution of each system listed in Figure 6.12 to the cardiovascular system.
3. In what way does the liver contribute to the functioning of the cardiovascular system?

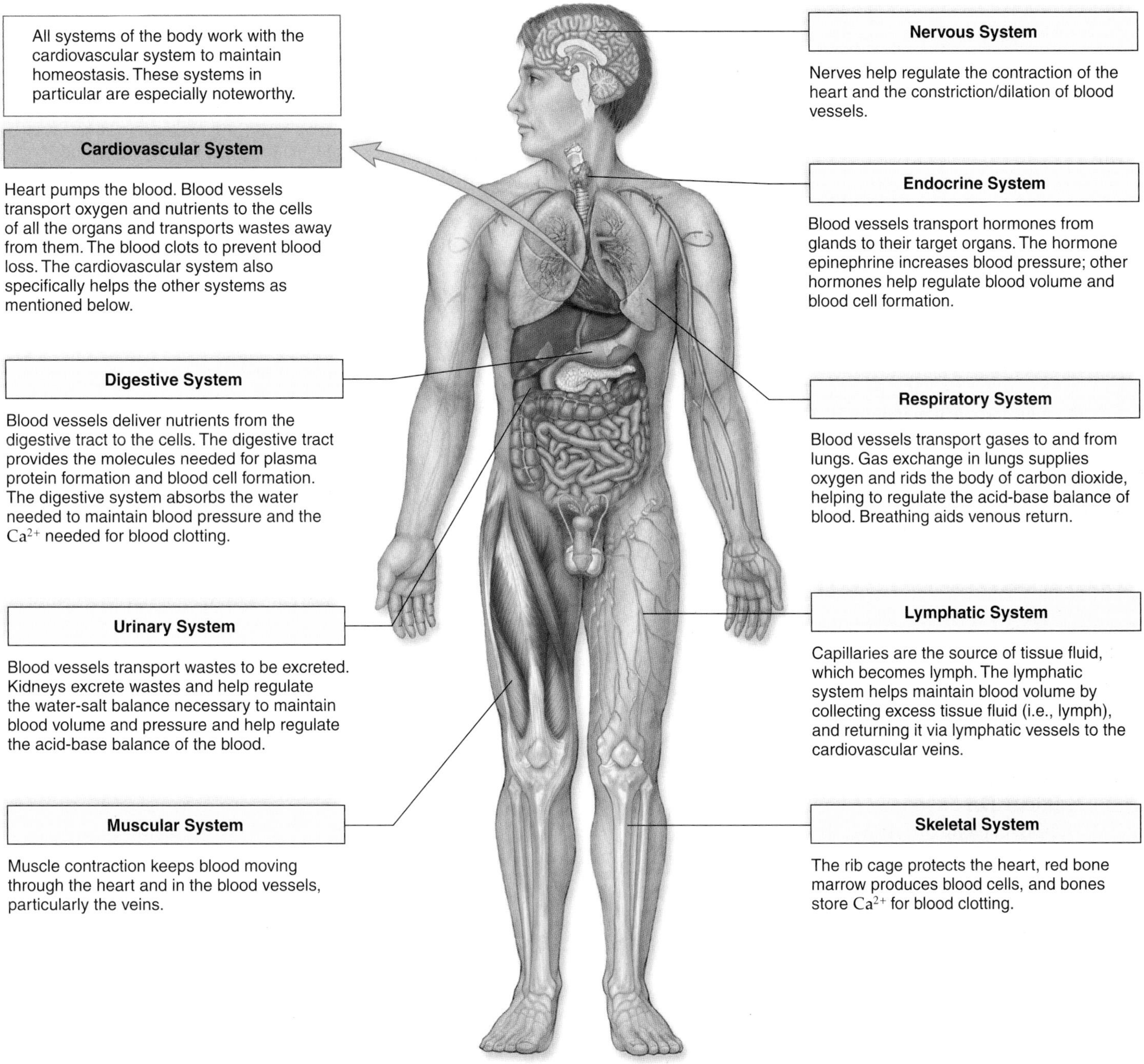

All systems of the body work with the cardiovascular system to maintain homeostasis. These systems in particular are especially noteworthy.

Cardiovascular System

Heart pumps the blood. Blood vessels transport oxygen and nutrients to the cells of all the organs and transports wastes away from them. The blood clots to prevent blood loss. The cardiovascular system also specifically helps the other systems as mentioned below.

Digestive System

Blood vessels deliver nutrients from the digestive tract to the cells. The digestive tract provides the molecules needed for plasma protein formation and blood cell formation. The digestive system absorbs the water needed to maintain blood pressure and the Ca^{2+} needed for blood clotting.

Urinary System

Blood vessels transport wastes to be excreted. Kidneys excrete wastes and help regulate the water-salt balance necessary to maintain blood volume and pressure and help regulate the acid-base balance of the blood.

Muscular System

Muscle contraction keeps blood moving through the heart and in the blood vessels, particularly the veins.

Nervous System

Nerves help regulate the contraction of the heart and the constriction/dilation of blood vessels.

Endocrine System

Blood vessels transport hormones from glands to their target organs. The hormone epinephrine increases blood pressure; other hormones help regulate blood volume and blood cell formation.

Respiratory System

Blood vessels transport gases to and from lungs. Gas exchange in lungs supplies oxygen and rids the body of carbon dioxide, helping to regulate the acid-base balance of blood. Breathing aids venous return.

Lymphatic System

Capillaries are the source of tissue fluid, which becomes lymph. The lymphatic system helps maintain blood volume by collecting excess tissue fluid (i.e., lymph), and returning it via lymphatic vessels to the cardiovascular veins.

Skeletal System

The rib cage protects the heart, red bone marrow produces blood cells, and bones store Ca^{2+} for blood clotting.

Figure 6.12 **How do body systems cooperate to ensure homeostasis?**
Each of these systems makes critical contributions to the functioning of the cardiovascular system and, therefore, to homeostasis. See if you can suggest contributions by each system before looking at the information given.

Summarizing the Concepts

6.1 Blood: An Overview

Blood functions to

- transport hormones, oxygen, and nutrients to cells;
- transport carbon dioxide and other wastes from cells;
- fight infections, and perform various regulatory functions;
- maintain blood pressure and regulate body temperature;
- keep the pH of body fluids within normal limits;

These functions help maintain homeostasis.

Blood has two main components: plasma and formed elements (red blood cells, white blood cells, and platelets)

Plasma

- 91% of blood plasma is water.
- Plasma proteins (albumins, globulins, and fibrinogen) are mostly produced by the liver.
- Plasma proteins maintain osmotic pressure, and help regulate pH. Albumins transport other molecules, globulins function in immunity, and prothrombin and fibrinogen enable blood clotting.

6.2 Red Blood Cells and Transport of Oxygen

Red blood cells lack a nucleus. Instead, they contain hemoglobin, which combines with oxygen and transports it to the tissues. Hemoglobin assists in carbon dioxide transport, as well.

red blood cells

Red blood cell production is controlled by the blood oxygen concentration . When oxygen concentration decreases, the kidneys increase production of the hormone erythropoietin. In response, more red blood cells are produced by the bone marrow.

6.3 White Blood Cells and Defense Against Disease

White blood cells are larger than red blood cells. They have a nucleus and are translucent unless stained. White blood cells are either granular leukocytes or agranular leukocytes.

- The granular leukocytes are eosinophils, basophils, and neutrophils. Neutrophils are abundant, respond first to infections, and phagocytize pathogens.

eosinophil   basophil   neutrophil

- The agranular leukocytes include monocytes and lymphocytes. Monocytes are the largest white blood cells. They can become macrophages that phagocytize pathogens and cellular debris. Lymphocytes (B cells and T cells) are responsible for specific immunity.

monocyte   lymphocyte

All blood cells are produced within red bone marrow from stem cells. Red blood cells live about 120 days and are eventually destroyed in the liver and spleen. Most white blood cells survive for only a few days, but some persist for months to years.

6.4 Platelets and Blood Clotting

Platelets result from fragmentation of megakaryocytes in the red bone marrow, and function in blood clotting.

Blood Clotting

Platelets and two plasma proteins, prothrombin and fibrinogen, function in blood clotting, an enzymatic process. Fibrin threads that trap red blood cells result from the enzymatic reaction.

6.5 Blood Typing and Transfusions

Blood typing usually involves determining the ABO blood group and whether the person is Rh⁻ or Rh⁺. Determining blood type is necessary for transfusions so that agglutination of red blood cells does not occur.

ABO Blood Groups

ABO blood typing determines the presence or absence of type A antigen and type B antigen on the surface of red blood cells.

- **Type A Blood**   Type A surface antigens; plasma has anti-B antibodies.
- **Type B Blood**   Type B surface antigens; plasma has anti-A antibodies.
- **Type AB Blood**   Both type A and type B surface antigens; plasma has neither anti-A nor anti-B antibodies (universal recipient).
- **Type O Blood**   Neither type A nor type B surface antigens. Plasma has both anti-A and anti-B antigens (universal donor).
- **Agglutination**   Agglutination occurs if the corresponding antigen and antibody are mixed (i.e., if the donor has type A blood and the recipient has type B blood).

Rh Blood Groups

The Rh antigen must also be considered when transfusing blood. It is very important during pregnancy because an Rh⁻ mother may form antibodies to the Rh antigen while carrying or after the birth of an Rh⁺ child. These antibodies can cross the placenta to destroy the red blood cells of an Rh⁺ child.

6.6 Homeostasis

Homeostasis depends upon the cardiovascular system because it serves the needs of the cells. Other body systems are also critical to cardiovascular system function:

- The digestive system supplies nutrients.
- The respiratory system supplies oxygen and removes carbon dioxide from the blood.
- The nervous and endocrine systems help maintain blood pressure. Endocrine hormones regulate red blood cell formation and calcium balance.
- The lymphatic system returns tissue fluid to the veins.
- Skeletal muscle contraction (skeletal system) and breathing movements (respiratory system) propel blood in the veins.

Understanding Key Terms

agglutination 126
agranular leukocyte 122
albumin 118
anemia 121
antibody 116
antigen 122
basophil 122
blood doping 120
blood transfusion 126
clotting 123
colony-stimulating factor
   (CSF) 121
deoxyhemoglobin 119
eosinophil 122
erythropoietin (EPO) 120
fibrin 125
fibrinogen 118
formed element 116
globulin 118
granular leukocyte 122
hemoglobin (Hb) 118
hemolysis 121
hemolytic disease of the
   newborn 121
hemophilia 125
immune system 121

infectious mononucleosis 123
leukemia 123
lymphocyte 122
mast cell 122
megakaryocyte 123
monocyte 122
neutrophil 122
osmotic pressure 117
oxyhemoglobin 119
plasma 116
plasma protein 117
platelet (thrombocyte) 123
prothrombin 123
prothrombin activator 123
red blood cell (erythrocyte) 118
serum 125
severe combined
   immunodeficiency disease
   (SCID) 123
sickle-cell disease 121
thrombin 123
thrombocytopenia 125
thromboembolism 125
thrombus 125
white blood cell
   (leukocyte) 121

Match the key terms to these definitions.

a. _____ Iron-containing protein in red blood cells that combines with and transports oxygen.

b. _____ Component of blood that is either cellular or derived from a cell.

c. _____ Liquid portion of blood.

d. _____ A group of cancerous conditions that involve uncontrolled white blood cell proliferation.

e. _____ Plasma protein converted to thrombin during the steps of blood clotting.

Testing Your Knowledge of the Concepts

1. The transport function of blood is dependent on what components? The defense function of blood is dependent on what components? The regulatory functions of blood are dependent on what components? (pages 116–18)

2. The osmotic pressure of blood is dependent on what components of blood? (pages 117–18)

3. Describe the structure of a red blood cell, including the molecule hemoglobin. What is the role of red blood cells in the blood? How is the production of red blood cells regulated? (pages 118–20)

4. What are the two main categories of white blood cells? Which of the white blood cells are phagocytes? Explain. (pages 121–22)

5. What formed element is crucial to blood clotting? What other substances are necessary for clotting? Explain the steps that take place when blood clots. (page 123)

6. List and describe three disorders involving RBCs, three disorders involving WBCs, and two disorders involving platelets. (pages 121, 123, and 125)

7. For each type of ABO blood, give the antigen(s) and antibody(ies) present. List which blood type each can receive and to which type each can be given. (pages 126–27)

8. Explain what occurs in hemolytic disease of the newborn. (pages 127–29)

9. Choose five body systems, and briefly explain how they are critical to cardiovascular system function. (pages 129–30)

In questions 10–14, match each description with a component of blood in the key. Answers can be used more than once.

Key:
  a. red blood cells
  b. white blood cells
  c. red and white blood cells and platelets
  d. plasma

10. Antigens in plasma membrane determine blood type

11. Transport oxygen

12. Includes monocytes

13. Contains plasma proteins

14. Formed elements

In questions 15–19, match each description with a white blood cell in the key.

Key:
  a. lymphocytes   d. monocytes
  b. neutrophils   e. eosinophils
  c. basophils

15. U-shaped nucleus, blue-stained granules that release histamine

16. Includes B cells and T cells that provide specific immunity

17. Bilobed nucleus, red-stained granules, allergic reactions, and parasitic worms

18. Largest, no granules, become macrophages

19. Most abundant, multilobed nucleus, first responders to invasion

20. Which of the following is not a formed element of blood?
  a. leukocyte
  b. eosinophil
  c. fibrinogen
  d. platelet

21. Which of the plasma proteins contributes most to osmotic pressure?
  a. albumin   c. erythrocytes
  b. globulins   d. fibrinogen

22. Stem cells are responsible for
  a. red blood cell production.
  b. white blood cell production.
  c. platelet production.
  d. the production of all formed elements.

23. Which hemoglobin component is recovered for reuse following red blood cell destruction?
    a. heme   c. iron
    b. globin   d. Both b and c are correct.

24. When the oxygen capacity of the blood is reduced,
    a. the liver produces more bile.
    b. the kidneys release erythropoietin.
    c. the bone marrow produces more red blood cells.
    d. sickle-cell disease occurs.
    e. Both b and c are correct.

25. Which of the following conditions can cause anemia?
    a. lack of iron, folic acid, or vitamin B_{12} in the diet
    b. red blood cell destruction
    c. lack of hemoglobin
    d. All of these are correct.
    e. All but a are correct.

26. Which of the following is not true of white blood cells?
    a. formed in red bone marrow
    b. carry oxygen and carbon dioxide
    c. can leave the bloodstream and enter tissues
    d. can fight disease and infection

27. Megakaryocytes give rise to
    a. basophils.   c. monocytes.
    b. lymphocytes.   d. platelets.

28. Which of the following is in the correct sequence for blood clotting?
    a. prothrombin activator, prothrombin, thrombin
    b. fibrin threads, prothrombin activator, thrombin
    c. thrombin, fibrinogen, fibrin threads
    d. prothrombin, clotting factors, fibrinogen
    e. Both a and c are correct.

29. Theoretically, a person with type AB blood should be able to receive
    a. type B and type AB blood.
    b. type O and type B blood.
    c. type A and type O blood.
    d. All of these are correct.

30. If a person has type B⁻ blood, it means there are
    a. anti-A antibodies in the plasma.
    b. no B antigens on the red blood cells.
    c. Rh antigens on the red blood cells.
    d. no Rh antigens on the red blood cells.
    e. Both a and d are correct.

Thinking Critically About the Concepts

After reading the homeostasis section of this chapter, you should better understand the interactions between different organ systems that are required for homeostasis. No system ever operates alone. Keep in mind what you've learned about the cardiovascular system as you continue your studies. You may be surprised by how often you hear about concepts from Chapters 5 and 6 in upcoming chapters.

Ben Monteiro, from the case study, suffers from leukemia. By causing huge numbers of abnormal white blood cells to be produced, the disease disrupts homeostasis in all organ systems. Thrombocytopenia, or platelet deficiency, may result in a fatal hemorrhage. Diseased white blood cells can't battle infections. Most important, when bone marrow produces leukemia cells instead of red blood cells, tissues cannot receive the oxygen they require. Without aggressive treatment, Ben's disease would be fatal.

1. Carbon monoxide (CO) is a deadly odorless, colorless gas. Hemoglobin in red blood cells binds much more closely to CO than it does to oxygen. Hemoglobin's ability to transport oxygen is severely compromised in the presence of CO.
   a. A malfunctioning furnace is one potential cause for CO poisoning. What are some other situations in which you could be exposed to carbon monoxide?
   b. In Chapter 3, you learned that oxygen is necessary for cellular metabolism. What is the molecule that serves as the source of cellular energy?
   c. If cells are deprived of oxygen, what is the effect on the production of cellular energy? Be specific.

2. There are three specific nutrients mentioned in this chapter that are necessary for the body to form red blood cells.
   a. Can you name all three?
   b. What are good food sources for these nutrients?

3. What type of organic molecule is hemoglobin? What nutrient is required to form hemoglobin? (Review Chap. 2 if necessary.)

4. The hormone erythropoietin is produced by the kidneys whenever more red blood cells are necessary. Think of several reasons when the body will need more red blood cells.

5. Athletes who abuse erythropoietin have many more red blood cells than usual.
   a. After examining Figure 6.3 and imagining many more red blood cells than usual in this capillary, explain why an athlete might die from having too many red blood cells.
   b. Why is death more likely at night when an athlete is sleeping, rather than during the day when he or she is active? Base your explanation on the heart rate.

6. After a trip to a country close to the equator, a student returns covered with mosquito bites. She has a low fever for a week, and blood tests show a very high level of eosinophils. Can you think of a disease that might cause these symptoms? Base your answer on what you know about tropical diseases.

7. Considering what you have learned from the "Homeostasis" section of this chapter, you should be able to think of at least two other organ systems that interact with the cardiovascular system (and the blood in particular) to regulate blood oxygen. List two organ systems and describe how they work with the cardiovascular system to make oxygen available to body cells.

CHAPTER

7

Lymphatic System and Immunity

The guys from Brown Hall were playing a game of football in front of the dorm. Nick threw his backpack on the front stoop of the dorm and joined them.

"Go long, Nick," his roommate Brent yelled. Nick ran to the lawn and turned, just in time to pluck the ball out of the air. He tucked it under his arm and started running. He didn't make it very far before being hit by one of the larger opponents. Then half of the team piled on top of him. There was a lot of yelling and joking, and it took several minutes before everyone climbed off of him.

"Hey, Nick, come on. Get up and give us the ball." One of the guys from his dorm grabbed Nick's arm and helped him stand up. A player from the opposing team scooped up the ball. Everyone started to line up for the next play. Nick stayed bent over with his hands on his thighs. Brent yelled from the line of scrimmage, "Come on Nick! What's the matter?"

Nick tried to answer, but he had a hard time catching his breath. Brent could tell that something was wrong. He started over to where Nick was standing. Just as Brent got to his side, Nick's knees buckled and he hit the ground in a heap.

"Nick! NICK!" Brent yelled. He grabbed Nick and shook him. "He passed out!" Brent yelled. "Dustin, call 911!" he screamed to a classmate who was standing on the porch. Dustin quickly reached for his cell phone.

"We need an ambulance right now," Dustin gasped. "Some guys were playing football, and this one guy just passed out."

"Where are you, sir?" the dispatcher inquired.

Dustin was so upset that he had to think for a moment. "Oh, um, we're in front of Brown Hall, at the University, on 3rd Street."

"The ambulance is on its way, sir," the dispatcher reassured. "Stay on the line and don't hang up. Is your friend breathing? Does he have a pulse?"

"Brent, make sure he's breathing! Has he got a pulse?" Dustin screamed as he ran towards his fallen friend. Brent nodded.

Soon they could hear the ambulance siren in the distance.

7.1 Microbes, Pathogens, and You
Bacteria and viruses are pathogens that cause human diseases.

7.2 The Lymphatic System
The lymphatic vessels return excess tissue fluid to cardiovascular veins. The lymphatic organs (red bone marrow, thymus, lymph nodes, and spleen) are important to immunity.

7.3 Innate Defenses
Innate defenses are barriers that prevent pathogens from entering the body and mechanisms able to deal with minor invasions.

7.4 Acquired Defenses
Acquired defenses specifically counteract an invasion in two ways: by producing antibodies and by outright killing abnormal cells.

7.5 Acquired Immunity
The two main types of acquired immunity are immunization by vaccines and the administration of prepared antibodies.

7.6 Hypersensitivity Reactions
The immune system is associated with allergies, tissue reaction, and autoimmune disorders. Treatment is available for these, but research goes forward into finding new and better cures.

7.1 Microbes, Pathogens, and You

Microbes (microscopic organisms, such as bacteria, viruses, and protists) are widely distributed in the environment. They cover inanimate objects and the surfaces of plants and animals. They are plentiful on and within our bodies. Many of the activities of microbes are useful to humans. We eat foods produced by bacteria every day. They contribute to the production of yogurt, cheese, bread, beer, wine, and many pickled foods. Today, drugs available through biotechnology are produced by bacteria. Microbes help us in still another way. Without the activity of decomposers, the biosphere (including ourselves) would cease to exist. When a tree falls to the forest floor, it eventually rots because decomposers, including bacteria and fungi, break down the remains of dead organisms to inorganic nutrients. Plants need these inorganic nutrients to make the many molecules that become food for us.

Unfortunately, human infectious diseases are typically caused by bacteria and viruses, collectively called **pathogens.** The body has three lines of defense against invasion:

1. Barriers to entry, such as the skin and mucous membranes of body cavities, act to prevent pathogens from gaining entrance into the body.
2. First responders, such as the phagocytic white blood cells, act to prevent an infection after an invasion has occurred.
3. Acquired defenses overcome an infection by killing the particular disease-causing agent that has entered the body. Acquired defenses also protect us against cancer.

Bacteria

Bacteria are single-celled prokaryotes that do not have a nucleus. Figure 7.1 illustrates the main features of bacterial anatomy and shows the three common shapes: **coccus** (spherical shaped), **bacillus** (rod shaped), and **spirillum** (plural **spirilla**). Spirilla may have spiral forms, or are comma-shaped. Bacteria have a cell wall that contains a unique amino-disaccharide. The "cillin" antibiotics, such as penicillin, interfere with the production of the cell wall. The cell wall of some bacteria is surrounded by a **capsule** that has a thick, gummy consistency. Capsules often allow bacteria to stick to surfaces such as teeth. They also prevent phagocytic white blood cells from taking them up and destroying them.

Motile bacteria usually have long, very thin appendages called flagella (sing., **flagellum**). The flagella rotate 360° and cause the bacterium to move backwards. Some bacteria have **fimbriae,** stiff fibers that allow the bacteria to adhere to surfaces such as host cells. Fimbriae allow a bacterium to cling to and gain access to the body. In contrast, a **pilus** is an elongated hollow appendage used to transfer DNA from one cell to another. Genes that allow bacteria to be resistant to antibiotics can be passed from one to the other through a pilus.

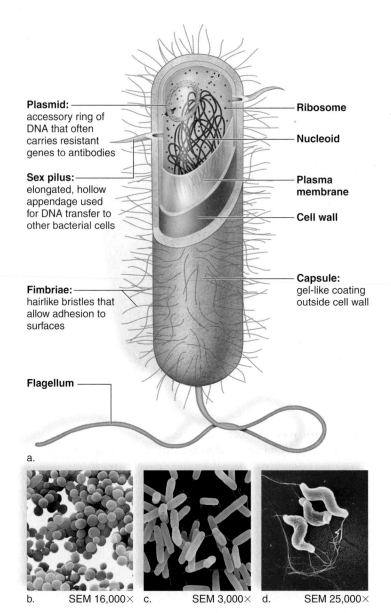

Plasmid: accessory ring of DNA that often carries resistant genes to antibodies

Sex pilus: elongated, hollow appendage used for DNA transfer to other bacterial cells

Fimbriae: hairlike bristles that allow adhesion to surfaces

Flagellum

Ribosome

Nucleoid

Plasma membrane

Cell wall

Capsule: gel-like coating outside cell wall

a.

b.   SEM 16,000×   c.   SEM 3,000×   d.   SEM 25,000×

Figure 7.1   What do bacteria look like?
a. Bacterial features, especially the four with definitions, contribute to the ability of bacteria to cause disease. Bacteria occur in three shapes.
b. *Staphylococcus aureus* is a sphere-shaped bacterium that causes toxic shock syndrome. **c.** *Pseudomonas aeruginosa* is a rod-shaped bacterium that causes urinary tract infections. **d.** *Campylobacter jejuni* is a curve-shaped bacterium that causes food poisoning.

Bacteria are independent cells that are metabolically competent. Their DNA is packaged in a chromosome that occupies the center of the cell. Many bacteria also have accessory rings of DNA called **plasmids.** Genes that allow bacteria to be resistant to antibiotics are often located in a plasmid. Abuse of antibiotic therapy increases the number of resistant bacterial strains that are difficult to kill, even with antibiotics.

Bacteria reproduce by a process called binary fission (Fig. 7.2). The single, circular chromosome attached to the plasma membrane is copied. Then the chromosomes are

Figure 7.2 How do bacteria reproduce?

When bacteria reproduce, division (fission) produces two (binary) cells identical to the original cell.

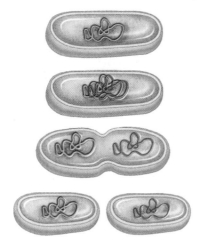

separated as the cell enlarges. The newly formed plasma membrane and cell wall separate the cell into two cells. Bacteria can reproduce rapidly under favorable conditions, with some species doubling their numbers every 12 minutes.

Strep throat, tuberculosis, gangrene, gonorrhea, and syphilis are well-known bacterial diseases. Not only does growth of bacteria cause disease but some bacteria release molecules called **toxins** that inhibit cellular metabolism. For example, it is important to have a tetanus shot because the bacteria that causes this disease, *Clostridium tetani*, produces a toxin that prevents relaxation of muscles. In time, the body contorts because all the muscles have contracted. Eventually, suffocation occurs.

Viruses

Viruses bridge the gap between the living and the nonliving. Outside a host, viruses are essentially chemicals that can be stored on a shelf. But when the opportunity arises, viruses replicate inside cells, and during this period, they clearly appear to be alive.

Viruses are acellular—not composed of cells. They are parasites and do not live independently. Viruses cause diseases such as colds, flu, measles, chicken pox, polio, rabies, AIDS, genital warts, and genital herpes.

Virus particles are about one quarter the size of a bacterium, about one hundredth the size of a eukaryotic cell (Fig. 7.3). A virus always has two parts: an outer capsid composed of protein units and an inner core of nucleic acid (Fig. 7.4). A virus carries the genetic information needed to reproduce itself. In contrast to cellular organisms, the viral genetic material need not be double-stranded DNA, nor even DNA. Indeed, some viruses, such as HIV and influenza, have RNA as their genetic material. A virus may also contain various enzymes that help it reproduce.

Viruses are microscopic pirates, commandeering the metabolic machinery of a host cell. Viruses gain entry into

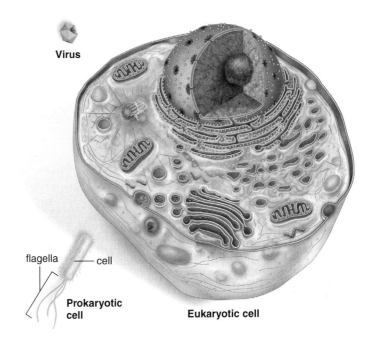

Figure 7.3 How do the sizes of viruses, bacteria, and eukaryotic cells compare?

Viruses are tiny noncellular particles while bacteria are small independent cells. Eukaryotic cells are more complex and larger because they contain a nucleus and many organelles.

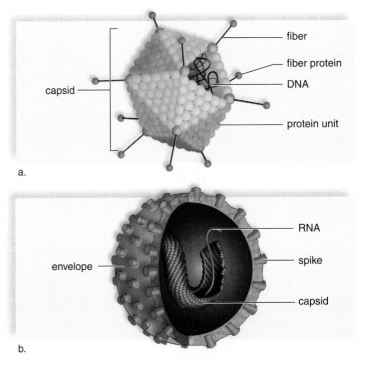

Figure 7.4 What do viruses look like?

Despite their diversity, all viruses have an outer capsid, composed of protein subunits, and a nucleic acid core, composed of either DNA or RNA, but not both. **a.** Adenoviruses cause colds and **(b)** influenza viruses cause the flu. Influenza viruses are surrounded by an envelope with spikes.

Mary Mallon: The Most Dangerous Woman in America

How would you care to be known as the most dangerous woman in America? This description was used to refer to Mary Mallon. What did Mary do?

Mary Mallon, also known as Typhoid Mary, was a cook in New York from 1900–15. But she was also a carrier of the bacteria that cause typhoid fever (often referred to as typhoid), even though she was healthy. She probably had a mild case of typhoid fever earlier in her life and recovered. However, in this age before antibiotics, Mary still carried the bacteria in her body. As a result, she infected those who ate her cooking. From 1900–07, Mary worked in seven different households. There, she caused 22 cases of typhoid fever and was responsible for one child's death. When Mary left the households, the typhoid fever cases stopped.

Typhoid is a bacterial disease spread via food or water contaminated by sewage. Infected patients shed the bacteria in their feces or urine. The disease is spread if an infected person prepares food or beverages with unwashed hands. High fever, headache, weakness, and severe diarrhea are symptoms of typhoid. These symptoms can persist for weeks. Further, the patient can tear the weakened intestinal wall during recovery, and then die from excessive bleeding. The fatality rate in the early 1900s was approximately 10% (Fig. 7A).

When public officials first suspected Mary was the cause of the outbreaks, they approached her to ask for urine and fecal samples. Enraged, her reply was to attack them with a kitchen knife. In 1907, public health officials took her into custody by force. She was sent to a quarantine island located

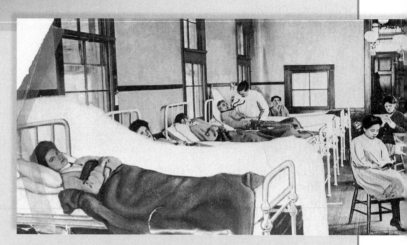

Figure 7A   **Why was Mary Mallon considered a menace to society?**
As a typhoid carrier, Mary spread typhoid to others. In an age before antibiotics typhoid had a high fatality rate.

in New York's East River. Mary never believed that she was a typhoid carrier—after all, she was perfectly healthy. Imprisoned against her will, she always maintained her innocence.

In 1910, she was released, with orders that she never again work in food preparation. She was trained to work as a laundress instead and directed to check in regularly with health officials. However, after her release, Mary seemed to disappear. In 1915, there was an outbreak of typhoid at the Sloane Maternity Hospital in Manhattan. Twenty-two people sickened and two died. The hospital had recently hired a cook named Mrs. Brown—Mary, using an alias. She was sent back to the same cottage on North Brother Island, where she was permanently imprisoned. Mary lived for another 23 years, dying in 1938.

Have You Ever Wondered...

Is "double-dipping" REALLY that bad?

You are standing around the reception table, eating chips and your favorite dip. Suddenly, you watch someone take a half-eaten chip and "double-dip"—dip the chip into the dip bowl again. Should you eat any more of that dip? Researchers from the Clemson University food science program studied the question. First, they "double-dipped" a snack cracker into sterile water, without taking a bite from the cracker. Then, after taking a bite from the cracker, the researchers "double-dipped" it into sterile water. "Double-dipping" increased the number of bacteria in the water threefold. In other words, if you see someone "double-dip," avoid the dip!

and are specific to a particular host cell because portions of the virus are specific for a receptor on the host cell's outer surface. Once the virus is attached, the viral genetic material (DNA or RNA) enters the cell. Inside the cell, the nucleic acid codes for the protein units in the capsid. In addition, the virus may have genes for special enzymes needed for the virus to reproduce and exit from the host cell. In large measure, however, a virus relies on the host's enzymes and ribosomes for its own reproduction.

Prions

Prions, proteinaceous infectious particles, cause a group of degenerative diseases of the nervous system, also called wasting diseases. Originally thought to be viral diseases, prions cause Creutzfeldt-Jakob disease (CJD) in humans; scrapie in sheep; and bovine spongiform encephalopathy (BSE), commonly called **mad cow disease.** These infections are apparently transmitted by ingestion of brain and nerve tissues

Figure 7.5 How do the normal prion protein and "rogue" prion protein compare?
The model on the left shows the normal prion protein composed of α-helices (coils). The model on the right shows the "rogue" prion protein with β-pleated sheets (flat arrows).

from infected animals. Prions are proteins of unknown function found normally in the brains of healthy individuals. The "rogue" form of the prion protein is folded into a new shape. The "rogue" protein is able to refold normal prion proteins into the new shape and thus cause disease. Two molecular models of a protein are shown in Figure 7.5. We can imagine the one on the left is the normal protein, and the one on the

right is the prion. Nervous tissue is lost and calcified plaques show up in the brain due to activity of prions. Thankfully, the incidence of prion diseases in humans is very low.

> **Check Your Progress 7.1**
> 1. Which anatomical structures in bacteria can be associated with virulence, the ability to cause disease?
> 2. Why are viruses considered obligate parasites and always cause disease?

7.2 The Lymphatic System

The lymphatic system consists of lymphatic vessels and the lymphatic organs. This system, closely associated with the cardiovascular system, has four main functions that contribute to homeostasis: (1) Lymphatic capillaries absorb excess tissue fluid and return it to the bloodstream; (2) in the small intestines, lymphatic capillaries called lacteals absorb fats in the form of lipoproteins and transport them to the bloodstream; (3) the lymphatic system is responsible for the production, maintenance, and distribution of lymphocytes; and (4) the lymphatic system helps defend the body against pathogens.

CASE STUDY THE AMBULANCE ARRIVES

The guys stepped back away from Nick when the ambulance arrived. Four paramedics leaped out. Two of the paramedics immediately went to Nick's side. One quickly took out a stethoscope and placed it against Nick's chest. The other checked for a pulse and peeled back an eyelid. A second team of paramedics opened the rear doors and began to unload a stretcher.

"Hey guys, what happened here?" the driver inquired. "What's his name?"

A circle of players was gathered around. "His name is Nick Williamson," Brent answered. "He's my roommate. We were playing football and he got tackled. Then when he tried to stand up, he fainted." The other players all began to talk.

"Someone must have tackled him too hard."

"Maybe he hit his head or something."

"We didn't hit him that hard. Really we didn't."

"But there were a lot of guys on top of him."

"Not that many. Besides, he stood up again before he passed out."

"He's got a MedicAlert® on him," a paramedic yelled. "Bee sting allergy. There's a big welt on his arm."

The driver quickly radioed to the hospital and described the situation. Meanwhile, Brent grabbed Nick's backpack from the stoop. "He told me about it! Here's his Epi-Pen®!" Within moments, the paramedics had injected the Epi-Pen® into Nick's thigh. The team then started an IV in Nick's arm.

"Nick told me about his allergy once," Brent told the driver. "He said he almost died from a bee sting when he was a little kid. He always kept that pen thing in his pocket or his backpack in case of emergencies. He must have passed out before he could get it." Brent sighed. "I guess I should have realized it faster, but I just thought he was hurt from the pile-up."

"No," the paramedic said. "He was having a severe allergic reaction to that bee sting. It happens pretty fast. We injected him with the epinephrine in the Epi-Pen®, and we'll get him to the hospital. We're giving him some fluids through the IV, and he should be all right. You guys did a good job by calling us so quickly."

"You can't ride with us, buddy, because of liability," he told Brent. "But you can follow us to University Hospital. That's only a few blocks away."

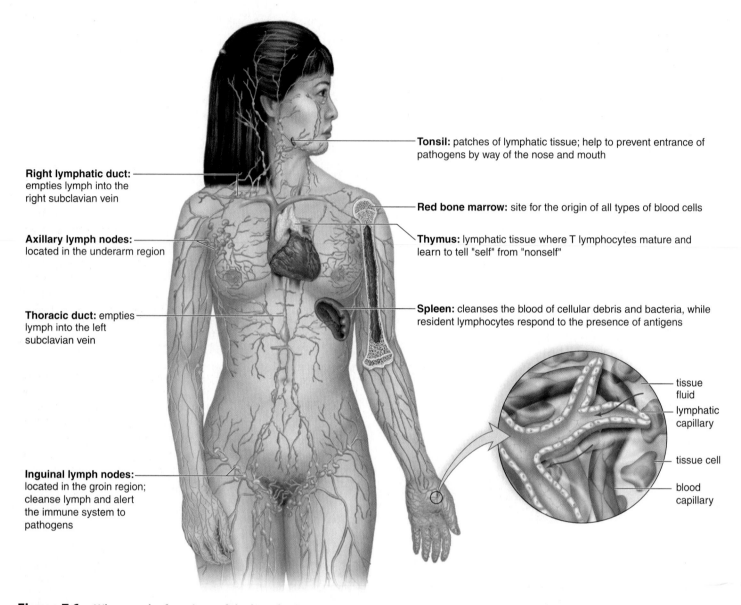

Tonsil: patches of lymphatic tissue; help to prevent entrance of pathogens by way of the nose and mouth

Right lymphatic duct: empties lymph into the right subclavian vein

Red bone marrow: site for the origin of all types of blood cells

Axillary lymph nodes: located in the underarm region

Thymus: lymphatic tissue where T lymphocytes mature and learn to tell "self" from "nonself"

Thoracic duct: empties lymph into the left subclavian vein

Spleen: cleanses the blood of cellular debris and bacteria, while resident lymphocytes respond to the presence of antigens

Inguinal lymph nodes: located in the groin region; cleanse lymph and alert the immune system to pathogens

tissue fluid

lymphatic capillary

tissue cell

blood capillary

Figure 7.6   **What are the functions of the lymphatic system components?**
Lymphatic vessels drain excess fluid from the tissues and return it to the cardiovascular system. The enlargement shows that lymphatic vessels, like cardiovascular veins, have valves to prevent backward flow. The lymph nodes, spleen, thymus gland, and red bone marrow are the main lymphatic organs that assist immunity.

Lymphatic Vessels

Lymphatic vessels form a one-way system of capillaries to vessels, and, finally, to ducts. These vessels take lymph to cardiovascular veins in the shoulders (Fig. 7.6). As you recall from Chapter 6, lymphatic capillaries take up excess tissue fluid. Tissue fluid is mostly water, but it also contains solutes (i.e., nutrients, electrolytes, and oxygen) derived from plasma. Tissue fluid also contains cellular products (i.e., hormones, enzymes, and wastes) secreted by cells. The fluid inside lymphatic vessels is called **lymph.** Lymph is usually a colorless liquid, but after a meal, it appears creamy because of its lipid content.

The lymphatic system has two ducts (highways): the thoracic duct and the right lymphatic duct. The larger thoracic duct returns lymph collected from the body below the thorax, the left arm, and left side of the head and neck. The thoracic duct drains fluid into the left subclavian vein. The right lymphatic duct returns lymph from the right arm and right side of the head and neck into the right subclavian vein.

The construction of the larger lymphatic vessels is similar to that of cardiovascular veins, including the presence of valves. The movement of lymph within lymphatic capillaries is largely dependent upon skeletal muscle contraction. Lymph forced through lymphatic vessels as a result of muscular compression is prevented from flowing backward by one-way valves.

Lymphatic Organs

Lymphatic organs are divided into primary: red bone marrow and the thymus gland; and secondary: lymph nodes and spleen (Fig. 7.7).

The Primary Lymphatic Organs

Red bone marrow produces all types of blood cells. In a child, most bones have red bone marrow, but in an adult, it is limited to the sternum, vertebrae, ribs, part of the pelvic girdle, and the upper ends of the humerus and femur. In addition to the red blood cells, bone marrow produces the various types of white blood cells: neutrophils, eosinophils, basophils, lymphocytes, and monocytes. Lymphocytes are either **B lymphocytes (B cells)** or **T lymphocytes (T cells).** B cells mature in the bone marrow, but the T cells mature in the thymus. Any B cell that reacts with cells of the body is removed in the bone marrow and does not enter the circulation. This ensures that the B cells do not harm normal cells of the body.

The soft, bilobed **thymus gland** is located in the thoracic cavity between the trachea and the sternum, superior to the heart. The thymus varies in size, but it is largest in children and shrinks as we get older.

The thymus has two functions: (1) The thymus gland produces thymic hormones, such as thymosin, thought to aid in the maturation of T lymphocytes. Thymosin may also have other functions in immunity. (2) Immature T lymphocytes migrate from the bone marrow through the bloodstream to the thymus, where they mature. Only about 5% of these cells ever leave the thymus. These T lymphocytes have survived a critical test: If any show the ability to react with the individual's cells, they die. If they have potential to attack a pathogen, they leave the thymus. The thymus is absolutely critical to immunity because without mature T cells, the body's response to specific pathogens is poor or absent.

Secondary Lymphatic Organs

The secondary lymphatic organs are the spleen, the lymph nodes, and other organs containing lymphoid tissue. These include the tonsils, Peyer's patches, and the appendix.

The **spleen** filters blood. The spleen, the largest lymphatic organ, is located in the upper left region of the abdominal cavity posterior to the stomach. Connective tissue divides the spleen into regions known as white pulp and red pulp (Fig. 7.7). The red pulp, which surrounds venous sinuses (cavities), is involved in filtering the blood. Blood entering the spleen must pass through the sinuses before exiting. Here, macrophages that are like powerful vacuum cleaners engulf pathogens and debris, such as worn-out red blood cells.

The spleen's outer capsule is relatively thin, and an infection or a blow can cause the spleen to burst. Although the spleen's functions are replaced by other organs, a person without a spleen is often slightly more susceptible to infections. They may have to receive antibiotic therapy indefinitely.

Lymph nodes, which occur along lymphatic vessels, filter lymph. Connective tissue forms a capsule and also divides

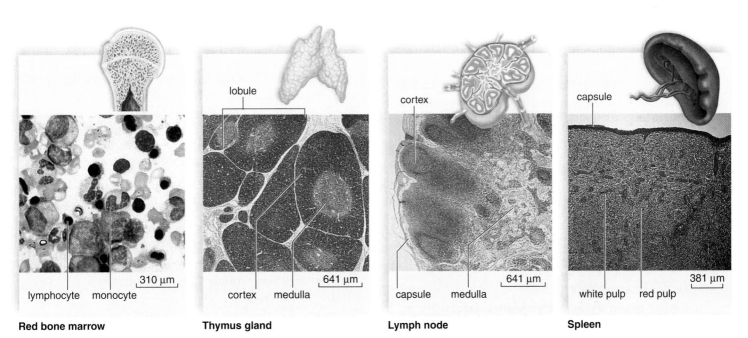

lymphocyte   monocyte
310 µm
Red bone marrow

lobule
cortex   medulla
641 µm
Thymus gland

cortex
capsule   medulla
641 µm
Lymph node

capsule
white pulp   red pulp
381 µm
Spleen

Figure 7.7 **What are the primary and secondary lymphatic organs?**
Left: Red bone marrow and the thymus gland are the primary lymphatic organs. Blood cells, including lymphocytes, are produced in red bone marrow. B cells mature in the bone marrow, but T cells mature in the thymus. *Right*: Lymph nodes and the spleen are secondary lymphatic organs. Lymph is cleansed in lymph nodes, while blood is cleansed in the spleen.

a lymph node into compartments. Each compartment contains a sinus that increases in size toward the center of the node. As lymph courses through the sinuses, it is exposed to macrophages, which engulf pathogens and debris. Lymphocytes, also present in sinuses, fight infections and attack cancer cells.

Lymph nodes are named for their location. For example, inguinal nodes are in the groin, and axillary nodes are in the armpits. Physicians often feel for the presence of swollen, tender lymph nodes in the neck as evidence that the body is fighting an infection. This is a noninvasive, preliminary way to help make such a diagnosis.

Lymphatic nodules are concentrations of lymphatic tissue not surrounded by a capsule. The **tonsils** are patches of lymphatic tissue located in a ring about the pharynx. The tonsils perform the same functions as lymph nodes, but because of their location, they are the first to encounter pathogens and antigens that enter the body by way of the nose and mouth.

Peyer's patches located in the intestinal wall and tissues within the appendix, a blind pouch of the large intestine, encounter pathogens that enter the body by way of the intestinal tract.

Check Your Progress 7.2

1. a. Which organs are primary lymphatic organs? b. Secondary lymphatic organs?
2. How do the primary and secondary lymphatic organs function in immunity?
3. Which organ produces blood cells, which filters blood, which filters lymph, and which causes T cells to mature?

7.3 Innate Defenses

Immunity, the ability to combat diseases and cancer, includes two innate lines of defense. These act indiscriminately against all pathogens:

- Barriers to entry. The body puts up barriers that can prevent the possible entry of pathogens.
- Phagocytic white blood cells, the neutrophils and macrophages. We will consider these two phagocytic white blood cells in the context of the **inflammatory response,** a special reaction of the body when first invaded. Protective proteins are also part of this line of defense.

Barriers to Entry

The body has built-in barriers, both physical and chemical, that serve as the first line of defense against an infection by pathogens.

Skin and Mucous Membranes   The intact skin is generally an effective physical barrier that prevents infection. Mucous membranes lining the respiratory, digestive, reproductive, and urinary tracts are also physical barriers to entry by pathogens. The ciliated cells that line the upper respiratory tract sweep mucus and trapped particles up into the throat, where they can be swallowed or expectorated (spit out).

Chemical Barriers   The chemical barriers to infection include the secretions of sebaceous (oil) glands of the skin. These secretions contain chemicals that weaken or kill certain bacteria on the skin.

Perspiration, saliva, and tears contain an antibacterial enzyme called **lysozyme.** Saliva also helps to wash microbes off the teeth and tongue, and tears wash the eyes. Similarly, as urine is voided from the body, it flushes bacteria from the urinary tract.

The acid pH of the stomach inhibits growth or kills many types of bacteria. At one time, it was thought that no bacterium could survive the acidity of the stomach. But now we know that ulcers are caused by the bacterium *Helicobacter pylori* (see pages 8–9). Similarly, the acidity of the vagina and its thick walls discourage the presence of pathogens.

Resident Bacteria   Finally, a significant chemical barrier to infection is created by the normal flora, microbes that usually reside in the mouth, intestine, and other areas. By using available nutrients and releasing their own waste, these resident bacteria prevent potential pathogens from taking up residence. For this reason, chronic use of antibiotics can make a person susceptible to pathogenic infection by killing off the normal flora.

Inflammatory Response

The inflammatory response exemplifies the second line of defense against invasion by a pathogen. Inflammation employs mainly neutrophils and macrophages to surround

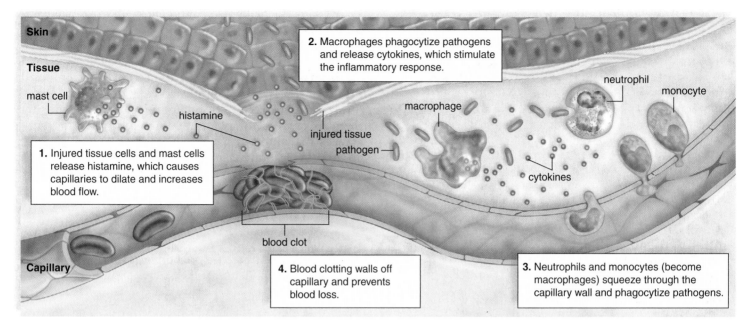

Figure 7.8   What are the steps of the inflammatory response?
1. Due to capillary changes in a damaged area and the release of chemical mediators, such as histamine by mast cells, an inflamed area exhibits redness, heat, swelling, and pain. The inflammatory response can be accompanied by other reactions to the injury. **2.** Macrophages release cytokines, which stimulate the inflammatory and other immune responses. **3.** Monocytes and neutrophils squeeze through capillary walls from the blood and phagocytize pathogens. **4.** A blood clot can form a seal in a break in a blood vessel.

and kill (engulf by phagocytosis) pathogens trying to get a foothold inside the body. Protective proteins are also involved. Inflammation is usually recognized by its four hallmark symptoms: redness, heat, swelling, and pain (Fig. 7.8).

The four signs of the inflammatory response are due to capillary changes in the damaged area, and all serve to protect the body. Chemical mediators, such as **histamine,** released by damaged tissue cells and **mast cells,** cause the capillaries to dilate and become more permeable. Excess blood flow due to enlarged capillaries causes the skin to redden and become warm. Increased temperature in an inflamed area tends to inhibit growth of some pathogens. Increased blood flow brings white blood cells to the area. Increased permeability of capillaries allow fluids and proteins, including blood clotting factors, to escape into the tissues. Clot formation in the injured area prevents blood loss. The excess fluid in the area presses on nerve endings, causing the familiar pain associated with swelling. Together, these events summon white blood cells to the area.

As soon as the white blood cells arrive, they move out of the bloodstream into the surrounding tissue. The neutrophils are first and actively devour debris, dead cells, and bacteria they encounter. The many neutrophils attracted to the area can usually localize any infection and keep it from spreading. If neutrophils die off in great quantity, they become a yellow-white substance we call **pus.**

When an injury is not serious, the inflammatory response is short-lived, and the healing process will quickly return the affected area to a normal state. Nearby cells secrete chemical factors to ensure the growth (and repair) of blood vessels and new cells to fill in the damaged area.

If, on the other hand, the neutrophils are overwhelmed, they call for reinforcements by secreting chemical mediators called **cytokines.** Cytokines attract more white blood cells to the area, including monocytes. Monocytes are longer-lived cells that become **macrophages,** even more powerful phagocytes than neutrophils. Macrophages can enlist the help of lymphocytes to carry out specific defense mechanisms.

Have You Ever Wondered...

How do antihistamines work?

Once histamine is released from mast cells, it binds to receptors on other body cells. There, the histamine causes the familiar symptoms associated with infections and allergies: sneezing, itching, runny nose, and watery eyes. Antihistamines work by blocking the receptors on the cells so that histamine can no longer bind. For allergy relief, antihistamines are most effective when taken before exposure to the allergen.

**Figure 7.9
How does the complement system kill a bacterium?**
When complement proteins in the blood plasma are activated by an immune response, they form a membrane attack complex that makes holes in bacterial cell walls and plasma membranes, allowing fluids to enter until the cell eventually bursts.

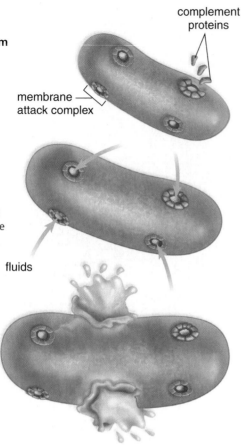

Inflammation is the body's natural response to an irritation or injury and serves an important role. Once the tide of the battle has turned, inflammation rapidly subsides. However, in some cases, chronic inflammation may last for weeks, months, or even years if an irritation or infection cannot be overcome. Inflammatory chemicals may cause collateral damage to the body, in addition to killing the invaders. Should an inflammation persist, anti-inflammatory medications, such as aspirin, ibuprofen, or cortisone, can minimize the effects of various chemical mediators.

Protective Proteins

The **complement system,** often simply called complement, is composed of a number of blood plasma proteins designated by the letter C and a number. The complement proteins "complement" certain immune responses, which accounts for their name. For example, they are involved in and amplify the inflammatory response because certain complement proteins can bind to mast cells and trigger histamine release. Others can attract phagocytes to the scene. Some complement proteins bind to the surface of pathogens already coated with antibodies, which ensures that the pathogens will be phagocytized by a neutrophil or macrophage.

Certain other complement proteins join to form a **membrane attack complex** that produces holes in the surface of bacteria. Fluids then enter the bacterial cell to the point that they burst (Fig. 7.9).

Interferons are proteins produced by virus-infected cells as a warning to noninfected cells in the area. Interferon binds to receptors of noninfected cells, causing them to prepare for possible attack by producing substances that interfere with viral replication. Interferons are used as treatment in certain viral infections, such as hepatitis C.

> ### Check Your Progress 7.3
> 1. What are some examples of the body's innate defenses?
> 2. Which blood cells, in particular, should you associate with innate defenses, and how do they function?
> 3. Why are the complement proteins so named?

7.4 Acquired Defenses

When innate defenses have failed to prevent an infection, acquired defenses come into play. Acquired defenses overcome an infection by doing away with the particular disease-causing agent that has entered the body. Acquired defenses also protect against cancer.

How Acquired Defense Works

Acquired defenses respond to **antigens,** molecules the immune system recognizes as foreign to the body. Antigens are typically large molecules, such as proteins. Fragments of bacteria, viruses, molds, or parasitic worms can all be antigenic. Further, abnormal plasma membrane proteins produced by cancer cells may also be antigens. We do not ordinarily become immune to our normal cells so it is said that the immune system is able to distinguish self from nonself.

Acquired defenses primarily depend on the action of lymphocytes, which differentiate as either B cells (B lymphocytes) or T cells (T lymphocytes). B cells and T cells are capable of recognizing antigens because they have specific antigen receptors. These antigen receptors are plasma membrane proteins whose shape allows them to combine with particular antigens. Each lymphocyte has only one type of receptor. It is often said that the receptor and the antigen fit together like a lock and a key. We encounter millions of different antigens during our lifetime, so we need a diversity of B cells and T cells to protect us against them. Remarkably, this diversification occurs during the maturation process. Millions of specific B cells and/or T cells are formed increasing the likelihood that at least one will recognize any possible antigen.

B cells differentiate into other types of cells, and so do T cells. The several types of cells we will be discussing are listed in Table 7.1 for easy reference.

Table 7.1	Immune Cell Type and Function
Cell	**Function**
B cells	Produce plasma cells and memory cells
Plasma cells	Produce specific antibodies
Memory cells	Ready to produce antibodies in the future
T cells	Regulate immune response; produce cytotoxic T cells and helper T cells
Cytotoxic T cells	Kill virus-infected cells and cancer cells
Helper T cells	Regulate immunity
Memory T cells	Ready to kill in the future

B Cells and Antibody-Mediated Immunity

The receptor on a B cell is called a **B-cell receptor (BCR).** The **clonal selection model** (Fig. 7.10) states that an antigen *selects*, then binds to the BCR of only one type B cell. Then this B cell produces multiple copies of itself. The resulting group of identical cells is called a *clone.* (Similarly, an antigen can bind to a T-cell receptor (TCR) and this T cell will clone.)

B Cells Become Plasma Cells and Memory B Cells   Note in Figure 7.10 that each B cell has a specific BCR represented by shape. Only the B cell with a BCR that has a shape that fits the antigen (green circle) undergoes clonal expansion. During clonal expansion, cytokines secreted by helper T cells (see page 149) stimulate B cells to clone. Most of the cloned B cells become **plasma cells,** which circulate in the blood and lymph. Plasma cells are larger than regular B cells because they have extensive rough endoplasmic reticulum. This is for the mass production and secretion of antibodies to a specific antigen. Antibodies identical to the BCR of the activated B cell are secreted from the plasma B cell. Some cloned B cells become memory cells, the means by which long-term immunity is possible. If the same antigen enters the system again, memory B cells quickly divide and transform into plasma cells. The correct type of antibody can quickly be produced by these plasma cells.

Once the threat of an infection has passed, the development of new plasma cells ceases, and those present undergo apoptosis. **Apoptosis** is the process of programmed cell death. It involves a cascade of specific cellular events leading to the death and destruction of the cell.

Defense by B cells is called humeral immunity or **antibody-mediated immunity** because activated B cells become plasma cells that produce antibodies. Collectively, plasma cells probably produce as many as 2 million different antibodies. A human being doesn't have 2 million genes, so there cannot be a separate gene for each type antibody. It has been found that scattered DNA segments can be shuffled

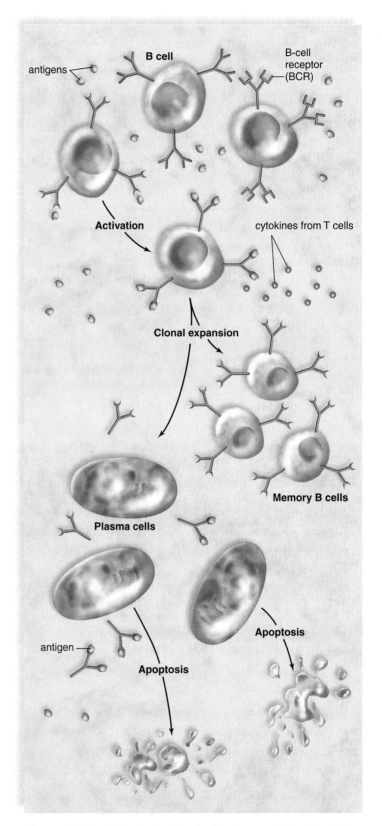

Figure 7.10   **What is the clonal selection model for B cells?** Activation of a B cell occurs when its B-cell receptor (BCR) combines with an antigen (colored green). In the presence of cytokines produced by T cells, the B cell undergoes clonal expansion, producing many plasma cells that secrete antibodies specific to the antigen.

and combined in various ways to produce the DNA sequence coding for the BCR unique to each type of B cell.

Characteristics of B Cells

- Antibody-mediated immunity against pathogens
- Produced and mature in bone marrow
- Directly recognize antigen and then undergo clonal selection
- Clonal expansion produces antibody-secreting plasma cells as well as memory B cells

---

Structure of an Antibody   The basic unit that composes antibody molecules is a Y-shaped protein molecule with two arms. Each arm has a "heavy" (long) polypeptide chain and a "light" (short) polypeptide chain (Fig. 7.11). These chains have constant regions, located at the trunk of the Y, where the sequence of amino acids is fixed. The class of antibody of each individual molecule is determined by the structure of the antibody's constant region. The variable regions form an antigen-binding site. Their shape is specific to a particular antigen. The antigen combines with the antibody at the antigen-binding site in a lock-and-key manner. Antibodies may consist of single Y-shaped molecules, called monomers, or may be paired together in a molecule termed a dimer. Very large antibodies belonging to the M class are pentamers—clusters of five Y-shaped molecules linked together.

Antigens can be part of a pathogen, such as a virus, or a toxin, like that produced by tetanus bacteria. Antibodies sometimes react with viruses and toxins by coating them completely, a process called neutralization. Often, the reaction produces a clump of antigens combined with antibodies, termed an immune complex. The antibodies in an immune complex are like a beacon that attracts white blood cells.

Classes of Antibodies   There are five different classes of circulating antibodies (Table 7.2). IgG antibodies are the major type in blood, and lesser amounts are found in lymph and tissue fluid. IgG antibodies bind to pathogens and their toxins. IgG antibodies can cross the placenta from a mother to her fetus, so the newborn has temporary, partial immune protection. As mentioned, IgM antibodies are pentamers. IgM antibodies are the first antibodies produced by a newborn's body. IgM antibodies are the first to appear in blood soon after an infection begins and the first to disappear before the infection is over. They are good activators of the complement system. IgA antibodies are monomers or dimers containing two Y-shaped structures. They are the main type of

---

CASE STUDY IN THE EMERGENCY ROOM

By the time they got to the hospital, Nick was conscious again. The paramedics put him in the emergency room and pulled the curtains around him. Brent soon joined him there. Someone had called Nick's parents, who were on their way. A nurse came in and checked his IV, blood pressure, and pulse. The emergency room doctor followed and talked to Brent and Nick.

"What you experienced, Nick, is called anaphylaxis, and it's your body's response to a bee sting," the doctor explained. "You've probably had this same sort of reaction before. Is that when you got the MedicAlert® bracelet?"

Nick nodded. "When I was younger. It was pretty bad."

The doctor nodded. "It's pretty scary. Usually a bee sting just affects the skin. Anaphylaxis is an allergic response that involves your whole body. It happens in seconds to minutes after you're exposed to the allergen. An allergen is anything your body decides to become allergic to. Your immune system severely overreacted to the bee venom—that's the allergen in your case. Your body released massive amounts of histamine."

Brent interrupted. "I've heard of antihistamines. You take those when you're all clogged up and can't breathe."

"Yes, that's right. Antihistamines block the action of histamine," the doctor said. "In this case, histamine caused your airways to narrow. You wheeze and can't catch your breath."

Nick nodded. "I remember being really short of breath just before I passed out."

"Histamine also causes your blood vessels to leak fluid, and your blood pressure falls. That's why you blacked out," the doctor asserted. "Your brain wasn't getting enough oxygen. If you don't get treated fast, you go into shock. Both the breathing difficulties and the shock can be fatal."

Nick looked up at Brent, "Thanks for calling the squad so fast."

The doctor agreed. "It was a good thing you did. The paramedics gave you that Epi-Pen®. Epinephrine counters the effects of the histamine. It raises your blood pressure and relaxes your airways so you can breathe."

The doctor turned to Brent. "The Epi-Pen® is for self-injection, but his allergic reaction came on so fast that he didn't have time to use it. He should probably teach you how to inject it, because you're his roommate. If this happens again, use it fast! And call 911 as fast as possible, just like you guys did. Tell the dispatcher that you're dealing with a severe allergic reaction as a result of a bee sting."

Nick's parents arrived, so Brent only stayed a few minutes longer. Some of the guys were gathered in the waiting room. "Nick's gonna be okay," Brent told them. "Let's go back to the dorm."

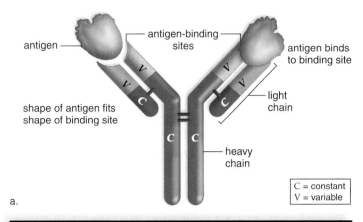

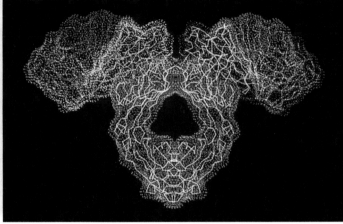

Figure 7.11 **What is the structure of an antibody?**
a. An antibody contains two heavy (long) polypeptide chains and two light (short) chains arranged so there are two variable regions, where a particular antigen is capable of binding with an antibody (*V* = variable region, *C* = constant region). **b.** Computer model of an antibody molecule. The antigen combines with the two side branches.

antibody found in body secretions: saliva, tears, mucus, and breast milk. IgA molecules bind to pathogens and prevent them from reaching the bloodstream. The main function of IgD molecules seems to be to serve as antigen receptors on immature B cells. IgE antibodies are responsible for prevention of parasitic worm infections, but they can also cause immediate allergic responses.

T Cells and Cell-Mediated Immunity

Cell-mediated immunity is named for the action of T cells that directly attack diseased cells and cancer cells. Other T cells, however, release cytokines that stimulate both nonspecific and specific defenses.

How T Cells Recognize an Antigen When a T cell leaves the thymus, it has a unique **T-cell receptor (TCR),** just as B cells have. Unlike B cells, however, T cells are unable to recognize an antigen without help. The antigen must be displayed to them by an **antigen-presenting cell (APC),** such as a macrophage. After phagocytizing a pathogen, such as a bacterium, APCs travel to a lymph node or spleen, where T cells also congregate. In the meantime, the APC has broken the pathogen apart in a lysosome. A piece of the pathogen is then displayed in the groove of an **MHC (major histocompatibility complex)** protein on the cell's surface.

Human MHC proteins are called **HLA (human leukocyte antigens).** These proteins are found on all of our body cells. More than 50 different proteins have been identified, and each human being has a unique combination—no two sets are exactly alike (with the exception of the set belonging to identical twins. Identical twins arise from division of a single zygote, so their HLA proteins are identical.) MHC antigens are self proteins because they mark the cell as belonging to a particular individual. The importance of self proteins in plasma membranes was first recognized when it was discovered that they contribute to the specificity of tissues and make it difficult to transplant tissue from one human to another. Comparison studies of the more than 50$^+$ MHC antigens must always be carried out before a transplant is attempted. The greater the number of these proteins that match, the more likely the transplant will be successful.

When an antigen-presenting cell links a foreign antigen to the self protein on its plasma membrane, it carries out an important safeguard for the rest of the body. The T cell to be

Table 7.2	**Antibodies**	
Class	**Presence**	**Function**
IgG	Main antibody type in circulation; crosses the placenta from mother to fetus	Binds to pathogens, activates complement, and enhances phagocytosis by white blood cells
IgM	Antibody type found in circulation; largest antibody; first antibody formed by a newborn; first antibody formed with any new infection	Activates complement; clumps cells
IgA	Main antibody type in secretions such as saliva and milk	Prevents pathogens from attaching to epithelial cells in digestive and respiratory tract
IgD	Antibody type found on surface of immature B cells	Presence signifies readiness of B cell
IgE	Antibody type found as antigen receptors on basophils in blood and on mast cells in tissues	Responsible for immediate allergic response and protection against certain parasitic worms

activated can compare the antigen and self protein side by side. The activated T cell and all of the daughter cells, which it will form, can recognize foreign from self. These T cells go on to destroy cells carrying foreign antigens, while leaving normal body cells unharmed.

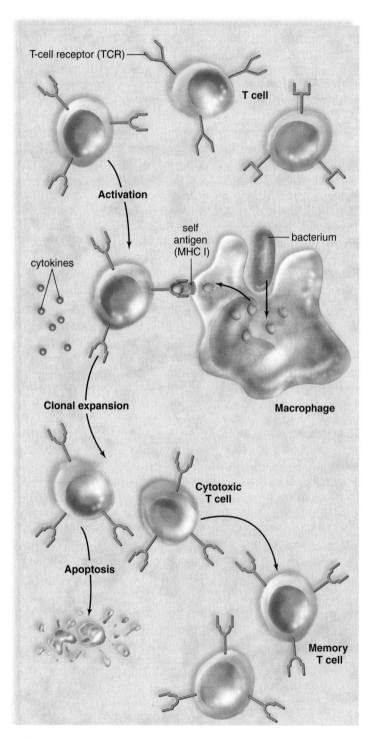

Figure 7.12 **What is the clonal selection model for T cells?** Activation of a T cell occurs when its T-cell receptor (TCR) can combine with an antigen presented by a macrophage. In this example, cytotoxic T cells are produced. When the immune response is finished, they undergo apoptosis. A small number of memory T cells remain.

Clonal Expansion In Figure 7.12, the T cells have specific TCRs, represented by their different shapes. A macrophage is presenting an antigen to a T cell that has the specific TCR that will combine with this particular antigen, represented by a green circle. The T cell is activated and undergoes clonal expansion. Many copies of the activated T cell are produced during clonal expansion. There are two classes of MHC proteins (called MHC I and MHC II). A subgroup of T cells recognize APC that display an antigen within the groove of an MHC I protein. These T cells will activate and become cytotoxic T cells. A subgroup of T cells recognize APC that display an antigen within the groove of an MHC II protein. These T cells will activate and become helper T cells. Helper T cells are necessary for regulating B cells.

As the illness disappears, the immune reaction wanes. Activated T cells become susceptible to apoptosis. As mentioned previously, apoptosis contributes to homeostasis by regulating the number of cells present in an organ, or in this case, in the immune system. When apoptosis does not occur as it should, the potential exists for an autoimmune response (see page 153) or for T-cell cancers (i.e., lymphomas and leukemias).

Cytotoxic T cells have storage vacuoles containing perforins and storage vacuoles containing enzymes called granzymes. After a cytotoxic T cell binds to a virus-infected cell or tumor cell, it releases perforin molecules, which punch

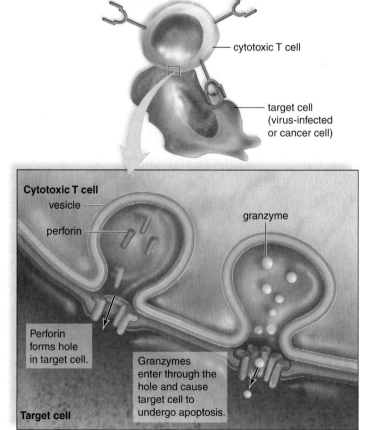

Figure 7.13 **How do cytotoxic T cells kill?** Cytotoxic T cells bind to the target cell, punch a hole in its membrane, and inject chemicals that cause the cell to die.

holes into the plasma membrane, forming a pore. Cytotoxic T cells then deliver granzymes into the pore. These cause the cell to undergo apoptosis and die. Once cytotoxic T cells have released the perforins and granzymes, they move on to the next target cell. Cytotoxic T cells are responsible for so-called **cell-mediated immunity** (Fig. 7.13).

Helper T cells regulate immunity by secreting cytokines, the chemicals that enhance the response of all types of immune cells. B cells cannot be activated without T cell help (see Fig 7.10). The human immunodeficiency virus (HIV), the virus that causes AIDS, infects helper T cells and other cells of the immune system. The virus thus inactivates the immune response and makes HIV-infected individuals susceptible to the opportunistic infections that eventually will kill them.

Notice in Figure 7.12 that a few of the clonally expanded T cells are **memory T cells.** They remain in the body and can jump-start an immune reaction to an antigen previously present in the body.

Characteristics of T Cells

- Cell-mediated immunity against virus-infected cells and cancer cells
- Produced in bone marrow, mature in thymus
- Antigen must be presented in groove of an HLA (MHC) molecule
- Cytotoxic T cells destroy nonself antigen-bearing cells
- Helper T cells secrete cytokines that control the immune response

---

Check Your Progress 7.4

1. How does innate defense differ from acquired defense?
2. What are some examples of the body's acquired defenses?
3. Which blood cells are mainly responsible for acquired defense, and how do they function?

7.5 Acquired Immunity

Immunity occurs naturally through infection or is brought about artificially by medical intervention. The two types of acquired immunity are active and passive. In **active immunity,** the individual alone produces antibodies against an antigen. In **passive immunity,** the individual is given prepared antibodies via an injection.

Active Immunity

Active immunity sometimes develops naturally after a person is infected with a pathogen. However, active immunity is often induced when a person is well to prevent future infection. Artificial exposure to an antigen through immunization can prevent future disease. The United States is committed to immunizing all children against the common types of childhood disease (Fig. 7.14).

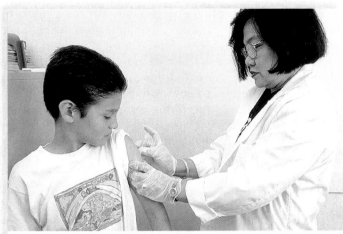

Suggested Immunization Schedule

Vaccine	Age (months)	Age (years)
Hepatitis B	Birth–18	11–12
Diphtheria, tetanus, pertussis (DTP)	2, 4, 6, 15–18	4–6
Tetanus and pertussis only		11–12
Haemophilus influenzae, type b	2, 4, 6, 12–15	
Polio	2, 4, 6–18	4–6, 11–12
Pneumococcal	2, 4, 6, 12–15	
Measles, mumps, rubella (MMR)	12–15	4–6, 11–12
Varicella (chicken pox)	12–18	11–12
Meningococcal		11–12
Hepatitis A (in selected areas)	12–23	11–12 2–18

a.

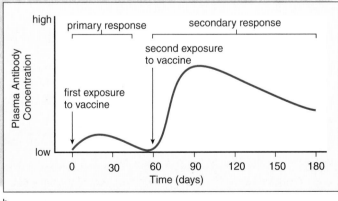

b.

Figure 7.14 **How do immunizations cause active immunity?**
a. Suggested immunization schedule for infants and children. **b.** During immunization, the primary response, after the first exposure to a vaccine, is minimal, but the secondary response, which may occur after the second exposure, shows a dramatic rise in the amount of antibody present in plasma.

Immunization involves the use of **vaccines,** substances that contain an antigen to which the immune system responds. Traditionally, vaccines are the pathogens themselves, or their products, that have been treated to be no longer virulent (able to cause disease). Today, it is possible to genetically engineer bacteria to mass-produce a protein from pathogens, and this protein can be used as a vaccine. This method has now produced a vaccine against hepatitis B, a viral-induced disease.

After a vaccine is given, it is possible to follow an immune response by determining the amount of antibody present in a sample of plasma—this is called the **antibody titer.** After the first exposure to a vaccine, a primary response occurs. For a period of several days, no antibodies are present. Then the titer rises slowly, levels off, and gradually declines as the antibodies bind to the antigen or simply break down (Fig. 7.14). After a second exposure to the vaccine, a secondary response is expected. The titer rises rapidly to a level much greater than before. Then it slowly declines. The second exposure is called a "booster" because it boosts the antibody titer to a high level. The high antibody titer now is expected to help prevent disease symptoms, even if the individual is exposed to the disease-causing antigen.

Active immunity depends upon the presence of memory B cells and memory T cells capable of responding to lower doses of antigen. Active immunity is usually long-lasting, although a booster may be required after many years.

Passive Immunity

Passive immunity occurs when an individual is given prepared antibodies or immune cells to combat a disease. These antibodies are not produced by the individual's plasma cells, so passive immunity is temporary. For example, newborn infants are passively immune to some diseases because IgG antibodies have crossed the placenta from the mother's blood (Fig. 7.15a). These antibodies soon disappear, and within a few months, infants become more susceptible to infections. Breast-feeding prolongs the natural passive immunity an infant receives from the mother because IgG and IgA antibodies are present in the mother's milk (Fig. 7.15b).

Even though passive immunity does not last, it is sometimes used to prevent illness in a patient who has been unexpectedly exposed to an infectious disease. Usually, the patient receives a **gamma globulin** injection (serum that contains antibodies), perhaps taken from individuals who have recovered from the illness (Fig. 7.15c). For example, a health-care worker who suffers an accidental needle stick may come into contact with the blood from a patient infected

CASE STUDY TREATMENT AND RELEASE

The doctor explained Nick's treatment to his parents. Nick's mother looked very nervous and held onto Nick's hand. "We are going to give Nick corticosteroids through his IV," the doctor said. "Nick got into trouble because his immune system went haywire when bee venom entered his circulation. Corticosteroids will help calm down his immune system."

"We'll also continue to give him extra fluids," the doctor continued. "That'll help bring his blood pressure back up. If he's stable later tonight, we'll let him go home." His parents both nodded.

Then the doctor turned to Nick. "Nick, you should go and see an allergist as soon as possible. Allergy shots can reduce or even eliminate insect venom allergies. There are also other treatments that will make a reaction less severe, too."

"You'll need to carry that Epi-Pen® with you all the time. But you already know that, right?" he smiled. "It's a good thing you

were wearing that MedicAlert® bracelet, too. It probably saved your life."

Six hours later, Nick was allowed to leave the emergency room. His parents took him home with them. He was still weak, but the swelling was gone, his blood pressure was normal, and he was breathing without difficulty. He called Brent from the car. "Hey, Brent, I guess I gave you a little scare."

"A little scare!" Brent exclaimed, "I thought you were going to die. Don't do that to me!"

"Sorry. I'm going home for a few days. I need to see the allergist," he explained. "Tell the guys I'm fine. Sorry I ruined the game."

Brent laughed. "Don't worry about the game. I'll let the guys know. Just get better, and we'll see you soon. And you need to show me how to use that Epi-Pen® thing in case this happens again. I'm just glad you're OK."

a. Antibodies (IgG) cross the placenta.

b. Antibodies (IgG, IgA) are secreted into breast milk.

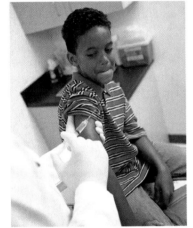

c. Antibodies can be injected by a physician.

Figure 7.15 **How are antibodies obtained in passive immunity?** During passive immunity, antibodies are received **(a)** by crossing the placenta, **(b)** in breast milk, or **(c)** by injection. The body is not producing the antibodies, so passive immunity is short-lived.

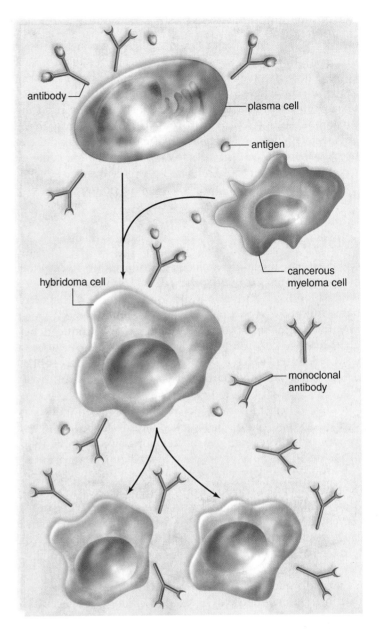

Figure 7.16 **How are monoclonal antibodies produced?** Plasma cells of the same type (derived from immunized mice) are fused with myeloma (cancerous) cells, producing hybridoma cells that are "immortal." Hybridoma cells divide and continue to produce the same type of antibody, called monoclonal antibodies.

with hepatitis virus. Immediate treatment with a gamma-globulin injection (along with simultaneous vaccination against the virus) can typically prevent the viruses from causing infection.

Monoclonal Antibodies

Every plasma cell derived from the same B cell secretes antibodies against a specific antigen. These are **monoclonal antibodies** because all of them are the same type and because they are produced by plasma cells derived from the same B cell. One method of producing monoclonal antibodies in vitro (outside the body in glassware) is depicted in Figure 7.16. B lymphocytes are removed from an animal (today, usually mice are used) and are exposed to a particular antigen. The resulting plasma cells are fused with myeloma cells (malignant plasma cells that live and divide

indefinitely; they are immortal cells). The fused cells are called hybridomas—*hybrid-* because they result from the fusion of two different cells, and *-oma* because one of the cells is a cancer cell.

At present, monoclonal antibodies are being used for quick and certain diagnosis of various conditions. For example, the hormone hCG (human chorionic gonadotropin) is present in the urine of a pregnant woman. A monoclonal antibody can be used to detect this hormone. If it is present, the woman knows she is pregnant. Monoclonal antibodies are also used to identify infections. And because they can distinguish between cancerous and normal tissue cells, they are used to carry radioisotopes or toxic drugs to tumors, which can then be selectively destroyed. Herceptin is a monoclonal antibody used in the treatment of breast cancer. It binds to a protein receptor on breast cancer cells and prevents the cancer cells from dividing as quickly. Antibodies that bind to cancer cells also can activate complement and can increase phagocytosis by macrophages and neutrophils.

Cytokines and Immunity

Cytokines are signaling molecules produced by T lymphocytes, macrophages, and other cells. Cytokines regulate white blood cell formation and/or function, so they are being investigated as a possible adjunct (or added on) therapy for cancer and AIDS. Both interferon, produced by virus-infected cells, and **interleukins,** produced by various white blood cells, have been used as immunotherapeutic drugs. These are used particularly to enhance the ability of the individual's T cells to fight cancer.

Most cancer cells carry an altered protein on their cell surface, so they should be attacked and destroyed by cytotoxic T cells. Whenever cancer develops, it is possible that cytotoxic T cells have not been activated. In that case, cytokines might awaken the immune system and lead to the destruction of the cancer. In one technique, researchers withdrew T cells from the patient, and presented cancer cell antigens to them. They then activated the cells by culturing them in the presence of an interleukin. The T cells were reinjected into the patient, who was given doses of interleukin to maintain the killer activity of the T cells.

Scientists actively engaged in interleukin research believe that interleukins soon will be used as adjuncts for vaccines. They may be used for the treatment of chronic infectious disease and perhaps for the treatment of cancer. Interleukin antagonists also may prove helpful in preventing skin and organ rejection, autoimmune diseases, and allergies.

> ### Check Your Progress 7.5
> 1.   a. What is acquired immunity? b. What are the two different types of acquired immunity? c. What are some examples of each?
> 2.   What are two types of immune therapies that can assist passive immunity?

7.6 Hypersensitivity Reactions

Sometimes, the immune system responds in a manner that harms the body, as when individuals develop allergies,

receive an incompatible blood type, suffer tissue rejection, or have an autoimmune disease.

Allergies

Allergies are hypersensitivities to substances, such as pollen, food, or animal hair, that ordinarily would do no harm to the body. The response to these antigens, called **allergens,** usually includes some degree of tissue damage.

An **immediate allergic response** can occur within seconds of contact with the antigen. The response is caused by antibodies known as IgE (see Table 7.2). IgE antibodies are attached to receptors on the plasma membrane of mast cells in the tissues and also to basophils in the blood. When an allergen attaches to the IgE antibodies on these cells, the cells release histamine and other substances that bring about the allergic symptoms. When pollen is an allergen, histamine stimulates the mucous membranes of the nose and eyes to release fluid. This causes the runny nose and watery eyes typical of **hay fever.** If a person has asthma, the airways leading to the lungs constrict, resulting in difficult breathing accompanied by wheezing. When food contains an allergen, nausea, vomiting, and diarrhea often result.

Anaphylactic shock is an immediate allergic response that occurs because the allergen has entered the bloodstream. Bee stings and penicillin shots are known to cause this reaction because both inject the allergen into the blood. Anaphylactic shock is characterized by a sudden and life-threatening drop in blood pressure due to increased permeability of the capillaries by histamine. Taking epinephrine can counteract this reaction until medical help is available.

People with allergies produce ten times more IgE than those people without allergies. A new treatment using injections of monoclonal IgG antibodies for IgEs is being tested in individuals with severe food allergies. More routinely, injections of the allergen are given so that the body will build up high quantities of IgG antibodies. The hope is that IgG antibodies will combine with allergens received from the environment before they have a chance to reach the IgE antibodies located in the membrane of mast cells and basophils.

A **delayed allergic response** is initiated by memory T cells at the site of allergen contact in the body. The allergic response is regulated by the cytokines secreted by both T cells and macrophages. A classic example of a delayed allergic response is the skin test for tuberculosis (TB). When the test result is positive, the tissue where the antigen was injected becomes red and hardened. This shows that there was prior exposure to the bacterium that causes TB. Contact dermatitis, which occurs when a person is allergic to poison ivy, jewelry, cosmetics, and many other substances that touch the skin, is also an example of a delayed allergic response.

Other Immune Problems

Certain organs, such as the skin, the heart, and the kidneys, could be transplanted easily from one person to another if the body did not attempt to reject them. Rejection of transplanted tissue results because the recipient's immune system recognizes that the transplanted tissue is not "self." Cytotoxic T cells respond by attacking the cells of the transplanted tissue.

Organ rejection can be controlled by carefully selecting the organ to be transplanted and administering **immunosuppressive** drugs. It is best if the transplanted organ has the same type of MHC antigens as those of the recipient, because cytotoxic T cells recognize foreign MHC antigens. Two well-known immunosuppressive drugs, cyclosporine and tacrolimus, act by inhibiting the production of certain T-cell cytokines.

Xenotransplantation is the use of animal organs instead of human organs in transplant patients. Scientists have chosen to use the pig because animal husbandry has long included the raising of pigs as a meat source, and pigs are prolific. Genetic engineering can make pig organs less antigenic. The ultimate goal is to make pig organs as widely accepted as type O blood.

An alternative to xenotransplantation exists because tissue engineering is making organs in the laboratory. Scientists have transplanted lab-grown urinary bladders into human patients. They hope that production of organs lacking HLA antigens will one day do away with the problem of rejection.

Immune system disorders occur when a patient has an immune deficiency, or when the immune system attacks the body's own cells.

When a person has an immune deficiency, the immune system is unable to protect the body against disease. Infrequently, a child may be born with an impaired immune system. For example, in **severe combined immunodeficiency disorder (SCID),** both antibody- and cell-mediated immunity are lacking or inadequate. Without treatment, even common infections can be fatal. Bone marrow transplants and gene therapy have been successful in SCID patients. Acquired immune deficiencies can be caused by infections, chemical exposure, or radiation. Acquired immunodeficiency syndrome (AIDS) is a result of an infection with the human immunodeficiency virus (HIV). As a result of a weakened immune system, AIDS patients show a greater susceptibility to infections and also have a higher risk of cancer (see "Infectious Diseases Supplement").

When cytotoxic T cells or antibodies mistakenly attack the body's own cells, the person has an **autoimmune disease.** The exact cause of autoimmune diseases is not known, although it appears to involve both genetic and environmental factors. People with certain HLA antigens are more

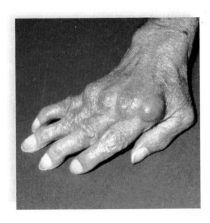

Figure 7.17   What causes the symptoms of rheumatoid arthritis? Rheumatoid arthritis is due to recurring inflammation in skeletal joints. Complement proteins, T cells, and B cells all participate in deterioration of the joints, which eventually become immobile.

susceptible. Women are more likely than men to develop an autoimmune disease.

Sometimes the autoimmune disease follows an infection. For example, in **rheumatic fever,** antibodies induced by a streptococcal (bacterial) infection of the throat also react with heart muscle. This causes an inflammatory response with damage to the heart muscle and valves. **Rheumatoid arthritis** (Fig. 7.17) is an autoimmune disease in which the joints are chronically inflamed. It is thought that antigen-antibody complexes, complement, neutrophils, activated T cells, and macrophages are all involved in the destruction of cartilage in the joints. A person with **systemic lupus erythematosus (SLE,** commonly just called lupus) has various symptoms including a facial rash, fever, and joint pain. In these patients, damage to the central nervous system, heart, and kidneys can be fatal. SLE patients produce high levels of anti-DNA antibodies. All human cells (except red blood cells) contain DNA, so the symptoms of lupus interfere with tissues throughout the body. **Myasthenia gravis** develops when antibodies attach to and interfere with the function of neuromuscular junctions. The result is severe muscle weakness, eventually resulting in death from respiratory failure. In **multiple sclerosis (MS),** T cells attack the myelin sheath covering nerve fibers, causing CNS dysfunction, double vision, and muscular weakness. Treatments for autoimmune diseases usually involve drugs designed to decrease the immune response.

> ### Check Your Progress 7.6
> 1. **What types of complications and disorders are associated with the functioning of the immune system? Explain each of these.**

Summarizing the Concepts

7.1   Microbes, Pathogens, and You

Microbes perform valuable services, but they also cause disease.

Bacteria

- are prokaryotic cells that live and reproduce independently of host cells; and
- cause disease by multiplying in hosts and also by producing toxins.

Viruses

- are noncellular particles, consisting of a protein coat and a nucleic acid core; and
- take over the machinery of the host to reproduce.

7.2   The Lymphatic System

The lymphatic system consists of lymphatic vessels that return lymph to cardiovascular veins.

The primary lymphatic organs are

- the red bone marrow, where all blood cells are made and the B lymphocytes mature; and
- the thymus gland, where T lymphocytes mature.

The secondary lymphatic organs are

- the spleen; lymph nodes; and other organs containing lymphoid tissue, such as the tonsils, Peyer's patches, and the appendix. Blood is cleansed of pathogens and debris in the spleen. Lymph is cleansed of pathogens and debris in the nodes.

7.3   Innate Defenses

Immunity involves innate and acquired defenses. The innate defenses include

- barriers to entry;
- the inflammatory reaction, which involves the phagocytic neutrophils and macrophages; and
- protective proteins.

7.4   Acquired Defenses

Acquired defenses require B cells and T cells, also called B lymphocytes and T lymphocytes.

B Cells and Antibody-Mediated Immunity

- Activated B cells undergo clonal selection with production of plasma cells and memory B cells, after their B-cell receptor combines with a specific antigen.
- Plasma cells secrete antibodies and eventually undergo apoptosis. Plasma cells are responsible for antibody-mediated immunity.
- An antibody is usually a Y-shaped molecule that has two binding sites for a specific antigen.
- Memory B cells remain in the body and produce antibodies if the same antigen enters the body at a later date.

T Cells and Cell-Mediated Immunity

- For a T cell to recognize an antigen, the antigen must be presented by an antigen-presenting macrophage, along with an HLA (human leukocyte antigen).

- Activated T cells undergo clonal expansion until the illness has been stemmed. Then, most of the activated T cells undergo apoptosis. A few cells remain, however, as memory T cells.
- The two main types of T cells are cytotoxic T cells and helper T cells.
- Cytotoxic T cells kill virus-infected cells or cancer cells on contact because they bear a nonself protein.
- Helper T cells produce cytokines and stimulate other immune cells.

7.5   Acquired Immunity

- Active (long-lived) immunity can be induced by vaccines when a person is well and in no immediate danger of contracting an infectious disease. Active immunity depends upon the presence of memory cells in the body.
- Passive immunity is needed when an individual is in immediate danger of succumbing to an infectious disease. Passive immunity is short-lived because the antibodies are administered to, and not made by, the individual.
- Monoclonal antibodies, produced by the same plasma cell, have various functions, from detecting infections to treating cancer.
- Cytokines, including interferon, are a form of passive immunity used to treat AIDS and to promote the body's ability to recover from cancer.

7.6   Hypersensitivity Reactions

Allergic responses occur when the immune system reacts vigorously to substances not normally recognized as foreign.

- Immediate allergic responses, usually consisting of coldlike symptoms, are due to the activity of antibodies.
- Delayed allergic responses, such as contact dermatitis, are due to the activity of T cells.
- Tissue rejection occurs when the immune system recognizes a tissue as foreign.
- Immune deficiencies can be inherited or can be caused by infection, chemical exposure, or radiation.
- Autoimmune disorders occur when the immune system reacts to tissues/organs of the individual as if they were foreign.

Understanding Key Terms

active immunity 149
allergen 152
allergy 152
anaphylactic shock 152
antibody-mediated
   immunity 145
antibody titer 150
antigen 144
antigen-presenting cell
   (APC) 147
apoptosis 145
autoimmune disease 153
bacillus 136
bacteria 136
B cells (B lymphocytes) 141
B-cell receptor (BCR) 145
capsule 136
cell-mediated immunity 149

clonal selection model 145
coccus 136
complement system 144
cytokine 143
cytotoxic T cell 148
delayed allergic response 152
fimbriae 136
flagellum 136
gamma globulin 150
hay fever 152
helper T cell 149
histamine 143
HLA (human leukocyte
   antigen) 147
immediate allergic
   response 152
immunity 142
immunization 150

immunosuppressive 153
inflammatory response 142
interferon 144
interleukin 152
lymph 140
lymphatic nodule 142
lymphatic organ 141
lymph node 141
lysozyme 142
macrophage 143
mad cow disease 138
mast cell 143
membrane attack complex 144
memory T cell 149
MHC (major histocompatibility complex) 147
monoclonal antibody 151
multiple sclerosis (MS) 153
myasthenia gravis 153
passive immunity 149
pathogen 136
Peyer's patches 142
pilus 136

plasma cell 145
plasmid 136
prion 138
pus 143
red bone marrow 141
rheumatic fever 153
rheumatoid arthritis 153
spirillum 136
severe combined immunodeficiency disease (SCID) 153
spleen 141
systemic lupus erythematosus (SLE) 153
T cells (T lymphocytes) 141
T-cell receptor (TCR) 147
thymus gland 141
tonsils 142
toxin 137
vaccine 150
virus 137
xenotransplantation 153

Match the key terms to these definitions.

a. _____ Antigens prepared in such a way that they can promote active immunity without causing disease.

b. _____ Series of proteins in plasma that form an innate defense mechanism against pathogen invasion.

c. _____ Foreign substance, usually a protein, that stimulates the immune system to react, such as by producing antibodies.

d. _____ Process of programmed cell death involving a cascade of specific cellular events leading to the death and destruction of the cell.

e. _____ Type of lymphocyte that matures in the thymus gland and exists in three varieties, one of which kills antigen-bearing cells outright.

Testing Your Knowledge of the Concepts

1. Describe the basic characteristics of bacteria. Explain how five particular features contribute to the ability of bacteria to cause disease. (pages 136–37)

2. What is the structure of a virus? Is a virus living? Explain how a virus is able to reproduce. (pages 137–38)

3. What are prions, and how do they cause disease? (pages 138–39)

4. Why are the red bone marrow and the thymus gland termed primary lymphatic organs? (page 141)

5. In what ways are the spleen and lymph nodes similar, and in what ways are they different? (pages 141–42)

6. How do innate defenses differ from acquired defenses? (pages 142, 144)

7. What is the first line of defense? Describe several methods of protection. (page 142)

8. What is the second line of defense? Describe the various cells and chemicals and their roles in protecting the body. (pages 142–44)

9. What are the four signs of inflammation, and what causes them? (pages 142–43)

10. What two types of cells are involved in providing acquired defenses against pathogens? What type of immunity does each provide? (pages 144–45)

11. What is the clonal selection model as it applies to B cells? What becomes of the clones that are produced? (page 145)

12. Describe the structure of an antibody. What are the five main classes, where are they found, and what are their functions? (pages 146–47)

13. Explain how T cells recognize an antigen. What are the types of T cells, and how do they function in immunity? (pages 147–49)

14. How is active immunity achieved? How is passive immunity achieved? (pages 149–51)

15. Discuss the production of monoclonal antibodies and their applications. (pages 151–52)

16. Discuss allergies, tissue rejection, and autoimmune diseases as they relate to the immune system. (pages 152–53)

17. Which of the following is a function of the spleen?
    a. produces T cells
    b. removes worn-out red blood cells
    c. produces immunoglobulins
    d. produces macrophages
    e. regulates the immune system

18. Which of the following is a function of the thymus gland?
    a. production of red blood cells
    b. secretion of antibodies
    c. production and maintenance of stem cells
    d. site for the maturation of T lymphocytes

For questions 19–23, match the lymphatic organs in the key to the description of its location.

Key:
    a. lymph nodes
    b. Peyer's patches
    c. spleen
    d. thymus gland
    e. tonsils

19. Arranged in a ring around the pharynx

20. In the intestinal wall

21. Occur along lymphatic vessels

22. Upper left abdominal cavity, posterior to the stomach

23. Superior to the heart, posterior to the sternum

24. Which of the following is a function of the secondary lymphatic organs?
    a. transport of lymph
    b. clonal selection of B cells
    c. located where lymphocytes encounter antigens
    d. All of these are correct.

25. Defense mechanisms that function to protect the body against many infectious agents are called
    a. acquired.
    b. innate.
    c. barriers to entry.
    d. immunity.

26. Which of the following is most directly responsible for the increase in capillary permeability during the inflammatory reaction?
    a. pain
    b. white blood cells
    c. histamine
    d. tissue damage

27. Which of the following is not a goal of the inflammatory reaction?
    a. bring more oxygen to damaged tissues
    b. decrease blood loss from a wound
    c. decrease the number of white blood cells in the damaged tissues
    d. prevent entry of pathogens into damaged tissues

28. Which of the following is not correct concerning interferon?
    a. Interferon is a protective protein.
    b. Virus-infected cells produce interferon.
    c. Interferon has no effect on viruses.
    d. Interferon can be used to treat certain viral infections.

29. Which one of these does not pertain to B cells?
    a. have passed through the thymus
    b. have specific receptors
    c. are responsible for antibody-mediated immunity
    d. synthesize and liberate antibodies

30. Which of these pertain(s) to T cells?
    a. have specific receptors
    b. are of more than one type
    c. are responsible for cell-mediated immunity
    d. stimulate antibody production by B cells
    e. All of these are correct.

31. During a secondary immune response,
    a. antibodies are made quickly and in great amounts.
    b. antibody production lasts longer than in a primary response.
    c. B cells become plasma cells.
    d. All of these are correct.

32. Active immunity may be produced by
    a. having a disease.
    b. receiving a vaccine.
    c. receiving gamma globulin injections.
    d. Both a and b are correct.
    e. Both b and c are correct.

33. Which of the following is not an innate body defense?
    a. complement
    b. the skin and mucous membranes
    c. antibodies
    d. lysozyme

Thinking Critically About the Concepts

An allergic response is an overreaction of the immune system in response to an antigen. Such responses always require a prior exposure to the antigen. The reaction can be immediate (within seconds to minutes) or delayed (within hours). Nick Williamson experienced an immediate response to a bee sting. His symptoms were present within seconds. Insect venom reactions often involve the development of hives and itching. Asthmalike symptoms include shortness of breath and wheezing. Decreased blood pressure will eventually cause loss of consciousness. This immediate, severe allergic effect, called anaphylaxis, can be fatal if untreated. In addition to insect venom, food allergens such as those in milk, peanut butter, and shellfish may also elicit life-threatening symptoms. In these cases, IgE antibodies, histamine, and other inflammatory chemicals are the culprits involved in making the reaction so severe.

1. How are B cells involved in immediate allergic responses?

2. Why did Nick require a prior exposure to bee venom to develop an allergic response?

3. The doctor suggested that Nick visit his allergist for treatment that will lessen the severity of Nick's allergic response. What might that treatment involve?

4. Lymphatic vessels transport tissue fluid back to the cardiovascular system. The pressure in the lymphatic vessels is low. Which blood vessels are lymphatic vessels most similar to with regard to structure?

5. Think of an analogy (something you're already familiar with) for the barrier defenses like your skin and mucous membranes.

6. Someone bitten by a poisonous snake should be given some antivenom (antibodies) to prevent them from dying. If they are bitten by the same type of snake three years after the initial bite, will they have immunity to the venom, or should they get another shot of antivenom? Justify your response with an explanation of the type of immunity someone gains from a shot of antibodies.

CHAPTER 8

Digestive System and Nutrition

Monica and her best friend Irene were checking out the latest fashions at the mall. "You have to be tall and a twig to wear these skinny jeans," Monica observed as she held a pair of jeans against her body. "What are those of us who are short and round supposed to wear?" she continued.

Irene poked her friend and joked, "It's probably too late to do anything about being short, but if you'd lay off the potato chips you might not be so round."

"Thanks a lot!" retorted Monica. "That's a nice thing for my best friend to say!"

That night, her mother heard Monica crying in her room. She entered the room cautiously. "Monica," her mother began softly, "what's wrong? Did you and Irene have a fight today?"

Between sniffs, Monica told her mom about the size 8 jeans. Monica sobbed. "Irene said that if I just quit eating potato chips, I wouldn't be so round and fat! I am so tired of being overweight!" Monica cried. "I'm still huge after several diets, fat camps and hypnosis. What else can I do?"

"Maybe it's time to make an appointment with your primary care physician," her mother replied. "She might have something else to suggest. I'll call her tomorrow."

"I'm really glad you came in," Dr. Emory commented several days later. "Your weight is not just out of the normal range, it has become dangerous."

She continued, "We use body mass index, BMI for short, to determine a normal weight range. You're five feet two inches and 228 pounds, which makes your BMI 41.7. People with a BMI of 40 or more are considered morbidly obese. That means you'll eventually die from complications of being overweight such as heart disease, type 2 diabetes and certain cancers. I know this will sound scary, but you need to consider bariatric surgery. I want to make an appointment for you to meet with Dr. Robert Hall who specializes in several bariatric procedures. We'll all be happier if we can get your BMI back into the normal range," Dr. Emory reassured Monica and her mother.

8.1 Overview of Digestion

The gastrointestinal tract (GI tract) contains a number of organs that carry out several processes as they digest food. The wall of the tract typically has four layers and is modified in each of the organs.

8.2 First Part of the Digestive Tract

The mouth receives food where it is chewed. Chemical digestion begins before food enters the pharynx and the esophagus, which take it to the stomach.

8.3 The Stomach and Small Intestine

The stomach stores food and continues chemical digestion, which is completed in the small intestine. The products of digestion are absorbed by the small intestine into the blood or into the lymph.

8.4 Three Accessory Organs and Regulation of Secretions

The pancreas produces pancreatic juice, which is sent to the small intestine for the chemical digestion of food. The pancreatic hormone insulin causes cells, including liver cells, to take up glucose from the blood. The liver is a major metabolic organ, which stores glucose as glycogen and produces bile. Bile is stored in the gallbladder before it is sent to the small intestine for emulsification of fats.

8.5 The Large Intestine and Defecation

The large intestine absorbs water and vitamins produced by bacteria that normally reside in the large intestine, and it carries out defecation.

8.6 Nutrition and Weight Control

Planning nutritious meals and snacks to control one's weight involves making healthy and informed food choices.

8.1 Overview of Digestion

The organs of the digestive system are located within a tube called the gastrointestinal tract (GI tract), depicted in Figure 8.1. Food, whether it is a salad or a cheeseburger, consists of the organic macromolecules you studied in Chapter 2: carbohydrates, fats, and proteins. These molecules are too big to cross plasma membranes. The purpose of digestion is to **hydrolyze** these macromolecules to their subunit molecules. The subunit molecules, namely sugars, amino acids, fatty acids, and glycerol, can cross plasma membranes. Our food also contains water, salts, vitamins, and minerals that help the body function normally. The nutrients made available by digestive processes are carried by the blood to our cells. The following processes are necessary to the digestive process.

- **Ingestion** occurs when the mouth takes in food. Ingestion can be associated with our diet. The expression "You are what you eat" recognizes that our diet is very important to our health. By learning about and developing good nutritional habits, you increase your likelihood of enjoying a longer, more active and productive life. Unfortunately, poor diet and lack of physical activity now rival smoking as a major cause of preventable death in the United States.

- **Digestion** can be mechanical or chemical. Mechanical digestion occurs when large pieces of food are divided into smaller pieces that can be acted on by the digestive enzymes. Cutting up our food prior to ingestion aids mechanical digestion. Mechanical digestion occurs primarily in the mouth and stomach.

  During chemical digestion digestive enzymes hydrolyze your food's macromolecules into absorbable subunits. Chemical digestion begins in the mouth, continues in the stomach, and is completed in the small intestine.

- **Movement** of GI tract contents along the digestive tract is very important for the tract to fulfill its other functions. For example, food must be passed along from one organ to the next, and indigestible remains must be expelled.

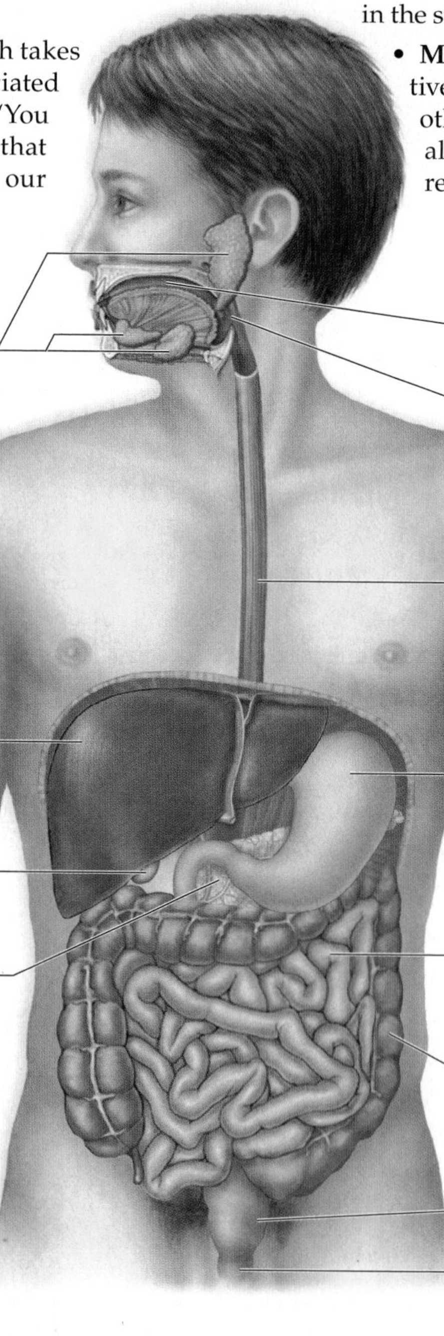

Accessory organs

Salivary glands
secrete saliva: contains digestive enzyme for carbohydrates

Liver
major metabolic organ: processes and stores nutrients; produces bile for emulsification of fats

Gallbladder
stores bile from liver; sends it to the small intestine

Pancreas
produces pancreatic juice: contains digestive enzymes, and sends it to the small intestine; produces insulin and secretes it into the blood after eating

Digestive tract organs

Mouth
teeth chew food; tongue tastes and pushes food for chewing and swallowing

Pharynx
passageway where food is swallowed

Esophagus
passageway where peristalsis pushes food to stomach

Stomach
secretes acid and digestive enzyme for protein; churns, mixing food with secretions, and sends chyme to small intestine

Small intestine
mixes chyme with digestive enzymes for final breakdown; absorbs nutrient molecules into body; secretes digestive hormones into blood

Large intestine
absorbs water and salt to form feces

Rectum
stores and regulates elimination of feces

Anus

Figure 8.1   How do the pancreas and liver assist digestion?
Food is digested within the organs of the GI tract. The liver and pancreas are accessory organs and produce chemicals that assist with digestion.

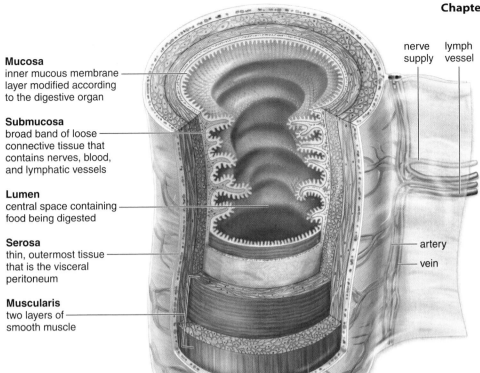

Mucosa
inner mucous membrane layer modified according to the digestive organ

Submucosa
broad band of loose connective tissue that contains nerves, blood, and lymphatic vessels

Lumen
central space containing food being digested

Serosa
thin, outermost tissue that is the visceral peritoneum

Muscularis
two layers of smooth muscle

nerve supply   lymph vessel

artery

vein

Figure 8.2 **What is the structure of the gastrointestinal tract wall?**
The wall of the gastrointestinal tract contains the four layers noted.

- **Absorption** occurs as subunit molecules produced by chemical digestion (i.e., nutrients) cross the wall of the GI tract and enter the cells lining the tract. From there, the nutrients enter the blood for delivery to the cells.

- **Elimination:** Molecules that cannot be digested need to be eliminated from the body. The removal of indigestible wastes through the anus is termed defecation.

Wall of the Digestive Tract

We can compare the GI tract to a garden hose that has a beginning (mouth) and an end (anus). The **lumen** is the central space that contains food being digested. The wall of the GI tract has four layers (Fig. 8.2). Each layer can be associated with a particular disorder.

The inner layer of the wall next to the lumen is called the **mucosa.** You would probably guess that the mucosa produces mucus, and you would be right. This mucus protects the other layers of the tract from the digestive enzymes inside the lumen. Glands in the mucosa of the mouth, stomach, and small intestine release digestive enzymes. Hydrochloric acid is produced by glands in the mucosa of the stomach.

Diverticulosis is a condition in which portions of the mucosa literally have pushed through the other layers and formed pouches, where food can collect. The pouches can be likened to an inner tube that pokes through weak places in a tire. When the pouches become infected or inflamed, the condition is called *diverticulitis*. This happens in 10–25% of people with diverticulosis.

The second layer in the GI wall is called the **submucosa.** The submucosal layer is a broad band of loose connective tissue that contains blood vessels, lymphatic vessels, and nerves. These are the vessels that will carry the nutrients absorbed by the mucosa. Lymph nodules, called Peyer's patches, are also in the submucosa. Like the tonsils, they help protect us from disease. The submucosa contains blood vessels, so it can be the site of an inflammatory response (see page 143) that leads to *inflammatory bowel disease (IBD)*. Chronic diarrhea, abdominal pain, fever, and weight loss are symptoms of IBD.

The third layer is termed the **muscularis,** and it contains two layers of smooth muscle. The inner, circular layer encircles the tract. The outer, longitudinal layer lies in the same direction as the tract. The contraction of these muscles, under nervous and hormonal control, accounts for movement of digested food from the esophagus to the anus. The muscularis can be associated with *irritable bowel syndrome (IBS)*, in which contractions of the wall cause abdominal pain, constipation, and/or diarrhea. The underlying cause of IBS is not known, although some suggest stress as an underlying cause.

The fourth layer of the tract is the **serosa** (serous membrane layer), which secretes a lubricating fluid. The serosa is a part of the peritoneum, the internal lining of the abdominal cavity. The **appendix** is a worm-shaped blind tube projecting from the first part of the large intestine on the right side of the abdomen. An inflamed appendix (*appendicitis*) has to be removed. Should the appendix burst, the result can be **peritonitis,** a life-threatening infection of the peritoneum.

> **Check Your Progress 8.1**
>
> 1. **Name and describe the processes that occur during the digestive process.**
> 2. **What are the four layers of the GI tract? Associate an illness with each of the layers.**

8.2 First Part of the Digestive Tract

The mouth, the pharynx, and the esophagus are in the first part of the GI tract.

The Mouth

The mouth receives food and begins the process of mechanical and chemical digestion. The mouth is bounded externally by the lips and cheeks. The lips extend from the base of the nose to the start of the chin. The red portion of the lips is poorly keratinized, and this allows blood to show through.

The roof of the mouth separates the nasal cavities from the oral cavity. The roof has two parts: an anterior (toward the front) **hard palate** and a posterior (toward the back) **soft palate** (Fig. 8.3a). The hard palate contains several bones, but the soft palate is composed entirely of muscle. The soft palate ends in a finger-shaped projection called the uvula. The tonsils are also in the back of the mouth on either side of the tongue. Tonsils are lymphatic tissue and help protect us from disease. In the nasopharynx, there is a single pharyngeal tonsil (by convention, it is called the adenoids).

Three pairs of **salivary glands** send secretions (saliva) by way of ducts to the mouth (see Fig. 8.1). One pair of salivary glands lies at the side of the face immediately below and in front of the ears. The ducts of these salivary glands open on the inner surface of the cheek just above the second upper molar. This pair of glands swells when a person has the mumps, a viral disease. The measles, mumps, and rubella (MMR) vaccination you likely had as a child prevents the mumps. Another pair of salivary glands lies beneath the tongue, and still another pair lies beneath the floor of the oral cavity. The ducts from these salivary glands open under the tongue. You can locate the openings if you use your tongue to feel for small flaps on the inside of your cheek and under your tongue. Saliva is a solution of mucus and water. Saliva also contains **salivary amylase,** an enzyme that begins the chemical digestion of starch, as well as bicarbonate.

The Teeth and Tongue

Mechanical digestion occurs when our teeth chew food into pieces convenient for swallowing. During the first two years of life, the 20 smaller deciduous, or baby, teeth appear. These are eventually replaced by 32 adult teeth (see Fig. 8.3a). The third pair of molars, called the wisdom teeth, sometimes fail to erupt. If they push on the other teeth and/or cause pain, they can be removed by a dentist or oral surgeon. Each tooth has two main divisions: a crown and a root (Fig. 8.3b). The crown has a layer of enamel, an extremely hard outer covering of calcium compounds; dentin, a thick layer of bonelike material; and an inner pulp, which contains the nerves and the blood vessels. Dentin and pulp are also found in the root.

Tooth decay, called **dental caries,** or cavities, occurs when bacteria within the mouth metabolize sugar. Acids produced during this metabolism erode the teeth. Tooth decay can be painful when it is severe enough to reach the nerves of the inner pulp. Two measures can prevent tooth decay: eating a limited amount of sweets and daily brushing and flossing of teeth. Fluoride treatments, particularly in children, can make the enamel stronger and more resistant to decay. Gum disease, known to be linked to cardiovascular disease, is more apt to occur with aging. Inflammation of the gums (gingivitis) can spread to the periodontal membrane, which lines the tooth socket. A person then has **periodontitis,** characterized by a loss of bone and loosening of the teeth. Extensive dental work may be required or teeth will be completely lost. Stimulation of the gums in a manner advised by your dentist is helpful in controlling this condition. Medications are also available.

Figure 8.3   What are the functions of the different types of teeth? What happens during a root canal?

a. The chisel-shaped incisors bite; the pointed canines tear; the fairly flat premolars grind; and the flattened molars crush food.
b. Longitudinal section of a tooth. The crown is the portion that projects above the gum line and can be replaced by a dentist if damaged. When a "root canal" is done, the nerves are removed. When the periodontal membrane is inflamed, the teeth can loosen.

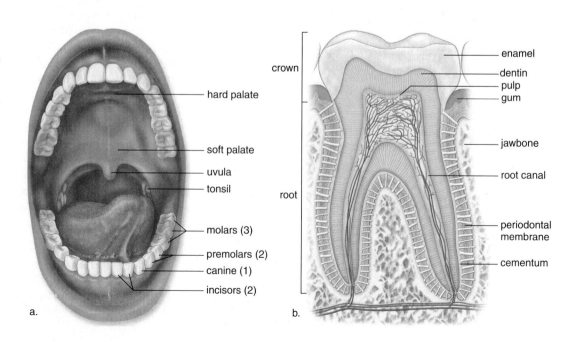

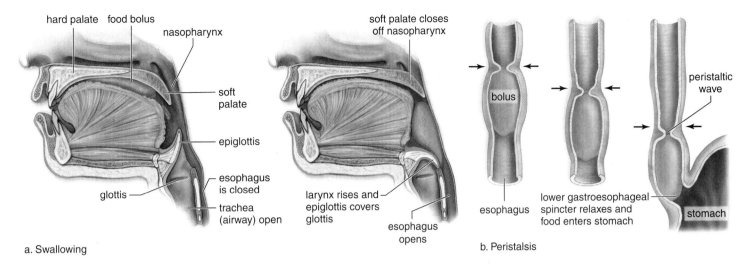

a. Swallowing

b. Peristalsis

Figure 8.4 **What happens to food during swallowing?**
a. When food is swallowed, the tongue pushes a bolus of food up against the soft palate *(left)*. Then, the soft palate closes off the nasal cavities, and the epiglottis closes off the larynx so the bolus of food enters the esophagus *(right)*. **b.** Peristalsis moves food through a sphincter into the stomach.

The tongue is covered by a mucous membrane, which contains the sensory receptors called taste buds (see page 308). When taste buds are activated by the presence of food, nerve impulses travel by way of nerves to the brain. The tongue is composed of skeletal muscle, and it assists the teeth in carrying out mechanical digestion by moving food around in the mouth. In preparation for swallowing, the tongue forms chewed food into a mass called a **bolus** that it pushes toward the pharynx.

The Pharynx and Esophagus

Both the mouth and the nasal passages lead to the **pharynx,** a hollow space at the back of the throat (Fig. 8.4). In turn, the pharynx opens into both the food passage (**esophagus**) and air passage (trachea, or windpipe). These two tubes are parallel to each other, with the trachea anterior (in front of) the esophagus.

Swallowing

Swallowing has a voluntary phase—from day-to-day living, you know that you can swallow voluntarily. (Try it!) However, once food or drink is pushed back into the pharynx, swallowing becomes a reflex action performed automatically (without you willing it). During swallowing, food normally enters the esophagus, a muscular tube that takes food to the stomach, because other possible avenues are blocked. The soft palate moves back to close off the nasal passages, and the trachea moves up under the **epiglottis** to cover the glottis. The **glottis** is the opening to the larynx (voice box), and therefore, the air passage. We do not breathe when we swallow. The up-and-down movement of the Adam's apple,

the front part of the larynx, is easy to observe when a person swallows (see Fig. 8.4*a*).

Unfortunately, we have all had the unpleasant experience of having food "go the wrong way." The wrong way may be either into the nasal cavities or into the trachea. If it is the latter, coughing will most likely force the food out of the trachea and into the pharynx again.

Peristalsis, rhythmic contractions of the smooth muscle, pushes food through the esophagus. These contractions continue in the stomach and intestines. The esophagus plays no role in the chemical digestion of food. Its sole purpose is to move the food bolus from the mouth to the stomach. A constriction, often called the *lower gastroesophageal sphincter*, marks the entrance of the esophagus to the stomach. **Sphincters** are muscles that encircle tubes and act as valves. The tubes close when the sphincters contract and they open when sphincters relax. When food or saliva is swallowed, the sphincter relaxes for a moment to allow the food or saliva to enter the stomach (see Fig. 8.4*b*). The sphincter then contracts, preventing the acidic stomach contents from backing up into the esophagus.

Heartburn, discussed in the Health Focus on page 162, occurs when some of the stomach's contents escape into the esophagus. When vomiting occurs, the abdominal muscles and the **diaphragm** (the muscle separating the thoracic and abdominal cavities) contract.

> **Check Your Progress 8.2**
> 1. Describe the mechanical digestion and the chemical digestion that occurs in the mouth.
> 2. What ordinarily prevents food from entering the nose or entering the trachea when you swallow?

Health Focus

Heartburn (GERD)

In the absence of other symptoms, that burning sensation in your chest may have nothing to do with your heart. Instead, it is likely due to acid reflux. The stomach contents are more acidic than those of the esophagus. When the stomach contents pass upward into the esophagus, the acidity begins to erode the lining of the esophagus, producing the burning sensation associated with heartburn.

Some people experience heartburn after eating a large meal that overfills their stomach. Women sometimes get heartburn during pregnancy when the developing fetus pushes the internal organs upwards. Pressure applied to the abdominal wall from obesity also causes heartburn.

Almost everyone has had acid reflux and heartburn at some time. The burning sensation occurs in the area of the esophagus that lies behind the heart, and that is why it is termed heartburn. We might think it means we have a heart problem.

When heartburn or acid reflux becomes a chronic condition, the patient is diagnosed with gastroesophageal reflux disease (GERD). The term signifies that the stomach (gastric refers to the stomach) and the esophagus are involved in the disease. In GERD, patients' reflux is more frequent, remains in the esophagus longer, and will often contain higher levels of acid than in a patient with typical heartburn or acid reflux. People diagnosed with GERD may also experience pain in their chest, feel like they are choking, and have trouble swallowing.

Patients with weak or abnormal esophageal contractions will have difficulty pushing food into the stomach. These weak contractions can also prevent reflux from being pushed back into the stomach after it has entered the esophagus, thus producing GERD. When patients are lying down, the effects of abnormal esophageal contractions become more severe. This is because gravity is not helping to return reflux to the stomach. Some people with GERD may have weaker than normal lower gastroesophageal sphincters. Their sphincters don't fully close after food is pushed into the stomach. Surgery to tighten the sphincter may alleviate GERD.

Many people take over-the-counter medications for acid reflux. These drugs have a basic pH, so they neutralize stomach acid. Some progress to taking prescribed medications, which

Figure 8A   **Which person is less likely to suffer from acid reflux?**

Diet and weight control help prevent acid reflux so the person riding the bike is less likely to suffer from acid reflux. A healthy diet and light exercise are helpful to someone who has GERD (gastroesophageal reflux disease).

shut down and reduce acid production. However, persons with acid reflux could try modifying their eating habits first.

Diet and Exercise

It has been found that diet and weight control can help control acid reflux. Here are some tips:

- Eat several small meals instead of three large meals a day. In the movie *Supersize Me*, Morgan Spurlock doesn't just have acid reflux after eating a meal consisting of a supersize fast-food meal. He throws up! Vomiting involves the same processes as does acid reflux.
- Avoid foods that lead to stomach acidity, such as tomato sauces and citrus fruits. Alcohol and caffeinated beverages can also increase acid production. High-fat meals (like those served at fast-food chains) contribute to acid generation as well. Instead, eat complex carbohydrates (such as multigrain bread, brown rice, and pasta) instead of foods high in refined sugar (like cakes and candy). Besides providing you with vitamins and minerals, complex carbohydrates can help you control acid reflux!
- Participate in light exercise such as riding a bike at a slow pace, walking, yoga, and light weight lifting. Both diet and exercise should help control weight.

8.3 The Stomach and Small Intestine

The stomach and small intestine complete the digestion of food, which began in the mouth.

The Stomach

The **stomach** (Fig. 8.5) is a thick-walled, J-shaped organ that lies on the left side of the body beneath the diaphragm. The stomach is continuous with the esophagus above and the duodenum of the small intestine below. The stomach stores food, initiates the digestion of protein, and controls the movement of food into the small intestine. Nutrients are not absorbed by the stomach. However, it does absorb alcohol, because alcohol is fat soluble and can pass through membranes easily.

The stomach wall has the usual four layers, but two of them are modified for particular functions. The muscularis contains three layers of smooth muscle (see

Have You Ever Wondered...

Can you drink through your nose?

It may be a good party trick if you can pull it off. You'll find videos of people drinking various liquids through their noses on YouTube, but it's not the socially acceptable method of fluid intake. In addition, fluids can damage the delicate membrane of the nasopharynx. There is also the danger of the fluid "going down the wrong way" and being sucked into lungs. However, the connection between the nasopharynx and the digestive system can be of medical use. The nasopharynx provides a route to the stomach when it needs to be pumped—for example, after the ingestion of a potentially lethal substance.

break down food into smaller fragments that are mixed with gastric juice.

The mucosa of the stomach has deep folds called **rugae.** These disappear as the stomach fills to an approximate capacity of 1 liter. The mucosa of the stomach has millions of gastric pits, which lead into **gastric glands** (Fig. 8.5b and c). The gastric glands produce gastric juice. Gastric juice contains an enzyme called **pepsin,** which digests protein, plus hydrochloric acid (HCl) and mucus. HCl causes the stomach to be very acidic with a pH of about 2. This acidity is beneficial because it kills most bacteria present in food. Although HCl does not digest food, it does break down the connective tissue of meat and activates pepsin.

Normally, the stomach empties in about 2–6 hours. When food leaves the stomach, it is a thick, soupy liquid of partially digested food called **chyme.** Chyme enters the small intestine in squirts. Peristaltic waves move the chyme toward the pyloric sphincter, which closes and squeezes most of the chyme back, letting only a small amount to enter the small intestine at one time (Fig. 8.5d).

Fig. 8.5a). In addition to the circular and longitudinal layers, the stomach also contains a layer of smooth muscle that runs obliquely to the other two. The oblique layer also allows the stomach to stretch and to mechanically

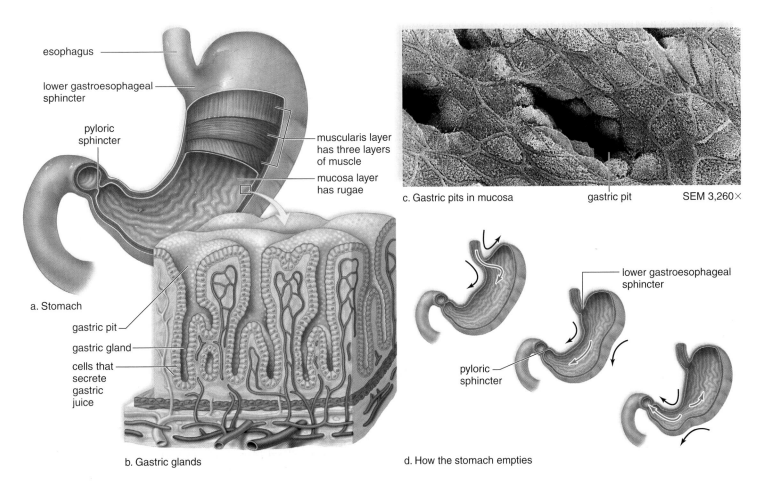

c. Gastric pits in mucosa   gastric pit   SEM 3,260×

a. Stomach

b. Gastric glands

d. How the stomach empties

Figure 8.5   How is the stomach layered? What substances are produced by the cells in the gastric glands?
a. Structure of the stomach showing that the muscularis has three muscle layers and the mucosa has folds called rugae. **b.** Gastric glands present in the mucosa secrete mucus, HCl, and pepsin, an enzyme that digests protein. **c.** Micrograph of gastric pits. **d.** Peristalsis in the stomach controls the secretion of chyme into the small intestine at the pyloric sphincter.

Historical Focus

A Window to the Stomach: Alexis St. Martin and Dr. William Beaumont

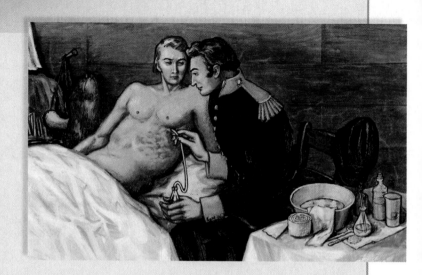

Not much was known about the digestive activity of the stomach, until a surgeon named William Beaumont was presented with a unique opportunity to study its workings. In 1822, Beaumont treated Alexis St. Martin for a gunshot wound to the abdomen. St. Martin's wound never fully healed. The injury left him with a gastric fistula—a permanent hole through the abdominal wall into his stomach.

Beaumont performed experiments on St. Martin in an effort to determine what happens to food in the stomach. Many of the experiments consisted of tying a piece of food to a string and inserting it through the hole in St. Martin's stomach. Every few hours, Beaumont would remove the food and examine the process of digestion. He was able to make observations about the breakdown process for different foods, such as meat and vegetables. Further, Beaumont studied the time required for digestion. Through careful observations, he discovered most of what we know about the stomach's digestive activities.

About 150 years later, in 1982, Drs. Marshall and Warren began their work on the cause of ulcers. As related in Chapter 1, they discovered that the bacterium *Helicobacter pylori* is able to overcome the protective function of mucus in the stomach. Erosion of the stomach lining results in a gastric (referring to the stomach) ulcer.

CASE STUDY AT THE SURGEON'S OFFICE

Several days after her appointment with Dr. Emory, Monica met with Dr. Hall, a specialist in bariatric surgery. Monica looked around nervously at the illustrations hanging on his office walls, while he reviewed her medical records and the results of her blood tests.

Dr. Hall noticed her worried look. "Don't worry, Monica," he reassured. "It looks scarier than it really is." He continued, "We still need the results of your psychological evaluation. You've got be ready to commit to a healthier lifestyle. If those results check out, we can schedule your surgery and get you on track."

Dr. Hall pointed to one of the illustrations. "It may help you understand what we'll do during your procedure. Check out these pictures. Most people don't realize that there are several different types of bariatric surgery. All are designed to promote weight loss in morbidly obese people."

"One of the most common procedures is gastric bypass surgery, which is what we'll be doing on you," he continued. "We'll reduce your stomach to the size of a golf ball. Then we'll reroute the digestive tract so that food bypasses the first two feet of the small intestine."

Dr. Hall warned, "There are always risks when you have surgery. You may develop an infection, and there will be pain afterwards.

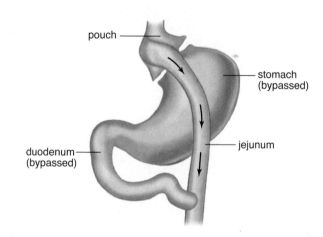

And if you don't change your eating habits, your weight will not change either."

He added, "You'll need to meet with a nutritionist to learn how and what to eat following your surgery. It's important to know how to eat, because if you overeat you may become very ill. The small intestines are bypassed, so some nutrients will not be absorbed as well. You need to know what to eat to avoid nutritional deficiencies."

Table 8.1	Major Digestive Enzymes			
Enzyme	**Produced By**	**Site of Action**	**Optimum pH**	**Digestion**
CARBOHYDRATE DIGESTION:				
Salivary amylase	Salivary glands	Mouth	Neutral	Starch + H_2O → maltose
Pancreatic amylase	Pancreas	Small intestine	Basic	Starch + H_2O → maltose
Maltase	Small intestine	Small intestine	Basic	Maltose + H_2O → glucose + glucose
Lactase	Small intestine	Small intestine	Basic	Lactose + H_2O → glucose + galactose
PROTEIN DIGESTION:				
Pepsin	Gastric glands	Stomach	Acidic	Protein + H_2O → peptides
Trypsin	Pancreas	Small intestine	Basic	Protein + H_2O → peptides
Peptidases	Small intestine	Small intestine	Basic	Peptide + H_2O → amino acids
NUCLEIC ACID DIGESTION:				
Nuclease	Pancreas	Small intestine	Basic	RNA and DNA + H_2O → nucleotides
Nucleosidases	Small intestine	Small intestine	Basic	Nucleotide + H_2O → base + sugar + phosphate
FAT DIGESTION:				
Lipase	Pancreas	Small intestine	Basic	Fat droplet + H_2O → monoglycerides + fatty acids

The Small Intestine

The **small intestine** is named for its small diameter (compared with that of the large intestine), but perhaps it should be called the long intestine. The small intestine averages about 6 m (18 ft) in length, compared with the large intestine, which is about 1.5 m (4½ ft) in length.

Digestion Is Completed in the Small Intestine

Notice in Table 8.1 that the small intestine contains enzymes to digest all types of foods, primarily carbohydrates, proteins, and fats. Most of these enzymes are secreted by the pancreas and enter via a duct at the **duodenum,** the name for the first 25 cm of the small intestine. Another duct brings bile from the liver and gallbladder into the duodenum (see Fig. 8.8). **Bile** emulsifies fat. Emulsification causes fat droplets to disperse in water. After fat is mechanically broken down to fat droplets by bile, it is hydrolyzed to glycerol and fatty acids by **lipase** present in pancreatic juice. Pancreatic amylase begins the digestion of carbohydrates. An intestinal enzyme completes the digestion of carbohydrates to glucose. Similarly, pancreatic trypsin begins and intestinal enzymes finish the digestion of proteins to amino acids. The intestine has a slightly basic pH because pancreatic juice contains sodium bicarbonate ($NaHCO_3$), which neutralizes chyme.

Nutrients Are Absorbed in the Small Intestine

The wall of the small intestine absorbs the molecules, namely sugars, amino acids, fatty acids, and glycerol, the products of the digestive process. The mucosa of the small intestine is modified for absorption. It has been suggested that the surface area of the small intestine is approximately that of a tennis court. What factors contribute to increasing its surface area? The mucosa of the small intestine contains fingerlike projections called villi (sing., **villus**), which give the intestinal wall a soft, velvety appearance (Fig. 8.6).

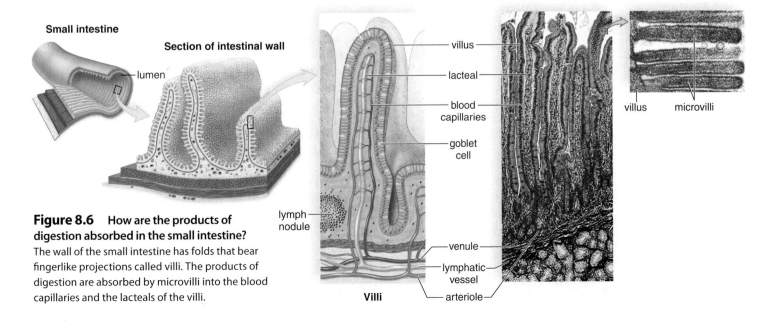

Figure 8.6 How are the products of digestion absorbed in the small intestine?
The wall of the small intestine has folds that bear fingerlike projections called villi. The products of digestion are absorbed by microvilli into the blood capillaries and the lacteals of the villi.

Figure 8.7 **What are the products of the digestion of carbohydrates, proteins, and fats? How are these products absorbed?** **a.** Carbohydrate is digested to glucose, which is actively transported into the cells of intestinal villi. From there, glucose moves into the bloodstream. **b.** Proteins are digested to amino acids, which are actively transported into the cells of intestinal villi. From there, amino acids move into the bloodstream. **c.** Fats are emulsified by bile and digested to monoglycerides and fatty acids. These diffuse into cells, where they recombine and join with proteins. These lipoproteins, called chylomicrons, enter a lacteal.

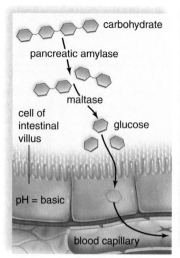

a. Carbohydrate digestion

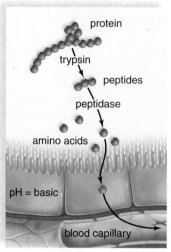

b. Protein digestion

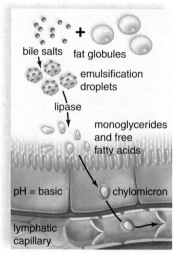

c. Fat digestion

A villus has an outer layer of columnar epithelial cells, and each of these cells has thousands of microscopic extensions called microvilli. Collectively, in electron micrographs, microvilli give the villi a fuzzy border known as a "brush border." The microvilli bear the intestinal enzymes, so these enzymes are called brush-border enzymes. The microvilli greatly increase the surface area of the villus for the absorption of nutrients.

Nutrients are absorbed into the vessels of a villus (Fig. 8.7). A villus contains blood capillaries and a small lymphatic capillary, called a **lacteal.** As you know, the lymphatic system is an adjunct to the cardiovascular system. Lymphatic vessels carry a fluid called lymph to the cardiovascular veins. Sugars (digested from carbohydrates) and amino acids (digested from proteins) enter the blood capillaries of a villus. Glycerol and fatty acids (digested from fats) enter the epithelial cells of the villi. Lipoprotein droplets, called chylomicrons, are formed when glycerol and fatty acids are rejoined in the villi epithelia cells. Chylomicrons then enter a lacteal. After nutrients are absorbed, they are eventually carried to all the cells of the body by the bloodstream.

Lactose Intolerance

Lactose is the primary sugar in milk. People who do not have the brush border enzyme called lactase cannot digest lactose. The result is a condition called **lactose intolerance,** characterized by diarrhea, gas, bloating, and abdominal cramps after ingesting milk and other dairy products. Diarrhea occurs because the indigestible lactose causes fluid retention in the small intestine. Gas, bloating, and cramps occur when bacteria break down the lactose anaerobically.

Persons with lactose intolerance can consume dairy products that are lactose-free or in which lactose has already been digested. These include lactose-free milk, cheese, and yogurt. A dietary supplement that aids in the digestion of lactose is available also.

> **Check Your Progress 8.3**
> 1. a. What are the functions of the stomach, and (b) how is the wall of the stomach modified to perform these functions?
> 2. a. What are the functions of the small intestine, and (b) how is the wall of the small intestine modified to perform these functions?

8.4 Three Accessory Organs and Regulation of Secretions

First, we will take a look at three accessory organs of digestion before considering how the secretions of these organs and those of the GI tract are regulated.

Three Accessory Organs

The **pancreas** is a fish-shaped, spongy, grayish pink organ that stretches across the back of the abdomen behind the stomach. Most pancreatic cells produce pancreatic juice, which enters the duodenum via the pancreatic duct (Fig. 8.8a). Pancreatic juice contains sodium bicarbonate ($NaHCO_3$) and digestive enzymes for all types of food. Sodium bicarbonate neutralizes acid chyme from the stomach. **Pancreatic amylase** digests starch, **trypsin** digests protein, and pancreatic lipase digests fat.

The pancreas is also an endocrine gland that secretes the hormone insulin into the blood. A **hormone** is a substance produced by one set of cells that affects a different set of cells, the so-called target cells. When the blood glucose level rises rapidly, the pancreas produces an overload of insulin to bring the level under control. Over the years, the body's cells can become insulin resistant and diabetes type 2 can occur.

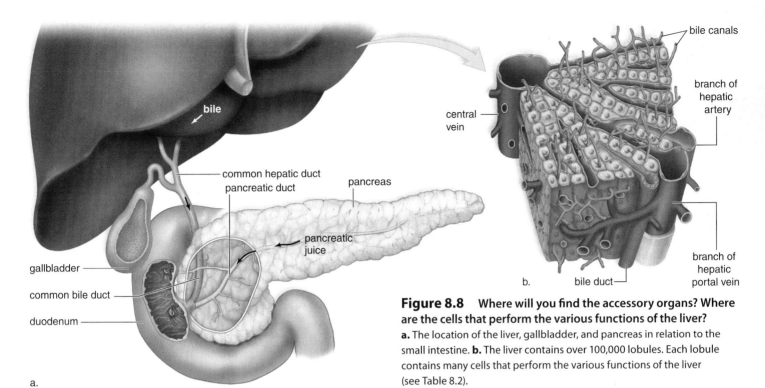

Figure 8.8 **Where will you find the accessory organs? Where are the cells that perform the various functions of the liver?**
a. The location of the liver, gallbladder, and pancreas in relation to the small intestine. **b.** The liver contains over 100,000 lobules. Each lobule contains many cells that perform the various functions of the liver (see Table 8.2).

The largest gland in the body, the **liver,** lies mainly in the upper-right section of the abdominal cavity, under the diaphragm (see Fig. 8.1). The liver is a major metabolic gland with approximately 100,000 lobules that serve as its structural and functional units (see Fig. 8.8b). The hepatic portal vein (see Fig. 8.8b) brings blood to the liver from the GI tract capillary bed. Capillaries of the lobules filter this blood. In a sense, the liver acts like a sewage treatment plant when it removes poisonous substances from the blood and detoxifies them (Table 8.2).

The liver is also a storage organ. It removes iron and the vitamins A, D, E, K, and B_{12} from blood and stores them. Blood glucose levels remain relatively constant after we eat. In the presence of insulin, the liver stores glucose as glycogen. When blood glucose becomes low, the liver releases glucose by breaking down glycogen. If need be, the liver converts glycerol (from fats) and amino acids to glucose molecules. As amino acids are converted to glucose, the liver combines their amino groups with carbon dioxide to form **urea,** the usual nitrogenous waste product in humans.

Plasma proteins are made by the liver. The liver also helps regulate blood **cholesterol** levels. Some cholesterol is converted to bile salts by the liver. **Bile** is a solution of bile salts, water, cholesterol, and bicarbonate. It has a yellowish-green color because it also contains bilirubin, formed during the breakdown of hemoglobin. Hemoglobin breakdown is yet another function of the liver. Bile is stored in the **gallbladder,** a pear-shaped organ just below the liver, until it is sent via the bile ducts to the duodenum. **Gallstones** form when liquid stored in the gallbladder hardens into pieces of stonelike material. In the small intestine, bile salts emulsify fat. When fat is emulsified, it breaks up into droplets. The droplets provide a large suface area that can be acted upon by digestive enzymes.

Liver Disorders

Hepatitis and cirrhosis are two serious diseases that affect the entire liver and hinder its ability to repair itself. Therefore, they are life-threatening diseases. When a person has a liver ailment, bile pigments may leak into the blood causing **jaundice.** Jaundice is a yellowish tint to the whites of the eyes and also to the skin of light-pigmented persons. Jaundice can result from **hepatitis,** inflammation of the liver. Viral hepatitis occurs in several forms. Hepatitis A is usually acquired from sewage-contaminated drinking water and food. Hepatitis B, which is usually spread by sexual contact, can also be spread by blood transfusions or contaminated needles. The hepatitis

Table 8.2	Functions of the Liver

1. Destroys old red blood cells; excretes bilirubin, a breakdown product of hemoglobin in bile, a liver product
2. Detoxifies blood by removing and metabolizing poisonous substances
3. Stores iron (Fe^{2+}) and the fat-soluble vitamins A, D, E, and K
4. Makes plasma proteins, such as albumins and fibrinogen, from amino acids
5. Stores glucose as glycogen after a meal, and breaks down glycogen to glucose to maintain the glucose concentration of blood between eating periods
6. Produces urea after breaking down amino acids
7. Helps regulate blood cholesterol level, converting some to bile salts

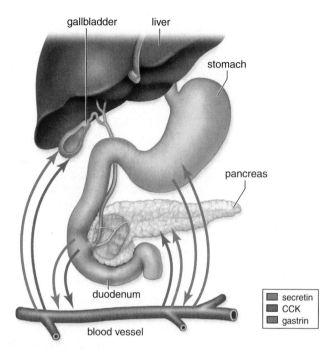

gallbladder liver

stomach

pancreas

duodenum

blood vessel

☐ secretin
☐ CCK
☐ gastrin

Figure 8.9 **What hormones control the secretions of digestive glands?**

Gastrin (blue), from the lower stomach, feeds back to stimulate the upper part of the stomach to produce digestive juice. Secretin (green) and CCK (purple) from the duodenal wall stimulate the pancreas to secrete digestive juice and the gallbladder to release bile.

B virus is more contagious than the AIDS virus, spread in the same way. Vaccines are available for hepatitis A and hepatitis B. Hepatitis C is usually acquired by contact with infected blood and can lead to chronic hepatitis, liver cancer, and death. There is no vaccine for hepatitis C.

Cirrhosis is another chronic disease of the liver. First, the organ becomes fatty, and then liver tissue is replaced by inactive fibrous scar tissue. Cirrhosis of the liver is often seen in alcoholics, due to malnutrition and to the excessive amounts of alcohol (a toxin) the liver is forced to break down. Physicians are beginning to see cirrhosis of the liver in obese people, who are overweight due to a diet high in fatty foods.

The liver has amazing regenerative powers and can recover if the rate of regeneration exceeds the rate of damage. During liver failure, however, there may not be enough time to let the liver heal itself. Liver transplantation is usually the preferred treatment for liver failure, but artificial livers have been developed and tried in a few cases. The liver is a vital organ, and its failure leads to death.

Regulation of Digestive Secretions

The secretions of digestive juices are controlled by the nervous system and by digestive hormones. When you look at or smell food, the parasympathetic nervous system automatically stimulates gastric secretion. Also, when a person has eaten a meal particularly rich in protein, the stomach pro-

duces the hormone gastrin. Gastrin enters the bloodstream, and soon the secretory activity of gastric glands increases.

Cells of the duodenal wall produce two other hormones of particular interest—secretin and cholecystokinin (CCK). Secretin release is stimulated by acid, especially the HCL present in chyme. Partially digested proteins and fat stimulate the release of CCK. Soon after these hormones enter the bloodstream, the pancreas increases its output of pancreatic juice. Pancreatic juice buffers the acidic chyme entering the intestine from the stomach and helps digest food. CCK also causes the liver to increase its production of bile and the gallbladder to contract and release stored bile. The bile then aids the digestion of fats that stimulated the release of CCK. Figure 8.9 summarizes the actions of gastrin, secretin, and CCK.

> **Check Your Progress 8.4**
> 1. **What are the three main accessory organs that assist with the digestive process?**
> 2. **How does each accessory organ contribute to the digestion of food?**
> 3. **How are digestive secretions regulated in the body?**

8.5 The Large Intestine and Defecation

The **large intestine** includes the cecum, the colon, the rectum, and the anal canal (Fig. 8.10). The large intestine is larger in diameter than the small intestine (6.5 cm compared with 2.5 cm), but it is shorter in length (see Fig. 8.1).

The **cecum,** which lies below the junction with the small intestine, is the blind end of the large intestine. The cecum usually has a small projection called the **vermiform appendix** (*vermiform* means wormlike) (see Fig. 8.10). In humans, the appendix also may play a role in fighting infections. Some scientists recently proposed that the appendix may contribute to the population of bacteria in the large intestine. As mentioned previously, the appendix can become infected, a condition called appendicitis.

The **colon** includes the ascending colon, which goes up the right side of the body to the level of the liver; the transverse colon, which crosses the abdominal cavity just below the liver and the stomach; the descending colon, which passes down the left side of the body; and the sigmoid colon, which enters the **rectum,** the last 20 cm of the large intestine. The rectum opens at the **anus,** where **defecation,** the expulsion of feces, occurs.

Functions of the Large Intestine

The large intestine absorbs water, an important function that prevents dehydration of the body. The large intestine does not produce any digestive enzymes and it does not absorb any nutrients.

The large intestine absorbs vitamins produced by bacteria called the intestinal flora. For many years, it was believed

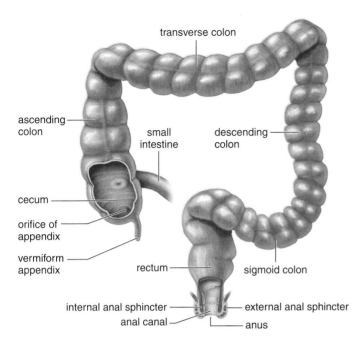

Figure 8.10 **What are the different regions of the large intestine?**

The regions of the large intestine are the cecum, colon, rectum, and anal canal.

Escherichia coli were the major inhabitants of the colon, but culture methods now show that over 99% of the colon bacteria are other types of bacteria. The bacteria in the large intestine break down indigestible material and also produce B complex vitamins and most of the vitamin K needed by our bodies. In this way, they perform a service for us.

The large intestine forms feces. Feces are normally three-quarters water and one-quarter solids. Bacteria, dietary **fiber** (indigestible remains), and other indigestible materials make up the solid portion. Bacterial action on indigestible materials causes the odor of feces and also accounts for the presence of gas. A breakdown product of bilirubin (see page 167) and the presence of oxidized iron cause the brown color of feces.

Defecation, ridding the body of feces, is also a function of the large intestine. Peristalsis occurs infrequently in the large intestine, but when it does, feces are forced into the rectum. Feces collect in the rectum until it is appropriate to defecate. At that time, stretching of the rectal wall initiates nerve impulses to the spinal cord. Shortly thereafter, the rectal muscles contract and the anal sphincters relax. This allows the feces to exit the body through the anus (see Fig. 8.10). A person can inhibit defecation by contracting the external anal sphincter. Ridding the body of indigestible remains is another way the digestive system helps maintain homeostasis.

Water is considered unsafe for swimming when the coliform (nonpathogenic intestinal) bacterial count reaches a certain number. A high count indicates that a significant amount of feces has entered the water. The more feces present, the greater the possibility that disease-causing bacteria are also present.

Disorders of the Colon and Rectum

The large intestine is subject to a number of disorders. Many of these can be prevented or minimized by a good diet and good bowel habits. Food poisoning caused by eating contaminated food causes an infection of the intestinal tract.

Diarrhea The major causes of **diarrhea** are infection of the lower intestinal tract and nervous stimulation. The intestinal wall becomes irritated, and peristalsis increases when an infection occurs. Water is not absorbed, and the diarrhea that results rids the body of the infectious organisms. In nervous diarrhea, the nervous system stimulates the intestinal wall, and diarrhea results. Prolonged diarrhea can lead to dehydration because of water loss. An imbalance of salts in the blood may affect heart muscle contraction and lead to death.

Constipation When a person is constipated, the feces are dry and hard, making it difficult for them to be expelled. Diets that lack whole grain foods, as well as ignoring the urge to defecate, are often the causes of **constipation.** When feces are not expelled regularly, additional water is absorbed from them. The material becomes drier, harder, and more difficult to eliminate. Adequate water and fiber intake can help regularity of defecation. The frequent use of laxatives is discouraged because dependence on their use can result. If, however, it is necessary to take a laxative, the most natural is a bulk laxative. Like fiber, it produces a soft mass of cellulose in the colon. Lubricants, such as mineral oil, make the colon slippery; saline laxatives, such as milk of magnesia, act osmotically—they prevent water from being absorbed. Some laxatives are irritants that increase peristalsis.

Chronic constipation is associated with the development of **hemorrhoids,** enlarged and inflamed blood vessels at the anus. Other contributing factors for hemorrhoid development include pregnancy, aging, and anal intercourse.

Diverticulosis As mentioned previously (see page 159), diverticulosis is the occurrence of little pouches of mucosa where food can collect. The pouches form when the mucosa pushes through weak spots in the muscularis. A frequent site is the last part of the descending colon.

Irritable Bowel Syndrome (IBS, spastic colon) Also mentioned previously (see page 159), IBS is a condition in which the muscularis contracts powerfully but without its normal coordination. The symptoms are abdominal cramps; gas; constipation; and urgent, explosive stools (feces discharge).

Inflammatory Bowel Disease (IBD, colitis) IBD is a collective term for a number of inflammatory disorders. Ulcerative colitis and Crohn's disease are the most common of these. Ulcerative colitis affects the large intestine and rectum, but in Crohn's disease, the inflammation can be more widespread and ulcers can penetrate more deeply. Ulcers are painful and cause bleeding because they erode the submucosal layer, where there are nerves and blood vessels. Colitis causes diarrhea, rectal bleeding, abdominal cramps, and urgency.

Health Focus

Swallowing a Camera

During a traditional endoscopy procedure, the doctor uses an endoscope (a retractable tubelike instrument with an embedded camera) to examine the patient's GI tract. PillCam™ has become a viable alternative to traditional endoscopy. With a gulp of water, PillCam™ is swallowed, and it travels through the digestive system. Instead of spending an uncomfortable half day or more at the doctor's office, a patient visits the doctor in the morning, swallows the camera, dons the recording device, and goes about his or her daily routine.

Propelled by the normal muscular movement of the digestive system, PillCam™ embarks on a four- to eight-hour journey through the digestive system. As it travels through the stomach, the twists and turns of the small intestine, and the large intestine, PillCam™ continuously captures high-quality, wide-angle film footage of its journey and beams this information to the recording device worn by the patient.

At the end of the day, PillCam™ reaches the end of its journey, and it is defecated. Later, the recording device is returned to the doctor's office so that the data may be retrieved. The doctor can view PillCam™'s journey as a 90-minute movie.

Like the food that we eat, PillCam™ traverses the numerous twists and turns of the intestine with ease. This provides a more accurate diagnosis, because a larger portion of the GI tract can be examined. PillCam™ does not create discomfort in the patient, so the considerable risks involved with using anaesthetics and painkillers are eliminated. And finally, the doctor does not need to be present for the entire procedure, saving valuable time and money for both doctor and patient.

Colonoscopy is a routine procedure used to examine and diagnose colon cancer. It employs a colonoscope, an instrument similar to an endoscope, inserted through the anus into the large intestine. It is capable of removing precancerous tissue before it becomes invasive. PillCam™ cannot be used to remove tissue samples for analysis, so its use as a replacement for colonoscopy is somewhat limited.

Figure 8B   How does a PillCam™ Work? PillCam™ traverses through the entire GI tract, taking pictures after it is swallowed.

Polyps and Cancer   The colon is subject to the development of **polyps,** small growths arising from the epithelial lining. Polyps, whether benign or cancerous, can be removed surgically. If colon cancer is detected while still confined to a polyp, the expected outcome is a complete cure. Some investigators believe that dietary fat increases the likelihood of colon cancer, because dietary fat causes an increase in bile secretion. It could be that intestinal bacteria convert bile salts to substances that promote the development of cancer. On the other hand, fiber in the diet seems to inhibit the development of colon cancer. Regular elimination reduces the time that the colon wall is exposed to any cancer-promoting agents in feces.

As discussed in the Health Focus, "Swallowing a Camera," endoscopy by means of a flexible tube (and camera) inserted into the GI tract, usually from the anus, is gradually being replaced by the PillCam™, a camera you swallow!

Check Your Progress 8.5

1. What are the different parts of the large intestine?
2. What are the functions of the large intestine?

Have You Ever Wondered...

What happens to things that kids swallow, like coins and toys?

Ideally, these items will pass through the GI tract, undigested of course. They'll be eliminated much like the PillCam™. However, foreign objects may perforate or block the tract somewhere between the mouth and anus. Another concern is that some metals, like nickel, are harmful when swallowed. So watch those little ones carefully!

8.6 Nutrition and Weight Control

Obesity—being grossly overweight—has doubled in the United States in only 20 years. Nearly one-third (33%) of adults are now obese. Seventeen percent of children and teens between the ages of 2–19 are estimated to be overweight or obese. These statistics are of great concern because excess body fat is associated with a higher risk for premature death, diabetes type 2, hypertension, cardiovascular disease, stroke, gallbladder disease, respiratory dysfunction, osteoarthritis, and certain types of cancer.

Obesity is also on the rise throughout the world. The term globesity has been coined in recognition of the worldwide problem. In all regions, obesity appears to escalate as income increases.

The Health Focus on page 172 tells us about the various ways people have tried to keep their weight under control. The conclusion is that while dieting, *"Eat a variety of foods, watch your weight, and exercise."* To reverse a trend toward obesity, eat fewer calories, be more active, and make wiser food choices.

How Obesity Is Defined

Today, obesity is often defined as a **body mass index (BMI)** of 32 or greater. If you know your height and weight, you can determine your BMI using the table in Figure 8.11. The BMI can also be calculated by dividing weight in pounds (lbs) by height in inches (in) squared and multiplying by a conversion factor of 703: [weight in pounds/height in inches)^2] $\times$ 703.

Most people find that using the table is a lot easier. As a general rule,

> healthy BMIs = 19.1 to 26.4;
> overweight BMIs = 26.5 to 31.1;
> obese BMIs = 32.3 to 39.9; and
> morbidly obese BMIs = 40 or more.

Your BMI gives you an idea of how much of your weight is due to adipose tissue—that is, fat. In general, the taller you are, the more you could weigh without it being due to fat. Using BMI in this way works for most people, especially if they tend to be sedentary. But your BMI number should be used only as a general guide. It does not take into account fitness, bone structure, or gender. For example, a weightlifter might have an obese BMI, not because he has body fat, but because of increased bone and muscle weight.

Classes of Nutrients

A **nutrient** can be defined as a component of food that performs a physiological function in the body. Nutrients

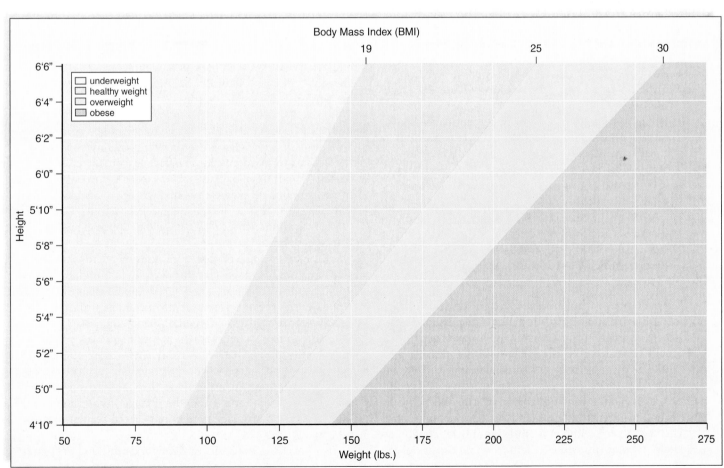

Source: U.S. Department of Agriculture: Dietary Guidelines for Americans 2005

Figure 8.11   What is your body mass index?
Match your weight with your height, then determine your body mass index (BMI). Healthy BMIs = 19.1 to 26.4; overweight BMIs = 26.5 to 31.1; obese BMIs = 32.3 to 39.9; morbidly obese BMIs = 40 or more.

Health Focus

Searching for the Magic Weight-Loss Bullet

"Eat a variety of foods, watch your weight, and exercise" doesn't sound like a very glamorous way to lose weight. Besides you can't sell the message to the public and make a lot of money. No wonder the public, always looking for the magic weight-loss bullet, is offered so many solutions to being overweight, most of which are not healthy. The solutions involve trendy diet programs, new prescription medications, and even surgery. The latter two options are for people who have tried a low-calorie diet and regular physical activity but have been unsuccessful in losing weight. Prescription medications should only be taken under a physician's supervision, and of course, surgeries are only done by physicians.

TRENDY DIET PROGRAMS: Various diets for the overweight have been around for many years, and here are some recently touted:

The Pritikin Diet   This diet encourages the consumption of large amounts of carbohydrates and fiber in the form of whole grains and vegetables. The diet is so low-fat that the dieter may not be able to consume a sufficient amount of "healthy" fats.

The Atkins Diet   This diet is just the opposite of the Pritikin Diet because it is a low-carbohydrate (carb) diet. It is based on the assumption that if we eat more protein and fat, our bodies lose weight by burning stored body fat. The Atkins diet is thought by many to be a serious threat to homeostasis. It puts a strain on the body to maintain the blood glucose level, the breakdown of fat lowers blood pH, and the excretion of nitrogen from protein breakdown stresses the kidneys.

The Zone Diet and the South Beach Diet   As a reaction to the Atkins diet, these diets recommend only "healthy" fats and permit low-sugar carbs. In other words, these diets are bringing us back, once again, to "Eat a variety of foods, maintain your weight, and exercise."

You may have heard of the caveman diet. The caveman diet mimics the diets of humans prior to agricultural pursuits. It promotes the consumption of meats, fish, fruits, and vegetables. If you would rather not change your lifestyle dramatically, you can "flush" fat away, according to some nutritionists. They assert that by consuming certain foods, such as cayenne pepper, mustard, cinnamon, green vegetables, and omega-3-rich fish, you can boost your metabolic rate and cleanse your body of fat.

Then again, according to a professor at Brigham Young University, the cure for endless dieting and the key to reaching a healthy weight is to listen to your body. Using a hunger scale of 1 to 10, with 1 being starving and 10 being very overfull, keep around 3 to 5 and you will eat less. Unfortunately, such sensible advice doesn't seem to have caught on yet.

PRESCRIPTION DRUGS: Once Medicare and Medicaid classified obesity a disease, drugmakers revved up the search for a magic weight-loss pill. Despite lawsuits against the manufacturers of the prescription drug fen-phen for causing heart problems and the belief that a drug called Meridia® elevates blood pressure, new drugs continue to enter clinical trials. Another one called Acomplia®, supposed to block pleasure receptors in the brain, is in late-stage clinical trials. Another idea is a weight-loss nasal spray. Orlistat, sold as Xenical® or Alli®, interferes with fat absorption in the small intestine.

BURNING CALORIES: Exercise should be part of any weight-loss effort. Schools are starting programs to increase the activity of students in an effort to combat childhood obesity. About three-quarters of U.S. teens fail to participate in the amount of exercise needed for good health. Many schools are trying to increase students' activity in novel ways, including incorporating the video dance game Dance Dance Revolution®.

Research suggests that a minimum of 10,000 (10K) steps per day is necessary for weight maintenance and good health. That number of steps per day is roughly equivalent to the recommended 30 minutes of daily exercise. To get started, you'll need a pedometer. You'll probably find you need to increase the amount of walking you do to reach the goal of 10K steps a day. There are a number of easy ways to add steps to your routine. Park a little farther away from your office or the store. Take the stairs instead of the elevator, or go for a walk after a meal. If your goal is to lose weight, 12–15K steps a day have been shown to promote weight loss.

We are facing an obesity epidemic and must do something about it. Are these weight-loss options the answer? There are signs that the public is no longer searching for the magic bullet and, instead, is focusing on "To achieve or maintain a healthy weight, eat a variety of healthy foods, and exercise."

provide us with energy, promote growth and development, and regulate cellular metabolism.

Carbohydrates

Carbohydrates are either simple or complex. Glucose is a simple sugar preferred by the body as an energy source. Complex carbohydrates with several sugar units are digested to glucose. While body cells can use fatty acids as an energy source, brain cells require glucose. For this reason alone, it is necessary to include carbohydrates in the diet because the body is unable to convert fatty acids to glucose.

Any product made from refined grains, such as white bread, cake, and cookies, should be minimized in the diet. During refinement, fiber and also vitamins and minerals are removed from grains, so primarily starch remains. In contrast, sources of complex carbohydrates, such as beans, peas, nuts, fruits, and whole-grain products, are recommended as a good source of vitamins, minerals, and also fiber (Fig. 8.12). Insoluble fiber adds bulk to fecal material and stimulates movements of the large intestine, preventing constipation. Soluble fiber combines with bile salts and cholesterol in the small intestine and prevents them from being absorbed.

Can Carbohydrates Be Harmful? Some nutritionists hypothesize that the high intake of refined carbohydrates and fructose sweeteners processed from cornstarch may be responsible for obesity in the United States. In addition, these foods are said to have a high **glycemic index (GI),** because they quickly increase blood glucose. When the blood glucose level rises rapidly, the pancreas produces an overload of insulin to bring the level under control. Investigators tell us that a chronically high insulin level may lead to insulin resistance, diabetes type 2, and increased fat deposition. Deposition of fat is associated with coronary heart disease, liver ailments, and several types of cancer.

Table 8.3 gives suggestions on how to reduce your intake of dietary sugars.

Table 8.3	Reducing High-Glycemic Index Carbohydrates

To reduce dietary sugar:

1. Eat fewer sweets, such as candy, soft drinks, ice cream, and pastries.
2. Eat fresh fruits or fruits canned without heavy syrup.
3. Use less sugar—white, brown, or raw—and less honey and syrups.
4. Avoid sweetened breakfast cereals.
5. Eat less jelly, jam, and preserves.
6. Eat fresh fruit; especially avoid artificial fruit juices.
7. When cooking, use spices, such as cinnamon, instead of sugar to flavor foods.
8. Do not put sugar in tea or coffee.
9. Avoid processed foods made from refined carbohydrates, such as white bread, rice, and pasta, and limit potato intake.

Carbohydrates are the preferred energy source for the body, but select complex carbohydrates in whole grains, beans, nuts, and fruits, contain fiber, vitamins, and minerals in addition to carbohydrate.

Proteins

Dietary proteins are digested to amino acids, which cells use to synthesize hundreds of cellular proteins. Of the 20 different amino acids, nine are **essential amino acids** that must be present in the diet. Children will not grow if their diets lack the essential amino acids. Eggs, milk products, meat, poultry, and most other foods derived from animals contain all nine essential amino acids and are "complete" or "high-quality" protein sources.

Legumes (beans and peas) (Fig. 8.13), other types of vegetables, seeds and nuts, and grains supply us with amino

Figure 8.13 Why are beans considered a good food?
Beans are a good source of complex carbohydrates and protein. But beans don't supply all nine of the essential amino acids. To ensure a complete source of protein in the diet, beans should be eaten in combination with a grain, such as rice.

Figure 8.12 What foods are good sources of fiber?
Plants provide a good sources of carbohydrates. They also provide a good source of vitamins, minerals, and fiber whey they are not processed (refined).

acids. However, each of these alone is an incomplete protein source, because of a deficiency in at least one of the essential amino acids. Absence of one essential amino acid prevents use of the other 19 amino acids. Therefore, vegetarians are counselled to combine two or more incomplete types of plant products to acquire all the essential amino acids. Tofu, soymilk, and other foods made from processed soybeans are complete protein sources. A balanced vegetarian diet is possible with a little knowledge and planning.

Amino acids are not stored in the body, and a daily supply is needed. However, it does not take very much protein to meet the daily requirement. Two servings of meat a day (one serving is equal in size to a deck of cards) is usually plenty.

Can Proteins Be Harmful?   The liver removes the nitrogen-containing compound from an amino acid. By converting this portion to urea, the liver enables potentially toxic nitrogen to be removed from our bodies. However, large amounts of water are needed to properly excrete urea. Therefore, dehydration may occur if protein consumption is excessive. High-protein diets, especially those rich in animal proteins, can also increase calcium loss in urine. Excretion of calcium may lead to kidney stones and bone loss.

Certain types of meat, especially red meat, are known to be high in saturated fats, while other sources of protein, such as chicken, fish, and eggs, are more likely to be low in saturated fats. Better protein sources include chicken, fish, and egg whites, which are more likely to be low in saturated fats. As you recall from Chapter 5, excessive dietary saturated fat is a risk factor for cardiovascular disease (page 111).

> Sufficient proteins are needed to supply the essential amino acids. Meat and dairy sources of protein may supply unwanted saturated fat, but vegetable sources do not.

Lipids

Fats, oils, and cholesterol are lipids. Saturated fats, which are solids at room temperature, usually have an animal origin. Two well-known exceptions are palm oil and coconut oil, which contain mostly saturated fats and come from the plants mentioned (Fig. 8.14). Butter and fats associated with meats (like the fat on steak and bacon) contain saturated fats.

Oils contain unsaturated fatty acids, which do not promote cardiovascular disease. Corn oil and safflower oil are high in polyunsaturated fatty acids. Polyunsaturated oils are the only type of fat that contains linoleic acid and linolenic acid, two fatty acids the body cannot make. These fatty acids must be supplied by diet, so they are called **essential fatty acids.**

Olive oil and canola oil are well known to contain a larger percentage of monounsaturated fatty acids than other types of cooking oils. Omega-3 fatty acids—with a double bond in the third position—are believed to preserve brain function and protect against heart disease. Flaxseed contains abundant omega-3 fatty acids. Cold-water fish like salmon, sardines, and trout are also excellent sources.

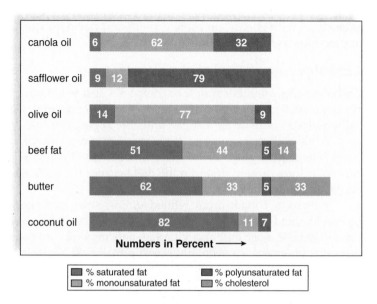

Figure 8.14   After observing this illustration, which fat would you judge to be the most healthy to use?
This illustration gives the percentages of saturated and unsaturated fatty acids in fats and oils.

Can Lipids Be Harmful?   The risk for cardiovascular disease is increased by a diet high in saturated fats and cholesterol. Saturated fats contribute to the formation of lesions associated with artherosclerosis inside the blood vessels. These lesions, termed atherosclerotic plaques, limit the flow of blood through these vessels. (See Chap. 5 for a review of the process.) Cholesterol is carried in the blood by the two transport proteins: high-density lipoprotein (HDL) and low-density lipoprotein (LDL). Cholesterol transported by HDL (the "good" lipoprotein) ends up in the liver where the cholesterol is metabolized. Cholesterol carried by LDL (the "bad" lipoprotein) ends up being deposited in the tissues. Artherosclerotic plaques form when levels of HDL are low and/or when levels of LDL are high. Recommended levels of HDL and LDL can be re-established by a diet low in saturated fats and cholesterol.

Trans-fatty acids (trans fats) arise when unsaturated fatty acids are hydrogenated to produce a solid fat. The function of the cell-membrane receptors that clear cholesterol from the bloodstream may be reduced by trans fats resulting in higher blood cholesterol level. Trans fats are found in commercially packaged goods, such as cookies and crackers. Unfortunately, other snacks such as microwave popcorn may be sources as well. Be aware that any packaged goods containing *"partially hydrogenated"* vegetable oils or shortening contain trans fats. Some margarines used for home cooking or baking incorporate hydrogenated vegetable oil. Commercially fried foods, such as French fries from some fast-food chains, should be strictly limited in a healthy diet. Though tasty, these are often full of trans fats.

Table 8.4 gives suggestions on how to reduce dietary saturated fat and cholesterol. It is not a good idea to rely on commercially produced low-fat foods. In some products, sugars have replaced the fat, and in others, protein is used.

Table 8.4	Reducing Certain Lipids

To reduce saturated fats and trans fats in the diet:

1. Choose poultry, fish, or dry beans and peas as a protein source.
2. Remove skin from poultry, and trim fat from red meats before cooking; place on a rack so that fat drains off.
3. Broil, boil, or bake rather than fry.
4. Limit your intake of butter, cream, trans fats, shortenings, and tropical oils (coconut and palm oils).
5. Use herbs and spices to season vegetables instead of butter, margarine, or sauces. Use lemon juice instead of salad dressing.
6. Drink skim milk instead of whole milk, and use skim milk in cooking and baking.

To reduce dietary cholesterol:

1. Avoid cheese, egg yolks, liver, and certain shellfish (shrimp and lobster). Preferably, eat white fish and poultry.
2. Substitute egg whites for egg yolks in both cooking and eating.
3. Include soluble fiber in the diet. Oat bran; oatmeal; beans; corn; and fruits, such as apples, citrus fruits, and cranberries, are high in soluble fiber.

Unsaturated fats such as those in oils do not lead to cardiovascular disease and are preferred. Fats and oils contain many more calories per gram than do carbohydrates and protein.

Minerals

Minerals are divided into major minerals and trace minerals. The body contains more than 5 grams of each major mineral and less than 5 grams of each trace mineral. Table 8.5 lists the selected minerals and gives their functions and food sources. The major minerals are constituents of cells and body fluids and are structural components of tissues.

The trace minerals are parts of larger molecules. For example, iron (Fe^{2+}) is present in hemoglobin, and iodine (I^-) is a part of hormones produced by the thyroid gland. Zinc (Zn^{2+}), copper (Cu^{2+}), and manganese (Mn^{2+}) are present in enzymes that catalyze a variety of reactions. As research continues, more and more elements are added to the list of trace minerals considered essential. During the past three decades, for example, very small amounts of selenium, molybdenum, chromium, nickel, vanadium, silicon, and even arsenic have

Table 8.5	Minerals

Mineral	Functions	Food Sources	Condition With	
			Too Little	**Too Much**
MAJOR (MORE THAN 100 MG/DAY NEEDED)				
Calcium (Ca^{2+})	Strong bones and teeth, nerve conduction, muscle contraction, blood clotting	Dairy products, leafy green vegetables	Stunted growth in children, low bone density in adults	Kidney stones; interferes with iron and zinc absorption
Phosphorus (PO_4^{3-})	Bone and soft tissue growth; part of phospholipids, ATP, and nucleic acids	Meat, dairy products, sunflower seeds, food additives	Weakness, confusion, pain in bones and joints	Low blood and bone calcium levels
Potassium (K^+)	Nerve conduction, muscle contraction	Many fruits and vegetables, bran	Paralysis, irregular heartbeat, eventual death	Vomiting, heart attack, death
Sulfur (S^{2-})	Stabilizes protein shape, neutralizes toxic substances	Meat, dairy products, legumes	Not likely	In animals, depresses growth
Sodium (Na^+)	Nerve conduction, pH and water balance	Table salt	Lethargy, muscle cramps, loss of appetite	Edema, high blood pressure
Chloride (Cl^-)	Water balance	Table salt	Not likely	Vomiting, dehydration
Magnesium (Mg^{2+})	Part of various enzymes for nerve and muscle contraction, protein synthesis	Whole grains, leafy green vegetables	Muscle spasm, irregular heartbeat, convulsions, confusion, personality changes	Diarrhea
TRACE (LESS THAN 20 MG/DAY NEEDED)				
Zinc (Zn^{2+})	Protein synthesis, wound healing, fetal development and growth, immune function	Meats, legumes, whole grains	Delayed wound healing, night blindness, diarrhea, mental lethargy	Anemia, diarrhea, vomiting, renal failure, abnormal cholesterol levels
Iron (Fe^{2+})	Hemoglobin synthesis	Whole grains, meats, prune juice	Anemia, physical and mental sluggishness	Iron toxicity disease, organ failure, eventual death
Copper (Cu^{2+})	Hemoglobin synthesis	Meat, nuts, legumes	Anemia, stunted growth in children	Damage to internal organs if not excreted
Iodine (I^-)	Thyroid hormone synthesis	Iodized table salt, seafood	Thyroid deficiency	Depressed thyroid function, anxiety
Selenium (SeO_4^{2-})	Part of antioxidant enzyme	Seafood, meats, eggs	Vascular collapse, possible cancer development	Hair and fingernail loss, discolored skin
Manganese (Mn^{2+})	Part of enzymes	Nuts, legumes, green vegetables	Weakness and confusion	Confusion, coma, death

been found to be essential to good health. Table 8.5 lists the functions of various minerals and gives their food sources and signs of deficiency and toxicity.

Occasionally, individuals do not receive enough iron (especially women), calcium, magnesium, or zinc in their diets. Adult females need more iron in their diet than males (18 mg compared with 10 mg) because they lose hemoglobin each month during menstruation. Stress can bring on a magnesium deficiency, and due to its high-fiber content, a vegetarian diet may make zinc less available to the body. However, a varied and complete diet usually supplies enough of each type of mineral.

Calcium

Calcium (Ca^{2+}) is a major mineral needed for the construction of bones and teeth. It's also necessary for nerve conduction, muscle contraction, and blood clotting. Many people take calcium supplements to prevent or counteract **osteoporosis,** a degenerative bone disease that afflicts an estimated one-fourth of older men and one-half of older women in the United States. Osteoporosis develops because bone-eating cells called osteoclasts are more active than bone-forming cells called osteoblasts. Therefore, the bones are porous, and they break easily because they lack sufficient calcium. Due to studies that show consuming more calcium does slow bone loss in elderly people, the guidelines were revised. A calcium intake of 1,000 mg a day is recommended for men and for women who are premenopausal, and 1,300 mg a day is recommended for postmenopausal women. To achieve this amount, supplemental calcium is most likely necessary.

Small-framed, Caucasian women with a family history of osteoporosis are at greatest risk of developing the disease. Smoking and drinking more than nine cups of caffienated drinks daily may also contribute. Vitamin D is an essential companion to calcium in preventing osteoporosis. Other vitamins may also be helpful; for example, magnesium has been found to suppress the cycle that leads to bone loss. In addition to adequate calcium and vitamin intake, exercise helps prevent osteoporosis. Medications are also available that slow bone loss while increasing skeletal mass. These are still being studied for their effectiveness and possible side effects.

Sodium

Sodium plays a major role in regulating the body's water balance, as does chloride (Cl^-). The recommended amount of sodium intake per day is 500 mg, although the average American takes in 4,000–4,700 mg every day. This imbalance has caused concern because sodium in the form of salt intensifies hypertension (high blood pressure) if you already have it. About one-third of the sodium we consume occurs naturally in foods. Another third is added during commercial processing; and we add the last third either during home cooking or at the table in the form of table salt.

Clearly, it is possible to cut down on the amount of sodium in the diet. Table 8.6 gives recommendations for doing so.

Health **Focus**

When Zero Is More Than Nothing

When the Food and Drug Administration required the addition of trans fat information to food labels in 2006, many food companies created labels touting their products as "trans-fat free." A check of the label details would list 0 grams of trans fat, in the area where fat grams are listed. But a more thorough check of the list of ingredients might reveal a bit of trans fat lurking in the food.

If you see "partially hydrogenated oil" listed with the ingredients, there *are* some trans fats in that particular food. Trans fats only have to be listed with the breakdown of fat grams when there are 0.5 gram or more per serving. Limiting trans fats to 1% of daily calories is recommended by the American Heart Association. Unfortunately, eating more than one serving of a food with "hidden" trans fats might push some people over the recommended daily intake of trans fats.

The absence of ingredient labeling for foods served at restaurants and bakeries is a concern as well. Trans fats were commonly used in foods to extend the food products' shelf life. Many companies have started to discontinue the use of trans fats, given the health risks associated with them. Several fast food companies have made very public announcements of their goal to eliminate the use of trans fats in their preparation of food. New York City and Philadelphia have both banned the use of trans fats in restaurant and bakery foods, and other cities are considering similar action.

Check the labels of foods in your house, and see if claims of "zero grams of trans fats" are true. When you discover hidden trans fats, be sure to watch how many servings you eat. Practice with the label pictured here and see how you do.

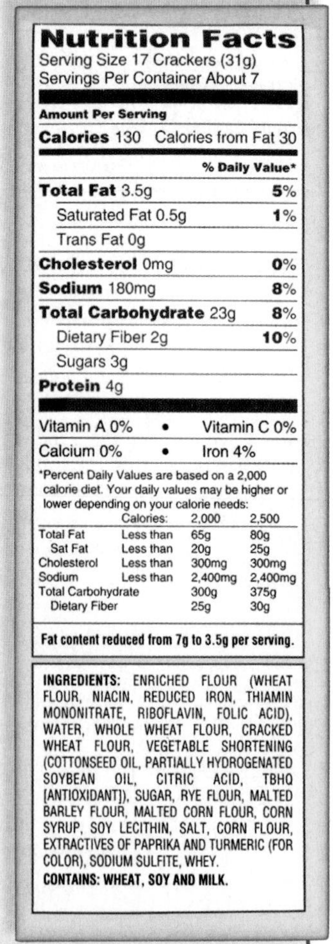

Nutrition Facts		
Serving Size 17 Crackers (31g)		
Servings Per Container About 7		
Amount Per Serving		
Calories 130　Calories from Fat 30		
		% Daily Value*
Total Fat 3.5g		**5%**
Saturated Fat 0.5g		**1%**
Trans Fat 0g		
Cholesterol 0mg		**0%**
Sodium 180mg		**8%**
Total Carbohydrate 23g		**8%**
Dietary Fiber 2g		**10%**
Sugars 3g		
Protein 4g		
Vitamin A 0%	•	Vitamin C 0%
Calcium 0%	•	Iron 4%

*Percent Daily Values are based on a 2,000 calorie diet. Your daily values may be higher or lower depending on your calorie needs:

	Calories:	2,000	2,500
Total Fat	Less than	65g	80g
Sat Fat	Less than	20g	25g
Cholesterol	Less than	300mg	300mg
Sodium	Less than	2,400mg	2,400mg
Total Carbohydrate		300g	375g
Dietary Fiber		25g	30g

Fat content reduced from 7g to 3.5g per serving.

INGREDIENTS: ENRICHED FLOUR (WHEAT FLOUR, NIACIN, REDUCED IRON, THIAMIN MONONITRATE, RIBOFLAVIN, FOLIC ACID), WATER, WHOLE WHEAT FLOUR, CRACKED WHEAT FLOUR, VEGETABLE SHORTENING (COTTONSEED OIL, PARTIALLY HYDROGENATED SOYBEAN OIL, CITRIC ACID, TBHQ [ANTIOXIDANT]), SUGAR, RYE FLOUR, MALTED BARLEY FLOUR, MALTED CORN FLOUR, CORN SYRUP, SOY LECITHIN, SALT, CORN FLOUR, EXTRACTIVES OF PAPRIKA AND TURMERIC (FOR COLOR), SODIUM SULFITE, WHEY.
CONTAINS: WHEAT, SOY AND MILK.

Figure 8C　How can you tell that a particular food isn't really free of trans fat?

Table 8.6	Reducing Dietary Sodium

To reduce dietary sodium:

1. Use spices instead of salt to flavor foods.
2. Add little or no salt to foods at the table, and add only small amounts of salt when you cook.
3. Eat unsalted crackers, pretzels, potato chips, nuts, and popcorn.
4. Avoid hot dogs, ham, bacon, luncheon meats, smoked salmon, sardines, and anchovies.
5. Avoid processed cheese and canned or dehydrated soups.
6. Avoid brine-soaked foods, such as pickles or olives.
7. Read nutrition labels to avoid high-salt products.

Vitamins

Vitamins are organic compounds (other than carbohydrate, fat, and protein) that the body uses for metabolic purposes but is unable to produce in adequate quantity. Many vitamins are portions of coenzymes, enzyme helpers. For example, niacin is part of the coenzyme NAD, and riboflavin is part of another dehydrogenase, FAD, discussed in Chapter 3. Coenzymes are needed in only small amounts because each can be used over and over. Not all vitamins are coenzymes. Vitamin A, for example, is a precursor for the visual pigment that prevents night blindness. If vitamins are lacking in the diet, various symptoms develop (Tables 8.7 and 8.8). There are 13 vitamins, divided into those that are fat soluble (see Table 8.7) and those that are water soluble (see Table 8.8).

Antioxidants

Over the past 20 years, numerous statistical studies have been done to determine whether a diet rich in fruits and vegetables can protect against cancer. Cellular metabolism generates free radicals, unstable molecules that carry an extra electron. The most common free radicals in cells are superoxide (O_2^-) and hydroxide (OH^-). To stabilize themselves, free radicals donate an electron to DNA, to proteins (including enzymes), or to li-

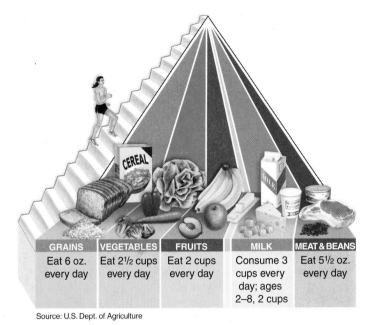

Source: U.S. Dept. of Agriculture

Figure 8.15 The 2005 USDA food guide pyramid. What is the recommended daily intake of fruits and vegetables according to this 2005 USDA food guide pyramid?
This food guide pyramid can be helpful when planning nutritious meals and snacks. Note the narrow yellow component that represents fats like butter and oils. They should be used sparingly.

pids, which can be found in plasma membranes. Such donations most likely damage these cellular molecules and thereby may lead to disorders, perhaps even cancer.

Vitamins C, E, and A are believed to defend the body against free radicals, and are therefore termed antioxidants. These vitamins are especially abundant in fruits and vegetables. The food pyramid in Figure 8.15 suggests we eat 2½ cups of vegetables and 2 cups of fruit a day. To achieve this goal, think in terms of salad greens, raw or cooked vegetables, dried fruit, and fruit juice, in addition to apples and oranges and other fresh fruits.

Table 8.7	Fat-Soluble Vitamins			
Mineral	**Functions**	**Food Sources**	**Conditions With**	
			Too Little	**Too Much**
Vitamin A	Antioxidant synthesized from beta-carotene; needed for healthy eyes, skin, hair, and mucous membranes, and for proper bone growth	Deep yellow/orange and leafy, dark-green vegetables; fruits; cheese; whole milk; butter; eggs	Night blindness, impaired growth of bones and teeth	Headache, dizziness, nausea, hair loss, abnormal development of fetus
Vitamin D	A group of steroids needed for development and maintenance of bones and teeth	Milk fortified with vitamin D, fish liver oil; also made in the skin when exposed to sunlight	Rickets, decalcification and weakening of bones	Calcification of soft tissues, diarrhea, possible renal damage
Vitamin E	Antioxidant that prevents oxidation of vitamin A and polyunsaturated fatty acids	Leafy green vegetables, fruits, vegetable oils, nuts, whole-grain breads and cereals	Unknown	Diarrhea, nausea, headaches, fatigue, muscle weakness
Vitamin K	Needed for synthesis of substances active in clotting of blood	Leafy green vegetables, cabbage, cauliflower	Easy bruising and bleeding	Can interfere with anticoagulant medication

Table 8.8	Water-Soluble Vitamins				
Vitamin	**Functions**	**Food Sources**	**Conditions With**		
			Too Little	**Too Much**	
Vitamin C	Antioxidant; needed for forming collagen; helps maintain capillaries, bones, and teeth	Citrus fruits, leafy green vegetables, tomatoes, potatoes, cabbage	Scurvy, delayed wound healing, infections	Gout, kidney stones, diarrhea, decreased copper	
Thiamine (vitamin B_1)	Part of coenzyme needed for cellular respiration; also promotes activity of the nervous system	Whole-grain cereals, dried beans and peas, sunflower seeds, nuts	Beriberi, muscular weakness, enlarged heart	Can interfere with absorption of other vitamins	
Riboflavin (vitamin B_2)	Part of coenzymes, such as FAD;* aids cellular respiration, including oxidation of protein and fat	Nuts, dairy products, whole-grain cereals, poultry, leafy green vegetables	Dermatitis, blurred vision, growth failure	Unknown	
Niacin (nicotinic acid)	Part of coenzymes NAD;+ needed for cellular respiration, including oxidation of protein and fat	Peanuts, poultry, whole-grain cereals, leafy green vegetables, beans	Pellagra, diarrhea, mental disorders	High blood sugar and uric acid, vasodilation, etc.	
Folacin (folic acid)	Coenzyme needed for production of hemoglobin and formation of DNA	Dark, leafy green vegetables; nuts; beans; whole-grain cereals	Megaloblastic anemia, spina bifida	May mask B_{12} deficiency	
Vitamin B_6	Coenzyme needed for synthesis of hormones and hemoglobin; CNS control	Whole-grain cereals, bananas, beans, poultry, nuts, leafy green vegetables	Rarely, convulsions, vomiting, seborrhea, muscular weakness	Insomnia, neuropathy	
Pantothenic acid	Part of coenzyme A needed for oxidation of carbohydrates and fats; aids in the formation of hormones and certain neurotransmitters	Nuts, beans, dark-green vegetables, poultry, fruits, milk	Rarely, loss of appetite, mental depression, numbness	Unknown	
Vitamin B_{12}	Complex, cobalt-containing compound; part of the coenzyme needed for synthesis of nucleic acids and myelin	Dairy products, fish, poultry, eggs, fortified cereals	Pernicious anemia	Unknown	
Biotin	Coenzyme needed for metabolism of amino acids and fatty acids	Generally in foods, especially eggs	Skin rash, nausea, fatigue	Unknown	

*FAD = flavin adenine dinucleotide
+NAD = nicotinamide adenine dinucleotide

Dietary supplements may provide a potential safeguard against cancer and cardiovascular disease. Nutritionists do not think people should take supplements instead of improving their intake of fruits and vegetables. There are many beneficial compounds in these foods that cannot be obtained from a vitamin pill. These compounds enhance one another's absorption or action and also perform independent biological functions.

Vitamin D

Skin cells contain a precursor cholesterol molecule converted to vitamin D after UV exposure. Vitamin D leaves the skin and is modified first in the kidneys and then in the liver until finally it becomes calcitriol. Calcitriol promotes the absorption of calcium by the intestines. When taking a calcium supplement, it's a good idea to get one with added vitamin D. The lack of vitamin D leads to rickets in chil-

dren. Rickets, characterized by bowing of the legs, is caused by defective mineralization of the skeleton. Most milk is fortified with vitamin D, which helps prevent the occurrence of rickets.

How to Plan Nutritious Meals

Many serious disorders in Americans are linked to a diet that results in excess body fat. While genetics is a factor in being overweight, a person cannot become fat without taking in more food energy (Calories, Kcal) than is needed. A person needs calories (energy) for their basal metabolism. Basal metabolism is the number of calories your body burns at rest to maintain normal body functions. A person also needs calories for exercise. The less exercise, the fewer calories needed beyond the basal metabolic rate. So, the first step in planning a diet is to limit the number

| Table 8.9 | Preparing Meals and Snacks Using the Food Guide Pyramid | | | | |

Breakfast	Snack	Lunch	Snack	Dinner	Dessert
½ cup oat flakes cereal	Whole-wheat muffin	Toasted American cheese sandwich (whole-grain bread)	Apple	3 oz. hamburger patty	½ cup frozen vanilla yogurt topped with sliced strawberries
8 oz. nonfat milk	2 Tbsp peanut butter mixed with 1 tsp sunflower seeds	1 cup vegetable salad	4 graham crackers	Whole-wheat bun	
½ cup blueberries	8 oz. nonfat milk	1 Tbsp olive oil and vinegar salad dressing	½ cup baked beans		
1 slice whole-grain toast		4 oz. tomato juice		1 cup steamed broccoli spears	
1 tsp soft margarine				1 tsp soft margarine	
4 oz. orange juice					

of calories to an amount that you will use each day. Let's say you do all the necessary calculations (beyond the scope of this book). You discover that, as a woman, the maximum number of calories you can take in each day is 2,000. If you're a man, you can afford 2,500 calories without gaining weight.

Now, the new food pyramid developed by the U.S. Department of Agriculture (USDA) can be used to help you decide how those calories should be distributed among the foods to eat (Fig. 8.15). The new pyramid emphasizes food that should be eaten often and omits foods that should not be eaten on a regular basis. Additionally, the USDA provides recommendations concerning the minimum quantity of foods in each group that should be eaten daily. In general, we should:

- eat a variety of foods. Foods from all food groups should be included in the diet.
- eat more of these foods: fruits, vegetables, whole grains, and fat-free or low-fat milk products. Choose dark-green vegetables, orange vegetables, and leafy vegetables. Dry beans and peas are good sources for fiber and a great protein source as well. Limit potatoes and corn. When eating grains, choose whole grains such as brown rice, oatmeal, whole-wheat bread. Choose fruit as a snack or a topping for foods, instead of sugar.
- choose lean meats such as poultry, fish high in omega-3 fatty acids, such as salmon, trout, and herring, in moderate sized portions. Include oils rich in monounsaturated and polyunsaturated fatty acids in the diet.
- eat less of foods high in saturated or trans fats, added sugars, cholesterol, salt, and alcohol.
- be physically active every day. If weight loss is needed, decrease calorie intake slowly, while maintaining adequate nutrient intake and increasing physical activity.

Table 8.9 gives you a sample menu, developed by a nutritionist, so you can see what a good diet for one day would look like. Table 8.9 also gives you an idea of what a nutritionist means by a "serving size."

Eating Disorders

People with eating disorders are dissatisfied with their body image. Social, cultural, emotional, and biological factors all contribute to the development of an eating disorder. Serious conditions such as obesity, anorexia nervosa, and bulimia nervosa can lead to malnutrition, disability, and death. Regardless of the eating disorder, early recognition and treatment are crucial.

Anorexia nervosa is a severe psychological disorder characterized by an irrational fear of getting fat. Victims refuse to eat enough food to maintain a healthy body weight (Fig. 8.16a). A self-imposed starvation diet is often accompanied by occasional binge eating, followed by purging and extreme physical activity to avoid weight gain. Binges usually include large amounts of high-calorie foods, and purging episodes involve self-induced vomiting and laxative abuse. About 90% of people suffering from anorexia nervosa are young women; an estimated 1 in 200 teenage girls is affected.

A person with **bulimia nervosa** binge-eats, and then purges to avoid gaining weight (Fig. 8.16b). The binge-purge cycle behavior can occur several times a day. People with bulimia nervosa can be difficult to identify because their body weights are often normal and they tend to conceal their binging and purging practices. Women are more likely than men to develop bulimia; an estimated 4% of young women suffer from this condition.

Other abnormal eating practices include binge-eating disorder and muscle dysmorphia. Many obese people suffer

a. Anorexia nervosa

b. Bulimia nervosa

c. Muscle dysmorphia

Figure 8.16 **What are the characteristics of different eating disorders?**
a. People with anorexia nervosa have a mistaken body image and think they are fat, even though they are thin. **b.** Those with bulimia nervosa overeat and then purge their bodies of the food they have eaten. **c.** People with muscle dysmorphia think their muscles are underdeveloped. They spend hours at the gym and are preoccupied with diet as a way to gain muscle mass.

CASE STUDY AFTER SURGERY

Prior to Monica's discharge from the hospital, she had to meet with a registered dietician to learn how to develop an acceptable postsurgery diet. A knock on her hospital room door told Monica that the dietician had arrived. "Hi Monica, I'm Holly," she said. "I hear you need some nutritional information before you can bail out of here."

Monica replied, "Come on in. I am *sooo* ready to get out of here and go home."

Holly put down her box of brochures. "Okay then, listen up," she began. "First, the most important thing to remember is that your stomach is now very small. If you eat too much, it can easily get too full, which will make you sick. Your stomach will stretch out a bit over time. Still, from now on you must be really careful about the amount of food you eat or liquids you drink." Monica took in this information, while thinking to herself, "I hope I know what I've gotten myself into!"

"For a few weeks, you need to puree anything solid you plan to eat. If you don't yet have a good blender, have someone get you an early birthday present," laughed Holly. "Once you're able to add solid foods to your diet, be sure to chew everything thoroughly."

She added, "Plan to have 5 to 7 small meals a day instead of 2 to 3 large ones. Don't eat food at the same time you have something to drink. You may need to start taking a multivitamin with added minerals to prevent deficiencies. Dr. Hall will advise you of that when you see him for follow up."

"The surgery won't work unless you change the food you eat," Holly warned. "For the rest of your life, you have to make every calorie count. You know what you should be eating, right? That's lean meat, fresh veggies and fruit, whole grains, and all that stuff." Monica nodded.

"All right then Monica, I think you're about ready to go home," Holly concluded. "I'll sign off on your paperwork, and we'll get you out of here as soon as possible. Remember, one of the most important things for you to do is exercise. Start with short walks as soon as you can. Build up to regular exercise 4 to 5 times a week."

She added, "Here's my card. Call me if you have any questions or want more information about joining our support group for folks who've had gastric bypass surgery."

"Good luck and take care," Holly smiled.

Have You Ever Wondered...

How you can easily determine the serving size of
a certain food?

The food pyramid (www.mypyramid.gov) provides good
information about the serving size for most foods you might
eat. For example, about 12 baby carrots is a serving (cup) of
carrots. Some good analogies for serving sizes will help. A
serving of meat (protein) should be the size of a deck of cards.
A potato should approximate the size of a light bulb or a
computer mouse, while an apple might be the size of a
baseball. A single serving of cheese should resemble the size
of a pair of dice.

from **binge-eating disorder,** a condition characterized
by episodes of overeating without purging. Stress,
anxiety, anger, and depression can trigger food binges.

A person suffering from **muscle dysmorphia** (Fig. 8.16*c*)
thinks his or her body is underdeveloped. Body-building
activities and a preoccupation with diet and body form accompany this condition. Each day, the person may spend
hours in the gym working out on muscle-strengthening
equipment. Unlike anorexia nervosa and bulimia, muscle
dysmorphia affects more men than women.

> ### Check Your Progress 8.6
>
> 1. **a.** What is the rationale behind using BMI to judge obesity?
>   **b.** What disorders are associated with obesity?
> 2. **Why might carbohydrates and why might fats be the cause of the obesity epidemic today?**
> 3. **a.** What fat types are dangerous to our health, and **(b)** what type is protective for cardiovascular disease?
> 4. **Proteins provide the essential amino acids. Why shouldn't they be eaten as an energy source?**
> 5. **In general, what does the new USDA food pyramid tell you?**
> 6. **Name and describe four eating disorders.**

CASE STUDY ONE YEAR LATER

Monica grinned when she realized the new text message on her cell phone was from Irene. Irene had been bugging her for days to go clothes shopping. After shedding 76 pounds following her surgery, Monica needed new clothes, especially some smaller jeans. She didn't plan to buy too many new items, because she hoped to lose another 20 pounds over the next several months. That would put her at 132 pounds and a BMI of 24. For a person five foot two, a BMI of 24 would put her in normal range.

"Normal," she thought. "I might not be tall and a twig, but I'll be normal." It had a nice ring to it.

Fortunately for Monica, recovery was quick. She would be left with some stretch marks, but she didn't have any loose folds of skin that would need to be removed surgically. Her mom recommended a concoction of shea butter, vitamin E, and collagen to minimize the appearance of the stretch marks. She found that her daily long walks and weight lifting were toning her muscles.

Her cell phone beeped to signal another incoming text from Irene. It read, "AYT" (Are You There?).

Monica laughed and texted back . . . "Meet U at my house 10 mins. B4N" (Bye for Now).

Summarizing the Concepts

8.1 Overview of Digestion

The organs of the digestive system are located within the GI tract. The processes of digestion require ingestion, digestion, movement, absorption, and elimination. All parts of the tract have four layers, called the mucosa, submucosa, muscularis, and serosa.

8.2 First Part of the Digestive Tract

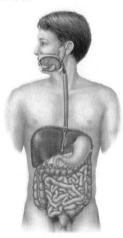

- In the mouth, teeth chew the food, saliva contains salivary amylase for digesting starch, and the tongue forms a bolus for swallowing.

- Both the mouth and the nose lead into the pharynx. The pharynx opens into both the food passage (esophagus) and air passage (trachea, or windpipe). During swallowing, the opening into the nose is blocked by the soft palate, and the epiglottis covers the trachea. Food enters the esophagus; and peristalsis begins. The esophagus moves food to the stomach.

8.3 The Stomach and Small Intestine

- The stomach expands and stores food and also churns, mixing food with the acidic gastric juices. This juice contains pepsin, an enzyme that digests protein.

- The duodenum of the small intestine receives bile from the liver and pancreatic juice from the pancreas. Bile emulsifies fat and readies it for digestion by lipase.

- The pancreas produces enzymes that digest starch (pancreatic amylase), protein (trypsin), and fat (lipase). The intestinal enzymes finish the process of chemical digestion.

- Small nutrient molecules are absorbed at the villi in the walls of the small intestine.

8.4 Three Accessory Organs and Regulation of Secretions

Three accessory organs of digestion send secretions to the duodenum via ducts. These organs are the pancreas, liver, and gallbladder.

- The pancreas produces pancreatic juice, which contains digestive enzymes for carbohydrate, protein, and fat.

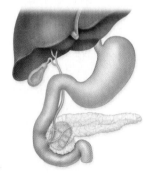

- The liver produces bile, destroys old blood cells, detoxifies blood, stores iron, makes plasma proteins, stores glucose as glycogen, breaks down glycogen to glucose, produces urea, and helps regulate blood cholesterol levels.

- The gallbladder stores bile, produced by the liver. The secretions of digestive juices are controlled by the nervous system and by hormones.

- Gastrin produced by the lower part of the stomach stimulates the upper part of the stomach to secrete pepsin.

- Secretin and CCK produced by the duodenal wall stimulate the pancreas to secrete its juices and the gallbladder to release bile.

8.5 The Large Intestine and Defecation

- The large intestine consists of the cecum; the colon (including the ascending, transverse, and descending colon); and the rectum, which ends at the anus.

- The large intestine absorbs water, salts, and some vitamins; forms the feces; and carries out defecation.

- Disorders of the large intestine include diverticulosis, irritable bowel syndrome, inflammatory bowel disease, polyps, and cancer.

8.6 Nutrition and Weight Control

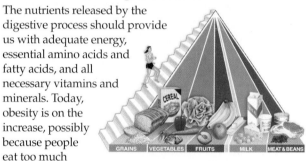

The nutrients released by the digestive process should provide us with adequate energy, essential amino acids and fatty acids, and all necessary vitamins and minerals. Today, obesity is on the increase, possibly because people eat too much food and make improper food choices. Obesity is associated with many illnesses, including diabetes type 2 and cardiovascular disease. The food guide pyramid shows foods to emphasize and foods to minimize for good health.

- Carbohydrates are necessary in the diet, but simple sugars and refined starches cause a rapid release of insulin that can lead to diabetes type 2.

- Proteins supply essential amino acids.

- Unsaturated fatty acids, particularly the omega-3 fatty acids, are protective against cardiovascular disease.

- Saturated fatty acids and trans fats contribute to heart disease.

- Vitamins and minerals are also required by the body in certain amounts.

Understanding Key Terms

absorption 159
anorexia nervosa 179
anus 168
appendix 159
bile 165, 167
binge-eating disorder 181
body mass index (BMI) 171
bolus 161
bulimia nervosa 179
cecum 168
cholesterol 167
chyme 163
cirrhosis 168
colon 168
constipation 169
defecation 168

dental caries 160
diaphragm 161
diarrhea 169
digestion 158
diverticulosis 159
duodenum 165
elimination 159
epiglottis 161
esophagus 161
essential amino acids 173
essential fatty acids 174
fiber 169
gallbladder 167
gallstone 167
gastric gland 163
glottis 161

glycemic index (GI) 173
hard palate 160
heartburn 161
hemorrhoid 169
hepatitis 167
hormone 166
hydrolyze 158
ingestion 158
jaundice 167
lacteal 166
lactose intolerance 166
large intestine 168
lipase 165
liver 167
lumen 159
mineral 175
movement 158
mucosa 159
muscle dysmorphia 181
muscularis 159
nutrient 171
obesity 170
osteoporosis 176

pancreas 166
pancreatic amylase 166
pepsin 163
periodontitis 160
peristalsis 161
peritonitis 159
pharynx 161
polyp 170
rectum 168
rugae 163
salivary amylase 160
salivary gland 160
serosa 159
small intestine 165
soft palate 160
sphincter 161
stomach 162
submuscosa 159
trypsin 166
urea 167
vermiform appendix 168
villus 165
vitamin 177

Match the key terms to these definitions.

a. _____ Essential requirement in the diet, needed in small amounts. Often a part of a coenzyme.

b. _____ Fat-digesting enzyme secreted by the pancreas.

c. _____ Lymphatic vessel in an intestinal villus; it aids in the absorption of fats.

d. _____ Muscular tube for moving swallowed food from the pharynx to the stomach.

e. _____ Organ attached to the liver that serves to store and concentrate bile.

Testing Your Knowledge of the Concepts

1. Argue that absorption is the most important of the five processes of digestion over the other four processes. (pages 158–59)

2. List the main organs of the digestive tract, and state the contribution of each to the digestive process. (pages 160–69)

3. Discuss the absorption of the products of digestion into the lymphatic and cardiovascular systems. (page 166)

4. Name the enzymes involved in the digestion of starch, protein, and fat, and tell where these enzymes are active and what they do. (page 165)

5. Why are the pancreas, liver, and gallbladder considered accessory organs of digestion and not an organ of digestion? (pages 166-67)

6. Name and state the functions of the hormones that assist the nervous system in regulating digestive secretions. (page 168)

7. What is the chief contribution of each of these in the body: carbohydrates, proteins, fats, fruits, and vegetables? (pages 173–78)

8. Which three eating disorders involve binge eating? How are these three disorders different from one another? (pages 179–80)

9. Tracing the path of food in the following list (a–f), which step is out of order first?
   a. mouth
   b. pharynx
   c. esophagus
   d. small intestine
   e. stomach
   f. large intestine

10. The appendix connects to the
    a. cecum.
    b. small intestine.
    c. esophagus.
    d. liver.
    e. All of these are correct.

11. Which association is incorrect?
    a. mouth—starch digestion
    b. esophagus—protein digestion
    c. small intestine—starch, lipid, protein digestion
    d. stomach—food storage
    e. liver—production of bile

12. Why can a person not swallow food and talk at the same time?
    a. To swallow, the epiglottis must close off the trachea.
    b. The brain cannot control two activities at once.
    c. To speak, air must come through the larynx to form sounds.
    d. A swallowing reflex is only initiated when the mouth is closed.
    e. Both a and c are correct.

13. Which association is incorrect?
    a. pancreas—produces alkaline secretions and enzymes
    b. salivary glands—produce saliva and amylase
    c. gallbladder—produces digestive enzymes
    d. liver—produces bile

14. Which of the following could be absorbed directly without need of digestion?
    a. glucose
    b. fat
    c. polysaccharides
    d. protein
    e. nucleic acid

15. Peristalsis occurs
    a. from the mouth to the small intestine.
    b. from the beginning of the esophagus to the anus.
    c. only in the stomach.
    d. only in the small and large intestine.
    e. only in the esophagus and stomach.

16. An organ is a structure made of two or more tissues performing a common function. Which of the four tissue types are present in the wall of the digestive tract?
    a. epithelium
    b. connective tissue
    c. nervous tissue
    d. muscle tissue
    e. All of these are correct.

17. Which association is incorrect?
    a. protein—trypsin
    b. fat—bile
    c. fat—lipase
    d. maltose—pepsin
    e. starch—amylase

18. Most of the products of digestion are absorbed across the
    a. squamous epithelium of the esophagus.
    b. striated walls of the trachea.
    c. convoluted walls of the stomach.
    d. fingerlike villi of the small intestine.
    e. smooth wall of the large intestine.

19. Bile
    a. is an important enzyme for the digestion of fats.
    b. cannot be stored.
    c. is made by the gallbladder.
    d. emulsifies fat.
    e. All of these are correct.

20. Which of the following is not a function of the liver in adults?
    a. produces bile
    b. detoxifies alcohol
    c. stores glucose
    d. produces urea
    e. makes red blood cells

21. The large intestine
    a. digests all types of food.
    b. is the longest part of the intestinal tract.
    c. absorbs water.
    d. is connected to the stomach.
    e. is subject to hepatitis.

In questions 22–26, match each function to an organ in the key.

Key:

    a. mouth
    b. esophagus
    c. stomach
    d. small intestine
    e. large intestine

22. Removes nondigestible remains

23. Serves as a passageway

24. Stores food

25. Absorbs nutrients

26. Receives food

27. The amino acids that must be consumed in the diet are called essential. Nonessential amino acids
    a. can be produced by the body.
    b. are only needed occasionally.
    c. are stored in the body until needed.
    d. can be taken in by supplements.

28. Which of the following are often organic portions of important coenzymes?
    a. minerals
    b. vitamins
    c. protein
    d. carbohydrates

29. The products of digestion are
    a. large macromolecules needed by the body.
    b. enzymes needed to digest food.
    c. small nutrient molecules that can be absorbed.
    d. regulatory hormones of various kinds.
    e. the food we eat.

In questions 30–34, match each statement to a layer of the wall of the esophagus in the key. Answers may be used more than once.

Key:

    a. mucosa
    b. submucosa
    c. muscularis
    d. serosa

30. Loose connective tissue that contains lymph nodules

31. Contains a layer of epithelium that lines the lumen

32. Very thin layer of squamous epithelium that secretes a fluid, keeping the organ moist

33. Contains digestive glands and mucus-secreting goblet cells

34. Two layers of smooth muscle

In questions 35–40, match each statement to an answer in the key. Answers may be used more than once. Some may have more than one answer.

Key:

    a. gastrin
    b. secretin
    c. CCK
    d. All of these are correct.
    e. None of these is correct.

35. Stimulates gallbladder to release bile

36. Hormone carried in bloodstream

37. Stimulates the stomach to digest protein

38. Enzyme that digests food

39. Secreted by duodenum

40. Secreted by the stomach

In questions 41–45, match each statement to a vitamin or mineral in the key.

Key:

    a. calcium
    b. vitamin K
    c. sodium
    d. iodine
    e. vitamin A

41. Needed to make thyroid hormone

42. Needed for night vision

43. Needed for bones, teeth, and muscle contraction

44. Needed for nerve conduction, pH, and water balance

45. Needed for making clotting proteins

Thinking Critically About the Concepts

Monica, from the case study, had bariatric surgery, which reduces the size of the stomach and enables food to bypass a section of the small intestine. The surgery is generally done when obese individuals have unsuccessfully tried numerous ways to lose weight and their health is compromised by their weight. There are many risks associated with the surgery, but it helps a number of people lose a considerable amount of weight. After people undergo the surgery, there are several lifestyle changes they must make to avoid nutritional deficiencies and to compensate for the small size of their stomach.

1. a. Why do some people who had bariatric surgery process their food in a blender (or have to chew thoroughly) before swallowing their food?
   b. Why should people who had bariatric surgery drink liquids between meals rather than with meals?

2. What risk is there to the esophagus after bariatric surgery?

3. The Historical Focus on page 164 describes how William Beaumont was able to observe the workings of the stomach through a fistula in St. Martin's stomach.
   a. What do you think Beaumont observed after he inserted a piece of meat into St. Martin's stomach?
   b. What do you think Beaumont observed after he inserted a piece of bread into St. Martin's stomach?

CHAPTER 9

Respiratory System

Justin stumbled down the stairs toward the kitchen where he could hear his roommate Chad making breakfast. "Ouch!" yelled Justin as he ran into the pantry door Chad left open.

Chad peered at Justin from behind the door and said, "Jeez, you look rough. The way you were snoring last night, I would have guessed you'd be ready to tackle anything today. Instead you look half dead."

"I am SO tired," complained Justin. "I even went to bed a little early last night, hoping that the extra sleep would make me feel better. Honestly, I think I may feel more tired than I did yesterday, which makes no sense!"

"Thanks for making a pot of coffee," he continued. "Hopefully a couple cups of java will do the trick. I've got a big physics exam today."

Later in the day, Justin woke with a start to find his physics professor shaking his shoulder. Dr. Edwards was saying, "Justin, wake up and finish your exam."

Justin blinked and tried to get his bearings. "I was sleeping during the exam?" he asked incredulously.

"Yes," replied Dr. Edwards. He continued, "This is not the first time you've fallen asleep in class. Are you not feeling well? Should I call someone to pick you up?"

Justin sighed and replied, "No, I don't feel sick. I'm just incredibly tired all the time, even when I get what should be plenty of sleep."

"I think you ought to go see the doctor at the student health service," Dr. Edwards responded. "There's got to be a good reason why a young healthy person always feels tired."

9.1 The Respiratory System

The organs of the respiratory system ensure that oxygen enters the body and carbon dioxide leaves the body (Fig. 9.1). During **inspiration,** or inhalation (breathing in), air is conducted from the atmosphere to the lungs by a series of cavities, tubes, and openings, illustrated in Figure 9.1. During **expiration,** or exhalation (breathing out), air is conducted from the lungs to the atmosphere by way of the same structures.

Ventilation is another term for breathing that includes both inspiration and expiration. Once ventilation has occurred, the respiratory system depends on the cardiovascular system to transport oxygen (O_2) from the lungs to the tissues and carbon dioxide (CO_2) from the tissues to the lungs.

Gas exchange is necessary because the cells of the body carry out cellular respiration to make energy in the form of ATP. During cellular respiration, cells use up O_2 and produce CO_2. The respiratory system provides these cells with O_2 and removes CO_2.

> **Check Your Progress 9.1**
> 1. Trace the path of air from the nasal cavities to the lungs.
> 2. What is the function of the respiratory system?

Figure 9.1 **How long is the respiratory tract?**
The respiratory tract extends from the nose to the lungs. Note the organs in the upper respiratory tract and the ones in the lower respiratory tract.

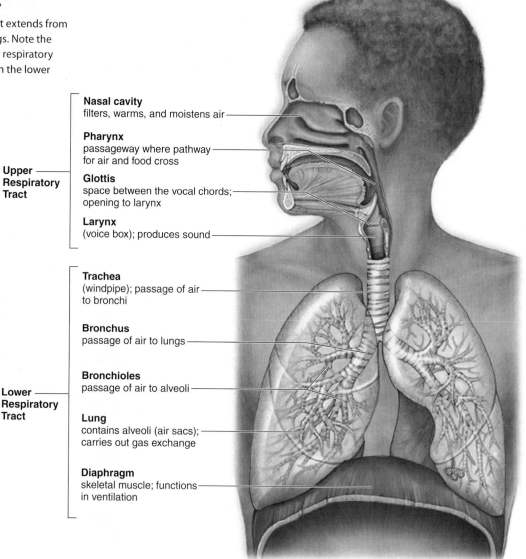

Upper Respiratory Tract

Nasal cavity
filters, warms, and moistens air

Pharynx
passageway where pathway for air and food cross

Glottis
space between the vocal chords; opening to larynx

Larynx
(voice box); produces sound

Lower Respiratory Tract

Trachea
(windpipe); passage of air to bronchi

Bronchus
passage of air to lungs

Bronchioles
passage of air to alveoli

Lung
contains alveoli (air sacs); carries out gas exchange

Diaphragm
skeletal muscle; functions in ventilation

9.2 The Upper Respiratory Tract

The nasal cavities, pharynx, and larynx are the organs of the upper respiratory tract (Fig. 9.2).

The Nose

The nose opens at the nares (nostrils) that lead to the **nasal cavities.** The nasal cavities are narrow canals separated from each other by a septum composed of bone and cartilage (Fig. 9.2).

Air entering the nasal cavities is met by large stiff hairs that act as a screening device. The hairs filter the air and trap small particles (dust, mold spores, pollen, etc.) so they don't enter air passages. The rest of the nasal cavities are lined by mucous membrane with a submucosa that has plentiful capillaries. These capillaries help warm and moisten the air, but they also make us more susceptible to nose bleeds if the nose suffers an injury. The mucus secreted by the mucous membrane helps trap dust and move it to the pharynx, where it can be swallowed or expectorated.

Moistened air leaves the nose, and when we breathe out on a cold day, it condenses so that we can see our breath.

Special ciliated cells in the narrow upper recesses of the nasal cavities act as odor receptors. Nerves lead from these cells to the brain, where the impulses generated by the odor receptors are interpreted as smell.

The tear (lacrimal) glands drain into the nasal cavities by way of tear ducts. When you cry, your nose runs as tears drain from the eye surface into the nose.

The nasal cavities also communicate with the cranial sinuses, and fluid may accumulate. Excess pressure from this fluid will cause a sinus headache from the increased pressure.

Air in the nasal cavities passes into the nasopharynx, the upper portion of the pharynx. The middle ears have tubes, called **auditory (Eustachian) tubes,** that empty into the nasopharynx. When air pressure inside the middle ears equalizes with the air pressure in the nasopharynx, the auditory tube openings may "pop." In an effort to move air from the ears and prevent the popping, some people chew gum or yawn when a plane takes off and lands.

Have You Ever Wondered...

Whether stifling a sneeze is harmful?

Scientists have found that air travels at 100 miles an hour during a sneeze. Further, sneeze droplets may travel up to 12 ft away from the person sneezing. If a sneeze is stifled, the air is forced into the Eustachian tube and middle ear. Damage to the ear could occur when a sneeze is stifled.

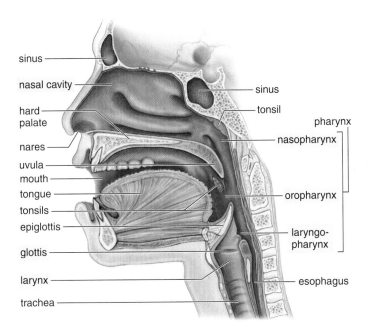

Figure 9.2   Can you trace the flow of air through the upper respiratory tract?
This drawing shows the path of air from the nasal cavities to the trachea, in the lower respiratory tract. The designated structures are in the upper respiratory tract.

The Pharynx

The **pharynx** is a funnel-shaped passageway that connects the nasal and oral cavities to the larynx. Therefore, the pharynx, commonly referred to as the "throat," has three parts: the nasopharynx, where the nasal cavities open above the soft palate; the oropharynx, where the oral cavity opens; and the laryngopharynx, which opens into the larynx.

The tonsils form a protective ring at the junction of the oral cavity and the pharynx. Inhaled air passes directly over the tonsils, making them the primary lymphatic defense during breathing. Being lymphatic tissue, the tonsils contain lymphocytes that protect against invasion of inhaled foreign antigens. In the tonsils, B cells and T cells are prepared to respond to antigens that may subsequently invade internal tissues and fluids. Therefore, the respiratory tract assists the immune system in maintaining homeostasis.

In the pharynx, the air passage and the food passage lie parallel to each other and share a common opening in the laryngopharynx. The larynx is normally open, allowing air to pass, but the esophagus is normally closed and opens only when a person swallows. If someone swallows and some of the food enters the larynx, coughing occurs in an

Figure 9.3 **When is the Heimlich maneuver used?**
The Heimlich maneuver is used when someone's ability to breathe is prevented by an obstruction blocking the airway. The steps associated with the Heimlich maneuver are shown in the illustration.

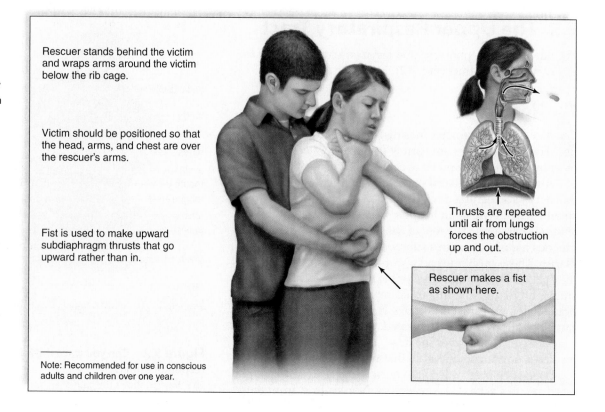

Rescuer stands behind the victim and wraps arms around the victim below the rib cage.

Victim should be positioned so that the head, arms, and chest are over the rescuer's arms.

Fist is used to make upward subdiaphragm thrusts that go upward rather than in.

Thrusts are repeated until air from lungs forces the obstruction up and out.

Rescuer makes a fist as shown here.

Note: Recommended for use in conscious adults and children over one year.

effort to dislodge the flood. A **cough** is a sudden, noisy explosion of air from the lungs. The Heimlich maneuver shown in Figure 9.3 may be used to dislodge food blocking the **epiglottis** and airway.

Larynx

The **larynx** is a cartilaginous structure that serves as a passageway for air between the pharynx and the trachea. The larynx can be pictured as a triangular box whose apex, the Adam's apple, is located at the front of the neck. The larynx is called the *voice box* because it houses the vocal cords. The **vocal cords** are mucosal folds supported by elastic ligaments, and the slit between the vocal cords is an opening called the **glottis** (Fig. 9.4). When air is expelled through the glottis, the vocal cords vibrate, producing sound. At the time of puberty, the growth of the larynx and the vocal cords is much more rapid and accentuated in the male than in the female, causing the male to have a more prominent Adam's apple and a deeper voice. The voice "breaks" in the young male due to his inability to control the longer vocal cords.

The high or low pitch of the voice is regulated when speaking and singing by changing the tension on the vocal cords. The greater the tension, as when the glottis becomes narrower, the higher the pitch. When the glottis is wider, the pitch is lower (Fig. 9.4, *right*). The loudness, or intensity, of the voice depends upon the amplitude of the vibrations—the degree to which the vocal cords vibrate.

Ordinarily when food is swallowed, the larynx moves upward against the epiglottis, a flap of tissue that prevents

food from passing into the larynx. You can detect the movement of the larynx by placing your hand gently on your larynx and swallowing.

Check Your Progress 9.2
1. **What organs are a part of the upper respiratory tract? What are the functions of each organ?**
2. **Several different organs have connections with the pharynx. What are they?**
3. **a. What two pathways cross in the pharynx? b. Why is it correct to say that they "cross"?**

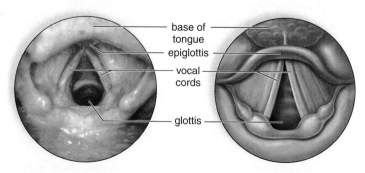

base of tongue
epiglottis
vocal cords
glottis

Figure 9.4 **How is sound produced?**
Viewed from above, the vocal cords can be seen to stretch across the glottis, the opening to the trachea. When air is expelled through the glottis, the vocal cords vibrate, producing sound. The glottis is narrow when we produce a high-pitched sound, and it widens as the pitch deepens.

9.3 The Lower Respiratory Tract

Whereas the nasal cavities, pharynx, and larynx are a part of the upper respiratory tract, the trachea and the rest of the respiratory system are in the lower respiratory tract.

The Trachea

The **trachea,** commonly called the windpipe, is a tube connecting the larynx to the primary bronchi. Its walls consist of connective tissue and smooth muscle reinforced by C-shaped cartilaginous rings. The rings prevent the trachea from collapsing.

The trachea lies anterior to the esophagus. It is separated from the esophagus by a flexible muscular wall. This orientation allows the esophagus to expand when swallowing. The mucous membrane that lines the trachea has an outer layer of pseudostratified ciliated columnar epithelium (see Fig. 4.7) and goblet cells (Fig. 9.5). The goblet cells produce mucus, which traps debris in the air as it passes through the trachea. The mucus is then swept toward the pharynx and away from the lungs by the cilia that project from the epithelium.

When one coughs, the tracheal wall contracts, narrowing its diameter. Therefore, coughing causes air to move more rapidly through the trachea, helping to expel mucus and foreign objects. Smoking is known to destroy the cilia, and consequently, the soot in cigarette smoke collects in the lungs. Smokers often develop heavy coughs as a result. Smoking cigarettes is discussed more fully in the Health Focus on pages 201–02.

If the trachea is blocked because of illness or the accidental swallowing of a foreign object, it is possible to insert a breathing tube by way of an incision made in the trachea.

CASE STUDY AT THE HEALTH CLINIC

Justin glanced around the waiting room, trying to guess why the various students were at the health clinic. It seemed obvious why some were there. They sniffled and wheezed so much that Justin moved a few chairs away from them, hoping to avoid getting whatever they had.

Moments later the nurse called Justin's name and showed him to a room. Justin's pulse, blood pressure, and body temperature were all taken and recorded. Shortly after that, the general practitioner walked through the door. "You look pretty healthy compared to a lot of the students who are waiting," observed the doctor. "Your blood pressure seems a little high for someone your age, but those extra pounds you're carrying around might be causing that. Is there a history of high blood pressure and heart disease in your family?"

Justin thought for a minute. "My grandfather takes medication for blood pressure, but he's the only person in my family who does. Or at least, he's the only one I know about."

"Have you ever had tests to check your thyroid function and blood sugar?" queried the doctor. Justin shook his head. "Then we'll definitely check that." The doctor made some notes on Justin's chart about getting blood work done.

"Tell me about your sleep habits, Justin," the doctor then asked. "Do you go to bed at a regular time? Is your bedroom comfortable—not too hot or cold?" Justin nodded. "Do you drink alcohol or use sedatives or sleeping pills?"

"I have a beer or two on the weekends, but that's it," Justin replied. "I don't take any sleeping pills or anything like that. I must snore a lot, though. My roommate complains all the time that I snore really loud. I guess I wake him up all the time."

"Do other people in your family snore or have a history of sleep disorders?" asked the doctor.

"Pretty much the entire bunch of us are heavy snorers," confirmed Justin.

"I think we should get you lined up for a visit to the sleep center," observed the doctor. "You may have obstructive sleep apnea, or OSA. Apnea means no breathing, and that's what happens with OSA. You stop breathing while you're asleep. Your breathing stops for about 20 seconds at a time, and then your brain jerks you awake. You gasp or gulp for breath. That may be why you don't feel rested after sleeping and why you're tired and fall asleep throughout the day. That may also explain the increased blood pressure we observed today."

"So I just go there and fall asleep?" Justin asked. "That ought to be a piece of cake for me."

"They'll be running tests on you while you're sawing logs," the doctor smiled. "They'll test for obstruction of your airway during sleep. You also get a polysomnography to assess what happens during sleeping."

"What's a polysom—or whatever that is?" Justin asked.

"It's a sleep test," the doctor replied. "They hook you up to monitors that track your heart, lungs, and brain while you sleep. The monitors also record your breathing, how much oxygen is in your blood, and your arm and leg movements. All of this goes on with you asleep. It helps to explain the quality of your sleep. If you have obstructive sleep apnea, the sleep center will make recommendations about how to treat it," the doctor finished.

"In the meantime, we'll get the blood work done to rule out problems with your thyroid. The thyroid gland is in your neck, and it controls your metabolism. If your thyroid isn't working, that can make your breathe too slowly."

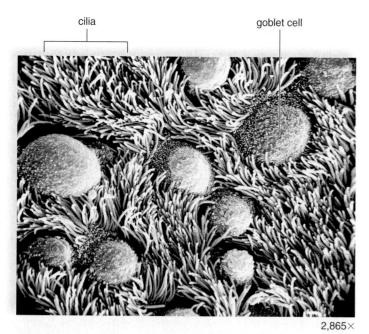

Figure 9.5   **How is inhaled debris prevented from reaching the lungs?**
Scanning electron micrograph of the surface of the mucous membrane lining the trachea consisting of goblet cells and ciliated cells. The cilia sweep mucus and debris embedded in it toward the pharynx, where it is swallowed or expectorated. Smoking causes the cilia to disappear; consequently, debris now enters the bronchi and lungs.
© Dr. Kessel & Dr. Kardon/Tissues & Organs/Visuals Unlimited

This tube acts as an artificial air intake and exhaust duct. The operation is called a **tracheostomy.**

The Bronchial Tree

The trachea divides into right and left primary bronchi (sing., **bronchus**), which lead into the right and left lungs (see Fig. 9.1). The bronchi branch into a few secondary bronchi that also branch, until the branches become about 1 mm in diameter and are called **bronchioles.** The bronchi resemble the trachea in structure. As the bronchial tubes divide and subdivide, their walls become thinner, and the small rings of cartilage are no longer present. During an asthma attack, the smooth muscle of the bronchioles contracts, causing bronchiolar constriction and characteristic wheezing. Each bronchiole leads to an elongated space enclosed by a multitude of air pockets, or sacs, called alveoli (sing., **alveolus**) (Fig. 9.6).

The Lungs

The **lungs** are paired, cone-shaped organs in the thoracic cavity. Located in the central area of the thoracic cavity are the trachea, heart, and esophagus. The lungs are on either side of this central area. The right lung has three lobes, and the left lung has two lobes, allowing room for the heart, which points left. A lobe is further divided into lobules, and each lobule has a bronchiole serving many alveoli.

The lungs follow the contours of the thoracic cavity, including the diaphragm, the muscle that separates the thoracic cavity from the abdominal cavity. Each lung is enclosed by **pleurae,** two layers of serous membrane that produce serous fluid. The parietal pleura adheres to the thoracic cavity wall, and the visceral pleura adheres to the surface of the lung. Surface tension is the tendency for water molecules to cling to one another due to hydrogen bonding between molecules. Surface tension holds the two pleural layers together. Therefore, the lungs must follow the movement of the thorax when breathing occurs. When someone has pleurisy, these layers are inflamed. Breathing, sneezing, and coughing are quite painful because the layers rub against each other.

The Alveoli

The lungs have about 300 million alveoli, with a total cross-sectional area of 50–70 m^2. That's about the size of a tennis court. Each alveolar sac is surrounded by blood capillaries. The walls of the sac and the capillaries are largely simple squamous epithelium (see Fig. 4.7). Gas exchange occurs

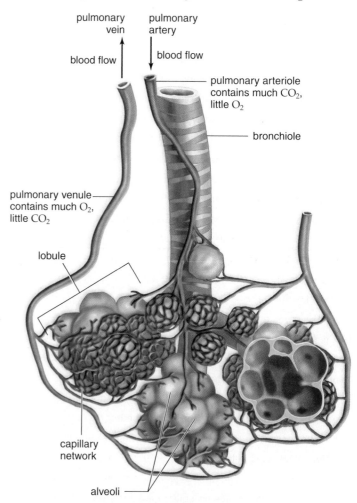

Figure 9.6   **Is blood in the pulmonary artery oxygenated? What about blood in a pulmonary vein?**
The lungs consist of alveoli surrounded by an extensive capillary network. The pulmonary artery carries O_2-poor (CO_2-rich) blood (colored blue), and the pulmonary vein carries O_2-rich blood (colored red).

between air in the alveoli and blood in the capillaries. Oxygen diffuses across the alveolar wall and enters the bloodstream, while carbon dioxide diffuses from the blood across the alveolar wall to enter the alveoli (see Fig. 9.6).

The alveoli of human lungs are lined with a **surfactant,** a film of lipoprotein that lowers the surface tension of water and prevents them from closing. The lungs collapse in some newborn babies, especially premature infants, who lack this film. The condition, called **infant respiratory distress syndrome,** is now treatable by surfactant replacement therapy.

> **Check Your Progress 9.3**
> 1. What organs are a part of the lower respiratory tract? What are the functions of each organ?
> 2. What part of the respiratory tract functions in gas exchange?

9.4 Mechanism of Breathing

Ventilation, or breathing, has two phases. The process of **inspiration,** also called inhalation, moves air into the lungs,

and the process of **expiration,** also called exhalation, moves air out of the lungs. To understand ventilation, the manner in which air enters and exits the lungs, it is necessary to remember the following facts:

1. Normally, there is a continuous column of air from the pharynx to the alveoli of the lungs.
2. The lungs lie within the sealed thoracic cavity. The rib cage, consisting of the ribs joined to the vertebral column posteriorly and to the sternum anteriorly, forms the top and sides of the thoracic cavity. The intercostal muscles lie between the ribs. The diaphragm and connective tissue form the floor of the thoracic cavity.
3. The lungs adhere to the thoracic wall by way of the pleura. Any space between the two pleurae is minimal due to the surface tension of the fluid between them.

Inspiration

Inspiration is the active phase of ventilation because this is the phase in which the diaphragm and the external intercostal muscles contract (Fig. 9.7a). In its relaxed state, the diaphragm is dome-shaped. During inspiration, it contracts

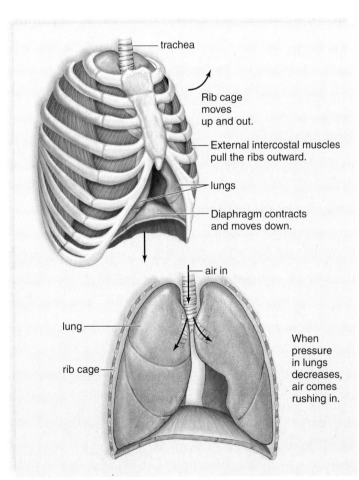

a. Inspiration

b. Expiration

Figure 9.7   What happens to the thoracic cavity and lungs during inspiration? What happens to them during expiration?
a. During inspiration, the thoracic cavity and lungs expand so that air is drawn in. **b.** During expiration, the thoracic cavity and lungs resume their original positions and pressures. Now air is forced out.

and becomes a flattened sheet of muscle. Also, the external intercostal muscles contract, causing the rib cage to move upward and outward.

Following contraction of the diaphragm and the external intercostal muscles, the volume of the thoracic cavity will be larger than it was before. As the thoracic volume increases, the lungs increase in volume as well because the lung adheres to the wall of the thoracic cavity. As the lung volume increases, the air pressure within the alveoli decreases, creating a partial vacuum. In other words, alveolar pressure is now less than atmospheric pressure (air pressure outside the lungs). Air will naturally flow from outside the body into the respiratory passages and into the alveoli, because a continuous column of air reaches into the lungs.

Air comes into the lungs because they have already opened up; air does not force the lungs open. This is why it is sometimes said that *humans inhale by negative pressure.* The creation of a partial vacuum in the alveoli causes air to enter the lungs. While inspiration is the active phase of breathing, the actual flow of air into the alveoli is passive. Just as with bellows used to fan a fire, when the handles of the bellows are pulled apart, air automatically flows into the bellows. When the pressure to pull apart the bellows is released, air flows out as the bellows close.

Expiration

Usually, expiration is the passive phase of breathing, and no effort is required to bring it about. During expiration, the diaphragm and external intercostal muscles relax. The rib cage returns to its resting position, moving down and inward (Fig. 9.7b). The elastic properties of the thoracic wall and lung tissue help them to recoil. In addition, the lungs recoil because the surface tension of the fluid lining the alveoli tends to draw them closed (see Chap. 2). If we continue the analogy of the bellows, imagine simply allowing the open bellows to fall shut. As the volume of the bellows decreases, the air pressure inside increases. Now air flows out (Fig. 9.8).

What keeps the alveoli from collapsing as a part of expiration? Recall that the presence of surfactant lowers the surface tension within the alveoli. Also, as the lungs recoil, the pressure between the pleura decreases, and this tends to make the alveoli stay open. The importance of the reduced intrapleural pressure is demonstrated when, by design or accident, air enters the intrapleural space. Now the lung collapses.

Maximum Inspiratory Effort and Forced Expiration

If you recall the last time you exercised vigorously—perhaps running in a race, or even just climbing all those stairs to your classroom—you probably remember that you were breathing a lot harder than normal, during and immediately after that heavy exercise. Maximum inspiratory effort involves muscles

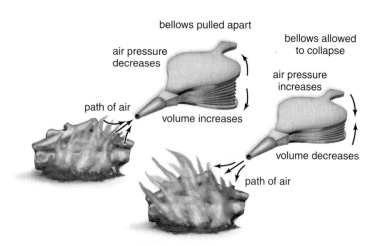

Figure 9.8 **What happens to air pressure when volume increases?**
When the bellows are pulled apart, the volume in the bellows increases and the air pressure decreases. When the bellows collapse, the volume decreases and air pressure increases.

of the back, chest, and neck. This increases the size of the thoracic cavity larger than normal, thus allowing maximum expansion of the lungs.

Expiration can also be forced. The maximum inspiratory efforts of heavy exercise are accompanied by forced expiration. Forced expiration is also necessary to sing, blow air into a trumpet, or blow out birthday candles. Contraction of the internal intercostal muscles can force the rib cage to move downward and inward. Also, when the abdominal wall muscles contract, they push on the abdominal organs. In turn, the organs push upward against the diaphragm, and the increased pressure in the thoracic cavity helps expel air.

Volumes of Air Exchanged During Ventilation

As ventilation occurs, air moves into the lungs from the nose or mouth during inspiration; and then moves out of the lungs during expiration. A free flow of air to and from the lungs is vitally important. Therefore, a technique has been developed that allows physicians to determine if there is a medical problem that prevents the lungs from filling with air upon inspiration and releasing it from the body upon expiration. This technique is illustrated in Figure 9.9, which shows the measurements recorded by a spirometer when a person breathes as directed by a technician. The actual numbers mentioned in the following discussion about lung volumes are averages. These numbers are affected by gender, height, and age, so your lung volumes may be different from the values stated here.

Tidal Volume   Normally when we are relaxed, only a small amount of air moves in and out with each breath, similar, perhaps, to the tide at the beach. This amount of air, called the **tidal volume,** is only about 500 mL.

Vital Capacity It is possible to increase the amount of air inhaled, and therefore the amount exhaled, by deep breathing. The maximum volume of air that can be moved in plus the maximum amount that can be moved out during a single breath is called the **vital capacity.** It is called vital capacity because your life depends on breathing, and the more air you can move, the better off you are. A number of different illnesses, discussed at the end of this chapter, can decrease vital capacity.

Inspiratory and Expiratory Reserve Volume As noted previously, we can increase inspiration by expanding the chest and also by lowering the diaphragm to the maximum extent possible. Forced inspiration **(inspiratory reserve volume)** usually adds another 2,900 mL of inhaled air, and that's quite a bit more than a tidal volume of only 500 mL!

We can increase expiration by contracting the abdominal and thoracic muscles. This so-called **expiratory reserve volume** is usually about 1,400 mL of air. You can see from Figure 9.9 that vital capacity is the sum of tidal, inspiratory reserve, and expiratory reserve volumes.

Residual Volume It is a curious fact that some of the inhaled air never reaches the lungs; instead, it fills the nasal cavities, trachea, bronchi, and bronchioles (see Fig. 9.1). These passages are not used for gas exchange, and therefore, they are said to contain **dead air space.** To ensure that newly inhaled air reaches the lungs, it is better to breathe slowly and deeply.

Also, note in Figure 9.9 that even after a very deep exhalation, some air (about 1,000 mL) remains in the lungs. This is called the **residual volume.** The residual volume is the amount of air that can't be exhaled from the lungs. In some

Have You Ever Wondered...

What happens when "the wind gets knocked out of you"?

The "wind may get knocked out of you" following a blow to the upper abdomen, in the area of the stomach. There is a network of nerves in that region called the solar plexus. Trauma to this area can cause the diaphragm to experience a sudden, involuntary, and painful contraction called a spasm. It's not possible to breathe while the diaphragm experiences this spasm; hence the feeling of being breathless. The pain and inability to breathe stop once the diaphragm relaxes.

lung diseases to be discussed later, the residual volume gradually increases because the individual has difficulty emptying the lungs. Increased residual volume will cause the expiratory reserve volume to be reduced. As a result, vital capacity is decreased as well.

> **Check Your Progress 9.4**
> 1. a. How is breathing achieved? b. How does the volume (size) of the thoracic cavity affect the pressure in the lungs?
> 2. a. Why do physicians measure our ability to move air into and out of the lungs? b. What measurements do they take?

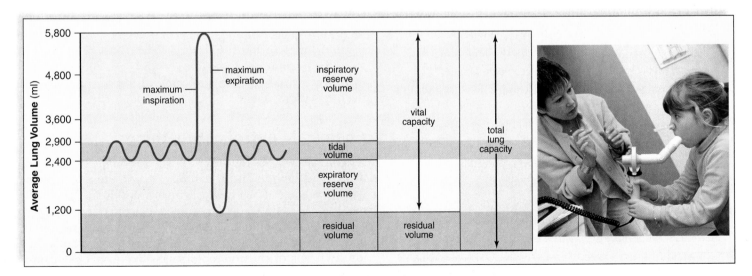

Figure 9.9 What instrument is used to measure vital capacity? How is vital capacity measured?
A spirometer is an instrument that measures the amount of air inhaled and exhaled with each breath. During inspiration, there is an upswing, and during expiration, there is a downswing. Vital capacity (red) is measured by taking the deepest breath and then exhaling as much as possible.

9.5 Control of Ventilation

Breathing is controlled in two ways. It is under nervous and chemical control.

Nervous Control of Breathing

Normally, adults have a breathing rate of 12 to 20 ventilations per minute. The rhythm of ventilation is controlled by a **respiratory control center** located in the medulla oblongata of the brain. The respiratory control center automatically sends out nerve signals to the diaphragm and the external intercostal muscles of the rib cage, causing inspiration to occur (Fig. 9.10). When the respiratory center stops sending nerve signals to the diaphragm and the rib cage, the muscles relax and expiration occurs.

Sudden infant death syndrome (SIDS), or crib death, claims the life of about 2,000 infants a year in the United States. An infant under one year of age is put to bed seemingly healthy, and sometime while sleeping, the child stops breathing. The cause of SIDS is not known, and it apparently is due to a respiratory center that stops sending out the impulses that cause us to breathe.

Although the respiratory center automatically controls the rate and depth of breathing, its activity can be influenced by nervous input. We can voluntarily change our breathing pattern to accommodate activities such as speaking, singing, eating, swimming under water, and so forth. Following forced inspiration, stretch receptors in the airway walls respond to increased pressure. These receptors initiate

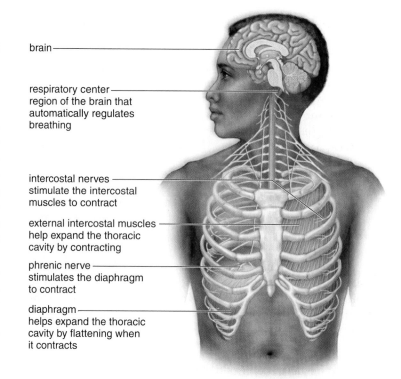

brain

respiratory center—
region of the brain that
automatically regulates
breathing

intercostal nerves —
stimulate the intercostal
muscles to contract

external intercostal muscles —
help expand the thoracic
cavity by contracting

phrenic nerve —
stimulates the diaphragm
to contract

diaphragm —
helps expand the thoracic
cavity by flattening when
it contracts

Figure 9.10 **Where is the respiratory center that controls breathing located? How does it work?**
During inspiration, the respiratory center, located in the medulla oblongata, stimulates the external intercostal (rib) muscles to contract via the intercostal nerves and stimulates the diaphragm to contract via the phrenic nerve. The thoracic cavity, and then the lungs, expand and air comes rushing in. Expiration occurs due to a lack of stimulation from the respiratory center to the diaphragm and intercostal muscles. Now, as the thoracic cavity, and then the lungs, resume their original size, air is pushed out.

Have You Ever Wondered...

How long can people hold their breath?

Most individuals can hold their breath for 1–2 minutes. With practice, many will be able to hold their breath for up to 3 minutes. Free divers are individuals who compete to see long they can hold their breath while diving in water. These divers can hold their breath for 5 minutes or more.

Researchers hope to more closely investigate divers' use of hyperventilation and lung "packing" to make free diving possible. Lung packing begins with very deep breathing. The diver then "packs," adding more air by additional breathing through the mouth. The use of hyperventilation or lung packing to prolong breath holding should never be tried by untrained individuals. People have drowned by attempting to do so.

inhibitory nerve impulses. The impulses travel from the inflated lungs to the respiratory center. This temporarily stops the respiratory center from sending out nerve signals. In this manner, excessive stretching of the elastic tissue of the lungs is prevented.

Chemical Control of Breathing

As you know, working cells produce carbon dioxide, which enters the blood. There, carbon dioxide combines with water, forming an acid, which breaks down and gives off hydrogen ions (H^+). These hydrogen ions can change the pH of the blood. **Chemoreceptors** are sensory receptors in the body that are sensitive to chemical composition of body fluids. Two sets of chemoreceptors sensitive to pH can cause breathing to speed up. A centrally placed set is located in the medulla oblongata of the brain stem, and a peripherally placed set is in the circulatory system. Carotid bodies, located in the carotid arteries, and aortic bodies, located in the aorta, are sensitive to blood pH. These chemoreceptors are not strongly affected by low oxygen (O_2) levels. Instead, they are stimulated when the carbon dioxide entering the blood is sufficient to change blood pH.

When the pH of the blood becomes more acidic (decreases), the respiratory center increases the rate and depth of breathing. With an increased breathing rate, more carbon dioxide is removed from the blood. The hydrogen ion concentration returns to normal, and the breathing rate returns to normal.

Most people are unable to hold their breath for more than one minute. When you hold your breath, metabolically produced carbon dioxide begins accumulating in the blood. As a result, H^+ accumulates and the blood becomes more acidic. The respiratory center, stimulated by the chemoreceptors, is able to override a person's voluntary inhibition of respiration. Breathing resumes, despite attempts to prevent it.

Check Your Progress 9.5

1. Explain why we automatically breath 12–20 times a minute.
2. Explain why it's not possible to hold your breath for more than a minute or so.

CASE STUDY AT THE SLEEP CENTER

Justin reported to the sleep center about a week after his visit to the health clinic. A complete physical exam of his nasal passages and throat didn't show any physical obstructions like nasal polyps (benign growths in the nose) or enlarged tonsils. Justin breathed a sigh of relief when he was told that news. He'd been dreading that surgery might be required to correct his symptoms. Tests showed that his thyroid gland was normal as well.

A polysomnography followed the physical exam. "This test monitors EEG, eye movements, heart rate, breathing, oxygen saturation, and your muscle tone and activity while you sleep," the technician explained. "We'll videotape any abnormal movements you make while you sleep."

"My doctor told me a little about the test. The EEG is brain waves, right?" Justin asked. "And oxygen saturation is how much oxygen you have in your blood?"

"You've got it," the technician replied as he completed the setup.

"I feel like I'm in a science fiction movie—hooked up to all these machines," Justin joked. "You really expect folks to be able to sleep like this?"

"Surprisingly, most people don't have a problem falling asleep here," replied the technician. Justin was no exception.

Soon after the sleep test was completed, Justin met with a sleep center physician. "The results of your sleep test suggest that you do have OSA, Justin," the doctor explained. "The muscle of your airways relaxes as you sleep. When that happens, it chokes off your breath for a few seconds. Your brain suddenly has too little oxygen, and it makes you wake up for a few seconds."

"So I'm waking up all night?" Justin asked, puzzled. "I don't remember doing that."

The doctor answered, "Most people don't. You aren't waking up completely, but you're not getting sound sleep either. And you're a heavy snorer, right?" Justin nodded. "That's because you gasp or choke to catch your breath. OSA explains that really loud snoring. It also explains your daytime sleepiness and the trouble you're having with focusing. Guys are most likely to have OSA, and it is seen more frequently if you're overweight."

"So if I lose weight, will it go away?" Justin inquired.

"Well, I definitely want you to lose weight," the doctor asserted. "You've got high blood pressure, and you're way too young for that. But I'm also going to get you fitted up with a CPAP machine."

"What's a see-pap machine?"

The doctor explained, "It's the abbreviation for Continuous Positive Airway Pressure. Basically, it gently pushes air into your airways all the time. That keeps the airways open while you're asleep. It means you can fall asleep and stay asleep, without gasping or choking. You'll get some real rest for a change."

He warned, "You can't quit using the thing, even if it takes a while to get used to it. Untreated OSA contributes to high blood pressure and heart failure, not to mention the risk of car accidents if you fall asleep while driving.

"The technician will set you up with a CPAP and make sure the mask is properly fitted. Come back for a followup visit after you've been sleeping with the CPAP for a month."

9.6 Gas Exchanges in the Body

Gas exchange is critical to homeostasis. Oxygen needed to produce energy must be supplied to all the cells, and carbon dioxide must be removed from the body during gas exchange. As mentioned previously, respiration includes not only the exchange of gases in the lungs, but also the exchange of gases in the tissues (Fig. 9.11).

The principles of diffusion govern whether O_2 or CO_2 enters or leaves the blood in the lungs and in the tissues. Gases exert pressure, and the amount of pressure each gas exerts is called its partial pressure, symbolized as P_{O_2} and P_{CO_2}. If the partial pressure of oxygen differs across a membrane, oxygen will diffuse from the higher to lower partial pressure.

External Respiration

External respiration refers to the exchange of gases between air in the alveoli and blood in the pulmonary capillaries (see Fig. 9.6 and Fig. 9.11a). Blood in the pulmonary capillaries has a higher P_{CO_2} than atmospheric air. Therefore, *CO_2 diffuses out of the plasma into the lungs.* Most of the CO_2 is carried in plasma as **bicarbonate ions** (HCO_3^-). In the low P_{CO_2} environment of the lungs, the reaction proceeds to the right:

$$H^+ + HCO_3^- \longrightarrow H_2CO_3 \xrightarrow{\text{carbonic anhydrase}} H_2O + CO_2$$

| hydrogen ion | bicarbonate ion | carbonic acid | water | carbon dioxide |

The enzyme **carbonic anhydrase** speeds the breakdown of carbonic acid (H_2CO_3) in red blood cells.

What happens if you hyperventilate (breathe at a high rate), and therefore push this reaction far to the right? The blood will have fewer hydrogen ions, and alkalosis, a high blood pH, results. In that case, breathing will be inhibited, and you may suffer from various symptoms ranging from dizziness to continuous contractions of the skeletal muscles. You may have heard that you should inhale and exhale from a paper bag after hyperventilating. Doing so increases the CO_2 in your blood, because you're inhaling the CO_2 you just exhaled into the bag. This restores a normal blood pH. What happens if you hypoventilate (breathe at a low rate) and this reaction does not occur? Hydrogen ions build up in the blood, and acidosis will occur. Buffers may compensate for the low pH, and breathing will most likely increase. Extreme changes in blood pH affect enzyme function; you may become comatose and die.

The pressure pattern for O_2 during external respiration is the reverse of that for CO_2. Blood in the pulmonary capillaries is low in oxygen, and alveolar air contains a higher partial pressure of oxygen. Therefore, *O_2 diffuses into plasma and then into red blood cells in the lungs.* Hemoglobin takes up this oxygen and becomes **oxyhemoglobin** (HbO_2):

$$Hb + O_2 \longrightarrow HbO_2$$

| deoxyhemoglobin | oxygen | oxyhemoglobin |

Internal Respiration

Internal respiration refers to the exchange of gases between the blood in systemic capillaries and the tissue cells. In Figure 9.11b, internal respiration is shown at the bottom. Blood entering systemic capillaries is a bright red color because red blood cells contain oxyhemoglobin. The temperature in the tissues is higher and the pH is slightly lower (more acidic), so oxyhemoglobin naturally gives up oxygen. After oxyhemoglobin gives up O_2, it diffuses out of the blood into the tissues:

$$HbO_2 \longrightarrow Hb + O_2$$

| oxyhemoglobin | deoxyhemoglobin | oxygen |

Oxygen diffuses out of the blood into the tissues because the P_{O_2} of tissue fluid is lower than that of blood. The lower P_{O_2} is due to cells continuously using up oxygen in cellular respiration (see Fig. 3.16). *Carbon dioxide diffuses into the blood from the tissues* because the P_{CO_2} of tissue fluid is higher than that of blood. Carbon dioxide is produced during cellular respiration and collects in tissue fluid.

After CO_2 diffuses into the blood, most enters the red blood cells, where a small amount is taken up by hemoglobin, forming **carbaminohemoglobin** ($HbCO_2$). In plasma, CO_2 combines with water, forming carbonic acid (H_2CO_3), which dissociates to hydrogen ions (H^+) and bicarbonate ions (HCO_3^-):

$$CO_2 + H_2O \xrightarrow{\text{carbonic anhydrase}} H_2CO_3 \longrightarrow H^+ + HCO_3^-$$

| carbon dioxide | water | carbonic acid | hydrogen ion | bicarbonate ion |

The enzyme carbonic anhydrase, mentioned previously, speeds the reaction in red blood cells. Bicarbonate ions (HCO_3^-) diffuse out of red blood cells and are carried in the plasma. The globin portion of hemoglobin combines with excess hydrogen ions produced by the overall reaction, and Hb becomes HHb, called **reduced hemoglobin.** In this way, the pH of blood remains fairly constant. Blood that leaves the systemic capillaries is a dark maroon color because red blood cells contain reduced hemoglobin.

> ### Check Your Progress 9.6
>
> 1. What are the differences between external respiration and internal respiration?
>
> 2. a. What respiratory pigment is found in human red blood cells? b. How does it function in the transport of oxygen; carbon dioxide?

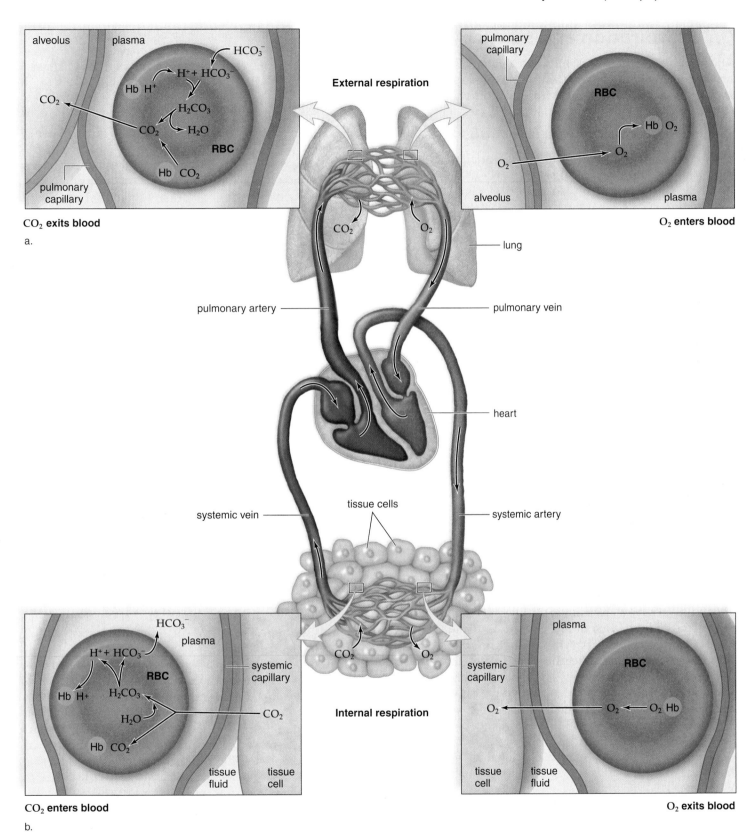

Figure 9.11   **What happens during external respiration in the lungs and internal respiration in the tissues?**
a. During external respiration in the lungs, HCO_3^{2} is converted to CO_2, which exits the blood. O_2 enters the blood and hemoglobin (Hb) carries O_2 to the tissues. **b.** During internal respiration in the tissues, O_2 exits the blood, and CO_2 enters the blood. Most of the CO_2 enters red blood cells, where it becomes the bicarbonate ion, carried in the plasma. Some hemoglobin combines with CO_2 and some combines with H^1.

9.7 Respiration and Health

The respiratory tract is constantly exposed to environmental air. The quality of this air, and whether it contains infectious pathogens such as bacteria and viruses, can affect our health.

Upper Respiratory Tract Infections

Upper respiratory infections (URI) can spread from the nasal cavities to the sinuses, middle ears, and larynx. What we call "strep throat" is a primary bacterial infection caused by *Streptococcus pyogenes* that can lead to a generalized URI and even a systemic (affecting the body as a whole) infection. The symptoms of strep throat are severe sore throat, high fever, and white patches on a dark red throat. Strep throat is bacterial so it can be treated successfully with antibiotics.

Sinusitis

Sinusitis develops when nasal congestion blocks the tiny openings leading to the sinuses (see Fig. 9.2). Symptoms include postnasal discharge, as well as facial pain that worsens when the patient bends forward. Pain and tenderness usually occur over the lower forehead or over the cheeks. If the latter, toothache is also a complaint. Successful treatment depends on restoring proper drainage of the sinuses. Even a hot shower and sleeping upright can be helpful. Nasal spray decongestants and oral antihistamines also alleviate the symptoms of sinusitis. Sprays may be preferred because they treat the symptoms without the side effects, such as drowsiness, of oral medicine. However, nasal sprays can become habit forming if used for long periods. Persistent sinusitis should be evaluated by a health-care professional.

Otitis Media

Otitis media is an infection of the middle ear. This infection is considered here because it is a complication often seen in children who have a nasal infection. Infection can spread by way of the auditory tube from the nasopharynx to the middle ear. Pain is the primary symptom of a middle ear infection. A sense of fullness, hearing loss, vertigo (dizziness), and fever may also be present. Antibiotics are prescribed if necessary, but physicians are aware today that overuse of antibiotics can lead to resistance of bacteria to antibiotics. Tubes (called tympanostomy tubes) are sometimes placed in the eardrums of children with multiple recurrences to help prevent the buildup of pressure in the middle ear and the possibility of hearing loss. Normally, the tubes fall out with time.

Tonsillitis

Tonsillitis occurs when the **tonsils** become inflamed and enlarged. The tonsil in the posterior wall of the nasopharynx is often called adenoids. If tonsillitis occurs frequently and enlargement makes breathing difficult, the tonsils can be removed surgically in a **tonsillectomy.** Fewer tonsillectomies are performed today than in the past because we now know that the tonsils trap many of the pathogens that enter the pharynx. Therefore, they are a first line of defense against invasion of the body.

Laryngitis

Laryngitis is an infection of the larynx with accompanying hoarseness, leading to the inability to talk in an audible voice. Usually, laryngitis disappears with treatment of the URI. Persistent hoarseness without the presence of a URI is one of the warning signs of cancer and should be looked into by a physician.

Lower Respiratory Tract Disorders

Lower respiratory tract disorders include infections, restrictive pulmonary disorders, obstructive pulmonary disorders, and lung cancer.

Lower Respiratory Infections

Acute bronchitis is an infection of the primary and secondary bronchi. Usually, it is preceded by a viral URI that has led to a secondary bacterial infection. Most likely, a nonproductive cough has become a deep cough that expectorates mucus and perhaps pus.

Pneumonia is a viral or bacterial infection of the lungs in which the bronchi and alveoli fill with thick fluid (Fig. 9.12). Most often, it is preceded by influenza. High fever and chills, with headache and chest pain, are symptoms of pneumonia. Rather than being a generalized lung infection, pneumonia may be localized in specific lobules of the lungs. Obviously, the more lobules involved, the more serious is the infection. Pneumonia can be caused by a bacterium that is usually held in check but has gained the upper hand due to stress and/or reduced immunity. AIDS patients are subject to a particularly rare form of pneumonia caused by the protozoan *Pneumocystis jiroveci* (formerly *Pneumocystis carinii*). Pneumonia of this type is almost never seen in individuals with a healthy immune system.

Pulmonary tuberculosis is a bacterial disease that was called *consumption* at one time. When the bacteria (*Mycobacterium tuberculosis*) invade the lung tissue, the cells build a protective capsule around the foreigners, isolating them from the rest of the body. This tiny capsule is called a tubercle. If the resistance of the body is high, the imprisoned organisms die, but if the resistance is low, the organisms eventually can be liberated. If a chest X-ray detects active tubercles, the individual is put on appropriate drug therapy to ensure the localization of the disease and the eventual destruction of any live bacteria. It is possible to tell if a person has ever been exposed to tuberculosis with a *tuberculin* test. This procedure uses a highly diluted bacterial extract injected into the patient's skin. If a tuberculin test is positive, an X-ray will be done to confirm an active disease. A person may have a positive test, but no active disease.

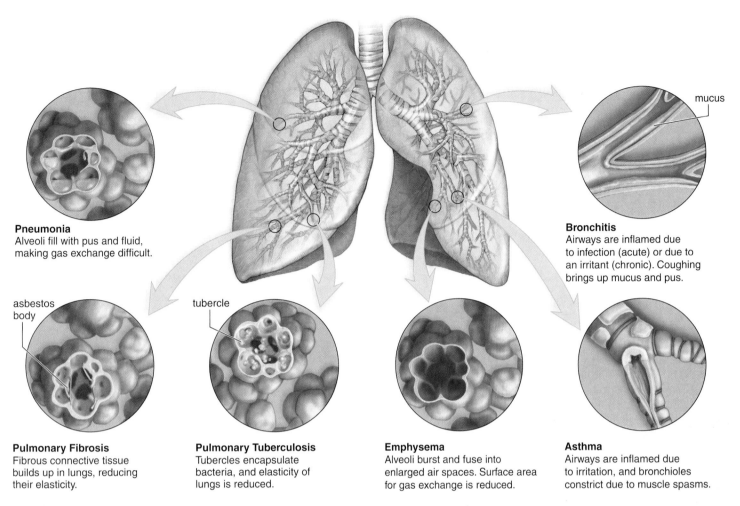

Pneumonia
Alveoli fill with pus and fluid, making gas exchange difficult.

Bronchitis
Airways are inflamed due to infection (acute) or due to an irritant (chronic). Coughing brings up mucus and pus.

mucus

asbestos body

tubercle

Pulmonary Fibrosis
Fibrous connective tissue builds up in lungs, reducing their elasticity.

Pulmonary Tuberculosis
Tubercles encapsulate bacteria, and elasticity of lungs is reduced.

Emphysema
Alveoli burst and fuse into enlarged air spaces. Surface area for gas exchange is reduced.

Asthma
Airways are inflamed due to irritation, and bronchioles constrict due to muscle spasms.

Figure 9.12 **What causes the diseases and disorders shown in this figure?**
Exposure to infectious pathogens and/or polluted air, including tobacco smoke, causes the diseases and disorders shown here.

Restrictive Pulmonary Disorders

In restrictive pulmonary disorders, vital capacity is reduced because the lungs have lost their elasticity. Inhaling particles such as silica (sand), coal dust, asbestos, and fiberglass can lead to **pulmonary fibrosis,** a condition in which fibrous connective tissue builds up in the lungs. The lungs cannot inflate properly and are always tending toward deflation. Breathing asbestos is also associated with the development of cancer. Asbestos was formerly used widely as a fireproofing and insulating agent, so unwarranted exposure has occurred. It has been projected that two million deaths caused by asbestos exposure—mostly in the workplace— will occur in the United States between 1990 and 2020.

Obstructive Pulmonary Disorders

In obstructive pulmonary disorders, air does not flow freely in the airways, and the time it takes to inhale or exhale maximally is greatly increased. Several disorders, including chronic bronchitis, emphysema, and asthma, are collectively referred to as chronic obstructive pulmonary disease (COPD) because they tend to recur.

In **chronic bronchitis,** the airways are inflamed and filled with mucus. A cough that brings up mucus is common. The bronchi have undergone degenerative changes, including the loss of cilia and their normal cleansing action. Under these conditions, an infection is more likely to occur. Smoking is the most frequent cause of chronic bronchitis. Exposure to other pollutants can also cause chronic bronchitis.

Emphysema is a chronic and incurable disorder in which the alveoli are distended and their walls damaged. As a result, the surface area available for gas exchange is reduced. Emphysema, most often caused by smoking, is often preceded by chronic bronchitis. Air trapped in the lungs leads to alveolar damage and a noticeable ballooning of the chest. The elastic recoil of the lungs is reduced, so not only are the airways narrowed, but the driving force behind expiration is also reduced. The victim is breathless and may have a cough. The surface area for gas exchange is reduced, so less oxygen reaches the heart and the brain. Even so, the heart works

Figure 9.13 How does the appearance of normal lungs differ from the lungs of a heavy smoker? **a.** Normal lung with heart in place. Note the healthy red color. **b.** Lungs of a heavy smoker. Notice how black the lungs are except where cancerous tumors have formed.

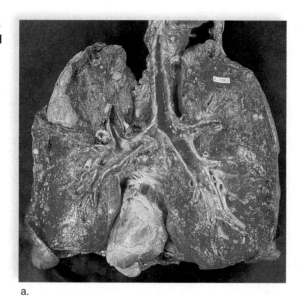

a.

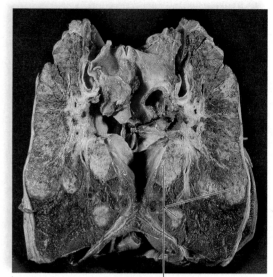

b.   cancerous tumors

furiously to force more blood through the lungs, and an increased workload on the heart can result. Lack of oxygen to the brain can make the person feel depressed, sluggish, and irritable. Exercise, drug therapy, supplemental oxygen, and giving up smoking may relieve the symptoms and possibly slow the progression of emphysema. Severe emphysema may be treated by lung transplantation or lung volume reduction surgery (LVRS). During LVRS, a third of the most diseased lung tissue is removed. The removal enables the remaining tissue to function better. The result is an increase in patients' breathing ability and lung capacity.

Asthma is a disease of the bronchi and bronchioles that is marked by wheezing, breathlessness, and sometimes a cough and expectoration of mucus. The airways are unusually sensitive to specific irritants, which can include a wide range of allergens such as pollen, animal dander, dust, tobacco smoke, and industrial fumes. Even cold air can be an irritant. When exposed to the irritant, the smooth muscle in the bronchioles undergoes spasms. It now appears that chemical mediators given off by immune cells in the bronchioles cause the spasms. Most asthma patients have some degree of bronchial inflammation that further reduces the diameter of the airways and contributes to the seriousness of an attack. Asthma is not curable, but it is treatable. Special inhalers can control the inflammation and possibly prevent an attack, while other types of inhalers can stop the muscle spasms should an attack occur.

Lung Cancer

Lung cancer is more prevalent in men than in women, but has surpassed breast cancer as a cause of death in women. The increase in the incidence of lung cancer in women is directly correlated to increased numbers of women who smoke. Autopsies on smokers have revealed the progressive steps by which the most common form of lung cancer develops. The first event appears to be thickening and callusing of the cells

lining the bronchi. (Callusing occurs whenever cells are exposed to irritants.) Then cilia are lost, making it impossible to prevent dust and dirt from settling in the lungs. Following this, cells with atypical nuclei appear in the callused lining. A tumor consisting of disordered cells with atypical nuclei is considered cancer *in situ* (at one location). A normal lung versus a lung with cancerous tumors is shown in Figure 9.13. A final step occurs when some of these cells break loose and penetrate other tissues, a process called metastasis. Now the cancer has spread. The original tumor may grow until a bronchus is blocked, cutting off the supply of air to that lung. The entire lung then collapses, the secretions trapped in the lung spaces become infected, and pneumonia or a lung abscess (localized area of pus) results. The only treatment that offers a possibility of cure is to remove a lobe or the whole lung before metastasis has had time to occur. This operation is called **pneumonectomy.** If the cancer has spread, chemotherapy and radiation are also required.

The Health Focus on pages 201–02 lists the various illnesses, including cancer, apt to occur when a person smokes. Research indicates that passive smoking (secondhand smoke)—exposure to smoke related by others who are smoking—can also cause lung cancer and other illnesses associated with smoking. If a person stops voluntary smoking and avoids passive smoking, and if the body tissues are not already cancerous, the lungs may return to normal over time.

> **Check Your Progress 9.7**
>
> 1. a. What are some common respiratory infections and disorders of the upper respiratory tract? b. Of the lower respiratory tract?
> 2. How can these common respiratory infections be treated?
> 3. What are three respiratory disorders commonly associated with smoking tobacco?

Health **Focus**

Questions About Smoking, Tobacco, and Health

Is cigarette smoking really addictive?

Yes. The nicotine in cigarette smoke causes addiction to smoking. Nicotine is an addictive drug (just like heroin and cocaine). Small amounts make the smoker want to smoke more. Smokers usually suffer withdrawal symptoms when they stop. Also, nicotine can affect the mood and nature of the smoker. The younger a person is when he or she begins to smoke, the more likely he or she is to develop an addiction to nicotine.

What are some of the short-term and long-term effects of smoking cigarettes?

Short-term effects include shortness of breath and nagging coughs, diminished ability to smell and taste, premature aging of the skin, and increased risk of sexual impotence in men. Smokers tend to tire easily during physical activity. Long-term effects include many types of cancer, heart disease, aneurysms, bronchitis, emphysema, and stroke. Smoking contributes to the severity of pneumonia and asthma.

Does smoking cause cancer?

Yes. Tobacco use accounts for about one-third of cancer deaths in the United States. Smoking causes almost 90% of lung cancers. Smoking also causes cancers of the oral cavity, pharynx, larynx (voice box), and esophagus. It contributes to the development of cancers of the bladder, pancreas, cervix, kidney, and stomach. Smoking is also linked to the development of some leukemias.

Why do smokers have "smoker's cough?"

Cigarette smoke contains chemicals that irritate the air passages and lungs. When a smoker inhales these substances, the body tries to protect itself by producing mucus and coughing. The nicotine in smoke decreases the sweeping action of cilia (hairlike formations lining the airways), so some of the poisons in the smoke remain in the lungs.

If you smoke but do not inhale, is there any danger?

Yes. Wherever smoke touches living cells, it does harm. Even if smokers don't inhale, they are breathing the smoke as second-hand smoke and are still at risk for lung cancer. Pipe and cigar smokers, who often do not inhale, are at an increased risk for lip, mouth, tongue, and several other cancers.

Does cigarette smoking affect the heart?

Yes. Smoking increases the risk of heart disease, the number one cause of death in the United States. Cigarette smoking is the biggest risk factor for sudden heart death. Smokers who have a heart attack are more likely to die within an hour of the heart attack than nonsmokers. Cigarette smoke at very low levels (much lower than the levels causing lung disease) can cause damage to the heart.

How does smoking affect pregnant women and their babies?

Smoking during pregnancy is linked with a greater chance of miscarriage, premature delivery, stillbirth, infant death, low birth weight, and sudden infant death syndrome (SIDS). Up to 10% of infant deaths would be prevented if pregnant women did not smoke. When a pregnant woman smokes, she really is smoking for two because the nicotine, carbon monoxide, and other dangerous chemicals in smoke enter her bloodstream and then pass into the baby's body. This prevents the baby from getting essential nutrients and oxygen for growth.

What are the dangers of environmental tobacco smoke (ETS)?

ETS causes about 3,000 lung cancer deaths and about 35,000 to 40,000 deaths from heart disease each year in healthy nonsmokers. Children whose parents smoke are more likely to suffer from asthma, pneumonia or bronchitis, ear infections, coughing, wheezing, and increased mucus production in the first two years of life.

Are chewing tobacco and snuff safe alternatives to cigarette smoking?

No. The juice from smokeless tobacco is absorbed directly through the lining of the mouth. This creates sores and white patches that often lead to cancer of the mouth. Smokeless tobacco users greatly increase their risk of other cancers, including those of the pharynx (throat). Other effects of smokeless tobacco include harm to teeth and gums.

How can people stop smoking?

There are a number of organizations, including the American Cancer Society and the American Lung Association, that offer suggestions for how to quit smoking. Both organizations also offer support groups to people interested in quitting. There is even advice about how people can help their friends quit smoking. Nicotine Anonymous offers a 12-step program for kicking a nicotine addiction, modeled after the 12-step Alcoholics Anonymous program. Recovering smokers have reported that a support system is very important.

What products are available to replace nicotine while attempting to quit?

Several over-the-counter products are available to replace nicotine. These include nicotine patches, gum, and

(continued)

Health Focus *(continued)*

lozenges. These products are designed to supply enough nicotine to alleviate the withdrawal symptoms people experience during attempts to stop smoking. A nicotine nasal spray and nicotine inhaler are also available by prescription. The delivery of nicotine to the brain by these prescription drugs is comparable to that of a cigarette. These drugs can gradually wean heavy smokers away from nicotine. However, nicotine is still a drug, with the same danger for overdose.

Are there new methods to help people stop smoking?

New smoking cessation drugs have recently been approved by the FDA, and others are in development phases. One such drug, Chantix®, affects areas in the brain stimulated by nicotine. Chantix® may lessen withdrawal symptoms by mimicking the effects of nicotine. Additional aids to smoking cessation include a vaccine that stimulates the synthesis of antibodies to nicotine. The antibodies bind to nicotine, preventing the stimulatory effects that result in addiction. This vaccine is being tested in clinical trials.

CASE STUDY TREATMENT

"Man, that is fierce," exclaimed Chad. "You look like a fighter pilot with that thing on." Chad had arrived home to find Justin trying on his CPAP mask.

"I think it looks a little crazy," Justin replied. "I'm a little worried that the tube that connects to the machine will get in my way while I'm sleeping."

"Hey, if it means you don't need surgery, I think it's something you can get used to," observed Chad.

A month later Justin returned to the sleep center for a followup appointment. "You look a lot more rested," commented the doctor.

"I feel so much better," Justin replied. "My CPAP has been a life saver. I haven't fallen asleep once in class since I started using the CPAP while I sleep. My roommate really likes it. I don't wake him up snoring like I did before."

"Well, just like I told you before, don't quit using it even though you feel better," the doctor warned. "You'll just go back to having obstructive sleep apnea again."

Justin asserted, "Now that I know what if feels like to be rested after sleeping, I'd be crazy not to use it. It used to be tough to finish reading and homework for my classes. Staying awake during lecture was practically impossible." He laughed. "My physics prof wants to suggest CPAP for everyone in his class!"

Bans on Smoking

In 1964, the surgeon general of the United States made it known to the general public that smoking was hazardous to our health, and thereafter, a health warning was placed on packs of cigarettes. At that time, 40.4% of adults smoked, but by 1990, only about 26% of adults smoked. In the meantime, however, the public became aware that passive smoking—just being in the vicinity of someone who is smoking—can also lead to cancer and other health problems. By now, many state and local governments have passed legislation that bans smoking in public places such as restaurants, elevators, public meeting rooms, and in the workplace.

Is legislation that restricts the freedom to smoke ethical? Or is such legislation akin to racism and creating a population of second-class citizens segregated from the majority on the basis of a habit? Are the desires of nonsmokers being allowed to infringe on the rights of smokers? Or is this legislation one way to help smokers become nonsmokers? One study showed that workplace bans on smoking reduce the daily consumption of cigarettes among smokers by 10%.

Bans on smoking have the potential for a significant economic impact. Legislation might ban smoking in family-style restaurants, but allow bars and restaurants associated with casinos to allow smoking. In certain states, a smoking ban in businesses has cost some underage employees their jobs. A bar can allow smoking if the age limit to enter is 21, but that means employees under 21 have to be let go. Is it fair to treat some businesses differently or to deny people their jobs? The selling of tobacco and even the increased need for health care it generates helps the economy. One smoker writes, "Smoking causes people to drink more, eat more, and leave larger tips. Smoking also powers the economy of Wall Street." Is this a reason to allow smoking to continue? Or should we require all places of business to put in improved air filtration systems? Would that do away with the dangers of passive smoking?

Does legislation that bans smoking in certain areas represent government invasion of our privacy? If yes, is reducing the chance of cancer a good enough reason to allow the government to invade our privacy? Some people are prone to cancer more than others. Should we all be regulated by the same legislation?

Decide Your Opinion

1. Is legislation that bans smoking in public places creating a group of second-class citizens whose rights are being denied?
2. Should we be concerned about passing and following legislation that possibly puts a damper on the economy, even if it does improve the health of people?
3. Are bans on smoking an invasion of our privacy? If so, is prevention of cancer in certain persons a good enough reason to risk a possible invasion of our privacy?

Summarizing the Concepts

9.1 The Respiratory System

The respiratory tract consists of the nose, the pharynx, the larynx, the trachea, the bronchi, the bronchioles, and the lungs.

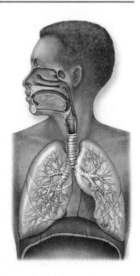

9.2 The Upper Respiratory Tract

Air from the nose enters the pharynx and passes through the glottis into the larynx:

- nose: filters and warms the air
- pharynx: opening into parallel air and food passageways
- larynx: the voice box that houses the vocal cords

9.3 The Lower Respiratory Tract

- The trachea (windpipe) is lined with goblet cells and ciliated cells.
- The bronchi, along with the pulmonary arteries and veins, enter the lungs.
- The lungs consist of the alveoli, air sacs surrounded by a capillary network.

9.4 Mechanism of Breathing

Breathing involves inspiration and expiration of air.

Inspiration

The diaphragm lowers, and the rib cage moves upward and outward; the lungs expand, and air rushes in.

Expiration

The diaphragm relaxes and moves up. The rib cage moves down and in; pressure in the lungs increases; air is pushed out of the lungs.

Respiratory volumes can be measured:

- Tidal volume: the amount of air that normally enters and exits with each breath.
- Vital capacity: the amount of air that moves in plus the amount that moves out with maximum effort.
- Inspiratory reserve volume and the expiratory reserve volume: the difference between normal amounts and the maximum effort amounts of air moved.
- Residual volume: the amount of air that stays in the lungs when we breathe.

9.5 Control of Ventilation

The respiratory center in the brain automatically causes us to breathe 12–20 times a minute. Extra carbon dioxide in the blood can decrease the pH, and if so, chemoreceptors alert the respiratory center, which increases the rate of breathing.

9.6 Gas Exchanges in the Body

Both external and internal respiration depend on diffusion. Hemoglobin activity is essential to the transport of gases and, therefore, to external and internal respiration.

External Respiration

- CO_2 diffuses out of plasma into lungs.
- O_2 diffuses into the plasma and then into red blood cells in the capillaries. O_2 is carried by hemoglobin.

Internal Respiration

- O_2 diffuses out of the blood into the tissues.
- CO_2 diffuses into the blood from the tissues. CO_2 is carried in the plasma as the bicarbonate ion (HCO_3^-).

9.7 Respiration and Health

A number of illnesses are associated with the respiratory tract.

Upper Respiratory Tract Infections

- Infections of the nasal cavities, sinuses, throat, tonsils, and larynx are all upper respiratory tract infections.
- These include sinusitis, otitis media, tonsillitis, and laryngitis.

Lower Respiratory Tract Disorders

- Lower respiratory tract infections include acute bronchitis, pneumonia, and pulmonary tuberculosis.
- Restrictive pulmonary disorders are exemplified by pulmonary fibrosis.
- Obstructive pulmonary disorders are exemplified by chronic bronchitis, emphysema, and asthma.
- Smoking can eventually lead to lung and other cancers.

Understanding Key Terms

acute bronchitis 199
alveolus 190
asthma 200
auditory (Eustachian) tube 187
bicarbonate ion 196
bronchiole 190
bronchus 190
carbaminohemoglobin 196
carbonic anhydrase 196
chemoreceptor 194
chronic bronchitis 199
cough 188
dead air space 193
emphysema 199
epiglottis 188
expiration 186, 191
expiratory reserve volume 193
external respiration 196
glottis 188
infant respiratory distress
   syndrome 191
inspiration 186, 191
inspiratory reserve volume 193
internal respiration 196
laryngitis 199
larynx 188
lung cancer 200

lungs 190
nasal cavity 187
otitis media 198
oxyhemoglobin 196
pharynx 187
pleurae 190
pneumonectomy 200
pneumonia 198
pulmonary fibrosis 199
pulmonary tuberculosis 198
reduced hemoglobin 196
residual volume 193
respiratory control center 194
sinusitis 198
sudden infant death syndrome
   (SIDS) 194
surfactant 191
tidal volume 192
tonsil 198
tonsillectomy 198
tonsillitis 198
trachea 189
tracheostomy 190
ventilation 186
vital capacity 193
vocal cord 188

Match the key terms to these definitions.

a. _____ Common passageway for both food intake and air movement, located between the mouth and the esophagus.

b. _____ Chemical in the lungs that reduces the surface tension of water to keep the alveoli from collapsing.

c. _____ Fold of tissue across the glottis within the larynx; creates vocal sounds when it vibrates.

d. _____ Form in which most of the carbon dioxide is transported in the bloodstream.

e. _____ Stage during breathing when air is pushed out of the lungs.

Testing Your Knowledge of the Concepts

1. Name as many structures as you can that connect to the pharynx. (page 187)

2. How is the structure of the trachea important for respiration, as well as digestion? (page 189)

3. Describe the structure of an alveolus, and explain how it is suited for gas exchange. (pages 190–91)

4. What are the steps of inspiration and expiration? How is breathing controlled? (pages 191–92, 194-95)

5. Using Figure 9.9, list and define the volumes and capacities of air movement. (pages 192-93)

6. Describe what occurs during external respiration, using two important equations. What is the driving force for the gas exchange? (pages 196–97)

7. Describe what occurs during internal respiration, using two important equations. What is the driving force for the gas exchange? (pages 196–97)

8. Describe several upper and lower respiratory tract disorders (other than cancer). If appropriate, explain why breathing is difficult with these conditions. (pages 198–200)

9. List the steps by which lung cancer develops. (page 200)

10. Which of these is anatomically incorrect?
    a. The nose has two nasal cavities.
    b. The pharynx connects the nasal and oral cavities to the larynx.
    c. The larynx contains the vocal cords.
    d. The trachea enters the lungs.
    e. The lungs contain many alveoli.

11. How is inhaled air modified before it reaches the lungs?
    a. It must be humidified.   c. It must be filtered.
    b. It must be warmed.   d. All of these are correct.

12. What is the name of the structure that prevents food from entering the trachea?
    a. glottis   c. epiglottis
    b. septum   d. Adam's apple

In questions 13–17, match each description with a structure in the key.

Key:
a. pharynx   d. trachea
b. glottis   e. bronchi
c. larynx   f. bronchioles

13. Branched tubes that lead from bronchi to the alveoli

14. Reinforced tube that connects larynx with bronchi

15. Chamber behind oral cavity and between nasal cavity and larynx

16. Opening into larynx

17. Divisions of the trachea that enter lungs

18. Which of these is incorrect concerning inspiration?
    a. Rib cage moves up and out.
    b. Diaphragm contracts and moves down.
    c. Pressure in lungs decreases, and air comes rushing in.
    d. The lungs expand because air comes rushing in.

19. Air enters the human lungs because
    a. atmospheric pressure is lower than the pressure inside the lungs.
    b. atmospheric pressure is greater than the pressure inside the lungs.
    c. although the pressures are the same inside and outside, the partial pressure of oxygen is lower within the lungs.
    d. the residual air in the lungs causes the partial pressure of oxygen to be lower than it is outside.

20. The maximum volume of air that can be moved in and out during a single breath is called the
    a. expiratory and inspira-   c. tidal volume.
       tory reserve volume.   d. vital capacity.
    b. residual volume.   e. functional residual capacity.

21. If air enters the intrapleural space (the space between the pleura),
    a. a lobe of the lung can collapse.
    b. the lungs could swell and burst.
    c. the diaphragm will contract.
    d. nothing will happen because air is needed in the intrapleural space.

22. Internal respiration refers to
    a. the exchange of gases between alveolar air and the blood in the lungs.
    b. the movement of air into the lungs.
    c. the exchange of gases between the blood and tissue fluid.
    d. cellular respiration, resulting in the production of ATP.

23. The chemical reaction that converts carbon dioxide to a bicarbonate ion takes place in
    a. the blood plasma.   c. the alveolus.
    b. red blood cells.   d. the hemoglobin molecule.

24. The enzyme carbonic anhydrase
    a. causes the blood to be more basic in the tissues.
    b. speeds up the conversion of carbonic acid to carbon dioxide and water, and the reverse.
    c. actively transports carbon dioxide out of capillaries.
    d. is active only at high altitudes.
    e. All of these are correct.

25. Hemoglobin assists transport of gases by
    a. combining with oxygen.
    b. combining with CO_2.
    c. combining with H^+.
    d. being present in red blood cells.
    e. All of these are correct.

26. In humans, the respiratory center
    a. is stimulated by carbon dioxide.
    b. is located in the medulla oblongata.
    c. controls the rate of breathing.
    d. All of these are correct.

27. Which of the following is not true of obstructive pulmonary disorders?
    a. Air does not flow freely in the airways.
    b. Vital capacity is reduced due to loss of lung elasticity.
    c. Disorders may include chronic bronchitis, emphysema, and asthma.
    d. Ventilation takes longer to occur.

28. Label this diagram of the human respiratory tract.

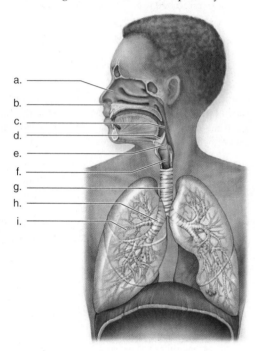

a. —
b. —
c. —
d. —
e. —
f. —
g. —
h. —
i. —

Thinking Critically About the Concepts

Children may also suffer from obstructive sleep apnea (OSA). Symptoms experienced during the daytime include breathing through the mouth and difficulty focusing. Children don't often have the excessive sleepiness during the day that adults like Justin have. Children with OSA often have enlarged tonsils or adenoids. Removal of the tonsils and adenoids often alleviates the problem. CPAP might be necessary if the OSA continues after surgery.

1. Why would enlarged tonsils or adenoids cause OSA?

2. Explain the expression, "the food went down the wrong way," by referring to structures along the path of air.

3. Long-term smokers often develop a chronic cough.
   a. What purpose does the smokers' cough serve?
   b. Why are nonsmokers less likely to develop a chronic cough?

4. Professional singers who need to hold a long note while singing must exert a great deal of control over their breathing. What muscle(s) needs/need conditioning to acquire better control over their breathing?

5. Why would someone who nearly drowned have a blue tint to their skin?

CHAPTER 10

Urinary System and Excretion

"**R**efills for everyone?" asked the waitress when she passed by the table. Casey and her friends were enjoying dessert and coffee after a movie.

"You can top off my cup," replied Casey. "Thanks."

Lia was sitting next to Casey. She commented, "I don't know how you do it. If I drank that much coffee, I'd be running to the bathroom every five minutes. Then I'd be awake for the rest of the night."

"Well, now that you mention it, I do need to hit the restroom before we head home. I'll be right back," responded Casey.

It took only a few minutes for Casey to drive from the restaurant to her apartment. Murphy wagged her tail when she realized that it was Casey walking through the door. Casey patted her dog on the head as she raced toward the bathroom. "Hang on, Murphy!" she called. "I'll take you for a walk as soon as I go to the bathroom." She thought, "I think I must have had too much coffee to drink." Minutes later, Casey and Murphy were walking down the street so that Murphy could be set for bedtime.

Casey woke with a start and realized she needed to go to the bathroom AGAIN! She'd already been to the bathroom right before going to bed. The urge had wakened her again at 11:30 PM and again at 1:30 AM. She squinted at her alarm clock after this bathroom trip, as she crawled back into bed. "Augh," she groaned. It was only 4 AM and this was her fourth trip that night.

Her final thought before drifting back to sleep was that she might need to make an appointment to see her physician, just in case she had a urinary tract infection brewing.

10.1 Urinary System

The kidneys are the primary organs of excretion. **Excretion** is the removal of metabolic wastes from the body. People sometimes confuse the terms defecation and excretion, but they do not refer to the same process. Defecation, the elimination of feces from the body, is a function of the digestive system. Excretion, on the other hand, is the elimination of metabolic wastes, the products of metabolism. For example, the undigested food and bacteria that make up feces have never been a part of the functioning of the body, while the substances excreted in urine were once metabolites in the body.

Organs of the Urinary System

The urinary system consists of the kidneys, ureters, urinary bladder, and urethra (Fig. 10.1).

Kidneys

The **kidneys** are paired organs located near the small of the back, on either side of the vertebral column. They lie in depressions beneath the peritoneum, where they receive some protection from the lower rib cage. A sharp blow to the back can dislodge a kidney from beneath the peritoneum.

After such an injury, the so-called **floating kidney** lacks its protective covering and is vulnerable to further damage.

The kidneys are bean-shaped and reddish brown in color. The fist-sized organs are covered by a tough capsule of fibrous connective tissue, called a renal capsule. Masses of adipose tissue adhere to each kidney. The concave side of a kidney has a depression, where a **renal artery** enters and a **renal vein** and a ureter exit the kidney. The renal artery transports blood to be filtered to the kidneys and the renal vein carries filtered blood away from the kidneys.

Ureters

The **ureters** conduct urine from the kidneys to the bladder. They are small, muscular tubes about 25 cm long and 5 mm in diameter. The wall of a ureter has three layers: an inner mucosa (mucous membrane), a smooth muscle layer, and an outer fibrous coat of connective tissue. Peristaltic contractions cause urine to enter the bladder even if a person is lying down. Urine enters the bladder in spurts that occur at the rate of one to five per minute.

Urinary Bladder

The **urinary bladder** stores urine until it is expelled from the body. The bladder has three openings: two for the ureters and one for the urethra, which drains the bladder (Fig. 10.2).

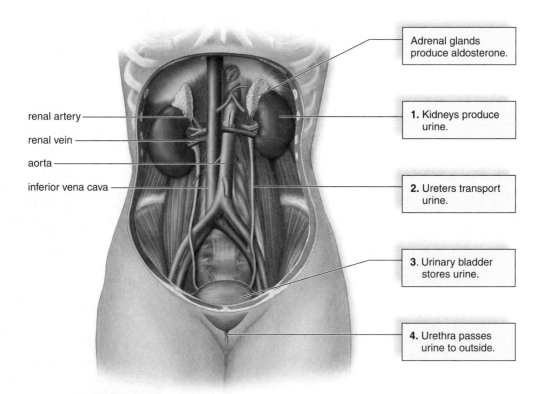

renal artery
renal vein
aorta
inferior vena cava

Adrenal glands produce aldosterone.

1. Kidneys produce urine.

2. Ureters transport urine.

3. Urinary bladder stores urine.

4. Urethra passes urine to outside.

Figure 10.1   **Which body structures contain urine?**
Urine is found only within the kidneys, the ureters, the urinary bladder, and the urethra.

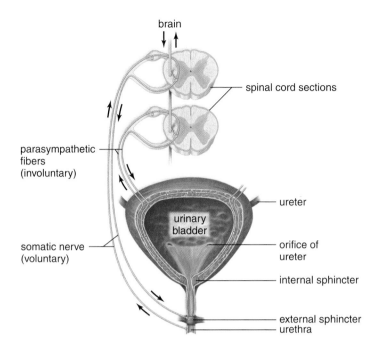

Figure 10.2   What happens to cause urination?
As the bladder fills with urine, sensory impulses go to the spinal cord and then to the brain. The brain can override the urge to urinate. When urination occurs, motor nerve impulses cause the bladder to contract and the sphincters to relax.

The bladder wall is expandable because it contains a middle layer of circular fibers of smooth muscle and two layers of longitudinal smooth muscle. The epithelium of the mucosa becomes thinner and folds in the mucosa called *rugae* disappear as the bladder enlarges. The bladder's rugae are similar to those of the stomach. A layer of transitional epithelium enables the bladder to stretch and contain an increased volume of urine.

The bladder has other features that allow it to retain urine. After urine enters the bladder from a ureter, small folds of bladder mucosa act like a valve to prevent backward flow. Two sphincters in close proximity are found where the urethra exits the bladder. The internal sphincter occurs around the opening to the urethra. It is composed of smooth muscle and is involuntarily controlled. An external sphincter is composed of skeletal muscle that can be voluntarily controlled.

When the urinary bladder fills to about 250 mL with urine, stretch receptors are activated by the enlargement of the bladder. These receptors send sensory nerve signals to the spinal cord. Subsequently, motor nerve impulses from the spinal cord cause the urinary bladder to contract and the sphincters to relax so that urination, also called **micturition,** is possible (Fig. 10.2).

Have You Ever Wondered...

Why do the people in commercials for Detrol LA® "gotta go, gotta go, gotta go right now"?

Detrol LA® is used to treat an overactive bladder. The muscles of the bladder contract, even though the bladder may not be full. Muscle contraction causes strong feelings of urgency to go to the bathroom. Use of the medication controls the symptoms by calming the bladder's muscle contractions.

Urethra

The **urethra** is a small tube that extends from the urinary bladder to an external opening. Therefore, its function is to remove urine from the body. The urethra has a different length in females than in males. In females, the urethra is only about 4 cm long. The short length of the female urethra makes bacterial invasion of the urinary tract easier. In males, the urethra averages 20 cm when the penis is flaccid (limp, nonerect). As the urethra leaves the male urinary bladder, it is encircled by the prostate gland. The prostate sometimes enlarges, restricting the flow of urine in the urethra. The Health Focus on page 211 discusses this problem in men.

In females, the reproductive and urinary systems are not connected. However, the urethra carries urine during urination and sperm during ejaculation in males.

Functions of the Urinary System

As the kidneys produce urine, they carry out the following four functions that contribute to homeostasis:

1. *Excretion of Metabolic Wastes*   The kidneys excrete metabolic wastes, notably nitrogenous wastes. Urea is the primary nitrogenous end product of metabolism in human beings, but humans also excrete some ammonium, creatinine, and uric acid.

    Urea is a by-product of amino acid metabolism. The breakdown of amino acids in the liver releases ammonia, which the liver rapidly combines with carbon dioxide to produce urea. Ammonia is very toxic to cells, but urea is much less toxic.

    Creatine phosphate is a high-energy phosphate reserve molecule in muscles. The metabolic breakdown of creatine phosphate results in **creatinine.**

    The breakdown of nucleotides, such as those containing adenine and thymine, produces **uric acid.** Uric acid is rather insoluble. If too much uric acid is present in blood, crystals form and precipitate out. Crystals of uric acid sometimes collect in the joints, producing a painful ailment called **gout.**

2. *Maintenance of Water-Salt Balance* A principal function of the kidneys is to maintain the appropriate water-salt balance of the blood. As we shall see, blood volume is intimately associated with the salt balance of the body. As you know, salts, such as NaCl, have the ability to cause osmosis, the diffusion of water—in this case, into the blood. The more salts there are in the blood, the greater the blood volume and the greater the blood pressure. In this way, the kidneys are involved in regulating blood pressure.

   The kidneys also maintain the appropriate level of other ions, such as potassium ions (K^+), bicarbonate ions (HCO_3^-), and calcium ions (Ca^{2+}), in the blood.

3. *Maintenance of Acid-Base Balance* The kidneys regulate the acid-base balance of the blood. For a person to remain healthy, the blood pH should be just about 7.4. The kidneys monitor and help control blood pH, mainly by excreting hydrogen ions (H^+) and reabsorbing the bicarbonate ions (HCO_3^-) as needed to keep blood pH at 7.4. Urine usually has a pH of 6 or lower because our diet often contains acidic foods.

4. *Secretion of Hormones* The kidneys assist the endocrine system in hormone secretion. The kidneys release renin, an enzyme that leads to aldosterone secretion.

Aldosterone is a hormone produced by the adrenal glands (see Fig. 10.1), which lie atop the kidneys. As described in Section 10.4, aldosterone is involved in regulating the water-salt balance of the blood.

Erythropoietin (EPO) is a hormone secreted by the kidneys. When blood oxygen decreases, EPO increases red blood cell synthesis by stem cells in the bone marrow (see Chap. 6, page 120). When the concentration of red blood cells increases, blood oxygen increases also. Genetically engineered EPO is sometimes prescribed for people in kidney failure (see Chap. 21 for a complete description of the genetic engineering process). Failing kidneys produce less EPO, resulting in fewer red blood cells and symptoms of fatigue. Supplemental EPO will increase red blood cell synthesis and energy levels.

5. The kidneys also reabsorb filtered nutrients and convert vitamin D. Vitamin D is a molecule that promotes calcium (Ca^{2+}) absorption from the digestive tract.

> **Check Your Progress 10.1**
>
> 1. **What are the organs of the urinary system, and what are their functions?**
>
> 2. **In what four ways do the kidneys help maintain homeostasis?**

CASE STUDY THE NEXT DAY

When the alarm clock finally rang at 6:30 AM, Casey groaned and dragged herself out of bed. Her mouth was so dry she could barely get out the words "good morning" to Murphy. The bottle of water she kept by her bedside was empty, and Casey frowned. She didn't remember drinking it all. "I guess I must have finished that water after one of my trips to the bathroom last night," she thought.

She headed to the kitchen to get something to drink. No sooner had she finished a glass of water than she felt the need to go to the bathroom again. "Poor Murphy," she muttered, as she exited the bathroom. She stroked her head. "Now I know what you must feel like when I'm late getting home from work, and you've been holding it all day. I'll definitely try harder to be home on time from now on."

After a quick breakfast of juice and peanut butter toast, Casey dashed off to work. There was a 9:00 AM staff meeting, and she needed a little time beforehand to get her file and look over the agenda. However, extra time was not to be had—once again, she had to make a stop at the ladies' room as soon as she arrived. Next, she had to buy a bottle of water because she felt absolutely parched. Casey raced by

her desk, picked up her file and slid into her chair at the far side of the room just as her manager began the meeting. The next 90 minutes were agonizing! She felt a strong need to urinate for almost the entire meeting.

Her colleague Anne noticed her crossing and uncrossing her legs in an effort to stave off the urge to go. "Are you okay?" whispered Anne.

"Is this meeting ever going to end?" Casey replied softly. "I've got to go to the bathroom SO bad!" When the meeting finally ended, Casey raced to the ladies room.

Anne followed Casey to the restroom. "What's up with you?" she queried.

"I think I may have a urinary tract infection," Casey fretted. "But my symptoms aren't the same as before. I don't see any blood in my urine, and it isn't painful to go. I just have to go constantly, and I'm so thirsty all the time. Right now, I feel really dizzy."

Anne frowned. "I'm going to drive you to the doctor, Casey," she asserted.

Health Focus

Urinary Difficulties Due to an Enlarged Prostate

The prostate gland, part of the male reproductive system, surrounds the urethra at the point where the urethra leaves the urinary bladder (Fig. 10A). The prostate gland produces and adds a fluid to semen as semen passes through the urethra within the penis. At about age 50, the prostate gland often begins to enlarge, growing from the size of a walnut to that of a lime or even a lemon. This condition is called benign prostatic hyperplasia (BPH). As it enlarges, the prostate squeezes the urethra, causing urine to back up—first into the bladder, then into the ureters, and finally, perhaps, into the kidneys.

Treatment Emphasis Is on Early Detection

The treatment for BPH can involve (1) a more invasive procedure to reduce the size of the prostate or (2) taking a drug that is expected to shrink the prostate and/or improve urine flow. Prostate tissue can be destroyed by applying microwaves to a specific portion of the prostate. In many cases, however, a physician may decide that prostate tissue should be removed surgically. In some cases, rather than performing abdominal surgery, which requires an incision of the abdomen, the physician gains access to the prostate via the urethra. This operation, called transurethral resection of the prostate (TURP), requires careful consideration because one study found that the death rate during the five years following TURP is much higher than that following abdominal surgery.

Some drug treatments recognize that prostate enlargement is due to a prostate enzyme (5a-reductase) that acts on the male sex hormone testosterone, converting it into a substance that promotes prostate growth. Growth is fine during puberty, but continued growth in an adult is undesirable. Three substances, one a nutrient supplement and two prescription drugs, interfere with the action of the enzyme that promotes growth. Saw palmetto, sold in tablet form as an over-the-counter nutrient supplement, is derived from a plant of the same name. This drug should not be taken unless the need for it is confirmed by a physician, but it is particularly effective during the early stages of prostate enlargement. The prescription drugs finasteride and dutasteride are more powerful inhibitors of the enzyme, but patients complain of erectile dysfunction and loss of libido while on the drugs.

Two other medications have a different mode of action. Nafarelin prevents the release of LH, which when present leads to testosterone production. When it is administered, approximately half of the patients report relief of urinary symptoms even after drug treatment is halted. However,

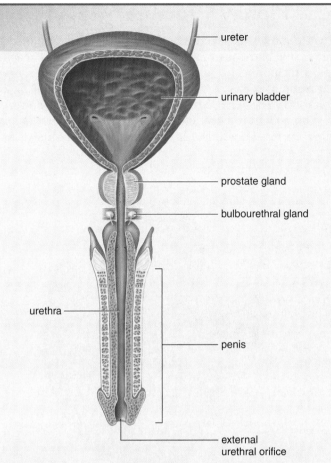

Figure 10A   What happens when the prostate enlarges?
Note the position of the prostate gland, which can enlarge to obstruct urine flow.

again the patients experience erectile dysfunction and other side effects, such as hot flashes. The drug terazosin, on the market for hypertension because it relaxes arterial walls, also relaxes muscle tissue in the prostate. Improved urine flow was experienced by 70% of the patients taking this drug. However, the drug has no effect on the prostate's overall size.

Many men are concerned that BPH may be associated with prostate cancer, but the two conditions are not necessarily related. BPH occurs in the inner zone of the prostate, whereas cancer tends to develop in the outer area. If prostate cancer is suspected, blood tests and a biopsy, in which a tiny sample of prostate tissue is surgically removed, will confirm the diagnosis.

Enlarged Prostate and Cancer

Although prostate cancer is the second most common cancer in men, it is not a major killer. Typically, prostate cancer is so slow growing that the survival rate is about 98% if the condition is detected early.

10.2 Kidney Structure

When a kidney is sliced lengthwise, it is possible to see that many branches of the renal artery and vein reach inside the kidney (Fig. 10.3*a*). If the blood vessels are removed, it is easier to identify the three regions of a kidney. (1) The **renal cortex** is an outer, granulated layer that dips down in between a radially striated inner layer called the renal medulla. (2) The **renal medulla** consists of cone-shaped tissue masses called renal pyramids. (3) The **renal pelvis** is a central space, or cavity, continuous with the ureter (Fig. 10.3*b* and *c*).

Microscopically, the kidney is composed of over one million **nephrons,** sometimes called renal, or kidney, tubules (Fig. 10.3*d*). The nephrons filter the blood and produce urine. Each nephron is positioned so that the urine flows into a collecting duct. Several nephrons enter the same collecting duct. The collecting ducts eventually enter the renal pelvis.

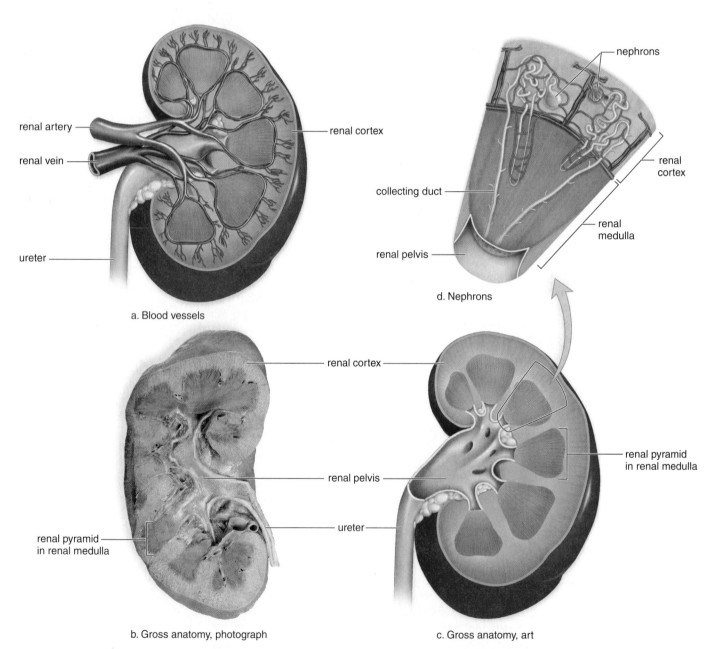

a. Blood vessels

d. Nephrons

b. Gross anatomy, photograph

c. Gross anatomy, art

Figure 10.3   **What blood vessels serve the kidneys? In what regions of the kidneys are the nephrons found?**
a. A longitudinal section of the kidney showing the blood supply. The renal artery divides into smaller arteries, and these divide into arterioles. Venules join to form small veins, which join to form the renal vein. **b.** and **c.** The same section without the blood supply. Now it is easier to distinguish the renal cortex; the renal medulla; and the renal pelvis, which connects with the ureter. The renal medulla consists of the renal pyramids. **d.** An enlargement showing the placement of nephrons.

Anatomy of a Nephron

Each nephron has its own blood supply, including two capillary regions (Fig. 10.4). From the renal artery, an afferent arteriole transports blood to the **glomerulus,** a knot of capillaries inside the glomerular capsule. Blood leaving the glomerulus is carried away by the efferent arteriole. Blood pressure is higher in the glomerulus because the efferent arteriole is narrower than the afferent arteriole. The efferent arteriole divides and forms the **peritubular capillary network,** which surrounds the rest of the nephron. Blood from the efferent arteriole travels through the peritubular

capillary network. Then the blood goes into a venule that carries blood into the renal vein.

Parts of a Nephron

Each nephron is made up of several parts (see Fig. 10.4). Some functions are shared by all parts of the nephron. However, the specific structure of each part is especially suited to a particular function.

First, the closed end of the nephron is pushed in on itself to form a cuplike structure called the **glomerular capsule** (Bowman's capsule). The outer layer of the glomerular

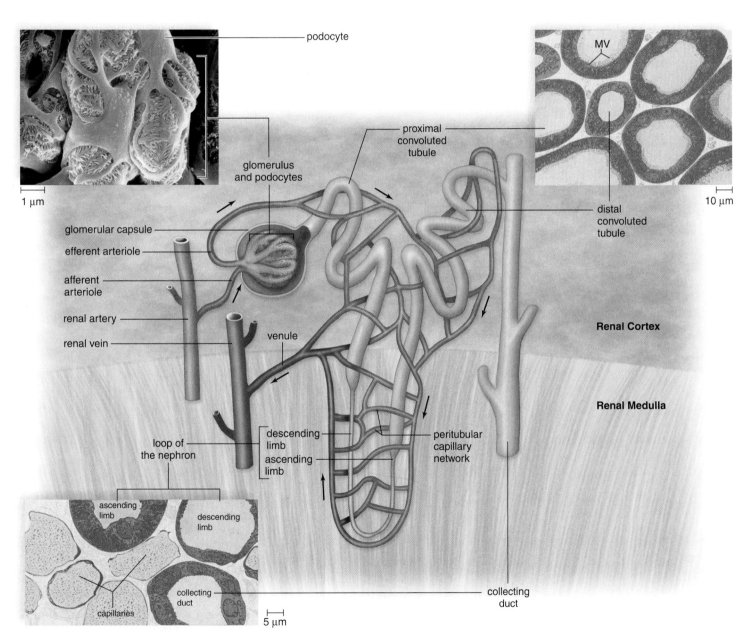

Figure 10.4 **What are the components of a nephron?**

A nephron is made up of a glomerular capsule, the proximal convoluted tubule, the loop of the nephron, the distal convoluted tubule, and the collecting duct. The photomicrographs show the microscopic anatomy of these structures. You can trace the path of blood about the nephron by following the arrows. (MV = microvilli)

capsule is composed of squamous epithelial cells. The inner layer is made up of podocytes that have long cytoplasmic extensions. The podocytes cling to the capillary walls of the glomerulus and leave pores that allow easy passage of small molecules from the glomerulus to the inside of the glomerular capsule. This process, called glomerular filtration, produces a filtrate of the blood.

Next, there is a **proximal convoluted tubule.** The cuboidal epithelial cells lining this part of the nephron have numerous microvilli, about 1 μm in length, that are tightly packed and form a brush border (Fig. 10.5). A brush border greatly increases the surface area for the tubular reabsorption of filtrate components. Each cell also has many mitochondria, which can supply energy for active transport of molecules from the lumen to the peritubular capillary network.

Simple squamous epithelium appears as the tube narrows and makes a U-turn called the **loop of the nephron** (loop of Henle). Each loop consists of a descending limb and an ascending limb. The descending limb of the loop allows water to diffuse into tissue surrounding the nephron. The ascending limb actively transports salt from its lumen to interstitial tissue. As we shall see, this activity facilitates the reabsorption of water by the nephron and collecting duct.

The cuboidal epithelial cells of the **distal convoluted tubule** have numerous mitochondria, but they lack microvilli. This means that the distal convoluted tubule is not spe-cialized for reabsorption. Instead, its primary function is ion exchange. During ion exchange, cells reabsorb certain ions, returning them to the blood. Other ions are secreted from the blood into the tubule. The distal convoluted tubules of several nephrons enter one collecting duct. Many **collecting ducts** carry urine to the renal pelvis.

As shown in Figure 10.4, the glomerular capsule and the convoluted tubules always lie within the renal cortex. The loop of the nephron dips down into the renal medulla. A few nephrons have a very long loop of the nephron, which penetrates deep into the renal medulla. Collecting ducts are also located in the renal medulla, and together they give the renal pyramids their appearance.

> **Check Your Progress 10.2**
>
> 1. What are the three major areas of a kidney?
> 2. What microscopic structure is responsible for the production of urine?
> 3. How do the parts of a nephron differ with regards to structure and function?

10.3 Urine Formation

Figure 10.6 gives an overview of urine formation, which is divided into three processes.

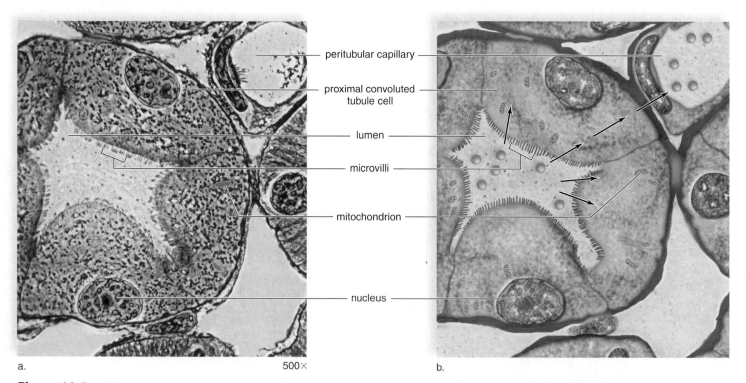

a.   500×   b.

Figure 10.5   What characteristics enable the cells of the proximal convoluted tubule to perform their function?
a. This photomicrograph shows that the cells lining the proximal convoluted tubule have a brushlike border composed of microvilli, which greatly increase the surface area exposed to the lumen. The peritubular capillary network surrounds the cells. **b.** Diagrammatic representation of (a) shows that each cell has many mitochondria, which supply the energy needed for active transport, the process that moves molecules (green) from the lumen of the tubule to the capillary, as indicated by the arrows.

Glomerular Filtration

Glomerular filtration occurs when whole blood enters the glomerulus by way of the afferent arteriole (recall that the glomerulus is a tuft of capillaries). Due to glomerular blood pressure, water and small molecules move from the glomerulus to the inside of the glomerular capsule. This is a filtration process because large molecules and formed elements

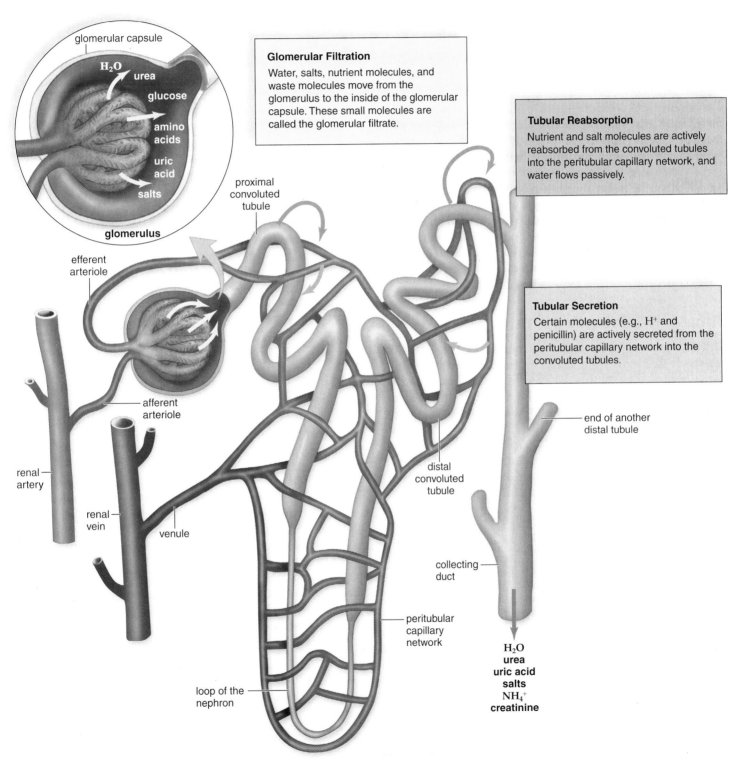

Glomerular Filtration

Water, salts, nutrient molecules, and waste molecules move from the glomerulus to the inside of the glomerular capsule. These small molecules are called the glomerular filtrate.

Tubular Reabsorption

Nutrient and salt molecules are actively reabsorbed from the convoluted tubules into the peritubular capillary network, and water flows passively.

Tubular Secretion

Certain molecules (e.g., H^+ and penicillin) are actively secreted from the peritubular capillary network into the convoluted tubules.

glomerular capsule

H_2O
urea
glucose
amino acids
uric acid
salts

glomerulus

proximal convoluted tubule

efferent arteriole

afferent arteriole

renal artery

renal vein

venule

end of another distal tubule

distal convoluted tubule

collecting duct

peritubular capillary network

loop of the nephron

H_2O
urea
uric acid
salts
NH_4^+
creatinine

Figure 10.6 How is urine produced?

The three main processes in urine formation are described in boxes and color coded to arrows that show the movement of molecules into or out of the nephron at specific locations. In the end, urine is composed of the substances within the collecting duct (see brown arrow).

are unable to pass through the capillary wall. In effect, then, blood in the glomerulus has two portions, the filterable components and the nonfilterable components:

Filterable Blood Components	Nonfilterable Blood Components
Water	Formed elements (blood cells and platelets)
Nitrogenous wastes	Plasma proteins
Nutrients	
Salts (ions)	

The nonfilterable components leave the glomerulus by way of the efferent arteriole. The **glomerular filtrate** inside the glomerular capsule now contains the filterable blood components in approximately the same concentration as plasma.

As indicated in Table 10.1, nephrons in the kidneys filter 180 liters of water per day, along with a considerable amount of small molecules (such as glucose) and ions (such as sodium). If the composition of urine were the same as that of the glomerular filtrate, the body would continually lose water, salts, and nutrients. Therefore, we can conclude that the composition of the filtrate must be altered as this fluid passes through the remainder of the tubule.

Tubular Reabsorption

Tubular reabsorption occurs as molecules and ions are passively and actively reabsorbed from the nephron into the blood of the peritubular capillary network. The osmolarity of the blood is maintained by the presence of plasma proteins and salt. When sodium ions (Na^+) are actively reabsorbed, chloride ions (Cl^-) follow passively. The reabsorption of salt (Na^+Cl^-) increases the osmolarity of the blood compared with the filtrate. Therefore, water moves passively from the tubule into the blood. About 65% of Na^+ is reabsorbed at the proximal convoluted tubule.

Table 10.1	Reabsorption from Nephrons		
Substance	**Amount Filtered (per day)**	**Amount Excreted (per day)**	**Reabsorption (%)**
Water, L	180	1.8	99.0
Sodium, g	630	3.2	99.5
Glucose, g	180	0.0	100.0
Urea, g	54	30.0	44.0

L = liters, g = grams

Nutrients such as glucose and amino acids return to the peritubular capillaries almost exclusively at the proximal convoluted tubule. This is a selective process because only molecules recognized by carrier proteins are actively reabsorbed. Glucose is an example of a molecule that ordinarily is completely reabsorbed because there is a plentiful supply of carrier proteins for it. However, every substance has a maximum rate of transport. After all its carriers are in use, any excess in the filtrate will appear in the urine. In **diabetes mellitus,** because the liver and muscles fail to store glucose as glycogen, the blood glucose level is above normal, and glucose appears in the urine. The presence of excess glucose in the filtrate raises its osmolarity. Therefore, less water is reabsorbed into the peritubular capillary network. The frequent urination and increased thirst experienced by untreated diabetics are due to less water being reabsorbed from the filtrate into the blood.

We have seen that the filtrate that enters the proximal convoluted tubule is divided into two portions, components reabsorbed from the tubule into blood, and components not reabsorbed that continue to pass through the nephron to be further processed into urine:

Reabsorbed Filtrate Components	Nonreabsorbed Filtrate Components
Most water	Some water
Nutrients	Much nitrogenous waste
Required salts (ions)	Excess salts (ions)

The substances not reabsorbed become the tubular fluid, which enters the loop of the nephron.

Tubular Secretion

Tubular secretion is a second way by which substances are removed from blood and added to the tubular fluid. Hydrogen ions (H^+), creatinine, and drugs such as penicillin are some of the substances moved by active transport from blood into the kidney tubule. In the end, urine contains substances that have undergone glomerular filtration but have not been reabsorbed and substances that have undergone tubular secretion. Tubular secretion is now known to occur along the length of the kidney tubule.

> **Check Your Progress 10.3**
> 1. What are the three major processes in urine formation?
> 2. How does the nephron carry out these processes?

Urinalysis

A routine urinalysis can detect abnormalities in urine color, concentration, and content. Significant information about the general state of one's health, particularly in the diagnosis of renal and metabolic diseases is provided by this analysis also. A complete urinalysis consists of three phases of examination: physical, chemical, and microscopic.

Physical Examination

During a physical examination, a clinician may examine urine color, clarity, and odor. The color of normal, fresh urine is usually pale yellow. However, the color may vary from almost colorless (dilute) to dark yellow (concentrated). Pink, red, or smoky brown urine is usually a sign of bleeding that may be due to a kidney, bladder, or urinary tract infection. Liver disorders can also produce dark brown urine. The clarity of normal urine may be clear or cloudy. However, a cloudy urine sample is also characteristic of abnormal levels of bacteria. Finally, urine odor is usually "nutty" or aromatic; but a foul-smelling odor is characteristic of a urinary tract infection. A sweet, fruity odor is characteristic of glucose in the urine or diabetes mellitus. Urine's odor is also affected by the consumption of garlic, curry, asparagus, and vitamin C.

Chemical Examination

The chemical examination is often done with a dipstick, a thin strip of plastic, impregnated with chemicals that change color upon reaction with certain substances present in urine (Fig. 10B). The color change on each segment of the dipstick is compared with a standardized color chart. Dipsticks can be used to determine urine's specific gravity; pH; and content of glucose, bilirubin, urobilinogen, ketone, protein, nitrite, blood, and white blood cells (WBCs) in the urine.

1. *Specific gravity* is an indicator of how well the kidneys are able to adjust tonicity in urine. Normal values for urine specific gravity range from 1.002 to 1.035. A high value (concentrated urine) may be a result of dehydration or diabetes mellitus.

2. Normal *urine pH* can be as low as 4.5 and as high as 8.0. In patients with kidney stone disease, urine pH has a direct effect on the type of stones formed.

3. *Glucose* is normally not present in urine. If present, diabetes mellitus is suspected.

4. *Bilirubin* (a by-product of hemoglobin degradation) is not normally present in the urine. *Urobilinogen* (a by-product of bilirubin degradation) is normally present in very small amounts. High levels of bilirubin or urobilinogen may indicate liver disease.

5. *Ketones* are not normally found in urine. Urine ketones are a by-product of fat metabolism. The presence of ketones in the urine may indicate diabetes mellitus or use of a low-carbohydrate diet, such as the Atkins diet.

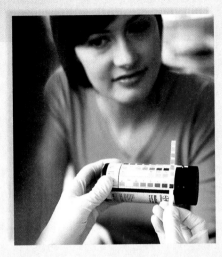

Figure 10B **What do the results of this person's urinalysis indicate?**
This patient's urine has been tested and the multiple test stick is being compared to a reference chart. The dark brown pad shows that the patient has glucose in her urine and this indicates she has diabetes mellitus.

6. Plasma *proteins* should not be present in urine. A significant amount of urine protein (proteinuria) is usually a sign of kidney damage.

7. Urine typically does not contain *nitrates*. The presence of nitrites is a sign of urinary tract infection.

8. It is not abnormal for blood to show up in the urine when a woman is menstruating. Blood present in the urine at other times may indicate a bacterial infection or kidney damage.

9. The chemical test for *WBCs* is normally negative. A high urine WBC count usually indicates a bacterial infection somewhere in the urinary tract.

Microscopic Examination

Finally, in a microscopic examination, the urine is centrifuged, and the sediment (solid material) is examined under a microscope. When renal disease is present, the urine will often contain an abnormal amount of cellular material. Urinary casts are sediments formed by the coagulation of protein material in the distal convoluted tubule or the collecting duct. Casts may be a sign of many different disorders, depending on the type present. The presence of crystals in the urine is characteristic of kidney stones, kidney damage, or problems with metabolism.

Forensic Analysis

Urinalysis is also used by the federal government to screen potential employees for the use of numerous illegal drugs. Business establishments conduct similar random testing. Drugs associated with date rape (such as Rohypnol®, known as "roofies," and GHB, or gamma hydroxybutyrate) can be detected in the urine of victims to determine if they were drugged.

There are more than 100 different tests that can be done on urine. Many different factors such as diet, kidney function, and other health disorders can affect what ends up in one's urine. Consequently, a urinalysis can provide a clinician with important information about a patient's overall health.

Can you drink urine if there is no other source of water available?

You're probably familiar with news reports of trapped individuals who survived for days without water by drinking their urine. Or if you're a fan of *Man vs. Wild*™ on the Discovery Channel, you may remember an episode in the Australian Outback when Bear Grylls drank his urine.

In some cultures, people drink urine as an alternative medicine or for cosmetic reasons. This practice is known as urine therapy. Urine from someone who is healthy should be sterile, and therefore drinkable. However, urine contains any number of other substances besides water, like salts and metabolites of various drugs. If these substances are highly concentrated, you may become dehydrated more quickly. Many survival guides advise against drinking your urine.

10.4 Regulatory Functions of the Kidneys

The kidneys maintain the water-salt balance of the blood within normal limits. In this way, they also maintain the blood volume and blood pressure. The kidneys are also involved in regulating the pH of the blood, so that it stays within normal limits.

Water-Salt Balance

Most of the water found in the filtrate is reabsorbed into the blood before urine leaves the body. All parts of a nephron and the collecting duct participate in the reabsorption of water. The reabsorption of salt always precedes the reabsorption of water. In other words, water is returned to the blood by the process of osmosis. During the process of reabsorption, water passes through water channels, called **aquaporins**, within a plasma membrane protein.

Sodium (Na^+) is an important ion in plasma. Usually more than 99% of the Na^+ filtered at the glomerulus is returned to the blood. The kidneys also excrete or reabsorb other ions, such as potassium ions (K^+), bicarbonate ions (HCO_3^-), and magnesium ions (Mg^{2+}) as needed.

Reabsorption of Salt and Water from Cortical Portions of the Nephron

The proximal convoluted tubule, the distal convoluted tubule, and the cortical portion of the collecting ducts are present in the renal cortex. Most (65%) of the water that enters the glomerular capsule is reabsorbed from the nephron into the blood at the proximal convoluted tubule. Na^+ is actively reabsorbed, and Cl^- follows passively. Aquaporins are always open and water is reabsorbed osmotically into the blood.

CASE STUDY AT THE DOCTOR'S OFFICE

"Casey O'Connor?" called the nurse. "Right here," Casey replied.

"Come this way," directed the nurse. "Do you think you can give me a urine sample?"

Casey laughed sadly. "That won't be a problem at all!"

In the exam room, the nurse took Casey's temperature, blood pressure, and pulse. "Your vital signs are all off, Casey. I'll hustle Dr. Taylor in right away," the nurse said as he pulled the door closed.

Minutes later Dr. Taylor entered the room with Casey's records in hand. "What seems to be the problem?" she asked. Casey began by describing how incredibly thirsty she'd been for the past few days and her need to urinate so often. "Right now, I'm really dizzy," she added.

"Does it sting, burn, or hurt when you go?" Dr. Taylor asked. She frowned a bit when Casey's reply was negative. "Well, let's see if your physical turns anything up. Does it hurt here?" She pushed gently just above Casey's pelvic bone. "How about here?" She continued as she pressed Casey's back, just below her ribs.

"No, not at all," Casey replied.

A knock at the door indicated the arrived of Casey's urinalysis. "No pain in your bladder or kidneys, no white or red blood cells, so no urinary tract infection," Dr. Taylor began as she examined the urinalysis test strip. "Glucose and ketones are absent, which rules out diabetes mellitus too." She continued, "The one test result that's way, way off is the urine specific gravity."

"Could you give that to me in English?" Casey asked.

"Sorry about that," Dr. Taylor continued. "It means your urine is too dilute. You said you've noticed excessive thirst and urination?"

"I am in the bathroom all day long and half the night," Casey replied in frustration.

"I'm sending you to the hospital right away for blood work and an MRI," Dr. Taylor ordered. "I'll get an ambulance. No way are you driving."

Hormones regulate the reabsorption of sodium and water in the distal convoluted tubule. **Aldosterone** is a hormone secreted by the adrenal glands, which sit atop the kidneys. This hormone promotes ion exchange at the distal convoluted tubule. Potassium (K^+) is excreted, and sodium (Na^+) is reabsorbed into the blood. The release of aldosterone is set into motion by the kidneys. The **juxtaglomerular apparatus** is a region of contact between the afferent arteriole and the distal convoluted tubule (Fig. 10.7). When blood volume (and therefore, blood pressure) falls too low for filtration to occur, the juxtaglomerular apparatus can respond to the decrease by secreting **renin.** Renin is an enzyme that ultimately leads to secretion of aldosterone by the adrenal glands. Research scientists speculate that excessive renin secretion—and thus, reabsorption of excess salt and water—might contribute to high blood pressure.

Aquaporins are not always open in the distal convoluted tubule. Another hormone called **antidiuretic hormone (ADH)** must be present. ADH is produced by the hypothalamus and secreted by the posterior pituitary according to the osmolarity of the blood. If our intake of water has been low, ADH is secreted by the posterior pituitary.

Water moves from the distal convoluted tubule and the collecting duct into the blood.

Atrial natriuretic hormone (ANH) is a hormone secreted by the atria of the heart when cardiac cells are stretched due to increased blood volume. ANH inhibits the secretion of renin by the juxtaglomerular apparatus and the secretion of aldosterone by the adrenal glands. Its effect, therefore, is to promote the excretion of Na^+, called *natriuresis.* Normally, salt reabsorption creates an osmotic gradient that causes water to be reabsorbed. Thus, by causing salt excretion, ANH causes water excretion, too. If ANH is present, less water will be reabsorbed, even if ADH is also present.

Reabsorption of Salt and Water from the Medulla Portions of the Nephron

The ability of humans to regulate the tonicity of their urine is dependent on the work of the medullary portions of the nephron (loop of the nephron) and the collecting duct.

The Loop of the Nephron A long loop of the nephron, which typically penetrates deep into the renal medulla, is

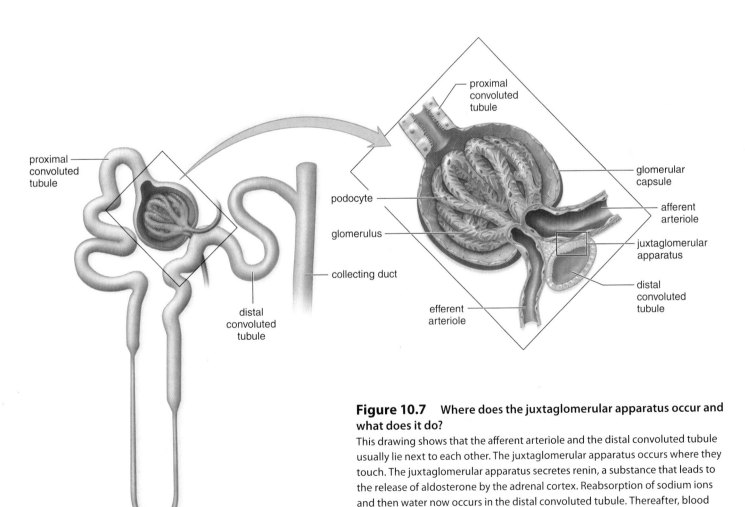

Figure 10.7 Where does the juxtaglomerular apparatus occur and what does it do?
This drawing shows that the afferent arteriole and the distal convoluted tubule usually lie next to each other. The juxtaglomerular apparatus occurs where they touch. The juxtaglomerular apparatus secretes renin, a substance that leads to the release of aldosterone by the adrenal cortex. Reabsorption of sodium ions and then water now occurs in the distal convoluted tubule. Thereafter, blood volume and blood pressure increase.

made up of a descending limb and an ascending limb. Salt (NaCl) passively diffuses out of the lower portion of the *ascending limb.* Any remaining salt is actively transported from the thick upper portion of the limb into the tissue of the outer medulla (Fig. 10.8). In the end, the concentration of salt is greater in the direction of the inner medulla. Surprisingly, however, the inner medulla has an even higher concentration of solutes than expected. It is believed that urea leaks from the lower portion of the collecting duct, and this molecule contributes to the high solute concentration of the inner medulla.

Water leaves the descending limb along its entire length via a countercurrent mechanism because of the osmotic gradient within the medulla. Although water is reabsorbed as soon as fluid enters the descending limb, the remaining fluid within the limb encounters an increasing osmotic concentration of solute. Therefore, water continues to be reabsorbed, even to the bottom of the descending limb. (The ascending limb does not reabsorb water—it has no aquaporins as indicated by the dark line in Fig. 10.8—its job is to help establish the solute concentration gradient.) Water is returned to the cardiovascular system when it is reabsorbed.

The Collecting Duct   Fluid within the collecting duct encounters the same osmotic gradient established by the ascending limb of the nephron. Therefore, water will diffuse from the entire length of the collecting duct into the blood if aquaporins are open, as they will be if ADH is present.

To understand the action of ADH, consider its name, antidiuretic hormone. Diuresis means increased amount of urine, and antidiuresis means decreased amount of urine. When ADH is present, more water is reabsorbed (blood volume and pressure rise) and a decreased amount of urine results. ADH is the ultimate fine tuner of the tonicity of urine according to the needs of the body. For example, ADH is secreted at night when we are not drinking water, and this explains why the first urine of the day is more concentrated.

Diuretics

Diuretics are chemicals that increase the flow of urine. Drinking alcohol causes diuresis because it inhibits the secretion of ADH. The dehydration that follows is believed to contribute to the symptoms of a hangover. Caffeine is a diuretic because it increases the glomerular filtration rate and decreases the tubular reabsorption of Na^+. Diuretic drugs developed to counteract high blood pressure also decrease the tubular reabsorption of Na^+. A decrease in water reabsorption and a decrease in blood volume and pressure follow.

Acid-Base Balance of Body Fluids

The pH scale, as discussed in Chapter 2, page 28, can be used to indicate the basicity (alkalinity) or the acidity of body fluids. A basic solution has a lesser hydrogen ion concentration [H^+] than the neutral pH of 7.0. An acidic solution has a greater [H^+] than neutral pH. The normal pH for body fluids is between 7.35 and 7.45. This is the pH at which our proteins, such as cellular enzymes, function properly. If the blood pH rises above 7.45, a person is said to have **alkalosis,** and if the blood pH decreases below 7.35, a person is said to have **acidosis.** Alkalosis and acidosis are abnormal conditions that may need medical attention.

The foods we eat add basic or acidic substances to the blood, and so does metabolism. For example, cellular respiration adds carbon dioxide that combines with water to form carbonic acid, and fermentation adds lactic acid. The pH of body fluids stays at just about 7.4 via several mechanisms, primarily acid-base buffer systems, the respiratory center, and the kidneys.

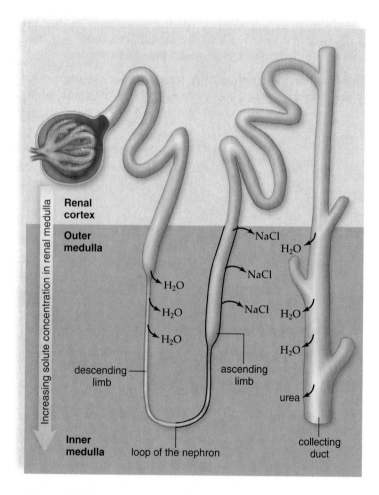

Figure 10.8   **What molecules are transported out of the ascending limb of the loop of Henle? How does the countercurrent mechanism allow urine to be concentrated?** Salt (NaCl) diffuses and is actively transported out of the ascending limb of the loop of the nephron into the renal medulla; also, urea is believed to leak from the collecting duct and to enter the tissues of the renal medulla. This creates a hypertonic environment, which draws water out of the descending limb and the collecting duct. This water is returned to the cardiovascular system. (The thick black outline of the ascending limb means that it is impermeable to water.)

Acid-Base Buffer Systems

The pH of the blood stays near 7.4 because the blood is buffered. A **buffer** is a chemical or a combination of chemicals that can take up excess hydrogen ions (H^+) or excess hydroxide ions (OH^-). One of the most important buffers in the blood is a combination of carbonic acid (H_2CO_3) and bicarbonate ions (HCO_3^-). When hydrogen ions (H^+) are added to blood, the following reaction occurs:

$$H^+ + HCO_3^- \rightarrow H_2CO_3$$

When hydroxide ions (OH^-) are added to blood, this reaction occurs:

$$OH^- + H_2CO_3 \rightarrow HCO_3^- + H_2O$$

These reactions temporarily prevent any significant change in blood pH. A blood buffer, however, can be overwhelmed unless some more permanent adjustment is made. The next adjustment to keep the pH of the blood constant occurs at pulmonary capillaries.

Respiratory Center

As discussed in Chapter 9, the respiratory center in the medulla oblongata increases the breathing rate if the hydrogen ion concentration of the blood rises. Increasing the breathing rate rids the body of hydrogen ions because the following reaction takes place in pulmonary capillaries:

$$H^+ + HCO_3^- \rightleftharpoons H_2CO_3 \rightleftharpoons H_2O + CO_2$$

In other words, when carbon dioxide is exhaled, this reaction shifts to the right, and the amount of hydrogen ions is reduced.

It is important to have the correct proportion of carbonic acid and bicarbonate ions in the blood. Breathing readjusts

---

CASE STUDY DIAGNOSIS AND TREATMENT

Casey's first day of testing had been a miserable experience. After being weighed, she was allowed no water to drink all morning, but the frequent urination persisted. Weighing, blood tests, and urine collection continued all morning. By early afternoon, she had lost several pounds due to water loss and was extremely dehydrated and dizzy. Finally, an IV was started and she was allowed to drink again. All day long hospital staff came and went from her room as X-rays, an MRI, and more blood and urine collection continued. Finally, Dr. Taylor stopped by to see her in the late afternoon. Casey smiled nervously. "Please tell me I'm going to be okay. Otherwise I may have to buy stock in a toilet paper company," she joked.

"Yes, you're going to be fine," Dr. Taylor reassured. "You have a disorder called **diabetes insipidus,** and it's very treatable."

"I thought you said I didn't have diabetes!" Casey exclaimed.

"You don't have the diabetes most folks are familiar with," Dr. Taylor explained. "*Diabetes* just means excessive urination. Diabetes insipidus means 'excessive watery urine.' It happens because your brain can't make a hormone."

"I thought hormones were from glands," Casey remarked.

"Think of your brain as a big hormone factory. The one your brain isn't making is called antidiuretic hormone, or ADH for short. Without it, your kidneys can't make your urine concentrated. You drink, but you urinate right away. So even if you're already dehydrated, like you were this morning when we wouldn't let you drink anything, your body still loses water. It's basically going right through you."

Casey asked, "What causes diabetes—whatever it is?"

"Sometimes it's a brain tumor, but that's not true in your case. The MRI came back perfectly normal. Sometimes it's a head injury—but you've never hit your head really hard, have you? Like in a car accident or something like that?" Dr. Taylor asked. Casey shook her head. "Then we may never know what caused it, but we can fix it just by giving you the missing hormone."

"So I can take pills for this, right?" Casey queried.

"The best way to control diabetes insipidus is with a nasal spray, so the hormone goes right into your blood. Unfortunately, we're going to need to keep you here for one more day, just to figure out how much hormone to use. If we don't use enough, you could get seriously dehydrated because you'll lose too much water. Or if the hormone spray works too well, your body will retain too much water. Either way, you'll be in bad shape." Dr. Taylor smiled reassuringly. "So stick around one more day, and let us keep an eye on you."

Casey returned to Dr. Taylor's office the week after being discharged from the hospital. She had monitored her water intake and urine output carefully all week. Blood work confirmed that the initial dose of ADH spray was working fine. "Are you feeling less thirst? How about fewer trips to the bathroom?" asked Dr. Taylor.

"No more late night bathroom visits," Casey smiled. "It's such a relief to sleep through the night. But will this diabetes insipidus thing affect me later in life?"

"It shouldn't," Dr. Taylor reassured. "But you'll need to keep a close eye on how much water you take in and how much urine you produce. Remember, water in and water out should be just about equal every single day."

capillary

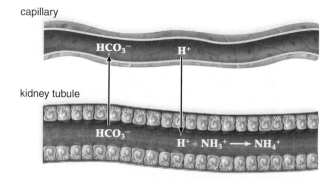

kidney tubule

Figure 10.9 How is blood pH maintained by the kidneys?
In the kidneys, bicarbonate ions (HCO_3^-) are reabsorbed, and hydrogen ions (H^+) are excreted as needed to maintain the pH of the blood. Excess hydrogen ions are buffered, for example, by ammonia (NH_3), which becomes ammonium (NH_4^+). Ammonia is produced in tubule cells by the deamination of amino acids.

this proportion so that this particular acid-base buffer system can continue to absorb H^+ and OH^- as needed.

The Kidneys

As powerful as the acid-base buffer and the respiratory center mechanisms are, only the kidneys can rid the body of a wide range of acidic and basic substances and, otherwise, adjust the pH. The kidneys are slower acting than the other two mechanisms, but they have a more powerful effect on pH. For the sake of simplicity, we can think of the kidneys as reabsorbing bicarbonate ions and excreting hydrogen ions as needed to maintain the normal pH of the blood (Fig. 10.9). If the blood is acidic, hydrogen ions are excreted, and bicarbonate ions are reabsorbed. If the blood is basic, hydrogen ions are not excreted, and bicarbonate ions are not reabsorbed. The urine is usually acidic, so it follows that an excess of hydrogen ions is usually excreted. Ammonia (NH_3) provides another means of buffering and removing the hydrogen ions in urine: ($H^+ + NH_3 \rightarrow NH_4^+$). Ammonia (whose presence is obvious in the diaper pail or kitty litter box) is produced in tubule cells by the deamination of amino acids. Phosphate provides another means of buffering hydrogen ions in urine.

The importance of the kidneys' ultimate control over the pH of the blood cannot be overemphasized. As mentioned, the enzymes of cells cannot continue to function if the internal environment does not have near-normal pH.

> ### Check Your Progress 10.4
> 1. a. Where is water reabsorbed, and (b) how is water reabsorbed so that urine is concentrated?
> 2. a. What three hormones influence urine production and kidney function? b. How do these hormones work together?
> 3. How do the kidneys regulate the pH of body fluids?

10.5 Disorders with Kidney Function

Many types of illnesses, especially diabetes, hypertension, and inherited conditions, cause progressive renal disease and renal failure. Infections are also contributory. If the infection is localized in the urethra, it is called **urethritis.** If the infection invades the urinary bladder, it is called **cystitis.** Finally, if the kidneys are affected, the infection is called **pyelonephritis.**

Urinary tract infections, an enlarged prostate gland, pH imbalances, or an intake of too much calcium can lead to kidney stones. Kidney stones are hard granules made of calcium, phosphate, uric acid, and protein. Kidney stones form in the renal pelvis and usually pass unnoticed in the urine flow. If they grow to several centimeters and block the renal pelvis or ureter, a reverse pressure builds up and destroys nephrons. When a large kidney stone passes, strong contractions within a ureter can be excruciatingly painful.

One of the first signs of nephron damage is albumin, white blood cells, or even red blood cells in the urine. As described on page 217, urinalysis can detect urine abnormalities rapidly. If damage is so extensive that more than two-thirds of the nephrons are inoperative, urea and other waste substances accumulate in the blood. This condition is called **uremia.** Although nitrogenous wastes can cause serious damage, the retention of water and salts is of even greater concern. The latter causes edema, fluid accumulation in the body tissues. Imbalance in the ionic composition of body fluids can lead to loss of consciousness and to heart failure.

Hemodialysis

Patients with renal failure can undergo **hemodialysis,** using either an artificial kidney machine or continuous ambulatory peritoneal dialysis (CAPD). *Dialysis* is defined as the diffusion of dissolved molecules through a semipermeable natural or synthetic membrane that has pore sizes that allow only small molecules to pass through. In an artificial kidney machine (Fig. 10.10), the patient's blood is passed through a membranous tube, which is in contact with a dialysis solution, or **dialysate.** Substances more concentrated in the blood diffuse into the dialysate, and substances more concentrated in the dialysate diffuse into the blood. The dialysate is continuously replaced to maintain favorable concentration gradients. In this way, the artificial kidney can be used either to extract substances from blood, including waste products or toxic chemicals and drugs, or to add substances to blood—for example, bicarbonate ions (HCO_3^-) if the blood is acidic. In the course of a 3–6 hour hemodialysis, from 50 to 250 g of urea can be removed from a patient, which greatly exceeds the amount excreted by normal kidneys. Therefore, a patient needs to undergo treatment only about twice a week.

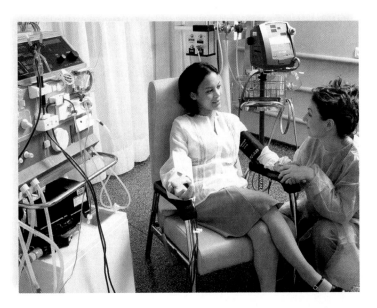

Figure 10.10 How is blood cleansed and adjusted by an artificial kidney machine?

As the patient's blood is pumped through dialysis tubing, it is exposed to a dialysate (dialysis solution). Wastes exit from blood into the solution because of a pre-established concentration gradient. In this way, blood is not only cleansed, but its water-salt and acid-base balances can also be adjusted.

CAPD is so named because the peritoneum is the dialysis membrane. A fresh amount of dialysate is introduced directly into the abdominal cavity from a bag that is temporarily attached to a permanently implanted plastic tube. The dialysate flows into the peritoneal cavity by gravity. Waste and salt molecules pass from the blood vessels in the abdominal wall into the dialysate before the fluid is collected 4–8 hours later. The solution is drained into a bag from the abdominal cavity by gravity, and then it is discarded. One advantage of CAPD over an artificial kidney machine is that the individual can go about his or her normal activities during CAPD.

Replacing a Kidney

Patients with renal failure sometimes undergo a kidney transplant operation, during which a functioning kidney from a donor is received. As with all organ transplants, there is the possibility of organ rejection. Receiving a kidney from a close relative has the highest chance of success. The current one-year survival rate is 97% if the kidney is received from a relative and 90% if it is received from a nonrelative. In the future, transplantable kidneys may be created in a laboratory. Another option could be to use kidneys from specially bred pigs whose organs would not be antigenic to humans.

> **Check Your Progress 10.5**
>
> 1. What are the most common causes of renal disease, and how can it be treated?

10.6 Homeostasis

Figure 10.11 tells us how the kidneys assist the work of our systems and/or how these systems help the urinary system carry out its functions. Our final discussion of the urinary system centers around these functions of the kidney.

The Kidneys Excrete Waste Molecules

In Chapter 8, we compared the liver to a sewage treatment plant because it removes poisonous substances from the blood and prepares them for excretion. Similarly, the liver produces urea, the primary nitrogenous end product of humans, which is excreted by the kidneys. If the liver is a sewage treatment plant, the tubules of the kidney are like the trucks that take the sludge, prepared waste, away from the town (the body).

Metabolic waste removal is absolutely necessary for maintaining homeostasis. The blood must constantly be cleansed of the nitrogenous wastes, end products of metabolism. The liver produces urea, and muscles make creatinine. These wastes, and also uric acid from the cells, are carried by the cardiovascular system to the kidneys (Fig. 10.12). The urine producing kidneys are responsible for the excretion of nitrogenous wastes. They are assisted to a limited degree by the sweat glands in the skin, which excrete perspiration, a mixture of water, salt, and some urea. In times of kidney failure, urea is excreted by the sweat glands and forms a so-called uremic frost on the skin.

Water-Salt Balance

If blood does not have the usual water-salt balance, blood volume and blood pressure are affected. Without adequate blood pressure, exchange across capillary walls cannot take place, nor is glomerular filtration possible in the kidneys.

What happens if you have insufficient Na^+ in your blood and tissue fluid? This can occur due to prolonged heavy sweating, as in athletes running a marathon. When blood Na^+ concentration falls too low, blood pressure falls and the renin-aldosterone sequence begins. Then the kidneys increase Na^+ reabsorption, conserving as much as possible. Subsequently, the osmolarity of the blood and the blood pressure return to normal.

The marathon runner should not drink too much water too fast. Quickly ingesting a large amount of pure water can dilute the body's remaining Na^+ and disrupt water-salt balance. Sports drinks preferred by athletes contain sodium and water, so both can be replaced simultaneously.

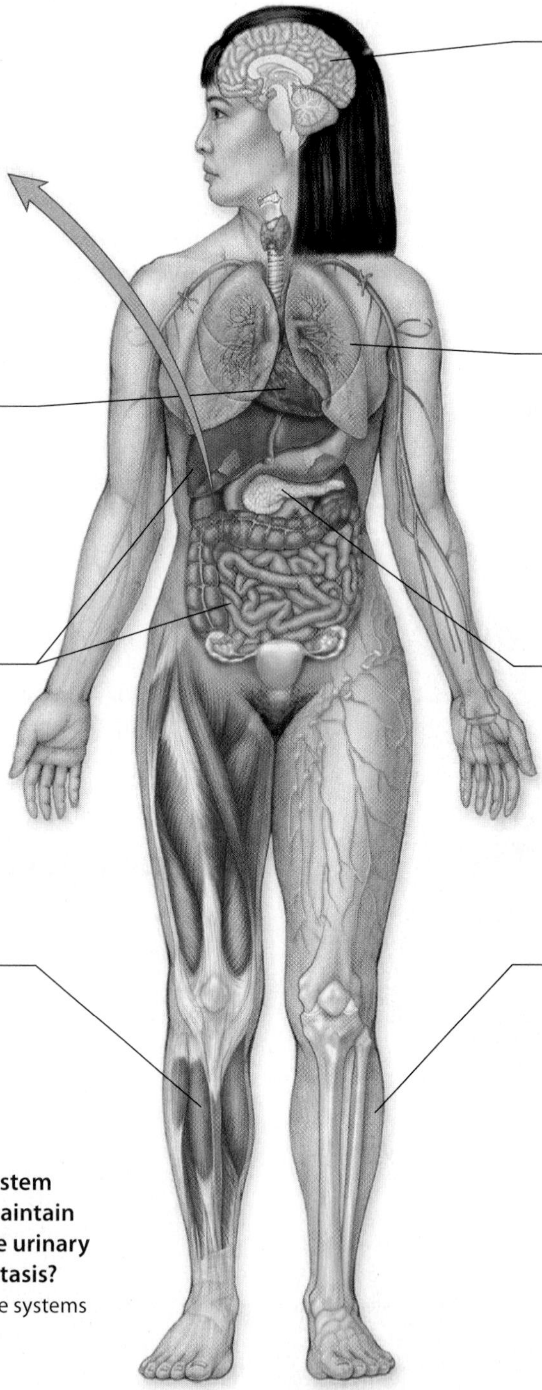

All systems of the body work with the urinary system to maintain homeostasis. These systems are especially noteworthy.

Urinary System

As an aid to all the systems, the kidneys excrete nitrogenous wastes and maintain the water-salt balance and the acid-base balance of the blood. The urinary system also specifically helps the other systems.

Cardiovascular System

Production of renin by the kidneys helps maintain blood pressure. Blood vessels transport nitrogenous wastes to the kidneys and carbon dioxide to the lungs. The buffering system of the blood helps the kidneys maintain the acid-base balance.

Digestive System

The liver produces urea excreted by the kidneys. The yellow pigment found in urine, called urochrome (breakdown product of hemoglobin), is produced by the liver. The digestive system absorbs nutrients, ions, and water. These help the kidneys maintain the proper level of ions and water in the blood.

Muscular System

The kidneys regulate the amount of ions in the blood. These ions are necessary to the contraction of muscles, including those that propel fluids in the ureters and urethra.

Nervous System

The kidneys regulate the amount of ions (e.g., K^+, Na^+, Ca^{2+}) in the blood. These ions are necessary for nerve impulse conduction. The nervous system controls urination.

Respiratory System

The kidneys help the lungs by exhaling carbon dioxide as bicarbonate ions, while the lungs help the kidneys maintain the acid-base balance of the blood by exhaling carbon dioxide.

Endocrine System

The kidneys produce renin, leading to the production of aldosterone, a hormone that helps the kidneys maintain the water-salt balance. The kidneys produce the hormone erythropoietin, and they change vitamin D to a hormone. The posterior pituitary secretes ADH, which regulates water retention by the kidneys.

Integumentary System

Sweat glands excrete perspiration, a solution of water, salt, and some urea.

Figure 10.11 How does the urinary system cooperate with other organ systems to maintain body equilibrium? What systems does the urinary system work with to bring about homeostasis?
The urinary system particularly works with these systems to bring about homeostasis.

Figure 10.12
What does the angiogram show about the kidneys?
Angiography involves injecting a contrast medium opaque to X-rays into the blood vessels, which highlights them when an X-ray of the area is taken.

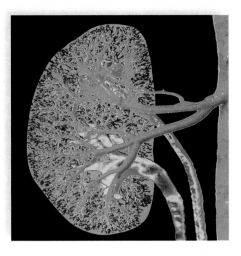

become more acidic. The chemoreceptors in the carotid bodies (located in the carotid arteries) and in the aortic bodies (located in the aorta) respond to the increased acid. The respiratory center is stimulated by the chemoreceptors to initiate exhalation.

As powerful as the combined bicarbonate buffer and breathing systems are, only the kidneys can rid the body of a wide range of acidic and basic substances—thus, the importance of the kidneys' ultimate control over blood pH cannot be overemphasized. The kidneys are slower acting than the buffer/breathing mechanism, but their effect on blood pH is stronger. They do this by adjusting the secretion of hydrogen ions and bicarbonate ions as needed.

By contrast, think what happens if you eat a big tub of salty popcorn at the movies. When salt (NaCl) is absorbed from the digestive tract, the Na^+ content of the blood increases above normal. This results in increased blood volume. The atria of the heart are stretched by this increased blood volume, and the stretch triggers the release of ANH by the heart. ANH inhibits sodium and water reabsorption by the proximal convoluted tubule and collecting duct. Blood volume then decreases because more sodium and water are excreted in the urine.

Acid-Base Balance of Blood

How is the acid-base balance of the blood maintained? Two closely associated mechanisms work to offset short-term challenges to the acid-base balance. These are the blood bicarbonate (HCO_3^-) buffering system and the process of breathing. Usually, the excretion of carbon dioxide (CO_2) by the lungs helps to keep blood pH within normal limits. You may remember from Chapter 9 that holding your breath increases blood carbon dioxide, which causes the blood to

The Kidneys Assist Other Systems

Aside from producing renin, the kidneys assist the endocrine system and also the cardiovascular system by producing erythropoietin (see Fig. 10.11). Erythropoietin is used to stimulate red bone marrow production in patients in renal failure or recovering from chemotherapy. The kidneys assist the skeletal, nervous, and muscular systems by helping to regulate the amount of calcium ions (Ca^{2+}) in the blood. The kidneys convert vitamin D to its active form needed for Ca^{2+} absorption by the digestive tract, and they regulate the excretion of electrolytes, including Ca^{2+}. The kidneys also regulate the sodium (Na^+) and potassium (K^+) content of the blood. These ions, needed for nerve conduction, are necessary to the contraction of the heart and other muscles in the body.

> **Check Your Progress 10.6**
> 1. What are three main functions of the kidneys, and how do they carry them out?
> 2. a. Why do we have to have at least one kidney to live without kidney dialysis? b. Why would one be enough?

⚛ Science **Focus**

Lab Grown Bladders

You're probably familiar with organ transplants done with organs harvested from people who've recently died or even living donors' organs. Did you know some organs can now be grown in a lab and used for transplantation?

Bladders grown in a lab have been successfully transplanted into a number of patients. Each patient in one particular study contributed cells from his/her own diseased bladder. Those cells were cultured in a lab and encouraged to form a new bladder by growing them on a collagen form shaped like a bladder. Eventually the new bladders were attached to each patient's diseased bladder. This alleviated the incontinence problems the patients had experienced. The risk of kidney damage was also decreased in these individuals by lowering the pressure inside their enlarged

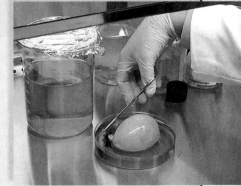

Figure 10C **What is the benefit of transplanting a lab grown bladder?**
There is less risk of rejection when a lab grown bladder is transplanted.

bladders. The cells for the new bladders were taken from each patient, so there was no risk of rejection.

Researchers hope this technique can be used to grow other types of tissues and organs that would be available for transplant. This could be the one potential solution to the limited number of organs available for transplant.

Summarizing the Concepts

10.1 Urinary System

Organs of the Urinary System

Only the urinary system contains urine.

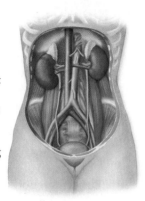

- Kidneys produce urine.
- Ureters take urine to the bladder.
- Urinary bladder stores urine.
- Urethra releases urine to the outside.

Functions of the Urinary System

- Excrete nitrogenous wastes, including urea, uric acid, and creatinine.
- Maintain the normal water-salt balance of the blood.
- Maintain the acid-base balance of the blood.
- Assist endocrine system in hormone secretion by secreting erythropoietin (stimulates red blood cell production) and releasing renin (leads to the secretion of aldosterone).

10.2 Kidney Structure

- Macroscopic structures are the renal cortex, renal medulla, and renal pelvis.
- Microscopic structures are the nephrons.

Anatomy of a Nephron

- Each nephron has its own blood supply: the afferent arteriole divides to become the glomerulus. The efferent arteriole branches into the peritubular capillary network.
- Each region of the nephron is anatomically suited to its task in urine formation.

10.3 Urine Formation

Urine is composed primarily of nitrogenous waste products and salts in water. The steps in urine formation include the following:

Glomerular Filtration

Water, salts, nutrients, and wastes move from the glomerulus to the inside of the glomerular capsule.

Tubular Reabsorption

Nutrients and salt molecules are reabsorbed from the convoluted tubules into the peritubular capillary network; water follows.

Tubular Secretion

Certain molecules are actively secreted from the peritubular capillary network into the convoluted tubules.

10.4 Regulatory Functions of the Kidneys

The kidneys maintain the water-salt balance of the blood and, therefore, participate in blood volume and pressure control.

Water-Salt Balance

- In the renal cortex, most of the reabsorption of salts and water occurs in the proximal convoluted tubule. Aldosterone controls the reabsorption of sodium and antidiuretic hormone (ADH) controls the reabsorption of water in the distal convoluted tubule. Atrial natriuretic hormone (ANH) acts contrary to aldosterone.
- In the renal medulla, the ascending limb of the loop of the nephron establishes a solute gradient that increases toward the inner medulla. The solute gradient draws water from the descending limb of the loop of the nephron and from the collecting duct. When ADH is present, more water is reabsorbed from the collecting duct, and a decreased amount of urine results.

Acid-Base Balance of Body Fluids

- The kidneys maintain the acid-base balance of blood (blood pH).
- They reabsorb HCO_3^- and excrete H^+ as needed to maintain the pH at about 7.4. Ammonia buffers H^+ in the urine.

10.5 Disorders with Kidney Function

Various types of problems, including diabetes, kidney stones, and infections, can lead to renal failure, which necessitates undergoing hemodialysis by using a kidney machine or CAPD, or by receiving a kidney transplant.

10.6 Homeostasis

- The kidneys maintain the water-salt balance of blood.
- The kidneys also keep blood pH within normal limits.
- The urinary system works with the other systems of the body to maintain homeostasis.

Understanding Key Terms

proximal convoluted tubule 214   tubular secretion 216
pyelonephritis 222   urea 209
renal artery 208   uremia 222
renal cortex 212   ureter 208
renal medulla 212   urethra 209
renal pelvis 212   urethritis 222
renal vein 208   uric acid 209
renin 219   urinary bladder 208
tubular reabsorption 216

Match the key terms to these definitions.

a. _____ Drug used to counteract hypertension by causing the excretion of water.

b. _____ Removal of metabolic wastes from the body.

c. _____ Tubular structure that receives urine from the bladder and carries it to the outside of the body.

d. _____ Hollow chamber in the kidney that lies inside the renal medulla and receives freshly prepared urine from the collecting ducts.

e. _____ Filtered portion of blood contained within the glomerular capsule.

Testing Your Knowledge of the Concepts

1. Describe the path of urine and the structure of each organ mentioned. (pages 208–09)

2. In what ways do the four functions of the urinary system contribute to homeostasis? (pages 209-10)

3. Describe the macroscopic structure of the kidney. (page 212)

4. Trace the path of blood into and out of the kidney. (pages 212–13)

5. Name the parts of a nephron, and describe the structure of each part. (pages 213–14)

6. Why would it be proper to associate glucose with the first two processes involved in urine formation but not the third process? (pages 215–16)

7. Explain how a hypertonic urine can be formed, detailing where and how salt and water move, and the influence of hormones on the process. (pages 218–20)

8. Describe three ways that the body maintains the acid-base balance of body fluids. (pages 220-22)

9. Explain how an artificial kidney machine and continuous ambulatory peritoneal dialysis work to cleanse the blood. (pages 222-23)

10. How do the kidneys assist other body systems? (pages 224–25)

11. Which of the following is a structural difference between the urinary systems of males and females?
    a. Males have a longer urethra than females.
    b. In males, the urethra passes through the prostate.
    c. In males, the urethra serves both the urinary and reproductive systems.
    d. All of these are correct.

12. Which of these is found in the renal medulla?
    a. loop of the nephron   c. peritubular capillaries
    b. collecting ducts   d. All of these are correct.

13. Which of these functions of the kidneys are mismatched?
    a. excretes metabolic wastes—rids the body of urea
    b. maintains the water-salt balance—helps regulate blood pressure
    c. maintains the acid-base balance—rids the body of uric acid
    d. secretes hormones—secretes erythropoietin
    e. All of these are correct.

14. Which of the following is not correct?
    a. Uric acid is produced from the breakdown of amino acids.
    b. Creatinine is produced from breakdown reactions in the muscles.
    c. Urea is the primary nitrogenous waste of humans.
    d. Ammonia results from the deamination of amino acids.

15. Which part of a nephron is out of sequence first?
    a. glomerular capsule
    b. proximal convoluted tubule
    c. distal convoluted tubule
    d. loop of the nephron
    e. collecting duct

16. Which portion of the nephron has cells with a brush border and many mitochondria?
    a. glomerular capsule
    b. proximal convoluted tubule
    c. loop of the nephron
    d. distal convoluted tubule
    e. collecting duct

17. When tracing the path of filtrate, the loop of the nephron follows which structure?
    a. collecting duct
    b. distal convoluted tubule
    c. proximal convoluted tubule
    d. glomerulus
    e. renal pelvis

18. When tracing the path of blood, the blood vessel that follows the renal artery is the
    a. peritubular capillary.   d. renal vein.
    b. efferent arteriole.   e. glomerulus.
    c. afferent arteriole.

19. The function of the descending limb of the loop of the nephron in the process of urine formation is
    a. reabsorption of water.   c. reabsorption of solutes.
    b. production of filtrate.   d. secretion of solutes.

20. Which of the following materials would not normally be filtered from the blood at the glomerulus?
    a. water   d. glucose
    b. urea   e. sodium ions
    c. protein

21. Which of the following materials would not be maximally reabsorbed from the filtrate?
    a. water   d. urea
    b. glucose   e. amino acids
    c. sodium ions

22. By what process are most molecules secreted from the blood into the tubule?
    a. osmosis   c. active transport
    b. diffusion   d. facilitated diffusion

23. Reabsorption of the glomerular filtrate occurs primarily at the
    a. proximal convoluted tubule.
    b. distal convoluted tubule.
    c. loop of the nephron.
    d. collecting duct.

24. A countercurrent mechanism draws water from the
    a. proximal convoluted tubule.
    b. descending limb of the loop of the nephron.
    c. distal convoluted tubule.
    d. collecting duct.
    e. Both b and d are correct.

25. Sodium is actively extruded from which part of the nephron?
    a. descending portion of the proximal convoluted tubule
    b. ascending portion of the loop of the nephron
    c. ascending portion of the distal convoluted tubule
    d. descending portion of the collecting duct

26. Excretion of a hypertonic urine in humans is best associated with the
    a. glomerular capsule and the tubules.
    b. proximal convoluted tubule only.
    c. loop of the nephron and collecting duct.
    d. distal convoluted tubule and peritubular capillary.

27. Which of these hormones is most likely to directly cause a drop in blood pressure?
    a. aldosterone
    b. antidiuretic hormone (ADH)
    c. erythropoietin
    d. atrial natriuretic hormone (ANH)

28. The presence of ADH (antidiuretic hormone) causes an individual to excrete
    a. sugars.   c. more water.
    b. less water.   d. Both a and c are correct.

29. To lower blood acidity,
    a. hydrogen ions are excreted, and bicarbonate ions are reabsorbed.
    b. hydrogen ions are reabsorbed, and bicarbonate ions are excreted.
    c. hydrogen ions and bicarbonate ions are reabsorbed.
    d. hydrogen ions and bicarbonate ions are excreted.
    e. urea, uric acid, and ammonia are excreted.

30. The function of erythropoietin is
    a. reabsorption of sodium ions.
    b. excretion of potassium ions.
    c. reabsorption of water.
    d. to stimulate red blood cell production.
    e. to increase blood pressure.

In questions 31–34, match the function of the urinary system to the human organ system in the key.

Key:

a. muscular system   e. respiratory system
b. nervous system   f. digestive system
c. endocrine system   g. reproductive system
d. cardiovascular system

31. Liver synthesizes urea.

32. Smooth muscular contraction assists voiding of urine.

33. ADH, aldosterone, and atrial natriuretic hormone regulate reabsorption of Na^+ by kidneys.

34. Blood vessels deliver waste to be excreted.

35. Label this diagram of a nephron.

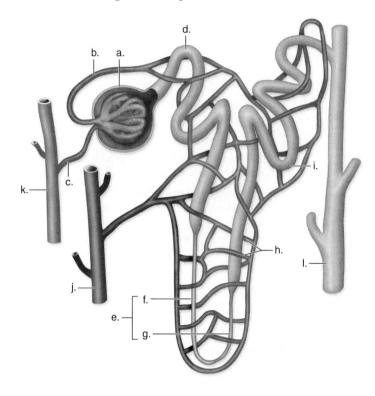

Thinking Critically About the Concepts

Casey was suffering from a form of diabetes that may be unfamiliar to most people. When the word "diabetes" is used, everyone tends to think of the type related to insulin. That form of diabetes is diabetes mellitus, which can be type 1 or type 2. Kidney function is affected in both types of diabetes. Several other hormones affect kidney function as well. Chapter 15 will provide you with more information about these and other hormones.

1. Describe the symptoms common to diabetes mellitus and diabetes insipidus. What symptom distinguishes diabetes mellitus from diabetes insipidus?

2. Why would increasing the production of red blood cells in people in renal failure alleviate their symptom of fatigue?

3. A side effect of using Detrol LA® to calm an overactive bladder is constipation. Look back to Chapter 8 and identify the type of muscle found throughout the digestive system. Speculate as to why constipation might occur after the use of Detrol LA®.

4. Men who have prostate cancer often have some or all of the prostate removed. The surgeon tries very hard not to damage the sphincter muscles associated with the bladder. What problem might a man experience if the bladder's sphincter muscles were damaged during the surgery?

5. Recall the anatomical position of the kidneys, and consider the ways they might be damaged in sports. Describe special equipment designed specifically to protect the kidneys.

CHAPTER **11**

Skeletal System

CASE STUDY EMILY HARAHAN

I t had taken some time to get used to, but Paul Harahan was beginning to enjoy his part-time duty as "Mr. Mom." He and Mary had worked their budget over, and it just made sense for her to return to work part-time. Her flexible schedule as a nurse allowed her to work only two evenings and on Saturdays. In turn, Paul was charged with 6-year-old Emily and 4-year-old Robby. Paul was becoming a regular at the nearby park, where a whole gang of kids considered him their giant playmate. The time passed quickly as he pushed swings, spun merry-go-rounds, and helped the littlest ones with their first time down the slide.

"Piggy-back ride, Dad!" Emily giggled. Yes, Paul thought, the Mr. Mom job had its benefits. He crouched and Emily climbed on his back, folding her arms around his neck. Bouncing her up and down, he growled fake growls, chasing Robby around the play yard. "My turn, Dad!" the little boy exclaimed.

"He's right, Emily, it's his turn for a piggyback ride," Paul said. He reached for her left forearm, swinging her to the ground. Suddenly, he heard a distinctive "pop" as he set her on the ground. She fell to the ground, screaming as she protected her left arm with her right. "What's wrong, honey?" he cried.

"You hurt my arm, Dad," Emily whimpered. Her forearm was cocked at an odd angle, as though a bone had broken, and her elbow was misshapen. "Oh, man, Emily, Daddy's sorry, sweetheart, it's going to be okay," he stammered.

Other parents gathered. "Is everything okay, Paul?" his neighbor asked. "Do you need me to call the squad? I don't think you should be driving right now."

"Yeah, I guess you'd better," Paul answered shakily.

"I'll take Robby home and give him some lunch with my boys," the neighbor offered. Paul nodded.

CHAPTER CONCEPTS

11.1 Overview of Skeletal System
Bones are the organs of the skeletal system. The tissues of the system are compact and spongy bone, various types of cartilage, and fibrous connective tissue in the ligaments that hold bones together.

11.2 Bone Growth, Remodeling, and Repair
Bone is a living tissue that grows, remodels, and repairs itself. In all of these processes, some bone cells break down bone, and some repair bone.

11.3 Bones of the Axial Skeleton
The axial skeleton lies in the midline of the body and consists of the skull, the hyoid bone, the vertebral column, and the rib cage.

11.4 Bones of the Appendicular Skeleton
The appendicular skeleton consists of the bones of the pectoral girdle, upper limbs, pelvic girdle, and lower limbs.

11.5 Articulations
Joints are classified according to their degree of movement. Synovial joints are freely moveable.

11.1 Overview of Skeletal System

The skeletal system consists of two types of connective tissue: bone and the cartilage found at **joints.** In addition, **ligaments** formed of fibrous connective tissue join the bones.

Functions of the Skeleton

The skeleton supports the body. The bones of the legs support the entire body when we are standing, and bones of the pelvic girdle support the abdominal cavity.

   The skeleton protects soft body parts. The bones of the skull protect the brain; the rib cage protects the heart and lungs. The vertebrae protect the spinal cord, which makes nervous connections to all the muscles of the limbs.

   The skeleton produces blood cells. All bones in the fetus have red bone marrow that produces blood cells. In the adult, only certain bones produce blood cells.

   The skeleton stores minerals and fat. All bones have a matrix that contains calcium phosphate, a source of calcium ions and phosphate ions in the blood. Fat is stored in yellow bone marrow.

   The skeleton, along with the muscles, permits flexible body movement. While articulations (joints) occur between all the bones, we associate body movement in particular with the bones of limbs.

Anatomy of a Long Bone

Figure 11.1 shows the anatomy of a long bone. The shaft, or main portion of the bone, is called the diaphysis. The diaphysis has a large **medullary cavity,** whose walls are composed of compact bone. The medullary cavity is lined with a thin, vascular membrane (the endosteum) and is filled with yellow bone marrow that stores fat.

   The expanded region at the end of a long bone is called an epiphysis (pl., epiphyses). The epiphyses are composed largely of spongy bone that contains red bone marrow, where blood cells are made. The epiphyses are coated with a thin layer of hyaline cartilage, called **articular cartilage** because it occurs at a joint.

Have You Ever Wondered...

How many bones are in a skeleton?

A newborn has nearly 300 bones, some of which fuse together as the child grows. The adult human skeleton has approximately 206 bones, but the number varies between individuals. Some people have extra bones, called Wormian bones, that help to fuse skull bones together. Others may have additional small bones in the ankles and feet.

   Except for the articular cartilage on its ends, a long bone is completely covered by a layer of fibrous connective tissue called the **periosteum.** This covering contains blood vessels, lymphatic vessels, and nerves. Note in Figure 11.1 how a blood vessel penetrates the periosteum and enters the bone. Branches of the blood vessel are found throughout the medullary cavity. Other branches can be found in hollow cylinders, called central canals, within the bone tissue. The periosteum is continuous with ligaments and tendons connected to a bone.

Bone

Compact bone is highly organized and composed of tubular units called osteons. In a cross section of an osteon, bone cells called **osteocytes** lie in lacunae, tiny chambers arranged in concentric circles around a central canal (see Fig. 11.1). Matrix fills the space between the rows of lacunae. Tiny canals called canaliculi (sing., canaliculus) run through the matrix. These canaliculi connect the lacunae with one another and with the central canal. The cells stay in contact by strands of cytoplasm that extend into the canaliculi. Osteocytes nearest the center of an osteon exchange nutrients and wastes with the blood vessels in the central canal. These cells then pass on nutrients and collect wastes from the other cells via gap junctions (see page 90).

   Compared with compact bone, **spongy bone** has an unorganized appearance (see Fig. 11.1). It contains numerous thin plates (called trabeculae) separated by unequal spaces. Although this makes spongy bone lighter than compact bone, spongy bone is still designed for strength. Just as braces are used for support in buildings, the trabeculae follow lines of stress. The spaces of spongy bone are often filled with **red bone marrow,** a specialized tissue that produces all types of blood cells. The osteocytes of spongy bone are irregularly placed within the trabeculae. Canaliculi bring them nutrients from the red bone marrow.

Cartilage

Cartilage is not as strong as bone, but it is more flexible. Its matrix is gel-like and contains many collagenous and elastic fibers. The cells, called **chondrocytes,** lie within lacunae that are irregularly grouped. Cartilage has no nerves, making it well suited for padding joints where the stresses of movement are intense. Cartilage also has no blood vessels and relies on neighboring tissues for nutrient and waste exchange. This makes it slow to heal.

   The three types of cartilage differ according to the type and arrangement of fibers in the matrix. *Hyaline cartilage* is firm and somewhat flexible. The matrix appears uniform and glassy, but actually it contains a generous supply of collagen fibers. Hyaline cartilage is found at the ends of long bones, in the nose, at the ends of the ribs, and in the larynx and trachea.

   Fibrocartilage is stronger than hyaline cartilage because the matrix contains wide rows of thick, collagen fibers. Fibrocartilage is able to withstand both tension and pressure and is found where support is of prime importance—in the disks located between the vertebrae and also in the cartilage of the knee.

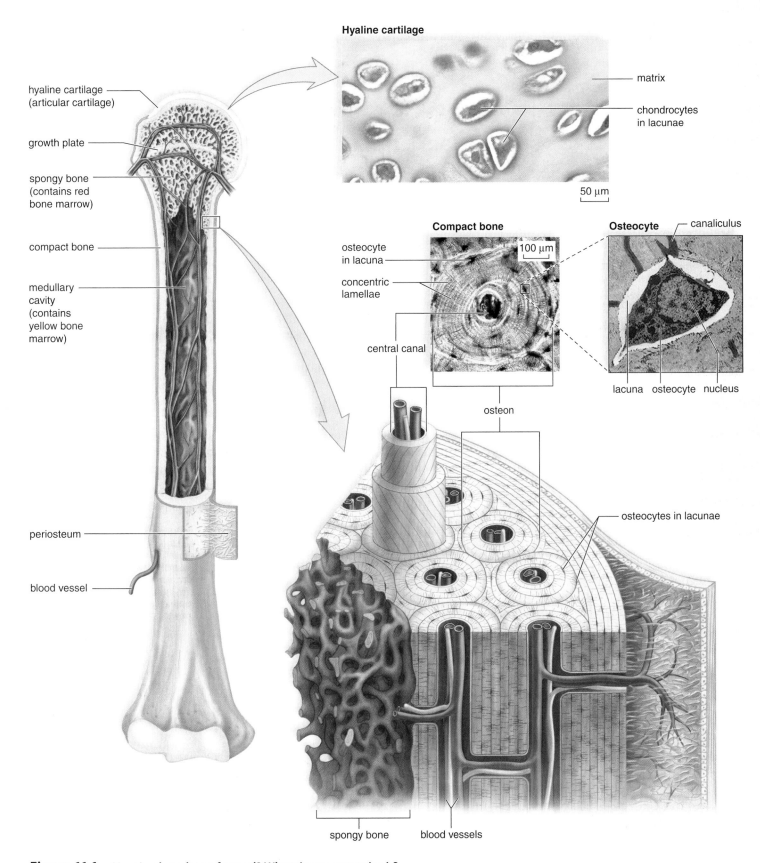

Hyaline cartilage

matrix

chondrocytes in lacunae

50 µm

hyaline cartilage (articular cartilage)

growth plate

spongy bone (contains red bone marrow)

compact bone

medullary cavity (contains yellow bone marrow)

periosteum

blood vessel

Compact bone

osteocyte in lacuna

concentric lamellae

central canal

100 µm

Osteocyte

canaliculus

lacuna   osteocyte   nucleus

osteon

osteocytes in lacunae

spongy bone   blood vessels

Figure 11.1 **How is a long bone formed? What tissues comprise it?**

Left: A long bone is formed of an outer layer of compact bone, shown in the enlargement and micrograph *(right)*. Spongy bone, which lies beneath compact bone, may contain red bone marrow. The central shaft of a long bone contains yellow marrow, a form of stored fat. Periosteum, a fibrous membrane, encases the bone except at its ends. Hyaline cartilage (see micrograph) covers the ends of bones.

Elastic cartilage is more flexible than hyaline cartilage because the matrix contains mostly elastin fibers. This type of cartilage is found in the ear flaps and the epiglottis.

Fibrous Connective Tissue

Fibrous connective tissue contains rows of cells called fibroblasts separated by bundles of collagenous fibers. This tissue makes up ligaments and tendons. Ligaments connect bone to bone. Tendons connect muscle to bone at a joint (also called an articulation).

> **Check Your Progress 11.1**
>
> 1. a. What is the anatomy of compact bone? b. Of spongy bone?
>
> 2. What are the three types of cartilage, and where are they found in the body?
>
> 3. What type of connective tissue makes up ligaments and tendons?
>
> 4. What is the structure of a typical bone, such as a long bone?

11.2 Bone Growth, Remodeling, and Repair

The importance of the skeleton to the human form is evident by its early appearance during development. The skeleton starts forming at about six weeks, when the embryo is only about 12 mm (0.5 in.) long. Most bones grow in length and width through adolescence, but some continue enlarging until about age 25. In a sense, bones can grow throughout a lifetime because they are able to respond to stress by changing size, shape, and strength. This process is called remodeling. If a bone fractures, it can heal by bone repair.

Bones are living tissues, as shown by their ability to grow, remodel, and undergo repair. Several different types of cells are involved in bone growth, remodeling, and repair:

Osteoblasts are bone-forming cells. They secrete the organic matrix of bone and promote the deposition of calcium salts into the matrix.

Osteocytes are mature bone cells derived from osteoblasts. They maintain the structure of bone.

Osteoclasts are bone-absorbing cells. They break down bone and assist in returning calcium and phosphate to the blood.

Throughout life, osteoclasts are removing the matrix of bone, and osteoblasts are building it up. When osteoblasts are surrounded by calcified matrix, they become the osteocytes within lacunae.

Bone Development and Growth

The term **ossification** refers to the formation of bone. The bones of the skeleton form during embryonic development in two distinctive ways: intramembranous ossification and endochondral ossification (Fig. 11.2).

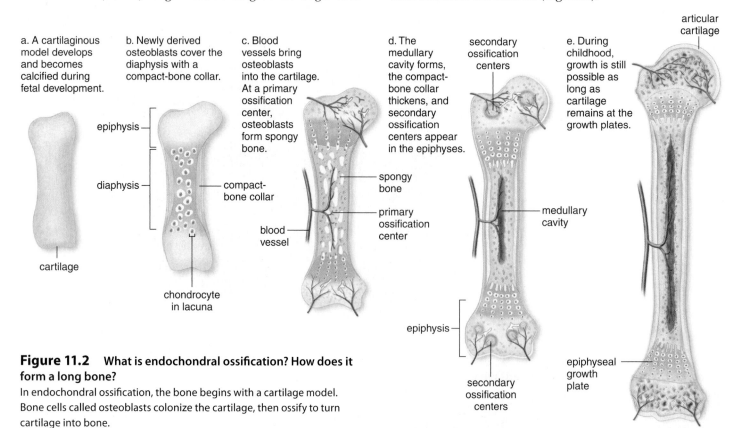

a. A cartilaginous model develops and becomes calcified during fetal development.

epiphysis

diaphysis

cartilage

b. Newly derived osteoblasts cover the diaphysis with a compact-bone collar.

compact-bone collar

chondrocyte in lacuna

c. Blood vessels bring osteoblasts into the cartilage. At a primary ossification center, osteoblasts form spongy bone.

blood vessel

spongy bone

primary ossification center

d. The medullary cavity forms, the compact-bone collar thickens, and secondary ossification centers appear in the epiphyses.

secondary ossification centers

medullary cavity

epiphysis

secondary ossification centers

e. During childhood, growth is still possible as long as cartilage remains at the growth plates.

articular cartilage

epiphyseal growth plate

Figure 11.2   What is endochondral ossification? How does it form a long bone?

In endochondral ossification, the bone begins with a cartilage model. Bone cells called osteoblasts colonize the cartilage, then ossify to turn cartilage into bone.

Intramembranous Ossification

Flat bones, such as the bones of the skull, are examples of intramembranous bones. In **intramembranous ossification,** bones develop between sheets of fibrous connective tissue. Here, cells derived from connective tissue cells become osteoblasts located in ossification centers. The osteoblasts secrete the organic matrix of bone. This matrix consists of mucopoly-saccharides and collagen fibrils. Calcification occurs when calcium salts are added to the organic matrix. The osteoblasts promote calcification, or ossification, of the matrix. Ossification results in the trabeculae of spongy bone. Spongy bone remains inside a flat bone. The spongy bone of flat bones, such as those of the skull and clavicles (collar-bones), contains red bone marrow.

A periosteum forms outside the spongy bone. Osteoblasts derived from the periosteum carry out further ossification. Trabeculae form and fuse to become compact bone. The compact bone forms a bone collar that surrounds the spongy bone on the inside.

Endochondral Ossification

Most of the bones of the human skeleton are formed by **endochondral ossification.** During endochondral ossification, bone replaces the cartilaginous models of the bones. Gradually, the cartilage is replaced by the calcified bone matrix that makes these bones capable of bearing weight.

Inside, bone formation spreads from the center to the ends, and this accounts for the term used for this type of ossification. (Endochondral literally means "within cartilage.") The long bones, such as the tibia, provide examples of endochondral ossification (see Fig. 11.2).

1. *The cartilage model.* In the embryo, chondrocytes lay down hyaline cartilage, which is shaped like the future bones. Therefore, they are called cartilage models of the future bones. As the cartilage models calcify, the chondrocytes die off.
2. *The bone collar.* Osteoblasts are derived from the newly formed periosteum. Osteoblasts secrete the organic bone matrix, and the matrix undergoes calcification. The result is a bone collar, which covers the diaphysis (see Fig. 11.2). The bone collar is composed of compact bone. In time, the bone collar thickens.
3. *The primary ossification center.* Blood vessels bring osteoblasts to the interior, and they begin to lay down spongy bone. This region is called a primary ossification center because it is the first center for bone formation.
4. *The medullary cavity and secondary ossification sites.* The spongy bone of the diaphysis is absorbed by osteoclasts, and the cavity created becomes the medullary cavity. Shortly after birth, secondary ossification centers form in the epiphyses. Spongy bone persists in the epiphyses, and they contain red bone marrow for quite some time. Cartilage is present at two locations: the epiphyseal (growth) plate and articular cartilage, which covers the ends of long bones.
5. *The epiphyseal (growth) plate.* A band of cartilage called a **growth plate** remains between the primary ossification center and each secondary center. The limbs keep increasing in length as long as growth plates are still present.

Figure 11.3 shows that the epiphyseal plate contains four layers. The layer nearest the epiphysis is the resting zone, where cartilage remains. The next layer is the proliferating

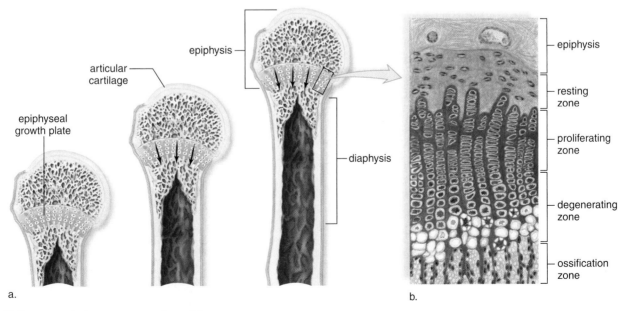

epiphysis

articular cartilage

epiphyseal growth plate

diaphysis

epiphysis

resting zone

proliferating zone

degenerating zone

ossification zone

a.

b.

Figure 11.3 **How do bones grow in length?**
a. Length of a bone increases when cartilage is replaced by bone at the growth plate. **b.** Chondrocytes produce new cartilage in the proliferating zone, and cartilage becomes bone in the ossification zone closest to the diaphysis. Arrows show the direction of ossification.

zone, in which chondrocytes are producing new cartilage cells. In the third layer, the degenerating zone, the cartilage cells are dying off; and in the fourth layer, the ossification zone, bone is forming. Bone formation here causes the length of the bone to increase. The inside layer of articular cartilage also undergoes ossification in the manner described.

The diameter of a bone enlarges as a bone lengthens. Osteoblasts derived from the periosteum are active in new bone deposition as osteoclasts enlarge the medullary cavity from inside.

Final Size of the Bones When the epiphyseal plates close, bone length can no longer occur. The epiphyseal plates in the arms and legs of women typically close between ages 16–18, while they do not close in men until about age 20. Portions of other types of bones may continue to grow until age 25. Hormones are chemical messengers secreted by the endocrine glands and distributed about the body by the bloodstream. Hormones control the activity of the epiphyseal plate, as is discussed next.

Hormones Affect Bone Growth

The importance of bone growth is signified by the involvement of several hormones in bone growth. A **hormone** is a chemical messenger, produced by one part of the body, which acts on a different part of the body.

Vitamin D is formed in the skin when it is exposed to sunlight, but it can also be consumed in the diet. Milk, in particular, is often fortified with vitamin D. In the kidneys, vitamin D is converted to a hormone that acts on the intestinal tract. The chief function of vitamin D is intestinal absorption of calcium. In the absence of vitamin D, children can develop rickets, a condition marked by bone deformities including bowed long bones (see Fig. 8.15a).

Growth hormone (GH) directly stimulates growth of the epiphyseal plate, as well as bone growth in general. However, growth hormone will be somewhat ineffective if the metabolic activity of cells is not promoted. Thyroid hormone, in particular, promotes the metabolic activity of cells. Too little growth hormone in childhood results in dwarfism. Too much growth hormone during childhood (prior to epiphyseal fusion) can produce excessive growth and even gigantism. Acromegaly results from excess GH in adults, following epiphyseal fusion. This condition produces excessive growth of bones in the hands and face (see Figs. 15.7 and 15.8).

Adolescents usually experience a dramatic increase in height, called the growth spurt, due to an increased level of sex hormones. These hormones apparently stimulate osteoblast activity. Rapid growth causes epiphyseal plates to become "paved over" by the faster growing bone tissue, within one or two years of the onset of puberty.

Bone Remodeling and Its Role in Homeostasis

Bone is constantly being broken down by osteoclasts and re-formed by osteoblasts in the adult. As much as 18% of bone is recycled each year. This process of bone renewal, often called **bone remodeling,** normally keeps bones strong (Fig. 11.4). In Paget's disease, new bone is generated at a faster-than-normal rate. This rapid remodeling produces bone that's softer and weaker than normal bone and can cause bone pain, deformities, and fractures.

Bone recycling allows the body to regulate the amount of calcium in the blood. To illustrate that the blood calcium level is critical, recall that calcium is required for blood to clot. Also, if the blood calcium concentration is too high, neurons and muscle cells no longer function. If calcium falls

Figure 11.4 What is bone remodeling?
Diameter of a bone increases as bone absorption occurs inside the shaft and is matched by bone formation outside the shaft.

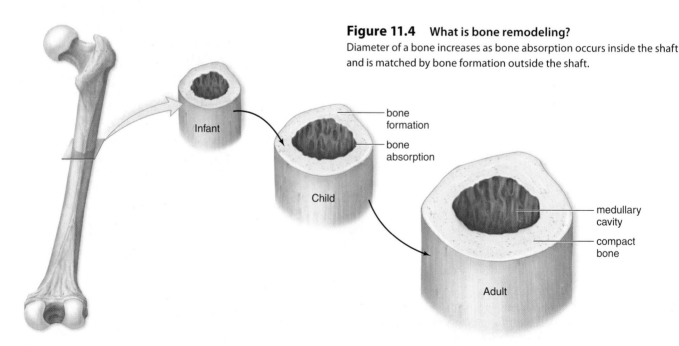

⚕ Health **Focus**

You Can Avoid Osteoporosis

Osteoporosis is a condition in which the bones are weakened due to a decrease in the bone mass that makes up the skeleton. The skeletal mass continues to increase until ages 20 to 30. After that, there is an equal rate of formation and breakdown of bone mass until ages 40 to 50. Then, reabsorption begins to exceed formation, and the total bone mass slowly decreases.

Over time, men are apt to lose 25% and women to lose 35% of their bone mass. But we have to consider that men, unless they have taken asthma medications that decrease bone formation, tend to have denser bones than women anyway. A man's testosterone (male sex hormone) level generally does not begin to decline significantly until after age 65. In contrast, the estrogen (female sex hormone) level in women begins to decline at about age 45. Sex hormones play an important role in maintaining bone strength, so this difference means that women are more likely than men to suffer a higher incidence of fractures, involving especially the hip, vertebrae, long bones, and pelvis. Although osteoporosis may at times be the result of various disease processes, it is essentially a disease that occurs as we age.

How to Avoid Osteoporosis

Everyone can take measures to avoid having osteoporosis when they get older. Adequate dietary calcium throughout life is an important protection against osteoporosis. The U.S. National Institutes of Health recommend a calcium intake of 1,200–1,500 mg per day during puberty. Males and females require 1,000 mg per day until the age of 65 and 1,500 mg per day after age 65. In older women, 1,500 mg per day is especially desirable.

A small daily amount of vitamin D is also necessary for the body to use calcium correctly. Exposure to sunlight is required to allow skin to synthesize a precursor to vitamin D.

If you reside on or north of a "line" drawn from Boston to Milwaukee, to Minneapolis, to Boise, chances are you're not getting enough vitamin D during the winter months. Therefore, you should take advantage of the vitamin D present in fortified foods such as low-fat milk and cereal.

Very inactive people, such as those confined to bed, lose bone mass 25 times faster than people who are moderately active. On the other hand, moderate weight-bearing exercise, such as regular walking or jogging, is another good way to maintain bone strength (Fig. 11A).

Diagnosis and Treatment

Postmenopausal women with any of the following risk factors should have an evaluation of their bone density:

- white or Asian race
- thin body type
- family history of osteoporosis
- early menopause (before age 45)
- smoking
- a diet low in calcium, or excessive alcohol consumption and caffeine intake
- sedentary lifestyle

Bone density is measured by a method called dual energy X-ray absorptiometry (DEXA). This test measures bone density based on the absorption of photons generated by an X-ray tube. Soon there may be blood and urine tests to detect the biochemical markers of bone loss. Then it will be made easier for physicians to screen older women and at-risk men for osteoporosis.

If the bones are thin, it is worthwhile to take all possible measures to gain bone density because even a slight increase can significantly reduce fracture risk. Although estrogen therapy does reduce the incidence of hip fractures, long-term estrogen therapy is rarely recommended for osteoporosis. Estrogen is known to increase the risk of breast cancer, heart disease, stroke, and blood clots. Other medications are available, however. Calcitonin, a thyroid hormone, has been shown to increase bone density and strength, while decreasing the rate of bone fractures. Also, the bisphosphonates are a family of nonhormonal drugs used to prevent and treat osteoporosis. To achieve optimal results with calcitonin or one of the bisphosphonates, patients should also receive adequate amounts of dietary calcium and vitamin D.

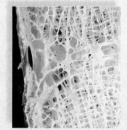

a. normal bone

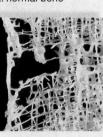

b. osteoporosis

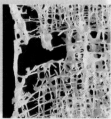

c.

Figure 11A **How can osteoporosis be prevented?**
Weight-lifting exercise, when we are young, can help prevent osteoporosis when we are older. **a.** Normal bone. **b.** Bone from a person with osteoporosis. **c.** An elderly person with osteoporosis.

too low, nerve and muscle cells become so excited that convulsions occur. The bones are the storage sites for calcium—if the blood calcium rises above normal, at least some of the excess is deposited in the bones. If the blood calcium dips too low, calcium is removed from the bones to bring it back up to the normal level.

Two hormones in particular are involved in regulating the blood calcium level. Parathyroid hormone (PTH) stimulates osteoclasts to dissolve the calcium matrix of bone. In addition, parathyroid hormone promotes calcium reabsorption in the small intestine and kidney. Thus, the blood calcium level increases. Calcitonin is a hormone that acts opposite to PTH. The female sex hormone estrogen can actually increase the number of osteoblasts, and the reduction of estrogen in older women is often given as reason for the development of weak bones, called osteoporosis. Osteoporosis is discussed in the Health Focus on page 235. In the young adult, the activity of osteoclasts is matched by the activity of osteoblasts, and bone mass remains stable until about age 45 years in women. After that age, bone mass starts to decrease. Men also experience osteoporosis later in life.

Bone remodeling also accounts for why bones can respond to stress (see Fig. 11.4). If you engage in an activity that calls upon the use of a particular bone, the bone will enlarge in diameter at the region most affected by the activity. During this process, osteoblasts in the periosteum form compact bone around the external bone surface, and osteoclasts break down bone on the internal bone surface, around the medullary cavity. Increasing the size of the medullary cavity prevents the bones from getting too heavy and thick. Today, exercises such as walking, jogging, and weight lifting are recommended. These exercises strengthen bone because they stimulate the work of osteoblasts instead of osteoclasts.

Bone Repair

Repair of a bone is required after it breaks or fractures. Fracture repair takes place over a span of several months in a series of four steps, shown in Figure 11.5 and listed here:

1. *Hematoma.* After a fracture, blood escapes from ruptured blood vessels and forms a hematoma (mass of clotted blood) in the space between the broken bones. The hematoma forms within 6–8 hours.
2. *Fibrocartilaginous callus.* Tissue repair begins, and a fibrocartilaginous callus fills the space between the ends of the broken bone for about three weeks.
3. *Bony callus.* Osteoblasts produce trabeculae of spongy bone and convert the fibrocartilage callus to a bony callus that joins the broken bones together. The bony callus lasts about three to four months.

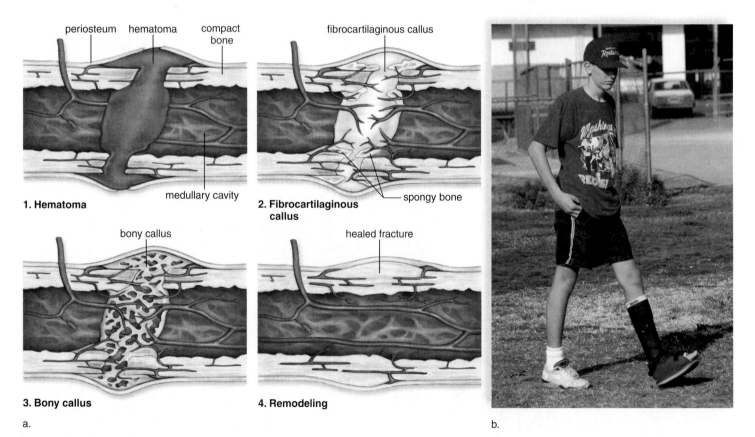

a.   b.

Figure 11.5   How is bone repaired after a fracture occurs?
a. Steps in the repair of a fracture. **b.** A cast helps stabilize the bones while repair takes place. Fiberglass casts are now replacing plaster of Paris as the usual material for a cast.

4. *Remodeling.* Osteoblasts build new compact bone at the periphery. Osteoclasts absorb the spongy bone, creating a new medullary cavity.

In some ways, bone repair parallels the development of a bone except that the first step, hematoma, indicates that injury has occurred. Further, a fibrocartilaginous callus precedes the production of compact bone.

The naming of fractures tells you what type of break occurred. A fracture is complete if the bone is broken clear through and incomplete if the bone is not separated into two parts. A fracture is simple if it does not pierce the skin and compound if it does pierce the skin. Impacted means that the broken ends are wedged into each other. A spiral fracture occurs when the break is ragged due to twisting of a bone.

Check Your Progress 11.2

1. **What are the types of cells involved in bone growth, remodeling, and repair?**
2. **How does bone growth occur during development?**
3. **What hormones are involved in bone growth?**
4. **How does remodeling affect the blood calcium concentration?**
5. **How does bone repair itself?**

11.3 Bones of the Axial Skeleton

The 206 bones of the skeleton are classified according to whether they occur in the axial skeleton or the appendicular skeleton (see Fig. 11.6). The **axial skeleton** lies in the midline of the body and consists of the skull, hyoid bone, vertebral column, and the rib cage.

The Skull

The **skull** is formed by the cranium (braincase) and the facial bones. However, some cranial bones contribute to the face.

The Cranium   The cranium protects the brain. In adults, it is composed of eight bones fitted tightly together. In newborns, certain cranial bones are not completely formed. Instead, these bones are joined by membranous regions called **fontanels.** The fontanels usually close by the age of 16 months by the process of intramembranous ossification.

Some of the bones of the cranium contain the **sinuses,** air spaces lined by mucous membrane. The sinuses reduce the weight of the skull and give a resonant sound to the voice. Two sinuses, called the mastoid sinuses, drain into the middle ear. **Mastoiditis,** a condition that can lead to deafness, is an inflammation of these sinuses.

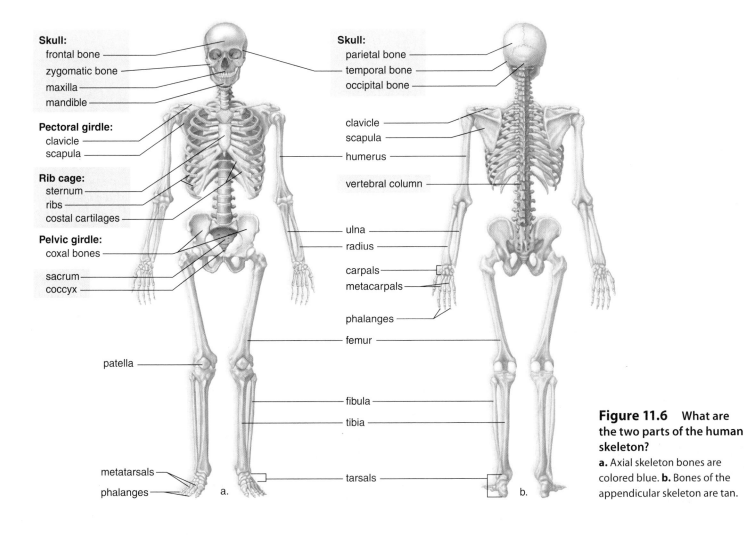

Figure 11.6 **What are the two parts of the human skeleton?**
a. Axial skeleton bones are colored blue. **b.** Bones of the appendicular skeleton are tan.

Science Focus

Identifying Skeletal Remains

Regardless of how, when, and where human bones are found unexpectedly, many questions must be answered. How old was this person at the time of death? Are these the bones of a male or female? What was the ethnicity? Are there any signs that this person was murdered?

Clues about the identity and history of the deceased person are available throughout the skeleton. Age is approximated by *dentition,* or the structure of the teeth in the upper jaw (maxilla) and lower jaw (mandible). For example, infants aged 0–4 months will have no teeth present; children aged approximately 6–10 years will have missing deciduous, or "baby teeth"; young adults acquire their last molars, or "wisdom teeth," around age 20. The age of older adults may be approximated by the number and location of missing or broken teeth. Studying areas of bone ossification also gives clues to the age of the deceased at the time of death. In older adults, signs of joint breakdown provide additional information about age. Hyaline cartilage becomes worn, yellowed, and brittle with age, and the hyaline cartilages covering bone ends wear down over time. The amount of yellowed, brittle, or missing cartilage helps scientists guess the person's age.

If skeletal remains include the individual's pelvic bones, these provide the best method for determining an adult's gender. The pelvis is more shallow and wider in the female than in the male. The long bones, particularly the humerus and femurs, give information about gender as well. Long bones will be thicker and denser in males, and points of muscle attachment will be bigger and more prominent. The skull of a male will have a square chin and more prominent ridges above the eye sockets or orbits.

Determining ethnic origin of skeletal remains can be difficult, because so many people have a mixed racial heritage. Forensic anatomists rely on observed racial characteristics of the skull. In general, individuals of African or African American descent will have a greater distance between the eyes, eye sockets that are roughly rectangular, and a jaw that is large and prominent. Skulls of Native Americans will typically have round eye sockets, prominent cheek (zygomatic) bones, and a rounded palate. Caucasian skulls will usually have a U-shaped palate, and a suture line between the frontal bones will often be visible. Additionally, the external ear canals in Caucasians are long and straight, so that the auditory ossicles can be seen.

Once the identity of the individual has been determined, the skeletal remains can be returned to the family for proper burial.

The major bones of the cranium have the same names as the lobes of the brain: frontal, parietal, occipital, and temporal. On the top of the cranium (Fig. 11.7*a*), the **frontal bone** forms the forehead, the **parietal bones** extend to the sides, and the **occipital bone** curves to form the base of the skull. Here there is a large opening, the **foramen magnum** (Fig. 11.7*b*), through which the spinal cord passes and becomes the brain stem. Below the much larger parietal bones, each **temporal bone** has an opening (external auditory canal) that leads to the middle ear.

The **sphenoid bone,** shaped like a bat with outstretched wings, extends across the floor of the cranium from one side to the other. The sphenoid is the keystone of the cranial bones because all the other bones articulate with it. The sphenoid completes the sides of the skull and also contributes to forming the orbits (eye sockets). The **ethmoid bone,** which lies in front of the sphenoid, also helps form the orbits and the nasal septum. The orbits are completed by various facial bones. The eye sockets are called orbits because we can rotate our eyes.

The Facial Bones The most prominent of the facial bones are the mandible, the maxillae (sing., maxilla), the zygomatic bones, and the nasal bones.

The mandible, or lower jaw, is the only movable portion of the skull, and it also forms the chin (Figs. 11.7, 11.8). The maxillae form the upper jaw and a portion of the eye socket. Further, the hard palate and the floor of the nose are formed by the maxillae (anterior) joined to the palatine bones (posterior). Tooth sockets are located on the mandible and on the maxillae. The grinding action of the mandible and maxillae allow us to chew our food.

The lips and cheeks have a core of skeletal muscle. The **zygomatic bones** are the cheekbone prominences, and the **nasal bones** form the bridge of the nose. Other bones (e.g., ethmoid and vomer) are a part of the nasal septum, which divides the interior of the nose into two nasal cavities. The lacrimal bone (see Fig. 11.7*a*) contains the opening for the nasolacrimal canal, which drains tears from the eyes to the nose.

Certain cranial bones contribute to the face. The temporal bone and the wings of the sphenoid bone account for the flattened areas we call the temples. The frontal bone forms the forehead and has supraorbital ridges, where the eyebrows are located. Glasses sit where the frontal bone joins the nasal bones.

The exterior portion of ears are formed only by cartilage and not by bone. The nose is a mixture of bones, cartilages,

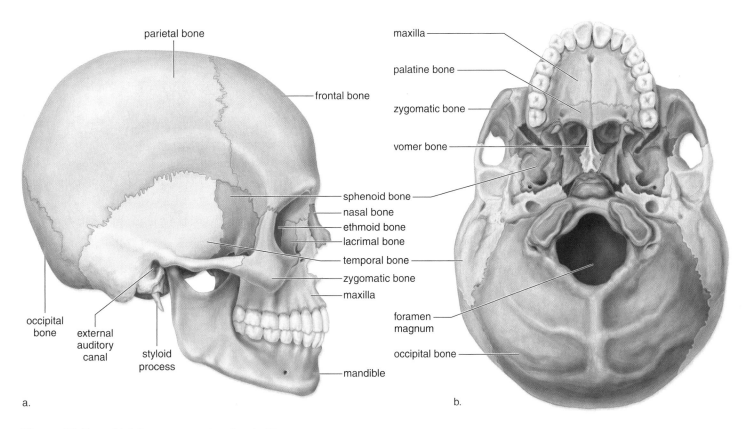

a. Lateral view. b. Inferior view.

Figure 11.7 Which bones comprise the skull?
a. Lateral view. b. Inferior view.

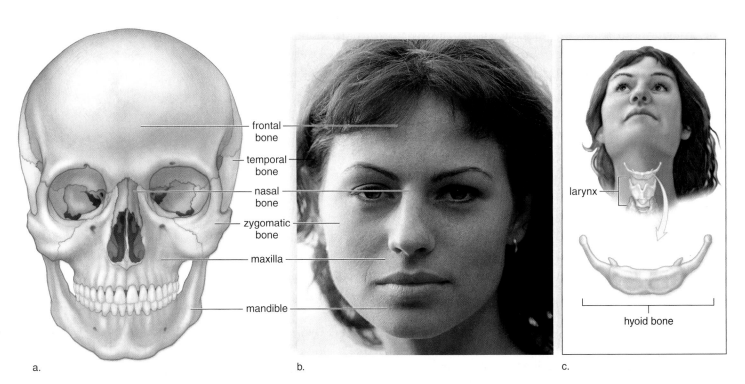

Figure 11.8 Which bones construct the face? Where is the hyoid bone located?
a. The frontal bone forms the forehead and eyebrow ridges; the zygomatic bones form the cheekbones; and the maxillae have numerous functions. They assist in the formation of the eye sockets and the nasal cavity. They form the upper jaw and contain sockets for the upper teeth. The mandible is the lower jaw with sockets for the lower teeth. The mandible has a projection we call the chin. b. The maxillae, frontal, and nasal bones help form the external nose. c. The hyoid bone is located as shown.

and connective tissues. The cartilages complete the tip of the nose, and fibrous connective tissue forms the flared sides of the nose.

The Hyoid Bone

Although the **hyoid bone** is not part of the skull, it will be mentioned here because it is a part of the axial skeleton. It is the only bone in the body that does not articulate with another bone (Fig. 11.8c). It is attached to the temporal bones by muscles and ligaments and to the larynx by a membrane. The larynx is the voice box at the top of the trachea in the neck region. The hyoid bone anchors the tongue and serves as the site for the attachment of muscles associated with swallowing. Due to its position, the hyoid bone is not usually easy to fracture in most situations. In cases of suspicious death, however, a fractured hyoid is a strong sign of *manual strangulation.*

The Vertebral Column

The **vertebral column** consists of 33 vertebrae (Fig. 11.9). Normally, the vertebral column has four curvatures that provide more resilience and strength for an upright posture than a straight column could provide. **Scoliosis** is an abnormal lateral (sideways) curvature of the spine. There are two other well-known abnormal curvatures. Kyphosis is an abnormal posterior curvature that often results in a "hunchback." An abnormal anterior curvature results in lordosis, or "swayback."

As the individual vertebrae are layered on top of one another, they form the vertebral column. The vertebral canal is in the center of the column, and the spinal cord passes through this canal (Fig. 11.10a). The intervertebral foramina (singular: foramen, a hole or opening) are found on either side of the column. Spinal nerves branch from the spinal cord and travel through the intervertebral foramina to locations throughout the body. Spinal nerves function to control skeletal muscle contraction, among other functions. If a vertebra is compressed, or slips out of position, the spinal cord and/or spinal nerves might be injured. The result can be paralysis or even death.

The spinous processes of the vertebrae can be felt as bony projections along the midline of the back. The transverse processes extend laterally. Both spinous and transverse processes serve as attachment sites for the muscles that move the vertebral column.

Types of Vertebrae   The various vertebrae are named according to their location in the vertebral column. The cervical vertebrae are located in the neck. The first cervical vertebra, called the **atlas,** holds up the head. It is so named because Atlas, of Greek mythology, held up the world. Movement of the atlas permits the "yes" motion of the head. It also allows the head to tilt from side to side. The second cervical vertebra is called the **axis** because it allows a degree of rotation, as when we shake the head "no." The thoracic vertebrae have long, thin, spinous processes

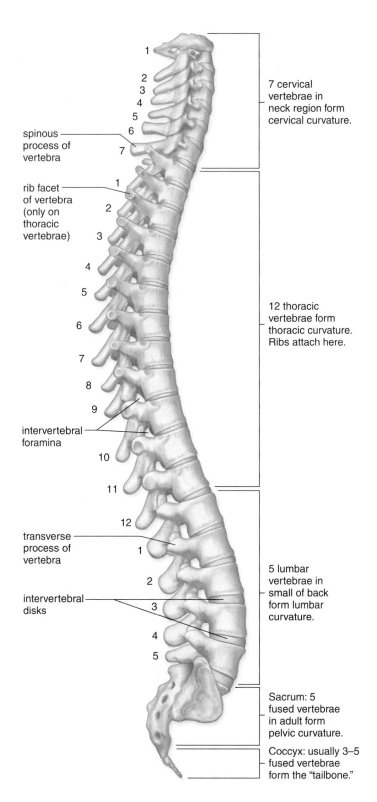

Figure 11.9   How many bones are in the vertebral column? The vertebral column is made up of 33 vertebrae separated by intervertebral disks. The intervertebral disks make the column flexible. The vertebrae are named for their location in the vertebral column. For example, the thoracic vertebrate are located in the thorax. Humans have a coccyx, which is also called a tailbone.

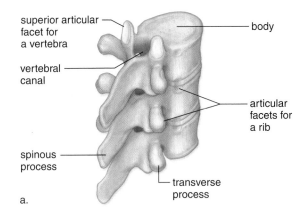

superior articular facet for a vertebra
vertebral canal
spinous process
body
articular facets for a rib
transverse process

a.

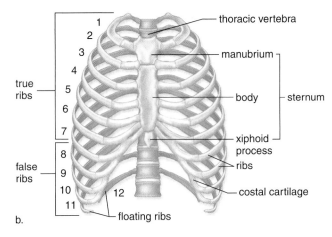

true ribs
false ribs
1 2 3 4 5 6 7 8 9 10 11 12
thoracic vertebra
manubrium
body — sternum
xiphoid process
ribs
costal cartilage
floating ribs

b.

Figure 11.10 How are the thoracic vertebrae, ribs, and sternum joined to form the rib cage?
a. The thoracic vertebrae articulate with one another and also with the ribs at articular facets. A thoracic vertebra has two facets for articulation with a rib; one is on the body, and the other is on the transverse process. **b.** The rib cage consists of the 12 thoracic vertebrae, the 12 pairs of ribs, the costal cartilages, and the sternum. The rib cage protects the lungs and the heart.

and articular facets for the attachment of the ribs (Fig. 11.10a). Lumbar vertebrae have a large body and thick processes. The five sacral vertebrae are fused together in the sacrum. The coccyx, or tailbone, is usually composed of four fused vertebrae.

Intervertebral Disks Between the vertebrae are **intervertebral disks** composed of fibrocartilage that acts as padding. They prevent the vertebrae from grinding against one another. The disks also absorb shock caused by movements such as running, jumping, and even walking. The presence of the disks allows the vertebrae to move as we bend forward, backward, and from side to side. Unfortunately, these disks become weakened with age and can herniate and rupture. Pain results if a disk presses against the spinal cord and/or spinal nerves. If that occurs, surgical removal of the disk may relieve the pain.

The Rib Cage

The rib cage, also called the thoracic cage, is composed of the thoracic vertebrae, the ribs and their associated cartilages, and the sternum (Fig. 11.10b). The rib cage is part of the axial skeleton.

The rib cage demonstrates how the skeleton is protective but also flexible. The rib cage protects the heart and lungs; yet it swings outward and upward upon inspiration and then downward and inward upon expiration.

The Ribs

A rib is a flattened bone that originates at the thoracic vertebrae and proceeds toward the anterior thoracic wall. There are 12 pairs of ribs. All 12 pairs connect directly to the thoracic vertebrae in the back. A rib articulates with the body and transverse process of its corresponding thoracic vertebra. Each rib curves outward and then forward and downward.

The upper seven pairs of ribs connect directly to the sternum by means of costal cartilages. These are called "true ribs." The "false ribs" are the next three pairs of ribs, which connect to the sternum by means of a common cartilage. The last two pairs are called "floating ribs" because they do not attach to the sternum.

The Sternum

The **sternum** lies in the midline of the body. Along with the ribs, it helps protect the heart and lungs. The sternum, or breastbone, is a flat bone that has the shape of a knife.

The sternum is composed of three bones. These bones are the manubrium (the handle), the body (the blade), and the xiphoid process (the point of blade). The manubrium articulates with the clavicles of the appendicular skeleton. Costal cartilages from the first pair of ribs also join to the manubrium. The manubrium joins with the body of the sternum at an angle. This is an important anatomical landmark because it occurs at the level of the second rib and therefore allows the ribs to be counted. Counting the ribs is sometimes done to determine where the apex of the heart is located—usually between the fifth and sixth ribs.

The xiphoid process is the third part of the sternum. The variably shaped xiphoid process serves as an attachment site for the diaphragm, which separates the thoracic cavity from the abdominal cavity.

Check Your Progress 11.3
1. What are the bones of the axial skeleton?
2. What are the bones of the cranium, and how are they situated?
3. What are the bones of the face, and how do they contribute to facial features?
4. a. Describe the various types of vertebrae and state their function. b. What is the structure and function of the rib cage?

11.4 Bones of the Appendicular Skeleton

The **appendicular skeleton** consists of the bones within the pectoral and pelvic girdles and their attached limbs. A pectoral (shoulder) girdle and upper limb are specialized for flexibility. The pelvic (hip) girdle and lower limbs are specialized for strength.

The Pectoral Girdle and Upper Limb

The body has left and right **pectoral girdles.** Each consists of a scapula (shoulder blade) and a clavicle (collarbone) (Fig. 11.11). The **clavicle** extends across the top of the thorax. It articulates with (joins with) the sternum and the acromion process of the **scapula,** a visible bone in the back. The muscles of the arm and chest attach to the coracoid process of the scapula. The **glenoid cavity** of the scapula articulates with, and is much smaller than, the head of the humerus. This allows the arm to move in almost any direction, but reduces stability. This is the joint most apt to dislocate. Ligaments, and also tendons, stabilize this joint. Tendons that extend to the humerus from four small muscles originating on the scapula form the **rotator cuff.** Vigorous circular movements of the arm can lead to rotator cuff injuries.

The components of a pectoral girdle freely follow the movements of the upper limb, which consists of the humerus within the arm and the radius and ulna within the forearm. The **humerus,** the single long bone in the arm, has a smoothly rounded head that fits into the glenoid cavity of the scapula as mentioned. The shaft of the humerus has a tuberosity (protuberance) where the deltoid, a shoulder muscle, attaches. You can determine, even after death, that the person did a lot of heavy lifting during their lifetime by the size of their deltoid tuberosity.

The far end of the humerus has two protuberances, called the capitulum and the trochlea, which articulate respectively with the **radius** and the **ulna** at the elbow. The bump at the back of the elbow is the olecranon process of the ulna.

When the upper limb is held so that the palm is turned forward, the radius and ulna are about parallel to each other. When the upper limb is turned so that the palm is turned backward, the radius crosses in front of the ulna, a feature that contributes to the easy twisting motion of the forearm.

The hand has many bones, and this increases its flexibility. The wrist has eight **carpal** bones, which look like small pebbles. From these, five **metacarpal** bones fan out to form a framework for the palm. The metacarpal bone that leads to the thumb is opposable to the other digits. An opposable thumb can touch each finger separately or cross the palm to grasp an object. (**Digits** is a term that refers to either fingers or toes.) The knuckles are the enlarged distal ends of the metacarpals. Beyond the metacarpals are the **phalanges,** the bones of the fingers and the thumb. The phalanges of the hand are long, slender, and lightweight.

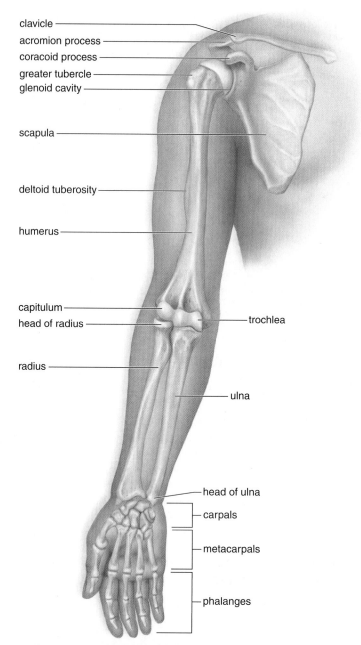

Figure 11.11 **Which two bones form the pectoral girdle? What are the bones of the upper limb?**

The pectoral girdle consists of the clavicle (collarbone) and the scapula (shoulder blade). The humerus is the single bone of the arm. The forearm is formed by the radius and ulna. A hand contains carpals, metacarpals, and phalanges.

Pectoral girdle
- scapula and clavicle

Upper limb
- arm contains humerus
- forearm contains radius and ulna
- hand contains carpals, metacarpals, and phalanges

The paramedics had splinted Emily's arm and loaded the little girl and her father into an ambulance. Paul was sure that the short drive to the pediatric hospital's emergency room had been the longest 8 minutes of his life. They waited in a curtained cubicle, with Emily crying and whimpering while Paul stroked her free hand. He felt like crying himself.

Mary rushed in, still wearing her scrubs and name badge. "I just got your call. What on earth happened?" Concerned, she asked, "Paul, are you okay? You're as white as a sheet."

"Yeah, I guess I'm okay," he answered thinly. "We were just playing in the park, and I swung her down from a piggyback ride. I think maybe I broke her arm or something."

An older gentleman entered just then. Dressed in a white lab coat, with thinning hair and bifocals, he looked like a doctor from TV. "You're Mary Harahan from the intensive care unit, right?" She nodded. "I'm Dave Mertz."

"Is this Emily?" he asked Emily. "What did you do to your arm, Emily?" He felt for a pulse and squeezed each tiny finger. "Can you feel my squeezes, Emily?" The little girl nodded. Removing the splint, he gently attempted to straighten her arm, but stopped when she cried out.

Paul repeated, "I swung her down from a piggyback ride. I think maybe I broke her arm or something."

"Well, she's going to be fine, sir. We'll take an X-ray and do an MRI to make sure, but I think she's just dislocated her elbow," the doctor reassured. "Mary, you go with Emily. Sir, I'd like you to come with me to the waiting area." He directed Paul to sit on the carpeted floor and put his head between his knees. "Stay there until you're not so light-headed."

Pelvic girdle
- coxal bones

Lower limb
- thigh contains femur
- leg contains tibia and fibula
- foot contains tarsals, metatarsals, and phalanges

The Pelvic Girdle and Lower Limb

Figure 11.12 shows how the lower limb is attached to the pelvic girdle. The **pelvic girdle** (hip girdle) consists of two heavy, large coxal bones (hip bones). The **pelvis** is a basin composed of the pelvic girdle, sacrum, and coccyx. The pelvis bears the

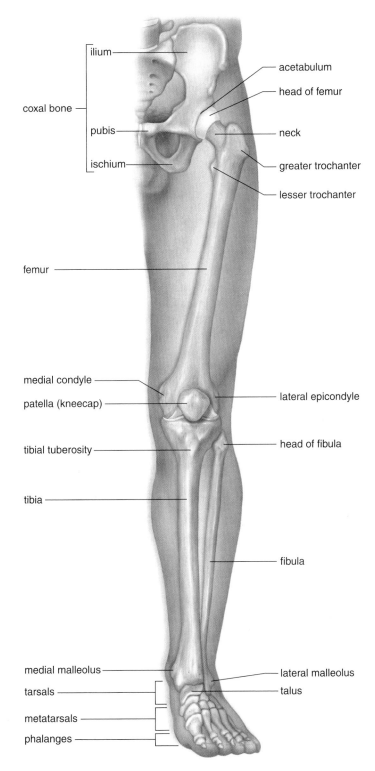

Figure 11.12 **Which three bones join to form a coxal bone? What bones must be added to complete the pelvis? What are the bones of the lower limb?**
The ilium, ischium, and pelvis join at the acetabulum (hip socket) to form a coxal bone. The pelvis is completed by addition of the sacrum and coccyx. The femur (thigh bone) and tibia and fibula (shin bones) form the leg. Tarsals, metatarsals, and phalanges construct the foot.

weight of the body, protects the organs within the pelvic cavity, and serves as the place of attachment for the legs.

Each **coxal bone** has three parts: the ilium, the ischium, and the pubis, which are fused in the adult (Fig. 11.12). The hip socket, called the acetabulum, occurs where these three bones meet. The ilium is the largest part of the coxal bones, and our hips form where it flares out. We sit on the ischium, which has a posterior spine called the ischial spine for muscle attachment. The pubis, from which the term pubic hair is derived, is the anterior part of a coxal bone. The two pubic bones are joined by a fibrocartilaginous joint, the **pubic symphysis.**

The male and female pelves differ from each other. In the female, the iliac bones are more flared; the pelvic cavity is more shallow, but the outlet is wider. These adaptations facilitate the birthing process during vaginal delivery.

The **femur** (thighbone) is the longest and strongest bone in the body. The head of the femur articulates with the coxal bones at the acetabulum, and the short neck better positions the legs for walking. The femur has two large processes, the greater and lesser trochanters, which are places of attachment for thigh muscles, buttock muscles, and hip flexors. At its distal end, the femur has medial and lateral condyles that articulate with the **tibia** of the leg. This is the region of the knee and the **patella,** or kneecap. The patella is held in place by the quadriceps tendon, which continues as a ligament that attaches to the tibial tuberosity. At the distal end, the medial malleolus of the tibia causes the inner bulge of the ankle. The **fibula** is the more slender bone in the leg. The fibula has a head that articulates with the tibia and a distal lateral malleolus that forms the outer bulge of the ankle.

Each foot has an ankle, an instep, and five toes. The many bones of the foot give it considerable flexibility, especially on rough surfaces. The ankle contains seven **tarsal** bones, one of which (the talus) can move freely where it joins the tibia and fibula. Strange to say, the calcaneus, or heel bone, is also considered part of the ankle. The talus and calcaneus support the weight of the body.

The instep has five elongated **metatarsal** bones. The distal ends of the metatarsals form the ball of the foot. If the ligaments that bind the metatarsals together become weakened, flat feet are apt to result. The bones of the toes are called **phalanges,** just like those of the fingers. In the foot, the phalanges are stout and extremely sturdy.

Check Your Progress 11.4

1. a. What are the bones of the pectoral girdle and (b) of the upper limb?
2. a. What are the bones of the pelvic girdle and (b) of the lower limb?

CASE STUDY DIAGNOSIS AND TREATMENT

Paul sat in the waiting area for almost three hours, drinking several cans of soda and downing two candy bars. He had read almost every tattered magazine before Dr. Mertz finally found him. "Mr. Harahan? You can come back this way. Emily would like to see you."

Paul jumped to his feel. "Is she all right?"

"Aside from being a little sleepy, she's just fine," the doctor replied. "She had just dislocated her elbow, but it took a while before the X-rays and MRI came back to prove that. I gave her a mild sedative to relax her, and then popped the bone right back into place. Took just a few minutes."

"Oh man," Paul groaned. The idea of popping bones made him light headed and queasy again. "I really didn't break her arm, then?"

"No, no, not at all," the doctor reassured. "And the medics checked her pulse at the scene. I rechecked it right away, just to make sure she didn't damage blood vessels. Squeezing her fingers showed she didn't damage nerves either. So no permanent damage was done.

"The elbow is a joint made of three bones," he explained. "The biggest two are the humerus in your upper arm, and the ulna in your forearm. These two bones are held in place with ligaments, and a fluid capsule around the bones helps them to slide freely. Are you with me, so far?" he asked. Paul nodded.

Dr. Mertz continued, "The third bone is the radius. It's largest at the wrist, but small at the elbow. It rides alongside the ulna, and it's held there by ligaments, too. In young kids, the ligaments aren't strong. When you lifted Emily by her arm, the weight of her body pulled the radius free from the ligaments that support it. This happens in little kids all the time. I'll bet I treat a dislocated elbow every other day—and it's for one of two causes. Either the kid has taken a tumble, or somebody lifted the kid by one arm."

"So she's got a sprained elbow?" Paul asked.

"Technically, no," Dr. Mertz replied. "A sprain is a tear to the ligament. A tear in muscle is called a strain. Emily's elbow ligaments aren't torn, and neither are the muscles. She'll be in a sling for a day or so, if you can just keep her in it," he smiled. "Her elbow joint is going to be weak for some time. As she gets older, the ligaments will get stronger, so this will probably never happen again."

"In the future," he warned, "if you have to lift Emily, lift under her armpits, never by her wrists and never by one arm. Now why don't you come back and see her?"

11.5 Articulations

Bones are joined at the joints, classified as fibrous, cartilaginous, or synovial. Many fibrous joints, such as the **sutures** between the cranial bones, are immovable. Cartilaginous joints may be connected by hyaline cartilage, as in the costal cartilages that join the ribs to the sternum. Other cartilaginous joints are formed by fibrocartilage, as in the intervertebral disks. Cartilaginous joints tend to be slightly movable. **Synovial joints** are freely movable.

Figure 11.13 illustrates the anatomy of a freely movable synovial joint. Ligaments connect bone to bone and support or strengthen the joint. A fibrous capsule formed by ligaments surrounds the bones at the joint. This capsule is lined with synovial membrane that secretes a small amount of synovial fluid to lubricate the joint. Fluid-filled sacs called bursae (sing., bursa) ease friction between bare areas of bone and overlapping muscles, or between skin and tendons. The full joint contains menisci (sing., meniscus), C-shaped pieces of hyaline cartilage between the bones. These give added stability and act as shock absorbers.

The ball-and-socket joints at the hips and shoulders allow movement in all planes, even rotational movement.

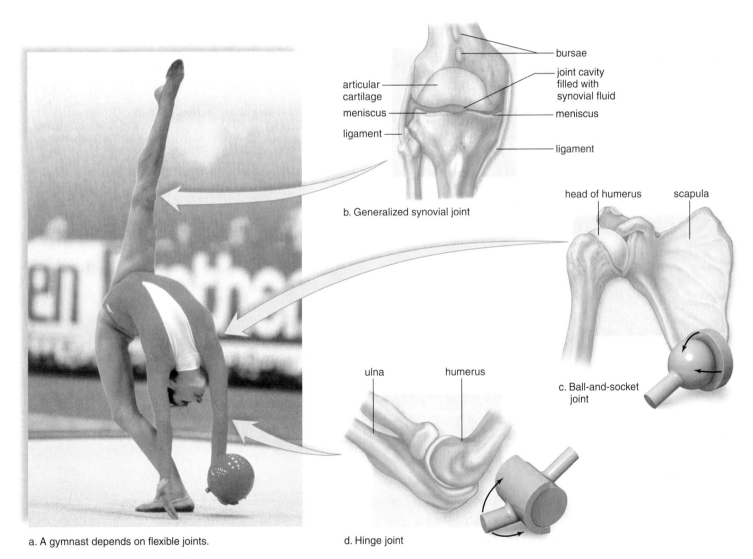

a. A gymnast depends on flexible joints.

b. Generalized synovial joint

c. Ball-and-socket joint

d. Hinge joint

Figure 11.13 What is the basic structure of a synovial joint? Are there different types of synovial joints?
a. Synovial joints are moveable and therefore flexible. **b.** The bones of joints are joined by ligaments that form a capsule. The capsule is lined with synovial membrane, which gives off synovial fluid as a lubricant. Bursae are fluid-filled sacs that reduce friction. Menisci (sing. meniscus), formed of cartilage, stabilize a joint, and articular cartilage caps the bones. **c.** Ball-and-socket joints form the hip and shoulder. **d.** Hinge joints construct the knee and elbow.

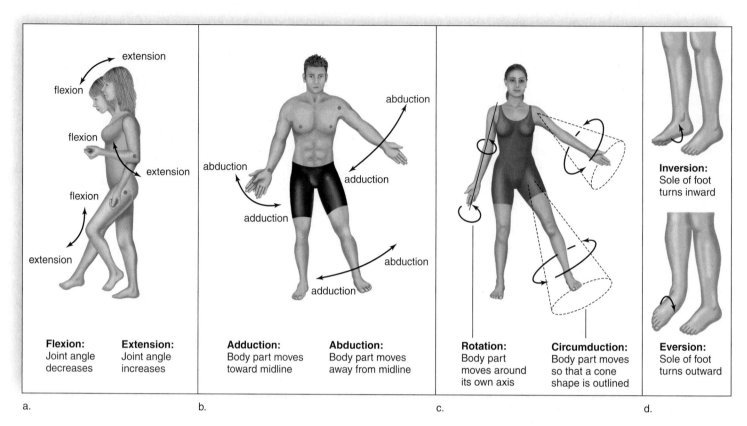

Figure 11.14   **What types of movement can occur at a synovial joint?**
a. Flexion and extension. **b.** Adduction and abduction. **c.** Rotation and circumduction. **d.** Inversion and eversion. Red dots indicate pivot points.

The elbow and knee joints are synovial joints called hinge joints. Like a hinged door, they largely permit movement in one direction only.

Movements Permitted by Synovial Joints

Intact skeletal muscles are attached to bones by tendons that span joints. When a muscle contracts, one bone moves in relation to another bone. The more common types of movements are described in Figure 11.14.

> **Check Your Progress 11.5**
>
> 1. a. What are the different types of articulations, and
>   (b) what types of movements are permitted by synovial joints?

Have You Ever Wondered...

Can I get arthritis from cracking my knuckles?

It's a common warning given to knuckle-crackers, and sometimes it gets them to stop this annoying habit. However, no reliable studies have proven this old wives' tale. A knuckle-crack occurs when the ligaments forming the capsule of a synovial joint are suddenly stretched. An air bubble forms in the synovial fluid, and when the bubble pops, the crack noise is heard. "Cracking" can occur at any synovial joint—so it's possible to crack toes and ankles, too.

Habitually "cracking" joints may not cause arthritis, but it does weaken the ligaments over time. The bones forming a joint are easily dislocated when the ligaments are weakened. When a joint seems stiff, don't "crack" it. Instead, gently stretch the muscles to relieve stiffness.

Historical Focus

Pioneer in Joint Replacement Surgery: Dr. John Charnley (1911–1982)

"It has rightly been said that every surgical operation is a biological experiment."

Osteoarthritis is a condition that will afflict nearly everyone, to a greater or lesser degree, as each person ages. The bones that unite to form joints, or articulations, are covered with a slippery cartilage. This articular cartilage wears down over time, as friction in the joint wears it away. The erosion process is accelerated in individuals who have subjected their joints to years of constant heavy activity: athletes, dancers, individuals who are obese, and so on. By age 80, people typically have osteoarthritis in one or more joints. By contrast, rheumatoid arthritis is an autoimmune disorder (see Chap. 7, page 153) that causes inflammation within the joint. Unlike osteoarthritis, which typically affects older people, rheumatoid arthritis can afflict a person of any age—even young children.

The result for both forms of arthritis is the loss of the joint's natural smoothness. This is what causes the pain and stiffness experienced by millions of arthritis sufferers in the United States. Arthritis will first be treated with medications for joint inflammation and pain and physical therapy to maintain and strengthen the joint. However, if these treatments fail, a total joint replacement is often performed. Successful replacement surgeries are now routine, thanks to the hard work and dedication of Dr. John Charnley.

Charnley was a British orthopedic surgeon who researched joint replacement. Early experimental surgeries by Charnley and others had been very disappointing. Fused joints were immobile, and fusion didn't always relieve the patient's pain. Postsurgical infection was common. The bones attached to the artificial joint eroded, and the supporting muscles wasted away because the joint wasn't useful. Charnley wanted to design a successful prosthetic hip, with the goal of replacing both parts of the diseased hip joint: the acetabulum, or "socket," as well as the ball-shaped head of the femur. First, though, he had to answer several key questions. Why did bone refuse to grow around the artificial joint? Why didn't the patient have a full range of motion after the surgery? Why were infection and swelling so common?

Charnley soon determined that surgical experimentation alone wasn't enough. He studied bone repair, persuading a colleague to operate on his own tibia, or shin bone, to see how repair occurred. He studied the mechan-

ics of the hip joint, testing different types of synthetic materials. He achieved his first success using a hip socket lined with Teflon, but soon discovered that the surrounding tissues became inflamed. Suspecting that Teflon caused inflammation, he injected himself with a solution containing microscopic bits of Teflon to prove his hypothesis. After multiple attempts, his perfected hip consisted of a socket of durable polyethylene, still used as the joint's plastic component. The head of his prosthetic femur was a small, highly polished metal ball. Stainless steel, cobalt, titanium, as well as chrome alloys, form the metal component today. Various techniques for cementing the polyethylene socket onto the pelvic bone had failed when bone pulled away from the cemented surface and refused to grow. Charnley's surgery used dental cement, slathered onto the bone surfaces. When the plastic components were attached, cement was squeezed into every pore of the bone. Now the bone could regenerate, growing around the plastic. Finally, Charnley devised a specialized surgical tent and instrument tray to minimize infection.

Charnley's ideas were innovative and unorthodox. As with many pioneers, Charnley's work was ridiculed by his colleagues. He was relegated to a former tuberculosis hospital, which he converted into a center for innovation in orthopedic surgery. His colleague, Dr. Frank Gunston, developed a prosthetic knee joint similar to the Charnley hip. In knee replacement surgery, the damaged ends of bones are removed and replaced with artificial components that resemble the original bone ends. Hip and knee replacement remain the most common joint replacement surgeries, but ankles, feet, shoulders, elbows, and fingers can also be replaced. Though many improvements on the procedure continue, the Charnley hip replacement remains

(continued)

Historical **Focus** *(continued)*

the "gold standard"—the technique after which all others are modeled. Diseased bone in the pelvis is surgically removed, and the plastic socket is inserted. The damaged head of the femur is detached, and a metal ball attached to a stem is inserted into the femur's hollow shaft. Bone cement is applied to all surfaces, then squeezed into the porous bone.

When a joint replacement is complete, the patient's hard work is vital to ensure the success of the procedure. Exercise and activity are critical to the recovery process. After surgery, the patient is encouraged to use the new joint as soon as possible. The day after a hip or knee replacement, patients will often begin standing and walking to start strengthening the replaced joint. The extent of im-

provement and range of motion of the joint depends on its stiffness before the surgery, as well as the amount of patient effort during therapy following surgery. Most patients can expect to experience some degree of pain in the replaced joint. This is due to the weakness of the muscles surrounding the joint, as well as healing of the damaged tissues. Complete recovery from pain varies from patient to patient, but most can expect recovery to take several months. Many older patients can expect their replacement to last about 10 years. Younger patients may need a second replacement later in life, if they happen to wear out their first prosthesis. Still, individuals who have joint replacement surgery can expect an improved quality of life and a bright future, with greater independence and healthier pain-free activity.

Summarizing the Concepts

11.1 Overview of Skeletal System

Functions of the skeletal system:

- Supports and protects the body.
- Produces blood cells.
- Stores mineral salts, particularly calcium phosphate. It also stores fat.
- Along with the muscles, permits flexible body movement.

The bones of the skeleton are composed of bone tissues and cartilage. Ligaments composed of fibrous connective tissue connect bones at joints.

In a long bone,

- hyaline cartilage covers the ends of a long bone.
- periosteum (fibrous connective tissue) covers the rest of the bone.
- spongy bone (containing red bone marrow) is in the epiphyses.
- yellow bone marrow is in the medullary cavity of the diaphysis.
- compact bone makes up the wall of the diaphysis.

11.2 Bone Growth, Remodeling, and Repair

Cells involved in growth, remodeling, and repair of bone are:

- osteoblasts, bone-forming cells;
- osteocytes, mature bone cells derived from osteoblasts; and
- osteoclasts, which break down and absorb bone.

Bone Development and Growth

- Intramembranous ossification: Bones develop between sheets of fibrous connective tissue. Examples are flat bones such as bones of the skull.
- Endochondral ossification: Cartilaginous models of the bones are replaced by calcified bone matrix.
- Bone growth is affected by vitamin D, growth hormone, and sex hormones.

Bone Remodeling and Its Role in Homeostasis

- Bone remodeling is the renewal of bone. Osteoclasts break down bone and osteoblasts re-form bone. Some bone is recycled each year.
- Bone recycling allows the body to regulate blood calcium.
- Two hormones, parathyroid hormone and calcitonin, direct bone remodeling and control blood calcium.

Bone Repair

Repair of a fracture requires four steps:

- hematoma formation,
- fibrocartilaginous callus,
- bony callus, and
- remodeling.

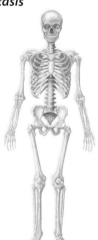

11.3 Bones of the Axial Skeleton

The axial skeleton consists of the skull, the hyoid bone, the vertebral column, and the rib cage.

- The skull is formed by the cranium, which protects the brain and the facial bones.
- The hyoid bone anchors the tongue and is the site of attachment of muscles involved with swallowing.
- The vertebral column is composed of vertebrae separated by shock-absorbing disks, which make the column flexible. It supports the head and truck, protects the spinal cord, and is a site for muscle attachment.
- The rib cage is composed of the thoracic vertebrae, ribs, costal cartilages, and sternum. It protects the heart and lungs.

11.4 Bones of the Appendicular Skeleton

The appendicular skeleton consists of the bones of the pectoral girdles, upper limbs, pelvic girdle, and lower limbs.

- The pectoral girdles and upper limbs are adapted for flexibility.
- The pelvic girdle and the lower limbs are adapted for supporting weight; the femur is the longest and strongest bone in the body.

11.5 Articulations

Bones are joined at joints, of which there are three types:

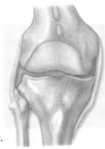

- Fibrous joints (such as the sutures of the cranium) are immovable.
- Cartilaginous joints (such as those between the ribs and sternum and the pubic symphysis) are slightly movable.
- Synovial joints (which have a synovial membrane) are freely movable.

Understanding Key Terms

appendicular skeleton 242	medullary cavity 230
articular cartilage 230	ossification 232
axial skeleton 237	osteoblast 232
bone remodeling 234	osteoclast 232
cartilage 230	osteocyte 230, 232
chondrocyte 230	pectoral girdle 242
compact bone 230	pelvic girdle 243
endochondral ossification 233	pelvis 243
fibrous connective tissue 232	periosteum 230
fontanel 237	red bone marrow 230
foramen magnum 238	rotator cuff 242
growth plate 233	scoliosis 240
hormone 234	sinus 237
intervertebral disk 241	skull 237
intramembranous ossification 233	spongy bone 230
joint 230	suture 245
ligament 230	synovial joint 245
mastoiditis 237	vertebral column 240

Match the key terms to these definitions.

a. _____ Fibrous connective tissue that connects bone to bone.

b. _____ Type of bone that contains osteons consisting of concentric layers of matrix and osteocytes in lacunae.

c. _____ Cavity or hollow space in cranial and facial bones.

d. _____ Membranous regions located between certain cranial bones in the skull of an infant or fetus.

e. _____ Mature bone cell located within the lacunae of bone.

Testing Your Knowledge of the Concepts

1. What are the five functions of the skeletal system? (page 230)
2. Where would you expect to find spongy bone in a long bone? What are its functions? (page 230)
3. Differentiate between compact and spongy bone. (page 230)
4. Describe the structure of hyaline cartilage. (page 230)
5. Why are osteoclasts needed, even though they break down bone? (page 232)
6. Explain how bone is formed by intramembranous ossification. (page 233)
7. Describe the process of endochondral ossification. (page 233)
8. Describe the actions of the hormones involved in bone growth and bone remodeling. (pages 234–36)
9. Provide the rationale for the four steps of bone repair. (page 236–37)
10. Explain the terms axial skeleton and appendicular skeleton. (pages 237–41)
11. List the bones that make up the cranium and the face. (pages 237–40)
12. What are the types of vertebrae, and how many of each are there? (pages 240–41)
13. What bones make up the rib cage? What are the functions of the rib cage? (page 241)
14. Name the bones of the pectoral girdle and upper limb. (page 242)
15. Name the bones of the pelvic girdle and lower limb. (pages 243–44)
16. What are the two types of cartilaginous joints, and what type of movement do cartilaginous joints have? (pages 245–46)
17. Describe the different types of movements made by synovial joints. (pages 245–46)

In questions 18–23, match each bone to a location in the key.

Key:

a. forehead	e. shoulder blade
b. chin	f. hip
c. cheekbone	g. arm
d. collarbone	

18. Zygomatic bone
19. Clavicle
20. Frontal bone
21. Humerus
22. Coxal bone
23. Scapula

In questions 24–29, match each bone to a feature in the key.

Key:

a. glenoid cavity	e. greater and lesser trochanters
b. trochlea	f. xiphoid process
c. acetabulum	
d. spinous process	

24. Femur
25. Scapula
26. Ulna
27. Coxal bone
28. Sternum
29. Vertebra
30. Spongy bone
    a. contains osteons.
    b. contains red bone marrow, where blood cells are formed.
    c. lends no strength to bones.
    d. takes up most of a leg bone.
    e. All of these are correct.
31. Which of these associations is mismatched?
    a. slightly movable joint—vertebrae
    b. hinge joint—hip
    c. synovial joint—elbow
    d. immovable joint—sutures in cranium

32. The bone cell responsible for breaking down bone tissue is the _____, while the bone cell that produces new bone tissue is the _____.
    a. osteoclast, osteoblast
    b. osteocyte, osteoclast
    c. osteoblast, osteocyte
    d. osteocyte, osteoblast
    e. osteoclast, osteocyte

33. All blood cells—red, white, and platelets—are produced by which of the following?
    a. yellow bone marrow
    b. red bone marrow
    c. periosteum
    d. medullary cavity

34. This bone is the only movable bone of the skull.
    a. sphenoid
    b. frontal
    c. mandible
    d. maxilla
    e. temporal

35. Which of the following is not a function of the skeletal system?
    a. production of blood cells
    b. storage of minerals
    c. involved in movement
    d. storage of fat
    e. production of body heat

36. When you bend your arm, the bump seen as your elbow is part of the
    a. humerus.
    b. radius.
    c. ulna.
    d. carpal.

37. The bump seen on the outside of the ankle is part of the
    a. femur.
    b. tibia.
    c. fibula.
    d. tarsal bones.

38. Which of the following is not a bone of the appendicular skeleton?
    a. the scapula
    b. a rib
    c. a metatarsal bone
    d. the patella

39. Which of the following statements is incorrect?
    a. A growth plate occurs between the primary ossification center and a secondary center.
    b. Each temporal bone has an opening that leads to the middle ear.
    c. Intervertebral disks are composed of fibrocartilage.
    d. Bone cells are rigid and hard because they are dead.
    e. The sternum is composed of three bones that fuse during fetal development.

40. The clavicle articulates with the
    a. scapula and humerus.
    b. humerus and manubrium.
    c. sternum and scapula.
    d. manubrium, scapula, and humerus.
    e. femur and tibia.

41. The vertebrae that articulate with the ribs are the
    a. lumbar vertebrae.
    b. sacral vertebrae.
    c. thoracic vertebrae.
    d. cervical vertebrae.
    e. coccyx.

In questions 42–46, indicate whether the statement is true (T) or false (F).

42. The pectoral girdle is specialized for weight-bearing, while the pelvic girdle is specialized for flexibility of movement. _____

43. The term phalanges refers to the bones in both the fingers and the toes. _____

44. Bones synthesize vitamin D for the body. _____

45. Bones store minerals and fat. _____

46. Most bones develop through endochondral ossification. _____

In questions 47–51, match each term with the correct characteristic given in the key.

Key:
    a. closing the angle at a joint
    b. movement in all planes
    c. made of cartilage
    d. movement toward the body
    e. fluid-filled sac

47. Ball-and-socket joints

48. Flexion

49. Bursa

50. Meniscus

51. Adduction

52. Label this diagram of a skeleton.

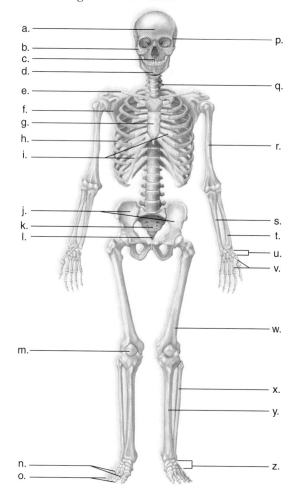

Thinking Critically About the Concepts

In the chapter case study, Paul Harahan accidentally dislocates his daughter Emily's elbow. This type of injury was historically termed "nursemaid's elbow," indicating the injury resulted from careless handling of a child by a caregiver. The ligaments that connect children's bones are flexible, yet weak. Dislocations of bones forming a synovial joint are common in children and are usually not serious. Likewise, bone fractures are common to children. Most strains, sprains, dislocations, and fractures result from a child's active lifestyle. Injuries like these are treated with pain management, maneuvers to return a bone/joint to normal, immobilization in a cast or splint, and surgery if necessary. The vast majority of these injuries heal quickly, with no lingering effects. The child is encouraged to be as active as possible during the healing process, because inactivity causes increased bone depletion. Balancing safety with the need for children to exercise and play is critical for normal growth.

1. What nutritional and personal habits would contribute to rapid bone repair?

2. Individuals who spend the majority of their day indoors (a person in a nursing home, for example) are more susceptible to fracture. Why?

3. Why do the broken bones of older people take much longer to mend than the broken bones of children and young adults?

4. Pediatricians are becoming increasingly concerned by the increased incidence of rickets (see Fig. 8.15) in children eating a typical fast-food diet. What nutrients are missing from the diet?

5. Two athletes show up in the emergency room following a college football game. One has a fracture of the fibula (see Fig. 11.2). The second has a severe ankle sprain. Which player is most likely to return to play first?

6. a. What stimulates the release of parathyroid hormone?
   b. What disease(s) is/are likely to result from hyperparathyroidism?

7. Why is the spinal cord almost always damaged when someone breaks one or more cervical vertebrae?

12

Muscular System

For about the hundredth time that month, Kayla Scott thanked heaven for work that paid well, but didn't require her full attention. Sure, she still worked very hard as a hotel housekeeper, but she had her routine down. Beds changed first, dust everything, bathroom next, mop the floor, vacuum everywhere—then on to the next room as fast as she could get there. Her manager was great. He really liked Kayla's work, so when she finished that day's set of rooms, she could leave. That meant she could pick up her two kids from school and spend her evenings with them. After their bedtime, she put in a couple of hours studying for her college classes. It was still hard after the divorce, and money was very tight, but she and the kids were getting by.

Her ear buds were stuck in her ears while she worked. Most of the time she listened to the podcasts of that week's class lectures, studying while she worked. Lately, though, the pain in her right arm distracted her from both the lecture and the work. Kayla popped two more ibuprofen, even though the package label said 6 tablets in 24 hours was the maximum dosage. She'd exceeded that dosage by midmorning and it hadn't really even touched the pain. She knew she might not finish that day and hoped Mr. Christiansen would understand. She pushed the cart back to the storage area that doubled as his office. She barely made it through the door before running to the trash can and getting sick. Mortified, she looked up at her manager and began apologizing.

"Kayla, would you stop it! I am not mad at you," Mr. Christiansen reassured. "Do you think you have the flu? Do you need to go home?"

"I think maybe I just took too much ibuprofen," she replied. "Maybe I could just sip some soda and eat lunch. That might settle my stomach. My arm's been killing me, but it's okay after I ice it.

"I'm really, really not a slacker," she continued. "I'll finish my shift. I promise."

"Look at your elbow!" Mr. Christiansen exclaimed. "You're going to the doctor right now. That thing is twice the normal size."

"I really can't, Mr. Christiansen," she replied. "My car's in the shop again. I don't think the bus runs out that way."

He pulled out his wallet. "This," he asserted, gently placing a $20 bill in her palm, "is to pay for a cab. And this is for the doctor's co-pay." He laid another bill in her palm. "You guys get medical insurance for a reason. Now get on out of here."

12.1 Overview of Muscular System

All muscles, regardless of their particular type, can contract—that is, shorten. When muscles contract, some part of the body or the entire body moves.

Types of Muscles

Humans have three types of muscle tissue: smooth, cardiac, and skeletal (Fig. 12.1). The cells of these tissues are called **muscle fibers.**

Smooth muscle fibers are shaped like narrow cylinders, with pointed ends. Each has a single nucleus (uninucleated). The cells are usually arranged in parallel lines, forming sheets. Striations (bands of light and dark) are seen in cardiac and skeletal muscle but not in smooth muscle. Smooth muscle is located in the walls of hollow internal organs and blood vessels, and it causes these walls to contract. Contraction of smooth muscle is involuntary, occurring without conscious control. Although smooth muscle is slower to contract than skeletal muscle, it can sustain prolonged contractions and does not fatigue easily.

Cardiac muscle forms the heart wall. Its fibers are generally uninucleated, striated, and tubular. Branching allows the fibers to interlock at **intercalated disks.** The plasma membranes at intercalated disks contain gap junctions (see page 76) that permit contractions to spread quickly throughout the heart wall. Cardiac fibers relax completely between contractions, which prevents fatigue. Contraction of cardiac muscle is rhythmic. It occurs without outside nervous stimulation and without conscious control. Thus, cardiac muscle contraction is involuntary.

Skeletal muscle fibers are tubular, multinucleated, and striated. They make up the skeletal muscles attached to the skeleton. Fibers run the length of the muscle and can be quite long. Skeletal muscle is voluntary because we can decide to move a particular part of the body, such as the arms and legs.

Functions of Skeletal Muscles

Skeletal muscles have numerous functions.

Skeletal muscles support the body. Skeletal muscle contraction opposes the force of gravity and allows us to remain upright.

Skeletal muscles cause movements of bones and other body structures. Muscle contraction accounts not only for the movement of arms and legs but also for movements of the eyes, facial expressions, and breathing.

Skeletal muscles help maintain a constant body temperature. Skeletal muscle contraction causes ATP to break down, releasing heat distributed about the body.

Skeletal muscle contraction assists movement in cardiovascular and lymphatic vessels. The pressure of skeletal muscle contraction keeps blood moving in cardiovascular veins and lymph moving in lymphatic vessels.

Skeletal muscles help protect internal organs and stabilize joints. Muscles pad the bones, and the muscular wall in the abdominal region protects the internal organs. Muscle tendons help hold bones together at joints.

Figure 12.1
What types of muscles can be found in the body?
Human muscles are of three types: smooth, cardiac, and skeletal. These muscles have different characteristics, as noted.

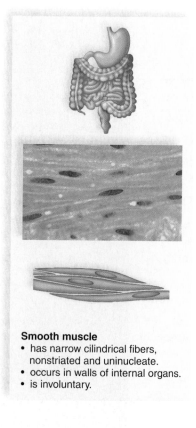

Smooth muscle
- has narrow cilindrical fibers, nonstriated and uninucleate.
- occurs in walls of internal organs.
- is involuntary.

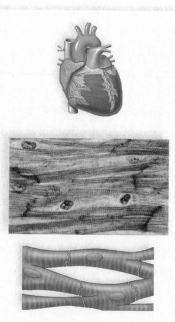

Cardiac muscle
- has striated, branched, generally uninucleated fibers.
- occurs in walls of heart.
- is involuntary.

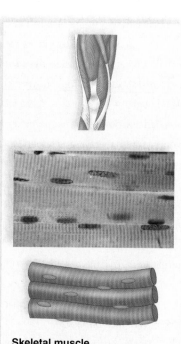

Skeletal muscle
- has striated, tubular, multinucleated fibers.
- is usually attached to skeleton.
- is voluntary.

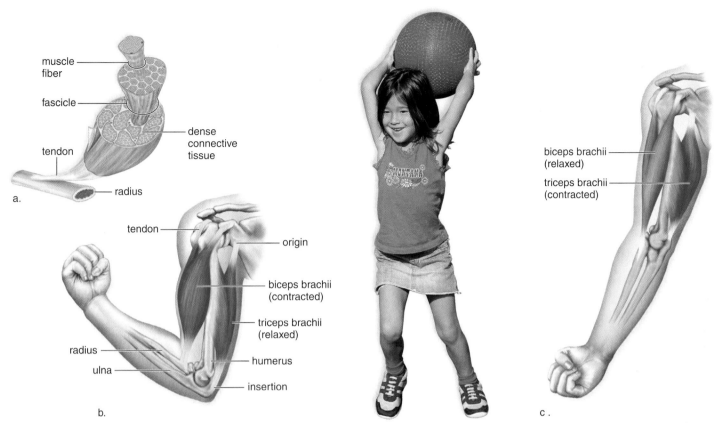

Figure 12.2 **How are muscles connected to bones? How do muscles cooperate during movement?**
a. Connective tissue separates bundles of muscle fibers that make up a skeletal muscle. A layer of connective tissue covering the muscle contributes to the tendon, which attaches muscle to bone. **b.** When the biceps brachii contracts, the forearm flexes. **c.** When the triceps brachii contracts, the forearm extends. Therefore, these two muscles are antagonistic. The origin of a skeletal muscle is on a bone that remains stationary, and the insertion of a muscle is on a bone that moves when the muscle contracts.

Skeletal Muscles of the Body

In the animal kingdom, humans are *vertebrates.* Vertebrate animals possess an internal vertebral column and a skeleton with jointed appendages. Our skeletal muscles are attached to the skeleton, and their contraction causes the movement of bones at a joint.

Basic Structure of Skeletal Muscles

Skeletal muscles are well organized. A whole muscle contains bundles of skeletal muscle fibers called fascicles (Fig. 12.2a). These are the strands of muscle that we see when we cut red meat and poultry. Within a fascicle, each fiber is surrounded by connective tissue, and the fascicle is also surrounded by connective tissue. Muscles are covered with fascia, a type of connective tissue that extends beyond the muscle and becomes its **tendon.** Tendons quite often extend past a joint before anchoring a muscle to a bone. Small, fluid filled sacs called bursae (sing. bursa) can often be found between tendons and bones. A **bursa** acts as a "cushion," allowing ease of movement.

Skeletal Muscles Work in Pairs

In general, each muscle is concerned with the movement of only one bone. For the sake of discussion, we can imagine the

movement of that bone and no others. The **origin** of a muscle is on a stationary bone, and the **insertion** of a muscle is on a bone that moves. When a muscle contracts, it pulls on the tendons at its insertion, and the bone moves. For example, when the biceps brachii contracts, it raises the forearm. To an anatomist, the upper limb consists of the arm above the elbow and the forearm below the elbow. Likewise, the lower limb is made up of the thigh above the knee and the leg below the knee.

Skeletal muscles usually function in groups. Consequently, to make a particular movement, your nervous system does not stimulate a single muscle. Rather, it stimulates an appropriate group of muscles. Even so, for any particular movement, one muscle does most of the work and is called the prime mover. While a prime mover is working, other muscles called synergists function as well. Synergists assist the prime mover and make its action more effective.

When muscles contract, they shorten. Therefore, muscles can only pull; they cannot push. This means that muscles work in opposite pairs. The muscle that acts opposite to a prime mover is called an antagonist. For example, the biceps brachii and the triceps brachii are antagonists. The biceps flexes the forearm (Fig. 12.2b), and the triceps extends the forearm (Fig. 12.2c). If both of these muscles contracted at once, the forearm would remain rigid. Smooth body movements depend on an antagonist relaxing when a prime mover is acting.

Names and Actions of Skeletal Muscles

Figure 12.3 shows that contraction of the facial muscles produce the facial expressions that tell us about the emotions and mood of a person. Figure 12.4a and b illustrates the location of some of the major skeletal muscles and gives their actions. When learning the names of muscles, considering what the name means will help you remember it. The names of the various skeletal muscles are often combinations of the following terms used to characterize muscles. Not all the muscles mentioned are featured in Figure 12.4, but most are.

1. **Size.** For example, the gluteus maximus is the largest muscle that makes up the buttocks. The gluteus minimus is the smallest of the gluteal muscles. Other

Figure 12.3 How can muscle contraction be used to convey emotion?
Our facial expressions and hand movements are due to muscle contractions.

terms used to indicate size are vastus (huge), longus (long), and brevis (short).
2. **Shape.** The deltoid is shaped like a triangle. (The Greek letter delta has this appearance: Δ.) The trapezius is

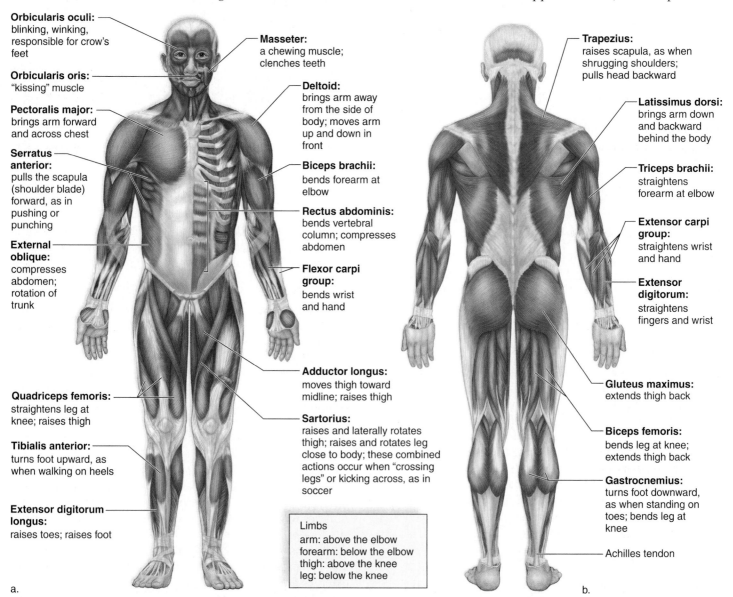

Figure 12.4 What are some of the major skeletal muscles?

Rigor Mortis

When a person dies, the physiological events that accompany death occur in an orderly progression. Respiration ceases, the heart ultimately stops beating, and tissue cells begin to die. The first tissues to die are those with the highest oxygen requirement. Brain and nervous tissue have an extremely high requirement for oxygen. Deprived of oxygen, these cells typically die after only 6 minutes because of a lack of ATP. However, tissues that can produce ATP by fermentation (which does not require oxygen) can "live" for an hour or more before ATP is completely depleted. Muscle is capable of generating ATP by fermentation. Therefore, muscle cells can survive for a time after clinical death occurs. Muscle death is signaled by a process termed **rigor mortis,** the "stiffness of death."

Stiffness occurs because, for biochemical reasons we will discuss later, muscles cannot relax unless they have a supply of ATP. Without ATP, the muscles remain fixed in their last state of contraction. If, for example, a murder victim dies while sitting at a desk, the body in rigor mortis will be frozen in the sitting position. Rigor mortis resolves approximately 24–36 hours after death. Muscles lose their stiffness because lysosomes rupture. The lysosomes release enzymes that break the bonds between the muscle proteins, actin and myosin.

Body temperature and the presence or absence of rigor mortis allows the time of death to be estimated. For example, the body of someone dead for 3 hours or less will still be warm (close to body temperature, 98.6°F, 37°C), and rigor mortis will be absent. After approximately 3 hours, the body will be significantly cooler than normal, and rigor mortis will begin to develop. The corpse of an individual dead at least 8 hours will be in full rigor mortis, and the temperature of the body will be the same as the surroundings. Forensic pathologists know that a person has been dead for more than 24 hours if the body temperature is the same as the environment, and there is no longer a trace of rigor mortis.

CASE STUDY IN THE SPECIALIST'S OFFICE

The longer she waited in this Dr. Kerner's office, the more anxious and upset Kayla felt. Her mother had picked up the kids the previous day, while Kayla waited in the primary-care doctor's office. Now it seemed to Kayla that Mom would have to pick them up again. Both parents had been happy to help after the divorce, but it really hurt her pride even to ask. The joint specialist had managed to fit Kayla into the schedule, yet the wait took forever—wait for the X-rays, wait for an exam room to free up, wait for the doctor. Kayla checked her watch, careful not to move the right elbow she was icing. An hour and fifteen minutes, just waiting for the doctor. Thirty minutes later, Kayla was ready to give up and go home when the exam room door opened. "Kayla Scott? Dr. Kerner. Let me take a look at that elbow."

"Wow," Kayla thought. "Some bedside manner. This one's all business."

She winced when Dr. Kerner attempted to straighten her elbow, then gasped when the doctor touched it. "How long has it bothered you?" Dr. Kerner asked.

"A couple of months, I guess," Kayla replied. "I noticed it one day at work, and it just got worse after that."

Dr. Kerner checked the records she carried. "You're a hotel housekeeper? So what does the job require? Vacuuming, mopping, wiping counters, that sort of thing, right?" Kayla nodded, smiling to herself. No one had ever asked for her job description before.

"So that kind of work will have to stop right away," the doctor remarked absently. "You have tendonitis and a serious bursitis. Your elbow is badly inflamed."

Kayla had had enough. "Will you slow down, please? What is tendonitis and bursitis? Why do I have to quit work?" she snapped at the doctor, who finally looked up.

"I'm sorry," Dr. Kerner replied. "It's just been one of those days for me. Let me start over. Doing repetitive arm motions like mopping, dusting, and vacuuming first irritated your muscle tendon. That's tendonitis. I'll bet you just kept right on working after it started to hurt, right?" Kayla nodded.

"That's when the bursa became inflamed," the doctor continued. "A bursa is like a plastic bag filled with a small amount of fluid. It sits between the muscle tendon and the bone of the elbow, and it gives the muscle and tendons a nice, smooth surface to glide over. Now that both the tendon and the bursa are affected, your elbow joint is seriously inflamed. Dr. Parker sent you over to me and we took X-rays, just to make sure you didn't have a small fracture in your elbow. The good news is there's no broken bone, so I can give you drugs like prescription ibuprofen or naproxen for pain and to reduce swelling. You can only take so much of that stuff, though, before it hurts your stomach." Kayla nodded. She'd had personal experience with that!

"We'll also start you out with a cortisone injection, so we can reduce the swelling in your elbow right away," Dr. Kerner reassured her, then warned, "But basically, the only way to treat this is just to rest the joint. If you don't, the bursa will develop scar tissue, and you'll lose movement permanently."

shaped like a trapezoid. Other terms used to indicate shape are latissimus (wide) and teres (round).

3. **Location.** For example, the external oblique muscles are located outside the internal obliques. The frontalis muscle overlies the frontal bone. Other terms used to indicate location are pectoralis (chest), gluteus (buttock), brachii (arm), and sub (beneath).

4. **Direction of muscle fibers.** For example, the rectus abdominis is a longitudinal muscle of the abdomen (rectus means straight). The orbicularis oculi is a circular muscle around the eye. Other terms used to indicate direction are transverse (across) and oblique (diagonal).

5. **Attachment.** For example, the sternocleidomastoid is attached to the sternum, clavicle, and mastoid process. The mastoid process is located on the temporal bone of the skull. The brachioradialis is attached to the brachium (arm) and the radius (forearm).

6. **Number of attachments**. For example, the biceps brachii has two attachments, or origins, and is located on the arm. The quadriceps femoris has four origins and is located on the femur.

7. **Action.** For example, the extensor digitorum extends the fingers or digits. The adductor longus is a large muscle that adducts the thigh. Adduction is the movement of a body part toward the midline. Other terms used to indicate action are flexor (to flex or bend), masseter (to chew), and levator (to lift).

With every muscle you learn, try to understand its name.

> **Check Your Progress 12.1**
>
> 1. What are the three types of muscles in the human body?
> 2. What are the functions of skeletal muscles?
> 3. How do skeletal muscles work together to cause bones to move?
> 4. What are the major skeletal muscles of the trunk? The arm? The thigh?

12.2 Skeletal Muscle Fiber Contraction

We have already examined the structure of skeletal muscle, as seen with the light microscope. As you know, skeletal muscle tissue has alternating light and dark bands, giving it a striated appearance. Now we will see that these bands are due to the arrangement of myofilaments in a muscle fiber.

Muscle Fibers and How They Slide

A muscle fiber is a cell containing the usual cellular components, but special names have been assigned to some of these components (Table 12.1 and Fig. 12.5). The plasma membrane is called the **sarcolemma;** the cytoplasm is the sarcoplasm; and the endoplasmic reticulum is the **sarcoplasmic reticulum.** A muscle fiber also has some unique anatomical characteristics. One feature is its T (for transverse) system. The sarcolemma forms **T (transverse) tubules** that penetrate, or dip down, into the cells. The transverse tubules come into contact—but do not fuse—with expanded portions of the sarcoplasmic reticulum. The expanded portions of the sarcoplasmic reticulum are calcium storage sites. Calcium ions (Ca^{2+}), as we shall see, are essential for muscle contraction.

The sarcolemma encases hundreds and sometimes even thousands of **myofibrils,** each about 1 μm in diameter. Myofibrils are the contractile portions of the muscle fibers. Any other organelles, such as mitochondria, are located in the sarcoplasm between the myofibrils. The sarcoplasm also contains glycogen, which provides stored energy for muscle contraction. In addition, sarcoplasm includes the red pigment myoglobin, which binds oxygen until it is needed for muscle contraction.

Myofibrils and Sarcomeres

As you see in Figure 12.5, a muscle cell or fiber, (a) is roughly cylindrical in shape. Grouped inside this larger cylinder are smaller cylinders called myofibrils (b). Myofibrils run the entire length of the muscle fiber. Myofibrils are composed of even smaller cylinders called myofilaments (c). Thus, the muscle cell is a set of small cylinders (myofilaments) assembled into larger cylinders (myofibrils) clustered within the largest cylinder (the muscle fiber). The light microscope shows that skeletal muscle fibers have light and dark bands called striations. At the higher magnification provided by an electron microscope, one can see that the striation of skeletal muscle fibers are formed by the placement of myofilaments within myofibrils. There are two types of myofilaments. Thick myofilaments are made up of a protein called **myosin,** and thin myofilaments are composed of a second protein

Table 12.1	**Microscopic Anatomy of a Muscle Fiber**
Name	**Function**
Sarcolemma	Plasma membrane of a muscle fiber that forms T tubules
Sarcoplasm	Cytoplasm of a muscle fiber that contains the organelles, including myofibrils
Glycogen	A polysaccharide that stores energy for muscle contraction
Myoglobin	A red pigment that stores oxygen for muscle contraction
T tubule	Extension of the sarcolemma that extends into the muscle fiber and conveys impulses that cause Ca^{2+} to be released from the sarcoplasmic reticulum
Sarcoplasmic reticulum	The smooth ER of a muscle fiber that stores Ca^{2+}
Myofibril	A bundle of myofilaments that contracts
Myofilament	Actin filaments or myosin filaments, whose structure and functions account for muscle striations and contractions

A muscle contains bundles of muscle fibers, and a muscle fiber has many myofibrils.

Figure 12.5 **What does a skeletal muscle fiber look like? How does it contract?**
A muscle fiber contains many myofibrils divided into sarcomeres, which are contractile. When the myofibrils of a muscle fiber contract, the sarcomeres shorten: The actin (thin) filaments slide past the myosin (thick) filaments toward the center. The Z lines have moved and the H zone has gotten smaller, to the point of disappearing.

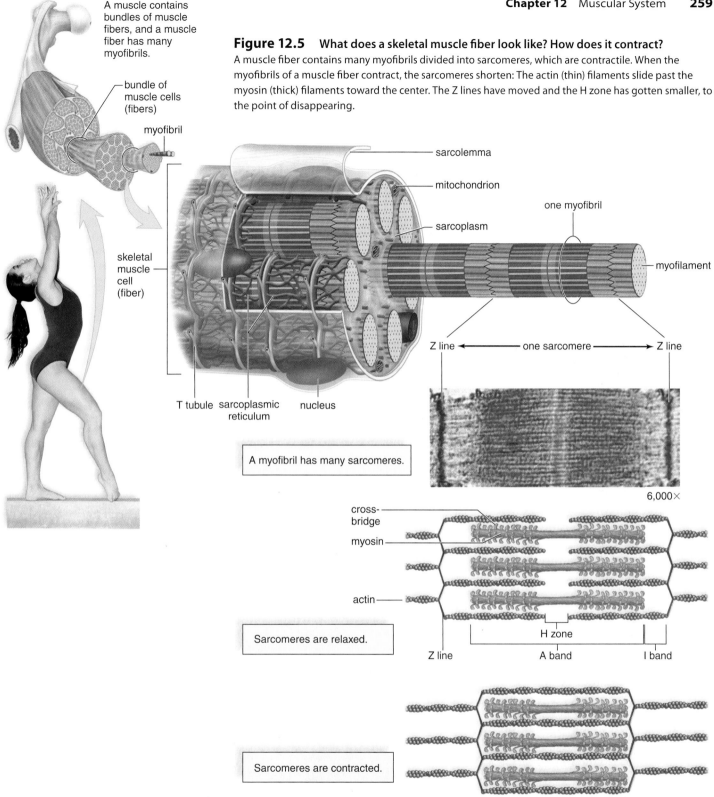

bundle of muscle cells (fibers)

myofibril

skeletal muscle cell (fiber)

sarcolemma

mitochondrion

sarcoplasm

one myofibril

myofilament

Z line ◄—— one sarcomere ——► Z line

T tubule sarcoplasmic reticulum nucleus

A myofibril has many sarcomeres.

6,000×

cross-bridge
myosin
actin

Sarcomeres are relaxed.

Z line

H zone

A band

I band

Sarcomeres are contracted.

termed **actin.** Myofibrils are further divided vertically into **sarcomeres** *(d)*. A sarcomere extends between two dark vertical lines called the Z lines. The I bands on either side of the Z line are light colored because each contains only the thin actin myofilaments. The dark central A band within the sarcomere is composed of overlapping actin and myosin myofilaments. Centered within the A band is a vertical H zone. This zone has only myosin myofilaments.

Myofilaments

The thick and thin filaments differ in the following ways:

Thick Filaments A thick filament is composed of several hundred molecules of the protein myosin. Each myosin molecule is shaped like a golf club, with the straight portion of the molecule ending in a globular head, or *cross-bridge*. The cross-bridges occur on each side of a sarcomere but not in the middle (Fig. 12.5*e*).

Thin Filaments Primarily, a thin filament consists of two intertwining strands of the protein actin. Two other proteins, called tropomyosin and troponin, also play a role, as we will discuss later in this section.

Sliding Filaments We will also see that when muscles are stimulated, electrical signals travel across the sarcolemma, and then down a T tubule. In turn, this signals calcium to be released from the sarcoplasmic reticulum. Now the muscle fiber contracts as the sarcomeres, within the myofibrils, shorten. As you compare the relaxed sarcomere (Fig. 12.5e) with the contracted sarcomere (Fig. 12.5f), note that the filaments themselves remain the same length. When a sarcomere shortens, the actin (thin) filaments approach one another as they slide past the myosin (thick) filaments. This causes the I band to shorten, the Z line to move

inward, and the H zone to almost or completely disappear (see Fig. 12.5f). The sarcomere changes from a rectangular shape to a square as it shortens. The movement of actin filaments in relation to myosin filaments is called the **sliding filament model** of muscle contraction. ATP supplies the energy for muscle contraction. Although the actin filaments slide past the myosin filaments, it is the myosin filaments that do the work. Myosin filaments break down ATP, and their cross-bridges pull the actin filament toward the center of the sarcomere.

As an analogy, think of yourself and a group of friends as myosin. Collectively, your hands are the cross-bridges, and you are pulling on a rope (actin) to get an object tied to the end of the rope (the Z line). As you pull the rope, you grab, pull, release, and then grab farther along on the rope.

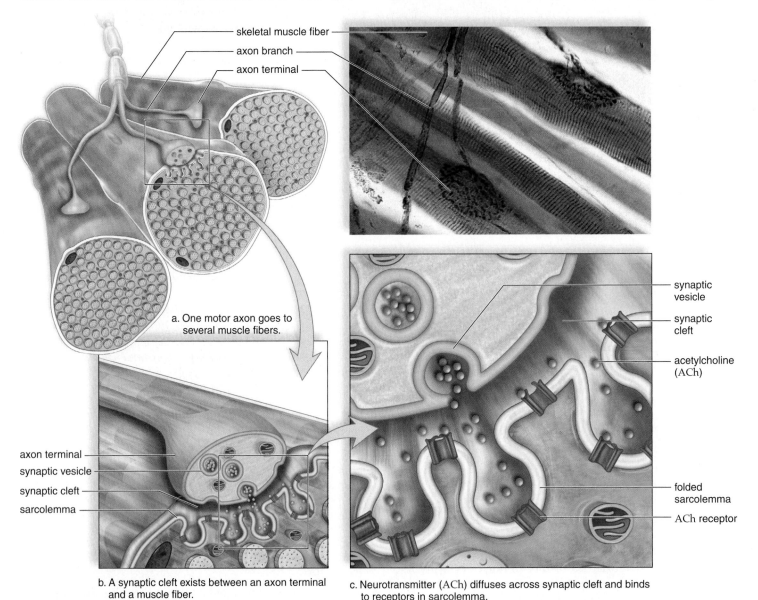

a. One motor axon goes to several muscle fibers.

b. A synaptic cleft exists between an axon terminal and a muscle fiber.

c. Neurotransmitter (ACh) diffuses across synaptic cleft and binds to receptors in sarcolemma.

Figure 12.6 How do a motor neuron and a skeletal muscle fiber join at a neuromuscular junction? What events occur at the neuromuscular junction?

a. The branch of a motor nerve fiber terminates in an axon terminal that meets. **b.** A synaptic cleft separates the axon terminal from the sarcolemma of the muscle fiber. **c.** Nerve impulses traveling down a motor fiber cause synaptic vesicles to discharge acetylcholine, which diffuses across the synaptic cleft and binds to ACh receptors. Impulses travel down the T system of a muscle fiber and the muscle fiber contracts.

Control of Muscle Fiber Contraction

Muscle fibers are stimulated to contract by motor neurons whose axons are grouped together to form nerves. The axon of one motor neuron can stimulate from a few to several muscle fibers of a muscle because each axon has several branches (Fig. 12.6*a*). Each branch of an axon ends in an axon terminal that lies in close proximity to the sarcolemma of a muscle fiber. A small gap, called a synaptic cleft, separates the axon terminal from the sarcolemma (Fig. 12.6*b*). This entire region is called a **neuromuscular junction.**

Axon terminals contain synaptic vesicles filled with the neurotransmitter acetylcholine (ACh). Nerve signals travel down the axons of motor neurons and arrive at an axon terminal. The signals trigger the synaptic vesicles to release ACh into the synaptic cleft (Fig. 12.6*c*). When ACh is released, it quickly diffuses across the cleft and binds to receptors in the sarcolemma. Now, the sarcolemma generates electrical signals that spread across the sarcolemma and down the T tubules. Recall that the T tubules lie adjacent to the sarcoplasmic reticulum, but the two structures are not connected. Nonetheless, signaling from the T tubules causes the release of Ca^{2+} from the sarcoplasmic reticulum, which leads to sarcomere contraction, as explained in Figure 12.7.

As described on page 82, sarcomere contraction can be prevented by the action of a potent neurotoxin. Botox® is the trade name for botulinum toxin A, produced by a bacterium. Botox® prevents wrinkling of the brow and skin about the eyes, because it blocks the release of ACh into the synaptic cleft. As a result, the sarcolemma is never activated and muscle contraction never occurs. Another point of interest concerns rigor mortis. We can well imagine that after death the synaptic vesicles lose their integrity and ACh pours into the synaptic cleft. The flood of ACh into the synaptic cleft would result in constant muscle signaling, and muscles would remain contracted. This might be the reason why rigor mortis is seen in muscles some time after a person is brain dead.

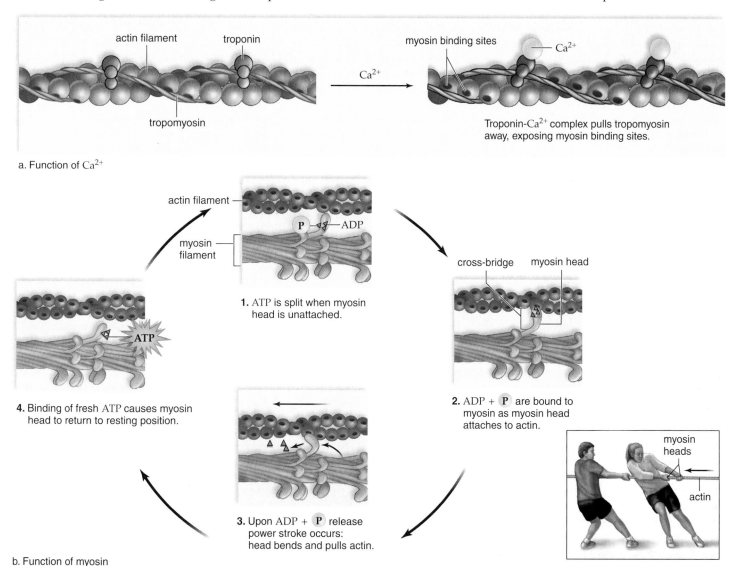

a. Function of Ca^{2+}

Troponin-Ca^{2+} complex pulls tropomyosin away, exposing myosin binding sites.

1. ATP is split when myosin head is unattached.

2. ADP + P are bound to myosin as myosin head attaches to actin.

3. Upon ADP + P release power stroke occurs: head bends and pulls actin.

4. Binding of fresh ATP causes myosin head to return to resting position.

b. Function of myosin

Figure 12.7 **What are the roles played by calcium ions and ATP during muscular contraction?**
a. Calcium (Ca^{2+}) binds to troponin, exposing myosin binding sites. **b.** Follow 1–4 to see how myosin uses ATP and does the work of pulling action toward the center of the sarcomere, much as a group of people pulling a rope *(right).*

Two other proteins are associated with an actin filament. Threads of **tropomyosin** wind about an actin filament, covering binding sites for myosin located on each actin molecule. **Troponin** occurs at intervals along the threads. When Ca^{2+} ions are released from the sarcoplasmic reticulum, they combine with troponin. This causes the tropomyosin threads to shift their position, exposing myosin binding sites. In other words, myosin can now bind to actin (see Fig.12.7a).

To fully understand muscle contraction, study Figure 12.7b. (1) The heads of a myosin filament have ATP binding sites. At this site, ATP is hydrolyzed, or split, to form ADP and Ⓟ. (2) The ADP and Ⓟ remain on the myosin heads, while the heads attach to an actin binding site. Joining myosin to actin forms temporary bonds called cross-bridges. (3) Now, ADP and Ⓟ are released and the cross-bridges bend sharply. This is the power stroke that pulls the actin filament toward the center of the sarcomere. (4) When ATP molecules again bind to the myosin heads, the cross-bridges are broken. Myosin heads detach from the actin filament. This is the step that does not happen during rigor mortis. Relaxing the muscle is impossible, because ATP is needed to break the bond between an actin binding site and the myosin cross-bridge.

In living muscle the cycle begins again, and myosin reattaches farther along the actin filament. The cycle recurs until calcium ions are actively returned to the calcium storages sites. This step also requires ATP.

Check Your Progress 12.2

1. What are the microscopic levels of structure in a skeletal muscle?
2. How does contraction of a skeletal muscle come about?
3. What is the role of ATP in muscle contraction?

12.3 Whole Muscle Contraction

Whole muscle contraction is dependent on muscle fiber contraction.

Muscles Have Motor Units

As mentioned, each axon within a nerve stimulates a number of muscle fibers. A nerve fiber with all of the muscle fibers it innervates, is called a **motor unit.** A motor unit obeys a principle called the **all-or-none law.** Why? Because all the muscle fibers in a motor unit are stimulated at once. They all either contract or do not contract. A variable of interest is the number of muscle fibers within a motor unit. For example, in the ocular muscles that move the eyes, the innervation ratio is one motor axon per 23 muscle fibers. By contrast, in the gastrocnemius muscle of the leg, the ratio is about one motor axon per 1,000 muscle fibers. Thus, moving the eyes requires finer control than moving the legs.

When a motor unit is stimulated by infrequent electrical impulses, a single contraction occurs. This response is called a **muscle twitch** and lasts only a fraction of a second. A muscle twitch is customarily divided into three stages. We can use our knowl-

edge of muscle fiber contraction to understand these events. The latent period is the time between stimulation and initiation of contraction (Fig. 12.8a). During this time, we can imagine the events that begin muscle contraction are occurring. The neurotransmitter ACh diffuses across the synaptic cleft, causing an electrical signal to spread across the sarcolemma and down the T tubules. The contraction period follows, as calcium leaves the sarcoplasmic reticulum and myosin-action cross-bridges form. As you know, the muscle shortens as it contracts. On the graph, the force increases as the muscle contracts. Finally, the relaxation period completes the muscle twitch. Myosin-action cross-bridges are broken, and calcium returns to the sarcoplasmic reticulum. Force diminishes as the muscle returns to its former length.

If a motor unit is given a rapid series of stimuli, it can respond to the next stimulus without relaxing completely. Summation is increased muscle contraction until maximal sustained contraction, called **tetanus,** is achieved (Fig. 12.8b). Tetanus continues until the muscle fatigues due to depletion of energy reserves. Fatigue is apparent when a muscle relaxes, even though stimulation continues. The tetanus of muscle cells is not the same as the infection called tetanus. The infection called *tetanus* is caused by the bacterium *Clostridium tetani*. Death occurs because the muscles, including the respiratory muscles, become fully contracted and do not relax.

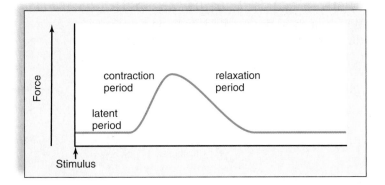

a.

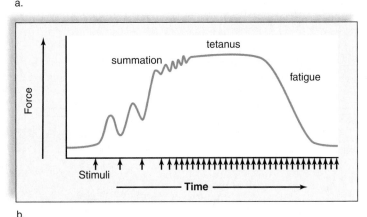

b.

Figure 12.8   What are the three phases of a single muscle twitch? Why do summation and tetanus create a greater force than that of a single twitch?
a. Stimulation of a muscle by a single electrical signal results in a simple muscle twitch: first, a latent period, followed by contraction and relaxation. **b.** Repeated stimulation results in summation and tetanus, which creates greater force because the motor unit cannot relax between stimuli.

A whole muscle typically contains many motor units. As the intensity of nervous stimulation increases, more motor units in a muscle are activated. This phenomenon is known as recruitment. Maximum contraction of a muscle would require that all motor units be undergoing tetanic contraction. This rarely happens because they could all fatigue at the same time. Instead, some motor units are contracting maximally, while others are resting, allowing sustained contractions to occur.

Muscle Tone

One desirable effect of exercise is to have good "**muscle tone**," dependent on muscle contraction. Some motor units are always contracted, but not enough to cause movement. As a result, the muscle is firm and solid—has good tone—as opposed to being soft and flabby.

Energy for Muscle Contraction

Muscles can use various fuel sources for energy, and they have various ways of producing ATP needed for muscle contraction.

Fuel Sources for Exercise

A muscle has four possible energy sources (Fig. 12.9). Two of these are stored in muscle, and two are acquired from blood. Both glycogen and fat (triglycerides) are stored in muscles. Which of these are used depends on exercise intensity and duration. Figure 12.9 shows the percentage of energy derived from these sources due to submaximal exercise (65–75% of effort) over time. As the length of the exercise period is increased, use of muscle glycogen decreases and use of muscle triglyceride (fat) increases.

Muscles also make use of blood glucose and plasma fatty acids as an energy source. Both of these are delivered to muscles by circulating blood. Some of us specifically exercise to lose weight, and we are particularly interested in the use of plasma fatty acids by exercising muscles. Adipose tissue is the source of plasma fatty acids for muscle contraction, but it also tends to make us look fat. Figure 12.9 shows that the amount of fat burned increases when more time is spent in exercise. If we are on a diet that restricts the amount of fat eaten, exercise

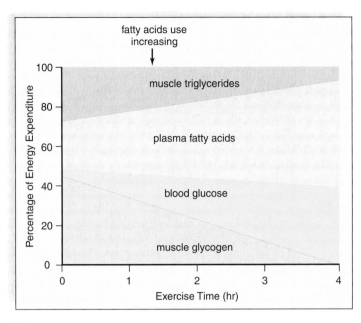

Figure 12.9 How does a muscle obtain the energy needed for contraction?
Percentage of energy derived from the four major fuel sources during submaximal exercise (65–75% of effort). The amount from plasma fatty acids increases during the time span shown.

will decrease body fat. Submaximal exercise burns fat better than maximal exercise, for reasons we will now explore.

Sources of ATP for Muscle Contraction

Muscle cells store limited amounts of ATP. Once stored ATP is used up, the cells have three ways to produce more ATP (Fig. 12.10). The three ways include (1) formation of ATP by the creatine phosphate (CP) pathway; (2) formation of ATP by fermentation (see page 59); and (3) formation of ATP by cellular respiration, which involves the use of oxygen by mitochondria. Aerobic exercising depends on cellular respiration to supply ATP. Neither formation of ATP by the CP pathway or by fermentation involves the need for oxygen. Both are anaerobic processes.

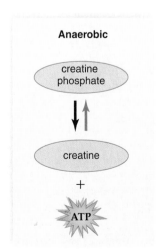
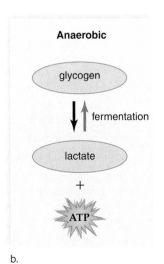
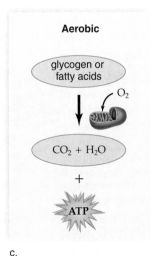

a.

b.

c.

Figure 12.10 How do muscle cells produce the ATP energy needed for contraction?
a. When contraction begins, muscle cells break down creatine phosphate to produce ATP. When resting, muscle cells rebuild their supply of creating phosphate (red arrow). **b.** Muscle cells also use fermentation to produce ATP quickly. When resting, muscle cells metabolize lactate, reforming as much glucose and then glycogen as possible (red arrow). **c.** For the long term, muscle cells switch to cellular respiration to produce ATP aerobically.

The CP Pathway The simplest and most rapid way for muscle to produce ATP is to use the CP pathway because it only consists of one reaction (Fig. 12.10a):

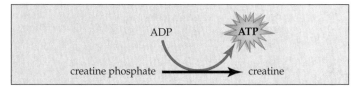

This reaction occurs in the midst of sliding filaments, and therefore, this method of supplying ATP is the speediest energy source available to muscles. Creatine phosphate is formed only when a muscle cell is resting, and only a limited amount is stored. The CP pathway is used at the beginning of submaximal exercise and during short-term, high-intensity exercise that lasts less than 5 seconds. The energy to complete a single play in a football game comes principally from the CP system. Intense activities lasting longer than 5 seconds will also make use of fermentation.

Fermentation Fermentation, as you know, produces two ATP molecules from the breakdown of glucose to lactate anaerobically. This pathway is the one most likely to begin with glycogen. Hormones provide the signal to muscle cells to break down glycogen, making glucose available as an energy source.

Fermentation, like the CP pathway, is fast-acting, but it results in the buildup of lactate (Fig. 12.10b). Formation of lactate is noticeable because it produces short-term muscle aches and fatigue upon exercising. We have all had the experience of needing to continue breathing following strenuous exercise. This continued intake of oxygen, called **oxygen debt**, is required, in part, to complete the metabolism of lactate and restore cells to their original energy state. The lactate is transported to the liver, where 20% of it is completely broken down to carbon dioxide and water. The ATP gained by this respiration is then used to reconvert 80% of the lactate to glucose and then glycogen. In persons who train, the number of mitochondria in individual muscles increases. There is a greater reliance on these additional mitochondria to produce ATP. Muscles rely less on fermentation as a result.

Cellular Respiration Cellular respiration is the slowest of all three mechanisms used to produce ATP. However, it is also the most efficient, typically producing several dozen molecules of ATP from each food molecule. As you will recall, cellular respiration occurs in the mitochondria. Thus, the process is aerobic, and oxygen is supplied by the respiratory system. In addition, a protein called **myoglobin** found within muscle cells delivers oxygen directly to the mitochondria. Cellular respiration can make use of glucose from the breakdown of stored muscle glycogen, glucose taken up from blood, and/or fatty acids from fat digestion (Fig. 12.10c). Also, cellular respiration is more likely to supply ATP when exercise is submaximal in intensity. According to Figure 12.9, if you are interested in exercising to lose weight, you should do so at a lower intensity and for a generous amount of time. This means that your aerobic exercise class at the local gym burns triglyceride, or fat, by making ATP using cellular respiration.

Fast-Twitch and Slow-Twitch Muscle Fibers

We have seen that all muscle fibers metabolize aerobically and anaerobically. However, some muscle fibers use one method more than the other to provide myofibrils with ATP. Fast-twitch fibers tend to rely on the creatine phosphate pathway and fermentation, anaerobic means of supplying ATP to muscles. Slow-twitch fibers tend to prefer cellular respiration, which is aerobic (Fig. 12.11).

Fast-Twitch Fibers

Fast-twitch fibers are usually anaerobic and seem to be designed for strength because their motor units contain many fibers. They provide explosions of energy and are most helpful in sports

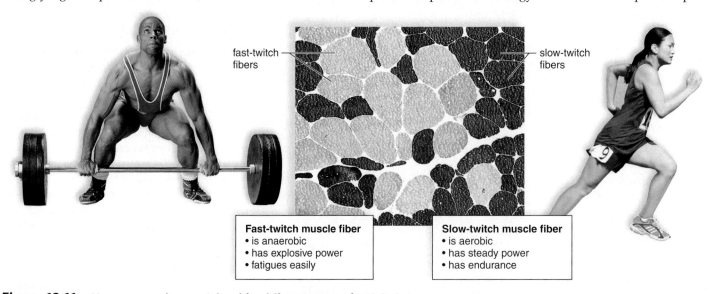

Figure 12.11 How are muscles specialized for different types of activity?
If your muscles contain many fast-twitch fibers (light color), you would probably do better at a sport like weight lifting. If your muscles contain many slow-twitch fibers (dark color), you would probably do better at a sport like cross-country running.

Exercise, Exercise, Exercise

Exercise programs improve muscular strength, muscular endurance, and flexibility. Muscular strength is the force a muscle group (or muscle) can exert against a resistance in one maximal effort. Muscular endurance is judged by the ability of a muscle to contract repeatedly or to sustain a contraction for an extended period. Flexibility is tested by observing the range of motion about a joint.

Exercise also improves cardiorespiratory endurance. The heart rate and capacity increase, and the air passages dilate so that the heart and lungs are able to support prolonged muscular activity. The blood level of high-density lipoprotein (HDL) increases. HDL is the molecule that slows the development of artherosclerotic plaques in blood vessels (see Chap. 5). Also, body composition—the proportion of protein to fat—changes favorably when you exercise.

Exercise also seems to help prevent certain types of cancer. Cancer prevention involves eating properly, not smoking, avoiding cancer-causing chemicals and radiation, undergoing appropriate medical screening tests, and knowing the early warning signs of cancer. However, studies show that people who exercise are less likely to develop colon, breast, cervical, uterine, and ovarian cancers.

Physical training with weights can improve the density and strength of bones and the strength and endurance of muscles in all adults, regardless of age. Even men and women in their eighties and nineties can make substantial gains in bone and muscle strength that help them lead more independent lives. Exercise helps prevent osteoporosis, a condition in which the bones are weak and tend to break

(see page 235). Exercise promotes the activity of osteoblasts in young as well as older people. The stronger the bones when a person is young, the less chance of osteoporosis as that person ages. Exercise helps prevent weight gain, not only because the level of activity increases but also because muscles metabolize faster than other tissues. As a person becomes more muscular, the body is less likely to accumulate fat.

Exercise relieves depression and enhances the mood. Some people report that exercise actually makes them feel more energetic. Further, after exercise, particularly in the late afternoon, people sleep better that night. Self-esteem rises because of improved appearance, as well as other factors that are not well understood. For example, vigorous exercise releases endorphins, hormonelike chemicals known to alleviate pain and provide a feeling of tranquility.

A sensible exercise program is one that provides all of these benefits without the detriments of a too-strenuous program. Overexertion can be harmful to the body and might result in sports injuries, such as lower back strains or torn ligaments of the knees. The beneficial programs suggested in Table 12A are tailored according to age.

Dr. Arthur Leon at the University of Minnesota performed a study involving 12,000 men, and the results showed that only moderate exercise is needed to lower the risk of a heart attack by one-third. In another study conducted by the Institute for Aerobics Research in Dallas, Texas, which included 10,000 men and more than 3,000 women, even a little exercise was found to lower the risk of death from cardiovascular diseases and cancer. Increasing daily activity by walking to the corner store instead of driving and by taking the stairs instead of the elevator can improve your health.

Table 12A	Staying Fit		
Exercise	**Children, 7–12**	**Teenagers, 13–18**	**Adults, 19–55**
Amount	Vigorous activity 1–2 hrs daily	Vigorous activity 1 hr 3–5 days a week; otherwise, ½ hr daily moderate activity	Vigorous activity 1 hr 3 days a week; otherwise, ½ hr daily moderate activity
Purpose	Free play	Build muscle with calisthenics	Exercise to prevent lower back pain: aerobics, stretching, or yoga
Organized	Build motor skills through team sports, dancing, or swimming	Continue team sports: dancing, hiking, or swimming	Do aerobic exercise to control buildup of fat cells
Group	Enjoy more exercise outside of physical education classes	Pursue sports that can be enjoyed for a lifetime: tennis, swimming, or horseback riding	Find exercise partners: join a running club, bicycle club, or outing group
Family	Participate in family outings: bowling, boating, camping, or hiking	Take active vacations: hike, bicycle, or cross-country ski	Initiate family outings: bowling, boating, camping, or hiking

activities such as sprinting, weight lifting, swinging a golf club, or throwing a shot. Fast-twitch fibers are light in color because they have fewer mitochondria, little or no myoglobin, and fewer blood vessels than slow-twitch fibers do. Fast-twitch fibers can develop maximum tension more rapidly than slow-twitch fibers can. In addition, their maximum tension is greater. However, their dependence on anaerobic energy leaves them vulnerable to an accumulation of lactate, which causes them to fatigue quickly.

Slow-Twitch Fibers

Despite having motor units with smaller numbers of muscle fibers, slow-twitch fibers have more stamina and a steadier "tug." These muscle fibers are most helpful in endurance sports, such as long distance running, biking, jogging, and swimming. They produce most of their energy aerobically, so they tire only when their fuel supply is gone. Slow-twitch fibers have many mitochondria and are dark in color because they contain myoglobin, the respiratory pigment found in muscles. They are also surrounded by dense capillary beds and draw more blood and oxygen than fast-twitch fibers. Slow-twitch fibers have a low maximum tension, which develops slowly, but the muscle fibers are highly resistant to fatigue. Slow-twitch fibers have a substantial reserve of glycogen and fat, so their abundant mitochondria can maintain a steady, prolonged production of ATP when oxygen is available.

Delayed Onset Muscle Soreness

Many of us have experienced delayed onset muscle soreness (DOMS), which generally appears some 24–48 hours after strenuous exercise. It is thought that DOMS is due to tissue injury that takes several days to heal. Any movement you aren't used to can lead to DOMS, but it is especially associated with any activity that causes muscles to contract while they are lengthening. Examples include walking down stairs, running downhill, lowering weights, and the downward motion of squats and push-ups. To prevent DOMS, try warming up thoroughly and cooling down completely. Stretch after exercising. When beginning a new activity, start gradually and build up your endurance gradually. Avoid making sudden major changes in your exercise routine.

> **Check Your Progress 12.3**
> 1. a. What is a motor unit, and (b) why does a motor unit obey the all-or-none law?
> 2. How does a whole muscle avoid fatigue caused by tetanus?
> 3. a. What are the sources of ATP for muscle contraction, and (b) which type of muscle fibers rely primarily on anaerobic ATP production?

12.4 Muscular Disorders

Muscular disorders are divided into those commonly seen and those more serious.

Common Muscular Conditions

Spasms are sudden and involuntary muscular contractions most often accompanied by pain. Spasms can occur in smooth and skeletal muscles. A spasm of the intestinal tract is a type

CASE STUDY THE NEXT DAY

"What else could go wrong today?" Kayla thought as she struggled into her uniform. Her car was trying to die for good, and was still in the shop. Her sick 8-year-old would have to stay at her mom and dad's house for another day. Now she was faced with a day spent cleaning rooms left-handed. She slipped into the storage area to retrieve her cart. She hoped to put the doctor's notice on Mr. Christiansen's desk, then get to work before he noticed that her right arm was stuck in a sling.

No such luck for her today—there he was at his desk, as usual. "Hi, Mr. Christiansen," she smiled weakly. "I'm sorry I was out yesterday. Here's a note from the doctor. I'll get right to work now."

"Wait a minute, Kayla," he replied. "Are you going to be able to use that arm?"

Her shoulders slumped. "Ummh . . . not exactly. The doctor says I can't move it at all. But I can work left-handed, honest." He shook his head. "Mr. Christiansen, I really need this job!" she cried. "I'm going to school and I have two kids! Please, please don't fire me!"

"Who said anything about firing anybody? You're the best worker I have. I'm not about to let you go," he reassured her. "You have disability insurance with this job, right? You're going to take some paid leave to rest that arm.

"And you can't clean when you get back, either," he continued. "No sense messing up that elbow again. So how about being the administrative assistant when Susan goes out on maternity leave? Deal?"

She sighed gratefully. "Deal."

Have You Ever Wondered...

Which muscles are best to use for intramuscular injections?

Health-care providers have to choose muscles that are sufficiently large and well-developed to tolerate intramuscular injections. At the same time, muscles containing large blood vessels or nerves must be avoided. An injection in these muscles could pierce a blood vessel or damage a nerve. Typically, one of three preferred injection sites is chosen. The deltoid muscle on the upper arm is usually well-developed in older children and adults. The vastus lateralis on the side of the thigh (part of the quadriceps group) is the best site for infants and young children. The gluteus medius is on the lower back, above the buttock. However, a clinician injecting into the gluteus medius must be careful to avoid the gluteus maximus (buttock) muscle. The body's largest nerve, the sciatic nerve, lies underneath and within the gluteus maximus.

Historical Focus

Iron Horse: Lou Gehrig (1903–1941)

It is truly unfortunate that over time, the name Lou Gehrig has become recognized first and foremost with a disease. Gehrig, the "Iron Horse" of baseball, was born in 1903. The son of German immigrants, Gehrig was a talented athlete who excelled in many sports. He was awarded a football scholarship by Columbia University in 1921. He played fullback for the Columbia football team in the fall of 1922, and then pitched and played first base for the Columbia baseball team in the spring of 1923. His goal was to complete an engineering degree. However, recruiters for the New York Yankees baseball team had other ideas. Over the objections of his parents, who wanted him to complete his college education, Gehrig signed with the Yankees in 1923.

Gehrig's dedication to baseball and love for the game were legendary. He won his nickname by playing 2,130 consecutive games. His record stood until 1995, when it was finally broken by Cal Ripken. Illness, back spasms, even broken bones in his hands—nothing stopped Gehrig from playing his game. Doctors who X-rayed his hands were amazed to discover 17 different fractures of the bones in Gehrig's hands. All had slowly healed as Gehrig continued to play baseball.

During his years of playing, Gehrig set numerous records that stood for decades. His career included 493 home runs and set the record for the most home runs hit by any first baseman in history. (Mark McGwire broke Gehrig's record in 1998, but McGwire is still being investigated for alleged steroid abuse. See Bioethical Focus, page 270.) He was the first American League player to hit four home runs in a single game. For 12 consecutive years, this quiet, humble man would have a batting average over .300. Gehrig would always play in the shadow of more famous players: first, home run slugger Babe Ruth, and later Joe DiMaggio. Yet he was always recognized and respected for his achievements. A *New York Times* sportswriter wrote of Gehrig:

> "…But his greatest record doesn't show in the book. It was the absolute reliability of Henry Louis Gehrig. He could be counted upon. He was there every day at the ballpark bending his back and ready to break his neck to win for his side. He was there day after day and year after year. He never sulked or whined or went into a pot or a huff. He was the answer to a manager's dream."

In 1938, Gehrig's performance went into a steady decline, and his consecutive game play ended on May 2, 1939. As the Yankee team captain, he presented the team lineup to the game umpire as always. But for the first time in fifteen years, his name wasn't included. He had voluntarily pulled himself out of the lineup, and his baseball career was over.

Gehrig was diagnosed with **amyotrophic lateral sclerosis (ALS),** a rare degenerative neuromuscular disease that now bears his name. The cause of ALS is unknown. ALS sufferers gradually lose the ability to walk, talk, chew, and swallow. Mental abilities and sensation are not affected, however. The patient dies of respiratory complications, usually within three years of diagnosis. Today, drugs that slow the progression of the disease are available, but there remains no cure.

Gehrig was arguably one of the best baseball players to have played the game, creating records that have stood until recently. More importantly, he modeled humility and hard work. His farewell speech to team and fans was given at Yankee Stadium on July 4, 1939. As he said goodbye to the game he loved, Gehrig showed his extraordinary courage:

> "Yet today, I consider myself the luckiest man on the face of this earth. . . . So I close in saying that I may have had a tough break, but I have an awful lot to live for."

Gehrig died on June 2, 1941.

Figure 12A July 4, 1939: Lou Gehrig says goodbye to baseball.

of colic sometimes called a bellyache. Multiple spasms of skeletal muscles are called a seizure, or **convulsion. Cramps** are strong, painful spasms, especially of the leg and foot, usually due to strenuous activity. Cramps can even occur when sleeping after a strenuous workout. **Facial tics,** such as periodic eye blinking, head turning, or grimacing, are spasms that can be controlled voluntarily, but only with great effort.

A **strain** is caused by stretching or tearing of a muscle. A **sprain** is a twisting of a joint leading to swelling and injury, not only of muscles but also of ligaments, tendons, blood vessels, and nerves. The ankle and knee are often subject to sprains. When a tendon is inflamed by a sprain, **tendinitis** results. Tendinitis may irritate the bursa underlyling the tendon, causing **bursitis.**

Muscular Diseases

These conditions are more serious and always require close medical care.

Myalgia and Fibromyalgia

Myalgia refers to achy muscles. The most common cause for myalgia is either overuse or overstretching of a muscle or group of muscles. Myalgia without a traumatic history is often due to viral infections. Myalgia may accompany myositis (inflammation of the muscles), either in response to viral infection or as an immune system disorder. **Fibromyalgia** is a chronic condition whose symptoms include achy pain, tenderness, and stiffness of muscles. Its precise cause is not known, but it may also be due to a underlying infection that is not obvious at first.

Muscular Dystrophy

Muscular dystrophy is a broad term applied to a group of disorders characterized by a progressive degeneration and weakening of muscles. As muscle fibers die, fat and connective tissue take their place. **Duchenne muscular dystrophy,** the most common type, is inherited through a flawed gene carried by the mother. It is now known that the lack of a protein called dystrophin causes the condition. When dystrophin is absent, calcium leaks into the cell and activates an enzyme that dissolves muscle fibers. In an attempt to treat the condition, muscles have been injected with immature muscle cells that do produce dystrophin.

Myasthenia Gravis

Myasthenia gravis is an autoimmune disease characterized by weakness that especially affects the muscles of the eyelids, face, neck, and extremities. Muscle contraction is impaired

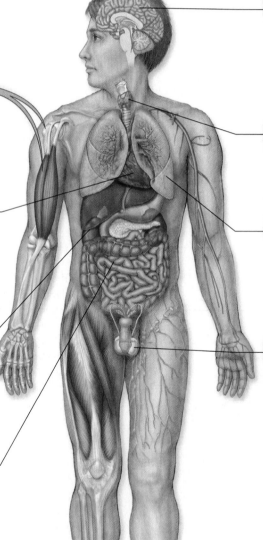

The muscular and skeletal systems work together to maintain homeostasis. The systems listed here in particular also work with these two systems.

Muscular and Skeletal Systems

These systems allow the body to move, and they provide support and protection for internal organs. Muscle contraction provides heat to warm the body; bones play a role in Ca^{2+} balance. These systems specifically help the other systems as mentioned below.

Cardiovascular System

Red bone marrow produces the blood cells. The rib cage protects the heart; red bone marrow stores Ca^{2+} for blood clotting. Muscle contraction keeps blood moving in the heart and blood vessels, particularly the veins.

Urinary System

Muscle contraction moves the fluid within ureters, bladder, and urethra. Kidneys activate vitamin D needed for Ca^{2+} absorption and help maintain the blood level of Ca^{2+} for bone growth and repair, and for muscle contraction.

Digestive System

Jaws contain teeth that chew food; the hyoid bone assists swallowing. Muscle contraction accounts for chewing of food and peristalsis to move food along digestive tract. The digestive tract absorbs ions needed for strong bones and muscle contraction.

Nervous System

Bones store Ca^{2+} needed for muscle contraction and nerve impulse conduction. The nervous system stimulates muscles and sends sensory input from joints to the brain. Muscle contraction moves eyes, permits speech, and creates facial expressions.

Endocrine System

Growth hormone and sex hormones regulate bone and muscle development; parathyroid hormone and calcitonin regulate Ca^{2+} content of bones.

Respiratory System

The rib cage protects lungs, and rib cage movement assists breathing, as does muscle contraction. Breathing provides the oxygen needed for ATP production so muscles can move.

Reproductive System

Muscle contraction moves gametes in oviducts, and uterine contraction occurs during childbirth. Sex hormones influence bone growth and density; androgens promote muscle growth.

Figure 12.12   Human systems work together.

because the immune system mistakenly produces antibodies that destroy acetylcholine (ACh) receptors. In many cases, the first sign of the disease is a drooping of the eyelids and double vision. Treatment includes drugs that inhibit the enzyme that digests acetylcholine so that ACh accumulates in neuromuscular junctions.

> **Check Your Progress 12.4**
> 1. What is the difference between a strain and a sprain?
> 2. As pertains to a potential cause, what could myalgia and myasthenia gravis have in common?

12.5 Homeostasis

In this section, our discussion centers on the contribution of the muscular and skeletal systems to homeostasis (Fig. 12.12).

Both Systems Produce Movement

Movement is essential to maintaining homeostasis. The skeletal and muscular systems work together to enable body movement (Fig. 12.13). This is most evidently illustrated by what happens when skeletal muscles contract and pull on the bones to which they are attached, causing movement at joints. Body movement of this sort allows us to respond to certain types of changes in the environment. For instance, if you are sitting in the sun and start to feel hot, you can get up and move to a shady spot.

The muscular and skeletal systems work for other types of movements that are just as important for maintaining homeostasis. Contraction of skeletal muscles associated with the jaw and tongue allow you to grind food with the teeth. The rhythmic smooth-muscle contractions of peristalsis move ingested materials through the digestive tract. These processes are necessary for supplying the body's cells with nutrients. The ceaseless beating of your heart, which propels blood into the arterial system, is caused by the contraction of cardiac muscle. Contractions of skeletal muscles in the body, especially those associated with breathing and leg movements, aid

Figure 12.13 How is movement possible?
Skeletal muscles and bones must cooperate to produce each dancer's elegant motion.

in the process of venous return by pushing blood back toward the heart. This is why soldiers and members of marching bands are cautioned not to lock their knees when standing at attention. The reduction in venous return causes a drop in blood pressure that can result in fainting. The pressure exerted by skeletal muscle contraction also helps to squeeze tissue fluid into the lymphatic capillaries, where it is referred to as lymph.

Both Systems Protect Body Parts

The skeletal system plays an important role by protecting the soft internal organs of your body. The brain, heart, lungs, spinal cord, kidneys, liver, and most of the endocrine glands are shielded by the skeleton. In particular, the nervous and endocrine organs must be defended so they can carry out activities necessary for homeostasis.

The skeletal muscles pad and protect the bones, while the tendons and bursae associated with skeletal muscles reinforce and cushion the joints. Muscles of the abdominal wall offer additional protection to the soft internal organs. Examples of these muscles include the rectus abdominis and external oblique muscles illustrated in Figure 12.4.

Bones Store and Release Calcium

Under the direction of the endocrine system, the skeletal system performs vital tasks for calcium homeostasis. Calcium ions (Ca^{2+}) are needed for a variety of processes in your body, such as muscle contraction and nerve conduction. They are also necessary for the regulation of cellular metabolism by acting in cellular messenger systems. Thus, it is important to always maintain an adequate level of Ca^{2+} in the blood. When you have plenty of Ca^{2+} in your blood, the hormone calcitonin from the thyroid gland ensures that calcium salts are deposited in bone tissue. Thus, the skeleton acts as a reservoir for storage of this important mineral. If your blood Ca^{2+} level starts to fall, parathyroid hormone secretion stimulates osteoclasts to break down bone tissue and thereby make Ca^{2+} available to the blood. Vitamin D is needed for the absorption of Ca^{2+} from the digestive tract, which is why vitamin D deficiency can result in weak bones. It is easy to get enough of this vitamin, because your skin produces it when exposed to sunlight, and the milk you buy at the grocery store is fortified with vitamin D.

Blood Cells Are Produced in Bones

The bones of your skeleton contain two types of marrow: yellow and red. Fat is stored in yellow bone marrow, thus making it part of the body's energy reserves.

Red bone marrow is the site of blood cell production. The red blood cells are the carriers of oxygen in the blood. Oxygen is necessary for the production of ATP by aerobic cellular respiration. White blood cells also originate in the red bone marrow. The white cells are involved in defending your body against pathogens and cancerous cells; without them, you would soon succumb to disease and die.

Anabolic Steroid Use

They're called "performance-enhancing steroids," and their use is alleged to be widespread in athletics, both amateur and professional. Whether the sport is baseball, football, professional cycling, or track and field events—no activity seems to be safe from drug abuse. Steroid abuse admitted by Marion Jones forever changed Olympic history. Jones was the first female athlete to win five medals, for track and field events during the 2000 Sydney Olympics. In 2008, she was stripped of all medals she had earned, as well as disqualified from a fifth-place finish in the 2004 Athens games. Future Olympic record books will not include her name. The records of her teammates in the relay events have also been tainted.

Baseball records will also likely require revisions. The exciting slugfest between Mark McGwire and Sammy Sosa in the summer of 1998 was largely credited with reviving national interest in baseball. Yet the great home run competition drew unwanted attention to the darker side of professional sports, when it was alleged that McGwire and Sosa were using anabolic steroids at the time. Similar charges of drug abuse may prevent baseball great Roger Clemens from entering the Baseball Hall of Fame, despite holding the record for Cy Young awards. Likewise, because controversy continues to surround baseball legend Barry Bonds, this talented athlete may never achieve Hall of Fame status. Though he scored a record 715 home runs and won more Most Valuable Player awards than anyone in history, Bonds remains accused of steroid abuse.

Use of anabolic steroids in professional sports continues to be denied by most athletes and officials. However, many people from both inside and outside the industry maintain that such abuse has been going on for many years, and that it continues despite the negative publicity. Congress continues to investigate the controversy, yet the finger-pointing and accusations steadily increase.

What Are Anabolic Steroids?

Steroids encompass a large category of substances, both beneficial and harmful. Anabolic steroids are a class of steroids that generally cause tissue growth by promoting protein production. They are naturally occurring hormones created by the body and commonly used to regulate many physiological processes, from growth to sexual function. Most anabolic steroids are closely related to male sex hormones, such as testosterone.

These metabolically potent drugs are controlled substances available only by prescription under the close supervision of a physician because of their many side effects and vast potential for abuse. However, a few anabolic steroids are still legal due to loopholes in drug laws, despite being banned by most professional sports organizations.

Robust muscle growth is not possible from prescription doses of anabolic steroids, so large doses must be used to obtain that effect. Athletes may take dangerously large amounts of anabolic steroids to enhance athletic performance or to increase strength, often with serious consequences. Steroid abusers may vary the type and quantity of drug taken (called "stacking") or may take the drugs and then stop for a time, only to resume later (called "cycling"). Stacking and cycling are done to minimize serious side effects while maximizing the desired effects of the drugs. Even so, dangerous health consequences can occur.

Dangerous Health Consequences

The most common health consequences include high blood pressure, jaundice (yellowing of the skin), acne, and a greatly increased risk of can-

Figure 12 B
Marion Jones was stripped of five gold medals from the 2000 Olympic games after admitting steroid use.

cer. In women, anabolic steroid abuse may cause masculinization, including a deepened voice, excessive facial and body hair, coarsening of the hair, menstrual cycle irregularities, and enlargement of the clitoris. Anabolic steroid abuse is even more dangerous during adolescence. When taken prior to or during the teenage growth spurt, steroids may result in permanently shortened height or early onset of puberty. Ironically, while proper use of anabolic steroids has been helpful in treating many cases of impotence in males, abuse of these drugs may cause impotence and even shrinking of the testicles.

Perhaps the most frightening aspect of anabolic steroid abuse are the reports of increased aggressive behavior and violent mood swings. Furthermore, many users have reported extremely severe withdrawal symptoms upon quitting. Also, many anabolic steroids have also been identified as a "gateway drug," leading abusers to escalate their drug habit to more dangerous drugs such as heroin and cocaine.

Widespread Use

Unfortunately, anabolic steroid use is not confined to professional sports. A 1999 study by the National Institute on Drug Abuse (NIDA) found that use of anabolic steroids has increased at an alarming rate among adolescents in middle and high school. The study also found that hundreds of thousands of adults abuse steroids on occasion. Nearly half of the individuals surveyed were unaware of the danger posed by these drugs, indicating the need for better education and awareness. In imitating their heroes, many young athletes are turning to anabolic steroids for the competitive edge that they offer, despite the many serious risks posed. Unfortunately, it may cause these young athletes to strike out.

Decide Your Opinion

1. Should Marion Jones' teammates be stripped of their gold medal status? Why or why not?
2. Should recognitions such as admission to the Hall of Fame be denied to athletes if steroid abuse is alleged but cannot be proven?
3. Do you believe the techniques athletes use to train and enhance performance should be regulated? If so, who can or should enforce regulation?
4. What regulations should be applied to middle-school and high-school athletes? Who should enforce these rules?
5. Is it acceptable for athletes to possibly endanger their health by practicing excessively? By gaining or losing weight? By taking drugs other than steroids?
6. Who, if anyone, should be in charge of regulating the behaviors of athletes to ensure that they do not harm themselves?

Muscles Help Maintain Body Temperature

The muscular system helps to regulate body temperature. When you are very cold, smooth muscle constricts inside the blood vessels supplying the skin. Thus, the amount of blood close to the surface of the body is reduced. This helps to conserve heat in the body's core, where vital organs lie. If you are cold enough, you may start to shiver. Shivering is caused by involuntary skeletal muscle contractions. This is initiated by temperature-sensitive neurons in the hypothalamus of the brain. Skeletal muscle contraction requires ATP, and using ATP generates heat. You may also notice that you get goose bumps when you are cold. This is because arrector pili muscles contract. These tiny bundles of smooth muscle are attached to the hair follicles and cause the hairs to stand up. This is not very helpful in keeping humans warm, but it is quite effective in our furrier fellow mammals. Think of a cat or dog outside on a cold winter day. Its fur is a better insulator when standing up than when lying flat. Goose bumps can also be a sign of fear. Although a human with goose bumps may not look very impressive, a frightened or aggressive animal whose fur is standing on end looks bigger and (it is hoped) more intimidating to a predator or rival.

> **Check Your Progress 12.5**
> 1. What functions important to homeostasis are performed by the skeletal and muscular systems working together?
> 2. a. What functions important to homeostasis are especially carried out by the skeletal system?
>   b. The muscular system?

Summarizing the Concepts

12.1 Overview of Muscular System

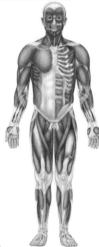

Humans have three types of muscle tissue:

- Smooth muscle is involuntary and occurs in walls of internal organs.
- Cardiac muscle is involuntary and occurs in walls of the heart.
- Skeletal muscle is voluntary, contains bundles of muscle fibers called fascicles, and is usually attached by tendons to the skeleton.
  Skeletal muscle functions:
- Helps maintain posture.
- Provide movement and heat.
- Protect underlying organs.

Skeletal Muscles of the Body

When achieving movement, some muscles are prime movers, some are synergists, and others are antagonists.

Names and Actions of Skeletal Muscles

Muscles are named for their size, shape, location, direction of fibers, number of attachments, and action.

12.2 Skeletal Muscle Fiber Contraction

Muscle fibers contain myofibrils, and myofibrils contain actin and myosin filaments. Muscle contraction occurs when sarcomeres shorten and actin filaments slide past myosin filaments.

- Nerve impulses travel down motor neurons and stimulate muscle fibers at neuromuscular junctions.
- The sarcolemma of a muscle fiber forms T tubules that almost touch the sarcoplasmic reticulum, which stores calcium ions.
- When calcium ions are released into muscle fibers, actin filaments slide past myosin filaments.
- At a neuromuscular junction, synaptic vesicles release acetylcholine (neurotransmitter), which diffuses across the synaptic cleft.

- When acetylcholine (ACh) is received by the sarcolemma, electrical signals begin and lead to the release of calcium.
- Calcium ions bind to troponin, exposing myosin binding sites.
- Myosin filaments break down ATP and attach to actin filaments, forming cross-bridges.
- When ADP and Ⓟ are released, cross-bridges change their positions.
- This pulls actin filaments to the center of a sarcomere.

12.3 Whole Muscle Contraction

Muscles Have Motor Units

- A muscle contains motor units: several fibers under the control of a single motor axon.
- Motor unit contraction is described in terms of a muscle twitch, summation, and tetanus.
- The strength of muscle contraction varies according to recruitment of motor units.
- In the body, a continuous slight tension (called muscle tone) is maintained by muscle motor units that take turns contracting.

Energy for Muscle Contraction

A muscle fiber has three ways to acquire ATP for muscle contraction.

- Creatine phosphate transfers a phosphate to ADP, and ATP results. This CP pathway is the most rapid.

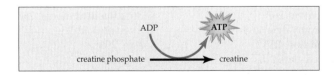

- Fermentation also produces ATP quickly. Fermentation is associated with an oxygen debt because oxygen is needed to metabolize the lactate that accumulates.

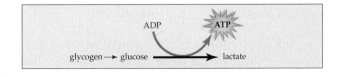

- Cellular respiration provides most of the muscle's ATP, but takes longer because much of the glucose and oxygen must be transported in blood to mitochondria. Cellular respiration occurs during aerobic exercise and burns fatty acids in addition to glucose.

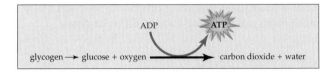

Fast-Twitch and Slow-Twitch Muscle Fibers

- Fast-twitch fibers, for sports like weight lifting, rely on an anaerobic means of acquiring ATP; have few mitochondria and myoglobin, but motor units contain more muscle fibers; and are known for explosive power, but fatigue quickly.
- Slow-twitch fibers, for sports like running and swimming, rely on aerobic respiration to acquire ATP and have a plentiful supply of mitochondria and myoglobin, which gives them a dark color.

12.4 Muscular Disorders

Muscular disorders include spasms and injuries, as well as diseases such as muscular dystrophy and myasthenia gravis.

12.5 Homeostasis

- The muscles and bones produce movement and protect body parts.
- The bones produce red blood cells and are involved in the regulation of blood calcium levels.
- The muscles produce the heat that gives us a constant body temperature.

Understanding Key Terms

actin 259
all-or-none law 262
amylotrophic lateral sclerosis
    (ALS) 267
bursa 255
bursitis 267
cardiac muscle 254
convulsion 267
cramp 267
Duchenne muscular
    dystrophy 268
facial tic 267
fibromyalgia 268
insertion 255
intercalated disk 254
motor unit 262
muscle fiber 254
muscle tone 263
muscle twitch 262
muscular dystrophy 268
myalgia 268
myasthenia gravis 268

myofibril 258
myoglobin 264
myosin 258
neuromuscular junction 261
origin 255
oxygen debt 264
rigor mortis 257
sarcolemma 258
sarcomere 259
sarcoplasmic reticulum 258
skeletal muscle 254
sliding filament model 260
smooth muscle 254
spasm 266
sprain 267
strain 267
T (transverse) tubule 258
tendinitis 267
tendon 255
tetanus 262
tropomyosin 262
troponin 262

Match the key terms to these definitions.

a. _____ Structural and functional unit of a myofibril; contains actin and myosin filaments.

b. _____ End of a muscle attached to a movable bone.

c. _____ Sustained maximal muscle contraction.

d. _____ Contraction of muscles at death due to lack of ATP.

e. _____ Stretching or tearing of a muscle.

Testing Your Knowledge of the Concepts

1. What are the characteristics of the three types of muscles in the human body? Where is each type found? (page 254)

2. Give an example to show that skeletal muscles work in antagonistic pairs. Explain. (page 255)

3. What criteria are used to name muscles? Give an example of each one. (pages 256, 258)

4. What are the functions of a muscle fiber's components? (pages 258–60)

5. Describe the sliding filament model of muscle contraction within a sarcomere. Begin with the nerve impulse and end with the relaxation of the muscle. (pages 261–63)

6. What are the four possible energy sources for a muscle and the three sources of ATP for muscle contraction? (pages 263–64)

7. Compare fast- and slow-twitch muscle fibers. (pages 264–66)

8. What are some common muscular disorders and some more serious muscular diseases? (pages 268–69)

9. How does the muscular system help maintain homeostasis? (pages 269, 271)

10. Impulses that move down the T system of a muscle fiber most directly cause
    a. movement of tropomyosin.
    b. attachment of the cross-bridges to myosin.
    c. release of Ca^{2+} from the sarcoplasmic reticulum.
    d. splitting of ATP.

11. Which of the following statements about cross-bridges is false?
    a. They are composed of myosin.
    b. They bind to ATP after they detach from actin.
    c. They contain an ATPase.
    d. They split ATP before they attach to actin.

12. Which statement about sarcomere contraction is incorrect?
    a. The A bands shorten.
    b. The H zones shorten.
    c. The I bands shorten.
    d. The sarcomeres shorten.

13. Which of the following muscles would have motor units with the lowest innervation ratio?
    a. leg muscles   c. muscles that move the fingers
    b. arm muscles   d. muscles of the trunk

14. The thick filaments of a muscle fiber are made up of
    a. actin.   c. fascia.
    b. troponin.   d. myosin.

15. As ADP and (P) are released from a myosin head,
    a. actin filaments move toward the H zone.
    b. myosin cross-bridges pull the thin filaments.
    c. a sarcomere shortens.
    d. Only a and c are correct.
    e. All of these are correct.

16. Which of these is a direct source of energy for muscle contraction?
    a. ATP   d. glycogen
    b. creatine phosphate   e. Both a and b are correct.
    c. lactic acid

17. When muscles contract,
    a. sarcomeres increase in length.
    b. actin breaks down ATP.
    c. myosin slides past actin.
    d. the H zone disappears.
    e. calcium is taken up by the sarcoplasmic reticulum.

18. Nervous stimulation of muscles
    a. occurs at a neuromuscular junction.
    b. results in an impulse that travels down the T system.
    c. causes calcium to be released from expanded regions of the sarcoplasmic reticulum.
    d. All of these are correct.

19. In a muscle fiber,
    a. the sarcolemma is connective tissue holding the myofibrils together.
    b. the T system causes release of Ca^{2+} from the sarcoplasmic reticulum.
    c. both actin and myosin filaments have cross-bridges.
    d. there is no endoplasmic reticulum.
    e. All of these are correct.

20. To increase the force of muscle contraction,
    a. individual muscle cells have to contract with greater force.
    b. motor units have to contract with greater force.
    c. motor units need to be recruited.
    d. All of these are correct.
    e. None of these is correct.

21. Lack of calcium in muscles will
    a. result in no contraction.
    b. cause weak contraction.
    c. cause strong contraction.
    d. will have no effect.
    e. None of these is correct.

22. Which of these energy relationships are mismatched?
    a. creatine phosphate—anaerobic
    b. cellular respiration—aerobic
    c. fermentation—anaerobic
    d. oxygen debt—anaerobic
    e. All of these are properly matched.

23. During muscle contraction,
    a. ATP is hydrolyzed when the myosin head is unattached.
    b. ADP and (P) are released as the myosin head attaches to actin.
    c. ADP and (P) release causes the head to change position and actin filaments to move.
    d. release of ATP causes the myosin head to return to resting position.

24. Proper functioning of a neuromuscular junction requires the
    a. presence of acetylcholine.
    b. presence of a synaptic cleft.
    c. presence of a motor terminal.
    d. sarcolemma of a muscle cell.
    e. All of these are correct.

In questions 25–28, match each muscle to a region of the body in the key.

Key:
    a. neck and back   c. arm
    b. abdomen   d. thigh

25. Biceps femoris

26. Trapezius

27. Rectus abdominis

28. Triceps brachii

In questions 29–32, match each function to a muscle of the buttocks and legs in the key.

Key:
    a. tibialis anterior   c. biceps femoris
    b. gluteus maximus   d. gastrocnemius

29. Bends leg at knee and extends thigh back

30. Bends ankle so that foot is upward

31. Bends leg at knee and bends sole of foot

32. Extends thigh back

In questions 33–35, match each function to a muscle of the head and neck in the key.

Key:
    a. trapezius   c. masseter
    b. orbicularis oculi

33. A chewing muscle

34. Moves head and scapula

35. Closes eye

In questions 36–38, match the functions to a muscle of the upper limb and trunk in the key.

Key:
    a. external oblique   c. pectoralis major
    b. triceps brachii

36. Compresses abdomen

37. Brings arm across chest

38. Straightens forearm at elbow

In questions 39–41, match each function to a muscle of the lower limb in the key.

Key:
    a. adductor longus   c. sartorius
    b. quadriceps femoris

39. Straightens leg at knee

40. Moves thigh toward the body

41. Lateral rotates thigh
42. Label this diagram of a muscle fiber, using these terms: myofibril, T tubule, sarcomere, sarcolemma, sarcoplasmic reticulum, Z line.

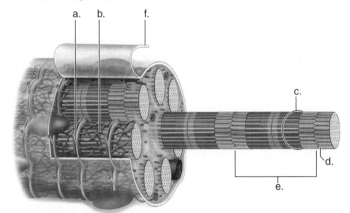

Thinking Critically About the Concepts

You've discovered from the case of Kayla Scott that overusing a joint can result in damage to tendons (tendinitis), ligaments, and bursae (bursitis). Similarly, abuse can cause damage to support muscles as well. Recall from Chapter 11 that long-term heavy exercise leads to the early development of osteoarthritis. These types of injuries at a joint are all difficult to treat. Pain-killing drugs such as ibuprofen or naproxen can upset the stomach and ultimately lead to ulcers. Kayla's cortisone shot would reduce inflammation and swelling, but this type of treatment has side effects. Complications of using cortisonetype steroids include decreased immune response and mood disturbances, among others. Resting the joint will ultimately be necessary for healing. However, when one's livelihood, like Kayla's, depends on using a joint over and over, resting can be very difficult!

1. Kayla Scott from the case study has bursitis and tendinitis affecting her elbow joint.
   a. What bones form the elbow joint?
   b. Why does tendinitis result in bursitis?

2. Overuse of a joint occurs in many professions. Where would the resulting tendinitis and bursitis occur in a
   a. baseball pitcher?
   b. cement contractor or carpet layer?
   c. ballet dancer?
   d. sprinter?

3. You learned about rigor mortis from the Science Focus reading on page 257. Perhaps you're also a fan of crime scene shows as well. If so, you know that the onset of rigor mortis in a deceased person can be influenced by a number of factors. Consider the following:
   a. If a body was rapidly cooled after death, how would this affect the timing of rigor mortis?
   b. Explain how rapid cooling affects the onset of rigor mortis.

4. Rigor mortis is usually complete within one to two days after death (depending on environmental variables). Why would rigor mortis diminish after several days?

5. What do you think causes a person with amyotrophic lateral sclerosis (Lou Gehrig's disease) to die? Hint: Think of the muscles used in respiration.

CHAPTER 13

Nervous System

"This is going to be pretty scary at first, Jill. You have multiple sclerosis—MS," the neurologist, Dr. Hansen, explained softly. "That explains the eye pain and double vision and your weakness and fatigue as well."

Jill felt breathless with fear. The news hit her almost like a physical blow. "Are you sure?"

"We saw patches of nerve cells on the MRI," the doctor answered. "The spinal tap showed that the nerve cells are being attacked by your white blood cells. There doesn't seem to be much doubt, unfortunately."

She took her husband's hand. He had started to sob, and turned his face away, embarrassed. "Hey Mike, it's gonna be okay," she reassured. "It's gonna be okay, honey," she repeated. "So now what?" she asked shakily.

"Most of the time, MS is called a relapsing-remitting disorder," Dr. Hansen replied. "That means it will get worse for a time, and then it will get better. Right now, you're in a relapse, a period when your symptoms get worse. A relapse usually lasts a few months, but it can persist for a year or more."

"Years?!" Now Jill was crying, too. "I have three kids, and the oldest is only 14! I have a job and a family to take care of! What's this going to do to the rest of my life? Am I going to die?"

"MS won't shorten your life, Jill," Dr. Hansen reassured. "We have some steroid medications to give that usually help to shorten a relapse. Medications can also make your symptoms less severe. And remissions can last for years, too. Your life may not be the same as before, but you'll still be able to do most of the things you've always done."

"There isn't any cure for this thing, is there?" Jill was afraid she already knew the answer. Dr. Hansen shook his head.

13.1 Overview of the Nervous System

The nervous system has two major divisions (Fig. 13.1a). The **central nervous system (CNS)** consists of the brain and spinal cord. The brain is completely surrounded and protected by the skull. It connects directly to the spinal cord, similarly protected by the vertebral column. The **peripheral nervous system (PNS)** consists of nerves. Nerves lie outside the CNS. The division between the CNS and the PNS is arbitrary. The two systems work together and are connected to each other (Fig. 13.1b).

The nervous system has three specific functions:

1. The nervous system receives sensory input. Sensory receptors in skin and other organs respond to external and internal stimuli by generating nerve signals that travel by way of the PNS to the CNS. For example, if you smell baking cookies, olfactory (smell) receptors in the nose will use the PNS to transmit that information to the CNS.
2. The CNS performs information processing and integration, summing up the input it receives from all over the body. The CNS reviews the information, stores the information as memories, and creates the appropriate motor responses. The smell of those baking cookies evokes pleasant memories of their taste, and you'll quickly conceive the idea of getting some for yourself. Other stored memories, from childhood lessons in proper behavior, will remind you that it's rude to grab those cookies without asking first!
3. The CNS generates motor output. Nerve signals from the CNS go by way of the PNS to the muscles, glands, and organs, all in response to the cookies. Signals to the salivary glands will make you salivate. Your stomach will generate the acid and enzymes you'll need to digest the cookies—even before you've had a bite. And you won't hesitate to grab those cookies as soon as you're allowed, using the appropriate muscles to do so.

Nervous Tissue

Nervous tissue contains two types of cells: neurons and neuroglia (neuroglial cells). **Neurons** are the cells that transmit nerve impulses between parts of the nervous system; **neuroglia** support and nourish neurons. Neuroglia were discussed in Chapter 4, and this section discusses the structure and function of neurons.

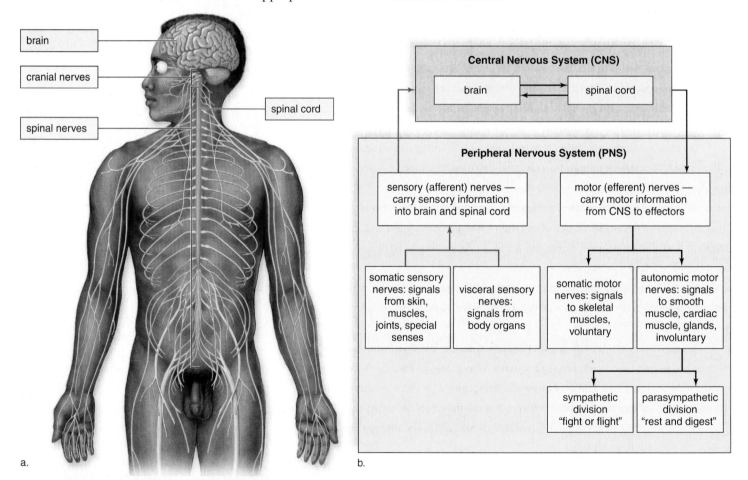

Figure 13.1   **What are the two divisions of the nervous system? How do these two divisions communicate?**
a. The central nervous system (CNS) consists of the brain and spinal cord. The peripheral nervous system (PNS) consists of the nerves, which lie outside the CNS. **b.** The red arrows are the pathway by which the CNS receives sensory information. The black arrows are the pathway by which the CNS communicates with the somatic nervous system and the autonomic nervous system, two divisions of the PNS.

Neuron Structure

Classified according to function, the three types of neurons are sensory neurons, interneurons, and motor neurons (Fig. 13.2). Their functions are best described relative to the CNS. A **sensory neuron** takes nerve signals from a sensory receptor to the CNS. **Sensory receptors** are special structures that detect changes in the environment. An **interneuron** lies entirely within the CNS. Interneurons can receive input from sensory neurons and also from other interneurons in the CNS. Thereafter, they sum up all the information received from other neurons before they communicate with motor neurons. A **motor neuron** takes nerve impulses away from the CNS to an effector (muscle fiber, organ, or gland). **Effectors** carry out our responses to environmental changes, whether these are external or internal.

Neurons vary in appearance, but all of them have just three parts: a cell body, dendrites, and an axon. The **cell body** contains the nucleus, as well as other organelles. **Dendrites** are short extensions that receive signals from sensory receptors or other neurons. Incoming signals from dendrites can result in nerve signals that are then conducted by an axon. The **axon** is the portion of a neuron that conducts nerve impulses. An axon can be quite long. Individual axons are termed nerve fibers. Collectively, they form a nerve.

In sensory neurons, a very long axon carries nerve signals from the dendrites associated with a sensory receptor to the CNS, and this axon is interrupted by the cell body. In interneurons and motor neurons, on the other hand, multiple dendrites take signals to the cell body, and then an axon conducts nerve signals away from the cell body.

Myelin Sheath

Many axons are covered by a protective **myelin sheath.** In the PNS, this covering is formed by a type of neuroglia called **Schwann cells,** which contain myelin (a lipid substance) in their plasma membranes. In the CNS, oligodendrocytes perform this function. The myelin sheath develops when these cells wrap themselves around an axon many times. Each neuroglia cell covers only a portion of an axon, so the myelin sheath is interrupted. The gaps where there is no myelin sheath are called **nodes of Ranvier** (Fig. 13.2).

Long axons tend to have a myelin sheath, but short axons do not. The gray matter of the CNS is gray because it contains no myelinated axons; the white matter of the CNS is white because it does. In the PNS, myelin gives nerve fibers their white, glistening appearance and serves as an excellent insulator. The myelin sheath also plays an important role in nerve regeneration within the PNS. If an axon is accidentally severed, the myelin sheath remains and serves as a passageway for new fiber growth.

The Nerve Signal

Nerve signals convey information within the nervous system. In the past, nerve signals could be studied only in excised neurons (i.e., neurons taken from the body). Sophisticated techniques now enable researchers to study nerve signals in single, intact nerve cells.

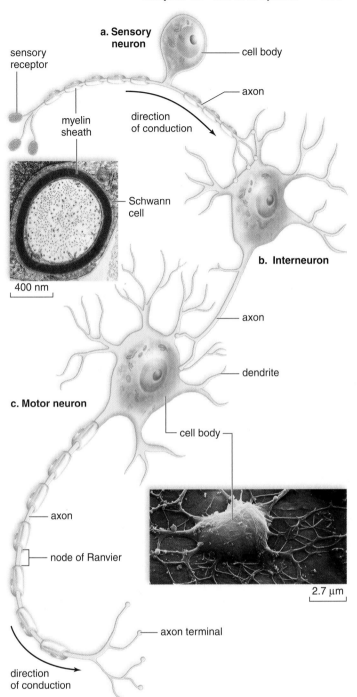

Figure 13.2 How do the three different types of neurons differ in structure?

a. A sensory neuron has a long axon covered by a myelin sheath that takes nerve impulses all the way from dendrites to the CNS. **b.** In the CNS, some interneurons, such as this one, have a short axon that is not covered by a myelin sheath. **c.** A motor neuron has a long axon covered by a myelin sheath that takes nerve impulses from the CNS to an effector.

Resting Potential

Think of all the devices you use that are powered by a battery: your calculator, your cell phone, a digital camera, a garage door opener, and countless others. Every battery is an energy source manufactured by separating positively charged ions

Figure 13.3   How does a stimulus translate resting potential energy into an action potential?
a. Resting potential occurs when a neuron is not conducting a nerve impulse. During an action potential, **(b)** the stimulus causes the cell to reach its threshold. **c.** Depolarization is followed by **(d)** repolarization. **e.** Visualizing an action potential.

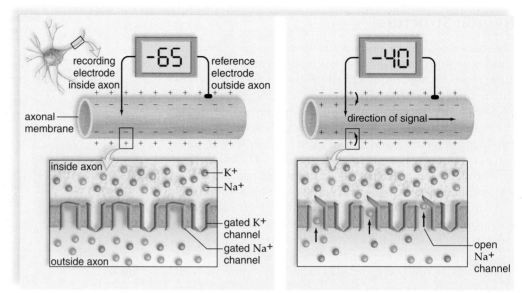

a. Resting potential: Na⁺ outside the axon, K⁺ and large anions inside the axon. Separation of charges polarizes the cell and causes the resting potential.

b. Stimulus causes the axon to reach its threshold; the axon potential increases from -65 to -40. The action potential has begun.

across a membrane from negative ions. The battery's *potential energy* can be used to perform work: running your calculator or lighting a flashlight, for example. A resting neuron also has potential energy, much like a fully charged battery. This energy, called the **resting potential,** exists because the cell membrane is *polarized:* positively charged ions are stashed outside the cell, negatively charged ions inside.

As Figure 13.3*a* shows, the outside of the cell is positive because positively charged sodium (Na^+) ions gather around the outside of the cell membrane. At rest, the neuron's cell membrane is permeable to potassium, but not to sodium. Thus, positively charged potassium ions contribute to the positive charge by diffusing out of the cell to join the sodium ions. The inside of the cell is negative because of the presence of large, negatively charged proteins and other molecules, stuck inside the cell because of their size.

Like a battery, the neuron's resting potential energy can be measured in volts. While a D-size flashlight battery has 1.5 volts, a nerve cell typically has 0.065 volts, or 65 millivolts (mV), of stored energy (see Fig. 13.3*a*). By convention, the voltage measurement is always a negative number. This is because scientists compare the inside of the cell (where negatively charged proteins and other large molecules cluster) to the outside of the cell (where positively charged sodium and potassium ions are gathered).

Just like rechargeable batteries, neurons must maintain their resting potential to be able to work. To do so, neurons actively transport sodium ions out of the cell and return potassium to the cytoplasm. A protein carrier in the membrane, called the **sodium-potassium pump,** pumps sodium (Na^+) out of the neuron and potassium (K^+) into the neuron. This action effectively "recharges" the cell, so like a fresh battery, it can perform work.

Action Potential

The resting potential energy of the neuron can be used to perform the work of the neuron: conduction of nerve signals. The process of conduction is termed an **action potential,** and it occurs in the axons of neurons. A **stimulus** activates the neuron and begins the action potential. For example, a stimulus for pain neurons in the skin would be pricking the skin with a sharp pin. However, the stimulus must be strong enough to cause the cell to reach **threshold,** the voltage that will result in an action potential. In Figure 13.3*b*, the threshold voltage is −40 mV. An action potential is an all-or-nothing event. Once threshold is reached, the action potential happens automatically and completely. On the other hand, if the threshold voltage is never reached, the action potential will not occur. Increasing the strength of a stimulus (like pressing harder with the pin) won't change the strength of an action potential. However, it may cause more action potentials to occur in a given period (and as a result, the person will perceive that pain has increased).

Sodium Gates Open   Protein channels specific for sodium ions are located in the cell membrane of the axon. When an action potential begins in response to a threshold stimulus, these protein channels open, and sodium ions rush into the cell. Adding positively charged sodium ions causes the inside of the axon to become positive, compared to the outside (see Fig. 13.3*c*). This change is called **depolarization** because the charge inside the axon (its polarity) changes from negative to positive.

Potassium Gates Open   Almost immediately after depolarization, the channels for sodium close and a separate set of potassium protein channels opens. Potassium flows rapidly from the cell. As positively charged potassium ions exit the cell, the inside of the cell becomes negative again (due to the presence of large negatively charged ions trapped inside the cell). This change in polarity is called **repolarization,** because the inside of the axon resumes a negative charge as potassium exits the axon (Fig. 13.3*d*). Finally, the sodium-potassium pump completes the action potential. Potassium is returned to the inside and sodium to the outside of the cell, and resting potential is restored.

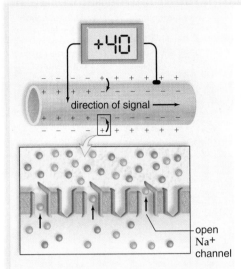 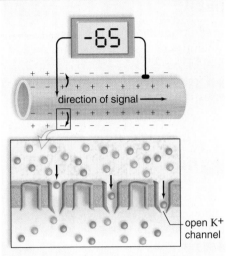 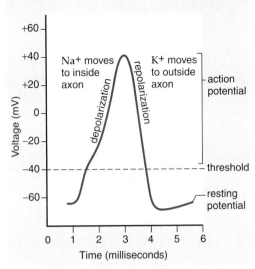

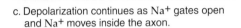

c. Depolarization continues as Na⁺ gates open and Na⁺ moves inside the axon.

d. Action potential ends: repolarization occurs when K⁺ gates open and K⁺ moves to outside the axon. The sodium-potassium pump returns the ions to their resting positions.

e. An action potential can be visualized if voltage changes are graphed over time.

Visualizing an Action Potential

To visualize such rapid fluctuations in voltage across the axonal membrane, researchers generally find it useful to plot the voltage changes over time (Fig. 13.3e). During depolarization the voltage increases from -65 mV to -40 mV to $+45$ mV, as Na⁺ moves to the inside of the axon. In repolarization, the opposite change occurs when potassium leaves the axon. The entire process is very rapid, requiring only 3–4 milliseconds to complete.

CASE STUDY PHYSICAL THERAPY

"Come on, Coach, one more rep! One more time for Doug! You go, girl!" Jill's physical therapist encouraged.

Jill extended her legs, raising the weights again. She reflected sourly that the job of high school math teacher/girl's soccer coach had its benefits, but it had some disadvantages too. It was exciting to see her former students thrive in technical programs, colleges, and universities, and then move on to graduate and professional schools. All over the community there were successful people like Doug Krasnewsky, who had studied algebra 2 and trigonometry in her classroom. Doug ended up being her physical therapist—a wonderful, supportive therapist, but one who was constantly, annoyingly cheerful and called her Coach.

She tried to control the weights as they lowered, but they clattered down. "Sorry, Doug, this hasn't been my day," she sighed.

The young man pulled up a chair and sat next to her weight bench. "This whole thing is pretty rough, Coach," he said sympathetically. "It stinks. I wish I could make it stop or go away for you."

Jill gulped hard. "Thank you, Doug. Wish I could, too." She tried hard to smile. "Okay, smart guy, I want you to think back to all those college classes and tell me how this happened."

"I have to tell you, I didn't remember much, so I surfed the Internet," he smiled. "Turns out this whole thing is an autoimmune response, where the immune system attacks nerves. Or maybe not so much the nerves themselves, but their coating. The coating is called the myelin sheath and acts like insulation around a nerve cell. If it's damaged, the nerve signal is slowed way, way down. That's what causes all of your symptoms.

"The nerve damage can happen in any part of your body, and it comes and goes—but you already know that. No one really knows why it happens," Doug continued. "When you're in relapse, we really have to work to maintain and build muscle so you can stay as mobile as possible. That's why I'm pestering you so much. How's your diet? Beth Olson over at University Hospital is your dietician, right? She was in the class the year behind me."

"All you kids are wonderful. Beth has me on the healthiest diet I've ever had in my life, along with some of the hugest vitamin pills I've ever had to choke down."

Doug studied her. "Your doctor put you on an antidepressant, right?"

"Yeah. I just wish it worked better," Jill smiled sadly. "Still, my family has been wonderful—even the kids have stayed strong for me. And I'm glad the MS hasn't hit my mind, and I can still work and coach. I can't show those soccer girls any moves, but they just need someone to yell at them anyway."

"I'm gonna get you off that power scooter, Coach," Doug reassured. "You can take it to the bank."

Propagation of an Action Potential

If an axon is unmyelinated, an action potential at one locale stimulates an adjacent part of the axon membrane to produce an action potential. Conduction along the entire axon in this fashion can be rather slow (approximately 1 meter/second in thin axons) because each section of the axon must be stimulated.

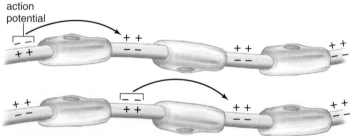

In myelinated fibers, an action potential at one node of Ranvier causes an action potential at the next node, jumping over the entire myelin-coated portion of the axon. This type of conduction is called **saltatory conduction** (*saltatio* is a Latin word that means "to jump") and is much faster. In thick, myelinated fibers, the rate of transmission is more than 100 m/sec. Regardless of whether an axon is myelinated or not, its action potentials are self-propagating. Each action potential generates another, along the entire length of the axon.

Like the action potential itself, conduction of an action potential is an all-or-none event—either an axon conducts its action potential or it does not. The intensity of a message is determined by how many action potentials are generated within a given time. An axon can conduct a volley of action potentials very quickly, because only a small number of ions are exchanged with each action potential. Once the action potential is complete, the ions are rapidly restored to their proper place through the action of the sodium-potassium pump.

As soon as the action potential has passed by each successive portion of an axon, that portion undergoes a short **refractory period** during which it is unable to conduct an action potential. This ensures the one-way direction of a signal, from the cell body down the length of the axon, to the axon terminal.

It is interesting to observe that all functions of the nervous system, from our deepest emotions to our highest reasoning abilities, are dependent on the conduction of nerve signals.

The Synapse

Every axon branches into many fine endings, each tipped by a small swelling called an **axon terminal.** Each terminal lies very close to either the dendrite or the cell body of another neuron. This region of close proximity is called a **synapse** (Fig. 13.4). At a synapse, a small gap called the **synaptic cleft** separates the sending neuron from the receiving neuron.

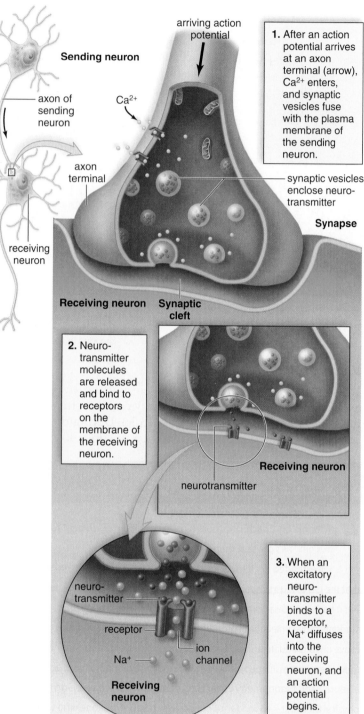

Figure 13.4   How does signal transmission occur at a synapse?

Transmission across a synapse from one neuron to another occurs when a neurotransmitter is released and diffuses across a synaptic cleft and binds to a receptor in the membrane of the receiving neuron.

The nerve signal is unable to jump the cleft. Therefore, another means is needed to pass the nerve signal from the sending neuron to the receiving neuron.

Transmission across a synapse is carried out by molecules called **neurotransmitters,** stored in synaptic vesicles in the axon terminals. (Recall from Chapter 3 that a vesicle is a membranous sac that stores and transports substances.) The events at a synapse are: (1) Nerve signals traveling along an axon to reach an axon terminal. (2) Calcium ions entering the terminal and stimulating synaptic vesicles to merge with the sending membrane. (3) Neurotransmitter molecules releasing into the synaptic cleft and diffusing across the cleft to the receiving membrane. There, neurotransmitter molecules bind with specific receptor proteins. Depending on the type of neurotransmitter, the response of the receiving neuron can be toward excitation or toward inhibition. In Figure 13.5, excitation occurs because the neurotransmitter, such as acetylcholine (ACh), has caused the sodium gate to open. Sodium ions diffuse into the receiving neuron. Inhibition would occur if a neurotransmitter caused K^+ to enter the receiving neuron.

Once a neurotransmitter has been released into a synaptic cleft and has initiated a response, it is removed from the cleft. In some synapses, the receiving membrane contains enzymes that rapidly inactivate the neurotransmitter. For example, the enzyme **acetylcholinesterase (AChE)** breaks down acetylcholine. In other synapses, the sending membrane rapidly reabsorbs the neurotransmitter, possibly for repackaging in synaptic vesicles or for molecular breakdown.

The short existence of neurotransmitters at a synapse prevents continuous stimulation (or inhibition) of receiving membranes. The receiving cell needs to be able to respond quickly to changing conditions. If the neurotransmitter were to linger in the cleft, the receiving cell would be unable to respond to a new signal from a sending cell.

Neurotransmitter Molecules

Among the more than 100 substances known or suspected to be neurotransmitters are **acetylcholine (ACh), norepinephrine (NE), dopamine, serotonin, glutamate,** and **GABA (gamma aminobutyric acid).** Neurotransmitters transmit signals between nerves. Nerve-muscle, nerve-organ, and nerve-gland synapses also communicate using neurotransmitters.

Acetylcholine and norepinephrine are active in both the CNS and PNS. In the PNS, these neurotransmitters act at synapses called neuromuscular junctions. Neuromuscular junctions are discussed on page 261.

In the PNS, ACh excites skeletal muscle but inhibits cardiac muscle. It has either an excitatory or inhibitory effect on smooth muscle or glands, depending on their location.

Norepinephrine generally excites smooth muscle. In the CNS, norepinephrine is important to dreaming, waking, and mood. Serotonin is involved in thermoregulation, sleeping, emotions, and perception. Many drugs that affect the nervous system act at the synapse. Some interfere with the actions of neurotransmitters, while other drugs prolong the effects of neurotransmitters (see page 295).

Synaptic Integration

A single neuron has a cell body and may have many dendrites (Fig. 13.5a). All can have synapses with many other neurons. Therefore, a neuron is on the receiving end of

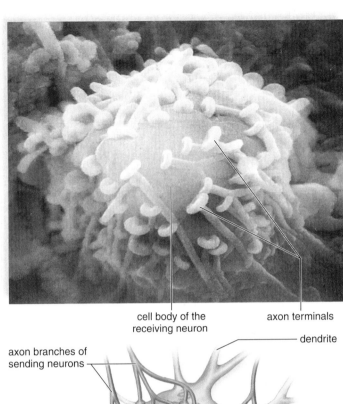
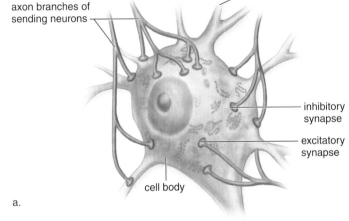

cell body of the receiving neuron / axon terminals / dendrite / axon branches of sending neurons / inhibitory synapse / excitatory synapse / cell body

a.

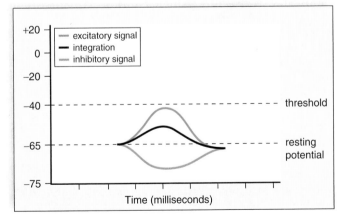

b.

Figure 13.5 How are excitatory and inhibitory signals integrated at the synapse?
a. Inhibitory signals and excitatory signals are summed up in the dendrite and cell body of the postsynaptic neuron. Only if the combined signals cause the membrane potential to rise above threshold does an action potential occur. **b.** In this example, threshold was not reached.

Artist and Scientist: Santiago Ramon y Cajal (1852–1934)

In any discipline—humanities, music and the arts, history, mathematics, education, social sciences—innovation and discovery happen because of the hard work of dedicated people. The contributions of the passionate scientist Dr. Santiago Ramon y Cajal established a foundation for ongoing studies of the nervous system.

In the late nineteenth century, the brain was believed to be a continuous network of "filaments," and scientists were not convinced that the filaments were even cells. Using a new technique, Cajal stained samples of brain tissue with a dye containing metallic silver. Careful microscopic studies showed Cajal that the brain was composed of individual cells. A later researcher named nerve cells *neurons*. Cajal then discovered that the neurons were not directly connected to one another. This discovery allowed later scientists to research this gap—the synapse—as well as the neurotransmitters that allow nerve cells to communicate across a synapse. An artist as well as a researcher, Cajal illustrated his microscopic discoveries. His drawings were reproduced in textbooks for decades.

Cajal's theory of "dynamic polarization" described the idea of the resting and action potential. He proposed that neu-

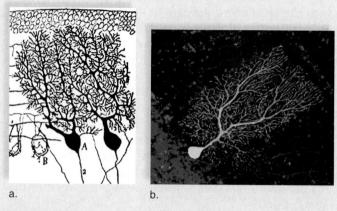

Figure 13B

a. Santiago Ramon y Cajal's sketch of neuron. **b.** Actual photomicrograph of the same neuron.

rons received signals at the cell body and dendrites and transmitted these signals via their axons to other neurons. Cajal described this basic principle of neuron function long before methods were devised to prove his theories.

Cajal was awarded the Nobel Prize in 1906 for his discoveries in the structure and function of the nervous system.

---

many signals, which can either be excitatory or inhibitory. Recall that an excitatory neurotransmitter produces an excitatory signal by opening sodium gates at a synapse. This drives the neuron closer to its threshold (illustrated by the green line on Fig. 13.5*b*). If threshold is reached, an action potential is inevitable. On the other hand, an inhibitory neurotransmitter drives the neuron farther from an action potential (red line on Fig. 13.5*b*) by opening the gates for potassium.

Neurons integrate these incoming signals. **Integration** is the summing up of excitatory and inhibitory signals. If a neuron receives enough excitatory signals (either from different synapses or at a rapid rate from a single synapse) to outweigh the inhibitory ones, chances are the axon will transmit a signal. On the other hand, if a neuron receives more inhibitory than excitatory signals, summing these signals may prohibit the axon from reaching threshold and then depolarizing (black solid line on Fig. 13.5*b*).

Check Your Progress 13.1

1. What are the two major divisions of the nervous system?
2. a. What are the three types of neurons, and (b) what are the three parts of a neuron?
3. a. What is a nerve impulse, and (b) how is it propagated?
4. Neurons don't physically touch, so how is an impulse transmitted from one neuron to the next?

13.2 The Central Nervous System

The spinal cord and the brain make up the CNS, where sensory information is received and motor control is initiated. As mentioned previously, both the spinal cord and the brain are protected by bone. The spinal cord is surrounded by vertebrae,

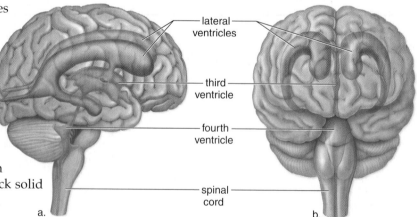

Figure 13.6 Where are the ventricles of the brain, and how do they interconnect?

The brain has four ventricles. A lateral ventricle is found on each side of the brain. They join at the third ventricle. The third ventricle connects with the fourth ventricle superiorly; the central canal of the spinal cord joins the fourth ventricle inferiorly. All structures are filled with cerebrospinal fluid. **a.** Lateral view of ventricles seen through a transparent brain. **b.** Anterior view of ventricles seen through a transparent brain.

and the brain is enclosed by the skull. Also, both the spinal cord and the brain are wrapped in protective membranes known as **meninges** (sing., meninx). *Meningitis* is an infection of the meninges. The spaces between the meninges are filled with **cerebrospinal fluid,** which cushions and protects the CNS. A small amount of this fluid is sometimes withdrawn from around the spinal cord for laboratory testing when a spinal tap (lumbar puncture) is performed.

Cerebrospinal fluid is also contained within the ventricles of the brain and in the central canal of the spinal cord. The brain has four **ventricles,** interconnecting chambers that produce and serve as a reservoir for cerebrospinal fluid (Fig. 13.6). Normally, any excess cerebrospinal fluid drains away into the cardiovascular system. However, blockages can occur. In an infant, the brain can enlarge due to cerebrospinal fluid accumulation, resulting in a condition called hydrocephalus ("water on the brain"). If cerebrospinal fluid collects in an adult, the brain cannot enlarge. Instead, it is pushed against the skull.

Such situations will cause severe brain damage and will be fatal unless quickly corrected.

The CNS is composed of two types of nervous tissue—gray matter and white matter. **Gray matter** contains cell bodies and short, nonmyelinated fibers. **White matter** contains myelinated axons that run together in bundles called **tracts.**

The Spinal Cord

The **spinal cord** extends from the base of the brain through a large opening in the skull called the foramen magnum (see pages 238–39). From the foramen magnum, the spinal cord proceeds inferiorly in the vertebral canal.

Structure of the Spinal Cord

A cross section of the spinal cord shows a central canal, gray matter, and white matter (Fig. 13.7*a*). Figure 13.7*b* shows how an individual vertebra protects the spinal cord. The spinal

Figure 13.7 How are white and gray matter organized to form the spinal cord and the spinal nerves?
a. Cross section of the spinal cord, showing arrangements of white and gray matter. **b.** Spinal nerves originating from the spinal cord. **c.** The spinal cord is protected by vertebrae. **d.** Spinal nerves emerging from the cord.

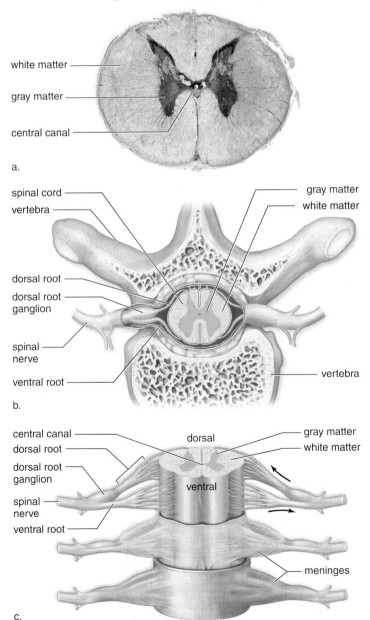

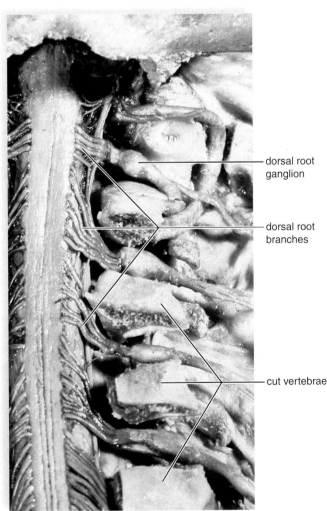

d. Dorsal view of spinal cord and dorsal roots of spinal nerves.

nerves project from the cord through intervertebral foramina (page 240). Fibrocartilage intervertebral disks separate the vertebrae. Should the disk rupture or herniate, the vertebra will compress a spinal nerve. Pain and loss of mobility result.

The central canal of the spinal cord contains cerebrospinal fluid, as do the meninges that protect the spinal cord. The gray matter is centrally located and shaped like the letter H (Fig. 13.7a, b, and c). Portions of sensory neurons and motor neurons are found in gray matter, as are interneurons that communicate with these two types of neurons. The dorsal root of a spinal nerve contains sensory fibers entering the gray matter. The ventral root of a spinal nerve contains motor fibers exiting the gray matter. The dorsal and ventral roots join before the spinal nerve leaves the vertebral canal, forming a mixed nerve (Fig. 13.7c, d). Spinal nerves are a part of the PNS.

The white matter of the spinal cord occurs in areas around the gray matter. The white matter contains ascending tracts taking information to the brain (primarily located posteriorly) and descending tracts taking information from the brain (primarily located anteriorly). Many tracts cross just after they enter and exit the brain, so the left side of the brain controls the right side of the body. Likewise, the right side of the brain controls the left side of the body.

Functions of the Spinal Cord

The spinal cord provides a means of communication between the brain and the peripheral nerves that leave the cord. When someone touches your hand, sensory receptors generate nerve signals that pass through sensory fibers to the spinal cord and up ascending tracts to the brain (see Fig. 13.1b, red arrows).

The gate control theory of pain proposes that the tracts in the spinal cord have "gates," and that these "gates" control the flow of pain messages from the peripheral nerves to the brain. Depending on how the gates process a pain signal, the pain message can be allowed to pass directly to the brain or can be prevented from reaching the brain. As mentioned earlier, endorphins can temporarily block pain messages and so can other messages, such as those received from touch receptors.

To touch the person back, the brain initiates voluntary control over our limbs. Motor signals originating in the brain pass down descending tracts to the spinal cord and out to our muscles by way of motor fibers (see Fig. 13.1b, black arrows). Therefore, if the spinal cord is severed, we suffer a loss of sensation and a loss of voluntary control—paralysis. If the cut occurs in the thoracic region, the lower body and legs are paralyzed, a condition known as paraplegia. If the injury is in the neck region, all four limbs are usually affected, a condition called quadriplegia.

Reflex Actions The spinal cord is the center for thousands of reflex arcs (see Fig. 13.16). A stimulus causes sensory receptors to generate signals that travel in sensory axons to the spinal cord. Interneurons integrate the incoming data and relay signals to motor neurons. A response to the stimulus occurs when motor axons cause skeletal muscles to contract.

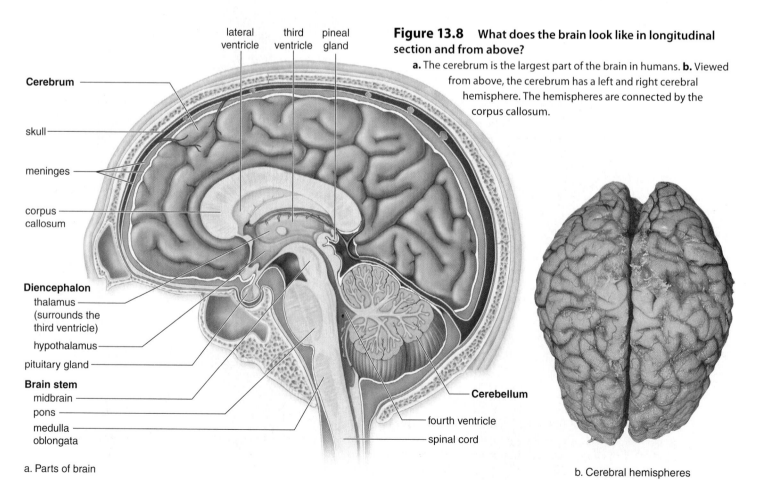

Figure 13.8 **What does the brain look like in longitudinal section and from above?**

a. The cerebrum is the largest part of the brain in humans. **b.** Viewed from above, the cerebrum has a left and right cerebral hemisphere. The hemispheres are connected by the corpus callosum.

lateral ventricle · third ventricle · pineal gland

Cerebrum

skull

meninges

corpus callosum

Diencephalon
thalamus (surrounds the third ventricle)
hypothalamus
pituitary gland

Brain stem
midbrain
pons
medulla oblongata

Cerebellum

fourth ventricle

spinal cord

a. Parts of brain

b. Cerebral hemispheres

Motor neurons in a reflex arc may also affect smooth muscle, organs, or glands. Each interneuron in the spinal cord has synapses with many other neurons. Therefore, interneurons send signals to other interneurons and motor neurons.

Similarly, the spinal cord creates reflex arcs for the internal organs. For example, when blood pressure falls, internal receptors in the carotid arteries and aorta generate nerve signals that pass through sensory fibers to the cord and then up an ascending tract to a cardiovascular center in the brain. Thereafter, nerve signals pass down a descending tract to the spinal cord. Motor signals then cause blood vessels to constrict so that the blood pressure rises.

The Brain

The human **brain** has been called the last great frontier of biology. The goal of modern neuroscience is to understand the structure and function of the brain's various parts so well that it will be possible to prevent or correct the thousands of mental disorders that rob human beings of a normal life. This section gives only a glimpse of what is known about the brain and the modern avenues of research.

We will discuss the parts of the brain with reference to the cerebrum, the diencephalon, the cerebellum, and the brain stem. The brain's four ventricles (described on pages 282–83) are called, in turn, the two lateral ventricles, the third ventricle, and the fourth ventricle. It may be helpful for you to associate the cerebrum with the two lateral ventricles, the diencephalon with the third ventricle, and the brain stem and the cerebellum with the fourth ventricle (Fig. 13.8*a*).

The Cerebrum

The **cerebrum,** also called the telencephalon, is the largest portion of the brain in humans. The cerebrum is the last center to receive sensory input and carry out integration before commanding voluntary motor responses. It communicates with and coordinates the activities of the other parts of the brain.

Cerebral Hemispheres   Just as the human body has two halves, so does the cerebrum. These halves are called the left and right **cerebral hemispheres** (Fig. 13.8*b*). A deep groove called the longitudinal fissure divides the left and right cerebral hemispheres. The two cerebral hemispheres communicate via the **corpus callosum,** an extensive bridge of nerve tracts.

Shallow grooves called sulci (sing., sulcus) divide each hemisphere into lobes (Fig. 13.9). The *frontal lobe* is the most anterior of the lobes (directly behind the forehead). The *parietal lobe* is posterior to the frontal lobe. The *occipital lobe* is

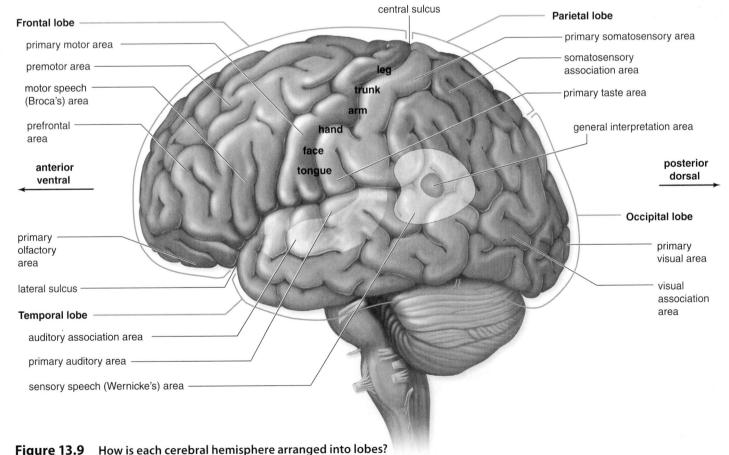

Figure 13.9 **How is each cerebral hemisphere arranged into lobes?**
Each cerebral hemisphere is divided into four lobes: frontal, parietal, temporal, and occipital. Centers in the frontal lobe control movement and higher reasoning, as well as the smell sensation. Somatic sensing is carried out by parietal lobe neurons, while those of the temporal lobe receive sound information. Visual information is received and processed in the occipital lobe.

posterior to the parietal lobe (at the rear of the head). The *temporal lobe* lies inferior to the frontal and parietal lobes (at the temple and the ear).

Each lobe is associated with particular functions, as indicated in Figure 13.9.

The Cerebral Cortex The **cerebral cortex** is a thin, highly convoluted outer layer of gray matter that covers the cerebral hemispheres. (Recall that gray matter consists of neurons whose axons are unmyelinated.) The cerebral cortex contains over 1 billion cell bodies and is the region of the brain that accounts for sensation, voluntary movement, and all the thought processes we associate with consciousness.

Primary Motor and Sensory Areas of the Cortex The cerebral cortex contains motor areas and sensory areas, as well as association areas. The **primary motor area** is in the frontal lobe just anterior to (before) the central sulcus. Voluntary commands to skeletal muscles begin in the primary motor area, and each part of the body is controlled by a certain section (Fig. 13.10*a*). Observe the illustration carefully. You'll see that large areas of cerebral cortex are devoted to controlling structures that carry out very fine, precise movements. Thus, the muscles that control facial movements—swallowing, salivation, expression—take up an especially large portion of the primary motor area. Likewise, hand movements require tremendous accuracy. Together, these two structures command nearly two-thirds of the primary motor area.

The **primary somatosensory area** is just posterior to the central sulcus in the parietal lobe. Sensory information from the skin and skeletal muscles arrives here, where each part of the body is sequentially represented (Fig. 13.10*b*). Like the primary motor cortex, large areas of the primary sensory cortex are dedicated to those body areas with acute sensation. Once again, the face and hands require the largest proportion of the sensory cortex.

Reception areas for the other primary sensations—taste, vision, hearing and smell—are located in other areas of the cerebral cortex (see Fig. 13.9). The primary taste area in the parietal lobe (pink) accounts for taste sensations. Visual information is received by the primary visual cortex (blue) in the occipital lobe. The primary auditory area in the temporal lobe (dark green) accepts information from our ears. Smell sensations travel to the primary olfactory area (yellow) found on the deep surface of the frontal lobe.

Association Areas Association areas are places where integration occurs and where memories are stored. Anterior to the primary motor area is a premotor area. The premotor area organizes motor functions for skilled motor activities, such as walking and talking at the same time. Next, the primary motor area sends signals to the cerebellum, which integrates them. A momentary lack of oxygen during birth can damage the motor areas of the cerebral cortex. Cerebral palsy, a condition characterized by a spastic weakness of the arms and legs, results. The *somatosensory association area*, located just posterior to the primary somatosensory area, processes and analyzes sensory information from the skin and muscles. The *visual association area* in the occipital lobe associates new visual information with stored visual memories. It might "decide," for example, if we have seen a face, scene, or symbol

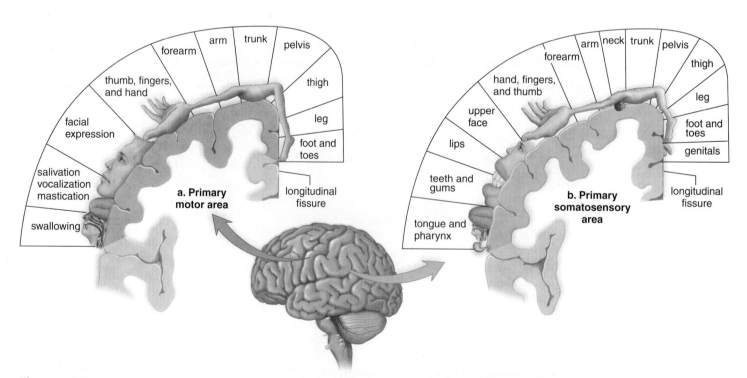

Figure 13.10 **Where are the primary motor and primary somatosensory areas located on the brain's surface?**
a. The primary motor area (blue) is located in the frontal lobe, adjacent to **(b)** the primary somatosensory area in the parietal lobe. The primary taste area is colored pink. The size of each body region shown indicates the relative amount of cortex devoted to control of that body region.

before. The *auditory association area* in the temporal lobe performs the same functions with regard to sounds.

Processing Centers Processing centers of the cortex receive information from the other association areas and perform higher-level analytical functions. The **prefrontal area,** an association area in the frontal lobe, receives information from the other association areas and uses this information to reason and plan our actions. Integration in this area accounts for our most cherished human abilities. Reasoning, critical thinking, and formulating appropriate behaviors are possible because of integration carried out in the prefrontal area.

The unique ability of humans to speak is partially dependent upon two processing centers found only in the left cerebral cortex. **Wernicke's area** is located in the posterior part of the left temporal lobe. **Broca's area** is located in the left frontal lobe. Broca's area is located just anterior to the portion of the primary motor area for speech musculature (lips, tongue, larynx, and so forth) (see Fig. 13.9). Wernicke's area helps us understand both the written and spoken word and sends the information to Broca's area. Broca's area adds grammatical refinements and directs the primary motor area to stimulate the appropriate muscles for speaking and writing.

Central White Matter Much of the rest of the cerebrum is composed of white matter. Myelination occurs and white matter develops as a child grows. Progressive myelination enables the brain to grow in size and complexity. For example, as neurons become myelinated within tracts designed for language development, children become more capable of speech. Descending tracts from the primary motor area communicate with lower brain centers, and ascending tracts from lower brain centers send sensory information up to the primary somatosensory area. Tracts within the cerebrum also take information between the different sensory, motor, and association areas pictured in Figure 13.9. As previously mentioned, the corpus callosum contains tracts that join the two cerebral hemispheres.

Basal Nuclei

Though the majority of each cerebral hemisphere is composed of tracts, there are masses of gray matter deep within the white matter. These **basal nuclei** integrate motor commands to ensure that the proper muscle groups are stimulated or inhibited. Integration ensures that movements are coordinated and smooth. **Parkinson disease** (see Chap. 17) is believed to be caused by degeneration of specific neurons in the basal nuclei.

The Diencephalon

The hypothalamus and the thalamus are in the **diencephalon,** a region that encircles the third ventricle. The **hypothalamus** forms the floor of the third ventricle. The hypothalamus is an integrating center that helps maintain homeostasis. It regulates hunger, sleep, thirst, body temperature, and water balance. The hypothalamus controls the

Have You Ever Wondered...

Why does a stroke on the right side of the brain cause weakness or paralysis, as well as decreased sensation, on the left side of the body?

Descending motor tracts (from the primary motor area) and ascending sensory tracts (from the primary somatosensory area) cross over in the spinal cord and medulla. Motor neurons in the right cerebral hemisphere control the left side of the body, and vice versa because of crossing over. Likewise, sensation from the left half of the body will travel to the right cerebral hemisphere. Destruction of brain tissue by a stroke interferes with outgoing motor signals to the opposite side of the body, as well as incoming sensory information.

pituitary gland and thereby, serves as a link between the nervous and endocrine systems.

The **thalamus** consists of two masses of gray matter located in the sides and roof of the third ventricle. The thalamus is on the receiving end for all sensory input except the sense of smell. Visual, auditory, and somatosensory information arrives at the thalamus via the cranial nerves and tracts from the spinal cord. The thalamus integrates this information and sends it on to the appropriate portions of the cerebrum. The thalamus is involved in arousal of the cerebrum, and it also participates in higher mental functions such as memory and emotions.

The pineal gland, which secretes the hormone melatonin, is located in the diencephalon. Presently there is much popular interest in the role of melatonin in our daily rhythms. Some researchers believe it can help alleviate jet lag or insomnia. Scientists are also interested in the possibility that the hormone may regulate the onset of puberty.

The Cerebellum

The **cerebellum** lies under the occipital lobe of the cerebrum and is separated from the brain stem by the fourth ventricle. The cerebellum has two portions joined by a narrow median portion. Each portion is primarily composed of white matter. In longitudinal section, the white matter has a treelike pattern called *arbor vitae*. Overlying the white matter is a thin layer of gray matter that forms a series of complex folds.

The cerebellum receives sensory input from the eyes, ears, joints, and muscles about the present position of body parts. It also receives motor output from the cerebral cortex about where these parts should be located. After integrating this information, the cerebellum sends motor signals by way of the brain stem to the skeletal muscles. In this way, the cerebellum maintains posture and balance. It also ensures that all of the muscles work together to produce smooth, coordinated, voluntary movements. The cerebellum assists the learning of new motor skills such as playing the piano or hitting a baseball.

The Brain Stem

The tracts cross in the **brain stem,** which contains the midbrain, the pons, and the medulla oblongata (see Fig. 13.8a). The **midbrain** acts as a relay station for tracts passing between the cerebrum and the spinal cord or cerebellum. It also has reflex centers for visual, auditory, and tactile responses. The word **pons** means "bridge" in Latin. True to its name, the pons contains bundles of axons traveling between the cerebellum and the rest of the CNS. In addition, the pons functions with the medulla oblongata to regulate breathing rate. Reflex centers in the pons coordinate head movements in response to visual and auditory stimuli.

The **medulla oblongata** contains a number of reflex centers for regulating heartbeat, breathing, and vasoconstriction (blood pressure). It also contains the reflex centers for vomiting, coughing, sneezing, hiccuping, and swallowing. The medulla oblongata lies just superior to the spinal cord, and it contains tracts that ascend or descend between the spinal cord and higher brain centers. Recall that tracts are groups of axons that travel together (page 283). Ascending tracts convey sensory information. Motor information is transmitted on descending tracts.

The Reticular Formation   The **reticular formation** is a complex network of **nuclei** (masses of gray matter) and fibers that extend the length of the brain stem (Fig. 13.11). The reticular formation is

a major component of the reticular activating system (RAS). The RAS receives sensory signals and sends them to higher centers. Motor signals received by the RAS are sent to the spinal cord.

The RAS arouses the cerebrum via the thalamus and causes a person to be alert. If you want to awaken the RAS, surprise it with sudden stimuli, such as an alarm clock ringing, bright lights, smelling salts, or splashing cold water on your face. The RAS can filter out unnecessary sensory stimuli, explaining why you can study with the TV on. Similarly, the RAS allows you to take a test without noticing the sounds of the people around you—unless the sounds are particularly distracting. To inactivate the RAS, remove visual or auditory stimuli, allowing yourself to become drowsy and drop off to sleep. General anesthetics function by artificially suppressing the RAS. A severe injury to the RAS can cause a person to be comatose, from which recovery may be impossible.

> **Check Your Progress 13.2**
>
> 1. What structures compose the central nervous system?
> 2. What are the two main functions of the spinal cord?
> 3. What are the four major parts of the brain and the general function of each?
> 4. What brain structure is responsible for a person being awake and alert, asleep, or comatose?

13.3 The Limbic System and Higher Mental Functions

The limbic system is intimately involved in our emotions and higher mental functions. After a short description, we will discuss the functions of the limbic system.

Limbic System

The **limbic system** is an evolutionary ancient group of linked structures deep within the cerebrum. It is a functional grouping rather than an anatomical one (Fig. 13.12). The limbic system blends primitive emotions and higher mental functions into a united whole. It accounts for why activities such as sexual behavior and eating seem pleasurable. Conversely, unpleasant sensations or emotions (pain, frustration, hatred, despair) are translated by the limbic system into a stress response.

Two significant structures within the limbic system are the amygdala and the hippocampus. The **amygdala,** in particular, can cause experiences to have emotional overtones, and it creates the sensation of fear. This center can use past knowledge fed to it by association areas to assess a current situation. If necessary, the amygdala can trigger the fight-or-flight reaction. So if you are out late at night, and you turn to see someone in a ski mask following you, the amygdala may immediately cause you to start running. The frontal cortex can override the limbic system and cause us to rethink the situation and prevent us from acting out strong reactions.

The **hippocampus** is believed to play a crucial role in learning and memory. The hippocampal region acts as an in-

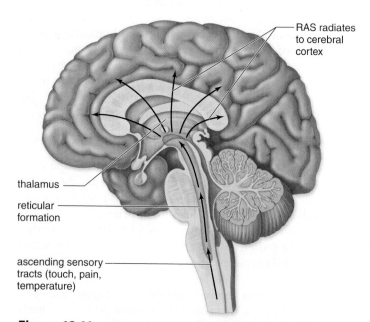

thalamus

reticular formation

ascending sensory tracts (touch, pain, temperature)

RAS radiates to cerebral cortex

Figure 13.11   **What is the function of the reticular formation?** The reticular formation receives and sends on motor and sensory information to various parts of the CNS. One portion, the reticular activating system (RAS; see arrows), arouses the cerebrum and, in this way, controls alertness versus sleep.

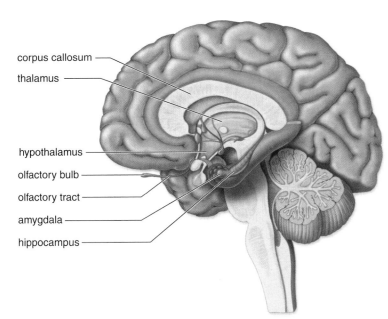

corpus callosum

thalamus

hypothalamus

olfactory bulb

olfactory tract

amygdala

hippocampus

Figure 13.12   How does the limbic system integrate emotion and reason?
In the limbic system (purple), structures deep within each cerebral hemisphere and surrounding the diencephalon join higher mental functions, such as reasoning, with more primitive feelings, such as fear and pleasure. Therefore, primitive feelings can influence our behavior, but reason can also keep them in check.

formation gateway during the learning process. It determines what information about the world is to be sent to memory and how this information is to be encoded and stored by other regions in the brain. Most likely, the hippocampus can communicate with the frontal cortex because we know that memories are an important part of our decision-making processes.

It's hoped that research into the functioning of the hippocampus will help in understanding **Alzheimer disease**, a brain disorder characterized by gradual loss of memory (see Chap. 17).

Higher Mental Functions

As in other areas of biological research, brain research has progressed due to technological breakthroughs. Neuroscientists now have a wide range of techniques at their disposal for studying the human brain, including modern technologies that allow us to record its functioning.

Memory and Learning

Just as the connecting tracts of the corpus callosum are evidence that the two cerebral hemispheres work together, so the limbic system indicates that cortical areas may work with lower centers to produce learning and memory. **Memory** is the ability to hold a thought in mind or to recall events from the past, ranging from a word we learned only yesterday to an early emotional experience that has shaped our lives. **Learning** takes place when we retain and use past memories.

Types of Memory   We have all tried to remember a seven-digit telephone number for a short time. If we say we are trying to keep it in the forefront of our brain, we are exactly correct. The prefrontal area, active during **short-term memory**, lies just posterior to our forehead! There are some telephone numbers that we have memorized. In other words, they have gone into **long-term memory**. Think of a telephone number you know by heart, and try to bring it to mind without also thinking about the place or person associated with that number. Most likely you cannot. Typically, long-term memory is a mixture of what is called **semantic memory** (numbers, words, etc.) and **episodic memory** (persons, events, etc.).

Skill memory is another type of memory that can exist independent of episodic memory. Skill memory is involved in performing motor activities such as riding a bike or playing ice hockey. When a person first learns a skill, more areas of the cerebral cortex are involved than after the skill is perfected. In other words, you have to think about what you are doing when you learn a skill, but later the actions become automatic. Skill memory involves all the motor areas of the cerebrum below the level of consciousness.

Long-Term Memory Storage and Retrieval   Our long-term memories are apparently stored in bits and pieces throughout the sensory association areas of the cerebral cortex. Visions are stored in the vision association area, sounds are stored in the auditory association area, and so forth. As previously mentioned, the hippocampus serves as a bridge between the sensory association areas (where memories are stored) and the prefrontal area (where memories are used). The prefrontal area communicates with the hippocampus, when memories are stored and when these memories are brought to mind. Why are some memories so emotionally charged? As you know, the amygdala seems to be responsible for fear conditioning and associating danger with sensory stimuli received from various parts of the brain.

Have You Ever Wondered...

What is amnesia?

Amnesia results from disruption of the memory pathways and can be temporary or permanent. In anterograde amnesia, injury to the limbic system separates long-term memories (those from events that occurred prior to the injury) from events that occur in the here-and-now. An affected person might carry on a conversation about past events (memories of a long-ago birthday) but cannot recall a breakfast menu from that morning. In retrograde amnesia, a blow to the head or similar injury abolishes all memories for a variable time before the injury. For example, a head injury occurring during a car accident may abolish all memories from hours to days prior to the accident.

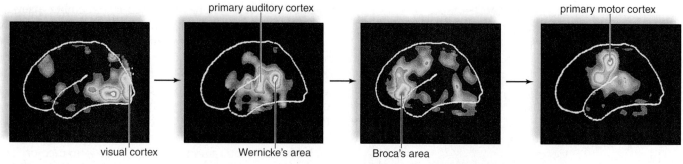

1. The word is seen in the visual cortex.

2. Information concerning the word is interpreted in Wernicke's area.

3. Information from Wernicke's area is transferred to Broca's area.

4. Information is transferred from Broca's area to the primary motor area.

Figure 13.13 **How are different areas of the brain sequentially activated so a person can read aloud?** These functional images were captured by a high-speed computer during PET (positron emission tomography) scanning of the brain. A radioactively labeled solution is injected into the subject, and then the subject is asked to perform certain activities. Cross-sectional images of the brain generated by the computer reveal where activity is occurring because the solution is preferentially taken up by active brain tissue and not by inactive brain tissue. These PET images show the cortical pathway for reading words and then speaking them. Red indicates the most active areas of the brain, and blue indicates the least active areas.

Long-Term Potentiation While it is helpful to know the memory functions of various portions of the brain, an important step toward curing mental disorders is understanding memory on the cellular level. After synapses have been used intensively for a short time, they release more neurotransmitters than before. This phenomenon, called **long-term potentiation (LTP),** may be involved in memory storage.

Language and Speech

Language depends on semantic memory. Therefore, we would expect some of the same areas in the brain to be involved in both memory and language. Any disruption of these pathways could contribute to an inability to comprehend our environment and use speech correctly.

Seeing and hearing words depends on sensory centers in the occipital and temporal lobes, respectively. Damage to Wernicke's area, discussed earlier, results in the inability to comprehend speech. Damage to Broca's area, on the other hand, results in the inability to speak and write. The functions of the visual cortex, Wernicke's area, and Broca's area are shown in Figure 13.13.

One interesting aside pertaining to language and speech is the recognition that the left brain and the right brain may have different functions. Recall that the left hemisphere contains both Broca's area and Wernicke's area. As you might expect, it appears that the left hemisphere plays a role of great importance in language functions. The role of the isolated left hemisphere can be studied in patients after surgery to sever the corpus callosum. This procedure is used for seizure control in patients with epilepsy. After surgery, the patient is termed "split brain," because there is no longer direct communication between the two cerebral hemispheres. If a split-brain individual views an object with only the right eye, its image will be sent only to the right hemisphere. This person will be able to choose the proper object for a particular use—scissors to cut paper, for example—but will be unable to name that object.

Research on the split brain is ongoing. In a very general way, the left brain can be contrasted with the right brain:

Left Hemisphere	Right Hemisphere
Verbal	Nonverbal, visuospatial
Logical, analytical	Intuitive
Rational	Creative

Researchers now believe that the hemispheres process the same information differently. The left hemisphere is more global, whereas the right hemisphere is more specific in its approach.

Check Your Progress 13.3

1. What is the function of the limbic system?
2. What limbic system structures are involved in the fight-or-flight reaction, learning, and long-term memory?
3. a. What areas of the cerebrum are involved in language and speech? b. Where are they located?

13.4 The Peripheral Nervous System

The peripheral nervous system (PNS), which lies outside the central nervous system, contains the nerves. Nerves are designated as cranial nerves when they arise from the brain and spinal nerves when they arise from the spinal cord. In any case, all nerves carry signals to and from the CNS. So right now, your eyes are sending messages by way of a cranial nerve to the brain, allowing you to read a page. When you're finished, your brain will direct the muscles in your fingers to turn the page by way of the spinal cord and a spinal nerve.

Figure 13.14 (*left*) gives the anatomy of a nerve. The cell body and the dendrites of neurons are in the CNS or ganglia.

Ganglia (sing. **ganglion**) are collections of nerve cell bodies outside the CNS. The axons of neurons project from the CNS and form the spinal cord. In other words, *nerves,* whether cranial or spinal, are composed of axons, the long part of neurons.

Humans have 12 pairs of **cranial nerves** attached to the brain. By convention, the pairs of cranial nerves are referred to by roman numerals (Fig. 13.14, *right*). Some cranial nerves are sensory nerves—they contain only sensory fibers; some are motor nerves that contain only motor fibers; and others are mixed nerves that contain both sensory and motor fibers. Cranial nerves are largely concerned with the head, neck, and facial regions of the body. However, the vagus nerve (X) has branches not only to the pharynx and larynx, but also to most of the internal organs. From which part of the brain do you think the vagus arises? It arises from the brain stem, specifically the medulla oblongata that communicates so well with the hypothalamus. These two parts of the brain control the internal organs.

As you know, the **spinal nerves** of humans emerge from either side of the spinal cord (see Fig. 13.7). There are 31 pairs of spinal nerves. The roots of a spinal nerve physically separate the axons of sensory neurons from the axons of motor neurons, forming an arrangement resembling a letter Y. The posterior root of a spinal nerve contains sensory fibers that direct sensory receptor information inward (toward the spinal cord). The cell body of a sensory neuron is in a posterior-root ganglion (also termed a **dorsal-root ganglion**). A ganglion is a collection of cell bodies outside the CNS. The anterior (also termed ventral) root of a spinal nerve contains motor fibers that conduct impulses outward (away from the cord) to the effectors. Observe in Figure 13.7 that the anterior and posterior roots rejoin to form a spinal nerve. All spinal nerves are called mixed nerves because they contain both sensory and motor fibers. Each spinal nerve serves the particular region of the body in which it is located. For example, the intercostal muscles of the rib cage are innervated by thoracic nerves.

Somatic System

The PNS has divisions, and right now we are going to consider the somatic system. The nerves in the **somatic system**

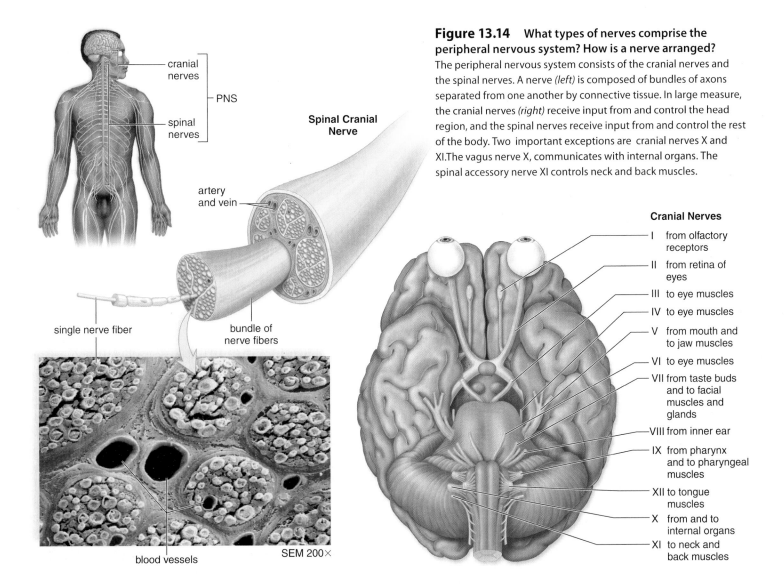

Figure 13.14   What types of nerves comprise the peripheral nervous system? How is a nerve arranged?
The peripheral nervous system consists of the cranial nerves and the spinal nerves. A nerve *(left)* is composed of bundles of axons separated from one another by connective tissue. In large measure, the cranial nerves *(right)* receive input from and control the head region, and the spinal nerves receive input from and control the rest of the body. Two important exceptions are cranial nerves X and XI.The vagus nerve X, communicates with internal organs. The spinal accessory nerve XI controls neck and back muscles.

cranial nerves

PNS

spinal nerves

Spinal Cranial Nerve

artery and vein

single nerve fiber

bundle of nerve fibers

blood vessels

SEM 200×

Cranial Nerves

I   from olfactory receptors

II   from retina of eyes

III   to eye muscles

IV   to eye muscles

V   from mouth and to jaw muscles

VI   to eye muscles

VII   from taste buds and to facial muscles and glands

VIII   from inner ear

IX   from pharynx and to pharyngeal muscles

XII   to tongue muscles

X   from and to internal organs

XI   to neck and back muscles

serve the skin, skeletal muscles, and tendons (see Fig. 13.1). The somatic system sensory nerves take sensory information from external sensory receptors to the CNS. Motor commands leaving the CNS travel to skeletal muscles via somatic motor nerves.

Not all somatic motor actions are voluntary. Some actions are automatic. Automatic responses to a stimulus in the somatic system are called **reflexes.** A reflex occurs quickly, without our even having to think about it. For example, a reflex may cause you to blink your eyes in response to a flash of light, without your willing it. We will study the path of a reflex because it allows us to study in detail the path of nerve signals to and from the CNS.

The Reflex Arc

Figure 13.15 illustrates the path of a reflex that involves only the spinal cord. If your hand touches a sharp pin, sensory receptors in the skin generate nerve signals that move along sensory fibers through the dorsal-root ganglia toward the spinal cord. Sensory neurons that enter the cord dorsally (posteriorly) pass signals on to many interneurons. Some of

these interneurons synapse with motor neurons whose short dendrites and cell bodies are in the spinal cord. Nerve signals travel along these motor fibers to an effector, which brings about a response to the stimulus. In this case, the effector is a muscle, which contracts so that you withdraw your hand from the pin. Various other reactions are also possible—you will most likely look at the pin, wince, and cry out in pain. This whole series of responses occurs because some of the interneurons involved carry nerve signals to the brain. The brain makes you aware of the stimulus and directs these other reactions to it. You don't feel pain until the brain receives the information and interprets it!

Autonomic System

The **autonomic system** is also in the PNS (see Fig. 13.1). The autonomic system regulates the activity of cardiac and smooth muscles, organs, and glands. The system is divided into the sympathetic and parasympathetic divisions (Fig. 13.16). Activation of these two systems generally causes opposite responses.

Figure 13.15 What happens in a spinal reflex?
A stimulus (e.g., sharp pin) causes sensory receptors in the skin to generate nerve signals that travel in sensory axons to the spinal cord. Interneurons integrate data from sensory neurons and then relay signals to motor neurons. Motor axons convey nerve signals from the spinal cord to a skeletal muscle, which contracts. Movement of the hand away from the pin is the response to the stimulus.

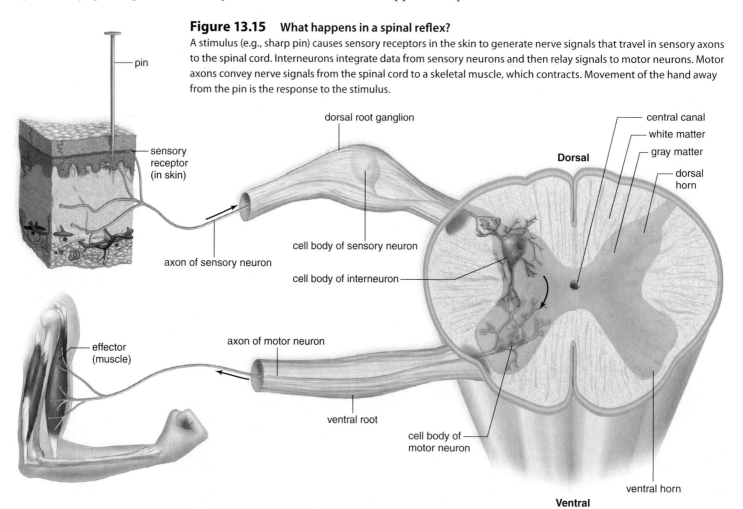

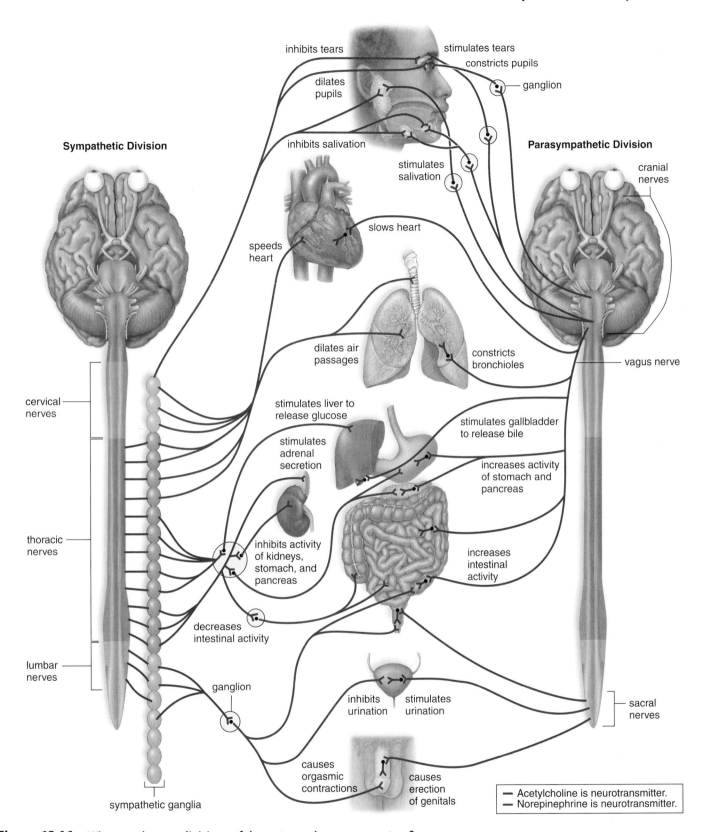

Sympathetic Division

inhibits tears

dilates pupils

inhibits salivation

speeds heart

stimulates tears

constricts pupils

ganglion

stimulates salivation

slows heart

Parasympathetic Division

cranial nerves

dilates air passages

constricts bronchioles

vagus nerve

cervical nerves

stimulates liver to release glucose

stimulates adrenal secretion

stimulates gallbladder to release bile

increases activity of stomach and pancreas

thoracic nerves

inhibits activity of kidneys, stomach, and pancreas

increases intestinal activity

lumbar nerves

decreases intestinal activity

ganglion

inhibits urination

stimulates urination

sacral nerves

causes orgasmic contractions

causes erection of genitals

sympathetic ganglia

— Acetylcholine is neurotransmitter.
— Norepinephrine is neurotransmitter.

Figure 13.16 **What are the two divisions of the autonomic nervous system?**
Sympathetic preganglionic fibers *(left)* arise from the thoracic and lumbar portions of the spinal cord; parasympathetic preganglionic fibers *(right)* arise from the cranial and sacral portions of the spinal cord. Each system innervates the same organs but has contrary effects.

Although their functions are different, the two divisions share some features: (1) They function automatically and usually in an involuntary manner; (2) they innervate all internal organs; and (3) they use two neurons and one ganglion for each impulse. The first neuron has a cell body within the CNS and a preganglionic fiber that enters the ganglion. The second neuron has a cell body within a ganglion and a postganglionic fiber that leaves the ganglion.

Reflex actions, such as those that regulate blood pressure and breathing rate, are especially important to the maintenance of homeostasis. These reflexes begin when the sensory neurons in contact with internal organs send messages to the CNS. They are completed by motor neurons within the autonomic system.

Sympathetic Division

Most preganglionic fibers of the **sympathetic division** arise from the middle, or thoracolumbar, portion of the spinal cord. They terminate almost immediately in ganglia that lie near the cord. Therefore, in this division, the preganglionic fiber is short, but the postganglionic fiber that contacts an organ is long.

The sympathetic division is especially important during emergency situations when you might be required to *fight or take flight.* It accelerates the heartbeat and dilates the bronchi—active muscles, after all, require a ready supply of glucose and oxygen. Sympathetic neurons inhibit the digestive organs, as well as the kidneys and urinary bladder. The activities of these organs—digestion, defecation, and urination—are not immediately necessary if you're under attack. The neurotransmitter released by the postganglionic axon is primarily norepinephrine (NE). The structure of NE is like that of epinephrine (adrenaline), an adrenal medulla hormone that usually increases heart rate and contractility.

Parasympathetic Division

The **parasympathetic division** includes a few cranial nerves (e.g., the vagus nerve) as well as fibers that arise from the sacral (bottom) portion of the spinal cord. Therefore, this division is often referred to as the craniosacral portion of the autonomic system. In the parasympathetic division, the preganglionic fiber is long, and the postganglionic fiber is short because the ganglia lie near or within the organ.

The parasympathetic division, sometimes called the housekeeper division, promotes all the internal responses we associate with a relaxed state. For example, it causes the pupil of the eye to contract, promotes digestion of food, and slows heart rate. It's been suggested that the parasympathetic system could be called the *rest and digest* system. The neurotransmitter used by the parasympathetic division is acetylcholine (ACh).

The Somatic Versus the Autonomic Systems

Recall that the PNS includes the somatic system and the autonomic system. Table 13.1 summarizes the features and functions of the somatic motor pathway with the motor pathways of the autonomic system.

> **Check Your Progress 13.4**
>
> 1. a. How many cranial nerves are there? b. How many spinal nerves are there?
> 2. What is the fastest way for you to react to a stimulus?
> 3. a. What is the autonomic system? b. How does it function?

CASE STUDY TEN YEARS LATER

Ten months of planning seemed to have flown by, Jill reflected. In a few minutes, her oldest girl Kirsten would promise "in sickness or in health, in good times or in bad" to Steve.

Jill was having a hard time not getting choked up, thinking of all the bad times that Kirsten had helped her though. During Jill's first relapse, Kirsten and the boys had seen their mother go from bad to worse. She'd needed Kirsten's help to button and zip her clothes and tie her shoes. When Jill was at her worst, Kirsten had even had to help her in the bathroom. Kirsten had cooked and cleaned, helped grade papers, and driven her two younger brothers around. Together, her children and her husband Mike had helped Jill through the blackest moods with laughter, practical jokes, and silliness. Slowly, Jill's strength had returned. She'd been able to move from a wheelchair to a power scooter, then to a walker. When she was finally able to walk

unassisted, the beautifully carved cane she'd used ever since had been a "graduation present" from Mike and her kids.

Then a blessedly long remission had freed Jill and Kirsten for a more normal mother-and-daughter relationship. Jill and Mike had encouraged her to move away for college, even though Kirsten was reluctant to leave. The relapses that had followed had all been brief, and she and Mike had insisted that Kirsten stay in school and enjoy her independence. *Married.* Jill smiled to herself, then sniffled. Kirsten was bound to be successful. She already understood the "good times, bad times" thing.

"Mom, you're not going to boo-hoo all over the place, are you?" her son Mark teased. She smiled back at him and shook her head. Then, with her sons supporting her arms, she grasped her cane and stepped off down the aisle.

Table 13.1	**Comparison of Somatic Motor and Autonomic Motor Pathway**		
	Somatic Motor Pathway	**Autonomic Motor Pathways**	
		Sympathetic	**Parasympathetic**
Type of control	Voluntary/involuntary	Involuntary	Involuntary
Number of neurons per message	One	Two (preganglionic shorter than postganglionic)	Two (preganglionic longer than postganglionic)
Location of motor fiber	Most cranial nerves and all spinal nerves	Thoracolumbar spinal nerves	Cranial (e.g., vagus) and sacral spinal nerves
Neurotransmitter	Acetylcholine	Norepinephrine	Acetylcholine
Effectors	Skeletal muscles	Smooth and cardiac muscle, glands and organs	Smooth and cardiac muscle, glands and organs

13.5 Drug Therapy and Drug Abuse

As you are reading these words, synapses throughout your brain are organizing, integrating, and cataloguing the information you take in. Neurotransmitters at these synapses control the firing of countless action potentials, thus creating a network of neural circuits. It is amazing to realize that all thoughts, feelings, and actions of a human being are dependent on neurotransmitters in the CNS and PNS. By modifying or controlling synaptic transmission, a wide variety of drugs with neurologic activity—both legal pharmaceuticals and illegal drugs of abuse—can alter mood, emotional state, behavior, and personality.

As mentioned previously (page 281), there are more than 100 known neurotransmitters. The most widely studied neurotransmitters to date are acetylcholine, norepinephrine, dopamine, serotonin, and GABA. Acetylcholine is an essential CNS neurotransmitter for memory circuits in the limbic system. Norepinephrine is important to dreaming, waking, and mood. The neurotransmitter dopamine plays a central role in the brain's regulation of mood. Dopamine is also the basal nuclei neurotransmitter that helps to organize coordinated movements. Serotonin is involved in thermoregulation, sleeping, emotions, and perception. GABA is an abundant inhibitory neurotransmitter in the CNS.

Neuromodulators are naturally occurring molecules that block the release of a neurotransmitter or modify a neuron's response to a neurotransmitter. Two well-known neuromodulators are substance P and endorphins. Substance P is released by sensory neurons when pain is present. Endorphins block the release of substance P and serve as natural painkillers. Endorphins are produced by the brain during times of physical and/or emotional stress. They are associated with the "runner's high" of joggers.

Both pharmaceuticals and illegal drugs have several basic modes of action:

- They promote the action of a neurotransmitter, usually by increasing the amount of neurotransmitter at a synapse. Examples include drugs such as Xanax® and Valium®, which increase GABA. These medications are used for panic attacks and anxiety. Reduced levels of norepinephrine and serotonin are linked to depression. Drugs such as Prozac®, Paxil®, and Cymbalta® allow norepinephrine and/or serotonin to accumulate at the synapse, which explains their effectiveness as antidepressants. **Alzheimer disease** causes a slow, progressive loss of memory, as you'll discover in Chapter 17. Drugs used for Alzheimer disease allow acetylcholine to accumulate at synapses in the limbic system.

- They interfere with or decrease the action of a neurotransmitter. For instance, antipsychotic drugs used for the treatment of schizophrenia decrease the activity of dopamine. The caffeine in coffee, chocolate, and tea keeps us awake by interfering with the effects of inhibitory neurotransmitters in the brain.

- They replace or mimic a neurotransmitter or neuromodulator. The opiates—namely, codeine, heroin, and morphine—bind to endorphin receptors and in this way reduce pain and produce a feeling of well-being.

Ongoing research into neurophysiology and neuropharmacology (the study of nervous system function and the way drugs work in the nervous system) continues to provide evidence that mental illnesses are caused by imbalances in neurotransmitters. These studies will undoubtedly improve treatments for mental illness, as well as provide insight into the problem of drug abuse.

Like mental illness, drug abuse is also linked to neurotransmitter levels. As mentioned previously, the neurotransmitter dopamine is essential for mood regulation. Dopamine plays a central role in the working of the brain's built-in *reward circuit*. The reward circuit is a collection of neurons that, under normal circumstances, promotes healthy, pleasurable activities, such as consuming food. It's possible to abuse behaviors, such as eating, spending, or gambling, because the behavior stimulates the reward circuit and make us feel good. Drug abusers take drugs that artificially affect the reward circuit to the point that they neglect their basic physical needs in favor of continued drug use.

Drug abuse is apparent when a person takes a drug at a dose level and under circumstances that increase the potential for a harmful effect. Drug abusers are apt to display a psychological and/or physical dependence on the drug. Psychological dependence is apparent when a person craves the drug, spends time seeking the drug, and takes it regularly. With physical dependence, formerly called "addiction," the person has become tolerant to the drug. More is needed to get the same effect, and withdrawal symptoms occur when he or she stops taking the drug. This not only is true for teenagers and adults, but it is also true for newborn babies of mothers who abuse and are addicted to drugs (Fig. 13.17). Alcohol, drugs, and tobacco can all adversely affect the developing embryo, fetus, or newborn.

Alcohol

Alcohol consumption is the most socially accepted form of drug use worldwide. Its widespread use probably correlates with its ancient origin and its derivation from grains or fruits. The approximate number of adults that consume alcohol in the United States on a regular basis is 65%. Of those who are drinkers, 5% are heavy drinkers. Notably, 80% of college-age young adults drink.

Alcohol (ethanol) has known harmful effects on the body and brain. Alcohol readily crosses cell membranes, including the blood-brain barrier. It denatures protein structures, causing damage to several tissues, including vital organs of the body such as the liver and brain. The liver is the major detoxification organ of the body and prolonged alcohol consumption scars the liver and impairs its function. Depending on the amount consumed, the effects of alcohol on the brain can lead to a feeling of relaxation, a lowering of inhibitions, impaired concentration and coordination,

slurred speech, and vomiting. If the alcohol level of blood becomes too high, coma or death can occur.

In the central nervous system (CNS), alcohol acts as a *depressant* (Table 13.2) and influences many brain regions and neurotransmitter systems. For example, alcohol increases the action of GABA and increases the release of endorphins in the hypothalamus. Chronic alcohol consumption can damage the frontal lobes, decrease overall brain size, and increase the size of the ventricles. Brain damage is manifested by permanent memory loss, amnesia, confusion, apathy, disorientation, or lack of motor coordination.

There is no effective treatment to cure alcoholism. However, some drugs have been approved by the FDA to help reduce cravings in those who seriously wish to quit drinking.

Nicotine

Nicotine is a small molecule, not normally produced by the body, that acts as a *stimulant*. During the smoking of tobacco, nicotine is rapidly delivered to the CNS, especially the midbrain. When nicotine binds to neurons in the CNS, dopamine is released. In the peripheral nervous system, nicotine mimics acetylcholine and increases skeletal muscle activity, heart rate, and blood pressure. Nicotine also promotes digestive tract motility, which may account for weight loss in smokers.

The physiological and psychological addictive nature of nicotine is well known. Approximately 70 million Americans smoke (Fig. 13.18). The addiction rate of smokers is about 70%. The failure rate in those who try to quit smoking is about 80–90% of smokers. Withdrawal symptoms include irritability, headache, insomnia, poor cognitive performance, the urge to smoke, and weight gain. The attempt to quit smoking can include the use of nicotine skin patches, nicotine gum, or the taking of drugs by mouth that block the actions of acetylcholine. The effectiveness of these therapies is variable. An experimental therapy involves "immunizing" the brain of smokers against nicotine. Injections cause the production of antibodies that bind to nicotine and prevent it from passing the blood-brain barrier. The effectiveness of this new therapy is not yet known.

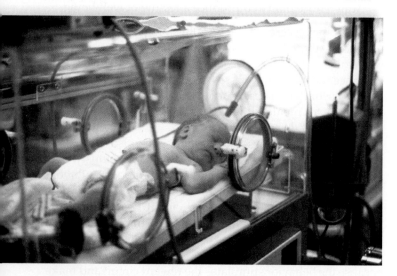

Figure 13.17   Can drug abuse affect the fetus of a pregnant woman?
This newborn is addicted to cocaine.

Table 13.2	Drug Influence on CNS and Route	
Substance	**Effect**	**Mode of Transmission**
Alcohol	Depressant	Drink
Nicotine	Stimulant	Smoked or smokeless tobacco
Cocaine	Stimulant	Sniffed/snorted, injected, or smoked
Methamphetamine	Stimulant	Smoked or pill form
Heroin	Depressant	Sniffed/snorted, injected, or smoked
Marijuana	Psychoactive	Smoked or consumed

Figure 13.18 Why do people use drugs?
Social motivations may be at play if a person smokes and drinks only in the presence of others. Other factors may be at work if the activity occurs to excess when the person is alone.

Cocaine

Cocaine is an alkaloid derived from the shrub *Erythroxylon coca*. Approximately 35 million Americans have used cocaine by sniffing/snorting, injecting, or smoking. Cocaine is a powerful *stimulant* in the CNS that interferes with the re-uptake of dopamine at synapses. The result is a rush of well-being that lasts from 5–30 minutes. Although it seemed at first that cocaine could relieve depression, it soon became apparent that cocaine is extremely addictive and its use is very harmful. During cocaine sprees (or binges), the drug is taken repeatedly and at ever-higher doses. The result is sleeplessness; lack of appetite; increased sex drive; tremors; and "cocaine psychosis," a condition that resembles paranoid schizophrenia. During the crash period, fatigue, depression, and irritability are common, along with memory loss and a confused state of cognition.

"Crack" is the street name given to cocaine processed for smoking. Approximately 8 million Americans use crack. The term *crack* refers to the crackling sound heard when smoking, allowing extremely high doses of the drug to reach the brain rapidly, so that there is an intense and immediate high or rush.

With continued use, the body makes less dopamine to compensate for the apparent excess at synapses. Drug tolerance leads to withdrawal symptoms, during which cravings for cocaine increase. Cocaine-related deaths are usually due to cardiac and/or respiratory arrest. The combination of cocaine and alcohol dramatically increase the risk of sudden death. Currently, there are no effective treatments for cocaine addiction, although cognitive-behavioral therapies have met with some success.

Methamphetamine

Methamphetamine is a synthetic drug made from amphetamine by the addition of a methyl group. Over 9 million United States residents have used methamphetamine at least once in their lifetime; teenagers and young adults represent approximately one-fourth of these. The addition of the methyl group is fairly simple, so methamphetamine is often produced from amphetamine in makeshift home laboratories. It is available as a powder (speed) or as crystals (crystal meth or ice). The crystals are smoked, and the effects are almost instantaneous and nearly as quick as when methamphetamine is snorted. The effects last 4 to 8 hours when smoked.

Methamphetamine has a structure similar to that of dopamine, and its *stimulatory* effect mimics cocaine. It reverses the effects of fatigue, maintains wakefulness, and temporarily elevates the mood of the user. After the initial rush, there is typically a state of high agitation that, in some individuals, leads to violent behavior. Chronic use can lead to what is called an amphetamine psychosis resulting in paranoia; auditory and visual hallucinations; self-absorption; irritability; and aggressive, erratic behavior. Drug tolerance, dependence, and addiction are common. Hyperthermia, convulsions, and death can occur.

Ecstasy is the street name for methylenedioxymethamphetamine (MDMA), a drug that has the same effects as methamphetamine, but without hallucinations. Ecstasy comes in a pill form, and it often contains other drugs besides MDMA, some of which are more dangerous.

There are no pharmacologic treatments for methamphetamine addiction. The most common therapy is cognitive-behavioral interventions.

Heroin

Heroin is derived from the resin or sap of the opium poppy plant, grown from Turkey to Southeast Asia and in parts of Latin America. Heroin is a highly addictive drug that acts as a *depressant* in the nervous system. The opiates, which also include morphine and codeine, have pain-killing effects.

Heroin is the most abused opiate because it is rapidly delivered to the brain, where it is converted to morphine. Morphine binds promptly to opioid receptors, and the result is a rush sensation and euphoric experience. Opiates depress breathing, activate the reward circuit, block pain pathways, cloud mental function, and sometimes cause nausea and vomiting. Long-term effects of heroin use are addiction, hepatitis, HIV/AIDS, and various bacterial infections because of shared needles. As with other drugs of abuse, tolerance and dependence are common, and heavy users may experience convulsions and death by respiratory arrest.

Heroin can be injected, snorted, or smoked. Abusers typically inject heroin up to four times a day. It is estimated that 4 million Americans have used heroin some time in

Medical Marijuana Use

The plant *Cannabis sativa* has been used in many cultures throughout history as a general-purpose medicine, pain reliever, and salve. In recent years, there has been a renewed interest in therapeutic applications of marijuana in the United States.

Marijuana Is Illegal

In 2005, the Supreme Court ruled that the constitutional authority of Congress to regulate interstate market in drugs extends to small quantities of doctor-recommended marijuana. So, patients using marijuana prescribed by a physician can be criminally prosecuted by federal law enforcement agencies.

Those opposed to medical marijuana (1) express reservations about its toxicity, possible interactions with other drugs, and potential for dependence; (2) believe that smoking marijuana might cause damage to the respiratory system; (3) feel that medical use of marijuana will make it more accessible to drug abusers and lead to use of other, more harmful drugs; (4) cite animal studies that show marijuana or its constituents may deteriorate motor coordination, negatively affect a fetus, and lower sperm count or motility.

Benefits of Marijuana Use

The goal of medical marijuana use is to relieve suffering in seriously ill patients, for whom other treatments have not worked. To this end, marijuana or its ingredients have been reported to reduce anxiety, relax muscles, increase appetite, and modulate the immune system and cardiovascular systems. All of these effects could be useful. Currently, some applications for medical marijuana include treating pain, preventing nausea and vomiting in patients undergoing chemotherapy, and lowering intraocular pressure in glaucoma patients.

The Food and Drug Administration (FDA) dismisses claims of marijuana's medical benefits. Ironically, however, the FDA has approved certain THC-based drug therapies, meant for use by cancer patients and AIDS patients experiencing extreme weight loss.

How to Make Medical Use of Marijuana Legal

Many researchers agree that more studies using doses of fixed purities and strength are needed to probe marijuana's safety and effectiveness. However, the legal status of marijuana poses significant hurdles to this research. Following the Supreme Court's 2005 ruling, advocates for medical marijuana use and continued research may need to bring their fight to Congress.

Decide Your Opinion

1. Should marijuana be available for medical use by all patients? Why or why not?
2. Do you agree with the Supreme Court's ruling that lets federal authorities prosecute medical marijuana users, even if such use is legal in the state in which they live?
3. How should the use of medical marijuana be regulated?

---

their lives, and over 300,000 people use heroin annually. The many available treatments for heroin addiction include synthetic opiate compounds, such as methadone or suboxone. They decrease withdrawal symptoms and block heroin's effects.

Marijuana

The dried flowering tops, leaves, and stems of the Indian hemp plant *Cannabis sativa* contain and are covered by a resin rich in tetrahydrocannabinol (THC). The names *cannabis* and *marijuana* apply to either the plant or THC. Usually, marijuana is smoked in a cigarette called a "joint," or it can be consumed. An estimated 22 million Americans use marijuana. Although the drug was banned in the United States in 1937, several states have legalized its use for medical purposes.

A neurotransmitter called anandamide was recently discovered. Both THC and anandamide belong to a class of chemicals called cannabinoids. It would seem, then, that THC mimics the actions of anadamide. Receptors that bind cannabinoids are located in the hippocampus, cerebellum, basal ganglia, and cerebral cortex. These brain areas are important for memory, orientation, balance, motor coordination, and perception.

When THC reaches the CNS by smoking a "joint," mild euphoria occurs, along with alterations in vision and judgment. Distortions of space and time can also occur in occasional users. In heavy users, hallucinations, anxiety, depression, rapid flow of ideas, body image distortions, paranoia, and psychotic symptoms can result. The terms cannabis psychosis and cannabis delirium describe such reactions of marijuana's influence on the brain. Regular use can cause cravings that make it difficult to stop marijuana use.

It is estimated that 100,000 people each year seek treatment for marijuana abuse. There is no "quick fix" for the problem of marijuana abuse. However, as with cocaine addiction, cognitive behavioral therapies are making inroads to address dependency problems for marijuana abusers.

> **Check Your Progress 13.5**
>
> 1. How can the abuse of drugs, including alcohol and nicotine, affect the nervous system?

Summarizing the Concepts

13.1 Overview of the Nervous System

The nervous system

- is divided into the central nervous system (CNS) and the peripheral nervous system (PNS).
- has three functions: (1) reception of input; (2) integration of data; and (3) generates motor output.

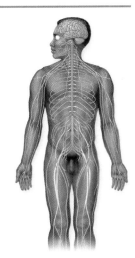

Nervous Tissue

Nervous tissue contains two types of cells: neurons and neuroglia.

- Neurons transmit nerve signals.
- Neuroglia nourish and support neurons.

Neuron Structure

A neuron is composed of dendrites, a cell body, and an axon. There are three types of neurons:

- Sensory neurons take nerve signals from sensory receptors to the CNS.
- Interneurons occur within the CNS.
- Motor neurons take nerve signals from the CNS to effectors (muscles or glands).

Myelin Sheath

- Long axons are covered by a myelin sheath.

The Nerve Signal

Resting Potential   More Na^+ outside the axon and more K^+ inside the axon. The axon does not conduct a signal.

Action Potential   A change in polarity across the axonal membrane as a nerve signal occurs: When Na^+ gates open, Na^+ moves to the inside of the axon, and a depolarization occurs. When K^+ gates open and K^+ moves to outside the axon, a repolarization occurs.

The Synapse

- When a neurotransmitter is released into a synaptic cleft, transmission of a nerve signal occurs.
- Binding of the neurotransmitter to receptors in the receiving membrane causes excitation or inhibition.
- Integration is the summing of excitatory and inhibitory signals.

13.2 The Central Nervous System

The CNS receives and integrates sensory input and formulates motor output. The CNS consists of the spinal cord and brain.

The Spinal Cord

- Gray matter of the spinal cord contains neuron cell bodies.
- White matter consists of myelinated axons that occur in tracts.
- Conduction to and from brain; carries out reflex actions.

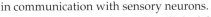

The Brain

Cerebrum   The cerebrum has two cerebral hemispheres connected by the corpus callosum.

- Sensation, reasoning, learning and memory, and language and speech take place in the cerebrum.
- The cerebral cortex covers the cerebrum. The cerebral cortex of each cerebral hemisphere has four lobes: frontal, parietal, occipital, and temporal.
- The primary motor area in the frontal lobe sends out motor commands to lower brain centers, which pass them on to motor neurons.

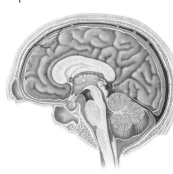

- The primary somatosensory area in the parietal lobe receives sensory information from lower brain centers in communication with sensory neurons.
- Association areas are located in all the lobes.

The Diencephalon   The hypothalamus controls homeostasis. The thalamus sends sensory input on to the cerebrum.

The Cerebellum   The cerebellum coordinates skeletal muscle contractions.

The Brain Stem   The medulla oblongata and the pons have centers for breathing and the heartbeat.

13.3 The Limbic System and Higher Mental Functions

- The limbic system lying deep in the brain is involved in determining emotions.
- The amygdala determines when a situation deserves the emotion we call "fear."
- The hippocampus is particularly involved in storing and retrieving memories.

13.4 The Peripheral Nervous System

- The PNS contains only nerves and ganglia.
- Cranial nerves take impulses to and from the brain.

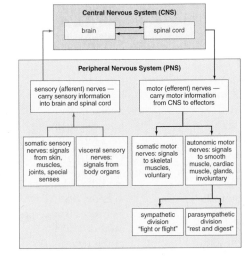

- Spinal nerves take impulses to and from the spinal cord.
- The PNS is divided into the somatic system and the autonomic system.

Somatic System

The somatic system serves the skin, skeletal muscles, and tendons.

- Some actions are due to reflexes, which are automatic and involuntary.
- Other actions are voluntary; originate in cerebral cortex.

Autonomic System

Two divisions in this system are the sympathetic division and the parasympathetic division.

Sympathetic Division Responses that occur during times of stress.

Parasympathetic Division Responses that occur during times of relaxation.

- Actions in these divisions are involuntary and automatic.
- These divisions innervate internal organs.
- Two neurons and one ganglion are used for each impulse.

13.5 Drug Therapy and Drug Abuse

- Neurological drugs promote, prevent, or mimic the action of a particular neurotransmitter.
- Dependency occurs when the body compensates for the presence of neurological drugs.

Understanding Key Terms

acetylcholine (ACh) 281	diencephalon 287
acetylcholinesterase (AChE) 281	dopamine 281
action potential 278	dorsal-root ganglion 291
Alzheimer disease 289, 295	drug abuse 296
amygdala 288	effector 277
association area 286	episodic memory 289
autonomic system 292	GABA (gamma aminobutyric acid) 281
axon 277	ganglia 291
axon terminal 280	ganglion 291
basal nuclei 287	glutamate 281
brain 285	gray matter 283
brain stem 288	hippocampus 288
Broca's area 287	hypothalamus 287
cell body 277	integration 282
central nervous system (CNS) 276	interneuron 277
	learning 289
cerebellum 287	limbic system 288
cerebral cortex 286	long-term memory 289
cerebral hemisphere 285	long-term potentiation (LTP) 290
cerebrospinal fluid 283	
cerebrum 285	medulla oblongata 288
corpus callosum 285	memory 289
cranial nerve 291	meninges (sing., meninx) 283
dendrite 277	midbrain 288
depolarization 278	motor neuron 277

myelin sheath 277	saltatory conduction 280
nerve signal 277	Schwann cell 277
neuroglia 276	semantic memory 289
neuron 276	sensory neuron 277
neurotransmitter 280	sensory receptor 277
node of Ranvier 277	serotonin 281
norepinephrine (NE) 281	short-term memory 289
nuclei 288	skill memory 289
parasympathetic division 294	sodium-potassium pump 278
Parkinson disease 287	somatic system 291
peripheral nervous system (PNS) 276	spinal cord 283
	spinal nerve 291
pons 288	stimulus 278
prefrontal area 287	sympathetic division 294
primary motor area 286	synapse 280
primary somatosensory area 286	synaptic cleft 280
	thalamus 287
reflex 292	threshold 278
refractory period 280	tract 283
repolarization 278	ventricle 283
resting potential 278	Wernicke's area 287
reticular formation 288	white matter 283

Match the key terms to these definitions.

a. _____ Automatic, involuntary response of an organism to a stimulus.

b. _____ Chemical stored at the ends of axons; responsible for transmission across a synapse.

c. _____ Part of the peripheral nervous system that regulates internal organs.

d. _____ Collection of neuron cell bodies, usually outside the central nervous system.

e. _____ Neurotransmitter active in the somatic system of the peripheral nervous system.

Testing Your Knowledge of the Concepts

1. What are the three functions of the nervous system? (page 276)

2. What are the functions performed by the three types of neurons? Describe the structure and functions of the three parts of a neuron. (page 277)

3. What is the resting potential, and how is it created and maintained? (page 278)

4. What is an action potential? Describe the two parts of the process. (pages 278–80)

5. Explain transmission of a nerve signal across a synapse. (page 281)

6. List and describe the functions of several neurotransmitters. (pages 281, 295)

7. Describe the structure and functions of the spinal cord. (pages 283–85)

8. Name the major parts of the brain, and give a function for each part. (pages 285–88)

9. What are the lobes of the cerebrum? Give the location and function of the following areas: primary motor area, primary somatosensory area, primary visual area, primary auditory area, primary taste area, and primary olfactory area. (pages 285–86)

10. What are association areas, and what roles do they play? (page 286)

11. What is the reticular formation, and what is the role of the RAS? (page 288)

12. What is the limbic system? What are the two main parts and their functions? (pages 288–89)

13. Describe the various types of memory. (pages 289–90)

14. What types of nerves make up the PNS? How many of each type are there, and what areas of the body do each serve? (pages 290–91)

15. Distinguish between the somatic and the autonomic nervous systems as to structure, effectors, neurotransmitters, and functions. (pages 291–94)

16. Trace the path of a reflex arc. (page 292)

17. Describe the physiological effects and mode of action of alcohol, nicotine, cocaine, methamphetamine, heroin, and marijuana. (pages 295–98)

18. What type of neuron lies completely in the CNS?
    a. motor neuron
    b. interneuron
    c. sensory neuron

19. Which of the following neuron parts receive(s) signals from sensory receptors of other neurons?
    a. cell body   c. dendrites
    b. axon   d. Both a and c are correct.

20. The neuroglial cells that form myelin sheaths in the PNS are
    a. oligodendrocytes.   d. astrocytes.
    b. ganglionic cells.   e. microglia.
    c. Schwann cells.

21. Which of these correctly describes the distribution of ions on either side of an axon when it is not conducting a nerve signal?
    a. more sodium ions (Na^+) outside and more potassium ions (K^+) inside
    b. more K^+ outside and less Na^+ inside
    c. charged protein outside; Na^+ and K^+ inside
    d. Na^+ and K^+ outside and water only inside
    e. chlorine ions (Cl^-) on outside and K^+ and Na^+ on inside

22. When the action potential begins, sodium gates open, allowing Na^+ to cross the membrane. Now the polarity changes to
    a. negative outside and positive inside.
    b. positive outside and negative inside.
    c. neutral outside and positive inside.
    d. All of these are correct.

23. Repolarization of an axon during an action potential is produced by
    a. inward diffusion of Na^+.
    b. outward diffusion of K^+.
    c. inward active transport of Na^+.
    d. active extrusion of K^+.

24. Transmission of the nerve signal across a synapse is accomplished by the
    a. movement of Na^+ and K^+.
    b. release of a neurotransmitter by a dendrite.
    c. release of a neurotransmitter by an axon.
    d. release of a neurotransmitter by a cell body.
    e. All of these are correct.

25. Synaptic vesicles are
    a. at the ends of dendrites and axons.
    b. at the ends of axons only.
    c. along the length of all long fibers.
    d. All of these are correct.

26. Which of the following cerebral areas is not correctly matched with its function?
    a. occipital lobe—vision
    b. parietal lobe—somatosensory area
    c. temporal lobe—primary motor area
    d. frontal lobe—Broca's motor speech area

27. The hypothalamus does not
    a. control skeletal muscles.
    b. regulate thirst.
    c. control the pituitary gland.
    d. regulate body temperature.

28. Which of the following brain regions is not correctly matched to its function?
    a. medulla oblongata—regulated heartbeat, breathing, and blood pressure
    b. cerebellum—coordinated voluntary muscle movements
    c. thalamus—secretes melatonin that regulates daily body rhythms
    d. midbrain—reflex centers for visual, auditory, and tactile responses

29. A spinal nerve takes nerve signals
    a. to the CNS.
    b. away from the CNS.
    c. both to and away from the CNS.
    d. from the CNS to the spinal cord.

30. Which of these are the first and last elements in a spinal reflex?
    a. axon and dendrite
    b. sensory receptor and muscle effector
    c. ventral horn and dorsal horn
    d. brain and skeletal muscle
    e. motor neuron and sensory neuron

31. The autonomic system has two divisions, called the
    a. CNS and PNS.
    b. somatic and skeletal divisions.
    c. efferent and afferent divisions.
    d. sympathetic and parasympathetic divisions.

32. The sympathetic division of the autonomic system does not cause
    a. the liver to release glycogen.
    b. dilation of bronchioles.
    c. the gastrointestinal tract to digest food.
    d. an increase in the heart rate.

33. Label this diagram.

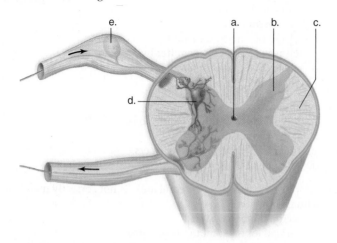

Thinking Critically About the Concepts

Demyelinating disorders like the multiple sclerosis that affected Jill Nielsen are the subject of numerous research projects. Many investigations focus on the cells that create myelin: the Schwann cells of the PNS and oligodendrocytes in the CNS. Other studies focus on immune system cells that attack this myelin sheath.

The goal of this research is to determine how to restore lost myelin, which might help (or possibly cure) folks living with MS and other demyelinating diseases. Investigations into the role played by the sheath in nerve regeneration may offer hope to victims of spinal cord injury.

1. Why are impulses transmitted more quickly down a myelinated axon than an unmyelinated axon?

2. A buildup of very long chain saturated fatty acids is believed to be the cause of myelin loss in adrenoleukodystrophy. This rare disease is a demyelinating disorder like MS. It is the subject of the film *Lorenzo's Oil*. This real-life drama focuses on Lorenzo Odone, whose parents successfully developed a diet that helped their son.
   a. From your study of chemistry in Chapter 2, a fatty acid is a part of what type of molecule?
   b. What distinguishes a saturated fatty acid from an unsaturated fatty acid?
   c. From your study of nutrition in Chapter 8, what types of foods would contain saturated fatty acids?

3. If you do some research on the Internet, you should be able to find more information about ALD. How does a child get ALD?

4. Why would you expect the motor skills of a child to improve as myelination continues during early childhood development?

Senses

The waitress in the pizza restaurant stood at their tableside, ready to take orders from Mike Fila and his daughter Lauren. "I'll have a medium double pepperoni and sausage pizza and a large cola. How about you, honey?"

"Just a side salad for me, thanks, Daddy," Lauren replied. "Dressing on the side and an iced tea. Unsweetened, please," she nodded to the waitress.

"Aw, honey, live a little," her father smiled. "Besides, you're eating for two right now."

"And I have to stay healthy, and you do too. I'm really worried about you, Dad. You aren't watching your weight like the doctor told you to. I'll bet you don't test your blood sugar twice a day, either," Lauren fretted. "Are you taking your insulin shots, at least?" she asked.

"I admit, I don't do as good a job without your mom around to remind me," Mike sighed. "It's been three years, and I miss her more than ever. It was really nice of you to come out and have supper with me."

"It's nice to get together with you when Mark's out of town like this. Mark and I don't like for you to be alone so much. You know you're always welcome for supper at our place," Lauren reassured. She knew that when her father ate dinner alone at home, his meals were enormous. She feared that he was not managing his diabetes well at all.

Mike made quick work of his pizza and beer, then entertained Lauren with corny jokes. She laughed until her sides hurt, and two hours passed quickly. "We'd better hit the road, Dad," Lauren said at last. "We both have to get up for work tomorrow. It'll be really nice to stop work when this baby comes."

Mike lurched heavily to his feet. He blinked several times, then rubbed his right eye. He blinked again, then shook his head, trying to clear his vision. "This is strange," he muttered.

"Are you all right?" Lauren asked anxiously.

"I'm not sure," her father replied. "It's like a curtain just closed over half of my right eye. I can't see a thing!"

"Sit back down, Dad. I'm calling 911." When he started to protest, Lauren cried shrilly, "Stop arguing with me, Dad!"

14.1 Sensory Receptors and Sensations

Sensory receptors are dendrites specialized to detect certain types of stimuli (sing., **stimulus**). **Exteroceptors** are sensory receptors that detect stimuli from outside the body, such as those that result in taste, smell, vision, hearing, and equilibrium (Table 14.1). **Interoceptors** receive stimuli from inside the body. Interoceptors include pressoreceptorors (sometimes referred to as baroreceptors) that respond to changes in blood pressure and osmoreceptors to monitor the body's water-salt balance. Chemoreceptors are interoceptors that monitor the pH of the blood.

Interoceptors are directly involved in homeostasis and are regulated by a negative feedback mechanism. For example, when blood pressure rises, pressoreceptors signal a regulatory center in the brain. The brain responds by sending out nerve signals to the arterial walls, causing their smooth muscle to relax. The blood pressure then falls. Once blood pressure is returned to normal, the pressoreceptors are no longer stimulated.

Exteroceptors such as those in the eye and ear continually send messages to the central nervous system. In this way, they keep us informed regarding the surrounding environment.

Types of Sensory Receptors

Sensory receptors in humans can be classified into just four categories: chemoreceptors, photoreceptors, mechanoreceptors, and thermoreceptors.

Chemoreceptors respond to chemical substances in the immediate vicinity. As Table 14.1 indicates, taste and smell, which detect external stimuli, use chemoreceptors. However, so do various other organs sensitive to internal stimuli. Chemoreceptors that monitor blood pH are located in the carotid arteries and aorta. If the pH lowers, the breathing rate increases. As more carbon dioxide is exhaled, the blood pH rises.

Pain receptors (nociceptors) are a type of chemoreceptor. They are naked dendrites that respond to chemicals released by damaged tissues. Pain receptors are protective because they alert us to possible danger. For example, without the pain of appendicitis, we might never seek the medical help needed to avoid a ruptured appendix.

Photoreceptors respond to light energy. Our eyes contain photoreceptors that are sensitive to light rays and thereby provide us with a sense of vision. Stimulation of the photoreceptors known as rod cells results in black-and-white vision. Stimulation of the photoreceptors known as cone cells results in color vision.

Mechanoreceptors are stimulated by mechanical forces, which most often result in pressure of some sort. When we hear, airborne sound waves are converted to fluid-borne pressure waves that can be detected by mechanoreceptors in the inner ear. Mechanoreceptors are responding to fluid-borne pressure waves when we detect changes in gravity and motion, helping us keep our balance. These receptors are in the vestibule and semicircular canals of the inner ear, respectively.

The sense of touch depends on pressure receptors sensitive to either strong or slight pressures. Pressoreceptors located in certain arteries detect changes in blood pressure, and stretch receptors in the lungs detect the degree of lung inflation. Proprioceptors respond to the stretching of muscle fibers, tendons, joints, and ligaments. Signals from proprioceptors make us aware of the position of our limbs.

Thermoreceptors located in the hypothalamus and skin are stimulated by changes in temperature. Those that respond when temperatures rise are called warmth receptors, and those that respond when temperatures lower are called cold receptors. On a hot summer day, rising body temperature will most likely have you seeking shade.

How Sensation Occurs

Sensory receptors respond to environmental stimuli by generating nerve signals. When the nerve signals arrive at the cerebral cortex of the brain, **sensation**, the conscious perception of stimuli, occurs.

As we discussed in Chapter 13, sensory receptors are the first element in a reflex arc. We are aware of a reflex action only when sensory information reaches the brain. At that time, the brain integrates this information with other

Table 14.1	**Exteroceptors**				
Sensory Receptor	**Stimulus**	**Category**	**Sense**	**Sensory Organ**	
Taste cells	Chemicals	Chemoreceptor	Taste	Taste buds	
Olfactory cells	Chemicals	Chemoreceptor	Smell	Olfactory epithelium	
Rod cells and cone cells in retina	Light rays	Photoreceptor	Vision	Eye	
Hair cells in spiral organ of the inner ear	Sound waves	Mechanoreceptor	Hearing	Ear	
Hair cells in semicircular canals of the inner ear	Motion	Mechanoreceptor	Rotational equilibrium	Ear	
Hair cells in vestibule of the inner ear	Gravity	Mechanoreceptor	Gravitational equilibrium	Ear	

information received from other sensory receptors. Consider what happens if you burn yourself and quickly remove your hand from a hot stove. The brain receives information not only from your skin, but also from your eyes, nose, and all sorts of sensory receptors.

Some sensory receptors are free nerve endings or encapsulated nerve endings, while others are specialized cells closely associated with neurons. Often the plasma membrane of a sensory receptor contains receptor proteins that react to the stimulus. For example, the receptor proteins in the plasma membrane of chemoreceptors bind to certain chemicals. When this happens, ion channels open, and ions flow across the plasma membrane. If the stimulus is sufficient, nerve signals begin and are carried by a sensory nerve fiber within the PNS to the CNS (Fig. 14.1). The stronger the stimulus, the greater the frequency of nerve signals. Nerve signals that reach the spinal cord first are conveyed to the brain by ascending tracts. If nerve signals finally reach the cerebral cortex, sensation and perception occur.

All sensory receptors initiate nerve signals. The sensation that results depends on the part of the brain receiving

Have You Ever Wondered...

Why does rubbing your closed eyes (as you might if you're tired or your eyes itch) produce a visual sensation?

As described in Section 14.4, the eye is a flexible container filled with fluid and a soft gelatinous material. Compressing the eyes by rubbing on them increases pressure in the eyes. In turn, the photoreceptors of the eye are stimulated by the increased eye pressure. When the nerve signals are conveyed to the brain, the brain senses "vision." We "see stars" because nerve signals from the eyes can only result in sight.

the nerve signals. Nerve signals that begin in the optic nerve eventually reach the visual areas of the cerebral cortex. Thereafter, we see objects. Nerve signals that begin in the auditory nerve eventually reach the auditory areas of the cerebral cortex. We hear sounds when the auditory cortex is stimulated. If it were possible to switch these nerves, stimulation of the eyes would result in hearing!

Before sensory receptors initiate nerve signals, they also carry out **integration,** the summing up of signals. One type of integration is called **sensory adaptation,** a decrease in response to a stimulus. We have all had the experience of smelling an odor when we first enter a room and then later not being aware of it. Some authorities believe that when sensory adaptation occurs, sensory receptors have stopped sending impulses to the brain. Others believe that the reticular activating system (RAS) has filtered out the ongoing stimuli. You will recall that sensory information is conveyed from the brain stem through the thalamus to the cerebral cortex by the RAS. The thalamus acts as a gatekeeper and passes on information only of immediate importance. Just as we gradually become unaware of particular environmental stimuli, we can suddenly become aware of stimuli that may have been present for some time. This can be attributed to the workings of the RAS, which has synapses with all the great ascending sensory tracts.

The functioning of our sensory receptors makes a significant contribution to homeostasis. Without sensory input, we would not receive information about our internal and external environment. This information leads to appropriate reflex and voluntary actions to keep the internal environment constant.

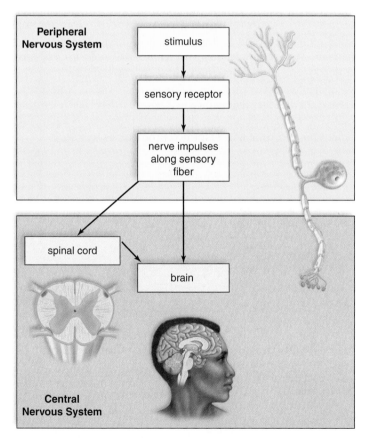

Figure 14.1 How does a sensation result in sensory perception? After detecting a stimulus, sensory receptors initiate nerve signals within the PNS. These signals give the central nervous system (CNS) information about the external and internal environment. The CNS integrates all incoming information, and then initiates a motor response to the stimulus.

Check Your Progress 14.1

1. What is the function of a sensory receptor?
2. What are the different types of sensory receptors?
3. Explain sensation.

14.2 Proprioceptors and Cutaneous Receptors

Sensory receptors in the muscles, joints and tendons, other internal organs, and skin send nerve signals to the spinal cord. From there, they travel up the spinal cord in tracts to the somatosensory areas of the cerebral cortex (see Fig. 13.10). These general sensory receptors can be categorized into three types: proprioceptors, cutaneous receptors, and pain receptors.

Proprioceptors

Proprioceptors are mechanoreceptors involved in reflex actions that maintain muscle tone, and thereby the body's equilibrium and posture. They help us know the position of our limbs in space by detecting the degree of muscle relaxation, the stretch of tendons, and the movement of ligaments. Muscle spindles, built into a muscle, act to increase muscle contraction. Golgi tendon organs, found in tendons, decrease muscle contraction. The result is a muscle that has the proper length and tension, or muscle tone.

Figure 14.2 illustrates the activity of a muscle spindle. In a muscle spindle, sensory nerve endings are wrapped around thin muscle cells within a connective tissue sheath. When the muscle relaxes and its length increases, the muscle spindle is stretched and nerve signals are generated. The more the muscle stretches, the faster the muscle spindle sends its signals. A reflex action then occurs, which results in contraction of muscle fibers adjoining the muscle spindle.

The knee-jerk reflex, which involves muscle spindles, offers an opportunity for physicians to test a reflex action. The information sent by muscle spindles to the CNS is used to maintain the body's equilibrium and posture. Proper balance and body position are maintained, despite the force of gravity always acting upon the skeleton and muscles.

Cutaneous Receptors

The skin is composed of two layers: the epidermis and the dermis. In Figure 14.3, the artist has dramatically indicated these two layers by separating the epidermis from the dermis in one location. The epidermis is stratified squamous epithelium (page 78). Cells become keratinized as they rise to the surface, where they are sloughed off. The dermis is a thick connective tissue layer. The dermis contains **cutaneous receptors,** which make the skin sensitive to touch, pressure, pain, and temperature (warmth and cold). The dermis is a mosaic of these tiny receptors, as you can determine by slowly passing a metal probe over your skin. At certain points, you will feel touch or pressure, and at others, you will feel heat or cold (depending on the probe's temperature).

Three types of cutaneous receptors are sensitive to fine touch. These receptors give a person specific information such as the location of the touch plus its shape, size, and texture. *Meissner corpuscles* and *Krause end bulbs* are concentrated in the fingertips, the palms, the lips, the tongue, the nipples, the penis, and the clitoris. *Merkel disks* are found where the epidermis meets the dermis. A free nerve ending called a *root hair plexus* winds around the base of a hair follicle. This receptor will respond if the hair is touched.

Figure 14.2 **How does a muscle spindle work?**
When a muscle is stretched, a muscle spindle sends ① sensory nerve impulses to the spinal cord. ② Motor nerve impulses from the spinal cord result in muscle fiber contraction so that muscle tone is maintained.

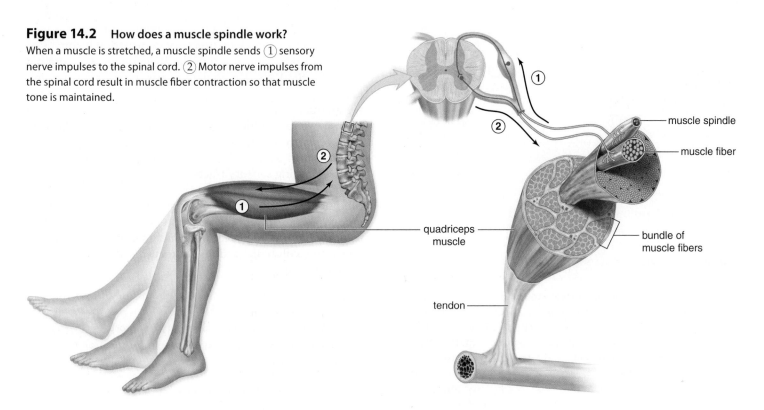

quadriceps muscle

tendon

muscle spindle

muscle fiber

bundle of muscle fibers

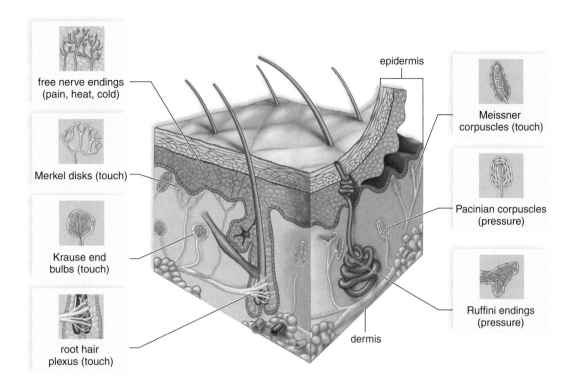

Figure 14.3 **Is each type of sensory receptor found in the skin specialized for a single function?**
The classical view is that each sensory receptor has the main function shown here. However, investigators report that matters are not so clear-cut. For example, microscopic examination of the skin of the ear shows only free nerve endings (pain receptors), and yet the skin of the ear is sensitive to all sensations. Therefore, it appears that the receptors of the skin are somewhat, but not completely, specialized.

Two types of cutaneous receptors sensitive to pressure are Pacinian corpuscles and Ruffini endings. *Pacinian corpuscles* are onion-shaped sensory receptors that lie deep inside the dermis. *Ruffini endings* are encapsulated by sheaths of connective tissue and contain lacy networks of nerve fibers.

Temperature receptors are simply free nerve endings in the epidermis. Some free nerve endings are responsive to cold. Others respond to warmth. Cold receptors are far more numerous than warmth receptors, but the two types have no known structural differences.

Pain Receptors

Like the skin, many internal organs have pain receptors, also called *nociceptors*. These receptors are sensitive to chemicals released by damaged tissues. When inflammation occurs, due to mechanical, thermal, or electrical stimuli or toxic substances, cells release chemicals that stimulate pain receptors. Aspirin and ibuprofen reduce pain by inhibiting the synthesis of one class of these chemicals.

Sometimes, stimulation of internal pain receptors is felt as pain from the skin, as well as the internal organs. This is called **referred pain.** Some internal organs have a referred pain relationship with areas located in the skin of the back, groin, and abdomen. For example, pain from the heart is often felt in the left shoulder and arm. This most likely happens when nerve impulses from the pain receptors of internal organs travel to the spinal cord and synapse with neurons also receiving impulses from the skin. As you recall

Have You Ever Wondered...

What are phantom sensation and phantom pain?

Suppose you've lost a foot and a leg due to an injury. In addition to dealing with loss of a limb, often an amputee must cope with the phenomena of phantom sensation and/or phantom pain. Phantom sensation is a painless awareness of the amputated limb. For example, a patient whose foot and lower leg have been removed may have an itchy or tingly sensation in the "foot," though the foot is no longer there. Similarly, pain can be sensed as originating from the absent body part. Researchers believe that any stimulus (such as a touch) to the stump will fool the brain into a perceived sensation, because the brain has received signals from the leg and foot for such a long time.

Phantom sensation may last for years, but usually disappears without treatment. Phantom pain must be treated with a combination of medication, massage, and physical therapy.

from the case study of Chapter 5, this type of referred pain is more common in men than in women. The nonspecific symptoms that women often experience during a heart attack may delay a diagnosis.

> ### Check Your Progress 14.2
> 1. What is the function of proprioceptors?
> 2. What are the functions of cutaneous receptors in the skin?

14.3 Senses of Taste and Smell

Taste and smell are called chemical senses because their receptors are sensitive to molecules in the food we eat and the air we breathe.

Taste cells and olfactory cells bear chemoreceptors. Chemoreceptors are also in the carotid arteries and in the aorta. Here, they are primarily sensitive to the pH of the blood. These chemoreceptors are called carotid and aortic bodies. They communicate via sensory nerve fibers with the respiratory center in the medulla oblongata. When the pH drops, they signal this center. Immediately thereafter, the breathing rate increases. Exhaling CO_2 raises the pH of the blood.

Chemoreceptors are plasma membrane receptors that bind to particular molecules. They are divided into two types: those that respond to distant stimuli and those that respond to direct stimuli. Olfactory cells act from a distance and taste cells act directly. pH receptors also respond to direct stimuli.

Sense of Taste

In adult humans, approximately 3,000 **taste buds** are located primarily on the tongue (Fig. 14.4b–e). Many taste buds lie along the walls of the papillae. These small elevations on the tongue are visible to the naked eye. Isolated taste buds are also present on the hard palate, the pharynx, and the epiglottis. There are at least four primary types of taste (sweet, sour, salty, and bitter). A fifth taste, called umami, allows us to enjoy the savory flavors of certain cheeses, beef, and mushrooms. Taste buds for each of these tastes are located throughout the tongue, although certain regions may be most sensitive to particular tastes. The tip of the tongue is most sensitive to sweet tastes, making it especially pleasurable to lick an ice cream cone (Fig. 14.4a). The margins of the tongue react to salty and sour tastes. The rear of the tongue responds to bitter tastes.

How the Brain Receives Taste Information

Taste buds open at a taste pore. They have supporting cells and a number of elongated taste cells that end in microvilli. When molecules bind to receptor proteins of the microvilli, nerve signals are generated in sensory nerve fibers that go to the brain. Signals reach the gustatory (taste) cortex, located

a.

Figure 14.4   How do you taste an ice cream cone?
a. Without a tongue, it would be impossible to lick an ice cream cone, and its taste would be minimal. **b.** Papillae on the tongue contain taste buds sensitive to sweet, sour, salty, bitter, and perhaps umami. **c.** Photomicrograph and enlargement of papilla. **d.** Taste buds occur along the walls of the papillae. **e.** Taste cells in microvilli that bear receptor proteins for certain molecules. When molecules bind to the receptor proteins, nerve signals are generated and go to the brain, where the sensation of taste occurs.

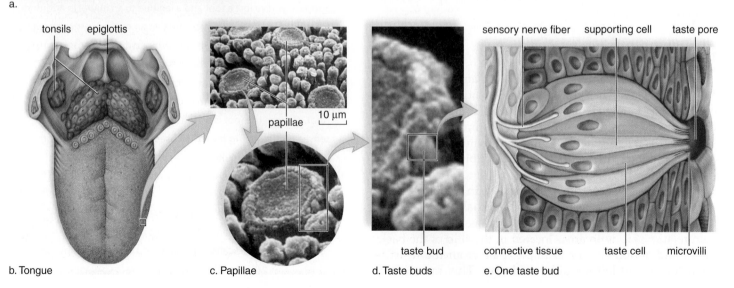

tonsils   epiglottis

papillae   10 μm

sensory nerve fiber   supporting cell   taste pore

b. Tongue   c. Papillae   taste bud   connective tissue   taste cell   microvilli

d. Taste buds   e. One taste bud

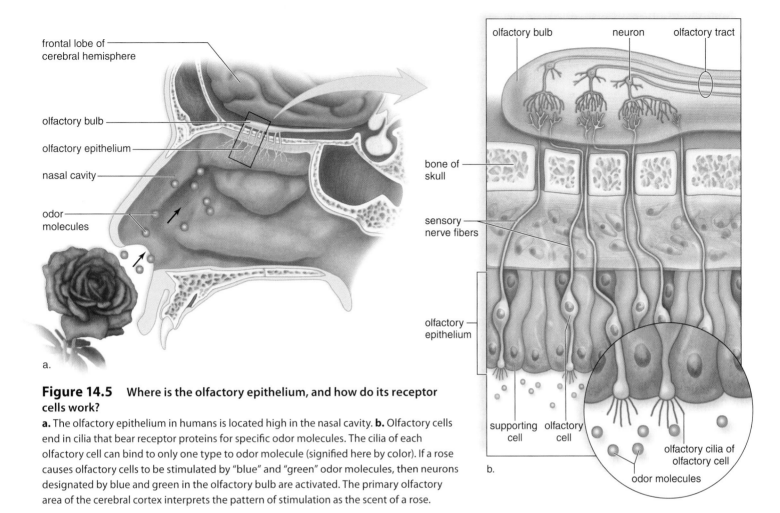

Figure 14.5 **Where is the olfactory epithelium, and how do its receptor cells work?**

a. The olfactory epithelium in humans is located high in the nasal cavity. **b.** Olfactory cells end in cilia that bear receptor proteins for specific odor molecules. The cilia of each olfactory cell can bind to only one type to odor molecule (signified here by color). If a rose causes olfactory cells to be stimulated by "blue" and "green" odor molecules, then neurons designated by blue and green in the olfactory bulb are activated. The primary olfactory area of the cerebral cortex interprets the pattern of stimulation as the scent of a rose.

primarily in the parietal lobe. There, they are interpreted as particular tastes.

Humans can respond to a range of sweet, sour, salty, and bitter tastes. As the signals are received, the brain appears to survey their overall pattern. A "weighted average" of all taste messages is used by the brain as the perceived taste. For example, a glass of lemonade transmits sour, sweet, and perhaps bitter sensations. If you remember the taste of lemonade, your brain will recognize this pattern, and you'll think, "ahh, lemonade." Again, we can note that even though our senses depend on sensory receptors, the cortex integrates the incoming information and gives us our sensations.

Sense of Smell

Approximately 80–90% of what we perceive as "taste" actually is due to the sense of smell. This accounts for how dull food tastes when we have a head cold or a stuffed-up nose. Our sense of smell depends on between 10 and 20 million **olfactory cells** located within olfactory epithelium high in the roof of the nasal cavity (Fig. 14.5*a* and *b*). Olfactory cells are modified neurons. Each cell ends in a tuft of about five olfactory cilia, which bear receptor proteins for odor molecules.

How the Brain Receives Odor Information

Each olfactory cell has only one out of several hundred different types of receptor proteins. Nerve fibers from similar olfactory cells lead to the same neuron in the olfactory bulb (an extension of the brain). An odor contains many odor molecules, which activate a characteristic combination of receptor proteins. For example, a rose might stimulate olfactory cells, designated by blue and green in Figure 14.5, while a carnation might stimulate a different combination. An odor's signature in the olfactory bulb is determined by which neurons are stimulated. When the neurons communicate this information via the olfactory tract to the olfactory areas of the cerebral cortex, we know we have smelled a rose or a carnation.

The olfactory cortex is located in the temporal lobe. Some areas of the olfactory cortex receive smell sensations, while other areas contain olfactory memories.

Have you ever noticed that a certain aroma vividly brings to mind a certain person or place and can recreate emotions you feel about that person or place? A person's cologne may depress you by reminding of a failed relationship, while the smell of boxwood might create happier emotions by recalling your grandfather's farm. The olfactory bulbs have direct connections with the limbic system and its centers for emotion

Figure 14.6   **What are the three layers, or coats, of a human eye?** The sclera (the outer layer of the eye) becomes the cornea and the choroid (the middle layer) is continuous with the ciliary body and the iris. The retina (the inner layer) contains the photoreceptors for vision. The fovea centralis is the region where vision is most acute.

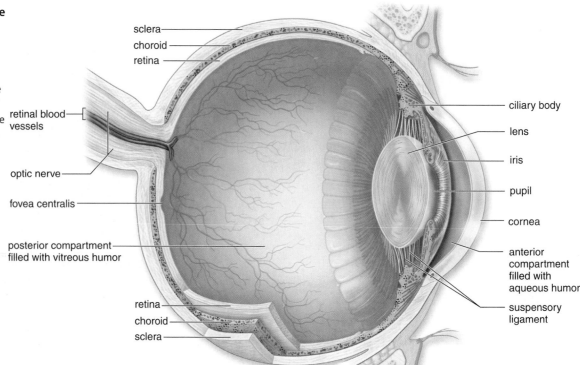

and memory. One investigator showed that when subjects smelled an orange while viewing a painting, memories of the painting were more vividly recalled, and the subjects also had many deep feelings about the painting.

The number of olfactory cells declines with age. This can be dangerous if an older person can't smell smoke or a gas leak. Older people also tend to apply excessive amounts of perfume or cologne before they can detect its smell.

Check Your Progress 14.3

1. a. Taste cells are what type of sensory receptor? b. What are the types of taste that result from stimulating these receptors?
2. a. What are the receptors responsible for smell called, and (b) where are they located?
3. Explain why a rose smells different than a carnation.

14.4 Sense of Vision

Vision requires the work of the eyes and the brain. As we shall see, much processing of stimuli occurs in the eyes before nerve signals are sent to the brain. Still, researchers estimate that at least a third of the cerebral cortex takes part in processing visual information.

Anatomy and Physiology of the Eye

The eyeball is an elongated sphere about 2.5 cm in diameter. It has three layers, or coats: the sclera, the choroid, and the retina (Fig. 14.6 and Table 14.2). The outer layer, the **sclera,** is white and fibrous except for the **cornea,** made of transparent collagen fibers. The cornea is the window of the eye.

The **choroid** is the thin middle coat. It has an extensive blood supply, and its dark pigment absorbs stray light rays that photoreceptors have not absorbed. This helps visual acuity. Toward the front, the choroid becomes the donut-shaped **iris.** The iris regulates the size of the **pupil,** a hole in the center of the iris through which light enters the eyeball. The color of the iris (and therefore the color of your eyes) correlates with its pigmentation. Heavily pigmented eyes are brown, while lightly pigmented eyes are green or blue. Behind the iris, the choroid thickens and forms the circular ciliary body. The **ciliary body** contains the ciliary muscle, which controls the shape of the lens for near and far vision.

Table 14.2	Structures of the Eye
Part	**Function**
SCLERA	Protects and supports eyeball
Cornea	Refracts light rays
Pupil	Admits light
CHOROID	Absorbs stray light
Ciliary body	Holds lens in place, accommodation
Iris	Regulates light entrance
RETINA	Contains sensory receptors for sight
Rod cells	Make black-and-white vision possible
Cone cells	Make color vision possible
Fovea centralis	Makes acute vision possible
OTHER	
Lens	Refracts and focuses light rays
Humors	Transmit light rays and support eyeball
Optic nerve	Transmits impulse to brain

The **lens** is attached to the ciliary body by suspensory ligaments and divides the eye into two compartments. The anterior compartment is in front of the lens and the posterior compartment is behind it. The anterior compartment is filled with a clear, watery fluid called the **aqueous humor.** A small amount of aqueous humor is continually produced each day. Normally, it leaves the anterior compartment by way of tiny ducts. When a person has **glaucoma,** these drainage ducts are blocked, and aqueous humor builds up. If glaucoma is not treated, the resulting pressure compresses the arteries that serve the nerve fibers of the retina, where photoreceptors are located. The nerve fibers begin to die due to lack of nutrients, and the person becomes partially blind. Eventually, total blindness can result.

The third layer of the eye, the **retina,** is located in the posterior compartment. This compartment is filled with a clear, gelatinous material called the **vitreous humor.** The vitreous humor holds the retina in place and supports the lens. The retina contains photoreceptors called rod cells and cone cells. The rods are very sensitive to light, but they do not see color. Therefore, at night or in a darkened room, we see only shades of gray. The cones, which require bright light, are sensitive to different wavelengths of light. This sensitivity gives us the ability to distinguish colors. The retina has a very special region called the **fovea centralis** where cone

cells are densely packed. Light is normally focused on the fovea when we look directly at an object. This is helpful because vision is most acute in the fovea centralis. Sensory fibers from the retina form the **optic nerve,** which takes nerve signals to the visual cortex.

Function of the Lens

The cornea, assisted by the lens and the humors, focuses images on the retina. Focusing starts with the cornea and continues as the rays pass through the lens and the humors. The image produced is much smaller than the object because light rays are bent (refracted) when they are brought into **focus.** If the eyeball is too long or too short, the person may need corrective lenses to bring the image into focus. The image on the retina is inverted (upside down) and reversed from left to right.

Visual accommodation occurs for close vision. During visual accommodation, the lens changes its shape to bring the image to focus on the retina. The shape of the lens is controlled by the ciliary muscle, within the ciliary body. When we view a distant object, the ciliary muscle is relaxed, causing the suspensory ligaments attached to the ciliary body to be taut. The ligaments put tension on the lens and cause it to remain relatively flat (Fig. 14.7a). When we view a near object, the ciliary muscle contracts, releasing the tension on the suspensory ligaments. The lens becomes round and thick

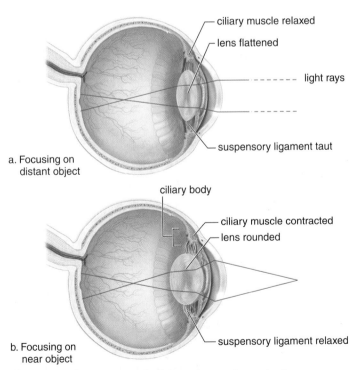

a. Focusing on distant object

— ciliary muscle relaxed
— lens flattened
— — — light rays
— suspensory ligament taut

b. Focusing on near object

ciliary body
— ciliary muscle contracted
— lens rounded
— suspensory ligament relaxed

Figure 14.7 **How is light focused onto the retina?**
Light rays from each point on an object are bent by the cornea and the lens in such a way that an inverted and reversed image of the object forms on the retina. **a.** When focusing on a distant object, the lens is flat because the ciliary muscle is relaxed and the suspensory ligament is taut. **b.** When focusing on a near object, the lens accommodates—it becomes rounded because the ciliary muscle contracts, causing the suspensory ligament to relax.

due to its natural elasticity (Fig. 14.7b). Thus, contraction or relaxation of the ciliary muscle allows the image to be focused on the retina. Close work requires contraction of the ciliary muscle, so it often causes muscle fatigue, known as eyestrain. Usually after the age of 40, the lens loses some of its elasticity and is unable to accommodate. For near vision, simple glasses with a magnifying lens solve this problem. Bifocal lenses may be necessary for those who already use corrective lenses.

Visual Pathway to the Brain

The pathway for vision begins once light has been focused on the photoreceptors in the retina. Some integration occurs in the retina, where nerve signals begin before the optic nerve transmits them to the brain.

Function of Photoreceptors Figure 14.8a illustrates the structure of the photoreceptors called **rod cells** and **cone cells.** Both rods and cones have an outer segment joined to an inner segment by a short stalk. Pigment molecules are embedded in the membrane of the many disks present in the outer segment. Synaptic vesicles are located at the synaptic endings of the inner segment.

The visual pigment in rods is a deep purple pigment called rhodopsin (Fig. 14.8b). **Rhodopsin** is a complex molecule made up of the protein opsin and a light-absorbing molecule called **retinal,** a derivative of vitamin A. When a rod absorbs light, rhodopsin splits into opsin and retinal. This leads to a cascade of reactions and the closure of ion channels in the rod cell's plasma membrane. The release of inhibitory transmitter molecules from the rod's synaptic vesicles ceases. Thereafter, signals go to other neurons in the retina. Rods are very sensitive to light and, therefore, are suited to night vision. (Carrots are rich in vitamin A, so it is true that eating carrots can improve your night vision.) Rod cells are plentiful throughout the entire retina, except the fovea. Therefore, rods also provide us with peripheral vision and perception of motion.

The cones, on the other hand, are located primarily in the fovea and are activated by bright light. They allow us to detect the fine detail and the color of an object. **Color vision** depends on three different types of cones, which contain pigments called the B (blue), G (green), and R (red) pigments. Each pigment is made up of retinal and opsin, but there is a slight difference in the opsin structure of each. This accounts for their individual absorption patterns. Various combinations of cones are believed to be stimulated by in-between shades of color.

Function of the Retina The retina has three layers of neurons (Fig. 14.9). The layer closest to the choroid contains

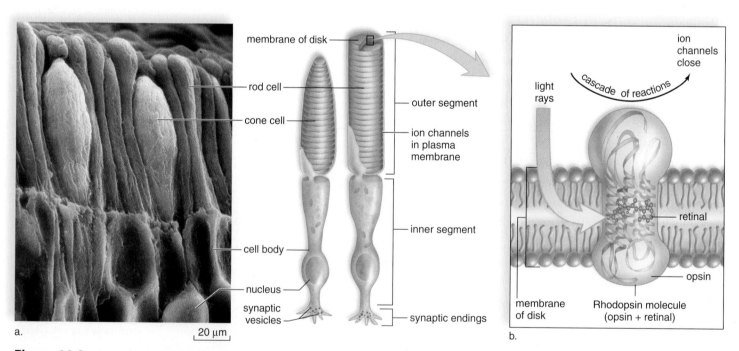

Figure 14.8 **How do the two types of photoreceptors generate nerve signals?**
a. The outer segment of rods and cones contains stacks of membranous disks, which contain visual pigments. **b.** In rods, the membrane of each disk contains rhodopsin, a complex molecule containing the protein opsin and the pigment retinal. When rhodopsin absorbs light energy, it splits, releasing retinal, which sets in motion a cascade of reactions that cause ion channels in the plasma membrane to close. Thereafter, nerve signals go to the brain.

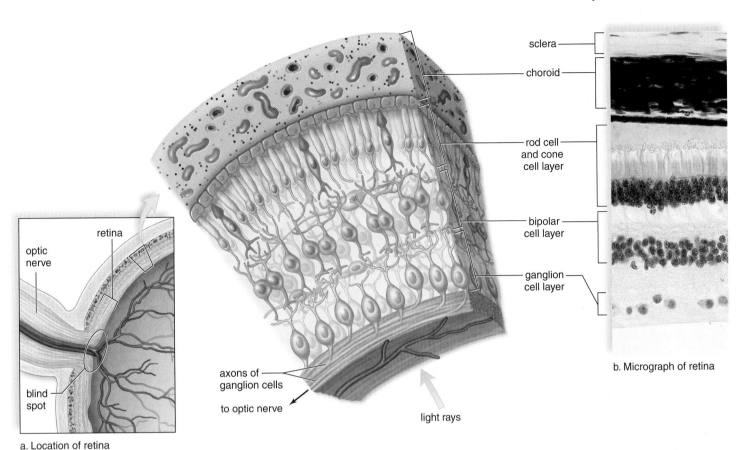

sclera

choroid

rod cell
and cone
cell layer

bipolar
cell layer

ganglion
cell layer

retina

optic
nerve

blind
spot

axons of
ganglion cells

to optic nerve

light rays

b. Micrograph of retina

a. Location of retina

Figure 14.9 **What is the structure of the retina? How does its structure suit the retina's function?**
a. The retina is the inner layer of the eyeball. Rod and cone cells, located at the back of the retina nearest the choroid, synapse with bipolar cells, which synapse with ganglion cells. Integration of signals occurs at these synapses; therefore, much processing occurs in bipolar and ganglion cells. Further, many rod cells share one bipolar cells, but cone cells do not. Certain cone cells synapse with only one ganglion cell. Cone cells, in general, distinguish more detail than do rod cells. **b.** Micrograph shows that the sclera and choroid are relatively thin compared to the retina, which has several layers of cells.

the rod cells and cone cells. A layer of bipolar cells covers the rods and cones. The innermost layer contains ganglion cells, whose sensory fibers become the optic nerve. Only the rod cells and the cone cells are sensitive to light, and therefore, light must penetrate to the back of the retina before they are stimulated.

The rod cells and the cone cells synapse with the bipolar cells. Next, signals from bipolar cells stimulate ganglion cells whose axons become the optic nerve. Notice in Figure 14.9 that there are many more rod cells and cone cells than ganglion cells. The retina has as many as 150 million rod cells and 6 million cone cells but only 1 million ganglion cells. The sensitivity of cones versus rods is mirrored by how directly they connect to ganglion cells. As many as 150 rods may activate the same ganglion cell. No wonder stimulation of rods results in vision that is blurred and indistinct. In contrast, some cone cells in the fovea centralis activate only one ganglion cell. This explains why cones, especially in the

fovea centralis, provide us with a sharper, more detailed image of an object.

As signals pass to bipolar cells and ganglion cells, integration occurs. Therefore, considerable processing occurs in the retina before ganglion cells generate nerve signals. Ganglion cells converge to form the optic nerve, which transmits information to the visual cortex. Additional integration occurs in the visual cortex.

Blind Spot Figure 14.9 also provides an opportunity to point out that there are no rods and cones where the optic nerve exits the retina. Therefore, no vision is possible in this area. You can prove this to yourself by putting a dot to the right of center on a piece of paper. Use your right hand to move the paper slowly toward your right eye, and make sure you look straight ahead. The dot will disappear at one point—this is your right eye's **blind spot.** The two eyes together provide complete vision because the blind spot for the right eye is not the same as the blind spot for the left eye. The blind spot for the right eye is right of center and the blind spot for the left eye is left of center.

From the Retina to the Visual Cortex To reach the visual cortex, the optic nerves carry nerve impulses from the eyes to the optic chiasma (Fig. 14.10). The **optic chiasma** has an X shape, formed by a crossing-over of optic nerve fibers. After exiting the optic chiasma, the optic nerves continue as **optic tracts.** Fibers from the right half of each retina converge and continue on together in the right optic tract. Similarly, the nerve fibers from the left half of each retina join to form the left optic tract, traveling together to the brain.

The optic tracts sweep around the hypothalamus, and most fibers synapse with neurons in nuclei (masses of neuron cell bodies) within the thalamus. Axons from the thalamic nuclei form optic radiations that take nerve impulses to the *visual cortex* within the occipital lobe. The image is split in the visual cortex. This division of incoming informa-

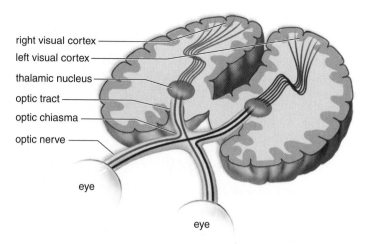

right visual cortex
left visual cortex
thalamic nucleus
optic tract
optic chiasma
optic nerve
eye
eye

Figure 14.10   **What is the function of the optic chiasma?** Because of the optic chiasma, data from the right half of each retina go to the right visual cortex, and data from the left half of the retina go to the left visual cortex. These data are then combined to allow us to see the entire visual field.

"Hey Mike, I got Dr. Kerry's report yesterday. Sounds like some pretty good news about your eye, huh?"

Mike had known his family doctor, Steve Cooke, forever. The two had been on a first name basis for a long, long time. "Yup, pretty good news, Stevie. My right eye is pretty well fixed up. I still can't see out of this corner." He pointed to an area close to his right temple.

"That's in the peripheral vision part of your eye. You can still see to read and watch the TV, right?" Dr. Cooke asked. "And the left eye is still good, according to this report."

Mike nodded. "Sure, I can still watch the tube. And I can read the sports page with reading glasses, same as always."

"Well, that means that the part of your retina that allows you to have close-up and color vision is still alive and getting its oxygen. You're a pretty lucky guy.

"You know what I'm gonna say next, don't you?" Dr. Cooke continued. Mike nodded to his friend, but wouldn't meet his gaze.

Dr. Cooke glared, but spoke softly. "Well, I'm going to repeat it all over again, and maybe this time you'll listen. You're about thirty pounds overweight, and your blood sugar is all over the map. You won't stick yourself to test your blood sugar twice a day like I asked

you to. You probably don't inject your insulin on a regular schedule either. The meds I gave you for glucose control and high blood pressure are gathering dust on the shelf, right?" He gave Mike's arm a gentle punch.

"Well, here's your really bad news, you big blockhead," the doctor continued. "You're starting to get some protein in your urine, and that means your kidneys are starting to fail. You mentioned tingly or numb sensations in your toes, so the blood vessels in your legs are getting blocked up. Bottom line is, you're either going to have a massive heart attack or stroke because of your diabetes and high blood pressure. Or maybe you'll go blind, or have to have a leg amputated. Or you'll have to go on kidney dialysis." Mike's shoulders slumped defeatedly.

"Pretty rough since Sally died, buddy?" Dr. Cooke asked gently. Mike nodded, then looked away again.

"Well, there's sure nothing like a new grandchild to cheer a guy up," his friend asserted. "This eye thing's your wake-up call. I'm going to write a prescription for a mild antidepressant, and I want you to go to a grief counselor. And if you promise to take care of yourself like I've been telling you, you'll be showing me pictures of Lauren's baby in a month or two!"

tion happens because the right visual cortex receives information from the right optic tract, and the left visual cortex receives information from the left optic tract. For good depth perception, the right and left visual cortices communicate with each other. Also, because the image is inverted and reversed, it must be righted in the brain for us to correctly perceive the visual field.

Abnormalities of the Eye

Color blindness and misshapen eyeballs are two common abnormalities of the eyes. Complete color blindness is extremely rare and is caused by a genetic mutation. In most instances, only one type of cone is defective or deficient in number. The most common mutation is the inability to see the colors red and green. This abnormality affects 5–8% of the male population. If the eye lacks cones that respond to red wavelengths, green colors are accentuated, and vice versa.

Distance Vision

If you can see from 20 feet what a person with normal vision can see from 20 feet, you are said to have 20/20 vision. Persons who can see close objects, but can't see the letters on an optometrist's chart from 20 feet, are said to be nearsighted. **Nearsighted** people can see close objects better than they can see objects at a distance. These individuals have an elongated eyeball, and when they attempt to look at a distant object, the image is brought to focus in front of the retina (Fig. 14.11a). They can see close objects because their lens can compensate for the long eyeball. To see distant objects, these people can wear concave lenses, which spread the light rays so that the image focuses on the retina.

Persons who can easily see the optometrist's chart but cannot see close objects well are **farsighted.** These individuals can see distant objects better than they can see close objects. They have a shortened eyeball, and when they try to see close objects, the image is focused behind the retina (Fig. 14.11b). When the object is distant, the lens can compensate for the short eyeball. When the object is close, these persons can wear convex lenses to increase the bending of light rays so that the image can be focused on the retina.

When the cornea or lens is uneven, the image is fuzzy. The light rays cannot be evenly focused on the retina. This condition, called **astigmatism,** can be corrected by an unevenly ground lens to compensate for the uneven cornea (Fig. 14.11c).

Many people today opt to have LASIK surgery instead of wearing lenses. LASIK surgery is discussed in the Health Focus on page 316.

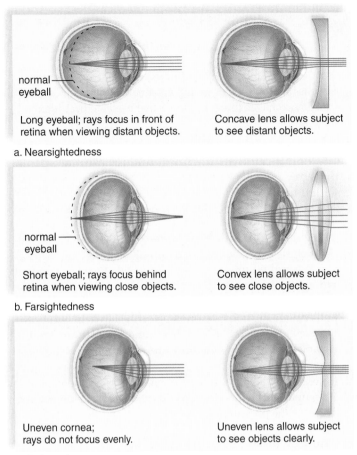

a. Nearsightedness

Long eyeball; rays focus in front of retina when viewing distant objects.

Concave lens allows subject to see distant objects.

b. Farsightedness

Short eyeball; rays focus behind retina when viewing close objects.

Convex lens allows subject to see close objects.

c. Astigmatism

Uneven cornea; rays do not focus evenly.

Uneven lens allows subject to see objects clearly.

Figure 14.11 What are the most common abnormalities of the eye? How do corrective lenses work to fix the problem?
a. A concave lens in nearsighted persons focuses light rays on the retina. **b.** A convex lens in farsighted persons focuses light rays on the retina. **c.** An uneven lens in persons with astigmatism focuses light rays on the retina.

> **Check Your Progress 14.4**
> 1. **What parts of the eye assist in focusing an image on the retina?**
> 2. **What are the two types of photoreceptors?**
> 3. **Once photoreceptors initiate a visual signal, what other cells in the eye integrate the signal and pass it on to the visual cortex?**

14.5 Sense of Hearing

The ear has two sensory functions: hearing and balance (equilibrium). The sensory receptors for both of these are located in the inner ear. Each consists of **hair cells** with

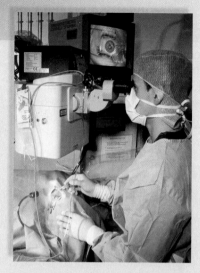

Figure 14A LASIK surgery.

Correcting Vision Problems

Poor vision can be due to a number of problems, some more serious than others. For example, retinal detachment, cataracts, and glaucoma are three conditions that need medical attention.

Cataracts and Glaucoma

Cataracts develop when the lens of the eye becomes cloudy. Normally the lens is clear and allows light to pass through easily. A cloudy lens will decrease light levels that reach the retina and slowly cause vision loss. Fortunately, a doctor can surgically remove the cloudy lens and replace it with a clear plastic lens. This will often restore the light level passing through the lens and improve the patients' vision.

When fluid pressure builds up inside the eye, a patient is diagnosed with glaucoma. Glaucoma can lead to a decrease in vision and eventually causes blindness. Special eye drops and oral medications are often prescribed to help reduce the intraocular pressure. If eye drops and medications are not capable of controlling the pressure, surgery may be the only option. During glaucoma surgery, the doctor will use a laser to create tiny holes in the eye where the cornea and iris meet. It is hoped this will increase fluid drainage from the eye and decrease the pressure inside the eye.

The Benefits of LASIK Surgery

For many people, a sign of aging is the slow and steady decrease of their ability to see close-up. A difficulty in focusing on small print is usually the first sign of presbyopia, a condition that usually begins in the late 30s. By age 55, nearly 100% of the population is affected. Reading in low light situations becomes more difficult, and letters begin to look fuzzy when reading close-up. Many people suffering from presbyopia will often experience headaches while reading. To accommodate for their deteriorating vision, presbyopia sufferers will hold reading materials at arm's length. One solution to presbyopia is to wear a simple magnifying lens (purchased at a pharmacy, for example). Those already wearing glasses may need bifocal lenses. Bifocals are designed to correct vision at a distance of 12" to 18". These lenses work well for most people while reading, but pose a problem for people who use a computer. Computer monitors are usually 19" to 24" away. This forces bifocal lens wearers to constantly move their head up and down in an attempt to switch between the close and distant viewing sections of the bifocals. Another solution is to wear contact lenses with one eye corrected to see close objects and the other eye corrected to see distant objects. The same type of correction can be done with LASIK surgery (Fig. 14A). LASIK, which stands for laser in-situ keratomileusis, is generally a safe and effective treatment option for a wide array of vision problems.

LASIK is a quick and painless procedure that involves the use of a laser to permanently change the shape of the cornea. For the majority of patients, LASIK will improve their vision and reduce their dependency on corrective lenses. The ideal LASIK candidate is over 18 years of age and has had a stable contact or glasses prescription for at least two years. Patients need to have a cornea thick enough to allow the surgeon to safely create a clean corneal flap of appropriate depth. Typically, patients affected by common vision problems (nearsightedness, astigmatism, or farsightedness) will respond well to LASIK. Anyone who suffers from any disease that will decrease their ability to heal properly after surgery will not be a candidate for LASIK. Candidates should thoroughly discuss the procedure with their eye-care professional before electing to have LASIK. People need to realize that the goal of LASIK is to reduce their dependency on glasses or contact lenses, not completely eliminate them.

Individuals who suffer from cataracts, advanced glaucoma, corneal or other eye diseases will not be considered for LASIK. Patients who expect LASIK to completely correct their visual problems and make them completely independent of their corrective lenses are not good candidates either.

The LASIK Procedure

During the LASIK procedure, a small flap of tissue (the conjunctiva) is cut away from the front of the eye. The flap is folded back exposing your cornea, allowing the surgeon to remove a defined amount of tissue from your cornea. Each pulse of the laser will remove a small amount of corneal tissue, allowing the surgeon to flatten or increase the steepness of the curve of your cornea. After the procedure, the flap of tissue is put back into place and allowed to heal on its own. LASIK patients receive eye drops or medications to help relieve the pain of the procedure. Improvements to your vision will begin as early as the day after the surgery, but typically take two to three months. Most patients will have vision close to 20/20, but your chances for improved vision are based in part on how good your eyes were before the surgery.

As with any surgery, complications are possible. Adverse effects include a sensation of having something in your eye or having blurred vision. You might also see halos around objects or be very sensitive to glare. In addition, dryness can cause eye irritation. Typically, these effects are temporary, and the rate of complications following surgery is very low. Always consult with your doctor before considering any type of surgery.

stereocilia (long microvilli) sensitive to mechanical stimulation. They are mechanoreceptors.

Anatomy and Physiology of the Ear

Figure 14.12 shows that the ear has three divisions: outer, middle, and inner. The **outer ear** consists of the **pinna** (external flap) and the **auditory canal.** The opening of the auditory canal is lined with fine hairs and sweat glands. Modified sweat glands are located in the upper wall of the canal. They secrete earwax, a substance that helps guard the ear against the entrance of foreign materials, such as air pollutants.

The **middle ear** begins at the **tympanic membrane** (eardrum) and ends at a bony wall containing two small openings covered by membranes. These openings are called the **oval window** and the **round window.** Three small bones are found between the tympanic membrane and the oval window. Collectively they are called the **ossicles.** Individually, they are the **malleus** (hammer), the **incus** (anvil), and the **stapes** (stirrup) because

their shapes resemble these objects. The malleus adheres to the tympanic membrane, and the stapes touches the oval window. An **auditory tube** (eustachian tube), which extends from the middle ear to the nasopharynx, permits equalization of air pressure. Chewing gum, yawning, and swallowing in elevators and airplanes help move air through the auditory tubes upon ascent and descent. As this occurs, we often feel the ears "pop."

Whereas the outer ear and the middle ear contain air, the inner ear is filled with fluid. The **inner ear** has three areas: The **semicircular canals** and the **vestibule** are concerned with equilibrium. The **cochlea** is concerned with hearing. The cochlea resembles the shell of a snail because it spirals.

Auditory Pathway to the Brain

The sound pathway begins with the auditory canal. Thereafter, hearing requires the other parts of the ear, the cochlear nerve, and the brain.

Through the Auditory Canal and Middle Ear The process of hearing begins when sound waves enter the auditory

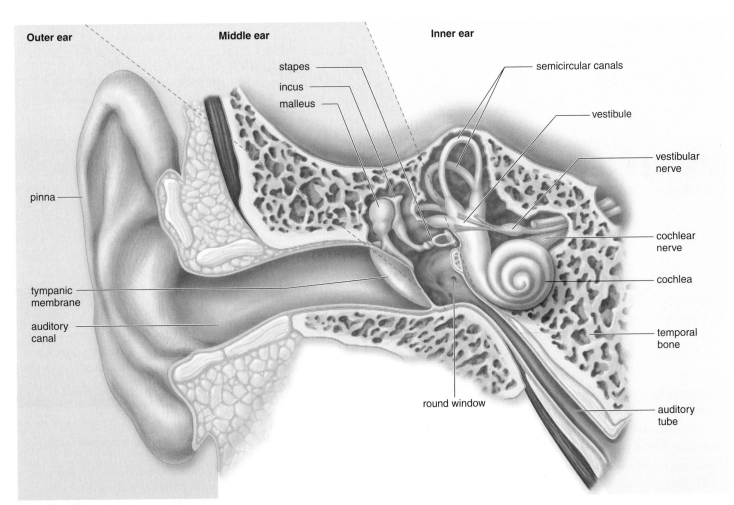

Figure 14.12 **What are the three divisions of the human ear?**
The external ear consists of the pinna (the structure commonly referred to as the "ear") and the auditory canal. The tympanic membrane separates the external ear from the middle ear. In the middle ear, the malleus (hammer), the incus (anvil), and the stapes (stirrup) amplify sound waves. In the inner ear, the mechanoreceptors for equilibrium are in the semicircular canals and the vestibule. The mechanoreceptors for hearing are in the cochlea.

canal. Just as ripples travel across the surface of a pond, sound waves travel by the successive vibrations of molecules. Ordinarily, sound waves do not carry much energy. However, when a large number of waves strike the tympanic membrane, it moves back and forth (vibrates) ever so slightly. As you know, the auditory ossicles attach to one another: malleus to incus, incus to stapes. The malleus is attached to the inner wall of the tympanic membrane. Thus, vibrations of the tympanic membrane will cause vibration of the malleus and in turn, the incus and stapes. The magnitude of the original pressure wave increases significantly as the vibrations move along the auditory ossicles. The pressure is multiplied about twenty times. Finally, the stapes strikes the membrane of the oval window, causing it to vibrate. In this way, the pressure is passed to the fluid within the cochlea.

From the Cochlea to the Auditory Cortex   If the cochlea is examined in cross section (Fig. 14.13), you can see that it has three canals. The sense organ for hearing, called the **spiral organ** (organ of Corti), is located in the cochlear canal. The spiral organ consists of little hair cells and a gelatinous material called the **tectorial membrane.** The hair cells sit on the basilar membrane, and their stereocilia are embedded in the tectorial membrane.

When the stapes strikes the membrane of the oval window, pressure waves move from the vestibular canal to the tympanic canal across the basilar membrane. The basilar membrane moves up and down, and the stereocilia of the hair cells embedded in the tectorial membrane bend. Then,

Have You Ever Wondered...

Why does that annoying song you hear seem to replay itself in your head all day?

Scientists who study the brain and special senses call such phenomena *earworms*. These are most likely caused by songs with repetitive, silly, or catchy lyrics. Their cause is not fully understood. Scientists have documented that the auditory cortex—the site for storing memories of sounds you've heard—becomes very active when earworms run through your mind. Earworms may be a way to keep the brain "idling," much as a car idles when you're at a stoplight. When you're ready for active thinking and problem solving, your brain is ready to go.

The good news is, you can make an earworm go away. Distract yourself with a complex task, such as reading or solving a puzzle. Eat something very spicy—the taste sensation will stimulate other areas of the brain. If all else fails, just think of a song that is equally annoying, and it will most likely replace the first!

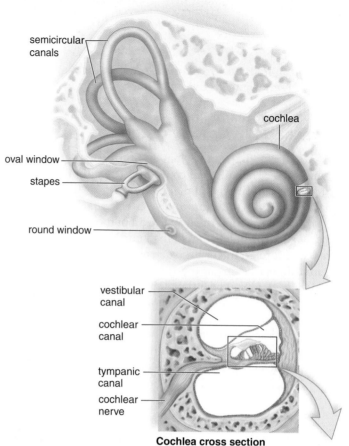

Cochlea cross section

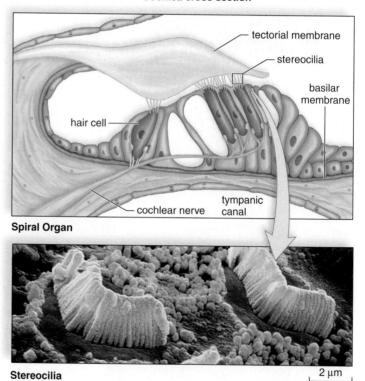

Spiral Organ

Stereocilia   2 μm

Figure 14.13   **How does the spiral organ (organ of Corti) translate sound waves into nerve signals?**
The spiral organ (organ of Corti) is located within the cochlea. The spiral organ consists of hair cells resting on the basilar membrane, with the tectorial membrane above. Pressure waves moving through the canals cause the basilar membrane to vibrate. This causes the stereocilia embedded in the tectorial membrane to bend. Nerve impulses traveling in the cochlear nerve result in hearing.

Noise Pollution

Though we can sometimes tune its presence out, unwanted noise is all around us. Noise pollution is noise from the environment that is annoying, distracting, and potentially harmful. It comes from airplanes, cars, lawn mowers, machinery, our own loud music, and that of our neighbors. It is present at our workplaces, in public spaces like amusement parks, and at home. Its prevalence allows loud noise to have a potentially high impact on our welfare.

How Does Noise Affect Us?

How does noise affect human health? Perhaps the greatest worry about noise pollution is that exposure to loud (>85 decibels) or chronic noises can damage cells of the inner ear and cause hearing loss (Fig. 14B). When we are young, we often do not consider the damage that noise may be doing to our spiral organ. The stimulation of loud music is often sought by young people at rock concerts without regard to the possibility that their hearing may be diminished as a result. Over the years, loud noises can bring deafness and accompanying depression when we are seniors.

Noise can affect well-being by other means, too. Data from studies of environmental noise can be difficult to interpret because of the presence of other confounding factors, including physical or chemical pollution. The tolerance level for noise also varies from person to person. Nonetheless, laboratory and field studies show that noise may be detrimental in nonauditory ways. Its effects on mental health include annoyance, inability to concentrate, and increased irritability. Long-term noise exposure from air or car traffic may impair cognitive ability, language learning, and memory in children.

(continued)

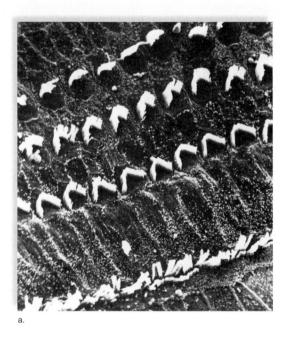

a.

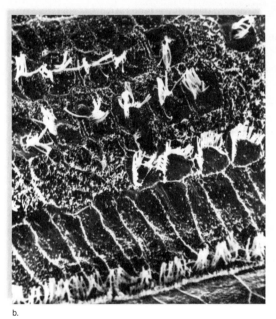

b.

Figure 14B How does loud noise damage the hair cells in the spiral organ and lead to hearing loss?
a. Normal hair cells in the spiral organ of a guinea pig. **b.** Damaged cells. This damage occurred after 24-hour exposure to a noise level equivalent to that at a heavy-metal rock concert (see Table 14A). Hearing is permanently impaired because lost cells will not be replaced, and damaged cells may also die.

Table 14A	Noises That Affect Hearing	
Type of Noise	**Sound Level (decibels)**	**Effect**
"Boom car," jet engine, shotgun, rock concert	Over 125	Beyond threshold of pain; potential for hearing loss high
Nightclub, thunderclap	Over 120	Hearing loss likely
Earbuds in external ear canal	110–120	Hearing loss likely
Chain saw, pneumatic drill, jackhammer, symphony orchestra, snowmobile, garbage truck, cement mixer	100–200	Regular exposure of more than 1 min risks permanent hearing loss
Farm tractor, newspaper press, subway, motorcycle	90–100	Fifteen minutes of unprotected exposure potentially harmful
Lawn mower, food blender	85–90	Continuous daily exposure for more than 8 hrs can cause hearing damage
Diesel truck, average city traffic noise	80–85	Annoying; constant exposure may cause hearing damage

(continued from p. 319)

Noise is often the grounds for losses in sleep and productivity and can induce stress. Additionally, several studies have suggested that it negatively affects cardiovascular health, though more research needs to be done to confirm this link.

Federal and Local Control

Noise pollution has been a concern for several decades. In 1972, the Noise Control Act was passed as a means for coordinating federal noise control and research and to develop noise emission standards. The aim was to protect Americans from "noise that jeopardizes their health or welfare." The Environmental Protection Agency (EPA) had federal authority to regulate noise pollution, and their Office of Noise Abatement and Control (ONAC) worked on establishing noise guidelines. However, funding for ONAC was cut off in 1981. Today, there is no national noise policy, though several federal governmental departments have some oversight.

Workplace noise exposure is controlled by the Occupational Safety and Health Administration (OSHA). OSHA has set guidelines for workplace noise. OSHA regulations require that protective gear be provided if sound levels exceed certain values. This may include noise-reducing ear muffs and other equipment for people who work around big equipment. However, OSHA guidelines don't cover things like telephone ringing and computer or typewriter noise that may be present in a nonindustrial environment such as an open-plan office. Aviation noise and traffic noise reduction plans are overseen by the Department of Transportation, the Federal Aviation Administration (FAA), and the Federal Highway Administration (FHA), respectively.

Some cities and local governments have taken steps to decrease unwanted or annoying noise. Many communities have noise ordinances and control noise levels during particular times of day. Persons throwing loud house parties might get a knock on their door from the police, and owners of barking dogs, a visit from animal control. Business enterprises may also be regulated. For example, concerts at outdoor venues may be required to finish by a certain time in the evening. However, there is often opposition to this type of control, on the basis that it impedes commerce.

Many people believe that much more should be done to curb noise pollution. Yet there are difficulties in regulating noise pollution generated by private persons or business owners. Individuals often feel that they should be free to do as they wish on private property, including playing music at full volume or running machinery. For example, club and café owners may believe that their livelihood depends on having reasonably loud sound systems. The same may hold true of sports fields and stadiums. There are also challenges to regulating airport and traffic noise. The main responsibilities of the FAA and FHA—promoting aviation industry growth and maintaining highway infrastructure—can be at odds with controlling noise for the public's sake. Therefore, noise abatement isn't necessarily their top priority.

Finding neighbor- and commerce-friendly solutions to noise problems will likely require individual citizens, communities, and governments to work together. Stopping the source of noise may not be the only answer to this issue. The implementation of new technology could also make things quieter by preventing sounds from traveling too far away from their sources.

Decide Your Opinion

1. Do you think that more should be done to curb noise pollution in your neighborhood? If so, why?
2. Do you believe that the government has the right to regulate what someone does on their own property? Why or why not?
3. Should regulation of airport and traffic noise fall to the Department of Transportation? If not, who may better control airport or highway noise?
4. If technology is available to reduce noise on highways, who do you think should pay for implementing it—tax payers, car owners, or others?

nerve signals begin in the **cochlear nerve** and travel to the brain. When they reach the auditory cortex in the temporal lobe, they are interpreted as a sound.

Each part of the spiral organ is sensitive to different wave frequencies, or pitch. Near the tip, the spiral organ responds to low pitches, such as a tuba. Near the base (beginning), it responds to higher pitches, such as a bell or a whistle. The nerve fibers from each region along the length of the spiral organ lead to slightly different areas in the auditory cortex. The pitch sensation we experience depends upon which region of the basilar membrane vibrates and which area of the auditory cortex is stimulated.

Volume is a function of the amplitude (strength) of sound waves. Loud noises cause the fluid within the vestibular canal to exert more pressure and the basilar membrane to vibrate to a greater extent. The resulting increased stimulation is interpreted by the brain as volume.

Check Your Progress 14.5

1. What parts in the ear assist in amplifying sound waves?
2. a. What receptors allow us to hear? b. Where are these receptors located? c. How do they function?

14.6 Sense of Equilibrium

The vestibular nerve originates in the semicircular canals, saccule, and utricle. It takes nerve signals to the brain stem and cerebellum (Fig. 14.14). Through its communication with the brain, the vestibular nerve helps us achieve

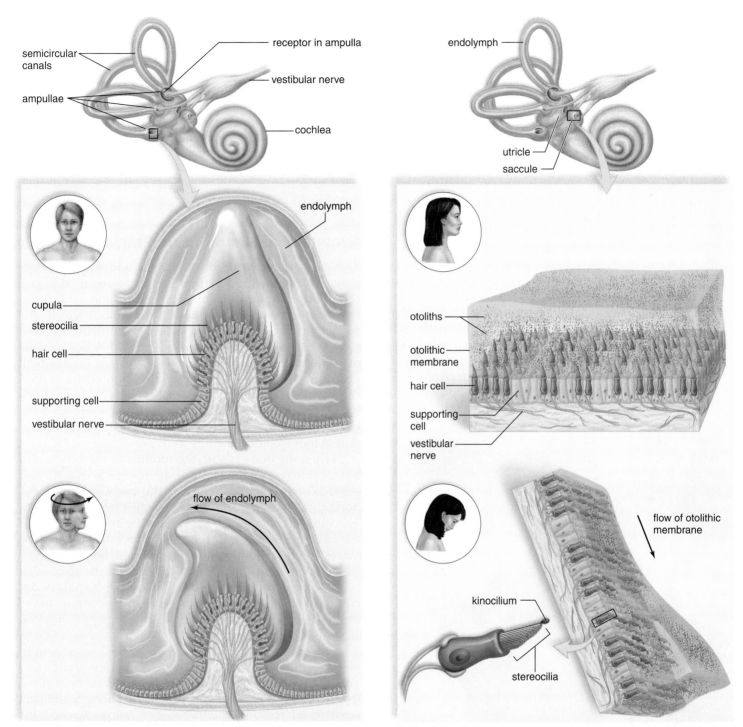

a. Rotational equilibrium: receptors in ampullae of semicircular canal

b. Gravitational equilibrium: receptors in utricle and saccule of vestibule

Figure 14.14   How do the mechanoreceptors of the inner ear sense body position?

a. Rotational equilibrium. The ampullae of the semicircular canals contain hair cells with stereocilia embedded in a cupula. When the head rotates, the cupula is displaced, bending the stereocilia. Thereafter, nerve impulses travel in the vestibular nerve to the brain. **b.** Gravitational equilibrium. The utricle and the saccule contain hair cells with stereocilia embedded in an otolithic membrane. When the head bends, otoliths are displaced, causing the membrane to sag and the stereocilia to bend. If the stereocilia bend toward the kinocilium, the longest of the stereocilia, nerve impulses increase in the vestibular nerve. If the stereocilia bend away from the kinocilium, nerve impulses decrease in the vestibular nerve. This difference tells the brain, by way of the vestibular nerve, in which direction the head moved.

equilibrium, but other structures in the body are also involved. For example, we already mentioned that proprioceptors are necessary for maintaining our equilibrium. Vision, if available, usually provides extremely helpful input the brain can act upon. To explain, let's take a look at the two sets of mechanoreceptors for equilibrium.

Rotational Equilibrium Pathway

Mechanoreceptors in the **semicircular canals** detect rotational and/or angular movement of the head **(rotational equilibrium)** (Fig. 14.14a). The three semicircular canals are arranged so that there is one in each dimension of space. The base of each of the three canals, called the **ampulla,** is slightly enlarged. Little hair cells, whose stereocilia are embedded within a gelatinous material called a cupula, are found within the ampullae. Each ampulla responds to head rotation in a different plane of space because of the way the semicircular canals are arranged. As fluid within a semicircular canal flows over and displaces a cupula, the stereocilia of the hair cells bend. This causes a change in the pattern of signals carried by the vestibular nerve to the brain. The brain uses information from the hair cells within each ampulla of the semicircular canals to maintain equilibrium. Appropri-

ate motor output to various skeletal muscles can correct our present position in space as needed.

Why does spinning around cause you to become dizzy? When we spin, the cupula slowly begins to move in the same direction we are spinning, and bending of the stereocilia causes hair cells to send messages to the brain. As time goes by, the cupula catches up to the rate we are spinning, and the hair cells no longer send messages to the brain. When we stop spinning, the slow-moving cupula continues to move in the direction of the spin and the stereocilia bend again, indicating that we are moving. Yet, the eyes know we have stopped. The mixed messages sent to the brain cause us to feel dizzy (Fig. 14.15).

Gravitational Equilibrium Pathway

The mechanoreceptors in the utricle and saccule detect movement of the head in the vertical or horizontal planes **(gravitational equilibrium).** The **utricle** and **saccule** are two membranous sacs located in the inner ear near the semicircular canals. Both of these sacs contain little hair cells, whose stereocilia are embedded within a gelatinous material called an otolithic membrane (see Fig. 14.14b). Calcium carbonate ($CaCO_3$) granules, or **otoliths,** rest on this membrane. The utricle is especially sensitive to horizontal (back-forth) movements and the bending of the head, while the saccule responds best to vertical (up-down) movements.

When the body is still, the otoliths in the utricle and the saccule rest on the otolithic membrane above the hair cells. When the head bends or the body moves in the horizontal and vertical planes, the otoliths are displaced. The otolithic membrane sags, bending the stereocilia of the hair cells beneath. If the stereocilia move toward the largest stereocilium, called the **kinocilium,** nerve impulses increase in the vestibular nerve. If the stereocilia move away from the kinocilium, nerve impulses decrease in the vestibular nerve. The frequency of nerve impulses in the vestibular nerve indicates whether you are moving up or down.

These data reach the cerebellum, which uses them to determine the direction of the movement of the head at that moment. It's important to remember that the cerebellum (page 287) is vital to maintain balance and gravitational equilibrium. The cerebellum processes information from the inner ear (the semicircular canals, utricle, and saccule) as well as visual and proprioceptive inputs. In addition, the motor cortex in the frontal lobe of the brain signals where the limbs should be located at any particular moment. After integrating all these nerve inputs, the cerebellum coordinates skeletal muscle contraction to correct our position in space if necessary.

Figure 14.15   **Is it still fun to spin around in circles, like you did as a child?**

Using playground equipment, children love to test their rotational equilibrium. Adults are not too interested.

Continuous stimulation of the stereocilia can contribute to motion sickness, especially when messages reaching the brain conflict with visual information from the eyes. Consider what might happen if you're standing inside a ship tossing up and down on the waves. Your visual inputs signal that you're standing still, because you see the wall in front of you and that wall isn't moving. However, the inputs from all three sensory areas of the inner ear tell your brain that you are moving up and down and from side to side. The result: you're seasick!

If you can match the two sets of information coming into the brain, you will begin to feel better. Thus, it makes sense to stand on deck if possible, so that visual signals and inner ear signals both tell your brain that you're moving. Prescription and over-the-counter medications can help alleviate motion sickness. Wristbands that apply pressure might also be helpful.

> **Check Your Progress 14.6**
>
> 1. a. Where are the receptors that detect rotation located?
>   b. How do they function?
> 2. a. Where are the receptors that detect gravity located?
>   b. How do they function?

Summarizing the Concepts

14.1 Sensory Receptors and Sensations

There are four types of sensory receptors: chemoreceptors; photoreceptors; mechanoreceptors; and thermoreceptors.

- Sensory receptors initiate nerve signals transmitted to the spinal cord and/or brain.
- Sensation occurs when nerve signals reach the cerebral cortex.
- Perception is an interpretation of sensations.

14.2 Proprioceptors and Cutaneous Receptors

Proprioceptors

- are mechanoreceptors involved in reflex actions.
- help maintain equilibrium and posture.

Cutaneous Receptors

- are found in the skin.
- sense touch, pressure, temperature, and pain.

14.3 Senses of Taste and Smell

Taste and smell are due to chemoreceptors stimulated by molecules in the environment.

Sense of Taste Microvilli of taste cells have receptor proteins for molecules that cause the brain to distinguish sweet, sour, salty, and bitter tastes.

Sense of Smell The cilia of olfactory cells have receptor proteins for molecules that cause the brain to distinguish odors.

14.4 Sense of Vision

Vision depends on the eye, the optic nerves, and the visual areas of the cerebral cortex.

Anatomy and Physiology of the Eye

The eye has three layers:

- The sclera (outer layer) protects and supports the eyeball.
- The choroid (middle, pigmented layer) absorbs stray light rays.

- The retina (inner layer) contains the rod cells (sensory receptors for dim light) and cone cells (sensory receptors for bright light and color).

  Function of the Lens The lens (assisted by the cornea and the humors) brings the light rays to focus on the retina. To see a close object, visual accommodation occurs as the lens becomes round and thick.

  Visual Pathway to the Brain The visual pathway begins when light strikes photoreceptors (rod cells and cone cells) in the retina. The optic nerves carry nerve impulses from the eyes to the optic chiasma, then pass through the thalamus before reaching the primary vision area in the occipital lobe of the brain.

Abnormalities of the Eye

- color blindness
- misshapen eyeballs (cause of nearsightedness, farsightedness, or astigmatism)

14.5 Sense of Hearing

Hearing depends on the ear, the cochlear nerve, and the auditory areas of the cerebral cortex.

Anatomy and Physiology of the Ear

The ear has three parts:

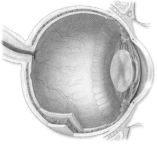

- In the outer ear, the pinna and the auditory canal direct sound waves to the middle ear.
- In the middle ear, the tympanic membrane and the ossicles (malleus, incus, and stapes) amplify sound waves.
- In the inner ear, the semicircular canals detect rotational equilibrium; the utricle and saccule detect gravitational equilibrium; and the cochlea houses the spiral organ, which contains mechanoreceptors for hearing.

  Auditory Pathway to the Brain The auditory pathway begins when the outer ear receives and the middle ear amplifies sound waves that then strike the oval window membrane.

- The mechanoreceptors for hearing are hair cells on the basilar membrane of the spiral organ.

- Nerve signals begin in the cochlear nerve and are carried to the primary auditory area in the temporal lobe of the cerebral cortex.

14.6 Sense of Equilibrium

The ear also contains mechanoreceptors for equilibrium.

Rotational Equilibrium Pathway

- Mechanoreceptors (hair cells) in the semicircular canals detect rotational and/or angular movement of the head.

Gravitational Equilibrium Pathway

- Mechanoreceptors (hair cells) in the utricle and saccule detect head movement in the vertical or horizontal planes.

Understanding Key Terms

ampulla 322	optic tracts 314
aqueous humor 311	ossicle 317
astigmatism 315	otolith 322
auditory canal 317	outer ear 317
auditory tube 317	oval window 317
blind spot 313	pain receptor 304
chemoreceptor 304	photoreceptor 304
choroid 310	pinna 317
ciliary body 310	proprioceptor 306
cochlea 317	pupil 310
cochlear nerve 320	referred pain 307
color vision 312	retina 311
cone cell 312	retinal 312
cornea 310	rhodopsin 312
cutaneous receptor 306	rod cell 312
exteroceptor 304	rotational equilibrium 322
farsighted 315	round window 317
focus 311	saccule 322
fovea centralis 311	sclera 310
glaucoma 311	semicircular canal 317, 322
gravitational equilibrium 322	sensation 304
hair cell 315	sensory adaptation 305
incus 317	sensory receptor 304
inner ear 317	spiral organ 318
integration 305	stapes 317
interoceptor 304	sterocilia 317
iris 310	stimulus 304
kinocilium 322	taste bud 308
lens 311	tectorial membrane 318
malleus 317	thermoreceptor 304
mechanoreceptor 304	tympanic membrane 317
middle ear 317	utricle 322
nearsighted 315	vestibule 317
olfactory cell 309	visual accommodation 311
optic chiasma 314	vitreous humor 311
optic nerve 311	

Match the key terms to these definitions.

a. _____ Structure that receives sensory stimuli and is a part of a sensory neuron or transmits signals to a sensory neuron.

b. _____ Inner layer of the eyeball containing the photoreceptors—rod cells and cone cells.

c. _____ Outer, white, fibrous layer of the eye that surrounds the eye except for the transparent cornea.

d. _____ Receptor sensitive to chemical stimulation—for example, receptors for taste and smell.

e. _____ Specialized region of the cochlea containing the hair cells for sound detection and discrimination.

Testing Your Knowledge of the Concepts

1. Contrast exteroceptors and interoceptors. (page 304)

2. What is sensory adaptation? (page 305)

3. What is proprioception, and what is the role of muscle spindles? (page 306)

4. List the cutaneous receptors and the type of stimulus to which each responds. (pages 306–08)

5. Describe the structure of a taste bud and explain how a taste cell functions. (pages 308–09)

6. Describe the structure and function of the olfactory epithelium. How does the sense of smell come about? (pages 309–10)

7. Describe the anatomy of the eye and the function of each part. (pages 310–14)

8. How does the eye respond when viewing an object far away? When viewing a close object? (pages 311–12)

9. Describe the sequence of events after rhodopsin absorbs light. (pages 312–14)

10. What are three abnormalities caused by a misshapen eyeball, and what type lens can be helpful? (page 315)

11. Describe the anatomy of the ear and the function of each part. (pages 315, 317)

12. Explain the pathway of sound and how sound is produced. (pages 317–18, 320)

13. What structures are involved in equilibrium, their structures, and functions? (pages 320–23)

14. A sensory receptor
    a. is the first portion of a reflex arc.
    b. initiates nerve impulses.
    c. can be internal or external.
    d. All of these are correct.

15. Receptors sensitive to changes in blood pressure are
    a. interoceptors.   c. proprioceptors.
    b. exteroceptors.   d. nociceptors.

16. Conscious interpretation of changes in the internal and external environment is called
    a. responsiveness.
    c. sensation.
    b. perception.
    d. accommodation.

17. Which of these is an incorrect difference between proprioceptors and cutaneous receptors?

Proprioceptors	**Cutaneous Receptors**
a. located in muscles and tendons	located in the skin
b. chemoreceptors	mechanoreceptors
c. respond to tension	respond to pain, hot, cold, touch, pressure
d. interoceptors	exteroceptors

    e. All of these contrasts are correct.

18. Pain perceived as coming from another location is known as
    a. intercepted pain.
    c. referred pain.
    b. phantom pain.
    d. parietal pain.

19. Tasting something "sweet" versus "salty" is a result of activating
    a. different sensory receptors.
    b. many versus few sensory receptors.
    c. no sensory receptors.
    d. None of these are correct.

20. Which structure of the eye is incorrectly matched with its function?
    a. lens—focusing
    b. cones—color vision
    c. iris—regulation of amount of light
    d. choroid—location of cones
    e. sclera—protection

21. Which of the following gives the correct path for light rays entering the human eye?
    a. sclera, retina, choroid, lens, cornea
    b. fovea centralis, pupil, aqueous humor, lens
    c. cornea, pupil, lens, vitreous humor, retina
    d. cornea, fovea centralis, lens, choroid, rods
    e. optic nerve, sclera, choroid, retina, humors

22. The thin, darkly pigmented layer that underlies most of the sclera is the
    a. conjunctiva.
    c. retina.
    b. cornea.
    d. choroid.

23. Adjustment of the lens to focus on objects close to the viewer is called
    a. convergence.
    c. focusing.
    b. visual accommodation.
    d. constriction.

24. Retinal is
    a. a derivative of vitamin A.
    b. sensitive to light energy.
    c. a part of rhodopsin.
    d. found in both rods and cones.
    e. All of these are correct.

25. To focus on objects that are close to the viewer, the
    a. suspensory ligaments must be pulled tight.
    b. lens needs to become more rounded.
    c. ciliary muscle will be relaxed.
    d. image must focus on the area of the optic nerve.

26. Which abnormality of the eye is incorrectly matched with its cause?
    a. astigmatism—either the lens or cornea is not even
    b. farsightedness—eyeball is shorter than usual
    c. nearsightedness—image focuses behind the retina
    d. color blindness—genetic disorder in which certain types of cones may be missing

27. Which of these associations is incorrectly matched?
    a. semicircular canals—inner ear
    b. utricle and saccule—outer ear
    c. auditory canal—outer ear
    d. cochlea—inner ear
    e. ossicles—middle ear

28. Which of the following wouldn't you mention if you were tracing the path of sound vibrations?
    a. auditory canal
    d. semicircular canals
    b. tympanic membrane
    e. cochlea
    c. ossicles

29. The middle ear is separated from the inner ear by the
    a. oval window.
    c. round window.
    b. tympanic membrane.
    d. Both a and c are correct.

30. Which one of these correctly describes the location of the spiral organ?
    a. between the tympanic membrane and the oval window in the inner ear
    b. in the utricle and saccule within the vestibule
    c. between the tectorial membrane and the basilar membrane in the cochlear canal
    d. between the nasal cavities and the throat
    e. between the outer and inner ear within the semicircular canals

31. Which of the following could result in hearing loss?
    a. symphony orchestra
    b. earphone use
    c. consistent use of loud equipment such as a jackhammer
    d. use of firearms
    e. All of these are correct.

32. Which of the following structures would allow you to know that you were upside down, even if you were in total darkness?
    a. utricle and saccule
    c. semicircular canals
    b. cochlea
    d. tectorial membrane

33. Which of these is an incorrect difference between olfactory receptors and equilibrium receptors?

Olfactory Receptors	**Equilibrium Receptors**
a. located in nasal cavities	located in the inner ear
b. chemoreceptors	mechanoreceptors
c. respond to molecules in the air	respond to movements of the body
d. communicate with brain via a tract	communicate with brain via vestibular nerve

    e. All of these contrasts are correct.

34. Both olfactory receptors and sound receptors
    a. are chemoreceptors.
    b. are a part of the brain.
    c. are mechanoreceptors.
    d. initiate nerve impulses.
    e. All of these are correct.

35. Label this diagram of a human eye.

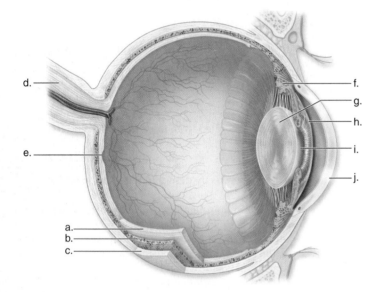

Thinking Critically About the Concepts

Both exteroreceptors and interoreceptors are essential to provide information to the central nervous system. In this chapter, Mike Fila faces loss of his vision because of retinal detachment. Loss of one of the special senses is immediately debilitating and can seriously affect homeostasis over time. Without vision, hearing, taste, or smell, we are vulnerable to dangerous conditions in our surroundings. Loss of cutaneous senses—touch, pressure, temperature, pain—can immediately impact survival. Imagine, for example, being unaware that you've been seriously cut because you can't feel pain. In that circumstance, it would be possible to bleed to death.

Keep in mind the hugely important role of interoreceptors in maintaining homeostasis. Any unacceptable change in internal homeostatic conditions is reported to the brain by some type of interoreceptor. Pressoreceptors of the cardiovascular system help to regulate blood pressure. Osmoreceptors in the hypothalamus help to control water-salt balance, and chemoreceptors monitor pH in the blood. Our lives depend on proper function of all components of the sensory system.

1. What type of receptors are activated when we enjoy supper in a pizza restaurant?

2. Besides the blood pH mentioned, what other homeostatic conditions are monitored by chemoreceptors?

3. If a person takes a blow to the back of the head, what sense is most likely to be affected?

4. Airport and construction workers are likely to be exposed to continuous, loud noises. What would you predict the long-term effect on their hearing to be? Why?

5. The acoustic and vestibular nerves travel together to the brain. If a tumor grows on this combined nerve, what sensations will be affected?

6. Besides vision, what other sensation will be affected if type 2 diabetes is poorly controlled?

CHAPTER 15

Endocrine System

Abbie glanced at her watch, and groaned when she realized only 20 minutes had passed. The exhaustion she felt had been occurring for several weeks. It was puzzling since she'd been sleeping eight hours a night and taking naps during her lunch hour.

"Maybe a snack will help," she thought. A sudden wave of dizziness came over her when she stood up. She grabbed onto the corner of her desk and steadied herself.

"Abbie, be careful!" her friend Simon yelled from the next cubicle.

"I'm fine," she reassured him. "Just a little dizzy. A candy bar should do the trick."

"Let me walk to the vending machines with you," Simon insisted. "I can scrape you up off the ground if you fall between here and there!" he joked.

Abbie peered at the thermostat as they walked by it on the way to the vending machines. "Wow—I can't believe it's 78 degrees in here! I've been freezing all day," she exclaimed as she pulled her sweater around her more tightly.

"Are you sure you're okay?" Simon asked. "We've been complaining all day about how hot it is in here. Someone is supposed to be coming to look at the air conditioning system because it's been blowing out warm air all day."

"Well, if it works any better, I'm gonna need an electric blanket," Abbie joked. She selected a chocolate bar with nuts hoping for a burst of energy to finish the workday and make it home without falling down. "I can't wait to get home and take a nap," she commented wistfully.

"I don't know, Abbie," observed Simon. "If you're that tired, you might want to go see your doctor."

"I'm just tired and a little chilly," Abbie insisted. "I don't have a fever and nothing hurts, so I don't think I've caught anything. I don't think it's anything that can't wait until next week. I get to do my annual 'visit-the-gynecologist' thing then."

"All right, I won't bug you about it anymore," Simon responded. "Just promise you'll mention what just happened today to your doctor. If you don't, I'll go *with* you to that appointment."

15.1 Endocrine Glands

The nervous system and the endocrine system work together to regulate the activities of the other systems. Both systems use chemical signals when they respond to changes that might threaten homeostasis. However, they have different means of delivering these signals (Fig. 15.1). As discussed, the nervous system is composed of neurons. In this system, sensory receptors (specialized dendrites) detect changes in the internal and external environment. The CNS integrates the information and responds by stimulating muscles and glands. Communication depends on nerve signals, conducted in axons, and neurotransmitters, which cross synapses. Axon conduction occurs rapidly and so does diffusion of a neurotransmitter across the short distance of a synapse. In other words, the nervous system is organized to respond rapidly to stimuli. This is particularly useful if the stimulus is an external event that endangers our safety—we can move quickly to avoid being hurt.

The endocrine system functions differently. The endocrine system is largely composed of glands (Fig. 15.2). These glands secrete **hormones,** carried by the bloodstream to target cells throughout the body. It takes time to deliver hormones, and it takes time for cells to respond. The effect initiated by the endocrine system is longer lasting. In other words, the endocrine system is organized for a slow but prolonged response.

Endocrine glands can be contrasted with **exocrine glands.** Exocrine glands have ducts and secrete their products into these ducts. The glands' products are carried to the lumens of other organs or outside the body. For example, the salivary glands send saliva into the mouth by way of the salivary ducts. Endocrine glands, as stated, secrete their products into the bloodstream, which delivers them throughout the body. It must be stressed that only certain cells, called target cells, can respond to certain hormones. A target cell for a particular hormone will have a receptor protein for that hormone. The receptor protein and hormone bind together like a key that fits a lock. The target cell then responds to that hormone.

Both the nervous system and the endocrine system make use of negative feedback mechanisms. If the blood pressure falls, sensory receptors signal a control center in the brain. This center sends out nerve signals to the arterial walls so that they constrict, and blood pressure rises. Now the sensory receptors are no longer stimulated, and the feedback mechanism is inactivated. Similarly, a rise in blood glucose level causes the pancreas to release insulin. This in turn, promotes glucose uptake by the liver, muscles, and other cells of the body (see Fig. 15.1). When the blood glucose level falls, the pancreas no longer secretes insulin.

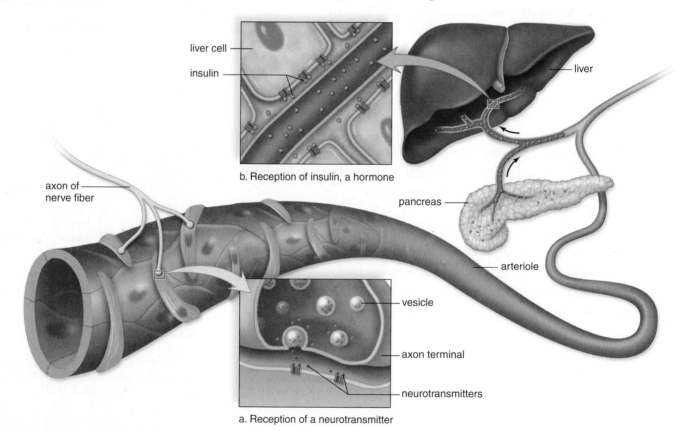

liver cell

insulin

liver

b. Reception of insulin, a hormone

axon of nerve fiber

pancreas

vesicle

arteriole

axon terminal

neurotransmitters

a. Reception of a neurotransmitter

Figure 15.1 **How does the action of a neurotransmitter differ from the action of a hormone?**
a. Nerve impulses passing along an axon cause the release of a neurotransmitter. The neurotransmitter, a chemical signal, causes the wall of an arteriole to constrict. **b.** The hormone insulin, a chemical signal, travels in the cardiovascular system from the pancreas to the liver, where it causes liver cells to store glucose as glycogen.

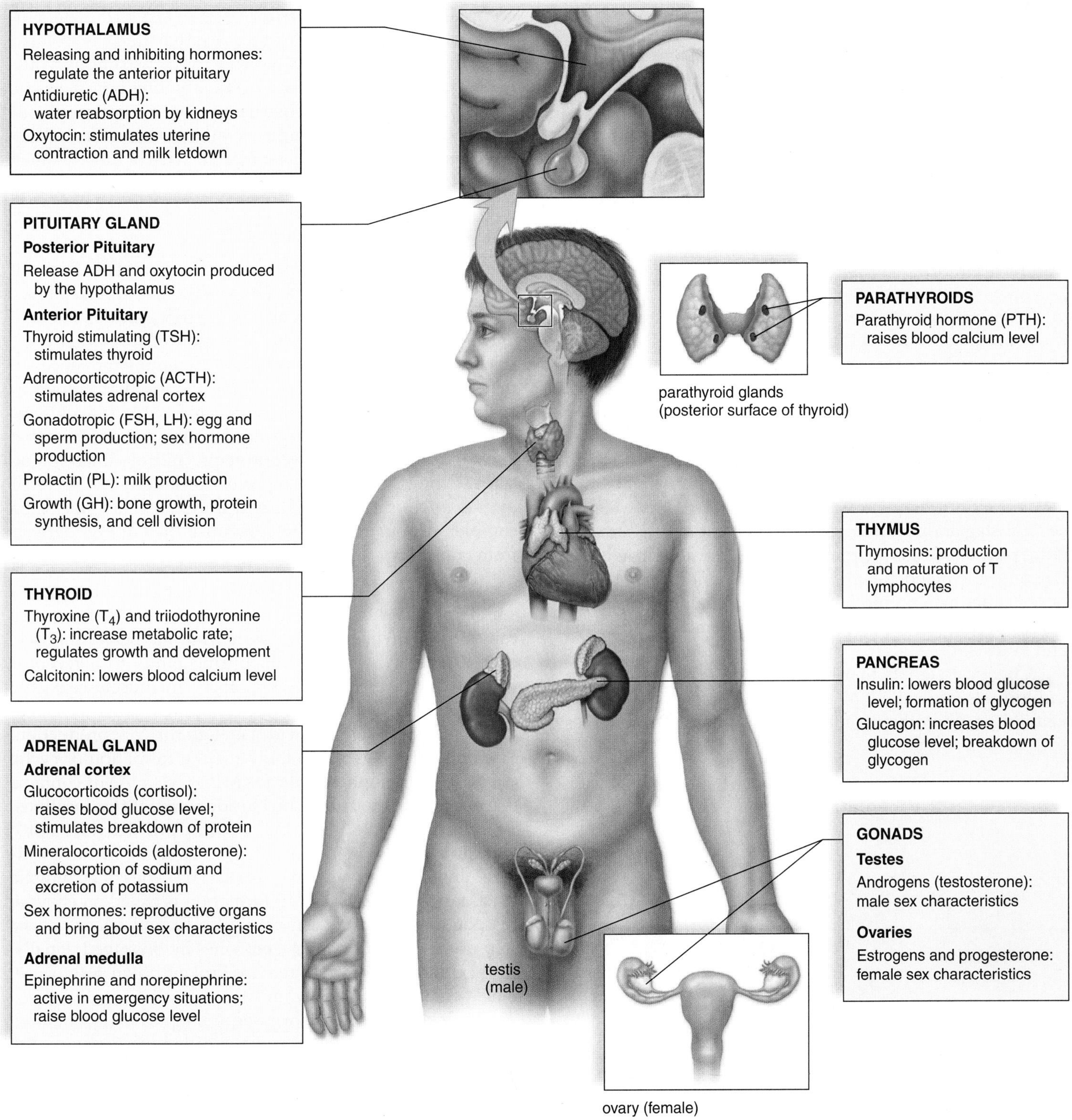

HYPOTHALAMUS
Releasing and inhibiting hormones:
 regulate the anterior pituitary
Antidiuretic (ADH):
 water reabsorption by kidneys
Oxytocin: stimulates uterine
 contraction and milk letdown

PITUITARY GLAND
Posterior Pituitary
Release ADH and oxytocin produced
 by the hypothalamus
Anterior Pituitary
Thyroid stimulating (TSH):
 stimulates thyroid
Adrenocorticotropic (ACTH):
 stimulates adrenal cortex
Gonadotropic (FSH, LH): egg and
 sperm production; sex hormone
 production
Prolactin (PL): milk production
Growth (GH): bone growth, protein
 synthesis, and cell division

THYROID
Thyroxine (T_4) and triiodothyronine
 (T_3): increase metabolic rate;
 regulates growth and development
Calcitonin: lowers blood calcium level

ADRENAL GLAND
Adrenal cortex
Glucocorticoids (cortisol):
 raises blood glucose level;
 stimulates breakdown of protein
Mineralocorticoids (aldosterone):
 reabsorption of sodium and
 excretion of potassium
Sex hormones: reproductive organs
 and bring about sex characteristics
Adrenal medulla
Epinephrine and norepinephrine:
 active in emergency situations;
 raise blood glucose level

PARATHYROIDS
Parathyroid hormone (PTH):
 raises blood calcium level

parathyroid glands
(posterior surface of thyroid)

THYMUS
Thymosins: production
 and maturation of T
 lymphocytes

PANCREAS
Insulin: lowers blood glucose
 level; formation of glycogen
Glucagon: increases blood
 glucose level; breakdown of
 glycogen

GONADS
Testes
Androgens (testosterone):
male sex characteristics
Ovaries
Estrogens and progesterone:
female sex characteristics

testis
(male)

ovary (female)

Figure 15.2 **What are some of the major glands of the endocrine system? What hormones do they produce?**
Major glands and the hormones they produce are depicted. Also, the endocrine sytem includes other organs such as the kidneys, gastrointestinal tract, and the heart, which also produce hormones but not as a primary function of these organs.

Hormones Are Chemical Signals

Like other **chemical signals,** hormones are a means of communication between cells, between body parts, and even between individuals. They affect the metabolism of cells that have receptors to receive them (Fig. 15.3). In a condition called *androgen insensitivity,* an individual has X and Y sex chromosomes. Their testes, which remain in the abdominal cavity, produce the sex hormone testosterone. However, the body cells lack receptors that are able to combine with testosterone. Therefore, the individual appears to be a normal female.

Like testosterone, most hormones act at a distance between body parts. They travel in the bloodstream from the gland that produced them to their target cells. Also considered to be hormones are the secretions produced by neurosecretory cells in the hypothalamus, a part of the brain. They travel in the capillary network that runs between the hypothalamus and the pituitary gland. Some of these secretions stimulate the pituitary to secrete its hormones, and others prevent it from doing so.

Not all hormones act between body parts. As we shall see, prostaglandins are a good example of a *local hormone.* After prostaglandins are produced, they are not carried elsewhere in the bloodstream. Instead, they affect neighboring cells, sometimes promoting pain and inflammation. Also, growth factors are local hormones that promote cell division and mitosis.

Chemical signals that influence the behavior of other individuals are called **pheromones.** Other animals rely heavily on pheromones for communication. Pheromones are used to mark one's territory and to attract a mate. Humans produce pheromones too. A researcher has isolated one released by men that reduces premenstrual nervousness and tension in women. Women who live in the same household often have menstrual cycles in synchrony. This is likely caused by the armpit secretions of a woman who is menstruating affecting the menstrual cycle of other women in the household.

The Action of Hormones

Hormones have a wide range of effects on cells. Some of these effects induce a target cell to increase its uptake of particular substances, such as glucose, or ions, such as calcium. Other effects bring about an alteration of the target cell's structure in some way. A few hormones simply influence cell metabolism. Growth hormone is a peptide that influences cell metabolism leading to a change in the structure of bone. The term **peptide hormone** is used to include hormones that are peptides, proteins, glycoproteins, and modified amino acids. Growth hormone is a protein produced and secreted by the anterior pituitary. **Steroid hormones** have the same complex of four carbon rings because they are all derived from cholesterol (see Fig. 2.17).

The Action of Peptide Hormones   Most endocrine glands secrete peptide hormones. The actions of peptide hormones can vary. We will concentrate on what happens in muscle cells after the hormone epinephrine binds to a receptor in the plasma membrane (Fig. 15.4). In muscle cells, the reception of epinephrine leads to the breakdown of glycogen to glucose, which provides energy for ATP production. The immediate result of binding is the formation of **cyclic adenosine monophosphate (cAMP).** Cyclic AMP contains one phosphate group attached to adenosine at two locations. Therefore, the molecule is cyclic. Cyclic AMP activates a protein kinase enzyme in the cell. This enzyme, in turn, activates another enzyme, and so forth. The series of enzymatic reactions that follows cAMP formation is called an enzyme cascade. Each enzyme can be used over and over at every step of the cascade, so more enzymes are involved. Finally, many molecules of glycogen are broken down to glucose, which enters the bloodstream.

Typical of a peptide hormone, epinephrine never enters the cell. Therefore, the hormone is called the **first messenger,** while cAMP, which sets the metabolic machinery in motion, is called the **second messenger.** To explain this terminology, let's imagine that the adrenal medulla, which produces epinephrine, is like the home office that sends out a courier (i.e., the hormone epinephrine is the first messenger) to a factory (the cell). The courier doesn't have a pass to enter the factory, so when he arrives at the factory, he tells a supervisor through the screen door that the home office wants the factory to produce a particular product. The supervisor (i.e., cAMP, the second messenger) walks over and flips a switch that starts the machinery (the enzymatic pathway), and a product is made.

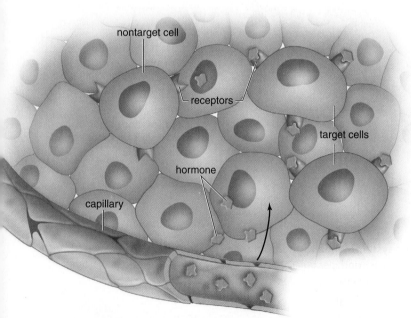

Figure 15.3   How do hormones recognize their target cells?
Most hormones are distributed by the bloodstream to target cells. Target cells have receptors for the hormone, and the hormone combines with the receptor as a key fits a lock.

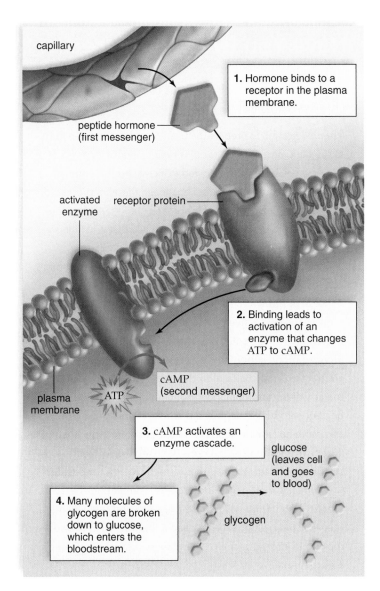

Figure 15.4 Where is the receptor for a peptide hormone? What happens when a peptide hormone binds to its receptor?
A peptide hormone (first messenger) binds to a receptor in the plasma membrane. Thereafter, cyclic AMP (second messenger) forms and activates an enzyme cascade.

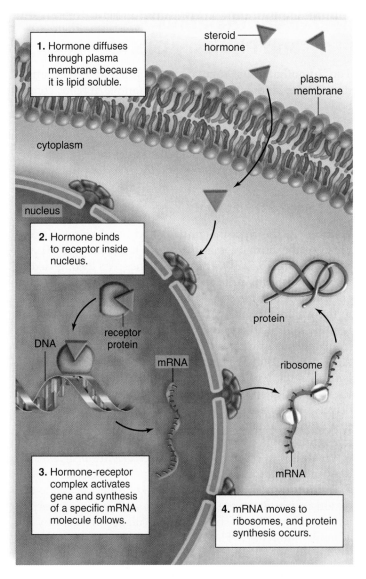

Figure 15.5 Where is the receptor for a steroid hormone? What happens when a steroid hormone binds to its receptor?
A steroid hormone passes directly through the target cell's plasma membrane before binding to a receptor in the nucleus or cytoplasm. The hormone-receptor complex binds to DNA, and gene expression follows.

The Action of Steroid Hormones Only the adrenal cortex, the ovaries, and the testes produce steroid hormones. Thyroid hormones are amines and act similarly to steroid hormones, even though they have a different structure. Steroid hormones do not bind to plasma membrane receptors. Instead they are able to enter the cell because they are lipids (Fig. 15.5). Once inside, a steroid hormone binds to a receptor, usually in the nucleus but sometimes in the cytoplasm. Inside the nucleus, the hormone-receptor complex binds with DNA and activates certain genes. Messenger RNA (mRNA) moves to the ribosomes in the cytoplasm and protein (e.g. enzyme) synthesis follows. To continue our analogy, a steroid hormone is like a courier that has a pass to enter the factory (the cell). Once

inside, it makes contact with the plant manager (DNA) who sees to it that the factory (cell) is ready to produce a product.

Steroids act more slowly than peptides because it takes more time to synthesize new proteins than to activate enzymes already present in cells. Their action lasts longer, however.

Check Your Progress 15.1
1. How can the nervous system be contrasted with the endocrine system?
2. What is a hormone?
3. How do hormones affect the metabolism of cells?

15.2 Hypothalamus and Pituitary Gland

The **hypothalamus** regulates the internal environment through the autonomic system. For example, it helps control body temperature and water balance. The hypothalamus also controls the glandular secretions of the **pituitary gland.** The pituitary, a small gland about 1 cm in diameter, is connected to the hypothalamus by a stalklike structure. The pituitary has two portions: the posterior and the anterior pituitary.

Posterior Pituitary

Neurons in the hypothalamus called neurosecretory cells produce the hormones **antidiuretic hormone (ADH)** and **oxytocin** (Fig. 15.6, *left*). These hormones pass through axons into the **posterior pituitary** where they are stored in axon endings. Certain neurons in the hypothalamus are sensitive to the water-salt balance of the blood. When these cells determine that the blood is too concentrated, ADH is released from the posterior pituitary. Upon reaching the kidneys, ADH causes more water to be reabsorbed into kidney capillaries. As the blood becomes dilute, ADH is no longer released. This is an example of control by negative feedback because the effect of the hormone (to dilute blood) acts to shut down the release of the hormone. Negative feedback maintains stable conditions and homeostasis.

Inability to produce ADH causes *diabetes insipidus,* the topic of the Chapter 10 case study. A person with this type of diabetes produces copious amounts of urine. Excessive urination results in severe dehydration and loss of important ions from the blood. The condition can be corrected by the administration of ADH.

Oxytocin, the other hormone made in the hypothalamus, causes uterine contraction during childbirth and milk let-down when a baby is nursing. The more the uterus contracts during labor, the more nerve signals reach the hypothalamus, causing oxytocin to be released. Similarly, as a baby suckles while being breast-fed, nerve signals from breast tissue reach the hypothalamus. As a result, oxytocin is produced by the hypothalamus and released from the posterior pituitary. The hormone causes the woman's breast milk to be released. The sound of a baby crying (even someone else's baby!) may also stimulate the release of oxytocin and milk let-down, much to the chagrin of women who are nursing. In both instances, the release of oxytocin from the posterior pituitary is controlled by **positive feedback.** The stimulus continues to bring about an effect that ever increases in intensity. Positive feedback terminates due to some external event as when a baby is full and stops suckling. Positive feedback is not a way to maintain stable conditions and homeostasis.

Anterior Pituitary

A portal system, consisting of two capillary systems connected by a vein, lies between the hypothalamus and the

anterior pituitary (Fig. 15.6, *right*). The hypothalamus controls the anterior pituitary by producing **hypothalamic-releasing** and **hypothalamic-inhibiting hormones,** which pass from the hypothalamus to the anterior pituitary by way of the portal system. For example, there is a thyroid-releasing hormone (TRH) and a thyroid-inhibiting hormone (TIH). TRH stimulates the anterior pituitary to secrete thyroid-stimulating hormone, and TIH inhibits the pituitary from secreting thyroid-stimulating hormone.

Three of the six hormones produced by the anterior pituitary have an effect on other glands: **Thyroid-stimulating hormone (TSH)** stimulates the thyroid to produce the thyroid hormones. **Adrenocorticotropic hormone (ACTH)** stimulates the adrenal cortex to produce cortisol. **Gonadotropic hormones** stimulate the gonads (the testes in males and the ovaries in females) to produce gametes and sex hormones. In each instance, the blood level of the last hormone in the sequence

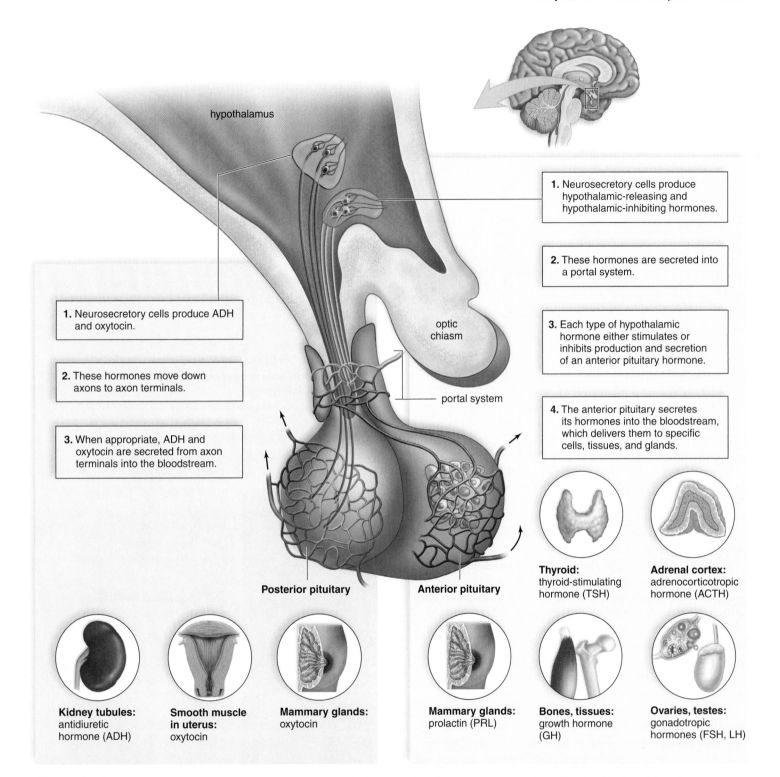

hypothalamus

1. Neurosecretory cells produce hypothalamic-releasing and hypothalamic-inhibiting hormones.

2. These hormones are secreted into a portal system.

1. Neurosecretory cells produce ADH and oxytocin.

optic chiasm

3. Each type of hypothalamic hormone either stimulates or inhibits production and secretion of an anterior pituitary hormone.

2. These hormones move down axons to axon terminals.

portal system

3. When appropriate, ADH and oxytocin are secreted from axon terminals into the bloodstream.

4. The anterior pituitary secretes its hormones into the bloodstream, which delivers them to specific cells, tissues, and glands.

Thyroid: thyroid-stimulating hormone (TSH)

Adrenal cortex: adrenocorticotropic hormone (ACTH)

Posterior pituitary

Anterior pituitary

Kidney tubules: antidiuretic hormone (ADH)

Smooth muscle in uterus: oxytocin

Mammary glands: oxytocin

Mammary glands: prolactin (PRL)

Bones, tissues: growth hormone (GH)

Ovaries, testes: gonadotropic hormones (FSH, LH)

Figure 15.6   **What hormones are produced by the hypothalamus? What hormones are produced by the pituitary gland?**
Left: The hypothalamus produces two hormones, ADH and oxytocin, stored and secreted by the posterior pituitary. *Right:* The hypothalamus controls the secretions of the anterior pituitary, and the anterior pituitary controls the secretions of the thyroid, adrenal cortex, and gonads, which are also endocrine glands.

exerts negative feedback control over the secretion of the first two hormones:

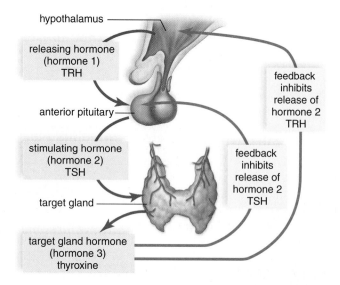

The other three hormones produced by the anterior pituitary do not affect other endocrine glands. **Prolactin (PRL)** is produced in quantity only after childbirth. It causes the mammary glands in the breasts to develop and produce milk. It also plays a role in carbohydrate and fat metabolism.

Melanocyte-stimulating hormone (MSH) causes skin-color changes in many fishes, amphibians, and reptiles having melanophores, special skin cells that produce color variations. The concentration of this hormone in humans is very low.

Growth hormone (GH), or somatotropic hormone, promotes skeletal and muscular growth. It stimulates the rate at which amino acids enter cells and protein synthesis occurs. It also promotes fat metabolism as opposed to glucose metabolism. The production of insulinlike growth factor 1 (IGF-1) by the liver is stimulated by growth hormone as well. IGF-1 is often measured as a means of determining GH level. Growth and development is also stimulated by IGF-1, and it may well be the means by which GH truly influences growth and development.

In the 1980s, growth hormone became a biotechnology product, and it was possible to treat short children and those diagnosed as pituitary dwarfs. A growth hormone blood test can be done to tell if a child is able to

produce the normal amount of growth hormone. If not, growth hormone can be injected as a medication.

Effects of Growth Hormone

GH is produced by the anterior pituitary. The quantity is greatest during childhood and adolescence, when most body growth is occurring. If too little GH is produced during childhood, the individual has **pituitary dwarfism,** characterized by perfect proportions but small stature. If too much GH is secreted, a person can become a giant (Fig. 15.7). Giants usually have poor health, primarily because GH has a secondary effect on the blood sugar level, promoting an illness called *diabetes mellitus* (see pages 341–42).

On occasion, GH is overproduced in the adult, and a condition called **acromegaly** results. Long bone growth is no longer possible in adults, so only the feet, hands, and face (particularly the chin, nose, and eyebrow ridges) can respond, and these portions of the body become overly large (Fig. 15.8).

> **Check Your Progress 15.2**
> 1. What role does the hypothalamus play in the endocrine system?
> 2. What hormones are produced by the anterior pituitary?

Figure 15.7 **How does growth hormone determine one's height? What happens in cases of GH excess or deficiency?**
a. The amount of growth hormone produced by the anterior pituitary during childhood affects the height of an individual. Plentiful growth hormone produces very tall basketball players. **b.** Too much growth hormone can lead to gigantism, while an insufficient amount results in limited stature and even pituitary dwarfism.

a.

b.

How Short Is Too Short?

Without treatment, children with a deficiency of growth hormone (GH) experience *pituitary dwarfism:* slow growth, short stature, and in some cases failure to begin puberty. Prior to the advent of biotechnology in the 1980s, treating these children was incredibly difficult and expensive. The GH needed to treat deficiencies had to be obtained from the cadaver pituitaries. While the treatment was generally very successful, the use of cadaveric GH caused Creutzfeldt-Jacob (a neurological disease similar to "mad cow" disease) in a small number of treated individuals.

Thanks to biotechnology, bacteria are now able to synthesize human GH (hGH). These bacteria have had the gene for hGH inserted into their genetic information. The altered bacteria are then grown in laboratories and make unlimited amounts of GH. Children with insufficient GH can be treated more safely and inexpensively with this GH. Recombinant hGH can also be used to treat other disorders such as the chromosomal deficiency known as Turner syndrome (discussed in Chap. 18). It may even be possible to slow or reverse the aging process with hGH treatments.

There is some controversy surrounding treating short children without hGH deficiency, for essentially cosmetic reasons. Unfortunately, Americans are obsessed with height. Shorter children are often bullied and teased by their peers. There are data to suggest that shorter individuals are discriminated against at their jobs. Their salaries are often lower than those of their taller counterparts with equivalent education and experience. Many people of short stature report having greater self-esteem problems than individuals of average to above average height. Treatment with hGH could be the solution to these problems.

Although the supply of hGH is seemingly unlimited, the cost of treatments is still quite high (though much cheaper than cadaveric GH). The cost of a year's treatment ranges from $13,000 to $30,000. Many insurance companies will not cover these costs. Of greater concern, however, are the potential side effects of supplemental hGH therapy, which are not well understood. Moreover, it is not clear whether hGH treatment will result in a significant increase in the final height of short children.

Decide Your Opinion

1. Now that hGH is easier to obtain, what potential abuses would you predict?
2. Do you think insurance companies should be expected to pay for hGH treatment if a child shows no hormone deficiency and is simply short?
3. Do you think our society will be somehow diminished, if pituitary dwarfs become "extinct" due to hGH therapy?

Age 9   Age 16   Age 33   Age 52

Figure 15.8 **What is the result of too much GH in an adult?**
Acromegaly is caused by overproduction of GH in the adult. It is characterized by enlargement of the bones in the face, the fingers, and the toes as a person ages.

15.3 Thyroid and Parathyroid Glands

The **thyroid gland** is a large gland located in the neck, where it is attached to the trachea just below the larynx (see Fig. 15.2). The parathyroid glands are embedded in the posterior surface of the thyroid gland.

Thyroid Gland

The thyroid gland is composed of a large number of follicles. Each follicle is a small spherical structure made of thyroid cells filled with triiodothyronine (T_3), which contains three iodine atoms, and **thyroxine (T_4),** which contains four.

Effects of Thyroid Hormones

To produce triiodothyronine and thyroxine, the thyroid gland actively acquires iodine. The concentration of iodine in the thyroid gland can increase to as much as 25 times that of the blood. If iodine is lacking in the diet, the thyroid gland is unable to produce the thyroid hormones. In response to constant stimulation by TSH from the anterior pituitary, the thyroid enlarges, resulting in a **simple goiter** (Fig. 15.9a). Some years ago, it was discovered that the use of iodized salt allows the thyroid to produce the thyroid hormones and, therefore, helps prevent simple goiter.

Thyroid hormones increase the metabolic rate. They do not have a target organ. Instead, they stimulate all cells of the body to metabolize at a faster rate. More glucose is broken down, and more energy is used.

If the thyroid fails to develop properly, a condition called **congenital hypothyroidism** results (Fig. 15.9b). Individuals with this condition are short and stocky and have had extreme hypothyroidism (undersecretion of thyroid hormone) since infancy or childhood. Thyroid hormone therapy can initiate growth, but unless treatment is begun within the first two months of life, mental retardation results. The occurrence of hypothyroidism in adults produces the condition known as **myxedema.** Lethargy, weight gain, loss of hair, slower pulse rate, lowered body temperature, and thickness and puffiness of the skin are characteristics of myxedema. The administration of adequate doses of thyroid hormones restores normal function and appearance.

In the case of hyperthyroidism (oversecretion of thyroid hormone), the thyroid gland is overactive, and a goiter forms. This type of goiter is called **exophthalmic goiter** (Fig. 15.9c). The eyes protrude because of edema in eye socket tissues and swelling of the muscles that move the eyes. The patient usually becomes hyperactive, nervous and irritable, and suffers from insomnia. Removal or destruction of a portion of the thyroid by means of radioactive iodine is sometimes effective in curing the condition. Hyperthyroidism can also be caused by a thyroid tumor, usually detected as a lump during physical examination. Again, the treatment is surgery in combination with administration of radioactive iodine. The prognosis for most patients is excellent.

Calcitonin

Calcium (Ca^{2+}) plays a significant role in both nervous conduction and muscle contraction. It is also necessary for blood clotting. The blood Ca^{2+} level is regulated in part by **calcitonin,** a hormone secreted by the thyroid gland when the blood Ca^{2+} level rises (Fig. 15.10). The primary effect of calcitonin is to bring about the deposit of Ca^{2+} in the bones. It also temporarily reduces the activity and number of osteoclasts. When the blood Ca^{2+} level lowers to normal, the release of calcitonin by the thyroid is inhibited.

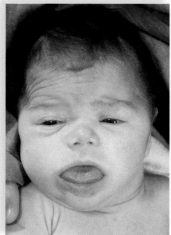
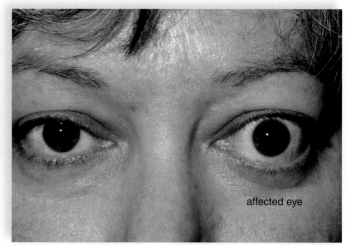

a. Simple goiter   b. Congenital hypothyroidism   c. Exophthalmic goiter

affected eye

Figure 15.9 **What happens to the thyroid if one's diet lacks iodine? What happens in thyroid hormone deficiency or excess?** **a.** An enlarged thyroid gland is often caused by a lack of iodine in the diet. Without iodine, the thyroid is unable to produce its hormones, and continued anterior pituitary stimulation causes the gland to enlarge. **b.** Individuals who develop hypothyroidism during infancy or childhood do not grow and develop as others do. Unless medical treatment is begun, the body is short and stocky; mental retardation is also likely. **c.** In exophthalmic goiter, a goiter is due to an overactive thyroid, and the eyes protrude because of edema in eye socket tissue.

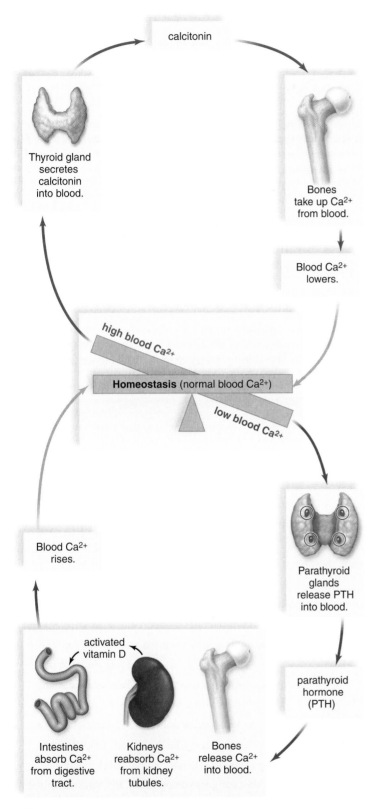

Figure 15.10 How is blood calcium regulated?
Top: When the blood calcium (Ca^{2+}) level is high, the thyroid gland secretes calcitonin. Calcitonin promotes the uptake of Ca^{2+} by the bones, and therefore, the blood Ca^{2+} level returns to normal. *Bottom:* When the blood Ca^{2+} level is low, the parathyroid glands release parathyroid hormone (PTH). PTH causes the bones to release Ca^{2+}. It also causes the kidneys to reabsorb Ca^{2+} and activate vitamin D; thereafter, the intestines absorb Ca^{2+}. Therefore, the blood Ca^{2+} level returns to normal.

Parathyroid Glands

Parathyroid hormone (PTH) produced by the **parathyroid glands,** causes blood Ca^{2+} level to increase. A low blood Ca^{2+} level stimulates the release of PTH. PTH promotes the activity of osteoclasts and the release of calcium from the bones. PTH also activates vitamin D in the kidneys. Activated vitamin D, a hormone sometimes called calcitriol, then promotes calcium reabsorption by the kidneys. The absorption of Ca^{2+} from the intestine is also stimulated by calcitriol. These effects bring the blood Ca^{2+} level back to the normal range, and PTH secretion stops.

Many years ago, the four parathyroid glands were sometimes mistakenly removed during thyroid surgery because of their size and location. Gland removal caused insufficient PTH production, called *hypoparathyroidism.* Hypoparathyroidism causes a dramatic drop in blood calcium, followed by excessive nerve excitability. Nerve signals happen spontaneously and without rest, causing a phenomenon called tetany. In **tetany,** the body shakes from continuous muscle contraction. Without treatment, severe hypoparathyroidism causes seizures, heart failure, and death.

Untreated *hyperparathyroidism* (oversecretion of PTH) can result in osteoporosis because of continuous calcium release from the bones. Hyperparathyroidism may also cause formation of calcium kidney stones.

When a bone is broken, homeostastis is disrupted. For the fracture to heal, osteoclasts will have to destroy old bone, and osteoblasts will have to lay down new bone. Many factors influence the formation of new bone, including parathyroid hormone, calcitonin, and vitamin D. Working together, the calcium needed to repair the fracture would be made readily available as new blood capillaries penetrate the fractured area.

> **Check Your Progress 15.3**
> 1. a. What hormones are produced by the thyroid gland?
>   b. How do thyroid hormones affect the metabolic rate?
> 2. a. What hormone is produced by the parathyroid gland? b. What is the function of this hormone? c. What hormone is secreted by the thyroid gland when the blood Ca^{2+} levels are high?

15.4 Adrenal Glands

The **adrenal glands** sit atop the kidneys (see Fig. 15.2). Each adrenal gland consists of an inner portion called the **adrenal medulla** and an outer portion called the **adrenal cortex.** These portions, like the anterior and the posterior pituitary, are two functionally distinct endocrine glands. The adrenal medulla is under nervous control. Portions of the adrenal cortex are under the control of corticotropin releasing hormone (CRH) from the hypothalamus and ACTH, an anterior pituitary hormone. Stress of all types, including emotional and physical trauma, prompts the hypothalamus to stimulate a portion of the adrenal glands (Fig. 15.11).

Adrenal Medulla

The hypothalamus initiates nerve signals that travel by way of the brain stem, spinal cord, and preganglionic sympathetic nerve fibers to the adrenal medulla. These signals stimulate the adrenal medulla to secrete its hormones. The cells of the adrenal medulla are thought to be modified postganglionic neurons.

   Epinephrine (adrenaline) and **norepinephrine** (noradrenaline) are the hormones produced by the adrenal medulla. These hormones rapidly bring about all the body changes that occur when an individual reacts to an emergency situation in a fight-or-flight manner. The effect of these hormones provide a short-term response to stress.

Adrenal Cortex

In contrast, the hormones produced by the adrenal cortex provide a long-term response to stress (see Fig. 15.11). The two major types of hormones produced by the adrenal cortex are the mineralocorticoids and the glucocorticoids. The **glucocorticoids,** whose secretion is controlled by ACTH, regulate carbohydrate, protein, and fat metabolism. The result of the actions of the glucocorticoids is an increase in blood glucose level. Glucocorticoids also suppress the body's inflammatory response. Cortisone, the medication often administered for inflammation of joints, is a glucocorticoid. The **mineralocorticoids** regulate water and salt balance, leading to increases in blood volume and blood pressure.

   The adrenal cortex also secretes a small amount of male sex hormones and a small amount of female sex hormones. This is the case in both males and females.

Glucocorticoids

ACTH stimulates those portions of the adrenal cortex that secrete the glucocorticoids. **Cortisol** and also cortisone are biologically significant glucocorticoids. Glucocorticoids raise the blood glucose level in at least two ways: (1) They promote the breakdown of muscle proteins to amino acids,

CASE STUDY AT THE DOCTOR'S OFFICE

The following week, Abbie headed to her doctor's appointment. Though she dreaded these annual exams, it was the one time of the year she actually went to see a doctor. It was quite a change from her childhood, when visits to her pediatrician happened regularly. Abbie told herself that it might be time to find a family physician, who could see her regularly for checkups and general health concerns.

   In the week following her dizzy spell, Abbie continued to feel extreme fatigue. She noticed she was often the only person at work wearing a sweater or a jacket. There had been several other times she'd felt lightheaded, but she'd managed to stay on her feet through all of them. Fortunately, Simon had not noticed her continued symptoms and didn't follow through on his threat to accompany her to the doctor. Simon's tendency to hover made her feel a little claustrophobic. Abbie hoped her gynecologist would be able to figure out a cause and treatment for her symptoms.

   While answering the litany of questions asked by Karen, Dr. Adair's nurse, Abbie described the symptoms she'd been having for several weeks. The nurse noted them on Abbie's chart. "You know Dr. Adair, Abbie. He's a pretty smart guy. I'll bet he can figure this one out, just like he always does.

   "Just come on into this room and put on this gown," Karen soothed. "Dr. Adair will be with you in a minute."

   Moments later, Dr. Adair tapped on the door, asking, "Are you decent in there?"

   "As decent as I can be in one of these crazy gowns!" Abbie laughed.

   "Now what's this I hear about you having dizzy spells and feeling tired all the time?" Dr. Adair inquired, while studying her chart.

   "Cold, too," Abbie added. "Everyone else at my work complains about how hot it is, and I'm bundled up in a sweater!"

   "Right," the doctor mused. "Could be you're anemic, not an uncommon thing in women your age. Guessing you don't eat much liver, which is a good source of iron," he smiled.

   Abbie blanched and made a face at the mention of liver. "I'll take that as a 'no,'" Dr. Adair laughed. "Let me feel your throat to see if there's anything unusual about your thyroid. Once we've finished the rest of your physical exam we'll send you off to the lab. They'll get some blood for tests I want to run," he finished. Dr. Adair made some notes in Abbie's chart after palpating her throat and completing the rest of her exam. His notes read, "Thyroid normal, check blood levels of TSH and T_3/T_4, and do a complete blood count." His notes described the remaining exam results as normal.

   "Okay, young lady. Get dressed and head on down to the lab," the doctor ordered. "Give them this paperwork, and they'll draw the blood samples we need.

   "Once we find out the results of your blood work, we'll give you a call and sort out what kind of treatment you need," Dr. Adair concluded as he closed up her file.

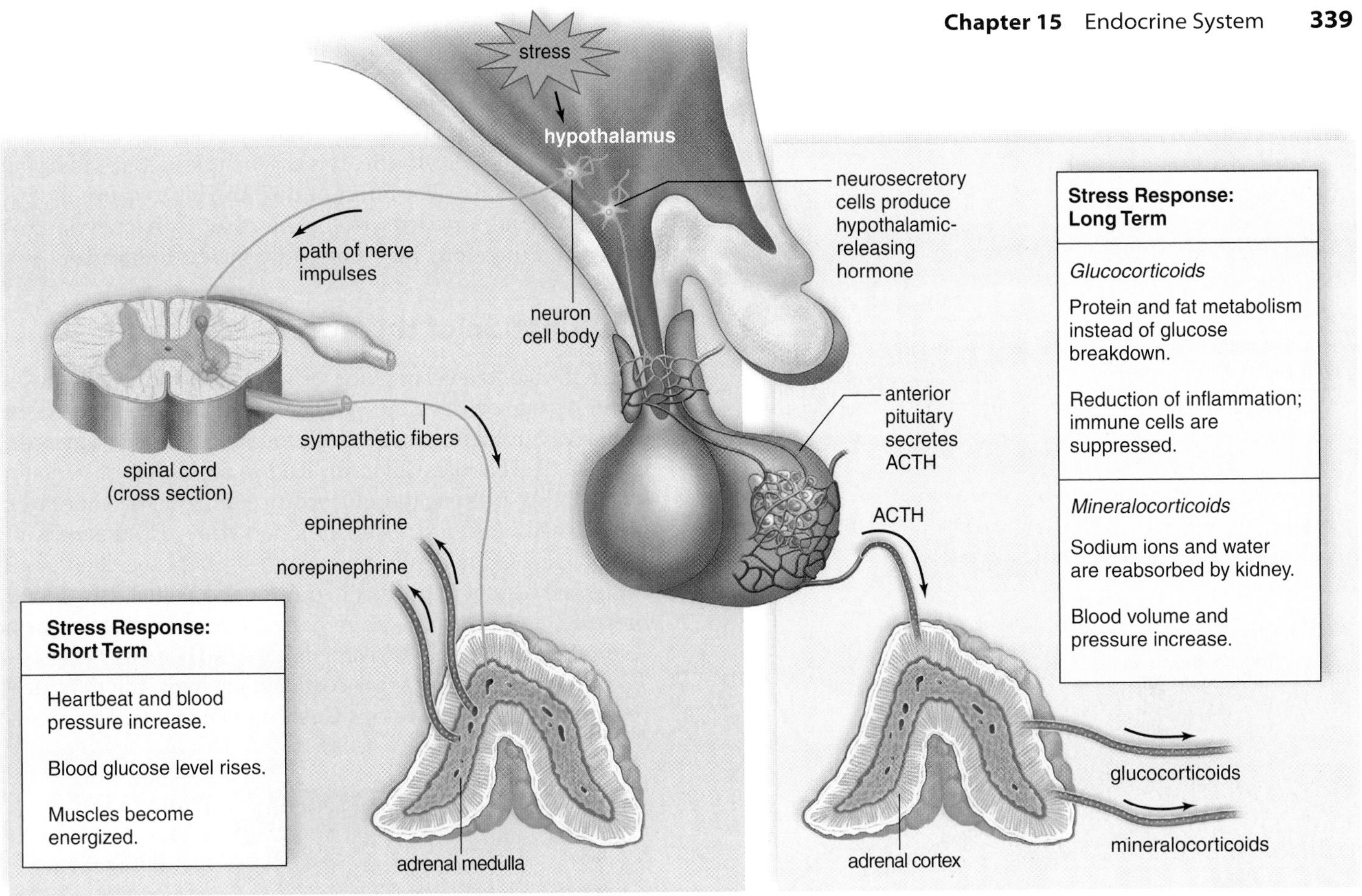

Figure 15.11   How do the adrenal medulla and the adrenal cortex respond differently to stress?
Both the adrenal medulla and the adrenal cortex are under the control of the hypothalamus when they help us respond to stress. *Left:* Nervous stimulation causes the adrenal medulla to provide a rapid, but short-term, stress response. *Right:* The adrenal cortex provides a slower, but long-term, stress response. ACTH causes the adrenal cortex to release glucocorticoids. Independently, the adrenal cortex releases mineralocorticoids.

taken up by the liver from the bloodstream. The liver then converts these excess amino acids to glucose, which enters the blood. (2) They promote the metabolism of fatty acids rather than carbohydrates, and this spares glucose.

The glucocorticoids also counteract the inflammatory response that leads to the pain and swelling. Very high levels of glucocorticoids in the blood can suppress the body's defense system, including the inflammatory response that occurs at infection sites. Cortisone and other glucocorticoids can relieve swelling and pain from inflammation. However, by suppressing pain and immunity, they can also make a person highly susceptible to injury and infection.

Mineralocorticoids

Aldosterone is the most important of the mineralocorticoids. Aldosterone primarily targets the kidney, where it promotes renal absorption of sodium (Na^+) and renal excretion of potassium (K^+).

The secretion of mineralocorticoids is not controlled by the anterior pituitary. When the blood Na^+ level and,

therefore, the blood pressure are low, the kidneys secrete **renin** (Fig. 15.12). Renin is an enzyme that converts the plasma protein angiotensinogen to angiotensin I. Angiotensin I is changed to angiotensin II by a converting enzyme found in lung capillaries. Angiotensin II stimulates the adrenal cortex to release aldosterone. The effect of this system, called the renin-angiotensin-aldosterone system, is to raise blood pressure in two ways. Angiotensin II constricts the arterioles, and aldosterone causes the kidneys to reabsorb Na^+. When the blood Na^+ level rises, water is reabsorbed, in part, because the hypothalamus secretes ADH (see page 332). Reabsorption means that water enters kidney capillaries and thus the blood. Then blood pressure increases to normal.

Recall that we studied the role of the kidneys in maintaining blood pressure on pages 218–20. At that time, we mentioned that if the blood pressure rises due to the reabsorption of Na^+, the atria of the heart are apt to stretch. Due to a great increase in blood volume, cardiac cells release a chemical called **atrial natriuretic hormone (ANH),** which inhibits the secretion of aldosterone from the adrenal cortex.

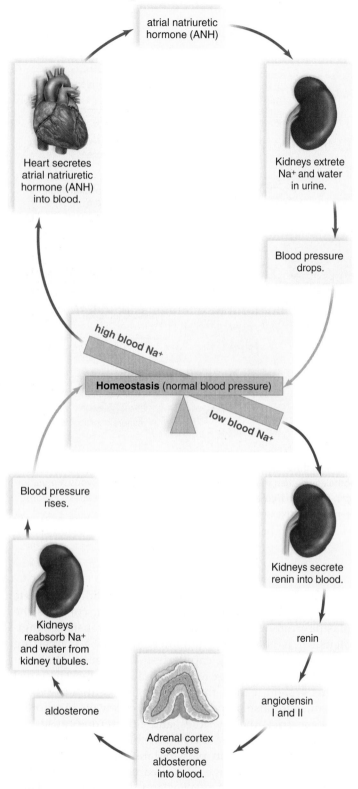

Figure 15.12 How is blood pressure regulated?

Bottom: When the blood sodium (Na^+) level is low, a low blood pressure causes the kidneys to secrete renin. Renin leads to the secretion of aldosterone from the adrenal cortex. Aldosterone causes the kidneys to reabsorb Na^+, and water follows, so that blood volume and pressure return to normal. *Top:* When a high blood Na^+ level accompanies a high blood volume, the heart secretes atrial natriuretic hormone (ANH). ANH causes the kidneys to excrete Na^+, and water follows. The blood volume and pressure return to normal.

In other words, the heart is among various organs in the body that releases a hormone, but obviously not as its major function. (Therefore, the heart is not included as an endocrine gland in Fig. 15.2.) The effect of this ANH is to cause the excretion of Na^+, or *natriuresis*. When Na^+ is excreted, so is water, and therefore, blood pressure lowers to normal.

Malfunction of the Adrenal Cortex

When the blood level of glucocorticoids is low due to hyposecretion, a person develops **Addison disease.** The presence of excessive but ineffective ACTH causes a bronzing of the skin because ACTH, like MSH, can lead to a buildup of melanin (Fig. 15.13). Without the glucocorticoids, glucose cannot be replenished when a stressful situation arises. Even a mild infection can lead to death. In some cases, hyposecretion of aldosterone results in a loss of sodium and water. Low blood pressure, and possibly severe dehydration can develop as a result. Left untreated, Addison disease can be fatal.

When the level of glucocorticoids is high due to hypersecretion, a person develops **Cushing syndrome.** The excess

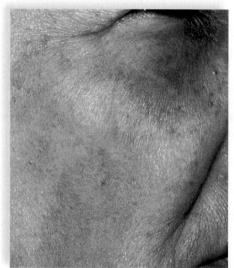

a.

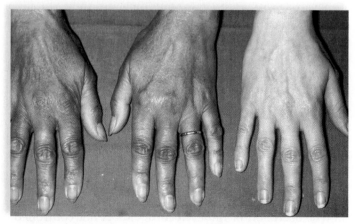

b.

Figure 15.13
What is a symptom of Addison disease?
Addison disease is characterized by a peculiar bronzing of the skin, particularly noticeable in these light-skinned individuals. Note the color of **(a)** the face and **(b)** the hands compared with the hand of an individual without the disease.

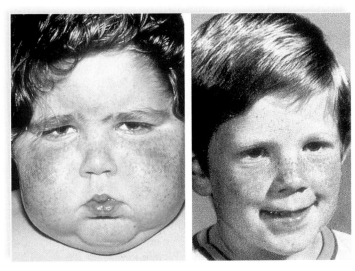

Figure 15.14 What causes Cushing syndrome?
Cushing syndrome results from hypersecretion of adrenal cortex hormones. *Left:* Patient first diagnosed with Cushing syndrome. *Right:* Four months later, after therapy.

Exocrine tissue produces digestive juice.

Pancreatic islet (islet of Langerhans)
Endocrine tissue produces insulin.

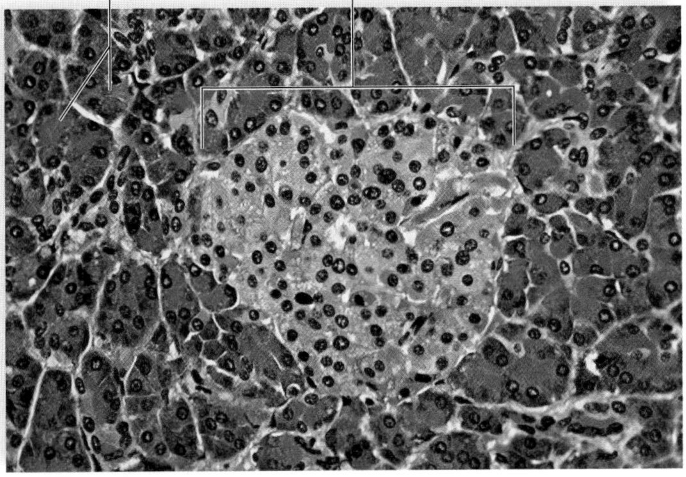

Figure 15.15 What are the products of the two types of endocrine cells?
This light micrograph shows that the pancreas has two types of cells. The exocrine tissue produces a digestive juice, and the endocrine tissue produces the hormones insulin and glucagon.

glucocorticoids result in a tendency toward diabetes mellitus, as muscle protein is metabolized and subcutaneous fat is deposited in the midsection. The result is a swollen "moon" face and an obese trunk, with arms and legs of normal size. Children will show obesity and poor growth in height (Fig. 15.14). Depending on the cause and duration of the Cushing syndrome, some people may have more dramatic changes. These include masculinization with increased blood pressure and weight gain.

> **Check Your Progress 15.4**
> 1. What are the two major hormones produced by the adrenal cortex, and what do they regulate?

15.5 Pancreas

The **pancreas** is a fish-shaped organ that stretches across the abdomen behind the stomach and near the duodenum of the small intestine. It is composed of two types of tissue. Exocrine tissue produces and secretes digestive juices that go by way of ducts to the small intestine. Endocrine tissue, called the **pancreatic islets** (islets of Langerhans), produces and secretes the hormones **insulin** and **glucagon** directly into the blood (Fig. 15.15).

The pancreas is not under pituitary control. Insulin is secreted when the blood glucose level is high, which usually occurs just after eating. Insulin stimulates the uptake of glucose by cells, especially liver cells, muscle cells, and adipose tissue cells. In liver and muscle cells, glucose is then stored as glycogen. In muscle cells, the glucose supplies energy for muscle contraction. Glucose enters the metabolic pool in fat cells and thereby sup-

plies glycerol for the formation of fat. In these various ways, insulin lowers the blood glucose level (Fig. 15.16, *top*).

Glucagon is secreted from the pancreas, usually between eating, when the blood glucose level is low. The major target tissues of glucagon are the liver and adipose tissue. Glucagon stimulates the liver to break down glycogen to glucose. It also promotes the use of fat and protein in preference to glucose as energy sources. Adipose tissue cells break down fat to glycerol and fatty acids. The liver takes these up and uses them as substrates for glucose formation. In these ways, glucagon raises the blood glucose level (Fig. 15.16, *bottom*).

Diabetes Mellitus

Over 20 million Americans (7% of the population) have **diabetes mellitus.** This disease is characterized by the inability of the body's cells to take up glucose, especially liver and muscle cells. This causes blood glucose to be higher than normal, and cells rely on other "fuels" like fatty acids for energy. A common symptom of diabetes mellitus is glucose in the urine (*mellitus,* from Greek, refers to honey or sweetness). As blood glucose increases, more glucose and water are excreted in the urine. This results in frequent urination and complaints of great thirst by affected people. Other symptoms include fatigue, unusual hunger, and weight changes. If untreated, diabetics develop serious and often fatal complications. Blurred vision and then vision loss rob the diabetic of sight. Sores that don't heal result in severe

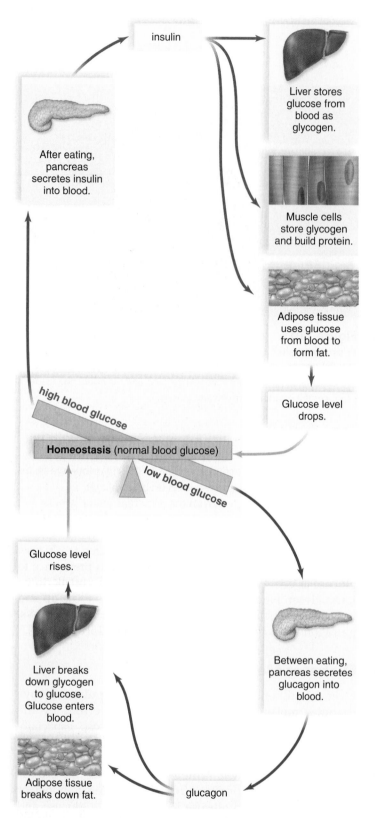

infections. Blood vessel changes cause kidney failure, nerve destruction, heart attack, or stroke.

The glucose tolerance test assists in the diagnosis of diabetes mellitus. After the patient is given 100 grams of glucose, the blood glucose concentration is measured at intervals. In a diabetic, the blood glucose level rises greatly and remains elevated for several hours (Fig. 15.17). In the meantime, glucose appears in the urine. In a nondiabetic, the blood glucose level rises somewhat and then returns to normal after about two hours.

Types of Diabetes

There are two types of diabetes mellitus. In *diabetes type 1,* the pancreas is not producing insulin. This condition is believed to be brought on by exposure to an environmental agent, most likely a virus, whose presence causes cytotoxic T cells to destroy the pancreatic islets. The body turns to the metabolism of fat, which leads to the buildup of ketones in the blood and, in turn, to acidosis (acid blood), which can lead to coma and death. As a result, the individual must have daily insulin injections. These injections control the diabetic symptoms but can still cause inconveniences because the blood sugar level may swing between hypoglycemia (low blood glucose level) and hyperglycemia (high blood glucose level). Without testing the blood glucose level, it is difficult to be certain which of these is present because the symptoms are similar. The symptoms include perspiration, pale skin, shallow breathing, and anxiety. Whenever these symptoms appear, immediate attention is required to bring the blood glucose level to its proper level. If the problem is hypoglycemia, the treatment is two

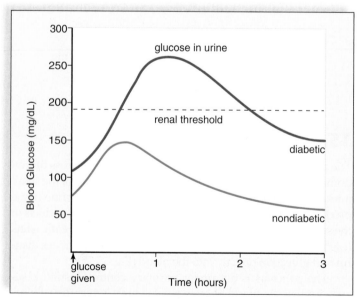

Figure 15.17 **What are the results of a glucose tolerance test performed on a diabetic?**
Following the administration of 100 grams of glucose, the blood glucose level rises dramatically in the diabetic, and glucose appears in the urine. Also, the blood glucose level at 2 hours is equal to more than 200 mg/dL.

Figure 15.16 **How is blood glucose regulated?**
Top: When the blood glucose level is high, the pancreas secretes insulin. Insulin promotes the storage of glucose as glycogen and the synthesis of proteins and fats. Therefore, insulin lowers the blood glucose level. *Bottom:* When the blood glucose level is low, the pancreas secretes glucagon. Glucagon acts opposite to insulin; therefore, glucagon raises the blood glucose level to normal.

Historical **Focus**

Surviving Diabetes Mellitus: Frederick Banting (1891–1941) and Charles Best (1899–1978)

Millions of diabetics owe an enormous debt of gratitude to two men that most have probably never heard of: Frederick Banting and Charles Best. Prior to their research, those unfortunates diagnosed with diabetes mellitus wasted away until death claimed them. Banting and Best were the team who first successfully isolated insulin, the hormone that allows diabetic patients to survive.

In 1920, Banting began work to isolate secretions from the pancreas that could treat the high blood sugar associated with diabetes. Earlier researchers had already determined that diabetes was caused by lack of a protein hormone produced by the Islets of Langerhans, clumps of cells found in the pancreas. Referring to its origin, the hormone was termed insulin (after the Latin *insula*, or island). Feeding diabetic patients pancreatic tissue failed to produce insulin activity. Attempts to extract insulin from the islets of Langerhans were also unsuccessful. Banting theorized that the unproductive attempts to recover insulin occurred because pancreatic enzymes digested the protein. If he could prevent insulin's destruction, Banting speculated, he might have a treatment for diabetes.

Banting presented his idea to a University of Toronto physiologist named John MacLeod, who offered Banting the use of lab space and experimental dogs. A medical student named Charles Best worked as Banting's assistant. The pair made amazing progress toward their goal of isolating insulin during the summer of 1921. Their achievements in such a short time are unparalleled. By fall, they were able to extend the lives of diabetic dogs with the material they'd isolated from the islets of Langerhans. Further experimentation allowed them to collect much larger insulin samples from cattle. Finally, here was a source that would provide enough insulin for human testing.

Early in 1922, they treated their first human patient, a 14-year-old boy whose life was saved by their extract. After

Figure 15A   **Charles Best and Frederick G. Banting.**
Best and Banting photographed at the University of Toronto in the summer of 1921. The dog was the first one to be kept alive with insulin treatment.

insulin treatment, the boy's blood sugar decreased, glucose was absent from his urine, and he was freed from other signs of diabetes. In February of 1922, Banting and Best published their first paper about the treatment of diabetes with insulin.

The Nobel Prize in Physiology or Medicine was awarded in 1923 to Banting and MacLeod, though MacLeod had never taken part in the original research. Best was not included in the award, because he was only a medical student. Banting was stung by the injustice to Best and divided his prize money with Best.

The work begun by Banting and Best continues to evolve. Today, human insulin is produced by genetically engineered bacteria, eliminating the need for an animal source. Several forms of human insulin are available, each with its own distinctive action. Combinations of insulins, delivered by high-tech insulin pumps, give diabetics stable insulin concentration throughout the day.

glucose tablets or two doses of glucose gel. Hard candy or orange juice would work too. If the problem is hyperglycemia, the treatment is insulin.

Some diabetics have learned to use an insulin pump to better regulate their blood sugar level. The pump is worn outside the body, usually attached to a belt or waistband. Insulin is pumped from a reservoir through a tube inserted under the skin of the abdominal wall. It is also possible to transplant a working pancreas into patients with diabetes type 1. To do away with the necessity of taking immunosuppressive drugs after the transplant, fetal pancreatic islet cells have been in-

jected into patients. Another experimental procedure is to place pancreatic islet cells in a capsule that allows insulin to get out but prevents antibodies and T lymphocytes from getting in. This artificial organ is implanted in the abdominal cavity.

Most of the diabetics in the United States have *diabetes type 2*. Often, the patient is obese. Usually, after insulin binds to a plasma membrane receptor, the number of protein carriers for glucose increases, and more glucose than usual enters the cell. In the case of diabetes type 2, glucose binds to the receptor, but the number of carriers does not increase. Therefore, the cell is said to be insulin resistant.

It is possible to prevent or at least control diabetes type 2 by adhering to a low-fat, low-sugar diet and exercising regularly. If this fails, oral drugs are available to treat diabetes type 2. These drugs stimulate the pancreas to secrete more insulin and enhance the metabolism of glucose in the liver and muscle cells. It is projected as many as 7 million Americans may have diabetes type 2 without being aware of it. Yet, the effects of untreated diabetes type 2 are as serious as those of diabetes type 1.

Long-term complications of both types of diabetes are blindness; kidney disease; and cardiovascular disorders, including atherosclerosis, heart disease, stroke, and reduced circulation. The latter can lead to gangrene in the arms and legs. Pregnancy carries an increased risk of diabetic coma, and the child of a diabetic is somewhat more likely to be stillborn or to die shortly after birth. These complications of diabetes are not expected to appear if the mother's blood glucose level is carefully regulated and kept within normal limits.

> **Check Your Progress 15.5**
>
>   1. **What hormones are produced by the pancreas, and how does each affect blood glucose?**
>   2. **What is diabetes mellitus?**

Have You Ever Wondered...

What is gestational diabetes, and what causes it?

Women who were not diabetic prior to pregnancy but with high blood glucose during pregnancy have *gestational* diabetes. Gestational diabetes affects a small percentage of pregnant women. This form of diabetes is caused by insulin resistance—body insulin concentration is normal, but cells fail to respond normally. Gestational diabetes and insulin resistance generally develop later in the pregnancy. Carefully planned meals and exercise often control this form of diabetes, but insulin injections may be necessary.

If the woman is not treated, additional glucose crosses the placenta, causing high blood glucose in the fetus. The extra energy in the fetus is stored as fat, resulting in macrosomia or a "fat" baby. Delivery of a very large baby can be dangerous for both the infant and the mother. Cesarean section is often necessary. Complications after birth are common for these babies. Further, there is a greater risk that the child will become obese and develop type 2 diabetes mellitus later in life.

Usually gestational diabetes goes away after the birth of the child. However, once a woman has experienced gestational diabetes, she has a greater chance of developing it again during future pregnancies. These women also tend to develop type 2 diabetes later in life.

15.6 Other Endocrine Glands

The **gonads** are the testes in males and the ovaries in females. The gonads are endocrine glands. Other lesser known glands and some tissues also produce hormones.

Testes and Ovaries

The activity of the testes and ovaries is controlled by the hypothalamus and pituitary. The **testes** are located in the scrotum, and the **ovaries** are located in the pelvic cavity. The testes produce **androgens** (e.g., **testosterone**), the male sex hormones. The ovaries produce **estrogens** and **progesterone,** the female sex hormones. These hormones feedback to control the hypothalmic secretion of gonadotropic releasing hormone (GnRH). The pituitary gland secretion of **follicle stimulating hormone** (FSH) and **luteinizing hormone** (LH), the gonadotropic hormones (Fig. 15.18) is controlled by feedback from the sex hormones, too. The activities of FSH and LH are discussed in Chapter 16.

Under the influence of the gonadotropic hormones, the testes begin to release increased amounts of testosterone at the time of puberty. Testosterone stimulates the growth of the penis and the testes. Testosterone also brings about and main-

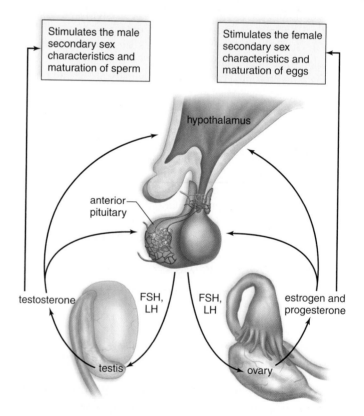

Figure 15.18   What hormones are produced by the testes and the ovaries?
The testes and ovaries secrete the sex hormones. The testes secrete testosterone, and the ovaries secrete estrogens and progesterone. In each sex, secretion of GnRH from the hypothalamus and secretion of FSH and LH from the pituitary are controlled by their respective hormones.

tains the male secondary sex characteristics that develop during puberty. These include the growth of facial, axillary (underarm) and pubic hair. It prompts the larynx and the vocal cords to enlarge, causing the voice to lower. Testosterone also stimulates oil and sweat glands in the skin. It is largely responsible for acne and body odor. Another side effect of testosterone is baldness. Although females, like males, do inherit genes for baldness, baldness is seen more often in males because of the presence of testosterone. Testosterone is partially responsible for the muscular strength of males, and this is why some athletes take supplemental amounts of **anabolic steroids,** which are either testosterone or related chemicals. The Bioethical Focus on page 270 discusses the detrimental effect anabolic steroids can have on the body.

The female sex hormones, estrogens (often referred to in the singular) and progesterone, have many effects on the body. In particular, estrogen secreted at the time of puberty stimulates the growth of the uterus and the vagina. Estrogen is necessary for egg maturation and is largely responsible for the secondary sex characteristics in females. These include female body hair and fat distribution. In general, females have a more rounded appearance than males because of a greater accumulation of fat beneath the skin. Also, the pelvic girdle is wider in females than in males, resulting in a larger pelvic cavity. Both estrogen and progesterone are required for breast development and for regulation of the uterine cycle. This includes monthly menstruation (discharge of blood and mucosal tissues from the uterus).

Thymus Gland

The lobular **thymus gland** lies just beneath the sternum (see Fig. 15.2). This organ reaches its largest size and is most active during childhood. With aging, the organ gets smaller and becomes fatty. Lymphocytes that originate in the bone marrow and then pass through the thymus are transformed into T lymphocytes. The lobules of the thymus are lined by epithelial cells that secrete hormones called **thymosins.** These hormones aid in the differentiation of lymphocytes packed inside the lobules. Although thymosins ordinarily work in the thymus, there is hope that these hormones could be injected into AIDS or cancer patients, where they would enhance T lymphocyte function.

Pineal Gland

The **pineal gland,** located in the brain (see Fig. 15.2), produces the hormone **melatonin,** primarily at night. Melatonin is involved in our daily sleep-wake cycle. Normally we grow sleepy at night when melatonin levels increase and awaken once daylight returns and melatonin levels are low (Fig. 15.19). Daily 24-hour cycles such as this are called **circadian rhythms.** These rhythms are controlled by a biological clock located in the hypothalamus.

Animal research suggests that melatonin also regulates sexual development. In keeping with these findings, it has been noted that children whose pineal gland has been destroyed due to a brain tumor experience early puberty.

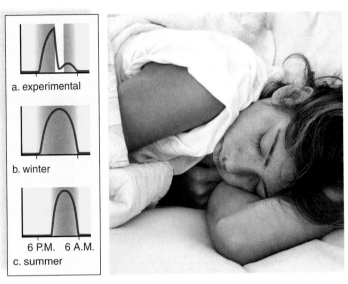

Figure 15.19 **What influences the production of melatonin?**
Melatonin production is greatest at night when we are sleeping. Light suppresses melatonin production **(a),** so it is secreted for a longer time in the winter **(b)** than in the summer **(c).**

Hormones from Other Organs/Tissues

Some organs not usually considered endocrine glands do secrete hormones. We have already mentioned that the kidneys secrete renin and that the heart produces atrial natriuretic hormone (see page 339). And you will recall that the stomach and the small intestine produce peptide hormones that regulate digestive secretions. A number of other types of tissues produce hormones.

Erythropoietin

In response to a low oxygen blood level, the kidneys secrete erythropoietin. Erythropoietin stimulates red blood cell formation in the red bone marrow. A greater number of red blood cells results in increased blood oxygen. A number of different types of organs and cells also produce peptide growth factors, which stimulate cell division and mitosis. Growth factors can be considered hormones because they act on cell types with specific receptors to receive them. Some are released into the blood; others diffuse to nearby cells.

Leptin

Leptin is a protein hormone produced by adipose tissue. Leptin acts on the hypothalamus, where it signals satiety or fullness. Strange to say, the blood of obese individuals may be rich in leptin. It is possible that the leptin they produce is ineffective because of a genetic mutation or else their hypothalamic cells lack a suitable number of receptors for leptin.

Prostaglandins

Prostaglandins are potent chemical signals produced within cells from arachidonate, a fatty acid. Prostaglandins are not distributed in the blood. They act locally, quite close to where they were produced. In the uterus, prostaglandins cause muscles to contract. Therefore, they are implicated in the pain and discomfort of menstruation in some women. Also, prostaglandins mediate the effects of pyrogens, chemicals believed to reset the temperature regulatory center in the brain. Aspirin reduces body temperature and controls pain because of its effect on prostaglandins.

Certain prostaglandins reduce gastric secretion and have been used to treat gastric reflux. Others lower blood pressure and have been used to treat hypertension. Still others inhibit platelet aggregation and have been used to prevent thrombosis. However, different prostaglandins have contrary effects, and it has been very difficult to successfully standardize their use. Therefore, prostaglandin therapy is still considered experimental.

> ### Check Your Progress 15.6
>
> 1. a. What other organs are considered major endocrine glands? b. Which of these are under the control of the anterior pituitary?
> 2. A number of other organs/tissues secrete hormones. Give some specific examples.
> 3. a. What is a local hormone? b. Give an example.

15.7 Homeostasis

The nervous and endocrine systems exert control over the other systems and thereby maintain homeostasis (Fig. 15.20).

Responding to External Changes

The nervous system is particularly able to respond to changes in the external environment. Some responses are automatic as you can testify by trying this: Take a piece of clear plastic and hold it just in front of your face. Get someone to gently toss a soft object, such as a wadded-up piece of paper, at the plastic. Can you prevent yourself from blinking? This reflex protects your eyes.

The eyes and other organs that have sensory receptors provide us with valuable information about the external environment. The central nervous system, on the receiving end of millions of bits of information, integrates information, compares it with previously stored memories, and "decides" on the proper course of action. The nervous system often responds to changes in the external environment through body movement. It gives us the ability to stay in as moderate an environment as possible. Otherwise, we test the ability of the nervous system to maintain homeostasis despite extreme conditions.

Responding to Internal Changes

The governance of internal organs usually requires that the nervous and endocrine systems work together. This usually occurs below the level of consciousness. Subconscious control often depends on reflex actions that involve the hypothalamus and the medulla oblongata. Let's take blood pressure as an example. You've just run three miles to raise money for hunger relief and decide to sit down under a tree to rest a bit. When you stand up to push off again, you feel faint. The feel-

CASE STUDY DIAGNOSIS AND TREATMENT

The blinking light on her answering machine caught Abbie's eye several days after her appointment with Dr. Adair. She pressed the play button and the voice of Dr. Adair's nurse filled the room. "Abbie, it's Karen from Dr. Adair's office. We have the results of your blood work. Please call me at your convenience so I can share them with you."

Abbie picked up the phone and dialed the number appearing on caller I.D. Her call was answered by the receptionist who immediately asked, "Can you hold, please?" Before Abbie had a chance to agree, classical musical began to play.

Minutes later, a voice asked, "Who are you holding for?"

"This is Abbie Phelps, I'm returning a call from Karen, Dr. Adair's nurse."

"One moment, please," the voice requested.

"Call waiting is SO annoying!" Abbie thought. She heaved a sigh of relief when she finally heard Karen's voice. "Hello, Abbie?"

"Yes," Abbie confirmed. "I got your message about the results of my blood work. Is everything okay?" she asked in a shaky voice.

"Relax," Karen calmly replied. "You're okay, you'll just need some medication. I'll get Dr. Adair on the phone. He can explain your test results much better than I can."

Moments later, Dr. Adair explained, "Your total blood count was normal, so you don't have anemia. However, your blood tests show that your TSH is elevated, and your thyroid hormones were low."

Abbie sank into a chair. "What does all that mean?" she asked weakly.

"It means you have hypothyroidism, or not enough thyroid hormone," the doctor replied. "Without it, your metabolism slows down, and you're tired and cold all the time. You could eventually gain weight, and you might have dry skin or lose your hair.

"But hypothyroidism is easy to treat," he reassured her. "I'm putting you on thyroid hormone medication. And you get to visit me in a month, and we'll recheck those hormones. We can fine-tune your dosage then, if we need to. Do you have any questions for me?" When Abbie declined, he replied, "Then I'll send you back to Karen."

"Now tell me what pharmacy to call about filling this prescription," Karen finished.

ing quickly passes because the medulla oblongata responds to input from the baroreceptors in the aortic arch and carotid arteries. The sympathetic system immediately acts to increase heart rate and constrict the blood vessels so that your blood pressure rises. Sweating may have upset the water-salt balance of your blood. If so, the hormone aldosterone from the adrenal cortex will act on the kidney tubules to conserve Na^+, and water reabsorption will follow. The hypothalamus can also help by sending antidiuretic hormone (ADH) to the posterior pituitary gland, which releases it into the blood. ADH actively promotes water reabsorption by the kidney tubules.

Recall from Chapter 13 that certain drugs, such as alcohol, can affect ADH secretion. When you consume alcohol, it is quickly absorbed across the stomach lining into the bloodstream, where it travels to the hypothalamus and inhibits ADH secretion. When ADH levels fall, the kidney tubules absorb less water. The result is increased production of dilute urine. Excessive water loss, or dehydration, is a disturbance of homeostasis. This is why drinking alcohol when you are exercising, or perspiring heavily on a hot day, is not a good idea. Instead of keeping you hydrated, an alcoholic beverage, such as beer, will have the opposite effect.

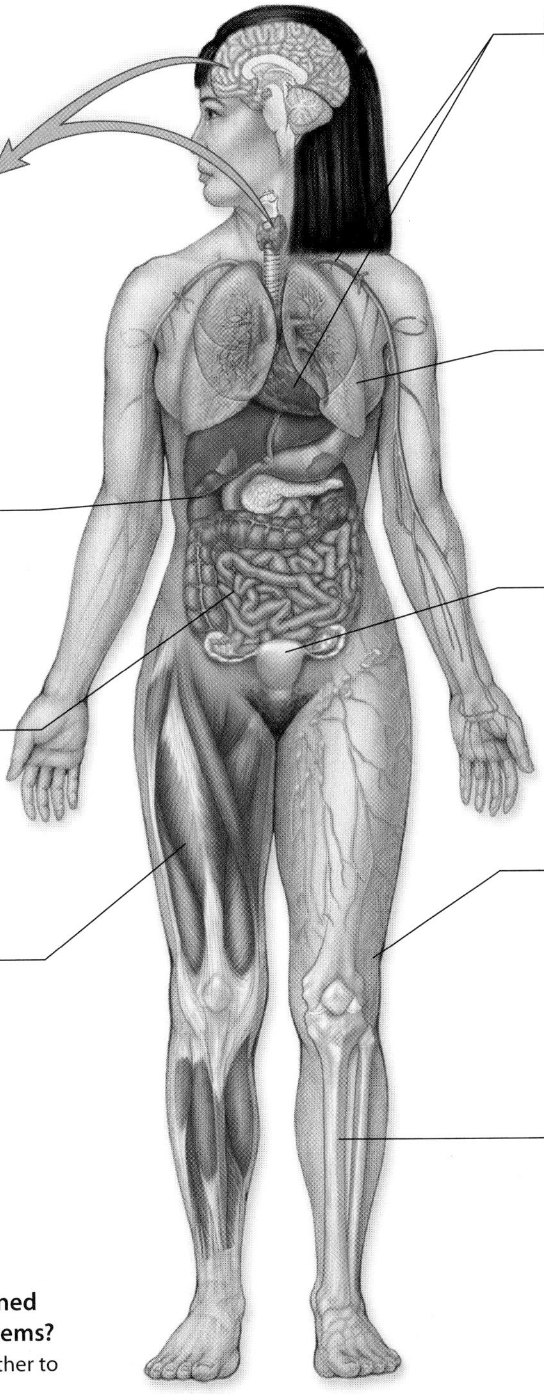

The nervous and endocrine systems work together to maintain homeostasis. The systems listed here in particular also work with these two systems.

Nervous and Endocrine Systems

The nervous and endocrine systems coordinate the activities of the other systems. The brain receives sensory input and controls the activity of muscles and various glands. The endocrine system secretes hormones that influence the metabolism of cells, the growth and development of body parts, and homeostasis.

Urinary System

Nerves stimulate muscles that permit urination. Hormones (ADH and aldosterone) help kidneys regulate the water-salt balance and the acid-base balance of the blood.

Digestive System

Nerves stimulate smooth muscle and permit digestive tract movements. Hormones help regulate digestive juices that break down food to nutrients for neurons and glands.

Muscular System

Nerves stimulate muscles, whose contractions allow us to move out of danger. Androgens promote growth of skeletal muscles. Sensory receptors in muscles and joints send information to the brain. Muscles protect neurons and glands.

Cardiovascular System

Nerves and epinephrine regulate contraction of the heart and constriction/dilation of blood vessels. Hormones regulate blood glucose and ion levels. Growth factors promote blood cell formation. Blood vessels transport hormones to target cells.

Respiratory System

The respiratory center in the brain regulates the breathing rate. The lungs carry on gas exchange for the benefit of all systems, including the nervous and endocrine systems.

Reproductive System

Nerves stimulate contractions that move gametes in ducts, and uterine contraction that occurs during childbirth. Sex hormones influence the development of the secondary sex characteristics.

Integumentary System

Nerves activate sweat glands and arrector pili muscles. Sensory receptors in skin send information to the brain about the external environment. Skin protects neurons and glands.

Skeletal System

Growth hormone and sex hormones regulate the size of the bones; parathyroid hormone and calcitonin regulate their Ca^{2+} content and therefore bone strength. Bones protect nerves and glands.

Figure 15.20 Homeostasis is maintained through cooperation of which organ systems? The nervous and endocrine systems work together to regulate and control the other systems.

Controlling the Reproductive System

Few systems intrigue us more than the reproductive system, which couldn't function without nervous and endocrine control. The hypothalamus controls the anterior pituitary, which controls the release of hormones from the testes and the ovaries and the production of their gametes. The nervous system directly controls the muscular contractions of the ducts that propel the sperm. Contractions of the oviducts, which move a developing embryo to the uterus where development continues, are stimulated by the nervous system, too. Without the positive feedback cycle involving oxytocin produced by the hypothalamus, birth might not occur.

The Neuroendocrine System

The nervous and endocrine systems work so closely together, they form what is sometimes called the neuroendocrine system. As we have seen, the hypothalamus certainly bridges the regulatory activities of the nervous and endocrine systems. In addition to producing the hormones released by the posterior pituitary, the hypothalamus produces hormones that control the anterior pituitary. The nerves of the autonomic system, which control other organs, are acted upon directly by the hypothalamus. The hypothalamus truly belongs to both the nervous and endocrine systems. Indeed, it is often and appropriately referred to as a neuroendocrine organ.

Check Your Progress 15.7

1. If sweating has caused you to lose Na$^+$ and water, how will your body restore its water-salt balance?

2. Give examples to show that the hypothalamus belongs to the nervous system and the endocrine system.

Summarizing the Concepts

15.1 Endocrine Glands

Endocrine glands secrete hormones into the bloodstream, and from there they are distributed to target organs or tissues.

- Hormones are a type of chemical signal that usually act at a distance between body parts.
- Hormones are either peptides or steroids.
- Reception of a peptide hormone at the plasma membrane activates an enzyme cascade inside the cell.
- Steroid hormones combine with a receptor, and the complex attaches to and activates DNA. Protein synthesis follows.

15.2 Hypothalamus and Pituitary Gland

Neurosecretory cells in the hypothalamus produce antidiuretic hormone and oxytocin, stored in axon endings in the posterior pituitary until they are released.

- The hypothalamus produces hypothalamic-releasing and hypothalamic-inhibiting hormones, which pass to the anterior pituitary by way of a portal system.
- The anterior pituitary produces at least six types of hormones, and some of these stimulate other hormonal glands to secrete hormones.

15.3 Thyroid and Parathyroid Glands

The thyroid gland requires iodine to produce triiodothyronine and thyroxine, which increase the metabolic rate.

- If iodine is available in limited quantities, a simple goiter develops.
- If the thyroid is overactive, an exophthalmic goiter develops.
- The thyroid gland produces calcitonin, which helps lower the blood calcium level.
- The parathyroid glands secrete parathyroid hormone, which raises the blood calcium level.

15.4 Adrenal Glands

The adrenal glands respond to stress:

- **Adrenal Medulla**   The adrenal medulla immediately secretes epinephrine and norepinephrine. Heartbeat and blood pressure increase, blood glucose level rises, and muscles become energized.
- **Adrenal Cortex**   The adrenal cortex produces the glucocorticoids (e.g., cortisol) and the mineralocorticoids (e.g., aldosterone). The glucocorticoids regulate carbohydrate, protein, and fat metabolism and also suppress the inflammatory response. Mineralocorticoids regulate water and salt balance, leading to increases in blood volume and blood pressure.

15.5 Pancreas

The pancreatic islets secrete the hormones insulin and glucagon.

- Insulin lowers the blood glucose level.
- Glucagon raises the blood glucose level.
- Diabetes mellitus is due to the failure of the pancreas to produce insulin or the failure of the cells to take it up.

15.6 Other Endocrine Glands

Other endocrine glands produce hormones.

- The testes and ovaries produce the sex hormones. Male sex hormones are the androgens (e.g., testosterone); female sex hormones are the estrogens and progesterone.
- The thymus gland secretes thymosins, which stimulate T- lymphocyte production and maturation.
- The pineal gland produces melatonin, which may be involved in circadian rhythms and the development of the reproductive organs.

Tissues also produce hormones.

- Kidneys produce erythropoietin.
- Adipose tissue produces leptin, which acts on the hypothalamus.
- Prostaglandins are produced within cells and act locally.

15.7 Homeostasis

The nervous and endocrine systems exert control over the other systems and thereby maintain homeostasis.

- The nervous system is able to respond to the external environment after receiving data from the sensory receptors. Sensory receptors are present in such organs as the eyes and ears.
- The nervous and endocrine systems work together to govern the subconscious control of internal organs. This

control often depends on reflex actions involving the hypothalamus and medulla oblongata.

- The nervous and endocrine systems work so closely together that they form what is sometimes called the neuroendocrine system.

Understanding Key Terms

acromegaly 334
Addison disease 340
adrenal cortex 337
adrenal gland 337
adrenal medulla 337
adrenocorticotropic hormone (ACTH) 332
aldosterone 339
anabolic steroid 345
androgen 344
anterior pituitary 332
antidiuretic hormone (ADH) 332
atrial natriuretic hormone (ANH) 339
calcitonin 336
chemical signal 330
circadian rhythm 345
congenital hypothyroidism 336
cortisol 338
Cushing syndrome 340
cyclic adenosine monophosphate (cAMP) 330
diabetes mellitus 341
endocrine gland 328
epinephrine 338
estrogen 344
exocrine gland 328
exophthalmic goiter 336
first messenger 330
follicle stimulating hormone (FSH) 344
glucagon 341
glucocorticoid 338
gonad 344
gonadotropic hormone 332
growth hormone (GH) 334
hormone 328
hypothalamic-inhibiting hormone 332
hypothalamic-releasing hormone 332

hypothalamus 332
insulin 341
leptin 345
luteinizing hormone (LH) 344
melanocyte-stimulating hormone (MSH) 334
melatonin 345
mineralocorticoid 338
myxedema 336
norepinephrine 338
ovary 344
oxytocin 332
pancreas 341
pancreatic islets 341
parathyroid gland 337
parathyroid hormone (PTH) 337
peptide hormone 330
pheromone 330
pineal gland 345
pituitary dwarfism 334
pituitary gland 332
positive feedback 332
posterior pituitary 332
progesterone 344
prolactin (PRL) 334
prostaglandin 345
renin 339
second messenger 330
simple goiter 336
steroid hormone 330
testes 344
testosterone 344
tetany 337
thymosin 345
thymus gland 345
thyroid gland 336
thyroid-stimulating hormone (TSH) 332
thyroxine (T_4) 336

Match the key terms to these definitions.

a. _____ Organ in the neck that secretes several important hormones, including thyroxine and calcitonin.

b. _____ Condition characterized by high blood glucose level and the appearance of glucose in the urine.

c. _____ Hormone secreted by the anterior pituitary that stimulates portions of the adrenal cortex.

d. _____ Type of hormone that causes the activation of an enzyme cascade in cells.

e. _____ Hormone released by the posterior pituitary that causes contraction of the uterus and milk letdown.

Testing Your Knowledge of the Concepts

1. Compare and contrast the nervous and endocrine systems. (page 328)
2. How does the action of a peptide hormone differ from that of a steroid hormone? (pages 330–31)
3. Explain the relationship between the hypothalamus and the posterior pituitary gland and to the anterior pituitary gland. List the hormones secreted by the anterior pituitary and the posterior pituitary glands and their actions. (pages 332–34)
4. Give an example of the negative feedback relationship among the hypothalamus, the anterior pituitary, and other endocrine glands. (page 332)
5. Discuss the action of growth hormone on the body. What occurs if there is too much or too little GH during the growing years? What occurs if there if too much GH in an adult? (pages 334–35)
6. What types of goiters and other conditions are associated with a malfunctioning thyroid gland? Explain each type. (page 336)
7. Explain how the thyroid and parathyroid glands work together to maintain blood calcium homeostasis. (pages 336–37)
8. What type of tissue is the adrenal medulla made of? What are its hormones and their actions? (page 338)
9. What hormones are secreted by the adrenal cortex, and what are their actions? (pages 338–40)
10. What are the causes and symptoms of Addison disease and Cushing syndrome? (pages 340–41)
11. Explain how insulin and glucagon maintain blood glucose homeostasis. What are the two types of diabetes mellitus, and what are the major symptoms? (pages 341–44)
12. Name the other endocrine glands and tissues mentioned in the chapter, and discuss the actions of the hormones they secrete. (pages 344–46)
13. How does the neuroendocrine system work with other systems to maintain homeostasis? (pages 346–48)
14. Hormones are never
    a. steroids.
    b. amino acids.
    c. glycoproteins.
    d. fats (triglycerides).
15. Which type of glands are ductless?
    a. exocrine
    b. endocrine
    c. Both a and b are correct.
    d. Neither a nor b is correct.
16. Which hormones can cross cell membranes?
    a. peptide hormones
    b. steroid hormones
    c. Both a and b are correct.
    d. Neither a nor b is correct.
17. The anterior pituitary controls the secretion(s) of both
    a. the adrenal medulla and the adrenal cortex.
    b. thyroid and adrenal cortex.
    c. ovaries and testes.
    d. Both b and c are correct.
18. Growth hormone is produced by the
    a. posterior adrenal gland.
    b. posterior pituitary.
    c. anterior pituitary.
    d. kidneys.
    e. None of these is correct.
19. _____ is released through positive feedback and causes _____ contractions.
    a. Insulin, stomach
    b. Oxytocin, stomach
    c. Oxytocin, uterine
    d. None of these is correct.

20. PTH causes the blood levels of calcium to _____, and calcitonin causes it to _____.
    a. increase, increase
    c. decrease, increase
    b. increase, decrease
    d. decrease, decrease

21. Bodily response to stress includes
    a. water reabsorption by the kidneys.
    b. blood pressure increase.
    c. increase in blood glucose levels.
    d. heart rate increase.
    e. All of these are correct.

22. Anabolic steroid use can cause
    a. liver damage.
    d. reduced testicular size.
    b. severe acne.
    e. All of these are correct.
    c. balding.

23. Lack of aldosterone will cause a blood imbalance of
    a. sodium.
    d. All of these are correct.
    b. potassium.
    e. None of these is correct.
    c. water.

24. Glucagon causes
    a. use of fat for energy.
    b. glycogen to be converted to glucose.
    c. use of amino acids to form fats.
    d. Both a and b are correct.
    e. None of these is correct.

25. Long-term complications of diabetes include
    a. blindness.
    d. All of these are correct.
    b. kidney disease.
    e. None of these is correct.
    c. circulatory disorders.

26. Diabetes mellitus is associated with
    a. too much insulin in the blood.
    b. too high a blood glucose level.
    c. blood that is too dilute.
    d. All of these are correct.

27. Which of these is not a pair of antagonistic hormones?
    a. insulin—glucagon
    b. calcitonin—parathyroid hormone
    c. cortisol—epinephrine
    d. aldosterone—atrial natriuretic hormone (ANH)

28. Which hormone and condition is mismatched?
    a. growth hormone—acromegaly
    b. thyroxine—goiter
    c. parathyroid hormone—tetany
    d. cortisol—myxedema
    e. insulin—diabetes

In questions 29–33, match the hormones to the correct gland in the key.

Key:

a. glucagon
d. insulin
b. prostaglandin
e. leptin
c. melatonin

29. Raises blood glucose levels

30. Conversion of glucose to glycogen

31. Hunger control

32. Controls circadian rhythms

33. Causes uterine contractions

In questions 34–38, match the hormones to the correct gland in the key.

Key:

a. pancreas
d. thyroid
b. anterior pituitary
e. adrenal medulla
c. posterior pituitary
f. adrenal cortex

34. Cortisol

35. Growth hormone (GH)

36. Oxytocin storage

37. Insulin

38. Epinephrine

39. Complete the following diagram.

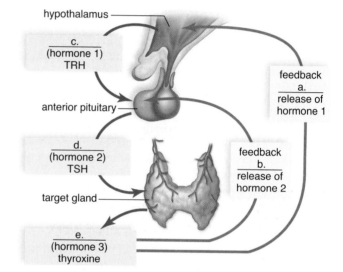

Thinking Critically About the Concepts

Blood tests are a way to diagnose any number of endocrine disorders because hormones are transported by the circulatory system. GH and IGF-1 can be checked to determine if deficiencies are the reason for a child's slow growth. Blood levels of TSH, T_3, and T_4 provide information about thyroid function.

Abbie's symptoms—fatigue and constantly feeling cold—are common to any number of ailments. The elevated TSH and low T_4 measurements from the blood tests ordered by Dr. Adair indicated that her symptoms were caused by hypothyroidism. Additional blood tests will allow Dr. Adair to fine-tune the amount of thyroid medication Abbie will need.

1. How is follicle stimulating hormone similar to growth hormone with regard to how their target cells respond to their signals?

2. Abbie's hypothyroidism was diagnosed by the high levels of TSH in her blood. Explain why her TSH was high. *Hint:* You might want to consider what happens to TSH when the activity of the thyroid is normal.

3. Why did Abbie's hypothyroidism cause her to feel tired and cold? *Hint:* Consider the effect of thyroxine on the body's cells.

4. Some people take supplemental thyroxine even when they don't have hypothyroidism. Why would someone want to do this? Can you think of some risks associated with this practice?

CHAPTER **16**

Reproductive System

CASE STUDY ANTHONY TODD

The door slammed, announcing Anthony's return from the telemarketing job he despised. Carter, his roommate, glanced up. "I guess you had a great time at work!"

"You know it," Anthony replied sarcastically. "One of my calls was monitored by my supervisor tonight. I got busted for not offering an alternate to an item that wasn't available. Never mind that I sold the customer $863 worth of stuff!"

"I've GOT to find another job!" He threw his hands up in frustration. "What did you do with the employment section of the newspaper?"

"Dude, I didn't do anything with the newspaper," Carter chastised. "You took it to work with you. What did YOU do with the employment section?"

"Augh! I left the paper at work," Anthony groaned. "I guess I better just study for the human reproduction quiz we have tomorrow. I'm sure Dr. Nicely will have dreamed up some killer questions. Remind me to check out tomorrow's employment section."

As predicted, Dr. Nicely's questions were tough. Anthony scowled while he tried to remember the pituitary hormones that controlled reproduction in males. "Why couldn't Dr. Nicely just ask about testosterone on the quiz?" he complained to Carter after the class. "I know all about how that controls male reproduction."

"You think so?" Carter replied. "Check out the flier posted on this bulletin board."

"Surely you jest," Anthony joked. He spun around to see what Carter was talking about. There on the board was a notice about a research project being conducted at the university. A trial study of a hormonal contraceptive gel needed male subjects.

"They're paying people to participate!" Anthony exclaimed. "I might be able to quit that telemarketing gig if it covers the rent. I'm going to see if I qualify to be a subject."

"Are you kidding?" Carter retorted. "What if it causes permanent damage? You want to have kids some day, right?"

"It's worth checking out, if I don't have to take those stupid phone orders anymore," Anthony replied. "I'll fill you in once I know more ."

CHAPTER CONCEPTS

16.1 Human Life Cycle
The male reproductive system produces sperm, and the female reproductive system produces eggs. Sperm and eggs have only 23 chromosomes each due to reduction division that takes place as they develop.

16.2 Male Reproductive System
The male reproductive system consists of the testes and a series of ducts that deliver sperm by way of the penis to a female sexual partner. The testes produce sperm and also the male sex hormones.

16.3 Female Reproductive System
In the female, the ovaries produce eggs and the sex hormones.

16.4 Female Hormone Levels
The sex hormones fluctuate in monthly cycles, resulting in ovulation once a month followed by menstruation if pregnancy does not occur. Pregnancy and the birth control pill prevent these events from occurring.

16.5 Control of Reproduction
Numerous birth control methods are available for those who wish to prevent pregnancy. Some couples are infertile, and if so, they may use assisted reproductive technologies to have a child.

16.6 Sexually Transmitted Diseases
Medications have been developed to control AIDS and genital herpes, but these STDs are not curable. A vaccine is now available for the most common form of genital warts and hepatitis A and B.

STDs caused by bacteria are curable with antibiotic therapy, but resistance is making this more and more difficult.

16.1 Human Life Cycle

Unlike the other systems of the body, the reproductive system is quite different in males and females. *Puberty* is the sequence of events by which a child becomes a sexually competent young adult. The reproductive system does not begin to fully function until puberty is complete. Sexual maturity occurs between the age of 10 and 14 in girls, and 12 and 16 in boys. At the completion of puberty, the individual is capable of producing children.

The reproductive organs (genitals) have the following functions:

1. Males produce sperm within testes, and females produce eggs within ovaries.
2. Males nurture and transport the sperm in ducts until they exit the penis. Females transport the eggs in uterine tubes to the uterus.
3. The male penis functions to deliver sperm to the female vagina, which functions to receive the sperm. The vagina also transports menstrual fluid to the exterior and is the birth canal.
4. The uterus of the female allows the fertilized egg to develop within her body. After birth, the female breast provides nourishment in the form of milk.
5. The testes and ovaries produce the sex hormones. The sex hormones have a profound effect on the body because they bring about masculinization or feminization of various features. In females, the sex hormones also allow a pregnancy to continue.

Mitosis and Meiosis

Our DNA is distributed among 46 chromosomes within the nucleus. Every cell in the body has 46 chromosomes. Ordinarily when a cell divides by a process called mitosis, the new cells also have 46 chromosomes. Mitosis is *duplication division*. (As an analogy, imagine the cell producing exact copies of itself during mitosis, much like a duplicating machine does with a page of notes.) In the life cycle of a human being, mitosis is the type of cell division that takes place during growth and repair of tissues (Fig. 16.1).

In addition to mitosis, human cells undergo a type of cell division called meiosis, which is *reduction division.* Meiosis takes place only in the testes of males during the production of sperm and in the ovaries of females during the production of eggs. During meiosis, the chromosome number is reduced from the normal 46 chromosomes, called the diploid, or 2n, number, down to 23 chromosomes, called the haploid, or n, number of chromosomes. Meiosis requires two successive divisions, called meiosis I and meiosis II.

The flagellated sperm is small compared to the egg. It is specialized to carry only chromosomes as it swims to the egg. The egg is specialized to await the arrival of a sperm and to provide the new individual with cytoplasm in addition to chromosomes. The first cell of a new human being is

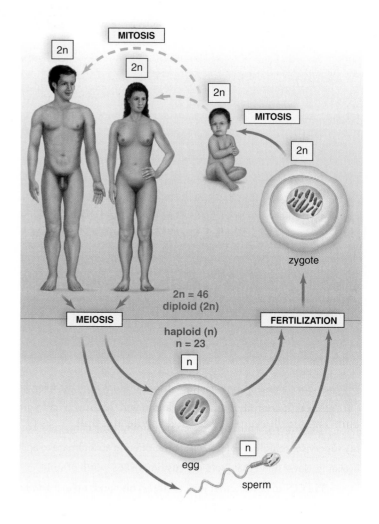

Figure 16.1 **Which type of cell division occurs during the human life cycle to produce sex cells?**
The human life cycle has two types of cell divisions: mitosis, in which the chromosome number stays constant, and meiosis, in which the chromosome number is reduced. During growth or cell repair, mitosis ensures that each new cell has 46 chromosomes. During production of sex cells, the chromosome number is reduced from 46 to 23. Therefore, an egg and a sperm each have 23 chromosomes so that when the sperm fertilizes the egg, the new cell, called a zygote, has 46 chromosomes.

called the zygote. A sperm has 23 chromosomes and the egg has 23 chromosomes, so the zygote has 46 chromosomes altogether. Without meiosis, the chromosome number in each generation of human beings would double, and the cells would no longer be able to function.

Check Your Progress 16.1

1. Contrast the two types of cell division in the human life cycle.
2. a. How many chromosomes does a mother contribute to the new individual? b. How many does the father contribute?
3. a. Where does meiosis occur in males? b. In females?

16.2 Male Reproductive System

The male reproductive system includes the organs depicted in Figure 16.2 and listed in Table 16.1. The male gonads, or primary sex organs, are paired **testes** (sing., testis), suspended within the sacs of the **scrotum.**

Sperm produced by the testes mature within the **epididymis** (pl., epididymides), a tightly coiled duct lying just outside each testis. Maturation seems to be required for sperm to swim to the egg. When sperm leave an epididymis, they enter a **vas deferens** (pl., vasa deferentia), also called the ductus deferens. The sperm may be stored for a time in the vas deferens. Each vas deferens passes into the abdominal cavity, where it curves around the bladder and empties into an ejaculatory duct. The ejaculatory ducts enter the **urethra.**

At the time of ejaculation, sperm leave the penis in a fluid called **semen.** The seminal vesicles, the prostate gland, and the bulbourethral glands (Cowper glands) add secretions to seminal fluid. The pair of **seminal vesicles** lie at the base of the bladder, and each has a duct that joins with a vas deferens. The **prostate gland** is a single, donut-shaped gland that surrounds the upper portion of the urethra just below the bladder. In older men, the prostate can enlarge and squeeze off the urethra, making urination painful and difficult. The condition can be treated medically. **Bulbourethral**

Table 16.1	Male Reproductive Organs
Organ	**Function**
Testes	Produce sperm and sex hormones
Epididymides	Ducts where sperm mature and some sperm are stored
Vasa deferentia	Conduct and store sperm
Seminal vesicles	Contribute nutrients and fluid to semen
Prostate gland	Contributes fluid to semen
Urethra	Conducts sperm
Bulbourethral glands	Contribute mucus-containing fluid to semen
Penis	Organ of sexual intercourse

glands are pea-sized organs that lie posterior to the prostate on either side of the urethra. Their secretion makes the seminal fluid gelatinous.

Each component of seminal fluid seems to have a particular function. Sperm are more viable in a basic solution, and seminal fluid, milky in appearance, has a slightly basic pH (about 7.5). Swimming sperm require energy, and seminal fluid contains the sugar fructose, which presumably serves as an energy source. Semen also contains prostaglandins, chemicals that cause the uterus to contract. Some investigators believe that uterine contractions help propel the sperm toward the egg.

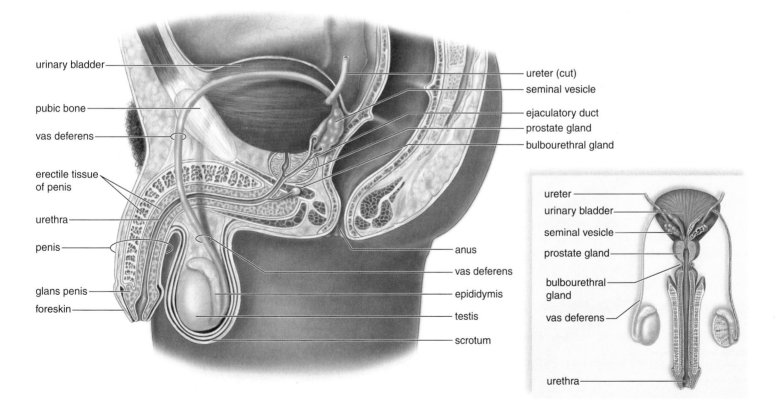

Figure 16.2 **Which organ produces each component of male semen?**
The testes produce sperm. The seminal vesicles, the prostate gland, and the bulbourethral glands provide a fluid medium for the sperm, which move from the vas deferens through the ejaculatory duct to the urethra in the penis. The foreskin (prepuce) is removed when a penis is circumcised.

Orgasm in Males

The **penis** (Fig. 16.3) is the male organ of sexual intercourse. The penis has a long shaft and an enlarged tip called the glans penis. The glans penis is normally covered by a layer of skin called the foreskin. **Circumcision,** the surgical removal of the foreskin, if done, is usually performed soon after birth.

Spongy, erectile tissue containing distensible blood spaces extends through the shaft of the penis. During sexual arousal, autonomic nerves release nitric oxide, NO. This stimulus leads to the production of cGMP (cyclic guanosine monophosphate), a high energy compound similar to ATP. The cGMP causes the smooth muscle of incoming arterial walls to relax and the erectile tissue to fill with blood. The veins that take blood away from the penis are compressed, and the penis becomes erect. **Erectile dysfunction** (formerly called impotency) exists when the erectile tissue doesn't expand enough to compress the veins. Medications for the treatment of erectile dysfunction inhibit the enzyme that breaks down cGMP, ensuring that a full erection will take place. Certain of these medications can cause vision problems because the same enzyme occurs in the retina. During an erection, a sphincter closes off the bladder so that no urine enters the urethra. (The urethra carries either urine or semen at different times.)

As sexual stimulation intensifies, sperm enter the urethra from each vas deferens, and the glands contribute secretions to the seminal fluid. Once seminal fluid is in the urethra, rhythmic muscle contractions cause it to be expelled from the penis in spurts (ejaculation).

The contractions that expel seminal fluid from the penis are a part of male orgasm, the physiological and psychological sensations that occur at the climax of sexual stimulation. The psychological sensation of pleasure is centered in the brain. However, the physiological reactions involve the genital (reproductive) organs and associated muscles, as well as the entire body. Marked muscular tension is followed by contraction and relaxation. Following ejaculation and/or loss of sexual arousal, the penis returns to its normal flaccid state. Usually a period of time, called the refractory period, follows during which stimulation does not bring about an erection. The length of the refractory period increases with age.

There may be in excess of 400 million sperm in the 3.5 ml of semen expelled during ejaculation. The sperm count can be much lower than this, however, and fertilization of the egg by a sperm can still take place.

Male Gonads, the Testes

The testes, which produce sperm and also the male sex hormones, lie outside the abdominal cavity of the male, within the scrotum. The testes begin their development inside the abdominal cavity. They descend into the scrotal sacs through the inguinal canal during the last two months of fetal development. If the testes do not descend and the male is not treated or operated on to place the testes in the scrotum, sterility (the inability to produce offspring) usually follows. This is because the internal temperature of the body is too high to produce viable sperm. The scrotum helps regulate the temperature of the testes by holding them closer to or farther away from the body.

Seminiferous Tubules and Interstitial Cells

A longitudinal section of a testis shows that it is composed of compartments called lobules, each of which contains one to three tightly coiled **seminiferous tubules** (Fig. 16.4a). A microscopic cross section of a seminiferous tubule reveals that it is packed with cells undergoing **spermatogenesis** (Fig. 16.4b), the production of sperm.

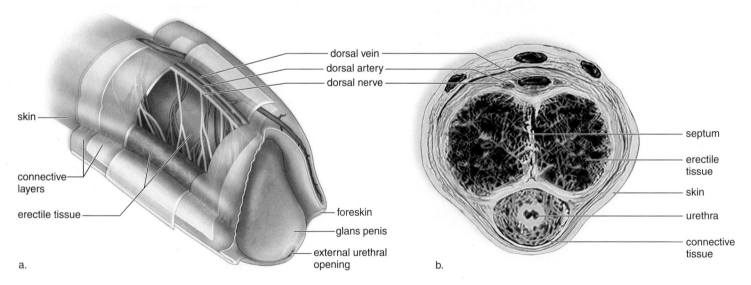

Figure 16.3 **How is the shaft of a male penis constructed?**
a. Penis shaft. The shaft of the penis ends in an enlarged tip called the glans penis, which, in uncircumsized males, is partially covered by a foreskin (prepuce). The penis contains columns of erectile tissue. **b.** Micrograph of shaft in cross section showing location of erectile tissue. One column surrounds the urethra.

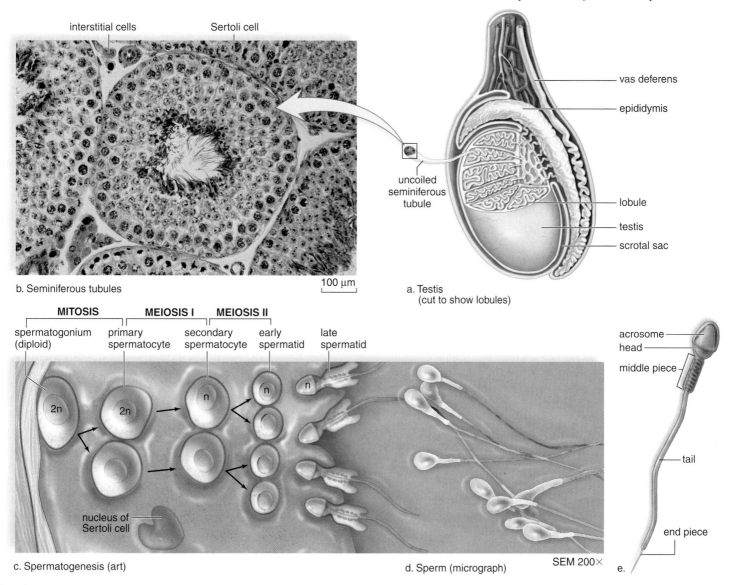

interstitial cells   Sertoli cell

vas deferens

epididymis

uncoiled seminiferous tubule

lobule

testis

scrotal sac

b. Seminiferous tubules

100 µm

a. Testis (cut to show lobules)

MITOSIS   **MEIOSIS I**   **MEIOSIS II**

spermatogonium (diploid) | primary spermatocyte | secondary spermatocyte | early spermatid | late spermatid

acrosome

head

middle piece

2n   2n   n   n   n

tail

nucleus of Sertoli cell

end piece

c. Spermatogenesis (art)

d. Sperm (micrograph)   SEM 200×   e.

Figure 16.4 **Where does spermatogenesis occur? What are the three parts of a sperm cell?**
a. The lobules of a testis contain seminiferous tubules. **b.** Electron micrograph of a cross section of the seminiferous tubules, where spermatogenesis occurs. Note the location of interstitial cell in clumps among the seminiferous tubules. **c.** Diagrammatic representation of spermatogenesis, which occurs in wall of tubules. **d.** Micrograph of sperm. **e.** A sperm has a head, a middle piece, and a tail. The nucleus is in the head, capped by the enzyme-containing acrosome.

Have You Ever Wondered...

Boxers or briefs?

Most people seem to know that the scrotum's role is to keep the temperature of the testes lower than body temperature. The lower temperature is necessary for normal sperm production. It might follow that the man's type of underwear might change that temperature, affecting sperm production. However, research has not supported this assumption. The style of underwear worn by a man, loose or close fitting, has not been shown to affect sperm count or fertility significantly. So, boxers or briefs? It's up to you!

During the production of sperm, spermatogonia divide to produce primary spermatocytes (2n). Primary spermatocytes move away from the outer wall, increase in size, and undergo meiosis I to produce secondary spermatocytes. Each secondary spermatocyte has only 23 chromosomes (Fig. 16.4c). Secondary spermatocytes (n) undergo meiosis II to produce four spermatids, each of which also has 23 chromosomes. Spermatids then develop into sperm (Fig. 16.4d). Note the presence of **Sertoli cells** (purple), which support, nourish, and regulate the process of spermatogenesis. It takes approximately 74 days for sperm to undergo development from spermatogonia to sperm.

Mature **sperm,** or spermatozoa, have three distinct parts: a head, a middle piece, and a tail (Fig. 16.4e). Mitochondria in the middle piece provide energy for the movement of the tail, which is a flagellum. The head contains a nucleus covered by a cap called the **acrosome,** which stores enzymes needed to penetrate the egg. The ejaculated semen of a normal human male contains several hundred million sperm, but only one sperm normally enters an egg. Sperm usually do not live more than 48 hours in the female genital tract.

Interstitial Cells

The male sex hormones, the androgens, are secreted by cells that lie between the seminiferous tubules. These cells are called **interstitial cells.** The most important of the androgens is testosterone, whose functions are discussed next.

Hormonal Regulation in Males

The hypothalamus has ultimate control of the testes' sexual function because it secretes a hormone called **gonadotropin-releasing hormone (GnRH).** GnRH stimulates the anterior pituitary to secrete the gonadotropic hormones. There are two gonadotropic hormones, **follicle-stimulating hormone (FSH)** and **luteinizing hormone (LH),** which are present in both males and females. In males, FSH promotes the production of sperm in the seminiferous tubules. LH in males controls the production of testosterone by the interstitial cells.

All these hormones are involved in a negative feedback relationship that maintains the fairly constant production of sperm and testosterone (Fig. 16.5). When the amount of testosterone in the blood rises to a certain level, it causes the hypothalamus and anterior pituitary to decrease their respective secretion of GnRH and LH. As the level of testosterone begins to fall, the hypothalamus increases its secretion of GnRH, and the anterior pituitary increases its secretion of LH. These stimulate the interstitial cells to produce testosterone. A similar feedback mechanism maintains the continuous production of sperm. The Sertoli cells in the wall of the seminiferous tubules produce a hormone called *inhibin* that blocks GnRH and FSH secretion when appropriate (Fig. 16.4).

Testosterone, the main sex hormone in males, is essential for the normal development and functioning of the organs listed in Table 16.1. Testosterone also brings about and maintains the male secondary sex characteristics that develop at the time of puberty. Males are generally taller than females and have broader shoulders and longer legs relative to trunk length. The deeper voices of males compared with those of females are due to a larger larynx with longer vocal cords. The so-called Adam's apple, part of the larynx, is usually more prominent in males than in females. Testosterone causes males to develop noticeable hair on the face; chest; and occasionally other regions of the body, such as the back. A related chemical also leads to the receding hairline and pattern baldness that occur in males.

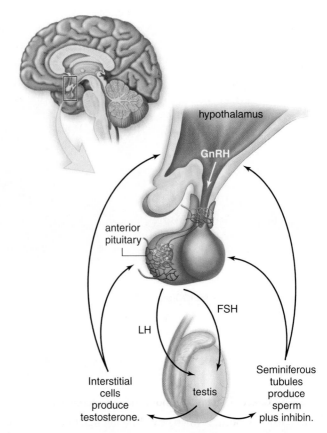

Figure 16.5   What hormones control the production of sperm and testosterone by the testes?
Gonadotropin-releasing hormone (GnRH) stimulates the anterior pituitary to secrete the gonadotropic hormones: Follicle-stimulating hormone (FSH) stimulates the production of sperm, and luteinizing hormone (LH) stimulates the production of testosterone. Testosterone and inhibin exert negative feedback control over the hypothalamus and the anterior pituitary, and this regulates the level of testosterone in the blood and the production of sperm by the testes.

Testosterone is responsible for the greater muscular development in males. Knowing this, both males and females sometimes take anabolic steroids, either testosterone or related steroid hormones resembling testosterone. Health problems involving the kidneys, the cardiovascular system, and hormonal imbalances can arise from such use. The testes shrink in size, and feminization of other male traits occurs (see page 270).

> *Check Your Progress 16.2*
>
> 1. a. What male structure functions to produce sperm?
>   b. What structures are involved in the production of seminal fluid? c. Which structure is involved in transport of semen out of the body?
>
> 2. a. Which endocrine glands are involved in promoting and maintaining the sex characteristics of males?
>   b. Which hormones are involved?

16.3 Female Reproductive System

The female reproductive system includes the organs depicted in Figure 16.6 and listed in Table 16.2. The female gonads are paired **ovaries** that lie in shallow depressions, one on each side of the upper pelvic cavity. The ovaries produce **eggs** (technically referred to as ova) and the female sex hormones, estrogen and progesterone.

The Genital Tract

The **oviducts,** also called the uterine or fallopian tubes, extend from the uterus to the ovaries. However, the oviducts are not attached to the ovaries. Instead, they have fingerlike projections called fimbriae (sing., **fimbria**) that sweep over the ovaries. When an egg (ovum) bursts from an ovary during ovulation, it usually is swept into an oviduct by the combined action of the fimbriae and the beating of cilia that line the oviducts.

Once in the oviduct, the egg is propelled slowly by ciliary movement and tubular muscle contraction toward the uterus.

Table 16.2	Female Reproductive Organs
Organ	**Function**
Ovaries	Produce eggs and sex hormones
Oviducts	Conduct eggs; location of fertilization (uterine or fallopian tubes)
Uterus (womb)	Houses developing fetus
Cervix	Contains opening to uterus
Vagina	Receives penis during sexual intercourse; serves as birth canal and as an exit for menstrual flow

An egg lives approximately 6–24 hours, unless fertilization occurs. Fertilization, and therefore **zygote** formation, usually takes place in the oviduct. A developing embryo normally arrives at the uterus after several days, and then **implantation** occurs. During implantation the embryo embeds in the uterine lining, which has been prepared to receive it.

CASE STUDY AT THE RESEARCH CLINIC

The clinic directory steered Anthony to the research office. He felt a twinge of nervousness when he walked through the door. "Steady, boy, we're only here to get information. It's going to be so great to quit that stupid job you hate!" he murmured softly, reassuring himself.

"Pardon me, did you say something?" the receptionist asked.

"Nothing, nothing," Anthony stammered. "I saw the notice about the research study thing. You know, the hormonal contraceptive gel for men? I'm interested in becoming a subject."

"Oh, sure," the receptionist replied. "Here's an informational packet that explains everything. The informed consent paperwork will have to be completed before you can enroll. Why don't you have a seat and start reading? I'll see if one of the investigators can come here to answer any questions you might have."

Anthony sat down and began to read. The introduction discussed hormonal control of human male reproduction, reminding Anthony of Dr. Nicely's quiz questions. A brief summary of research on male hormonal contraception (MHC) followed. Anthony's studies paid off—he was able to figure out the technical jargon.

The earliest versions of MHC had involved weekly shots of androgens. Anthony shivered at the thought of weekly shots; he hated needles. He remembered that testosterone, the hormone controlling male secondary sex characteristics, was an androgen. But he was surprised to learn that the best results had been achieved by combining testosterone and a synthetic form of progesterone called progestin. He was relieved that this study was investigating using a gel instead of injections.

As he continued to read, Anthony learned about the ways that scientists hoped to develop a safe and effective MHC. The ideal MHC, he discovered, would result in azoospermia, the complete absence of sperm in the ejaculate. Severe oligospermia, fewer than 1 million sperm per millimeter of semen, was considered infertile as well. An effective contraceptive would produce dramatic reductions in sperm count like these, but fertility would be restored soon after discontinuing use of MHC. Twenty million sperm per milliliter is considered fertile, and Anthony was reassured by prior research studies. It took only three to four months for fertility to return. His plans to have a "basketball team" of his own kids someday shouldn't be affected if he was accepted into the study.

Anthony flipped to the section about side effects. Some men had experienced increased muscle mass, acne, and mood swings. Researchers predicted being able to minimize these effects, however. Anthony was more concerned with the results about HDL. From Dr. Nicely's class, Anthony remembered that HDL is the good cholesterol transport molecule associated with preventing heart disease. Decreased levels of HDL had happened in earlier studies, and that didn't sound good.

Suddenly the receptionist interrupted his thoughts. "I'm sorry, but all the investigators are in a meeting right now. No one is available to talk to you about the study. Why don't you finish reading those materials at home and come back tomorrow? I'll make an appointment for you with the chief investigator, Dr. Robbins. They can probably do all the preliminary blood work then, too."

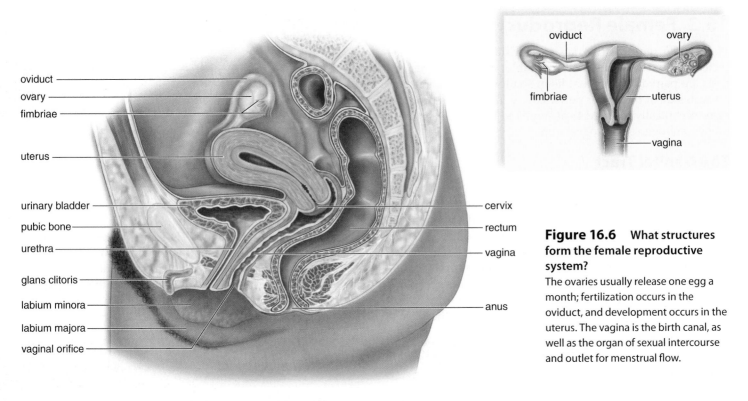

oviduct
ovary
fimbriae
uterus
urinary bladder
pubic bone
urethra
glans clitoris
labium minora
labium majora
vaginal orifice

cervix
rectum
vagina
anus

oviduct   ovary
fimbriae   uterus
vagina

Figure 16.6   What structures form the female reproductive system?
The ovaries usually release one egg a month; fertilization occurs in the oviduct, and development occurs in the uterus. The vagina is the birth canal, as well as the organ of sexual intercourse and outlet for menstrual flow.

The **uterus** is a thick-walled, muscular organ about the size and shape of an inverted pear. Normally, it lies above and is tipped over the urinary bladder. The oviducts join the uterus at its upper end, while at its lower end, the **cervix** enters the vagina nearly at a right angle.

Cancer of the cervix is a common form of cancer in women. Early detection is possible by means of a **Pap test,** which requires the removal of a few cells from the region of the cervix for microscopic examination. If the cells are cancerous, a physician may recommend a hysterectomy. A hysterectomy is the removal of the uterus, including the cervix. Removal of the ovaries in addition to the uterus is technically termed an ovariohysterectomy (radical hysterectomy). The vagina remains, so the woman can still engage in sexual intercourse.

Development of the embryo and fetus normally takes place in the uterus. This organ, sometimes called the womb, is approximately 5 cm wide in its usual state. It is capable of stretching to over 30 cm wide to accommodate a growing fetus. The lining of the uterus, called the **endometrium,** participates in the formation of the placenta (see page 364). The endometrium supplies nutrients needed for embryonic and fetal development. The endometrium has two layers: one layer is a basal layer of reproducing cells. The innermost endometrial lining is the functional layer. In the nonpregnant female, the functional layer of the endometrium varies in thickness according to a monthly reproductive cycle called the uterine cycle.

A small opening in the cervix leads to the vaginal canal. The **vagina** is a tube that lies at a 45° angle to the small of the back. The mucosal lining of the vagina lies in folds and can extend. This is especially important when the vagina serves as the birth canal, and it facilitates sexual intercourse, when the vagina receives the penis. The vagina also acts as an exit for menstrual flow. Several different types of bacteria normally reside in the vagina and create an acidic environment. This environment is protective against the possible growth of pathogenic bacteria, but sperm prefer the basic environment provided by seminal fluid.

External Genitals

The external genital organs of the female are known collectively as the **vulva** (Fig. 16.7). The vulva includes two large, hair-covered folds of skin called the labia majora.

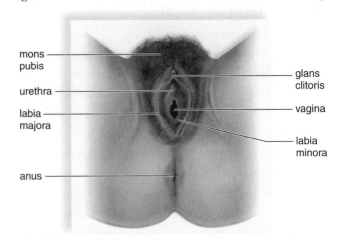

mons pubis
urethra
labia majora
anus

glans clitoris
vagina
labia minora

Figure 16.7   What structures form the external genitals of a female?
The external genitals of the female include the labia majora, labia minora, and glans clitoris. These organs are also referred to as the vulva.

Male and Female Circumcision: Medical Option, Cultural Practice, or Child Abuse?

At birth a layer of skin (the foreskin) covers the end of a male baby's penis. In the United States, more than 50% of infant males are circumcised shortly after birth. During circumcision, the glans penis is exposed when the foreskin is removed during a surgical procedure. This procedure is done before babies go home from the hospital or during religious ceremonies in the home.

The decision to circumcise a baby boy is made by the parents and is often based on the religious or cultural beliefs of the parents. Circumcision may be done so that male children will resemble their father. Some choose circumcision because of concerns about cleanliness. Claims that circumcision increases or decreases sexual pleasure later in life have not been supported by research.

There is some evidence that suggests urinary tract infections are less common in circumcised infants. Research also shows that circumcision reduces the spread of HIV during heterosexual contact. In areas of the world where HIV infections are prevalent, circumcision may become an important means of limiting the spread of AIDS.

As with any type of surgery, there are risks associated with circumcision. The most common complications are minor bleeding and localized infections that can be treated easily. One of the biggest concerns is about the pain experienced by the baby during circumcision. The American Academy of Pediatrics (AAP) now recommends using a form of local anesthesia during the procedure. The AAP does not recommend nor argue against circumcision of male babies.

However, the circumcision of females is a highly controversial topic. Female circumcision (also referred to as female genital cutting [FGC], or female genital mutilation) is done strictly for cultural or religious reasons, though no religion specifically calls for its practice. The procedure involves partially or totally cutting away the external genitalia of a female. Cultures that practice FGC believe it to be a necessary rite of passage for girls. In the views of these cultures, FGC must be done to preserve the virginity of females and to prevent promiscuity. It is also done for aesthetic reasons, because the clitoris is thought to be an unhealthy and unattractive organ. Moreover, FGC is seen as an essential prerequisite for marriage. Females with an intact clitoris are believed to be unclean. Such women are considered to be potentially harmful to a man during intercourse or to a baby during childbirth, if either is touched by the clitoris. Many believe that FGC enhances a husband's sexual pleasure and a woman's fertility.

Many girls die from infection after FGC. FGC also causes lifelong urinary and reproductive tract infections, infertility, and pelvic pain. Victims report an absent or greatly diminished pleasurable response to sexual intercourse.

FCG is most commonly performed on girls between the ages of 4 and 12 and in countries in central Africa. It is also performed in some Middle Eastern countries and among Muslim groups in various other locations. With increasing immigration from these countries, there are also greater numbers of women who have been subjected to FGC. Likewise, there are more girls in the United States who are at risk for FGC.

Figure 16A Waris Dirie, Somalian-born supermodel, victim of FGC, advocates against female genital circumcision in her book *Desert Flower.*

Thanks to the efforts of mutilation victim Waris Dirie and others like her, the need to eliminate FGC is now discussed openly, and action is being taken in many countries to outlaw the practice. FGC is considered to be a violation of Human Rights by the United Nations, UNICEF, and the World Health Organization. It is illegal to perform FGC in many African and Middle Eastern countries, but the practice continues because the laws are not enforced. In the United States, FGC is a criminal practice. In 1996, the United States granted asylum to a woman from Togo, who was trying to escape an arranged marriage and the FGC that would accompany it. Unfortunately, many immigrants to the United States continue the practice of FGC, by sending their daughters abroad for the procedure, or by importing someone to perform it. A number of educational approaches to eliminate FGC have been tried. These include community education that teaches about the harm done by FGC, and substituting alternative rituals for the rite of passage to womanhood. Education may do even more to halt FGC, because more highly educated women are less likely to support having their daughters mutilated in this fashion.

Decide Your Opinion

1. In your view, is male circumcision unjustifiable? Why or why not?
2. Should families who accept the idea of FGC be allowed to immigrate?
3. How should the United States prosecute parents who have subjected their daughters to FGC?

The labia majora extend backward from the mons pubis, a fatty prominence underlying the pubic hair. The labia minora are two small folds lying just inside the labia majora. They extend forward from the vaginal opening to encircle and form a foreskin for the glans clitoris. The glans clitoris is the organ of sexual arousal in females and, like the penis, contains a shaft of erectile tissue that becomes engorged with blood during sexual stimulation.

The cleft between the labia minora contains the openings of the urethra and the vagina. The vagina may be partially closed by a ring of tissue called the hymen. The hymen is ordinarily ruptured by sexual intercourse or by other types of physical activities. If remnants of the hymen persist after sexual intercourse, they can be surgically removed.

The urinary and reproductive systems in the female are entirely separate. For example, the urethra carries only urine, and the vagina serves only as the birth canal and the organ for sexual intercourse.

Orgasm in Females

Upon sexual stimulation, the labia minora, the vaginal wall, and the clitoris become engorged with blood. The breasts also swell, and the nipples become erect. The labia majora enlarge, redden, and spread away from the vaginal opening.

The vagina expands and elongates. Blood vessels in the vaginal wall release small droplets of fluid that seep into the vagina and lubricate it. Mucus-secreting glands beneath the labia minora on either side of the vagina also provide lubrication for entry of the penis into the vagina. Although the vagina is the organ of sexual intercourse in females, the clitoris plays a significant role in the female sexual response. The extremely sensitive clitoris can swell to two or three times its usual size. The thrusting of the penis and the pressure of the pubic symphyses of the partners act to stimulate the clitoris.

Orgasm occurs at the height of the sexual response. Blood pressure and pulse rate rise, breathing quickens, and the walls of the uterus and oviducts contract rhythmically. A sensation of intense pleasure is followed by relaxation when organs return to their normal size. Females have no refractory period, and multiple orgasms can occur during a single sexual experience.

Check Your Progress 16.3

1. Among the female structures, which (a) produce the egg, (b) transport the egg, (c) house a developing embryo, and (d) serve as the birth canal?
2. Among the external genitalia, which plays a significant role in female sexual response?

16.4 Female Hormone Levels

Hormone levels cycle in the female on a monthly basis, and the ovarian cycle drives the uterine cycle, as discussed in this section.

Ovarian Cycle: Nonpregnant

An ovary contains many **follicles,** and each one contains an immature egg, called an oocyte. A female is born with as many as two million follicles, but the number is reduced to 300,000–400,000 by the time of puberty. Only a small number of follicles (about 400) ever mature because a female usually produces only one egg per month during her reproductive years. As the follicle matures during the **ovarian cycle,** it changes from a primary to a secondary to a vesicular (Graafian) follicle (Fig. 16.8). Epithelial cells of a primary follicle surround a primary oocyte. Pools of follicular fluid bathe the oocyte in a secondary follicle. In a vesicular follicle, the fluid-filled cavity increases to the point that the follicle wall balloons out on the surface of the ovary.

Figure 16.8*b* traces the steps of **oogenesis.** A primary oocyte undergoes meiosis I, and the resulting cells are haploid with 23 chromosomes each. One of these cells is called a polar body. A polar body is a sort of cellular "trash can" because its function is simply to hold discarded chromosomes. The secondary oocyte undergoes meiosis II, but only if it is first fertilized by a sperm cell. If the secondary oocyte remains unfertilized, it never completes meiosis and will die shortly after being released from the ovary.

When appropriate, the vesicular follicle bursts, releasing the oocyte (often called an egg) surrounded by a clear membrane. This process is referred to as **ovulation.** Once a vesicular follicle has lost the oocyte, it develops into a **corpus luteum,** a glandlike structure. If the egg is not fertilized, the corpus luteum disintegrates.

As mentioned previously, the ovaries produce eggs and also the female sex hormones estrogen and progesterone. A primary follicle produces estrogen, and a secondary follicle produces estrogen and some progesterone. The corpus luteum produces progesterone and some estrogen.

Phases of the Ovarian Cycle

Similar to the testes, the hypothalamus has ultimate control of the ovaries' sexual function because it secretes gonadotropin-releasing hormone, or GnRH. GnRH stimulates the anterior pituitary to produce FSH and LH, and these hormones control the ovarian cycle. The gonado-

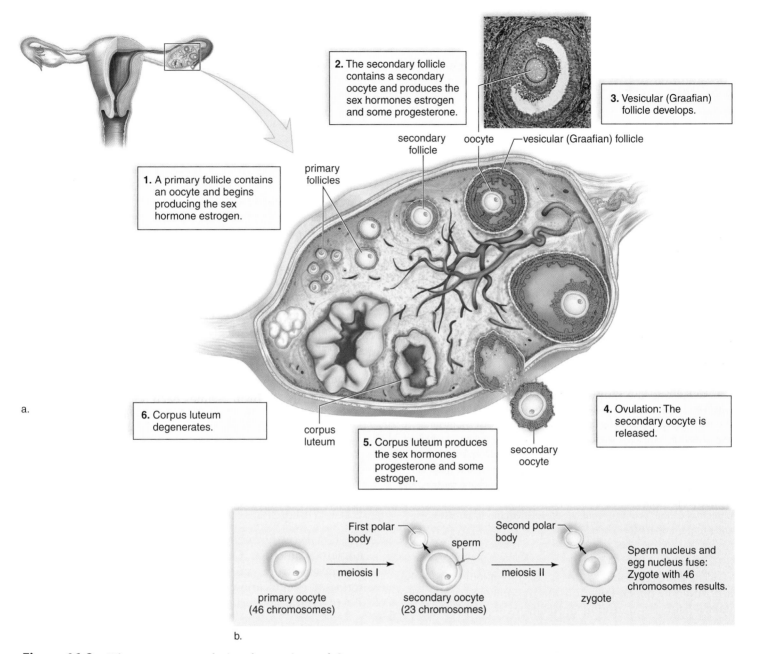

2. The secondary follicle contains a secondary oocyte and produces the sex hormones estrogen and some progesterone.

3. Vesicular (Graafian) follicle develops.

secondary follicle   oocyte   vesicular (Graafian) follicle

primary follicles

1. A primary follicle contains an oocyte and begins producing the sex hormone estrogen.

a.

6. Corpus luteum degenerates.

corpus luteum

5. Corpus luteum produces the sex hormones progesterone and some estrogen.

secondary oocyte

4. Ovulation: The secondary oocyte is released.

First polar body —

meiosis I

primary oocyte (46 chromosomes)

sperm

secondary oocyte (23 chromosomes)

Second polar body —

meiosis II

zygote

Sperm nucleus and egg nucleus fuse: Zygote with 46 chromosomes results.

b.

Figure 16.8   **What events occur during the ovarian cycle?**
a. A single follicle goes through all stages (1–6) in one place within the ovary. As the follicle matures, layers of follicle cells surround a secondary oocyte. Eventually, the mature follicle ruptures, and the secondary oocyte is released. The follicle then becomes the corpus luteum, which eventually disintegrates. **b.** During oogenesis, the chromosome number is reduced from 46 to 23. Fertilization restores the full number of chromosomes.

tropic hormones are not present in constant amounts. Instead they are secreted at different rates during the cycle. For simplicity's sake, it is convenient to emphasize that during the first half, or *follicular phase*, FSH promotes the development of follicles that primarily secrete estrogen (Fig. 16.9). As the estrogen level in the blood rises, it exerts negative feedback control over the anterior pituitary secretion of FSH. The follicular phase now comes to an end.

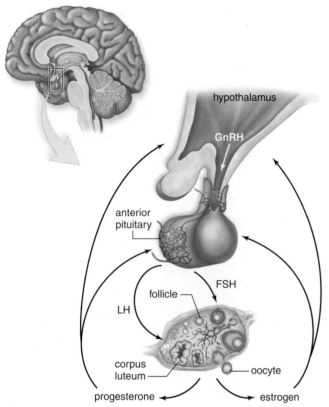

Figure 16.9 **What hormones control the production of estrogen and progesterone by the ovaries?**
The hypothalamus produces gonadotropin-releasing hormone (GnRH). GnRH stimulates the anterior pituitary to produce follicle-stimulating hormone (FSH) and luteinizing hormone (LH). FSH stimulates the follicle to produce primarily estrogen, and LH stimulates the corpus luteum to produce primarily progesterone. Estrogen and progesterone maintain the sexual organs (e.g., uterus) and the secondary sex characteristics, and they exert feedback control over the hypothalamus and the anterior pituitary. Feedback control regulates the relative amounts of estrogen and progesterone in the blood.

The estrogen spike at the end of the follicular phase has a positive feedback effect on the hypothalamus and pituitary gland. As a result, GnRH from the hypothalamus increases. There is a corresponding surge of LH released from the anterior pituitary. The LH surge triggers ovulation at about day 14 of a 28-day cycle.

Now, the *luteal phase* begins. During the luteal phase of the ovarian cycle, LH promotes the development of the corpus luteum. The corpus luteum secretes high levels of progesterone and some estrogen. When pregnancy does not occur, the corpus luteum regresses, and a new cycle begins with menstruation (Fig. 16.10).

Estrogen and Progesterone

Estrogen and progesterone affect not only the uterus but other parts of the body as well. Estrogen is largely responsible for the secondary sex characteristics in females, including body hair and fat distribution. In general, females have a more rounded appearance than males because of a greater accumulation of fat beneath the skin. Like males, females develop axillary and pubic hair during puberty. In females, the upper border of pubic hair is horizontal, but in males, it tapers toward the navel. Both estrogen and progesterone are also required for breast development. Other hormones are involved in milk production (prolactin) following pregnancy and milk letdown (oxytocin) when a baby begins to nurse.

The pelvic girdle is wider and deeper in females, so the pelvic cavity usually has a larger relative size compared with that of males. This means that females have wider hips than males and their thighs converge at a greater angle toward the knees. The female pelvis tilts forward, so females tend to have more of a lower back curve than males, an abdominal bulge, and protruding buttocks.

Menopause, the period in a woman's life during which the ovarian cycle ceases, is likely to occur between ages 45 and 55. The ovaries are no longer responsive to the gonadotropic hormones produced by the anterior pituitary, and the ovaries no longer secrete estrogen or progesterone. At the onset of menopause, menstruation becomes irregular, but as long as it occurs, it is still possible for a woman to conceive. Therefore, a woman is usually not considered to have completed menopause until menstruation is absent for a year.

Until recently, many women took combined estrogen-progestin drugs to ease menopausal symptoms. However, a new study conducted by the Women's Health Initiative (WHI) found that the long-term use of the combined drugs by most menopausal women caused increases in breast cancer, heart attacks, strokes, and blood clots. Those risks outweigh the drugs' actual benefits of a small decrease in hip fractures and a decrease in cases of colorectal cancer.

Uterine Cycle: Nonpregnant

The female sex hormones, **estrogen** and **progesterone,** have numerous functions. One function of these hormones affects the endometrium, causing the uterus to undergo a cyclical series of events known as the **uterine cycle** (Fig. 16.9). Twenty-eight-day cycles are divided as follows:

During *days 1–5,* a low level of estrogen and progesterone in the body causes the endometrium to disintegrate and its blood vessels to rupture. On day 1 of the cycle, a flow of blood

Have You Ever Wondered...

Do women make testosterone?

The adrenal glands and ovaries of women make small amounts of testosterone. Womens' small testosterone levels may affect libido, or sex drive. The use of supplemental testosterone to restore a woman's libido has not been well researched.

By the way, men make estrogen, too. Androgens are converted to estrogen by an enzyme in the gonads and peripheral tissues. Estrogen may prevent osteoporosis in males.

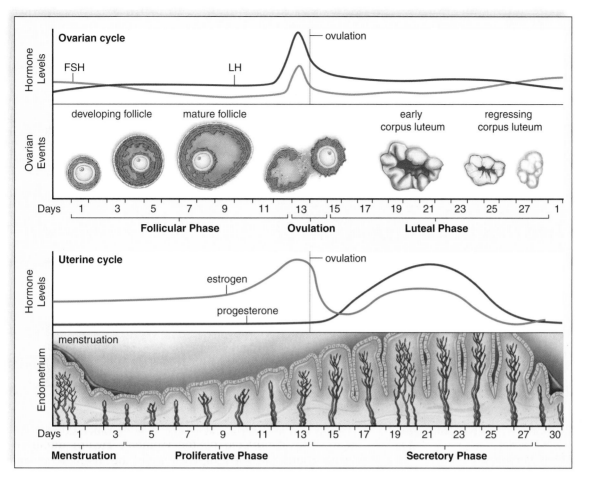

Figure 16.10 How do estrogen and progesterone affect the endometrium throughout the uterine cycle?

During the follicular phase, FSH released by the anterior pituitary promotes the maturation of a follicle in the ovary. The ovarian follicle produces increasing levels of estrogen, which causes the endometrium to thicken during the proliferative phase of the uterine cycle. After ovulation and during the luteal phase of the ovarian cycle, LH promotes the development of the corpus luteum. Progesterone in particular causes the endometrial lining to become secretory. Menses, due to the breakdown of the endometrium, begins when progesterone production declines to a low level.

and tissues, known as the menses, passes out of the vagina during **menstruation,** also called the menstrual period.

During *days 6–13,* increased production of estrogen by a new ovarian follicle in the ovary causes the endometrium to thicken and become vascular and glandular. This is called the proliferative phase of the uterine cycle.

On day 14 of a 28-day cycle, ovulation usually occurs.

During *days 15–28,* increased production of progesterone by the corpus luteum in the ovary causes the endometrium of the uterus to double or triple in thickness (from 1 mm to 2–3 mm). The uterine glands mature and produce a thick mucoid secretion in response to increased progesterone. This is called the secretory phase of the uterine cycle. The endometrium is now prepared to receive the developing embryo. If this does not occur, the corpus luteum in the ovary regresses.

The low level of progesterone in the female body results in the endometrium breaking down during menstruation.

Table 16.3 compares the stages of the uterine cycle with those of the ovarian cycle when pregnancy does not occur.

Fertilization and Pregnancy

Following unprotected sexual intercourse, many sperm will make their way into the oviduct, where the egg is located following ovulation. Only one sperm is needed to fertilize the egg, which is then called a zygote. Development begins even as the zygote travels down the oviduct to the uterus. The endometrium is now prepared to receive the developing embryo. The embryo implants in the endometrial lining several days following fertilization. Implantation signals the

Table 16.3	Ovarian and Uterine Cycles: Nonpregnant		
Ovarian Cycle	**Events**	**Uterine Cycle**	**Events**
Follicular phase—Days 1–13	FSH secretion begins.	Menstruation—Days 1–5	Endometrium breaks down.
	Follicle maturation occurs.	Proliferative phase—Days 6–13	Endometrium rebuilds.
	Estrogen secretion is prominent.		
Ovulation—Day 14*	LH spike occurs.		
Luteal phase—Days 15–28	LH secretion continues.	Secretory phase—Days 15–28	Endometrium thickens, and glands are secretory.
	Corpus luteum forms.		
	Progesterone secretion is prominent.		

*Assuming a 28-day cycle.

363

Figure 16.11   **What is the effect of pregnancy on the corpus luteum and endometrium?**
If pregnancy occurs, the corpus luteum does not regress. Instead, the corpus luteum is maintained and secretes increasing amounts of progesterone. Therefore, menstruation does not occur, and the uterine lining, where the embryo resides, is maintained.

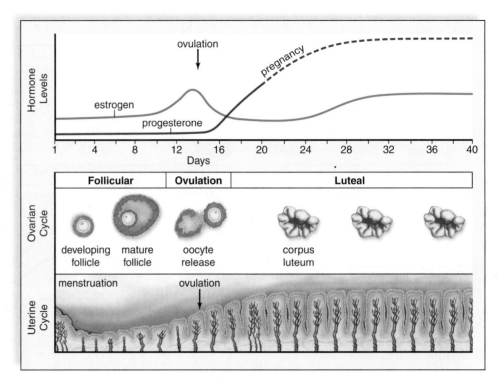

beginning of a pregnancy. An abortion may be spontaneous (referred to as a miscarriage) or induced. Each type of abortion ends with the loss of the embryo or fetus.

The **placenta,** which sustains the developing embryo and later the fetus, originates from both maternal and fetal tissues. It is the region of exchange of molecules between fetal and maternal blood, although the two rarely mix. At first, the placenta produces **human chorionic gonadotropin (hCG),** which maintains the corpus luteum in the ovary. (A pregnancy test detects the presence of hCG in the blood or urine.) Rising amounts of hCG stimulate the corpus luteum to produce increasing amounts of progesterone. This progesterone shuts down the hypothalamus and anterior pituitary so that no new follicles begin in the ovary. The progesterone maintains the uterine lining where the embryo now resides. The absence of menstruation is a signal to the woman that she may be pregnant (Fig. 16.11).

Eventually, the placenta produces progesterone and some estrogen. The corpus luteum is no longer needed and it regresses.

Many women use birth control pills to prevent pregnancy. The most commonly used pills include active pills, containing a synthetic estrogen and progesterone, taken for 21 days. Seven days of inactive pills that lack estrogen and progesterone are also provided. Negative feedback from hormones in the active pills inhibits release of the hypothalamus and pituitary gland hormones. As a result, no follicles develop in the ovary, and ovulation does not occur. In effect, the birth control pills fool the body into acting as if pregnancy had occurred (Fig. 16.12).

Birth control pills also thicken the cervical mucus, preventing sperm from entering the uterus. The uterine lining builds up to some degree while the active pills are being taken. Progesterone decreases when the last of the active pills are taken,

causing a minimenstruation to occur. Some women skip taking the inactive pills and start taking a new pack of active pills right away to skip menstruation (a period). A new form of birth control pills consists of three months of active pills. Women taking them have only four menstrual periods a year.

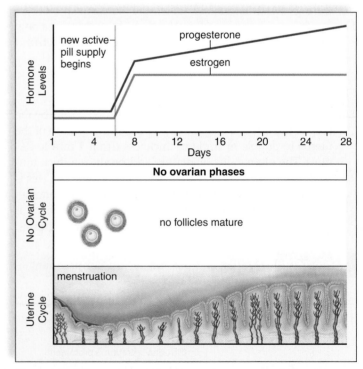

Figure 16.12   **What is the effect of birth control pills on the ovarian cycle?**
Active pills cause the uterine lining to build up, and this lining is shed when inactive pills are taken. Feedback inhibition of the hypothalamus and anterior pituitary means that the ovarian cycle does not occur.

Check Your Progress 16.4

1. **a.** Which hormones control the ovarian cycle? **b.** Which hormones control the uterine cycle?
2. **a.** What changes occur in these cycles when a woman becomes pregnant? **b.** When she takes the birth control pill?

16.5 **Control of Reproduction**

Several means are available to dampen or enhance our reproductive potential. **Birth control methods** are used to regulate the number of children an individual or couple will have.

Birth Control Methods

The most reliable method of birth control is abstinence—not engaging in sexual intercourse. This form of birth control has the added advantage of preventing transmission of a sexually transmitted disease. Table 16.4 lists other means of birth control used in the United States and rates their effectiveness. For example, with the birth control pill, we expect 98% effectiveness. Only 2% of sexually active women will get pregnant within the year. On the other hand, with the withdrawal method, we expect that 75% of women will not get pregnant. That makes the withdrawal method one of the least effective methods of contraception.

Figure 16.13 features some of the most effective and commonly used means of birth control. **Contraceptives** are

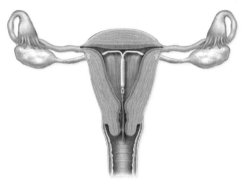

a. Intrauterine device

b. Hormone skin patch

c. Depo-Provera®

d. Diaphragm and spermicidal jelly

e. Female condom

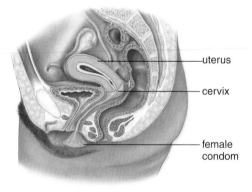

uterus

cervix

female condom

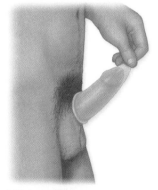

f. Male condom

g. Implant

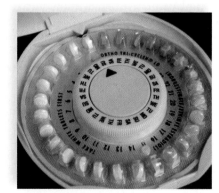

h. Oral contraception

Figure 16.13 What are some of the different forms of birth control now available?
a. Intrauterine device. **b.** Hormone skin patch. **c.** Depo-Provera® (a progesterone injection). **d.** Diaphragm and spermicidal jelly. **e.** Female condom.
f. Male condom. **g.** Implant. **h.** Oral contraception (birth control pills).

medications and devices that reduce the chance of pregnancy. Oral contraception (**birth control pills**) contains a combination of estrogen and progesterone for the first 21 days, followed by seven inactive pills (Fig. 16.13*h*). The estrogen and progesterone in the birth control pill or a patch (Fig. 16.13*b*) applied to the skin effectively shuts down the pituitary production of both FSH and LH. Follicle development in the ovary is prevented. Since ovulation does not

Table 16.4	Common Methods of Contraception			
Name	**Procedure**	**How Does It Work?**	**Effectiveness?**	**Health Risk**
Abstinence	Refrain from sexual intercourse	No sperm in vagina	100%	None; also protects against STDs
Natural family planning	Determine day of ovulation by keeping records	Intercourse is avoided during the time while ovum is viable	80%	None
Withdrawal method	Penis withdrawn from vagina just before ejaculation	Ejaculation outside the woman's body; no sperm in vagina	75%	None
Douching	Vagina cleansed after intercourse	Washes sperm out of vagina	≥ 70%	May cause inflammation
Male condom	Sheath of latex, polyurethane, or natural material fitted over erect penis	Prevents entry of sperm into vagina; latex and polyurethane forms protect against STDs	89%	Latex allergy with latex forms; no protection against STDs with natural material condoms
Female condom	Polyurethane liner fitted inside vagina	Prevents entry of sperm into vagina; some protection against STDs	79%	Possible allergy or irritation, urinary tract infection
Spermicide: jellies, foams, creams	Spermicidal products inserted into vagina before intercourse	Spermicide nonoxynol-9 kills large numbers of sperm cells	50–80%	Irritation, allergic reaction, urinary tract infection
Contraceptive sponges	Sponge containing spermicide inserted into vagina and placed against cervix	Spermicide nonoxynol-9 kills large numbers of sperm cells	72–86%	Irritation, allergic reaction, urinary tract infection, toxic shock syndrome
Combined hormone: vaginal ring	Flexible plastic ring inserted into vagina; releases hormones absorbed into the bloodstream	Combined hormonal methods suppress ovulation by the combined actions of the hormones estrogen and progestin	98%	Combined hormonal method's can cause dizziness; nausea; changes in menstruation, mood, and weight; rarely, cardiovascular disease, including high blood pressure, blood clots, heart attack, and strokes
Combined hormone pill	Pills are swallowed daily; chewable form also available		98%	
Combined hormone 91-day regimen	Pills are swallowed daily; user has 3-4 menstrual periods a year		98%	
Combined hormones injection (Lunelle®)	Injection of long-acting hormone given once a month		99%	
Combined hormone patch	Patch is applied to skin and left in place for 1 week; new patch applied		98%	
Progestin-only mini-pill	Pills are swallowed daily	Thickens cervical mucus, preventing sperm from contacting egg	98%	Irregular bleeding, weight gain, breast tenderness
Progesterone-only injection (Depo-Provera®)	Injection of progestin once every three months	Inhibits ovulation; prevents sperm from reaching the egg; prevents implantation	99%	Irregular bleeding, weight gain, breast tenderness, osteoporosis possible
Emergency contraception	Must be taken within 72 hours after unprotected intercourse	Suppresses ovulation by the combined actions of the hormones estrogen and progestin; prevents implantation	80%	Nausea, vomiting, abdominal pain, fatigue, headache
Diaphragm	Latex cup, placed into vagina to cover cervix before intercourse	Blocks entrance of sperm into uterus, spermicide kills sperm	90% with spermicide	Irritation, allergic reaction, urinary tract inflection, toxic shock syndrome
Cervical cap	Latex cap held over cervix	Blocks entrance of sperm into uterus, spermicide kills sperm	90% with spermicide	Irritation, allergic reaction, toxic shock syndrome, abnormal Pap smear
Cervical shield	Latex cap in upper vagina, held in place by suction	Blocks entrance of sperm into uterus, spermicide kills sperm	90% with spermicide	Irritation, allergic reaction, urinary tract infection, toxic shock syndrome
Intrauterine device Copper T	Placed in uterus	Causes cervical mucus to thicken; fertilized embryo cannot implant	99%	Cramps, bleeding, infertility, perforation of uterus
Intrauterine device progesterone-releasing type	Placed in uterus	Prevents ovulation; causes cervical mucus to thicken; fertilized embryo cannot implant	99%	Cramps, bleeding, infertility, perforation of uterus

occur, pregnancy cannot take place. Women taking birth control pills or using a patch should see a physician regularly, because of possible side effects.

An **intrauterine device (IUD)** is a small piece of molded plastic inserted into the uterus by a physician (Fig. 16.13*a*). IUDs are believed to alter the environment of the uterus and oviducts so that fertilization probably will not occur. If fertilization should occur, implantation cannot take place.

The **diaphragm** is a soft latex cup with a flexible rim that lodges behind the pubic bone and fits over the cervix (Fig. 16.13*d*). Each woman must be properly fitted by a physician, and the diaphragm can be inserted into the vagina no more than 2 hours before sexual relations. Also, it must be used with spermicidal jelly or cream and should be left in place at least 6 hours after sexual relations. The cervical cap is a minidiaphragm.

There has been a renewal of interest in barrier methods of birth control, because these methods offer some protection against sexually transmitted diseases. A **female condom,** now available, consists of a large polyurethane tube with a flexible ring that fits onto the cervix (Fig. 16.13*e*). The open end of the tube has a ring that covers the external genitals. A **male condom** is most often a latex sheath that fits over the erect penis (Fig. 16.13*f*). The ejaculate is trapped inside the sheath, and thus does not enter the vagina. When used in conjunction with a spermicide, the protection is better than with the condom alone.

Contraceptive implants use a synthetic progesterone to prevent ovulation by disrupting the ovarian cycle. The older version of the implant consists of six match-sized, time-release capsules surgically implanted under the skin of a woman's upper arm. The newest version consists of a single capsule that remains effective for about three years (Fig. 16.13*g*).

Contraceptive injections are available as progesterone only (Fig. 16.13*c*) or a combination of estrogen and progesterone. The length of time between injections can vary from one to several months.

Contraceptive vaccines are now being developed. For example, a vaccine intended to immunize women against hCG, the hormone so necessary to maintaining the implantation of the embryo, was successful in a limited clinical trial. Since hCG is not normally present in the body, no autoimmune reaction is expected, but the immunization does wear off with time. Others believe that it would also be possible to develop a safe antisperm vaccine that could be used in women.

Vasectomy and Tubal Ligation

Vasectomy and tubal ligation are two methods to bring about sterilization, the inability to reproduce (Fig. 16.14). **Vasectomy** consists of cutting and sealing the vas deferens from each testis so that the sperm are unable to reach the seminal fluid ejected at the time of orgasm. The sperm are then largely reabsorbed. Following this operation, which can be done in a doctor's office, the amount of ejaculate remains normal because sperm account for only about 1% of the volume of semen. Also, there is no effect on the secondary sex characteristics because testosterone continues to be produced by the testes.

Tubal ligation consists of cutting and sealing the oviducts. Pregnancy rarely occurs because the passage of the egg through the oviducts has been blocked. Using a method called laparoscopy, which requires only two small incisions, the surgeon inserts a small, lighted telescope to view the oviducts and a small surgical blade to sever them.

It is best to view a vasectomy or tubal ligation as permanent. Even following successful reconnection, fertility is usually reduced by about 50%.

Morning-After Pills

A morning-after pill, or emergency contraception, refers to a medication that will prevent pregnancy after unprotected intercourse. The expression "morning-after pill" is a misnomer in that the medication can begin one to several days after unprotected intercourse.

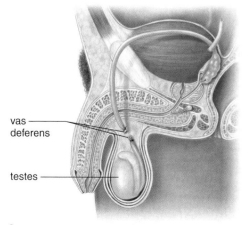

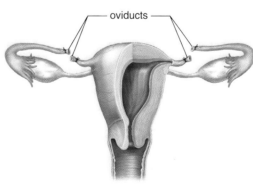

oviducts

a. b.

Figure 16.14 **What occurs during sterilization?**

a. Vasectomy involves making two small cuts in the skin of the scrotum. Each vas deferens is lifted out and cut. The cut ends are tied or sealed with an electric current. The openings in the scrotum are closed with stitches. **b.** During tubal ligation, one or two small incisions are made in the abdomen. Using instruments inserted through the incisions, the oviducts are coagulated (burned), sealed shut with cautery, or cut and tied. The skin incision is then stitched closed.

vas deferens

testes

One type, a kit called Preven®, contains four synthetic progesterone pills; two are taken up to 72 hours after unprotected intercourse, and two more are taken 12 hours later. The medication upsets the normal uterine cycle, making it difficult for an embryo to implant in the endometrium. In one study, it was estimated that the medication was 85% effective in preventing unintended pregnancies.

Mifepristone, better known as RU-486®, is a pill presently used to cause the loss of an implanted embryo by blocking the progesterone receptor proteins of endometrial cells. Without functioning receptors for progesterone, the endometrium sloughs off, carrying the embryo with it. When taken in conjunction with a prostaglandin to induce uterine contractions, RU-486® is 95% effective. It is possible that some day this medication will also be a "morning-after pill," taken when menstruation is late without evidence that pregnancy has occurred.

Infertility

Infertility is the failure of a couple to achieve pregnancy after one year of regular, unprotected intercourse. The American Medical Association (AMA) estimates that 15% of all couples are infertile. The cause of infertility can be attributed to the male (40%), the female (40%), or both (20%).

Causes of Infertility

The most frequent cause of infertility in males is low sperm count and/or a large proportion of abnormal sperm, which can be due to environmental influences. It appears that a sedentary lifestyle coupled with smoking and alcohol consumption is most often the cause of male infertility. When males spend most of the day sitting in front of a computer or the TV or driving, the testes temperature remains too high for adequate sperm production.

CASE STUDY PART OF THE STUDY

Anthony arrived early for his appointment at the clinic the next day. He felt the same twinge of nervousness as he walked through the clinic's door. The same receptionist greeted him and escorted him to a private office. "I'll let Dr. Robbins know you are here," she directed.

Anthony decided to skim through the informational packet one more time. As he began to unfold the brochures, a very tall gentleman entered the office. "Anthony?"

"Yes," Anthony replied, quickly standing.

"Hi, I'm Dr. Robbins." The two men shook hands. "Have a seat. I understand you might be interested in being a research subject for the male hormonal contraception gel project," Dr. Robbins continued. "Do you have any questions I can answer?"

Anthony glanced down at the large envelope in his lap. "Well, I guess one question. How do I get chosen to be in the study?"

Dr. Robbins replied, "You look pretty healthy. You're between the ages of 18 and 50, right?" Anthony nodded. "First, you get a very thorough physical exam, and it's free. Then, we'll have to do some blood work to check your reproductive hormones."

"Let me guess," Anthony asked. "You want to test for GnRH, FSH, LH, and testosterone, right?"

Dr. Robbins grinned. "Are you, by any chance, in Dr. Nicely's Bio 313 class, the one where he slams you with all of those hormones?" Anthony nodded and smiled back.

"Small world! He was one of my PhD advisors," Dr. Robbins continued. "Good. We'll also check for your normal sperm

concentration and your prostate-specific antigen. If that all checks out, you're in the study," Dr. Robbins finished.

"And I'd get paid for my participation?" Anthony asked hesitantly. Dr. Robbins smiled and nodded. Anthony sighed in relief, explaining, "I'd really like to quit my telemarketing job. I'll be able to, if the money from the study is enough to pay my rent."

Dr. Robbins warned, "Just one more thing—if you have sex during the study, you or your partner will need to use a reliable form of contraception." Anthony nodded his agreement. "Let's go on over to the clinic and go over the required paperwork. We can get your physical exam and blood tests scheduled, too."

A few days later, Anthony applied his first dose of the gel containing testosterone and progestin to his upper arm. "What does it feel like?" Carter asked. "Nothing unusual," reported Anthony. "It feels like I'm putting on sunscreen or body lotion."

After three weeks of daily gel applications, Anthony returned to the clinic for more blood tests. "What's this blood work tell you?" he asked.

Dr. Robbins explained, "We compare today's FSH, LH, and testosterone levels to your first set of measurements. If FSH and LH are lower, that means your sperm production is being inhibited. The rest of your physical exam results check out, so the gel isn't hurting you. That includes your blood HDL levels. We'll check the prostate-specific antigen again, too, to be sure the treatment hasn't done anything to your prostate."

"Thanks again for volunteering for this study," Dr. Robbins smiled. "Ideally, our results will get us that much closer to MHC that works and is safe. That'll be another contraceptive option."

Should Infertility Be Treated

Every day, couples make plans to start or expand their families. Yet for many, their dreams might not be realized because conception is difficult or impossible. Before seeking medical treatment for infertility, a couple might want to decide how far they are willing to go to have a child. Here are some of the possible risks.

Some of the Procedures Used

If a man has low sperm count or motility, artificial or intrauterine insemination of his partner with a large number of specially selected sperm may be done to stimulate pregnancy. Although there are dangers in all medical procedures, artificial insemination is generally safe.

If a woman is infertile because of physical abnormalities in her reproductive system, she may be treated surgically. While surgeries are now very sophisticated, they nonetheless have risks, including bleeding, infection, organ damage, and adverse reactions to anesthesia. Similar risks are associated with collecting eggs for in vitro fertilization (IVF). To ensure the collection of several eggs, a woman may be placed on hormone-based medications that stimulate egg production. Such medications may cause ovarian hyperstimulation syndrome—enlarged ovaries and abdominal fluid accumulation. In mild cases, the only symptom is discomfort, but in severe cases (though rare), a woman's life may be endangered, and in any case, the fluid has to be drained.

Usually, IVF involves the creation of many embryos; the healthiest-looking ones are transferred into the woman's body. Others may be frozen for future attempts at establishing pregnancy, given to other infertile couples, donated for research, or destroyed. Of those that are transferred, none, one, or all might develop into fetuses. The significant increase of multifetal pregnancies in the United States in the last 15 years has been largely attributed to fertility treatment (Fig. 16B). While the number of triplet and higher number multiple pregnancies started to level off in 1999, twin pregnancies continue to climb.

Figure 16B   Are assisted reproductive technologies a boon to society?
What about to individuals who wish to have children? Or are they a detriment to both?

While they might seem like a dream come true, multifetal pregnancies are difficult. The mother is more likely to develop complications such as gestational diabetes and high blood pressure than are women carrying single babies. Positioning of the babies in the uterus may make vaginal delivery less likely, and there is likely a chance of preterm labor. Babies born prematurely face numerous hardships. Infant death and long-term disabilities are also more common with multiple births. This is true even of twins. Even if all babies are healthy, parenting multiples poses unique challenges.

What Happens to Frozen Embryos?

Despite potential trials, thousands of people undergo fertility treatment every year. Its popularity has brought a number of ethical issues to light. For example, the estimated high numbers of stored frozen embryos (a few hundred thousand in the United States) has generated debate about their fate, complicated by the fact that the long-term viability of frozen embryos is not well understood. Scientists may worry that embryos donated to other couples are ones screened out from one implantation and not be likely to survive. Some religious groups strongly oppose destruction of these embryos or their use in research. Patients for whom the embryos were created generally feel that they should have sole rights to make decisions about their fate. However, a fertility clinic may no longer be receiving monetary compensation for the storage of frozen embryos and unable to contact the couples for whom they were produced. The question then becomes whether the clinic now has the right to determine their fate.

Who Should Be Treated?

Additionally, because fertility treatment is voluntary, is it ever acceptable to turn some people away? What if the prospect of a satisfactory outcome is very slim or almost nonexistent? This may happen when one of the partners is ill or the woman is at an advanced age. Should a physician go ahead with treatment even if it might endanger a woman's (or baby's) health? Those in favor of limiting treatment argue that a physician has a responsibility to prevent potential harm to a patient. On the other hand, there is concern that if certain people are denied fertility for medical reasons, might they be denied for other reasons also, such as race, religion, sexuality, or income?

Decide Your Opinion

1. Should couples go to all lengths to have children even if it could endanger the life of one or both spouses?
2. Should couples with multiples due to infertility treatment receive assistant from private and public services?
3. To what lengths should society go to protect frozen embryos?
4. Do you think that anyone should be denied fertility treatment? If so, what factors do you think a doctor should take into consideration when deciding whether to provide someone fertility treatment?

Body weight appears to be the most significant factor in causing female infertility. In women of normal weight, fat cells produce a hormone called leptin that stimulates the hypothalamus to release GnRH. FSH release and normal follicle development follow. In overweight women, leptin levels are higher, which impacts GnRH and FSH. The ovaries of overweight women often contain many small follicles that fail to ovulate. Other causes of infertility in females are blocked oviducts due to pelvic inflammatory disease (see page 373) and endometriosis. Endometriosis is the presence of uterine tissue outside the uterus, particularly in the oviducts and on the abdominal organs. Backward flow of menstrual fluid allows living uterine cells to establish themselves in the abdominal cavity. The cells go through the usual uterine cycle, causing pain and structural abnormalities that make it more difficult for a woman to conceive.

Sometimes the causes of infertility can be corrected by medical intervention so that couples can have children. If no obstruction is apparent and body weight is normal, it is possible to give females fertility drugs. These drugs are gonadotropic hormones that stimulate the ovaries and bring about ovulation. Such hormone treatments may cause multiple ovulations and multiple births.

When reproduction does not occur in the usual manner, many couples adopt a child. Others sometimes try one of the assisted reproductive technologies discussed in the following paragraphs.

Assisted Reproductive Technologies

Assisted reproductive technologies (ART) consist of techniques used to increase the chances of pregnancy. Often, sperm and/or eggs are retrieved from the testes and ovaries, and fertilization takes place in a clinical or laboratory setting.

Artificial Insemination by Donor (AID) During artificial insemination, sperm are placed in the vagina by a physician. Sometimes a woman is artificially inseminated by her partner's sperm. This is especially helpful if the partner has a low sperm count, because the sperm can be collected over time and concentrated so that the sperm count is sufficient to result in fertilization. Often, however, a woman is inseminated by sperm acquired from a donor who is a complete stranger to her. At times, a combination of partner and donor sperm is used.

A variation of AID is *intrauterine insemination (IUI)*. In IUI, fertility drugs are given to stimulate the ovaries. Then the donor's sperm are placed in the uterus, rather than in the vagina.

If the prospective parents wish, sperm can be sorted into those believed to be X-bearing or Y-bearing to increase the chances of having a child of the desired sex. Fertilization of an egg with an X-bearing sperm results in a female child. Fertilization by a Y-bearing sperm yields a male child.

In Vitro Fertilization (IVF) During IVF, conception occurs in laboratory glassware. Ultrasound machines can now spot follicles in the ovaries that hold immature eggs; therefore, the latest method is to forgo the administration of fertility drugs and retrieve immature eggs by using a needle. The

Figure 16.15   **How is a sperm introduced into an egg?**
A microscope connected to a television screen is used to carry out intracytoplasmic sperm injection. A pipette holds the egg steady while a needle (not visible) introduces the sperm into the egg.

immature eggs are then brought to maturity in glassware, and then concentrated sperm are added (Fig. 16.15). After about two to four days, the embryos are ready to be transferred to the uterus of the woman, who is now in the secretory phase of her uterine cycle. If desired, the embryos can be tested for a genetic disease, and only those found to be free of disease will be used. If implantation is successful, development is normal and continues to term.

Gamete Intrafallopian Transfer (GIFT) The term **gamete** refers to a sex cell, either a sperm or an egg. GIFT was devised to overcome the low success rate (15–20%) of in vitro fertilization. The method is exactly the same as in vitro fertilization, except the eggs and the sperm are placed in the oviducts immediately after they have been brought together. GIFT has the advantage of being a one-step procedure for the woman—the eggs are removed and reintroduced all at the same time. A variation on this procedure is to fertilize the eggs in the laboratory and then place the zygotes in the oviducts.

Surrogate Mothers In some instances, women are contracted and paid to have babies. These women are called surrogate mothers. The sperm and even the egg can be contributed by the contracting parents.

Intracytoplasmic Sperm Injection (ICSI) In this highly sophisticated procedure, a single sperm is injected into an egg. It is used effectively when a man has severe infertility problems.

> **Check Your Progress 16.5**
> 1. What are some common birth control methods available to men and women today?
> 2. If a couple has trouble conceiving a child, what options can they explore?

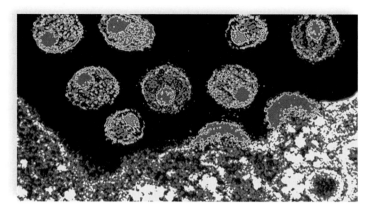

Figure 16.16 What types of cells are infected by HIV, the AIDS virus?
False-colored micrographs showing HIV particles budding from an infected helper T cell. These viruses can infect helper T cells and also macrophages, which work with helper T cells to stem the infection.

16.6 Sexually Transmitted Diseases

Sexually transmitted diseases (STDs) are caused by viruses, bacteria, fungi, and parasites.

STDs Caused by Viruses

Among those STDs caused by viruses, effective treatment is available for AIDS (acquired immunodeficiency syndrome) and genital herpes. However, treatment for HIV/AIDS and genital herpes cannot presently eliminate the virus from the person's body. Drugs used for treatment can merely slow replication of the viruses. Thus, neither viral disease is presently curable. Further, antiviral drugs have serious, debilitating side effects on the body.

HIV Infections

The Infectious Diseases supplement, which begins on page 379, discusses HIV infections at greater length than this brief summary. At present there is no vaccine to prevent an HIV infection nor is there a cure for AIDS. The best course of action is to follow the guidelines for preventing transmission outlined in the Health Focus on page 375.

The primary host for HIV is a helper T lymphocyte (Fig. 16.16). These are the very cells that stimulate an immune response, so the immune system becomes severely impaired in persons with AIDS. During the first stage of an HIV infection, symptoms are few, but the individual is highly contagious. Several months to several years after infection, the helper T lymphocyte count falls. Following this decrease, infections, such as other sexually transmitted diseases, begin to appear. In the last stage of infection, called AIDS, the helper T cell count falls way below normal. At least one opportunistic infection is present. Such diseases have the opportunity to occur only because the immune system is severely weakened. Persons with AIDS typically die from an opportunistic disease, such as *Pneumocystis* pneumonia.

There is no cure for AIDS. A treatment called highly active antiretroviral therapy (HAART) is usually able to stop HIV reproduction to the extent that the virus becomes undetectable in the blood. The medications must be continued indefinitely because as soon as HAART is discontinued, the virus rebounds.

Genital Warts

Genital warts are caused by the human papillomaviruses (HPVs). Many times, carriers either do not have any sign of warts or merely have flat lesions. When present, the warts commonly are seen on the penis and foreskin of men and near the vaginal opening in women. A newborn can become infected while passing through the birth canal.

Individuals currently infected with visible growth may have those growths removed by surgery, freezing, or burning with lasers or acids. However, visible warts that are removed may recur. A vaccine has been released for the human papillomaviruses that most commonly cause genital warts. This development is an extremely important step in the prevention of cancer, as well as in the prevention of warts themselves. Genital warts are associated with cancer of the cervix, as well as tumors of the vulva, vagina, anus, and penis. Researchers believe that these viruses may be involved in up to 90% of all cases of cancer of the cervix. Vaccination might make such cancers a thing of the past.

Genital Herpes

Genital herpes is caused by herpes simplex virus. Type 1 usually causes cold sores and fever blisters, while type 2 more often causes genital herpes (Fig. 16.17).

Persons usually get infected with herpes simplex virus type 2 when they are adults. Some people exhibit no symptoms. Others may experience a tingling or itching sensation before blisters appear on the genitals. Once the blisters rupture, they leave painful ulcers that may take as long as three weeks or as little as five days to heal. The blisters may be accompanied by fever; pain on urination; swollen lymph

Have You Ever Wondered...

Can you catch an STD from a toilet seat?

When HIV/AIDS was first identified in the mid-1980s, many people were concerned about being infected by the virus on toilet seats. Toilet seats are plastic and inert, so they're not very hospitable to disease-causing organisms. So if you're deciding whether to hover or sit, remember sitting on a toilet seat will not give you an STD.

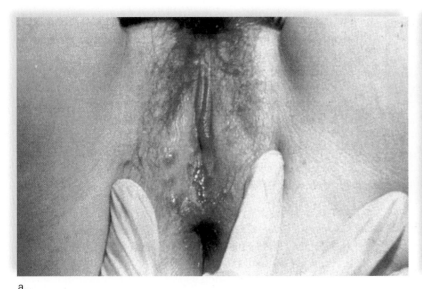

a.

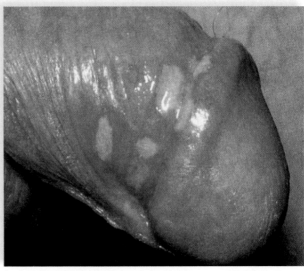

b.

Figure 16.17 **What virus causes genital herpes? What causes the symptoms of genital herpes?**
There are several types of herpes viruses, and usually the ones called herpes simplex-2 cause genital herpes. About one million persons become infected each year, most of them teens and young adults. Symptoms of genital herpes are due to an outbreak of blisters, which can be present on the labia of females **(a)** or on the penis of males **(b).**

nodes in the groin; and in women, a copious discharge. At this time, the individual has an increased risk of acquiring an HIV infection.

After the ulcers heal, the disease is only latent, and blisters can recur, although usually at less frequent intervals and with milder symptoms. Fever, stress, sunlight, and menstruation are associated with recurrence of symptoms. Exposure to herpes in the birth canal can cause an infection in the newborn, which leads to neurological disorders and even death. Birth by cesarean section prevents this possibility. There are antiviral drugs available that reduce the number and length of outbreaks. However, these drugs are not a cure for genital herpes. Latex or polyurethane condoms are recommended by the FDA to prevent the transmission of the virus to sexual partners.

Hepatitis

Hepatitis infects the liver and can lead to liver failure, liver cancer, and death. There are six known viruses that cause hepatitis, designated A-B-C-D-E-G. Hepatitis A is usually acquired from sewage-contaminated drinking water, but this infection can also be sexually transmitted through oral/anal contact. Hepatitis B is spread through sexual contact and by blood-borne transmission (accidental needle stick on the job; receiving a contaminated blood transfusion; a drug abuser sharing infected needles while injecting drugs; from mother to fetus, etc.). Simultaneous infection with hepatitis B and HIV is common, because both share the same routes of transmission. Fortunately, a combined vaccine is

available for hepatitis A and B. It is recommended that all children receive the vaccine to prevent infection (see page 149). Hepatitis C (also called non-A, non-B hepatitis) causes most cases of posttransfusion hepatitis. Hepatitis D and G are sexually transmitted, while hepatitis E is acquired from contaminated water. Screening of blood and blood products can prevent transmission of hepatitis viruses during a transfusion. Proper water-treatment techniques can prevent contamination of drinking water.

STDs Caused by Bacteria

Only STDs caused by bacteria are curable with antibiotics. Antibiotic resistance acquired by these bacteria may require treatment with extremely strong drugs for an extended period to achieve a cure.

Chlamydia

Chlamydia is named for the tiny bacterium that causes it *(Chlamydia trachomatis)*. The incidence of new chlamydia infections has steadily increased since 1984.

Chlamydia infections of the lower reproductive tract are usually mild or asymptomatic, especially in women. About 18 to 21 days after infection, men may experience a mild burning sensation on urination and a mucoid discharge. Women may have a vaginal discharge along with the symptoms of a urinary tract infection. Chlamydia also causes cervical ulcerations, which increase the risk of acquiring HIV.

If the infection is misdiagnosed or if a woman does not seek medical help, there is a particular risk of the infection spreading from the cervix to the uterine tubes so that pelvic inflammatory disease (PID) results. This very painful condition can result in blockage of the uterine tubes with the possibility of sterility and infertility. If a baby comes in contact with chlamydia during birth, inflammation of the eyes or pneumonia can result.

Gonorrhea

Gonorrhea is caused by the bacterium *Neisseria gonorrhoeae.* Diagnosis in the male is not difficult, because typical symptoms are pain upon urination and a thick, greenish yellow urethral discharge. In males and females, a latent infection leads to pelvic inflammatory disease (PID), which can also cause sterility in males. If a baby is exposed during birth, an eye infection leading to blindness can result. All newborns are given eyedrops to prevent this possibility.

Gonorrhea proctitis, an infection of the anus characterized by anal pain and blood or pus in the feces, also occurs in patients. Oral/genital contact can cause infection of the mouth, throat, and tonsils. Gonorrhea can spread to internal parts of the body, causing heart damage or arthritis. If, by chance, the person touches infected genitals and then touches his or her eyes, a severe eye infection can result. Up to now, gonorrhea was curable by antibiotic therapy. However, resistance to antibiotics is becoming more and more common, and 40% of all strains are now known to be resistant to therapy.

Syphilis

Syphilis is caused by a bacterium called *Treponema pallidum* (Fig. 16.18). As with many other bacterial diseases, penicillin is an effective antibiotic. Syphilis has three stages, often separated by latent periods, during which the bacteria are resting before multiplying again. During the primary stage, a hard **chancre** (ulcerated sore with hard edges) indicates the site of infection. The chancre usually heals spontaneously, leaving little scarring. During the secondary stage, the victim breaks out in a rash that does not itch and is seen even on the palms of the hands and the soles of the feet. Hair loss and infectious gray patches on the mucous membranes may also occur. These symptoms disappear of their own accord.

The tertiary stage lasts until the patient dies. During this stage, syphilis may affect the cardiovascular system by causing aneurysms, particularly in the aorta. In other instances, the disease may affect the nervous system, resulting in psychological disturbances. Also, gummas, large destructive ulcers, may develop on the skin or within the internal organs.

Congenital syphilis is caused by syphilitic bacteria crossing the placenta. The child is born blind and/or with numerous anatomical malformations. Control of syphilis depends on prompt and adequate treatment of all new cases. Therefore, it is crucial for all sexual contacts to be traced so they can be treated. Diagnosis of syphilis can be made by blood tests or by microscopic examination of fluids from lesions.

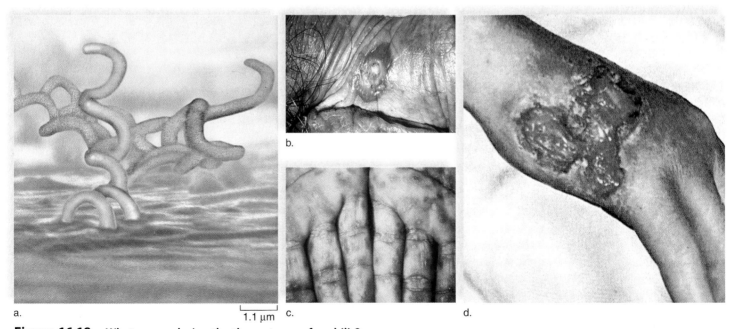

a. 1.1 μm   c.   d.

Figure 16.18 **What occurs during the three stages of syphilis?**
a. Scanning electron micrograph of *Treponema pallidum,* the cause of syphilis. **b.** The three stages of syphilis. The primary stage of syphilis is a chancre at the site where the bacterium enters the body. **c.** The secondary stage is a body rash that occurs even on the palms of the hands and soles of the feet. **d.** In the tertiary stage, gummas may appear on the skin or internal organs.

Vaginal Infections

The term vaginitis is used to describe any vaginal infection or inflammation. It is the most commonly diagnosed gynecological condition. Bacterial vaginosis (BV) is believed to cause 40–50% of the cases of vaginitis in the United States. Overgrowth of certain bacteria inhabiting the vagina causes vaginosis. A common culprit is the bacterium *Gardnerella vaginosis*. Overgrowth of this organism and subsequent symptoms can occur for nonsexual reasons. However, symptomless males may pass on the bacterium to women, who do experience symptoms.

The symptoms of BV are vaginal discharge that has a strong odor, a burning sensation during urination, and/or itching or pain in the vulva. Some women with BV have no signs of the infection. How women acquire these infections is not well understood. Having a new sex partner or multiple sex partners seems to increase the risk of getting BV, but females who are not sexually active get BV as well. Douching also appears to increase the incidence of BV. Women with BV are more susceptible to infection by other STDs, including HIV, herpes, chlamydia, and gonorrhea. Pregnant women with BV are at greater risk of premature delivery.

The yeast *Candida albicans*, and a protozoan, *Trichomonas vaginalis*, are two other causes of vaginitis. *Candida albicans* is normally found living in the vagina. Under certain circumstances, its growth increases above normal, causing vaginitis. For example, women taking birth control pills or antibiotics may be prone to yeast infections. Both can alter the normal balance of vaginal organisms, causing a yeast infection. A yeast infection causes a thick, white, curdlike vaginal discharge and is accompanied by itching of the vulva and/or vagina. Antifungal medications inserted into the vagina are used to treat yeast infections. Trichomoniasis caused by *Trichomonas vaginalis* affects both males and females. The urethra is usually the site of infection in males. Infected males are often asymptomatic and pass the parasite to their partner during sexual intercourse. Symptoms of trichomoniasis in females are a foul-smelling, yellow-green frothy discharge and itching of the vulva/vagina. Having trichomoniasis greatly increases the risk of infection by HIV. Prescription drugs are used to treat trichomoniasis but if one partner remains infected, reinfection will occur. It is recommended that both partners in a sexual relationship be treated and abstain from having sex until the treatment is completed.

Check Your Progress 16.6

1. What condition can occur due to both a chlamydial and gonorrheal infection?
2. Genital warts are associated with what other medical condition in women?

CASE STUDY THREE YEARS LATER

"**W**hirrrrrrrrrr." The cell phone on Carter's desk vibrated, indicating an incoming call. Distractedly Carter glanced at the caller I.D., then grinned. Quickly, he answered the call. "Anthony!" he yelled into the phone. "Long time, no hear from you! What's the occasion?"

"Dude, how do you feel about being a godfather?" Anthony replied.

"W-W-What? Are you kidding?" Carter stammered.

"No joke. Tina and I have a brand new baby boy named Carter. We figured since he's named for you, maybe you should be his godfather," Anthony explained.

"Absolutely. I'd be honored to be a godfather for my namesake," Carter agreed. "Well, what do you know? You're a dad! I was a little worried."

"What do you mean?" Anthony asked.

"Remember that research project you participated in to pay the rent? Guess it didn't do any long-term damage after all," Carter joked. "Congratulations!"

Health **Focus**

Preventing Transmission of STDs

Sexual Activities Transmit STDs

Abstain from sexual intercourse or develop a long-term monogamous (always the same partner) sexual relationship with a partner who is free of STDs (Fig. 16C).

Refrain from multiple sex partners or having relations with someone who has multiple sex partners. If you have sex with two other people and each of these has sex with two people, and so forth, the number of people who are relating is quite large.

Be aware that having relations with an intravenous drug user is risky because the behavior of this group risks AIDS and hepatitis B. Be aware that anyone who already has another sexually transmitted disease is more susceptible to an HIV infection.

Avoid anal-rectal intercourse (in which the penis is inserted into the rectum) because this behavior increases the risk of an HIV infection. The lining of the rectum is thin, and infected CD4 T cells can easily enter the body there. Also, the rectum is supplied with many blood vessels, and insertion of the penis into the rectum is likely to cause tearing and bleeding that facilitate the entrance of HIV. The vaginal lining is thick and difficult to penetrate, but the lining of the uterus is only one cell thick at certain times of the month, and does allow CD4 T cells to enter.

Uncircumcised males are more likely to become infected than circumcised males because vaginal secretions can remain under the foreskin for a long time.

Practice Safer Sex

Always use a latex condom during sexual intercourse if you are not in a monogamous relationship. Be sure to follow the directions supplied by the manufacturer for the use of a condom. At one time, condom users were advised to use nonoxynol-9 in conjunction with a condom, but testing shows that this spermicide has no effect on viruses, including HIV.

Avoid fellatio (kissing and insertion of the penis into a partner's mouth) **and cunnilingus** (kissing and insertion of the tongue into the vagina) because they may be a means of transmission. The mouth and gums often have cuts and sores that facilitate catching an STD.

Practice penile, vaginal, oral, and hand cleanliness. Be aware that hormonal contraceptives make the female genital tract receptive to the transmission of sexually transmitted diseases, including HIV.

Be cautious about using alcohol or any drug that may prevent you from being able to control your behavior.

Drug Use Transmits HIV

Stop, if necessary, or do not start the habit of injecting drugs into your veins. Be aware that HIV and hepatitis B can be spread by blood-to-blood contact.

Always use a new sterile needle for injection or one that has been cleaned in bleach if you are a drug user and cannot stop your behavior (Fig. 16D).

Figure 16C Sexual activities transmit STDs.

Figure 16D Sharing needles transmits STDs.

Summarizing the Concepts

16.1 Human Life Cycle

The life cycle of higher organisms requires two types of cell division: mitosis and meiosis.

- **Mitosis:** growth and repair of tissues.
- **Meiosis:** gamete production.

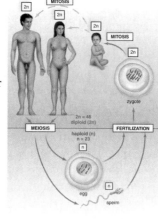

16.2 Male Reproductive System

The external genitals of males are

- the penis (organ of sexual intercourse).
- the scrotum (contains the testes).

Spermatogenesis, occurring in seminiferous tubules of the testes, produces sperm.

- Mature sperm are stored in the epididymides.
- Sperm pass from the vasa deferentia to the urethra.
- The seminal vesicles, prostate gland, and bulbourethral glands add fluids (by secretion) to sperm.
- Sperm and secretions are called semen or seminal fluid.

Orgasm in males results in ejaculation of semen from the penis.

Hormonal Regulation in Males

- Hormonal regulation, involving secretions from the hypothalamus, the anterior pituitary, and the testes, maintains a fairly constant level of testosterone.
- FSH from the anterior pituitary promotes spermatogenesis.
- LH from the anterior pituitary promotes testosterone production by interstitial cells.

16.3 Female Reproductive System

Oogenesis occurring within the ovaries typically produces one mature follicle each month.

- This follicle balloons out of the ovary and bursts, releasing an egg that enters an oviduct.
- The oviducts lead to the uterus, where implantation and development occur.

The female external genital area includes the vaginal opening, the clitoris, the labia minora, and the labia majora.

- The vagina is the organ of sexual intercourse and the birth canal in females.

Orgasm in females culminates in uterine and oviduct contractions.

16.4 Female Hormone Levels

Ovarian Cycle: Nonpregnant

- The ovarian cycle is under the hormonal control of the hypothalamus and the anterior pituitary.
- During the cycle's first half, FSH from the anterior pituitary causes maturation of a follicle that secretes estrogen and some progesterone.
- After ovulation and during the cycle's second half, LH from the anterior pituitary converts the follicle into the corpus luteum.
- The corpus luteum secretes progesterone and some estrogen.

Uterine Cycle: Nonpregnant

Estrogen and progesterone regulate the uterine cycle.

- Estrogen causes the endometrium to rebuild.
- Ovulation usually occurs on day 14 of a 28-day cycle.
- Progesterone produced by the corpus luteum causes the endometrium to thicken and become secretory.
- A low level of hormones causes the endometrium to break down as menstruation occurs.

Fertilization and Pregnancy

If fertilization takes place, the embryo implants in the thickened endometrium.

- The corpus luteum is maintained because of hCG production by the placenta, and therefore, progesterone production does not cease.
- Menstruation usually does not occur during pregnancy.

16.5 Control of Reproduction

Numerous birth control methods and devices are available.

- A few of these are the birth control pill, diaphragm, and condom.
- Effectiveness varies.

Assisted reproductive technologies may help infertile couples to have children. Some of these technologies are

- artificial insemination by donor (AID).
- in vitro fertilization (IVF).
- gamete intrafallopian transfer (GIFT).
- intracytoplasmic sperm injection (ICSI).

16.6 Sexually Transmitted Diseases

STDs are caused by viruses, bacteria, fungi, and parasites.

STDs Caused by Viruses

- AIDS is caused by HIV (human immunodeficiency virus).
- Genital warts are caused by human papillomaviruses; these viruses cause warts or lesions on genitals and are associated with certain cancers.
- Genital herpes is caused by herpes simplex virus type 2; causes blisters on genitals.
- Hepatitis is caused by hepatitis viruses A, B, C, D, E, and G. A and E are usually acquired from contaminated water; hepatitis B and C from bloodborne transmission; and B, D, and G are sexually transmitted.

STDs Caused by Bacteria

- Chlamydia is caused by *Chlamydia trachomatis;* PID can result.
- Gonorrhea is caused by *Neisseria gonorrhoeae;* PID can result.
- Syphilis is caused by *Treponema pallidum.* It has three stages, with the third stage resulting in death.

Vaginal Infections

- Bacterial vaginosis commonly results from bacterial overgrowth. *Gardnerella vaginosis* often causes such infections.
- Infection with the yeast *Candida albicans* also occurs because of overgrowth, and antibiotics or hormonal contraceptives trigger this condition.
- The parasite *Trichomonas vaginalis* also causes vaginosis. This type affects both men and women, though men are often asymptomatic.

Understanding Key Terms

acrosome 356
birth control method 365
birth control pill 366
bulbourethral gland 353
cervix 358
chancre 373
chlamydia 372
circumcision 354
contraceptive 365
contraceptive implant 367
contraceptive injection 367
contraceptive vaccine 367
corpus luteum 360
diaphragm 367
egg 357
endometrium 358
epididymis 353
erectile dysfunction 354
estrogen 362
female condom 367
fimbria 357
follicle 360
follicle-stimulating hormone (FSH) 356
gamete 370
gonadotropin-releasing hormone (GnRH) 356
human chorionic gonadotropin (hCG) 364
implantation 357
infertility 368
interstitial cell 356
intrauterine device (IUD) 367

luteinizing hormone (LH) 356
male condom 367
menopause 362
menstruation 363
oogenesis 360
ovarian cycle 360
ovary 357
oviduct 357
ovulation 360
Pap test 358
penis 354
placenta 364
progesterone 362
prostate gland 353
scrotum 353
semen 353
seminal vesicle 353
seminiferous tubule 354
Sertoli cell 355
sperm 355
spermatogenesis 354
testes 353
testosterone 356
tubal ligation 367
urethra 353
uterine cycle 362
uterus 358
vagina 358
vas deferens 353
vasectomy 367
vulva 358
zygote 357

Match the key terms to these definitions.

a. _____ Release of an oocyte from the ovary.

b. _____ Female sex hormone that causes the endometrium of the uterus to become secretory during the uterine cycle; along with estrogen, it maintains secondary sex characteristics in females.

c. _____ Thick, whitish fluid consisting of sperm and secretions from several glands of the male reproductive tract.

d. _____ Narrow end of the uterus, which projects into the vagina.

e. _____ Cap at the anterior end of a sperm that partially covers the nucleus and contains enzymes that help the sperm penetrate the egg.

Testing Your Knowledge of the Concepts

1. What type of cell division produces the gametes? Why is this type of cell division necessary? (page 352)

2. Trace the path of sperm. What glands contribute fluids to semen? (page 353)

3. Where are sperm produced in the testes? What is the process called? Where is testosterone produced in the testes? (pages 354–56)

4. Name the hormones involved in maintaining the sex characteristics of the male, and tell what each does. (page 356)

5. What are the organs of the female reproductive tract and their functions? (pages 357–58)

6. Describe the ovarian cycle in a nonpregnant female and the hormones involved. (pages 360–62)

7. Describe the uterine cycle in a nonpregnant female, and relate it to the ovarian cycle. (page 364)

8. Describe the hormonal role of the placenta. (pages 363–64)

9. Briefly describe various birth control methods, along with their effectiveness. (pages 365–68)

10. Briefly describe various types of assisted reproductive technologies. (page 370)

11. What STDs are caused by viruses? List the causative agent, symptoms, and treatments. What STDs are caused by bacteria? List the causative agent, symptoms, and treatments. (pages 371–74)

12. Label this diagram of the male reproductive system, and trace the path of sperm.

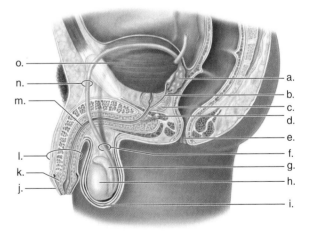

13. Which of these associations is mismatched?
    a. interstitial cells—testosterone
    b. seminiferous tubules—sperm production
    c. vasa deferentia—seminal fluid production
    d. urethra—conducts sperm

14. Follicle-stimulating hormone (FSH)
    a. is secreted by females but not by males.
    b. stimulates the seminiferous tubules to produce sperm.
    c. secretion is controlled by gonadotropin-releasing hormone (GnRH).
    d. Both b and c are correct.

15. In tracing the path of sperm, you would mention the vasa deferentia before the
    a. testes.
    c. urethra.
    b. epididymides.
    d. uterus.

16. Semen does not contain
    a. prostate fluid.
    d. prostaglandins.
    b. urine.
    e. Both b and d are correct.
    c. fructose.

17. The scrotum is
    a. part of the primary sex organ.
    b. important in regulating the temperature of the testes.
    c. poorly innervated.
    d. an extension of the spermatic cord.

18. Testosterone is produced and secreted by
    a. spermatogonia.
    c. seminiferous tubules.
    b. sustentacular cells.
    d. interstitial cells.

19. Luteinizing hormone in males
    a. stimulates sperm development.
    b. triggers ovulation.
    c. is responsible for secondary sex characteristics.
    d. controls testosterone production by interstitial cells.

20. Secondary sex characteristics are
    a. the only body features that regress as we age.
    b. the same in males and females.
    c. those characteristics that develop only after puberty.
    d. the only reproductive structures affected by hormones.

21. The release of the oocyte from the follicle is caused by
    a. a decreasing level of estrogen.
    b. a surge in the level of follicle-stimulating hormone.
    c. a surge in the level of luteinizing hormone.
    d. progesterone released from the corpus luteum.

22. Which of the following is not an event of the ovarian cycle?
    a. FSH promotes the development of a follicle.
    b. The endometrium thickens.
    c. The corpus luteum secretes progesterone.
    d. Ovulation of an egg occurs.

23. An oocyte is fertilized in the
    a. vagina.
    c. oviduct.
    b. uterus.
    d. ovary.

24. Following implantation, the corpus luteum is maintained by
    a. estrogen.
    b. progesterone.
    c. follicle-stimulating hormone.
    d. human chorionic gonadotropin.

25. During pregnancy,
    a. the ovarian and uterine cycles occur more quickly than before.
    b. GnRH is produced at a higher level than before.
    c. the ovarian and uterine cycles do not occur.
    d. the female secondary sex characteristics are not maintained.

26. Female oral contraceptives prevent pregnancy because
    a. the pill inhibits the release of luteinizing hormone.
    b. oral contraceptives prevent the release of an egg.
    c. follicle-stimulating hormone is not released.
    d. All of these are correct.

27. Which contraceptive method is most effective?
    a. abstinence
    c. IUD
    b. diaphragm
    d. Depo-Provera® injection

In questions 28–30, match each method of protection with a means of birth control in the key.

Key:
a. vasectomy
d. diaphragm
b. oral contraception
e. male condom
c. intrauterine device (IUD)

28. Blocks entrance of sperm to uterus

29. Traps sperm and also prevents STDs

30. Prevents implantation of an embryo

31. Which of the following may cause an eye infection in newborns delivered to infected mothers?
    a. chlamydia
    c. AIDS
    b. gonorrhea
    d. Both a and b are correct.

32. For which of the following infections is there currently no cure?
    a. chlamydia
    c. syphilis
    b. gonorrhea
    d. herpes

33. Which of the following is indicative of the third stage of a syphilis infection?
    a. chancre
    c. gumma
    b. rash
    d. generalized edema

Thinking Critically About the Concepts

Male hormonal contraception (MHC) would be similar to hormonal contraception used by women. The primary goal of hormonal contraception is to inhibit or lower the gonadotropins, which in turn prevents gamete production or maturation. A successful product must have minimal side effects, and fertility should be restored shortly after the contraception is discontinued. There are numerous clinical trials being conducted to determine hormone dosages and an effective delivery method for MHC. Ideally, an MHC will be available to the general public in the next decade. That is good news for couples who would like an option for male contraception in addition to condoms and vasectomies.

1. Review Figures 16.5 and 16.12. See if you can figure out how giving males androgens and progestin (a synthetic progesterone) would inhibit the gonadotropins.

2. Refer to Figure 16.10.
   a. Redraw the illustration and add progesterone early in the cycle (to represent use of birth control pills) and draw new lines to show how FSH/LH will be affected by progesterone's presence early in the cycle.
   b. Redraw the illustration and show the effects of hCG (produced when the female gets pregnant) on the corpus luteum, progesterone levels, and endometrium.

3. Women who use birth control pills appear to have a lower risk of developing ovarian cancer. Women who use fertility enhancing drugs (which increase the number of follicles that develop) may increase their risk of developing ovarian cancer. Speculate about how these therapies may affect a woman's risk of developing ovarian cancer.

4. It is fairly common for very serious female athletes to not have their periods. These athletes might have a 10–15% body fat composition. Explain the connection between low body fat composition and the absence of a menstrual cycle in these athletes.

Infectious Diseases Supplement

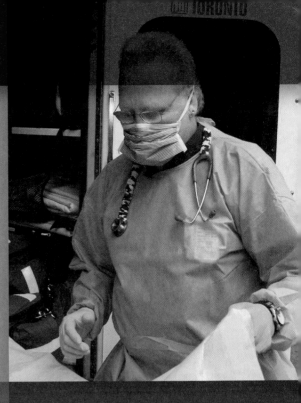

Epidemiology is the study of diseases in populations, including the causes, distribution, and control of these diseases. Humans have a long history of struggle with **infectious diseases**—diseases caused by **pathogens,** which include certain bacteria, viruses, fungi, parasites, protozoans, and prions. The term pathogen comes from the Greek language, meaning "suffering."

Infectious disease outbreaks are described in the earliest written documents. Smallpox outbreaks were described 10,000 years ago in the Nile Valley. Malaria was described 4,000 years ago in China. The Chinese treated it with the Qinghao plant—the source of the drug currently used to treat malaria. Outbreaks of bubonic plague—the so-called "Black Death"—in Europe decimated the continent's population. The spread of the disease, as well as the public response to it, literally determined European history.

In more modern times, the battle between infectious agents and humans includes the development of vaccinations. The Chinese first developed vaccines around 200 B.C.E. In 1796, the smallpox vaccine was introduced to the Western world. Death rates from this virus dropped dramatically. The disease was eradicated from the human population in 1977. A rabies vaccine was developed in 1885, and vaccines for diphtheria, typhus, tetanus, and polio soon followed. Child mortality rates plummeted and remain low in countries where vaccination is routine.

The death rates from infectious diseases dropped still further in the 1940s with the development of antibiotics. With the developments of vaccines and antibiotics, scientists hoped they could eradicate other pathogens or, at least, significantly limit the suffering caused by infectious diseases. However, infectious diseases still kill large numbers of people. According to the World Health Organization (WHO), infectious diseases kill over 13 million people annually. In developing countries, infectious diseases account for 54% of deaths. Lower respiratory infections, followed by HIV/AIDS, diarrheal diseases, malaria, and tuberculosis are the leading causes of infectious disease deaths.

Further, as globalization increases, diseases can now spread quickly from one section of the world to another. Global epidemics, called **pandemics,** now exist (Fig. S.1). Emerging diseases, those never before seen, must now be confronted, as well. In addition, some well-known pathogens are developing resistance to antibiotics and are undergoing resurgence. Organizations such as the Center for Disease Control and Prevention (CDC) and the WHO monitor and respond to the growing infectious disease threats.

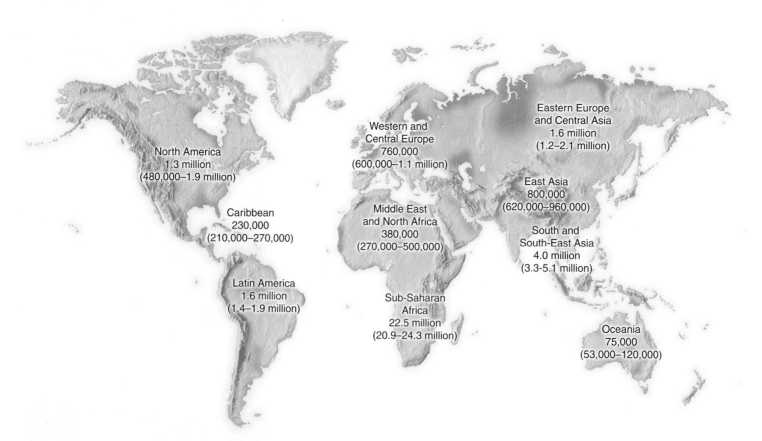

Figure S.1   How prevalent is HIV worldwide?
HIV/AIDS occurs in all continents of the globe. Approximately 33.2 million people are infected according to 2007 statistics.

S.1 AIDS and Other Pandemics

A disease is classified as an **epidemic** if there are more cases of the disease than expected in a certain area for a certain period. The number of cases that constitute an epidemic depends on what is expected. For example, a few cases of a very rare disease may constitute an epidemic, whereas a larger number of a very common disease may not be. If the epidemic is confined to a local area, it is usually called an **outbreak.** As mentioned previously, a pandemic could potentially affect the global population. In 2003, the General Assembly of the United Nations passed a resolution to fight the pandemics of HIV/AIDS, tuberculosis, malaria, and other infectious diseases for the good of the people of the world.

HIV/AIDS

Acquired immunodeficiency syndrome (AIDS) is caused by a virus known as a **human immunodeficiency virus (HIV).** There are two main types of HIV: HIV-1 and HIV-2. HIV-1 is the more widespread, virulent form of HIV.

HIV can infect cells with particular surface receptors. Most importantly HIV infects and destroys cells of the immune system, particularly helper T cells and macrophages. As the number of helper T cells declines, the body's ability to fight an infection also declines. As a result, the person becomes ill with various diseases. AIDS is the advanced stage of HIV infection, when a person develops one or more of a number of opportunistic infections. An **opportunistic infection** is one that has the *opportunity* to occur only because the immune system is severely weakened.

Origin of and Prevalence of HIV

It is generally accepted that HIV originated in Africa and then spread to the United States and Europe, by way of the Caribbean. HIV has been found in a preserved 1959 blood sample taken from a man who lived in an African country now called the Democratic Republic of the Congo. Even before this discovery, scientists speculated that an immunodeficiency virus may have evolved into HIV during the late 1950s.

Of the two types of HIV, HIV-2 corresponds to a type of immunodeficiency virus found in the green monkey, which lives in western Africa. In addition, researchers have found a virus identical to HIV-1 in a subgroup of chimpanzees once common in west-central Africa. Perhaps HIV viruses were originally found only in nonhuman primates. They could have mutated to HIV after humans ate nonhuman primates for meat.

British scientists have been able to show that AIDS came to their country perhaps as early as 1959. They examined the preserved tissues of a Manchester seaman who died that year and concluded that he most likely died of AIDS. Similarly, it is thought that HIV entered the United States on numerous occasions as early as the 1950s. But the first documented case is a 15-year-old male who died in Missouri in 1969, with skin lesions now known to be characteristic of an AIDS-related cancer. Doctors froze some of his tissues because they could not identify the cause of death. Researchers also want to test the preserved tissue samples of a 49-year-old Haitian who died in New York in 1959, of the type of pneumonia now known to be AIDS-related.

Throughout the 1960s, it was customary in the United States to list leukemia as the cause of death in immunodeficient patients. Most likely, some of these people actually died of AIDS. HIV is not extremely infectious. Thus, it took several decades for the number of AIDS cases to increase to the point that AIDS became recognizable as a specific and separate disease. The name AIDS was coined in 1982, and HIV was found to be the cause of AIDS in 1983–1984.

Worldwide, as shown at the bottom of Table S.1, an estimated 33.2 million people are now living with HIV infection. Among the 2.5 million new HIV infections, nearly 20% are under the age of 15. Table S.1 also tells us that 2.1 million people died during 2007, but since the beginning of the epidemic,

Table S.1	HIV Global Statistics, 2007			
	People Living with HIV	**New Infection**	**AIDS Deaths**	**Adult Prevalence (percentage)**
Sub-Saharan Africa	22.5 million	1.7 million	1.6 million	5.0
Asia	4.8 million	432,000	302,000	0.4
Latin America	1.6 million	100,000	58,000	0.5
Caribbean	230,000	17,000	11,000	1.0
North American, Western and Central Europe	2 million	77,000	33,000	0.9
Eatern Europe and Central Asia	1.6 million	150,000	55,000	0.9
North Africa and the Middle East	380,000	35,000	25,000	0.3
Oceania	75,000	14,000	1,200	0.4
Total	**33.2 million**	**2.5 million**	**2.1 million**	**0.8%**

over 25 million people have died of AIDS. Today, at least 0.8% of the adults in the world have an HIV infection.

As we can deduce from studying Figure S.1 and Table S.1, most people infected with HIV live in the developing (poor, low- to middle-income) countries. The few of the hardest hit regions are as follows:

- **Sub-Saharan Africa and Asia:** Of all the HIV-positive people in the world, 68% live in sub-Saharan Africa. In 2007, 76% of all deaths due to AIDS occurred in this region. Most of the Southern region is seriously affected, but there was a significant decline in HIV prevalence in Zimbabwe. Other countries have either reached a plateau or are approaching one.

   In Asia, the HIV prevalence is highest in Southeast Asia. Cambodia, Myanmar, and Thailand all showed declines. However, the epidemic is growing in Indonesia and Vietnam.

- **Latin America and the Caribbean:** In Latin America, the population numbers for those affected by HIV remain fairly stable. Transmission is prevalent among sex workers and men who have sex with men. These modes of infection account for three-fourths of adult HIV cases in the Caribbean, Haiti, and the Dominican Republic. In these countries, AIDS is the leading cause of death among persons aged 25 to 44 years.

- **North America, Europe, and Central Asia:** In North America and Western and Central Europe, the number of people living with HIV is increasing. This is primarily due to the advances in medication available to prolong the life of HIV patients. Approximately 2.0 million people in these regions are living with HIV. In 2007, 33,000 died. In Eastern Europe and Central Asia, nearly 90% of the newly reported cases of HIV were in the Russian Federation and Ukraine. IV drug abuse is the most common reason for HIV transmission in the Russian Federation.

- **North Africa and the Middle East:** The reported numbers of HIV cases in these areas remain small due to limited epidemiological surveillance. In the Sudan, heterosexual sex transmits HIV in the majority of cases. In other countries, mostly men are infected.

- **Oceania:** The epidemic is still expanding in Papua, New Guinea, where over 70% of the HIV-positive people in this region live. In Australia and New Zealand, the main risk factor for acquiring HIV appears to be unprotected sex between men.

Phases of an HIV Infection

HIV occurs as several subtypes. HIV-1C is prominent in Africa, while HIV-1B causes most infections in the United States. The following description of the phases of HIV infection pertains to an HIV-1B infection. The helper T cells and macrophages infected by HIV are called CD4 cells because they display a molecule called CD4 on their surface. With the destruction of CD4 cells, the immune system is significantly impaired. After all, macrophages present the antigen to helper T cells. In turn, helper T lymphocytes coordinate the immune response. B lymphocytes are stimulated to produce antibodies and cytotoxic T cells destroy cells infected with a virus. In the United States, one of the most common causes of AIDS deaths is from *Pneumocystis jiroveci* pneumonia (PCP), while tuberculosis kills more HIV-infected people in Africa than any other AIDS-related illness.

In 1993, the Centers for Disease Control (CDC) in the United States issued clinical guidelines for the classification of HIV to help clinicians track the status, progression, and phases of HIV infection. The classification of HIV infection will be discussed in three categories (or phases), and the system is based on two aspects of a person's health—the CD4 T cell count and the history of AIDS-defining illnesses.

Category A: Acute Phase

A person in category A typically has no apparent symptoms (asymptomatic), is highly infectious, and has a CD4 T cell count that has never fallen below 500 cells per mm^3 (500 cells/mm^3) of blood, which is sufficient for the immune system to function normally (Fig. S.2). A normal CD4 T cell count is at least 800 cells/mm^3.

Today, investigators are able to track not only the blood level of CD4 T cells, but also the viral load. The viral load is the number of HIV particles in the blood. At the start of an HIV-1B infection, the virus replicates ferociously, and the killing of CD4 T cells is evident because the blood level of these cells drops dramatically. During the first few weeks of infection, some people (1–2%) develop flulike symptoms (fever, chills, aches, swollen lymph nodes) that may last an average of two weeks. After this, a person may remain "symptom free" for years. At the beginning of this acute phase of infection, an HIV antibody test is usually negative because it generally takes an average of 25 days before there are detectable levels of HIV antibodies in body fluids.

After time, the body responds to the infection by increased activity of immune cells, and the HIV blood test becomes positive. During this phase, the number of CD4 T cells is greater than the viral load (Fig. S.2). But some investigators believe that a great unseen battle is going on. The body is staying ahead of the hordes of viruses entering the blood by producing as many as one to two billion new helper T lymphocytes each day. This is called the "kitchen sink model" for CD4 T cell loss. The sink's faucet (production of new CD4 T cells) and the sink's drain (destruction of CD4 T cells) are wide open. As long as the body can produce enough new CD4 T cells to keep pace with the destruction of these cells by HIV and by cytotoxic T cells, the person has a healthy immune system that can deal with the infection. In other words, a person in category A would have no history of conditions listed in categories B and C.

Category B: Chronic Phase A person in category B would have a CD4 count between 499 and 200 cells/mm^3, and one or more of a variety of symptoms related to an impaired immune system. The symptoms might include yeast infections of the

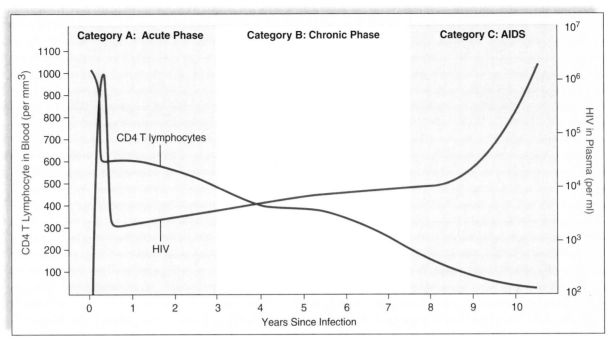

Figure S.2   **What are the stages of an HIV infection?**

In category A individuals, the number of HIV particles in plasma rises upon infection and then falls. The number of CD4 T lymphocytes falls, but stays above 400/mm³. In category B individuals, the number of HIV particles in plasma is slowly rising, and the number of T lymphocytes is decreasing. In category C individuals, the number of HIV particles in plasma rises dramatically as the number of T lymphocytes falls below 200/mm³.

mouth or vagina, cervical dysplasia (precancerous abnormal growth), prolonged diarrhea, thick sores on the tongue (hairy leukoplakia), or shingles (to list a few). Swollen lymph nodes, unexplained persistent or recurrent fevers, fatigue, coughs, or diarrhea are often seen as well. During this chronic stage of infection, the number of HIV particles is on the rise (Fig. S.2). However, they do not as yet have any of the conditions listed for category C.

Category C: AIDS   A person in category C is diagnosed with AIDS. When a person has AIDS, the CD4 T cell count has fallen below 200 cells/mm³, or the person has developed one or more of the 25 AIDS-defining illnesses (or opportunistic infections) described by the CDC's list of conditions in the 1993 AIDS surveillance case definition. Persons with AIDS die from one or more opportunistic diseases rather than from the HIV infection. Recall that an opportunistic illness occurs only when the immune system is weakened. Examples of these diseases include:

- *Pneumocystis jiroveci* pneumonia—a fungal infection of the lungs.
- *Mycobacterium tuberculosis*—a bacterial infection usually of lymph nodes or lungs but may be spread to other organs.
- Toxoplasmic encephalitis—a protozoan parasitic infection, often seen in the brain of AIDS patients.
- Kaposi's sarcoma—an unusual cancer of the blood vessels, which gives rise to reddish purple, coin-sized spots and lesions on the skin.
- Invasive cervical cancer—a cancer of the cervix, which spreads to nearby tissues.

Once one or more of these opportunistic infections has occurred, the person will remain in category C. Newly developed drugs can treat opportunistic diseases. Still, most AIDS patients are repeatedly hospitalized due to weight

Have You Ever Wondered...

What is shingles?

Shingles is a reactivation of the same virus that causes chicken pox. Once you have been infected with chicken pox, the *Varicella zoster* virus does not leave your body when you recover. Instead, the virus hides within your nervous system, burying itself deep inside a sensory nerve. Later, the virus can reactivate. It then travels down the nerve axons to the skin, where it causes a chicken pox-like rash. The rash is generally found on the chest and lower back, though it can occur anywhere. Unlike the chicken pox rash, the shingles rash is usually found in a broad stripe on only one side of the body.

Generally, a shingles outbreak is much more severe than chicken pox. The virus can affect the face, spinal cord, and brain. The rash can be very painful, and the pain can persist for years. The best treatment for shingles is prevention. Immunization can protect children from chicken pox, and a booster immunization reduces the risk of shingles in adults.

loss, constant fatigue, and multiple infections. Death usually follows in two to four years. Although there is still no cure for AIDS, many people with HIV infection are living longer, healthier lives due to the expanding use of antiretroviral therapy.

HIV Structure and Life Cycle

HIV consists of two single strands of RNA (its nucleic acid genome); various proteins; and an envelope, which it acquires from its host cell (Fig. S.3). The virus's genetic material is protected by a series of three protein coats: the nucleocapsid, capsid, and the matrix. Within the matrix, there are three very important enzymes, reverse transcriptase, integrase, and protease.

- **Reverse transcriptase** catalyzes reverse transcription, the conversion of the viral RNA to viral DNA.
- **Integrase** catalyzes the integration of viral DNA into the DNA of the host cell.
- **Protease** catalyzes the breakdown of the newly synthesized viral polypeptides into functional viral proteins.

Embedded in HIV's envelope are protein spikes referred to as Gp120. These spikes must be present for HIV to gain entry into its target immune cells. The genome for HIV consists of RNA instead of DNA, which classifies HIV as a retrovirus. A **retrovirus** must use reverse transcription to convert its RNA into viral DNA. Then, it can insert its genome into the host's genome (DNA).

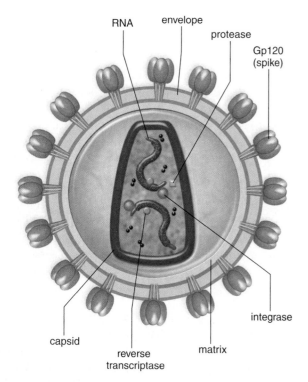

Figure S.3   **What are the components of human immunodeficiency virus?**
HIV is composed of two strands of RNA, a capsid, and an envelope containing spike proteins.

HIV Life Cycle   The events that occur in the reproductive cycle of an HIV virus (Fig. S.4) are these:

1. *Attachment.* During attachment, HIV binds to the plasma membrane of its target cell. Gp120, the spike located on the surface of HIV, binds to a CD4 receptor on the surface of a helper T cell or macrophage.
2. *Fusion.* After attachment occurs, HIV fuses with the plasma membrane, and the virus enters the cell.
3. *Entry.* During a process called uncoating, the capsid and protein coats are removed, releasing RNA and viral proteins into the cytoplasm of the host cell.
4. *Reverse transcription.* This event in the reproductive cycle is unique to retroviruses. During this phase, the enzyme called reverse transcriptase catalyzes the conversion of HIV's single-stranded RNA into double-stranded viral DNA. Usually in cells, DNA is transcribed into RNA. Retroviruses can do the opposite only because they have a unique enzyme from which they take their name. (*Retro* in Latin means reverse.)
5. *Integration.* The newly synthesized viral DNA, along with the viral enzyme integrase, migrates into the nucleus of the host cell. Then, with the help of integrase, the host cell's DNA is spliced. Double-stranded viral DNA is then integrated into the host cell's DNA (chromosome). Once viral DNA has integrated into the host cell's DNA, HIV is referred to as a **provirus,** meaning it is now a part of the cell's genetic material. HIV is usually transmitted to another person by means of cells that contain proviruses. Also, proviruses serve as a latent reservoir for HIV during drug treatment. Even if drug therapy results in an undetectable viral load, investigators know that there are still proviruses inside infected lymphocytes.
6. *Biosynthesis and cleavage.* When the provirus is activated, perhaps by a new and different infection, the normal cell machinery directs the production of more viral RNA. Some of this RNA becomes the genetic material for new viral particles. The rest of viral RNA brings about the synthesis of very long polypeptides. These polypeptides have to be cut up into smaller pieces. This cutting process, called cleavage, is catalyzed by a third HIV enzyme called protease.
7. *Assembly.* Capsid proteins, viral enzymes, and RNA can now be assembled to form new viral particles.
8. *Budding.* During budding, the virus gets its envelope and envelope marker coded for by the viral genetic material. The envelope is actually host plasma membrane.

The life cycle of an HIV virus includes transmission to a new host. Body secretions, such as semen from an infected male, contain proviruses inside CD4 T cells. When this semen is discharged into the vagina, rectum, or mouth, infected CD4 T cells migrate through the organ's lining and enter the body. The receptive partner in anal-rectal intercourse appears to be most at risk, because the lining of the

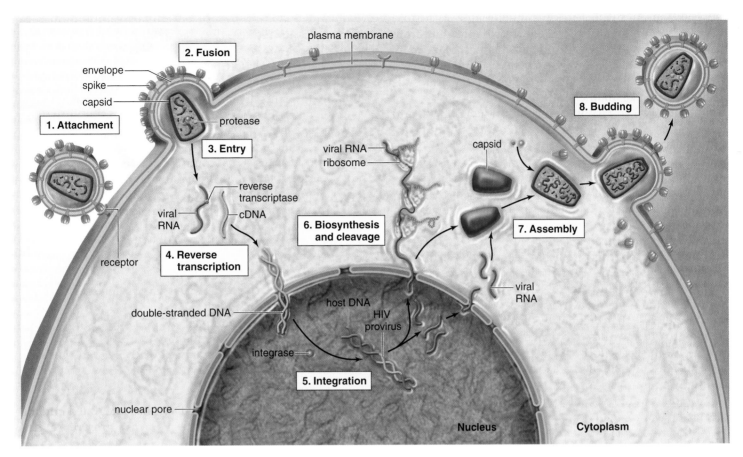

Figure S.4 **How does HIV reproduce?**

HIV is a retrovirus that uses reverse transcription to produce viral DNA. Viral DNA integrates into the cell's chromosomes before it reproduces and buds from the cell.

rectum is apparently prone to penetration. CD4 macrophages present in tissues are believed to be the first infected when proviruses enter the body. When these macrophages move to the lymph nodes, HIV begins to infect CD4 T cells. HIV can hide out in local lymph nodes for some time, but eventually the lymph nodes degenerate. Large numbers of HIV particles can then enter the bloodstream. Now the viral load begins to increase; when it exceeds the CD4 T-cell count, the individual progresses to the final phase of an HIV infection.

Transmission and Prevention of HIV HIV is transmitted by sexual contact with an infected person, including vaginal or rectal intercourse and oral/genital contact. Also, needle-sharing among intravenous drug users is high-risk behavior. A less common mode of transmission (and now rare in countries where blood is screened for HIV) is through transfusions of infected blood or blood-clotting factors. Babies born to HIV-infected women may become infected before or during birth, or through breast-feeding after birth. From a global perspective, heterosexual sex is

the main mode of HIV transmission. In some nations, however, men who have sex with men, IV drug abusers, and sex industry (prostitution) workers are the most common transmitters of HIV. Differences in cultures, sexual practices, and belief systems around the world influence the type of HIV prevention strategies needed to fight the spread of the disease.

Blood, semen, vaginal fluid, and breast milk are the body fluids known to have the highest concentrations of HIV. However, HIV is not transmitted through casual contact in the workplace, schools, or social settings. Casual kissing, hugging, or shaking hands won't spread the virus. Likewise, you can't be infected by touching toilet seats, door knobs, dishes, drinking glasses, food, or pets. The general message of HIV prevention across the globe is abstinence, sex with only one uninfected partner, and accurate and consistent condom use during each sexual encounter.

HIV Testing and Treatment for HIV Generally, an HIV test does not test for the virus. Instead, initial HIV tests are designed to detect the presence of HIV antibodies in the body. Most people will develop antibodies to HIV within two to eight weeks (average of 25 days), but it can take from three to six months. Test results for conventional blood, oral, and urine HIV antibody tests may not be available for weeks. Rapid HIV antibody tests are now being used, with results available within 20 minutes.

At one time, an HIV infection almost invariably led to AIDS and an early death, because there were no drugs for controlling the progression of HIV disease in infected people. But since late 1995, scientists have gained a much better understanding of the structure of HIV and its life cycle. Now, therapy is available that successfully controls HIV replication. Patients remain in the chronic phase of infection for a variable number of years, so that the development of AIDS is postponed. However, this is not the case for those in poor nations where funds for HIV drugs are limited. Only 20% of those worldwide who need antiretroviral therapy are receiving this treatment.

AIDS in the United States is presently caused by HIV-1B, and drug therapy has brought the condition under control. But the drug therapy has two dangers. Infected people may become lax in their efforts to avoid infection because they know that drug therapy is available. Further, drug use leads to drug-resistant viruses. Even now, some HIV-1B viruses have become drug-resistant when patients have failed to adhere to their drug regimens.

Another possible problem is that the present treatment is designed for HIV-1B. It's possible that in 5–20 years, the more-developed countries, including the United States, will experience a new epidemic of AIDS caused by HIV-1C. Therefore, it behooves the more-developed countries to do all they can to help African countries aggressively seek a solution to the HIV-1C epidemic.

There is no cure for AIDS, but a treatment called highly active antiretroviral therapy (HAART) is usually able to stop HIV replication to such an extent that the viral load becomes undetectable. HAART uses a combination of drugs that interfere with the life cycle of HIV. Entry inhibitors stop HIV from entering a cell. The virus is prevented from binding to a receptor in the plasma membrane. Reverse transcriptase inhibitors, such as zidovudine (AZT), interfere with the operation of the reverse transcriptase enzyme. Integrase inhibitors prevent HIV from inserting its genetic material into that of the host cells. Protease inhibitors prevent protease from cutting up newly created polypeptides. Assembly and budding inhibitors are in the experimental stage, and none are available as yet. Ideally, a drug combination will make the virus less likely to replicate or successfully mutate. If reproduction can be suppressed, viral resistance to this therapy will be less likely to occur, and the drugs won't lose their effectiveness.

Investigators have found that when HAART is discontinued, the virus rebounds. Therefore, therapy must be continued indefinitely. An HIV-positive pregnant woman who takes reverse transcriptase inhibitors during her pregnancy reduces the chances of HIV transmission to her newborn. If possible, drug therapy should be delayed until the tenth to twelfth week of pregnancy to minimize any adverse effects of AZT on fetal development. But, if treatment begins at this time, the chance of transmission is reduced by 66%. If treatment is delayed until the last few weeks of a pregnancy, transmission of HIV to an offspring is still cut by 50%.

The consensus is that control of the AIDS pandemic will not occur until a vaccine preventing HIV infection is developed. Unfortunately, none of these would be a cure. Instead, the vaccine would be a preventive measure that helps people who are not yet infected escape infection (preventive vaccine). Alternatively, a therapeutic vaccine could slow the progression of the disease upon future infection. Scientists have studied more than 50 different preventive vaccines and over 30 therapeutic vaccines. Scientists do not expect to have results until the year 2010.

The Science Focus on page 387 reviews the difficulties in developing an effective AIDS vaccine so that you can understand why we do not have an AIDS vaccine yet and, indeed, why we may never have an ideal one.

TUBERCULOSIS

In 1882, Robert Koch was the first to see the causative agent of tuberculosis (TB) with a microscope (Fig. S.5). At that time, TB caused 14% of deaths in Europe. The disease was called *consumption,* because it seemed to consume the patients from the inside until they wasted away. In the 1940s, with the advent of effective antibiotics to fight TB, it was

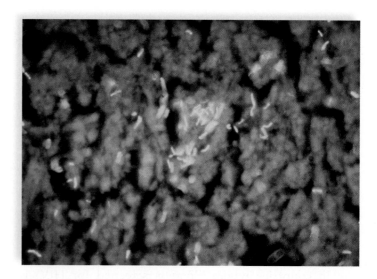

Figure S.5   **What is the causative agent of tuberculosis?**
Tuberculosis is caused by *Mycobacterium tuberculosis,* shown here in a sputum sample stained with a fluorescent dye.

Science **Focus**

Is an AIDS Vaccine Possible?

An ideal AIDS vaccine would be inexpensive and able to provide lifelong protection against all strains of HIV (Fig. SA). Is this possible? Effective vaccines have been developed against diseases such as hepatitis B, smallpox, polio, tetanus, influenza, and the measles. But what about AIDS? The top 10 "setbacks" in AIDS vaccine development are these:

1. The ideal AIDS vaccine is one that would prevent HIV entry into human cells and prevent the progression and transmission of the disease. However, no vaccine has ever proven to be 100% effective at blocking a virus from entry into cells. Instead, most vaccines prevent, modify, or weaken the disease caused by the infection.
2. Due to HIV's high rate of mutation, there are several genetically different types and subtypes of HIV. HIV strains may differ by 10% within one person and by 35% in people across the globe. Viruses that are genetically different may have different surface proteins, and HIV surface proteins are the focus of many AIDS vaccines. The question is, will scientists need a vaccine for each HIV subtype, or will one vaccine provide protection for all HIV variants?
3. The vaccine may produce only short-term protection, meaning people would need to continue to get booster shots similar to shots given yearly for the flu.
4. There are fears that an AIDS vaccine would make people more vulnerable to HIV infection. It is believed that in some diseases such as yellow fever and Rift Valley fever, the antibodies produced after receiving the vaccine helped the virus infect more cells.
5. HIV infects and destroys immune cells, particularly T cells. It would be hard to create an effective vaccine to stimulate the immune cells that HIV is targeting for destruction.
6. HIV can be transmitted as a free virus and in infected cells, so perhaps a vaccine would need to stimulate both cellular and antibody-mediated responses. To date, most of the successful vaccines stimulate only antibody production.
7. Most vaccines in use today against other diseases are prepared from live-attenuated (weakened) forms of the infectious virus. There are concerns that an AIDS vaccine made using weakened forms of the live virus will cause HIV disease (AIDS).
8. No person has ever completely recovered from HIV infection.
9. HIV inserts its genetic material into human cells, where it can hide from the immune system.
10. Scientists do not have an ideal animal model for AIDS vaccine testing.

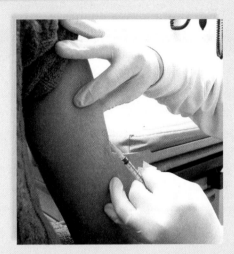

Figure SA   **What problems complicate the development of an HIV vaccine?**
Different strains of HIV as well as a high viral mutation rate are just two of the reasons it has been difficult to develop a vaccine against HIV.

Although there are many difficulties in vaccine development, AIDS vaccine trials are underway. The process can take many years. After a vaccine has been tested in animals, it must pass through three phases of clinical trials before it is marketed or administered to the public. In Phases I and II, the vaccine is tested from one to two years in a small number of HIV-uninfected volunteers. The most effective vaccines move into Phase III. In Phase III, the vaccine is tested three to four years in thousands of HIV-uninfected people.

There are no approved preventive or therapeutic vaccines for HIV infection. So, is an AIDS vaccine possible? Most scientists have many reasons to be optimistic that an AIDS vaccine can and will be developed. One reason is that scientists have had some success with vaccinating monkeys. But the most compelling reason for optimism is the human body's ability to suppress the infection. The immune system is able to successfully and effectively decrease the HIV viral load in the body, helping to delay the onset of AIDS an average of ten years in 60% of people who are HIV-infected in the United States. Studies have shown a small number of people remain HIV-uninfected after repeated exposure to the virus, and a few HIV-infected individuals maintain a healthy immune system for over 15 years. It is these stories of the human body's ability to fight HIV infection that keep scientists hopeful that there is a way to help the body fight HIV infection. Most scientists agree that, despite the obstacles or the "setbacks," an AIDS vaccine is possible somewhere in our future.

Have You Ever Wondered...

What is a sanatorium?

A sanatorium was a hospital for long-term care of those with untreatable diseases. When tuberculosis was ravaging a world without antibiotics, rest and relaxation in areas with clean air was considered the best way to allow the body to fight the disease. Sufferers were sent to sanatoriums to either heal or die. The word sanatorium came from the Latin verb *sano*, meaning to heal. A sanatorium is often distinguished from a sanitarium. Sanitariums, based on the Latin noun *sanitas*, meaning health, are health resorts.

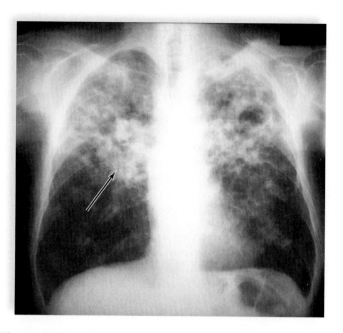

Figure S.6   **What does TB in the lungs look like?**
TB causes tubercles in the lungs. In the active state, these liquify, and cavities are seen on X-ray.

thought that the disease could be eliminated. However, control measures were not implemented consistently, and tuberculosis cases began to rise in the 1980s. It is estimated that one-third of the world's population has been exposed to TB. Each year, approximately 9 million people are infected, and 2 million die. HIV infections are a contributing factor to the increase in TB cases. Tuberculosis is currently the number one cause of death in AIDS patients.

Causative Agent and Transmission

Tuberculosis is caused by the rod-shaped bacterium *Mycobacterium tuberculosis*. In nature, it is a very slow-growing bacterium. The cells have a thick, waxy coating and can exist for weeks in a dehydrated state. The organism is spread by air-borne droplets, introduced into the air when an infected person coughs, sings, or sneezes. The bacteria can float in the air for several hours and still be infectious. The likelihood of infection increases with the length and frequency of exposure to an individual with active TB. This makes TB especially contagious on airplane flights longer than eight hours. It also makes the caregivers for those with TB at risk.

Disease

The incubation period is four to 12 weeks, and the disease develops very slowly. Once the bacteria reach the lungs, they are consumed by macrophages. Other white blood cells rush to the infected area. Together, they wall off the original infection site, producing small, hard nodules, or *tubercles,* in the lungs. It is these tubercles that give the disease its name. In most patients, the bacteria remain alive within the tubercles, but the disease does not progress. These patients are said to have latent TB. They do not feel sick, and they are not contagious. A person with latent TB will test positively on a tuberculosis skin test. Tubercles

often calcify and can be seen on a chest X-rays (Fig. S.6). The combination of skin testing and X-ray findings confirms the diagnosis of tuberculosis.

If the immune system fails to control *Mycobacterium tuberculosis* in the lungs, active disease may occur. A person with active disease is contagious. The tubercle liquefies and forms a cavity. The bacteria can then spread from these cavities throughout the body, especially to the kidney, spine, and brain. It may be fatal. Symptoms of active TB include a bad cough, chest pain, and coughing up blood or sputum. As the disease progresses, symptoms include fatigue, loss of appetite, chills, fevers, and night sweats. The patient begins to lose weight and wastes away.

Treatment and Prevention

Due to the resurgence of antibiotic-resistant strains, multiple anti-TB drugs are given simultaneously for 12 to 24 months. The most common drugs are isoniazid (INH), rifampin (RIF), ethambutol, and pyrazinamide. It is unlikely that bacteria can develop resistance to multiple drugs at the same time. It takes at least six months to kill all of the *Mycobacterium tuberculosis* in the body, so the length of drug treatment is long.

Public health officials try to prevent the spread of TB by identifying and treating all cases of active TB. Active TB patients are isolated for at least two weeks at the beginning of drug treatment to prevent transmission. Thereafter, they are monitored for drug compliance and reappearance of symptoms. Anyone exposed to an active case of TB is treated.

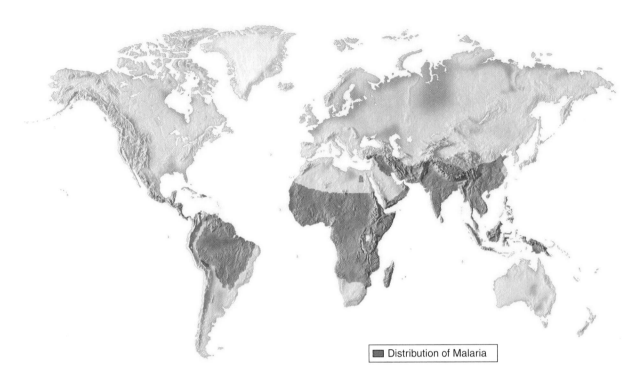

■ Distribution of Malaria

Figure S.7 **What areas of the world are hardest hit by malaria?**
Malaria is found in subtropical areas where its mosquito vector lives.

MALARIA

Malaria is called the world's invisible pandemic. Most people in the West do not even consider malaria a major health threat, yet there are 350–500 million cases per year with over 1 million deaths, mostly in sub-Saharan Africa (Fig. S.7). Key to the geographic distribution of malaria is transmission of the disease by a mosquito vector that depends on temperature and rainfall, and thus survives well in tropical areas. A **vector** is a living organism, usually an insect or animal, which transfers the pathogen from one host to another.

Causative Agent and Transmission

The parasites that cause malaria belong to the genus *Plasmodium*. These are protists (see Fig. 1.5). There are four species that infect humans: *P. malariae, P. falciparum, P. vivax,* and *P. ovale. P. falciparum* causes more disease and death than the other species. The parasite is spread by the female *Anopheles gambiae* mosquito. Half of the life cycle occurs in the human, while the remainder happens in the mosquito. As the female mosquito feeds on human blood, she injects saliva containing an anticoagulant, along with the parasite. The parasites travel to the liver, where they undergo asexual reproduction. The parasites are released from the liver, to infect more liver cells and erythrocytes (red blood cells). Inside the erythrocyte, the *Plasmodium* enlarges and divides, until it bursts the erythrocytes. This red blood cell stage is cyclic and repeats every 48 to 72 hours. Some parasites within the erythrocytes don't destroy their host cells. Instead, they develop into the sexual form of the parasite. When these are ingested by another mosquito during a blood meal, they develop into male and female gametes within the gut of the mosquito. The gametes fuse, undergo mitosis, and form the parasites that migrate to the salivary glands of the mosquito to continue the cycle.

Disease People at significant risk for malaria include those who have little or no immunity to the parasite. Children, pregnant women, and travelers are most likely to fall victim to the disease. Diagnosis of malaria depends upon the presence of parasites in the blood. The incubation period from time of bite to onset of symptoms varies from seven to 30 days. The symptoms of malaria range from very mild to fatal. Most infected people develop a flulike illness with chills and fevers, interrupted by sweating. These symptoms exhibit a cyclical pattern every 48 to 72 hours corresponding to bursting of the red blood cells in the body. Milder cases of malaria are often confused with influenza or a cold, and therefore treatment is delayed. More severe cases cause severe anemia due to destruction of red blood cells, cerebral malaria, acute kidney failure, cardiovascular collapse, shock, and death.

Figure S.8   **How can malaria transmission be prevented?**
Studies show that the use of insecticide-treated mosquito nets reduces the risk of malaria transmission.

Treatment and Prevention   Malaria is a curable disease if it is diagnosed and treated correctly in a short period. If a person exhibits malaria symptoms, they should be treated within 24 hours of onset. Common antimalarial drugs include quinine and artesunate. Treatment helps reduce symptoms and breaks the transmission pattern for the disease. Antimalarial drugs can also be used prophylactically, or before infection. They do not prevent the initial infection following the mosquito bite. Instead, they prevent the development of the parasites in the blood.

Health organizations are working on the prevention of infection through vector control. Strategies include eliminating the mosquito by removing its breeding sites and by insecticide fogging of large areas. Additional efforts are aimed at preventing humans from being bitten by the mosquito, using simple mosquito nets. The use of insecticide-treated mosquito nets for children has reduced the incidence of malaria (Fig. S.8).

Drug-resistant *Plasmodium* and insecticide-resistant *Anopheles* are becoming significant problems. *P. falciparum* and *P. vivax* have developed strains that are resistant to the antimalarial drugs. Efforts to develop a malaria vaccine are ongoing.

> **Check Your Progress S.1**
> 1. Define a pandemic and give examples.
> 2. Describe the three stages of an HIV infection.
> 3. Describe how *Mycobacterium* causes tuberculosis.
> 4. Why does a malaria infection involve both a mosquito and a parasite?

S.2 Emerging Diseases

Since 2003, avian influenza and severe acute respiratory syndrome (SARS) have generated a lot of press. These are considered new or **emerging diseases.** The National Institute of Allergy and Infectious Diseases (NIAID) lists 16 pathogens that are newly recognized in the last two decades. Re-emerging diseases are ones that have reappeared after a significant decline in incidence. Tuberculosis would be an example of a re-emerging disease. Finally, there are diseases that have been known throughout human history, but the disease was not known to be caused by an infectious agent or the pathogen had never been identified. Ulcers caused by *Helicobacter pylori* (recognized in 1983) would be an example.

Where do emerging diseases come from? Some of these diseases may result from new and/or increased exposure to animals or insect populations that act as vectors for disease. Changes in human behavior and use of technology can result in new diseases. SARS is thought to have arisen in Guandong, China, due to consumption of civets, a type of exotic cat considered a delicacy (Fig. S.9). The civets were possibly infected by exposure to horseshoe bats sold in open markets. Legionnaires' disease emerged in 1976 due to contamination of a large air conditioning system in a hotel. The bacteria thrived in the cooling tower used as the water source for the air conditioning system. In addition, globalization results in the transport of diseases all over the world that were previously restricted to isolated communities. The first SARS cases were reported in southern China the week of November 16, 2002. By the end of February 2003, SARS had reached nine countries/provinces, mostly through airline travel (Fig. S.10). Some pathogens mutate and change hosts, jumping from birds to humans, for example. Before 1997, avian flu was thought to only affect birds. A mutated strain jumped

Figure S.9   **What was the source of the recent SARS epidemic?**
It appears that civets may have been the source of the SARS epidemic. Civet meat is considered a delicacy in some parts of China.

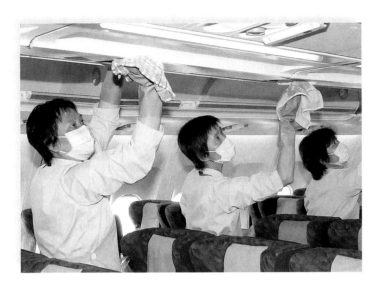

Figure S.10   How does globalization affect the spread of infectious diseases?
Airline travel was responsible for spreading SARS from China to other parts of the world. Workers clean an airplane to prevent further transmission of SARS.

to humans in the 1997 outbreak. To control that epidemic, officials killed 1.5 million chickens to remove the source of the virus.

NIAID also monitors re-emerging diseases. Re-emerging diseases have been known in the past but were thought to have been controlled. Diseases in this category include known diseases that are spreading from their original geographical location or diseases that have suddenly increased in incidence.

NIAID lists five diseases in this category. A change in geographical location could be due to global warming allowing expansion of habitats for insect vectors. Re-emerging diseases can also be due to human carelessness, as in the abuse of antibiotics or poorly implemented vaccination programs. This allows previously controlled diseases to resurge.

> **Check Your Progress S.2**
> 1. Define an emerging disease, and give an example.
> 2. How do emerging diseases arise?
> 3. What is responsible for re-emerging diseases?

S.3 Antibiotic Resistance

Some well-known pathogens are becoming more difficult to fight due to the advent of **antibiotic resistance.** Just four years after penicillin was introduced in 1943, bacteria began developing resistance to it. The use of antibiotics does not cause humans to become resistant to the drugs. Instead, pathogens become resistant. There are some organisms in a population naturally resistant to the drug (Fig. S.11). They have acquired this resistance through mutations or interactions with other organisms. The drug regimen kills the susceptible ones while leaving naturally resistant ones to multiply and repopulate the patient's body. The new population is then resistant to the drug. Tuberculosis, malaria, gonorrhea, *Staphylococcus aureus,* and enterococci (or group D *Streptococcus*) are a few of the diseases and organisms connected to antibiotic resistance. Unfortunately, more and more organisms are becoming

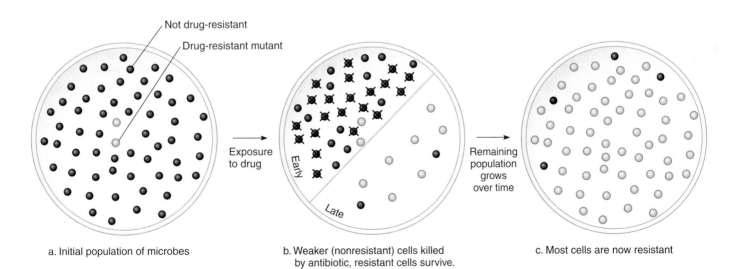

a. Initial population of microbes

b. Weaker (nonresistant) cells killed by antibiotic, resistant cells survive.

c. Most cells are now resistant

Figure S.11   How does antibiotic resistance develop?
a. In a microbe population, random mutation can result in cells with drug resistance. **b.** Drug use kills all nonresistant microbes, leaving only the stronger, resistant cells to survive. **c.** The new microbe population is now mostly resistant to the antibiotic.

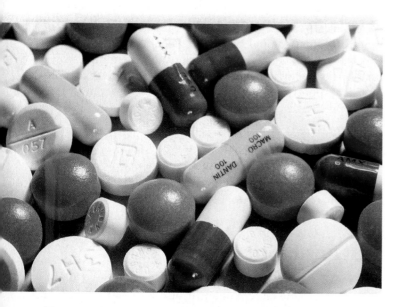

Figure S.12   **How has our ability to treat infections been compromised?**
Misuse of antibiotics has resulted in increased drug resistance among bacteria. For example, MRSA is becoming a serious health threat.

multidrug resistant leaving health care facilities with little choice for treatment of infection.

What is being done about this problem? The CDC, U.S. Food and Drug Administration, and U.S. Department of Agriculture have formed a cooperative organization charged with monitoring antibiotic-resistant organisms. Pharmaceutical companies are developing new antibiotics. However, the best way to fight antibiotic resistance is to prevent it from happening in the first place by using antibiotics wisely. Take all of the antibiotics prescribed as directed. Do not skip doses or discontinue treatment when you feel better. Do not expect a doctor to prescribe antibiotics for all infections. For example, antibiotics are ineffective in treating virus infections such as colds. Do not save unused antibiotics or take antibiotics prescribed for a different infection. Antibiotics are important drugs for fighting infection and improper usage lessens their effectiveness for all of us (Fig. S.12).

XDR TB

XDR TB stands for extensively drug-resistant tuberculosis. It is resistant to almost all of the drugs used to treat TB. This includes the first-line antibiotics, the older and cheaper ones, as well as the second-line antibiotics, the newer and more expensive drugs. Therefore the treatment options are very limited. Fortunately, this is still relatively rare. There have been only 49 cases in the United States between 1993 and 2006. MDR TB, multidrug resistant TB, is more common. These organisms are resistant to the first-line antibiotics. Eastern Europe and Southeast Asia have the highest rates of MDR TB.

As described in the Bioethical Box on page 16, a patient diagnosed with XDR TB traveled to Europe and Canada by air, and then tried to enter the United States by car. The CDC put him under federal quarantine in Atlanta. Patients with XDR TB should not be traveling on airlines and the CDC recommended testing for passengers and crew on those airline flights.

MRSA

Methicillin-resistant *Staphylococcus aureus,* or **MRSA,** is another antibiotic-resistant bacterium. It causes "staph" infections such as boils and infection of the hair follicles. In 1974, only 2% of staph infections were caused by MRSA. In 1995, the number had risen to 22%. In 2004, it was 63%. It is especially common in athletes who share equipment. MRSA is resistant to methicillin and other common antibiotics such as penicillin and amoxicillin. MRSA is prevalent in nursing homes and hospitals where patients have already reduced immune responses. Patients infected with MRSA generally have longer hospital stays with poorer outcomes. It can be fatal. It is passed from nonsymptomatic carriers to patients, usually through hand contact. Hand washing is critical for preventing transmission.

> ### Check Your Progress S.3
> 1. How do antibiotic-resistant bacteria arise?
> 2. How should antibiotics be taken?
> 3. What are some examples of antibiotic-resistant bacteria?

CHAPTER

17

Development and Aging

CASE STUDY AMBER AND KENT FORREST

Amber and Kent Forrest studied the blue lines on the test stick in Amber's hand. Amber compared the lines that appeared on the test stick with the kit's directions. She broke out in a grin. "Woo Hoo!" she yelled, "We're pregnant!" Kent hugged her, and they both jumped up and down. They had been hoping for this!

A week later, Amber and Kent were in the office of her obstetrician, Dr. Davis. "Well, Amber and Kent, congratulations! You are about six weeks pregnant."

Dr. Davis scanned her chart. "You're healthy, and you've started taking the prenatal vitamins. Those are important throughout your pregnancy, but especially during these first three months. It's amazing how much growth and development your little person will do during the first trimester. The folic acid those big 'horse pill' vitamins contain helps prevent birth defects."

"They're so huge," Amber smiled. "I can't take them in the morning because I'm a little sick, so I take them at night instead."

Dr. Davis smiled back, then continued, "That's fine. You've already been vaccinated for rubella. It used to be called German measles way back when. The disease causes birth defects, but you're immune. No worries there.

"And it looks like you're both doing all the right things already. Neither of you smokes, right?" he asked. Both shook their heads. "No drug use?" Again, both shook their heads, this time vehemently.

"Well, you probably know all the rest of my orders: No alcohol. No drugs, even prescription drugs or over-the-counter drugs, unless you check with my office first," he warned, then added, "Help yourselves to all the brochures in the waiting room. They'll help explain what's going on with your body during the first trimester. They've got great illustrations of the baby's development, too. If anything seems unusual or out of place, give the office a call," Dr. Davis concluded, shaking hands with Amber and Kent. "Congratulations, you two! We'll be seeing each other a lot during the next nine months!"

CHAPTER CONCEPTS

17.1 Fertilization
During fertilization, a sperm nucleus fuses with the egg nucleus. Once one sperm penetrates the plasma membrane, the egg undergoes changes that prevent any more sperm from entering.

17.2 Pre-Embryonic and Embryonic Development
Pre-embryonic development occurs between the time of fertilization and implantation in the uterine lining. By the end of embryonic development, all organ systems are established, and there is a mature and functioning placenta. The embryo is only about 38 mm (1½ in.) long.

17.3 Fetal Development
During fetal development, the gender becomes obvious, the skeleton continues to ossify, fetal movement begins, and the fetus gains weight.

17.4 Pregnancy and Birth
During pregnancy, the mother gains weight as the uterus comes to occupy most of the abdominal cavity. A positive feedback mechanism that involves uterine contractions and oxytocin explains the onset and continuation of labor so that the child is born.

17.5 Development After Birth
Development after birth consists of infancy, childhood, adolescence, and adulthood. Aging is influenced by our genes, whole body changes, and extrinsic factors.

17.1 Fertilization

Fertilization is the union of a sperm and egg to form a **zygote,** the first cell of the new individual (Fig. 17.1).

Steps of Fertilization

The tail of a sperm is a flagellum, which allows it to swim toward the egg. The middle piece contains energy-producing mitochondria. The head contains a nucleus capped by a membrane-bound acrosome. The acrosome is an organelle containing digestive enzymes. Only the nucleus from the sperm head fuses with the egg nucleus. Therefore, the zygote receives cytoplasm and organelles only from the mother.

The plasma membrane of the egg is surrounded by an extracellular matrix termed the zona pellucida. In turn, the zona pellucida is surrounded by a few layers of adhering follicular cells, collectively called the corona radiata. These cells nourished the egg when it was in a follicle of the ovary.

During fertilization, several sperm penetrate the corona radiata. Several sperm attempt to penetrate the zona pellucida, but only one sperm enters the egg. The acrosome plays a role in allowing sperm to penetrate the zona pellucida. After a sperm head binds tightly to the zona pellucida, the acrosome releases digestive enzymes that forge a pathway for the sperm through the zona pellucida. When a sperm binds to the egg, their plasma membranes fuse. This sperm (the head, the middle piece, and usually the tail) enters the egg. Fusion of the sperm nucleus and the egg nucleus follows.

To ensure proper development, only one sperm should enter an egg. Prevention of polyspermy (entrance of more than one sperm) depends on changes in the egg's plasma membrane and in the zona pellucida. As soon as a sperm touches an egg, the egg's plasma membrane depolarizes (from 265 mV to 10 mV), and this prevents the binding of any other sperm. Then, vesicles called cortical granules release enzymes that cause the

Have You Ever Wondered...

How many sperm compete to fertilize the egg?

Studies indicate that there are between 40 and 150 million sperm in a man's ejaculation. Of that number, only 100 or so will make it near the egg. Out of those 100, only a single sperm will fertilize the egg.

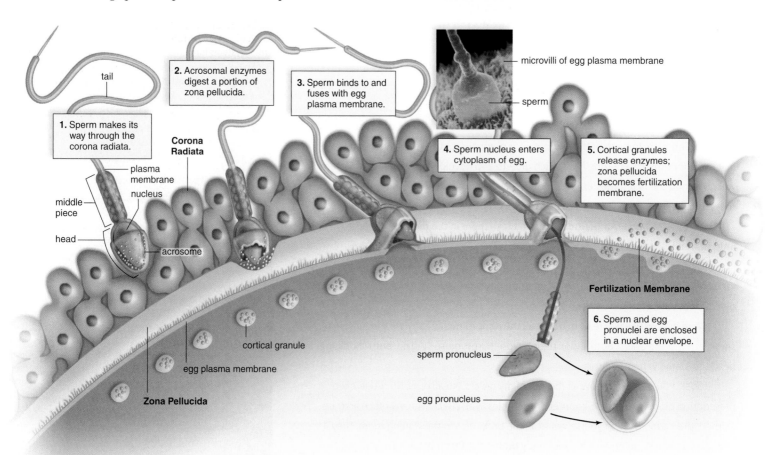

Figure 17.1   How does fertilization occur?
During fertilization, a single sperm is drawn into the egg by microvilli of its plasma membrane (micrograph). With the help of enzymes from the acrosome, a sperm makes its way through the zona pellucida. After a sperm binds to the plasma membrane of the egg, changes occur that prevent other sperm from entering the egg. Fertilization is complete when the sperm pronucleus and the egg pronucleus contribute chromosomes to the zygote.

zona pellucida to become an impenetrable fertilization membrane. Now sperm cannot bind to the zona pellucida either.

> **Check Your Progress 17.1**
> 1. **How does fertilization occur?**

17.2 Pre-Embryonic and Embryonic Development

Human development proceeds from pre-embryonic to embryonic development, and then through fetal development. Table 17.1 outlines the major events during development.

Table 17.1	Human Development	
Time	**Events for Mother**	**Events for Baby**
Pre-Embryonic Development		
First week	Ovulation occurs.	Fertilization occurs. Cell division begins and continues. Chorion appears.
Embryonic Development		
Second week	Symptoms of early pregnancy (nausea, breast swelling, and fatigue) are present. Blood pregnancy test is positive.	Implantation occurs. Amnion and yolk sac appear. Embryo has tissues. Placenta begins to form.
Third week	First menstruation is missed. Urine pregnancy test is positive. Symptoms of early pregnancy continue.	Nervous system begins to develop. Allantois and blood vessels are present. Placenta is well formed.
Fourth week		Limb buds form. Heart is noticeable and beating. Nervous system is prominent. Embryo has tail. Other systems form.
Fifth week	Uterus is the size of a hen's egg. Mother feels frequent need to urinate due to pressure of growing uterus on bladder.	Embryo is curved. Head is large. Limb buds show divisions. Nose, eyes, and ears are noticeable.
Sixth week	Uterus is the size of an orange.	Fingers and toes are present. Skeleton is cartilaginous.
Two months	Uterus can be felt above the pubic bone.	All systems are developing. Bone is replacing cartilage. Facial features are becoming refined. Embryo is about 38 mm (1½ in.) long.
Fetal Development		
Third month	Uterus is the size of a grapefruit.	Gender can be distinguished by ultrasound. Fingernails appear.
Fourth month	Fetal movement is felt by a mother who has previously been pregnant.	Skeleton is visible. Hair begins to appear. Fetus is about 150 mm (6 in.) long and weighs about 170 grams (6 oz).
Fifth month	Fetal movement is felt by a mother who has not previously been pregnant. Uterus reaches up to level of umbilicus, and pregnancy is obvious.	Protective cheesy coating, called vernix caseosa, begins to be deposited. Heartbeat can be heard.
Sixth month	Doctor can tell where baby's head, back, and limbs are. Breasts have enlarged, nipples and areolae are darkly pigmented, and colostrum is produced.	Body is covered with fine hair called lanugo. Skin is wrinkled and reddish.
Seventh month	Uterus reaches halfway between umbilicus and rib cage.	Testes descend into scrotum. Eyes are open. Fetus is about 300 mm (12 in.) long and weighs about 1,350 grams (3 lb).
Eighth month	Weight gain is averaging about a pound a week. Standing and walking are difficult for the mother because her center of gravity is thrown forward.	Body hair begins to disappear. Subcutaneous fat begins to be deposited.
Ninth month	Uterus is up to rib cage, causing shortness of breath and heartburn. Sleeping becomes difficult.	Fetus is ready for birth. It is about 530 mm (20½ in.) long and weighs about 3,400 grams (7½ lb).

Processes of Development

As a human being develops, these processes occur:

Cleavage Immediately after fertilization, the zygote begins to divide so that there are first 2, then 4, 8, 16, and 32 cells, and so forth. Increase in size does not accompany these divisions (see Fig. 17.3). Cell division during cleavage is mitotic, and each cell receives a full complement of chromosomes and genes.

Growth During embryonic development, cell division is accompanied by an increase in size of the daughter cells.

Morphogenesis Morphogenesis refers to the shaping of the embryo and is first evident when certain cells are seen to move, or migrate, in relation to other cells. By these movements, the embryo begins to assume various shapes.

Differentiation When cells take on a specific structure and function, differentiation occurs. The first system to become visibly differentiated is the nervous system.

Extraembryonic Membranes

The **extraembryonic membranes** are not part of the embryo and fetus. Instead, as implied by their name, they are outside the embryo (Fig. 17.2). The names of the extraembryonic membranes in humans are strange to us because they are named for their function in shelled animals! In shelled animals, the chorion lies next to the shell and carries on gas exchange. The amnion contains the protective amniotic fluid, which bathes the developing embryo. The allantois collects nitrogenous wastes. The yolk sac surrounds the yolk, which provides nourishment.

The functions of the extraembryonic membranes are different in humans because humans develop inside the uterus. The extraembryonic membranes have these functions in humans:

1. **Chorion.** The chorion develops into the fetal half of the **placenta,** the organ that provides the embryo/fetus with nourishment and oxygen and takes away its waste. The chorionic villi are fingerlike projections of the chorion that increase the absorptive area of the chorion. Blood vessels within the chorionic villi are continuous with the umbilical blood vessels.

2. **Allantois.** The allantois, like the yolk sac, extends away from the embryo. It accumulates the small amount of urine produced by the fetal kidneys and later gives rise to the urinary bladder. For now, its blood vessels become the umbilical blood vessels, which take blood to and from the fetus. The umbilical arteries carry O_2-poor blood from the fetus to the placenta, and the umbilical veins carry O_2-rich blood from the placenta to the fetus.

3. **Yolk sac.** The yolk sac is the first embryonic membrane to appear. In shelled animals such as birds, the yolk sac contains yolk, food for the developing embryo. In mammals such as humans, this function is taken over by the placenta, and the yolk sac contains little yolk. But the yolk sac contains plentiful blood vessels. It is the first site of blood cell formation.

4. **Amnion.** The amnion enlarges as the embryo and then the fetus enlarges. It contains fluid to cushion and protect the embryo, which develops into a fetus.

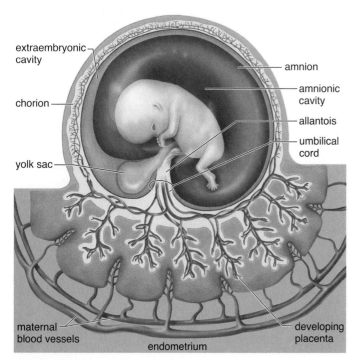

Figure 17.2 **What are the extraembryonic membranes? What are their functions?**
The chorion and amnion surround the embryo. The two other extraembryonic membranes, the yolk sac and allantois, contribute to the umbilical cord.

Have You Ever Wondered...

How is a baby's due date calculated?

The due date for arrival of the baby is calculated from the first day of the woman's last menstrual cycle before pregnancy. From conception to birth is approximately 266 days. Conception occurs approximately 14 days after the menstrual cycle begins (assuming ovulation occurs in the middle of the menstrual cycle). This gives a total of 280 days until the due date, or approximately 40 weeks.

To estimate the actual date, a calculation called Naegele's rule is often used:

1. Use the first day of the last period as a starting point.
2. Subtract 3 months from the month in which the period occurred.
3. Add seven days to the first day of the last period.

For example, if a woman's last period started on January 1, her baby's approximate due date will be October 8.

Stages of Development

Pre-embryonic development encompasses the events of the first week, while **embryonic development** begins with the second week and lasts until the end of the second month.

Pre-Embryonic Development

The events of the first week of development are shown in Figure 17.3.

Immediately after fertilization, the zygote divides repeatedly as it passes down the oviduct to the uterus. A **morula** is a compact ball of embryonic cells that becomes a **blastocyst.** The many cells of the blastocyst arrange themselves so that there is an **inner cell mass** surrounded by an outer layer of cells. The inner cell mass will become the embryo, and the layer of cells will become the chorion. The early appearance of the chorion emphasizes the complete dependence of the developing embryo on this extraembryonic membrane.

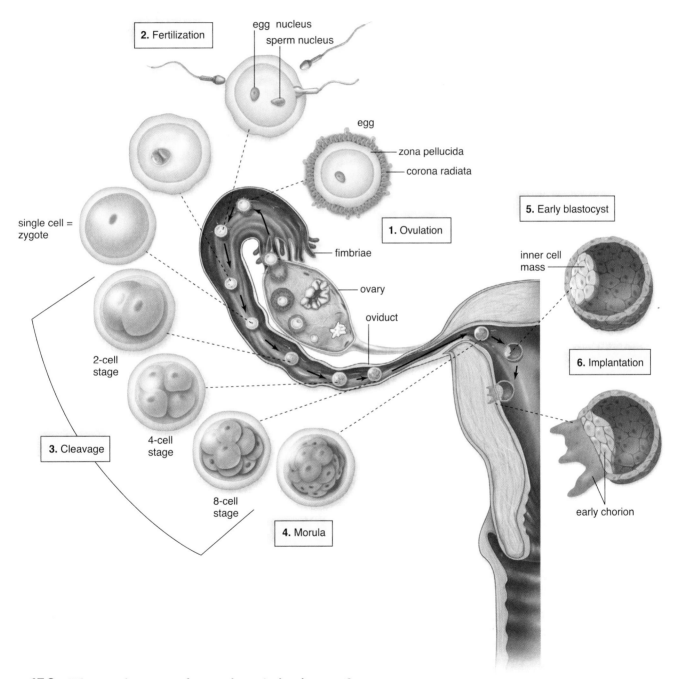

Figure 17.3   What are the stages of pre-embryonic development?

Structures and events proceed counter-clockwise. (1) At ovulation, the secondary oocyte leaves the ovary. A single sperm nucleus enters the egg, and (2) fertilization occurs in the oviduct. As the zygote moves along the oviduct, it undergoes (3) cleavage to produce (4) a morula. (5) The blastocyst forms and (6) implants in the uterine lining.

Each cell within the inner cell mass has the genetic capability of becoming any type of tissue. Sometimes during development, the cells of the morula separate, or the inner cell mass splits, and two pre-embryos are present rather than one. If all goes well, these two pre-embryos will be identical twins because they have inherited exactly the same chromosomes. Fraternal twins arise when two different eggs are fertilized by two different sperm. They do not have identical chromosomes.

Embryonic Development

The second week begins the process of implantation. Embryonic development lasts until the end of the second month of development. At the end of embryonic development the embryo is easily recognized as human.

Second Week   At the end of the first week, the **embryo** usually begins the process of implanting itself in the wall of the uterus. When **implantation** is successful, a woman is clinically pregnant. On occasion, it happens that the embryo implants itself in a location other than the uterus—most likely, the oviduct. Such a so-called **ectopic pregnancy** cannot succeed because an oviduct is unable to support it.

During implantation, the chorion secretes enzymes to digest away some of the tissue and blood vessels of the endometrium of the uterus. The chorion also begins to secrete **human chorionic gonadotropin (hCG),** the hormone that is the basis for the pregnancy test. hCG acts like luteinizing hormone (LH) in that it serves to maintain the corpus luteum past the time it normally disintegrates. It is being stimulated, so the corpus luteum secretes progesterone, which maintains

the endometrial wall. The endometrium is maintained, so the expected menstruation does not occur.

The embryo is now about the size of the period at the end of this sentence. As the week progresses, the inner cell mass becomes the **embryonic disk,** and two more extraembryonic membranes form (Fig. 17.4a,b). The yolk sac is the first site of blood cell formation. The amniotic

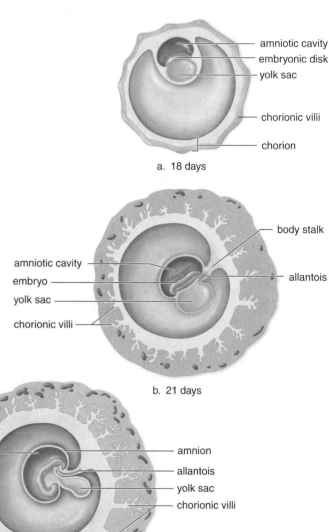

a. 18 days

b. 21 days

c. 25 days

d. 35+ days

Figure 17.4   **How does the embryo develop?**
a. At first, no organs are present in the embryo, only tissues. The amniotic cavity is above the embryonic disk, and the yolk sac is below. The chorionic villi are present. **b, c.** The allantois and yolk sac, two more extraembryonic membranes, are positioned inside the body stalk as it becomes the umbilical cord. **d.** At 35+days, the embryo has a head region and a tail region. The umbilical cord takes blood vessels between the embryo and the chorion (placenta).

cavity surrounds the embryo (and then the fetus) as it develops. In humans, amniotic fluid acts as an insulator against cold and heat. It also absorbs shock, such as that caused by the mother exercising.

The start of the major event called **gastrulation** turns the inner cell mass into the embryonic disk. Gastrulation is an example of morphogenesis during which cells move or migrate. In this case, cells migrate to become tissue layers called the **primary germ layers.** By the time gastrulation is complete, the embryonic disk has become an embryo with three primary germ layers: ectoderm, mesoderm, and endoderm. Figure 17.5 shows the significance of the primary germ layers—all the organs of an individual can be traced back to one of the primary germ layers.

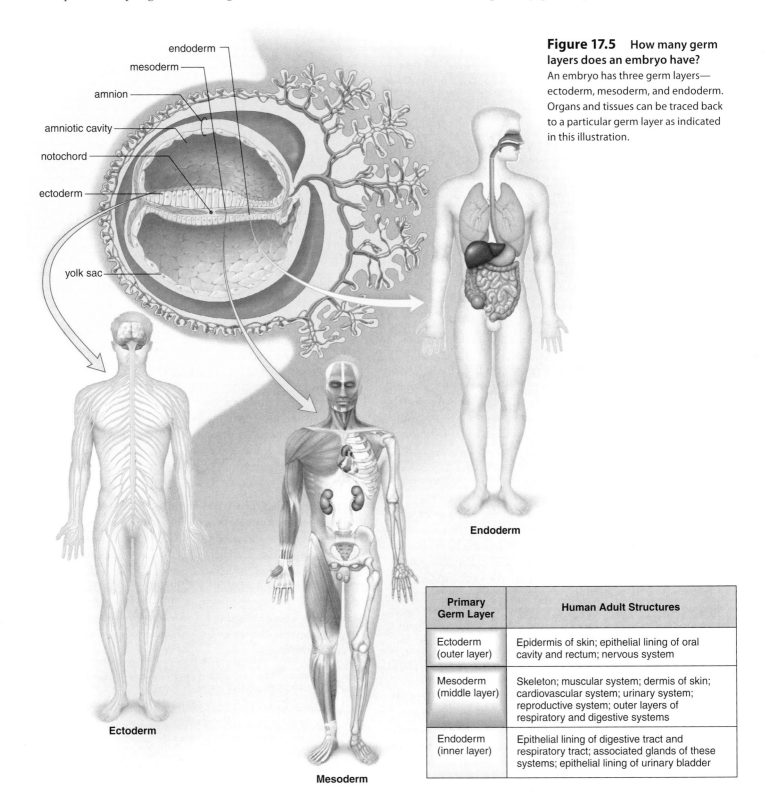

Figure 17.5 How many germ layers does an embryo have?
An embryo has three germ layers—ectoderm, mesoderm, and endoderm. Organs and tissues can be traced back to a particular germ layer as indicated in this illustration.

endoderm
mesoderm
amnion
amniotic cavity
notochord
ectoderm
yolk sac

Ectoderm

Mesoderm

Endoderm

Primary Germ Layer	Human Adult Structures
Ectoderm (outer layer)	Epidermis of skin; epithelial lining of oral cavity and rectum; nervous system
Mesoderm (middle layer)	Skeleton; muscular system; dermis of skin; cardiovascular system; urinary system; reproductive system; outer layers of respiratory and digestive systems
Endoderm (inner layer)	Epithelial lining of digestive tract and respiratory tract; associated glands of these systems; epithelial lining of urinary bladder

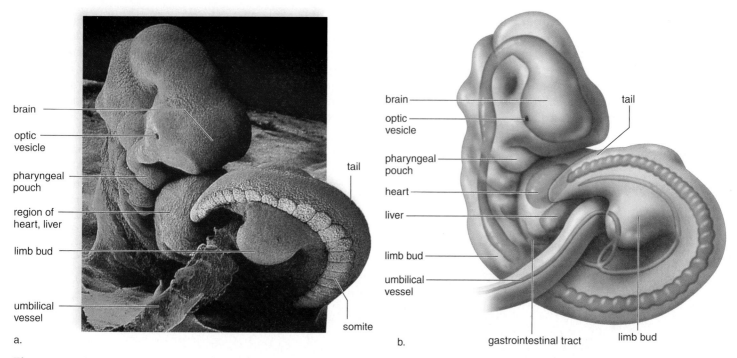

brain
optic vesicle
pharyngeal pouch
region of heart, liver
limb bud
umbilical vessel
a.

tail

brain
optic vesicle
pharyngeal pouch
heart
liver
limb bud
umbilical vessel
b.

tail

somite

gastrointestinal tract   limb bud

Figure 17.6   **What does a human embryo look like after five weeks of development?**
a. Scanning electron micrograph. **b.** The embryo is curled so that the head touches the heart and liver, the two organs whose development is farther along than the rest of the body. The organs of the gastrointestinal tract are forming, and the arms and the legs develop from the bulges called limb buds. The tail is an evolutionary remnant; its bones regress and become those of the coccyx (tailbone). The pharyngeal pouches become functioning gills in fishes and amphibian larvae; in humans, the first pair of pharyngeal pouches becomes the auditory tubes. The second pair becomes the tonsils, while the third and fourth become the thymus gland and the parathyroid glands.

Third Week   Two important organ systems make their appearance during the third week. The nervous system is the first organ system to be visually evident. At first, a thickening appears along the entire posterior length of the embryo. Then the center begins to fold inward forming a pocket. The edges are called neural folds. When the neural folds meet at the midline, the pocket becomes a tube called the neural tube. The neural tube later develops into the brain and the spinal cord.

Development of the heart begins in the third week and continues into the fourth week. At first, there are cells from both sides of the body that migrate to form the heart. When these fuse to form a continuous tube, the heart begins pumping blood, even though the chambers of the heart are not fully formed. The veins enter posteriorly, and the arteries exit anteriorly from this largely tubular heart. Later the heart twists so that all major blood vessels are located anteriorly.

Fourth and Fifth Weeks   At four weeks, the embryo is barely larger than the height of this print. A body stalk (future umbilical cord) connects the embryo to the chorion, which has treelike projections called **chorionic villi** (see Fig. 17.4c,d). The fourth extraembryonic membrane, the allantois, lies within the body stalk. Its blood vessels become the umbilical blood vessels. The head and the tail then lift up, and the body stalk moves anteriorly by constriction. Once this process is complete, the **umbilical cord** is fully formed (see Fig. 17.4d). The umbilical cord connects the developing embryo to the placenta.

Soon limb buds appear (Fig. 17.6). Later, the arms and the legs develop from the limb buds, and even the hands and the feet become apparent. At the same time—during the fifth week—the head enlarges, and the sense organs become more prominent. It is possible to make out the developing eyes and ears, and even the nose.

Sixth Through Eighth Weeks   During the sixth through eighth weeks of development, the embryo changes to a form easily recognized as a human being. Concurrent with brain development, the head achieves its normal relationship with the body as a neck region develops. The nervous system is developed well enough to permit reflex actions, such as a startle response to touch. At the end of this period, the embryo is about 38 mm (1.5 in.) long and weighs no more than an aspirin tablet, even though all organ systems have been established.

Check Your Progress 17.2

1. Development involves what processes?
2. a. What are the extraembryonic membranes? b. What are their functions?
3. a. What happens during pre-embryonic development? b. During embryonic development?

17.3 Fetal Development

The placenta is the source of progesterone and estrogen during pregnancy. These hormones have two functions: (1) because of negative feedback on the hypothalamus and anterior pituitary, they prevent any new follicles from maturing, and (2) they maintain the endometrium. Menstruation does not usually occur during pregnancy.

The placenta has a fetal side contributed by the chorion and a maternal side consisting of uterine tissues (Fig. 17.7).

The blood of the mother and the fetus never mix because exchange always takes place across the villi. Exchange occurs via diffusion. Carbon dioxide and other wastes move from the fetal side to the maternal side. Nutrients and oxygen move from the maternal side to the fetal side of the placenta. Harmful chemicals can also cross the placenta, and this is of particular concern during the embryonic period, when various structures are first forming. Each organ or part seems to have a sensitive period during which a substance can alter its normal function.

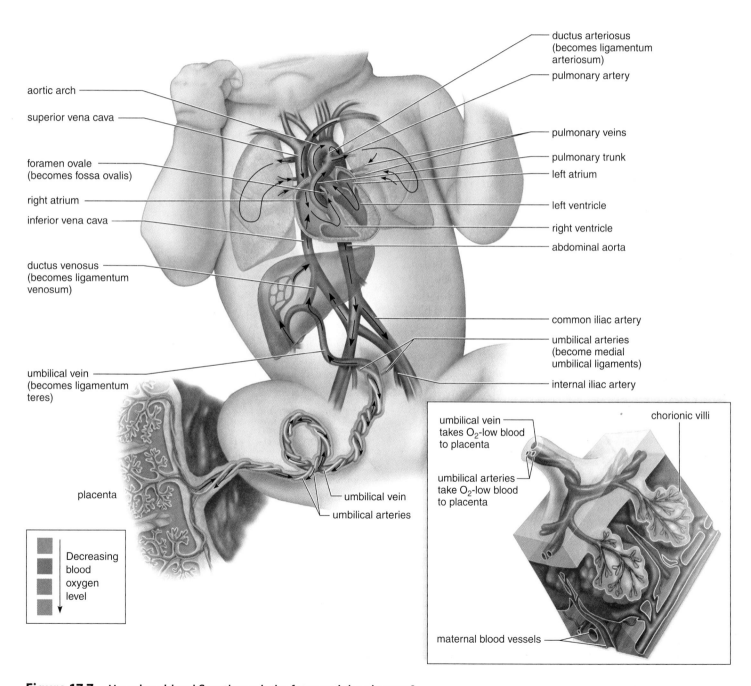

Figure 17.7 How does blood flow through the fetus and the placenta?

a. Trace the path of blood by following the arrows. **b.** At the placenta, an exchange of molecules between fetal and maternal blood takes place across the walls of the chorionic villi.

Preventing Birth Defects

Most birth defects are not due to inheritance of an abnormal number of chromosomes or any other genetic abnormality. Tragically, congenital defects (those the baby has at birth) are often the result of poor health practices during pregnancy. While it's certainly true that everyone's physical condition depends on maintaining good health habits, an expectant mother must remember that her fitness determines the well-being of both herself and her unborn baby. Pregnant women must follow certain sensible guidelines to help ensure the birth of a healthy baby.

If possible, a woman should receive a complete physical exam before becoming pregnant. Chicken pox, mumps, measles, and rubella can be very harmful to the embryo or fetus, and are prevented just by updating vaccinations. The woman can also be treated for dangerous sexually transmitted diseases, such as syphilis. If she is HIV-positive (see the Infectious Diseases supplement), therapies are available to improve maternal and fetal health.

Basic nutrients—protein, carbohydrate, and fat—are required in adequate amounts to meet the demands of the pregnant woman and fetus. Further, it is crucial that certain vitamins and minerals be supplemented. A pregnant woman needs more folate (folic acid) to meet demand for the increased cell division rate and DNA synthesis in her body, as well as that of the developing child. Without proper folate, the baby is at risk for a *neural tube defect*. The spinal cord will not close completely, or the brain will not fully develop. The B vitamins must be supplemented for proper metabolism. Calcium is necessary for bone growth, and iron for red blood cell formation. A health-care professional can help the pregnant woman to plan a balanced diet.

"You're eating for two right now," is a saying that a pregnant woman might hear. It's true, to a point. The pregnant woman needs to consume more calories so that her baby will have a normal weight. However, if her food intake becomes excessive, the mother will gain too much weight and so will her baby. A baby that is too large is at risk for physical injury during delivery. Moderate exercise can usually continue throughout pregnancy, and hopefully will contribute to ease of delivery.

At one time, it was believed that the placenta was an effective barrier between the mother and her unborn baby. Now we know that this is definitely not the case. Harmful substances can cross the placenta and damage the fetus. Smoking must stop during pregnancy. Recall (from Chap. 6) that carbon monoxide found in cigarette smoke binds to hemoglobin. Oxygen transport in the baby's blood is prevented as a result. Even second-hand smoke contains fetotoxic chemicals and must be avoided. Alcohol easily crosses the placenta, and even one drink a day appears to increase the chance of a miscarriage. The more alcohol consumed, the greater the chances of physical abnormalities if the pregnancy continues. Heavy consumption puts a fetus at risk for mental defects, because alcohol enters the brain. Illegal drugs, such as marijuana, cocaine, and heroin should be completely avoided.

Medications and supplements may also affect normal development. Excessive vitamin A, sometimes used to treat acne, may damage an embryo. Sex hormones, including

---

CASE STUDY SIX WEEKS LATER: THE HEARTBEAT

Amber woke up excited. Today was her 12-week appointment. They were going to hear the baby's heartbeat today! She shook Kent. "Wake up, sleepy head. Today's the day."

Kent grinned. "I know. You've been telling me for weeks." He switched to a sing-song voice. "The baby's heart begins to beat at five weeks and can be seen on a sonogram at five to six weeks. With a Doppler instrument using sound waves and a microphone for amplification, the heartbeat can be heard around 12–13 weeks." Amber laughed and hit him with her pillow. Perhaps she had played that DVD from the library one too many times.

At the doctor's office, Amber lay on her back on the exam table. The technician explained, "I'm going to put some gel on the end of this probe and scan it across your belly. I'll need to change the angle and the amount of pressure to find the heartbeat clearly. It may feel slightly cold, and you'll feel some pressure but it won't hurt. Okay?" Amber nodded. "So just relax and let's listen."

At first, there were regular "lub-dup, lub-dup" noises. "That's your heartbeat, Amber," the technician explained. But then the nurse changed the angle of the probe and pressed harder, and the couple could clearly hear the baby's racing heartbeat. Amber held her breath and listened. "I'm listening to my baby's heartbeat?"

Kent looked puzzled. "It's beating so much faster than Amber's heart."

The technician nodded. "Yes, it's beating at about 150 beats per minute. That's how we know that we're listening to the baby's heart, not Amber's. A rate this fast is normal for this stage of development. Heart rate slows down as the baby develops."

"My girlfriend told me that you could figure out the baby's sex by the heart rate," Amber asked. "Is that really true?"

"I hear that one all the time," the technician responded, and shook her head. "I looked it up. There was a university study, and it showed that heart rate can't predict sex. So don't go buying all blue or pink clothes yet," she concluded with a smile.

birth control pills, can cause abnormalities of the sex organs. Some antibiotics and antidepressants are dangerous as well. Health-care professionals are generally very cautious about prescribing drugs during pregnancy, and no pregnant woman should take any drug—even ordinary cold remedies or aspirin—without checking first with her care provider.

A pregnant woman should avoid having X-rays, if at all possible. Diagnostic X-rays that expose her to a small amount of radiation (such as dental X-rays) are probably safe. Still, the expectant mother should be sure her care provider knows that she is, or may possibly be, pregnant. Many organic industrial chemicals, such as vinyl chloride, formaldehyde, asbestos, benzenes, and pesticides, cause mutations when they enter the baby's bloodstream. These compounds should be strictly avoided. Someone who works with hazardous chemicals should inform her employer that she is pregnant.

Now that physicians and laypeople are aware of the various ways birth defects can be prevented, it is hoped that the incidence of birth defects will decrease in the future.

Path of Fetal Blood

The umbilical cord stretches between the placenta and the fetus. It is the lifeline of the fetus because it contains the umbilical arteries and vein. Blood within the fetal aorta travels to its various branches, including the iliac arteries. The iliac arteries connect to the *umbilical arteries* carrying O_2-poor blood to the placenta. The *umbilical vein* carries blood rich in nutrients and O_2 from the placenta to the fetus. The umbilical vein enters the liver and then joins the *ductus venosus* (venous duct). This merges with the inferior vena cava, a vessel that returns blood to the right atrium. This mixed blood enters the heart and most of it is shunted to the left atrium through the *foramen ovale* (oval opening). The left ventricle pumps this blood into the aorta. O_2-poor blood that enters the right atrium is pumped into the pulmonary trunk. It then joins the aorta by way of the *ductus arteriosus* (arterial duct). Therefore, most blood entering the right atrium by-passes the lungs.

Various circulatory changes occur at birth due to the tying of the cord and the expansion of the lungs. These include:

1. Inflation of the lungs. This reduces the resistance to blood flow through the lungs. This allows an increased amount of blood flow from the right atrium to the right ventricle and into the pulmonary arteries. Now gas exchange occurs in the lungs, not at the placenta. Oxygen-rich blood returns to the left side of the heart through the pulmonary veins.

2. An increase in blood flow from the pulmonary veins to the left atrium. This increases the pressure in the left atrium causing a flap to cover the foramen ovale. Even if this mechanism fails, passage of blood from the right atrium to the left atrium rarely occurs because either the opening is small or it closes when the atria contract.

Figure 17.8 What does a five- to seven-month old fetus look like? Wrinkled skin is covered by fine hair.

3. The ductus arteriosus closes at birth because endothelial cells divide and block off the duct.

4. Remains of the ductus arteriosus and parts of the umbilical arteries and vein later are transformed into connective tissue.

Events of Fetal Development

Fetal development includes the third through the ninth months of development. At this time, the fetus is recognizably human (Fig. 17.8), but many refinements still need to be added. The fetus usually increases in size and gains the weight that will be needed to allow it to live as an independent individual.

Third and Fourth Months

At the beginning of the third month, the fetal head is still very large relative to the rest of the body. The nose is flat, the eyes are far apart, and the ears are well formed. Head growth now begins to slow down as the rest of the body increases in length. Fingernails, nipples, eyelashes, eyebrows, and hair on the head appear.

Cartilage begins to be replaced by bone as ossification centers appear in most of the bones. Cartilage remains at the ends of the long bones, and ossification is not complete until age 18 or 20 years. The skull has six large membranous areas called **fontanels.** These permit a certain amount of flexibility as the head passes through the birth canal and allow rapid growth of the brain during infancy. Progressive fusion of the skull bones causes the fontanels to close, usually by 2 years of age.

Human Cloning: Should It Be Done?

In March 1997, Scottish investigators announced they had cloned a sheep called Dolly, and their procedure is now routinely used (Fig. 17A). A donor 2n nucleus is substituted for the n nucleus of an egg. A stimulus is applied that triggers cell division, and the resulting embryo is implanted into a surrogate mother where it develops to term. Using the procedure developed by the Scottish researchers, it is now common practice to clone all sorts of farm animals (horses, cows, sheep, goats, pigs) and also cats and monkeys.

Success of Cloning

Even so, cloning of animals is still in its infancy, and many problems still exist. (1) The vast majority of pregnancies involving clones are not successful. To clone Dolly the sheep, it took 247 tries before one was successful. In many cases, the clone grows abnormally large, and the uterus enlarges with fluid to the point it can rip apart. Almost all clone pregnancies spontaneously abort. (2) Of the small number of animal clones born, most have severe abnormalities: malfunctioning livers, abnormal blood vessels and heart problems, underdeveloped lungs, diabetes, and immune system deficiencies are seen in newborns. Several cow clones had head deformities—none survived very long. (3) Even if the newborn clone appears healthy, it usually soon develops diseases seen in older animals. Dolly was euthanized in 2003 because she was suffering from lung cancer and crippling arthritis. She had lived only half the normal life span for a Dorset sheep.

Reproductive Cloning Versus Therapeutic Cloning

Animal cloning is a form of reproductive cloning. In **reproductive cloning,** the desired end is to create an individual. No reputable scientist or scientific institution is attempting to clone a living human being. Even if such a feat were possible, a clone would not be identical to the person being cloned. Recall that mitochondria have genes, and these genes would be contributed by the egg even if the egg nucleus is removed. The clone could be identical to its genetic parent only if the donor nucleus and the donor egg came from the same woman. Further, the clone would be subject to different environmental factors and a different upbringing from his/her genetic parent.

In **therapeutic cloning,** the desired end is *not* an individual. Rather, it is the embryonic stem cells that could possibly be "coaxed" into becoming other types of cells. So far, therapeutic cloning of humans is experimental and has not been perfected. (Claims of success by a Korean researcher in autumn 2005 were later proven to be a hoax.)

However, many teams are attempting to clone human embryos to recover stem cells. The goal is two-fold. Stem cell research will undoubtedly yield useful information about how cells develop and become specialized. However, the ultimate goal of therapeutic cloning is to provide cells and tissues that could treat human illnesses: insulin-secreting cells for diabetics, nerve cells for stroke patients or those with Parkinson's disease, cardiac cells for those with heart disease, and so forth. Yet, ethical concerns about therapeutic cloning remain—after all, any pre-embryo is potentially a living, breathing human being!

Anticipating intense interest in therapeutic cloning, the U.S. National Academy of Science proposed strict new guidelines for federally funded research in 2005. As a result of the Academy's recommendations, Embryonic Stem Cell Research Oversight (or ESCRO) committees must approve embryonic stem cell research before it is begun. Thus, this type of research is subject to two reviews: an ESCRO committee analysis in addition to reviews already required by an Institutional Review Board. Why scrutinize research so carefully? You don't need to look any farther than the Tuskegee scandal (see page 15) to understand.

ESCRO committees consist of bioethicists and legal experts, as well as members of the general public. All research requires informed consent from the donors of ova or sperm prior to beginning the research. Stem cells created for therapeutic cloning can never be used for reproductive purposes under the proposed guidelines. The guidelines also require that the embryos created cannot be grown in culture for longer than 14 days. At that point, the primitive streak of the developing nervous system begins to form.

Decide Your Opinion

1. Do you approve of the restrictions currently in place for therapeutic cloning? If not, how would you see these restrictions changed?
2. Scientific researchers have made great strides in stem cell research, successfully converting both adult cells and umbilical cord cells back to stem cell forms. In light of these successes, should therapeutic cloning be funded at all?
3. A commercial company, now out of business, once offered the opportunity to clone people's cats—for around $50,000. Should the cloning of pets be allowed when there are so many unwanted pets in the United States?
4. Should scientists be allowed to bring extinct species back to life, similar to *Jurassic Park,* using frozen tissue samples and reproductive cloning?

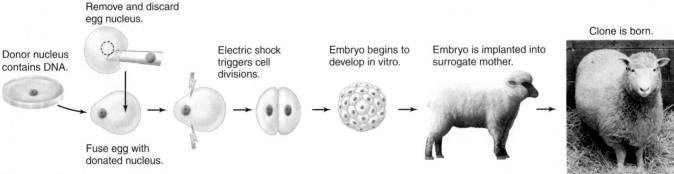

Donor nucleus contains DNA.

Remove and discard egg nucleus.

Fuse egg with donated nucleus.

Electric shock triggers cell divisions.

Embryo begins to develop in vitro.

Embryo is implanted into surrogate mother.

Clone is born.

Figure 17A   How is reproductive cloning carried out?

Have You Ever Wondered...

Can a baby hear in the womb?

A baby's ears begin to develop around eight weeks, but are not fully formed. At 18 weeks, the bones of the skull and middle ear, as well as their nerve connections to the brain, have developed so that a baby can hear. At this time, sounds such as the mother's heartbeat are heard. At 25 weeks the baby listens to voices. The mother's voice, in particular, is recognized. The baby's heart rate will slow slightly—a sign that her voice comforts her baby.

For the rest of pregnancy, the baby's sense of hearing transmits information from the outside world. Soft music will lull him to sleep. Loud sounds will startle the baby and wake him. However, all sounds reaching the baby in the womb are muffled. The baby's ears are filled with amniotic fluid, and the outer ear is covered with the waxy coating that protects the baby's skin.

Sometime during the third month, it is possible to distinguish males from females. As discussed on page 407, the presence of an *SRY* gene, usually on the Y chromosome, leads to the development of testes and male genitals. Otherwise, ovaries and females genitals develop. At this time, either testes or ovaries are located within the abdominal cavity. Later, in the last trimester of fetal development, the testes descend into the scrotal sacs (scrotum). Sometimes the testes fail to descend. In that case, an operation may be done later to place them in their proper location.

During the fourth month, the fetal heartbeat is loud enough to be heard when a physician applies a stethoscope to the mother's abdomen. By the end of this month, the fetus is about 152 mm (6 in.) in length and weighs about 171 g (6 oz).

Fifth Through Seventh Months

During the fifth through seventh months (Fig. 17.8), the mother begins to feel movement. At first, there is only a fluttering sensation, but as the fetal legs grow and develop, kicks and jabs are felt. The fetus, though, is in the fetal position, with the head bent down and in contact with the flexed knees.

CASE STUDY THE ULTRASOUND

"Amber?" Amber nodded and started to stand. The technician walked over. "Hi! My name is Joan. I'm going to be doing your ultrasound today. Please come with me."

Amber asked, "Can my husband come with me?"

Joan replied, "Of course. Follow me." Amber and Kent followed her to the ultrasound room. Amber was now 18 weeks pregnant and Dr. Davis had ordered an ultrasound. It was routine for him to check on the baby's growth and determine whether the pregnancy was proceeding normally. Amber changed into the gown Joan gave her, and Joan helped her onto the table.

Amber groaned. "My bladder is so full. I think it might burst! Why did they make me drink so much water? If you push on my belly I'm going to have a problem."

Joan said, "We need to have your bladder full because then it will descend out of the way so we can see the baby.

"Besides," Joan continued, "the liquid transmits sound waves well. Ultrasounds bounce high frequency sound waves off the baby, and we see patterns of dark and light on the screen. It's a little uncomfortable, I know, but it won't last long. Try to relax."

The ultrasound screen was turned away from Amber and Kent so they couldn't see anything. They studied Joan's face but there was no expression on it. Both Amber and Kent felt apprehensive. Was everything okay? Joan pressed buttons on the computer keyboard and moved the probe around Amber's abdomen. About 15 minutes later, Joan put the probe away and said, "Just rest for a minute. I'll be right back."

In a few minutes, Joan was back. "Why don't you empty your bladder? Then, I'll show you some pictures." Joan helped Amber off the table.

When Amber returned from the restroom, a second person had joined Joan. "Hello, Amber. I'm Dr. Bratton. I'll be reading the ultrasound and sending the report to Dr. Davis. Everything looks fine."

Amber was finally able to relax. Dr. Bratton turned the screen around and began showing ultrasound images to Amber and Kent. "Here you can see the head, from the top. And now from the side. Joan took measurements so that we could determine the baby's approximate gestational age. Your baby is the right size for its age," she reassured.

Kent looked at the blobs of light and dark on the screen. It took him awhile to figure out what was what, but suddenly he saw the profile of a nose! He looked at Amber. She had already figured it out and was grinning. Dr. Bratton showed them the arms and hands.

Amber gasped. "You can see the fingers!"

Dr. Bratton laughed. "Yes, and there are five." Then she asked, "Do you want to know if it's a boy or girl?" Amber and Kent shook their heads.

"We'd like to be surprised," the couple replied together, and then both laughed.

"Okay. We'll skip that part." Dr. Bratton continued to show them details of the ultrasound and printed out a picture of the baby. "You have a healthy baby."

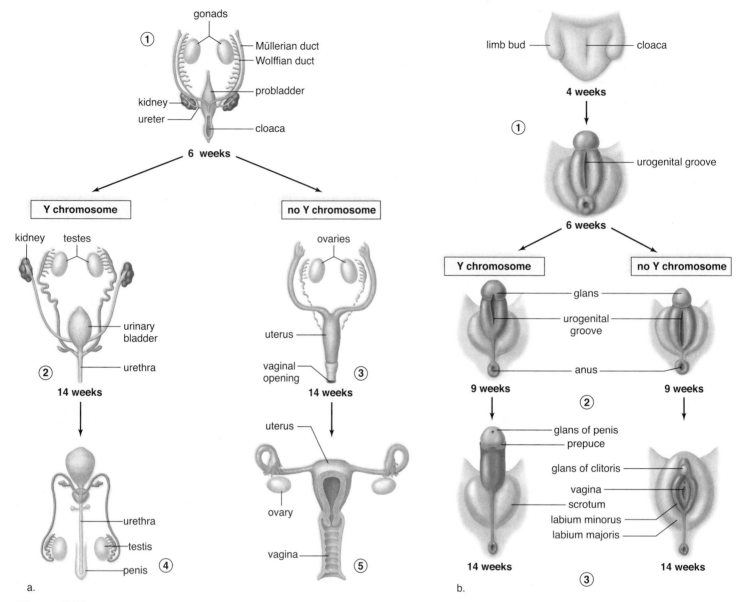

Figure 17.9 How do the male and female organs develop?
a. Development of gonads and ducts. **b.** Development of external genitals.

The wrinkled, translucent skin is covered by a fine down called **lanugo.** This, in turn, is coated with a white, greasy, cheeselike substance called **vernix caseosa,** which probably protects the delicate skin from the amniotic fluid. The eyelids are now fully open.

At the end of this period, the fetus's length has increased to about 300 mm (12 in.), and it weighs about 1,380 g (3 lb). It is possible that, if born now, the baby will survive.

Eighth Through Ninth Months

At the end of nine months, the fetus is about 530 mm (20½ in.) long and weighs about 3,400 g (7½ lb). Weight gain is due largely to an accumulation of fat beneath the skin. Full-term babies have the best chance of survival. Premature babies are subject to various challenges, such as respiratory distress syndrome because their lungs are underdeveloped (see page 191), jaundice (see page 167), and infections.

As the end of development approaches, the fetus usually rotates so that the head is pointed toward the cervix. However, if the fetus does not turn, a **breech birth** (rump first) is likely. It is very difficult for the cervix to expand enough to accommodate this form of birth, and asphyxiation of the baby is more likely to occur. Thus, a **cesarean section** (incision through the abdominal and uterine walls) may be prescribed for delivery of the fetus.

Development of Male and Female Genitals

The sex of an individual is determined at the moment of fertilization. Males have a pair of chromosomes designated as X and Y, and females have two X chromosomes.

Normal Development of the Genitals

Development of the internal and external genitals is shown in Figure 17.9.

Internal Genitals During the first several weeks of development, it is impossible to tell by external inspection whether the unborn child is a boy or a girl. Gonads don't start developing until the seventh week of development. The tissue that gives rise to the gonads is called *indifferent* because it can become testes or ovaries, depending on the action of hormones.

In Figure 17.9a, notice that ① at six weeks, both males and females have the same types of ducts. During this indifferent stage, an embryo has the potential to develop into a male or a female. If a gene called *SRY* is present, testes develop and **testosterone** produced by the testes stimulate the Wolffian ducts to become male genital ducts. ② The Wolffian ducts enter the urethra, which belongs to both the urinary and reproductive systems in males. An anti-Müllerian hormone causes the Müllerian ducts to regress. In the absence of an *SRY* gene, ovaries develop instead of testes from the same indifferent tissue. ③ Now the Wolffian ducts regress, and under the influence of estrogen from the ovaries, the Müllerian ducts develop into the uterus and oviducts. Estrogen has no effect on the Wolffian duct, which degenerates in females. A developing vagina also extends from the uterus. There is no connection between the urinary and genital systems in females.

At 14 weeks, both the primitive testes and ovaries are located deep inside the abdominal cavity. An inspection of the interior of the testes would show that sperm are even now starting to develop, and similarly, the ovaries already contain large numbers of tiny follicles, each having an ovum. ④ Toward the end of development, the testes descend into the scrotal sac; ⑤ the ovaries remain in the abdominal cavity.

External Genitals Figure 17.9b shows the development of the external genitals. These tissues are also indifferent at first—they can develop into either male or female genitals. ① At six weeks, a small bud appears between the legs; this can develop into the male penis or the female clitoris. ② At nine weeks, a urogenital groove bordered by two swellings appears. ③ By 14 weeks, this groove has disappeared in males, and the scrotum has formed from the original swellings. In females, the groove persists and becomes the vaginal opening. Labia majora and labia minora are present instead of a scrotum. These changes are due to the presence or absence of another hormone produced by the testes. It is called dihydrotestosterone (DHT), derived from testosterone.

Abnormal Development of Genitals

It's not correct to say that all XY individuals develop into males. Some XY individuals become females (XY female syndrome). Similarly, some XX individuals develop into males (XX male syndrome). In individuals with the XY female syndrome, a piece of the Y chromosome is missing. In individuals with the XX male syndrome, this same small piece is present on an X chromosome. The piece of a Y chromosome that causes male genitals to develop is called the *SRY* (sex determining region of the Y) gene. The *SRY* gene causes testes to form, and then the testes secrete these hormones: (1) Testosterone stimulates

Figure 17.10
How does androgen insensitivity affect sexual development? This individual has a female appearance but the XY chromosomes of a male. She developed as a female because her plasma membrane receptors for testosterone are ineffective. Underdeveloped testes are in the abdominal cavity, instead of a uterus and ovaries.

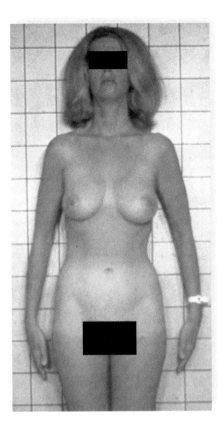

development of the epididymides, vasa deferentia, seminal vesicles, and ejaculatory duct. (2) Anti-Müllerian hormone prevents further development of female structures and instead causes them to degenerate. (3) Dihydrotestosterone, derived from testosterone, directs the development of the urethra, prostate gland, penis, and scrotum.

Ambiguous Sex Determination The absence of any one or more of these hormones results in ambiguous sex determination. The individual has the external appearance of a female, although the gonads of a female are absent.

In *androgen insensitivity syndrome*, these three types of hormones are produced by testes during development, but the individual develops as a female because the plasma membrane receptors for testosterone are ineffective (Fig. 17.10). The external genitalia develop as female, and the Wolffian duct degenerates internally. The individual does not develop a scrotum, so the testes fail to descend and instead remain deep within the body. The individual develops the secondary sex characteristics of a female, and no abnormality is suspected until the individual fails to menstruate.

> **Check Your Progress 17.3**
> 1. a. Trace the path of blood in the fetus starting with the placenta; (b) name the structures unique to fetal circulation.
> 2. What are the major events during fetal development?
> 3. How does the development of the genitals differ in males and females?

17.4 Pregnancy and Birth

Major changes that take place in the mother's body during pregnancy are due to placental hormones.

The Energy Level Fluctuates

When first pregnant, the mother may experience nausea and vomiting, loss of appetite, and fatigue. These symptoms subside, and some mothers report increased energy levels and a general sense of well-being despite an increase in weight. During pregnancy, the mother gains weight due to breast and uterine enlargement; weight of the fetus; amount of amniotic fluid; size of the placenta; her own increase in total body fluid; and an increase in storage of proteins, fats, and minerals. The increased weight can lead to lordosis (swayback) and lower back pain.

The Uterus Relaxes

Aside from an increase in weight, many of the physiological changes in the mother are due to the presence of the placental hormones that support fetal development (Table 17.2). Progesterone decreases uterine motility by relaxing smooth muscle, including the smooth muscle in the walls of arteries. The arteries expand, and this leads to a low blood pressure that sets in motion the renin-angiotensin-aldosterone mechanism, promoted by estrogen. Aldosterone activity promotes sodium and water retention, and blood volume increases until it reaches its peak sometime during weeks 28–32 of pregnancy. Altogether, blood volume increases from 5 L to 7 L—a 40% rise. An increase in the number of red blood cells follows. With the rise in blood volume, cardiac output increases by 20–30%. Blood flow to the kidneys, placenta, skin, and breasts rises significantly. Smooth muscle relaxation also explains the common gastrointestinal effects of pregnancy. The heartburn experienced by many is due to relaxation of the esophageal sphincter and reflux of stomach contents into the esophagus. Constipation is caused by a decrease in intestinal tract motility.

The Pulmonary Values Increase

Of interest is the increase in pulmonary values in a pregnant woman. The bronchial tubes relax, but this alone cannot explain the typical 40% increase in vital capacity and tidal volume. The increasing size of the uterus from a nonpregnant weight of 60–80 g to 900–1,200 g contributes to an improvement in respiratory functions. The uterus comes to occupy most of the abdominal cavity, reaching nearly to the xiphoid process of the sternum. This increase in size not only pushes the intestines, liver, stomach, and diaphragm superiorly, but it also widens the thoracic cavity. Compared with nonpregnant values, the maternal oxygen level changes little. Blood carbon dioxide levels fall by 20%, creating a concentration gradient favorable to

Table 17.2	Effects of Placental Hormones on Mother
Hormone	**Chief Effects**
Progesterone	Relaxation of smooth muscle; reduced uterine motility; reduced maternal immune response to fetus
Estrogen	Increased uterine blood flow; increased renin-angiotensin-aldosterone activity; increased protein biosynthesis by the liver
Peptide hormones	Increased insulin resistance

Source: Moore, Thomas R., *Gestation Encyclopedia of Human Biology,* Vol. 7, 7th edition. Copyright © 1997 Academic Press.

the flow of carbon dioxide from fetal blood to maternal blood at the placenta.

Still Other Effects

The enlargement of the uterus does result in some problems. In the pelvic cavity, compression of the ureters and urinary bladder can result in stress incontinence. Compression of the inferior vena cava, especially when lying down, decreases venous return, and the result is edema and varicose veins.

Aside from the steroid hormones progesterone and estrogen, the placenta also produces some peptide hormones. One of these makes cells resistant to insulin, and the result can be pregnancy-induced diabetes. Some of the integumentary changes observed during pregnancy are also due to placental hormones. **Striae gravidarum,** commonly called "stretch marks," typically form over the abdomen and lower breasts in response to increased steroid hormone levels rather than stretching of the skin. Melanocyte activity also increases during pregnancy. Darkening of certain areas of the skin, including the face, neck, and breast areolae, is common.

Birth

The uterus has contractions throughout pregnancy. At first, these are light, lasting about 20–30 seconds and occurring every 15–20 minutes. Near the end of pregnancy, the contractions may become stronger and more frequent so that a woman thinks she is in labor. "False-labor" contractions are called **Braxton Hicks contractions.** However, the onset of true labor is marked by uterine contractions that occur regularly every 15–20 minutes and last for 40 seconds or longer.

A positive feedback mechanism can explain the onset and continuation of labor. Uterine contractions are induced by a stretching of the cervix, which also brings about the release of oxytocin from the posterior pituitary gland. Oxytocin stimulates the uterine muscles, both directly and through the action of prostaglandins. Uterine contractions push the fetus downward, and the cervix stretches even more. This cycle keeps repeating itself until birth occurs.

Prior to or at the first stage of **parturition,** the process of giving birth to an offspring, there can be a "bloody show" caused by expulsion of a mucous plug from the cervical canal. This plug prevented bacteria and sperm from entering the uterus during pregnancy.

Stage 1

During the first stage of labor, the uterine contractions of labor occur in such a way that the cervical canal slowly disappears as the lower part of the uterus is pulled upward toward the baby's head. This process is called **effacement,** or "taking up the cervix." With further contractions, the baby's head acts as a wedge to assist cervical dilation (Fig. 17.11a). If the amniotic membrane has not already ruptured, it is apt to do so during this stage, releasing the amniotic fluid, which leaks out of the vagina (an event sometimes referred to as "breaking water"). The first stage of parturition ends once the cervix is dilated completely.

Stage 2

During the second stage of parturition, the uterine contractions occur every 1–2 minutes and last about 1 minute each. They are accompanied by a desire to push, or bear down. As the baby's head gradually descends into the vagina, the desire to push becomes greater. When the baby's head reaches the exterior, it turns so that the back of the head is uppermost (Fig. 17.11b). To enlarge the vaginal orifice, an **episiotomy** is often performed. This incision, which enlarges the opening, is sewn together later. As soon as the head is

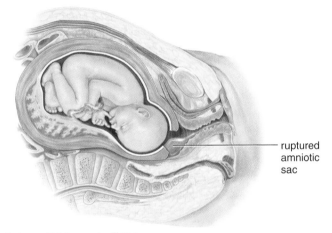

a. First stage of birth: cervix dilates

ruptured amniotic sac

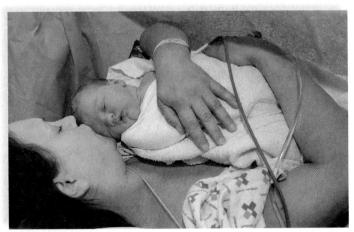

c. Baby has arrived

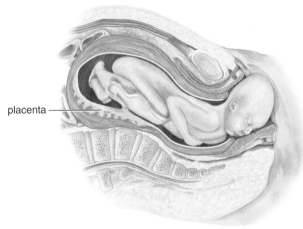

placenta

b. Second stage of birth: baby emerges

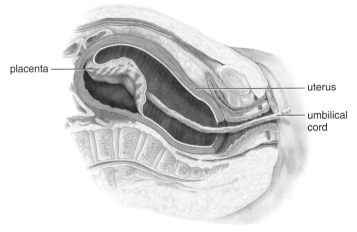

placenta

uterus

umbilical cord

d. Third stage of birth: expelling afterbirth

Figure 17.11 **What are the stages of birth?**
Birth proceeds in three stages. The baby emerges in stage two.

delivered, the physician may hold the head and guide it downward, while one shoulder and then the other emerges. The rest of the baby follows easily (Fig. 17.11c).

Once the baby is breathing normally, the umbilical cord is cut and tied, severing the child from the placenta. The stump of the cord shrivels and leaves a scar, the umbilicus (belly button).

Stage 3

The placenta, or **afterbirth,** is delivered during the third stage of parturition (Fig. 17.11d). About 15 minutes after delivery of the baby, uterine muscular contractions shrink the uterus and dislodge the placenta. The placenta then is expelled into the vagina. As soon as the placenta and its membranes are delivered, the third stage of parturition is complete.

> ### Check Your Progress 17.4
> 1. **What chemical factors are responsible for the many physiological changes in a pregnant woman?**
> 2. **Maternal blood carbon dioxide levels fall by 20% during pregnancy. How does this benefit the fetus?**
> 3. **Describe the three stages of labor.**

17.5 Development After Birth

Development does not cease once birth has occurred, but continues throughout the stages of life: infancy, childhood, adolescence, and adulthood. Infancy, the toddler years, and preschool years are times of remarkable growth. During the birth to 5-year-old stage, humans acquire gross motor and fine motor skills. These include the ability to sit up and then to walk, as well as being able to hold a spoon and manipulate small objects. Language usage begins during this time and will become increasingly sophisticated throughout childhood. As infants and toddlers explore their environment, their senses—vision, taste, hearing, smell, and touch—mature dramatically. Socialization is very important, as a child forms emotional ties with its caregivers and learns to separate self from others. Babies do not all develop at the same rate, and there is a large variation in what is considered normal.

The preadolescent years, from 6 to 12 years of age, are a time of continued rapid growth and learning. Preadolescents form identities apart from parents, and peer approval becomes very important. Adolescence begins with the onset of puberty, as the young person achieves sexual maturity. For girls, puberty begins between 10 and 14 years of age, whereas for boys it generally occurs between ages 12 and 16. During this time, the sex-specific hormones (see page 345) cause the secondary sexual characteristics to appear.

CASE STUDY   THE DAY ARRIVES

Amber was miserable. "I am two days past my due date. Why do they give you a due date if they don't inform the baby? I feel like a cow." She was propped up on pillows in bed. "Kent, can you please help me up? I'm sorry but I have to go to the bathroom again."

He helped her up. Then he knocked on her swollen abdomen as though it was a door. "Hello in there! Hey baby! It's time to come out." Kent grinned at her. Amber tried to grin back, but she just felt uncomfortable.

At 2 A.M., Amber awoke when she felt a gush of liquid. "Kent!" She reached over and grabbed his arm. He was instantly awake. "I think my membranes just ruptured."

"Okay, now what do we do?" Kent asked.

"I'm already packed. I think I'll take a shower and wait for the contractions. When they're regular and about ten minutes apart, we'll go to the hospital." While Amber showered, Kent paced around the house. Then both waited and timed contractions. Amber walked slowly around the house, with Kent at her arm to support her.

Finally Kent said, "I can't stand this. It's 6 A.M. I think we should go to the hospital, regular contractions or not.

"Who says this kid is following the book?" he fretted. "I don't want to deliver this baby. I'd feel much better if a nurse looked at you. The worst they can do is send us home." Amber didn't argue with him.

But the examining nurse didn't send them home. Amber was admitted to the birthing unit. By 9 A.M., her contractions were strong, painful, and regular. Dr. Davis entered the room dressed in surgical scrubs, and the couple was pleased to see him.

"Do you feel an urge to push, Amber?" Dr. Davis asked. She nodded. He instructed her to push strongly during each contraction and to rest in between. Giving birth seemed to happen so quickly—20 minutes of intense hard work—and Jared Michael was born; 7 pounds, 8 ounces with a head full of black hair. As soon as he was breathing normally, Dr. Davis clamped the umbilical cord and handed scissors to Kent. "Want to cut the cord, Dad?"

The nurse handed her son to Amber, and Amber started to cry. Kent carefully hugged his wife and their new son. "Congratulations, Amber!" he grinned. "We're parents!"

 Historical **Focus**

An End to "Laudable Pus": Ignaz Semmelweis (1818–1865)

Imagine that it's the nineteenth century. Now imagine that you or your loved one is expecting a baby. You're a little down on your luck, and you can't seem to scrape together enough money for private childbirth at home. Instead, you'd be forced to go to a hospital, and you're terrified. You would fear for your life.

This scenario occurred all across Europe. In 1847, almost 20% of women whose babies were delivered by a doctor in a hospital died of so-called "childbed fever." A day or so after delivery, the woman began to show signs of infection: high fever, low blood pressure, and discharge of pus from her reproductive tract. Physicians of the time accepted these almost-routine infections as part of the delivery process. Even the discharge was thought to be beneficial and an indication of healing—and thus, referred to as "laudable (praiseworthy) pus."

By contrast, a woman whose delivery was performed at home by a midwife had a very low incidence of childbed fever, and the Hungarian physician Ignaz Semmelweis wondered why. At the time there was no concept of germs or infection, but Semmelweis began to suspect that the doctors were killing their patients. The increasing urbanization of society had resulted in more and more women going to hospitals built to care for the poor. Urban hospitals were training grounds for physicians, and part of their education involved performing autopsies. Semmelweis began to understand the source of childbed fever when one of his mentors died, following an accidental stab wound inflicted during an autopsy. When Semmelweis autopsied his mentor, the body looked like those of women with childbed fever.

Semmelweis proposed that something contagious was being carried from the autopsies to the delivery wards.

He placed basins of chlorinated lime at the entrance to each ward and required that all who entered wash their hands. He also instituted the use of nailbrushes for cleaning the fingernails. The incidence of childbed fever dropped from 18% to 1.3% during the first seven months after the hand washing protocol was instituted and continued to drop thereafter.

Unfortunately, despite his dramatic success, Semmelweis was neither liked nor respected by his colleagues. Suggesting that dirty, careless doctors were to blame for the mothers' deaths earned only hostility. Presentations and publications of his work were largely ignored. In 1865, he began to deteriorate mentally and was committed to a mental institution. He died two weeks later at the age of 47. Ironically, his own death was most likely from an infection similar to "childbed fever."

Profound social and psychological changes are also associated with the transition from childhood to adulthood.

Aging encompasses the progressive changes from infancy until eventual death. Today, **gerontology,** the study of aging, is of great interest because there are now more older individuals in our society than ever before, and the number is expected to rise dramatically. In the next half-century, the number of people over age 65 will increase 147%. The human life span is judged to be a maximum of 120–125 years. The present goal of gerontology is not necessarily to increase the life span, but to increase the health span, the number of years that an individual enjoys the full functions of all body parts and processes (Fig. 17.12).

Hypotheses of Aging

Of the many hypotheses about the cause of aging, three are considered here.

Genetic in Origin

Several lines of research indicate that aging has a genetic basis. Researchers working with simple organisms, such as

Figure 17.12   What is aging?
Aging is a slow process during which the body undergoes changes that eventually bring about death, even if no marked disease or disorder is present. Medical science is trying to extend the human life span and the health span, the length of time the body functions normally.

yeast and roundworms, have identified a host of genes whose expression decreases the life span. If these genes are silenced through mutations or restricted food intake, the organism lives longer. What do these genes have in common? Apparently, when these genes are inactive, mitochondria do not produce energy—the cell uses alternative pathways. The current *mitochondrial hypothesis of aging* has been supported by engineering mice that have a defective DNA polymerase. (Recall that mitochondria have their own DNA.) These mice aged much faster than their peers. Why? Possibly because their defective mitochondria produced more free radicals than usual. Free radicals (see page 177) are unstable molecules that carry an extra electron. To become stable, free radicals donate an electron to another molecule, such as DNA or proteins (e.g., enzymes) or lipids, found in plasma membranes. Eventually, these molecules are unable to function, and the cell is destroyed. The well-known observation that a low-calorie diet can expand the life span is consistent with the mitochondrial hypothesis of aging. Caloric restriction also shuts down the genes that decrease the life span—the genes that turn on the activity of mitochondria!

Whole-Body Process

A decline in the hormonal system can affect many different organs of the body. For example, diabetes type 2 is common in older individuals. The pancreas makes insulin, but the cell's receptors are ineffective. Menopause in women and andropause in men occurs for similar reasons. The bloodstream contains adequate amounts of anterior pituitary hormones, but the ovaries and testes do not respond. The ovaries do not produce adequate amounts of estrogen and progesterone. Likewise, testosterone secretion diminishes. Perhaps aging results from the loss of hormonal activities and a decline in the functions they control.

The immune system, too, no longer performs as it once did, and this can affect the body as a whole. The thymus gland gradually decreases in size, and eventually most of it is replaced by fat and connective tissue. The incidence of cancer increases among the elderly, which may signify that the immune system is no longer functioning as it should. This idea is also substantiated by the increased incidence of autoimmune diseases in older individuals.

It is possible, though, that aging is not due to the failure of a particular system that can affect the body as a whole, but to a specific type of tissue change that affects all organs. It has been noticed for some time that proteins such as the collagen fibers present in many support tissues become increasingly cross-linked as people age. Undoubtedly, this cross-linking contributes to the stiffening and loss of elasticity characteristic of aging tendons and ligaments. It may also account for the inability of such organs as the blood vessels, the heart, and the lungs to function as they once did.

Extrinsic Factors

The current data about the effects of aging are often based on comparisons of the elderly to younger age groups. When today's elderly were young people, they may not have been aware of the benefits of such things as diet and exercise to general health. It is possible, then, that much of what we attribute to aging is instead due to years of poor health habits.

Consider osteoporosis. This condition is associated with a progressive decline in bone density in both males and females, so fractures are more likely to occur after only minimal trauma. Osteoporosis is common in the elderly. By age 65, one-third of women will have vertebral fractures. By age 81, one-third of women and one-sixth of men will have suffered a hip fracture. While there is no denying that a decline in bone mass occurs as a result of aging, certain extrinsic factors are also important. The occurrence of osteoporosis is associated with cigarette smoking, heavy alcohol intake, and inadequate calcium intake. Not only is it possible to eliminate these negative factors by personal choice, but it is also possible to add a positive factor. A moderate exercise program has been found to slow down the progressive loss of bone mass.

Even more important, a sensible exercise program and a proper diet that includes at least five servings of fruits and vegetables a day will most likely help eliminate cardiovascular disease. Experts no longer believe that the cardiovascular system necessarily suffers a large decrease in functioning ability with age. Persons 65 years of age and older can have well-functioning hearts and open coronary arteries if their health habits are good and they continue to exercise regularly.

Effect of Age on Body Systems

Data about how aging affects body systems are necessarily based on past events. It is possible that, in the future, age will not have these effects or at least not to the same degree as those described here.

Skin

As aging occurs, skin becomes thinner and less elastic because the number of elastic fibers decreases. The collagen fibers undergo cross-linking, as discussed previously. Also, there is less adipose tissue in the subcutaneous layer. Therefore, older people are more likely to feel cold. The loss of thickness partially accounts for sagging and wrinkling of the skin.

Homeostatic adjustment to heat is also limited because there are fewer sweat glands for sweating to occur. The hair on the scalp and the extremities thins out because of fewer hair follicles. The number of oil (sebaceous) glands is reduced, and the skin tends to crack. Older people also experience a decrease in the number of melanocytes, making their hair gray and their skin pale. In contrast, some of the remaining pigment cells are larger, and pigmented blotches appear on the skin.

Processing and Transporting

Cardiovascular disorders are the leading cause of death today. The heart shrinks because of a reduction in cardiac muscle cell size. This leads to loss of cardiac muscle strength and reduced cardiac output. Still, the heart, in the absence of disease, is able to meet the demands of increased activity. It can double its rate or triple the amount of blood pumped each minute even though the maximum possible output declines.

The middle layer of arteries contains elastic fibers, most likely subject to cross-linking, so the arteries become more rigid with time. Their size is further reduced by plaque, a buildup of fatty material. Therefore, blood pressure readings gradually rise. Such changes are common in individuals living in Western industrialized countries but not in agricultural societies. A diet low in cholesterol and saturated fatty acids has been suggested as a way to control degenerative changes in the cardiovascular system.

Blood flow to the liver is reduced, and this organ does not metabolize drugs as efficiently as before. This means that, as a person gets older, less medication is needed to maintain the same level of a drug in the bloodstream.

Cardiovascular problems are often accompanied by respiratory disorders, and vice versa. Growing inelasticity of lung tissue means that ventilation is reduced. We rarely use the entire vital capacity, so these effects are not noticed unless the demand for oxygen increases.

Blood supply to the kidneys is also reduced. The kidneys become smaller and less efficient at filtering wastes. Salt and water balance is difficult to maintain, and the elderly dehydrate faster than young people do. Difficulties involving urination include incontinence (lack of bladder control) and the inability to urinate. In men, the prostate gland may enlarge and reduce the diameter of the urethra. This can make urination so difficult that surgery is often needed.

Integration and Coordination

While most tissues of the body regularly replace their cells, some at a faster rate than others, the brain and the muscles ordinarily do not. However, contrary to previous opinion, studies show that few neural cells of the cerebral cortex are lost during the normal aging process. This means that cognitive skills remain unchanged even though a loss in short-term memory characteristically occurs. Although the elderly learn more slowly than the young, they can acquire and remember new material. The results of tests indicate that when more time is given for the subject to respond, age differences in learning decrease.

Neurons are extremely sensitive to oxygen deficiency, and if neuron death does occur, it may be due not to aging but to reduced blood flow in narrowed blood vessels. Reaction time, however, does slow, and more stimulation is needed for hearing, taste, and smell receptors to function as before. After age 50, the ability to hear tones at higher frequencies decreases gradually, and this can make it difficult to identify individual voices and to understand conversation in a group. The lens of the eye does not accommodate as well and also may develop a cataract. Glaucoma, the buildup of pressure due to increased fluid, is more likely to develop because of a reduction in the size of the anterior cavity of the eye.

Loss of skeletal muscle mass is not uncommon, but it can be controlled by following a regular exercise program.

The capacity to do heavy labor decreases, but routine physical work should be no problem. Lung function may decrease, due to decreases in both rib cage flexibility and respiratory muscle strength. Still, many healthy seniors will be able to continue everyday activities well into the eighth or ninth decade. Reduced muscularity of the urinary bladder contributes to an inability to empty the bladder completely during urination. Occasional urinary incontinence may result. Risk of urinary tract infection is also increased.

The Reproductive System

As mentioned, females undergo menopause, and thereafter, the level of female sex hormones in the blood falls markedly. The uterus and the cervix decrease in size, and the walls of the oviducts and the vagina become thinner. The external genitals become less pronounced. Males undergo andropause, and the level of androgens falls gradually over the age span of 50–90, but some sperm production continues until death.

As a group, females live longer than males. Males suffer a marked increase in heart disease in their forties. An increase is not noted in females until after menopause, when women lead men in the incidence of stroke. Men are still more likely than women to have a heart attack, however. At one time it was thought that estrogen offers women some protection against cardiovascular disorders. This hypothesis is called into question because postmenopausal administration of estrogen has been shown to increase the risk of cardiovascular disorders in women.

Conclusion

We have listed many adverse effects of aging, but although such effects are seen, they are not inevitable (Fig. 17.13). We must discover any extrinsic factors that precipitate these adverse effects and guard against them. Just as it is wise to make the proper preparations to remain financially independent when older, it is also wise to realize that, biologically, successful old age begins with the health habits developed when we are younger.

Figure 17.13 **What steps can an individual take to increase health span?**
Gerontology research has shown that regular physical exercise, as well as staying engaged both mentally and socially, can slow the progress of aging and lengthen the health span.

Check Your Progress 17.5

1. What are the different hypotheses of aging?
2. What is the effect of aging on the various body systems?
3. What is the best way to keep healthy, even though aging occurs?

Health **Focus**

Degenerative Brain Disorders

In 1900, the average life span in the United States was 47 years of age. Today it is 75 years of age. Normal aging does involve some changes in mental faculties, but many of the changes we associate with old age are related to disease and not aging. Two of the more common diseases are Alzheimer disease (AD) and Parkinson disease.

AD is characterized by the presence of abnormally structured neurons and a reduced amount of acetylcholine (ACh). The AD neuron has two characteristic features. Bundles of fibrous protein, called neurofibrillary tangles, extend from the axon to surround the nucleus of the neuron. Tangles form when the supporting protein called tau becomes malformed (Fig. 17B) and twists the neurofibrils, which are normally straight. In addition, protein-rich accumulations called amyloid plaques envelop branches of the axon. The plaques grow so dense that they trigger an inflammatory reaction that causes neuron death.

Treatment for Alzheimer Disease

Treatment for AD involves using one of two categories of drugs. Cholinesterase inhibitors (Aricept®, Cognex®, Exelon®, Reminyl®) work at neuron synapses in the brain, slowing the activity of acetylcholinesterase, the enzyme that breaks down acetylcholine (ACh). Allowing ACh to accumulate in synapses keeps memory pathways in the brain functional for a longer period. A second drug, memantine (Namenda®) blocks *excito-toxicity:* the tendency of diseased neurons to self-destruct. This recently approved medication is used only in moderately to severely affected patients. Using the drug allows neurons involved in memory pathways to survive longer in affected patients. Successes with these medications indicates that treatment for AD patients should begin as soon as possible after diagnosis and continue indefinitely. However, neither type of medication cures AD—both merely slow the progress of disease symptoms, allowing the patient to function independently for a longer time. Additional research is currently underway to test the effectiveness of anticholesterol *statin* drugs, as well as anti-inflammatory medications, in slowing the progress of the disease.

Much of current research on AD focuses on the prevention and cure of the disease. Scientists believe that curing AD

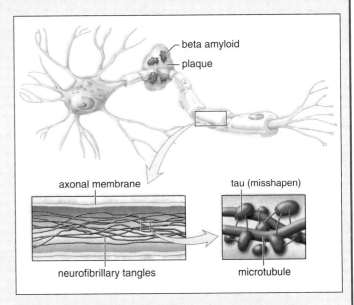

Figure 17B   What causes Alzheimer disease?
Some of the neurons of Alzheimer disease (AD) patients have beta amyloid plaques and neurofibrillary tangles. AD neurons are present throughout the brain but concentrated in the hippocampus and amygdala.

will require an early diagnosis, because it is thought that the disease may begin in the brain 15–20 years before symptoms ever develop. Currently, diagnosis can't be made with absolute certainty until the brain is examined at autopsy. A new test on the cerebrospinal fluid may allow early detection of amyloid proteins and a much earlier diagnosis of the disease. Researchers are also testing vaccines for AD, which would target the patient's immune system to destroy amyloid plaques.

Early findings have shown that risk factors for cardiovascular disease—heart attacks and stroke—also contribute to an increased incidence of AD. Risk factors for cardiovascular disease include elevated blood cholesterol and blood pressure, smoking, obesity, sedentary lifestyle, and diabetes mellitus. Thus, evidence suggests that a lifestyle tailored for good cardiovascular health may also prevent AD.

Summarizing the Concepts

17.1 Fertilization

The acrosome of a sperm releases enzymes that digest a pathway for the sperm through the zona pellucida. The sperm nucleus enters the egg and fuses with the egg nucleus.

17.2 Pre-Embryonic and Embryonic Development

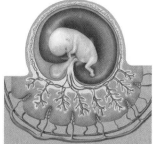

- Cleavage, growth, morphogenesis, and differentiation are the processes of development.
- The extraembryonic membranes (chorion, allantois, yolk sac, and amnion) function in internal development.

17.3 Fetal Development

- At the end of the embryonic period, all organ systems are established, and there is a mature and functioning placenta. The umbilical arteries and umbilical vein take blood to and from the placenta, where exchanges take place.
- Exchanges supply the fetus with oxygen and nutrients and rid the fetus of carbon dioxide and wastes.
- The venous duct joins the umbilical vein to the inferior vena cava.
- The oval duct and arterial duct allow the blood to pass through the heart without going to the lungs. Fetal development extends from the third through the ninth months.
- During the third and fourth months, the skeleton is becoming ossified.
- The sex of the fetus becomes distinguishable. If an *SRY* gene is present, testes and male genitals develop. Otherwise, ovaries and female genitals develop.

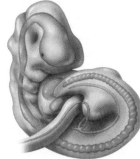

- During the fifth through the ninth months, the fetus continues to grow and to gain weight.

17.4 Pregnancy and Birth

Major changes take place in the mother's body during pregnancy.

- Weight gain occurs as the uterus occupies most of the abdominal cavity.
- Many complaints, such as constipation, heartburn, darkening of certain skin areas, and pregnancy-induced diabetes, are due to the presence of placental hormones.

Birth

- A positive feedback mechanism that involves uterine contractions and oxytocin explains the onset and continuation of labor.
- During stage 1 of parturition (birth), the cervix dilates.

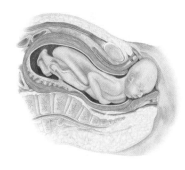

- During stage 2, the child is born.
- During stage 3, the afterbirth is expelled.

17.5 Development After Birth

Development after birth consists of infancy, childhood, adolescence, and adulthood.

- Aging encompasses progressive changes from about age 20 on that contribute to an increased risk of infirmity, disease, and death.

Hypotheses of Aging

- Aging may have a genetic basis.
- Aging may be due to changes that affect the whole body (e.g., decline of hormonal system).
- Aging may be due to extrinsic factors (e.g., diet and exercise).

Effect of Age on Body Systems

- Deterioration of organ systems can possibly be prevented or reduced in part by using good health habits.

Understanding Key Terms

afterbirth 410
aging 411
allantois 396
amnion 396
blastocyst 397
Braxton Hicks contraction 408
breech birth 406
cesarean section 406
chorion 396
chorionic villi 400
cleavage 396
differentiation 396
ectopic pregnancy 398
effacement 409
embryo 398
embryonic development 397
embryonic disk 398
episiotomy 409
extraembryonic membrane 396
fertilization 394
fetal development 403
fontanel 403
gastrulation 399

gerontology 411
growth 396
human chorionic gonadotropin (hCG) 398
implantation 398
inner cell mass 397
lanugo 406
morphogenesis 396
morula 397
parturition 409
placenta 396
pre-embryonic development 397
primary germ layer 399
reproductive cloning 404
striae gravidarum 408
testosterone 407
therapeutic cloning 404
umbilical cord 400
vernix caseosa 406
yolk sac 396
zygote 394

Match the key terms to these definitions.

a. _____ Short, fine hair present during the later portion of fetal development.

b. _____ The placenta delivered during the third stage of parturition.

c. _____ The study of aging.

d. _____ Union of a sperm nucleus and an egg nucleus, which creates a zygote with the diploid number of chromosomes.

e. _____ Mitotic cell division of the zygote with no increase in cell size.

Testing Your Knowledge of the Concepts

1. Describe how polyspermy is prevented during fertilization. (page 394)

2. Name the four embryonic membranes and give a human function for each one. (page 396)

3. Justify the division of development into pre-embryonic, embryonic, and fetal development. (pages 395–407)

4. What are the three primary germ layers, and what body structures come from each germ layer? (page 399)

5. Briefly summarize the weekly events of embryonic development. (pages 398–400)

6. Briefly summarize the monthly events of fetal development. (pages 401–06)

7. Explain how blood circulates to and from the placenta and the fetus. How is blood shunted away from the lungs? (pages 401, 403)

8. List the hormones involved in the development of the male and female internal and external sex organs and state their functions. (pages 406–07)

9. Describe some of the changes that occur in the mother during pregnancy. (page 408)

10. What event marks the end of each stage of birth? (pages 409–10)

11. Discuss three hypotheses concerning aging. How can you prevent the major changes that can occur in the body as we age? (pages 411–13)

12. Only one sperm enters an egg because
    a. sperm have an acrosome.
    b. the corona radiata gets larger.
    c. changes occur in the zona pellucida.
    d. the cytoplasm hardens.
    e. All of these are correct.

13. Which of these statements is correct?
    a. All major organs are formed during embryonic development.
    b. The hands and feet begin as paddlelike structures.
    c. The heart is at first tubular.
    d. The placenta functions until birth occurs.
    e. All of these are correct statements.

14. When all three germ layers are present (ectoderm, endoderm, and mesoderm), what event has occurred?
    a. blastulation
    b. limb formation
    c. gastrulation
    d. morulation

15. Which of these is not a process of development?
    a. cleavage
    b. parturition
    c. growth
    d. morphogenesis
    e. differentiation

16. Which of these is mismatched?
    a. chorion—sense perception
    b. yolk sac—first site of blood cell formation
    c. allantois—umbilical blood vessels
    d. amnion—contains fluid that protects embryo

17. In human development, which part of the blastocyst will develop into a embryo?
    a. trophoblast
    b. inner cell mass
    c. chorion
    d. yolk sac

18. Which primary germ layer is not correctly matched to an organ system or organ that develops from it?
    a. ectoderm—the nervous system
    b. endoderm—lining of the digestive tract
    c. mesoderm—skeletal system
    d. endoderm—cardiovascular system

19. Human chorionic gonadotropin is a
    a. hormone.
    b. basis of pregnancy test.
    c. cause of ectopic pregnancy.
    d. Both a and b are correct.

20. Which is a correct sequence that ends with the stage that implants?
    a. morula, blastocyst, embryonic disk, gastrula
    b. ovulation, fertilization, cleavage, morula, early blastocyst
    c. embryonic disk, gastrula, primitive streak, neurula
    d. primitive streak, neurula, extraembryonic membranes, chorion

21. Differentiation is equivalent to which term?
    a. morphogenesis
    b. growth
    c. specialization
    d. gastrulation

22. Which process refers to the shaping of the embryo and involves cell migration?
    a. cleavage
    b. differentiation
    c. growth
    d. morphogenesis

23. Which association is not correct?
    a. third and fourth months—fetal heart has formed, but it does not beat
    b. fifth through seventh months—mother feels movement
    c. eighth through ninth months—usually head is now pointed toward the cervix
    d. All of these are correct.

24. At three months, the embryo has
    a. become a fetus.
    b. body systems already.
    c. a head, arms, and legs.
    d. ears and eyes, which don't function.
    e. All but b are correct.

25. Which of these structures is not a circulatory feature unique to the fetus?
    a. arterial duct
    b. oval opening
    c. umbilical vein
    d. pulmonary trunk

26. Which of these statements is correct?
    a. Fetal circulation, like adult circulation, takes blood equally to a pulmonary circuit and a systemic circuit.
    b. Fetal circulation shunts blood away from the lungs but makes full use of the systemic circuit.
    c. Fetal circulation includes exchange of substances between fetal blood and maternal blood at the placenta.
    d. Unlike adult circulation, fetal blood always carries O_2-rich blood and therefore has no need for the pulmonary circuit.
    e. Both b and c are correct.

27. Which of these is a hormone involved in development of male and female sex organs?
    a. estrogen
    b. anti-Müllerian hormone
    c. dihydrotestosterone
    d. testosterone
    e. All of these hormones are involved.

28. Which hormone can be administered to begin the process of childbirth?
    a. estrogen
    b. oxytocin
    c. prolactin
    d. testosterone
    e. Both b and d are correct.

29. Label this diagram illustrating the placement of the extraembryonic membranes, and give a function for each membrane in humans.

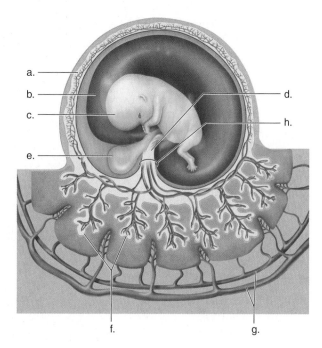

Thinking Critically About the Concepts

Amber and Kent used a home pregnancy test to determine if she was pregnant. These tests detect the level of hCG (human chorionic gonadotropin; see page 398) in the urine. This hormone is released following implantation of the embryo into the uterus, usually around six days after fertilization. Some tests claim that they are sensitive enough to detect hCG on the date that menstruation is expected to begin. However, doctors recommend waiting until menstruation is one week late. If pregnant, a woman's level of hCG rises with each passing day, and testing is more likely to be accurate. However, even with a negative test result, the woman may still be pregnant if hCG levels are too low to be detected at the time of the first test. The test should be repeated later if menstruation doesn't begin. The home pregnancy tests contain a positive control. This is a visual sign (usually a line or a +) that appears if the test is working correctly. If this line does not appear, the test is not valid and must be repeated.

1. At home, pregnancy tests check for the presence of hCG in a female's urine. Where does hCG come from? Why is hCG found in a pregnant woman's urine?

2. A blood test at a doctor's office can also check for the presence of hCG in a female's blood.
   a. Why would you expect to find hCG circulating in a pregnant female's blood?
   b. hCG is a protein, so how does hCG affect its target cells?

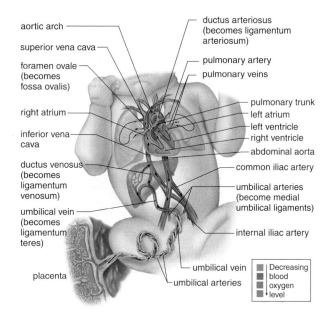

3. a. What recommendations would be made to a female (or male) who hopes to prevent osteoporosis?
   b. Why is it better/easier to prevent osteoporosis than treat it once someone has it?

4. a. What pituitary hormone is checked with a blood test to diagnose menopause?
   b. Will levels of this hormone be increased or decreased if the female is in menopause?
   c. How does the changed (increased or decreased) level of this pituitary hormone cause the onset of menopause (cessation of menses)?

CHAPTER 18

Patterns of Chromosome Inheritance

"That was awful! I didn't study nearly enough."

"Ugh. I think I'm depressed."

"Let's go do something fun. Do you guys want to do something tonight?"

Mechelle Wilson, her roommate Tamika, and several other students were hanging around in the sociology department lounge. It was 4:00 P.M. on Friday, and they had just finished a particularly exhausting essay exam in one of their sociology classes. The dining hall wasn't open for dinner yet, and no one wanted to think about starting on homework. Besides, there was still some venting to do about that exam!

"I don't know," Mechelle answered. "How 'bout a chick-flick at the mall?"

Marcus, one of the guys, moaned aloud. "Do we have to do a chick-flick? How 'bout something with explosions?" A girl from the group playfully hit him with a pillow from the couch.

"Ok, what about a comedy instead?" Tamika suggested.

"That sounds good," Marcus agreed. "We can eat in the food court, too."

The group decided to go to the mall to eat and catch a movie. They piled into a few cars and headed out. By the time they made it to the movie theater and argued about which movie to see, they had just missed the last matinee showing. The next showing wouldn't start for two hours. After dinner in the food court, there was still an hour before the movie started. Across from the movie theater was a popular jewelry store, specializing in in-store ear piercing.

"Let's get our ears pierced while we wait!" Tamika suggested. Two of the guys opted out, but the rest stood in line while the salesperson pierced their ears. Mechelle had her ears pierced as a baby, but she decided to go ahead and get a second piercing. This one would be slightly above the first earring on her right ear lobe. She picked out some really nice gold hoop earrings that would look good on her.

They went into the movie and laughed themselves silly. It was a fun evening with friends. Mechelle had almost forgotten how awful the test was!

18.1 Chromosomes and the Cell Cycle
A karyotype is a picture of chromosomes about to divide. Cell division, called mitosis, is a part of the cell cycle. The cell cycle consists of interphase and mitosis.

18.2 Mitosis
Mitosis is duplication division in which the daughter cells have the same number and types of chromosomes as the mother cell.

18.3 Meiosis
Meiosis is reduction division in which the daughter cells have half the number of chromosomes as the parent cell. In the end, the daughter cells also have a different combination of chromosomes and genes than the parent cell.

18.4 Comparison of Meiosis and Mitosis
Meiosis I uniquely pairs and separates the paired chromosomes so that the daughter cells have half the number of chromosomes. Meiosis II is exactly like mitosis, except the cells have half the number of chromosomes.

18.5 Chromosome Inheritance
Abnormalities in chromosome inheritance occur due to changes in chromosome number and changes in chromosome structure. Changes in chromosome number in humans leads to one less than normal and one more than normal. Changes in chromosome structure can include deletions of some segments and duplications of other segments.

18.1 Chromosomes and the Cell Cycle

A human nucleus is only about 5–8 μm, yet it holds all the chromatin that condenses to form the chromosomes when cells divide. Humans have 46 chromosomes that occur in 23 pairs. Twenty-two of these pairs are called autosomes. All of these chromosomes are found in both males and females. One pair of chromosomes is called the sex chromosomes because they do contain the genes that control gender. Males have the sex chromosomes X and Y, and females have two X chromosomes. (Recall that a Y chromosome contains the *SRY* gene that causes testes to develop.)

Suppose you wanted to see the chromosomes of an individual. What would you do? Any cell in the body except red blood cells, which lack a nucleus, can be a source of chromosomes for examination. In adults, it is easiest to obtain and use white blood cells separated from a blood sample for the purpose of looking at the chromosomes. After a cell sample has been obtained, the cells are stimulated to divide in a culture medium. When a cell divides, chromatin condenses to form chromosomes. The nuclear envelope fragments, liberating the chromosomes. Next, a chemical is used to stop the division process when the chromosomes are most compacted and visible microscopically. Stains are applied to the slides, and the cells can be photographed with a camera attached to a microscope. Staining causes the chromosomes to have dark and light cross-bands of varying widths, and these can be used in addition to size and shape to help pair up the chromosomes. A computer is used to arrange the chromosomes in pairs (Fig. 18.1). The display is called a karyotype. The karyotype in Figure 18.1 is that of a normal male.

Have You Ever Wondered...

How many chromosomes do other organisms have? Is the number of chromosomes related to how complex an organism is?

In eukaryotes, like humans, the number of chromosomes varies considerably. A fruit fly has 8 chromosomes, while yeasts have 32. Humans have 46, and horses have 64. The largest number of chromosomes appears to be found in a particular type of fern. It has 1,252 chromosomes. The number of chromosomes doesn't seem to determine an organism's complexity.

A Karyotype

A normal karyotype tells us a lot about a body cell. First, we should notice that a normal body cell is diploid—it has the full complement of 46 chromosomes. How did it happen that every body cell you ever had or will have has 46 chromosomes? **Mitosis**—duplication division—which began when the fertilized egg started dividing, ensures that every cell has 46 chromosomes.

The enlargement of a pair of chromosomes shows that in dividing cells each chromosome is composed of two identical parts, called **sister chromatids** (see Fig. 18.1). These chromosomes are said to be replicated or duplicated chromosomes because the two sister chromatids contain the

Figure 18.1   What do the chromosomes in our body look like?
In body cells, the chromosomes occur in pairs. In a karyotype, the pairs have been numbered. These chromosomes are duplicated, and each one is composed of two sister chromatids.

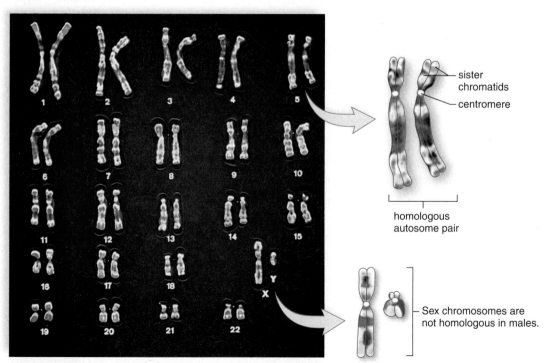

The 46 chromosomes of a male

sister chromatids

centromere

homologous autosome pair

Sex chromosomes are not homologous in males.

same genes. Genes are the units of heredity that control the cell. This is possible only because each chromatid contains a DNA double helix.

The chromatids are held together at a region called the centromere. A **centromere** has the function of holding the chromatids together until a certain phase of mitosis when the centromere splits. Once separated, each sister chromatid is a chromosome. In this way, a duplicated chromosome gives rise to two individual daughter chromosomes. When daughter chromosomes separate, the new cell gets one of each type and, therefore, a full complement of chromosomes.

Obtaining Fetal Chromosomes

Physicians and prospective parents sometimes want to view an unborn child's chromosomes to determine whether a chromosomal abnormality exists. These abnormalities can cause diseases such as Fragile X syndrome or Down syndrome. A **syndrome** is a group of symptoms that always occur together.

Chorionic villi sampling (CVS) is usually performed from the eighth to the twelfth week of pregnancy. The doctor inserts a long, thin tube through the vagina into the uterus. With the help of ultrasound, which gives a picture of the uterine contents, the tube is placed between the uterine lining and the chorionic villi. Fetal cells are obtained from the villi by suction (Fig. 18.2*a*). Enough cells are obtained to allow karyotyping to be done right away. CVS carries a greater risk of spontaneous abortion than amniocentesis—0.8% compared with 0.3%. The advantage of CVS is getting the results of karyotyping at an earlier date.

Amniocentesis is usually performed from the fifteenth to the seventeenth week of pregnancy. A long needle is passed through the abdominal wall to withdraw a small amount of amniotic fluid, along with a few fetal cells (Fig. 18.2*b*). These few cells are allowed to undergo mitosis in the laboratory until there are enough for karyotyping to be done. This may take about four weeks.

The Cell Cycle

The **cell cycle** is an orderly process that has two parts: interphase and cell division. To understand the cell cycle, it is necessary to recall the structure of a cell (see Fig. 3.4). A human cell has a plasma membrane, which encloses the cytoplasm, the content of the cell outside the nucleus. In the cytoplasm are various organelles, which carry on various functions necessary to the life of the cell. When a cell is not undergoing division, the DNA (and associated proteins) within a nucleus is a tangled mass of thin threads called chromatin.

Interphase

As Figure 18.3 shows, most of the cell cycle is spent in **interphase.** This is the time when the organelles carry on their usual functions. Also, the cell gets ready to divide. It grows larger, the number of organelles doubles, and the amount of chromatin doubles as DNA synthesis occurs.

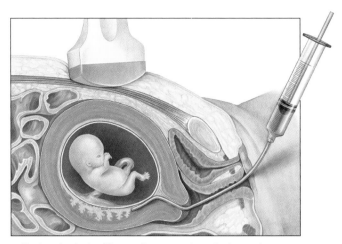

a. During chorionic villi sampling, a suction tube is used to remove cells from the chorion, where the placenta will develop.

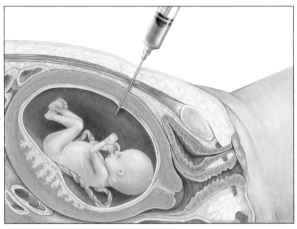

b. During amniocentesis, a long needle is used to withdraw amniotic fluid containing fetal cells.

Figure 18.2 How are fetal chromosomes obtained for karyotyping?

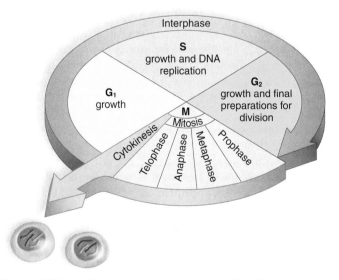

Figure 18.3 What are the stages of the cell cycle?
The cell cycle has four stages. During interphase, which consists of G$_1$, S, and G$_2$, the cell gets ready to divide, and during the mitotic stage, nuclear division and cytokinesis (cytoplasmic division) occur.

Interphase is divided into three stages: the G_1 stage occurs before DNA synthesis, the S stage includes DNA synthesis, and the G_2 stage occurs after DNA synthesis. Originally, G stood for the "gaps"—those times during interphase when DNA synthesis was not occurring. But now that we know growth happens during these stages, the G can be thought of as standing for growth. Let us see what specifically happens during these stages.

G_1 *stage.* The cell returns to normal size and resumes its function within the body. A cell doubles its organelles (e.g., mitochondria and ribosomes), and it accumulates the materials needed for DNA synthesis.

S stage. DNA replication occurs. A copy is made of all the DNA in the cell. DNA replication occurs, so each chromosome consists of two identical DNA double helix molecules. These molecules occur in the strands called sister chromatids.

G_2 *stage.* The cell synthesizes the proteins needed for cell division, such as the protein found in microtubules. The role of microtubules in cell division is described in a later section.

The amount of time the cell takes for interphase varies widely. Some cells, such as nerve and muscle cells, typically do not complete the cell cycle and are permanently arrested in G_1. These cells are said to have entered a G_0 stage. Embryonic cells spend very little time in G_1 and complete the cell cycle in a few hours.

Cell Division

Following interphase, the cell enters the cell division part of the cell cycle. Cell division has two stages: M (for mitotic) stage and cytokinesis. Mitosis is a type of nuclear division. Recall that mitosis is called *duplication division* because each new nucleus contains the same number and type of chromosomes as the former cell. **Cytokinesis** is division of the cytoplasm.

During mitosis, the sister chromatids of each chromosome separate, becoming chromosomes distributed to two daughter nuclei. When cytokinesis is complete, two daughter cells are now present. Mammalian cells usually require only about 4 hours to complete the mitotic stage.

The cell cycle, including interphase and cell division occurs continuously in certain tissues. Right now your body is producing thousands of new red blood cells, skin cells, and cells that line your respiratory and digestive tracts. The process of **apoptosis,** programmed cell death, also occurs to do away with any cells dividing when they shouldn't. The control of the cell cycle is discussed on page 426.

Check Your Progress 18.1

1. What are the three stages of interphase?
2. How does interphase prepare a cell for cell division?

18.2 Mitosis

Mitosis is *duplication division. The nuclei of the two new cells have the same number and types of chromosomes as the cell that divides.* The cell that divides is called the **parent cell,** and the new cells are called the **daughter cells.** The parent cell and daughter cell have the same number and types of chromosomes, so they are genetically identical.

Overview of Mitosis

As mentioned, when mitosis is going to occur, chromatin in the nucleus becomes highly condensed. The chromosomes become visible. Replication of DNA occurred, so each chromosome is now duplicated. Each is composed of two identical parts, called sister chromatids, held together at a centromere. They are called sister chromatids because they contain the same genes:

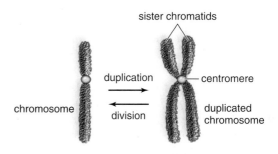

Figure 18.4 gives an overview of mitosis. For simplicity, only four chromosomes are depicted. (In determining the number of chromosomes, it is necessary to count only the number of independent centromeres.) As you know, the complete number of chromosomes is called the **diploid (2n)** number.

During mitosis, the centromeres divide and the sister chromatids separate. (Following separation, each chromatid is called a chromosome.) Each daughter cell gets a complete set of chromosomes and is 2n. Therefore, each daughter cell receives the same number and types of chromosomes as the parent cell. Each daughter cell is genetically identical to the other and to the parent cell.

The Spindle

Another event of importance during mitosis is the duplication of the **centrosome,** the microtubule organizing center of the cell. After centrosomes duplicate, they separate and form the poles of the **mitotic spindle,** where they assemble the microtubules that make up the spindle fibers. The chromosomes are attached to the spindle fibers at their centromeres (Fig. 18.5). An array of microtubules, called an **aster** (because it looks like a star), is also at the poles.

The **centrioles** are short cylinders of microtubules that are present in centrosomes. The centrioles lie at right angles to one another. Although it has not been shown, they could possibly assist in the formation of the spindle that separates the chromatids during mitosis.

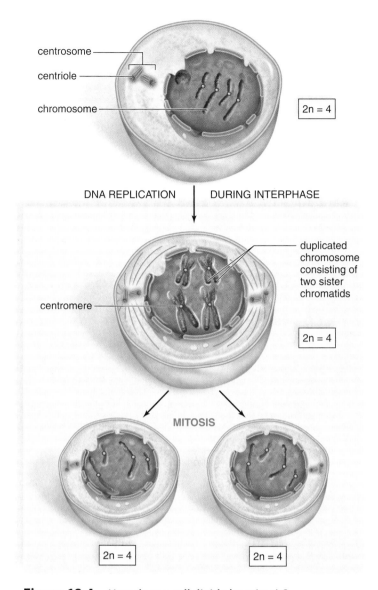

centrosome
centriole
chromosome

2n = 4

DNA REPLICATION DURING INTERPHASE

duplicated chromosome consisting of two sister chromatids

centromere

2n = 4

MITOSIS

2n = 4   2n = 4

Figure 18.4   How does a cell divide by mitosis?
Following DNA replication, each chromosome is duplicated. When the centromeres split, the sister chromatids, now called chromosomes, move into daughter nuclei. (The blue and red chromosomes were inherited from different parents.)

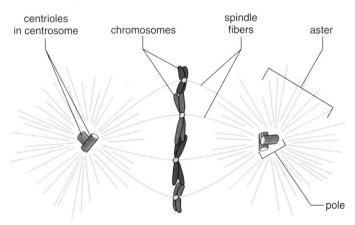

centrioles in centrosome   chromosomes   spindle fibers   aster

pole

Figure 18.5   How is the mitotic spindle involved in mitosis?
Before mitosis begins, the centrosome and the centrioles duplicate. During mitosis, they separate, and the mitotic spindle, composed of microtubules, forms between them.

Phases of Mitosis

As an aid in describing the events of mitosis, the process is divided into four phases: prophase, metaphase, anaphase, and telophase (Fig. 18.6). Although the stages of mitosis are depicted as if they were separate, they are continuous. One stage flows from the other with no noticeable interruption.

Prophase

Several events occur during **prophase** that visibly indicate the cell is preparing to divide. The centrosomes outside the nucleus have duplicated, and they begin moving away from one another toward opposite ends of the nucleus. Spindle fibers appear between the separating centrosomes. The nuclear envelope begins to fragment. The nucleolus, a special region of DNA, disappears as the chromosomes coil and become condensed.

The chromosomes are now visible. Each is composed of two sister chromatids held together at a centromere. Spindle fibers attach to the centromeres as the chromosomes continue to shorten and to thicken. During prophase, chromosomes are randomly placed in the nucleus (Fig. 18.6).

Metaphase

During **metaphase,** the nuclear envelope is fragmented, and the spindle occupies the region formerly occupied by the nucleus. The chromosomes are now at the equator (center) of the spindle. Metaphase is characterized by a fully formed spindle. The chromosomes, each with two sister chromatids, are aligned at the equator (Fig. 18.6).

Anaphase

At the start of **anaphase,** the centromeres uniting the sister chromatids divide. Then the sister chromatids separate, becoming chromosomes that move toward opposite poles of the spindle. Separation of the sister chromatids ensures that each cell receives a copy of each type of chromosome, and thereby has a full complement of genes. Anaphase is characterized by the 2n (diploid) number of chromosomes moving toward each pole.

Remember that counting the number of centromeres indicates the number of chromosomes. Therefore, in Figure 18.6, each pole receives four chromosomes: two are red and two are blue.

Function of the Spindle   The spindle brings about chromosomal movement. Two types of spindle fibers are involved in the movement of chromosomes during anaphase. One type extends from the poles to the equator of the spindle. There, they overlap. As mitosis proceeds, these fibers increase in length. This helps push the chromosomes apart. The chromosomes themselves are attached to other spindle fibers that extend from their centromeres to the poles. These fibers (composed of microtubules that can disassemble) become shorter as the chromosomes

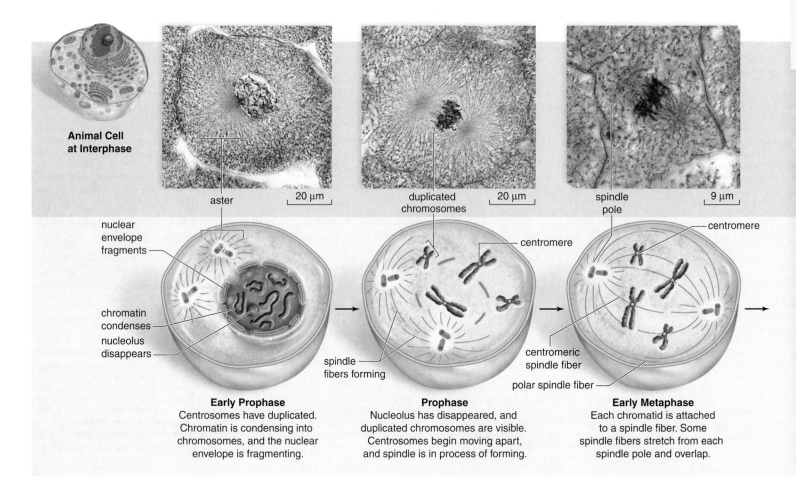

Animal Cell at Interphase

aster | 20 μm

duplicated chromosomes | 20 μm

spindle pole | 9 μm

nuclear envelope fragments

chromatin condenses

nucleolus disappears

spindle fibers forming

centromere

centromere

centromeric spindle fiber

polar spindle fiber

Early Prophase
Centrosomes have duplicated. Chromatin is condensing into chromosomes, and the nuclear envelope is fragmenting.

Prophase
Nucleolus has disappeared, and duplicated chromosomes are visible. Centrosomes begin moving apart, and spindle is in process of forming.

Early Metaphase
Each chromatid is attached to a spindle fiber. Some spindle fibers stretch from each spindle pole and overlap.

Figure 18.6 What occurs during the phases of mitosis?

move toward the poles. Therefore, they pull the chromosomes apart.

Spindle fibers, as stated earlier, are composed of microtubules. Microtubules can assemble and disassemble by the addition or subtraction of tubulin (protein) subunits. This is what enables spindle fibers to lengthen and shorten and what ultimately causes the movement of the chromosomes.

Telophase

Telophase begins when the chromosomes arrive at the poles. During telophase, the chromosomes become indistinct chromatin again. The spindle disappears as the nuclear envelope components reassemble in each cell. Each nucleus has a nucleolus because each has a region of the DNA where ribosomal subunits are produced. Telophase is characterized by the presence of two daughter nuclei.

Cytokinesis

Cytokinesis is the division of the cytoplasm and organelles. In human cells, a slight indentation, called a **cleav-**

age furrow, passes around the circumference of the cell. Actin filaments form a contractile ring, and as the ring becomes smaller, the cleavage furrow pinches the cell in half. As a result, each cell becomes enclosed by its own plasma membrane.

The Importance of the Cell Cycle and Mitosis

The cell cycle, which includes mitosis, is very important to the well-being of humans. Mitosis is responsible for new cells in the developing embryo, fetus, and child. It is also responsible for replacement cells in an adult (Fig. 18.7a).

Ordinarily, a cell cycle control system works perfectly to produce more cells only to the extent necessary (Fig. 18.7c). A benign tumor or a cancerous tumor develops when the cell cycle control system is not functioning properly. A benign tumor is confined to a particular location.

In the case study, Mechelle formed a keloid on her ear. A keloid is a benign, noncancerous tumor that forms when a wound heals (Fig. 18.7b). When a keloid forms, the cell cycle keeps occurring and doesn't stop when it should.

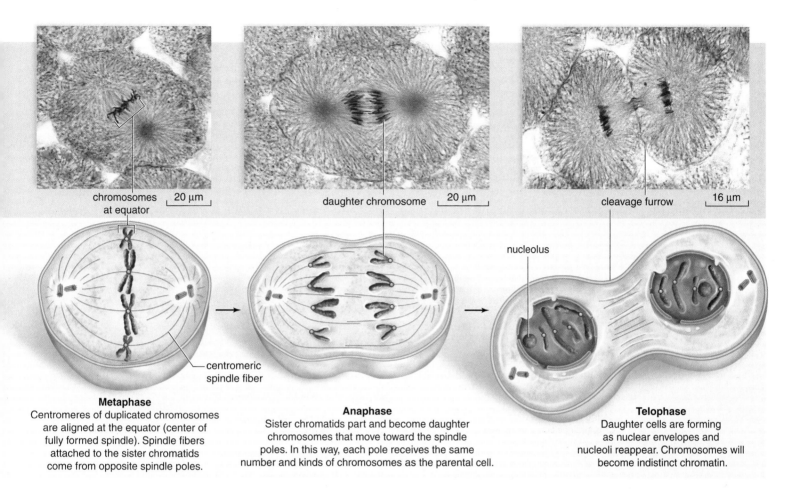

Metaphase
Centromeres of duplicated chromosomes are aligned at the equator (center of fully formed spindle). Spindle fibers attached to the sister chromatids come from opposite spindle poles.

Anaphase
Sister chromatids part and become daughter chromosomes that move toward the spindle poles. In this way, each pole receives the same number and kinds of chromosomes as the parental cell.

Telophase
Daughter cells are forming as nuclear envelopes and nucleoli reappear. Chromosomes will become indistinct chromatin.

chromosomes at equator 20 µm

daughter chromosome 20 µm

cleavage furrow 16 µm

centromeric spindle fiber

nucleolus

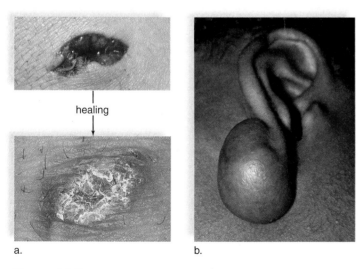

healing

a.

b.

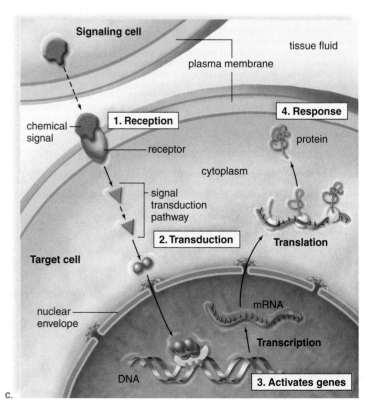

Signaling cell

tissue fluid

plasma membrane

chemical signal

1. Reception

receptor

4. Response

protein

cytoplasm

signal transduction pathway

2. Transduction

Translation

Target cell

nuclear envelope

mRNA

Transcription

DNA

3. Activates genes

c.

Figure 18.7 **What is the role of mitosis in our bodies? How is the cell cycle controlled?**
a. The cell cycle, including mitosis, occurs when humans grow and when tissues undergo repair. **b.** A keloid occurs when mitosis continues abnormally in response to an injury. **c.** Growth factors stimulate a cell-signaling pathway that stretches from the plasma membrane to the genes that regulate the occurrence of the cell cycle.

A cancerous tumor is not confined. Cancer cells have a tendency to leave the original tumor by way of the blood and lymphatic vessels and start new tumors in other parts of the body.

Cell Cycle Control System

The cell cycle control system extends from the plasma membrane to particular genes in the nucleus. Some external signals such as hormones and growth factors can stimulate a cell to go through the cell cycle. At a certain time in the menstrual cycle of females, the hormone progesterone stimulates cells lining the uterus to prepare the lining for implantation of a fertilized egg. Epidermal growth factor stimulates skin in the vicinity of an injury to finish the cell cycle, thereby repairing damage.

As shown in Figure 18.7c, ① during reception, an external signal delivers a message to a specific receptor embedded in the plasma membrane of a receiving cell. ② The receptor relays the signal to proteins inside the cell's cytoplasm. The proteins form a pathway called the signal transduction pathway because they pass the signal from one to the other. ③ The last signal activates genes whose protein product ④ stimulates or inhibits the cell cycle. Genes called proto-oncogenes stimulate the cell cycle, and genes called tumor-suppressor genes inhibit the cell cycle. These genes are discussed more in Chapter 19, which is about cancer.

CASE STUDY AT THE DERMATOLOGIST

Mechelle took one last look at herself in the mirror before she walked out the door. She made sure that her hair was pulled down over her earlobe. *"Why did I ever get that ear pierced again?"* she thought to herself. It had seemed like a good idea at the time. Everybody had multiple ear piercings. Why did hers end up in this big round mass on her earlobe?

Mechelle was relieved to finally visit her dermatologist. She hoped something could be done about the big, ugly lump on her ear. But in the doctor's exam room, Mechelle was a little nervous. She felt stupid because she had done this to herself. She didn't really want to explain to the dermatologist that she'd gotten her ear pierced again on a whim, just for something fun to do. She was also hoping that he could help her get rid of the big ugly lump. She didn't want to go around for the rest of her life with this thing on her earlobe.

Dr. Kendrig entered. "Hi, Mechelle. I understand you have a problem with an ear piercing?"

Mechelle nodded and pulled back her hair. Dr. Kendrig examined the earlobe, turning it over and looking at both sides. He examined her other ear and then looked at her neck, face, and arms. Then he pulled up a chair and sat opposite her.

"When did you get your ear pierced?" he asked.

"About six months ago," Mechelle answered. "At one of those mall jewelry stores."

"I noticed you have other ear piercings," Dr. Kendrig observed. "When did you get those other piercings?"

"When I was a baby," she replied. "One in each ear lobe. Those seemed fine so I didn't expect to have any problems with another one."

"No, I can see that you wouldn't have expected it," he responded thoughtfully. "Does it hurt?"

"Sometimes," Mechelle nodded. "It itches a lot."

"This is called a keloid," Dr. Kendrig explained. "It's a form of scar tissue. Instead of healing normally, the healing process gets out of control. Fibrous tissue just keeps dividing and multiplying, and a big mass like this grows.

"Does anyone else in your family have this sort of mass due to some type of skin injury?" he asked. "A piercing? A cut of some kind?"

"My mom has a really big scar on her arm where she fell once," Mechelle answered. "It's not pretty, but it isn't as awful as this lump on my ear."

"Uh-huh," Dr. Kendrig nodded. "That's called hypertrophic scarring, which means too much scar tissue. Hypertrophic scars and keloids are similar. The tendency to grow them runs in families, and that's why I asked about your family.

"We aren't sure what causes keloids to form," he continued. "African Americans and young women of all races are more likely to get them. Some people get them with their first piercing. Others don't grow one until the second piercing, and some lucky folks don't get one at all. Babies and kids just don't seem to grow them."

"So that's why getting my ears pierced the first time wasn't a problem, just the second time," Mechelle responded. "But this lump is much bigger than the original piercing. Why is that?"

"The closer you get to the cartilage on the ear, the more risk there is of forming a keloid," Dr. Kendrig answered. "This second piercing is just below the ear cartilage. I strongly suggest that you don't get any more piercings. When it comes to keloids, prevention is the best medicine.

"And, if you ever need to have surgery," he warned, "the surgeon needs to know about this keloid. There are procedures done before surgery that can help keep hypertrophic scars and keloids from developing."

Check Your Progress 18.2

1. Following mitosis, how does the chromosome number of the daughter cell compare with the chromosome number of the parent cell?

2. What are the phases of mitosis, and what happens during each phase?

3. How is the cytoplasm divided between the daughter cells following mitosis?

18.3 Meiosis

Meiosis is *reduction division. Meiosis involves two divisions, so there are four daughter cells. Each daughter cell has one of each type of chromosome and, therefore, half as many chromosomes as the parent cell.* The parent cell has the diploid (2n) number of chromosomes, while the daughter cells have half this number, called the **haploid (n)** number of chromosomes. The daughter cells that result from meiosis go on to become the gametes.

In Figure 18.8, the diploid (2n) number of chromosomes is four chromosomes. The parent cell has the 2n number of chromosomes, while the daughter cells have the haploid (n) number of chromosomes, equal to two chromosomes.

Overview of Meiosis

At the start of meiosis, the parent cell is 2n, or diploid, and the chromosomes occur in pairs. For simplicity's sake, there are only two pairs of chromosomes (Fig. 18.8). The short chromosomes are one pair, and the long chromosomes are another. The members of a pair are called **homologous chromosomes,** or **homologues,** because they look alike and carry genes for the same traits, such as type of hair or color of eyes.

Meiosis I

The two cell divisions of meiosis are called meiosis I and meiosis II. Prior to meiosis I, DNA replication has occurred, and the chromosomes are duplicated. Each chromosome consists of two chromatids held together at a centromere. During meiosis I the homologous chromosomes come together and line up side by side. This so-called **synapsis** results in an association of four chromatids that stay in close proximity during the first two phases of meiosis I. Synapsis is significant because its occurrence leads to a reduction of the chromosome number.

There are pairs of homologous chromosomes at the equator during meiosis I because of synapsis. Only during meiosis I is it possible to observe paired chromosomes at the equator. When the members of these pairs separate, each daughter nucleus receives one member of each pair. Therefore, each daughter cell now has the haploid (n) number of chromosomes, as you can verify by counting its centromeres. Each chromosome, however, is still duplicated. No replication of DNA occurs between meiosis I

and meiosis II. The time between meiosis I and meiosis II is called **interkinesis.**

Meiosis II and Fertilization

During meiosis II, the centromeres divide. The sister chromatids separate, becoming chromosomes that are distributed to daughter nuclei. In the end, each of four daughter cells has the n, or haploid, number of chromosomes. Each chromosome consists of one chromatid.

In humans, the daughter cells mature into **gametes** (sex cells—sperm and egg) that fuse during fertilization. **Fertilization** restores the diploid number of chromosomes in the **zygote,** the first cell of the new individual. If the

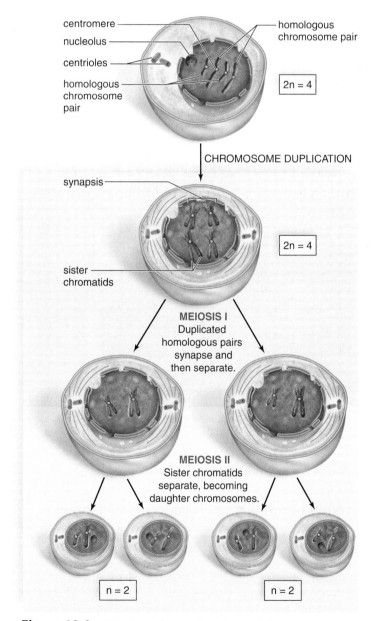

Figure 18.8 What are the results of meiosis?
DNA replication is followed by meiosis I when homologous chromosomes pair and then separate. During meiosis II, the sister chromatids become chromosomes that move into daughter nuclei.

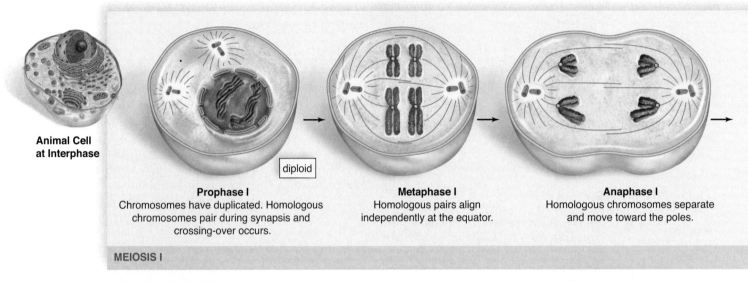

Animal Cell at Interphase

diploid

MEIOSIS I

Prophase I
Chromosomes have duplicated. Homologous chromosomes pair during synapsis and crossing-over occurs.

Metaphase I
Homologous pairs align independently at the equator.

Anaphase I
Homologous chromosomes separate and move toward the poles.

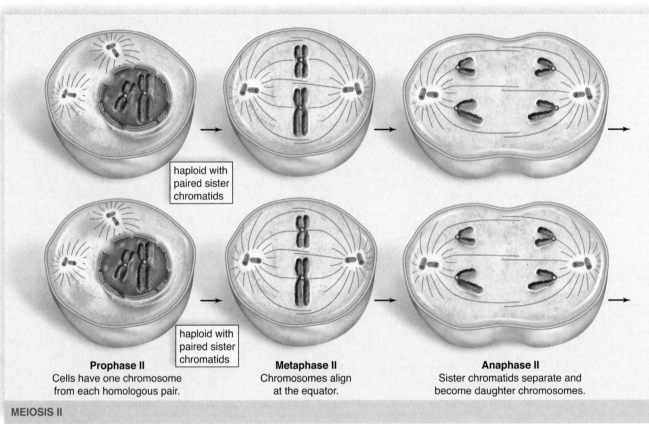

haploid with paired sister chromatids

haploid with paired sister chromatids

MEIOSIS II

Prophase II
Cells have one chromosome from each homologous pair.

Metaphase II
Chromosomes align at the equator.

Anaphase II
Sister chromatids separate and become daughter chromosomes.

Figure 18.9   What occurs during the phases of meiosis?
Homologous chromosomes pair and then separate during meiosis I. Crossing-over, which occurs during meiosis I, is discussed more fully on page 430. Chromatids separate, becoming daughter chromosomes during meiosis II. Following meiosis II, there are four haploid daughter cells.

gametes carried the diploid instead of the haploid number of chromosomes, the chromosome number would double with each fertilization. After several generations, the zygote would be nothing but chromosomes.

Figure 18.9 shows the phases of meiosis I and meiosis II. Meiosis I has four phases: prophase I, metaphase I, anaphase I, and telophase I. Some of the meiosis I stages are discussed more fully on pages 429–30. The stages of meiosis II are named similarly to those of meiosis I, except the name is followed by a II.

Stages of Meiosis

Meiosis is a part of sexual reproduction. The process of meiosis ensures that the next generation of individuals

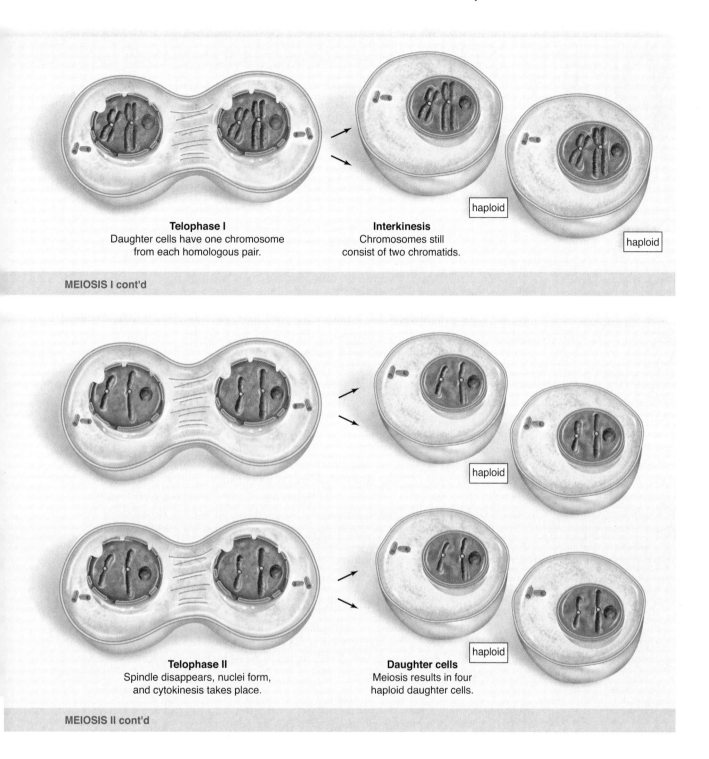

Telophase I
Daughter cells have one chromosome
from each homologous pair.

Interkinesis
Chromosomes still
consist of two chromatids.

haploid

haploid

MEIOSIS I cont'd

haploid

haploid

Telophase II
Spindle disappears, nuclei form,
and cytokinesis takes place.

Daughter cells
Meiosis results in four
haploid daughter cells.

haploid

MEIOSIS II cont'd

will have the diploid number of chromosomes and a combination of characteristics different from that of either
parent. Both meiosis I and meiosis II have the same four
stages of nuclear division as did mitosis—prophase, metaphase, anaphase, and telophase. Here we discuss only
prophase I and metaphase I because special events occur
during these phases.

Prophase I

In prophase I, synapsis occurs, and then the spindle appears.
The nuclear envelope fragments and the nucleolus disappears. During synapsis, the homologous chromosomes
come together and line up side by side. Now, an exchange of
genetic material may occur between the nonsister chromatids

sister chromatids

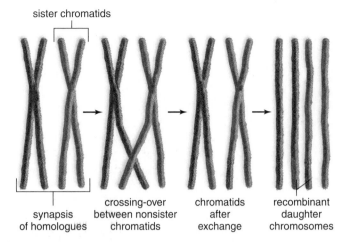

synapsis of homologues

crossing-over between nonsister chromatids

chromatids after exchange

recombinant daughter chromosomes

Figure 18.10 How do synapsis and crossing-over increase variability?

During meiosis I, *from left to right*, duplicated homologous chromosomes undergo synapsis and line up with each other. During crossing-over, nonsister chromatids break and then rejoin in the manner shown. Two of the resulting chromosomes will have a different combination of genes than they had before.

of the homologous pair (Fig. 18.10). This exchange is called **crossing-over.** Crossing-over means that the chromatids held together by a centromere are no longer identical. When the chromatids separate during meiosis II, the daughter cells will receive chromosomes with recombined genetic material.

To appreciate the significance of crossing-over, it is necessary to realize that the members of a homologous pair can carry slightly different instructions for the same genetic trait. For example, one homologue may carry instructions for brown eyes and blond hair, while the corresponding homologue may carry instructions for blue eyes and red hair. Crossing-over causes the offspring to receive a different combination of instructions than the mother or the father received. Therefore, offspring could receive brown eyes and red hair or blue eyes and blond hair.

Metaphase I

During metaphase I, the homologous pairs align independently at the equator. This means that the maternal or

paternal member may be oriented toward either pole. Figure 18.11 shows the eight possible orientations for a cell that contains only three pairs of chromosomes. The first four orientations will result in gametes that have different combinations of maternal and paternal chromosomes. The next four will result in the same types of gametes as the first four. For example, the first cell and the last cell will both produce gametes with either three red or three blue chromosomes.

Once all possible orientations are considered, the result will be 2^3, or 8, possible combinations of maternal and paternal chromosomes in the resulting gametes from this cell. In humans, where there are 23 pairs of chromosomes, the number of possible chromosomal combinations in the gametes is a staggering 2^{23}, or 8,388,608. And this does not even consider the genetic variations introduced due to crossing-over.

The events of prophase I and metaphase I help ensure that gametes will not have the same combination of chromosomes and genes.

Significance of Meiosis

Meiosis is a part of gametogenesis, production of the sperm and egg. One function of meiosis is to keep the chromosome number constant from generation to generation. The gam-

Figure 18.11 How does independent alignment increase variability?

When a parent cell has three pairs of homologous chromosomes, there are eight possible chromosome alignments at the equator due to independent assortment. Among the 16 daughter nuclei resulting from these alignments, there are eight different combinations of chromosomes.

etes are haploid, so the zygote has only the diploid number of chromosomes.

An easier way to keep the chromosome number constant is to reproduce asexually. Unicellular organisms such as bacteria, protozoans, and yeasts reproduce by binary fission:

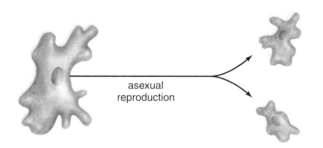

asexual
reproduction

This is a form of asexual reproduction because one parent has produced identical offspring. Binary fission is a quick and easy way to asexually reproduce many organisms within a short time. A bacterium can increase to over 1 million cells in about 7 hours, for example. Then why do organisms expend the energy to reproduce sexually? It takes energy to find a mate, carry out a courtship, and produce eggs or sperm that may never be used for reproductive purposes. A human male produces 300 million sperm per day, and very few of these will fertilize an egg.

Most likely, humans and other animals practice sexual reproduction that includes meiosis because it results in genetic recombination. Genetic recombination ensures that offspring will be genetically different compared to each other and to their parents. Genetic recombination occurs because of crossing-over and independent alignment of chromosomes. Also, at the time of fertilization, parents contribute genetically different chromosomes to the offspring.

All environments are subject to a change in conditions. Those individuals able to survive in a new environment are able to pass on their genes. Environments are subject to change, so sexual reproduction is advantageous. It generates the diversity needed so that at least a few will be suited to new and different environmental circumstances.

> **Check Your Progress 18.3**
>
> 1. Following meiosis, how does the chromosome number of the daughter cells compare to the chromosome number of the parent cell?
> 2. How does meiosis reduce the likelihood that gametes will have the same combination of chromosomes and genes?
> 3. What happens during the two cell divisions of meiosis?

CASE STUDY TREATMENT OPTIONS

Mechelle listened attentively to Dr. Kendrig. Although she understood what he was saying, she was really most interested in what could be done about the keloid. She was hoping he could make it disappear completely.

Dr. Kendrig continued, "But now that you have a keloid, we can talk about a few things to maybe help this one go away." Mechelle looked up at Dr. Kendrig hopefully. "There are numerous ways to treat keloids, but none of them work consistently."

Mechelle looked crushed. Did that mean this couldn't be fixed? Dr. Kendrig noticed her dejected expression. "That just means that there is no single treatment that helps everyone," he reassured. "There are several different things we can try, and some treatments work better for some people than for others. But you need to understand that your ear probably won't look exactly like it looked before, and it probably won't look just like your other earlobe.

"We can give you steroid creams or injections to shrink the keloid. We can also try radiation treatment and pressure treatment. Laser surgery or traditional surgery are options, too."

Dr. Kendrig went on to describe the various options in greater detail. He gave her several flyers that summarized these treatments and their success rates. Then he recommended, "First, let's try surgery, apply a tight pressure dressing after surgery, and follow up with steroid injections. A surgery might make a keloid return—even bigger than before—so that's why we'll also use pressure therapy and the steroid shots. Patients that follow my directions after surgery usually have good results, but I can't make any promises," he warned.

It wasn't exactly what Mechelle had hoped to hear. "Does insurance cover this?" she asked.

"I'm afraid it doesn't," Dr. Kendrig replied. "Unless the keloid is affecting your ability to function in some way, it's considered cosmetic surgery and isn't covered by insurance. In your case, I would consider this cosmetic surgery. You'll have to pay for the treatment yourself." He handed her a sheet that listed the costs of the various procedures.

"We have payment plans, and I can have our office manager talk with you about the best way to proceed if you're interested. You may want to think about it and talk to your parents," he concluded.

Before leaving the office, Mechelle gathered all of the information she'd need to discuss her treatment with her parents. Dr. Kendrig had been very kind and hadn't made her feel stupid for getting her ear pierced again. She didn't think her parents would be as understanding—especially when they saw the costs of the treatment!

18.4 Comparison of Meiosis and Mitosis

Meiosis and mitosis are both nuclear divisions, but there are several differences between them. You will want to be able to easily recognize how they differ. To this end, Figure 18.12 compares meiosis to mitosis. You will want to study each of the differences listed here until you are familiar with them.

- DNA replication takes place only once prior to both meiosis and mitosis. However, meiosis requires two nuclear divisions, but mitosis requires only one.
- Four daughter nuclei are produced by meiosis, and following cytokinesis, there are four daughter cells. Mitosis followed by cytokinesis results in two daughter cells.
- The four daughter cells following meiosis are haploid (n) and have half the chromosome number of the parent cell (2n). The daughter cells following mitosis have the same chromosome number as the parent cell—the 2n, or diploid, number.
- The daughter cells from meiosis are not genetically identical to each other or to the parent cell. The daughter cells from mitosis are genetically identical to each other and to the parent cell.

The specific differences between these nuclear divisions can be categorized according to occurrence and process.

Occurrence

Meiosis occurs only at certain times in the life cycle of sexually reproducing organisms. In humans, meiosis occurs only in the reproductive organs and produces the gametes. Mitosis is more common because it occurs in all tissues during growth and repair. Which type of cell division can lead to cancer? Mitosis can result in a proliferation of body cells. Abnormal mitosis can lead to cancer.

Process

Comparison of Meiosis I with Mitosis

These events distinguish meiosis I from mitosis:

- Homologous chromosomes pair and undergo crossing-over during prophase I of meiosis, but not during mitosis.

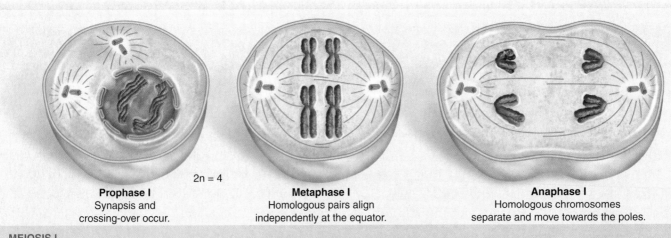

2n = 4

Prophase I
Synapsis and
crossing-over occur.

Metaphase I
Homologous pairs align
independently at the equator.

Anaphase I
Homologous chromosomes
separate and move towards the poles.

MEIOSIS I

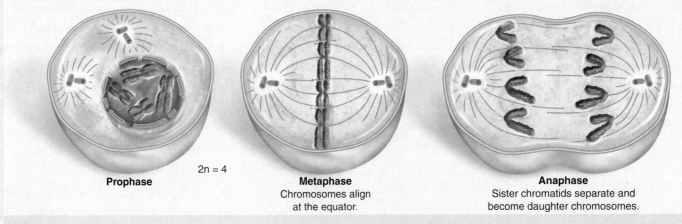

2n = 4

Prophase

Metaphase
Chromosomes align
at the equator.

Anaphase
Sister chromatids separate and
become daughter chromosomes.

MITOSIS

- Paired homologous chromosomes align at the equator during metaphase I in meiosis. These paired chromosomes have four chromatids altogether. Individual chromosomes align at the equator during metaphase in mitosis. They each have two chromatids.

  This difference makes it easy to tell whether you are looking at mitosis, meiosis I, or meiosis II. For example, if a cell has 16 chromosomes, then 16 chromosomes are at the equator during mitosis but only 8 chromosomes during meiosis II. Only meiosis I has paired duplicated chromosomes at the equator.

- Homologous chromosomes (with centromeres intact) separate and move to opposite poles during anaphase I of meiosis. Centromeres split, and sister chromatids, now called chromosomes, move to opposite poles during anaphase in mitosis.

Comparison of Meiosis II with Mitosis

The events of meiosis II are just like those of mitosis except in meiosis II, the nuclei contain the haploid number of chromosomes. If the parent cell has 16 chromosomes, then the cells undergoing meiosis II have 8 chromosomes, and the daughter cells have 8 chromosomes, for example.

Summary

To summarize the process, Tables 18.1 and 18.2 separately compare meiosis I and meiosis II with mitosis.

Table 18.1	**Comparison of Meiosis I with Mitosis**
Meiosis I	**Mitosis**
Prophase I	*Prophase*
Pairing of homologous chromosomes	No pairing of chromosomes
Metaphase I	*Metaphase*
Homologous duplicated chromosomes at equator	Duplicated chromosomes at equator
Anaphase I	*Anaphase*
Homologous chromosomes separate	Sister chromatids separate, becoming daughter chromosomes that move to the poles
Telophase I	*Telophase*
Two haploid daughter cells	Two daughter cells, identical to the parent cell

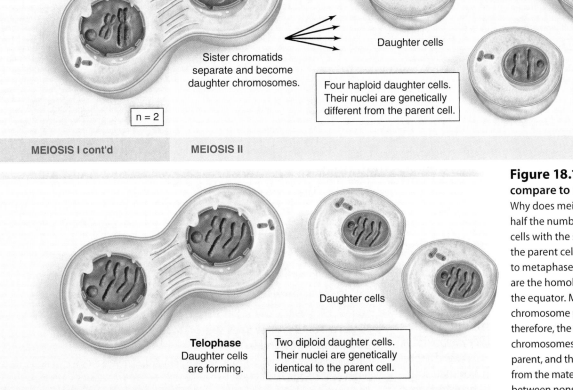

Telophase I
Daughter cells are forming and will go on to divide again.

n = 2

Sister chromatids separate and become daughter chromosomes.

Daughter cells

n = 2

n = 2

Four haploid daughter cells. Their nuclei are genetically different from the parent cell.

MEIOSIS I cont'd   **MEIOSIS II**

Telophase
Daughter cells are forming.

Daughter cells

Two diploid daughter cells. Their nuclei are genetically identical to the parent cell.

MITOSIS cont'd

Figure 18.12   How does meiosis compare to mitosis?

Why does meiosis produce daughter cells with half the number while mitosis produces daughter cells with the same number of chromosomes as the parent cell? Compare metaphase I of meiosis to metaphase of mitosis. Only in metaphase I are the homologous chromosomes paired at the equator. Members of homologous chromosome pairs separate during anaphase I, and therefore, the daughter cells are haploid. The blue chromosomes were inherited from the paternal parent, and the red chromosomes were inherited from the maternal parent. The exchange of color between nonsister chromatids represents the crossing-over that occurs during meiosis I.

Table 18.2	Comparison of Meiosis II with Mitosis
Meiosis II	**Mitosis**
Prophase II	*Prophase*
No pairing of chromosomes	No pairing of chromosomes
Metaphase II	*Metaphase*
Haploid number of duplicated chromosomes at equator	Duplicated chromosomes at equator
Anaphase II	*Anaphase*
Sister chromatids separate, becoming daughter chromosomes that move to the poles	Sister chromatids separate, becoming daughter chromosomes that move to the poles
Telophase II	*Telophase*
Four haploid daughter cells	Two daughter cells, identical to the parent cell

Check Your Progress 18.4

Key:

    a. mitosis   c. neither

    b. meiosis   d. both

1. Which one results in four daughter cells?
2. Which one has no pairing of chromosomes?
3. Which one is out of control when cancer occurs?
4. a. Which one is duplication division? b. Reduction division?
5. Which one results in cells exactly like the parent cell?
6. Which one results in cells genetically different from the parent cell?
7. Which one is a type of nuclear division?
8. Which one is involved in gametogenesis?
9. Which one is part of the cell cycle?

Spermatogenesis and Oogenesis

Meiosis is a part of **spermatogenesis,** the production of sperm in males, and **oogenesis,** the production of eggs in females (Fig. 18.13). Following meiosis, the daughter cells mature to become the gametes.

Spermatogenesis

After puberty, the time of life when the sex organs mature, spermatogenesis is continual in the testes of human males. As many as 300,000 sperm are produced per minute, or 400 million per day.

Spermatogenesis is shown in Figure 18.13, *top.* The *primary spermatocytes,* which are diploid (2n), divide during meiosis I to form two *secondary spermatocytes,* which are haploid (n). Secondary spermatocytes divide during meiosis II to produce four *spermatids,* which are also haploid (n). What's the difference between the chromosomes in haploid secondary spermatocytes and those in haploid

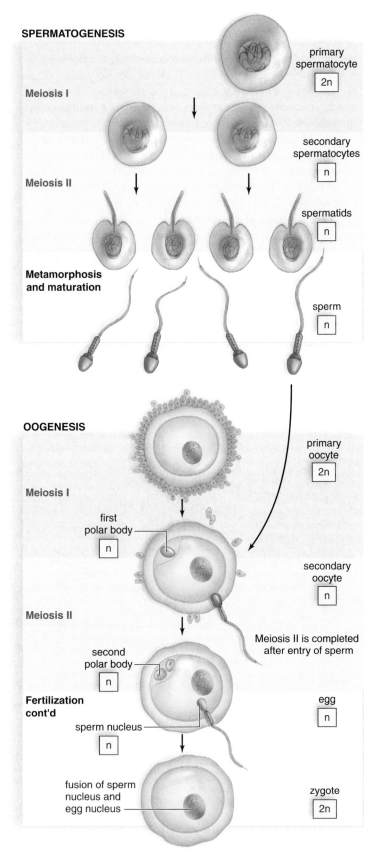

Figure 18.13 **How does spermatogenesis compare with oogenesis in mammals?**

Spermatogenesis produces four viable sperm, whereas oogenesis produces one egg and at least two polar bodies. In humans, both sperm and egg have 23 chromosomes each; therefore, following fertilization, the zygote has 46 chromosomes.

spermatids? The chromosomes in secondary spermatocytes are duplicated and consist of two chromatids, while those in spermatids consist of only one. Spermatids mature into sperm (spermatozoa). In human males, sperm have 23 chromosomes, the haploid number. The process of meiosis in males always results in four cells that become sperm. In other words, all four daughter cells—the spermatids—become sperm.

Oogenesis

As you know, the ovary of a female contains many immature follicles (see Fig. 16.8). Each of these follicles contains a primary oocyte arrested in prophase I. As shown in Figure 18.13 *bottom*, a primary oocyte, which is diploid (2n), divides during meiosis I into two cells, each of which is haploid. The chromosomes are duplicated. One of these cells, termed the **secondary oocyte,** receives almost all the cytoplasm. The other is the first polar body. A **polar body** acts like a trash can to hold discarded chromosomes. The first polar body contains duplicated chromosomes and completes meiosis II occasionally. The secondary oocyte begins meiosis II but stops at metaphase II and doesn't complete it unless a sperm enters during the fertilization process.

The secondary oocyte (for convenience, called the egg) leaves the ovary during ovulation and enters an oviduct, where it may be fertilized by a sperm. If so, the oocyte is activated to complete the second meiotic division. Following meiosis II, there is one egg and two or possibly three polar bodies. The mature egg has 23 chromosomes. The polar bodies disintegrate. They are a way to discard unnecessary chromosomes, while retaining much of the cytoplasm in the egg.

One egg can be the source of identical twins if after one division of the fertilized egg during development, the cells separate and each one becomes a complete individual. On the other hand, the occurrence of fraternal twins requires that two eggs be ovulated and then fertilized separately.

18.5 Chromosome Inheritance

Normally, an individual receives 22 pairs of autosomes and two sex chromosomes. Each pair of autosomes carries alleles for particular traits. The alleles can be different as when one calls for freckles and one does not.

Changes in Chromosome Number

Sometimes individuals are born with either too many or too few autosomes or sex chromosomes, most likely due to nondisjunction during meiosis. **Nondisjunction** occurs during meiosis I, when both members of a homologous pair go into the same daughter cell. It can also occur during meiosis II, when the sister chromatids fail to separate and both daughter chromosomes go into the same gamete. Figure 18.14 assumes that nondisjunction has occurred during oogenesis. Some

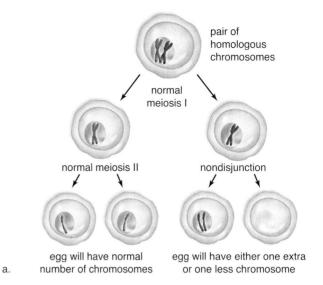

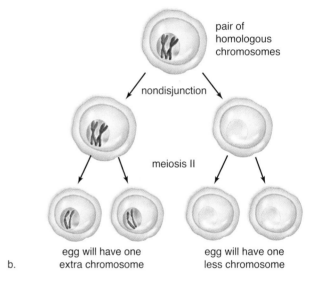

Figure 18.14 **What happens when there is nondisjunction of chromosomes during oogenesis, followed by fertilization with normal sperm?**

a. Nondisjunction can occur during meiosis II if the sister chromatids separate, but the resulting chromosomes go into the same daughter cell. Then the egg will have one more (24) or one less (22) then the usual number of chromosomes. Fertilization of these abnormal eggs with normal sperm produces an abnormal zygote with 47 or 45 chromosomes. **b.** Nondisjunction can also occur during meiosis I and result in abnormal eggs that also have one more or one less than the normal number of chromosomes. Fertilization of these abnormal eggs with normal sperm results in a zygote with an abnormal chromosome number.

abnormal eggs have 24 chromosomes, while others have only 22 chromosomes. If an egg with 24 chromosomes is fertilized with a normal sperm, the result is a **trisomy,** so called because one type of chromosome is present in three copies. If an egg with 22 chromosomes is fertilized with a normal sperm, the result is a **monosomy,** so called because one type of chromosome is present in a single copy.

Have You Ever Wondered...

Down syndrome is a trisomy of chromosome 21. Are there trisomies of the other chromosomes?

There are other chromosomal trisomies. However, because most chromosomes are much larger than chromosome 21, the abnormalities associated with three copies of these other chromosomes are much more severe than those found in Down syndrome. The extra genetic material causes profound congenital defects, resulting in fatality. Trisomies of the X and Y chromosomes appear to be exceptions to this, as noted in the text.

Chromosome 8 trisomy occurs rarely. Affected fetuses generally do not survive to birth or die shortly after birth. There are also trisomies of chromosomes 13 and 18. Again, these babies usually die within the first few days of life.

Normal development depends on the presence of exactly two of each type of chromosome. An abnormal number of autosomes will cause developmental abnormality. Monosomy of all but the X chromosome is fatal. The affected infant rarely develops to full term. Trisomy is usually fatal, though there are some exceptions. Among autosomal trisomies, only trisomy 21 (Down syndrome) has a reasonable chance of survival after birth.

The chances of survival are greater when trisomy or monosomy involves the sex chromosomes. In normal XX females, one of the X chromosomes becomes a darkly staining mass of chromatin called a **Barr body** (named after the person who discovered it). A Barr body is an inactive X chromosome. Therefore, we now know that the cells of females function with a single X chromosome just as those of males do. This is most likely the reason why a zygote with one X chromosome (Turner syndrome) can survive. Then, too, all extra X chromosomes beyond a single one become Barr bodies, and this explains why poly-X females and XXY males are seen fairly frequently. An extra Y chromosome, called Jacobs syndrome, is tolerated in humans, most likely because the Y chromosome carries few genes. Jacobs syndrome (XYY) is due to nondisjunction during meiosis II of spermatogenesis. We know this because two Ys are present only during meiosis II in males.

Down Syndrome, an Autosomal Trisomy

The most common autosomal trisomy seen among humans is Down syndrome, also called trisomy 21. Persons with Down syndrome usually have three copies of chromosome 21 because the egg had two copies instead of one. (In 23% of the cases studied, however, the sperm had the extra chromosome 21.) The chances of a woman having a Down syndrome child increase rapidly with age, starting at about age 40. The reasons for this are still being determined.

Although an older woman is more likely to have a Down syndrome child, most babies with Down syndrome are born to women younger than age 40 because this is the age group having the most babies. Karyotyping can detect a Down syndrome child. However, young women are not routinely encouraged to undergo the procedures necessary to get a sample of fetal cells (i.e., amniocentesis or chorionic villi sampling) because the risk of complications is greater than the risk of having a Down syndrome child. Fortunately, a test based on substances in maternal blood can help identify fetuses who may need to be karyotyped.

Down syndrome is easily recognized by these common characteristics: short stature; an eyelid fold; a flat face; stubby fingers; a wide gap between the first and second toes; a large, fissured tongue; a round head; and a palm crease, the so-called simian line. Unfortunately, mental retardation, which can vary in intensity, is also a characteristic. Chris Burke (Fig. 18.15a) was born with Down syndrome, and his parents were advised to put him in an institution. But Chris's parents didn't do that. They gave him the same loving care and attention they gave their other children, and it paid off. Chris is remarkably talented. He is a playwright, actor, and musician. He starred in *Life Goes On,* a TV series written just for him, and he is sometimes asked to be a guest star in other TV shows. His love of music and collaboration with other musicians has led to the release of several albums—like Chris, the songs are uplifting and inspirational. You can read more about this remarkable young man in his autobiography, *A Special Kind of Hero.*

The genes that cause Down syndrome are located on the bottom third of chromosome 21 (Fig. 18.15b). Extensive investigative work has been directed toward discovering the specific genes responsible for the characteristics of the syndrome. Thus far, investigators have discovered several genes that may account for various conditions seen in persons with Down syndrome. For example, they have located genes most likely responsible for the increased tendency toward leukemia, cataracts, accelerated rate of aging, and mental retardation. The gene for mental retardation, dubbed the *Gart* gene, causes an increased level of purines in the blood, a finding associated with mental retardation. One day, it may be possible to control the expression of the *Gart* gene even before birth so that at least this symptom of Down syndrome does not appear.

Changes in Sex Chromosome Number

An abnormal sex chromosome number is the result of inheriting too many or too few X or Y chromosomes. Figure 18.14 can be used to illustrate nondisjunction of the sex chromosomes during oogenesis, if you assume that the chromosomes shown represent X chromosomes. Nondisjunction during oogenesis or spermatogenesis can result in gametes that have too few or too many X or Y chromosomes.

A person with Turner syndrome (XO) is a female, and a person with Klinefelter syndrome (XXY) is a male. This shows that in humans, the presence of a Y chromosome, not the number of X chromosomes, determines maleness. The *SRY*

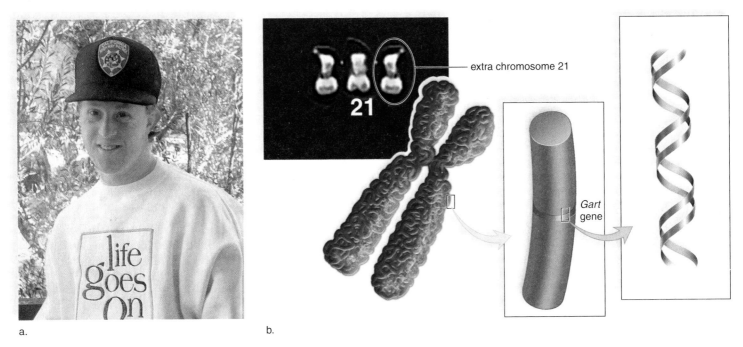

a.   b.

Figure 18.15   What causes Down syndrome?
a. Chris Burke was born with Down syndrome. Common characteristics of the syndrome include a wide, rounded face and a fold on the upper eyelids. Mental retardation, along with an enlarged tongue, makes it difficult for a person with Down syndrome to speak distinctly. **b.** Karotype of an individual with Down syndrome shows an extra chromosome 21. More sophisticated technologies allow investigators to pinpoint the location of specific genes associated with the syndrome. An extra copy of the *Gart* gene, which leads to a high level of purines in the blood, may account for the mental retardation seen in persons with Down syndrome.

(sex-determining region of Y) gene, on the short arm of the Y chromosome, produces a hormone called testis-determining factor. This hormone plays a critical role in the development of male genitals.

Turner Syndrome   From birth, an individual with Turner syndrome has only one sex chromosome, an X. As adults, Turner females are short, with a broad chest and folds of skin on the back of the neck. The ovaries, oviducts, and uterus are very small and underdeveloped. Turner females do not undergo puberty or menstruate, and their breasts do not develop. However, some have given birth following in vitro fertilization using donor eggs. They usually are of normal intelligence and can lead fairly normal lives if they receive hormone supplements.

Klinefelter Syndrome   About 1 in 650 live males are born with two X chromosomes and one Y chromosome. The symptoms of this condition (referred to as "47, XXY") are often so subtle that only 25% are ever diagnosed, and those are usually not diagnosed until after age 15. Earlier diagnosis opens the possibility for educational accommodations and other interventions that can help mitigate common symptoms, which include speech and language delays. Those 47, XXY males who develop more severe symptoms as adults are referred to as having "Klinefelter syndrome." All 47, XXY adults will require assisted reproduction to father children. Affected individuals commonly receive testosterone supplementation beginning at puberty.

Poly-X Females   A poly-X female has more than two X chromosomes and extra Barr bodies in the nucleus. Females with three X chromosomes have no distinctive phenotype, aside from a tendency to be tall and thin. Although some have delayed motor and language development, most poly-X females are not mentally retarded. Some may have menstrual difficulties, but many menstruate regularly and are fertile. Their children usually have a normal karyotype.

Females with more than three X chromosomes occur rarely. Unlike XXX females, XXXX females are more likely to be retarded. Various physical abnormalities are seen, but they may menstruate normally.

Jacobs Syndrome   XYY males with Jacobs syndrome can only result from nondisjunction during spermatogenesis. Affected males are usually taller than average, suffer from persistent acne, and tend to have speech and reading problems. At one time, it was suggested that these men were likely to be criminally aggressive, but it has been shown that the incidence of such behavior among them may be no greater than among XY males.

Changes in Chromosome Structure

Changes in chromosome structure are another type of chromosomal mutation. Various agents in the environment, such as radiation, certain organic chemicals, or even viruses, can cause chromosomes to break. Ordinarily, when breaks occur

When Your Child Is Disabled: Getting Help

If you are a parent who has just discovered that your child has a genetic disability, you are probably feeling confused and afraid. Most parents in this situation suffer from overwhelming feelings of loss and grief. This is normal. Though further genetic testing may determine if one parent carries the disease, this information can also lead to feelings of guilt or blame. You'll find that mothers and fathers often respond differently to the diagnosis in their child. This can lead to increased stress in a relationship. And you may be so overwhelmed in caring for this child that you feel as if you are neglecting your other children.

Fortunately, there are support groups and information sources to help you raise your disabled child. Major medical centers often have a genetics education center with medical specialists to help you. A vast amount of reliable information is available on the Internet. Knowing what to expect alleviates some confusion and fear. Meeting and talking with other parents of disabled children—those who are not only coping, but thriving—will enable you to regain your sense of hope. Marriage and family counselling may be advisable.

Your doctor should refer you to the proper agencies to get you started. Depending on your child's needs, government aid may be available to help pay for medical costs. A case worker will help you apply for those programs for which you qualify. There are waiting lists in some states so it is wise to apply as soon as possible.

Early intervention is important to allow your child to be as healthy and functional as possible. Regular doctor visits with a specialist in the area of the child's disability will enable you to take advantage of proper medications and therapies to help your child. Many states have free intervention services for children from birth to three years of age. Your case worker will know about these services.

The Individuals with Disabilities Education Act (IDEA) of 2004 requires that public schools develop an Individualized Education Plan (IEP) designed to meet the educational needs of qualifying disabled children. The special education experts at your local school will assist you in providing the appropriate education for your child. Some special education services, such as speech, occupational, and physical therapy, may also be available.

Although raising a child with a disability is not easy, help is available. You are not alone. Educate yourself and take advantage of groups and programs available in your area.

---

in chromosomes, the two broken ends reunite to give the same sequence of genes. Sometimes, however, the broken ends of one or more chromosomes do not rejoin in the same pattern as before. The result is various types of chromosomal mutation.

Changes in chromosome structure include deletions, translocations, duplications, and inversions of chromosome segments. A **deletion** occurs when an end of a chromosome breaks off or when two simultaneous breaks lead to the loss of an internal segment (Fig. 18.16*a*). Even when only one member of a pair of chromosomes is affected, a deletion often causes abnormalities.

A **duplication** is the presence of a chromosomal segment more than once in the same chromosome (Fig. 18.16*b*).

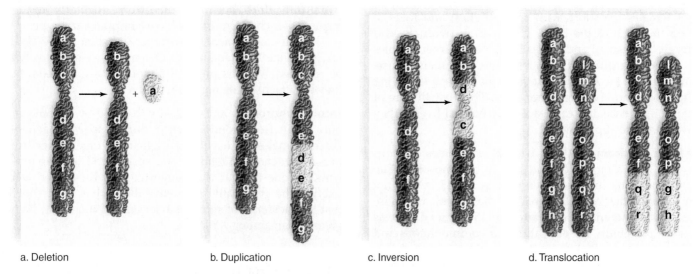

a. Deletion   b. Duplication   c. Inversion   d. Translocation

Figure 18.16 **What are the various types of chromosomal mutations?**
a. Deletion is the loss of a chromosome piece. **b.** Duplication occurs when the same piece is repeated within the chromosome. **c.** Inversion occurs when a piece of chromosome breaks loose and then rejoins in the reversed direction. **d.** Translocation is the exchange of chromosome pieces between nonhomologous pairs.

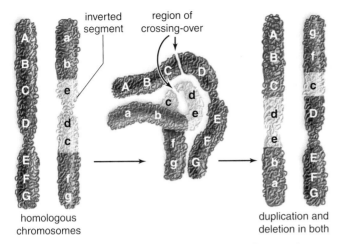

inverted segment

region of crossing-over

homologous chromosomes

duplication and deletion in both

Figure 18.17 How does an inversion cause changes in chromosome structure?

Left: A segment of one homologue is inverted. In the shaded segment, *edc* occurs instead of *cde. Middle:* The two homologues can pair only when the inverted sequence forms an internal loop. After crossing-over, a duplication and a deletion can occur. *Right:* The homologue on the left has *AB* and *ab* sequences and neither *fg* nor *FG* genes. The homologue on the right has *gf* and *FG* sequences and neither *AB* nor *ab* genes.

An **inversion** has occurred when a segment of a chromosome is turned around 180° (Fig. 18.16c). This reversed sequence of genes can lead to altered gene activities and to deletions and duplications as described in Figure 18.17.

A **translocation** is the movement of a chromosome segment from one chromosome to another nonhomologous chromosome (Fig. 18.16d). In 5% of cases, a translocation that occurred in a previous generation between chromosomes 21 and 14 is the cause of Down syndrome. In other words, because a portion of chromosome 21 is now attached to a portion of chromosome 14, the individual has three copies of the alleles that bring about Down syndrome when they are present in triplet copy. In these cases, Down syndrome is not related to the age of the mother but instead tends to run in the family of either the father or the mother.

Human Syndromes

Changes in chromosome structure occur in humans and lead to various syndromes, many of which are just now being discovered.

Deletion Syndromes Williams syndrome occurs when chromosome 7 loses a tiny end piece (Fig. 18.18). Children who have this syndrome look like pixies, with turned-up noses, wide mouths, a small chin, and large ears. Although their academic skills are poor, they exhibit excellent verbal and musical abilities. The gene that governs the production of the protein elastin is missing. This affects the health of the cardiovascular system and causes their skin to age prematurely. Such individuals are very friendly but need an ordered life, perhaps because of the loss of a gene for a protein normally active in the brain.

Cri du chat (cat's cry) syndrome is seen when chromosome 5 is missing an end piece. The affected individual has a small head, is mentally retarded, and has facial abnormalities. Abnormal development of the glottis and larynx results in the most characteristic symptom—the infant's cry resembles that of a cat.

Translocation Syndromes A person who has both of the chromosomes involved in a translocation has the normal amount of genetic material and is healthy, unless the chromosome exchange broke an allele into two pieces. The person who inherits only one of the translocated chromosomes will no doubt have only one copy of certain alleles and three copies of certain other alleles. A genetic counselor begins to suspect a translocation has occurred when spontaneous abortions are commonplace and family members suffer from various syndromes.

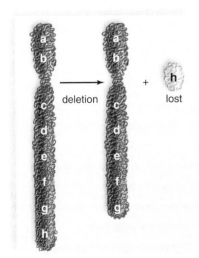

deletion

lost

a.

b.

Figure 18.18
What is an example of a chromosomal deletion and its result?
a. When chromosome 7 loses an end piece, the result is Williams syndrome. **b.** These children, although unrelated, have the same appearance, health, and behavioral problems.

Figure 18.19

What is an example of a chromosomal translocation and its result?

a. When chromosomes 2 and 20 exchange segments, **(b)** Alagille syndrome, with distinctive facial features, sometimes results because the translocation disrupts an allele on chromosome 20.

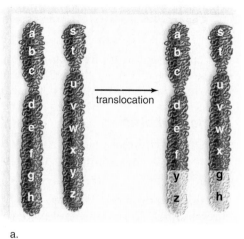

translocation

a.

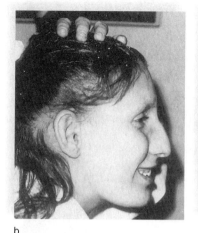

b.

Figure 18.19 shows a daughter and father who have a translocation between chromosomes 2 and 20. Although they have the normal amount of genetic material, they have the distinctive face (broad, prominent forehead and small, pointed chin), abnormalities of the eyes and internal organs, and severe itching characteristic of Alagille syndrome. People with this syndrome ordinarily have a deletion on chromosome 20. Therefore, it can be deduced that the translocation disrupted an allele on chromosome 20 in the father. The symptoms of Alagille syndrome range from mild to severe, so some people may not be aware they have the syndrome. This father did not realize it until he had a child with the syndrome.

Translocations can also be responsible for a variety of other disorders including certain types of cancer. In the 1970s, new staining techniques identified that a translocation from a portion of chromosome 22 to chromosome 9 was responsible for chronic myelogenous leukemia. This translocated chromosome was called Philadelphia chromosome. In Burkett lymphoma, a cancer common in children in equatorial Africa, a large tumor develops from lymph glands in the region of the jaw. This disorder involves a translocation from a portion of chromosome 8 to chromosome 14.

> **Check Your Progress 18.5**
>
> 1. What process usually causes an individual to have an abnormal number of chromosomes?
> 2. What is the specific chromosome abnormality of a person with Down syndrome?
> 3. What are some syndromes that result from inheritance of an abnormal sex chromosome number?
> 4. What are other types of chromosome mutations, aside from abnormal chromosome number?

CASE STUDY TEN MONTHS LATER

Mechelle checked her hair in the mirror before she left the house for her follow-up appointment with Dr. Kendrig. She could wear her hair behind her ears now. The earlobe was healing nicely, and she no longer had to hide it.

Her parents had not been terribly pleased with her. But in the end, they agreed that it was important to have the keloid removed, even if insurance wouldn't cover the procedure. Her parents had paid for half of the cost before treatment began, and Mechelle had to pay for the rest. Ten months later, her treatment was paid off and the ear looked good. She'd been very careful to follow Dr. Kendrig's instructions completely. She wanted to maximize the possibility of success. Not only did she not want to have that ugly lump on her ear, she didn't want to have to pay for any more procedures!

Dr. Kendrig's nurse greeted her and took her to an exam room. The surgery had been followed by cortisone injections around the edges of the wound every six weeks for six months. She had to wear a special pressure earring 24 hours a day for an entire year. It wasn't a hoop like she normally wore, but it was still very attractive. It looked almost like two fancy buttons, joined by a hinge, on both sides of her ear. When she put it on, it felt tight, like her earlobe was being pinched. But after a few minutes, she got used to the pressure and it wasn't noticeable.

Dr. Kendrig entered the room. "Hi, Mechelle. How's the earlobe?" Mechelle removed the earring and let Dr. Kendrig examine it. He looked pleased.

"Good job, Mechelle. This is healing nicely. Keep wearing those pressure earrings. It's important to constantly keep the pressure on the earlobe while it heals to prevent another keloid from forming. Do you have any questions or problems?"

Mechelle replaced her earring and shook her head. "Nope. I'm pretty happy." Her earlobe didn't look exactly like it had before she'd had it pierced, and it didn't look like the other earlobe. But she was probably the only one who would notice the difference.

"Good," Dr. Kendrig replied. "I'll see you again in two months, about a year post-op, and then again only if you have a problem."

Selecting Children

Human beings have always attempted to influence the characteristics of their children. For example, couples have attempted to determine the sex of their children for centuries through a variety of methods. Amniocentesis has allowed us to test fetuses for chromosomal abnormalities and debilitating developmental defects before birth. Modern genetic testing technology enables parents to directly select children bearing desired traits, even at the very earliest stages of development. See the Health Focus in Chapter 20, *Preimplantation Genetic Diagnosis,* for an explanation of this technology.

Fanconi's Anemia

Recently, preimplantation genetic diagnosis selected an embryo for a couple because the newborn could save the life of his sister (Fig. 18A). The couple, Jack and Lisa Nash, had a daughter with Fanconi's anemia, a rare inherited disorder in which affected persons cannot properly repair DNA damage that results from certain toxins. The disease primarily afflicts the bone marrow and, therefore, results in a reduction of all types of blood cells. Anemia occurs, due to a deficiency of red blood cells. Patients are also at high risk of infection, because of low white blood cell numbers, and of leukemia, because white blood cells cannot properly repair any damage to their DNA.

Fanconi's anemia may be treated by a traditional bone marrow transplant or by an adult stem cell transplant. The donor should be preferably a parent or sibling, because the risk of rejection is lower. Adult stem cells are almost always the preferred treatment option because stem cells are hardier and much less likely to be rejected than a bone marrow transplant. Recall that the umbilical cord of a newborn is a rich source of adult stem cells for all types of blood cells. (These are called adult because they are not embryonic stem cells.)

Testing the Embryo

During preimplantation genetic diagnosis, an embryonic cell is removed and tested for a disorder that runs in the family. In this case, doctors wanted to find an embryo that would become a healthy individual and also who would be able to benefit his/her sister. The parents underwent in vitro fertilization, and the 15 resulting embryos were screened to see which ones would not produce an individual with Fanconi's anemia and also had similar recognition proteins as their daughter. Two embryos met these requirements, but only one implanted in the uterus, and it developed into a healthy baby boy. Adult stem cells were harvested from the umbilical cord of the newborn and were successfully used to treat his sister's anemia. The physician who performed the genetic screening stated that he has received numerous inquiries about performing the procedure for other couples with diseased children. Molly Nash, now eleven years old, is still doing well.

Ethical Dilemma

This case, and other related cases, has raised a number of ethical issues surrounding prenatal selection of children based on genetic traits. The American Medical Association (AMA) insists that selection based on traits not related to disease is unethical. However, the AMA's chair of the Council on Ethical and Judicial Affairs made an exception for this case because the child was selected for medical reasons. Dr. Jacques Montagut, who helped develop in vitro fertilization, believes that it is dangerous to bear children for the purpose of curing others, and compared it with "a new form of biological slavery." Still others think that children will be selected for less altruistic reasons, such as for their height, physical prowess, or intellectual abilities.

Decide Your Opinion

1. In general, do you think it is ethical to have children to cure medically related conditions, regardless of how fertilization occurred? If not, do you agree with the AMA that this case is an acceptable exception?

2. The brother was created ostensibly as a treatment for his sister's disease. Do you believe that there is a moral obligation to provide him with compensation?

3. Would embryonic stem cells, derived from an aborted fetus and cultured in the laboratory, be an acceptable substitute?

4. Would you willingly donate sperm or eggs for in vitro fertilization to produce a healthy child for a couple who could not have one because of the risk of an inherited disease, such as Fanconi's anemia?

Figure 18A   Why did the Nash family's new baby raise such ethical issues?

Jack, Molly holding baby Adam, and Lisa Nash. Adam was genetically selected as an embryo because the stem cells of his umbilical cord would save the life of his sister, Molly.

Summarizing the Concepts

18.1 Chromosomes and the Cell Cycle

- Chromosomes occur in pairs in body cells.
- A karyotype is a visual display of a person's chromosomes.

  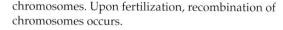

  A normal human karyotype shows 22 homologous pairs of autosomes and one pair of sex chromosomes.
- Normal sex chromosomes in males: XY
- Normal sex chromosomes in females: XX

The Cell Cycle

- The cell cycle occurs continuously and has four stages: G₁, S, G₂, (the interphase stages), and M (the mitotic stage, which includes cytokinesis and the stages of mitosis).
- In G₁, a cell doubles organelles and accumulates materials for DNA synthesis.
- In S, DNA replication occurs.
- In G₂, a cell synthesizes proteins needed for cell division.
- Apoptosis also occurs during the cell cycle.

18.2 Mitosis

Mitosis is duplication division that assures that all body cells have the diploid number and the same types of chromosomes as the cell that divides.

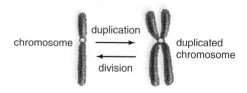

The phases of mitosis are prophase, metaphase, anaphase, and telophase.

- **Prophase** Chromosomes attach to spindle fibers.
- **Metaphase** Chromosomes align at the equator.
- **Anaphase** Chromatids separate, becoming chromosomes that move toward the poles.
- **Telophase** Nuclear envelopes form around chromosomes; cytokinesis begins.

  Cytokinesis is the division of cytoplasm and organelles following mitosis.
- The proper workings of the cell cycle and mitosis are critical to growth and tissue repair.

18.3 Meiosis

Meiosis involves two cell divisions: meiosis I and meiosis II.

- **Meiosis I** Homologous chromosomes pair and then separate.
- **Meiosis II** Sister chromatids separate, resulting in four cells with the haploid number of chromosomes that move into daughter nuclei.

  Meiosis results in genetic recombination due to crossing-over; gametes have all possible combinations of chromosomes. Upon fertilization, recombination of chromosomes occurs.

18.4 Comparison of Meiosis and Mitosis

- In prophase I, homologous chromosomes pair; there is no pairing in mitosis.
- In metaphase I, homologous duplicated chromosomes align at equator.
- In anaphase I, homologous chromosomes separate.

Spermatogenesis and Oogenesis

- **Spermatogenesis** In males, produces four viable sperm.
- **Oogenesis** In females, produces one egg and two or three polar bodies. Oogenesis goes to completion if the sperm fertilizes the developing egg.

18.5 Chromosome Inheritance

Meiosis is a part of gametogenesis (spermatogenesis in males and oogenesis in females) and contributes to genetic diversity.

Changes in Chromosome Number

- Nondisjunction changes the chromosome number in gametes, resulting in trisomy or monosomy.
- Autosomal syndromes include trisomy and Down syndrome.

Changes in Sex Chromosome Number

- Nondisjunction during oogenesis or spermatogenesis can result in gametes that have too few or too many X or Y chromosomes.
- Syndromes include Turner, Klinefelter, poly-X, and Jacobs.

Changes in Chromosome Structure

- Chromosomal mutations can produce chromosomes with deleted, duplicated, inverted, or translocated segments.
- These result in various syndromes such as Williams and cri du chat (deletion) and Alagille and certain cancers (translocation).

Understanding Key Terms

anaphase 423	interphase 421
apoptosis 422	inversion 439
aster 422	meiosis 427
Barr body 436	metaphase 423
cell cycle 421	mitosis 420
centriole 422	mitotic spindle 422
centromere 421	monosomy 435
centrosome 422	nondisjunction 435
cleavage furrow 424	oogenesis 434
crossing-over 430	parent cell 422
cytokinesis 422	polar body 435
daughter cell 422	prophase 423
deletion 438	secondary oocyte 435
diploid (2n) 422	sister chromatids 420
duplication 438	spermatogenesis 434
fertilization 427	synapsis 427
gamete 427	syndrome 421
haploid (n) 427	telophase 424
homologous chromosome 427	translocation 439
homologue 427	trisomy 435
interkinesis 427	zygote 427

Match the key terms to these definitions.

a. _____ Dark-staining nuclei of females that contains a condensed, inactive X chromosome.

b. _____ Member of a pair of chromosomes that are alike and come together in synapsis of prophase of meiosis I.

c. _____ Having one less chromosome than usual.

d. _____ A change in chromosome structure in which a segment is turned around 180°.

e. _____ Having the n number of chromosomes; for humans, n=23.

Testing Your Knowledge of the Concepts

1. Describe the two parts of the cell cycle. (pages 421–22)

2. What is the function of the mitotic spindle? (pages 423–24)

3. What is the importance of mitosis, and how is the process controlled? (pages 424, 426)

4. How do the terms diploid (2n) and haploid (n) relate to meiosis? (page 427)

5. List and describe the events in meiosis I and meiosis II. (pages 427–30)

6. What is the significance of meiosis? (pages 430–31)

7. Contrast mitosis and meiosis I and meiosis II. (pages 432–34)

8. How does spermatogenesis differ from oogenesis? (pages 434–35)

9. What is nondisjunction, when can it occur, and what are the results? (page 435)

10. What are some syndromes caused by changes in chromosome number, and what is the change? (pages 435–37)

11. Describe three changes that can occur in chromosome structure and some syndromes caused by such changes. (pages 437–40)

12. The point of attachment for two sister chromatids is the
    a. centriole.
    b. centromere.
    c. chromosome.
    d. karyotype.

13. Label the drawing of the cell cycle; then tell the main event of each stage.

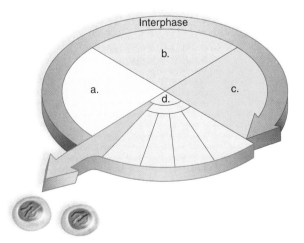

Interphase

In questions 14–20, match the statement to interphase or the phase of mitosis in the key.

Key:
  a. metaphase
  b. interphase
  c. telophase
  d. prophase
  e. anaphase

14. Spindle fibers begin to appear.

15. DNA synthesis occurs.

16. Chromosomes line up at the equator.

17. Duplicated chromosomes become visible.

18. Centromere splits and sister chromosomes move to opposite poles.

19. Cytokinesis occurs during this phase.

20. Chromosomes duplicate..

21. If a parent cell has 18 chromosomes before mitosis, how many chromosomes will the daughter cells have?
    a. 18   c. 9
    b. 36   d. 27

22. In humans, mitosis is necessary to
    a. grow and repair damaged tissue.
    b. form the gametes.
    c. maintain the same chromosome number in body cells.
    d. Both a and c are correct.

23. If a parent cell has 18 chromosomes, how are they arranged at the equator during metaphase of mitosis?
    a. single file of 18 replicated chromosomes in random order
    b. double file of 9 replicated chromosomes in random order
    c. single file of 18 replicated chromosomes from number 1 to 18
    d. double file of 9 replicated chromosomes with 1–9 from the mother on one side and 1–9 from the father on the other side

24. Crossing-over occurs between
    a. sister chromatids of the same chromosome.
    b. chromatids of nonhomologous chromosomes.
    c. nonsister chromatids of a homologous pair.
    d. Both b and c are correct.

25. At the equator in metaphase of meiosis I, there are
    a. single chromosomes in random order.
    b. unpaired duplicated chromosomes in random order.
    c. homologous pairs in random order.
    d. homologous pairs with those from the mother on one side and from the father on the other side.

26. The products of _____ are _____ cells.
    a. mitosis, diploid
    b. meiosis, haploid
    c. meiosis, diploid
    d. Both a and b are correct.

27. If a parent cell has 22 chromosomes, the daughter cells following meiosis II will have
    a. 22 chromosomes.
    b. 44 chromosomes.
    c. 11 chromosomes.
    d. Any of these could be correct.

28. Which of these drawings represents metaphase of meiosis I?

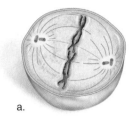 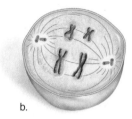

a.   b.

In questions 29–33, match the part of the diagram to the correct label.

29. Meiosis II

30. Primary spermatocyte

31. Sperm

32. Secondary spermatocyte

33. Spermatids

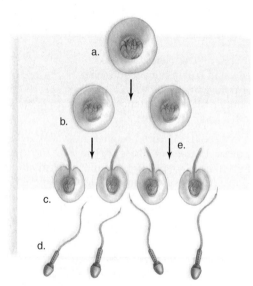

34. Which of these helps to ensure that genetic diversity will be maintained?
    a. independent alignment during metaphase I
    b. crossing-over during prophase I
    c. fusion of sperm and egg nuclei during fertilization
    d. All of these are correct.

35. How many viable cells are produced by oogenesis?
    a. four
    b. three
    c. one
    d. two

36. Monosomy or trisomy occurs because of
    a. crossing-over.
    b. inversion.
    c. translocation.
    d. nondisjunction.

37. Trisomy of chromosome 21 is called
    a. Down syndrome.
    b. Klinefelter syndrome.
    c. Turner syndrome.
    d. Jacobs syndrome.

38. A person with Klinefelter syndrome is _____ and has _____ sex chromosomes.
    a. male, XYY
    b. male, XXY
    c. female, XXY
    d. female, XO

Thinking Critically About the Concepts

Mechelle Wilson, the patient in the case study, developed a keloid due to an ear piercing. A keloid is a form of abnormal wound healing resulting in a benign tumor. Affected cells undergo increased mitosis in response to an injury. Keloid formation tends to run in families, so it may have a genetic component. Mechelle's mother had a bad scar on her arm, suggesting that she, too, had this tendency to heal abnormally.

1. Benign and cancerous tumors occur when the cell cycle control mechanisms no longer operate correctly. What types of genes may be involved in these cell cycle control mechanisms? (*Hint:* Reexamine Fig. 18.7.)

2. Mitosis goes on continually in your body: in your blood cells, skin cells, and the cells that line the respiratory and digestive tracts. Why might Mechelle have problems with wound healing, but not with these other sites where mitosis occurs continuously?

3. Explain how the separation of homologous chromosomes during meiosis affects the appearance of siblings (that some resemble each other and others look very different from one another).

4. a. What would you conclude about the ability of nervous and muscle tissue to repair themselves if nerve and muscle cells are typically arrested in G_1 of interphase?
   b. What are the implications of this arrested state to someone who suffers a spinal cord injury or heart attack?

S omething about cute little Cody Keenan didn't strike Mary Jean Evans quite right. After the diaper change that followed naptime at the day care center, she studied his diaper carefully. She wasn't imagining this—there was blood with the urine in that diaper. And after she'd raised her kids and grandkids, and then dozens of babies at the center, she could always settle a baby. None of her tricks was working this time, however, and that worried her. Cody fussed and wouldn't take his bottle.

Kim Keenan, his mother, was always on time for pick up, and she always wanted a full report on her baby's day. Mary Jean described the fussiness, then showed Kim the pinkish stain on Cody's diaper. "See that, Kim?" Mary Jean pointed out. "And he's giving us the 'my tummy hurts' signal. He's pulling up his legs and crossing his arms over his belly."

"He seems warm," Kim answered worriedly. "He's teething, so he might have a little fever. Or maybe he's got some kind of stomach bug."

"Maybe," Mary Jean answered. "But that blood came out with his urine. I think you need to take him to the doctor."

"He's due into the pediatrician on Wednesday for another round of shots and a well-baby checkup," Kim replied absently. Cody continued to wail. "Maybe the pediatrician can just move his appointment up. I think the office is open until seven o'clock.

"I don't think we'll be in tomorrow," she told Mary Jean. "I'll just stay home with him."

"You really do need to take him to the doctor right away. Oh, and don't forget to give him those acetaminophen drops before his shots," Mary Jean reminded Kim. "That will keep him from being so fussy later. It always worked with my grandbabies, but you know, you can't use aspirin any more because they found out it might hurt babies. . . ."

"Thanks, Mary Jean," Kim interrupted with a tired smile as she left. "I won't forget and I'll keep you posted."

Mary Jean dreaded the call she had to make next. Kim was a friendly person and seemed like such a loving mother … but blood in a baby's wet diaper was *never* normal. Most likely it was a sign of illness, but it could also signal child abuse. As a child care provider, Mary Jean knew she was obligated by law to call the state Children's Services department any time that abuse might have occurred.

CHAPTER CONCEPTS

19.1 Cancer Cells
Cancer cells have a number of abnormal characteristics that prevent them from functioning in the same manner as normal cells. They divide repeatedly and form tumors in the place of origin and in other parts of the body.

19.2 Causes and Prevention of Cancer
Whether cancer develops is partially due to inherited genes, but exposure to carcinogens such as UV radiation, tobacco smoke, pollutants, industrial chemicals, and certain viruses play a significant role also.

19.3 Diagnosis of Cancer
Cancer is usually diagnosed by certain screening procedures and by imaging the body and tissues, using various techniques.

19.4 Treatment of Cancer
Surgery followed by radiation and/or chemotherapy has now become fairly routine. Immune therapy, bone marrow transplants, and other methods are under investigation.

19.1 Cancer Cells

Cancer is over a hundred different diseases and each type of cancer can vary from another. However, some characteristics are common to cancer cells.

Characteristics of Cancer Cells

Cancer is a cellular disease, and cancer cells share traits that distinguish them from normal cells.

Cancer Cells Lack Differentiation

Cancer cells are nonspecialized and do not contribute to the functioning of a body part. A cancer cell does not look like a differentiated epithelial, muscle, nervous, or connective tissue cell. Instead, it looks distinctly abnormal.

Cancer Cells Have Abnormal Nuclei

In Figure 19.1, you can compare the appearance of normal cervical cells (Fig. 19.1a) with that of cancerous cervical cells. The nuclei of cancer cells are enlarged and may contain an abnormal number of chromosomes. The nuclei of the cervical cancer cells shown in Figure 19.1b, c have increased to the point that they take up most of the cell.

In addition to nuclear abnormalities, cancer cells have defective chromosomes. Some portions of the chromosomes may be duplicated, and/or some may be deleted. In addition, gene amplification (extra copies of specific genes) is seen much more frequently than in normal cells. Ordinarily, cells with damaged DNA undergo **apoptosis,** or programmed cell death. Cancer cells fail to undergo apoptosis, even though they are abnormal cells.

Tissues that divide frequently, such as those that line the respiratory and digestive tracts, are more likely to become cancerous. Cell division gives them the opportunity to undergo genetic mutations, each one making the cell more abnormal and giving it the ability to produce more of its own type.

Cancer Cells Have Unlimited Potential to Replicate

Ordinarily, cells divide about 60–70 times and then just stop dividing and die. Cancer cells are immortal and keep on dividing for an unlimited number of times.

Just as shoelaces are capped by small pieces of plastic, chromosomes in human cells end with special repetitive DNA sequences called **telomeres.** Specific proteins bind to telomeres in both normal and cancerous cells. These telomere proteins protect the ends of chromosomes from DNA repair enzymes. Though the enzymes effectively repair DNA in the center of the chromosome, they always tend to bind together the naked ends of chromosomes. In a normal cell, the telomeres get shorter after each cell cycle and protective telomere proteins gradually decrease. In turn, repair enzymes eventually cause the chromosomes' ends to bind together, causing the cell to undergo apoptosis and die. Telomerase is an enzyme that can rebuild telomere sequences and in that way prevent a cell from ever losing its potential to divide. The gene that codes for telomerase is constantly turned on in cancer cells, and telomeres are continuously rebuilt. The telomeres remain at a constant length, and the cell can keep diving over and over.

Cancer Cells Form Tumors

Normal cells anchor themselves to a substratum and/or adhere to their neighbors. They exhibit *contact inhibition*—when they come in contact with a neighbor, they stop dividing. Cancer cells have lost all restraint. They pile on top of one another and grow in multiple layers, forming a **tumor.** As cancer develops, the most aggressive cell becomes the dominant cell of the tumor (Fig. 19.2).

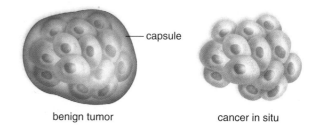

benign tumor   cancer in situ

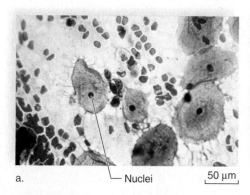

a.   Nuclei   50 μm

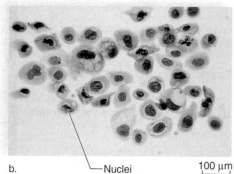

b.   Nuclei   100 μm

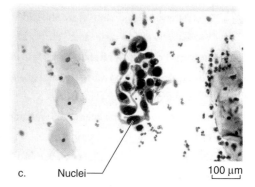

c.   Nuclei   100 μm

Figure 19.1   **How are normal tissue cells structurally different from cancer cells?**
a. Normal cervical cells. **b.** Precancerous cervical cells. **c.** Cancerous cervical cells.

Have You Ever Wondered...

What's the difference between a benign tumor and a malignant tumor?

A **benign** tumor is usually surrounded by a connective tissue capsule. A benign tumor will not invade adjacent tissue because of the capsule. The cells of a benign tumor resemble normal cells fairly closely. Chances are you have one of these tumors. If you have a mole, or *nevus,* on your body, you have a benign tumor of the skin melanocytes. Cells from a mole closely resemble other melanocytes.

However, despite the term benign, this type of tumor isn't necessarily harmless. If it presses on normal tissue, or restricts the normal tissue's blood supply, a benign tumor can be fatal. For example, a benign neuroma (nerve cell tumor) growing near the brainstem eventually affects control centers for heartbeat and respiration.

On the other hand, a **malignant** tumor is able to invade surrounding tissues, and its cells don't resemble normal cells. By traveling in blood or lymph vessels, the tumor can spread throughout the body. As a general rule, badly deformed cells are most likely to spread. Thus, they are the most malignant.

Cancer *in situ* (in place) is a malignant tumor found in its place of origin. As yet, the cancer has not spread beyond the basement membrane, the nonliving material that anchors tissues to one another. If recognized and treated early, cancer *in situ* is usually curable.

Cancer Cells Have No Need for Growth Factors

Chemical signals between cells tell them whether they should be dividing or not dividing. These chemical signals, called growth factors, are of two types: stimulatory growth factors and inhibitory growth factors. Cancer cells keep on dividing, even when stimulatory growth factors are absent, and they do not respond to inhibitory growth factors.

Cancer Cells Gradually Become Abnormal

Figure 19.2 illustrates that **carcinogenesis,** the development of cancer, is a multistage process that can be divided into these three phases:

Initiation: A single cell undergoes a mutation that causes it to begin to divide repeatedly (Fig. 19.2*a*).
Promotion: A tumor develops, and the tumor cells continue to divide. As they divide, they undergo mutations (Fig. 19.2*b,c*).
Progression: One cell undergoes a mutation that gives it a selective advantage over the other cells (Fig. 19.2*c*). This process is repeated several times, and, eventually, there is a cell that has the ability to invade surrounding tissues (Fig. 19.2*d-e*).

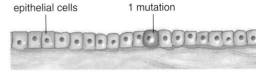

a. Cell (dark pink) acquires a mutation for repeated cell division.

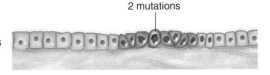

b. New mutations arise, and one cell (brown) has the ability to start a tumor.

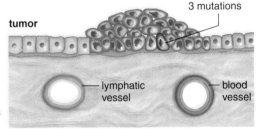

c. Cancer in situ. The tumor is at its place of origin. One cell (purple) mutates further.

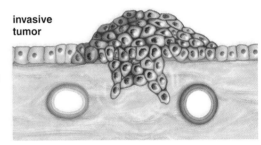

d. Cells have gained the ability to invade underlying tissues by producing a proteinase enzyme.

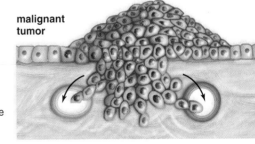

e. Cancer cells now have the ability to invade lymphatic and blood vessels.

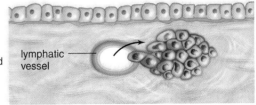

f. New metastatic tumors are found some distance from the original tumor.

Figure 19.2 **How does cancer progress from a single mutation into a full-fledged tumor?**
a. One cell (dark pink) in a tissue mutates. **b.** This mutated cell divides repeatedly and a cell (brown) with two mutations appears. **c.** A tumor forms and a cell with three mutations (purple) appears. **d.** This cell (purple), which takes over the tumor, can invade underlying tissue. **e.** Tumor cells invade lymphatic and blood vessels. **f.** A new tumor forms at a distant location.

Cancer Cells Undergo Angiogenesis and Metastasis

To grow larger than about a billion cells (about the size of a pea), a tumor must have a well-developed capillary network to bring it nutrients and oxygen. **Angiogenesis** is the formation of new blood vessels. The low oxygen content in the middle of a tumor may turn on genes coding for angiogenic growth factors that diffuse into the nearby tissues and cause new vessels to form.

Due to mutations, cancer cells tend to be motile. They have a disorganized internal cytoskeleton and lack intact actin filament bundles. To metastasize, cancer cells must make their way across the basement membrane and invade a blood vessel or lymphatic vessel. Invasive cancer cells are sperm-shaped (see Fig. 19.1c) and don't look at all like the normal cell seen nearby. Cancer cells produce proteinase enzymes that degrade the basement membrane and allow them to invade underlying tissues. Malignancy is present when cancer cells are found in nearby lymph nodes. When these cells begin new tumors far from the primary tumor, **metastasis** has occurred (Fig. 19.2f). Not many cancer cells achieve this feat (maybe 1 in 10,000), but those that successfully metastasize to various parts of the body lower the prognosis (the predicted outcome of the disease) for recovery.

Cancer Results from Gene Mutation

Recall that the cell cycle, as discussed on pages 421–26, consists of interphase, followed by mitosis. **Cyclin** is a protein molecule that has to be present for a cell to proceed from interphase to mitosis. When cancer develops, the cell cycle occurs repeatedly, in large part, due to mutations in two types of genes. Figure 19.3 shows that:

1. **Proto-oncogenes** code for proteins that promote the cell cycle and prevent apoptosis. They are often likened to the gas pedal of a car because they cause acceleration of the cell cycle.
2. **Tumor-suppressor genes** code for proteins that inhibit the cell cycle and promote apoptosis. They are often likened to the brakes of a car because they inhibit acceleration.

Proto-Oncogenes Become Oncogenes

When proto-oncogenes mutate, they become cancer-causing genes called **oncogenes.** These mutations can be called "gain-of-function" mutations because overexpression is the result (Fig. 19.4). Whatever a proto-oncogene does, an oncogene does it better.

A **growth factor** is a signal that activates a cell-signaling pathway, resulting in cell division. Some proto-oncogenes code for a growth factor or for a receptor protein that receives a growth factor. When these proto-oncogenes become oncogenes, receptor proteins are easy to activate and may even be stimulated by a growth factor produced by the receiving cell. Several proto-oncogenes code for Ras

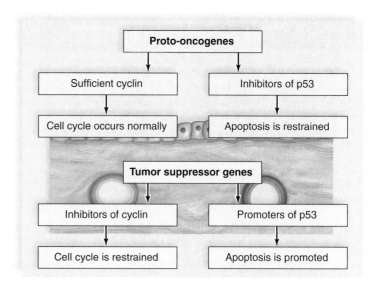

Figure 19.3 How do proto-oncogenes and tumor suppressor genes function in a normal cell?
In normal cells, proto-oncogenes code for sufficient cyclin to keep the cell cycle going normally, and they code for proteins that inhibit p53 and apoptosis. Tumor-suppressor genes code for proteins that inhibit cyclin and promote p53 and apoptosis.

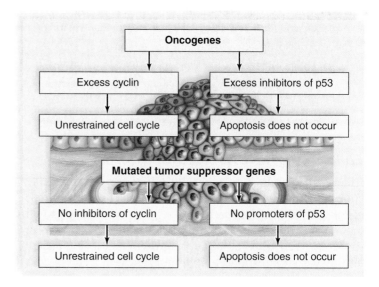

Figure 19.4 What happens after mutation to proto-oncogenes or tumor suppressor genes?
Oncogenes and mutated tumor-suppressor genes have the same end effects: an unrestrained cell cycle, and apoptosis does not occur.

proteins that promote mitosis by activating cyclin. *Ras* oncogenes are typically found in many different types of cancers. *Cyclin D* is a proto-oncogene that codes for cyclin directly. When this gene becomes an oncogene, cyclin is readily available all the time.*

*By convention, the gene's name is italicized. Its product protein is not.

p53 is a protein that activates repair enzymes. At the same time, p53 turns on genes that stop the cell cycle from proceeding. If repair is impossible, the p53 protein promotes *apoptosis,* programmed cell death. Apoptosis is an important way for carcinogenesis to be prevented. One proto-oncogene codes for a protein that functions to make p53 unavailable. When this proto-oncogene becomes an oncogene, no p53 will be available, regardless of how much is made. Many tumors are lacking in p53 activity.

Tumor-Suppressor Genes Become Inactive

When tumor-suppressor genes mutate, their products no longer inhibit the cell cycle nor promote apoptosis. Therefore, these mutations can be called "loss of function" mutations (Fig. 19.4).

Mutation of the tumor suppressor gene *Bax* is a good example. Its product, the protein Bax, promotes apoptosis. When *Bax* mutates, Bax protein is not present, and apoptosis is less likely to occur. The *Bax* gene contains a line of eight consecutive G bases in its DNA (see page 38). When the same base molecules are lined up in this fashion, the gene is more likely to be subject to mutation.

Types of Cancer

Statistics indicate that one in three Americans will deal with cancer in their lifetime. Therefore, this is a topic of considerable importance to the health and well-being of every individual. **Oncology** is the study of cancer. A medical specialist in cancer is, therefore, known as an **oncologist.** The patient's prognosis (probable outcome) depends on (1) whether the tumor has invaded surrounding tissues and (2) whether there are metastatic tumors in distant parts of the body.

Tumors are classified according to their place of origin. **Carcinomas** are cancers of the epithelial tissues and adenocarcinomas are cancers of glandular epithelial cells. Carcinomas include cancer of the skin, breast, liver, pancreas, intestines, lung, prostate, and thyroid. **Sarcomas** are cancers

CASE STUDY CENTRAL CHILDREN'S HOSPITAL

"What do you mean, we can't take Cody home?" Scott Keenan growled. "He's our son."

"Yes, Mr. Keenan," the social worker replied gently, "but in any situation with a potential for child abuse, the infant is placed in protective hospital custody. As soon as the test results come back, Child Protective Services will re-evaluate your situation. You or your wife can stay with him, but you can't take him home." The social worker left the waiting room and closed the door softly behind him.

Kim entered soon afterwards, and sat next to her husband. "They're doing the CT scan right now, so I can't be in there. They gave him something to make him drowsy, so at least he isn't crying anymore. Their pediatrician said he felt something in Cody's abdomen. Oh, Scott, I just want all this to be over!"

Scott clasped her hand. "They won't let us take him home until all the tests are finished. The social worker said they have to check him for child abuse, because of the blood in his urine." He scrubbed his face with his free hand. "This is like some kind of nightmare."

The X-ray technician entered an hour later, cradling their sleepy son. She was accompanied by two doctors, both dressed in surgical scrubs. The young woman offered her hand. "Mr. and Mrs. Keenan? I'm Beth Morgan, and I'm the doctor who first examined Cody. This is Dr. Stanislawski."

"Call me Stan, okay?" the older doctor smiled. "I'm a pediatric urologist—a kidney and urinary specialist. After your pediatrician sent you to the hospital, Beth called me in for a consult on your son. We found no evidence at all of child abuse," Stan asserted. "No bruising externally, no damage to any of the internal organs."

"We could have told you that already!" Scott exclaimed in disgust. "We wouldn't hurt our baby."

"We have to check; it's the law. But we're still going to admit Cody to the hospital right away," Stan continued. "We think Cody has a nephroblastoma. It used to be called a Wilms' tumor. It's a cancer in his kidney."

"Cancer in his kidney?" Kim echoed. "I didn't know you could get cancer there!"

"You can get cancer in any part of the body," Stan explained. "It just happens whenever and wherever a group of cells grows out of control. Cody is lucky in a way, because this type of cancer sometimes doesn't show symptoms at all. Blood in the urine and pain in his belly tipped us off, and the tumor is still small. We know that blood in the urine means the kidney's blood vessels have been damaged, and removing the kidney is Cody's best chance for a cure.

"Don't forget that Cody has two kidneys," Stan reassured. "Once we make sure his left kidney is working properly, we'll remove the right kidney and put him on chemotherapy. Just for a little while," he comforted the distraught couple. "And this cancer is one of the most curable there is, if you catch it early and the child is little."

that arise in muscles and connective tissue, such as bone and fibrous connective tissue. **Leukemias** are cancers of the blood, and **lymphomas** are cancers of lymphatic tissue. A blastoma is a cancer composed of immature cells. Recall that the embryo is formed from three primary germ layers: ectoderm, mesoderm, and endoderm (see Chap. 17). Each blastoma cell resembles the cells in its original primary germ layer. For example, a nephroblastoma (like Cody's in the case study) has cells similar to mesoderm cells, because the kidney grows from mesoderm.

Common Cancers

Cancer occurs in all parts of the body, but some organs are more susceptible than others (Fig. 19.5). In the respiratory system, lung cancer is the most common type. Overall, this is one of the most common types of cancer, and smoking is known to increase a person's risk for this disease. Smoking also increases the risk for cancer in the oral cavity.

In the digestive system, colorectal cancer (colon/rectum) is another common tumor. Other cancers of the digestive system include those of the pancreas, stomach, esophagus, and other organs. In the cardiovascular system, cancers include leukemia and plasma cell tumors. In the lymphatic

system, cancers are classified as either Hodgkin or non-Hodgkin lymphoma. Hodgkin lymphomas develop from mutated B-cells. Non-Hodgkin lymphomas can arise from B-cells or T-cells. Thyroid cancer is the most common type of tumor in the endocrine system, whereas brain and spinal tumors are found in the central nervous system.

Breast cancer is one of the most common types of cancer. Although predominantly found in women, occasionally it is also found in men. Cancers of the cervix, ovaries, and other reproductive structures also occur in women. In males, prostate cancer is one of the most common cancers. Other cancers of the male reproductive system include cancer of the testis and the penis. Bladder and kidney cancers are associated with the urinary system. Skin cancers include melanoma and basal cell carcinoma.

Check Your Progress 19.1

1. What characteristics of cancer cells allow them to grow uncontrollably?
2. Mutations in what two types of genes lead to uncontrollable growth?
3. What are the most common types of cancer?

Male	Female
Prostate 218,890 (29%)	Breast 178,480 (26%)
Lung and bronchus 114,760 (15%)	Lung and bronchus 98,620 (15%)
Colon and rectum 79,130 (10%)	Colon and rectum 74,630 (11%)
Urinary bladder 54,040 (7%)	Uterine corpus 39,080 (6%)
Melanoma of the skin 33,190 (4%)	Non-Hodgkin lymphoma 28,990 (4%)
Non-Hodgkin lymphoma 34,200 (4%)	Melanoma of the skin 26,030 (4%)
Kidney and renal pelvis 31,590 (4%)	Thyroid 25,480 (4%)
Oral cavity and pharynx 24,180 (3%)	Ovary 22,430 (3%)
Leukemia 24,800 (3%)	Urinary bladder 17,120 (3%)
Pancreas 18,830 (2%)	Pancreas 18,340 (3%)
All sites 766,860 (100%)	All sites 678,060 (100%)

a. **Cancer cases by site and sex**

Male	Female
Lung and bronchus 89,510 (31%)	Lung and bronchus 70,880 (26%)
Colon and rectum 26,000 (10%)	Breast 40,460 (15%)
Prostate 27,050 (9%)	Colon and rectum 26,180 (10%)
Pancreas 16,840 (6%)	Pancreas 16,530 (6%)
Leukemia 12,320 (4%)	Ovary 15,280 (6%)
Liver and intrahepatic bile duct 11,280 (4%)	Leukemia 9,470 (4%)
Esophagus 10,900 (4%)	Non-Hodgkin lymphoma 9,060 (3%)
Non-Hodgkin lymphoma 9,600 (3%)	Uterine corpus 7,400 (3%)
Urinary bladder 9,630 (3%)	Multiple myeloma 5,240 (2%)
Kidney and renal pelvis 8,080 (3%)	Brain and other nervous system 5,590 (2%)
All sites 289,550 (100%)	All sites 270,100 (100%)

b. **Cancer deaths by site and sex**

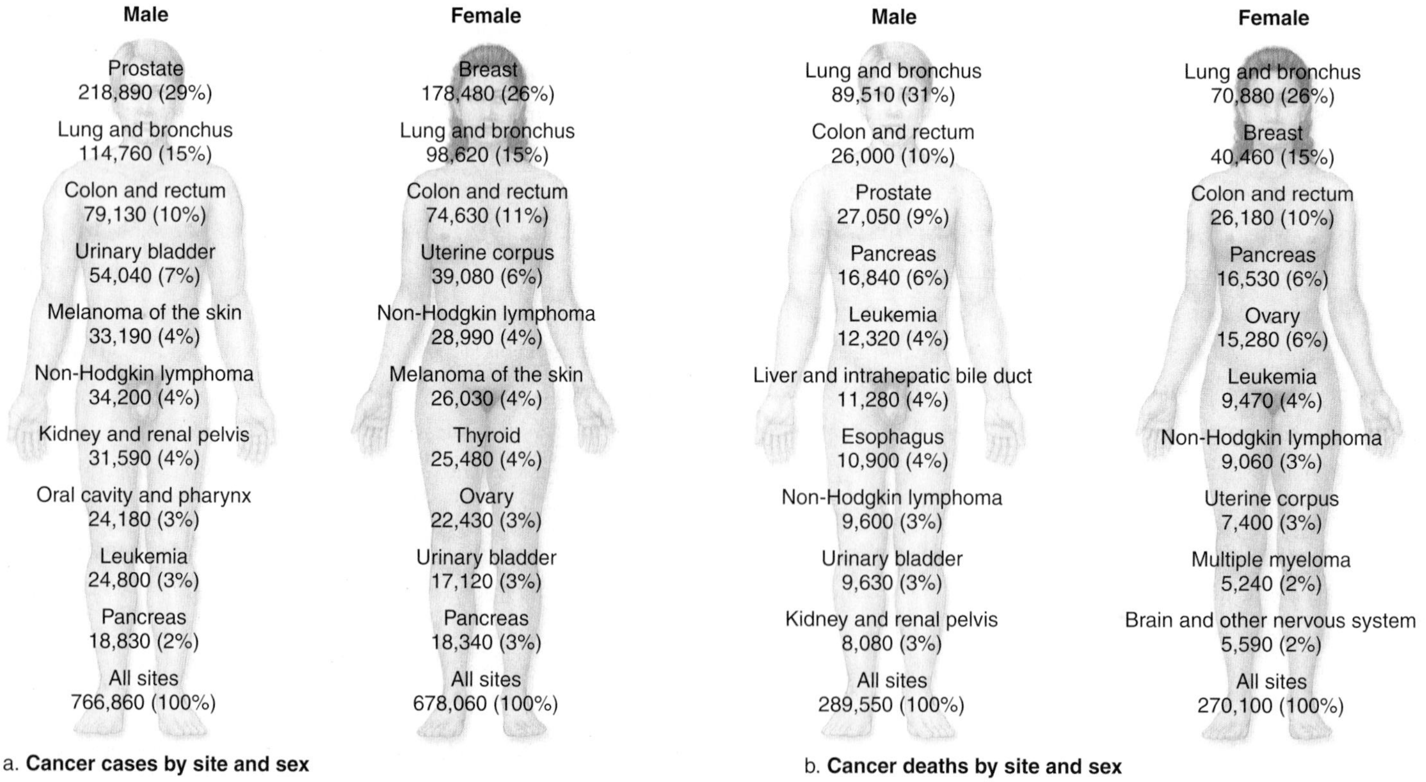

Figure 19.5 Which types of cancers are most likely to occur in the United States? Which types are most likely to be fatal?
a. Estimated new cancer cases; percentages for leading sites in 2007. **b.** Cancer deaths; percentages of leading sites in 2007.

19.2 Causes and Prevention of Cancer

Our current understanding of the causes of cancer are incomplete. However, by studying patterns of cancer development in populations, scientists have determined that both heredity and environmental risk factors come into play. Obviously, one's genetic inheritance can't be changed. But avoiding risk factors and following dietary guidelines can help prevent cancer.

Heredity

The first gene associated with breast cancer was discovered in 1990. Scientists named this gene *breast cancer 1* or *BRCA1* (pronounced brak-uh). Later, they found that breast cancer could also be due to another breast cancer gene they called *BRCA2*. These genes are tumor-suppressor genes that follow a particular inheritance pattern. We inherit two copies of every gene; one from each parent. If a mutated copy of *BRCA1* or *2* is inherited from either parent, a mutation in the other copy is required before predisposition to cancer is increased. Each cell in the body already has the single mutated copy, but cancer develops wherever the second mutation occurs. If the second mutation occurs in the breast, breast cancer may develop. Ovarian cancer may develop if a second cancer-causing mutation occurs there, instead.

The *RB* gene is also a tumor-suppressor gene. It takes its name from its association with retinoblastoma, a rare eye cancer that occurs almost exclusively in early childhood. In an affected child, a single mutated copy of the *RB* gene is inherited. A second mutation, this time to the other copy of the *RB* gene, causes the cancer to develop (Fig. 19.6). Retinoblastoma affecting a single eye is treated by removal of the eye, because total re-moval is more likely to result in a cure. Radiation, laser, and chemotherapy are options when both eyes are involved, but these treatments are less likely to produce a cure.

An abnormal *RET* gene, which predisposes an individual to thyroid cancer, can be passed from parent to child. *RET* is a proto-oncogene known to be inherited in an autosomal dominant manner—only one mutation is needed to increase a predisposition to cancer (see Chap. 20). The remainder of the mutations necessary for a thyroid cancer to develop are acquired (not inherited).

Environmental Carcinogens

A **mutagen** is an agent that causes mutations. A simple laboratory test, called an Ames test, is capable of testing if a substance is mutagenic. A **carcinogen** is a chemical that causes cancer by being mutagenic. Some carcinogens cause only initiation. Others cause initiation and promotion.

Heredity can predispose a person to cancer, but whether it develops or not depends on environmental mutagens, such as the ones we will be discussing.

Radiation

Ionizing radiation, such as in radon gas, nuclear fuel, and X-rays, is capable of affecting DNA and causing mutations. Though not a form of ionizing radiation, ultraviolet light also causes mutations. Ultraviolet radiation in sunlight and tanning lamps is most likely responsible for the dramatic increases seen in skin cancer in the past several years. Today, at least six cases of skin cancer occur for every one case of lung cancer. Nonmelanoma skin cancers are usually curable through surgery. However, melanoma skin cancer tends to metastasize and is responsible for 1–2% of total cancer deaths in the United States.

Another natural source of ionizing radiation is radon gas, which comes from the natural (radioactive) breakdown of uranium in soil, rock, and water. The Environmental Protection Agency recommends that every home be tested for radon because it is the second-leading cause of lung cancer in the United States. The combination of radon gas and smoking cigarettes can be particularly dangerous. A vent pipe system and fan, which pulls radon from beneath the house and vents it to the outside, is the most common way to rid a house of radon after the house is constructed.

We are well aware of the damaging effects of a nuclear bomb explosion or accidental emissions from nuclear power plants. For example, more cancer deaths have occurred in the vicinity of the Chernobyl Power Station (in Ukraine), which suffered a terrible accident in 1986. Usually, however, diagnostic X-rays account for most of our exposure to artificial sources of radiation. The benefits of these procedures can far outweigh the possible risk, but it is still wise to avoid X-ray procedures that are not medically warranted. When X-rays are necessary for therapy, nearby noncancerous tissues should be shielded carefully.

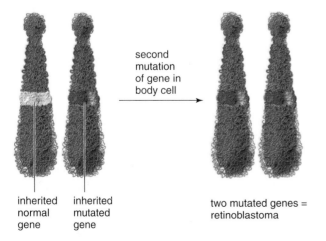

second mutation of gene in body cell

inherited normal gene

inherited mutated gene

two mutated genes = retinoblastoma

Figure 19.6 **How is retinoblastoma inherited?**
A child is at risk for an eye tumor when a mutated copy of the *RB* gene is inherited, even though a second mutation in the normal copy is required before the tumor develops.

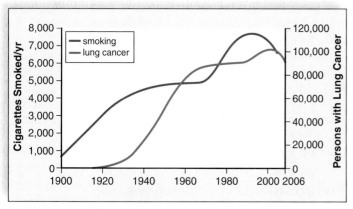

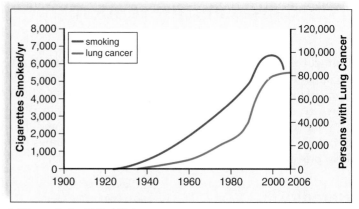

a. Men in the United States

b. Women in the United States

Figure 19.7   Does smoking really cause cancer?

a. As the incidence of smoking in men grew, so did the incidence of lung cancer. Lung cancer was rare in 1920, but as more men took up smoking, lung cancer became more common. **b.** As late as 1960, the incidence of lung cancer in women was low and so was the incidence of smoking. As women became smokers, the incidence of lung cancer in women also rose.

Recently, there has been a great deal of public concern regarding the presumed danger of nonionizing radiation. This energy form is given off by cell phones, electric lines, and appliances. No evidence has been found that links nonionizing radiation to cancer, however.

Organic Chemicals

Certain chemicals, particularly synthetic organic chemicals, have been found to be risk factors for cancer. We will mention only two examples: the organic chemicals in tobacco and those pollutants in the environment. There are many other chemical carcinogens.

Tobacco Smoke   Tobacco smoke contains a number of organic chemicals that are known mutagens, including nitroso-nor-nicotine, vinyl chloride, and benzo[a]pyrenes (a known suppressor of p53). The greater the number of cigarettes smoked per day and the earlier the habit starts, the more likely it is that cancer will develop. On the basis of data, such as shown in Figure 19.7, scientists estimate that about 80% of all cancers, including oral cancer and cancers of the larynx, esophagus, pancreas, bladder, kidney, and cervix, are related to the use of tobacco products. When smoking is combined with drinking alcohol, the risk of these cancers increases.

Passive smoking, or inhalation of someone else's tobacco smoke, is also dangerous. Researchers continue to collect evidence that confirms the link between passive smoking and cancer.

Pollutants   Being exposed to substances such as metals, dust, chemicals, or pesticides at work can increase the risk of cancer. Asbestos, nickel, cadmium, uranium, radon, vinyl chloride, benzidine, and benzene are well-known examples of carcinogens in the workplace. For example, inhaling asbestos fibers increases the risk of lung diseases, including cancer. The cancer risk is especially high for asbestos workers who smoke.

Data show the incidence in soft tissue sarcomas (STS), malignant lymphomas, and non-Hodgkin lymphomas is increased in farmers living in Nebraska and Kansas. All used 2,4-D (a commonly used herbicidal agent) on crops and to clear weeds along railroad tracks.

Viruses

At least four types of DNA viruses—hepatitis B and C viruses, Epstein-Barr virus, and human papillomavirus—are directly believed to cause human cancers.

In China, almost all the people have been infected with the hepatitis B virus. This correlates with the high incidence of liver cancer in that country. A combined vaccine is now available for hepatitis A and B. For a long time, circumstances suggested that cervical cancer was a sexually transmitted disease. Now, human papillomaviruses (HPVs) are routinely isolated from cervical cancers. Burkitt lymphoma occurs frequently in Africa, where virtually all children are infected with the Epstein-Barr virus. In China, the Epstein-Barr virus is isolated in nearly all nasopharyngeal cancer specimens.

RNA-containing retroviruses, in particular, are known to cause cancers in animals (see the Infectious Diseases supplement). In humans, the retrovirus HTLV-1 (human T-cell lymphotropic virus, type 1) has been shown to cause hairy cell leukemia. This disease occurs frequently in parts of Japan, the Caribbean, and Africa, particularly in regions where people are known to be infected with the virus. HIV, the virus that causes AIDS, and also Kaposi's sarcoma-associated herpesvirus (KSHV) are responsible for the development of Kaposi's sarcoma and certain lymphomas. This occurs due to the suppression of proper immune system functions.

Health Focus

Prevention of Cancer

Protective Behaviors

These behaviors help prevent cancer:

Don't use tobacco Cigarette smoking accounts for about 30% of cancer deaths. Smoking is responsible for 90% of lung cancer cases among men and 79% among women—about 87% on average. People who smoke two or more packs of cigarettes a day have lung cancer mortality rates 15–25 times greater than those of nonsmokers. Smokeless tobacco (chewing tobacco or snuff) increases the risk of cancers of the mouth, larynx, throat, and esophagus.

Don't sunbathe or use a tanning booth Almost all cases of basal-cell and squamous-cell skin cancers are considered sun-related. Further, sun exposure is a major factor in the development of melanoma, and the incidence of this cancer increases for people living near the equator.

Avoid radiation Excessive exposure to ionizing radiation can increase cancer risk. Even though most medical and dental X-rays are adjusted to deliver the lowest dose possible, unnecessary X-rays should be avoided. Excessive radon exposure in homes increases the risk of lung cancer, especially for cigarette smokers. It is best to test your home for radon and take the proper remedial actions.

Be tested for cancer Do the shower check for breast cancer or testicular cancer. Have other exams done regularly by a physician. See Table 19.1.

Be aware of occupational hazards Exposure to several different industrial agents (nickel, chromate, asbestos, vinyl chloride, etc.) and/or radiation increases the risk of various cancers. Risk from asbestos is greatly increased when combined with cigarette smoking.

Be aware of postmenopausal hormone therapy A new study conducted by the Women's Health Initiative found that estrogen-progestin combined therapy prescribed to ease the symptoms of menopause increased the incidence of breast cancer. Further, that risk outweighed the possible decrease in the number of colorectal cancer cases.

Get vaccinated Get vaccinated for HPV and hepatitis A and B. Consult with your physician or other health professional for advice regarding these vaccines and whether they are appropriate for you.

The Right Diet

Statistical studies have suggested that people who follow certain dietary guidelines are less likely to have cancer.

The following dietary guidelines greatly reduce your risk of developing cancer:

Avoid obesity Obesity increases the risk for many different types of cancer, especially those related to the reproductive systems in both men and women. Cancers of the colon, rectum, esophagus, breast, kidney, and prostate are all associated with being overweight.

Eat plenty of high-fiber foods Studies have indicated that a high-fiber diet (whole-grain cereals, fruits, and vegetables) may protect against colon cancer, a frequent cause of cancer deaths. Foods high in fiber also tend to be low in fat!

Increase consumption of foods that are rich in vitamins A and C Beta-carotene, a precursor of vitamin A, is found in dark green, leafy vegetables; carrots; and various fruits. Vitamin C is present in citrus fruits. These vitamins are called antioxidants because in cells they prevent the formation of free radicals (organic ions having an unpaired electron) that can possibly damage DNA. Vitamin C also prevents the conversion of nitrates and nitrites into carcinogenic nitrosamines in the digestive tract.

Reduce consumption of salt-cured, smoked, or nitrite-cured foods Salt-cured or pickled foods may increase the risk of stomach and esophageal cancers. Smoked foods, such as ham and sausage, contain chemical carcinogens similar to those in tobacco smoke. Nitrites are sometimes added to processed meats (e.g., hot dogs and cold cuts) and other foods to protect them from spoilage. As mentioned previously, nitrites are converted to nitrosamines in the digestive tract.

Include vegetables from the cabbage family in the diet The cabbage family includes cabbage, broccoli, brussels sprouts, kohlrabi, and cauliflower. These vegetables may reduce the risk of gastrointestinal and respiratory tract cancers.

Drink alcohol in moderation Risks of cancer development rise as the level of alcohol intake increases. The risk increases still further for those who smoke or chew tobacco while drinking. Heavy drinkers face a much greater risk for oral, pharyngeal, esophageal, and laryngeal cancer. Higher rates for cancers of the breast and liver are also linked to alcohol abuse. Men should limit daily alcohol consumption to two drinks or fewer. Women should have only one alcoholic drink daily. (A drink is defined as 12 oz of beer, 5 oz of wine, or 1.5 oz of distilled spirits.)

Dietary Choices

Nutrition is emerging as a way to help prevent cancer. The incidence of breast and prostate cancer parallels a high-fat diet, as does obesity. The Health Focus on page 453 discusses how to protect yourself from cancer. The American Cancer Society recommends consumption of fruits and vegetables, whole grains instead of processed (refined) grains, and limited consumption of red meats (especially high-fat and processed meats). Moderate to vigorous activity for 30–45 minutes a day, five or more days a week, is also recommended.

> **Check Your Progress 19.2**
> 1. **What evidence is there that you can inherit genes that lead to cancer?**
> 2. **What environmental carcinogens, in particular, are known to play a role in the development of cancer?**
> 3. **What protective steps can you take to reduce your risk of cancer?**

19.3 Diagnosis of Cancer

The earlier a cancer is detected, the more likely it can be effectively treated. At present, physicians have ways to detect several types of cancer before they become malignant. Researchers are always looking for new and better detection methods. A growing number of researchers now believe the future of early detection lies in testing for the "fingerprints"—molecular changes caused by cancer. Several teams of scientists are working on blood, saliva, and urine tests to catch cancerous gene and protein patterns in these bodily fluids before a tumor develops. In the meantime, cancer is usually diagnosed by the methods discussed here.

Seven Warning Signs

At present, diagnosis of cancer before metastasis is difficult, although treatment at this stage is usually more successful. The American Cancer Society publicizes seven warning signals, which spell out the word CAUTION and of which everyone should be aware:

C hange in bowel or bladder habits
A sore that does not heal
U nusual bleeding or discharge
T hickening or lump in breast or elsewhere
I ndigestion or difficulty in swallowing
O bvious change in wart or mole
N agging cough or hoarseness

Keep in mind that these signs do not necessarily mean that you have cancer. However, they are an indication that something is wrong and a medical professional should be consulted. Unfortunately, some of these symptoms are not obvious until cancer has progressed to one of its later stages.

Routine Screening Tests

Self-examination, followed by examination by a physician, can help detect the presence of cancer. For example, the letters ABCDE can help your self-exam of the skin for melanoma, the most serious form of skin cancer (Fig. 19.8). Breast cancer in women and testicular cancer in men can often be detected during a monthly shower check. The technique is discussed in the Health Focus on page 455.

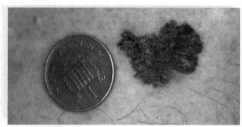

A = Asymmetry, one half the mole does not look like the other half.

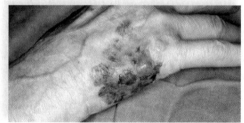

B = Border, irregular scalloped or poorly circumscribed border.

C = Color, varied from one area to another; shades of tan, brown, black, or sometimes white, red, or blue.

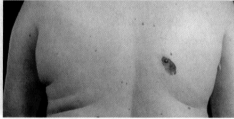

D = Diameter, larger than 6 mm (the diameter of a pencil eraser).

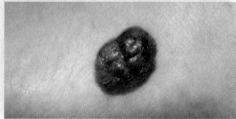

E = Elevated above skin surface, and **Evolving,** or changing over time

Figure 19.8   How can a melanoma be detected?
Suspicion of melanoma can begin by discovering a mole that has one or more of these characteristics.

☤ Health **Focus**

Shower Check for Cancer

The American Cancer Society urges women to do a breast self-exam and men to do a testicle self-exam every month. Breast cancer and testicular cancer are far more curable if found early, and we must all take on the responsibility of checking for one or the other.

Shower Self-Exam for Women

1. Check your breasts for any lumps, knots, or changes about one week after your period.
2. Place your right hand behind your head. Move your *left* hand over your *right* breast in a circle. Press firmly with the pads of your fingers (Fig. 19A). Also check the armpit.
3. Now place your left hand behind your head and check your *left* breast with your *right* hand in the same manner as before. Also check the armpit.
4. Check your breasts while standing in front of a mirror right after you do your shower check. First, put your hands on your hips and then raise your arms above your head (Fig. 19B). Look for any changes in the way your breasts look: dimpling of the skin, changes in the nipple, or redness or swelling.
5. If you find any changes during your shower or mirror check, see your doctor right away.

You should know that the best check for breast cancer is a mammogram. When your doctor checks your breasts, ask about getting a mammogram.

Shower Self-Exam for Men

1. Check your testicles once a month.
2. Roll each testicle between your thumb and finger as shown in Figure 19C. Feel for hard lumps or bumps.
3. If you notice a change or have aches or lumps, tell your doctor right away so he or she can recommend proper treatment.

Cancer of the testicles can be cured if you find it early. You should also know that prostate cancer is the most common cancer in men. Men over age 50 should have an annual health checkup that includes a prostate examination.

Information provided by the American Cancer Society. Used by permission.

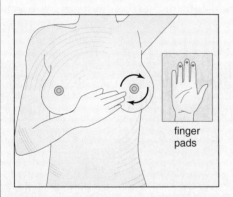

Figure 19A Shower check for breast cancer.

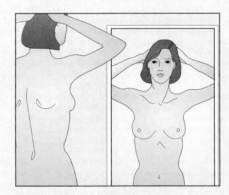

Figure 19B Mirror check for breast cancer.

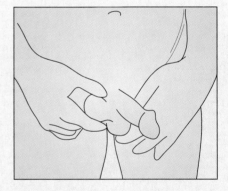

Figure 19C Shower check for testicular cancer.

The ideal tests for cancer are relatively easy to do, cost little, and are fairly accurate. The Pap test for cervical cancer fulfills these three requirements. A physician merely takes a sample of cells from the cervix, which are then examined microscopically for signs of abnormality (see Fig. 19.1). Any woman who chooses to receive the new HPV vaccine should still get regular Pap tests because (1) the vaccine will not protect against all types of HPV that cause cervical center, and (2) the vaccine does not protect against any HPV infections she may already have if she is sexually active. Regular Pap tests are credited with preventing over 90% of deaths from cervical cancer. Screening for colon cancer also depends

upon three types of testing. A digital rectal examination performed by a physician is of limited value because only a portion of the rectum can be reached. With flexible sigmoidoscopy, the second procedure, a much larger portion of the colon can be examined by using a thin, pliable, lighted tube. Finally, a stool blood test (fecal occult blood test) consists of examining a stool sample to detect any hidden blood. The sample is smeared on a slide, and a chemical is added that changes color in the presence of hemoglobin. This procedure is based on the supposition that a cancerous polyp bleeds, although some polyps do not bleed. Moreover, bleeding is not always due to a polyp. Therefore, the percentage of false negatives and false positives is high. All positive tests are followed up by a colonoscopy, an examination of the entire colon, or by X-ray after a barium enema. If the colonoscopy detects polyps, they can be destroyed by laser therapy. Blood tests are used to detect leukemia. Urinalysis aids in the diagnosis of bladder cancer.

Breast cancer is not as easily detected, but three procedures are recommended. First, every woman should do a monthly breast self-examination. But this is not a sufficient screen for breast cancer. Therefore, all women should have an annual physical examination, especially women over 40, when a physician will do the same procedure. Even then, this type of examination may not detect lumps before metastasis has already taken place. That is the goal of the third recommended procedure, *mammography,* an X-ray study of the breast (Fig. 19.9). However, mammograms do not show all cancers, and new tumors may develop in the interval between mammograms. The hope is that a mammogram will reveal a lump too small to be felt, at a time when the cancer is still highly curable. Table 19.1 outlines when routine screening tests should be done for various cancers, including breast cancer.

The diagnosis of cancer in other parts of the body may involve other types of imaging. Computerized axial tomography (or CAT scan) uses computer analysis of scanning X-ray images to create cross-sectional pictures that portray a tumor's size and location (see Fig. 2.3). Magnetic resonance imaging (MRI) is another type of imaging technique that depends on computer analysis. MRI is particularly useful for analyzing tumors in tissues surrounded by bone, such as tumors of the

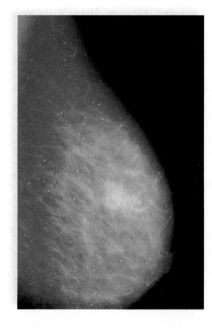

Figure 19.9
Why should women have regular mammograms?
An X-ray image of the breast can find tumors too small to be felt.

brain or spinal cord. A radioactive scan obtained after a radioactive isotope is administered can reveal any abnormal isotope accumulation due to a tumor. During ultrasound, echoes of high-frequency sound waves directed at a part of the body are used to reveal the size, shape, and location of tissue masses. Ultrasound can confirm tumors of the stomach, prostate, pancreas, kidney, uterus, and ovary.

Aside from various imaging procedures, a diagnosis of cancer can be confirmed without major surgery by performing

Table 19.1	**Recommendations for the Early Detection of Cancer in Average-Risk Asymptomatic People**		
Cancer	**Population**	**Test or Procedure**	**Frequency**
Breast	Women, age ≥ 20 years	Breast self-examination Clinical breast examination and mammography	Monthly, starting at age 20 years; see Health Focus, p. 455. For women in 20s and 30s, every 3 years; aged 40 years and over, preferably annually.
Colorectal	Men and women, age ≥ 50 years	Fecal occult blood test (FOBT) or fecal immunochemical test (FIT)*, *or* FOBT (or FIT) and flexible sigmoidoscopy *or* double contrast barium enema *or* colonoscopy	Annually, starting at age 50 years. Annually, starting at age 50 years. Every 5 years, starting at age 50 years. Every 10 years, starting at age 50 years.
Prostate	Men, age ≥ 50 years	Digital rectal examination and prostate-specific antigen test (PSA)	Annually, starting at age 50 years, for men who have a life expectancy of at least 10 more years.
Cervical	Women, age ≥ 18 years	Pap test	3 years after vaginal intercourse begins but no later than 18 years; after age 30 years, every 2 to 3 years if three normal pap tests in a row. Women age 70 years may choose to stop cervical cancer screening if three normal pap tests in a row.
Endometrial	Women, after menopause	Report any unexpected bleeding or spotting to a physician	
Testicular	Men, age 20 years	Testicle, self-examination	Monthly, starting at age 20 years, see Health Focus, p. 455.
Other cancers	Men and women, age 20 years	On the occasion of a periodic health examination, a cancer-related checkup should include examination for other cancers. See Health Focus, p. 453, for counseling about tobacco, sun exposure, diet and nutrition, risk factors, and environmental and occupational exposures.	

*FOBT as it is sometimes done in physicians' offices, with the single stool sample collected on a fingertip during a digital rectal examination, is not an adequate substitute for the recommended at-home procedure of collecting two samples from three consecutive specimens. Toilet-bowl FOBT tests also are not recommended. FIT is more patient-friendly and is likely to be equal or better in sensitivity and specificity. *2006 ACS Guidelines for Early Cancer Detection*

a *biopsy* or viewing body parts. Needle biopsies allow removal of a few cells for examination. Sophisticated techniques, such as laparoscopy, permit viewing of body parts.

Tumor Marker Tests

Tumor marker tests are blood tests for antigens and/or antibodies. Blood tests are possible because tumors release substances that provoke an antibody response in the body. For example, if an individual has already had colon cancer, it is possible to use the presence of an antigen called *CEA* (for *carcinoembryonic antigen*) to detect any relapses. When the CEA level rises, additional tumor growth has occurred.

There are also tumor marker tests that can be used for early cancer diagnosis. They are not reliable enough to count on solely, but in conjunction with physical examination (see pages 455–56) and ultrasound, they are considered useful. For example, there is a *prostate-specific antigen (PSA) test* for prostate cancer, a *CA-125* test for ovarian cancer, and an *alpha-fetoprotein (AFP) test* for liver tumors.

Genetic Tests

Tests for genetic mutations in proto-oncogenes and tumor-suppressor genes are making it possible to detect the likelihood of cancer before the development of a tumor. Tests are available that signal the possibility of colon, bladder, breast, and thyroid cancers, as well as melanoma. Physicians now believe that a mutated *RET* gene means that thyroid cancer is present or may occur in the future, and a mutated *p16* gene appears to be associated with melanoma. Genetic testing can also be used to determine if cancer cells still remain after a tumor has been removed.

A genetic test is also available for the presence of *BRCA1* (breast cancer gene 1). A woman who has inherited this gene can choose to either have prophylactic surgery or to be frequently examined for signs of breast cancer. Physicians can use microsatellites to detect chromosomal deletions that accompany bladder cancer. Microsatellites are small regions of DNA that always have di-, tri-, or tetranucleotide repeats. They compare the number of nucleotide DNA repeats in a lymphocyte microsatellite with the number in a microsatellite of a cell found in urine. When the number of repeats is less in the cell from urine, a bladder tumor is suspected.

Telomerase, you will recall, is the enzyme that keeps telomeres a constant length in cells. The gene that codes for telomerase is turned off in normal cells but is active in cancer cells. Therefore, if the test for the presence of telomerase is positive, the cell is cancerous.

Check Your Progress 19.3

1. What routine screening tests are available to detect and diagnose cancer?
2. What other types of tests are available to detect and diagnose cancer even before the routine tests indicate the presence of cancer?

19.4 Treatment of Cancer

Certain therapies for cancer have been available for some time. Other methods of therapy are in clinical trials. If successful, these new therapies will become more generally available in the future.

Standard Therapies

Surgery, radiation therapy, and chemotherapy are the standard methods of cancer therapy.

Surgery

Surgery alone is sufficient for cancer in situ. But because there is always the danger that some cancer cells have been left behind, surgery is often preceded by and/or followed by radiation therapy.

Radiation Therapy

Ionizing radiation causes chromosomal breakage and cell cycle disruption. Therefore, rapidly dividing cells, such as cancer cells, are more susceptible to its effects than other cells. Powerful X-rays or gamma rays can be administered through an externally applied beam. In some instances, tiny radioactive sources can be implanted directly into the patient's body. Cancer of the cervix and larynx, early stages of prostate cancer, and Hodgkin disease are often treated with radiation therapy alone.

Although X-rays and gamma rays are the mainstays of radiation therapy, protons and neutrons also work well. Proton beams can be aimed at the tumor like a rifle bullet hitting the bull's-eye of a target.

Side effects of radiation therapy greatly depend upon which part of the body is being irradiated and how much radiation is used. Weakness and fatigue are very typical. Dry mouth, nausea, and diarrhea often affect the digestive tract. Dry, red, or irritated skin, or even blistering burns may occur at the treatment site. Hair loss at the treatment site, which in some situations can be permanent, can be very depressing to the patient. Fortunately, most side effects of radiation treatment are temporary.

Chemotherapy

Radiation is localized therapy, while chemotherapy is a way to catch cancer cells that have spread (Figs. 19.10 and 19.11). Unlike radiation, which treats only the part of the body exposed to the radiation, chemotherapy treats the entire body. As a result, any cells that may have escaped from where the cancer originated are treated. Most chemotherapeutic drugs kill cells by damaging their DNA or interfering with DNA synthesis. The hope is that all cancer cells will be killed, while leaving enough normal cells untouched to allow the body to keep functioning. Combining drugs that have different actions at the cellular level may help destroy a greater number of cancer cells. Combinations might also reduce the

Figure 19.10   **How is a cancer patient treated with radiation?**
Most people who receive radiation therapy for cancer have external radiation on an outpatient basis. The patient's head is immobilized so that radiation can be delived to a very precise area. .

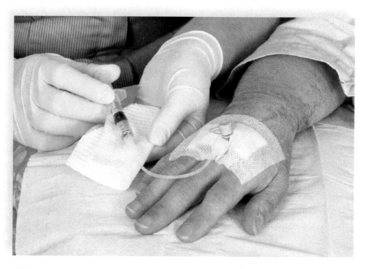

Figure 19.11   **How is a cancer patient treated using chemotherapy?**
The intravenous route is the most common, allowing chemotherapy drugs to spread quickly throughout the entire body by way of the bloodstream.

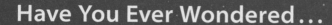

Have You Ever Wondered…

What can be done to lessen the side effects of radiation and chemotherapy?

Victims of cancer will often describe the side effects of treatment as the worst part of the disease. Nausea, vomiting, diarrhea, weight loss and hair loss, anxiety and/or depression, and extreme fatigue are common symptoms that may be caused by chemotherapy or radiation therapy. However, strategies now exist to help sufferers deal with treatment side effects, and others continue to be devised. Antiemetic drugs alleviate nausea and vomiting. Genetically engineered erythropoietin will help to stimulate red blood cell production and reduce fatigue (see Chap. 6). Marinol®, a medication closely related to marijuana, stimulates appetite and helps with weight loss. Hair loss can be minimized by cryotherapy, which involves applying cold packs on the scalp during a treatment. Cold temperatures slow the metabolic rate of hair follicles, and more follicles survive treatment. Antidepressants and antianxiety medications will help the patient to deal with the psychological effects of the disease.

Increasingly, cancer patients are turning to alternative therapies as well, for help in coping with the disease. Calming techniques, including yoga, meditation, and tai chi, help the sufferer to relax. Hypnosis can have the same restful effect. Aromatherapy, massage therapy, and music therapy seem to ease tension and stress, perhaps by stimulating pleasure centers in the brain. Exercise is known to produce endorphins, the brain's pain relieving neurotransmitters (see Chap. 13).

risk of the cancer developing resistance to one particular drug. What chemicals are used is generally based on the type of cancer and the patient's age, general health, and perceived ability to tolerate potential side effects. Some of the types of chemotherapy medications commonly used to treat cancer include

Alkylating agents. These medications interfere with the growth of cancer cells by blocking the replication of DNA.

Antimetabolites. These drugs block the enzymes needed by cancer cells to live and grow.

Antitumor antibiotics. These antibiotics—different from those used to treat bacterial infections—interfere with DNA, blocking certain enzymes and cell division and changing cell membranes.

Mitotic inhibitors. These drugs inhibit cell division or hinder certain enzymes necessary in the cell reproduction process.

Nitrosoureas. These medications impede the enzymes that help repair DNA.

Whenever possible, chemotherapy is specifically designed for the particular cancer. For example, in some cancers, a small portion of chromosome 9 is missing. Therefore, DNA metabolism differs in the cancerous cells compared with normal cells. Specific chemotherapy for the cancer can exploit this metabolic difference and destroy the cancerous cells.

One drug, *taxol*, extracted from the bark of the Pacific yew tree, was found to be particularly effective against advanced ovarian cancers, as well as breast, head, and neck tumors. Taxol interferes with microtubules needed for cell division. Now, chemists have synthesized a family of related drugs, called taxoids, which may be more powerful and have fewer side effects than taxol.

Certain types of cancer, such as leukemias, lymphomas, and testicular cancer, are now successfully treated by combination chemotherapy alone. The survival rate for children with childhood leukemia is 80%. Hodgkin disease, a lymphoma, once killed two out of three patients. Combination therapy, using four different drugs, can now wipe out the disease in a matter of months. Three out of four patients achieve a cure, even when the cancer is not diagnosed immediately. In other cancers—most notably, breast and colon cancer—chemotherapy can reduce the chance of recurrence after surgery has removed all detectable traces of the disease.

Chemotherapy sometimes fails because cancer cells become resistant to one or several chemotherapeutic drugs, a phenomenon called multidrug resistance. This occurs because a plasma membrane carrier pumps the drug (or drugs) out of the cancer cell before it can be harmed. Researchers are testing drugs known to poison the pump in an effort to restore the effectiveness of the drugs. Another possibility is to use combinations of drugs with different toxic activities, because cancer cells can't become resistant to many different types at once.

Bone marrow transplants are sometimes done in conjunction with chemotherapy. The red bone marrow contains large populations of dividing cells. Therefore, red bone marrow is particularly prone to destruction by chemotherapeutic drugs. In bone marrow autotransplantation, a patient's stem cells are harvested and stored before chemotherapy begins. High doses of radiation or chemotherapeutic drugs are then given within a relatively short time. This prevents multidrug resistance from occurring, and the treatment is more likely to catch every cancer cell. The stored stem cells can then be returned to the patient by injection. They automatically make their way to bony cavities, where they resume blood cell formation.

Newer Therapies

Several therapies are now in clinical trials and are expected to be increasingly used to treat cancer.

Immunotherapy

When cancer develops, the immune system has failed to dispose of cancer cells, even though they bear antigens that make them different from the body's normal cells. A vaccine, called Melacine, which contains broken melanoma cells from two different sources is under investigation for use against melanoma.

Another idea is to use immune cells, genetically engineered to bear the tumor's antigens (Fig. 19.12). When these cells are returned to the body, they produce cytokines. Recall that cytokines are chemical messengers for immune cells (see Chap. 7). Cytokines stimulate the body's immune cells to attack the tumor. Further, the altered immune cells present the tumor antigen to cytotoxic T cells, which then go forth and destroy tumor cells in the body.

Passive immunotherapy is also possible. Monoclonal antibodies have the same structure because they are produced by the same plasma cell (see Fig. 7.17). Some monoclonal antibodies are designed to zero in on the receptor proteins of cancer cells. To increase the killing power of monoclonal

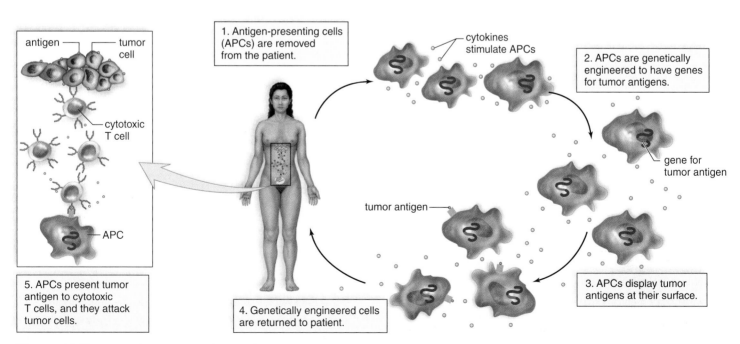

Figure 19.12 How does immunotherapy for cancer work?
1. Antigen-presenting cells (APCs) are removed and (2) are genetically engineered to (3) display tumor antigens. 4. After these cells are returned to the patient, (5) they present the antigen to cytotoxic T cells, which then kill tumor cells.

"I'm pretty upset with you, Kim," Dr. Phipps, their family doctor, scolded gently. "You look terrible. You've lost at least ten pounds, and you were slender to begin with. Are you sleeping okay? I'm glad you quit your job."

"I didn't have any more personal time or sick time left at my job," Kim explained. "I wasn't in love with it in the first place, and Scott makes more than I do right now. We just tightened our budget. We're getting by.

"Cody's therapy is very, very tough for me," she sighed. "I hate to see him cry after he gets stuck for chemotherapy. Then for the next two days, he cries constantly, vomits, and has diarrhea. Watching all of that is tearing me apart."

"Well, in between those sessions, he does fine, right?" the doctor asked. "Look at him now." The little boy smiled at him, his drooly grin

showing four little teeth. "He's walking a few steps, isn't he? So he's doing what other kids his age do. He's a little underweight, but that's to be expected.

"If you worry yourself sick, you won't help him at all," Dr. Phipps continued. "There's a support group at the hospital for people whose children have or had cancer. It might help you to talk with those other parents. But remember, his type of cancer has a 95% cure rate if found early, and they did. You were lucky. Did you say Cody's symptoms were first noticed by a day care worker?"

She nodded. "Thank goodness for her. With us working as many hours as we were back then, we might not have caught it right away like she did."

antibodies, they are linked to radioactive isotopes or chemotherapeutic drugs. It is expected that soon they will be used as initial therapies, in addition to chemotherapy.

p53 Gene Therapy

Researchers believe that *p53* gene expression is needed for only 19 hours to trigger apoptosis, programmed cell death. And the *p53* gene seems to trigger cell death only in cancer cells—elevating the *p53* level in a normal cell doesn't do any harm, possibly because apoptosis requires extensive DNA damage.

Ordinarily, when adenoviruses infect a cell, they first produce a protein that inactivates *p53*. In a cleverly designed procedure, investigators genetically engineered an adenovirus that lacks the gene for this protein. Now, the adenovirus can infect and kill only cells that lack a *p53* gene. Which cells are those? Tumor cells, of course. Another plus to this procedure is that the injected adenovirus spreads through the cancer, killing tumor cells as it goes. This genetically engineered virus is now in clinical trials.

Other Therapies

Many other therapies are now being investigated. Among them, drugs that inhibit angiogenesis are a proposed therapy under investigation. Antiangiogenic drugs confine and reduce tumors by breaking up the network of new capillaries in the vicinity of a tumor. A number of antiangiogenic compounds are currently being tested in clinical trials. Two highly effective drugs, called angiostatin and endostatin, have been shown to inhibit angiogenesis in laboratory animals and are expected to do the same in humans.

Check Your Progress 19.4

1. What three types of therapy are presently the standard ways to treat cancer?
2. a. What is an active form of immunotherapy? b. Describe a passive form of immunotherapy for cancer.
3. What type of *p53* therapy is available?
4. What is the rationale for antiangiogenesis therapy?

By the time the doctors allowed Cody to return to day care, Kim had landed a different job. This time, she hoped, the job might be interesting and fulfilling. She and Scott needed the extra income to deal with Cody's medical expenses, which were considerable despite their health insurance. Now that there was an opening in the toddler room, it seemed like a great time for Kim to return to the workforce.

That first morning back, she had another stop to make before the toddler room. Mary Jean Evans sat in her rocking chair, holding

an infant while four others sat propped in baby seats in front of her. She sang to all while rocking the baby in her lap. Kim waited for her to pause, then interrupted. "Mary Jean? Do you remember us?"

"I think so," the older woman replied, smiling hesitantly. "No—wait—that's Cody, right? He's gotten so big! And you're Kathy—no, wait—Kim. It's great to see you, after all this time! How are you both?"

"We're cancer survivors," Kim replied with a smile, presenting Mary Jean a small flower bouquet.

The Immortal Henrietta Lacks

If you've ever had food spoil in your refrigerator, you already know that it's very easy to grow mold and even easier to grow bacteria. Not so for healthy human tissue cells. These cells can be coaxed to reproduce using a technique called tissue culture. Unfortunately, healthy cells typically all die very rapidly, usually after dividing only a few times. However, metastatic cells—cancer cells—can survive and thrive in tissue culture. The techniques used to grow human cells successfully in tissue culture are part of the legacy of Henrietta Lacks (Fig. 19D). The cancer that claimed her life ensured that a part of Henrietta will live forever.

Lacks was a young African American woman and the mother of five children. In February of 1951, she experienced unexplained vaginal bleeding and sought help from a Baltimore hospital. Physicians found a quarter-sized tumor on Henrietta's cervix. Samples of the tumor were quickly obtained, then sent to the tissue culture laboratories at Johns Hopkins University. Though radiation treatment was attempted, the tumor ravaged Henrietta's body, appearing on all her major organs within months. After eight months of suffering, she died in October 1951. She was survived by her husband and five children, three of whom were still in diapers.

Henrietta's tumor cells ended up in the laboratory of George and Margaret Gey at Johns Hopkins. The couple had been trying to culture human cells, with little success, for more than two decades. The cells from Henrietta Lacks' tumor, now termed HeLa cells (Fig. 19E), not only lived but multiplied like wildfire. At last, here was a source of human cells that grew rapidly and seemed almost indestructible.

Using these cells, the Geys directed their research toward curing polio. Infantile paralysis, or polio, occurred in epidemics that damaged the brain and spinal cord in a small percentage of patients. Paralysis of major muscle groups resulted. If the diaphragm and other respiratory muscles were affected, the disease could be fatal. For the first time, the virus could be grown in human cells—HeLa cells—so that its characteristics could be studied. The results of these studies enabled Dr. Jonas Salk to develop a vaccine for polio. Today, polio is almost unheard of in the Western world.

Research with HeLa cells did not stop with the Geys. For over 50 years, these durable cells have been used worldwide to study many types of viruses, as well as leukemia and other cancers. The cells are human in origin, so they have also been the test subjects to determine the harmful effects caused by drugs or radiation. Effective testing was developed for chromosome abnormalities and hereditary diseases using HeLa. Samples of HeLa cells have even been launched in the space shuttle, for experiments involving a zero gravity environment. HeLa cell colonies can be found around the world. The cells can even be purchased from biological supply companies.

Figure 19D Henrietta Lacks 1920–1951.

HeLa cells are almost too sturdy. In 1974, researchers were shocked and dismayed to learn that cells thought to originate from other body tissues were HeLa. Countless hours and millions of dollars had been wasted by scientists who thought they were studying the brain, for example. HeLa cells, like bacteria, could be transmitted through the air, on a researcher's glove, or on contaminated glassware. This discovery led to better sterile technique not only in the laboratory, but also in operating rooms where cancerous tumors were removed. Once again, the cells of a poor woman, buried in an unmarked grave, served humanity by living on.

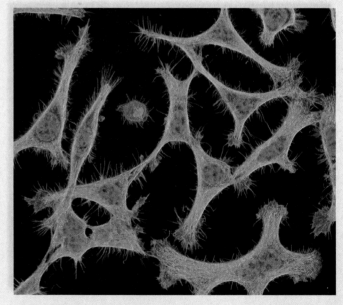

Figure 19E SEM of HeLa cell.

Summarizing the Concepts

19.1 Cancer Cells

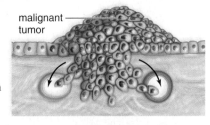

Certain characteristics are common to cancer cells. Cancer cells

- lack differentiation and do not contribute to function.
- do not undergo apoptosis—they enter the cell cycle an unlimited number of times.
- form tumors and do not need growth factors to signal them to divide.
- gradually become abnormal—carcinogenesis is comprised of initiation, promotion, and progression.
- undergo angiogenesis (the growth of blood vessels to support them) and can spread throughout the body (metastasis).

Cancer Is a Genetic Disease

Cells become increasingly abnormal due to mutations in proto-oncogenes and tumor-suppressor genes.

In normal cells: The cell cycle functions normally.

- **Proto-oncogenes** promote cell cycle activity and restrain apoptosis. Proto-oncogenes can mutate into oncogenes.
- **Tumor-suppressor genes** restrain the cell cycle and promote apoptosis.

In cancer cells: The cell cycle is accelerated and occurs repeatedly.

- **Oncogenes** cause an unrestrained cell cycle and prevent apoptosis.
- **Mutated tumor-suppressor genes** cause an unrestrained cell cycle and prevent apoptosis.

Types of Cancer

Cancers are classified according to their places of origin.

- Carcinomas originate in epithelial tissues.
- Sarcomas originate in muscle and connective tissues.
- Leukemias originate in blood.
- Lymphomas originate in lymphatic tissue.

Certain body organs are more susceptible to cancer than others.

19.2 Causes and Prevention of Cancer

Development of cancer is determined by a person's genetic profile, plus exposure to environmental carcinogens.

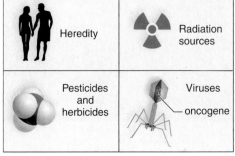

Some sources contributing to development of cancer

- Cancers that run in families are most likely due to the inheritance of mutated genes (e.g., breast cancer, retinoblastoma tumor).
- Certain environmental factors are carcinogens (e.g., UV radiation, tobacco smoke, pollutants).
- Industrial chemicals (e.g., pesticides and herbicides) are carcinogenic.
- Certain viruses (e.g., hepatitis B and C, human papillomavirus, and Epstein-Barr virus) cause specific cancers.

19.3 Diagnosis of Cancer

The earlier a cancer is diagnosed, the more likely it can be effectively treated. Tests for cancer include

- Pap test for cervical cancer.
- mammogram for breast cancer.
- tumor marker tests—blood tests that detect tumor antigens/antibodies.
- tests for genetic mutations of oncogenes and tumor-suppressor genes.
- biopsy and imaging—used to confirm the diagnosis of cancer.

C	hange in bowel or bladder habits
A	sore that does not heal
U	nusual bleeding or discharge
T	hickening or lump in breast or elsewhere
I	ndigestion or difficulty in swallowing
O	bvious change in wart or mole
N	agging cough or hoarseness

19.4 Treatment of Cancer

Surgery, radiation, and chemotherapy are traditional methods of treating cancer. Other methods include

- chemotherapy involving bone marrow transplants.
- immunotherapy.
- *p53* gene therapy (one type of *p53* gene therapy ensures that cancer cells undergo apoptosis).
- other therapies, such as inhibitory drugs for angiogenesis and metastasis, which are being investigated.

Understanding Key Terms

angiogenesis 448
apoptosis 446
benign 447
bone marrow transplant 459
cancer 446
carcinogen 451
carcinogenesis 447
carcinoma 449
cyclin 448
growth factor 448
initiation 447
leukemia 450
lymphoma 450
malignant 447

metastasis 448
mutagen 451
oncogene 448
oncologist 449
oncology 449
progression 447
promotion 447
proto-oncogene 448
sarcoma 449
telomere 446
tumor 446
tumor marker test 457
tumor-suppressor gene 448

Match the key terms to these definitions.

a. _____ End of chromosome that prevents it from binding to another chromosome; normally get shorter with each cell division.

b. _____ Environmental agent that contributes to the development of cancer.

c. _____ Normal gene involved in cell growth and differentiation that becomes an oncogene through mutation.

d. _____ Formation of new blood vessels, such as a capillary network.

e. _____ Spread of cancer from the place of origin to throughout the body; caused by the ability of cancer cells to migrate and invade tissues.

Testing Your Knowledge of the Concepts

1. List and briefly discuss the seven characteristics of cancer cells that cause them to be abnormal. (pages 446–48)

2. What are the roles of the two genes that mutate, causing cancer to develop? (pages 448–49)

3. What are the four types of cancer based on place of origin? (pages 449–50)

4. What are the top four types of new cancer cases in males? In females? (page 450)

5. What are the general causes of cancer? What are several types of environmental mutagens? (pages 451–52)

6. What are the seven warning signals of cancer? (page 454)

7. List and describe three tests designed to detect cancer. (pages 454–57)

8. Describe three standard therapies for cancer treatment. (pages 457–59)

9. Describe some newer therapies that may be successful in cancer treatment. (pages 459–60)

10. Whereas _____ stimulate the cell cycle, _____ inhibit the cell cycle.
   a. tumor-suppressor genes, oncogenes
   b. oncogenes, tumor-suppressor genes
   c. proto-oncogenes, oncogenes
   d. proto-oncogenes, tumor-suppressor genes

11. Growth factors lead to
   a. increased cell division.
   b. the functioning of cyclin proteins.
   c. progression through the cell cycle.
   d. All of these are correct.

12. Which of these is not true of the gene *p53*?
   a. Mutations of both proto-oncogenes and tumor-suppressor genes lead to inactivity of *p53*.
   b. Normally, *p53* functions to stop the cell cycle and initiate repair enzymes, when necessary.
   c. Normally, *p53* restores the length of telomeres.
   d. Cancer cells shut down the activity of *p53*, even though it may be present.

13. Which association is incorrect?
   a. proto-oncogenes—code for cyclin and proteins that inhibit the activity of *p53*
   b. oncogenes—"grain of function" genes
   c. mutated tumor-suppressor genes—code for cyclin and proteins that inhibit the activity of *p53*
   d. mutated tumor-suppressor genes—"loss of function" mutations
   e. Both a and c are incorrect.

14. To stimulate the immune system to fight cancer, which type of cells is genetically engineered?
   a. Macrophages because they will devour cancer cells.
   b. Tumor cells because they can be engineered to display more antigens than normal.
   c. Antigen-presenting cells because they can present tumor antigens to cytotoxic T cells.
   d. Any type of cell can present tumor antigens to cytotoxic T cells.
   e. All of these are correct.

15. Following each cell cycle, telomeres
   a. get longer.
   b. get shorter.
   c. return to the same length.
   d. bind to cyclin proteins.

16. Bone marrow transplants are done in conjunction with chemotherapy because
   a. Smoking and pollutants can kill bone marrow stem cells.
   b. Blood cells will be engineered to have cancer antigens.
   c. This procedure is no longer done.
   d. Chemotherapy sometimes kills bone marrow stem cells.
   e. All of these are correct.

17. Angiogenic growth factors function to
   a. stimulate the development of new blood vessels.
   b. activate tumor-suppressor genes.
   c. change proto-oncogenes into oncogenes.
   d. promote metastasis.

18. A tumor with cells that spread to secondary locations is referred to as
   a. benign.
   b. cancer in situ.
   c. malignant.
   d. encapsulated.

19. Which of the following is not a type of carcinogen?
   a. tobacco smoke
   b. radiation
   c. pollutants
   d. viruses
   e. All of these are carcinogens.

20. What type of cancer is associated with human papillomaviruses?
   a. breast
   b. cervical
   c. kidney
   d. lymphatic

21. Why is cancer called a genetic disease?
   a. Cancer is always inherited.
   b. Carcinogenesis is accompanied by mutations.
   c. Cancer causes mutations that are passed on to offspring.
   d. All of these are correct.

22. Leukemia is a form of cancer that affects
    a. lymphatic tissue.
    b. bone tissue.
    c. blood-forming cells.
    d. nervous system structures.

23. Which is the name of the tumor-suppressor gene that causes retinoblastoma?
    a. *p21*   c. *ras*
    b. *RB*   d. *TGF-b*

24. What is the most common form of cancer in men?
    a. colon   c. prostate
    b. lung   d. testicular

25. Concerning the causes of cancer, which one is incorrect?
    a. Genetic mutations cause cancer.
    b. Genetic mutations can be caused by environmental influences, such as radiation, organic chemicals, and viruses.
    c. An active immune system, diet, and exercise can help prevent cancer.
    d. Heredity cannot be a cause of cancer.

26. Which of the following is not a warning signal for cancer?
    a. sore that does not heal
    b. change in bowel or bladder habits
    c. nagging cough or hoarseness
    d. shortness of breath or fatigue

27. Which of these tests for the particular cancer is mismatched?
    a. breast cancer—mammogram
    b. lung cancer—X-ray
    c. cervical cancer—Pap test
    d. prostate cancer—CA-125 test

28. Following a biopsy, what does a doctor look for to diagnose cancer?
    a. He/She looks for abnormal-appearing cells.
    b. He/She sees if the cells can divide.
    c. He/She sees if the cells will respond to growth factors.
    d. He/She sees if the chromosomes have telomeres.

29. Most chemotherapeutic drugs kill cells by
    a. producing pores in plasma membranes.
    b. interfering with protein synthesis.
    c. interfering with cellular respiration.
    d. interfering with DNA and/or enzymes.

30. Multidrug resistance to chemotherapeutic drugs occurs because
    a. cancer cells can use plasma membrane carriers to pump drugs out of the cell.
    b. one drug may interfere with the activity of another drug.
    c. using several drugs at once will overtax the patient's immune system.
    d. using several drugs at once decreases the effectiveness of radiation therapy.

31. *p53* gene therapy
    a. triggers cytotoxic T cells to destroy tumor cells.
    b. triggers apoptosis in cancer cells.
    c. produces monoclonal antibodies against the tumor cells.
    d. reduces tumors by breaking up their blood vessels.

Thinking Critically About the Concepts

Both Ben Monteiro from Chapter 6 and little Cody Keenan were diagnosed with cancers most often seen in children and young adults. Ben's cancer could only be treated with chemotherapy. In Cody's case, cancer was treated with surgery and chemotherapy. Both survived their ordeals.

1. After their initial treatments in the hospital, both Ben and Cody had to avoid public exposure. Based on what you know about the immune system, explain.

2. Cody's cancer was called a nephroblastoma, and Ben's blood was crowded with immature white blood cells, or lymphoblasts. The suffix –*blast* refers to an immature cell. Would these types of cancer be more or less likely to spread throughout the body?

3. Cody's nephroblastoma caused blood to be found in his urine. What large blood vessels supplying the kidney might have been damaged by cancer? (See Chapter 5.)

4. In addition to blood in his urine, what other abnormal symptoms did Cody show?

5. Many types of cancer (ovarian cancer, for example) show no symptoms until they are well advanced. What might be the consequence for the patient?

6. The vaccine for HPV is recommended for females ages 11–26. There is some controversy over this vaccine. Some believe vaccination for sexually transmitted disease might lull these young women into a false sense of security regarding safety during sexual intercourse. Others fear that vaccination will encourage girls to engage in sex.
    a. Why would the vaccine be recommended for young girls who may not be sexually active yet?
    b. What type of immunity would women develop from getting vaccinated for HPV? You may need to revisit Chapter 7 on immunity to answer this question.
    c. If women are not vaccinated against HPV, how else could they protect themselves from infection?

7. Why are lymph nodes surrounding a breast with an invasive cancer often removed when a mastectomy is performed?

8. What are some healthy lifestyle choices you could make that might prevent cancer later in your life?

CHAPTER 20

Patterns of Genetic Inheritance

CASE STUDY JEREMY CALLEN

Jeremy Callen was sitting in the back seat of the car, looking out the window as he and his mother drove through the country. He was extremely excited and just couldn't stop talking. "The sun is very bright. The weatherman said that it would be sunny today. I like sunny. No thunderstorms. Do you like boom, Mom? I don't like boom. But no booms today. The weatherman said no. Going to college today. See Jackie. No booms."

Every now and then, he would ask his mother Teresa a question, but she didn't answer. Jeremy didn't seem to notice and just kept up a steady stream of conversation with himself. Teresa was lost in thought. Actually, she was very worried. Jeremy was going to visit his big sister, Jackie, at college for "Little Sib's Weekend." It would be his first time away from home without his mother.

Teresa glanced in the rearview mirror for a moment, studying her handsome 17-year-old son, with his dark eyes and long, expressive face. It has been such a blow, she reflected, to discover that Jeremy had fragile X syndrome. As a baby, he had learned to crawl and then to walk just a little more slowly than his brothers and sister had. Of greater concern was his long delay in being able to speak. He didn't like to be held or touched and often refused to make eye contact. An alert pediatric genetic specialist had helped the Callens to obtain a diagnosis for Jeremy. From then on, the entire family was able to rally around their special son and brother. Teresa and her husband John were extremely proud of each of Jeremy's hard-won accomplishments.

However, Teresa was worried about how he was going to handle sleeping in a dorm. But Jeremy missed his sister so much when she went off to college and wanted so badly to visit her. So Teresa had agreed to let him spend one night with Jackie, while she stayed in a nearby hotel.

Teresa pulled up in the visitor's parking lot. Jackie and two of her friends were waiting for them. When Jeremy got out of the car, he was grinning from ear to ear. "Jackie! How are you? How was your day?"

Jackie grabbed him and gave him a big hug. "I've missed you so much, Jeremy!

CHAPTER CONCEPTS

20.1 Genotype and Phenotype
The genotype refers to the genes for a particular trait. The phenotype refers to physical characteristics, such as hairline, blood type, color blindness, and even any cellular disorder.

20.2 One- and Two-Trait Inheritance
Each gene is represented by two alleles. Some alleles are dominant and some are recessive. It is not possible to tell genotype by the phenotype when a person has one dominant and one recessive allele. Many disorders, including cellular disorders, are inherited in the same manner as normal dominant/recessive traits.

20.3 Beyond Simple Inheritance Patterns
There are other inheritance patterns beyond simple dominant/recessive ones. Some are especially affected by the environment, in addition to the genotype.

20.4 Sex-Linked Inheritance
Some traits are controlled by alleles on the X chromosome that have nothing to do with the gender of the person. The Y is blank for X-linked recessive alleles, and any on the X chromosome will be expressed. Therefore, males are more apt to have an X-linked disorder than are females.

20.1 Genotype and Phenotype

Genotype refers to the genes of the individual. Alternative forms of a gene having the same position (locus) on a pair of chromosomes and affecting the same trait are called **alleles.** It is customary to designate an allele by a letter, which represents the specific trait (characteristic) it controls. A **dominant allele** is assigned an uppercase (capital) letter, while a **recessive allele** is given the same letter but in lowercase. A dominant gene "dominates" its recessive partner, which "recedes" into the background. In humans, for example, unattached (free) earlobes are dominant over attached earlobes. A suitable key would be *E* for unattached earlobes and *e* for attached earlobes. Free earlobes dominate over attached ones.

Alleles occur in pairs. Therefore, the individual normally has two alleles for a trait. Just as one of each pair of chromosomes is inherited from each parent, so too is one of each pair of alleles inherited from each parent. Indeed, the alleles occur in the same location, called their **locus** (plural, loci), on a pair of homologous chromosomes.

Figure 20.1 shows three possible fertilizations, the resulting genetic makeup of the zygote and, therefore, the individual. In the first instance, the chromosomes of both the sperm and the egg carry an *E*. Consequently, the zygote and subsequent individual have the alleles *EE*, called a **homozygous dominant** genotype. A person with genotype *EE* has unattached earlobes. The physical appearance of the individual—in this case, unattached earlobes—is called the **phenotype.** Notice in Figure 20.1 that the genotype (letters) and then the phenotype (description) are given after the drawing.

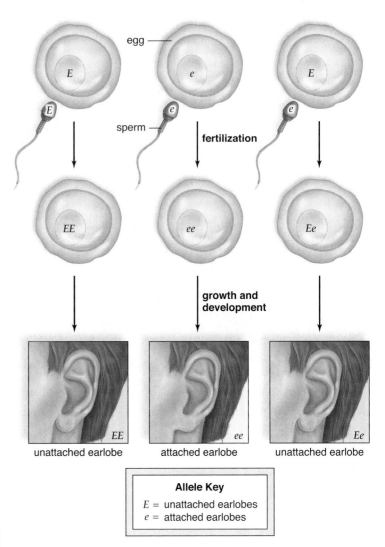

Figure 20.1   How does genetic inheritance affect our characteristics?
Individuals inherit a minimum of two alleles for every characteristic of their anatomy and physiology. The inheritance of a single dominant allele (*E*) causes an individual to have unattached earlobes; two recessive alleles (*ee*) cause an individual to have attached earlobes. Each individual receives one allele from the father (by way of a sperm) and one allele from the mother (by way of an egg).

Have You Ever Wondered...

How can some alleles be dominant over other alleles?

In a simple example, the dominant allele (*A*) codes for a particular protein. Let's assume this gene codes for an enzyme (protein) responsible for brown eye color. The recessive allele (*a*) is a mutated allele that no longer codes for the enzyme. At the molecular level, having the enzyme is dominant to the lack of the enzyme. In other words, having brown in the eye is dominant over not having dark pigment in the eye. Using this example, a homozygous dominant individual (*AA*) may have twice the amount of enzyme as a heterozygous individual (*Aa*). However, it may not be possible to detect a difference between homozygous and heterozygous, as long as there is enough enzyme to bring about the dominant phenotype. Half the amount of enzyme may still turn the eyes completely brown. So a person with the genotype *Aa* would have eyes just as brown as a person with the genotype *AA*. A homozygous recessive individual (*aa*) makes no enzyme, so the brown phenotype is absent. Gene expression is discussed further in Chapter 21.

In the second fertilization, the zygote has received two recessive alleles (*ee*), and the genotype is called **homozygous recessive.** An individual with this genotype has the recessive phenotype, attached earlobes. In the third fertilization, the resulting individual has the alleles *Ee*, called a **heterozygous** genotype. A heterozygote shows the dominant phenotype. Therefore, this individual has unattached earlobes.

How many dominant alleles must an individual inherit to have the dominant phenotype? These examples show that a dominant allele contributed from only one parent can bring about a particular dominant phenotype. How many recessive alleles must an individual inherit to have the recessive phenotype? A recessive allele must be received from both parents to bring about the recessive phenotype.

Phenotype

In our example, the phenotype was attached or unattached earlobes. From this you might get the impression that the phenotype has to be an easily observable trait. However, the phenotype can be any characteristic of the individual. This could include color blindness or a metabolic disorder such as the lack of an enzyme to metabolize the amino acid phenylalanine.

> **Check Your Progress 20.1**
> 1. What is the difference between genotype and phenotype?
> 2. What are the three possible genotypes and the two possible phenotypes for a characteristic controlled by two alleles, one being dominant and the other recessive?

20.2 One- and Two-Trait Inheritance

In one-trait (one characteristic) crosses, the inheritance of only one set of alleles is being considered. In two-trait (two characteristics) crosses, the inheritance of two sets of alleles is being considered. For both types of crosses, it will be necessary to determine the gametes of both individuals who are reproducing.

Forming the Gametes

When sperm and eggs are formed, the chromosome number is divided in half. An individual has 46 chromosomes, but the sperm or egg only has 23 chromosomes. If reduction didn't happen, the new individual would have twice as many chromosomes after fertilization. This division of chromosome numbers occurs during meiosis, when homologous chromosomes separate. One of these homologous chromosomes was originally from the mother, while the second was donated by the father. The alleles are on the homologous chromosomes, so they also separate during meiosis.

For example, let's say that the gene for unattached or attached earlobes is on a particular chromosome. On the homologous chromosome originally from the mother, the allele is an *E*. On the father's paired chromosome, the allele is also an *E*. *E* is the only option for alleles on either chromosome, so every gamete formed will carry an *E*. This occurs whether the gamete gets the mother's or the father's original chromosome. Similarly, if the homologous chromosome originally from the mother carries an *e* and the father's chromosome also bears an *e*, then all gametes will have the *e* allele. What if the mother's homologous chromosome has an *E* allele and the father's allele is *e*? Then, half of the gametes formed will receive *E* allele from the mother's homologue. The other half will get the father's homologue with the *e*.

Figure 20.2 shows the genotypes and phenotypes for certain other traits in humans. You can practice deciding what alleles the gametes would carry in order to produce

a. Widow's peak: *WW* or *Ww*

b. Straight hairline: *ww*

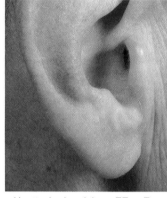
c. Unattached earlobes: *EE* or *Ee*

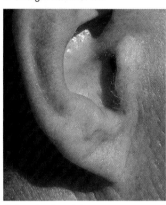
d. Attached earlobes: *ee*

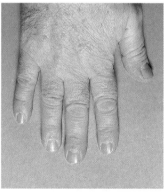
e. Short fingers: *SS* or *Ss*

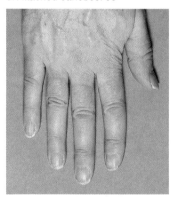
f. Long fingers: *ss*

g. Freckles: *FF* or *Ff*

h. No freckles: *ff*

Figure 20.2 What are some common inherited traits in human beings?
The allele keys indicate which traits are dominant and which are recessive.

"Are you ready to meet my friends?" Jackie asked. Jeremy looked down at his shoes. "Sure," he said. He stuck out his hand to shake hands, but he wouldn't look up. Making eye contact with others was still very difficult for Jeremy. He whispered, "I'm okay, I'm okay." Teresa felt herself growing tense. When Jeremy was nervous, he used "self-talk" to convince himself that he was fine. Most of the time, this form of coping worked very well for Jeremy. After Jeremy shook hands with Jackie's friends he looked at his mother and reported, "I said hello to everyone."

Teresa laughed. They had been practicing how to meet other people and shake hands for several weeks. "Yes, you did, Jeremy. Just like we practiced. Good job." Jeremy beamed. Teresa said to Jackie, "I'm going to the hotel. I have my cell phone on. If there is any problem, Jackie, you call me immediately. I can be here in 5 minutes."

Teresa handed Jackie the duffel bag from the back seat. Then she turned to Jeremy. "And you, young man. Be good and listen to your sister. Okay?"

Jeremy nodded and replied, "I'm always good, Mom."

When Jackie got to her room, Jeremy sat on the desk chair. He wasn't talking anymore. His eyes were very wide, and he was rubbing his upper lip. This was a sign of stress for Jeremy that Jackie knew well. Jackie sat on the bed next to Jeremy. She knew that he didn't like to be touched, so she just talked quietly instead.

"Okay, Jeremy. I typed up your schedule for today and tomorrow, so you'll know exactly what we're going to do. I made a copy for you." She handed the schedule to Jeremy, who read it quickly. Folding the paper very neatly, Jeremy carefully put it in his pocket.

Jackie was glad she'd taken the time to put the list together. Jeremy could relax when his day was organized with a set schedule. Without knowing what was to happen next, he became very fearful. Already he'd stopped rubbing his lip.

"Right now, we're going to go visit Steve," Jackie explained. "You know Steve. You are going to sleep in Steve's room tonight. He'll show you your bed and where you will brush your teeth. Is that okay?"

"Where are your clothes?" Jeremy asked. Jackie smiled ruefully. She knew that a short attention span was characteristic of fragile X syndrome, and that it was hard to get Jeremy to concentrate on any one topic for any length of time. Sometimes, coping with Jeremy's distractibility was difficult for her. But she also knew that Jeremy picked up far more information than others thought he did. He simply hadn't seen her clothes and was wondering where they were.

"In the closet, silly, just like at home," Jackie replied. She opened the closet door and showed him, and he smiled and

nodded. "Now listen. After we meet Steve, then we are going to the dining hall to eat. . . . "

"Can I have chicken nuggets?" Jeremy interrupted. He loved food and enjoyed cooking. He really loved salad with ranch dressing.

"Chicken nuggets and all the salad you want," Jackie reassured him, and Jeremy beamed. "Then we're going to watch a movie."

"What movie? I wanna watch *Scooby-Doo®*."

Jackie laughed. "No *Scooby-Doo®* at college." She was secretly grateful for that. Jeremy had memorized every episode of *Scooby-Doo®*. That meant Jackie had watched the show over and over again, too. Jackie was familiar with all the stories and dialogue, and she was very tired of the program.

"Sorry, but we're going to watch one of your other favorites— *Finding Nemo®*," she replied.

"Okay. I like that one, too," Jeremy replied.

"And then it will be time to brush your teeth and put on your pj's," Jackie explained. "Then you can go to sleep."

"What's for breakfast?" Jeremy asked quickly. Jackie had been anticipating that question. "Anything you want!" she answered.

She continued, describing the schedule for the following morning. Besides the meals, which Jeremy thought sounded great, there was a single planned activity. The siblings were supposed to tie-dye T-shirts together. Jeremy didn't understand what that meant, but Jackie told him that she'd help him. He was relieved. Once again, understanding his planned schedule reassured him. After lunch, Teresa would pick him up to go home. Jackie and her mother had agreed that one night in the dorm was plenty for Jeremy.

"Okay, Jeremy," Jackie concluded. "So, what's first?"

"We go see Steve!" he responded excitedly. "And see my bed."

Steve was one of Jackie's best friends. They had gone to the same high school that Jeremy still attended. Jeremy took academic classes in the morning, followed by vocational training in the afternoon. Most of Jeremy's classes were with other developmentally delayed students, but he was "mainstreamed" with the other students for some of his classes. Steve had tutored Jeremy in a biology class, where Jeremy had done very well.

Through his work with Jeremy, Steve had come to respect the splintered abilities of those with fragile X. He knew that these individuals are very good at some things and can't do other tasks at all. Steve had found his work with Jeremy to be enormously rewarding, and he'd decided to major in special education to help other people like Jeremy.

these genotypes. If the genotype is *EeSs*, all combinations of any two different letters can be present. Therefore, *ES, eS, Es,* and *es* are all possible gametes.

> ### Check Your Progress 20.2
>
> 1. For each of the following genotypes, give all possible gametes.
>   a. *WW*
>   b. *WWSs*
>   c. *Tt*
>   d. *Ttgg*
>   e. *AaBb*
> 2. For each of the following, state whether a genotype or a gamete is represented.
>   a. *D*
>   b. *Ll*
>   c. *Pw*
>   d. *LlGg*
> 3. What is the genotype of the individual from the following crosses?
>   a. *ES* × *es*
>   b. *eS* × *eS*

One-Trait Crosses

Many times, parents would like to know the chances of having a child with a certain genotype and, therefore, a certain phenotype. To illustrate, let us consider a cross involving freckles. What happens when two parents without freckles have children? Will the children of this couple have freckles? In solving the problem, we will: (1) use the key provided in Figure 20.3 to indicate the genotype of each parent; (2) determine the possible gametes for each parent; (3) combine all possible gametes; and (4) finally, determine the genotypes and the phenotypes of all the offspring. If both parents do not have freckles, then their genotypes are both *ff*. The only gametes they can produce contain the *f* allele. All of the children will therefore be *ff* and will not have freckles. In the following diagram, the letters in the first row give the genotypes of the parents. Each parent has only one type of gamete with regard to freckles, and therefore, all the children have a similar genotype and phenotype.

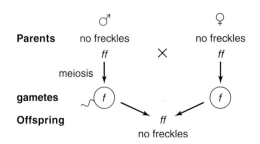

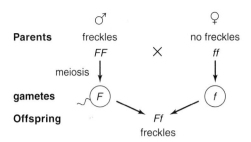

To challenge you a bit more, let's consider the results when a homozygous dominant man with freckles has children with a woman with no freckles. Will the children of this couple have freckles? The children are heterozygous (*Ff*) and have freckles. When writing a heterozygous genotype, always put the capital letter first to avoid confusion.

These children are **monohybrids.** They are heterozygous with regard to one pair of alleles. If they have children with someone else of the same genotype, will their children have freckles? In this problem (*Ff* × *Ff*), each parent has two possible types of gametes (*F* or *f*), and we must ensure that all types of sperm have an equal chance to fertilize all possible types of eggs. One way to do this is to use a **Punnett square** (Fig. 20.3), in which all possible types of sperm are lined up vertically and all possible types of eggs are lined up horizontally (or vice versa). Every possible combination of gametes occurs within the squares.

After we determine the genotypes and the phenotypes of the offspring, we can first determine the genotypic

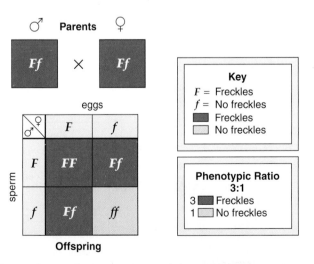

Figure 20.3 **What are the expected results of a monohybrid cross?**
A Punnett square diagrams the results of a cross. When the parents are heterozygous, each child has a 75% chance of having the dominant phenotype and a 25% chance of having the recessive phenotype.

and then the phenotypic ratio. The genotypic ratio is 1 *FF*: 2 *Ff*: 1 *ff* or simply 1:2:1, but the phenotypic ratio is 3:1. Why? Three individuals will have freckles and one will not have freckles.

This 3:1 phenotypic ratio is always expected for a monohybrid cross when one allele is completely dominant over the other. The exact ratio is more likely to be observed if a large number of matings take place and if a large number of offspring result. Only then do all possible types of sperm have an equal chance of fertilizing all possible types of eggs. Naturally, we do not routinely observe hundreds of offspring from a single type of cross in humans. The best interpretation of Figure 20.3, in humans, is to say that each child has three chances out of four to have freckles, or one chance out of four to not have freckles.

Chance has no memory. For example, if two heterozygous parents already have three children with freckles and are expecting a fourth child, this child still has a 75% chance of having freckles and a 25% chance of not having freckles. Suppose a man and a women did not have freckles. What chance would their children have for freckles? The answer would be 0% chance. How about the child's chance of not having freckles? The answer would be 100% chance. How about the cross *Ff* × *ff*? What are the chances of freckles? No freckles? In each instance, the child has a 50% chance for freckles and a 50% chance for no freckles.

Other One-Trait Crosses

It is not possible to tell by inspection if a person expressing a dominant allele is homozygous dominant or heterozygous. However, it is sometimes possible to tell by the results of a cross. For example, Figure 20.4 shows two possible results when a man with a widow's peak reproduces with a woman who has a straight hairline. If the man is homozygous dominant, all his children will have a widow's peak. If the man is heterozygous, each child has a 50% chance of having a straight hairline. The birth of just one child with a straight hairline indicates that the man is heterozygous.

Consider, also, that a person's parentage sometimes tells you that he is heterozygous. In Figure 20.4, each of the offspring with a widow's peak has to be heterozygous. Why? One of the parents was homozygous recessive, and therefore had to give each offspring a *w*. You will want to do all the practice problems in Check Your Progress 20.3 to ensure that you can do one-trait genetics problems.

The Punnett Square and Probability

Two laws of probability apply to genetics. The first is the product rule. According to this rule, the chance of two different events occurring simultaneously is equal to the multiplied probabilities of each event occurring separately. For example, what is the probability that a coin toss will be "heads"? There are only two options, so the probability of "heads" is 1 out of 2 or 50%. If we wanted to know what the probability was of a first coin toss being "heads" *and* a second coin toss being "heads," we use the product rule. (The product rule is often applied to cases where the word "and" is used.) This would be ½ × ½ = ¼, or 25%. The Punnett square allows you to determine the probability that an offspring will have a particular genotype/phenotype. When you bring the allele(s) donated by the sperm and egg together into the same square, you are using the product rule. Both father and mother each give a chance of having a particular allele, so the probability for that allele in an offspring is the multiple of these two separate chances.

The second law of probability is the sum rule. Using this rule, individual probabilities can be added to determine total probability for an event. If we toss a coin and we want to know the probability that it will be either heads *or* tails, we use the sum rule. (The sum rule is often applied to cases where the word "or" is used.) The probability of heads is ½.

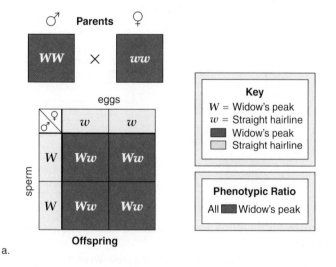

a.

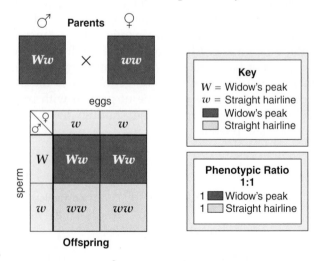

b.

Figure 20.4   How can a one-trait cross determine the genotype of a dominant phenotype?
This cross will determine if an individual with the dominant phenotype is homozygous or heterozygous. **a.** All offspring show the dominant characteristic, so the individual is most likely homozygous, as shown. **b.** The offspring show a 1:1 phenotypic ratio, so the individual is heterozygous, as shown.

The probability of tails is ½. Therefore, the probability of the coin toss giving heads or tails is ½ + ½ = 1 (meaning it's a sure thing). In the Punnett square, you use the sum rule when you add up the results of each square to determine the final genotype/phenotype ratios.

Two-Trait Crosses

Figure 20.5 allows you to relate the events of meiosis to the formation of gametes when a cross involves two traits. In the example given, a cell has two pairs of homologues, recognized by length. One pair of homologues is short, and the other is long. (The color signifies that we inherit chromosomes from our parents. One homologue of each pair is the "paternal" chromosome, and the other is the "maternal" chromosome.)

The homologues separate during meiosis I, so each gamete receives one member from each pair of homologues. The homologues separate independently so it does not matter which member of a pair goes into which gamete. All possible combinations of alleles occur in the gametes. In the simplest of terms, a gamete in Figure 20.5 will *receive one short and one long chromosome of either color.* Therefore, all possible combinations of chromosomes and alleles are in the gametes.

Specifically, assume that the alleles for two genes are on these homologues. The alleles *S* and *s* are on one pair of homologues, and the alleles *W* and *w* are on the other pair

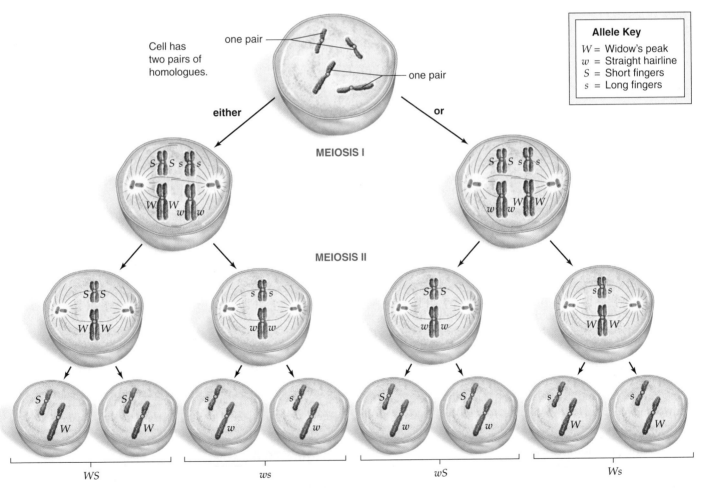

Figure 20.5 **How does meiosis result in genetic diversity among gametes?**

A cell has two pairs of homologous chromosomes (homologues), recognized by length, not color. The long pair of homologues carries alleles for type of hairline and the short pair of homologues carries alleles for finger length. The homologues, and the alleles they carry, align independently during meiosis. Therefore, all possible combinations of chromosomes and alleles occur in the gametes as shown in the last row of cells.

of homologues. The homologues separate, so a gamete will have either an *S* or an *s* and either a *W* or a *w*. They will never have two of the same letter of the alphabet. Also, because the homologues align independently at the equator, either the paternal or maternal chromosome of each pair can face either pole.

Therefore, there are no restrictions as to which homologue goes into which gamete. A gamete can receive either an *S* or an *s* and either a *W* or a *w* in any combination. In the end, the gametes will collectively have all possible combinations of alleles. You should be able to transfer this information to any cross that involves two traits. In other words, the process of meiosis explains why a person with the genotype *EeFf* would produce the gametes *EF, ef, Ef,* and *eF* in equal number.

The Dihybrid Cross

In the two-trait cross depicted in Figure 20.6, a person homozygous for widow's peak and short fingers (*WWSS*) reproduces with one who has a straight hairline and long fingers (*wwss*). The gametes for the *WWSS* parent must be *WS* and the gametes for the *wwss* parent must be *ws*. Therefore, the offspring will all have the genotype *WwSs* and the same phenotype (widow's peak with short fingers). This genotype is called a **dihybrid** because the individual is heterozygous in two regards: hairline and fingers.

When a dihybrid *WwSs* has children with another dihybrid that is *WwSs*, what gametes are possible? Each gamete can have only one letter of each type in all possible combinations. Therefore, these are the gametes for both dihybrids: *WS, Ws, wS,* and *ws.*

A Punnett square makes sure that all possible sperm fertilize all possible eggs. If so, these are the expected phenotypic results:

9 widow's peak and short fingers:
3 widow's peak and long fingers:
3 straight hairline and short fingers:
1 straight hairline and long fingers.

This 9:3:3:1 phenotypic ratio is always expected for a dihybrid cross when simple dominance is present. We can use this expected ratio to predict the chances of each child receiving a certain phenotype. For example, the chance of getting the two dominant phenotypes together is 9 out of 16. The chance of getting the two recessive phenotypes together is 1 out of 16.

Two-Trait Crosses and Probability

It is also possible to use the rules of probability we discussed on page 470 to predict the results of a dihybrid cross. For example, we know the probable results for two separate monohybrid crosses are as follows:

1. Probability of widow's peak = ¾
   Probability of straight hairline = ¼
2. Probability of short fingers = ¾
   Probability of long fingers = ¼

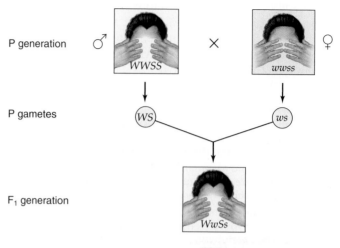

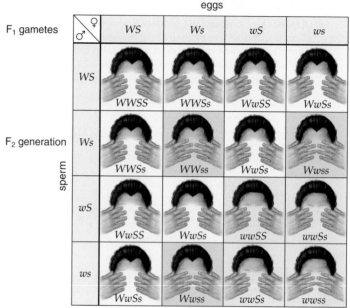

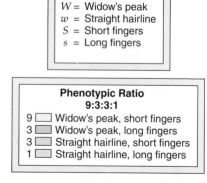

Allele Key

W = Widow's peak
w = Straight hairline
S = Short fingers
s = Long fingers

Phenotypic Ratio
9:3:3:1

9 ☐ Widow's peak, short fingers
3 ☐ Widow's peak, long fingers
3 ☐ Straight hairline, short fingers
1 ☐ Straight hairline, long fingers

Figure 20.6 **What are the expected results of a dihybrid cross?**
Each dihybrid can form four possible types of gametes, so four different phenotypes occur among the offspring in the proportions shown.

Using the product rule, we can calculate the probable outcome of a dihybrid cross as follows:

Probability of widow's peak and short fingers =
$$¾ × ¾ = ⁹⁄₁₆$$

Probability of widow's peak and long fingers =
$$¾ × ¼ = ³⁄₁₆$$

Probability of straight hairline and short fingers =
$$¼ × ¾ = ³⁄₁₆$$

Probability of straight hairline and long fingers =
$$¼ × ¼ = ¹⁄₁₆$$

In this way, the rules of probability tell us that the expected phenotypic ratio when all possible sperm fertilize all possible eggs is 9:3:3:1.

Other Two-Trait Crosses

It is not possible to tell by inspection whether an individual expressing the dominant alleles for two traits is homozygous

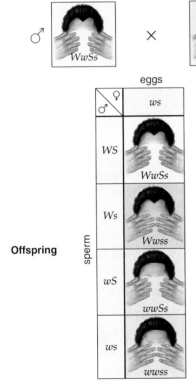

Figure 20.7 **What are the expected results of this two-trait cross?** The results of this cross indicate that the individual with the dominant phenotypes is heterozygous for both traits because some of the children are homozygous recessive for one or both traits. The chance of receiving any possible phenotype is 25%.

Table 20.1	Phenotypic Ratios of Common Crosses
Genotypes	**Phenotypes**
Monohybrid Gg × monohybrid Gg	3:1 (dominant to recessive)
Monohybrid Gg × recessive gg	1:1 (dominant to recessive)
Dihybrid GgRr × dihybrid GgRr	9:3:3:1 (9 both dominant, 3 one dominant, 3 other dominant, 1 both recessive)
Dihybrid GgRr × recessive ggrr	1:1:1:1 (all possible combinations in equal number)

dominant or heterozygous. But, if the individual has children with the homozygous recessive, it may be possible to tell. For example, if a man homozygous dominant for widow's peak and short fingers reproduces with a female homozygous recessive for both traits, then all his children will have the dominant phenotypes. However, if a man is heterozygous for both traits, then each child has a 25% chance of showing either one or both recessive traits. A Punnett square (Fig. 20.7) shows that the expected ratio is 1 widow's peak with short fingers: 1 widow's peak with long fingers: 1 straight hairline with short fingers: 1 straight hairline with long fingers, or 1:1:1:1.

For practical purposes, if a parent with the dominant phenotype, in either trait, has an offspring with the recessive phenotype, the parent has to be heterozygous for that trait. Also, it is possible to tell if a person is heterozygous by knowing the parentage. In Figure 20.7, no offspring showing a dominant phenotype is homozygous dominant for either trait. Why? The mother is homozygous recessive for that trait.

Table 20.1 gives the phenotypic results for certain crosses we have been studying. These crosses always give these phenotypic results. Therefore, it is not necessary to do a Punnett square to arrive at the results. To facilitate doing crosses, you will want to study Table 20.1 and understand why these are the results expected for these crosses.

> ### Check Your Progress 20.4
>
> 1. Attached earlobes and straight hairline are recessive. What genotype does a man with unattached earlobes and a widow's peak have if his mother has attached earlobes and a straight hairline?
>
> 2. What genotype do children have if one parent is homozygous recessive for earlobes and homozygous dominant for hairline and the other is homozygous dominant for unattached earlobes and homozygous recessive for hairline?
>
> 3. If an individual from this cross reproduces with another of the same genotype, what are the chances that they will have a child with a straight hairline and attached earlobes?
>
> 4. A child who does not have dimples or freckles is born to a man who has dimples and freckles (both dominant) and a woman who does not. What are the genotypes of all persons concerned?

Family Pedigrees for Inheritance of Genetic Disorders

We inherit many different traits from our parents. Some we may like, while others we don't care for. But we can also inherit serious diseases from our parents. Many of these diseases occur as a result of changes, or mutations, in our parents' genetic code. The abnormal gene could be present in each of your parent's cells, and thus passed down in the sperm or egg. Your parent may or may not have been affected by this genetic mutation. Alternatively, the genetic mutation might have occurred only in the sperm or egg that became a part of you. Some genetic diseases require two damaged alleles for the disease to manifest itself. Others need only one. When a genetic disorder is autosomal dominant, an individual with the alleles *AA* or *Aa* will have the disorder. When a genetic disorder is autosomal recessive, only individuals with the alleles *aa* will have the disorder. Genetic counselors often construct pedigrees to determine whether a condition that runs in the family is dominant or recessive. A pedigree shows the pattern of inheritance for a particular condition. Consider these two possible patterns of inheritance:

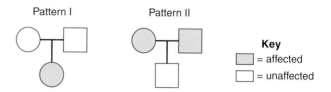

In both patterns, males are designated by squares and females by circles. Shaded circles and squares are affected individuals. A line between a square and a circle represents a union. A vertical line going downward leads, in these patterns, to a single child. (If there are more children, they are placed off a horizontal line.) Which pattern of inheritance do you suppose represents an autosomal dominant characteristic? Which represents an autosomal recessive characteristic?

CASE STUDY THE DINING HALL

As Jackie and Jeremy walked into his room, Steve stood up. "Hey, Jeremy, good to see you." Jeremy looked down at his shoes and stuck out his hand just like he'd practiced. "Look at me with your eyes, Jeremy," Steve reassured. "You know me." Jeremy glanced up and back down very quickly. His face broke out in a grin, and he stuck out his hand.

"Hi, Steve. How are you? How was your day?" It was Jeremy's standard greeting.

"Great!" Steve answered. "Wanna see your bed? It's all ready for you."

After a tour of the dorm room and the bathrooms, Jackie, Steve, and Jeremy walked over to the dining hall. Along the way, Jeremy started to chatter again. He didn't particularly talk to anyone. He just liked to talk. He talked through the schedule twice and described the sheets on the bed. Jackie liked to hear him chatter. That meant that he was calm and happy.

"WOW!" Jeremy shouted. He'd caught sight of the salad bar with all the fixings. Jackie and Steve laughed.

"Use your inside voice, Jeremy," Jackie reminded him. She handed him a plate and told him to help himself to salad.

"Where are the chicken nuggets?" Jeremy asked. "Next is *Finding Nemo*®. Right? *Finding Nemo*®?"

"One thing at a time, Jeremy," Jackie responded patiently. "Let's get your salad. Then we'll go get your chicken nuggets." They found a seat with Jackie's friends and sat down to eat.

Jeremy talked all through dinner. Sometimes he had trouble keeping his voice down because he was so excited. He asked Jackie repeatedly if the movie was next. Jackie had already told her friends all about Jeremy. They'd asked a lot of questions, and she'd described the family's life together. When Jackie and Jeremy were younger, the entire family had gone to a genetic counselor. There, they'd received Jeremy's diagnosis of fragile X syndrome. When she was in high school, Jackie had read all the information she could find about the condition. She had learned that fragile X is an inherited chromosomal abnormality, and that it is a syndrome because a variety of symptoms accompany the condition. She now understood that the type of symptoms, as well as their severity, varied from person to person.

Jeremy had worked with speech, physical, and occupational therapists for most of his life to develop to his full potential. Jackie knew how hard Jeremy worked, and she was very proud of him. She'd wanted her friends to meet him.

When they had finished dinner, the three cleaned and stacked their food trays. "Is the movie next NOW?" Jeremy demanded.

"Now," Jackie smiled.

Autosomal Recessive Disorder

In pattern I, the child is affected, but neither parent is. This can happen if the condition is recessive and the parents are *Aa*. The parents are **carriers** because they appear to be normal but are capable of having a child with a genetic disorder. Figure 20.8 shows a typical pedigree chart for a recessive genetic disorder. Other ways to recognize an autosomal recessive pattern of inheritance are also listed in the figure. If both parents are affected, all the children are affected. Why? The parents can pass on only recessive alleles for this condition. All children will be homozygous recessive.

Chance has no memory. Therefore, each child born to heterozygous parents has a 25% chance of having the disorder. In other words, it is possible that if a heterozygous couple has four children, each child might have the condition.

Autosomal Dominant Disorder

In pattern II, the child is unaffected, but the parents are affected. Of the two patterns, this one shows a dominant pattern of inheritance. The condition is dominant, so the parents can be *Aa* (heterozygous). The child inherited a recessive allele from each parent and, therefore, is unaffected. Figure 20.9 shows a typical pedigree for a dominant disorder. Other ways to recognize an autosomal dominant pattern of inheritance are also listed. When a disorder is dominant, an affected child must have at least one affected parent.

Genetic Disorders of Interest

Medical genetics has traditionally focused on disorders caused by single gene mutations, and we will discuss a few of the better-known disorders.

Autosomal Recessive Disorders

Inheritance of two recessive alleles is required before an autosomal recessive disorder will appear.

Tay-Sachs **Tay-Sachs disease** is a well-known autosomal recessive disorder that occurs usually among Jewish people in the United States, most of whom are of central and eastern European descent. Tay-Sachs disease results from a lack of an enzyme and the subsequent storage of its substrate in lysosomes. Lysosomes build up in many body cells, but the primary sites of storage are the cells of the brain. This accounts for the onset of symptoms and the progressive deterioration of psychomotor functions (Fig. 20.10).

At first, it is not apparent that a baby has Tay-Sachs disease. However, development begins to slow between four and eight months of age, and neurological impairment and psychomotor difficulties then become apparent. The child gradually becomes blind and helpless, develops uncontrollable seizures, and eventually becomes paralyzed.

Cystic Fibrosis **Cystic fibrosis** is an autosomal recessive disorder that occurs among all ethnic groups. It is the most common lethal genetic disorder among Caucasians in the

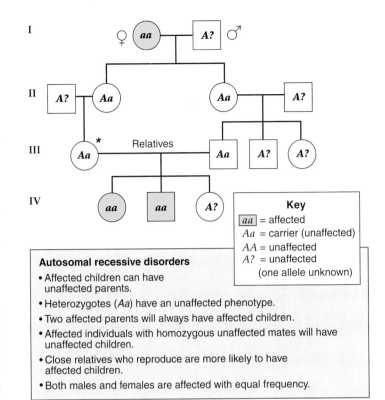

Autosomal recessive disorders
- Affected children can have unaffected parents.
- Heterozygotes (*Aa*) have an unaffected phenotype.
- Two affected parents will always have affected children.
- Affected individuals with homozygous unaffected mates will have unaffected children.
- Close relatives who reproduce are more likely to have affected children.
- Both males and females are affected with equal frequency.

Figure 20.8 **What does a pedigree for an autosomal recessive disorder reveal?**

The list gives ways to recognize an autosomal recessive disorder. How would you know that the individual at the asterisk is heterozygous?[1]

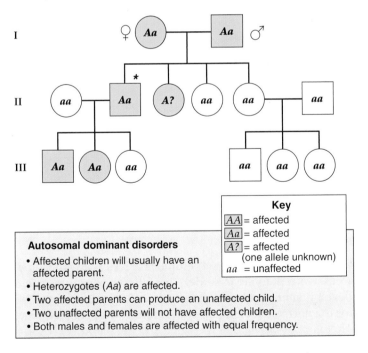

Autosomal dominant disorders
- Affected children will usually have an affected parent.
- Heterozygotes (*Aa*) are affected.
- Two affected parents can produce an unaffected child.
- Two unaffected parents will not have affected children.
- Both males and females are affected with equal frequency.

Figure 20.9 **What does a pedigree for an autosomal dominant disorder reveal?**

The list gives ways to recognize an autosomal dominant disorder. How would you know that the individual at the asterisk is heterozygous?[1]

[1]See Appendix B for answers.

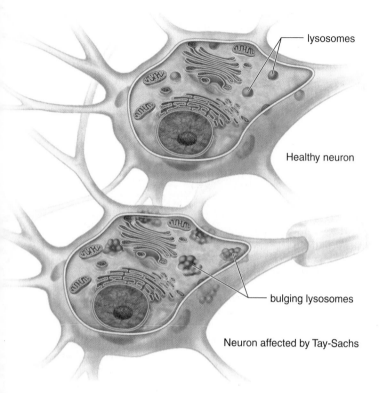

Figure 20.10   What causes Tay-Sachs disease?
In Tay-Sachs disease a lysosomal enzyme is missing. This causes the substrate of that enzyme to accumulate within the lysosomes.

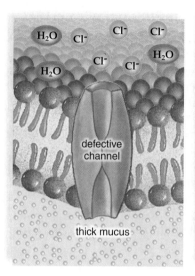
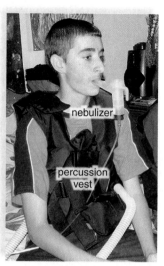

Figure 20.11   Why does cystic fibrosis occur? How is this disease treated?
Cystic fibrosis is due to a faulty protein that is supposed to regulate the flow of chloride ions into and out of cells through a channel protein. The nebulizer delivers drugs in aerosol form to the bronchial passages. The percussion vest pounds the chest to loosen mucus.

United States. Research has demonstrated that chloride ions (Cl⁻) fail to pass through a plasma membrane channel protein in the cells of these patients (Fig. 20.11). Ordinarily, after chloride ions have passed through the membrane, sodium ions (Na⁺) and water follow. It is believed that lack of water is the cause of abnormally thick mucus in bronchial tubes and pancreatic ducts. In these children, the mucus in the bronchial tubes and pancreatic ducts is particularly thick and viscous, interfering with the function of the lungs and pancreas. To ease breathing, the thick mucus in the lungs has to be manually loosened periodically, but still the lungs become infected frequently. Clogged pancreatic ducts prevent digestive enzymes from reaching the small intestine. To improve digestion, patients take digestive enzymes before every meal.

Phenylketonuria   Phenylketonuria (PKU) is an autosomal recessive metabolic disorder that affects nervous system development. Affected individuals lack an enzyme needed for the normal metabolism of the amino acid phenylalanine. Therefore, it appears in the urine and the blood. Newborns are routinely tested in the hospital for elevated levels of phenylalanine in the blood. If elevated levels are detected, newborns will develop normally if they are placed on a diet

low in phenylalanine. This diet must be continued until the brain is fully developed, around the age of seven or else severe mental retardation develops. Some doctors recommend that the diet continue for life. A pregnant woman with phenylketonuria must be on the diet to protect her unborn child from harm.

Sickle-Cell Disease   Sickle-cell disease is an autosomal recessive disorder in which the red blood cells are not biconcave disks like normal red blood cells. Many are sickle-shaped or shaped like a boomerang. The defect is caused by an abnormal hemoglobin that differs from normal

Have You Ever Wondered...

Why do diet sodas carry the warning: "Phenylketonurics: Contains Phenylalanine"?

The sweetener used in diet sodas is aspartame. Aspartame is formed by the combination of two amino acids: aspartic acid and phenylalanine. When aspartame is broken down by the body, phenylalanine is released. Phenylalanine is toxic for those with PKU and must be avoided. Therefore, those with PKU should avoid all diet products that may contain aspartame.

hemoglobin by one amino acid in the protein globin. The single amino acid change causes hemoglobin molecules to stack up and form insoluble rods. This causes the red blood cells to become sickle-shaped.

Sickle-shaped cells can't pass along narrow capillary passageways as disk-shaped cells can, so they clog the vessels and break down. This is why people with sickle-cell disease suffer from poor circulation, anemia, and low resistance to infection. Internal hemorrhaging leads to further complications, such as jaundice, episodic pain in the abdomen and joints, and damage to internal organs.

Sickle-cell heterozygotes have sickle-cell traits in which the blood cells are normal unless they experience dehydration or mild oxygen deprivation. Intense exertion may cause sickling and athletes should be aware of this possibility.

Autosomal Dominant Disorders

Inheritance of only one dominant allele is necessary for an autosomal dominant genetic disorder to appear. We discuss only two of the many autosomal dominant disorders here.

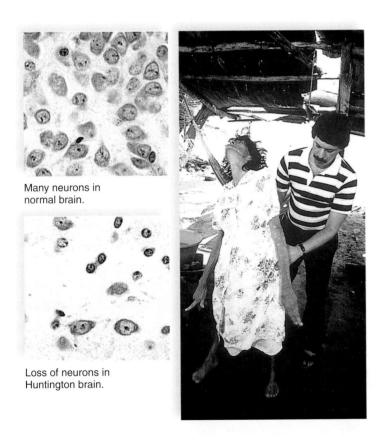

Many neurons in normal brain.

Loss of neurons in Huntington brain.

Figure 20.12 What causes Huntington disease? What effects does it have on the body?

Huntington disease is characterized by increasingly serious psychomotor and mental disturbances because of a loss of nerve cells.

Marfan Syndrome   **Marfan syndrome,** an autosomal dominant disorder, is caused by a defect in an elastic connective tissue protein, called fibrillin. This protein is normally abundant in the lens of the eye; the bones of limbs, fingers, and ribs; and also in the wall of the aorta. This explains why the affected person often has a dislocated lens, long limbs and fingers, and a caved-in chest. The wall of the aorta is weak and can possibly burst without warning. A tissue graft can strengthen the aorta, but Marfan patients with aortic symptoms still should not overexert themselves.

Huntington Disease   **Huntington disease** is an autosomal dominant neurological disorder that leads to progressive degeneration of brain cells (Fig. 20.12). The disease is caused by a mutated copy of the gene for a protein, called huntingtin. Most patients appear normal until they are of middle age and have already had children, who may later also be stricken. Occasionally, the first sign of the disease will appear during the teen years or even earlier. There is no effective treatment, and death comes 10 to 15 years after the onset of symptoms.

Several years ago, researchers found that the gene for Huntington disease was located on chromosome 4. A test was developed for the presence of the gene, but few people want to know if they have inherited the gene because there is no cure. The defective gene contains segments of DNA in which the base sequence CAG repeats again and again. This type of structure, called a trinucleotide repeat, causes the huntingtin protein to have too many copies of the amino acid glutamine. The normal version of huntingtin has stretches of between 10 and 25 glutamines. If huntingtin has more than 36 glutamines, it changes shape and forms large clumps inside neurons. Even worse, it attracts and causes other proteins to clump with it. One of these proteins, called CBP, helps nerve cells survive. Researchers hope they may be able to combat the disease by boosting CBP levels.

Check Your Progress 20.5

1. In a pedigree, all the members of one family are affected. a. If the trait is recessive, what are their genotypes? b. If the trait is dominant, what are their genotypes?

2. A baby has Tay-Sachs, but the parents are normal. What are the genotypes of the parents and the child?

3. What are the chances that homozygous normal parents for cystic fibrosis will have a child with cystic fibrosis?

4. What causes the sickle-shaped red blood cells in sickle-cell disease?

5. A child has Marfan syndrome. Why would you expect one of the parents to also have Marfan syndrome?

Preimplantation Genetic Diagnosis

I f prospective parents are heterozygous for one of the genetic disorders discussed on pages 475–77, they may want the assurance that their offspring will be free of the disorder. Determining the genotype of the embryo will provide this assurance. For example, if both parents are *Aa* for a recessive disorder, the embryo will develop normally if it has the genotype *AA* or *Aa*. On the other hand, if one of the parents is *Aa* for a dominant disorder, the embryo will develop normally only if it has the genotype *aa*.

Following in vitro fertilization (IVF), the zygote (fertilized egg) divides. When the embryo has eight cells (Fig. 20A*a*), removal of one of these cells for testing purposes has no effect on normal development. Only embryos that will not have the genetic disorders of interest are placed in the uterus to continue developing.

It is estimated that over 10,000 children have been born worldwide with normal genotypes following preimplantation embryo analysis for genetic disorders that run in their families. No American agency currently tracks these statistics, however. In the future, it's possible that embryos who test positive for a disorder could be treated by gene therapy, so that they, too, would be allowed to continue to term.

Testing the egg is possible if the condition of concern is recessive. Recall that meiosis in females results in a single egg and at least two polar bodies (see page 435). Polar bodies later disintegrate. They receive very little cytoplasm, but they do receive a haploid number of chromosomes. When a woman is heterozygous for a recessive genetic disorder, about half the first polar bodies will have received the mutated allele. In these instances, the egg received the normal allele. Therefore, if a polar body tests positive for a recessive mutated allele, the egg received the normal dominant allele. Only normal eggs are then used for IVF. Even if the sperm should happen to carry the mutation, the zygote will, at worst, be heterozygous. But the phenotype will appear normal.

Decide Your Opinion

1. Of the two diagnostic procedures described, does either seem more ethically responsible? Why?
2. Caring for an individual with a genetic condition can be very costly. Should society require preimplantation studies for the carriers of a genetic disease?

Figure 20A How is preimplantation genetic diagnosis carried out?
a. Following IVF and cleavage, genetic analysis is performed on one cell removed from an eight-cell embryo. If it is found to be free of the genetic defect of concern, the seven-cell embryo is implanted in the uterus and develops into a newborn with a normal phenotype. **b.** Chromosomal and genetic analysis is performed on a polar body attached to an egg. If the egg is free of a genetic defect, it is used for IVF, and the embryo is implanted in the uterus for further development.

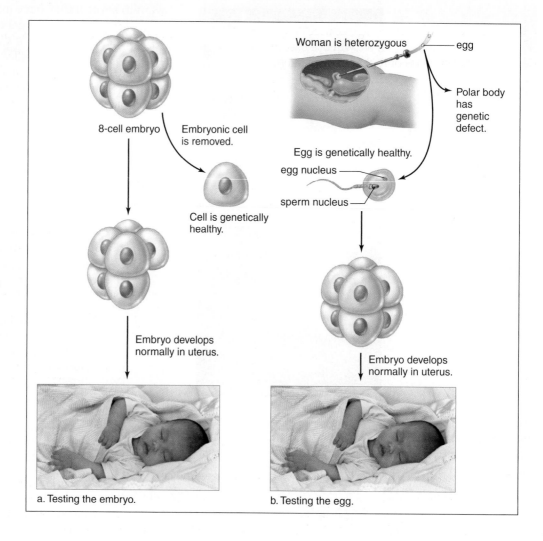

a. Testing the embryo.

b. Testing the egg.

20.3 Beyond Simple Inheritance Patterns

Certain traits, such as those studied in Section 20.1, are controlled by one set of alleles that follows a simple dominant or recessive inheritance. We now know of many other types of inheritance patterns.

Polygenic Inheritance

Polygenic traits, such skin color and height, are governed by several sets of alleles. The individual has a copy of all allelic pairs, possibly located on many different pairs of chromosomes. Each dominant allele codes for a product, and therefore, the dominant alleles have a quantitative effect on the phenotype. These effects are additive. The result is a *continuous variation* of phenotypes, resulting in a distribution of these phenotypes that resembles a bell-shaped curve. The more genes involved, the more continuous the variations and distribution of the phenotypes. Also, environmental effects cause many intervening phenotypes. In the case of height, differences in nutrition bring about a bell-shaped curve (Fig. 20.13).

Skin Color

Skin color is an example of a polygenic trait likely controlled by many pairs of alleles. Even so, we will use the simplest model and assume that skin has only two pairs of alleles (*Aa* and *Bb*) and that each capital letter contributes pigment to the skin. When a very dark person has children with a very light person, the children have medium-brown skin. When two people with the genotype *AaBb* have children with one another, the children may range in skin color from very dark to very light:

Genotypes	Phenotypes
AABB	Very dark
AABb or *AaBB*	Dark
AaBb or *AAbb* or *aaBB*	Medium brown
Aabb or *aaBb*	Light
aabb	Very light

Figure 20.14 How does temperature affect coat color in Himalayan rabbits? The dark ears, nose, and feet of this rabbit are believed to be due to a lower body temperature in these areas.

A range of phenotypes exists and several possible phenotypes fall between the two extremes. Therefore, the distribution of these phenotypes is expected to follow a bell-shaped curve. Few people have the extreme phenotypes and most people have the phenotype that lies in the middle. Skin color is also a **multifactorial trait,** a polygenic trait particularly influenced by the environment. After all, skin color is influenced by sun exposure.

Multifactorial Disorders

Many human disorders, such as cleft lip and/or palate, clubfoot, congenital dislocations of the hip, hypertension, diabetes, schizophrenia, and even allergies and cancers, are most likely controlled by polygenes subject to environmental influences. The coats of Siamese cats and Himalayan rabbits are darker in color at the ears, nose, paws, and tail (Fig. 20.14). The metabolic cause for this phenomena in Himalayan rabbits is known. Himalayan rabbits are homozygous for the allele *ch*, involved in the production of melanin. Experimental evidence suggests that the enzyme coded for by this gene is active only at a low temperature. Therefore, black fur occurs only at the extremities where body heat is lost to the environment.

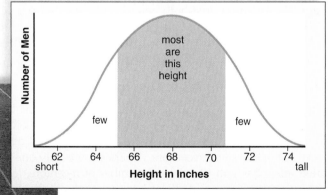

Figure 20.13 Why is height considered a polygenic trait? When you record the heights of a large group of people chosen at random, the values follow a bell-shaped curve. Such a continuous distribution is due to control of a trait by several sets of alleles. Environmental effects are also involved.

Reports have surfaced that all sorts of behavioral traits, such as alcoholism, phobias, and even suicide, can be associated with particular genes. No doubt behavioral traits are to a degree controlled by genes, but it has not been possible to determine to what degree. Very few scientists would support the idea that these behavioral traits are predetermined by our genes. Therefore, they must be multifactorial traits. Researchers are engaged in trying to determine what percentage of the trait is due to nature (inheritance) and what percentage is due to nurture (the environment). Some studies use identical and fraternal twins separated from birth. Then it's known that the twins have a different environment. The supposition is that, if identical twins in different environments share the same trait, that trait is most likely inherited. Identical twins are more similar in their intellectual talents, personality traits, and levels of lifelong happiness than are fraternal twins separated from birth. This substantiates the belief that behavioral traits are partly heritable. It also supports the belief that genes exert their effects by acting together in complex combinations susceptible to environmental influences.

Incomplete Dominance and Codominance

Incomplete dominance occurs when the heterozygote is intermediate between the two homozygotes. For example, when a curly-haired individual has children with a straight-haired individual, their children have wavy hair. When two wavy-haired persons have children, the expected phenotypic ratio among the offspring is 1:2:1—one curly-haired child to two with wavy hair to one with straight hair. We can explain incomplete dominance by assuming that only one allele codes for a product and the single dose of the product gives the intermediate result.

Codominance occurs when alleles are equally expressed in a heterozygote. A familiar example is the human blood type AB, in which the red blood cells have the characteristics of both type A and type B blood. We can explain codominance by assuming that both genes code for a product, and we observe the results of both products being present. Blood type inheritance is said to be an example of multiple alleles, described in the next section.

Incompletely Dominant Disorders

The prognosis in **familial hypercholesterolemia (FH)** parallels the number of LDL-cholesterol receptor proteins in the plasma membrane. A person with two mutated alleles lacks LDL-cholesterol receptors. A person with only one mutated allele has half the normal number of receptors, and a person with two normal alleles has the usual number of receptors. People with the full number of receptors do not have familial hypercholesterolemia. When receptors are completely absent, excessive cholesterol is deposited in various places in the body, including under the skin (Fig. 20.15).

The presence of excessive cholesterol in the blood causes cardiovascular disease. Therefore, those with no receptors die of cardiovascular disease as children. Individuals with

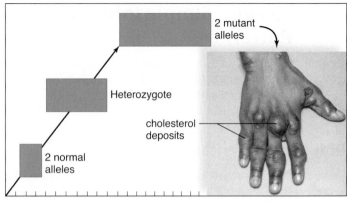

Figure 20.15   How is familial hypercholesterolemia inherited?
Familial hypercholesterolemia is incompletely dominant. Persons with one mutated allele have an abnormally high level of cholesterol in the blood, and those with two mutated alleles have a higher level still.

half the number of receptors may die when young or after they have reached middle age.

Multiple Allele Inheritance

When a trait is controlled by **multiple alleles,** the gene exists in several allelic forms. But each person can only have two of the possible alleles.

ABO Blood Types

Three alleles for the same gene control the inheritance of ABO blood types. These alleles determine the presence or absence of antigens on red blood cells:

$$I^A = \text{A antigen on red blood cells}$$
$$I^B = \text{B antigen on red blood cells}$$
$$i = \text{Neither A nor B antigen on red blood cells}$$

Each person has only two of the three possible alleles, and both I^A and I^B are dominant over i. Therefore, there are two possible genotypes for type A blood and two possible genotypes for type B blood. On the other hand, I^A and I^B are fully expressed in the presence of the other. Therefore, if a person inherits one of each of these alleles, that person will have type AB blood. Type O blood can result only from the inheritance of two i alleles.

The possible genotypes and phenotypes for blood type are as follows:

Phenotype	Genotype
A	$I^A I^A$, $I^A i$
B	$I^B I^B$, $I^B i$
AB	$I^A I^B$
O	ii

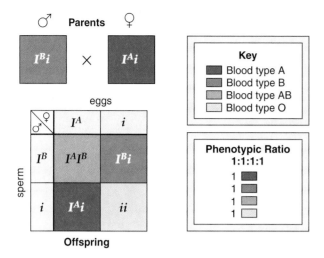

Figure 20.16 How are blood types inherited?
Blood type exemplifies multiple allele inheritance. The *I* gene has two codominant alleles, designated as *I^A* and *I^B*, and one recessive allele, designated by *i*. Therefore, a mating between individuals with type A blood and type B blood can result in any one of the four blood types. Why? The parents are *I^A i* and *I^B i*. If both parents were type AB blood, no child would have what blood type? See Appendix B for answers.

Blood typing can sometimes aid in paternity suits. However, a blood test of a supposed father can only suggest that he *might* be the father, not that he definitely *is* the father. For example, it is possible, but not definite, that a man with type A blood (genotype $I^A i$) is the father of a child with type O blood. On the other hand, a blood test sometimes can definitely prove that a man is not the father. For example, a man with type AB blood cannot possibly be the father of a child with type O blood. Therefore, blood tests can be used in legal cases only to try to exclude a man from possible paternity.

Figure 20.16 shows that matings between certain genotypes can have surprising results in terms of blood type. Parents with type A and type B blood can have offspring with all four possible blood types.

As a point of interest, the Rh factor is inherited separately from A, B, AB, or O blood types. When you are Rh positive, your red blood cells have a particular antigen, and when you are Rh negative, that antigen is absent. There are multiple recessive alleles for Rh⁻, but they are all recessive to Rh⁺.

Check Your Progress 20.6

1. A polygenic trait is controlled by several sets of alleles. Using the example of skin color, what are the two extreme genotypes for this trait?

2. What are some examples of human multifactorial traits?

3. What is the genotype of the lightest child that could result from a mating between two medium-brown individuals?

4. What is an example of incomplete dominance in humans?

5. How is ABO blood type an example of codominance and multiple allele inheritance?

6. A child with type O blood is born to a mother with type A blood. What is the genotype of the child? The mother? What are the possible genotypes of the father?

20.4 Sex-Linked Inheritance

Normally, both males and females have 23 pairs of chromosomes; 22 pairs are called **autosomes**, and one pair is the sex chromosomes. These are called the **sex chromosomes** because they differ between the sexes. In humans, males have the sex chromosomes X and Y, and females have two X chromosomes. The Y chromosome contains the gene responsible for determining male gender.

Traits controlled by genes on the sex chromosomes are said to be **sex-linked.** An allele on an X chromosome is **X-linked,** and an allele on the Y chromosome is Y-linked. Most sex-linked genes are only on the X chromosomes. The Y chromosome is lacking these. Very few Y-linked alleles have been found on the much smaller Y chromosome.

Many of the genes on the X chromosomes, such as those that determine normal as opposed to red-green color blindness, are unrelated to the gender of the individual. In other words, the X chromosome carries genes that affect both males and females. It would be logical to suppose that a sex-linked trait is passed from father to son or from mother to daughter, but this is not the case. A male always receives an X-linked allele from his mother, from whom he inherited an X chromosome. *The Y chromosome from the father does not carry an allele for the trait.* Usually, a sex-linked genetic disorder is recessive. Therefore, a female must receive two alleles, one from each parent, before she has the disorder.

X-Linked Alleles

When considering X-linked traits, the allele on the X chromosome is shown as a letter attached to the X chromosome. For example, this is the key for red-green color blindness, a well-known X-linked recessive disorder:

$$X^B = \text{normal vision}$$
$$X^b = \text{color blindness}$$

The possible genotypes and penotypes in both males and females are

Genotypes	Phenotypes
$X^B X^B$	Female who has normal color vision
$X^B X^b$	Carrier female who has normal color vision
$X^b X^b$	Female who is color-blind
$X^B Y$	Male who has normal vision
$X^b Y$	Male who is color-blind

The second genotype is a carrier female. Although a female with this genotype appears normal, she is capable of passing on an allele for color blindness. Color-blind females are rare because they must receive the allele from both parents. Color-blind males are more common because they need only one recessive allele to be color-blind. The allele for color blindness must be inherited from their mother because it is on the X chromosome. Males only inherit the Y chromosome from their father (Fig. 20.17).

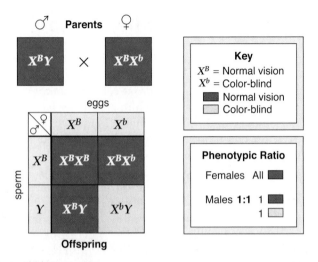

Figure 20.17 **What result can be predicted from crossing an X-linked allele?**
The male parent is normal, but the female parent is a carrier—an allele for color blindness is located on one of her X chromosomes. Therefore, each son has a 50% chance of being color-blind. The daughters will appear normal, but each one has a 50% chance of being a carrier.

Now let us consider a mating between a man with normal vision and a heterozygous woman (Fig. 20.17). What is the chance that this couple will have a color-blind daughter? A color-blind son? All daughters will have normal color vision because they all receive an X^B from their father. The sons, however, have a 50% chance of being color blind, depending on whether they receive an X^B or an X^b from their mother. The inheritance of a Y chromosome from their father cannot offset the inheritance of an X^b from their mother. The Y chromosome doesn't have an allele for the trait, so it can't possibly prevent color blindness in a son. Notice in Figure 20.17 that the phenotypic results for sex-linked traits are given separately for males and females.

Pedigree for X-Linked Disorders

Like color blindness, most sex-linked disorders are usually carried on the X chromosome. Figure 20.18 gives a pedigree for an *X-linked recessive disorder*. More males than females have the disorder because recessive alleles on the X chromosome are always expressed in males. The Y chromosome

---

CASE STUDY JEREMY'S VISIT CONCLUDES

The rest of the evening flew by. Jeremy fell asleep during the movie. It had been an exciting day for him. Jackie and Steve walked him back to Steve's dorm room, where Jeremy put on his *Scooby-Doo*® pajamas and brushed his teeth. Jackie tucked him into bed, just like she used to do at home. He looked pretty contented. She stayed in the dorm lounge until she knew that he'd gone to sleep, and then went back to her room.

Jackie called her mom on her cell phone. "Hi, Mom. He's fine. He's asleep in Steve's room. No problems at all."

Teresa breathed a sigh of relief. "He knows Steve. He should be fine." Then she added, "You need to go to sleep now—the two of you have a big day tomorrow!"

The next day, breakfast went well. Jeremy knew where the dining hall was and how to find everything he needed. To Katie, it seemed that he was feeling pretty comfortable with his surroundings and the bustle going on around him. He said hello to the workers and even shook hands with the cashier.

The tie-dye activity didn't go as smoothly. The event was loud and chaotic, and dozens of younger siblings ran around. The children's noise and activity upset Jeremy. He didn't like the looks of all the jars of dye and was afraid of getting his hands messy. He started flapping his hands—a sign that he was really stressed.

"Come on, Jeremy," Jackie reassured. "We can go now. Would you just like your T-shirt to be white? Is that okay?"

"Yeah, I want a white T-shirt," he answered. "I wanna go to the library."

"Okay, Jeremy," she answered. "The library is great. That's where we'll go next."

Jeremy was impressed with the library. He had never seen so many books. There weren't many at his reading level, but Jeremy liked sitting at the long table with a book and turning the pages. Best of all, it was quiet. "I'm studying, just like you, Jackie," he whispered.

After lunch, Teresa met them in the parking lot. "How'd it go, sweetheart?" she asked. Both Jeremy and Jackie started talking at the same time.

"Great! We had fun," Jackie replied.

"I had chicken nuggets! I studied in the library. And I made salad and brushed my teeth and we watched *Nemo* with Steve…." The chattering continued as Jeremy climbed in the back seat. Teresa kissed her daughter goodbye, inquiring, "What are you going to do with the rest of your weekend?"

Jackie laughed. "Go to bed! I'm worn out!"

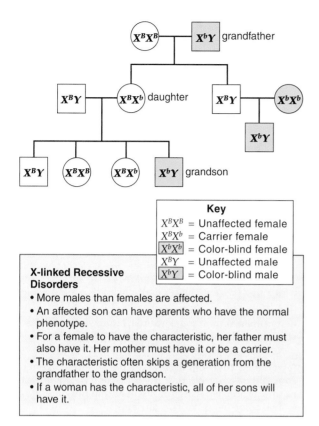

Key	
$X^B X^B$	= Unaffected female
$X^B X^b$	= Carrier female
$X^b X^b$	= Color-blind female
$X^B Y$	= Unaffected male
$X^b Y$	= Color-blind male

X-linked Recessive Disorders

- More males than females are affected.
- An affected son can have parents who have the normal phenotype.
- For a female to have the characteristic, her father must also have it. Her mother must have it or be a carrier.
- The characteristic often skips a generation from the grandfather to the grandson.
- If a woman has the characteristic, all of her sons will have it.

Figure 20.18 What does the pedigree for an X-linked recessive disorder reveal?

This pedigree for color blindness exemplifies the inheritance pattern of an X-linked recessive disorder. The list gives various ways of recognizing the X-linked recessive pattern of inheritance.

lacks an allele for the disorder. X-linked recessive conditions often pass from grandfather to grandson because the daughters of a male with the disorder are carriers. Figure 20.18 lists various ways to recognize a recessive X-linked disorder.

Only a few known traits are *X-linked dominant*. If a disorder is X-linked dominant, affected males pass the trait *only* to daughters, who have a 100% chance of having the condition. Females can pass an X-linked dominant allele to both sons and daughters. If a female is heterozygous and her partner is normal, each child has a 50% chance of escaping an X-linked dominant disorder. This depends on the maternal X chromosome which is inherited.

X-Linked Recessive Disorders of Interest

Color blindness, an X-linked recessive disorder, does not prevent males from leading a normal life. About 8% of Caucasian men have red-green color blindness. Most of these see brighter greens as tans, olive greens as browns, and reds as reddish browns. A few cannot tell reds from greens at all. They see only yellows, blues, blacks, whites, and grays.

Duchenne muscular dystrophy is an X-linked recessive disorder characterized by a wasting away of the muscles. Symptoms, such as waddling gait, toe walking, frequent

falls, and difficulty in rising, may appear as soon as the child starts to walk. Muscle weakness intensifies until the individual is confined to a wheelchair. Death usually occurs by age 20. Therefore, affected males are rarely fathers. The recessive allele remains in the population by passage from carrier mother to carrier daughter.

The absence of a protein, now called dystrophin, is the cause of Duchenne muscular dystrophy. Much investigative work determined that dystrophin is involved in the release of calcium from the sarcoplasmic reticulum in muscle fibers. The lack of dystrophin causes calcium to leak into the cell, which promotes the action of an enzyme that dissolves muscle fibers. (Refer to Chap. 12 to review this process.) When the body attempts to repair the tissue, fibrous tissue forms (Fig. 20.19). This cuts off the blood supply so that more and more cells die. Immature muscle cells can be injected into muscles, but it takes 100,000 cells for dystrophin production to increase by 30–40%.

Fragile X syndrome is the most common cause of inherited mental impairment. These impairments can range from mild learning disabilities to more severe intellectual disabilities. It is also the most common known cause of *autism,* a class of social, behavioral, and communication disorders. Males with full symptoms of the condition have characteristic physical abnormalities. A long face, prominent jaw, and large ears are facial features often seen in fragile X males. Joint laxity (excessively flexible joints) and genital abnormalities are common as well. In the case study, Jeremy Callen exhibits many of the behavioral patterns typical of his condition: tactile defensiveness (dislike of being touched), poor eye contact, repetitive speech patterns, hand flapping, and distractibility. Females with the condition present with variable symptoms. Most of the same traits seen in males with fragile X have been reported in females as well, but often the symptoms are milder in females and present with lower frequency.

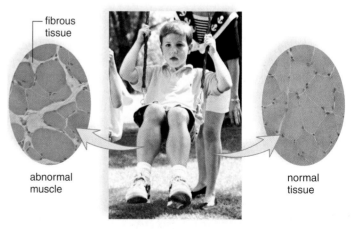

Figure 20.19 What are the characteristics of muscular dystrophy?

In muscular dystrophy, an X-linked recessive disorder, calves enlarge because fibrous tissue develops as muscles waste away, due to lack of the protein dystrophin.

A person with fragile X does not make the protein FMRP (fragile X mental retardation protein). Lack of this protein in the brain results in the various manifestations of the disease. As the name implies, a gene defect is found on the X chromosome. The genetic basis of fragile X is similar to that seen in Huntington disease (see page 477). In both cases, the gene in question has too many repeated copies of a DNA sequence containing three nucleotides (called a trinucleotide repeat). In fragile X, the DNA sequence is CGG. (Recall that in Huntington disease, the repeated sequence is CAG.) Fewer than 59 copies of the repeated sequence is considered normal. Between 59 and 200 copies is considered "premutation." Generally, both males and females with the premutation genotype have normal intellect and appearance. They may have subtle intellectual or behavioral symptoms, however. Persons whose DNA has over 200 copies of the repeat have "full mutation" and show the physical and behavioral traits of fragile X.

The trinucleotide repeat disorders, like Huntington disease and fragile X syndrome, exhibit what is called *anticipation*. This means that the number of repeats in the gene can increase in each successive generation. For example, a female with 100 copies of the repeat is considered to have a premutation. When this female passes her X chromosome with the premutation to her offspring, the number of repeats may expand to over 200. This would result in a full mutation in her child. In a recent study, maternal premutations of between 90 and 200 repeats resulted in an expansion to full mutation in 80–100% of the offspring.

Hemophilia is an X-linked recessive disorder. There are two common types. Hemophilia A is due to the absence or minimal presence of a clotting factor known as factor VIII, and hemophilia B is due to the absence of clotting factor IX. Hemophilia is called the bleeder's disease because the affected person's blood either does not clot or clots very slowly. Although hemophiliacs bleed externally after an injury, they also bleed internally, particularly around joints. Hemorrhages can be stopped with transfusions of fresh blood (or plasma) or concentrates of the clotting protein. Also, factors VIII and IX are now available as a biotechnology product.

> **Check Your Progress 20.7**
>
> 1. Why are more males than females color-blind?
>
> 2. a. What phenotypic ratio is expected for a cross in which both parents have one X-linked recessive allele? b. In which both parents have one X-linked dominant allele?
>
> 3. Both the mother and the father of a son with hemophilia appear to be normal. From whom did the son inherit the allele for hemophilia? What are the genotypes of the mother, the father, and the son?
>
> 4. A woman is color-blind. What are the chances that her sons will be color-blind? If she is married to a man with normal vision, what are the chances that her daughters will be color-blind? Will be carriers?
>
> 5. Both the husband and wife have normal vision. The wife gives birth to a color-blind daughter. Is it more likely the father had normal vision or was color-blind? What does this lead you to deduce about the girl's parentage?

Historical **Focus**

Hemophilia: The Royal Disease

The pedigree in Figure 20B shows why hemophilia is often referred to as "The Royal Disease." Queen Victoria of England, who reigned from 1837 to 1901, was the first of the royals to carry the gene. From her, the disease eventually spread to the Prussian, Spanish, and Russian royal families. In that era, monarchs arranged marriages between their children to consolidate political alliances. This practice allowed the gene for hemophilia to spread throughout the royal families. It is assumed that a spontaneous mutation arose either in Queen Victoria after her conception or in one of the gametes of her parents. However, in the book *Queen Victoria's Gene* by D. M. Potts, the author postulates that Edward Augustus, Duke of Kent, may not have been Queen Victoria's father. Potts suggests that Victoria may have instead been the illegitimate child of a hemophiliac male. Regardless of her parentage, had Victoria not been crowned, the fate of the various royal households may have been very different. Further, the history of Europe could have been changed dramatically as well.

However, Victoria did become queen. Queen Victoria and her husband Prince Albert had nine children. Fortunately, only one son, Leopold, suffered from hemophilia. He experienced severe hemorrhages and died in 1884 at the age of 31 as the result of a minor fall. He left behind a daughter, Alice, a carrier for the disease. Her son, Rupert, also suffered from hemophilia and in 1928 died of a brain hemorrhage as a result of a car accident. Queen Victoria's eldest son, Edward VII, and the heir to the throne, did not have the disease and so the current British royal family is free of the disease.

Two of Queen Victoria's daughters, Alice and Beatrice, were carriers of the disease. Alice married Louis IV, the Grand Duke of Hesse. Of her six children, three were affected by hemophilia. Her son, Frederick, died of internal bleeding from a fall. Alice's daughter Irene married Prince Henry of Prussia, her first cousin. Two of their three sons suffered from hemophilia. One of these sons, Waldemar, died at age 56 due to the lack of blood transfusion supplies during World War I. Henry, Alice's other son, bled to death at the age of four.

Alice's daughter, Alexandra, married Nicholas II of Russia. Alexandra gave birth to four daughters before giving birth to Alexei, the heir to the Russian throne. It was obvious almost from birth that Alexei had hemophilia. Every fall caused bleeding into his joints, which led to crippling of his limbs and excruciating pain. The best medical doctors could not help Alexei. Desperate to relieve his suffering, his parents turned to the monk Rasputin. Rasputin was able to relieve some of Alexei's suffering by hypnotizing him and putting him to sleep. Alexandra and Nicholas, the Tsar and Tsarina, put unlimited trust in Rasputin. The illness of the only heir to the Tsar's throne, the strain Alexei's illness placed on the Tsar and Tsarina, and the power of Rasputin were all factors leading to the Russian Revolution of 1917. The Tsar and Tsarina, as well as their children, were all murdered during the revolution.

Queen Victoria's other carrier daughter, Beatrice, married Prince Henry of Battenberg. Her son, Leopold, was a hemophiliac, dying at 32 during a knee operation. Beatrice's daughter, Victoria Eugenie, married Alfonso XII of Spain. Queen Ena, as Victoria Eugenie came to be known, was not popular with the Spanish people. Her firstborn son and the heir to the Spanish throne, Alfonso, did not stop bleeding upon his circumcision. When it became obvious that she had given her son hemophilia, it is alleged that her husband never forgave her. Like his cousin Rupert, Alfonso died in 1938 from internal bleeding after a car accident. Victoria's youngest son, Gonzalo, was also a hemophiliac whose life was claimed by a car accident in 1934.

Today, no members of any European royal family are known to have hemophilia. Individuals with the disease gene born in the late 1800s and early 1900s have all died, eliminating the gene from the current royal houses.

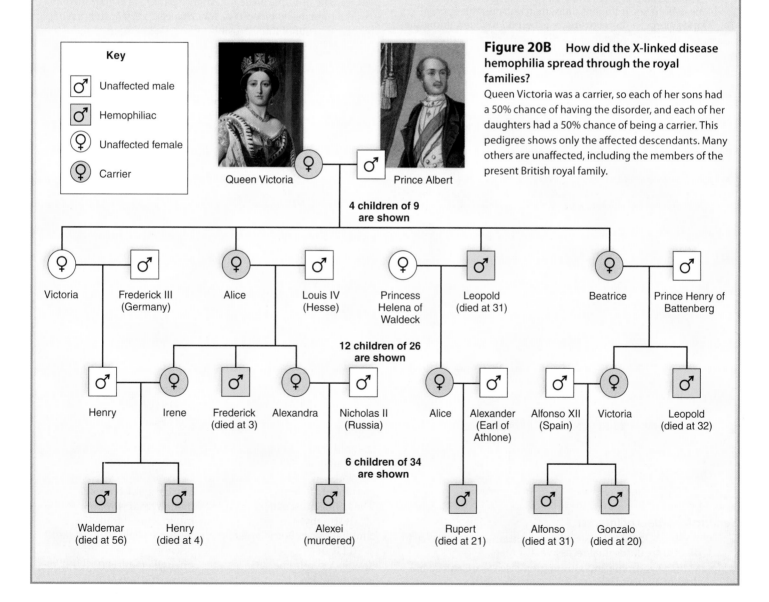

Figure 20B How did the X-linked disease hemophilia spread through the royal families?

Queen Victoria was a carrier, so each of her sons had a 50% chance of having the disorder, and each of her daughters had a 50% chance of being a carrier. This pedigree shows only the affected descendants. Many others are unaffected, including the members of the present British royal family.

Summarizing the Concepts

20.1 Genotype and Phenotype

Genotype refers to the alleles of the individual, and phenotype refers to the physical characteristics associated with these alleles.

- Homozygous dominant individuals (*EE*) have the dominant phenotype (i.e., unattached earlobes).
- Homozygous recessive individuals (*ee*) have the recessive phenotype (i.e., attached earlobes).
- Heterozygous individuals (*Ee*) have the dominant phenotype (i.e., unattached earlobes).

20.2 One- and Two-Trait Inheritance

One-Trait Crosses

The first step in doing one-trait problems is to determine the genotype and then the gametes.

- An individual has two alleles for every trait, but a gamete has one allele for every trait.

  The next step is to combine all possible sperm with all possible eggs. If there is more than one possible sperm and/or egg, a Punnett square is helpful in determining the genotypic and phenotypic ratio among the offspring.
- For a monohybrid × monohybrid cross, a 3:1 ratio is expected among the offspring.
- For a monohybrid × recessive cross, a 1:1 ratio is expected among the offspring.
- The expected ratio can be converted to the chance of a particular genotype/phenotype. For example, a 3:1 ratio = a 75% chance of the dominant phenotype and a 25% chance of the recessive phenotype.

Two-Trait Crosses

- If an individual is heterozygous for two traits, 4 gamete types are possible as can be substantiated by knowledge of meiosis.
- For a dihybrid × dihybrid cross (*AaBb* × *AaBb*), a 9:3:3:1 ratio is expected among the offspring.
- For a dihybrid × recessive cross (*AaBb* × *aabb*), a 1:1:1:1 ratio is expected among the offspring.

Family Pedigrees for Inheritance of Genetic Disorders

A pedigree shows the pattern of inheritance for a trait from generation to generation of a family. This first pattern appears in a family pedigree for a recessive disorder—both parents are carriers. The second pattern appears in a family pedigree for a dominant disorder. Both parents are again heterozygous.

Trait is recessive.

Trait is dominant.

Genetic Disorders of Interest

- Tay-Sachs, cystic fibrosis, phenylketonuria, and sickle-cell disease are autosomal recessive disorders.
- Marfan syndrome and Huntington disease are autosomal dominant disorders.

20.3 Beyond Simple Inheritance Patterns

In some patterns of inheritance, the alleles are not just dominant or recessive.

Polygenic Inheritance

Polygenic traits, such as skin color and height, are controlled by more than one set of alleles. The dominant alleles have an additive effect on the phenotype.

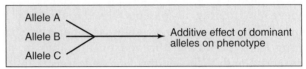

Incomplete Dominance and Codominance

In incomplete dominance (e.g., familial hypercholesterolemia), the heterozygote is intermediate between the two homozygotes. In codominance (e.g., blood type AB), both dominant alleles are expressed equally.

Multiple Allele Inheritance

The multiple allele inheritance pattern is exemplified in humans by blood type inheritance. Every individual has two out of three possible alleles: I^A, I^B, i. Both I^A and I^B are expressed. Therefore, this is also a case of codominance.

20.4 Sex-Linked Inheritance

Many genes on the X chromosomes, such as those that determine normal vision as opposed to color blindness, are unrelated to the gender of the individual. Common X-linked genetic crosses are

$X^B X^b \times X^B Y$ All daughters will be normal, even though they have a 50% chance of being carriers, but sons have a 50% chance of being color-blind.

$X^B X^B \times X^b Y$ All children are normal (daughters will be carriers).

Pedigree for X-Linked Disorders

- A pedigree for an X-linked recessive disorder shows that the trait often passes from grandfather to grandson by way of a carrier daughter. Also, more males than females have the characteristic.
- Like most X-linked disorders, color blindness, muscular dystrophy, fragile X, and hemophilia are recessive.

Understanding Key Terms

allele 466
autosome 481
carrier 475
codominance 480
color blindness 483
cystic fibrosis 475
dihybrid 472
dominant allele 466
Duchenne muscular dystrophy 483
familial hypercholesterolemia (FH) 480
fragile X syndrome 483
genotype 466

hemophilia 484
heterozygous 466
homozygous dominant 466
homozygous recessive 466
Huntington disease 477
incomplete dominance 480
locus 466
Marfan syndrome 477
monohybrid 469
multifactorial trait 479
multiple allele 480
phenotype 466
phenylketonuria (PKU) 476
polygenic trait 479

Match the key terms to these definitions.

a. _____ Gridlike device used to calculate the expected results of simple genetic crosses.

b. _____ Alternate forms of a gene that occur at the same site on homologous chromosomes.

c. _____ Allele that exerts its phenotype effect in the heterozygote; it masks the expression of the recessive allele.

d. _____ Particular site where a gene is found on a chromosome.

e. _____ Alleles of an individual for a particular trait or traits, expressed such as *BB*, *Aa*, or *BBAa*.

Testing Your Knowledge of the Concepts

1. Parents both have unattached earlobes, but some of their children have attached earlobes. Explain, in terms of phenotype and genotype. (pages 466–67)

2. Explain why the gametes have only one allele for a trait. (page 467)

3. What is the chance of producing a child with the dominant phenotype from each of the following crosses? (pages 469–70)
   a. *AA* × *AA*   c. *Aa* × *Aa*
   b. *Aa* × *AA*   d. *aa* × *aa*

4. Which of the crosses in question 3 can result in an offspring with the recessive phenotype? Explain. (pages 469–70)

5. What are the expected results of the following crosses? (pages 469–73)
   a. monohybrid × monohybrid
   b. monohybrid × recessive
   c. dihybrid × dihybrid
   d. dihybrid × recessive in both traits

6. Which of these crosses would best allow you to determine that an individual with the dominant phenotype is homozygous or heterozygous? (pages 470, 473)

7. What is the genotype of a heterozygote with a widow's peak and short fingers? Give all possible gametes for this individual. (page 472)

8. Using a pedigree, show an autosomal recessive disorder pattern and an autosomal dominant disorder pattern. (pages 474–75)

9. What is required for an autosomal recessive disorder to appear? Name four autosomal recessive disorders. What is required for an autosomal dominant disorder to appear? Name two autosomal dominant disorders. (pages 474–77)

10. Define and give an example of each of the following inheritance patterns: polygenic inheritance, multifactorial trait, incomplete dominance, codominance, and multiple alleles. (pages 479–81)

11. What are the phenotypes and genotypes for the ABO blood groups? If both parents are type AB blood, what are the possible genotypes and phenotypes for their children, including the expected percentages? (pages 480–81)

12. How is an X-linked trait different from an autosomal trait? (page 481)

13. A woman with normal vision reproduces with a man with normal vision. Three sons are color-blind, and one son has normal vision. Explain how this occurred, using the genotypes and gametes of all individuals. (pages 481–82)

14. Which of these is a correct statement?
    a. Each gamete contains two alleles for each trait.
    b. Each individual has one allele for each trait.
    c. Fertilization gives each new individual one allele for each trait.
    d. All of these are correct.
    e. None of these is correct.

15. Which of the following indicates a heterozygous individual?
    a. *AB*   c. *Aa*
    b. *AA*   d. *aa*

16. What possible gametes can be produced by *AaBb*.
    a. *Aa, Bb*   c. *AB, ab*
    b. *A, a, B, b*   d. *AB, Ab, aB, ab*

17. In humans, pointed eyebrows (*P*) are dominant over smooth eyebrows (*p*). Mary's father has pointed eyebrows, but she and her mother have smooth. What is the genotype of the father?
    a. *pp*   c. *PPpp*
    b. *Pp*   d. *pPpP*

18. The genotypic ratio from a monohybrid cross is
    a. 1:1.   d. 9:3:3:1.
    b. 3:1.   e. 1:1:1:1.
    c. 1:2:1.

19. A straight hairline is recessive. If two parents with a widow's peak have a child with a straight hairline, then what is the chance that their next child will have a straight hairline?
    a. no chance   d. ½
    b. ¼   e. ¹⁄₁₆
    c. ³⁄₁₆

20. What is the chance that an *Aa* individual will be produced from an *Aa* × *Aa* cross?
    a. 50%   d. 25%
    b. 75%   e. 100%
    c. 0%

21. The genotype of an individual with the dominant phenotype can be determined best by reproduction with
    a. the recessive genotype or phenotype.
    b. a heterozygote.
    c. the dominant phenotype.
    d. the homozygous dominant.
    e. Both a and b are correct.

22. The homologous chromosomes align independently at the equator during meiosis so
    a. all possible combinations of alleles can occur in the gametes.
    b. only the parental combinations of gametes can occur in the gametes.
    c. only the nonparental combinations of gametes can occur in the gametes.

23. What is the chance that a dihybrid cross will produce a homozygous recessive in both traits?
    a. $\frac{9}{16}$
    b. $\frac{1}{4}$
    c. $\frac{1}{16}$
    d. $\frac{3}{16}$
    e. $\frac{1}{8}$

24. Which of the following is not a feature of multifactorial inheritance?
    a. Effects of dominant alleles are additive.
    b. Genes affecting the trait may be on multiple chromosomes.
    c. Environment influences phenotype.
    d. Recessive alleles are harmful.

25. The ABO blood system exhibits
    a. codominance.
    b. multiple alleles.
    c. incomplete dominance.
    d. Both a and b are correct.

26. Assume two normal parents have a color-blind son. Which parent is responsible for color blindness in the son?
    a. the mother
    b. the father
    c. either parent
    d. Neither parent—two normal parents cannot have a color-blind son.

27. If a child has type O blood and the mother is type A, then which of the following could be the blood type of the child's father?
    a. A only
    b. B only
    c. O only
    d. A or O
    e. A, B, or O

28. Under what condition(s) will an autosomal recessive disorder appear?
    a. inheritance of a single recessive allele
    b. inheritance of one dominant and one recessive allele
    c. inheritance of two recessive alleles
    d. Both a and c are correct.

29. Alice and Henry are at the opposite extremes for a multifactorial trait. Their children will
    a. be bell-shaped.
    b. be a phenotype typical of a 3:1 ratio.
    c. have the middle phenotype between their two parents.
    d. look like one parent or the other.

30. Two wavy-haired individuals (neither curly hair nor straight hair are completely dominant) reproduce. What are the chances that their children will have wavy hair?
    a. 0%
    b. 25%
    c. 50%
    d. 100%

Thinking Critically About the Concepts

Jeremy Callen, in the case study, has a full mutation resulting in fragile X syndrome. But Jeremy's mother is not considered a carrier for fragile X syndrome. How is that possible? Due to the phenomenon of anticipation, Jeremy's mother does not have to carry a full mutation. She carries a premutation that expanded to a full mutation, resulting in Jeremy's condition. In the case of fragile X, even males can carry premutations. However, anticipation seems to be found predominantly in maternal transmission of the X chromosome to the offspring.

1. If Jeremy had a different X-linked disorder, such as hemophilia, draw the family pedigree (Mom, Dad, Jeremy, and Jackie) that would result. Use X^A and X^a for the hemophilia alleles. Can you determine whether Jackie is a carrier?

2. Draw the family pedigree for fragile X syndrome. Use X^+ for the normal allele, X^P for a premutation, and X^F for a full mutation. What assumptions must you make about Mom's genotype? Can you determine Jackie's risk of having a son with fragile X syndrome?

3. Why are the best matches for organ/tissue transplants often a sibling of the person needing the transplant?

21

DNA Biology and Technology

Bianca Ramirez threw her backpack on the bed in her dorm room. "Yay!" she exclaimed. "Done for the weekend."

Her roommate, Becky, looked up. "So you've already finished your assignment for Bio 127?" she inquired. "You know, the one due Monday."

A cloud of dismay shadowed Bianca's face. "Isn't it something I can do Sunday night?"

Becky laughed and handed Bianca her notebook. There on the front page, the syllabus stated:

> Due Monday: Interview a person who has a disease treated by a
> product made by genetic technology. Write a 1–2 page, single-spaced,
> typed report including the disease history of this person. Explain
> the specifics of how the gene product is made and how it helps this
> patient. Possible diseases will be discussed in class.

"Oh, no," Bianca groaned. "If I didn't have to talk to someone, I could do it Sunday night. But now I have to find someone and talk to them." Bianca looked at Becky. "What did you do?"

Becky said, "I have a cousin with cystic fibrosis. She takes Pulmozyme®. I called my aunt to get all the details. Pulmozyme® contains an enzyme that digests the DNA found in the mucus that blocks her airways and her lungs. She inhales it, and it helps her to breathe easier."

Bianca sighed, "I don't know anybody with cystic fibrosis. What other diseases could I do?"

Becky grinned. "You were in class when we talked about it. Remember?"

Bianca opened her notebook and flipped through the pages to the lecture on recombinant DNA pharmaceuticals. She ran her finger down the list. "Cystic fibrosis, kidney failure, cancer, hepatitis B and C infections, leukemia, lymphoma, multiple sclerosis," she read aloud. "Wow. This is a long list."

She kept reading. "Awesome! Here's one. Diabetes. Cameron Kelly has diabetes. He lives over in Dunn Hall. I can talk to him." Bianca flipped open her cell phone and dialed Cameron's number. "Hey, Cameron," she began. "I need to do a report on diabetes for biology. Can you meet me sometime this weekend to talk? Sure, I'll meet you in the dining hall! See you in 15 minutes." Bianca laughed in relief. "Maybe I'll get this assignment done on time for a change!"

21.1 DNA and RNA Structure and Function

DNA is a double helix composed of two poly-nucleotide strands. When DNA replicates, each strand serves as a template for a new strand. RNA is also made off a DNA template.

21.2 Gene Expression

Gene expression results in an RNA or protein product. Each protein has a sequence of amino acids according to the blueprint provided by the sequence of nucleotides in DNA.

21.3 Genomics

Genomics is the study of genetic information in a particular cell or organism, including humans and other organisms.

21.4 DNA Technology

DNA technology allows us to clone a portion of DNA for various purposes, including DNA fingerprinting and transfer of DNA to other organisms. Transgenic (genetically modified) organisms receive and express foreign DNA. During gene therapy, humans receive foreign DNA to cure some particular condition.

21.1 DNA and RNA Structure and Function

DNA (deoxyribonucleic acid) is the genetic material. DNA is largely found in the chromosomes, located in the nucleus of a cell. (Recall from page 56 that mitochondria also contain DNA. For the purpose of this discussion, we will consider nuclear DNA.) Any genetic material has to be able to do three things: (1) replicate so that it can be transmitted to the next generation, (2) store information, and (3) undergo mutations that provide genetic variability.

Structure of DNA

DNA is a **double helix.** It is composed of two strands that spiral about each other (Fig. 21.1*a*). Each strand is a polynucleotide because it is composed of a series of nucleotides. A nucleotide is a molecule composed of three subunits—phosphoric acid (phosphate), a pentose sugar (deoxyribose),

and a nitrogen-containing base (either A, C, G, or T, page 38). Looking at just one strand of DNA, notice that the phosphate and sugar molecules make up a backbone and the bases project to one side. Put the two strands together, and DNA resembles a ladder (Fig. 21.1*b*). The phosphate-sugar backbones make up the supports of the ladder. The rungs of the ladder are the paired bases. The bases are held together by hydrogen bonding: A pairs with T, by forming two hydrogen bonds, and G pairs with C, by forming three hydrogen bonds, or vice versa. This is called **complementary paired bases** (Fig. 21.1*c*).

The bases are important to the functioning of DNA. It will be helpful to remember that a purine (has two rings) is always paired with a pyrimidine (has one ring), like this:

Purines	Pyrimidines
Adenine (A)	Thymine (T)
Guanine (G)	Cytosine (C)

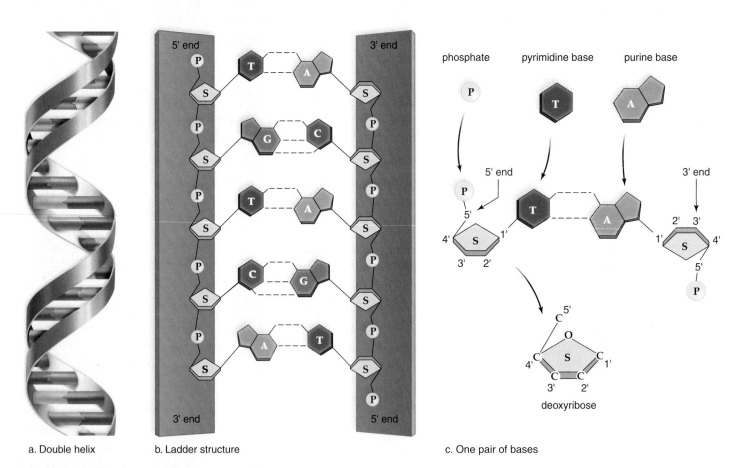

a. Double helix   b. Ladder structure   c. One pair of bases

Figure 21.1   **What is the structure of DNA?**
a. DNA double helix. **b.** When the helix is unwound, a ladder configuration shows that the supports are composed of sugar (S) and phosphate (P) molecules and the rungs are complementary bases. The bases in DNA pair in such a way that the phosphate-sugar backbones are oriented in different directions. **c.** The DNA strands are antiparallel, which is apparent by numbering the carbon atoms in deoxyribose.

Historical **Focus**

Overlooked Genius: Rosalind Franklin

A brilliant scientist and researcher, Dr. Rosalind Franklin made one of the most significant contributions to understanding the structure of DNA. Her X-ray "Photograph 51" was crucial in revealing DNA's makeup. But when the Nobel Prize for the discovery of the double helix was awarded in 1962, the honor went to James Watson, Francis Crick, and Maurice Wilkins. Rosalind Franklin was not listed as one of the recipients.

Rosalind Franklin was born in 1920 in London, England. At the age of 15 she decided to become a scientist, against her father's wishes. She went on to earn a doctorate in physical chemistry from Cambridge University by the age of 26. Her first job was in X-ray diffraction in a cutting-edge laboratory in Paris. In this process, a substance is first formed into crystals. X-rays shone through the crystals provide information about the molecule's structure. In 1951, Rosalind joined a laboratory at King's College, London. There, she began working on the X-ray diffraction of DNA. She refined an X-ray machine and perfected her photographs of crystallized DNA, some of which required 100-hour exposures to complete. By studying these photographs, she discovered that the linked sugar-phosphate strands in DNA were located on the outside of the molecule. She also discovered that the DNA helix was composed of two separate strands.

Yet Rosalind faced intense gender discrimination in her work at King's College. Women were not allowed to eat in the male-only dining rooms at the university. Her colleague, Maurice Wilkins, treated her as a research assistant, not as a peer. There was friction between the two. The atmosphere was so uncomfortable that Franklin left King's College to accept a position at Brikbeck College in London.

At that time, there was intense competition worldwide to be the first to determine the structure of DNA. In America, Linus Pauling had discovered the alpha helix structure of

Figure 21A Rosalind Franklin (1920–1958).

proteins and was using these data to investigate DNA. Francis Crick and James Watson at Cambridge University were also studying DNA. Wilkins showed Watson one of Franklin's X-ray photographs, without her permission. With the information from the photograph, Watson and Crick were finally able to determine the structure of DNA. They published their findings in the journal *Nature* in April, 1953. Although Franklin's work was published in the same issue, she was not given credit on the publication describing the arrangement of DNA.

Would Rosalind Franklin's work ever have received the acclaim she earned? Might she have been included in the 1962 Nobel Prize? The answer will never be known. Tragically, Franklin developed ovarian cancer in 1956, and she died in 1958 at the age of 37. The Nobel Prize is not awarded posthumously, and so Franklin was ineligible.

The two strands of DNA are antiparallel—they run in opposite directions. In one strand, the sugar molecules appear right-side up, and in the other, they appear upside down. While important, this need not concern us here.

Replication of DNA

When cells divide, each new cell gets an exact copy of DNA. The process of copying a DNA helix is called **DNA**

replication. During replication, the double-stranded structure of DNA allows each original strand to serve as a **template** (mold) for the formation of a complementary new strand. DNA replication is termed *semiconservative* because each new double helix has one original strand and one new strand. In other words, one of the original strands is conserved, or present, in each new double helix. Each original strand has produced a new strand through complementary base pairing, so there are now

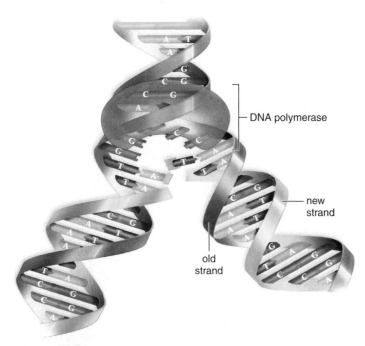

Figure 21.2   Why is DNA replication "semiconservative"?
Replication is called semiconservative because each new double helix
is composed of an original strand and a new strand.

Parental DNA molecule contains
so-called old strands hydrogen-bonded
by complementary base pairing.

Region of replication. Parental DNA
is unwound and unzipped. New
nucleotides are pairing with those in
old strands.

Replication is complete. Each double
helix is composed of an old (parental)
strand and a new (daughter) strand.

Figure 21.3   How is DNA replicated?
Use of the ladder configuration better illustrates how complementary
nucleotides, available in the cell, pair with those of each old strand before
they are joined together to form a daughter strand.

two DNA helices identical to each other and to the original molecule (Fig. 21.2).

Figure 21.3 shows how complementary nucleotides pair
in the daughter strand.

1. Before replication begins, the two strands that make up
   parental DNA are hydrogen-bonded to each other.
2. An enzyme unwinds and "unzips" double-stranded
   DNA (i.e., the weak hydrogen bonds between the paired
   bases break).
3. New complementary DNA nucleotides, always present in
   the nucleus, fit into place by the process of complementary
   base pairing. These are positioned and joined by the
   enzyme *DNA polymerase.*
4. To complete replication, an enzyme seals any breaks in
   the sugar-phosphate backbone. The DNA returns to its
   coiled structure once again.

Have You Ever Wondered...

How long does it take to copy the DNA in one human cell?

The enzyme DNA polymerase in humans can copy
approximately 50 bases per second. If only one DNA polymerase
were used to copy human DNA, it would take almost three
weeks! However, multiple DNA polymerases copy the human
genome by starting at many different places. The entire 3 billion
base pairs can be copied in 8 hours in a rapidly dividing cell.

5. The two double-helix molecules are identical to each
   other and to the original DNA molecule.

Rarely, a replication error occurs making the sequence of
the bases in the new strand different from the parental strand.
But, if an error does occur, the cell has repair enzymes that
usually fix it. A replication error that persists is a **mutation,** a
permanent change in the sequence of bases. A mutation can
possibly cause a change in the phenotype and introduce
variability. Such variabilities make you different from your
neighbor and humans different from other animals.

The Structure and Function of RNA

RNA (ribonucleic acid) is made up of nucleotides containing the sugar ribose. This sugar accounts for the scientific
name of this polynucleotide. The four nucleotides that make
up the RNA molecule have the following bases: adenine (A),
uracil (U), cytosine (C), and guanine (G) (Fig. 21.4). In RNA,
the base uracil replaces the base thymine.

RNA, unlike DNA, is single-stranded (Fig. 21.4). The
single RNA strand sometimes doubles back on itself, and
complementary base pairing still occurs. Similarities and
differences between these two nucleic acid molecules are
listed in Table 21.1.

In general, RNA is divided into coding and noncoding
RNAs. The coding RNA is messenger RNA (mRNA), translated into protein. Noncoding RNAs are divided into ribosomal RNA (rRNA), transfer RNA (tRNA), and the small
RNAs. The small RNAs are involved in the expression of
mRNA and rRNA.

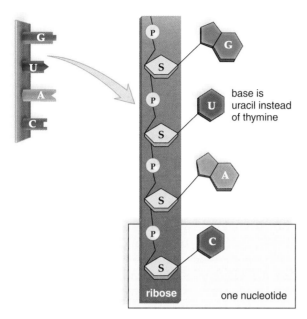

Figure 21.4 **What is the structure of RNA?**
Like DNA, RNA is a polymer of nucleotides. In an RNA nucleotide, the sugar ribose is attached to a phosphate molecule and to a base: G, U, A, or C. In RNA, the base uracil replaces thymine as one of the pyrimidine bases. RNA is single-stranded, whereas DNA is double-stranded.

Messenger RNA

Messenger RNA (mRNA) is produced in the nucleus where DNA serves as a template for its formation. This type of RNA carries genetic information from DNA to the ribosomes in the cytoplasm, where protein synthesis occurs. Messenger RNA is a linear molecule as shown in Figure 21.4.

Table 21.1	DNA-RNA Similarities and Differences

DNA-RNA SIMILARITIES

Nucleic acids
Composed of nucleotides
Sugar-phosphate backbone
Four different types of bases

DNA-RNA DIFFERENCES

DNA	RNA
Found in nucleus and mitochondria	Found in nucleus and cytoplasm
The genetic material	Helper to DNA
Sugar is deoxyribose	Sugar is ribose
Bases are A, T, C, G	Bases are A, U, C, G
Double-stranded	Single-stranded
Is transcribed (to give RNA)	Can be translated (to give proteins)

Ribosomal RNA

Ribosomal RNA (rRNA) is produced in the nucleolus of a nucleus where a portion of DNA serves as a template for its formation. Ribosomal RNA joins with proteins made in the cytoplasm to form the subunits of ribosomes. The subunits then leave the nucleus. Together, they assemble in the cytoplasm and form a ribosome when protein synthesis is about to begin. Proteins are synthesized at the ribosomes. In low-power electron micrographs ribosomes look like granules arranged along the endoplasmic reticulum. The endoplasmic reticulum is a system of tubules and saccules within the cytoplasm. Some ribosomes appear free in the cytoplasm or in clusters called polyribosomes.

Transfer RNA

Transfer RNA (tRNA) is produced in the nucleus, and a portion of DNA also serves as a template for its production. Appropriate to its name, tRNA transfers amino acids to the ribosomes. At the ribosomes, the amino acids are joined, forming a protein. There are 20 different types of amino acids in proteins. Therefore, at least 20 tRNAs must be functioning in the cell. Each type of tRNA carries only one type of amino acid.

Small RNAs

Small RNAs are divided into several classes. Small nuclear RNAS (snRNAs) are involved in splicing the mRNA (see page 496) before it is exported from the nucleus to the cytoplasm for translation. Small nucleolar RNAs (snoRNAs) modify the ribosomal RNAs within the nucleolus of the cell. MicroRNAs (miRNAs) attach to mRNAs in the cytoplasm. mRNAs are thus prevented from being translated unnecessarily. Small interfering RNAs (siRNAs) also bind to mRNAs. Attachment of an siRNA prepares the mRNA for degradation.

> **Check Your Progress 21.1**
> 1. How does the structure of DNA allow it to be replicated?
> 2. How is RNA structure similar to, but also different from, that of DNA?
> 3. What are the different types of RNA?

21.2 Gene Expression

As we shall see, DNA provides the cell with a blueprint for synthesizing proteins. DNA resides in the nucleus, and protein synthesis occurs in the cytoplasm. First, mRNA carries a copy of DNA's blueprint into the cytoplasm. Second, the other RNA molecules we just discussed are involved in bringing about protein synthesis.

Before discussing the mechanics of gene expression, let's review the structure of proteins.

Structure and Function of Proteins

Proteins are composed of subunits called amino acids (Table 21.2). Twenty different amino acids are commonly found in proteins, synthesized at the ribosomes in the cytoplasm of cells. Proteins differ because the number and order of their amino acids differ. Figure 21.5 shows that the sequence of amino acids in a protein leads to its particular shape. Proteins are found in all parts of the body. Some are structural proteins, and some are enzymes. The protein hemoglobin is responsible for the red color of red blood cells. Albumins and globulins (antibodies) are well-known plasma proteins. Muscle cells contain the proteins actin and myosin, which give muscles substance and the ability to contract.

Enzymes are organic catalysts that speed reactions in cells. The reactions in cells form metabolic or chemical pathways. A pathway can be represented as follows:

$$\begin{array}{ccccc} & E_A & E_B & E_C & E_D \\ A & \rightarrow & B \rightarrow & C \rightarrow & D \rightarrow E \end{array}$$

In this pathway, the letters are molecules, and the notations over the arrows are enzymes. Molecule A becomes molecule B, and enzyme E_A speeds the reaction. Molecule B becomes molecule C, and enzyme E_B speeds the reaction; and so forth. Enzymes are *specific*. Enzyme E_A can only convert A to B, enzyme E_B can only convert B to C, and so forth.

Proteins determine the structure and function of the various cells in the body.

Gene Expression: An Overview

The first step in gene expression is called transcription. The second step is called translation (Fig. 21.6). During **transcription,** a strand of mRNA forms that is complementary to a portion of DNA. The mRNA molecule that forms is a *transcript* of a gene. Transcription means to make a faithful copy. In this case, a sequence of nucleotides in DNA is copied to a sequence of nucleotides in mRNA.

We can liken the DNA in the nucleus to a cookbook that contains recipes for making proteins. The original recipe is valuable and must be preserved. A copy (mRNA) is made, and the copy goes to the production center (ribosomes) to prepare the food (protein).

Table 21.2	Amino Acids
Amino Acid	**Abbreviation**
alanine	ala
arginine	arg
asparagine	asn
aspartic acid	asp
cysteine	cys
glutamine	gln
glutamic acid	glu
glycine	gly
histidine	his
isoleucine	ile
leucine	leu
lysine	lys
methionine	met
phenylalanine	phe
proline	pro
serine	ser
threonine	thr
tryptophan	trp
tyrosine	tyr
valine	val

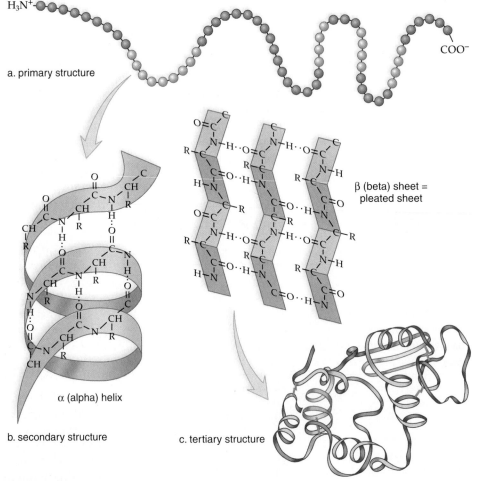

a. primary structure

α (alpha) helix

b. secondary structure

β (beta) sheet = pleated sheet

c. tertiary structure

Figure 21.5 **What are the three levels of protein structure?**
a. The primary structure of a protein is the sequence of its amino acids. **b.** The secondary structure can be either a helix or pleated sheet. **c.** The tertiary structure is the final three-dimensional shape.

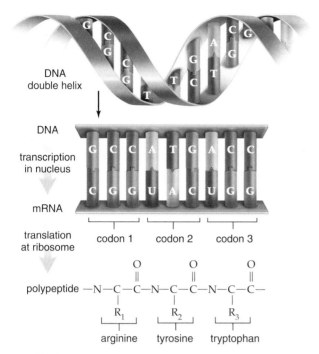

Figure 21.6 **How is a gene (DNA) expressed as a protein?** Transcription occurs when DNA acts as a template for RNA (e.g., mRNA) synthesis. (Uracil [U], in RNA, takes the place of thymine [T], in DNA.) Translation occurs when the sequence of codons of mRNA specify the sequence of amino acids in a polypeptide.

First Base	Second Base				Third Base
	U	**C**	**A**	**G**	
U	UUU phenylalanine	UCU serine	UAU tyrosine	UGU cysteine	**U**
	UUC phenylalanine	UCC serine	UAC tyrosine	UGC cysteine	**C**
	UUA leucine	UCA serine	UAA *stop*	UGA *stop*	**A**
	UUG leucine	UCG serine	UAG *stop*	UGG tryptophan	**G**
C	CUU leucine	CCU proline	CAU histidine	CGU arginine	**U**
	CUC leucine	CCC proline	CAC histidine	CGC arginine	**C**
	CUA leucine	CCA proline	CAA glutamine	CGA arginine	**A**
	CUG leucine	CCG proline	CAG glutamine	CGG arginine	**G**
A	AUU isoleucine	ACU threonine	AAU asparagine	AGU serine	**U**
	AUC isoleucine	ACC threonine	AAC asparagine	AGC serine	**C**
	AUA isoleucine	ACA threonine	AAA lysine	AGA arginine	**A**
	AUG (start) methionine	ACG threonine	AAG lysine	AGG arginine	**G**
G	GUU valine	GCU alanine	GAU aspartate	GGU glycine	**U**
	GUC valine	GCC alanine	GAC aspartate	GGC glycine	**C**
	GUA valine	GCA alanine	GAA glutamate	GGA glycine	**A**
	GUG valine	GCG alanine	GAG glutamate	GGG glycine	**G**

Figure 21.7 **How are the messenger RNA codons decoded?** In this chart, each of the codons (white rectangles) is composed of three letters representing the first base, second base, and third base. For example, find the rectangle where C for the first base and A for the second base intersect. You will see that U, C, A, or G can be the third base. CAU and CAC are codons for histidine; CAA and CAG are codons for glutamine.

Protein synthesis requires the process of **translation.** Translation means to put information into a different language. In this case, a sequence of *nucleotides* is translated into the sequence of *amino acids.* This is possible only if the bases in DNA and mRNA code for amino acids. This code is called the genetic code.

The Genetic Code

Recognizing that there must be a genetic code, investigators wanted to know how four bases (A, C, G, U) could provide enough combinations to code for 20 amino acids. If the code were a singlet code (only one base stands for an amino acid), only four amino acids could be encoded. If the code were a doublet (any two bases stand for one amino acid), it would still not be possible to code for 20 amino acids. But if the code were a triplet, then the four bases could supply 64 different triplets. This is far more than needed to code for 20 different amino acids. It should come as no surprise, then, to learn that the code is a **triplet code.**

Each three-letter (base) unit of an mRNA molecule is called a **codon.** The translation of all 64 mRNA codons has been determined (Fig. 21.7). Sixty-one triplets correspond to a particular amino acid. The remaining three are stop codons, which signal polypeptide termination. The one codon that stands for the amino acid methionine is also a start codon signaling polypeptide initiation. Most amino acids have more than one codon. Leucine, serine, and arginine have six different codons, for example. This offers some protection against possibly harmful mutations that change the sequence of the bases.

To crack the code, a cell-free experiment was done. Artificial RNA was added to a medium containing bacterial ribosomes and a mixture of amino acids. Comparison of the bases in the RNA with the resulting polypeptide, allowed investigators to decipher the code. For example, an mRNA with a sequence of repeating guanines (GGG'GGG'...) would encode a string of glycine amino acids.

The genetic code is just about universal in living things. This suggests that the code dates back to the very first organisms on Earth and that all living things are related.

Transcription

During transcription, a segment of the DNA serves as a template for the production of an RNA molecule. Although all classes of RNA are formed by transcription, we will focus on transcription to form mRNA.

Forming mRNA

Transcription begins when the enzyme **RNA polymerase** opens up the DNA helix just in front of it so that complementary base pairing can occur. Then, RNA polymerase joins the RNA nucleotides, and an mRNA molecule results. When mRNA forms, it has a sequence of bases complementary to DNA. Wherever A, T, G, or C is present in the DNA template, U, A, C, or G is incorporated into the mRNA molecule (Fig. 21.8). Now, mRNA is a faithful copy of the sequence of bases in DNA with U in place of T.

Processing mRNA

After the mRNA is transcribed in human cells, it must be *processed* before entering the cytoplasm.

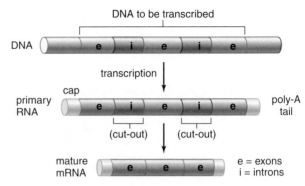

Figure 21.9 **What happens to an mRNA during posttranscriptional processing?**
During processing, a cap and tail are added to mRNA, and the introns are removed so that only exons remain.

The newly synthesized *primary mRNA* molecule becomes a *mature mRNA* molecule after processing (Fig. 21.9). Most genes in humans are interrupted by segments of DNA that are not part of the gene. These portions are called *introns* because they are intragene segments. The other portions of the gene are called *exons* because they are ultimately expressed. Only exons result in a protein product.

Primary mRNA contains bases complementary to both exons and introns, but during processing, (1) one end of the mRNA is capped, by the addition of an altered guanine nucleotide. The other end is given a tail, by the addition of adenosine nucleotides. (2) The introns are removed, and the exons are joined to form a mature mRNA molecule consisting of continuous exons. This *splicing* of mRNA is done by a complex composed of both RNA and protein. Surprisingly, RNA, not the protein, is the enzyme, and so it is called a *ribozyme*. One of the small RNAs is involved in this process.

Ordinarily, processing brings together all the exons of a gene. In some instances, cells use only certain exons rather than all of them to form a mature RNA transcript. Alternate mRNA splicing is believed to account for the ability of a single gene to result in different proteins in a cell. Increasingly, small RNA molecules have been found that regulate not only mRNA processing but also transcription and translation. DNA codes for proteins but RNA orchestrates the outcome.

Translation

During translation, transfer RNA (tRNA) molecules bring amino acids to the ribosomes (Fig. 21.10), where polypeptide synthesis occurs. A ribosome consists of a large and a

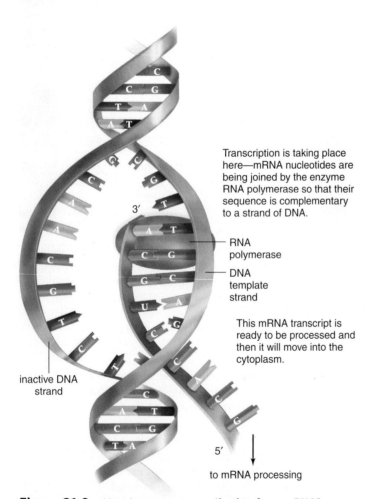

Transcription is taking place here—mRNA nucleotides are being joined by the enzyme RNA polymerase so that their sequence is complementary to a strand of DNA.

RNA polymerase

DNA template strand

This mRNA transcript is ready to be processed and then it will move into the cytoplasm.

inactive DNA strand

to mRNA processing

Figure 21.8 **How is a gene transcribed to form mRNA?**
During transcription, complementary RNA is made from a DNA template. A portion of DNA unwinds and unzips at the point of attachment of RNA polymerase. A strand of mRNA is produced when complementary bases join in the order dictated by the sequence of bases in template DNA.

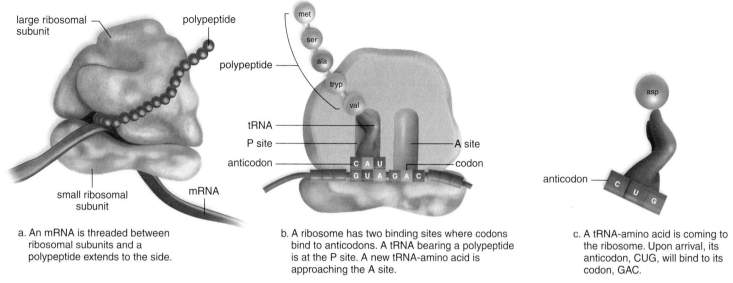

a. An mRNA is threaded between ribosomal subunits and a polypeptide extends to the side.

b. A ribosome has two binding sites where codons bind to anticodons. A tRNA bearing a polypeptide is at the P site. A new tRNA-amino acid is approaching the A site.

c. A tRNA-amino acid is coming to the ribosome. Upon arrival, its anticodon, CUG, will bind to its codon, GAC.

Figure 21.10   How does a ribosome function in translation?
Protein synthesis occurs at a ribosome. **a.** Side view of a ribosome showing mRNA and a growing polypeptide. **b.** The large ribosomal subunit contains two binding sites for tRNAs. **c.** tRNA structure and function.

small subunit that join together in the cytoplasm just as protein synthesis begins. Each ribosome has a binding site for mRNA, as well as binding sites for two tRNA molecules at a time. Usually, there are many tRNA molecules for each of the 20 amino acids found in proteins. The amino acid binds to one end of the tRNA molecule. Therefore, the entire complex is designated as tRNA–amino acid. At the other end of each tRNA is a specific **anticodon,** a group of three bases complementary to an mRNA codon. The tRNA molecules attach to the ribosome at the A (for amino acid) site where each anticodon pairs by hydrogen bonding with a codon.

The order in which tRNA–amino acids link to the ribosome is directed by the sequence of the mRNA codons. In this way, the order of codons in mRNA brings about a particular order of amino acids in a protein. The tRNA attached to the growing polypeptide moves to the P site of a ribosome, as shown in Figure 21.10.

If the codon sequence in a portion of the mRNA is ACC, GUA, and AAA, what will be the sequence of amino acids in a portion of the polypeptide? Inspection of Figure 21.7 (see page 495) allows us to determine this:

Codon	Anticodon	Amino Acid
ACC	UGG	Threonine
GUA	CAU	Valine
AAA	UUU	Lysine

Polypeptide synthesis requires three steps: initiation, elongation, and termination (Fig. 21.11).

1. During *initiation,* mRNA binds to the smaller of the two ribosomal subunits. Then the larger subunit associates with the smaller one.
2. During *elongation,* the polypeptide lengthens, one amino acid at a time. An incoming tRNA–amino acid complex arrives at the A site and then receives the peptide from the outgoing tRNA. The ribosome moves laterally so that again the P site is filled by a tRNA-peptide complex. The A site is now available to receive another incoming tRNA–amino acid complex. In this manner, the peptide grows, and the linear structure of a polypeptide comes about. (The particular shape of a polypeptide is formed later.)
3. Then *termination* of synthesis occurs at a codon that means stop and does not code for an amino acid. The ribosome dissociates into its two subunits and falls off the mRNA molecule.

During elongation, about five amino acids are added to a polypeptide every second. However, many ribosomes are at work forming the same polypeptide. As soon as the initial portion of mRNA has been translated by one ribosome and the ribosome has begun to move down the mRNA, another ribosome attaches to the mRNA. Therefore, several ribosomes, collectively called a **polyribosome,** can move along one mRNA at a time. Several

Figure 21.11   How are polypeptides assembled?

Polypeptide synthesis takes place at a ribosome and has three steps:
(1) initiation,
(2) elongation, and
(3) termination.

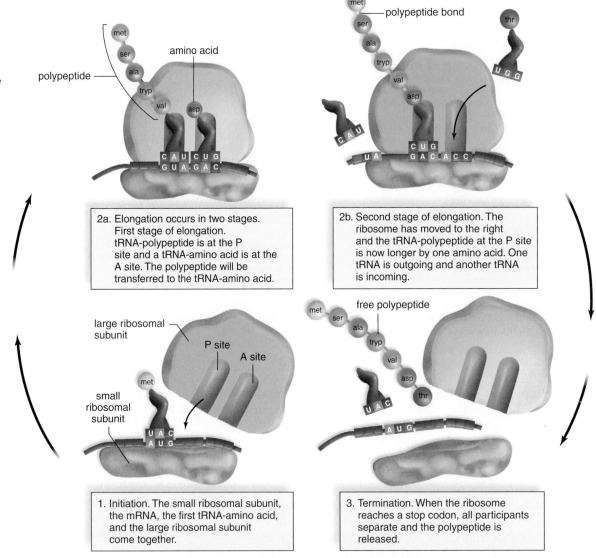

2a. Elongation occurs in two stages. First stage of elongation. tRNA-polypeptide is at the P site and a tRNA-amino acid is at the A site. The polypeptide will be transferred to the tRNA-amino acid.

2b. Second stage of elongation. The ribosome has moved to the right and the tRNA-polypeptide at the P site is now longer by one amino acid. One tRNA is outgoing and another tRNA is incoming.

1. Initiation. The small ribosomal subunit, the mRNA, the first tRNA-amino acid, and the large ribosomal subunit come together.

3. Termination. When the ribosome reaches a stop codon, all participants separate and the polypeptide is released.

Figure 21.12   What is the structure of a polyribosome? How does a polyribosome function?

Several ribosomes, collectively called a polyribosome, move along an mRNA molecule at one time. They function independently of one another; therefore, several polypeptides can be made simultaneously.

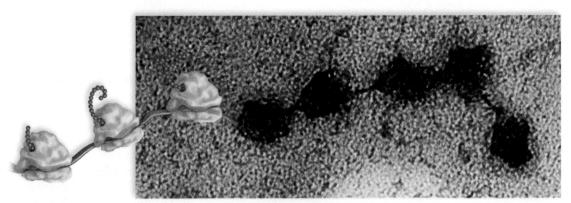

polypeptides of the same type can be synthesized using one mRNA molecule (Fig. 21.12).

Review of Gene Expression

DNA in the nucleus contains genes that are transcribed into RNAs. Some of these RNAs are mRNAs that will then be translated into proteins. During transcription, a segment of a DNA strand serves as a template for the formation of RNA. The bases in RNA are complementary to those in DNA. In mRNA, every three bases is a *codon* for a certain amino acid (Fig. 21.13 and Table 21.3). Messenger RNA is processed before it leaves the nucleus. During processing the introns are removed and the ends are modified. Messenger RNA carries

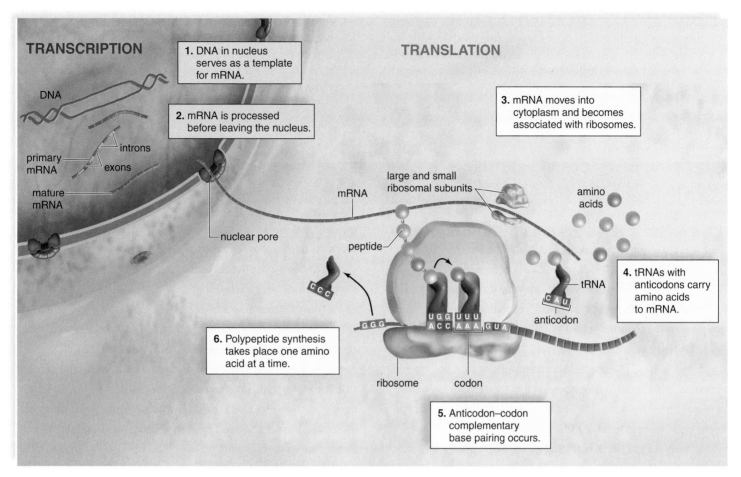

Figure 21.13 **How do all the steps in gene expression function together?**
Gene expression leads to the formation of a product, most often a protein. The two steps required for gene expression are transcription, which occurs in the nucleus, and translation, which occurs in the cytoplasm at the ribosomes.

a sequence of codons to the *ribosomes,* composed of rRNA and proteins. A tRNA bonded to a particular amino acid has an *anticodon* that pairs with a codon in mRNA. During translation, tRNAs and their attached amino acids arrive at the ribosomes. The linear sequence of codons of mRNA determines the order in which amino acids become incorporated into a protein.

The Regulation of Gene Expression

All cells receive a copy of all genes. However, cells differ as to which genes are actively expressed. Muscle cells, for example, have a different set of genes that are turned on in the nucleus and different proteins that are active in the cytoplasm than do nerve cells. A variety of mechanisms regulate gene expression, from transcription to protein activity in our cells. These mechanisms can be grouped under five primary levels of control—three that pertain to the nucleus and two that pertain to the cytoplasm:

1. *Pretranscriptional control* In the nucleus, the DNA must be available to the enzymes necessary for transcription. The chromosome in the region must decondense, or uncoil. Proteins and chemical modifications that protect the DNA must be removed before transcription can begin.

Table 21.3	Participants in Gene Expression	
Name of Molecule	**Special Significance**	**Definition**
DNA	Genetic information	Sequence of DNA bases
mRNA	Codons	Sequence of three RNA bases complementary to DNA
tRNA	Anticodon	Sequence of three RNA bases complementary to codon
rRNA	Ribosome	Site of protein synthesis
Amino acid	Building block for protein	Transported to ribosome by tRNA
Protein	Enzyme, structural protein, or secretory product	Amino acids joined in a predetermined order

2. *Transcriptional control* In the nucleus, a number of mechanisms regulate which genes are transcribed and/or the rate at which transcription of genes occurs. These include the use of transcription factors that initiate transcription, the first step in gene expression.

3. *Posttranscriptional control* Posttranscriptional control occurs in the nucleus after DNA is transcribed and mRNA is formed. How mRNA is processed before it leaves the nucleus and also how fast mature mRNA leaves the nucleus can affect the amount of gene expression.

CASE STUDY IN THE DINING HALL

Bianca joined Cameron at the table in the dining hall. He was already finished eating and was waiting for her. "Hi Cameron! How's it going?" she greeted him. "I really appreciate your doing this for me."

Cameron smiled at her. "No problem. What do you need to know?"

"Well, I need to know a little history of your diabetes, what drugs you take, and how the drugs help you," Bianca replied.

"That's not too bad," Cameron said. "When I was little I was diagnosed with type 1 diabetes." Bianca looked confused, so Cameron explained. "That's the kind of diabetes where I don't make any insulin at all, so I have to take it. Insulin lets the blood sugar into my cells. You know, blood glucose to make ATP energy and all that."

"Yeah, Bio 127 lecture stuff," Bianca nodded. "I pretty much remember that."

"If I have too much sugar in my blood, that's bad," Cameron continued. "If I have too little sugar in my blood, that's bad too. So about 6 or 8 times a day, I measure my blood sugar with my sugar or glucose meter. The blood glucose level determines how much insulin I should take."

Bianca asked, "How do you use the glucose meter?"

"I stick my finger with a sterile lancet and get a drop of blood on a test strip," Cameron replied. "The strip goes into the meter, and I get a number for the amount of glucose in my blood."

"You stick yourself 6 or 8 times a day?" Bianca exclaimed with a shudder.

Cameron laughed. "It's not that bad. You get used to it."

"So what kind of insulin do you take?" Bianca asked.

Cameron explained, "I take Humalog® because it works really fast."

"So, you just take a pill before you eat, right?" Bianca interjected.

"I wish," Cameron replied ruefully. "You don't swallow insulin, you inject it. If you swallowed it, your stomach would just digest it and it wouldn't work."

Bianca looked sick. "Inject?"

Cameron grinned. "Yes, with the aid of my mighty cell phone." Bianca looked puzzled. He explained, "Just kidding. See this thing on my waist that looks like a cell phone? It's an insulin pump. It really keeps my blood sugar controlled. Besides, I don't have to give myself shots four times a day like I used to. This thing holds the insulin and feeds it into my body all day long."

"How?" Bianca asked.

"You put this tiny little tube right into the fat under your skin." He showed her a neat white circular patch anchoring the tube just above his waist. Bianca winced and he repeated, smiling, "You get used to it. The pump feeds small amounts of the fast-acting insulin into my body all day long. It also lets me adjust a dosage and give myself extra if I need to, like right before a meal. Its got computers and alarms on it, so the insulin won't run out."

"What do you do if you want to go in the water or something?" Bianca wondered. "You're on the swim team, right?"

"Yup," Cameron nodded. "I went scuba diving last month, too. I was in the water all day long. So I take off the pump and inject insulin instead. The other kind of insulin I use is called Humalin® N. It acts slower and keeps my blood sugar levels stable when I'm not using the pump."

"It seems like so much trouble," Bianca said.

"Yeah, well, if I don't do it, I'm gonna be really messed up," Cameron asserted. "You get all this training when you're first diagnosed. That's great, but it's pretty scary, too. If you don't control your diabetes, you go blind, lose your fingers or toes or maybe your whole leg. You can have a heart attack or a stroke, or even die. So I'm really careful to take good care of myself. I test my blood all day long, write down what I eat, stuff like that," Cameron concluded.

"Wow. I never knew it was that complicated," Bianca replied. "Do you know anything about how the insulin is made?" Bianca asked.

Cameron laughed. "Sorry, Bianca. You're going to have to look that one up. I have no idea. I can give you the drug inserts from the packages, but I can't tell you anything else."

"Oh well. Thanks so much, Cameron. You've been a big help!"

4. *Translational control* Translational control occurs in the cytoplasm after mRNA leaves the nucleus and before there is a protein product. The life expectancy of mRNA molecules (how long they exist in the cytoplasm) can vary, as can their ability to bind ribosomes. It is also possible that some mRNAs may need additional changes before they are translated. Two of the small RNA classes, microRNAs and small interfering RNAs, are involved at this level of gene expression. MicroRNAs bind to mRNAs and block the target mRNA from being translated. Small interfering RNAs bind to mRNAs, marking the mRNA for destruction by nucleases.

5. *Posttranslational control* Posttranslational control, which also occurs in the cytoplasm, occurs after protein synthesis. The polypeptide product may have to undergo additional changes before it is biologically functional. Also, a functional enzyme is subject to feedback control—the binding of an enzyme's product can change its shape so that it is no longer able to carry out its reaction.

Activated Chromatin

For a gene to be transcribed in human cells, the chromosome in that region must first decondense. The chromosomes within the developing egg cells of many vertebrates are called lampbrush chromosomes because they have many loops that appear to be bristles (Fig. 21.14). Here, mRNA is

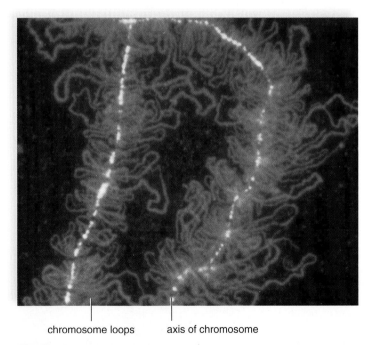

chromosome loops   axis of chromosome

Figure 21.14   What is a lampbrush chromosome?
These chromosomes, seen in amphibian egg cells, give evidence that when mRNA is being synthesized, chromosomes decondense. Each chromosome has many loops extending from its axis (white on the micrograph). Many mRNA transcripts are being made off these DNA loops.

being synthesized in great quantity so that adequate protein synthesis can be carried out.

Transcription Factors

In human cells, **transcription factors** are DNA-binding proteins. Every cell contains many different types of transcription factors. A specific combination of transcription factors and modifiers is believed to regulate the activity of any particular gene. After the right combination binds to DNA, an RNA polymerase attaches to the DNA and begins the process of transcription.

As cells mature, they differentiate and become specialized. Specialization is determined by which genes are active and, therefore, perhaps, by which transcription factors are active in that cell. Signals received from inside and outside the cell could turn on or off genes that code for certain transcription factors. For example, in an embryo, the hand is flattened and paddle-shaped. The five fingers are joined by webs of skin and connective tissue. The gene for apoptosis (programmed cell death) is turned on in the embryo's hand. Separated fingers are formed when the webs of tissue die off.

> **Check Your Progress 21.2**
>
> **1.** **a.** What are the levels of structure of a protein, and **(b)** what function do proteins perform in a metabolic pathway?
>
> **2.** What are the two steps of gene expression, and what has been formed at the end of each step?
>
> **3.** **a.** What are the various levels of genetic control in human cells? **b.** How is transcriptional control achieved?

21.3 Genomics

The Human Genome Has Been Sequenced

Genetics in the twenty-first century concerns genomics, the study of genomes—our genes and the genes of other organisms. Due to the Human Genome Project (HGP), a 13-year effort that involved both university and private laboratories around the globe, we now know the order of the 3 billion A, T, C, and G bases in our genome.

How was this feat accomplished? First, investigators developed a laboratory procedure that would allow them to decipher a short sequence of base pairs. Then instruments became available that would carry out this procedure automatically. Sperm DNA was the material of choice for study, because sperm DNA has a much higher ratio of DNA to protein than other type of cells. White blood cells from the blood of female donors were also used to include female-originated samples. The male and female donors were of European, African, American (both North and South), and Asian ancestry.

Have You Ever Wondered...

How big is a human gene?

The average gene in the human genome consists of 3,000 bases. Sizes can vary greatly, however. The dystrophin gene, coding for a muscle protein, consists of 2.4 million bases. Recognized genes make up less than 5% of the approximately 3 billion bases in the human genome. The rest of the genome consists of nonprotein coding sequences.

As you can imagine, many small regions of DNA that vary among individuals—polymorphisms—were identified during the HGP. Most of these are single nucleotide polymorphisms (SNPs). Many SNPs have no physiological effect. Others may contribute to the diversity of human beings or possibly increase an individual's susceptibility to disease and response to medical treatments.

A surprising finding has been that genome size is not proportionate to the number of genes and does not correlate to complexity of the organism. When scientists first completed the DNA sequence of the human genome, they made an estimate of the number of genes based on sequences that appeared to encode proteins. Researchers were surprised to find less than 25,000 genes in the human genome. Based on comparison with the genomes of other organisms, that number has now been reduced to 20,500 genes. The plant *Arabidopsis thaliana*, the mouse-ear cress, with a much smaller number of bases, has approximately the same number of genes as a human being! A gene in this context is a sequence of DNA bases transcribed into any type of RNA molecule. It is not yet known what each of the 20,500 human genes do specifically.

Today, laboratory instruments are commercially available that can automatically analyze up to 2 million base pairs of DNA in a 24-hour period. Improvements in DNA sequencers are still being designed, however. Researchers are now working on systems that can read as many as 1,700 bases a second! This means that a DNA sequencer might be available by 2014 that could read a genome for only $1,000. The Personal Genome Project is underway. This would not only produce faster sequences, but also determine the possible benefits and drawbacks of every person having their genome sequenced.

Functional and Comparative Genomics

Goals of continuing research in the Human Genome Project are to determine how our 20,500 genes function, and how together they form a human being. It is hoped that this type of study, called **functional genomics,** will also discover the role of noncoding regions. Genes make up less than 5% of the human genome, and scientists in the field of genomics are working hard to find the function of the many intergenic DNA sequences.

Comparing genomes is one way to determine how species have evolved and how genes and noncoding regions of the genome function. In one study, researchers compared our genome to that of chromosome 22 in chimpanzees (Fig. 21.15). They found three types of genes of particular interest: a gene for proper speech development, several for hearing, and several for smell. Genes necessary for speech development are thought to have played an important role in human evolution. You can suppose that changes in hearing may also have facilitated using language for communication between people. To explain differences in smell genes, investigators speculated that the olfaction genes may have affected dietary changes or sexual selection. Or, they may have been involved in other traits, rather than just smell. The researchers who did this study were surprised to find that many of the other genes they located and studied are known to cause human diseases. They wondered if comparing genomes would be a way of finding other genes associated with human diseases. Investigators are taking all sorts of avenues to link human base sequence differences to illnesses.

Another surprising discovery has been how very similar the genomes of all vertebrates are. Researchers weren't surprised to learn that the base sequence of chimpanzees and humans is 95–98% alike. However, they didn't expect to find

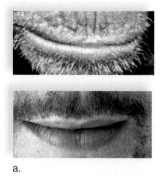
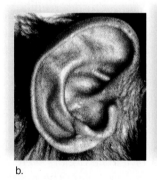
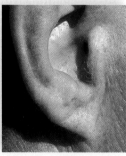
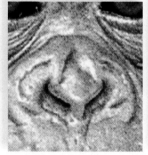

a.   b.   c.

Figure 21.15   How does the study of functional genomics help us understand human-chimpanzee differences?
Did changes in the genes for **(a)** speech, **(b)** hearing, and **(c)** smell influence the evolution of humans?

that our sequence is also 85% similar to that of a mouse, for example. So, the quest is on to discover how possibly the regulation of genes explains why we have one set of traits and mice have another set, despite the similarity of our base sequences. One possibility is alternative gene splicing. We may differ from mice by what types of proteins we manufacture and/or by when and where certain proteins are present.

Proteomics and Bioinformatics Are New Endeavors

Proteomics is the study of the structure, function, and interaction of cellular proteins. Many of our genes are translated into proteins at some time, in some of our cells. The translation of all coding genes results in a collection of proteins, called the human proteome. The analysis of proteomes is more challenging than the analysis of genomes. Protein concentrations differ widely in cells. Researchers have to be able to identify proteins, no matter whether there is one or thousands of copies of a protein in a cell. Any particular protein differs minute by minute in concentration, interactions, cellular location, and chemical modifications among other features. Yet, to understand a protein, all these features must be analyzed. Computer modeling of the three-dimensional shape of cellular proteins is an important part of proteomics. The study of cellular proteins and how they function is essential to the discovery of better drugs. Most drugs are proteins or molecules that affect the function of proteins.

Bioinformatics is the application of computer technologies to the study of the genome. Genomics and proteomics produce raw data. These fields depend on computer analysis to find significant patterns in the data. As a result of bioinformatics, scientists are hopeful of finding cause-and-effect relationships between various genetic profiles and genetic disorders caused by multifactorial genes. By correlating any sequence changes with resulting phenotypes, bioinformatics research may find that noncoding regions of the genome do have functions. New computational tools will, most likely, be needed to accomplish these goals.

A Person's Genome Can Be Modified

Gene therapy is the insertion of genetic material into human cells for the treatment of a disorder. Gene therapy has been used to cure inborn errors of metabolism and also to treat more generalized disorders such as cardiovascular disease and cancer. Most recently, a 2008 clinical trial showed that gene therapy could successfully treat a type of inherited blindness. Both ex vivo (outside the body) and in vivo (inside the body) gene therapy methods are used.

Ex Vivo Gene Therapy

Figure 21.16 describes the methodology for treating children who have severe combined immunodeficiency (SCID, page 153). These children lack the enzyme ADA (adenosine

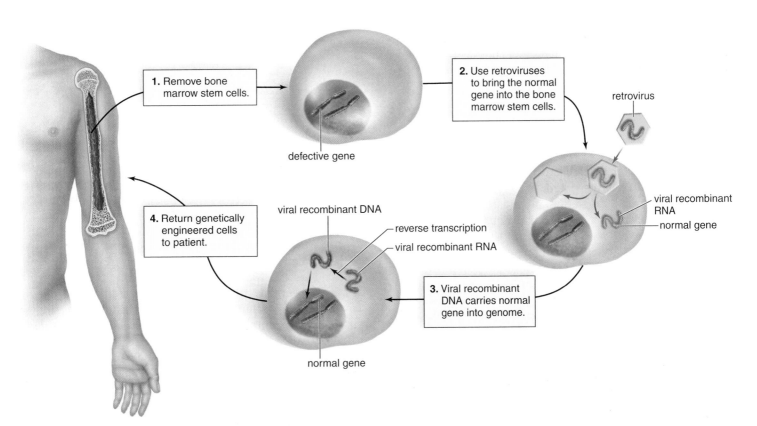

Figure 21.16 What are the steps for ex vivo gene therapy in humans?

deaminase). ADA is involved in the maturation of T and B cells. To carry out gene therapy, bone marrow stem cells are removed from bone marrow. These cells are infected with an RNA retrovirus that carries a normal gene for the enzyme. Then the cells are returned to the patient. Bone marrow stem cells are preferred for this procedure because they divide to produce more cells with the same genes. Patients who have undergone this procedure show significantly improved immune function. This improvement is associated with a sustained rise in the level of ADA enzyme activity in the blood.

In Vivo Gene Therapy

Cystic fibrosis patients lack a gene that codes for the transmembrane carrier of the chloride ion. They often suffer from numerous and potentially deadly infections of the respiratory tract. In gene therapy trials, the gene needed to cure cystic fibrosis is sprayed into nose. The corrective gene may also be the delivered to the lower respiratory tract by adenoviruses or by the use of liposomes, microscopic vesicles that spontaneously form when lipoproteins are put into a solution. Investigators are trying to improve uptake, and they are also hypothesizing that a combination of all three methods might be more successful.

Genes are being used to treat medical conditions such as poor coronary circulation. It has been known for some time that vascular endothelial growth factor (VEGF) can cause the growth of new blood vessels. The gene that codes for this growth factor can be injected alone or within a virus into the heart to stimulate branching of coronary blood vessels. Patients report that they have less chest pain and can run longer on a treadmill.

Problems

Gene therapy is still considered experimental. It has not yet been approved by the U.S. Food and Drug Administration (FDA). After tragedies occurring in clinical trials, research was severely restricted. In 1999, 18-year-old Jesse Gelsinger died, four days after starting gene therapy for a genetic disorder that prevented protein digestion. His death was attributed to a severe immune response against the virus vector used to carry the gene into his body. Likewise, gene therapy for severe combined immunodeficiency (SCID) was discontinued in 2002, even though it had been successful in producing cures. The procedure was stopped when a child developed a leukemialike condition after treatment. Following these incidents, researchers continue to modify vectors to make gene therapy safe and effective.

> **Check Your Progress 21.3**
> 1. **What are genomics and proteomics?**
> 2. **How can a person's genome be modified?**

21.4 DNA Technology

Genes Can Be Isolated and Cloned

In biology, **cloning** is the production of genetically identical copies of DNA, cells, or organisms through an asexual means. **Gene cloning** can be done to produce many identical copies of the same gene. **Recombinant DNA (rDNA),** which contains DNA from two or more different sources, allows genes to be cloned. To create recombinant DNA, a technician needs a **vector,** by which the gene of interest will be introduced into a host cell, such as a bacterium. One common vector is a plasmid. Recall that **plasmids** are small accessory rings of DNA found in bacteria. The ring is not part of the bacterial chromosome and replicates on its own.

The steps for gene cloning include (Fig. 21.17):

① A **restriction enzyme** is used to cleave human DNA and plasmid DNA. Hundreds of restriction enzymes occur naturally in bacteria, where they cut up any viral DNA that enters the cell. They are called restriction enzymes because they restrict the growth of viruses. They also act as molecular scissors to cleave any piece of DNA at a specific site. For example, the restriction enzyme called *Eco*RI always cuts double-stranded DNA at this sequence of bases and in this manner:

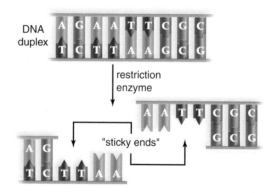

The restriction enzyme creates a gap in plasmid DNA in which foreign DNA (e.g., the insulin gene) can be placed if it ends in bases complementary to those exposed by the restriction enzyme. To ensure this, it is only necessary to use the same type of restriction enzyme to cleave both human DNA containing the gene for insulin and plasmid DNA.

② The enzyme called **DNA ligase** is used to seal foreign DNA into the opening created in the plasmid. The single-stranded, but complementary, ends of a cleaved DNA molecule are called "sticky ends." This is because they can bind a piece of DNA by complementary base pairing. Sticky ends facilitate the sealing of the plasmid DNA with human DNA for the insulin gene. Now the vector is complete, and an rDNA molecule has been prepared.

③ Some of the bacterial cells take up a recombinant plasmid, especially if the bacteria have been treated to make them more permeable.

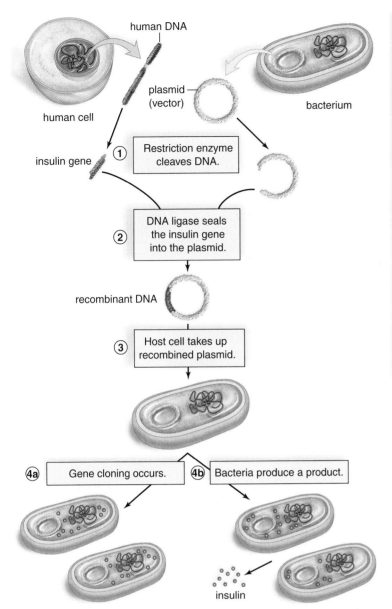

Figure 21.17 What are the steps for cloning a human gene?

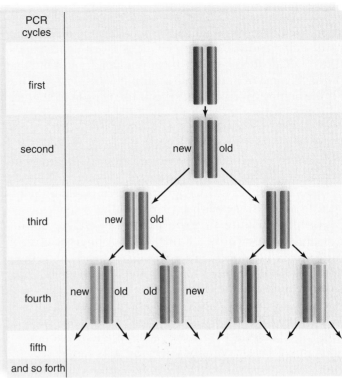

Figure 21.18 How does the polymerase chain reaction (PCR) result in multiple copies of a gene?

Copies produced per PCR cycle per one original strand.

(4a) Gene cloning occurs as the plasmid replicates on its own. Scientists clone genes for a number of reasons. They might want to determine the difference in base sequence between a normal gene and a mutated gene. Or, they might use the genes to genetically modify organisms in a beneficial way.

(4b) The bacterium can also make a product (e.g., insulin) that it could not make before. For a human gene to express itself in a bacterium, the gene has to be accompanied by regulatory regions unique to bacteria. Also, the gene should not contain introns because bacteria don't have introns. However, it is possible to make a human gene that lacks introns. The enzyme called reverse transcriptase can be used to make a DNA copy of mRNA. The DNA molecule, called **complementary DNA (cDNA),** does not contain introns.

Specific DNA Sequences Can Be Cloned

If only small pieces of identical DNA are needed, the **polymerase chain reaction (PCR),** developed by Kary Mullis in 1985, can create copies of a segment of DNA quickly in a test tube. PCR is very specific. It amplifies (makes copies of) a targeted DNA sequence. The targeted sequence can be less than one part in a million of the total DNA sample!

PCR requires the use of DNA polymerase, the enzyme that carries out DNA replication. It also needs a supply of nucleotides for the new DNA strands. PCR is a chain reaction because the targeted DNA is repeatedly replicated as long as the process continues. The colors in Figure 21.18 distinguish old DNA from new DNA. The amount of DNA doubles with each replication cycle.

DNA amplified by PCR is often analyzed for various purposes. Mitochondrial DNA base sequences in modern living populations were used to decipher the evolutionary history of human populations. Little DNA is required for PCR to be effective, so it has even been possible to sequence DNA taken from mummified human brains.

Also, following PCR, DNA can be subjected to DNA fingerprinting. Today, DNA fingerprinting is often carried out by detecting how many times a short sequence (two to five bases) is repeated. Organisms differ by how many repeats they have at particular locations. Recall that PCR amplifies only particular portions of the DNA. Therefore, the greater the number of repeats at a location, the longer the section of

DNA amplified by PCR. During a process called gel electrophoresis, these DNA fragments can be separated according to their size, and the result is a pattern of distinctive bands. If two DNA patterns match, there is a high probability that the DNA came from the same source. It is customary to test for the number of repeats at several locations to further define the source.

DNA fingerprinting has many uses. When the DNA matches that of a virus or mutated gene, it is known that a viral infection, genetic disorder, or cancer is present. Fingerprinting DNA from a single sperm is enough to identify a suspected rapist. DNA fingerprinted from blood or tissues at a crime scene has been successfully used in convicting criminals. Figure 21.19 shows how DNA fingerprinting can be used to identify a criminal. DNA fingerprinting is used to identify the remains of bodies. It was extensively used in identifying the victims of the September 11, 2001, terrorist attacks in the United States.

Applications of PCR and DNA fingerprinting are limited only by our imagination. PCR analysis has been used to identify unknown soldiers and members of the royal Russian family. Paternity suits can be settled. Environmental law enforcement authorities can identify illegally poached ivory and whale meat using these technologies. These applications have shed new light on evolutionary studies by comparing DNA from fossils with that of living organisms.

Biotechnology Products

Today, bacteria, plants, and animals are genetically engineered to produce **biotechnology products.** Organisms that have had a foreign gene inserted into them are called **transgenic organisms.**

Figure 21.20   How are transgenic organisms used?
a. Transgenic bacteria, grown industrially, are used to produce medicines. Transgenic animals and plants increase the food supply. **b.** These salmon (*right*) are much heavier and reproduce much sooner because of the growth hormone genes they received as embryos. **c.** Pests didn't consume unblemished peas because the plants that produced them received genes for pest inhibitors. **d.** Gene guns are often used to insert genes in plant embryos while they are in tissue culture. **e.** Transgenic bacteria can also be used for environmental cleanup. These bacteria are able to break down oil.

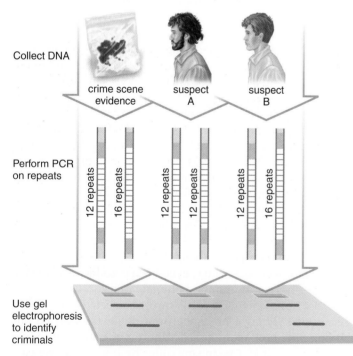

Figure 21.19   How can DNA fingerprinting be used in a criminal investigation?
DNA fingerprinting reveals that suspect A could not be the criminal.

Have You Ever Wondered...

How many convicted criminals have been released due to new DNA evidence?

The first exoneration based on DNA evidence occurred in 1989. According to the Innocence Project, a national public policy and litigation group, 218 wrongfully convicted persons had been exonerated through August 2008. There is a significant backlog of cases needing DNA analysis, but not enough criminal forensics laboratories to keep up with the demand.

From Bacteria

Recombinant DNA technology is used to produce transgenic bacteria, grown in huge vats called bioreactors (Fig. 21.20*a*). The gene product is collected from the medium. Biotechnology products on the market produced by bacteria include insulin, clotting factor VIII, human growth hormone, tissue plasminogen activator (t-PA), and hepatitis B vaccine. Transgenic bacteria have many other uses as well. Some have been produced to promote the health of plants. For example, bacteria that normally live on plants and encourage the formation of ice crystals have been changed from frost-plus to frost-minus bacteria. As a result, new crops such as frost-resistant strawberries are being developed.

Bacteria can be selected for their ability to degrade a particular substance, and this ability can then be enhanced by genetic engineering. For instance, naturally occurring bacteria that eat oil can be genetically engineered to do an even better job of cleaning up beaches after oil spills (Fig. 21.20*e*). Further, these bacteria were given "suicide" genes that caused them to self-destruct when the job had been accomplished.

From Plants

Corn, potato, soybean, and cotton plants have been engineered to be resistant to either insect predation or herbicides that are widely used (Figs. 21.20*c* and 21.21). Some corn and cotton plants have been developed that are both insect- and herbicide-resistant. In 2008, 92% of the soybeans and 80% of the corn planted in the United States had been genetically engineered. If crops are resistant to a broad-spectrum herbicide and weeds are not, then the herbicide can be used to kill the weeds. When herbicide-resistant plants were planted, weeds were easily controlled, less tillage was needed, and soil erosion was minimized.

Crops with other improved agricultural and food quality traits are desirable. A salt-tolerant tomato has already been developed (Fig. 21.22). First, scientists identified a gene coding for a protein that transports Na^+ across the vacuole membrane. Sequestering the Na^+ in a vacuole prevents it from interfering with plant metabolism. Then, the scientists cloned the gene and used it to genetically engineer plants that overproduce the channel protein. The modified plants thrived when watered with a salty solution. Today, crop production is limited by the effects of salinization on about 50% of irrigated lands. Salt-tolerant crops would increase yield on this land. Salt- and also drought- and cold-tolerant crops might help provide enough food for a world population that may nearly double by 2050.

a. Herbicide-resistant soybean plants

b. Nonresistant potato plant

c. Pest-resistant potato plant

Figure 21.21 **What characteristics are being genetically engineered in plants?**
Genetic engineering can produce plants resistant to herbicides **(a)** and pests **(b, c)**. Drought- and salt-resistant plants can be used in dry desert areas. Plants that are richer in nutrients can also be engineered.

Figure 21.22
What types of traits are desirable for genetic engineering in crop plants?

Transgenic Crops of the Future	
Improved Agricultural Traits	
Disease-protected	Wheat, corn, potatoes
Herbicide-resistant	Wheat, rice, sugar beets, canola
Salt-tolerant	Cereals, rice, sugarcane
Drought-tolerant	Cereals, rice, sugarcane
Cold-tolerant	Cereals, rice, sugarcane
Improved yield	Cereals, rice, corn, cotton
Modified wood pulp	Trees
Improved Food Quality Traits	
Fatty acid/oil content	Corn, soybeans
Protein/starch content	Cereals, potatoes, soybeans, rice, corn
Amino acid content	Corn, soybeans

a. Desirable traits

b. Salt-intolerant   Salt-tolerant

Potato blight is the most serious potato disease in the world. About 150 years ago, it was responsible for the Irish potato famine, which caused the death of millions of people. By placing a gene from a naturally blight-resistant wild potato into a farmed variety, researchers have now made potato plants that are no longer vulnerable to a range of blight strains. Some progress has also been made to increase the food quality of crops. Soybeans have been developed that mainly produce the monounsaturated fatty acid oleic acid, a change that may improve human health.

CASE STUDY AT THE LIBRARY

Bianca logged onto the Internet in the library and brought up the search engine. She entered "Humalog" into the address bar, then hit enter. *"Hmmm … 610,000 hits. Well, I'll have plenty of places to start looking,"* she thought. *"I think I'll start with the first one."*

Bianca spent several minutes wandering through the various links for Humalog® within the manufacturer's website. She couldn't find anything about how Humalog® was made using recombinant DNA. *"I think I'll send this link to Cameron, though,"* she thought. *"This is really interesting. If he's taking this drug, he probably ought to know this information."*

Bianca returned to the original search page and checked out several other sites. There was more medical information, but still no recombinant DNA information. Then she found a drug index site. "Jackpot!" Bianca typed notes:

Insulin—two chains: A chain and B chain

A chain is 21 amino acids long, B chain is 30 amino acids long

Two chains are held together by disulfide bonds

Humalog® is also called insulin lispro because its B chain has a lysine at amino acid #28 and a proline at amino acid #29 (reverse of normal)

Change in amino acids responsible for rapid action

Recombinant DNA (rDNA) origin

Made in nonpathogenic *E. coli* bacteria genetically modified with insulin gene

Bianca returned to the search engine and typed in "recombinant DNA insulin." She browsed through several sites. Insulin was the first recombinant drug to be marketed, beginning in 1982. *"Before Cameron was even born,"* she thought. She continued reading. The recombinant DNA version is human insulin because it's made from the human gene. It eventually replaced insulin produced from the pancreases of cows and pigs. Now pharmaceutical companies have modified the basic structure of the insulin to alter some of its properties. Reversing the two amino acids in the B chain makes Humalog® much faster acting. *"Ahh,"* Bianca thought. *"That's why Cameron's insulin pump injects into his body all day long. He said he could give himself extra right before meals. It works right away."*

Bianca felt a tap on her shoulder. She looked up. "Hi, Becky."

"What are you so engrossed in, Bianca?" Becky asked.

"Insulin!"

Becky looked confused. "Really?"

"Yes. This Bio 127 assignment is pretty interesting!" Bianca replied.

Are Genetically Engineered Foods Safe?

The genetically engineered foods debate is raging around the world, even though a large number of consumers in the United States seems to know nothing about it. In a 2008 *CBS News/New York Times* poll, 44% of those surveyed knew "not much" or "nothing at all" about foods containing genetically modified ingredients. On one side of the debate are the people who argue that genetically engineered foods are not only safe, but also the answer to eliminating pesticides and ending world hunger. However, opponents believe that there isn't enough evidence about genetically engineered food to support the safety claims. There are also fears that the public is being kept in the dark when it comes to the prevalence of these foods in our diet. This group cites the StarLink controversy to support its concerns.

The discovery by activists that a genetically engineered corn called StarLink had inadvertently made it into the food supply in 2000 triggered the recall of taco shells, tortillas, and many other corn-based foodstuffs from supermarkets. Further, the makers of StarLink were forced to buy back StarLink from farmers and to compensate food producers at an estimated cost of several hundred million dollars. StarLink is a type of "BT" corn. It contains a foreign gene taken from a common soil organism, *Bacillus thuringiensis,* whose insecticidal properties have been long known. About a dozen BT varieties, including corn, potato, and even a tomato, have now been approved for human consumption. These strains contain a gene for an insecticidal protein called CrylA. The makers of StarLink decided to use a gene for a related protein called Cry9C. They thought that using this molecule might slow down the chances of pest resistance to BT corn. To obtain FDA approval for use in foods, the makers of StarLink performed the required tests. Like the other now-approved strains, StarLink wasn't poisonous to rodents, and its biochemical structure is not similar to those of most food allergens. But the Cry9C protein resisted digestion longer than the other BT proteins when it was put in simulated stomach acid and subjected to heat. Most food allergens are stable like this, so StarLink was not approved for human consumption.

The scientific community is now trying to devise more tests for allergens because it has not been possible to determine conclusively whether Cry9C is or is not an allergen. Also, at this point, it is unclear how resistant to digestion a protein must be to be an allergen, and it is also unclear what degree of sequence similarity a potential allergen must have to a known allergen to raise concern. Dean D. Metcalfe, chief of the Laboratory of Allergic Diseases at the National Institute of Allergy and Infectious Diseases, said, "We need to understand thresholds for sensitization to food allergens and thresholds for elicitation of a reaction with food allergens."

Other scientists are concerned about the following potential drawbacks to the planting of BT corn: (1) resistance among populations of the target pest, (2) exchange of genetic material between the transgenic crop and related plant species, and (3) BT crops' impact on nontarget species. They feel that many more studies are needed before it can be said for certain that BT corn has no ecological drawbacks.

Despite controversy, the planting of genetically engineered crops continues. According to United States Department of Agriculture (USDA) statistics for 2008, U.S. farmers planted genetically engineered corn on 80% of corn acreage. Likewise, genetically engineered soybeans made up 92% of the U.S. crop, and cotton growers used engineered seed for 86% of their crop (Fig. 21B). In the same 2008 *CBS News/New York Times* poll cited previously, 87% of respondents believed that foods containing genetically modified ingredients should be labeled as such. This may not be easy to accomplish, however. For example, most cornmeal is derived from both conventional and genetically engineered corn. So far, there has been no attempt to sort out one type of food product from the other.

a.

b.

Figure 21B What are some common genetically engineered crops?
Genetically engineered **(a)** corn, **(b)** soybeans, and **(c)** cotton crops are increasingly being planted by today's farmers.

c.

Other types of genetically engineered plants are also expected to increase productivity. Leaves might be engineered to lose less water and take in more carbon dioxide. To accomplish the latter, the efficiency of the enzyme that captures carbon dioxide in plants could be improved. A team of Japanese scientists is working on altering the process of photosynthesis in rice so that rice plants can do well in hot, dry weather. These modifications would require a more complete reengineering of plant cells than the single-gene transfers that have been done so far.

Single-gene transfers have allowed plants to produce various products, including human hormones, clotting factors, and antibodies. One type of antibody made by corn can deliver radioisotopes to tumor cells, and another made by soybeans may be developed to treat genital herpes.

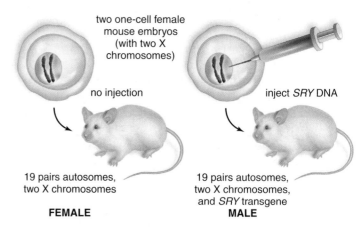

Figure 21.24   **How were transgenic mice used to explain the genetic basis of maleness?**
Transgenic mice showed that maleness is due to *SRY* DNA.

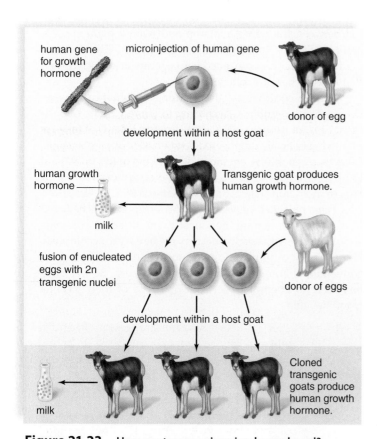

Figure 21.23   **How are transgenic animals produced?**
Once the desired human gene is introduced into a fertilized ovum, a single transgenic goat is produced. Using the first goat's cell nuclei and ova, a herd of transgenic goats can be developed. Each produces human growth hormone in her milk.

From Animals

Techniques have been developed to insert genes into the eggs of animals. It is possible to microinject foreign genes into eggs by hand. Another method uses vortex mixing. The eggs are placed in an agitator with DNA and silicon-carbide needles, and the needles make tiny holes through which the DNA can enter. When these eggs are fertilized, the resulting offspring are transgenic animals. Using this technique, many types of animal eggs have taken up the gene for bovine growth hormone (bGH). The procedure has been used to produce larger fish, cows, pigs, rabbits, and sheep. Pigs being bred to supply transplant organs for humans can be genetically modified in this manner also.

Gene pharming, the use of transgenic farm animals to produce pharmaceuticals, is being pursued by a number of firms. Genes that code for therapeutic and diagnostic proteins are incorporated into an animal's DNA, and the proteins appear in the animal's milk (Fig. 21.23). Plans are under way to produce drugs for the treatment of cystic fibrosis, cancer, blood diseases, and other disorders by this method. Figure 21.23 outlines the procedure for producing transgenic mammals. The gene of interest is microinjected into donor eggs. Following in vitro fertilization, the zygotes are placed in host females, where they develop. After transgenic female offspring mature, the product is secreted in their milk. Then, cloning can be used to produce many animals that produce the same product. Female clones produce the same product in their milk.

DNA Fingerprinting and the Criminal Justice System

Traditional fingerprinting has been used for years to identify criminals and to exonerate those wrongly accused of crimes. The opportunity now arises to use DNA fingerprinting in the same way. DNA fingerprinting requires only a small DNA sample. This sample can come from blood left at the scene of the crime, semen from a rape case, even a single hair root!

Advocates of DNA fingerprinting claim that identification is "beyond a reasonable doubt." But how can investigators be certain? Much of the forensic DNA fingerprinting done today uses short tandem repeats (STRs). These are stretches of noncoding DNA in our genome that contain repeated DNA sequences. Most commonly, these repeats are four bases in length; for example, CATG. You may have 11 copies of this repeat on a particular chromosome inherited from your father and only three copies on the homologous chromosome from your mother. When analyzed by electrophoresis, greater numbers of repeats correspond to increasing lengths on the DNA. Different people have unique repeat patterns, so these STRs can be used to discriminate between individuals. A particular STR pattern on a single chromosome may be shared by a number of people. However, by studying multiple STR sites, a statistically unique pattern can be developed for everyone—unless you share your DNA with an identical twin! In the United States, CODIS, the FBI's Combined DNA Index System, uses 13 STR sites to identify individuals.

Opponents of this technology, however, point out that it is not without its problems. Police or laboratory negligence can invalidate the evidence. For example, during the O. J. Simpson trial, the defense claimed that the DNA evidence was inadmissible because it could not be proven that the police had not "planted" O. J.'s blood at the crime scene. There have also been reported problems with sloppy laboratory procedures and the credibility of forensic experts. In one particular case, Curtis McCarty had been placed on death row three times by the same team of prosecutor and police lab analysts. After 21 years in prison, he was exonerated. The prosecutor has been accused of misconduct, while the police lab analyst was fired for falsifying laboratory data to obtain convictions.

In addition to identifying criminals, DNA fingerprinting can be used to establish paternity and maternity; determine nationality for immigration purposes; and identify victims of a national disaster, such as the terrorist attacks of September 11, 2001.

Considering the usefulness of DNA fingerprints, perhaps everyone should be required to contribute blood to create a national DNA fingerprint databank. Some say, however, this would constitute an unreasonable search, which is unconstitutional.

Decide Your Opinion

1. Would you be willing to provide your DNA for a national DNA databank? Why or why not? What types of privacy restrictions would you want on your DNA?
2. If not everyone, do you think that convicted felons, at least, should be required to provide DNA for a databank?
3. Should all defendants have access to DNA fingerprinting (at government expense) to prove they didn't commit a crime? Should this include those already convicted of crimes who want to reopen their cases using new DNA evidence?

Many researchers are using transgenic mice technology for various research projects in the laboratory. Figure 21.24 shows how this technology demonstrated that a section of DNA called *SRY* (sex-determining region of the Y chromosome) produces a male animal. This *SRY* DNA was cloned, and then one copy was injected into one-celled mouse embryos with two X chromosomes. Injected embryos developed into males, but any that were not injected developed into females. Mouse models have also been created to study human diseases. An allele such as the one that causes cystic fibrosis can be cloned and inserted into mice embryonic stem cells. Occasionally, a mouse embryo homozygous for cystic fibrosis will result. This embryo develops into a mutant mouse that has a phenotype similar to a human with cystic fibrosis. New drugs for the treatment of cystic fibrosis can then be tested in these mice.

Xenotransplantation is the use of animal organs, instead of human organs, in transplant patients. Scientists have chosen to use the pig because animal husbandry has long included the raising of pigs as a meat source, and pigs are prolific. The ability of pigs to express genes for human recognition proteins means that pig organs will be more readily accepted by humans and rejection will be less likely to occur.

Check Your Progress 21.4

1. What two techniques allow scientists to clone a portion of DNA?
2. a. How and (b) why is DNA fingerprinting done?
3. In what ways does genetic engineering improve agricultural plants and farm animals?

Summarizing the Concepts

21.1   DNA and RNA Structure and Function

DNA is the genetic material found in the chromosomes. It replicates, stores information, and mutates for genetic variability.

Structure of DNA

- Double helix composed of two polynucleotide strands. Each nucleotide is composed of a deoxyribose sugar, a phosphate, and a nitrogen-containing base (A, T, C, G).
- The base A is bonded to T, and G is bonded to C.

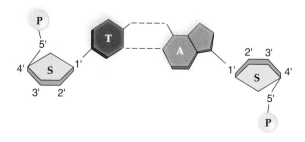

Replication of DNA

- DNA strands unzip, and a new complementary strand forms opposite each old strand, resulting in two identical DNA molecules:

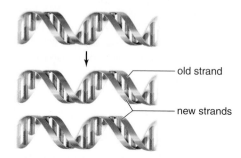

old strand

new strands

The Structure and Function of RNA

- RNA is a single-stranded nucleic acid in which the base U (uracil) occurs instead of T (thymine).
- The four forms of RNA are rRNA (found in the ribosomes); mRNA (carries the DNA message to the ribosomes); tRNA (transfers amino acids to the ribosomes where protein synthesis occurs); and small RNAs (control of gene expression).

21.2   Gene Expression

Gene expression leads to the formation of a product, either an RNA or a protein. Proteins differ by the sequence of their amino acids. Gene expression for proteins requires transcription and translation.

- **Transcription** Occurs in the nucleus. The DNA triplet code is passed to an mRNA that contains codons. Introns are removed from mRNA during mRNA processing.
- **Translation** Occurs in the cytoplasm at the ribosomes. tRNA molecules bind to their amino acids, and then their anticodons pair with mRNA codons.

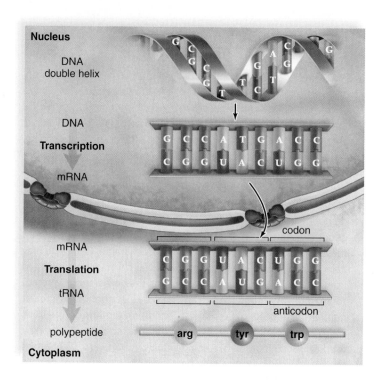

The Regulation of Gene Expression

Regulation of gene expression occurs at five levels in a human cell:

- **Pretranscriptional Control** In the nucleus; the DNA is made available to transcription factors and enzymes.
- **Transcriptional Control** In the nucleus; the degree to which a gene is transcribed into mRNA determines the amount of gene product.
- **Posttranscriptional Control** In the nucleus; involves mRNA processing and how fast mRNA leaves the nucleus.
- **Translational Control** In the cytoplasm; affects when translation begins and how long it continues. Includes inactivation and degradation of mRNA.
- **Posttranslational Control** In the cytoplasm; occurs after protein synthesis.

21.3   Genomics

The human genome has now been sequenced via the 13-year-long Human Genome Project.

- Genomes of other organisms have also been sequenced.
- The Personal Genome Project is underway, which would allow individuals to have their genome sequenced.

Functional and Comparative Genomics

- Functional genomics is the study of how the 20,500 genes in a human genome function.
- Comparative genomics is a way to determine how species have evolved and how genes and noncoding regions of the genome function.

Proteomics and Bioinformatics Are New Endeavors

- Proteomics is the study of the structure, function, and interaction of cellular proteins.
- Bioinformatics is the application of computer technologies to the study of the genome.

A Person's Genome Can Be Modified

- Gene therapies such as ex vivo gene therapy or in vivo gene therapy can treat various medical conditions.

21.4 DNA Technology

- Recombinant DNA contains DNA from two different sources. The foreign gene and vector DNA are cut by the same restriction enzyme and then the foreign gene is sealed into vector DNA. Bacteria take up recombinant plasmids.
- PCR uses DNA polymerase to make multiple copies of a specific piece of DNA. Following PCR, DNA can be subjected to DNA fingerprinting.
- Transgenic organisms (bacteria, plants, and animals that have had a foreign gene inserted into them) can produce biotechnology products, such as hormones and vaccines.
- Transgenic bacteria can promote plant health, remove sulfur from coal, clean up toxic waste and oil spills, extract minerals, and produce chemicals.
- Transgenic crops can resist herbicides and pests.
- Transgenic animals can be given growth hormone to produce larger animals; can supply transplant organs; and can produce pharmaceuticals.

Understanding Key Terms

anticodon 497
bioinformatics 503
biotechnology product 506
cloning 504
codon 495
complementary DNA (cDNA) 505
complementary paired bases 490
DNA (deoxyribonucleic acid) 490
DNA ligase 504
DNA replication 491
double helix 490
functional genomics 502
gene cloning 504
genetic engineering 507
messenger RNA (mRNA) 493
mutation 492
plasmid 504

polymerase chain reaction (PCR) 505
polyribosome 497
proteomics 503
recombinant DNA (rDNA) 504
restriction enzyme 504
ribosomal RNA (rRNA) 493
RNA (ribonucleic acid) 492
RNA polymerase 496
small RNAs 493
template 491
transcription 494
transcription factor 501
transfer RNA (tRNA) 493
transgenic organism 506
translation 495
triplet code 495
uracil (U) 492
vector 504
xenotransplantation 511

Match the key terms to these definitions.

a. _____ Cluster of ribosomes attached to the same mRNA molecule; each ribosome is producing a copy of the same polypeptide.

b. _____ Free-living organism in the environment that has had a foreign gene inserted into it.

c. _____ Pattern or guide used to make copies.

d. _____ The study of the structure, function, and interactions of cellular proteins.

e. _____ A means to transfer foreign genetic material into a cell.

Testing Your Knowledge of the Concepts

1. Explain why each new DNA double helix is like the parental DNA helix. (pages 491–92)
2. Describe the types of RNA and how they function. (page 493)
3. Explain why the terms transcription and translation are appropriate for these processes. (pages 494–95)
4. If a DNA strand is TAC AAT AAA CGT GTC ATT, what are the codons of mRNA, the anticodons of tRNA, and the amino acid sequence? (pages 495–97)
5. Tell where you would expect each level of genetic control to occur in the cell. Explain. (pages 499–501)
6. Why do the fields of proteomics and bioinformatics depend on genomics? (page 503)
7. Is ex vivo or in vivo gene therapy harder to perform? Why? (pages 503–04)
8. What is the methodology for producing transgenic bacteria? (pages 504–05)
9. What is the polymerase chain reaction (PCR), and how is it used to produce multiple copies of a DNA segment? (page 505)
10. What are some biotechnology products from bacteria, plants, and animals, and what are they used for? (pages 507–08, 510)
11. The double-helix model of DNA resembles a twisted ladder in which the rungs of the ladder are
    a. complementary base pairs.
    b. A paired with G and C paired with T.
    c. A paired with T and G paired with C.
    d. a sugar-phosphate paired with a sugar-phosphate.
    e. Both a and c are correct.
12. The enzyme responsible for adding new nucleotides to a growing DNA chain during DNA replication is
    a. helicase.
    b. RNA polymerase.
    c. DNA polymerase.
    d. ribozymes.
13. RNA processing
    a. removes the introns, leaving only the exons.
    b. is the same as transcription.
    c. is an event that occurs after RNA is transcribed.
    d. is the rejection of old, worn-out RNA.
    e. Both a and c are correct.
14. During protein synthesis, an anticodon of a tRNA pairs with
    a. amino acids in the polypeptide.
    b. DNA nucleotide bases.
    c. rRNA nucleotide bases.
    d. mRNA nucleotide bases.
15. Which of these associations does not correctly compare DNA and RNA?

DNA	RNA
a. Contains the base thymine.	Contains the base uracil.
b. Is double-stranded.	Is a single-stranded helix.
c. Is found only in the nucleus.	Is found in the nucleus and cytoplasm.
d. The sugar is deoxyribose.	The sugar is ribose.

16. The process of converting the information contained in the nucleotide sequence of RNA into a sequence of amino acids is called
    a. transcription.
    b. translation.
    c. translocation.
    d. replication.

17. Which of the following is involved in controlling gene expression?
    a. the occurrence of transcription
    b. activity of the polypeptide product
    c. life expectancy of the mRNA molecule in the cell
    d. All of these are involved.

18. Complementary base pairing
    a. involves T, A, G, C.
    b. is necessary to replication.
    c. uses hydrogen bonds.
    d. All of these are correct.

19. PCR
    a. uses RNA polymerase.
    b. takes place in huge bioreactors.
    c. uses DNA replication in vitro.
    d. makes many nonidentical copies of DNA.
    e. All of these are correct.

20. Restriction enzymes found in bacterial cells are ordinarily used
    a. during DNA replication.
    b. to degrade the bacterial cell's DNA.
    c. to degrade viral DNA that enters the cell.
    d. to attach pieces of DNA together.

21. Which of the following is a benefit to having insulin produced by biotechnology?
    a. It is just as effective.
    b. It can be mass-produced.
    c. It is nonallergenic.
    d. It is less expensive.
    e. All of these are correct.

22. Following is a segment of a DNA molecule. (Remember that only one strand is transcribed.) What are (a) the mRNA codons, (b) the tRNA anticodons, and (c) the sequence of amino acids?

template strand

noncoding strand

Thinking Critically About the Concepts

When Bianca looked up the possible diseases treated with pharmaceutical products produced using recombinant DNA, there was quite a list. In the case of Cameron's diabetes, recombinant DNA-produced insulin replaces that which his body can no longer produce. With careful management, Cameron functions normally and enjoys a high quality of life. Becky's cousin suffers from cystic fibrosis. The recombinant DNA-produced enzyme she inhales digests excessive amounts of DNA in the mucous clogging her airways. This dilutes the mucus and allows her to clear it from her lungs. This treatment doesn't cure cystic fibrosis, but it does provide relief for its symptoms.

1. What are the advantages of using a recombinant DNA human product instead of a product isolated from another organism? (For example, what advantage would there be to using human recombinant insulin versus insulin produced from cows or pigs?)

2. Are there any disadvantages to using a recombinant DNA product? (*Hint:* You may want to think about how the product is produced and then purified.)

3. Recombinant human growth hormone is available for children and adults with growth hormone deficiency.
   a. When would use of the hormone be appropriate? Under what circumstances would hormone use be improper?
   b. How would a physician know that the hormone treatment was completely effective? What further information might you need to answer this question?

4. In general, diseases caused by a single protein deficiency are more easily treated with a recombinant DNA-produced product. Yet cancer is on the list that Bianca read. What aspects of cancer could be treated with a recombinant DNA-produced product?

CHAPTER

22

Human Evolution

Andrew removed his baseball cap and wiped the sweat off his forehead with the back of his hand. He looked up at the sun. "How hot do you think it is out here?" he asked.

Jamie laughed but didn't look up. "Not so bad. Don't think about it, and you won't feel so hot."

"Right," Andrew sighed. "Like my not thinking about the heat will make it cooler out here."

Jamie continued to shake the sieve. It was a square plywood form approximately 4 in. deep with a metal mesh stretched across the bottom. It sat on legs that allowed it to be rocked back and forth over a large tarp laid on the ground. Anything smaller than ¼ in. in diameter would go through the sieve, while anything larger would remain on top of the mesh. Eventually everything in the sieve had shaken onto the tarp except a few rocks. Jamie examined the rocks and then threw them into the bucket.

Jamie looked up. "You'll never get done if you keep stopping. You complained that you were bored when we were in lecture. You said you couldn't wait to get out into the field. Now we're in the field. Quit complaining and give me some more dirt."

Andrew leaned over and sunk the small skimming shovel into the area he was digging up. He filled it with dirt and dumped it into the sieve. "I know, I know. I just didn't expect it to be so hot. I guess that's what happens when you take an archaeology class in the summer."

Andrew and Jamie continued to work for another hour until the site was completely excavated. The final hole was about 18 in. in diameter and 2 ft deep. Nothing of any value had been larger than ¼ in. in diameter. Jamie gathered the supplies while Andrew shoveled the dirt back into the hole. Jamie checked the label on the orange flag in the ground where Andrew had dug the hole. She recorded the information on the clipboard and marked the artifacts column on the worksheet with a zero: No results. That was discouraging.

"If we finish one more site, we can quit for today," Jamie said.

Andrew groaned. "Then you get to shovel the next one."

Jamie laughed. "It isn't any cooler if you shake than if you dig!"

CHAPTER CONCEPTS

22.1 Origin of Life
Data suggest that chemical evolution produced the first cell.

22.2 Biological Evolution
Descent from a common ancestor explains the unity of living things. For example, all living things have a cellular structure and a common chemistry from their common ancestor. Adaptation to different environments explains the great diversity of living things.

22.3 Classification of Humans
The classification of humans can be used to trace the ancestry of humans. Humans are hominids in the order primates, a type of mammal in the domain Eukarya.

22.4 Evolution of Hominids
The evolutionary tree of hominids resembles a bush. This means that there is not a straight line of fossils leading to modern humans.

22.5 Evolution of Humans
Among the members of the genus *Homo*, *Homo habilis* was the first to make tools. *Homo erectus* was the first to have the use of fire. Neandertals were heavily muscled. Modern humans are believed to have evolved in Africa. Cro-Magnons resembled modern humans in appearance and most likely in behavior as well.

22.1 **Origin of Life**

Our study of evolution begins with the origin of life. A fundamental principle of biology states that all living things are made of cells and that every cell comes from a preexisting cell (see page 44). But if this is so, how did the first cell come about? It was the very first living thing, so it had to come from nonliving chemicals. Could there have been a slow increase in the complexity of chemicals? Could a **chemical evolution** have produced the first cell(s) on the primitive Earth?

The Primitive Earth

The sun and the planets, including Earth, probably formed over a 10-billion-year period from aggregates of dust particles and debris. At 4.6 billion years ago (BYA), the solar system was in place. Dense silicate minerals became the semiliquid mantle.

The Earth's mass is such that the gravitational field is strong enough to have an atmosphere. If the Earth had had less mass, atmospheric gases would have escaped into outer space. The early Earth's atmosphere was not the same as today's atmosphere. Most likely, the first atmosphere was formed by gases escaping from volcanoes. If so, the primitive atmosphere would have consisted mostly of water vapor (H_2O), nitrogen (N_2), and carbon dioxide (CO_2), with only small amounts of hydrogen (H_2) and carbon monoxide (CO). The primitive atmosphere had little, if any, free oxygen.

At first, the Earth and its atmosphere were extremely hot. Water, existing only as a gas, formed dense, thick clouds. Then, as the Earth cooled, water vapor condensed to liquid water, and rain began to fall. Rain fell in such enormous quantities over hundreds of millions of years that the oceans of the world were produced.

Small Organic Molecules

The rain washed the other gases, such as N_2 and CO_2, into the oceans (Fig. 22.1a). The primitive Earth had many sources of energy. These included volcanoes, meteorites, radioactive isotopes, lightning, and ultraviolet radiation. In the presence of so much available energy, the primitive gases may have reacted with one another. This may have produced small organic compounds, such as nucleotides and amino acids (Fig. 22.1b).

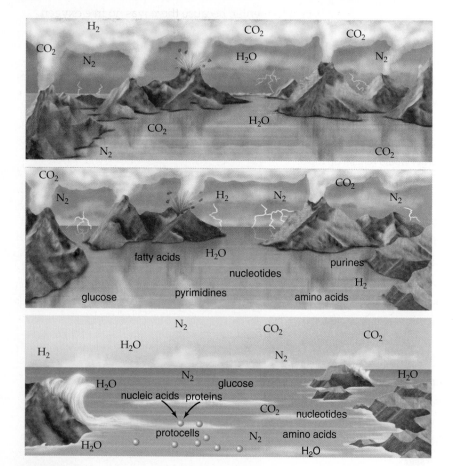

a. The primitive atmosphere contained gases, including H_2O, CO_2, and N_2, that escaped from volcanoes. As the water vapor cooled, some gases were washed into the oceans by rain.

b. The availability of energy from volcanic eruption and lightning allowed gases to form small organic molecules, such as nucleotides and amino acids.

c. Small organic molecules could have joined to form proteins and nucleic acids, which became incorporated into membrane-bound spheres. The spheres became the first cells, called protocells. Later protocells became true cells that could reproduce.

Figure 22.1   How could life have originated on planet Earth?
A chemical evolution could have produced the protocell, which became a true cell once it had genes composed of DNA and could reproduce.

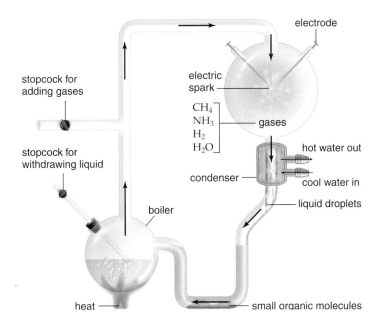

Figure 22.2 **How did Stanley Miller attempt to make organic molecules?**

Gases thought to be present in the early Earth's atmosphere were admitted to the apparatus, circulated past an energy source (electric spark), and cooled to produce a liquid that could be withdrawn. Upon chemical analysis, the liquid was found to contain various small organic molecules.

In 1953, Stanley Miller performed an experiment. He placed a mixture of gases resembling the Earth's early atmosphere in a closed system. He heated the mixture, and circulated it past an electric spark. After cooling, Miller discovered a variety of small organic molecules in the resulting liquid (Fig. 22.2). Later researchers do not believe Miller used the correct mixture of gases although other investigators have achieved the same results using various mixtures of gases.

Macromolecules

The newly formed small organic molecules likely joined to produce organic macromolecules (Fig. 22.1c). There are two hypotheses of special interest concerning this stage in the origin of life. One is the **RNA-first hypothesis.** This hypothesis suggests that only the macromolecule RNA (ribonucleic acid) was needed at this time to progress toward formation of the first cell(s). This hypothesis was formulated after the discovery that RNA can sometimes be both a substrate and an enzyme during RNA processing (see Fig. 21.9). At that time the splicing of mRNA to remove introns was done by a complex composed of both RNA and protein. The RNA and not the protein is the enzyme. RNA enzymes are called ribozymes. Then, too, ribosomes where protein synthesis occurs contain rRNA. Perhaps RNA, then, could have carried out the processes of life commonly associated today with DNA (deoxyribonucleic acid) and proteins. Scientists who sup-

port this hypothesis are fond of saying that it was an "RNA world" some 3.5 BYA. DNA, being a double helix, is more stable than RNA. Perhaps this accounts for the fact that DNA, not RNA, became the first genetic material.

Another hypothesis is termed the **protein-first hypothesis.** Sidney Fox has shown that amino acids join together when exposed to dry heat. He suggests that amino acids collected in shallow puddles along the rocky shore. The heat of the sun caused them to form proteinoids, small polypeptides that have some catalytic properties. When proteinoids are returned to water, they form microspheres. Microspheres are structures composed only of protein that have many of the properties of a cell.

The Protocell

A cell has a lipid-protein membrane. Fox has shown that if lipids are made available to microspheres, the two tend to become associated, producing a lipid-protein membrane. A **protocell,** which could carry on metabolism but could not reproduce, could have come into existence in this manner.

The protocell would have been able to use the still-abundant small organic molecules in the ocean as food. Therefore, the protocell was, most likely, a **heterotroph,** an organism that takes in preformed food. Further, the protocell would have been a fermenter, because there was no free oxygen.

The True Cell

A true cell can reproduce. In today's cells, DNA replicates before cell division occurs. Enzymatic proteins carry out the replication process.

How did the first cell acquire both DNA and enzymatic proteins? Scientists who support the RNA-first hypothesis propose a series of steps. According to this hypothesis, the first cell had RNA genes that, like messenger RNA, could have specified protein synthesis. Some of the proteins formed would have been enzymes. Perhaps one of these enzymes, such as reverse transcriptase found in retroviruses, could use RNA as a template to form DNA. Replication of DNA would then proceed normally.

By contrast, supporters of the protein-first hypothesis suggest that some of the proteins in the protocell would have evolved the enzymatic ability to synthesize DNA from nucleotides in the ocean. Then DNA would have gone on to specify protein synthesis, and in this way, the cell could have acquired all its enzymes, even the ones that replicate DNA.

> **Check Your Progress 22.1**
> 1. What type of evolution produced the first cells?
> 2. What evidence is there to suggest that the gases of early Earth could have become the first organic molecules?
> 3. What evidence is there that RNA could have been the original genetic material?

22.2 Biological Evolution

The first true cells were the simplest of life-forms. Therefore, they must have been **prokaryotic cells,** which lack a nucleus. Later, **eukaryotic cells** (protists), which have nuclei, evolved. Then, multicellularity and the other kingdoms (fungi, plants, and animals) evolved (see Fig. 1.5). Obviously, all these types of organisms—even prokaryotic cells—are alive today. Each type of organism has its own evolutionary history that is traceable back to the first cell(s).

Biological evolution is the process by which a species changes through time. Biological evolution has two important aspects: descent from a common ancestor and adaptation to the environment. Descent from the original cell(s) explains why all living things have a common chemistry and a cellular structure. An **adaptation** is a characteristic that makes an organism able to survive and reproduce in its environment. Adaptations to different environments help explain the diversity of life—why there are so many different types of living things.

Common Descent

Charles Darwin was an English naturalist who first formulated the theory of evolution that has since been supported by so much independent data. At the age of 22, Darwin sailed around the world as the naturalist on board the HMS *Beagle.* Between 1831 and 1836, the ship sailed in the tropics of the Southern Hemisphere. There, life-forms are more abundant and varied than in Darwin's native England.

Even though it was not his original intent, Darwin began to gather evidence that life-forms change over time and from place to place. The types of evidence that convinced Darwin that common descent occurs were fossil, anatomical, and biogeographical.

Fossil Evidence Supports Evolution

Fossils are the best evidence for evolution. They are the actual remains of species that lived on Earth at least 10,000 years ago and up to billions of years ago. Fossils can be the traces of past life or any other direct evidence that past life existed. Traces include trails, footprints, burrows, worm casts, or even preserved droppings. Fossils can also be such items as pieces of bone, impressions of plants pressed into shale, and even insects trapped in tree resin (which we know as amber). Most fossils, however, are found embedded in or recently eroded from sedimentary rock. Sedimentation, a process that has been going on since the Earth was formed, can take place on land or in bodies of water. Weathering and erosion of rocks produces an accumulation of particles. These particles vary in size and nature and are called sediment. Sediment becomes a stratum (pl., strata), a recognizable layer in a sequence of layers. Any given stratum is older than the one above it and younger than the one immediately below it (Fig. 22.3). This allows fossils to be dated.

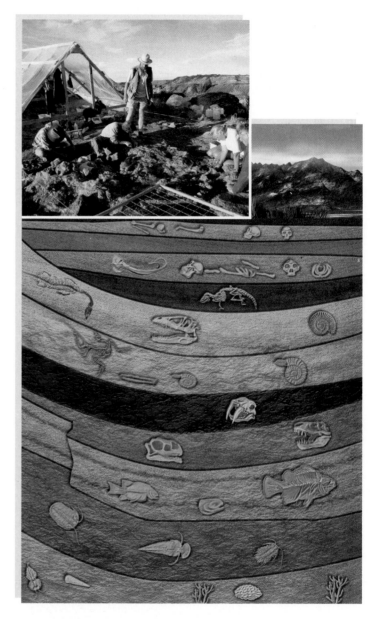

Figure 22.3   How are fossils recovered?
Paleontologists carefully remove fossils from strata.

Usually when an organism dies, the soft parts are either consumed by scavengers or decomposed by bacteria. This means that most fossils consist only of hard parts such as shells, bones, or teeth. These are usually not consumed or destroyed. When a fossil is found encased by rock, the remains were first buried in sediment. The hard parts were preserved by a process called mineralization. Finally, the surrounding sediment hardened to form rock. Subsequently, the fossil has to be found by a human. Most estimates suggest that less than 1% of past species have been preserved as fossils. Only a small fraction of these have been found.

More and more fossils have been found because researchers, called paleontologists, and their assistants

Figure 22.4 Why is *Archaeopteryx* considered a transitional fossil?

Archaeopteryx had a combination of reptilian and bird characteristics.

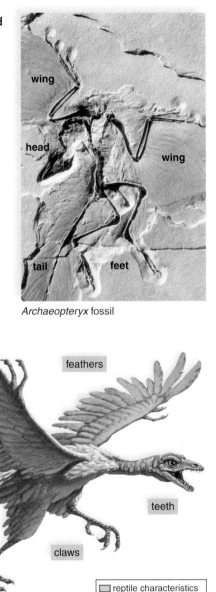

Archaeopteryx fossil

Artist depiction of *Archaeopteryx*

feathers

tail with vertebrae

claws

teeth

☐ reptile characteristics
☐ bird characteristics

that, in general, life has progressed from the simple to the complex. Unicellular prokaryotes are the first signs of life in the fossil record. Unicellular eukaryotes and then multicellular eukaryotes followed. Among the latter, fishes evolved before terrestrial plants and animals. On land, nonflowering plants preceded the flowering plants. Amphibians preceded the reptiles, including the dinosaurs. Dinosaurs are directly linked to the birds, but only indirectly linked to the evolution of mammals, including humans.

Transitional fossils are those that have characteristics of two different groups. In particular, they tell us who is related to whom and how evolution occurred. Even in Darwin's day, scientists knew of the *Archaeopteryx* fossils, which are intermediate between reptiles and birds. The dinosaurlike skeleton of these fossils had reptilian features, including jaws with teeth, and a long, jointed tail. But *Archaeopteryx* also had feathers and wings. Figure 22.4 shows not only a fossil of *Archaeopteryx*, it also gives us an artist's representation of the animal based on the fossil remains. Many more semibird fossils have been discovered in China.

It had always been thought that whales had terrestrial ancestors. Now, fossils have been discovered that support this hypothesis. *Ambulocetus natans* (meaning the walking whale that swims) was the size of a large sea lion, with broad webbed feet on both fore- and hindlimbs. This animal could both walk and swim. It also had tiny hooves on its toes and the primitive skull and teeth of early whales. Figure 22.5 is an artist's recreation based on fossil remains. It depicts *Ambulocetus* as a predator that patrolled freshwater streams looking for prey.

The origin of land mammals is also well documented. The synapsids are mammal-like reptiles whose descendants were wolflike and bearlike predators, as well as several types of piglike herbivores. Slowly, mammalian-like fossils acquired features such as a palate, which would have enabled them to breathe and eat at the same time. They also acquired a muscular diaphragm and rib cage that would have helped them breathe efficiently. The earliest true mammals were shrew-size creatures found in fossil beds about 200 million years old.

Other Evidence Supports Evolution

Many different types of evidence support the hypothesis that organisms are related through common descent.

have been out in the field looking for them (Fig. 22.3). Usually, paleontologists remove fossils from the strata to study them in the laboratory. Then they may decide to exhibit them. The **fossil record** is the history of life recorded by fossils. Paleontology is the science of discovering the fossil record. Decisions about the history of life, ancient climates, and environments can be made using the fossil record. The fossil record is the most direct evidence we have that evolution has occurred. The species found in ancient sedimentary rock are not the species we see today.

Darwin relied on fossils to formulate his theory of evolution. Today, we have a far more complete record than was available to Darwin. The record is complete enough to tell us

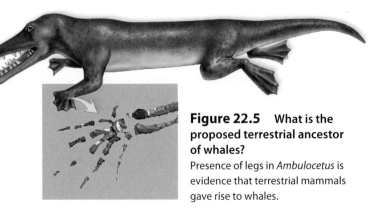

Figure 22.5 What is the proposed terrestrial ancestor of whales?

Presence of legs in *Ambulocetus* is evidence that terrestrial mammals gave rise to whales.

Darwin cited much of the evidence to support his theory of evolution. However, he had no knowledge of the genetic and biochemical data that became available after his time.

Biogeographical Evidence   **Biogeography** is the study of the distribution of plants and animals in different places throughout the world. Such distributions are consistent with the hypothesis that life-forms evolved in a particular locale before they spread out. Therefore, you would expect a different mix of plants and animals whenever geography separates continents, islands, or seas. For example, Darwin noted that South America lacked rabbits, even though the environment was suitable for them. He concluded that no rabbits lived in South America because rabbits evolved somewhere else and had no means of reaching South America. Instead, the Patagonian hare lives in South America. The Patagonian hare resembles a rabbit in anatomy and behavior. But it has the face of a guinea pig, from which it probably evolved.

As another example, both cactuses and euphorbia are plants adapted to a hot, dry environment. Both are succulent, spiny, flowering plants. Why do cactuses grow in North American deserts and euphorbia grow in African deserts, when each would do well on the other continent? They just happened to evolve on their respective continents.

The islands of the world are home to many unique species of animals and plants found no place else, even when the soil and climate are the same. Why do so many species of finches live on the Galápagos Islands, when these same species are not on the mainland? The reasonable explanation is that finches from the ancestral species

CASE STUDY AT THE DIG

Jamie looked at the map on the clipboard and then scouted the area around them. "The next flag should be this way."

The archaeology class had mapped the area earlier in the summer, setting orange flags at appropriate intervals. This was part of a project to determine whether there were other historical sites on the college property that could be included in the National Register of Historic Places. This week the various members of the class had been divided into groups of two to conduct a systematic shovel survey. At the previously mapped sites, the team would sample the soil to determine if any historic Native American artifacts were present.

She and Andrew carried their equipment toward the next site. "It's kind of spooky to think about the Native Americans who lived here, isn't it? It's sort of like spying on them. We find where they lived and see what they left behind. And from that we try to determine how they lived." She smiled ruefully. "Makes you wonder who'll be going through your trash a thousand years from now."

Andrew grimaced. "That's *real* nice. I'm collecting other people's trash! When I used to think about archaeology, I thought about pharaohs in Egypt 5,000 years ago. And I'd be collecting treasure—not trash. I never really thought about Native Americans living right here in the United States 1,200 years ago."

"Well, I think it's great that we can have an archaeology dig right on college land," Jamie commented. "Of course, it'd be really neat to go to Egypt—but can you imagine how hot that would be?"

She laughed. "And just think how much you'd complain!"

Andrew gave her a nasty look. "You're so funny. Look, here's the orange flag. It's your turn to shovel."

They didn't talk much while they were getting set up. Once Jamie began to shovel dirt into the sieve, Andrew started the shaking process.

Jamie started to talk while she was waiting for Andrew to empty the sieve. "Besides, archaeologists don't just find treasures. They're real scientists. We could be like those scientists who found the bones of Lucy, the first human, and be famous!"

"Probably not," Andrew replied sourly. "We're looking for animal bones, shells, burned wood, or seeds. And if we are really, really lucky, arrowheads, flint flakes, ceramic, glass, or metal.

"Yeah, that's treasure all right," he remarked sarcastically. "I don't care about being famous—I just want to pass this class."

As they neared the bottom of the hole, Jamie heard the shovel clank softly against something that wasn't dirt. She got down on her hands and knees and used a smaller trowel to carefully remove the dirt from the hole. She dumped it into the sieve and held her breath as Andrew shook the dirt out. Lying on top of the mesh was a wire nail and a shard of clear glass.

Jamie let out a yell. "JACKPOT!" Andrew placed the nail and the glass in separate plastic bags. He carefully labeled them with the site location and the depth at which they had been uncovered. Carefully Jamie began to remove more dirt from the hole with the trowel. It was much slower going now, as Jamie removed the soil in the hole in layers. It took almost another hour to finish excavating the hole, but when they were finished even Andrew was pleased.

"Wow, this is great!" Andrew said. "We've got a nail, two pieces of clear glass, a piece of brown glass, one arrowhead of some kind, and 17 flint flakes. I love archaeology!"

Jamie just looked at Andrew and shook her head.

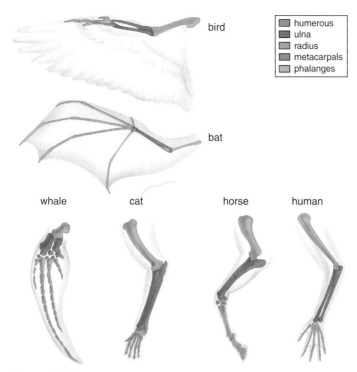

Figure 22.6 **Why are vertebrate forelimbs called homologous?**
Despite differences in function, vertebrate forelimbs have the same bones.

groups as each continued along its own evolutionary pathway. Structures that are anatomically similar because they are inherited from a common ancestor are called **homologous structures.** In contrast, **analogous structures** serve the same function, but they are not constructed similarly, nor do they share a common ancestry. The wings of birds and insects and the jointed appendages of a lobster and humans are analogous structures. The presence of homology, not analogy, is evidence that organisms are related.

Vestigial structures are anatomical features that are fully developed in one group of organisms but that are reduced and may have no function in similar groups. Modern whales have a vestigial pelvic girdle and legs. The ancestors of whales walked on land, but whales are totally aquatic animals today. Most birds have well-developed wings used for flight. Some bird species (e.g., ostrich), however, have greatly reduced wings and do not fly. Similarly, snakes have no use for hindlimbs, and yet some have remnants of a pelvic girdle and legs. Humans have a tailbone but no tail. The presence of vestigial structures can be explained by the common descent. Vestigial structures occur because organisms inherit their anatomy from their ancestors. They are traces of an organism's evolutionary history.

The homology shared by vertebrates extends to their embryological development (Fig. 22.7). At some time

migrated to all the different islands. Then, geographic isolation allowed the ancestral finches to evolve into a different species on each island.

In the history of the Earth, South America, Antarctica, and Australia were originally connected. Marsupials (pouched mammals) arose at this time and today are found in both South America and Australia. But when Australia separated and drifted away, the marsupials diversified into many different forms suited to various environments of Australia. They were free to do so because there were few, if any, placental mammals in Australia. In South America, where there are placental mammals, marsupials are not as diverse. This supports the hypothesis that evolution is influenced by the mix of plants and animals in a particular continent—by biogeography.

Anatomical Evidence Darwin was able to show that a common descent hypothesis offers a plausible explanation for anatomical similarities among organisms. Vertebrate forelimbs are used for flight (birds and bats), orientation during swimming (whales and seals), running (horses), climbing (arboreal lizards), or swinging from tree branches (monkeys). Yet all vertebrate forelimbs contain the same sets of bones organized in similar ways, despite their dissimilar functions (Fig. 22.6). The most plausible explanation for this unity is that the basic forelimb plan belonged to a common ancestor. The basic plan was then modified in the succeeding

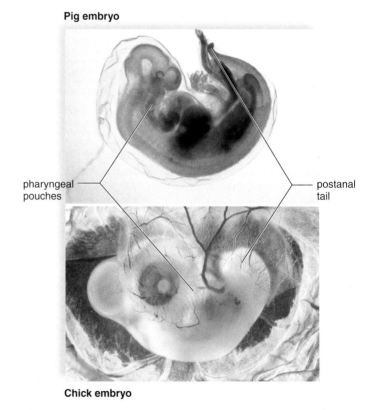

Figure 22.7 **What homologous structures do embryos share?**
Vertebrate embryos have features in common, such as pharygeal pouches, despite different ways of life as adults.

during development, all vertebrates have a postanal tail and exhibit paired pharyngeal pouches. In fish and amphibian larvae, these pouches develop into functioning gills. In humans, the first pair of pouches becomes the cavity of the middle ear and the auditory tube. The second pair becomes the tonsils. The third and fourth pairs become the thymus and parathyroid glands, respectively. Why should terrestrial vertebrates develop and then modify structures like pharyngeal pouches that have lost their original function? The most likely explanation is that fish are ancestral to other vertebrate groups.

Biochemical Evidence   Almost all living organisms use the same basic biochemical molecules, including DNA (deoxyribonucleic acid), ATP (adenosine triphosphate), and many identical or nearly identical enzymes. Further, organisms use the same DNA triplet code and the same 20 amino acids in their proteins. The sequences of DNA bases in the genomes of many organisms are now known, so it has become clear that humans share a large number of genes with much simpler organisms. Evolutionists who study development have also found that many developmental genes are shared in animals ranging from worms to humans. It appears that life's vast diversity has come about by only a slight difference in the regulation of genes. The result has been widely divergent types of bodies.

When the degree of similarity in DNA base sequences or the degree of similarity in amino acid sequences of proteins is examined, the data are as expected, assuming common descent. Cytochrome *c* is a molecule used in the electron transport chain of many organisms. Data regarding differences in the amino acid sequence of cytochrome *c* show that the sequence in a human differs from that in a monkey by only two amino acids. The human sequence differs from that in a duck by 11 amino acids, and from that in a yeast by 51 amino acids (Fig. 22.8). These data are consistent with other data regarding the anatomical similarities of these organisms and, therefore, their relatedness.

Intelligent Design

Evolution is a scientific theory. Sometimes we use the word theory when we mean a hunch or a guess. But in science, the term theory is reserved for those ideas that scientists have found to be all-encompassing because they are based on evidence (data) collected in a number of different fields, as just discussed. In other words, evolutionary theory has been supported by repeated scientific experiments and observations.

Some persons advocate the teaching of ideas that run contrary to the theory of evolution in schools. The emphasis is currently placed on Intelligent Design, a belief system that maintains that the diversity of life could never have arisen without the involvement of an "intelligent agent." Many scientists, and even religions, argue that intelligent design is

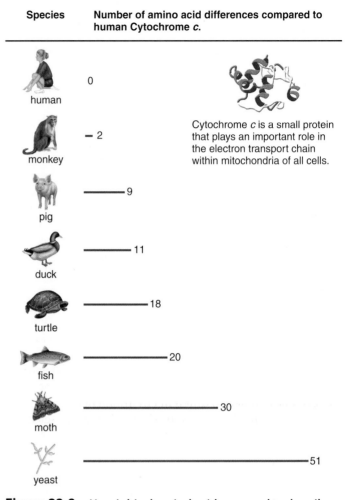

Cytochrome *c* is a small protein that plays an important role in the electron transport chain within mitochondria of all cells.

Figure 22.8   **How is biochemical evidence used to describe evolutionary relationships?**
Closer biochemical structures suggest a closer evolutionary relationship. The number of amino acid differences in cytochrome *c* between humans and other species is indicated.

faith based and not science. It would not be possible to test in a scientific way whether an intelligent agent exists. If it were possible to structure such an experiment, scientists would be the first to do it.

Natural Selection

When Darwin returned home, he spent the next 20 years gathering data to support the principle of biological evolution. Darwin's most significant contribution was to describe a mechanism for adaptation—**natural selection.** During adaptation, a species becomes suited to its environment. On his trip, Darwin visited the Galápagos Islands. He saw a number of finches that resembled one another but had different ways of life. Some were seed-eating ground finches, some cactus-eating ground finches, and some insect-eating tree finches. A warbler-type finch had a beak that could take

Have You Ever Wondered...

When did Darwin publish his book on natural selection? What was the reaction to his work?

Darwin published his book on natural selection in 1859. It was entitled *On the Origin of Species by Means of Natural Selection, or The Preservation of Favoured Races in the Struggle for Life*. The title is usually shortened to *On the Origin of Species*. It was instantly controversial, even though he avoided the use of the word "evolution." He only alludes to humans with the statement that by his theory "... light will be thrown on the origin of man and his history."

The book is still in print and you can obtain a copy.

Lamarck's proposal	Darwin's proposal

Originally, giraffes had short necks.	Originally, giraffe neck length varied.

Giraffes stretched their necks in order to reach food.	Competition for resources causes long-necked giraffes to have the most offspring.

With continual stretching, most giraffes now have long necks.	Due to natural selection, most giraffes now have long necks.

Figure 22.9 **What were the two mechanisms for evolution proposed in the nineteenth century?**
This diagram contrasts Jean-Baptiste Lamarck's process of acquired characteristics with Charles Darwin's process of natural selection.

honey from a flower. A woodpecker-type finch lacked the long tongue of a woodpecker but could use a cactus spine or twig to pull insects from cracks in the bark of a tree. Darwin thought the finches were all descended from a mainland ancestor whose offspring had spread out among the islands and become adapted to different environments.

To emphasize the nature of Darwin's natural selection process, it is often contrasted with a process espoused by Jean-Baptiste Lamarck, another nineteenth-century naturalist. Lamarck's explanation for the long neck of the giraffe was based on the assumption that the ancestors of the modern giraffe were trying to reach into the trees to browse on high-growing vegetation (Fig. 22.9). Continual stretching of the neck caused it to become longer, and this acquired characteristic was passed on to the next generation. Lamarck's mechanism will not work because acquired characteristics cannot be inherited (Fig. 22.9).

The critical elements of the natural selection process are:

- *Variation.* Individual members of a species vary in physical characteristics. Physical variations can be passed from generation to generation. (Darwin was never aware of genes, but we know today that the inheritance of the genotype determines the phenotype.)
- *Competition for limited resources.* Even though each individual could eventually produce many descendants, the number in each generation usually stays about the same. Why? Resources are limited and competition for resources results in unequal reproduction among members of a population.
- *Adaptation.* Those members of a population with advantageous traits capture more resources and are more likely to reproduce and pass on these traits. Thus, over time the environment "selects" for the better-adapted traits. Each subsequent generation includes more individuals adapted in the same way to the environment.

Natural selection can account for the great diversity of life. Environments differ widely, and, therefore, adaptations are varied. From vampire bats, to sea turtles, to the many finches observed by Darwin, all the different organisms are adapted to their way of life.

> **Check Your Progress 22.2**
> 1. **a.** What is biological evolution, and **(b)** what are its two most important aspects?
> 2. **What type of evidence convinced Charles Darwin that biological evolution does occur?**
> 3. **What process accounts for the great diversity of living things?**

22.3 Classification of Humans

To begin a study of human evolution, we turn to the classification of humans because biologists classify organisms according to their evolutionary relatedness. The **binomial name** of an organism gives its genus and species. Organisms in the same domain have only general characteristics in common. Those in the same genus have specific characteristics in common. Table 22.1 lists some of the characteristics that help classify humans. The dates in the first column of the table tell us when these groups of animals first appear in the fossil record.

DNA Data and Human Evolution

We are accustomed to using characteristics given in Table 22.1 to determine evolutionary relationships, but researchers are just as apt to use DNA data. Researchers are depending more and more on DNA data to trace the history of life. DNA data is particularly useful when anatomical differences are unavailable.

For example, in the late 1970s, Carl Woese and his colleagues at the University of Illinois decided to use rRNA sequence data to decide how prokaryotes are related. They knew that the DNA coding for ribosomal RNA (rRNA) changes slowly during evolution. rRNA genes may change only when there is a major evolutionary event. Woese reported that on the basis of rRNA sequence data, there are three domains of life and the archaea are more closely related to eukaryotes than to bacteria (Fig. 22.10). (See page 6 for a description of these domains.) In other words, major decisions regarding the history of life are now being made on the basis of DNA/rRNA/protein sequencing data. (Fig. 22.8 gave an example of the use of protein data

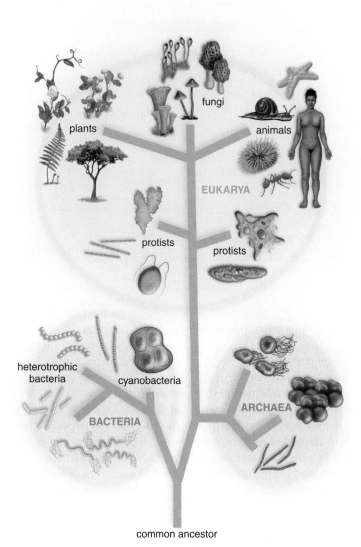

Figure 22.10   What are the three domains of life?
Representatives of each domain are depicted in the ovals. The evolutionary tree of life shows that domain Archaea is more closely related to domain Eukarya than either is to domain Bacteria.

to show evolutionary relationships.) For example, studies of rRNA sequences indicate that among the major groups of eukaryotes, animals are more closely related to fungi than they are to plants.

Later in this chapter, you will learn that DNA sequencing data were used to decide when the last common ancestor for the apes and humans must have existed. As yet, the fossil record has not revealed this ancestor. Comparative DNA data between apes and humans tell us this ancestor must have existed about 7 MYA. Mitochondrial DNA (mtDNA) is used to decide the timing of recent evolutionary events because mtDNA changes occur frequently. Mitochondrial DNA data indicate that modern humans arose in Africa and later migrated to Eurasia. (See page 533 for a discussion of this study.)

Table 22.1	Evolution and Classification of Humans	
BYA/MYA*	**Classification Category**	**Characteristics**
2 BYA	Domain Eukarya	Membrane-bound nucleus
600 MYA	Kingdom Animalia	Multicellular, motile, heterotrophic
540 MYA	Phylum Chordata	Sometime in life history: dorsal tubular nerve cord, notochord, pharyngeal pouches
120 MYA	Class Mammalia	Vertebrates with hair, mammary glands
60 MYA	Order Primates	Well-developed brain, adapted to live in trees
7 MYA	Family Hominidae	Adapted to upright stance and bipedal locomotion
3 MYA	Genus *Homo*	Most developed brain, made and used tools
0.1 MYA	Species *Homo sapiens*†	Modern humans; speech centers of brain well-developed

*BYA = billions of years ago; MYA = millions of years ago.
†To specify an organism, you must use the full binomial name, such as *Homo sapiens*.

Humans Are Primates

In contrast to the other orders of placental mammals, **primates** are adapted to an arboreal life—for living in trees. Primates have mobile limbs; grasping hands; a flattened face; binocular vision; a large, complex brain; and a reduced reproductive rate. The order Primates has two suborders. The **prosimians** include the lemurs, tarsiers, and lorises. The **anthropoids** include the monkeys, apes (Fig. 22.11), and humans. This classification tells us that humans are more closely related to the monkeys and apes than they are to the prosimians. Remarkably, there is more than 95 to 98% similarity between related genes in humans and in apes. This difference still results in a number of major changes.

Mobile Forelimbs and Hindlimbs

Primate limbs are mobile, and the hands and feet both have five digits each. Many primates, such as chimpanzees, have both an opposable big toe and thumb. The big toe or thumb can touch each of the other toes or fingers. (Humans don't have an opposable big toe, but the thumb is opposable. This results in a grip that is both powerful and precise.) The opposable thumb allows a primate to easily reach out and bring food, such as fruit, to the mouth. When locomoting, primates grasp and release tree limbs freely because nails have replaced claws.

Binocular Vision

In chimps, like other primates, the snout is shortened considerably, allowing the eyes to move to the front of the head. The stereoscopic vision (or depth perception) that results permits primates to make accurate judgments about the distance and position of adjoining tree limbs. Humans and the apes have three different cone cells, which are able to discriminate between greens, blues, and reds. (See Chap. 14 for a review.) Cone cells require bright light, but the image is sharp and in color. The lens of the eye focuses light directly on the fovea, a region of the retina, where cone cells are concentrated.

Large, Complex Brain

The evolutionary trend among primates is toward a larger and more complex brain. The brain size is smallest in prosimians and largest in modern humans. The cerebral cortex, with many association areas, expands so much that it becomes extensively folded in humans. The portion of the brain devoted to smell gets smaller. The portions devoted to sight increase in size and complexity during primate evolution. Also, more of the brain is involved in controlling and processing information received from the hands and the thumb. The result is good hand-eye coordination in chimpanzees and humans.

Reduced Reproductive Rate

It is difficult to care for several offspring while moving from limb to limb, and one birth at a time is the norm in primates. The juvenile period of dependency is extended, and there is an emphasis on learned behavior and complex social interactions.

Asian Apes

African Apes

White-handed gibbon,
Hylobates lar

Orangutan, *Pongo pygmaeus*

Chimpanzee, *Pan troglodytes*

Western lowland gorilla,
Gorilla gorilla

Figure 22.11 **What apes exist today?**
The apes can be divided into the Asian apes (gibbons and orangutans) and the African apes (chimpanzees and gorillas). Molecular data and the location of early hominid fossil remains tell us that we are more closely related to the African than to the Asian apes.

Comparing Human Skeleton to the Chimpanzee Skeleton

Figure 22.12 compares anatomical differences between chimpanzees and humans, which relate to upright stance of humans when they walk compared to the chimpanzees practice of knuckle-walking. When chimpanzees walk, the forearms rest on their knuckles.

These differences in anatomy between chimpanzees and humans determine that humans, but not chimps, are adapted for an upright stance: (1) In humans, the spine exits inferior to the center of the skull, and this places the skull in the midline of the body; (2) the longer S-shaped spine of humans places the trunk's center of gravity squarely over the feet; (3) the broader pelvis and hip joint of humans keep them from swaying when they walk; (4) the longer neck of the femur in humans causes the femur to angle inward at the knees; (5) the human knee joint is modified to support the body's weight—the femur is larger at the bottom, and the tibia is larger at the top; and (6) the human toe is not opposable;

instead, the foot has an arch. The arch enables humans to walk long distances and run with less chance of injury.

> **Check Your Progress 22.3**
> 1. How are humans classified—from domain to species?
> 2. What are the characteristics of primates?
> 3. What major differences exist between the chimpanzee skeleton and the human skeleton?

22.4 Evolution of Hominids

Once biologists have studied the characteristics of a group of organisms, they can construct an **evolutionary tree** that is a working hypothesis of their past history. The evolutionary tree in Figure 22.13 shows that all primates share one common ancestor and that the other types of primates diverged from the human line of descent over time. ① The common ancestor for all primates may have resembled a tree shrew.

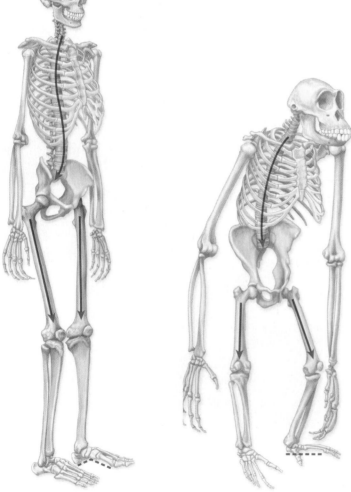

Figure 22.12   How is the human skeleton adapted for standing?
a. Human skeleton compared to **(b)** chimpanzee skeleton.

Human spine exits from the skull's center; ape spine exits from rear of skull.

Human spine is S-shaped; ape spine has a slight curve.

Human pelvis is bowl-shaped; ape pelvis is longer and more narrow.

Human femurs angle inward to the knees; ape femurs angle out a bit.

Human knee can support more weight than ape knee.

Human foot has an arch; ape foot has no arch.

a.   b.

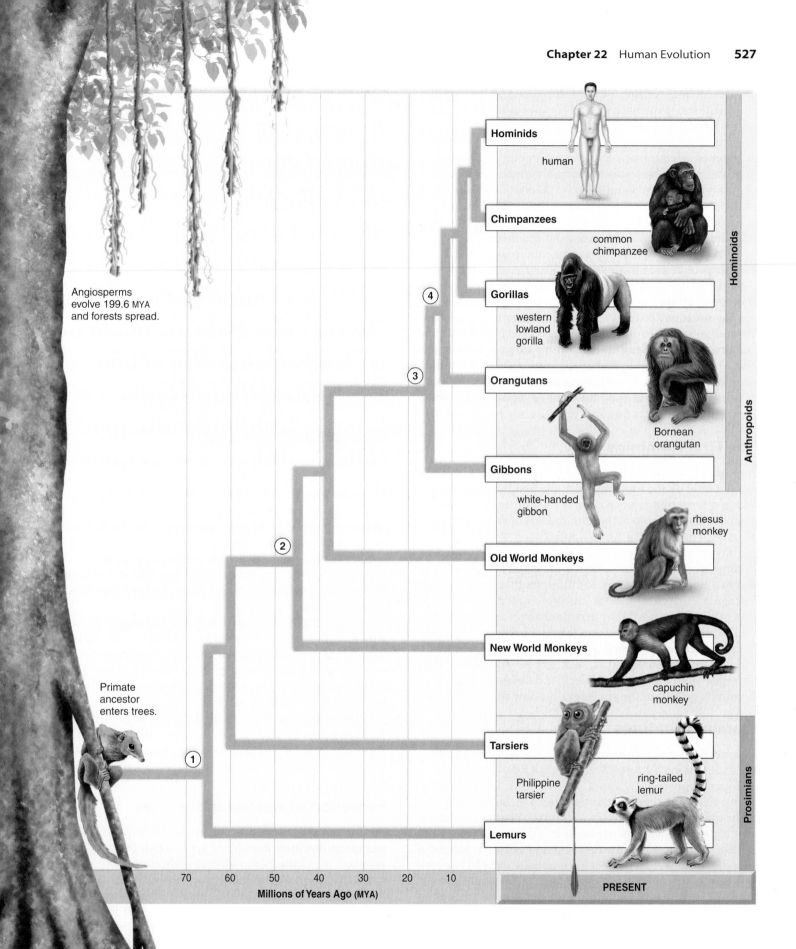

Angiosperms evolve 199.6 MYA and forests spread.

Primate ancestor enters trees.

Hominids

human

Chimpanzees

common chimpanzee

Gorillas

western lowland gorilla

Orangutans

Bornean orangutan

Gibbons

white-handed gibbon

Old World Monkeys

rhesus monkey

New World Monkeys

capuchin monkey

Tarsiers

Philippine tarsier

ring-tailed lemur

Lemurs

Hominoids

Anthropoids

Prosimians

70   60   50   40   30   20   10

Millions of Years Ago (MYA)

PRESENT

Figure 22.13 How are primates thought to have evolved?
See text for explanation of circled numbers.

The descendants of this ancestor developed traits such as a shortened snout and nails instead of claws as they adapted to life in trees. The time when each type of primate diverged from the main line of descent is known from the fossil record. A common ancestor was living at each point of divergence. For example, ② there was a common ancestor for monkeys, apes, and hominids about 45 MYA; ③ one for all apes and hominids about 15 MYA; and ④ another for just African apes and hominids about 7 MYA. The split between the ape and human lineage may have occurred about this time.

One of the most unfortunate misconceptions concerning human evolution is the belief that Darwin and others suggested that humans evolved from apes. On the contrary, humans and apes are thought to have shared a common apelike ancestor. Today's apes are our distant cousins, and we couldn't have evolved from our cousins because we are contemporaries—living on Earth at the same time. Humans and apes have been evolving separately from a common ancestor for about 7 million years. Following the split between humans and apes, different environments selected for the different traits that apes and humans have now.

The First Hominids

Biologists have not been able to agree on which extinct form known only by the fossil record is the first hominid. **Hominid** is a term that refers to our branch of the evolutionary tree. Any fossil placed in the hominid line of descent is closer to us than to one of the African apes.

When any two lines of descent, called a **lineage,** first diverge from a common ancestor, the genes and proteins of the two lineages are nearly identical. As time goes by, each lineage accumulates genetic changes, which lead to RNA and protein changes. Many genetic changes are neutral (not tied to adaptation) and accumulate at a fairly constant rate. Such changes can be used as a type of **molecular clock** to indicate the relatedness of two groups and when they diverged from each other. Molecular data also suggest that hominids split from the ape line of descent about 7 MYA.

Hominid Features

Paleontologists use certain anatomical features when they try to determine if a fossil is a hominid. One of these features is **bipedal posture** (walking on two feet). Until recently, many scientists thought that hominids began to walk upright on two feet because of a dramatic change in climate that caused forests to be replaced by grassland. Now, some biologists suggest that the first hominid began to assume a bipedal posture even while it lived in trees. Why? They cannot find evidence of a dramatic shift in vegetation about 7 MYA. The first hominid's environment is now thought to have included some forest, some woodland, and some grassland. While still living in trees, the first hominids may have walked upright on large branches as they collected fruit from overhead. Then, when they began to forage on the ground among bushes, it would have been easier to shuffle along on their hindlimbs. Bipedalism would also have prevented them from getting heatstroke because an upright stance exposes more of the body to breezes. Bipedalism may have been an advantage in still another way. Males may have acquired food far afield. If they could carry it back to females, they would have been more assured of having sexual intercourse.

Two other hominid features of importance are the shape of the face and brain size. Today's humans have a flatter face and a more pronounced chin than do the apes because the human jaw is shorter than that of the apes. Then, too, our teeth are generally smaller and less specialized. We don't have the sharp canines of an ape, for example. Chimpanzees have a brain size of about 400 cm^3, and modern humans have a brain size of about 1,300 cm^3.

It's hard to decide which fossils are hominids because human features evolved gradually and they didn't evolve at the same rate. Most investigators rely first and foremost on bipedal posture as the hallmark of a hominid, regardless of the size of the brain.

Earliest Fossil Hominids

Fossils have been found that can be dated at the time the ape and human lineages split. The oldest of these fossils, called *Sahelanthropus tchadensis,* dated at 7 MYA, was found in Chad, located in central Africa, far from eastern and southern Africa where other hominid fossils were excavated. The only find, a skull, appears to be that of a hominid because it has smaller canines and thicker tooth enamel than an ape. The braincase, however, is very apelike. It is impossible to tell if this hominid walked upright. Some suggest this fossil is ancestral to the gorilla.

Orrorin tugenensis, dated at 6 MYA and found in eastern Africa, is thought to be another early hominid, especially because the limb anatomy suggests a bipedal posture. However, the canine teeth are large and pointed, and the arm and finger bones retain adaptations for climbing. Some suggest this fossil is ancestral to the chimpanzee.

Ardipithecus kadabba, found in eastern Africa and dated between 5.8 and 5.2 MYA, is closely related to the later-appearing *Ardipithecus ramidus.* This ardipithecine is thought to be closely related to the australopithecines, discussed next.

Evolution of Australopithecines

The hominid line of descent begins in earnest with the **australopithecines,** a group of species that evolved and diversified in Africa. Some australopithecines were slight of frame and termed *gracile* (slender) types. Some were *robust* (powerful) and tended to have strong upper bodies and especially massive jaws. The gracile types most likely fed on soft fruits and leaves, while the robust types had a more fibrous diet that may have included hard nuts. In other words, the skull structure of australopithecines was suited to their particular diets.

Southern Africa

The first australopithecine to be discovered was unearthed in southern Africa by Raymond Dart in the 1920s. This hominid, named *Australopithecus africanus*, is a gracile type dated about 2.8 MYA. *A. robustus*, dated from 2 to 1.5 MYA, is a robust type from southern Africa. Both *A. africanus* and *A. robustus* had a brain size of about 500 cm³. Their skull differences are essentially due to dental and facial adaptations to different diets.

Limb anatomy suggests these hominids walked upright. However, the proportions of the limbs are apelike. The forelimbs are longer than the hindlimbs. Some argue that *A. africanus*, with its relatively large brain, is a possible ancestral candidate for early *Homo*, whose limb proportions are similar to those of this fossil.

Eastern Africa

More than 20 years ago, a team led by Donald Johanson unearthed nearly 250 fossils of a hominid called *A. afarensis*. A now-famous female skeleton dated at 3.18 MYA is known worldwide by its field name, Lucy. Although her brain was small (400 cm³), the shapes and relative proportions of her limbs indicate that Lucy stood upright and walked bipedally (Fig. 22.14*a*). Even better evidence of bipedal loco-

motion comes from a trail of footprints in Laetoli dated about 3.7 MYA. The larger prints are double, as though a smaller-sized being was stepping in the footfalls of another. There are additional small prints off to the side, within hand-holding distance (Fig. 22.14*b*).

Since the australopithecines were apelike above the waist (small brain) and humanlike below the waist (walked erect), it shows that human characteristics did not evolve all at one time. The term **mosaic evolution** is applied when different body parts change at different rates and, therefore, at different times.

A. afarensis, a gracile type, is most likely ancestral to the robust types found in eastern Africa: *A. aethiopicus* and *A. boisei*. *A. boisei* had a powerful upper body and the largest molars of any hominid. These robust types died out, and therefore, it is possible that *A. afarensis* is ancestral to both *A. africanus* and early *Homo*.

> **Check Your Progress 22.4**
> 1. Name three features characteristic of hominids.
> 2. Name a gracile and a robust australopithecine.

a.

b.

Figure 22.14 Who was *Australopithecus afarensis*?
a. A reconstruction of Lucy on display at the St. Louis Zoo. **b.** These fossilized footprints occur in ash from a volcanic eruption some 3.7 MYA. The larger footprints are double (one followed behind the other), and a third, smaller individual was walking to the side. (A female holding the hand of a youngster may have been walking in the footprints of a male.) The footprints suggest that *A. afarensis* walked bipedally.

22.5 Evolution of Humans

Fossils are assigned to the genus *Homo* if (1) the brain size is 600 cm³ or greater, (2) the jaw and teeth resemble those of humans, and (3) tool use is evident (Fig. 22.15). In this section, we will discuss early *Homo: Homo habilis* and *Homo erectus;* and later *Homo:* the Neandertals and the Cro-Magnons, the first modern humans.

Early *Homo*

Homo habilis, dated between 2.0 and 1.9 MYA, may be ancestral to modern humans. Some of these fossils have a brain size as large as 775 cm³, about 45% larger than that of *A. afarensis.* The cheek teeth are smaller than even those of the gracile australopithecines. Therefore, it is likely that these early members of the genus *Homo* were omnivores who ate meat in addition to plant material. Bones at their campsites bear cut marks, indicating that they used tools to strip them of meat.

The stone tools made by *H. habilis,* whose name means "handy man," are rather crude. It's possible that these are the cores from which they took flakes sharp enough to scrape away hide, cut tendons, and easily remove meat from bones.

Early *Homo* skulls suggest that the portions of the brain associated with speech areas were enlarged. We can speculate that the ability to speak may have led to hunting cooperatively. Other members of the group may have remained plant gatherers. If so, both hunters and gatherers most likely ate together and shared their food. In this way, society and culture could have begun.

Culture, which encompasses human behavior and products (e.g., technology and the arts), depends upon the capacity to speak and transmit knowledge. We can further

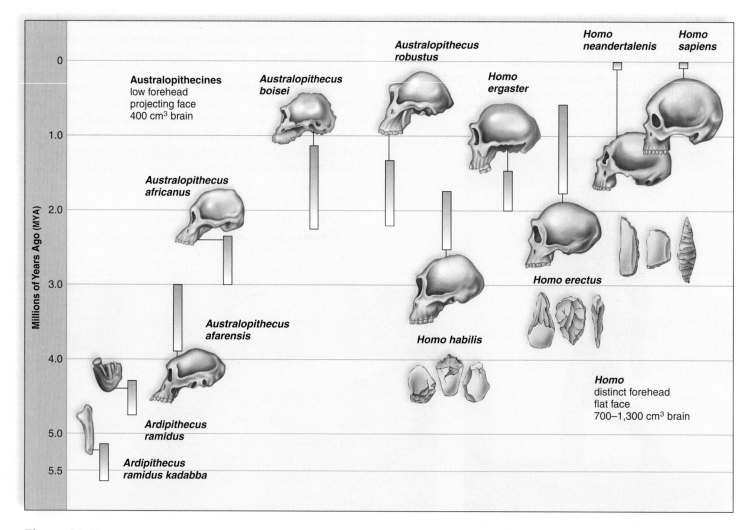

Figure 22.15   How are humans thought to have evolved?
The length of time each species existed is indicated by the vertical blue bars. There have been times when two or more hominids existed at the same time. Therefore, human evolution resembles a "bush" rather than a single branch. *(Left)* Australopithecines have a low forehead, projecting face, and a brain size of about 400 cm³. *(Right) Homos* have an increasingly high forehead, a flat face, and a brain size of 700 to 1,300 cm³.

speculate that the advantages of a culture to *H. habilis* may have hastened the extinction of the australopithecines.

Homo erectus

Homo erectus and like fossils are found in Africa, Asia, and Europe and dated between 1.9 and 0.3 MYA. A Dutch anatomist named Eugene Dubois was the first to unearth *H. erectus* bones in Java in 1891. Since that time, many other fossils have been found in the same area. Although all fossils assigned the name *H. erectus* are similar in appearance, enough discrepancy exists to suggest that several different species have been included in this group. In particular, some experts suggest that the Asian form is *Homo erectus* and the African form is *Homo ergaster* (Fig. 22.16).

Compared with *H. habilis*, *H. erectus* had a larger brain (about 1,000 cm³) and a flatter face. The nose projected, however. This type of nose is adaptive for a hot, dry climate because it permits water to be removed before air leaves the body. The recovery of an almost complete skeleton of a ten-year-old boy indicates that *H. ergaster* was much taller than the hominids discussed thus far. Males were 1.8 meters tall (about 6 feet), and females were 1.55 meters (approaching 5 feet). Indeed, these hominids were erect and most likely had a striding gait like ours. The robust and most likely heavily muscled skeleton still retained some australopithecine features. Even so, the size of the birth canal indicates that infants were born in an immature state that required an extended period of care.

H. ergaster may have first appeared in Africa and then migrated into Asia and Europe. At one time, the migration was thought to have occurred about 1 MYA. Recently, *H. erectus* fossil remains in Java and the Republic of Georgia have been dated at 1.9 and 1.6 MYA, respectively. These remains push the evolution of *H. erectus* in Africa to an earlier date than has yet been determined. In any case, such an extensive population movement is a first in the history of humankind and a tribute to the intellectual and physical skills of the species.

H. erectus was the first hominid to use fire and also fashioned more advanced tools than early *Homos*. These hominids used heavy, teardrop-shaped axes and cleavers. Flakes were probably used for cutting and scraping. It could be that *H. ergaster* was a systematic hunter and brought kills to the same site over and over. In one location, researchers have found over 40,000-bones and 2,647 stones. These sites could have been "home bases," where social interaction occurred and a prolonged childhood allowed time for learning. Perhaps a language evolved and a culture more like our own developed.

> **Check Your Progress 22.5**
>
> 1. Which fossil hominid is the oldest of the australopithecines?
> 2. Which hominid is the first to have culture? Explain.
> 3. Which hominid was the first to migrate out of Africa?

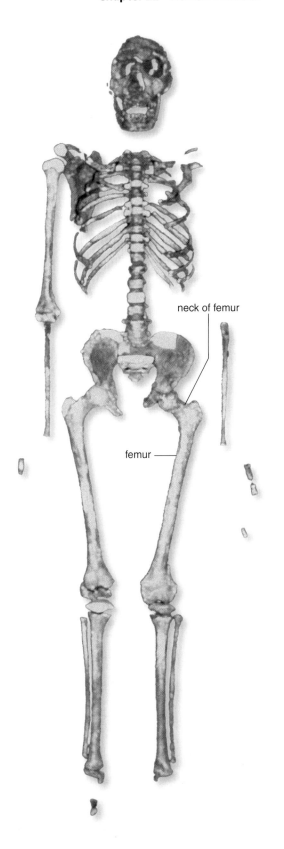

neck of femur

femur

Figure 22.16 Who was *Homo ergaster*?
This skeleton of a ten-year-old boy who lived 1.6 MYA in eastern Africa shows angled femurs because the femur neck is quite long.

⚛ Science **Focus**

Homo floresiensis

In 2003, scientists made one of the most spectacular discoveries in evolutionary history. Nine skeletons were discovered in a cave on the isle of Flores. Flores is an Indonesian island east of Bali, located midway between Asia and Australia. This new species of humans grew no taller than a modern three-year-old child (Fig. 22A). One skeleton was that of an adult female who died when she was approximately 30 years old. She stood 1 meter tall (3.3 ft) and weighed approximately 25 kilograms (55 lbs). Scientists estimate that she died around 18,000 years ago.

After examining the first skeleton, the research team concluded that they had discovered a new human species. They named the species *Homo floresiensis* after the island where it was found. The workers at the excavation site nicknamed the tiny creatures "hobbits" after the fictional creatures in the *Lord of the Rings* books by J. R. R. Tolkien.

a. *Homo floresiensis*, artist's impression

b. Comparison of skulls

Figure 22A   **Who was *Homo floresiensis*?**
a. Artist's recreation of *H. floresiensis*. **b.** The *H. floresiensis* skull is smaller than that of *H. erectus* and that of *H. sapiens*.

Classification of *Homo floresiensis*

Homo floresiensis have skulls the size of a grapefruit and a brain size of approximately 417 cc. The teeth are humanlike, the eyebrow ridges are thick, the forehead slopes sharply, and the face lacks a chin. Even though the hobbits stand only 3 ft tall, they are not classified as pygmies. Despite the small body size, small brain size, and a mixture of primitive and advanced anatomical features, *H. floresiensis* is distinctly a member of genus *Homo*. The researchers believe that *H. floresiensis* possibly evolved from a population of *Homo erectus* that reached Flores approximately 840,000 years ago.

Culture of *Homo floresiensis*

Many of the habits exhibited by *H. floresiensis* are remarkably similar to those of other *Homo* species. Archaeological evidence indicates that *H. floresiensis* had the use of fire. The skeletons discovered on Flores were found in sediment deposits that also contained stone tools and the bones of dwarf elephants, giant rodents, and Komodo dragons. The dwarf elephants, or stegodons, weighed about 1,000 kilograms (2,200 lbs) and would have posed a serious challenge to men who were only 1 meter tall. Successful hunting would have required communication among members of the hunting party. The Flores diet also included fish, frogs, birds, rodents, snakes, and tortoises.

The hobbits produced sophisticated stone tools, hunted successfully in groups, and crossed at least two bodies of water to reach Flores from mainland Asia. And yet, their brain was about one-third the size of modern humans. *H. floresiensis* is the smallest species of human ever discovered. Pound for pound, they outcompete every other member of genus *Homo*.

Further Research Needed

Researchers are interested in determining why the hobbits were so small. The first-discovered skeleton was believed to be that of a small child. There is no evidence of any other 1 meter tall adults in genus *Homo*. Modern pygmies are 1.4 to 1.5 meters (4.6 to nearly 5 ft) tall. Over thousands of years, it is possible that a population of *Homo erectus* evolved into *H. floresiensis*. If so, a smaller body size would have been favored by natural selection. The members of each generation could have been smaller than the previous generation. Dwarfing of mammals on islands is a well-known process that can be seen worldwide. Islands generally have a limited food supply, few predators, and at least a few species competing for the same ecological niche. It behooves species living on islands to minimize their daily energy requirements. The smaller the body size, the fewer the calories per day required for survival. At this point, there is no substantiation for this hypothesis, but continued research may produce an answer as to why hobbits are so small.

The Extinction of *Homo floresiensis*

It appears that many of Flores' inhabitants became extinct approximately 12,000 years ago due to a major volcanic eruption. Researchers found *Homo floresiensis* and pygmy stegodon remains below a 12,000-year-old volcanic ash layer. Hobbits reached the island approximately 11,000 years ago and possibly intermingled with modern humans. Rumors, myths, and legends among the indigenous tribes of Flores about "the tiny people who lived in the forest" have persisted.

Evolution of Modern Humans

Most researchers accept the idea that *Homo sapiens* (modern humans) evolved from *H. erectus,* but they differ as to the details. Perhaps *Homo sapiens* evolved from *H. erectus* separately in Asia, Africa, and Europe. The hypothesis that *Homo sapiens* evolved in several different locations is called the **multiregional continuity hypothesis** (Fig. 22.17*a*). This hypothesis proposes that evolution to modern humans was essentially similar in several different places. If so, each region should show a continuity of its own anatomical characteristics from the time when *H. erectus* first arrived in Europe and Asia.

Opponents argue that it seems highly unlikely that evolution would have produced essentially the same result in these different places. They suggest, instead, the **out-of-Africa hypothesis,** which proposes that *H. sapiens* evolved from *H. erectus* only in Africa, and thereafter *H. sapiens* migrated to Europe and Asia about 100,000 years BP (before present) (Fig. 22.17*b*). If so, there would be no continuity of characteristics between fossils dated 200,000 years BP and 100,000 years BP in Europe and Asia.

According to which hypothesis would modern humans be most genetically alike? The multiregional continuity hypothesis states that human populations have been evolving separately for a long time. Therefore, genetic differences are expected. The out-of-Africa hypothesis states that we are all descended from a few individuals from about 100,000-years BP. Therefore, the out-of-Africa hypothesis suggests that we are more genetically similar.

A few years ago, a study attempted to show that all the people of Europe (and the world, for that matter) have essentially the same mitochondrial DNA. Called the "mitochondrial Eve" hypothesis by the press (this is a misnomer because no single ancestor is proposed), the statistics that calculated the date of the African migration were found to be flawed. Still, the raw data—which indicate a close genetic relationship among all Europeans—support the out-of-Africa hypothesis.

These opposing hypotheses have sparked many other innovative studies to test them. Most scientists have come to the conclusion that the out-of-Africa hypothesis is supported.

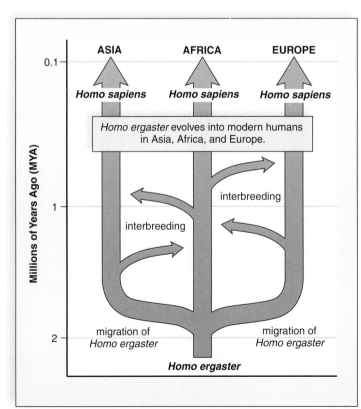

a. Multiregional continuity

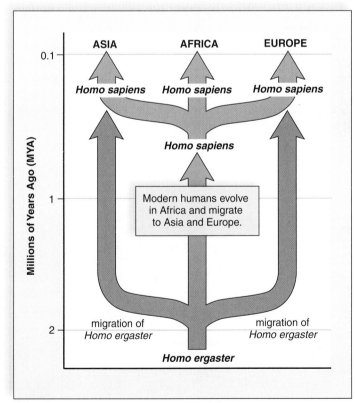

b. Out of Africa

Figure 22.17 **How do the two theories of human evolution compare? Which theory best explains the evolution of modern humans?**

a. The multiregional continuity hypothesis proposes that *Homo sapiens* evolved separately in at least three different places: Asia, Africa, and Europe. Therefore, continuity of genotypes and phenotypes is expected in each region but not between regions. **b.** The out-of-Africa hypothesis proposes that *Homo sapiens* evolved only in Africa and then migrated out of Africa, as *H. ergaster* did many years before. *H. sapiens* would have supplanted populations of *Homo* in Asia and Europe about 100,000 years BP.

Neandertals

Neandertals *(H. neanderthalensis)* take their name from Germany's Neander Valley, where one of the first Neandertal skeletons, dated some 200,000 years BP, was discovered. The Neandertals had massive brow ridges, and their nose, jaws, and teeth protruded far forward. The forehead was low and sloping, and the lower jaw lacked a chin. New fossils show that the pubic bone was long compared with ours.

According to the out-of-Africa hypothesis, Neandertals were eventually supplanted by modern humans. Surprisingly, however, the Neandertal brain was, on the average, slightly larger than that of *Homo sapiens* (1,400 cm^3, compared with 1,360 cm^3 in most modern humans). The Neandertals were heavily muscled, especially in the shoulders and neck (Fig. 22.18). The bones of the limbs were shorter and thicker than those of modern humans. It is hypothesized that a larger brain than that of modern humans was required to control the extra musculature. The Neandertals lived in Europe and Asia during the last ice age, and their sturdy build could have helped conserve heat.

The Neandertals give evidence of being culturally advanced. Most lived in caves, but those living in the open may have built houses. They manufactured a variety of stone tools, including spear points, which could have been used for hunting. Scrapers and knives could have helped in food preparation. They most likely successfully hunted bears, woolly mammoths, rhinoceroses, reindeer, and other contemporary animals. They used and could control fire, which probably helped them cook meat and keep themselves warm. They even buried their dead with flowers and tools and may have had a religion. Perhaps they believed in life after death. If so, they were capable of thinking symbolically.

Cro-Magnons

Cro-Magnons are the oldest fossils to be designated *Homo sapiens.* Cro-Magnons are named for a fossil location in France. In keeping with the out-of-Africa hypothesis, the Cro-Magnons were the modern humans who entered Asia and Europe from Africa 100,000 years BP or even earlier. Cro-Magnons had a thoroughly modern appearance (Fig. 22.19). Analysis of Neandertal DNA indicates that it is so different from Cro-Magnon DNA, these two groups of people did not interbreed. Instead Cro-Magnons seem to have replaced the Neandertals in the Middle East and then spread to Europe 40,000 years ago. There, they lived side by side with the Neandertals for several thousand years. If so, the Neandertals are cousins and not ancestors to us.

Cro-Magnons made advanced stone tools, including compound tools, as when stone flakes were fitted to a wooden handle. They may have been the first to make knife-like blades and throw spears, enabling them to kill animals from a distance. They were such accomplished hunters that some researchers suggest they were responsible for the extinction of many larger mammals, such as the giant sloth, the mammoth, the saber-toothed tiger, and the giant ox, during the late Pleistocene epoch.

Cro-Magnons hunted cooperatively, and perhaps they were the first to have a language. Most likely, they lived in small groups, with the men hunting by day, while the women remained at home with the children. It's possible that this hunting way of life among prehistoric people influences our behavior today. The Cro-Magnon culture included art. They sculpted small figurines out of reindeer bones and antlers. They also painted beautiful drawings of animals on cave walls in Spain and France (Fig. 22.19).

Figure 22.18 Who were Neandertals?
This drawing shows that the nose and the mouth of the Neandertals protruded from their faces, and their muscles were massive. They made stone tools and were most likely excellent hunters.

Figure 22.19 Who were Cro-Magnons?
Cro-Magnon people are one of the earliest groups to be designated *Homo sapiens.* Their tool-making ability and other cultural attributes, including artistic talents, are legendary.

Human Variation

Human beings have been widely distributed about the globe since they evolved. As with any other species that has a wide geographical distribution, phenotypic and genotypic variations are noticeable between populations. Today, we say that people have different ethnicities (Fig. 22.20*a*).

It has been hypothesized that human variations evolved as adaptations to local environmental conditions. One obvious difference among people is skin color. A darker skin is protective against the high ultraviolet (UV) intensity of bright sunlight. On the other hand, a white skin ensures vitamin D production in the skin when the UV intensity is low. Harvard University geneticist Richard Lewontin points out, however, that this hypothesis concerning the survival value of dark and light skin has never been tested.

Two correlations between body shape and environmental conditions have been noted since the nineteenth century. The first, Bergmann's rule, states that animals in colder regions of their range have a bulkier body build. The second, Allen's rule, states that animals in colder regions of their range have shorter limbs, digits, and ears. Both of these effects help regulate body temperature by increasing the surface-area-to-volume ratio in hot climates and decreasing the ratio in cold climates. For example, Figure 22.20*b,c* shows that the Massai of East Africa tend to be very tall and slender, with elongated limbs. By contrast, the Eskimos, who live in northern regions, are bulky with short limbs.

Other anatomical differences among ethnic groups, such as hair texture, a fold on the upper eyelid (common in Asian

Figure 22.20 **Why are humans of varying ethnicities different?**
a. Some of the differences between the three prevalent ethnic groups in the United States may be due to adaptations to the original environment. **b.** The Massai live in East Africa. **c.** Eskimos live near the Arctic circle.

CASE STUDY IN THE LABORATORY

Andrew and Jamie bent over the lab table. Together they studied the artifacts collected from the area they had worked on. Of the 87 survey sites on the map, 15 had yielded artifacts. They were now identifying and categorizing them. There were a total of 223 artifacts, so it was going to take some time.

Professor Carr stopped by their table and picked up the wire nail. "Jamie, tell me about this artifact."

Jamie explained. "This is a wire nail from a later Euro-American settlement. It would date after the time period we're interested in. Probably 1800s. It's evidence that there were people on the site after the Native Americans. It helps that we have historical records of when nails of this type were first used."

Professor Carr nodded and then picked up one of the stone flakes. "Andrew, what's this?"

Andrew began, "It's a flint flake."

Professor Carr interrupted, "Flint?"

Andrew hesitated. "Well, actually chert. Low-quality flint."

Professor Carr nodded, so Andrew continued. "Chert nodules were roughly hammered into flat shapes. Eventually, they'd be finely worked into tools like arrowheads or knives. When they were hammered, these chert flakes broke off. This is evidence that tools were made at this site."

"So, what's your conclusion about this plot?" Professor Carr asked the pair.

Andrew looked nervously at Jamie, who replied. "Well, this is a relatively large area with a fairly high density of Native American artifacts. From the historical records, it appears that this area was used as a pasture and an orchard. There are some mature trees in the area, so that suggests that the soil may be undisturbed by modern farmers. So this is a good area for further surveys, and maybe we can see if it is eligible for the National Register."

"Very good, Jamie—oh, and you too, Andrew," Professor Carr laughed. "You'll need to get all this written up. You can't really call yourself an archaeologist until the paperwork is done!"

Effects of Biocultural Evolution on Population Growth

Human beings today undergo **biocultural evolution** because culture has developed to the point that adaptation to the environment is not dependent on genes but on the passage of culture from one generation to the next.

Tool Use and Language Began

The first step toward biocultural evolution began when *Homo habilis* made primitive stone tools. *Homo erectus* continued the tradition and most likely was a hunter of sorts. It's possible that the campsites of *H. erectus* were "home bases," where the women stayed behind with the children while the men went out to hunt. Hunting was an important event in the development of culture, especially because it encourages the development of language. If *Homo erectus* didn't have the use of language, certainly Cro-Magnon did. People who have the ability to speak a language would have been able to cooperate better as they hunted and even as they sought places to gather plants. Among animals, only humans have a complex language that allows them to communicate their experiences. Words are not objects and events. They stand for objects and events that can be pictured in the mind.

Agriculture Began

About 10,000 years ago, people gave up being full-time hunter-gatherers and became at least part-time farmers. What accounts for the rise of agriculture? The answer is not known, but several explanations have been put forth. About 12,000 years ago, a warming trend occurred as the ice age came to a close. A variety of big game animals became extinct, including the saber-toothed cats, mammoths, and mastodons. This may have made hunting less productive. As the weather warmed, the glaciers retreated and left fertile valleys, where rivers and streams were full of fish and the soil was good. The Fertile Crescent in Mesopotamia is one such example. Here, fishing villages may have sprung up, causing people to settle down.

The people were probably already knowledgeable about what crops to plant. Most likely, as hunter-gatherers, people had already selected seeds with desirable characteristics for propagation. Then, a chance mutation may have made these plants particularly suitable as a source of food. So now they began to till the good soil where they had settled, and they began to systematically plant crops (Fig. 22B).

As people became more sedentary, they may have had more children, especially because the men were home more often. Population increases may have tipped the scales and caused them to adopt agriculture full time, especially if agriculture could be counted on to provide food for hungry mouths. The availability of agricultural tools must have contributed to making agriculture worthwhile. The digging stick, the hoe, the sickle, and the plow were improved when iron tools replaced bronze in the stone-bronze-iron sequence of ancient tools. Irrigation began as a way to control water supply, especially in semiarid areas and regions of periodic rainfall.

If evolutionary success is judged by population size, agriculture was extremely beneficial to our success because it caused a rapid increase of human numbers all over the Earth. Also, agriculture ushered in civilization as we know it. When crops became bountiful, some people were freed from raising their own food. They began to specialize for other ways of life in towns and then cities. Some people became traders, shopkeepers, bakers, and teachers, to name a few possibilities. Others became the nobility, priests, and soldiers. Today, farming is highly mechanized and cities are extremely large. However, we are on a treadmill. As the human population increases, we need new innovations to produce greater amounts of food. As soon as food production increases, populations grow once again, and the demand for food becomes still greater. Will there be a point when the population is greater than the food capacity? Is that time already upon us?

Industrial Revolution Began

The industrial revolution began in England during the eighteenth century and with it a demand for energy in the form of coal and oil that today seems unlimited. Our ability to construct any number of tools, including high-tech computers, is not stored in our genes. We learn it from the previous generation. Our modern civilization that began due to the advent of agriculture is now altering the global environment in a way that affects the evolution of other species. Species are becoming extinct, unless they are able to adapt to the presence of our civilization. It could be that biocultural evolution will be so harmful to the biosphere that the human species will eventually be driven to extinction also.

Decide Your Opinion

1. Can technology be used to help us not pollute the environment? How?
2. Should the extinction of other species be prevented? How can it be prevented?
3. Should the human population be reduced in size? How could this occur?

Figure 22B   **When did primitive agriculture begin?**
A primitive form of agriculture began in several locations on Earth about 10,000 years ago.

peoples), or the shape of lips, cannot be explained as adaptations to the environment. Perhaps these features became fixed in different populations due to genetic drift. As far as intelligence is concerned, no significant disparities have been found among different ethnic groups.

Genetic Evidence for a Common Ancestry

The two hypotheses regarding the evolution of humans, discussed on page 533, pertain to the origin of ethnic groups. The multiregional hypothesis suggests that different human populations came into existence as long as a million years ago, giving time for significant ethnic differences to accumulate despite some gene flow. The out-of-Africa hypothesis, on the other hand, proposes that all modern humans have a relatively recent common ancestor who evolved in Africa and then spread into other regions. Paleontologists tell us that the variation among modern populations is considerably less than among human populations some 250,000 years ago. This would mean that all ethnic groups evolved from the same single, ancestral population.

A comparative study of mitochondrial DNA shows that the differences among human populations are consistent with their having a common ancestor no more than a million years ago. Lewontin, mentioned previously, found that the genotypes of different modern populations are extremely similar. He examined variations in 17 genes, including blood groups and various enzymes, among seven major geographic groups: Caucasians, black Africans, mongoloids, south Asian Aborigines, Amerinds, Oceanians, and Australian Aborigines. He found that the great majority of genetic variation—85%—occurs within ethnic groups, not among them. In other words, the amount of genetic variation between individuals of the same ethnic group is greater than the variation between ethnic groups.

> **Check Your Progress 22.6**
> 1. How might *Homo habilis,* living about 2 MYA, have differed from the australopithecines?
> 2. How might *Homo erectus,* living about 1.9–1.6 MYA, have differed from *Homo habilis*?
> 3. a. What is the out-of-Africa hypothesis, and (b) what bearing does it have on the evolution of humans?
> 4. How do Cro-Magnons, the first of the modern humans, differ from the other species in the genus *Homo*?

Summarizing the Concepts

22.1 Origin of Life

A chemical evolution could have produced the protocell.

- Using an outside energy source, small organic molecules were produced by reactions between early Earth's atmospheric gases.
- Macromolecules evolved and interacted.
- The RNA-first hypothesis—only macromolecule RNA was needed for the first cell(s).
- The protein-first hypothesis—amino acids join to form polypeptides when exposed to dry heat.
- The protocell, a heterotrophic fermenter, lived on preformed organic molecules in the ocean.

The protocell eventually became a true cell once it had genes composed of DNA and could reproduce.

22.2 Biological Evolution

Biological evolution explains both the unity and diversity of life.

- Descent from a common ancestor explains the unity (sameness) of living things.
- Adaptation to different environments explains the great diversity of living things.
- **Fossil evidence supports evolution** The fossil record gives us the history of life in general and allows us to trace the descent of a particular group.

Darwin discovered much evidence for common descent.

- **Biogeographical evidence** The distribution of organisms on Earth is explainable by assuming that organisms evolved in one locale.
- **Anatomical evidence** The common anatomies and development of a group of organisms are explainable by descent from a common ancestor.
- **Biochemical evidence** All organisms have similar biochemical molecules.

Darwin developed a mechanism for adaptation known as natural selection:

Observation	Result	Conclusion
1 a. Organisms have variations.	b. New adaptations to the environment arise.	Organisms become more adapted with each generation.
2 a. Organisms struggle to exist.	b. More organisms are present than can survive.	
3 a. Organisms differ in fitness.	b. Organisms best suited to the environment survive and reproduce.	

- The result of natural selection is a population adapted to its local environment.

22.3 Classification of Humans

The classification of humans can be used to trace their ancestry.

- Humans are primates.
- A primate evolutionary tree shows that humans share a common ancestor with African apes.

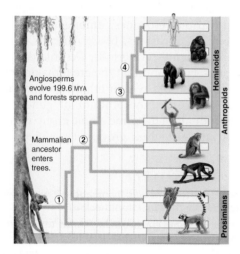

Angiosperms evolve 199.6 MYA and forests spread.

Mammalian ancestor enters trees.

22.4 Evolution of Hominids
- The first hominid (includes humans) most likely lived about 6–7 MYA.
- Certain features (bipedal posture, flat face, and brain) identify fossil hominids.
- Ardipithecines were most likely hominids.

Evolution of Australopithecines
The evolutionary tree of hominids resembles a bush (not a straight line of fossils leading to modern humans).
- Australopithecines (a hominid) lived about 3 MYA.
- They could walk erect, but they had a small brain.
- This testifies to a mosaic evolution for humans (not all advanced features evolved at the same time).

22.5 Evolution of Humans
Fossils are classified as *Homo* with regard to brain size (over 600 cm³), jaws and teeth (resemble modern humans), and evidence of tool use.
- *H. habilis* made and used tools.
- *H. erectus* was the first *Homo* to have a brain size of more than 1,000 cm³.
- *H. erectus* migrated from Africa into Europe and Asia.
- *H. erectus* used fire and may have been big-game hunters.

Evolution of Modern Humans
Two hypotheses of modern human evolution are being tested.
- The multiregional continuity hypothesis suggests that modern humans evolved separately in Europe, Africa, and Asia.
- The out-of-Africa hypothesis says that *H. sapiens* evolved in Africa but then migrated to Asia and Europe.

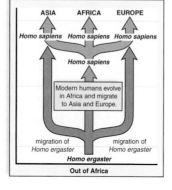

Neandertals and Cro-Magnons
- The Neandertals were already living in Europe and Asia before modern humans arrived.
- They had a culture, but did not have the physical traits of modern humans.
- Cro-Magnons are the oldest fossil to be designated *H. sapiens*. Their tools were sophisticated, and they had a culture.

Understanding Key Terms

adaptation 518
analogous structure 521
anthropoid 525
australopithecine 528
binomial name 524
biocultural evolution 536
biogeography 520
biological evolution 518
bipedal posture 528
chemical evolution 516
Cro-Magnon 534
culture 530
eukaryotic cell 518
evolutionary tree 526
fossil 518
fossil record 519
heterotroph 517
hominid 528
Homo erectus 531

Homo habilis 530
homologous structure 521
Homo sapiens 533
lineage 528
molecular clock 528
mosaic evolution 529
multiregional continuity hypothesis 533
natural selection 522
Neandertal 534
out-of-Africa hypothesis 533
primate 525
prokaryotic cell 518
prosimian 525
protein-first hypothesis 517
protocell 517
RNA-first hypothesis 517
vestigial structure 521

Match the key terms to these definitions.

a. _____ Process by which populations become adapted to their environment.

b. _____ Organism's modification in structure, function, or behavior suitable to the environment.

c. _____ Structure that is similar in two or more species because of common ancestry.

d. _____ Increase in the complexity of chemicals over time that could have led to the first cells.

e. _____ Any remains of an organism that have been preserved in the Earth's crust.

Testing Your Knowledge of the Concepts

1. List and discuss the steps by which a chemical evolution could have produced a protocell. (pages 516–17)
2. You are studying finches on the Galápagos Islands. What evidence do you need to show common descent and adaptation to the environment? (pages 518–23)
3. Describe the evidences that support descent from a common ancestor. (pages 518–22)
4. Describe the critical elements of Darwin's natural selection process. (pages 522–23)
5. List a characteristic of each category of human classification. (page 524)

6. What type of life are primates adapted to? Discuss several primate characteristics that aid them in their adaptation. (page 525)

7. How is the human skeleton like that of a chimpanzee skeleton? What's the major difference? (page 526)

8. Why do most investigators use bipedal locomotion, and not size of brain, to designate a hominid? (page 528)

9. Describe the characteristics of australopithecines, early *Homo*, *Homo erectus*, Neandertals, and Cro-Magnons. (pages 528–34)

10. Contrast the multiregional continuity hypothesis with the out-of-Africa hypothesis for the evolution of *Homo sapiens*. (page 533)

11. Which of these did Stanley Miller place in his experimental system to show that organic molecules could have arisen from inorganic molecules on the early Earth?
    a. microspheres
    b. purines and pyrimidines
    c. gases in the atmosphere of early Earth
    d. only RNA
    e. All of these are correct.

12. Which of these is the chief reason the protocell was probably a fermenter?
    a. The protocell didn't have any enzymes.
    b. The atmosphere didn't have any oxygen.
    c. Fermentation provides the most energy.
    d. There was no ATP yet.
    e. All of these are correct.

13. Evolution of the DNA → RNA → protein system was a milestone because the protocell could now
    a. be a heterotrophic fermenter.
    b. pass on genetic information.
    c. use energy to grow.
    d. take in preformed molecules.
    e. All of these are correct.

14. According to Darwin,
    a. the adapted individual is the one who survives and passes on its genes to offspring.
    b. changes in phenotype are passed on by way of the genotype to the next generation.
    c. organisms are able to bring about a change in their phenotype.
    d. evolution is striving toward particular traits.
    e. All of these are correct.

15. Organisms
    a. compete with other members of their species.
    b. vary in physical characteristics.
    c. are adapted to their environment.
    d. are related by descent from common ancestors.
    e. All of these are correct.

16. If evolution occurs, we would expect different biogeographical regions with similar environments to
    a. contain the same mix of plants and animals.
    b. each have its own specific mix of plants and animals.
    c. have plants and animals with similar adaptations.
    d. have plants and animals with different adaptations.
    e. Both b and c are correct.

17. The fossil record offers direct evidence for evolution because you can
    a. see that the types of fossils change over time.
    b. sometimes find common ancestors.
    c. trace the ancestry of a particular group.
    d. trace the biological history of living things.
    e. All of these are correct.

18. Organisms such as whales and sea turtles adapted to an aquatic way of life
    a. will probably have homologous structures.
    b. will have similar adaptations but not necessarily homologous structures.
    c. may very well have analogous structures.
    d. will have the same degree of fitness.
    e. Both b and c are correct.

19. Which of these gives the correct order of divergence from the main line of descent leading to humans?
    a. prosimians, monkeys, Asian apes, African apes, humans
    b. gibbons, baboons, prosimians, monkeys, African apes, humans
    c. monkeys, gibbons, prosimians, African apes, baboons, humans
    d. African apes, gibbons, monkeys, baboons, prosimians, humans
    e. *H. habilis*, *H. erectus*, *H. neanderthalensis*, Cro-Magnon

20. Lucy is a member of what species?
    a. *Homo erectus*
    b. *Australopithecus afarensis*
    c. *H. habilis*
    d. *A. robustus*
    e. *A. anamensis* and *A. afarensis* are alternative forms of Lucy.

21. What possibly may have influenced the evolution of bipedalism?
    a. A larger brain developed.
    b. Food gathering was easier.
    c. The climate became colder.
    d. Both b and c are correct.
    e. Both a and c are correct.

22. *H. ergaster* could have been the first to
    a. use and control fire.
    b. migrate out of Africa.
    c. make tools.
    d. Both a and b are correct.
    e. a, b, and c are correct.

23. Which of these characteristics is not consistent with the others?
    a. opposable thumb   d. well-developed brain
    b. learned behavior   e. stereoscopic vision
    c. multiple births

24. The last common ancestor for African apes and hominids
    a. has been found, and it resembles a gibbon.
    b. has not yet been identified, but it is expected to be dated from about 6–7 MYA.
    c. has been found, and it has been dated at 30 MYA.
    d. is not expected to be found because there was no such common ancestor.
    e. most likely lived in Asia, not Africa.

25. If the multiregional continuity hypothesis is correct, then
    a. hominid fossils in China after 100,000 BP are not expected to resemble earlier fossils.
    b. hominid fossils in China after 100,000 BP are expected to resemble earlier fossils.
    c. the mitochondrial Eve study must be invalid.
    d. Both a and c are correct.
    e. Both b and c are correct.

26. A primate evolutionary tree
    a. exists only for humans.
    b. shows the evolutionary relationship among the different types of primates.
    c. should not include extinct forms.
    d. indicates that the ape lineage and human lineage are still joined.
    e. All of these are correct.

27. Classify humans by filling in the missing lines.

    Domain   Eukarya
    Kingdom   Animalia
    Phylum   a. _____
    b. _____   Mammalia
    c. _____   Primates
    Family   d. _____
    e. _____   *Homo*
    Species   f. _____

In questions 28–31, match each description to a type of evolutionary evidence in the key.

Key:
   a. biogeography
   b. fossil record
   c. comparative biochemistry
   d. comparative anatomy

28. Species change over time.

29. Forms of life are variously distributed.

30. A group of related species have homologous structures.

31. The same types of molecules are found in all living things.

Thinking Critically About the Concepts

Much of the work done in archaeology is very tedious, as Andrew and Jamie discovered. For example, the Science Focus on page 532 describes the discovery of *H. floresiensis* in a limestone cave on the island of Flores. Researchers Michael Morwood, of the University of New England in Australia, and Radien Soejono, of the Indonesian Center for Archaeology in Jakarta, were in charge of the dig. They began the work in July 2001. It was not until the end of three seasons of field work that they found evidence of hominids—a single tooth. Once the first female skeleton had been found, colleague Peter Brown spent three months analyzing it.

1. In archaeology, why is it critically important to know where to dig? What clues would you look for to determine where to dig?

2. Andrew and Jamie did a systematic shovel survey. What is the purpose of this technique?

3. What artifacts do archaeologists use to determine a population's bioculture?

4. What types of information can scientists derive about an ancient people, using only a single human skull from that group?

Global Ecology and Human Interferences

CASE STUDY ALEX AND ARTHUR KEELING

"There's a good chance of some rain later in the week. Stayed tuned for traffic news."

Alex reached over to turn off the radio that signaled the start of his day. His dad would be hollering if Alex wasn't at the barn pronto to help milk the cows. He dressed quickly and made a quick stop by the coffee pot. He called, "Morning, Mom! Thanks for making coffee," before pulling on his boots and coveralls and heading out the door.

He paused to let his eyes adjust to the darkness before walking toward the glow coming from the barn. All was quiet except for the sound of the cows coming into the barn and the birds calling to one another. Alex loved everything about the dairy farm where he'd spent his entire life. It had been established by his great-grandfather Keeling in the late 1800s. Someday he would take over running the operation when his dad retired.

"Morning, Alex," his dad spoke sternly when Alex arrived at the barn. "You're late. The cows don't appreciate when you're late."

"Sorry," Alex mumbled. "I spent too long enjoying the morning on my way here."

"I used to tell your grandfather the same thing," Alex's dad replied with a fleeting smile. "Now stop making excuses and let's get these cows milked. I've got to make myself presentable before the awards ceremony at the university later today."

Alex joked with his dad about his fear of public speaking. "Do you want me to help you write an acceptance speech?" he teased. "I just finished my college speech class."

In reality, Alex was proud of his dad for being the recipient of the annual Governor's Environmental Stewardship Award. They'd made so many improvements to the farm's water quality over the past few years. It was nice to be recognized for all their efforts.

"Stop jabbering or we'll never get these cows milked," his dad retorted. "They might decide to give the award to another farmer who has better time management skills!

"You can give me all kinds of advice after you finish that Conservation Farming Practices course next semester," his dad grinned. "Until then, all I want to hear is the sound of those milking machines getting our girls milked."

CHAPTER CONCEPTS

23.1 The Nature of Ecosystems
The biosphere encompasses that part of Earth where living things live in ecosystems. Populations interact among themselves and with the physical environment in ecosystems. Ecosystems are characterized by energy flow and chemical cycling.

23.2 Energy Flow
Ecosystems contain food webs, in which the various populations are connected by predator and prey. Energy and chemicals are passed from one population to the next, as one population feeds on another. Food chains have a limited length. As demonstrated by food pyramids, only about 10% of energy is passed from one feeding level to the other. Eventually all the energy dissipates but the chemicals cycle back to the photosynthesizers.

23.3 Global Biogeochemical Cycles
Biogeochemical cycles contain reservoirs, which retain nutrients; exchange pools, where nutrients are readily available; and the biotic community, which passes nutrients from one population to the next. The water, carbon, and nitrogen cycles are gaseous because the exchange pool is the atmosphere. The phosphorus cycle is a sedimentary cycle. Particular ecological problems arise because human activities alter the normal transfer rates within each cycle.

Figure 23.1 **What are the temperature and rainfall characteristics of the major terrestrial ecosystems?**
Temperate forests have moderate temperatures and occur where rainfall is moderate, yet sufficient to support trees. Deserts have changeable temperatures with minimal rainfall. Tropical rain forests, which generally occur near the equator, have a high average temperature and the greatest amount of rainfall of all the terrestrial ecosystems. A tropical grassland (savanna) has high temperatures and moderate/seasonal rainfall. A temperate grassland (prairie) has low to high temperatures, with low annual rainfall. The taiga, a coniferous forest that encircles the globe, has a low average temperature, but moderate rainfall. The tundra is the northernmost terrestrial ecosystem and has the lowest average temperature of all the terrestrial ecosystems, with minimal to moderate rainfall.

23.1 The Nature of Ecosystems

The **biosphere** is where organisms are found on planet Earth, from the atmosphere above to the depths of the oceans below and everything in between. Specific areas of the biosphere where organisms interact among themselves and the physical and chemical environment are called **ecosystems.** The interactions that occur in an ecosystem maintain balance in that specific area, which in turn affects the balance of the biosphere. Human activities can alter the interactions between organisms and their environments in ways that reduce the abundance and diversity of life in an ecosystem. It is important to understand how ecosystems function so that we can repair past damage and predict how human activities might change normal conditions.

Ecosystems

Scientists recognize several distinctive major types of terrestrial ecosystems, also called biomes (Fig. 23.1). Temperature and rainfall define the biomes. A variety of organisms adapted to the regional climate characterize each biome. The tropical rain forest, which occurs at the equator, is dominated by large evergreen, broad-leaved trees. The savanna is a tropical grassland that supports many types of grazing animals. Temperate grasslands receive less rainfall than temperate forests (where many trees lose their leaves during the winter). Deserts receive scant rainfall, and therefore lack trees. The taiga is a very cold northern forest of conifers such as pine, spruce, hemlock, and fir. Bordering the North Pole is the frigid tundra, which has long winters and a short growing season. A permafrost persists even during the summer in the tundra and prevents large plants from becoming established.

Aquatic ecosystems are divided into those composed of freshwater and those composed of salt water (marine ecosystems). The ocean is a marine ecosystem that covers 70% of the Earth's surface. Two types of freshwater ecosystems are those with standing water, such as lakes and ponds, and those with running water, such as rivers and streams. The richest marine ecosystems lie near the coasts. Coral reefs are located offshore, while salt marshes occur where rivers meet the sea (Fig. 23.2).

a.

b.

c.

d.

Figure 23.2 **What are examples of freshwater and saltwater ecosystems?**
Aquatic ecosystems are divided into those that have salt water, such as the ocean **(a)** and those that have freshwater, such as a river **(b).** Saltwater, or marine, ecosystems also include coral reefs **(c)** and salt marshes **(d).**

Biotic Components of an Ecosystem

The abiotic components of an ecosystem are the nonliving components, such as soil type, water, and weather. The biotic components are living things that can be categorized according to their food source (Fig. 23.3). Some species are autotrophs, and some are heterotrophs.

Autotrophs

Autotrophs require only inorganic nutrients and an outside energy source to produce organic nutrients for their own use and for all the other members of a community. Therefore, they are called **producers,** meaning they produce food (Fig. 23.3*a*). Photosynthetic organisms produce most of the organic nutrients for the biosphere. Algae of all types possess chlorophyll and carry on photosynthesis in freshwater and marine habitats. Green plants are the dominant photosynthesizers on land.

Heterotrophs

Heterotrophs need a source of organic nutrients. They are **consumers,** meaning they consume food. **Herbivores** are organisms that graze directly on plants or algae (Fig. 23.3*b*). You're probably familiar with deer, rabbits, and cows that are herbivores in terrestrial habitats. Some insects are herbivores, too. In aquatic ecosystems, some protists are herbivores. **Carnivores** feed on other animals. Spiders that feed on insects are carnivores, as are osprey that feed on fish (Fig. 23.3*c*). This example allows us to mention that there are primary consumers (e.g., plant-eating insects), secondary consumers (e.g., insect-eating spiders), and tertiary consumers (e.g., osprey, hawks). Sometimes tertiary consumers are called top predators. **Omnivores** are animals that feed on plants and animals. Humans are omnivores.

　　Detritus feeders are organisms that feed on detritus, decomposing particles of organic matter. Marine fan worms take

Have You Ever Wondered...

How do producers at the bottom of the ocean floor make food without sunlight?

Producers at the bottom of the ocean floor use chemical energy instead of sunlight to make food. They are called *chemoautotrophs*. Volcanoes at the bottom of the ocean floor release hydrogen sulfide gas through cracks called hydrothermal vents. (Hydrogen sulfide is the nasty smelling gas we associate with rotten eggs.) Some chemoautotrophs split hydrogen sulfide to obtain the energy needed to link carbon atoms together to form glucose. The glucose contained in these chemoautotrophs sustains a variety of bizarre organisms such as giant tube worms, anglerfish, and giant clams.

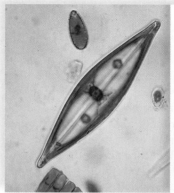

a. Producers

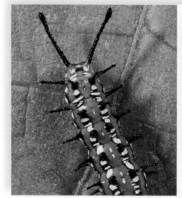

b. Herbivores

c. Carnivores

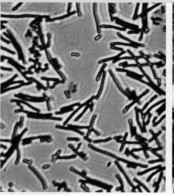

d. Decomposers

Figure 23.3 **What are the biotic components of an ecosystem?** **a.** Diatoms and green plants are autotrophs. **b.** Caterpillars and rabbits are herbivores. **c.** Spiders and osprey are carnivores. **d.** Some bacteria and some mushrooms are decomposers.

detritus from the water, while clams take it from the substratum. Earthworms and some beetles, termites, and ants are terrestrial detritus feeders. Bacteria and fungi, including mushrooms, are decomposers. They acquire nutrients by breaking down dead organic matter, including animal wastes. Decomposers perform a valuable service because they release inorganic substances that are taken up by plants once more (Fig. 23.3*d*). Otherwise, plants would be completely dependent only on physical processes, such as the release of minerals from rocks, to supply them with inorganic nutrients.

Niche

A **niche** is the role of an organism in an ecosystem. Descriptions of a niche include how an organism gets its food and what eats it, and how it interacts with other populations in the same community. Human beings are also part of the cycle of life. The chemicals making up our bodies must also return to nature.

Energy Flow and Chemical Cycling

When we diagram the interactions of all the populations in an ecosystem, it is possible to illustrate two phenomena that characterize every ecosystem. One of these phenomena is energy flow, which begins when producers absorb solar (and in some cases, chemical) energy. The second, nutrient cycling,

occurs when producers take in inorganic chemicals from the physical environment. Thereafter, producers make organic nutrients (food) directly for themselves and indirectly for the other populations of the ecosystem. Energy flow occurs because as nutrients pass from one population to another, all the energy content is eventually converted to heat. The heat then dissipates in the environment. Therefore, most ecosystems cannot exist without a continual supply of solar energy. Chemicals cycle when inorganic nutrients are returned to the producers from the atmosphere or soil (Fig. 23.4).

Only a portion of the organic nutrients made by autotrophs is passed on to heterotrophs because plants use organic molecules to fuel their cellular respiration (see Fig. 3.18 on page 57). Similarly, only a small percentage of nutrients taken in by heterotrophs is available to higher-level consumers. Figure 23.5 shows why. Some of the food eaten by a herbivore is never digested and is eliminated as feces. Metabolic wastes are excreted in urine. Of the assimilated energy, a large portion is used during cellular respiration and thereafter becomes heat. Only the remaining food energy, converted into increased body weight (or additional offspring), becomes available to carnivores. Plants carry on cellular respiration too, so only about 55% of the original energy absorbed by plants is available to an ecosystem. Further, as organisms feed on one another, less and less of this 55% is available in a usable form.

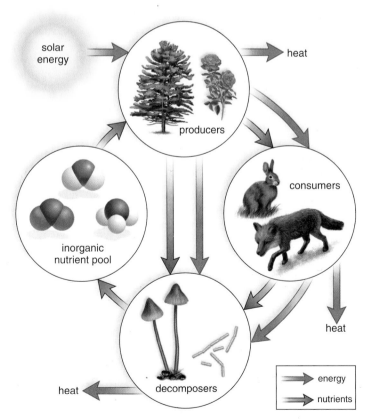

Figure 23.4 **Why does energy flow through an ecosystem?**
Chemicals cycle, but energy flows through an ecosystem. As energy transformations repeatedly occur, all the energy derived from the sun eventually dissipates as heat.

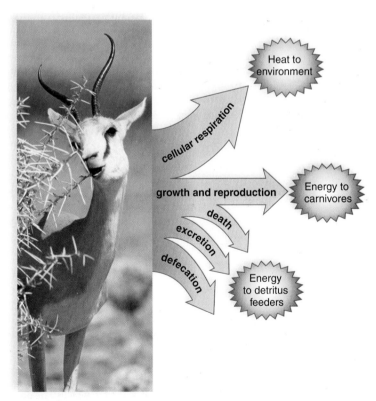

Figure 23.5 **What happens to the food energy taken in by an herbivore?**
Only about 10% of the food energy taken in by an herbivore is passed on to carnivores. A large portion goes to detritus feeders via defecation, excretion, and death, and another large portion is used for cellular respiration.

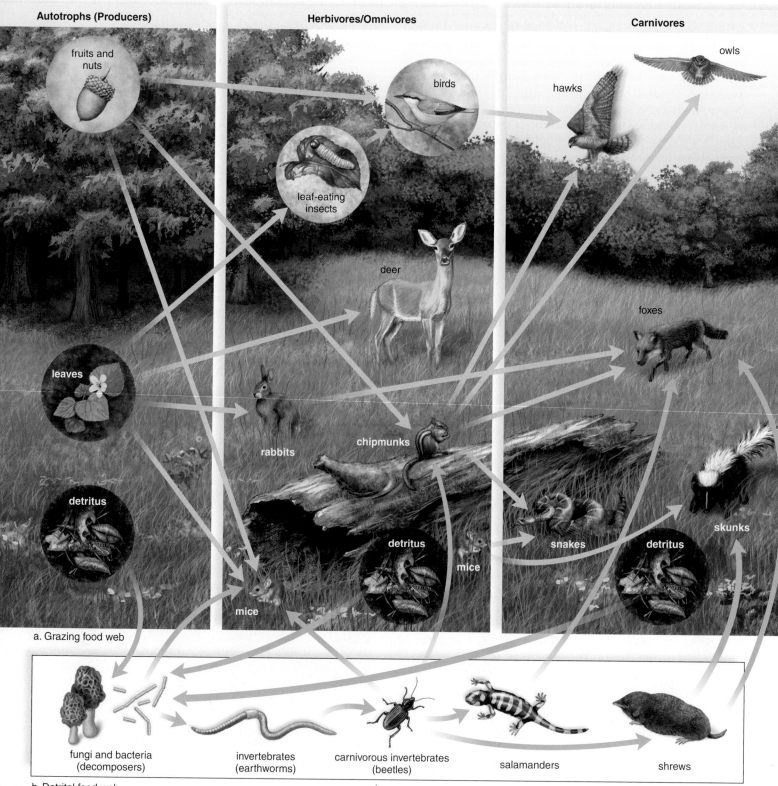

Autotrophs (Producers)

fruits and nuts

leaves

detritus

a. Grazing food web

Herbivores/Omnivores

birds

leaf-eating insects

deer

rabbits

chipmunks

detritus

mice

Carnivores

owls

hawks

foxes

snakes

detritus

mice

skunks

b. Detrital food web

fungi and bacteria (decomposers)

invertebrates (earthworms)

carnivorous invertebrates (beetles)

salamanders

shrews

Figure 23.6 What is a food web?

Food webs are descriptions of who eats whom. **a.** Tan arrows illustrate possible grazing food webs. The tree and other organisms that convert sun energy into food energy are producers (first trophic level). Mice and other animals that eat the producers are primary consumers (second trophic level). Carnivores that rely on primary consumer animals for energy (the hawk, fox, skunk, snake, and owl) are secondary consumers (third trophic level). **b.** Green arrows illustrate possible detrital food webs. These begin with detritus: organic waste and remains of dead organisms. Decomposers and detritus feeders recycle these organic nutrients. The organisms in the detrital food web may be prey for animals in the grazing food web, as when chipmunks feed on bugs. Thus, the grazing food web and detrital food web are interconnected.

The elimination of wastes and the deaths of all organisms does not mean that substances are lost to an ecosystem. Instead they are nutrients made available to decomposers. Decomposers convert organic nutrients, such as glucose, back into inorganic chemicals, such as carbon dioxide and water. The inorganic chemicals are then released to the soil or atmosphere. Chemicals complete their cycle within an ecosystem when inorganic chemicals are absorbed by the producers from the atmosphere or soil.

> ### Check Your Progress 23.1
> 1. Why could it be said that the biosphere is a giant ecosystem?
> 2. a. Why are autotrophs called producers and (b) heterotrophs called consumers?
> 3. What are the different types of consumers in an ecosystem?
> 4. What two processes characterize an ecosystem? Explain.

23.2 Energy Flow

The principles we have been discussing can now be applied using a forest ecosystem. The various interconnecting paths of energy flow are represented by a **food web.** This diagram describes **trophic (feeding) relationships.** Figure 23.6a is a **grazing food web** because it begins with an oak tree and grass. Caterpillars feed on oak leaves, while mice, rabbits, and deer feed on leaves and grass at or near the ground. Birds, chipmunks, and mice feed on seeds and nuts, but they are omnivores because they also feed on caterpillars. These herbivores and omnivores are then food for a number of different carnivores.

Figure 23.6b is a **detrital food web.** This type of food web begins with wastes and the remains of dead organisms. Detritus is food for decomposers and soil organisms like earthworms. Earthworms may be food for carnivorous invertebrates. In turn, salamanders and shrews may consume the carnivorous insects. The members of detrital food webs may become food for above ground carnivores, so the detrital and grazing food webs are connected.

We naturally tend to think that aboveground plants, such as trees, are the largest storage form of organic matter and energy. This is not necessarily the case. In this particular forest, the organic matter lying on the forest floor and mixed into the soil contains over twice as much energy as the leaves of living trees. Therefore, more energy in a forest may be funneling through the detrital food web than through the grazing food web.

Trophic Levels

The arrangement of the species in Figure 23.6 suggests that organisms are linked to one another in a straight line, according to feeding or predator-prey relationships. Diagrams that show a single path of energy flow are called **food chains.** For example, in the grazing food web, we could find this **grazing food chain:**

leaves → caterpillars → birds → hawks

And in the detrital food web (Fig. 23.6b), we could find this **detrital food chain:**

detritus → earthworms → beetles → shrews

A **trophic level** is composed of all the organisms that feed at a particular link in a food chain. In the grazing food web in Figure 23.6a, going from left to right, the trees are producers (first trophic level). The first series of animals are primary consumers (second trophic level). The next group of animals are secondary consumers (third trophic level).

Ecological Pyramids

The shortness of food chains can be attributed to the loss of energy between trophic levels. As mentioned, only about 10% of the energy of one trophic level is available to the next trophic level. Therefore, if an herbivore population consumes 1,000 kg of plant material, only about 100 kg is converted to herbivore tissue, 10 kg to first-level carnivores, and 1 kg to second-level carnivores. The so-called 10% rule of thumb explains why so few carnivores can be supported in a food web. The flow of energy with large losses between successive trophic levels is sometimes depicted as an **ecological pyramid** (Fig. 23.7).

Figure 23.7 **Why is there a sharp drop in biomass from the producer level to the herbivore level?**
The biomass, or dry weight (g/m²), for trophic levels in a grazing food web in a bog at Silver Springs, Florida. There is a sharp drop in biomass between the producer level and herbivore level. This is consistent with the common knowledge that the detrital food web plays a significant role in bogs.

top carnivores
1.5 g/m²

carnivores
11 g/m²

herbivores
37 g/m²

producers
809 g/m²

Energy losses between trophic levels also result in pyramids based on the number of organisms in each trophic level. When constructing a pyramid based on number of organisms, problems arise, however. For example, in Figure 23.6a, each tree would contain numerous caterpillars. Therefore, there would be more herbivores than autotrophs. The explanation has to do with size. An autotroph can be as tiny as a microscopic alga or as big as a beech tree. Similarly, an herbivore can be as small as a caterpillar or as large as an elephant.

Pyramids of biomass eliminate size as a factor because **biomass** is the number of organisms multiplied by the weight of organic matter contained in one organism. You would certainly expect the biomass of the producers to be greater than the biomass of the herbivores and that of the herbivores to be greater than that of the carnivores. In aquatic ecosystems, such as lakes and open seas, the herbivores may have a greater biomass than the producers. This is because algae are the only producers. Over time, algae reproduce rapidly, but are also consumed at a high rate.

These types of problems are making some ecologists hesitant about using pyramids to describe ecological relationships.

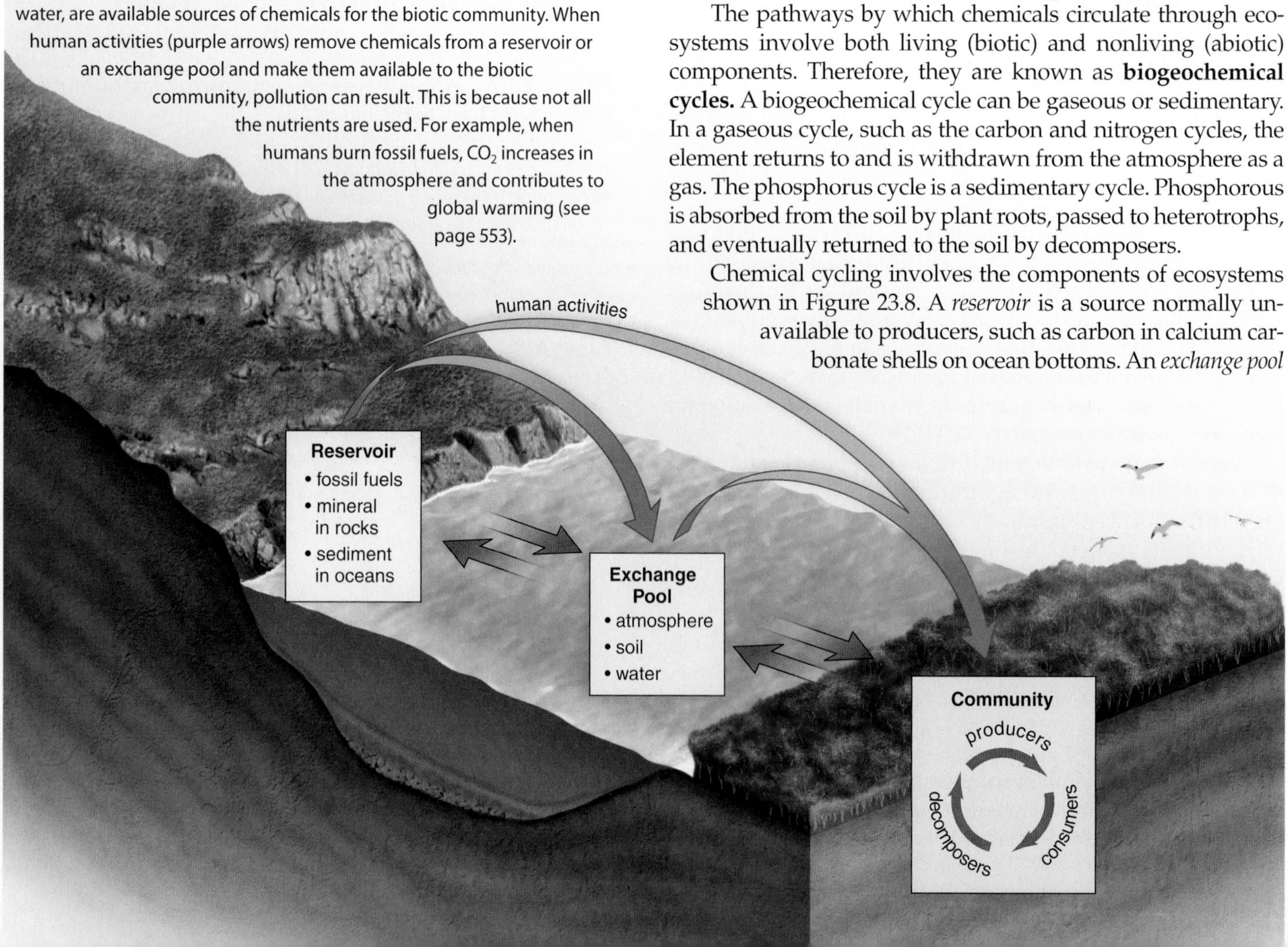

Figure 23.8   Where are nutrients found that are readily available for the biotic community?

Reservoirs, such as fossil fuels, minerals in rocks, and sediments in oceans, are normally relatively unavailable sources of nutrients for the biotic community. Nutrients in exchange pools, such the atmosphere, soil, and water, are available sources of chemicals for the biotic community. When human activities (purple arrows) remove chemicals from a reservoir or an exchange pool and make them available to the biotic community, pollution can result. This is because not all the nutrients are used. For example, when humans burn fossil fuels, CO_2 increases in the atmosphere and contributes to global warming (see page 553).

human activities

Reservoir
- fossil fuels
- mineral in rocks
- sediment in oceans

Exchange Pool
- atmosphere
- soil
- water

Community
producers
consumers
decomposers

Another issue concerns the role played by decomposers. These organisms are rarely included in pyramids, even though a large portion of energy becomes detritus in many ecosystems.

> **Check Your Progress 23.2**
> 1. What type of diagram represents the various paths of energy flow in an ecosystem?
> 2. What is the difference between a grazing food web and a detrital food web?
> 3. What type of diagram represents a single path of energy flow in an ecosystem?
> 4. What type of diagram illustrates that usable energy is lost in ecosystems?

23.3 Global Biogeochemical Cycles

In this section, we will examine in more detail how chemicals cycle through ecosystems. All organisms require a variety of organic and/or inorganic nutrients. For example, carbon dioxide and water are necessary nutrients for photosynthesizers. Nitrogen is a component of all the structural and functional proteins and nucleic acids that sustain living tissues. Phosphorus is essential for ATP and nucleotide production.

The pathways by which chemicals circulate through ecosystems involve both living (biotic) and nonliving (abiotic) components. Therefore, they are known as **biogeochemical cycles.** A biogeochemical cycle can be gaseous or sedimentary. In a gaseous cycle, such as the carbon and nitrogen cycles, the element returns to and is withdrawn from the atmosphere as a gas. The phosphorus cycle is a sedimentary cycle. Phosphorous is absorbed from the soil by plant roots, passed to heterotrophs, and eventually returned to the soil by decomposers.

Chemical cycling involves the components of ecosystems shown in Figure 23.8. A *reservoir* is a source normally unavailable to producers, such as carbon in calcium carbonate shells on ocean bottoms. An *exchange pool* is a source from which organisms do generally take chemicals, such as the atmosphere or soil. Chemicals move along food chains in a *biotic community.*

Human activities (purple arrows) remove chemicals from reservoirs and exchange pools and make them available to the biotic community. In this way, human activities result in pollution because it upsets the normal balance of nutrients for producers in the environment.

The Water Cycle

The **water (hydrologic) cycle** is described in Figure 23.9. The width of the arrows in this and the other cycles that will be examined indicate the transfer rate of water between components of an ecosystem

① During **evaporation** in the water cycle, the sun's rays cause freshwater to evaporate from seawater, leaving the salts behind. Net condensation occurs. During condensation, a gas is changed into a liquid. ② Vaporized freshwater rises into the atmosphere, condenses, and then falls as **precipitation** (e.g., rain, snow, sleet, hail, and fog) over the oceans and the land. ③ Water also evaporates from land and from plants (evaporation from plants is called transpiration). ④ Land lies above sea level, so gravity eventually returns all freshwater to the sea. In the meantime, much water is contained within standing waters (lakes and ponds), flowing water (streams and rivers), and groundwater. ⑤ **Runoff** is water

that flows directly into nearby streams, lakes, wetlands, or the ocean.

Instead of running off, some precipitation sinks, or percolates, into the ground. This saturates the earth to a certain level. The top of the saturation zone is called the groundwater table, or the water table. ⑥ Sometimes, groundwater is also located in **aquifers,** rock layers that contain water and release it in appreciable quantities to wells or springs. Aquifers are recharged when rainfall and melted snow percolate into the soil.

Human Activities

Humans interfere with the water cycle in three ways. First, they withdraw water from aquifers. Second, they clear vegetation from land and build roads and buildings that

Figure 23.9 What happens to water as it moves through the hydrologic (water) cycle?
Evaporation from the ocean exceeds precipitation, so there is a net movement of water vapor onto land. There precipitation results in surface water and groundwater that flow back to the sea. On land, transpiration by plants contributes to evaporation.

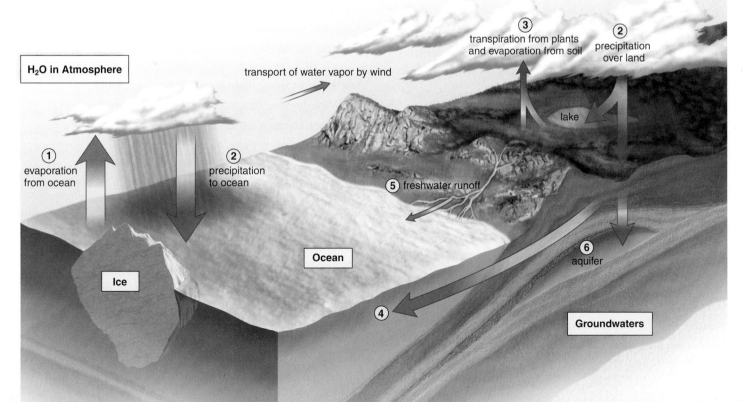

prevent percolation and increase runoff. Third, they interfere with the natural processes that purify water and instead add pollutants like sewage and chemicals to water.

In some parts of the United States, especially the arid West and southern Florida, withdrawals from aquifers exceed any possibility of recharge. This is called "groundwater mining." In these locations, the groundwater is dropping. Residents may run out of groundwater, at least for irrigation purposes, within a few short years. Freshwater, which makes up only about 3% of the world's supply of water, is called a renewable resource because a new supply is always being produced. However, it is possible to run out of freshwater when the available supply runs off instead of entering bodies of freshwater and aquifers. Freshwater may also become so polluted that it is not usable.

CASE STUDY THE AWARDS CEREMONY

Alex grinned as he read about his dad's accomplishments in the awards program later that day.

> "Arthur Keeling is a third-generation dairy farmer from McRay County. His commitment to improving the water quality of Third Creek has turned the formerly muddy and murky creek into a glistening watershed. With the help of government cost-share programs, Mr. Keeling built fencing to prevent the farm's dairy cattle from entering the creek. In addition, he added several thousand feet of piping to create a permanent water supply from a natural spring for his animals."

Alex thought back to the summer when installing that fence and pipe was their shared project. Hot and miserable, he'd whined constantly about being his dad's unpaid labor. Now he was able to see the project's value to the surrounding environment.

> "His commitment to animal health has served as an example for all in the farming community."

Keeping those cows out of the creek turned out to be both humane and profitable. The cattle no longer had to walk 2 miles from the barn to get water from the creek. They didn't stand around in the mud in the creek as they had before. As a result, hoof diseases—a problem the Keelings had always dealt with in the past—almost disappeared from the herd. Milk production improved significantly once the cows' feet no longer bothered them. Maybe Alex had imagined it, but "the girls" even *looked* happier.

> "As a result of his efforts, fouling of the creek by sediment and livestock waste has been halted. Yearly cleanup efforts begun by Mr. Keeling have organized neighbors along the creek's entire length. Thousands of pounds of trash have been removed from the creek and its surroundings. Natural wetlands have been restored on that portion of the creek bordered by Keeling Farm. Aquatic life has re-established and several species of water birds have been observed in the wetland area. Third Creek has now been established as a state scenic waterway."

Now you could walk the entire distance of the creek, as Alex often did. From the Wagner farm on the north to the Settledge farm on the south, the creek was clear. In the deepest spots, there were even some small-mouth bass. Alex had learned that bass couldn't tolerate pollution. Finding them was a great indicator that the water was clean enough to support life.

> "All of Third Creek will benefit from what Mr. Keeling has done to improve water quality. While many farmers have begun using more environmentally friendly farming techniques, Mr. Keeling has shown great initiative and has been a leader in McRay County and the state."

"Awesome," Alex whispered to his mom. "I'll bet Grandpa Keeling would be pleased as punch to read this. Wasn't he into the soil conservation thing back in the day?"

His mom nodded and beamed with pride. She pointed across the auditorium and whispered back, "There's the governor now. They should be starting any minute." She squeezed Alex's hand. "Someday, you might get your own award when the farm passes on to you. That would make all of us proud."

"Good afternoon, everyone," the governor began. "It gives me great pleasure to welcome you all here to help us acknowledge the fine work of the people selected to receive the Environmental Stewardship Award.

"First, in the area of environmental education, we present the teachers and students from Knox County High School. . . ."

The governor droned on, and Alex's attention wandered as he imagined his own future." … I'd like to share with you some details about the work done by Mr. Alex Keeling. His dad, Arthur, was honored sixteen years ago for improvements to the water quality of Third Creek. Alex is following in his dad's footsteps, using conservation farming techniques he learned at this very university. . . ."

Suddenly Alex felt his mom's elbow connect with his arm. "What are you daydreaming about now?" she scolded in a whisper. "Pay attention to your father's big moment!"

"Our farmers are important caretakers of our state's natural beauty," continued the governor. "I am very pleased to call forward the recipient of the award for Environmental Stewardship in Farming, Mr. Arthur Keeling from McRay County."

The Carbon Cycle

The carbon dioxide (CO_2) in the atmosphere is the exchange pool for the carbon cycle. In this cycle, organisms in both terrestrial and aquatic ecosystems exchange carbon dioxide with the atmosphere (Fig. 23.10). ① On land, plants take up carbon dioxide from the air. Through photosynthesis, they incorporate carbon into nutrients used by autotrophs and heterotrophs. ② When organisms, including plants, respire, carbon is returned to the atmosphere as carbon dioxide. Therefore, carbon dioxide recycles to plants by way of the atmosphere.

In aquatic ecosystems, the exchange of carbon dioxide with the atmosphere is indirect. ③ Carbon dioxide from the air combines with water to produce bicarbonate ion (HCO_3^-). This is a source of carbon for algae that produce food for themselves and for heterotrophs. Similarly, when aquatic organisms respire, the carbon dioxide they give off becomes bicarbonate ion. ④ The amount of bicarbonate in the water is in equilibrium with the amount of carbon dioxide in the air.

Reservoirs Hold Carbon

⑤ Living and dead organisms contain organic carbon and serve as one of the reservoirs for the carbon cycle. The world's biotic components, particularly trees, contain 800 billion tons of organic carbon. An additional 1,000–3,000 billion metric tons are estimated to be held in the remains of plants and animals in the soil. Ordinarily, decomposition of animals returns CO_2 to the atmosphere.

⑥ In the history of the Earth, some 300 MYA, plant and animal remains were transformed into coal, oil, and natural

Figure 23.10 **How is carbon transferred into the atmosphere during the carbon cycle?**

The carbon cycle is a gaseous biogeochemical cycle. Producers take in carbon dioxide from the atmosphere and convert it to organic molecules that feed all organisms. The transfer rate of carbon into the atmosphere due to respiration approximately matches the rate due to withdrawal by plants for photosynthesis. Fossil fuels arise when organisms die but do not decompose. When humans burn fossil fuels and destroy vegetation (purple arrows), more carbon dioxide is added to the atmosphere than is withdrawn. This causes environmental pollution.

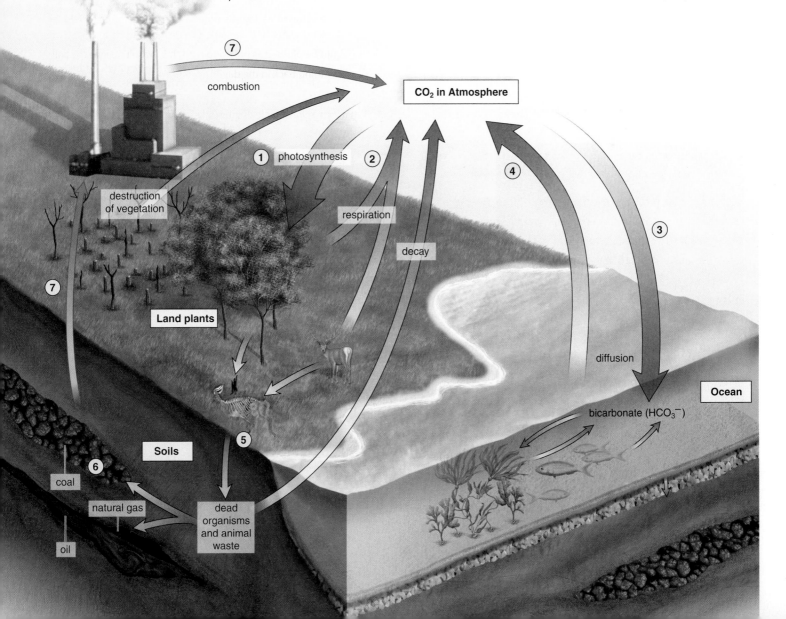

Guaranteeing Access to Safe Drinking Water

The ability to get safe drinking water may not be something you've ever thought about. In the United States and other developed countries, safe, clean water usually pours out of your home faucets every time you turn them on. When water is plentiful, it's taken for granted and a great deal of it is wasted. However, water is a precious and often scarce resource in many parts of the world.

Almost 20% of the Earth's population doesn't have access to safe drinking water (Fig. 23A). Further, basic sanitation (proper disposal of human and animal waste) is unavailable to 40%. Most people lacking safe drinking water and/or basic sanitation live in China, the Middle East, or Africa. These are some of the world's poorest people. If they have to pay for water, they are often charged much more than wealthy people living in the same area. They may have to walk miles to collect water and then return to their homes.

Having safe drinking water and basic sanitation would impact people's lives and health more than any other intervention. More than 80% of diseases, including typhoid fever and cholera, are associated with foul water and improper sanitation. Each year, more than 5 million people die from water pollution-related illnesses. This is the leading cause of death for children under five.

Solving the problem of making safe water available to more people is no easy task. One of the Millennium Development Goals (agreed upon by members of the United Nations in 2000) is to halve the number of people without access to safe water by 2015. The cost has been projected to be between $4 billion and $11.3 billion per year. An unusual twist to the water crisis is the interest in water as an investment and commodity to be traded or sold. The Goldman Sachs Group, Inc. (often referred to as *Goldman Sachs;* on the New York Stock Exchange, GS), is a large global banking company involved in investment banking. Goldman Sachs estimates the value of the water industry at $425 billion. At a recent conference hosted by the company, the water shortage was listed among the world's top threats to global stability.

Each day an average American uses 100–176 gallons of water. By contrast, the typical African family uses about 5 gallons. We may need to consider how much of the water we use daily is necessary. More effort will be needed to ensure all people have the ability to get safe drinking water. A more stable global community may depend on it.

Decide Your Opinion

1. Should developed nations assume responsibility for providing safe drinking water and sanitation to developing countries? If so, to what extent?
2. Water conservation is mandated by law in many states. Should legislation requiring water conservation be extended to include all states?
3. What form might this legislation take? How should it be applied?

Figure 23A Many people in developing nations lack access to safe drinking water.
Individuals in developing countries may have little or no access to safe driwnking water. Providing a universal clean water supply for all is a global necessity.

gas. We call these materials the **fossil fuels.** Another reservoir for carbon is the inorganic carbonate that accumulates in limestone and in calcium carbonate shells. Many marine organisms have calcium carbonate shells that remain in bottom sediments long after the organisms have died. Geological forces change these sediments into limestone.

Human Activities

The transfer rates of carbon dioxide due to photosynthesis and cellular respiration, which includes the work of decomposers, are just about even. However, more carbon dioxide is being deposited in the atmosphere than is being removed. ⑦ This increase is largely due to the burning of fossil fuels and the destruction of forests to make way for farmland and pasture. When we do away with forests, we reduce a reservoir and lose the organisms that take up excess carbon dioxide. Today, the amount of carbon dioxide released into the atmosphere is about twice the amount that remains in the atmosphere. It's believed that much of this has been dissolving into the ocean.

CO_2 and Global Warming Carbon dioxide and other gases are being emitted due to human activities. The other gases include nitrous oxide (N_2O) and methane (CH_4). Fertilizers and animal wastes are sources of nitrous oxide. In the digestive tracts of animals, bacterial decomposition produces methane. Decaying sediments and flooded rice paddies are methane sources, as well. In the atmosphere, nitrous oxide and methane are known as **greenhouse gases** because they act just like the panes of a greenhouse. They allow solar radiation to penetrate to Earth, but hinder the escape of infrared rays (heat) back into space—a phenomenon called the **greenhouse effect.** The greenhouse gases are contributing significantly to an overall rise in the Earth's ambient temperature. This phenomenon is called **global warming.**

Figure 23.11 shows the Earth's radiation balances. One thing to be learned from this diagram is that water vapor is a greenhouse gas. Clouds (composed of water vapor) also reradiate heat back to Earth. If the Earth's temperature rises, more water will evaporate, forming more clouds. This sets up a positive feedback effect that could increase global warming still more. The global climate has already warmed about 0.6°C since the Industrial Revolution. Computer models are unable to consider all possible variables, but the Earth's temperature may rise 1.5–4.5°C by 2100 if greenhouse emissions continue at the current rates.

Global warming will bring about other effects, which computer models attempt to forecast. It is predicted that, as the oceans warm, temperatures in the polar regions will rise to a greater degree than in other regions. As a result, sea levels will rise because glaciers will melt, and water expands as it warms. Water evaporation will increase, and most likely there will be increased rainfall along the coasts and dryer conditions inland. The occurrence of droughts will reduce agricultural yields and also cause trees to die off. Expansion of forests into arctic areas might not offset the loss of forests in the temperate zones. Coastal agricultural lands, such as the deltas of Bangladesh and China, will be inundated with water. Billions of dollars will have to be spent to keep coastal cities such as New Orleans, New York, Boston, Miami, and Galveston from disappearing into the sea.

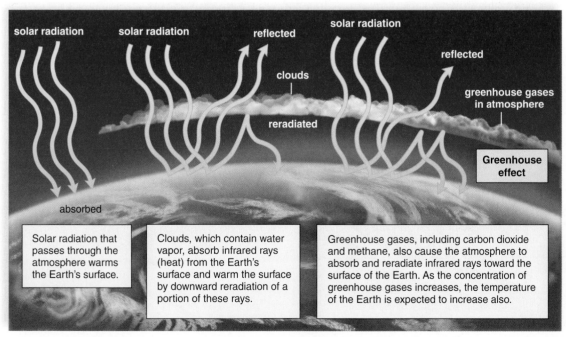

Figure 23.11 How do the greenhouse gases affect the temperature of the Earth?

The effect of the greenhouse gases *(far right)* is contributing to global warming. A positive feedback cycle is predicted. As the Earth's temperature rises, more water will evaporate, causing clouds to thicken and absorb more solar radiation. Water vapor in clouds and the other greenhouse gases prevent the escape of heat (infrared rays), and it is reradiated back to the Earth, causing even more evaporation, and so forth.

Solar radiation that passes through the atmosphere warms the Earth's surface.

Clouds, which contain water vapor, absorb infrared rays (heat) from the Earth's surface and warm the surface by downward reradiation of a portion of these rays.

Greenhouse gases, including carbon dioxide and methane, also cause the atmosphere to absorb and reradiate infrared rays toward the surface of the Earth. As the concentration of greenhouse gases increases, the temperature of the Earth is expected to increase also.

The Nitrogen Cycle

Nitrogen gas (N_2) makes up about 78% of the atmosphere, but plants cannot make use of nitrogen in this form. Therefore, nitrogen can be a nutrient that limits the amount of growth in an ecosystem.

Ammonium (NH_4^+) Formation and Use

① In the nitrogen cycle, **nitrogen fixation** occurs when nitrogen gas (N_2) is converted to ammonium (NH_4^+), a form plants can use (Fig. 23.12). Some cyanobacteria in aquatic ecosystems and some free-living bacteria in soil are able to fix atmospheric nitrogen in this way. Other nitrogen-fixing bacteria live in nodules on the roots of legumes, such as beans, peas, and clover. They make organic compounds containing nitrogen available to the host plants so that the plant can form proteins and nucleic acids.

Nitrate (NO_3^-) Formation and Use

Plants can also use nitrates (NO_3^-) as a source of nitrogen. The production of nitrates during the nitrogen cycle is called **nitrification.** Nitrification can occur in various ways: ② Nitrogen gas (N_2) is converted to nitrate (NO_3^-) in the atmosphere, where high energy is available to react nitrogen with oxygen. This energy may be supplied by cosmic radiation, meteor trails, and lightning. ③ Ammonium (NH_4^+) in the soil from various sources, including decomposition of organisms and animal wastes, is converted to nitrate by soil

Figure 23.12   How is the nitrogen cycle affected by human activities?
Nitrogen is primarily made available to biotic communities by internal cycling of the element. Without human activities, the amount of nitrogen returned to the atmosphere (denitrification in terrestrial and aquatic communities) exceeds withdrawal from the atmosphere (N_2 fixation and nitrification). Human activities (dark purple arrow) result in an increased amount of NO_3^- in terrestrial communities, with resultant runoff to aquatic biotic communities.

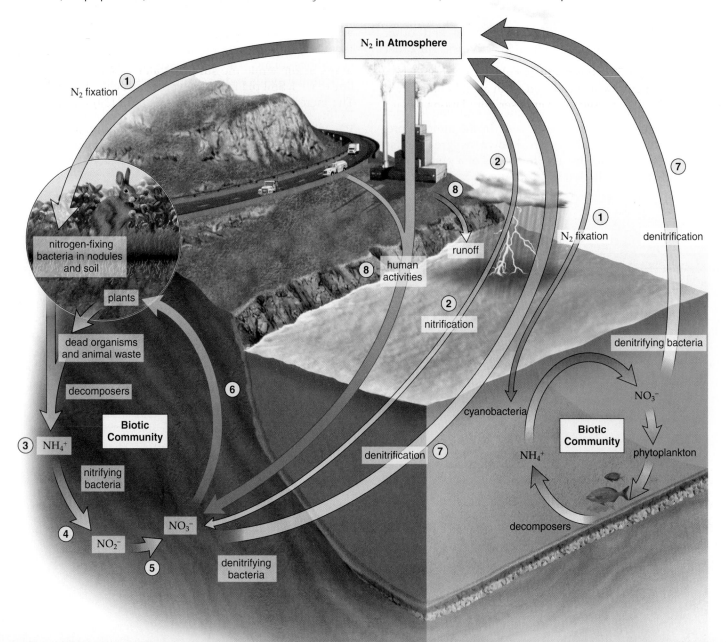

bacteria. ④ Nitrite-producing bacteria convert ammonium to nitrite (NO_2^-). ⑤ Nitrate-producing bacteria convert nitrite to nitrate. ⑥ During the process of **assimilation,** plants take up ammonia and nitrate from the soil and use these ions to produce proteins and nucleic acids.

In Figure 23.12, observe the biotic community subcycles occurring on land and in water. Those subcycles do not depend on the presence of nitrogen gas.

Formation of Nitrogen Gas from Nitrate

⑦ **Denitrification** is the conversion of nitrate back to nitrogen gas, which enters the atmosphere. Denitrifying bacteria living in the anaerobic mud of lakes, bogs, and estuaries carry out this process during their metabolism. In the nitrogen cycle, denitrification would counterbalance nitrogen fixation except for human activities.

Human Activities

⑧ Human activities significantly alter the transfer rates in the nitrogen cycle by producing fertilizers from N_2. They nearly double the fixation rate. Fertilizer, which also contains phosphate, runs off into lakes and rivers. This results in an overgrowth of algae and rooted aquatic plants. As discussed on page 557, the result is cultural eutrophication (over-enrichment), which can lead to an algal boom. When the algae die off, enlarged decomposer populations use up all the oxygen in the water. The result is a massive fish kill.

Acid deposition occurs because nitrogen oxides (NO_x) and sulfur dioxide (SO_2) enter the atmosphere from the

Have You Ever Wondered . . .

How acidic is acid rain?

Refer to Figure 2.9 (page 28). The pH of normal rain varies from 4.5 to 5.6. That's about the acidity of tomatoes or black coffee, and neither of these is particularly corrosive or harmful to the body. Rain collected in the Eastern United States during the summer has an average pH of 3.6. Vinegar has a pH similar to this rain—can you imagine watering trees or house plants with vinegar? Fog with a pH of 2 has been measured in some locations. A solution with pH 2 is only slightly less acid than stomach acid or battery acid. It's easy to imagine the damage such a solution could do to the environment.

burning of fossil fuels (Fig. 23.13*a*). Both these gases combine with water vapor to form acids that eventually return to the Earth in acid rain. Acid deposition has drastically affected forests and lakes in northern Europe, Canada, and northeastern United States. The soil in these forests is naturally acidic while the surface water is only mildly alkaline (basic). Soil pH is made even more acidic by acid rain. The increased acidity kills trees (Fig. 23.13*b*) and reduces agricultural yields. Marble, metal, and stonework are corroded by acid deposition, which results in the loss of architectural features (Fig. 23.13*c*).

a.

b.

c.

Figure 23.13
What causes acid deposition, and what harm does it cause?
a. Many forests in higher elevations of northeastern North America and northern Europe are dying due to acid deposition.
b. Air pollution due to fossil-fuel burning in factories and modes of transportation is the major cause of acid deposition.
c. Acid deposition damages architectural features.

 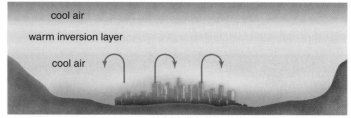

a. Normal pattern

b. Thermal inversion

Figure 23.14　**How does a thermal inversion trap pollutants?**
a. Normally, pollutants escape into the atmosphere when warm air rises. **b.** During a thermal inversion, a layer of warm air (warm inversion layer) overlies and traps pollutants in cool air below.

Nitrogen oxides and hydrocarbons (HC) from the burning of fossil fuels react with one another in the presence of sunlight to produce smog. Smog contains dangerous pollutants. Warm air near the Earth usually escapes into the atmosphere, taking pollutants with it. However, during a thermal inversion, pollutants are trapped near the Earth beneath a layer of warm, stagnant air. The air does not circulate, so pollutants can build up to dangerous levels. Areas surrounded by hills are particularly susceptible to the effects of a thermal inversion. This is because the air tends to stagnate and little turbulent mixing can occur (Fig. 23.14).

The Phosphorus Cycle

In the phosphorus cycle (Fig. 23.15), ① phosphorus trapped in oceanic sediments moves onto land after a geological

Figure 23.15　**What process supplies phosphorous to terrestrial and aquatic organisms? How do humans upset the balance of the phosphorous cycle?**
The weathering of rocks provides phosphorus, which cycles locally in both terrestrial and aquatic biota. Human activities (purple arrows) produce fertilizers, which add to the amount of phosphorus available to biotic communities. Eventually, fertilizers become a part of the runoff that enriches waters. Sewage treatment plants directly add phosphorus to local waters. When phosphorus becomes a part of oceanic sediments, it is lost to biotic communities for many years.

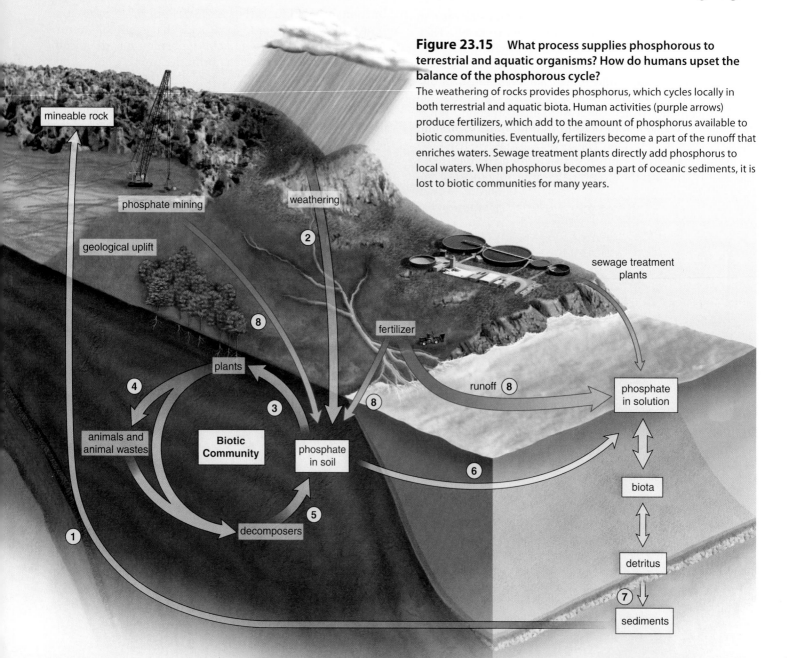

upheaval. ② On land, the very slow weathering of rocks places phosphate ions (PO_4^{3-} and HPO_4^{2-}) in the soil. ③ Some of this becomes available to plants, which use phosphate in a variety of molecules. Molecules requiring phosphate include phospholipids, ATP, and the nucleotides that become a part of DNA and RNA. ④ Animals eat producers and incorporate some of the phosphate into teeth, bones, and shells, which take many years to decompose. ⑤ However, the eventual death and decomposition of all organisms and their wastes do make phosphate ions available to producers once again. The available amount of phosphate is already being used within food chains, so phosphate is usually a limiting inorganic nutrient for plants. In other words, the lack of it limits the size of populations in ecosystems.

⑥ Some phosphate naturally runs off into aquatic ecosystems. There algae acquire phosphate from the water before it becomes trapped in sediments. ⑦ Phosphate in marine sediments does not become available to producers on land again until a geological upheaval exposes sedimentary rocks on land. Now, the cycle begins again. Phosphorus does not enter the atmosphere. Therefore, the phosphorus cycle is called a sedimentary cycle.

Phosphorus and Water Pollution

⑧ Human beings boost the supply of phosphate by mining phosphate ores for fertilizer and detergent production. As mentioned previously, runoff of phosphate and nitrogen into water occurs due to fertilizer use. This contamination, as well as that from animal wastes in livestock feedlots and discharge from sewage treatment plants, results in **cultural eutrophication** (overenrichment) of waterways.

Figure 23.16 lists the various sources of water pollution. Point sources of pollution are specific, and nonpoint sources are those caused by runoff from the land. Industrial wastes can include heavy metals and organochlorides, such as DDT and PCBs. These materials are not readily degraded under natural conditions or in conventional sewage treatment plants. **Biological magnification** occurs as these toxic chemicals pass along a food chain. Pollutants become increasingly concentrated with each higher consumer level, because they remain in the body and are not excreted. Aquatic food chains are more likely to experience biological magnification because there are more links than in a terrestrial food chain. Mercury levels can be high in some of the fish we eat (top consumers like sharks, swordfish, and tuna) for this reason.

Coastal regions are the immediate receptors for local pollutants and the final receptors for pollutants carried by rivers that empty at a coast. Waste dumping occurs at sea. However, ocean currents sometimes transport both trash and pollutants back to shore. Offshore mining and shipping add pollutants to the oceans. Some 5 million metric tons of oil a year end up in the oceans. Large oil spills kill plankton, fish, and shellfishes, as well as birds and marine mammals.

Sources of Water Pollution	
Leading to Cultural Eutrophication	
oxygen-demanding waste	Biodegradable organic compounds (e.g., sewage, wastes from food-processing plants, paper mills, and tanneries)
plant nutrients	Nitrates and phosphates from detergents, fertilizers, and sewage treatment plants
sediments	Enriched soil in water due to soil erosion
thermal discharges	Heated water from power plants
Health Hazards	
disease-causing agents	Bacteria and viruses from sewage and barnyard waste (causing, for example, cholera, food poisoning, and hepatitis)
synthetic organic compounds	Pesticides, industrial chemicals (e.g., PCBs)
inorganic chemicals and minerals	Acids from mines and air pollution; dissolved salts; heavy metals (e.g., mercury) from industry
radiation	Radioactive substances from nuclear power plants, medical and research facilities

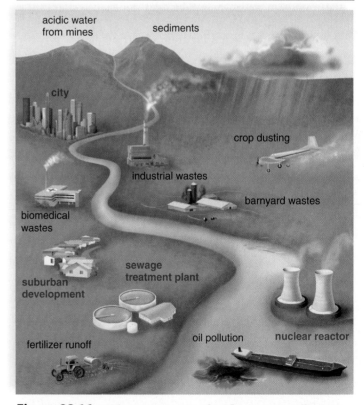

Figure 23.16 **What are sources of surface water pollution?**
Many bodies of water are dying due to the introduction of pollutants from point sources, which are easily identifiable, and nonpoint sources, which cannot be specifically identified.

Science Focus

Ozone Shield Depletion

Ozone is a gas that can be found in the troposphere (the layer of atmosphere closest to the ground) and in the stratosphere (a layer of the Earth's upper atmosphere). Exhaust from our cars and emissions from various industries contribute to the formation of ozone in the troposphere. It's more likely to form during the hot, sunny days associated with late spring through early fall. This type of ozone is "bad" ozone and is considered a pollutant. A large component of smog is ozone. You may have heard ozone levels being discussed in your area during daily air quality reports. When ozone levels around us are high, it adversely affects our ability to breathe and may cause permanent lung damage. Children and people who already suffer from lung diseases or conditions should stay indoors when tropospheric ozone levels are high.

In the stratosphere, ozone forms when ultraviolet (UV) radiation from the sun splits molecules of oxygen gas (O_2). Single oxygen atoms (O) combine with molecules of O_2 to form ozone (O_3). This ozone (often called the **ozone shield**) is considered "good" ozone because of its ability to absorb most of the sun's UV rays. The absorption of UV radiation by the ozone shield is critical for living things. In humans, UV radiation causes mutations that can lead to skin cancer and can make the lens of the eye develop cataracts. A United Nations Environmental Program report predicts a 26% rise in cataracts and nonmelanoma skin cancers for every 10% drop in the ozone level. Further, the immune system is damaged by UV radiation, making us more susceptible to infectious diseases. The growth of crops and trees is also adversely affected by UV radiation. In water ecosystems, algae and krill (tiny shrimplike animals) that sustain other aquatic life are killed by UV radiation. Therefore, an inadequate ozone shield that allows more UV rays to strike the Earth would threaten both our health and our food sources.

In the 1980s, it became apparent that loss of ozone in the shield had occurred worldwide. The depletion was most severe over Antarctica in the spring. There, the so-called **ozone hole** covered an area two and a half times the size of Europe (Fig. 23B). This bare area exposed not just Antarctica to harmful UV rays. The southern tip of South America and vast areas of the Pacific and Atlantic oceans were affected as well. Australia is also impacted by the Antarctica hole. A second ozone hole formed above the Arctic. Other holes have been detected over areas where large populations of people live.

Scientists discovered that chlorine atoms are a major cause of ozone depletion. These chlorine atoms come primarily from the breakdown of **chlorofluorocarbons (CFCs)**. The CFC you may be most familiar with is Freon®, a coolant found in refrigerators and air conditioners. CFCs are also used as a foaming agent during the production of Styrofoam®

CASE STUDY AT SEMESTER'S END

Alex sighed and logged onto his account in student services to find out about his final grades. He squinted at the monitor, expecting the worst and hoping for the best. It was critical to keep his GPA above a 3.0 so he'd continue to receive the state scholarship. Without that money, it would be tough for him to attend school full time.

He was thrilled to see a host of Bs listed for the courses he'd taken. There was even an A in the Conservation Farming Practices class. His dad would be pleased to hear that news.

Alex was excited to return to the farm to share with his dad all that he'd learned in that course. One of the coolest things he'd learned about was fly management. Insects known as fly parasitoids attack the pupal stage of the pest fly. Employing parasitoids could mean the elimination of chemical fly control. Further, insect use would save the farm some money and be more environmentally sound. Alex already knew that making the dairy cows more comfortable would translate into higher milk production. That, too, would mean additional profit for the farm.

College graduation for Alex and his friends was just around the corner. Their minds were occupied with resumes, interviews, rejection letters, job offers, and future careers. Alex smiled. His career would begin in earnest—just as soon as he got back home.

coffee cups, egg cartons, insulation, and padding. At one time CFCs were used at propellants in spray cans, but this use is now banned in the United States and several European countries. A pesticide called methyl bromide significantly decreases ozone levels as well.

The Montreal Protocol was drafted in 1987 and was initially signed by 43 countries. Agreement to phase-out the use of CFCs by 1995 was reached (though developing countries have until 2010 to complete their phase-out). In 2007, China closed five of its six remaining plants that produced CFCs, ahead of schedule. Methyl bromide was also later included in the Montreal Protocol. Its phase-out in developed countries occurred in 2005. The Montreal Protocol has now been signed by at least 183 countries.

Recent satellite measurements indicate that the amount of harmful chlorine pollution in the stratosphere has started to decline. However, recovery of the ozone shield may take many more years (the Antarctic hole is expected to be present until 2050 even if the Montreal Protocol is followed). Other pollution-fighting approaches may be required in addition to lowering chlorine pollution. The global warming phenomenon may also adversely impact ozone restoration. As greenhouse gases increase, less heat is reflected from the Earth's surface, and the stratosphere becomes colder. Ozone depletion increases when the stratosphere is very cold.

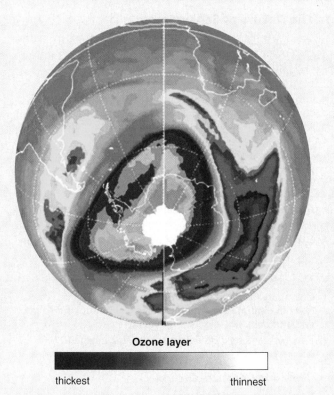

Ozone layer

thickest   thinnest

Figure 23B Ozone shield depletion.
Map of ozone levels in the atmosphere of the Southern Hemisphere during September, 2007. The ozone depletion, often called an ozone hole, is larger than the size of Europe.

In the last 50 years, humans have polluted the seas and exploited their resources to the point that many species are on the brink of extinction. Fisheries once rich and diverse, such as George's Bank off the coast of New England, are in severe decline. Haddock was once the most abundant species in this fishery, but now it accounts for less than 2% of the total catch. Cod and bluefin tuna have suffered a 90% reduction in population size. In warm, tropical regions, many areas of coral reefs are now overgrown with algae. This is because the fish that normally keep the algae under control have been killed off.

Check Your Progress 23.3

1. Chemical cycling in the biosphere may involve what three components?
2. a. What are two examples of gaseous biogeochemical cycles? b. What is an example of a sedimentary biogeochemical cycle?
3. What roles do bacteria play in the nitrogen cycle?
4. What ecological problems are associated with water, carbon, nitrogen, and phosphorus cycles?

Summarizing the Concepts

23.1 The Nature of Ecosystems

Ecology is the study of the interactions of organisms with each other and with the physical environment.

- Organisms interact with the physical and chemical environment, and the result is an ecosystem.
- Terrestrial ecosystems are forests (tropical rain forests, coniferous, temperate deciduous); grasslands (savanna and prairie); and deserts, which includes the tundra.
- Aquatic ecosystems are either salt water (i.e., seashores, oceans, coral reefs, estuaries) or freshwater (i.e., lakes, ponds, rivers, and streams).

Biotic Components of an Ecosystem

- In a community, each population has a habitat (residence) and a niche (its role in the community).
- Autotrophs (producers) produce organic nutrients for themselves and others from inorganic nutrients and an outside energy source.
- Heterotrophs (consumers) consume organic nutrients.
- Consumers are herbivores (eat plants/algae), carnivores (eat other animals), and omnivores (eat both plants/algae and animals).
- Decomposers feed on detritus, releasing inorganic substances back into the ecosystem.

Energy Flow and Chemical Cycling

Ecosystems are characterized by energy flow and chemical cycling.

- Energy flows through the populations of an ecosystem.
- Chemicals cycle within and among ecosystems.

23.2 Energy Flow

Various interconnecting paths of energy flow are called a food web.

- A food web is a diagram showing how various organisms are connected by eating relationships.
- Grazing food webs begin with vegetation eaten by a herbivore that becomes food for a carnivore.
- Detrital food webs begin with detritus, food for decomposers and for detritivores.
- Members of detrital food webs can be eaten by aboveground carnivores, joining the two food webs.

Trophic Levels

A trophic level is all the organisms that feed at a particular link in a food chain.

- Ecological pyramids illustrate that biomass and energy content decrease from one trophic level to the next because of energy loss.

23.3 Global Biogeochemical Cycles

Chemicals circulate through ecosystems via biogeochemical cycles, pathways involving both biotic and geological components. Biogeochemical cycles:

- can be gaseous or sedimentary.
- have reservoirs (e.g., ocean sediments, the atmosphere, and organic matter) that contain inorganic nutrients available to living things on a limited basis.

Exchange pools are sources of inorganic nutrients.

- Nutrients cycle among the biotic communities (producers, consumers, decomposers) of an ecosystem.

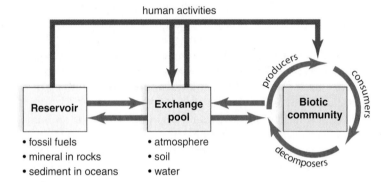

The Water Cycle

- The reservoir of the water cycle is freshwater that evaporates from the ocean.
- Water that falls on land enters the ground, surface waters, or aquifers and evaporates again.
- All water returns to the ocean.

The Carbon Cycle

- The reservoirs of the carbon cycle are organic matter (e.g., forests and dead organisms for fossil fuels), limestone, and the ocean (e.g., calcium carbonate shells).
- The exchange pool is the atmosphere.
- Photosynthesis removes carbon dioxide from the atmosphere.
- Respiration and combustion add carbon dioxide to the atmosphere.

The Nitrogen Cycle

- The reservoir of the nitrogen cycle is the atmosphere.
- Nitrogen gas must be converted to a form usable by plants (producers).
- Nitrogen-fixing bacteria (in root nodules) convert nitrogen gas to ammonium, a form producers can use.
- Nitrifying bacteria convert ammonium to nitrate.
- Denitrifying bacteria convert nitrate back to nitrogen gas.

The Phosphorus Cycle

- The reservoir of the phosphorus cycle is ocean sediments.
- Phosphate in ocean sediments becomes available through geological upheaval, which exposes sedimentary rocks to weathering.
- Weathering slowly makes phosphate available to the biotic community.
- Phosphate is a limiting nutrient in ecosystems.

Understanding Key Terms

acid deposition 555	food web 547
aquifer 549	fossil fuel 553
assimilation 555	global warming 553
autotroph 544	grazing food chain 547
biogeochemical cycle 548	grazing food web 547
biological magnification 557	greenhouse effect 553
biomass 548	greenhouse gases 553
biosphere 543	herbivore 544
carnivore 544	heterotroph 544
chlorofluorocarbon	niche 545
(CFC) 558	nitrification 554
consumer 544	nitrogen fixation 554
cultural eutrophication 557	omnivore 544
denitrification 555	ozone hole 558
detrital food chain 547	ozone shield 558
detrital food web 547	precipitation 549
detritus feeder 544	producer 544
ecological pyramid 547	runoff 549
ecosystem 543	trophic level 547
evaporation 549	trophic relationship 547
food chain 547	water (hydrologic) cycle 549

Match the key terms to these definitions.

a. _____ Animals that feed on both plants and animals.

b. _____ The organisms that feed at a particular link in a food chain.

c. _____ Remains of once-living organisms that are burned to release energy, such as coal, oil, and natural gas.

d. _____ Process by which atmospheric nitrogen gas is changed to forms that plants can use.

e. _____ Photosynthetic organism at the start of a grazing food chain that makes its own food.

Testing Your Knowledge of the Concepts

1. What are the major types of terrestrial ecosystems? Describe each with its temperature, rainfall, and type of vegetation. (pages 542–43)

2. What are the major types of aquatic ecosystems? (page 543)

3. What is a niche? (page 545)

4. Name the four different types of consumers (heterotrophs) found in ecosystems. (page 544)

5. Explain why energy flows but chemicals cycle through an ecosystem. (page 545)

6. Describe the two types of food webs and two types of food chains in terrestrial ecosystems. Which of these typically moves more energy through an ecosystem? (pages 546–47)

7. What is a trophic level? An ecological pyramid? (pages 547–48)

8. What is a biochemical cycle? What is a reservoir and an exchange pool? Give an example of each. (page 548)

9. Draw a diagram to illustrate the water, carbon, nitrogen, and phosphorus biochemical cycles. Include how human activities can change a particular cycle's balance. (pages 549–57)

In questions 10–13, match each description to a population in the key.

Key:

a. producer   c. decomposer
b. consumer   d. herbivore

10. Heterotroph that feeds on plant material.

11. Autotroph that manufactures organic nutrients.

12. Any type of heterotroph that feeds on plant material or on other animals.

13. Heterotroph that breaks down detritus as a source of nutrients.

14. Of the total amount of energy that passes from one trophic level to another, about 10% is
    a. respired and becomes heat.
    b. passed out as feces or urine.
    c. stored as body tissue.
    d. recycled to autotrophs.
    e. All of these are correct.

15. Compare this food chain:
    algae → water fleas → fish → green herons
    with this food chain:
    trees → tent caterpillars → red-eyed vireos → hawks

    Both water fleas and tent caterpillars are
    a. carnivores.
    b. primary consumers.
    c. detritus feeders.
    d. present in grazing and detrital food webs.
    e. Both a and b are correct.

16. In what way are decomposers like producers?
    a. Either may be the first member of a grazing or a detrital food chain.
    b. Both produce oxygen for other forms of life.
    c. Both require nutrient molecules and energy.
    d. Both are present only on land.
    e. Both produce organic nutrients for other members of ecosystems.

17. Why are ecosystems dependent on a continual supply of solar energy?
    a. Carnivores have a greater biomass than producers.
    b. Decomposers process the greatest amount of energy in an ecosystem.
    c. Energy transformation results in a loss of usable energy to the environment.
    d. Energy cycles within and between ecosystems.

18. Nutrient cycles always involve
    a. rocks as a reservoir.
    b. movement of nutrients through the biotic community.
    c. the atmosphere as an exchange pool.
    d. loss of the nutrients from the biosphere.

19. Which of the following contribute(s) to the carbon cycle?
    a. respiration
    b. photosynthesis
    c. fossil fuel combustion
    d. decomposition of dead organisms
    e. All of these are correct.

20. How do plants contribute to the carbon cycle?
    a. When plants respire, they release CO_2 into the atmosphere.
    b. When plants photosynthesize, they consume CO_2 from the atmosphere.
    c. When plants photosynthesize, they provide oxygen to heterotrophs.
    d. When plants emigrate, they transport carbon molecules between ecosystems.
    e. Both a and b are correct.

21. How do nitrogen-fixing bacteria in the soil contribute to the nitrogen cycle?
    a. They return nitrogen to the atmosphere.
    b. They change ammonium to nitrate.
    c. They change nitrogen to ammonium.
    d. They withdraw nitrate from the soil.
    e. They decompose and return nitrogen to autotrophs.

22. What is the reservoir in the phosphorus cycle?
    a. oceans
    b. marine sediments
    c. plants
    d. animals

For questions 23–26, match each human activity with one or more of the cycles listed in the key. More than one answer can be used, and answers can be used more than once.

Key:

a. water cycle   d. phosphorus cycle
b. carbon cycle   e. none of these
c. nitrogen cycle   f. all of these

23. Drive cars

24. Use fertilizers

25. Take showers

26. Grow crops

For questions 27–31, match each characteristic to the cycles listed in the key for questions 23–26. More than one answer can be used, and answers can be used more than once.

27. Occurs on land but not in the water

28. Can occur without the participation of humans

29. Always involves the participation of decomposers

30. The atmosphere is involved

31. Rocks are the reservoir in this cycle

32. Label the following diagram of an ecosystem.

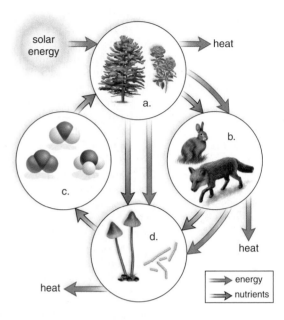

Thinking Critically About the Concepts

Alex and his dad are making a concerted effort to practice farming in an environmentally sound manner. Recognition from the state in the form of stewardship awards may be an incentive for agencies, businesses, organizations, individuals, and educational institutions to consider beginning environmental projects and/or conservation methods. Many colleges and universities have initiatives that make the campuses more environmentally friendly. Our own contributions help, too. Water conservation, recycling trash, replacing paper or plastic containers with cloth, conserving electricity, starting a compost pile, collecting rain water for plants in rain barrels—these are just a few of many ideas for you to consider.

1a. What impact might fungicides and pesticides have on detrital food webs?

b. How do you think the nutrient cycles would be affected by the use of fungicides and pesticides?

2. Why would an agricultural extension agent recommend planting a legume such as clover in a pasture?

3a. What types of things could you do at home to facilitate the cycling of nutrients such as carbon and nitrogen?

b. What types of things could you do at home to conserve water or improve the quality of a nearby body of water?

4. If Alex and his dad use fly parasitoids to control flies on the farm, they may be able to stop using pesticides that contain organochlorides. What impact would this have on food chains involving wildlife around the farm?

Human Population, Planetary Resources, and Conservation

CASE STUDY **JACOB HOPEMAN, MEGAN DONNELLY, EVA PHIPPS, ADAM RAY**

The Earth Day Festival, where Jacob, Megan, Eva, and Adam had met during their freshman year, was the most important activity sponsored by the university's Outdoor Club. The four friends, now all seniors, recognized the importance of teaching environmental awareness to children, as well as to their parents and guardians. This year, club members were gratified that over 500 children and adults had attended. Megan's group helped younger children plant seeds in paper cups they'd decorated. Their adult chaperones received free tree seedlings, along with plans to construct a simple compost bin. Alongside her area, Jacob and Eva had hosted their "fishing pond"—a plastic swimming pool filled with floating fish shapes, with fishing rods constructed from long sticks. By requiring the children to toss rare "fish" back into the pool, the little ones were introduced to the problem of overfishing. Grownups were provided with recipes and advice for making better choices of fish to eat. Adam's group taught about biodiversity using insects from the entomology lab. Harmless yet fascinating insects, such as hissing cockroaches, easily kept everyone's attention.

When the last participants finally departed, the four friends finished cleanup and sat around an empty picnic table. Megan smiled ruefully. "I've been waiting for graduation all year, and now that it's less than a month away, I don't want to leave. Doing this Earth Day thing is a blast. I'll miss it."

"Aren't you still doing Teach for America?" Eva asked. "You'll be around kids all the time."

"First, I'm going on a church mission trip to Mexico City to improve my Spanish," Megan replied. "Then I'm going to the Rio Grande valley. It's really poor in that area, and it's mostly a farming community. I'm guessing that I'll be able teach the kids about ecology and Earth-friendly farming."

"You're still going to China to teach English, right? *Ni hao!*"

Eva nodded, and laughed. "Hi to you too!" Then she sighed. "I'll be in China, Adam's in Togo with the Peace Corps, and Jacob's globe-trotting all over Europe. I'm really gonna miss you guys."

"If you send me your news, I can make sure everyone else gets it," Megan offered. "Adam, you can write, and everyone else can email."

CHAPTER CONCEPTS

24.1 Human Population Growth
The present growth rate for the world's population has decreased to 1.2%, but still 78 million more people are expected within a year because the human population is so large.

24.2 Human Use of Resources and Pollution
Human beings use land, water, food, energy, and minerals to meet their basic needs. Use of these resources leads to pollution.

24.3 Biodiversity
A biodiversity crisis is upon us because of habitat loss. The introduction of alien species, pollution, overexploitation, and disease all contribute to the crisis. Yet, wildlife has both a direct value and an indirect value for us.

24.4 Working Toward a Sustainable Society
Our present day society is not sustainable, and ways are given to make it more sustainable.

24.1 Human Population Growth

The world's population has risen steadily to a present size of close to 7 billion people (Fig. 24.1). Prior to 1750, the growth of the human population was relatively slow. As more reproducing individuals were added, population growth increased. The curve began to slope steeply upward, indicating that the population was undergoing **exponential growth.** The number of people added annually to the world population peaked at about 87 million around 1990. Currently it is a little over 78 million per year.

The **growth rate** of a population is determined by considering the difference between the number of persons born per year (birthrate, or natality) and the number who die per year (death rate, or mortality). It is customary to record these rates per 1,000 persons. For example, the world at the present time has a birthrate of 21 per 1,000 per year, but it has a death rate of 9 per 1,000 per year. This means that the world's population growth, or its growth rate, is

$$\frac{21 - 9}{1,000} \ = \ \frac{12}{1,000} \ = \ 0.012 \ \times \ 100 \ = \ 1.2\%$$

(While the birthrate and death rate are expressed in terms of 1,000 persons, the growth rate is expressed per 100 persons, or as a percentage.) After 1750, the world population growth rate steadily increased, until it peaked at 2% in 1965. It has since fallen to its present 1.2%. Yet, the world population is still steadily growing because of its past exponential growth.

In the wild, exponential growth indicates that a population is enjoying its **biotic potential.** This is the maximum growth rate under ideal conditions. Growth begins to decline because of limiting factors such as food and space. Finally, the population levels off at the carrying capacity. The **carrying capacity** is the maximum population that the environment can support for an indefinite period. The carrying capacity of the Earth for humans has not been determined. Some authorities think the Earth may be able to sustain 50–100 billion people. Others think we already have more humans than the Earth can adequately support.

The MDCs Versus the LDCs

The countries of the world can be divided into two groups. The more-developed countries (MDCs), typified by countries in North America and Europe, are those in which population growth is modest. The people in these countries enjoy a good standard of living. The less-developed countries (LDCs), typified by some countries in Asia, Africa, and Latin America, are those in which population growth is dramatic. The majority of people in these countries live in poverty.

The MDCs

The MDCs did not always have low population increases. Between 1850 and 1950, they doubled their populations. This was largely because of a decline in the death rate due to development of modern medicine and improvements in public health and socioeconomic conditions. The decline in the death rate was followed shortly thereafter by a decline in the birthrate. As a result, the MDCs have experienced only modest growth since 1950 (Fig. 24.1).

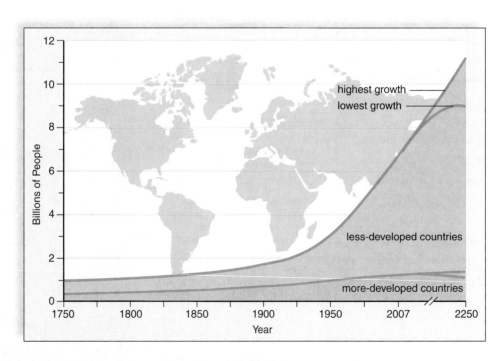

Figure 24.1   **Based on current statistics, what can be predicted for human population by 2050?**
The world's population of humans is now close to 7 billion. It is predicted that the world's population size may level off at 9 billion or increase to more than 11 billion by 2250, depending on the speed with which the growth rate declines.

The growth rate for the MDCs as a whole is about 0.1%. In some countries, population is not increasing or is decreasing in size. The MDCs are expected to increase by 52 million between 2002 and 2050, but this amount will still keep their total population at just about 1.2 billion. In contrast to the other MDCs, growth in the United States has not leveled off. The population of the United States is now greater than 300 million and continues to increase. While the birth rate in the United States has increased slightly, much of the continued population growth is due to immigration.

The LDCs

The death rate began to decline steeply in the LDCs following World War II with the introduction of modern medicine. However, the birthrate remained high. The growth rate of the LDCs peaked at 2.5% between 1960 and 1965. Since that time, the collective growth rate for the LDCs has declined. However, the growth rate has not declined in all LDCs. In many countries in sub-Saharan Africa, women give birth to more than five children each.

Between 2002 and 2050, the population of the LDCs may jump from 5 billion to at least 8 billion. Some of this increase will occur in Africa, but most will occur in Asia. Many deaths from AIDS are slowing the growth of the African population. Continued growth in Asia is expected to cause acute water scarcity, a significant loss of biodiversity, and more urban pollution. Twelve of the world's 15 most polluted cities are in Asia.

Comparing Age Structure

The LDCs are experiencing a population momentum because they have more women entering the reproductive years than older women leaving them. Populations have three age groups: prereproductive, reproductive, and postreproductive. This is best visualized by plotting the proportion of individuals in each group on a bar graph. This produces an age-structure diagram (Fig. 24.2).

Laypeople are sometimes under the impression that if each couple has two children, zero population growth will take place immediately. However, **replacement reproduction,** as this practice is called, will still cause most LDCs today to have a positive growth rate. This is because there are more young women entering the reproductive years than older women leaving them.

Most MDCs have a stabilized age-structure diagram. Therefore, their populations are expected to remain just about the same or decline if couples are having fewer than two children each.

Check Your Progress 24.1

1. Are the MDCs or the LDCs undergoing a larger population growth?
2. The growth rate of the world's population is decreasing. Why, then, is the population increasing so wildly?

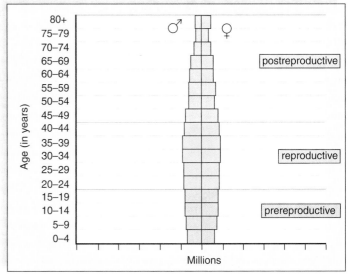

a. More-developed countries (MDCs)

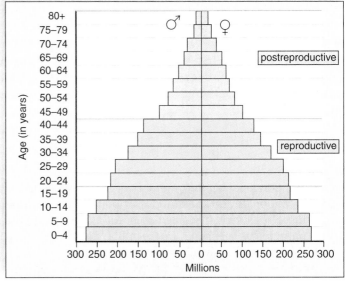

b. Less-developed countries (LDCs)

c.

Figure 24.2   What do these age-structure diagrams indicate about populations of MDCs and LDCs?
The shape of these age-structure diagrams allows us to predict that **(a)** the populations of MDCs are approaching stabilization, and **(b)** the populations of LDCs will continue to increase for some time. **c.** Improved women's rights and increasing contraceptive use could change this scenario. Here a community health worker is instructing women in Bangladesh about the use of contraceptives.

24.2 Human Use of Resources and Pollution

Human beings have certain basic needs. A resource is anything from the biotic or abiotic environment that helps meet these needs. Land, water, food, energy, and minerals are the maximally used resources that will be discussed in this chapter (Fig. 24.3). The total amount of resources used by an individual to meet his/her needs is sometimes referred to as an ecological footprint. A person can make their ecological footprint smaller by driving an energy-efficient car, living in a smaller house, owning fewer possessions, eating vegetables as opposed to meat, and so forth.

Some resources are nonrenewable, and some are renewable. **Nonrenewable resources** are limited in supply. For example, the amount of land, fossil fuels, and minerals is finite and can be exhausted. Efficient use, recycling, or substitution can make the supply last longer, but eventually these resources will run out.

Renewable resources are capable of being naturally replenished. We can use water and certain forms of energy (e.g., solar energy) or harvest plants and animals for food. A new supply will always be forthcoming. Even with renewable resources, though, we have to be careful not to squander them.

Unfortunately, a side effect of resource consumption can be pollution. **Pollution** is any undesired alteration of the environment and is often caused by human activities. The effect of humans on the environment is proportional to the population size. As the population grows, so does the need for resources and the amount of pollution caused by using these resources. Seven people adding waste to the ocean may not be alarming, but 7 billion people doing so would certainly affect its cleanliness. In modern times, the consumption of mineral and energy resources has grown faster than population size. This has occurred as people in LDCs have increased their use of resources.

Land

People need a place to live. Naturally, land is also needed for a variety of uses aside from homes. Land is used for agriculture, electric power plants, manufacturing plants, highways, hospitals, schools, and so on.

Beaches and Human Habitation

At least 40% of the world population lives within 100 km (60 mi) of a coastline, and this number is expected to increase. Living right on the coast is an unfortunate choice because it leads to beach erosion. Loss of habitat for marine organisms and loss of a buffer zone for storms also occur. Figure 24.4 shows how severe the problem can be in the United States. Coastal wetlands are also impacted when people fill them in. One reason to protect coastal wetlands is that they are spawning areas for fish and other forms of marine life. They are also habitats for certain terrestrial species, including many types of birds. Wetlands also protect coastal areas from

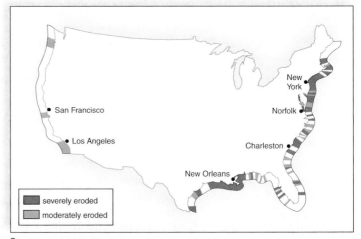

a.

b.

Figure 24.4   As beach erosion occurs, what happens to homes built right on the coast?

a. Most of the United States coastline is subject to beach erosion. **b.** Therefore, people who choose to live near the coast may eventually lose their homes.

Human population

| land | water | food | energy | minerals |

Figure 24.3   What resources are used by humans to meet their basic needs?

Human beings use land, water, food, energy, and minerals to meet their basic needs, such as a place to live, food to eat, and products that make their lives easier.

storms. The coast is particularly subject to pollution because toxic substances placed in freshwater lakes, rivers, and streams may eventually find their way to the coast.

Semiarid Lands and Human Habitation

Forty percent of the Earth's lands are already deserts. Land adjacent to a desert is in danger of becoming unable to support human life if it is improperly managed by humans (Fig. 24.5). **Desertification** is the conversion of semiarid land to desertlike conditions.

Often, desertification begins when humans allow animals to overgraze the land. The soil can no longer hold rainwater, and it runs off instead of keeping the remaining plants alive or replenishing wells. Humans then remove whatever vegetation they can find to use as fuel or fodder for their animals. The result is a lifeless desert. That area is then abandoned as people move on to continue the process someplace else. Some estimate that nearly three-quarters of all rangelands worldwide are in danger of desertification. Many

famines are due, at least in part, to degradation of the land to the point that it can no longer support human beings and their livestock.

Tropical Rain Forest and Human Habitation

Deforestation, the removal of trees, has long allowed humans to live in areas where forests once covered the land (Fig. 24.6). This land, too, is subject to desertification. Soil in the tropics is often thin and nutrient-poor because all the nutrients are tied up in the trees and other vegetation. When the trees are felled and the land is used for agriculture or grazing, it quickly loses its fertility. Then it is subject to desertification.

Water

In the water-poor areas of the world, people may not have ready access to drinking water, and if they do, the water may be impure. It's considered a human right for people to

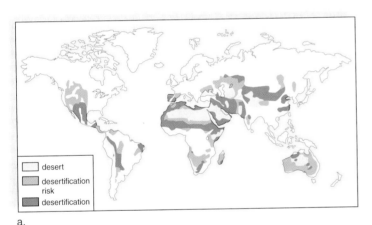

a.

b.

Figure 24.5 How does desertification affect humans?
a. Desertification is a worldwide occurrence that **(b)** reduces the amount of land suitable for human habitation.

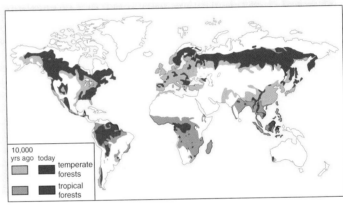

a.

b.

Figure 24.6 Does deforestation yield land suitable for long-term farming?
a. Nearly half of the world's forest lands have been cleared for farming, logging, and urbanization. **b.** The soil of tropical rain forests is not suitable for long-term farming.

a. Agriculture uses most of the freshwater consumed.

b. Industrial use of water is about half that of agricultural use.

c. Domestic use of water is about half that of industrial use.

Figure 24.7   How is water used by agriculture, industries, and households?
a. Agriculture primarily uses water for irrigation. **b.** Industry uses water variously. **c.** Households use water to drink, shower, flush toilets, and water lawns.

have clean drinking water. In reality, most freshwater is used by industry and agriculture (Fig. 24.7). Worldwide, 70% of freshwater is used to irrigate crops! Much of a recent surge in demand for water stems from increased industrial activ-

ity and irrigation-intensive agriculture. This type of agriculture now supplies about 40% of the world's food crops. In the MDCs, more water is usually used for bathing, flushing toilets, and watering lawns than for drinking and cooking.

Increasing Water Supplies

The needs of the human population overall do not exceed the renewable supply. However, this is not the case in certain regions of the United States and the world. About 40% of the world's land is desert, and deserts are bordered by semiarid land. When needed, humans increase the supply of freshwater by damming rivers and withdrawing water from aquifers.

Dams   The world's 45,000 large dams catch 14% of all precipitation runoff and provide water for up to 40% of irrigated land. They also give some 65 countries more than half their electricity. Damming of certain rivers has been so extensive that they no longer flow as they once did. The Yellow River in China fails to reach the sea most years. The Colorado River barely makes it to the Gulf of California. Even the Rio Grande dries up before it can merge with the Gulf of Mexico. The Nile in Egypt and the Ganges in India are also so overexploited they hardly make it to the ocean at some times of the year.

There are other drawbacks to dams. They lose water due to evaporation and seepage into underlying rock beds. The amount of water lost sometimes equals the amount made available! The salt left behind by evaporation and agricultural runoff increases salinity and can make a river's water unusable farther downstream. Dams hold back less water with time because of sediment buildup. Sometimes a reservoir becomes so full of silt that it is no longer useful for storing water.

Aquifers   To meet their freshwater needs, people are pumping vast amounts of water from **aquifers.** These are reservoirs found just below or as much as 1 km below the surface. Aquifers hold about 1,000 times the amount of water that falls on land as precipitation each year. This water accumu-

Figure 24.8 **What causes sinkholes?**
Sinkholes occur when an underground cavern collapses after ground-water has been withdrawn.

a.

b.

c.

Figure 24.9 **What measures can be used to conserve water?**
a. Planting drought-resistant crops in the field and drought-resistant plants in parks and gardens cuts down on the need to irrigate. **b.** When irrigation is necessary, drip irrigation is preferable to using sprinklers. **c.** Wastewater can be treated and reused instead of withdrawing more water from a river or aquifer.

lates from rain that fell in far-off regions. In the past 50 years, groundwater depletion has become a problem in many areas of the world. In substantial portions of the High Plains Aquifer (stretching from South Dakota to Texas) more than half of the water has been pumped out.

Consequences of Groundwater Depletion Removal of water from aquifers is causing land **subsidence,** a settling of the soil as it dries out. An area in California's San Joaquin valley has subsided at least 30 cm due to groundwater depletion. In the worst spot, the surface of the ground has dropped more than 9 meters! Subsidence damages canals, buildings, and underground pipes. Withdrawal of groundwater can cause **sinkholes.** These form when an underground cavern collapses because water no longer holds up its roof (Fig. 24.8).

Saltwater intrusion is another consequence of aquifer depletion. The flow of water from streams and aquifers usually keeps them fairly free of seawater. As water is withdrawn, the water table can lower to the point that seawater backs up into streams and aquifers. Saltwater intrusion reduces the supply of freshwater along the coast.

Conservation of Water

By 2025, two-thirds of the world's population may be living in countries facing serious water shortages. Some solutions for expanding water supplies have been suggested. Planting drought- and salt-tolerant crops would help a lot. Using drip irrigation delivers more water to crops and increases crop yields as well (Fig. 24.9). Although the first drip systems were developed in 1960, they're used on only less than

1% of irrigated land. Most governments subsidize irrigation so heavily that farmers have little incentive to invest in drip systems or other water-saving methods. Reusing water and adopting conservation measures could help the world's industries cut their water demands by more than half.

Food

In 1950, the human population numbered 2.5 billion. There was only enough food produced to provide less than 2,000 calories per person per day. Now, with almost 7 billion people on Earth, the world food supply provides more calories per person per day. Generally speaking, food comes from three activities: growing crops, raising animals, and fishing the seas. The increase in the food supply has largely been possible because of modern farming methods. Unfortunately, many of these methods include some harmful practices:

1. Planting of a few genetic varieties. The majority of farmers practice monoculture. Wheat farmers plant the same type of wheat, and corn farmers plant the same type of corn. Monoculture means that a single type of parasite can destroy entire crops.

2. Heavy use of fertilizers, pesticides, and herbicides. Fertilizer production is energy intensive, and fertilizer runoff contributes to water pollution. Pesticides reduce soil fertility because they kill off beneficial soil organisms as well as pests. Some pesticides and herbicides are linked to the development of cancer. **Agricultural runoff** places these chemicals in our water supply.

3. Generous irrigation. As already discussed, water is sometimes taken from aquifers. In the future, the water content of these aquifers may become so reduced that it could be too expensive to pump out any more.

4. Excessive fuel consumption. Irrigation pumps remove water from aquifers. Large farming machines are used to spread fertilizers, pesticides, and herbicides, as well as to sow and harvest the crops. In effect, modern farming methods transform fossil fuel energy into food energy.

Figure 24.10 shows ways to minimize the harmful effects of modern farming practices. Polyculture is the planting of two or more different crops in the same area. In Figure 24.10a, a farmer has planted alfalfa in between strips of corn. The alfalfa replenishes the nitrogen content of the soil so that fertilizer doesn't have to be added. In Figure 24.10b, contour farming with no-till conserves topsoil because it reduces

agricultural runoff. Contour farming is planting and plowing according to the slope of the land. No-till farming allows the previous crop to remain on the land, recycling nutrients and preventing soil erosion. Biological control (Fig. 24.10c) relies on the use of natural predators to destroy organisms that harm crops. This reduces the need for pesticides.

Soil Loss

Land suitable for farming and grazing animals is being degraded worldwide. Topsoil is the richest in organic matter and the most capable of supporting grass and crops. When bare soil is acted on by water and wind, soil erosion occurs and topsoil is lost. As a result, marginal rangeland becomes desertized, and farmland loses its productivity.

The custom of planting the same crop in straight rows that facilitates the use of large farming machines has caused the United States and Canada to have one of the highest rates of soil erosion in the world. Conserving the nutrients now being lost could save farmers billions of dollars annually in fertilizer costs. Much of the eroded sediment ends up in lakes and streams, where it reduces the ability of aquatic species to survive.

Green Revolutions

About 50 years ago, researchers began to breed tropical wheat and rice varieties specifically for farmers in the LDCs. The dramatic increase in yield due to the introduction of these new varieties around the world was called "the green revolution." These plants helped the world food supply keep pace with the rapid increase in world population. Unfortunately, most green revolution plants are called "high

a. Polyculture

b. Contour farming

c. Biological pest control

Figure 24.10   **What methods make farming more friendly to the environment?**
a. Polyculture reduces the ability of one parasite to wipe out an entire crop and reduces the need to use a herbicide to kill weeds. This farmer has planted alfalfa in between strips of corn, which also replenishes the nitrogen content of the soil (instead of adding fertilizers). Alfalfa, a legume, has root nodules that contain nitrogen-fixing bacteria. **b.** Contour farming with no-till conserves topsoil because water has less tendency to run off. **c.** Instead of pesticides, it is possible to use a natural predator. Here, ladybugs are feeding on cottony-cushion scale insects on citrus trees.

Figure 24.11 How does raising livestock such as these pigs affect the environment?
Raising livestock requires the use of more fossil fuels and water than raising crops. Livestock waste often washes into nearby bodies of water creating water pollution.

responders" because they need high levels of fertilizer, water, and pesticides to produce a high yield. They require the same subsidies and create the same ecological problems as do modern farming methods.

Genetic Engineering As discussed in Chapter 21 (see page 507), genetic engineering can produce transgenic plants with new and different traits. For example, resistance to both insects and herbicides are traits that can be introduced into plant DNA. When herbicide-resistant crops are planted, weeds are easily controlled, less tillage is needed, and soil erosion is minimized. Researchers also want to produce crops that tolerate salt, drought, and cold. Some progress has also been made in increasing the food quality of crops so that they will supply more of the proteins, vitamins, and minerals people need. Genetically engineered crops could result in still another green revolution.

Some are opposed to the use of genetically engineered crops. It is feared that these crops will damage the environment and lead to health problems in humans. The Health Focus on page 509 discusses this issue.

Domestic Livestock

A low-protein, high-carbohydrate diet consisting only of grains such as wheat, rice, or corn can lead to malnutrition. In the LDCs, kwashiorkor, caused by a severe protein deficiency, is seen in infants and children ages 1–3. It usually occurs after a new arrival in the family and the older children are no longer fed milk. The diet then consists of protein-poor starches. Such children are lethargic, irritable, and have bloated abdomens. Mental retardation is expected.

In the MDCs, many people tend to have more than enough protein in their diet. Almost two-thirds of United States cropland is devoted to producing livestock feed. This means that a large percentage of the fossil fuel, fertilizer, water, herbicides, and pesticides used are for the purpose of raising livestock. Typically, cattle are range-fed for about four months. Then they are brought to crowded feedlots where they may receive growth hormone and antibiotics. At the feedlots, they feed on

grain or corn. Most pigs and chickens spend their entire lives cooped up in crowded pens and cages (Fig. 24.11).

If livestock eat a large proportion of the crops in the United States, then raising livestock accounts for much of the pollution associated with farming. Fossil fuel energy is needed not just to produce herbicides and pesticides and to grow food, but also to make the food available to the livestock. Raising livestock is extremely energy-intensive in the MDCs. In addition, water is used to wash livestock wastes into nearby bodies of water, where they add significantly to water pollution. Whereas human wastes are sent to sewage treatment plants, raw animal wastes are not.

For these reasons, it is prudent to recall the ecological energy pyramid (see Fig. 23.7), which shows that as you move up the food chain, energy is lost. As a rule of thumb, for every 10 calories of energy from a plant, only 1 calorie is available for the production of animal tissue in a herbivore. A great deal of energy is wasted when the human diet contains more protein than is needed to maintain good health. It is possible to feed ten times as many people on grain as on meat.

> **Check Your Progress 24.2**
> 1. What five resources are maximally used by humans?
> 2. Which ecosystems suffer the most due to human habitation?
> 3. a. How do humans increase the supply of freshwater, and (b) what are the consequences of doing so?
> 4. a. What farming methods are in use today, and (b) what are the possible environmental consequences?

Energy

Modern society runs on various sources of energy. Some of these energy sources are nonrenewable, and others are renewable. The consumption of nonrenewable energy supplies results in environmental degradation. Renewable energy is expected to be used more in the future.

Nonrenewable Sources

Presently, about 6% of the world's energy supply comes from nuclear power, and 75% comes from fossil fuels. Both of these are finite, nonrenewable sources. It was once predicted that the nuclear power industry would fulfill a significant portion of the world's energy needs. However, this has not happened for two reasons. One is that people are very concerned about nuclear power dangers, such as the meltdown at the Chernobyl nuclear power plant in Russia in 1986. Further, radioactive wastes from nuclear power plants remain a threat to the environment for thousands of years. We still have not decided how best to safely store them.

Fossil fuels (oil, natural gas, and coal) are derived from the compressed remains of plants and animals that died thousands of years ago. Of the fossil fuels, oil burns more cleanly than coal, which may contain a lot of sulfur. When the use of coal releases sulfur, acid rain forms. Thus, despite that coal is plentiful in the United States, imported oil is our preferred fossil fuel. Regardless of which fossil fuel is used, all contribute to environmental problems because of the pollutants released when they're burned. Current research into clean-coal technology may make U.S. coal less polluting.

Fossil Fuels and Global Climate Change   In 1850, the level of carbon dioxide in the atmosphere was about 280 parts per million (ppm), and today it is about 350 ppm. This increase is largely due to the burning of fossil fuels and the burning and clearing of forests. Human activities are causing the emission of other gases as well. These gases are known as **greenhouse gases.** Just like the panes of a greenhouse, they allow solar radiation to pass through but hinder the escape of infrared heat back into space.

Computer models predict the Earth may warm to temperatures never before experienced by living things. The global climate has already warmed about 0.6°C, and it may rise as much as 1.5–4.5°C by 2100. If so, sea levels will rise as glaciers melt and warm water expands. Major coastal cities of the United States could eventually be threatened. The present wetlands will be inundated. Great losses of aquatic habitat will occur wherever wetlands cannot move inward because of coastal development and levees. Coral reefs, which prefer shallow waters, will most likely "drown" as the waters rise.

On land, regions of suitable climate for various species will shift toward the poles and higher elevations. Plants migrate when seeds disperse and growth occurs in a new locale. The present assemblages of species in ecosystems will be disrupted as some species migrate northward faster than others. Trees, for example, cannot migrate as fast as nonwoody plants. Also, too many species of organisms are confined to relatively small habitat patches surrounded by agricultural or urban areas. Even if such species have the capacity to disperse to new sites, suitable habitats may not be available.

Renewable Energy Sources

Renewable types of energy include hydropower, geothermal, wind, and solar.

Hydropower   Hydroelectric plants convert the energy of falling water into electricity (Fig. 24.12a). Hydropower accounts for about 10% of the electric power generated in the United States and almost 98% of the total renewable energy used. Worldwide, hydropower presently generates 19% of all electricity used. This percentage is expected to rise because of increased use in certain countries.

Much of the hydropower development in recent years has been due to the construction of enormous dams. These are known to have detrimental environmental effects (see page 568). The better choice is believed to be small-scale dams that generate less power per dam but do not have the same environmental impact.

Geothermal Energy   Elements such as uranium, thorium, radium, and plutonium undergo radioactive decay below the Earth's surface. This heats the surrounding rocks to hundreds of degrees Celsius. When the rocks are in contact with underground streams or lakes, huge amounts of steam and hot water are produced. This steam can be piped up to the surface to supply hot water for home heating or to run steam-driven turbogenerators. The California's Geysers project is the world's largest geothermal electricity-generating complex.

Wind Power   Wind power is expected to account for a significant percentage of our energy needs in the future. A common belief is that a huge amount of land is required for the "wind farms" that produce commercial electricity. The amount of land needed for a wind farm compares favorably with the amount of land required by a coal-fired power plant or a solar thermal energy system (Fig. 24.12b).

A community generating its own electricity by using wind power can solve the problem of uneven energy production. Electricity can be sold to a local public utility when an excess is available. Then electricity is bought from the same facility when wind power is in short supply.

Energy and the Solar-Hydrogen Revolution   Solar energy is diffuse energy that must be collected, converted to another form, and stored if it is to compete with other available forms of energy. Passive solar heating of a house is successful when the windows of the house face the sun, and the building is well insulated. Successful heating also requires that heat can be stored in water tanks, rocks, bricks, or some other suitable material.

In a **photovoltaic (solar) cell,** a wafer of the electron-emitting metal is in contact with another metal that collects the electrons. Electrons are then passed along into wires in a steady stream. Spurred by the oil shocks of the 1970s, the United States government has been supporting the development of photovoltaics ever since. As a result, the price of buying one has dropped from about $100 per watt to around $4. Photovoltaic cells placed on roofs generate electricity that can be used inside a building and/or sold back to a power company (Fig. 24.12c).

Several types of solar power plants are now operational in California. In one type, huge reflectors focus sunlight on a pipe containing oil. The heated pipes boil water, generating

a.

b.

c.

d.

Figure 24.12   **What are some sources of renewable energy?**
a. Hydropower dams provide a clean form of energy but can be ecologically disastrous in other ways. **b.** Wind power requires land on which to palce enough windmills to generate energy. **c.** Photovoltaic cells on rooftops and **(d)** sun-tracking mirrors on land can collect diffuse solar energy more cheaply than could be done formerly.

steam that drives a conventional turbogenerator. In another type, 1,800 sun-tracking mirrors focus sunlight onto a molten salt receiver mounted on a tower. The hot salt generates steam that drives a turbogenerator (Fig. 24.12d).

Scientists are working on the possibility of using solar energy to extract hydrogen from water via electrolysis. The hydrogen can then be used as a clean-burning fuel. When it burns, water is produced. Presently, cars have internal combustion engines that run on gasoline. In the future, vehicles are expected to be powered by fuel cells, which use hydrogen

to produce electricity (Fig. 24.13). The electricity runs a motor that propels the vehicle. Fuel cells are now powering buses in Vancouver and Chicago. Additional buses are planned.

Hydrogen fuel can be produced locally or in central locations, using energy from photovoltaic cells. The fuel produced in central locations can be piped to filling stations using the natural gas pipes already plentiful in the United States. However, two major hurdles are storage and production. Advantages of a solar-hydrogen revolution are decreased dependence on oil and fewer environmental problems.

Figure 24.13 How can hydrogen fuel cells reduce air pollution and fossil fuel consumption?
a. Hydrogen fuel cells. **b.** This bus is powered by hydrogen fuel. **c.** The use of fuel-cell hybrid vehicles, such as this prototype, will reduce air pollution and dependence on fossil fuels.

Minerals

Minerals are nonrenewable raw materials in the Earth's crust that can be mined (extracted) and used by humans. Nonmetallic raw materials, such as sand, gravel, and phosphate are considered minerals. Metals, such as aluminum, copper, iron, lead, and gold fall into this category, as well.

One of the greatest threats to the maintenance of ecosystems and biodiversity is surface mining, called strip mining. In the United States, huge machines can go as far as removing mountain tops to reach a mineral. The land devoid of vegetation takes on a surreal appearance, and rain washes toxic waste deposits into nearby streams and rivers.

The most dangerous metals to human health are the heavy metals. These include lead, mercury, arsenic, cadmium, tin, chromium, zinc, and copper. They are used to produce batteries, electronics, pesticides, medicines, paints, inks, and dyes. In the ionic form, they enter the body and inhibit vital enzymes. That's why these items should be discarded carefully and taken to hazardous waste sites.

Hazardous Wastes

The consumption of minerals and use of synthetic organic chemicals contribute to the build up of hazardous waste in the environment. Every year, countries around the world discard billions of tons of solid waste on land and in freshwater and salt water. The Environmental Protection Agency (EPA) oversees the cleanup of hazardous waste disposal sites in the United States. An allocation of monies called the Superfund helps pay for this cleanup. Commonly found contaminants include heavy metals (such as lead, mercury, and arsenic) and chlorine-containing organic chemicals (such as chloroform and polychlorinated biphenyls, or PCBs). Some of these contaminants interfere with hormone activity and proper endocrine system functioning.

Humanmade organic chemicals play a role in the production of plastics, pesticides, herbicides, cosmetics, and hundreds of other products. For example, the so-called *halogenated hydrocarbons* are compounds made from carbon and hydrogen, and also include halogen atoms such as chlorine and fluorine. These compounds, **chloroflourocarbons (CFCs),** have been shown to damage the Earth's ozone shield. Recall from Chapter 23 that

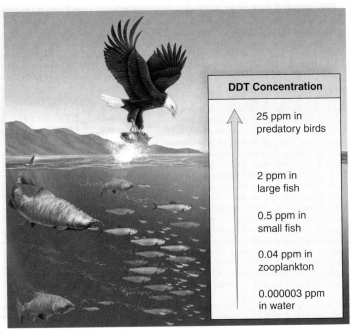

Figure 24.14 **What occurs during biological magnification?** Various synthetic organic chemicals, such as DDT, accumulate in animal fat. Therefore, the chemical becomes increasingly concentrated at higher trophic levels. By the time DDT was banned in the United States, it had interfered with predatory bird reproduction by causing egg-shell thinning.

this shield protects terrestrial life from harmful UV radiation and has recently been depleted by the use of CFCs.

Further, these types of synthetic compounds pose a threat to the health of living things, including humans, because they undergo **biological magnification.** Such chemicals accumulate in fat and are not excreted, so they remain in an organism's body. Therefore, they become more concentrated as they pass from organism to organism along a food chain. Biological magnification is more apt to occur in aquatic food chains, which have more links than terrestrial food chains. Rachel Carson's book *Silent Spring*, published in 1962, made the public aware of the harmful effects of pesticides such as DDT. These substances accumulate in the mud of deltas and estuaries of highly polluted rivers and cause environmental problems if disturbed. After working its way up the food chain, high concentrations of DDT in predatory birds like bald eagles and pelicans interfered with their ability to reproduce (Fig. 24.14). There are health advisories about eating certain types of fish due to the high levels of mercury they contain. Humans are often the final consumers in a variety of food chains and are affected by biological magnification as well. The breast milk of humans has been found to contain significant levels of DDT, PCBs, solvents, and heavy metals.

Raw sewage causes oxygen depletion in lakes and rivers. As the oxygen level decreases, the diversity of life is greatly reduced. Also, human feces can contain pathogenic microorganisms that cause cholera, typhoid fever, and dysentery. In regions of the LDCs where sewage treatment is practically nonexistent, many children die each year from these diseases. Typically, sewage treatment plants use bacteria to break down organic matter to inorganic nutrients, such as nitrates and

phosphates, which then enter surface waters. The result can be cultural eutrophication discussed on page 557.

> **Check Your Progress 24.3**
> 1. Why is it better for humans to use renewable rather than nonrenewable sources of energy?
> 2. The consumption of mineral resources has what environmental consequences?
> 3. Name some hazardous wastes present in today's environment.

24.3 Biodiversity

Biodiversity can be defined as the variety of life on Earth and described in terms of the number of different species. We are presently in a biodiversity crisis. The number of extinctions (loss of species) expected to occur in the near future is unparalleled in the history of the Earth.

Loss of Biodiversity

Figure 24.15 identifies the major causes of extinction.

Figure 24.15 **What causes the loss of biodiversity?**
a. Habitat loss, alien species, pollution, overexploitation, and disease have been identified as causes of extinction of organisms. **b.** Macaws that reside in South American tropical rain forests are endangered for the reasons listed in the graph.

Habitat Loss

Human occupation of the coastline, semiarid lands, tropical rain forests (see pages 566–67), and other areas have contributed to the loss of biodiversity. Scientists are especially concerned about the tropical rain forests and coral reefs because they are particularly rich in species. Already, tropical rain forests have been reduced from their original 14% of landmass to the present 6%. Also, 60% of coral reefs have been destroyed or are on the verge of destruction. It's possible that all coral reefs may disappear during the next 40 years.

Alien Species

Alien species, sometimes called exotics, are nonnative members of an ecosystem. Humans have introduced alien species into new ecosystems via colonization, horticulture and agriculture, and accidental transport. For example, the pilgrims brought the dandelion to the United States as a familiar salad green. Kudzu is a vine from Japan that the U.S. Department of Agriculture thought would help prevent soil erosion. The plant now covers much landscape in the South. The zebra mussel from the Caspian Sea was accidentally introduced into the Great Lakes in 1988. It is now found in many U.S. rivers where it forms dense beds that squeeze out native mussels. Alien species that crowd out native species are termed **invasive.** One way to counteract alien species is to replant native (original to the area) species.

Pollution

Pollution brings about environmental change that adversely affects the lives and health of living things. Acid deposition weakens trees and increases their susceptibility to disease and insects. This can decimate a forest. Global warming is predicted to be the cause of habitat loss due to temperature shifts. For example, coral reefs die off as ocean temperatures increase. Coastal wetlands may be lost as sea levels increase.

The depletion of the ozone shield limits crop and tree growth and kills plankton that sustain ocean life. Human-made organic chemicals released into the environment may interfere with hormone function and affect the reproductive ability of different species.

Overexploitation

Overexploitation occurs when the number of individuals taken from a wild population is so great that the population becomes severely reduced. A positive feedback cycle explains overexploitation. The smaller the population, the more valuable its members, and the greater the incentive to exploit the few remaining organisms.

Markets for decorative plants and exotic pets support both legal and illegal trade in wild species. Rustlers dig up rare cacti, such as the single-crested saguaro, and sell them to gardeners. Parakeets and macaws are among the birds taken from the wild for sale to pet owners. For every bird delivered alive, many more have died in the process. The same holds true for tropical fish, which often come from the coral reefs of Indonesia and the Philippines. Divers dynamite reefs or use plastic squeeze-bottles of cyanide to stun them. In the process, many fish die.

Declining species of mammals are still hunted for their hides, tusks, horns, or bones. A single Siberian tiger is now worth more than $500,000 because of its rarity. Its bones are pulverized and used as a medicinal powder. The horns of rhinoceroses become ornate carved daggers, and their bones are ground up to sell as a medicine. The ivory of an elephant's tusk is used to make art objects, jewelry, or piano keys. The fur of a Bengal tiger sells for as much as $100,000 in Tokyo.

Fish are a renewable resource if harvesting does not exceed the ability of the fish to reproduce. Today, larger and more efficient fishing fleets decimate fishing stocks (Fig. 24.16). Tuna and similar fish are captured by purse-seine fishing. A very large net surrounds a school of fish, and then the net is closed in the same manner as a drawstring purse. Dolphins that accompany the tuna are killed by this type of net. Other fishing boats drag huge trawling nets, large enough to accommodate 12 jumbo jets, along the seafloor to capture bottom-dwelling fish. Trawling has been called the marine equivalent of clear-cutting trees because after the net goes by, the sea bottom is devastated (Fig. 24.16b). Only large fish are kept. Undesirable small fish and sea turtles are discarded, dying, back into the ocean. Cod and haddock were once the most abundant bottom-dwelling fish along the Northeast coast. Now they are often outnumbered by dogfish and skate.

A marine ecosystem can be disrupted by overfishing, as exemplified on the United States West coast. When sea otters began to decline in numbers, investigators found that they were being eaten by orcas (killer whales). Usually, orcas prefer seals and sea lions to sea otters, but they began eating sea otters when seals and sea lions could not be found. What caused a decline in seals and sea lions? Their

Have You Ever Wondered...

Why are there beagle dogs at the Customs area in an international airport?

The beagles working at the airport's Customs checkpoint are trained to detect agricultural items (foodstuffs, plants, animals, and the like) that may not be brought into the United States. Importing foreign species is prohibited by law. The goal of this law is to prevent the introduction of exotic life and/or diseases that might affect crops or animals in the United States. About $2 million worth of illegal articles are seized yearly. Dogs assist in about 10% of those cases. Beagles are used because of their keen sense of smell and good-natured temperament.

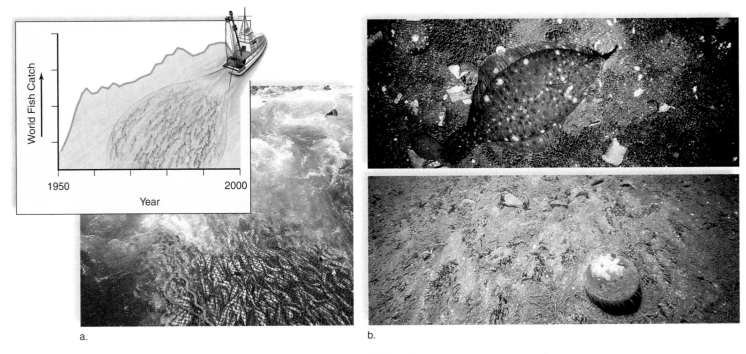

Figure 24.16 **How have modern fishing methods impacted the fish catch?**
a. The world fish catch has declined in recent years (insert). **b.** Devastation of the seafloor after trawling (lower photo).

preferred food sources (perch and herring) were no longer plentiful due to overfishing. Ordinarily, sea otters keep the population of sea urchins, which feed on kelp, under control. But with fewer sea otters around, the sea urchin population exploded and decimated the kelp beds. Thus, overfishing set in motion a chain of events that adversely affected the food web of an ecosystem.

Disease

Wildlife is subject to emerging diseases just as humans are. Exposure to domestic animals and their pathogens occurs due to the encroachment of humans on wildlife habitats. Wildlife can also be infected by animals not ordinarily encountered. For example, African elephants carry a strain of herpes virus that is fatal to Asian elephants. Deaths can result if the two types of elephants are housed together.

The significant effect of diseases on biodiversity is underscored by a National Wildlife Health Center study. The study found that almost half of sea otter deaths along the coast of California are due to infectious diseases. Scientists tell us that the number of pathogens that cause disease are on the rise. Just as human health is threatened, so is that of wildlife. Extinctions due to disease may occur.

Direct Value of Biodiversity

Various individual species perform useful services for human beings and contribute greatly to the value of biodiversity. The direct value of wildlife is related to their medicinal value,

Have You Ever Wondered...

How to choose the best fish to eat?

Before you place your order for fish, take a moment to think about your choice. You may want to consider if it's a sustainable species of fish, one that's being overfished, or one that contains high levels of mercury. Pacific halibut is a wild fish that is a good choice. The Pacific halibut is usually caught with bottom long lines, rather than trawling nets. This fishing technique won't damage the surrounding environment or catch unwanted fish or animals. Tilapia that is farm grown in the United States is another good choice. It's grown in closed inland systems that prevent exposure of the fish to pollutants. But avoid tilapia from Chile or Taiwan—fish there are raised in open systems where pollutants, especially mercury, can affect the fish. You might also want to consider the levels of healthy omega-3 fatty acids in the fish you eat. Farm-raised salmon or trout supply high concentrations of omega-3 acids (thought to protect against heart disease) and contain little or no mercury or other pollutants.

Shark, sole, haddock, and swordfish are poor choices. These species have been severely overfished and/or contain high levels of mercury.

agricultural value, and consumptive use value. These are the most obvious values that are discussed here and illustrated in Figure 24.17.

Medicinal Value

Most of the prescription drugs used in the United States were originally derived from living organisms. The rosy periwinkle from Madagascar is a tropical plant that has provided us with useful medicines. Potent chemicals from this plant are now used to treat leukemia and Hodgkin disease. The survival rate for childhood leukemia has gone from 10% to 90%, and Hodgkin disease is usually curable because of these drugs. Although the value of saving a life cannot be calculated, it is still sometimes easier for us to appreciate the worth of a resource if it is explained in monetary terms. Based on past success, it has been estimated that more than 300 types of drugs may yet be found in tropical rain forests. The value of this resource could be in excess of $140 billion.

You may already know that the antibiotic penicillin is derived from a fungus. Certain species of bacteria pro-duce the antibiotics tetracycline and streptomycin. These drugs have proven to be indispensable in the treatment of diseases.

Leprosy is a disease for which there is no cure as of yet. The bacterium that causes leprosy will not grow in the laboratory. However, scientists discovered that it grows naturally in the nine-banded armadillo. Having a source for the bacterium may make it possible to find a cure for leprosy. The blood of horseshoe crabs contains a substance called limulus amoebocyte lysate. This chemical is used to ensure that medical devices, such as pacemakers, surgical implants, and prosthetic devices, are free of bacteria. Blood is taken from 250,000 crabs a year, and then they are returned to the sea unharmed.

Agricultural Value

Crops, such as wheat, corn, and rice, are derived from wild plants that have been modified to be high producers. The same high-yield, genetically similar strains tend to be grown worldwide. At one time rice crops in Africa were being dev-

Figure 24.17   What is the direct value of wildlife?
The direct services of wild species benefit human beings immensely, and it is sometimes possible to calculate the monetary value, which is always surprisingly large.

Wild species, like the rosy periwinkle, are sources of many medicines.

Wild species, like many marine species, provide us with food.

Wild species, like the nine-banded armadillo, play a role in medical research.

astated by a virus. Researchers grew wild rice plants from thousands of seed samples until they found one that contained a gene for resistance to the virus. These wild plants were then used in a breeding program to transfer the gene into high-yield rice plants. If this variety of wild rice had become extinct before being discovered, rice cultivation in Africa might have collapsed.

Biological pest controls (natural predators and parasites) are often preferable to using chemical pesticides. When a rice pest, called the brown planthopper, became resistant to pesticides, farmers began to use natural brown planthopper enemies instead. The economic savings were calculated at well over $1 billion. Similarly, cotton growers in Cañete Valley, Peru, found that the cotton aphid was resistant to the pesticides being used. Research identified natural predators that are now being used to an ever greater degree by cotton farmers. Again, savings have been enormous.

Most flowering plants are pollinated by animals, such as bees, wasps, butterflies, beetles, birds, and bats. The honeybee, *Apis mellifera,* has been domesticated. It now pollinates almost $10 billion worth of food crops annually in the United States. The value of wild bee pollinators to the U.S. agricultural economy has been calculated at $4.1 to $6.7 billion a year. And yet, modern agriculture often kills wild bees by spraying fields with pesticides.

Consumptive Use Value

We have had much success cultivating crops, keeping domesticated animals, growing trees in plantations, and so forth. However, aquaculture, the growing of fish and shellfish for human consumption, has contributed only minimally to human welfare. Instead, most freshwater and marine harvests depend on the catching of wild fish (e.g., trout and tuna), crustaceans (e.g., shrimps and crabs), and mammals (e.g., whales). These aquatic organisms are an invaluable biodiversity resource.

The environment provides all sorts of other products that are sold in the marketplace worldwide. Wild fruits and vegetables, skins, fibers, beeswax, and seaweed are only a few examples. Also, some people obtain their meat directly

Wild species, like the lesser long-nosed bat, are pollinators of agricultural and other plants.

Wild species, like ladybugs, play a role in biological control of agricultural pests.

Wild species, like rubber trees, can provide a product indefinitely if the forest is not destroyed.

✿ Science **Focus**

Mystery of the Vanishing Bees

Imagine standing in the produce section of your supermarket. You're shocked to see that there are no apples, cucumbers, broccoli, onions, pumpkins, squash, carrots, blueberries, avocados, almonds, or cherries. This could happen at grocery stores in the future. All the crops mentioned, as well as many others, are dependent on honey bees for pollination. Your diet and $15 billion worth of crops could suffer if the honey bees aren't available to perform their important job of moving pollen. Although there are wild bee populations that pollinate crops, domestic honey bees are easily managed and transported from place to place when their pollination services are needed.

Colonies of honey bees have experienced a number of health problems since the 1980s. Mites—animals similar to ticks—have always been a danger for bees. Varroa mites and tracheal mites were early causes of colony stress and bee deaths (Fig. 24A). However, beekeepers were very alarmed in 2006 when entire colonies of bees began to vanish. Researchers started referring to the phenomenon as Colony Collapse Disorder (CCD). There doesn't appear to be one single factor that causes seemingly healthy bees to vanish from their hives. Scientists now believe multiple factors may stress bees, causing them to be vulnerable to infection by a parasite or pathogen. The indiscriminate use of pesticides, the strain of being moved from place to place to pollinate crops, and/or poor nutrition (because genetically engineered plants don't provide as much food for the bees) may contribute to CCD.

CCD is also occurring in the honey bee populations in other countries. Ideally, a cause and effective treatment for CCD will be found soon, thanks to worldwide research dedicated to solving the problem, as well as improved funding

Figure 24A　Parasitic mites can destroy entire bee colonies.

from agricultural agencies. Until then, there are things you can do to keep bees in your area healthy. Research and then plant native plants in your yard and garden. These typically require less fertilizer and water than other plants, and they will provide more pollen and nectar for the bees. In the Southern and Midwest regions of the United States, bees enjoy red clover, foxglove, bee balm, and joe-pye weed. Desert willow and manzanita will attract desert bees. Choose palms for tropical areas. In addition, native plants that flower at different times of the year will provide a constant food source. Midday is typically when bees are out foraging, so if you have to use pesticides apply them late in the day. Plants that rely on the honey bees for pollination—as well as your body—will thank you!

from the environment. The economic value of wild pig in the diet of native hunters in Sarawak, East Malaysia, has been calculated to be approximately $40 million per year.

Similarly, many trees are still felled in the natural environment for their wood. Researchers have calculated that a species-rich forest in the Peruvian Amazon is worth far more if the forest is used for fruit and rubber production than for timber production. Fruit and the latex needed to produce rubber can be brought to market for an unlimited number of years. Once the trees are gone, no timber can be harvested until new trees replace those that have been taken.

Indirect Value of Biodiversity

To bring about the preservation of wildlife, it is necessary to make all people aware that biodiversity is a resource of immense value. If we want to preserve wildlife, it is more

economical to save ecosystems than individual species. Ecosystems perform many useful services for modern humans, who increasingly live in cities. These services are said to be indirect because they are pervasive and not easily discernible. Our very survival depends on the functions that ecosystems perform for us. The indirect value of biodiversity can be associated with the following services.

Waste Disposal

Decomposers break down dead organic matter and other types of wastes to inorganic nutrients. The nutrients are then used by the producers within ecosystems. This function aids humans immensely because we dump millions of tons of waste material into natural ecosystems each year. Waste would soon cover the entire surface of our planet without decomposition. We can build expensive sewage treatment plants, but few of them break down solid wastes completely to inorganic nutrients. It is less expensive and more efficient to water plants and trees with partially treated wastewater and let soil bacteria cleanse it completely.

Biological communities are also capable of breaking down and immobilizing pollutants. These include heavy metals and pesticides, that humans release into the environment.

Provision of Freshwater

Few terrestrial organisms are adapted to living in a salty environment. They need freshwater. The water cycle continually supplies freshwater to terrestrial ecosystems. Humans use freshwater in innumerable ways, including drinking it and irrigating their crops. Freshwater ecosystems, such as rivers and lakes, also provide us with fish and other types of organisms for food.

Unlike other commodities, there is no substitute for freshwater. We can remove salt from seawater to obtain freshwater. However, the cost of desalination is about four to eight times the average cost of freshwater acquired via the water cycle.

Forests and other natural ecosystems exert a "sponge effect." They soak up water and then release it at a regular rate. When rain falls in a natural area, plant foliage and dead leaves lessen its impact. The soil then slowly absorbs it, especially if the soil has been aerated by organisms. The water-holding capacity of forests reduces the possibility of flooding. Forests release water slowly for days or weeks after the rains have ceased. Rivers flowing through forests in West Africa release between three and five times as much at the end of the dry season, as do rivers from coffee plantations.

Prevention of Soil Erosion

Intact ecosystems naturally retain soil and prevent soil erosion. The importance of this ecosystem attribute is especially observed following deforestation. In Pakistan, the world's largest dam, the Tarbela Dam, is losing its storage capacity of 12 billion cubic meters many years sooner than expected. Deforestation is causing silt to build up behind the dam, decreasing its storage capacity. At one time, the Philippines was exporting $100 million worth of oysters, mussels, clams, and cockles each year. Now, silt carried down rivers following deforestation is smothering the mangrove ecosystem that serves as a nursery for these shellfish. Most coastal ecosystems are not as bountiful as they once were because of deforestation and a myriad of other assaults.

Biogeochemical Cycles

You'll recall from Chapter 23 that ecosystems are characterized by energy flow and chemical cycling. The biodiversity within ecosystems contributes to the workings of the water, phosphorus, nitrogen, carbon, and other biogeochemical cycles. We depend on these cycles for freshwater, provision of phosphate, uptake of excess soil nitrogen, and removal of carbon dioxide from the atmosphere. When human activities upset the usual workings of biogeochemical cycles, the dire environmental consequences include the release of excess pollutants that are harmful to us. Technology is unable to substitute for any of the biogeochemical cycles.

Regulation of Climate

At the local level, trees provide shade and reduce the need for fans and air conditioners during the summer. In a rainforest, trees maintain area rainfall. Forests would become arid without the trees.

Globally, forests stabilize the climate because they take up carbon dioxide. The leaves of trees use carbon dioxide when they photosynthesize, and the bodies of the trees store carbon. When trees are cut and burned, carbon dioxide is released into the atmosphere. Carbon dioxide makes a significant contribution to global warming, which is expected to be stressful for many plants and animals. Only a small percentage of wildlife may be able to move northward, where the weather will be suitable for them.

Ecotourism

Almost everyone prefers to vacation in the natural beauty of an ecosystem. In the United States, nearly 100 million people enjoy vacationing in a natural setting. To do so, they spend $4 billion each year on fees, travel, lodging, and food. Many tourists want to go sport fishing, whale watching, boat riding, hiking, bird watching, and the like.

> **Check Your Progress 24.4**
> 1. **What are the chief causes of species extinctions?**
> 2. **What are the direct and indirect benefits of preserving wildlife?**

24.4 Working Toward a Sustainable Society

A **sustainable** society would always be able to provide the same amount of goods and services for future generations, as it does at present. At the same time, biodiversity would be preserved.

To achieve a sustainable society, resources cannot be depleted and must be preserved. In particular, future generations need clean air, water, an adequate amount of food, and enough space in which to live. This goal is not possible unless we carefully regulate our consumption of resources today, taking into consideration that the human population is still increasing.

Today's Unsustainable Society

We are quick to realize that population growth in the LDCs creates an environmental burden. However, we also need to consider that the excessive resource consumption of the MDCs also stresses the environment. Sustainability is incompatible with the current level of consumption plus the

To: senoritadonnelly@gmail.com; jhope132@aol.com

From: phippsygurl@yahoo.com

Attachment: summerpalace.jpg forbiddencity.jpg

Hi Megan, Jacob and Adam,

I'm writing in a really sleep-deprived state. The construction starts at 6 A.M. outside of my apartment and doesn't quit until around 11 P.M. No, I'm really not kidding. They are building so fast that is seems like it takes a week to throw up a 20-story high rise. I read someplace that China uses 50% of the world's concrete—and people in Togo are ruining the rainforest digging it all up.

This is a great country. It's so diverse and so beautiful, and the people are very friendly. But they've got real problems with air and water pollution. Meg, I think my pictures will remind you of Mexico City because of the haze. What this country needs is huanbao xiehui—that's a Chinese conservation society, like our Outdoor Club. Miss you all!

generation of wastes practiced by the MDCs. Overpopulation of the LDCs and overconsumption by the MDCs accounts for the increasing amount of worldwide pollution and the extinction of wildlife observable today (Fig. 24.18).

At present, a considerable proportion of land is being used for human purposes (homes, agriculture, factories, etc.). Agriculture uses large inputs of fossil fuels, fertilizer, and pesticides. These create much pollution. More freshwater is used for agriculture than in homes. Almost half of the agricultural yield in the United States goes toward feeding animals. According to the ten-to-one rule of thumb, it takes 10 lb of grain to grow 1 lb of meat. Therefore, it is wasteful for citizens in MDCs to eat as much meat as they do.

Farm animals and crops require freshwater from surface water and groundwater, and so do humans. Available supplies are dwindling, and what groundwater remains is in danger of being contaminated. Sewage and animal wastes wash into bodies of surface water and cause overenrichment, which robs aquatic animals of the oxygen they need to survive.

Our society primarily uses nonrenewable fossil fuel energy, which leads to global warming, acid deposition, and smog. The result is weakened ecosystems. The demand for goods has increased to the point that facilities to meet the demand are strained. Construction of improved infrastructure to support increased transportation needs only increases the use of nonrenewable energy resources. LDCs have increased needs for energy making it imperative for the MDCs to develop renewable energy sources.

The human population is expanding into all regions on the face of the planet, so habitats for other species are being lost, and a wildlife extinction crisis is expected.

Figure 24.18   What activities are characteristic of an unsustainable society?
Arrows point outward to signify that these types of activities reduce the carrying capacity of the Earth.

Characteristic of a Sustainable Society

A natural ecosystem can offer clues about how to make today's society sustainable. A natural ecosystem makes use of only renewable solar energy. Its materials cycle through the various populations back to the producer once again. For example, coral reefs have been sustaining themselves for millions of years. At the same time, the reefs have provided sustenance to the LDCs. The value of coral reefs has been assessed at over $300 billion a year. Their aesthetic value is immeasurable.

It is clear that if we want to develop a sustainable society, we too should use renewable energy sources and recycle materials. We should protect natural ecosystems that help sustain our modern society. At least a quarter of the coral reefs exist close to the shores of an MDC country, and the chances are good that these coral reefs will be protected. Unfortunately, other coral reefs are threatened by unsustainable practices. The good news is that reefs are remarkably regenerative and will return to their former condition if left alone. The message of today's environmentalists is about what can be done to improve matters and make the environment sustainable (Fig. 24.19). There is still time to make changes and improvements.

Sustainability should be practiced in various areas of human endeavor, from agriculture to business enterprises. Efficiency is the key to sustainability. For example, an efficient car would be ultralight and gas thrifty. Efficient cars could be just as durable and speedy as the inefficient ones of today. Only through efficiency can we meet the challenges of limited resources and finances in the future.

People generally live in either the country or the city, but the two regions depend on one another. Achieving sustainability will require that we understand how the two regions are interdependent. It would be impossible to have one sustainable and not the other because the two regions are linked. What happens within one will ultimately affect the other. Let's consider, therefore, the importance of both rural and urban sustainability.

Rural Sustainability

In rural areas, we must put the emphasis on preservation. We need to preserve ecosystems, including terrestrial ecosystems (such as forests and prairies) and aquatic ecosystems (freshwater and brackish ones along the coast). We should also preserve agricultural land, groves of fruit trees, and other areas that provide us with renewable resources.

It is imperative that we take all possible steps to preserve what remains of our topsoil and replant areas with native plants. Native grasses stabilize the soil and rebuild soil nutrients, while also serving as a source of renewable biofuel. Native trees can be planted to break the wind and protect the soil from erosion, while providing a product as well. Creative solutions to today's ecological problems are very much needed.

Here are some other possible ways to help make rural areas sustainable:

- Plant *cover crops*, which often are a mixture of legumes and grasses, to stabilize the soil between rows of cash crops or between seasonal plantings of cash crops.

Figure 24.19 What activities are characteristic of a sustainable society?
Arrows point inward to signify that these types of activities increase the carrying capacity of the Earth.

- Use *multiuse farming* by planting a variety of crops and use a variety of farming techniques to increase the amount of organic matter in the soil.
- Replenish soil nutrients through composting, organic gardening, or other self-renewable methods.
- Use low flow or trickle irrigation, retention ponds, and other water-conserving methods.
- Increase the planting of *cultivars* (plants propagated vegetatively), which are resistant to blight, rust, insect damage, salt, drought, and encroachment by noxious weeds.
- Use *precision farming (PF)* techniques that rely on accumulated knowledge to reduce habitat destruction, while improving crop yields.
- Use *integrated pest management (IPM)*, which encourages the growth of competitive beneficial insects and uses biological controls to reduce the abundance of a pest.
- Plant a variety of species, including native plants, to reduce our dependence on traditional crops.
- Plant *multipurpose trees*—trees with the ability to provide numerous products and perform a variety of functions, in addition to serving as windbreakers (Fig. 24.20). Remember that mature trees can provide many different types of products. For example, mature rubber trees provide us with rubber, and tagua nuts are an excellent substitute for ivory.
- Maintain and restore wetlands, especially in hurricane or tsunami-prone areas. Protect deltas from storm damage. By protecting wetlands, we protect the spawning ground for many valuable fish nurseries.
- Use renewable forms of energy, such as wind and biofuel.
- Support local farmers, those who fish, and feed stores by buying food products produced close to home.

Urban Sustainability

More and more people are moving to a city. Much thought needs to be given about how to serve the needs of new arrivals, without overexpansion of the city. Resources need to be shared in a way that will allow urban sustainability.

Here are some other possible ways to help make a city sustainable:

- Design an energy-efficient transportation system to rapidly move people about.
- Use solar or geothermal energy to heat buildings. Cool them with an air-conditioning system that uses seawater. In general, use conservation methods to regulate the temperature of buildings.
- Use *green roofs*. Grow a wild garden of grasses, herbs, and vegetables on the tops of buildings. This will assist temperature control, supply food, reduce the amount of rainwater runoff, and be visually appealing (Fig. 24.21).
- Improve storm-water management by using sediment traps for storm drains, artificial wetlands, and holding ponds. Increase use of porous surfaces for walking paths, parking lots, and roads. These surfaces reflect less heat, while soaking up rainwater runoff.
- Instead of traditional grasses, plant native species that attract bees and butterflies. These require less water and fertilizers.
- Create *greenbelts* that suit the particular urban setting. Include plentiful walking and bicycle paths.
- Revitalize old sections of a city, before developing new sections.
- Use lighting fixtures that hug the walls or ground and send light down. Control noise levels by designing quiet motors.
- Promote sustainability by encouraging recycling of business equipment. Use low-maintenance building materials, rather than wood.

Figure 24.20 **What purposes do trees serve in a sustainable society?**
Trees planted by a farmer to break the wind and prevent soil erosion can also have other purposes, such as supplying nuts and fruits.

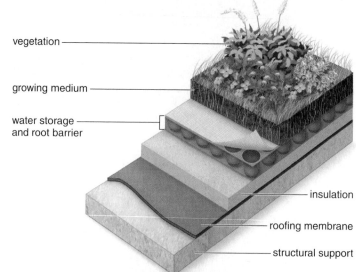

Figure 24.21 **What is a green roof?**
A green roof has plants growing on it that help control temperature, supply food, and reduce water runoff.

Have You Ever Wondered...

What are some simple things YOU can do to conserve energy and/or water and help solve environmental problems like global warming?

A few things you could easily do include:

- Change the light bulbs in your home to compact fluorescent bulbs. They use 75–80% less electricity than incandescent bulbs.

- Walk, ride your bike, carpool, or use mass transit. It will save you a lot of money, too!

- Get cloth or mesh bags for groceries and other purchases. Plastic bags may take 10–20 years to degrade. They're also dangerous to wildlife if mistaken for food and consumed.

- Turn off the water while you brush your teeth. If you don't finish a bottle of water, use it to water your plants. In dry climates, plant native plants that won't require frequent watering.

Assessing Economic Well-Being and Quality of Life

The gross national product (GNP) is a measure of the flow of money from consumers to businesses, in the form of goods and services purchased. It can also be considered the total costs of all manufacturing, production, and services. Costs include salaries and wages, mortgage and rent, interest and loans, taxes, and profit within and outside the country. In other words, GNP pertains solely to economic activities.

When calculating GNP, economists do not necessarily consider whether an activity is environmentally or socially harmful. For example, destruction of forests due to clear-cutting, strip mining, or land development is not a part of the GNP. In the same way, the cost of medical services does not include the pain or suffering caused by illness, for example.

Measures that include noneconomic indicators are most likely better at revealing our quality of life than is the GNP. The Index of Sustainable Economic Welfare (ISEW) includes real per capita income, distributional equity, natural resources depletion, environmental damage, and the value of unpaid labor. The ISEW *does* take into account other forms of value, beyond the purely monetary value of goods and services. Another such index is called the Genuine Progress Indicator (GPI). This indicator attempts to consider the quality of life, an attribute that does not necessarily depend on worldly goods. For example, the quality

of life might depend on how much respect we give other human beings. The Grameen Bank in Bangladesh decided that if women were loaned small amounts of money, they would pay it back after starting up small businesses. The loans give women the opportunity to make choices that can improve the quality of their lives. For these women, a loan is a way to sustain their lives while, in part, fulfilling their dreams. It is difficult to assign a value to well-being or happiness. However, economists are trying to devise a way to measure these values. The following criteria, among others, can be used.

Use value: actual price we pay to use or consume a resource, such as the entrance fees into national parks.

Option value: preserving options for the future, such as saving a wetland or a forest.

Existence value: saving things we might not realize exist yet. This might be flora and fauna in a tropical rain forest that, one day, could be the source of new drugs.

Aesthetic value: appreciating an area or creature for its beauty and/or contribution to biodiversity.

Cultural value: factors such as language, mythology, and history that are important for cultural identity.

Scientific and educational value: valuing the knowledge of naturalists, or even an experience of nature, as types of rational facts.

Development of the environment will always continue. Still, we can use these values to help us direct future development. Growth creates increases in demand, but development includes the direction of growth. If we permit unbridled growth, resources will become depleted. However, if development restrains resource consumption, while still promoting economic growth, perhaps a balance can be reached. We can then retain sources for future generations.

Each person has a particular comfort level, and human beings do not like to make sacrifices that reduce their particular comfort level. So, despite our knowledge of the need to protect fisheries and forests, we continue to exploit them. People from LDCs directly depend on these resources to survive and so have much to lose. Even so, it is difficult for them to sacrifice today for the sake of the future. Yet, there is still hope because nature is incredibly resilient. One solution to deforestation is reforestation. Costa Rica has been successfully reforesting since the early 1980s. Also, declining fisheries can be restocked and then managed for sustainability. What will it take to move toward sustainability? It will take an informed citizenry, creativity, and a desire to bring about change for the better.

> **Check Your Progress 24.5**
> 1. **What are the characteristics of today's unsustainable society?**
> 2. **How can we change today's society to one that is sustainable?**
> 3. **How can we assess how well we are doing?**

Summarizing the Concepts

24.1 Human Population Growth
- Populations have a biotic potential for increase in size.
- Biotic potential is normally held in check by environmental resistance.
- Population size usually levels off at carrying capacity.

The MDCs Versus the LDCs
- The MDCs have a 0.1% growth rate since 1950.
- The LDC growth rate is presently 1.6% after peaking at 2.5% in the 1960s.

  Age-structure diagrams can be used to predict population growth.
- MDCs are approaching a stable population size.
- LDC populations will continue to increase in size.

24.2 Human Use of Resources and Pollution

Five resources are maximally used by humans:

Human population

land   water   food   energy   minerals

Resources are either nonrenewable or renewable.
- Nonrenewable resources are not replenished and are limited in quantity (e.g., land, fossil fuels, minerals).
- Renewable resources are replenished but still are limited in quantity (e.g., water, solar energy, food).

Land
Human activities, such as habitation, farming, and mining, contribute to erosion, pollution, desertification, deforestation, and loss of biodiversity.

Water
Industry and agriculture use most of the freshwater supply. Water supplies are increased by damming rivers and drawing from aquifers. As aquifers are depleted, subsidence, sinkhole formation, and saltwater intrusion can occur. If used by industries, water conservation methods could cut world water consumption by half.

Food
Food comes from growing crops, raising animals, and fishing.
- Modern farming methods increase the food supply, but some methods harm the land, pollute water, and consume fossil fuels excessively.
- Genetically engineered plants increase the food supply and reduce the need for chemicals.
- Raising livestock contributes to water pollution and uses fossil fuel energy.
- The increased number and high efficiency of fishing boats have caused the world fish catch to decline.

Energy
Fossil fuels (oil, natural gas, coal) are nonrenewable sources. Burning fossil fuels and burning to clear land for farming cause pollutants and gases to enter the air.
- Greenhouse gases include CO_2 and other gases. Greenhouse gases cause global warming because solar radiation can pass through, but infrared heat cannot escape back into space.
- Renewable resources include hydropower, geothermal, wind, and solar power.

Minerals
Minerals are nonrenewable resources that can be mined. These raw materials include sand, gravel, phosphate, and metals. Mining causes destruction of the land by erosion, loss of vegetation, and toxic runoff into bodies of water. Some metals are dangerous to health. Land ruined by mining can take years to recover.

Hazardous Wastes   Billions of tons of solid waste are discarded on land and in water.
- Heavy metals (lead, arsenic, cadmium, chromium).
- Synthetic organic chemicals include chlorofluorocarbons (CFCs), which are involved in the production of plastics, pesticides, herbicides, and other products.
- Ozone shield destruction is associated with CFCs.
- Other synthetic organic chemicals enter the aquatic food chain, where the toxins become more concentrated (biological magnification).

24.3 Biodiversity
Biodiversity is the variety of life on Earth. The five major causes of biodiversity loss and extinction are
- habitat loss,
- introduction of alien species,
- pollution,
- overexploitation of plant and animals, and
- disease.

Direct Value of Biodiversity
Direct values of biodiversity are
- medicinal value (medicines derived from living organisms),
- agricultural value (crops derived from wild plants; biological pest controls and animals pollinators), and
- consumptive use values (food production).

Indirect Value of Biodiversity
Biodiversity in ecosystems contributes to
- waste disposal (through the action of decomposers and the ability of natural communities to purify water and take up pollutants),
- freshwater provision through the water biogeochemical cycle,
- prevention of soil erosion, which occurs naturally in intact ecosystems,
- function of biogeochemical cycles,
- climate regulation (plants take up carbon dioxide), and
- ecotourism (human enjoyment of a beautiful ecosystem).

24.4 Working Toward a Sustainable Society

A sustainable society would use only renewable energy sources, would reuse heat and waste materials, and would recycle almost everything. It would also provide the same goods and services presently provided and would preserve biodiversity.

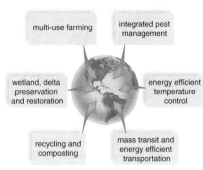

multi-use farming

integrated pest management

wetland, delta preservation and restoration

energy efficient temperature control

recycling and composting

mass transit and energy efficient transportation

Understanding Key Terms

agricultural runoff 570
alien species 576
aquifer 568
biodiversity 575
biological magnification 575
biotic potential 564
carrying capacity 564
chlorofluorocarbon (CFC) 574
deforestation 567
desertification 567
exponential growth 564
fossil fuel 572
greenhouse gases 572

growth rate 564
invasive 576
mineral 574
nonrenewable resource 566
photovoltaic (solar) cell 572
pollution 566
renewable resource 566
replacement reproduction 565
saltwater intrusion 569
sinkhole 569
subsidence 569
sustainable 581

Match the key terms to these definitions.

a. _____ The ability of a society or ecosystem to maintain itself while also providing services to human beings.

b. _____ Largest number of organisms of a particular species that can be maintained indefinitely in an ecosystem.

c. _____ Gases such as carbon dioxide and methane in the atmosphere that trap heat.

d. _____ Concentration of a synthetic organic chemical as it passes along a food chain.

e. _____ Water reservoir below the Earth's surface.

Testing Your Knowledge of the Concepts

1. Define carrying capacity and discuss the relevancy of this concept to the size of today's human population. (page 564)

2. Distinguish between MDCs and LDCs. Why are most LDCs, but not most MDCs, increasing in size? (pages 564–65)

3. Name three locales where humans have settled with unfortunate environmental consequences. What are those consequences? (pages 566–67)

4. What steps can be taken to conserve water? (page 569)

5. What are the environmental benefits of reducing the amount of meat in the American diet? (pages 569–71)

6. What are the types of fossil fuels, and what environmental problems are associated with burning fossil fuels? (page 572)

7. What renewable energy sources are available? What are the benefits of the solar-hydrogen revolution? (pages 572–73)

8. What are minerals, and what are the drawbacks of mining them? (pages 574–75)

9. Exponential growth is best described by
   a. steep unrestricted growth.
   b. an S-shaped growth curve.
   c. a constant rate of growth.
   d. growth that levels off after rapid growth.
   e. Both b and d are correct.

10. When the carrying capacity of the environment is exceeded, the population will typically
   a. increase, but at a slower rate.
   b. stabilize at the highest level reached.
   c. decrease.
   d. die off entirely.

11. Decreased death rate followed by decreased birthrate has occurred in
   a. MDCs.   c. MDCs and LDCs.
   b. LDCs.   d. neither MDCs nor LDCs.

12. Renewable resources
   a. are in limited supply compared to nonrenewable resources.
   b. are always forthcoming but still may be inadequate for human needs.
   c. include such energy sources as wind, solar, and biomass.
   d. All of these are correct.

13. Desertification is often caused by
   a. overuse of aquifers.   c. air pollution.
   b. urban sprawl.   d. overgrazing.

14. Soil in the tropics is often nutrient-poor because
   a. nutrients are tied up in plants.
   b. it is mostly sand.
   c. it has a high pH.
   d. rainfall leaches out minerals.

15. Most freshwater is used for
   a. domestic purposes, such as bathing, flushing toilets, and watering lawns.
   b. domestic purposes, such as cooking and drinking.
   c. agriculture.
   d. industry.

16. Which of the following is not a major problem with the damming of rivers?
   a. increase in salinity
   b. loss of water through evaporation
   c. change in the level of the water table
   d. buildup of sediment

17. Removal of groundwater from aquifers may cause
   a. pollution.   d. soil erosion.
   b. subsidence.   e. All of the above are correct.
   c. mineral depletion.

18. Which of the following is not a component of modern agriculture?
    a. dependency on chemical inputs
    b. frequent irrigation
    c. high fuel consumption
    d. high diversity of cultivars planted

19. The raising of domestic livestock
    a. consumes large amounts of fossil fuels.
    b. leads to water pollution.
    c. is energetically wasteful.
    d. All of the above are correct.

20. The best way to maintain fish supplies is to
    a. limit harvesting to the ability of fish to reproduce.
    b. do away with all the other animals that feed on fish.
    c. use larger and better types of nets
    d. All of these are good ways to maintain fish supplies.

For questions 21–25, match the description to the type of fuel in the key. Each answer can be used more than once. Each question may have more than one answer.

Key:

a. nuclear power
b. fossil fuels
c. hydropower
d. solar power
e. wind power
f. geothermal power

21. Renewable source of power

22. Waste products are harmful

23. Detrimental environmental effects

24. More available in certain geographic locations

25. Contribute(s) to global warming

26. Heavy metals are dangerous to humans because they
    a. inhibit important enzymes.
    b. cause apoptosis.
    c. reduce the body's ability to carry oxygen.
    d. break down mitochondria.

27. The Earth's ozone shield has been damaged by
    a. heavy metals.
    b. strip mining.
    c. chlorofluorocarbons.
    d. fossil fuels.

28. The preservation of ecosystems indirectly provides freshwater because
    a. trees produce water as a result of photosynthesis.
    b. animals excrete water-based products.
    c. forests soak up water and release it slowly.
    d. ecosystems promote the growth of bacteria that release water into the environment.

29. Which of these are indirect values of species?
    a. participation in biogeochemical cycles
    b. participation in waste disposal
    c. provision of freshwater
    d. prevention of soil erosion
    e. All of these are indirect values.

30. Which of the following is not a function that ecosystems can perform for humans?
    a. purification of water
    b. immobilization of pollutants
    c. reduction of soil erosion
    d. removal of excess soil nitrogen
    e. breakdown of heavy metals

31. A transition to hydrogen fuel technology will
    a. be long in coming and not likely to be of major significance.
    b. lessen many current environmental problems.
    c. not be likely because it will always be expensive and as polluting as natural gas.
    d. be of major consequence, but resource limitations for obtaining hydrogen will hinder its progress.

32. In which of the following is biological magnification most pronounced?
    a. aquatic food chains
    b. terrestrial food chains
    c. long food chains
    d. energy pyramids
    e. Both a and c are correct.

33. Complete the following graph by labeling each bar with a cause of extinction, from the most influential to the least.

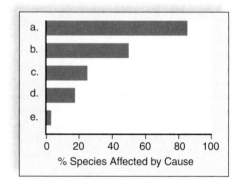

Thinking Critically About the Concepts

1. What environmental reasons would you give a friend for becoming a vegetarian?

2. a. What size footprint do you currently have on Earth? Visit http://www.earthday.net/footprint/index.asp and determine the size of your footprint.
   b. What types of things could you do to decrease the size of your footprint on Earth?

3. How would failure to recycle items that can be recycled (they end up in a landfill) affect the nutrient cycles, such as the carbon cycle covered in Chapter 23?

4. a. What type of environmental activities/initiatives (Earth Day festivities, recycling program, etc.) are in place at your school?
   b. How could you increase awareness of environmental issues on your campus?

Appendix A

Periodic Table of the Elements

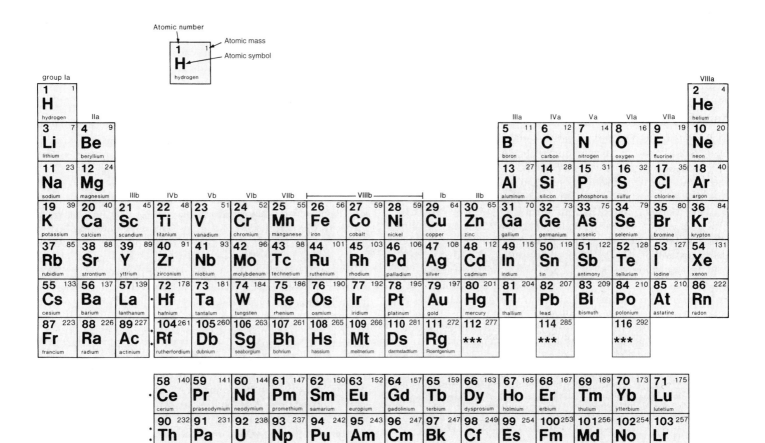

Appendix B

This appendix contains the answers to the Understanding Key Terms, Testing Your Knowledge of the Concepts, and Thinking Critically About the Concepts questions, which appear at the end of each chapter, and the Check Your Progress questions, which appear within each chapter.

Chapter 1

Check Your Progress

(1.1) 1. Organization, acquiring materials and energy, reproducing, growing and developing, being homeostatic, responding to stimuli, evolution; 2. Ancestry can be traced to the first cells. (1.2) 1. Presence of vertebrae tells us humans are vertebrates; hair and mammary glands tell us humans are mammals; 2. Brains, upright stance, language, usage of tools; 3. Continuation of our species. (1.3) 1. Observation, hypothesis, experiment, conclusion; 2. A control group and one or more test groups; 3. Review process and the investigator usually is the author. (1.4) 1a. Not reliable because not based on studies involving large numbers of subjects; 1b. Not all correlations are causations; 2a. Conclusion is an interpretation of data (results); 2b. conclusion; 3. Graphs summarize data, statistics evaluate data. (1.5). 1. Nuclear physics (cancer therapy, nuclear bomb), biotechnology (GM plants/bacteria, possibly endangering biosphere), gene technology (stem cells/destroying embryos); 2. All members; 3. Yes.

Understanding Key Terms

a. biosphere; b. cell; c. scientific theory; d. homeostasis; e. experiment

Testing Your Knowledge of the Concepts

9. c; 10. b; 11. e; 12. a; 13. a; 14. c; 15. b; 16. c; 17. a; 18. c; 19. a; 20. d; 21. a; 22. c; 23. b; 24. e; 25. d

Thinking Critically About the Concepts

1. Currently the smallest living things are single-celled organisms. These organisms have a cell membrane, genetic information, and cytoplasm (cell contents). Viruses lack cell membranes and contents and are considered by most people to be nonliving. Evolution or genetic change is the only characteristic of living things displayed by viruses. 2. It would be unethical to deny people a treatment that produces great results in order to continue a research project. The study should be discontinued after such a finding and everyone should receive the treatment. 3. Data collected by surveying classmates about their knowledge and successful use of the 5-second rule would be anecdotal data. This is not typically reliable data. 4. Salmonella and specific strains of *E. coli* have been implicated in many recent cases of illnesses associated with the consumption of tomatoes and spinach. In 1993, four children died after consuming hamburgers contaminated by a strain of *E. coli*.

Chapter 2

Check Your Progress

(2.1) 1. An atom is composed of a single nucleus that contains its protons and neutrons. Electrons circle the atoms in orbitals. Orbitals are then arranged into groups called *shells*. 2. The first isotope has

$40 - 20 = 20$ neutrons. The second isotope has $48 - 20 = 28$ neutrons. 3. Isotopes are the forms of an element that differ in the number of neutrons. A radioisotope decays over time, releasing rays and subatomic particles. Radioisotopes can be used for medical purposes and for sterilizing objects, including food. 4. The two basic types of bonding found between atoms are the ionic bond and the covalent bond. In ionic bonding, one or more atoms lose electrons and other atoms gain electrons. In this way, ions are formed. In covalent bonding, the ions are shared between atoms. (2.2) 1. Water is a liquid at room temperature, its temperature changes very slowly, and it requires a great deal of heat to become vapor. These characteristics keep water available for supporting life and enable it to help in cooling. Frozen water is less dense than water, so ice floats. Ice helps to insulate the water underneath, allowing marine life to survive the cold. Finally, water molecules are cohesive yet flow freely, allowing water to be the perfect solvent. 2. If the solution contains 0.001 moles/liter [H^+], it can be written in scientific notation: 1×10^{-3} moles/liter. Its pH $= -(-3)$, or pH 3. The solution is an acid, because its pH is less than 7. 3. Water contains equal amounts of acid and basic ions. Its pH is 7. Acids contain greater numbers of acid ions than basic ions. Their pH will be less than 7. Bases contain greater numbers of basic ions than acid ions. Their pH will be greater than 7. 4. The pH of soil and the body remain relatively constant because of the action of buffers. (2.3) 1. carbohydrates, lipids, proteins, and nucleic acids; 2. dehydration reaction. (2.4) 1. quick and short-term energy; 2. Simple carbohydrates contain low numbers of carbon atoms (3–7), complex carbohydrates contain high numbers of carbon atoms in chains of sugar (glucose) units; 3. Insoluble fiber stimulates movements of the large intestine, soluble fiber prevents the absorption of cholesterol. (2.5) 1a. energy storage; 1b. one glycerol molecule and three fatty acids; 2. phospholipids—cellular membranes, steroids—sex hormones. (2.6) 1. support, enzymes, transport, defense, hormones, motion; 2. —NH_2 (amino group) and the —COOH (acid group); 3. Once a protein loses its normal shape, it is no longer able to perform its usual function. (2.7) 1a. DNA contains genes, RNA conveys DNA's instructions regarding the amino acid sequence in a protein; 1b. DNA contains deoxyribose and the bases A, T, C, and G; RNA contains ribose, and the bases A, U, C, G (see Table 2.1); 2. hydrogen; 3. adenosine and three phosphate groups, high energy molecule.

Understanding Key Terms

a. emulsification; b. ion; c. covalent bond; d. hydrophilic; e. hydrogen bond

Testing Your Knowledge of the Concepts

15. b; 16. c; 17. b; 18. c; 19. d; 20. a; 21. b; 22. d; 23. c; 24. b; 25. a; 26. d; 27. b; 28. d; 29. a; 30. c; 31. a; 32. c; 33. d; 34. b; 35. a. subunits; b. dehydration reaction; c. macromolecule; d. hydrolysis reaction

Thinking Critically About the Concepts

1. As you know from the hemoglobin discussion, each hemoglobin molecule contains a molecule of iron-containing heme. In lead poisoning, iron is replaced by lead. When lead substitutes for iron, oxygen transport is decreased because lead can't bind irreversibly

to lead as it does to iron. **2.** In anemia, blood does not transport enough oxygen for the cells. Without oxygen, body cells cannot form the ATP energy needed for body systems. Thus, one symptom of anemia is feeling tired and lethargic. **3.** Sickle-cell anemia is an inherited form of anemia in which the hemoglobin protein is abnormal. Red blood cells containing the abnormal hemoglobin are misshapen and easily broken. Red blood cell count is decreased, as is oxygen transport. **4.** In this form, the gold is a stable element that will safely pass out of your body after you eat it. Gold compounds could react with body chemicals and might be dangerous. **5.** Nickel can be covered by silver, gold, platinum, and titanium. **6.** Iron, magnesium, manganese, zinc, copper, chromium, molybdenum, and vanadium are transition metals found in vitamins.

Chapter 3
Check Your Progress

(3.1) 1. A cell is the basic unit of life, all living things are made up of cells, new cells arise only from preexisting cells; **2.** Small cells have a greater surface-area-to-volume ratio, thus a greater ability to get material in and out of the cell; **3.** Light microscopes use light rays to magnify objects and can be use to view living specimens; electron microscopes use a stream of electrons to magnify objects and have a higher resolving power than light microscopes. **(3.2) 1a.** nucleus, plasma membrane, cytoplasm; **1b.** plasma membrane, cytoplasm; **1c.** nuclues; **2.** Nucleus arose from invagination of the plasma membrane; mitochondria and chloroplast evolved by engulfing prokaryotic cells. **(3.3) 1.** Structure is a phospholipid bilayer with attached or embedded proteins (fluid-mosaic model), function is to keep cells intact and selectively allow passage of molecules and ions; **2a.** isotonic solutions have the same solute concentration as in the cell, hypotonic solutions have a lower solute concentration than in the cell, hypertonic solutions have a higher solute concentration than in the cell; **2b.** isotonic solution—cell's shape stays the same, hypotonic solution—cell gains water, hypertonic solution—cell loses water; **3a.** diffusion, osmosis, facilitated diffusion, active transport; **3b.** passive—diffusion, osmosis, facilitated diffusion; active—active transport; **(3.4) 1.** nucleus—stores genetic information that specifies the proteins in cells, ribosomes—site of protein synthesis, rough endoplasmic reticulum—contains ribosomes, the sites of protein synthesis; **2.** endoplasmic reticulum—processing, modification and formation of transport vesicles, Golgi apparatus—processing, packaging, and secretion; **3.** smooth endoplasmic reticulum—synthesizes the phospholipids that occur in membranes, lysosomes—fuse with vesicles and digest their contents; **4.** move molecules from the ER to the Golgi. **(3.5) 1.** microtubules, actin filaments, intermediate filaments; **2.** microtubules; **3.** The cytoskeleton helps maintain a cell's shape and either anchors organelles or assists in the movement of organelles, cilia are involved in movement (sweeping of debris in respiratory tract, moving of egg along oviduct). **(3.6) 1.** So the energy within a glucose molecule can be released slowly and ATP can be produced gradually; **2a.** citric acid cycle; **2b.** electron transport chain; **3.** Enzymes are protein molecules with specific active sites that speed up metabolic reactions, coenzymes are nonprotein molecules that assist the activity of an enzyme; **4a.** to receive electrons at the end of the electron transport chain; **4b.** glucose; **5a.** burst of energy for a short time; **5b.** only produces two ATP per glucose molecule.

Understanding Key Terms

a. fluid-mosaic model; **b.** osmosis; **c.** selectively permeable; **d.** fermentation; **e.** cellular respiration

Testing Your Knowledge of the Concepts

13. c; **14.** b; **15.** e; **16.** a; **17.** c; **18.** c; **19.** a; **20.** d; **21.** b; **22.** c; **23.** c; **24.** a; **25.** e; **26.** c; **27.** b; **28. a.** carbohydrate chain; **b.** hydrophilic heads; **c.** phospholipid bilayer; **d.** hydrophobic tails; **e.** filaments of cytoskeleton; **f.** membrane protein; **g.** protein; **29. a.** active site; **b.** substrate; **c.** product; **d.** enzyme; **e.** enzyme-substrate complex; **f.** enzyme.

Thinking Critically About the Concepts

1. LHON is inherited through mitochondria and only the egg contributes mitochondria to the next generation, so only the mother can pass on a mitochondrial defect to her children. The father contributes sperm but no mitochondria to the next generation. **2.** LHON affects tissues that have high energy demand, such as eyes, heart, and muscle. The failure of the heart to obtain enough energy to pump blood can result in death.

Chapter 4
Check Your Progress

(4.1) 1. connective tissue, muscular tissue, nervous tissue, and epithelial tissue. **(4.2) 1.** fibrous connective tissue—loose fibrous tissue, adipose tissue, dense fibrous connective tissue; supportive connective tissue—cartilage, bone; fluid connective tissue—blood, lymph; **2.** Loose fibrous tissue supports epithelium and forms a protective covering enclosing many internal organs allowing them to expand, dense fibrous tissue contains many collagen fibers packed together and has more specific functions than loose connective tissue does. Both loose and dense fibrous connective tissues have cells called fibroblasts located some distance from one another separated by a jellylike ground substance. **3.** Compact bone makes up the shaft of a long bone and consists of osteons composed of rings of hard matrix, spongy bone is contained in the ends of a long bone and has an open bony latticework. Both consist of inorganic salts deposited around protein fibers. **4.** Blood is located in the blood vessels and consists of formed elements (4.3) and plasma, lymph is carried by lymphatic vessels and is derived from tissue fluid and contains white blood cells. Both are fluid connective tissues. **(4.3) 1.** skeletal muscle—voluntary movement of body, striated cells with multiple nuclei; smooth muscle—movement of substances in lumens of body, spindle-shaped cells each with a single nuclei; cardiac muscle—pumping blood, branching striated cells with a single nucleus; **2.** skeletal muscles—attached to the skeleton, smooth muscle—blood vessels and walls of the digestive tract, cardiac muscle—walls of the heart. **(4.4) 1a,b.** Dendrites—receive signals from sensory receptors or other neurons; cell body—contains most of the cell's cytoplasm and the nucleus; axon—conducts the nerve impulses; **2.** support and nourish neurons, engulf bacterial and cellular debris, produce hormones, form myelin sheaths. **(4.5) 1.** Squamous epithelium is composed of flattened cells—lining the air sacs of lungs and walls of blood vessels; cuboidal epithelium is composed of cube-shaped cells—glands, covering ovaries and lining kidney tubules; columnar epithelium is composed of rectangular cells—lining the digestive tract; **2a.** Appears to be layered, irregularly placed nuclei; **2b.** upward motion of the cilia carries mucus to the back of the throat; **3a.** Stratified epithelia have layers of cells piled one on top of the other; **3b.** and lines the nose, mouth, esophagus, anal canal, the outer portion of the cervix, and the vagina. **(4.6) 1a,b.** tight junctions—hold epithelial cells close together creating an impermeable barrier; adhesion junctions—firmly attach cytoskeletal fibers of one cell to that of another cell and allow the layer of cells to bend; gap junctions—occur when adjacent plasma membranes come together and leave a tiny channel between them allowing the passage of small molecules; **2.** Epithelial cells must create a layer that prevents leakage from one side of the sheet to the other because they cover the surface of organs and line the body cavities; **3a.** gap junctions; **b.** they allow the heart to beat as a coordinated whole. **(4.7) 1.** The epidermis is made up of stratified epithelium, forms a protective covering over the entire body, contains Langerhans cells and melanocytes, and produces vitamin D; the dermis (located beneath the epidermis) consists of dense fibrous connective tissue, which prevents the skin from being torn, contains blood vessels and sensory receptors; **2a.** melanin, carotene, and hemoglobin; **2b.** broad-spectrum sunscreen, protective clothing, sunglasses, and avoid tanning machines; **3a.** Sweat glands are present in all regions of the skin and play a role in modifying body temperature, oil glands are located in hair follicles and secrete sebum; **3b.** Secretions are acidic and retard bacterial growth; **4a.** beneath the dermis; **4b.** stores fat. **(4.8) 1.** integumentary system—homeostatic functions, cardiovascular system—circulates blood, lymphatic and immune systems—transports lymph and defends against disease, digestive system—digests food and eliminates waste, respiratory system—brings oxygen into the body and removes carbon dioxide, urinary system—rids the body of metabolic wastes, skeletal system—protects body parts, muscular system—maintains posture and accounts for movement, nervous system—conducts nerve impulses, endocrine system—secretes hormones, reproductive system—produces sex cells; **2a,b.** Ventral cavity—thoracic and abdominal cavities, Dorsal cavity—cranial cavity and vertebral canal; **3.** mucous, serous, synovial, and meninges. **(4.9) 1a.** The body's ability to maintain a relative constancy of its internal environment by adjusting its physiological processes; **1b.** Fluctuation in internal conditions can result in illness; **2.** Issue electrochemical signals, release hormones, supply oxygen, maintain body temperature, remove waste, maintain adequate nutrient levels, adjust the water-salt and acid-base balance of the blood (4.14); **3.** negative feedback keeps a variable close to a particular value, positive feedback brings about an ever greater change in the same direction, which assists the body in completing a process.

Understanding Key Terms

a. ligament; **b.** epidermis; **c.** carcinoma; **d.** homeostasis; **e.** spongy bone

Testing Your Knowledge of the Concepts

12. a; **13.** d; **14.** d; **15.** d; **16.** c; **17.** c; **18.** b; **19.** b; **20.** d; **21.** c; **22.** d; **23.** b; **24.** d; **25.** d; **26.** b; **27.** a; **28.** b; **29.** b; **30. a.** cranial cavity; **b.** vertebral cavity; **c.** dorsal cavity; **d.** diaphragm; **e.** pelvic cavity; **f.** abdominal cavity; **g.** thoracic cavity; **h.** ventral cavity

Thinking Critically About the Concepts

1a. The epidermis, the dermis, and the subcutaneous tissues under the dermis; **1b.** Only the epidermis is involved in a superficial burn. Redness and blistering are present but deeper tissues are not destroyed; **1c.** In a full burn, the epidermis, dermis, and some of the underlying tissues are involved. **2.** Blood vessels, nerves, and sweat glands are all destroyed. Muscles may also be destroyed. **3.** The wound was moist, meaning she had sweat glands, and she still had pain, meaning that the nerves were not destroyed. **4a.** Enzymes; **4b.** Blood pH; **4c.** Respiratory and urinary systems; **4a.** Shivering involves muscle contractions, causing energy loss, which can increase body temperature.; **4b.** No, as a component of a negative feedback mechanism, shivering would cease. **5a.** Nervous and endocrine; **5b.** The nervous system uses neurotransmitters and the endocrine system uses hormones.**6a.** Cartilaginous rings, mucus producing cells, and ciliated epithelium; **6b.** The bladder needs to increase in size as the volume of urine being stored increases, which transitional epithelial allows it to do. **7a.** enzymes; **7b.** they are denatured and no longer function; **7c.** change in pH, lungs, and kidneys.

Chapter 5
Check Your Progress

(5.1) 1. heart and blood vessels; **2.** generate blood pressure, transport blood, create exchange at the capillaries, and regulate blood flow; **3.** lymphatic vessels collect excess tissue fluid and return it to the cardiovascular system. **(5.2) 1.** arteries, capillaries, veins; **2.** Arteries carry blood away from the heart and have strong walls; capillaries have thin walls, where exchange can occur; veins carry blood to the heart, have thinner walls than arteries, and contain valves. **(5.3) 1.** The right ventricle of the heart sends blood through the lungs, and the left ventricle sends blood throughout the body; **2.** "Lub" occurs when increasing pressure of blood inside a ventricle forces the cusps of the AV valve to slam shut, "dup" occurs when the ventricles relax and blood in the arteries pushes back, causing the semilunar valves to close; **3.** cardiac control center of the medulla oblongata. **(5.4) 1.** heart rate; **2.** blood pressure; **3.** skeletal muscle pump, respiratory pump, and valves. **(5.5) 1.** O_2-poor blood in pulmonary artery is on its way to the lungs and O_2-rich blood in the pulmonary vein is coming from the lungs. **2.** Right ventricle→pulmonary trunk→pulmonary arteries→pulmonary capillaries→pulmonary veins→left atrium. **3.** left ventricle→aorta→mesenteric arteries→digestive tract capillary bed→hepatic portal vein→liver capillary bed→hepatic vein→inferior vena cava→right atrium; **(5.6) 1.** Blood pressure causes water to exit at the arterial end and osmotic pressure causes water to enter at the venous end. Solutes diffuse according to their concentration gradient; **2.** It is collected by the lymphatic capillaries and eventually returned to the systemic venous blood. **(5.7) 1a.** Hypertension, stroke, heart attack, aneurysm, heart failure; **1b.** drugs to lower blood pressure, nitroglycerin given at onset of heart attack, replace diseased/damaged portion of the vessel, open clogged arteries, dissolve clots, surgery, heart transplant.

Figure Question

Fig. 5.11: Heart

Understanding Key Terms

a. diastole; **b.** inferior vena cava; **c.** pulse; **d.** venule; **e.** systemic circuit

✓ Testing Your Knowledge of the Concepts

15. b; **16.** a; **17.** c; **18.** b; **19.** a; **20.** e; **21.** b; **22.** d; **23.** c; **24.** c; **25.** c; **26.** e; **27.** b; **28.** d; **29.** e; **30.** d; **31.** c; **32. a.** jugular vein; **b.** pulmonary artery; **c.** superior vena cava; **d.** inferior vena cava; **e.** hepatic vein; **f.** hepatic portal vein; **g.** renal vein; **h.** iliac vein; **i.** carotid artery; **j.** pulmonary vein; **k.** aorta; **l.** mesenteric arteries; **m.** renal artery; **n.** iliac artery. **33. a.** blood pressure; **b.** osmotic pressure; **c.** blood pressure; **d.** osmotic pressure

Thinking Critically About the Concepts

1. Extreme fatigue and weakness, nausea and vomiting, as well as nonspecific chest pain are some of the heart attack symptoms more common in women. These can cause a woman to mistake her heart attack for the "flu." **2.** Smoking, obesity, poorly controlled diabetes, high blood pressure, and a sedentary lifestyle are all risk factors for heart attack. In addition, Mrs. Hairston's ethnicity—African American—increased her risk. **3.** When heart muscle cells die during a heart attack, proteins from the cells spill into the blood. Blood tests detect these proteins and signal that a heart attack has occurred. **4a.** The aorta has a very thick wall, as do other arteries. It's rare for these blood vessels to rupture, because they are very strong. **4b.** If the aorta tears open as in an aneurysm, bleeding is extremely severe and blood pressure will fall very rapidly. Without immediate medical attention, the victim dies from internal bleeding. **5a.** In hypertrophic cardiomyopathy, the walls of the heart become too thick. **5b.** Believe it or not, if the heart walls are too thick, the heart is a less efficient pump because the walls are less flexible. **6.** Like cardiovascular veins, lymphatic vessels have one-way valves that keep lymph moving toward the heart. Muscle contraction compresses lymphatic vessels, ultimately returning lymph to the circulatory system. **7.** Many possible answers.

Chapter 6
Check Your Progress

(6.1) 1a. transport, defense and regulation; **1b.** formed elements and plasma; **2a.** 92% water, other 8% consists of various salts and organic molecules; **2b.** help maintain homeostasis. **(6.2) 1.** hemoglobin; **2.** they lose a nucleus during maturation; **3.** anemia, sickle-cell disease, hemolytic disease of the newborn. **(6.3) 1.** granular leukocytes (neutrophils, eosinophils, basophils), agranular leukocytes (lymphocytes, monocytes); **2.** neutrophils—granular with a multilobed nucleus, first responders in bacterial infection; eosinophils—granular with a bilobed nucleus, increase in number during a parasitic worm infection or allergic reaction; basophils—granular with a U-shaped nucleus, release histamine; lymphocytes—do not have granules and have nonlobular nuclei, responsible for specific immunity to particular pathogens and their toxins; monocytes—do not have granules and are the largest of the white blood cells, phagocytize pathogens, and stimulate other white blood cells; **3.** severe combined immunodeficiency—stem cells of white blood cells lack adenosine deaminase; leukemia—uncontrolled white blood cell proliferation; infectious mononucleosis—EBV infection of lymphocytes. **(6.4) 1.** platelets—clump at the site of puncture; thrombin—activates fibrinogen, forming long threads of fibrin; fibrin threads—wind around platelets, plasmin—destroys the fibrin network; serum (released after blood clots)—contains all the components of plasma. **(6.5) 1a.** A, B, AB, and O; **1b.** based on the presence or absence of type A antigen and type B antigen; **2.** Type O is the universal donor because it has neither type A nor type B antigens on the red blood cells; type AB is the universal acceptor because it has neither anti-A nor anti-B antibodies in plasma; **3.** when an RH⁻ mother is pregnant with her second RH⁺ baby. **(6.6) 1.** delivers oxygen from the lungs and nutrients from the digestive system, and removes metabolic wastes; **2.** digestive system—provides molecules for plasma protein and blood cell formation, urinary system—helps maintain water-salt balance of blood, muscular system—moves blood, nervous system—regulates heart contraction and constriction/dilation of blood vessels, endocrine system—produces hormones that regulate blood pressure and volume, respiratory system—gas exchange, lymphatic system—helps maintain blood volume, skeletal system—protects heart and red bone marrow; **3.** Glycogen stored in the liver can be broken down to make glucose available to the blood.

Understanding Key Terms
a. hemoglobin; **b.** formed element; **c.** plasma; **d.** leukemia; **e.** prothrombin

Testing Your Knowledge of the Concepts
10. a; **11.** a; **12.** b; **13.** d; **14.** c; **15.** c; **16.** a; **17.** e; **18.** d; **19.** b; **20.** c; **21.** a; **22.** d; **23.** d; **24.** e; **25.** d; **26.** b; **27.** d; **28.** e; **29.** d; **30.** e

Thinking Critically About the Concepts
1a. Carbon monoxide is found in automobile exhaust. Burning charcoal, or wood in some cases, will also give off CO. Fumes from all of these sources must be properly vented to the outside. **1b.** ATP is the source of cellular energy (compounds which are very similar in structure, like GTP, also supply very small amounts of cellular energy). **1c.** Without oxygen, the cell mitochondria can't metabolize food to form ATP. **2a.** protein, iron, vitamin B_{12}, and folic acid **2b.** Good sources of protein include lean meats, eggs, milk, nuts, and soy. Foods rich in B_{12} include organ meats, fish, shellfish, and dairy products. Folic acid can be found in cereals, baked goods, leafy vegetables, fruits (bananas, melons, lemons), and organ meat. **3.** Hemoglobin is a protein, and you might recall from Chapter 2 that it is composed of smaller subunits. Iron is necessary for hemoglobin synthesis. **4.** After blood loss for any reason, erythropoietin will be needed to grow erythrocytes. At high altitude, erythropoietin will help a person to acclimate to the lower concentration of oxygen in the air. Remember that it will take several weeks to replace lost red blood cells. **5a.** Blood packed with too many red blood cells becomes too dense for the heart to pump. **5b.** The athlete is more likely to die at night because the person isn't drinking fluids, and blood pressure falls during sleep. Blood becomes increasingly dense during sleep, further increasing the workload on the heart. **6.** Any parasitic disease can cause the student's symptoms. The most common parasitic disease is malaria. **7.** The respiratory system, the skeletal system, and the urinary system are examples. Lungs oxygenate the blood, bone marrow within the bones produces blood cells, and the kidneys produce erythropoietin to stimulate erythrocyte production.

Chapter 7
Check Your Progress

(7.1) 1. capsule, flagella, fimbriae, pilus, plasmids, and toxins; **2.** Viruses must replicate inside a living cell. **(7.2) 1a.** red bone marrow, thymus; **1b.** lymph nodes, spleen; **2.** Primary lymphatic organs are the sites where white blood cells are produced and mature, secondary lymphatic organs are the sites where white blood cells react to pathogens and blood and lymph are purified. **3.** red bone marrow, spleen, lymph nodes, thymus gland. **(7.3) 1.** skin and mucous membranes, chemical barriers, resident bacteria, inflammatory response, and protective proteins; **2.** neutrophils and macrophages engulf pathogens by phagocytosis; **3.** They "complement" certain immune responses. **(7.4) 1.** Nonspecific defenses act indiscriminately against all pathogens, specific defenses respond to antigens; **2.** Antibody-mediated immunity, cell-mediated immunity; **3.** B cells produce plasma and memory cells, plasma cells produce antibodies, memory cells produce antibodies in the future, T cells regulate immune responses and produce cytotoxic T cells and helper T cells, cytotoxic T cells kill virus infected cells and cancer cells, helper T cells regulate immunity, and memory T cells kill in the future. **(7.5) 1a.** Immunity that occurs naturally through infection or is brought about artificially by medical intervention; **1b,c.** active immunity—infection with a pathogen, immunization; passive immunity—transfer of IgG antibodies across the placenta, breast feeding, gamma globulin injection; **2.** monoclonal antibodies, cytokine therapy. **(7.6) 1.** allergies—hypersensitivities to substances that ordinarily would do no harm to the body, tissue rejection—the recipient's immune system recognizes that the transplanted tissue is not self, autoimmune disease—cytotoxic T cells or antibodies mistakenly attack the body's own cells, as if they bear foreign antigens.

Understanding Key Terms

a. vaccine; **b.** complement system; **c.** antigen; **d.** apoptosis; **e.** T cell

Testing Your Knowledge of the Concepts

17. b; **18.** d; **19.** e; **20.** b; **21.** a; **22.** c; **23.** d; **24.** d; **25.** b; **26.** c; **27.** c; **28.** c; **29.** a; **30.** e; **31.** d; **32.** d; **33.** c

Thinking Critically About the Concepts

1. B cells produce the IgE necessary for the allergic reaction. **2.** Upon the first exposure to bee venom, Nick's B cells produced antibodies against the venom but this took some time. Upon the second exposure, the reaction of the memory B cells was immediate. **3.** Often an allergist will inject very small doses of the allergen to help a patient build immunity to the allergen. **4.** Veins; **5.** Barrier defenses are supposed to prevent the entrance of pathogens to someone's body, similar as a fence keeping intruders off your property; **6.** They should get another shot of anti-venom, because the first shot gave them passive immunity.

Chapter 8

Check Your Progress

(8.1) 1. ingestion—the mouth takes in food, digestion—divides food into pieces and hydrolyzes food to molecular nutrients, movement—food is passed along from one organ to the next and indigestible remains are expelled, absorption—unit molecules produced by digestion cross the wall of the GI tract and enter the blood for delivery to cells, elimination—removal of indigestible wastes through the anus; **2.** mucosa—diverticulitis, submucosa—inflammatory bowel disease, muscularis—irritable bowel syndrome, serosa—appendicitis. **(8.2) 1.** mechanical digestion—teeth chew food into pieces convenient for swallowing and the tongue moves food around the mouth; chemical digestion—salivary amylase begins the process of digesting starch; **2.** The soft palate moves back to close off the nasal passages, and the trachea moves up under the epiglottis to cover the glottis. **(8.3) 1a.** store food, initiate the digestion of protein, and control the movement of chyme into the small intestine; **1b.** The muscularis contains an oblique layer that allows the stomach to stretch and mechanically break food down, the mucosa has millions of gastric pits, which lead into gastric glands that produce gastric juice; **2a.** Complete digestion using enzymes, which digest all types of food and absorb the products of the digestive process; **2b.** It contains villi that have an outer layer of columnar epithelial cells, each containing thousands of microvilli. **(8.4) 1.** pancreas, liver, gallbladder; **2.** Pancreas produces pancreatic juice, liver produces bile, gallbladder stores bile; **3.** nervous system and hormones. **(8.5) 1.** cecum, colon, rectum, and anal canal; **2.** absorb water and vitamins, form feces, defecation. **(8.6) 1a.** Gives an idea of how much weight is due to adipose tissue; **1b.** premature death, diabetes type 2, hypertension, cardiovascular disease, stroke, gallbladder disease, respiratory disfunction, osteoarthritis, and certain cancers; **2.** High intake of refined carbohydrates and fructose sweeteners as well as a high intake of fat leads to increased deposition of fat; **3a.** saturated

fats and trans-fatty acids; **3b.** unsaturated, particularly omega-3 fatty acids; **4.** Brain cells require glucose and excess nitrogen excretion stresses the kidneys; **5.** Which foods should be eaten often and which foods should not be eaten on a regular basis; **6.** anorexia nervosa—severe psychological disorder characterized by an irrational fear of getting fat, bulimia nervosa—binge eating and then purging to avoid gaining weight, binge eating disorder—episodes of overeating that are not followed by purging, muscle dysmorphia—thinking one's body is underdeveloped.

Understanding Key Terms

a. vitamin; **b.** lipase; **c.** lacteal; **d.** esophagus; **e.** gallbladder

Testing Your Knowledge of the Concepts

9. d; **10.** a; **11.** b; **12.** e; **13.** c; **14.** a; **15.** b; **16.** e; **17.** d; **18.** d; **19.** d; **20.** e; **21.** c; **22.** e; **23.** b; **24.** c; **25.** d; **26.** a; **27.** a; **28.** b; **29.** c; **30.** b; **31.** a; **32.** d; **33.** a; **34.** c; **35.** c; **36.** d; **37.** a; **38.** e; **39.** b, c; **40.** a; **41.** d; **42.** e; **43.** a; **44.** c; **45.** b

Thinking Critically About the Concepts

1a. To prepare the food for enzymes to perform chemical digestion, such as chewing does. The smaller stomach from bariatric surgery no longer performs a significant amount of mechanical digestion; **1b.** Because the stomach now holds only a few ounces at a time; **2.** If stomach is overfilled, stomach contents would flow into the esophagus causing heartburn; if chronic, serious problems can result; **3a.** It would have been digested chemically; **3b.** It would not have been digested chemically, carbohydrates are digested in the mouth and small intestine.

Chapter 9

Check Your Progress

(9.1) 1. nasal cavity→pharynx→glottis→larynx→trachea→bronchus →bronchioles→lungs; **2.** ensure that oxygen enters the body and carbon dioxide leaves the body. **(9.2) 1.** nose—filter, warm, and moisten the air; pharynx—connect nasal and oral cavities to the larynx; larynx—sound production; **2.** nasal cavity, oral cavity, and larynx; **3a.** Air passage and food passage; **3b.** The larynx, which receives air, is in front of the esophagus, which receives food. **(9.3) 1.** trachea—keeps lungs clean by sweeping mucus upwards and connects the larynx to the primary bronchi; bronchial tree—passage of air to the lungs; lungs—site of gas exchange between air in the alveoli and blood in the capillaries; **2.** alveoli. **(9.4) 1a.** During inspiration, the diaphragm contracts expanding the lungs, a partial vacuum is created, and air moves into the lungs. During expiration, the diaphragm relaxes; the lungs recoil, and abdominal organs press up against the diaphragm, forcing air out of the lungs; **1b.** As the volume (size) of the thoracic cavity increases (when you inhale), the pressure in the lungs decreases. As the volume decreases (when you exhale), the pressure in the lungs increases. **2a.** A free flow of air from the nose or mouth to the lungs and from the lungs to the nose or mouth is vitally important; **2b.** tidal volume, vital capacity, inspiratory and expiratory reserve volume, and residual volume. **(9.5) 1.** The rhythm of ventilation is controlled by a respiratory control center, located in the medulla oblongata; **2.** As you hold your breath, blood CO_2 increases, which makes the blood more acidic. The respiratory center initiated exhalation in response to the increased acidity. **(9.6) 1.** External respiration refers to the exchange of gases between air in the alveoli and blood in the pulmonary capillaries; internal respiration refers to the exchange of gases between the blood in systemic capillaries and the tissue fluid; **2a.** hemoglobin; **2b.** Takes up oxygen and becomes oxyhemoglobin, takes up carbon dioxide and becomes carbaminohemoglobin. **(9.7) 1a.** strep throat, sinusitis, otitis media, tonsillitis, laryngitis; **1b.** lower respiratory infections, restrictive pulmonary disorders, obstructive pulmonary disorders; **2.** When the common respiratory infections are caused by bacteria, they are treated with antibiotics. **3.** chronic bronchitis, emphysema, lung cancer.

Understanding Key Terms

a. pharynx; **b.** surfactant; **c.** vocal cord; **d.** bicarbonate ion; **e.** expiration

Testing Your Knowledge of the Concepts

10. d; **11.** d; **12.** c; **13.** f; **14.** d; **15.** a; **16.** b; **17.** e; **18.** d; **19.** b; **20.** d; **21.** a; **22.** c; **23.** b; **24.** b; **25.** e; **26.** d; **27.** b; **28. a.** nasal cavity; **b.** nostril; **c.** pharynx; **d.** epiglottis; **e.** glottis; **f.** larynx; **g.** trachea; **h.** bronchus; **i.** bronchiole

Thinking Critically About the Concepts

1. The enlarged tonsils or adenoids physically obstruct the airway, inhibiting airflow. Surgical removal may be required to restore sufficient passage of air. **2.** The expression describes the movement of food particles past the epiglottis, through the glottis, and into the trachea; **2.** Holding her breath increased the amount of CO_2 in the blood. Some of which combines with water and forms carbonic acid, which dissociates and forms H^+ and the bicarbonate ion that influence blood pH. This triggers involuntary exhalation; **3a.** Ciliated cells are damaged from smoking, so coughing prevents dust, bacteria, and other airborne contaminants from reaching the lungs; **3b.** Their ciliated epithelium is fully functioning; **4.** diaphragm and abdominal muscles; **5.** The blood O_2 of someone who has nearly drowned is low. When hemoglobin is not bound to O_2, blood is a much darker color and appears bluish because of the diffusion of light by the skin.

Chapter 10

Check Your Progress

(10.1) 1. kidneys—primary organs of excretion, ureters—conduct urine from the kidneys to the bladder, urinary bladder—stores urine until it is expelled, urethra—small tube that extends from the urinary bladder to an external opening; **2.** excretion of metabolic wastes, maintenance of water-salt balance, maintenance of acid-base balance, and secretion of hormones. **(10.2) 1.** renal cortex, renal medulla, and renal pelvis; **2.** nephrons; **3.** The glomerulus structure facilitates easy passage of small molecules to the glomerular capsule. The proximal convoluted tubule has a large surface area for the reabsorption of filtrate components. **(10.3) 1.** glomerular filtration, tubular reabsorption, and tubular secretion; **2.** Due to glomerular blood pressure, water and small blood molecules move from the glomerulus to the inside of the glomerular capsule; tubular reabsorption occurs as molecules and ions are both passively and actively reabsorbed from the nephron into the blood of the peritubular capillary network; certain molecules are then actively secreted from the peritubular capillary network into the convoluted tubules. **(10.4) 1a.** loop of the nephron and the collecting duct; **1b.** Salt is reabsorbed, the solute gradient is established, and water leaves the descending limb of the nephron and the collecting duct because of the osmotic gradient within the renal medulla; **2a.** aldosterone, atrial natriuretic hormone, and antidiuretic hormone; **2b.** Aldosterone promotes the excretion of potassium ions and the reabsorption of sodium ions, leading to a reabsorption of water; atrial natriuretic hormone promotes the excretion of sodium, leading to a decrease

in water reabsorption; antidiuretic hormone increases water reabsorption; **3.** If the blood is acidic, hydrogen ions are excreted and bicarbonate ions are reabsorbed; if the blood is basic, hydrogen ions are not excreted and bicarbonate ions are not reabsorbed. **(10.5) 1.** infections, diabetes, hypertension, and inherited conditions cause renal disease, which can be treated using hemodialysis or replacing the kidney. **(10.6) 1.** excrete waste molecules—cleanse blood of nitrogenous wastes and excrete them in urine, water-salt balance—increase or decrease Na^+ and water absorption, acid-base balance—remove a wide range of acidic and basic substances. **2a.** Only kidneys excrete nitrogenous waste, maintain water-salt balance, and sufficiently keep pH constant. **2b.** A single kidney is sufficient to perform the required functions.

Understanding Key Terms

a. diuretic; **b.** excretion; **c.** urethra; **d.** renal pelvis; **e.** glomerular filtrate

Testing Your Knowledge of the Concepts

11. d; **12.** d **13.** c; **14.** a; **15.** c; **16.** b; **17.** c; **18.** c; **19.** a; **20.** c; **21.** d; **22.** c; **23.** a; **24.** b; **25.** b; **26.** c; **27.** d; **28.** b; **29.** a; **30.** d; **31.** f; **32.** a; **33.** c; **34.** d; **35. a.** glomerular capsule; **b.** efferent arteriole; **c.** afferent arteriole; **d.** proximal convoluted tubule; **e.** loop of the nephron; **f.** descending limb; **g.** ascending limb; **h.** peritubular capillaries; **i.** distal convoluted tubule; **j.** renal vein; **k.** renal artery; **l.** collecting duct

Thinking Critically About the Concepts

1. Symptoms common to diabetes mellitus and diabetes insipidus are great thirst and frequent urination. The symptom that distinguishes the two types of diabetes is the presence of glucose in urine with diabetes mellitus (type 1 or 2). **2.** If there are more red blood cells, more oxygen will be delivered to the cells that will use the oxygen to make more ATP; **3.** It calms the smooth muscle of the bladder, which the large intestine also has. When the large intestine's contractions are lessened, constipation may result; **4.** The man is likely to be incontinent; **5.** Damage to the kidneys may result from football tackles or blows from a hard ball (like one used in lacrosse). Players often wear pads designed to protect the ribs and kidneys from damage.

Chapter 11

Check Your Progress

(11.1) 1a. compact bone is highly organized and composed of tubular units called osteons; **1b.** spongy bone has an unorganized appearance and is composed of trabeculae; **2.** hyaline cartilage—at the ends of long bones, in the nose, at the ends of the ribs, and in the larynx and trachea; fibrocartilage—the disks located between the vertebrae and also in the cartilage of the knee; elastic cartilage—the ear flaps and the epiglottis; **3.** fibrous connective tissue; **4.** The main portion of the bone (diaphysis) is composed of compact bone, the expanded region at each end of a long bone (epiphysis) is composed of spongy bone, and the long bone is covered by a layer of fibrous connective tissue (periosteum). **(11.2) 1.** osteoblasts, osteocytes, and osteoclasts; **2.** Through intramembranous ossification, in which bone develops between sheets of fibrous connective tissue, and endochondral ossification, in which bone replaces a cartilage model; **3.** A vitamin D converted hormone, growth hormone, and thyroid hormone; **4.** Allows the body to regulate the amount of calcium in the blood; **5.** A hematoma is formed, next tissue repair begins, and a fibrocartilaginous callus is formed between the ends of the broken bone, then the fibrocartilaginous callus is converted into a bony callus and remodeled. **(11.3) 1.** skull, hyoid bone, vertebral column, rib cage, and the ear ossicles; **2.** The frontal bone forms the forehead, the parietal bones extend to the sides, and the occipital bone curves to form the base of the skull, each temporal bone is located below the parietal bones, the sphenoid bone extends across the floor of the cranium from one side to the other, and the ethmoid bone lies in front of the sphenoid; **3.** mandible forms the lower jaw and chin, maxillae forms the upper jaw and the anterior portion of the hard palate, zygomatic bones are the cheekbone prominences, and the nasal bones form the bridge of the nose; **4a.** cervical vertebrae—located in the neck and allow movement of the head; thoracic vertebrae—form the thoracic curvature and have long, thin, spinous processes and articular facets for the attachment of the ribs; lumbar vertebrae—form the lumbar curvature and have a large body and thick processes; sacral vertebrae—fused together, forming the pelvic curvature; coccyx—fused vertebrae that form the tailbone; **4b.** Structure consists of thoracic vertebrae, the ribs, and the sternum; function is to protect the heart and lungs and to move during inspiration and expiration. **(11.4) 1a.** scapula, clavicle; **1b.** humerus, radius, ulna, carpals, metacarpals, phalanges; **2a.** two coxal bones; **2b.** femur, patella, tibia, fibula, tarsals, metatarsals, phalanges. **(11.5) 1a.** fibrous, cartilaginous, synovial; **1b.** flexion and extension, adduction and abduction, rotation and circumduction, inversion and eversion.

Understanding Key Terms

a. ligament; **b.** compact bone; **c.** sinus; **d.** fontanel; **e.** osteocyte

Testing Your Knowledge of the Concepts

18. c; **19.** d; **20.** a; **21.** g; **22.** f; **23.** e; **24.** e; **25.** a; **26.** b; **27.** c; **28.** f; **29.** d; **30.** b; **31.** b; **32.** a; **33.** b; **34.** c; **35.** e; **36.** c; **37.** c; **38.** b; **39.** d; **40.** c; **41.** c; **42.** F; **43.** T; **44.** F; **45.** T; **46.** T; **47.** b; **48.** a; **49.** e; **50.** c; **51.** d; **52. a.** frontal bone; **b.** zygomatic bone; **c.** maxilla; **d.** mandible; **e.** clavicle; **f.** scapula; **g.** sternum; **h.** ribs; **i.** costal cartilages; **j.** coxal bones; **k.** sacrum; **l.** coccyx; **m.** patella; **n.** metatarsals; **o.** phalanges; **p.** temporal bone; **q.** vertebral column; **r.** humerus; **s.** ulna; **t.** radius; **u.** carpals; **v.** metacarpals; **w.** femur; **x.** fibula; **y.** tibia; **z.** tarsals

Thinking Critically About the Concepts

1. A balanced diet rich in calcium and containing the proper amount of protein will speed bone repair. Weight-bearing exercise stimulates bond growth at any age. **2.** Without sunlight, a person produces less vitamin D. Thus, less calcium can be absorbed, and bones become weakened. Supplemental vitamin D can help to prevent this. **3.** The growth rate for all cells becomes slower as a person ages. **4.** The typical fast-food diet is deficient in calcium, vitamin D, and other key nutrients. It may also be low in protein. **5.** Believe it or not, it is the player whose fibula is fractured. Bone cells reproduce faster than cells found in cartilage, such as the ligaments that may be torn in a sprain. **6a.** Low blood calcium triggers parathyroid hormone release. **6b.** In hyperparathyroidism, the bones lose too much calcium and become fragile. Calcium deposits form elsewhere in the body, especially in the kidneys. **7.** If the cervical vertebrae are fractured, they can no longer support the weight of the head. If the head falls forward or backward, the spinal cord will stretch and be damaged.

Chapter 12
Check Your Progress

(12.1) 1. smooth, cardiac, and skeletal; **2.** make bones move, help maintain a constant body temperature, assist movement in cardiovascular and lymphatic vessels, protect internal organs and stabilize joints; **3.** In opposite pairs, for example, one flexes and the other extends; **4.** Major trunk muscles include the pectoralis major, trapezius, latissimus dorsi, external oblique, and rectus abdominis. The deltoid, biceps brachii, triceps brachii, and the extensor and flexor carpi muscle are the major muscles of the arm and forearm. The quadriceps group, the adductor longus, sartorius, gluteus maximus, and biceps femoris are some of the thigh muscles. **(12.2) 1.** Muscle fiber consists of many myofibrils, myofibril has many sarcomeres, sarcomeres contain two types of protein myofilaments; **2.** Muscle fibers are stimulated to contract by motor neurons whose axons are in nerves.; **3.** ATP supplies the energy for muscle contraction. **(12.3) 1a.** A nerve fiber together with all of the muscle fibers it innervates; **1b.** Because all the muscle fibers in a motor unit are stimulated at once; **2.** Not all motor units contract at the same time; **3a.** creatine phosphate pathway, fermentation, and cellular respiration; **3b.** fast-twitch. **(12.4) 1.** In a strain, the muscle at a joint is stretched or torn. A sprain results in stretching and tearing of tendons and ligaments at a joint, as well as possible blood vessel and nerve damage. **2.** Both myalgia and myasthenia gravis are thought to result from disorder of the immune system. **(12.5) 1.** body movement and protection of body parts; **2a.** store and release calcium, and production of blood cells; **2b.** provides heat.

Understanding Key Terms

a. sarcomere; **b.** insertion; **c.** tetanus; **d.** rigor mortis; **e.** strain

Testing Your Knowledge of the Concepts

10. c; **11.** b; **12.** a; **13.** c; **14.** d; **15.** e; **16** a; **17.** d; **18.** d; **19.** b; **20.** c; **21.** a; **22.** e; **23.** c; **24.** e; **25.** d; **26.** a; **27.** b; **28.** c; **29.** c; **30.** a; **31.** d; **32.** b; **33.** c; **34.** a; **35.** b; **36.** a; **37.** c; **38.** b; **39.** b; **40.** a; **41.** c; **42. a.** T tubule; **b.** sarcoplasmic reticulum; **c.** myofibril; **d.** Z line; **e.** sarcomere; **f.** sarcolemma

Thinking Critically About the Concepts

1a. The humerus, radius, and ulna form the elbow joint. **1b.** A tendon is cushioned by a bursa. If the tendon is overused and becomes inflamed, the inflammation will spread to the bursa, too. **2a.** in the elbow or shoulder **2b.** in the knee **2c.** in the hip, knee, ankle, or toe joints **2d.** in the knee and ankle joints **3a.** Rapid cooling would slow the progression of rigor mortis, because it would slow cellular metabolism. **3b.** Likewise, rapid cooling would affect the onset of rigor mortis, because cell metabolism is slowed. **4.** Rigor mortis diminishes because the decay process will cause the body to lose its stiffness. **5.** If the diaphragm and external intercostal muscles fail, a person can no longer breathe.

Chapter 13
Check Your Progress

(13.1) 1. central nervous system and peripheral nervous system; **2a.** sensory neurons, interneurons, and motor neurons; **2b.** cell body, dendrites, and axon; **3a.** action potential traveling along a neuron; **3b.** exchange of ions generated along the length of an axon; **4.** neurotransmitters. **(13.2) 1.** spinal cord and brain; **2.** provide a means of communication between the brain and the peripheral nerves, the center for reflex actions; **3.** cerebrum—main part of the brain, communicates and coordinates the activities of the other parts of the brain; diencephalon—contains the hypothalamus and thalamus, maintains homeostasis, receives sensory input; cerebellum—sends out motor impulses by way of the brain stem to the skeletal muscles, produces smooth, coordinated voluntary movements; brain stem—contains the midbrain, pons, and medulla oblongata; acts as a relay station and medulla has reflex centers; **4.** reticular activating system. **(13.3) 1.** blend primitive emotions and higher mental functions into a united whole; **2.** amygdala and hippocampus; **3a.** Wernicke's area and Broca's area; **3b.** left hemisphere. **(13.4) 1a.** 12 pairs; **1b.** 31 pairs; **2.** reflex; **3a.** a system that regulates the activity of cardiac and smooth muscles and glands; **3b.** Functions automatically, has two systems that generally cause opposite responses. **(13.5) 1.** affects the limbic system and either promotes or decreases the action of a particular neurotransmitter.

Understanding Key Terms

a. reflex; **b.** neurotransmitter; **c.** autonomic system; **d.** ganglion; **e.** acetylcholine

Testing Your Knowledge of the Concepts

18. b; **19.** d; **20.** c; **21.** a; **22.** a; **23.** b; **24.** c; **25.** b; **26.** c; **27.** a; **28.** c; **29.** c; **30.** b; **31.** d; **32.** c; **33. a.** central canal, **b.** dorsal horn; **c.** white matter; **d.** dorsal root; **e.** dorsal root ganglion; **f.** spinal nerve; **g.** cell body of motor neuron; **h.** cell body of interneuron.

Thinking Critically About the Concepts

1. Myelin enables the signal to jump from node to node quickly, because the depolarization process occurs only at the node of Ranvier. **2a.** triglyceride (fat or oil); **2b.** Unsaturated fatty acids are characterized by one or more double bonds between carbons, while saturated fatty acids are all single bonds; **2c.** Animal fat, such as butter and fatty cuts of meat; **3.** by inheriting a recessive allele from each parent. **4.** myelination enables signals to travel quickly down an axon, which helps motor skills.

Chapter 14
Check Your Progress

(14.1) 1. To detect certain types of stimuli; **2.** chemoreceptors, photoreceptors, mechanoreceptors, and thermoreceptors; **3.** Sensation is the conscious perception of stimuli that occurs after sensory receptors generate a nerve impulse that arrives at the cerebral cortex. **(14.2) 1.** Assist the brain in knowing the position of the limbs in space; **2.** To make the skin sensitive to touch, pressure, pain, and temperature. **(14.3) 1a.** chemoreceptors; **1b.** sweet, sour, salty, bitter, and perhaps umami; **2a.** olfactory cells; **2b.** within olfactory epithelium, high in the roof of the nasal cavity; **3.** Because it activates a different combination of receptor proteins on olfactory cells, thus stimulating different neurons. **(14.4) 1.** cornea, assisted by the lens and the humors; **2.** rod cells, cone cells; **3.** Rods and cones synapse with bipolar cells, which synapse with ganglion cells whose axons become the optic nerve. **(14.5) 1.** tympanic membrane, malleus, incus, stapes; **2a.** mechanoreceptors; **2b.** inner ear; **2c.** sensitive to mechanical stimulation. **(14.6) 1a.** semicircular canals; **1b.** Ampullae of the semicircular canals contain hair cells with stereocilia embedded in a cupula; when the head rotates, the cupula is displaced, bending the stereocilia.; **2a.** utricle and saccule; **2b.** Utricle and saccule contain hair cells with stereocilia embedded in an otolithic

membrane; when the head bends, otoliths are displaced causing the stereocilia to bend.

Understanding Key Terms

a. sensory receptor; **b.** retina; **c.** sclera; **d.** chemoreceptor; **e.** spiral organ

Testing Your Knowledge of the Concepts

14. d; **15.** a; **16.** c; **17.** b; **18.** c; **19.** a; **20.** d; **21.** c; **22.** d; **23.** b; **24.** e; **25.** b; **26.** c; **27.** b; **28.** d; **29.** d; **30.** c; **31.** e; **32.** a; **33.** e; **34.** d; **35. a.** retina; **b.** choroid; **c.** sclera; **d.** optic nerve; **e.** fovea centralis; **f.** ciliary body; **g.** lens; **h.** iris; **i.** pupil; **j.** cornea

Thinking Critically About the Concepts

1. Just about the entire sensory system: taste, smell, vision (seeing your pizza!), as well as receptors for temperature and texture in your mouth; **2.** Chemoreceptors also monitor the oxygen and carbon dioxide in the blood. Taste and smell receptors are also chemoreceptors; **3.** Vision, because the visual sensory and memory areas are found in the occipital lobe; **4.** Hearing receptors are severely damaged by continual loud noise. Without ear protection, the workers will go deaf; **5.** Both hearing and balance will be affected, sometimes severely; **6.** All cutaneous, or skin sensations, will be affected. The person will lose sensitivity to pain and temperature and can be injured as a result.

Chapter 15
Check Your Progress

(15.1) 1. To respond to stimuli, the nervous system uses nerve impulses to produce a quick response; the endocrine system uses hormones deposited in the bloodstream for a response that is slower but lasts longer; **2.** A chemical signal, a means of communication between cells, between body parts, and even between individuals.; **3.** Peptide hormones cause the formation of cyclic AMP, which results in a series of enzymatic reactions, steroid hormones promote the synthesis of a specific enzyme. **(15.2) 1.** Controls the glandular secretions of the pituitary gland; **2.** thyroid-stimulating hormone, adrenocorticotropic hormone, gonadotropic hormone, prolactin hormone, melanocyte-stimulating hormone, and growth hormone. **(15.3) 1a.** triiodothyronine and thyroxine; **1b.** increase the metabolic rate; **2a.** parathyroid hormone; **2b.** Causes the blood Ca^{2+} level to increase; **2c.** calcitonin. **(15.4) 1.** mineralocorticoids—regulate salt and water balance, glucocorticoids—regulate carbohydrate, protein, and fat metabolism. **(15.5) 1.** insulin—lowers the blood glucose level, glucagon—raises the blood glucose level; **2.** A hormonal disease in which liver cells, and most body cells, are unable to take up glucose as they should. **(15.6) 1a.** testes, ovaries, thymus gland, and pineal gland; **1b.** testes and ovaries; **2.** Kidneys secrete renin and erythropoietin, heart produces atrial natriuretic hormone, adipose tissue produces leptin, tissue cells secrete prostaglandins; **3a.** A hormone that acts where it is produced; **3b.** prostaglandins. **(15.7) 1.** The hormone aldosterone from the adrenal cortex will act on the kidney tubules to conserve Na^+ and water reabsorption will follow; **2.** Hypothalamus acts directly through the nerves of the autonomic system to control *internal* organs, produces hormones released by the posterior pituitary, and produces hormones that control the anterior pituitary.

Understanding Key Terms

a. thyroid gland; **b.** diabetes mellitus; **c.** adrenocorticotropic hormone (ACTH); **d.** peptide hormone; **e.** oxytocin

Testing Your Knowledge of the Concepts

14. d; **15.** b; **16.** b; **17.** d; **18.** c; **19.** c; **20.** b; **21.** e; **22.** e; **23.** d; **24.** d; **25.** d; **26.** b; **27.** c; **28.** d; **29.** a; **30.** d; **31.** e; **32.** c; **33.** b; **34.** f; **35.** b; **36.** c; **37.** a; **38.** e; **39.** d; **40. a.** inhibits; **b.** inhibits; **c.** releasing hormone; **d.** stimulating hormone; **e.** target gland hormone

Thinking Critically About the Concept

1. Follicle stimulating hormone and growth hormone are both protein hormones. They both bind to a receptor in the plasma membrane and activate the cAMP second messenger system; **2.** When thyroxine is produced, negative feedback occurs to stop TSH. Without thyroxine, there is no negative feedback to stop TSH. Therefore TSH levels remain high in someone with hypothyroidism; **3.** Thyroxine increases cellular metabolism (making energy). If thyroxine is low, cell metabolism will be low, causing the person to feel tired. When energy is made, some is lost as heat and that helps maintain normal body temperature. If metabolism is low and less energy is being made, less energy will be lost as heat, causing a person to feel cold; **4.** People who take thyroxine even though they don't have hypothyroidism hope to increase their metabolism enough to lose weight. Some risks associated with this practice include increased heart rate and blood pressure, which could damage the heart and cause high blood pressure.

Chapter 16
Check Your Progress

(16.1) 1. Mitosis is duplication division (number of chromosomes stays the same), meiosis is reduction division (number of chromosomes is reduced); **2a.** 23; **2b.** 23; **3a.** testes; **3b.** ovaries. **(16.2) 1a.** testes; **1b.** seminal vesicles, prostate gland, and bulbourethral gland; **1c.** urethra; **2a.** hypothalamus, anterior pituitary, testes; **2b.** gonadotropin-releasing hormone, follicle-stimulating hormone, luteinizing hormone, testosterone. **(16.3) 1a.** ovary; **1b.** oviducts; **1c.** uterus; **1d.** vagina; **2.** glans clitoris. **(16.4) 1a.** follicle-stimulating hormone, luteinizing hormone; **1b.** estrogen, progesterone; **2a.** Corpeus luteum is maintained in the ovary and produces increasing concentrations of progesterone; progesterone shuts down the hypothalamus and anterior pituitary so that no new follicles begin in the ovary; **2b.** The hormones in birth control pills feedback to inhibit the hypothalamus and the anterior pituitary; therefore, no new follicles begin in the ovary. **(16.5) 1.** abstinence, birth control pills, intrauterine devices, diaphragm, condom, contraceptive implants, contraceptive injections, contraceptive vaccines, vasectomy, tubal ligation, morning-after pills; **2.** artificial insemination by donor, in vitro fertilization, gamete intrafallopian transfer, surrogate mothers, intracytoplasmic sperm injection. **(16.6) 1.** pelvic inflammatory disease; **2.** cancer of the cervix.

Understanding Key Terms

a. ovulation; **b.** progesterone; **c.** semen; **d.** cervix; **e.** acrosome

Testing Your Knowledge of the Concepts

12a. seminal vesicle; **b.** ejaculatory duct; **c.** prostate gland; **d.** bulbourethral gland; **e.** anus; **f.** vas deferens; **g.** epididymis; **h.** testis; **i.** scrotum; **j.** foreskin; **k.** glans penis; **l.** penis; **m.** urethra; **n.** vas deferens; **o.** urinary bladder. Path of sperm: h, g, f, n, b, m; **13.** c; **14.** d; **15.** c; **16.** b; **17.** b; **18.** d; **19.** d; **20.** c; **21.** c; **22.** b; **23.** c; **24.** d; **25.** c; **26.** d; **27.** a; **28.** d; **29.** e; **30.** c; **31.** d; **32.** d; **33.** c

Thinking Critically About the Concepts

1. Increased testosterone and progesterone inhibit GnRH. The inhibition of GnRH inhibits FSH production. Lower FSH decreases sperm production; **2a.** Drawing should show the introduction of progesterone early in the cycle, inhibiting FSh and LH; **2b.** Drawing should show that HCH prevents the corpus luteum from regressing, which means progesterone remains high and the uterine lining remains thick; **3.** Birth control pills inhibit FSH, which in turn inhibit follicle development. Fewer follicles may be the cause of a lower risk of ovarian cancer. Fertility drugs increase the number of follicles, which may be the cause of the higher risk of ovarian cancer; **4.** A low body fat composition may prevent the female athlete from producing estrogen and progesterone, which prevents the occurrence of a menstrual cycle.

Infectious Diseases Supplement
Check Your Progress

(S.1) 1. A global epidemic, AIDS, tuberculosis, malaria; **2.** Acute phase—no apparent symptoms but highly infectious, CD4 T cell count above 500 cells/mm^3, chronic phase—CD4 count between 200 and 499 cells/mm^3, one or more symptoms of an impaired immune system, AIDS—CD4 cell count below 200 cells/mm^3, one or more of the 25 AIDS-defining illnesses; **3.** The bacteria is inhaled into the lungs where it is eaten by macrophages. This causes a hypersensitivity reaction resulting in tubercles in the lungs. The bacteria remain alive within the tubercles; **4.** The mosquito carries the parasite to the human. The parasite completes the sexual part of its life cycle within the mosquito and the asexual part within the human. **(S.2) 1.** Diseases that are newly recognized or that have reappeared after a significant decline in incidence, avian influenza, SARS, tuberculosis; **2.** New or increased exposure to animals/insects, changes in human behavior, mutations in pathogens; **3.** Mutations in pathogens that allow them to be resistant to antibiotics, human carelessness, global warming. **(S.3) 1.** Through mutations in their genetic material, often as a result of the misuse of antibiotics; **2.** Take the entire dose of antibiotics for the entire time prescribed, do not skip doses or discontinue treatment early, do not share antibiotics, do not treat viral infections with antibiotics; **3.** the causative agent of tuberculosis (XDR TB), methicillin resistant *Staphylococcus aureus* (MRSA).

Chapter 17
Check Your Progress

(17.1) 1. A sperm makes its way through the corona radiata, acrosome releases digestive enzymes to digest zona pellucida, sperm binds to the egg, their plasma membranes fuse, sperm enters the egg, egg and sperm nuclei fuse. **(17.2) 1.** cleavage, growth, morphogenesis, differentiation; **2a.** Membranes that are outside of the embryo, not part of the embryo and fetus; **2b.** chorion—develops into fetal half of the placenta, allantois—forms umbilical blood vessels, yolk sac—produces blood cells, amnion—contains fluid to cushion and protect embyo; **3a.** Zygote divides repeatedly, develops into a morula, and then a blastocyst; **3b.** Embryo implants in the uterine wall, gastrulation occurs and primary germ layers are formed, organ systems appear and develop; **(17.3) 1a.** *chorionic villi of placenta→umbilical vein→venous ducts→*inferior vena cava→heart→blood is shunted into the left atrium by way of the *oval opening* and some enters the aorta by way of the *arterial duct→aorta→umbilical arteries*; **1b.** structures unique to fetus in italics; **2.** skeleton becomes ossified, sex of fetus is distinguishable, fetus grows and gains weight; **3.** The presence of an *SRY* gene, on the Y chromosome, leads to the development of testes and male genitals; otherwise, ovaries and female genitals develop; **(17.4) 1.** placental hormones; **2.** creates a concentration gradient favorable to the flow of carbon dioxide from fetal blood to maternal blood at the placenta; **3.** stage 1—cervix dilates; stage 2—baby is born; stage 3—placenta/ afterbirth is delivered. **(17.5) 1.** genetic in origin, whole-body process, extrinsic factors; **2.** skin—becomes thinner and less elastic, homeostatic adjustment to heat is limited; processing and transporting—heart shrinks because of a reduction in cardiac muscle size, arteries become rigid, cardiovascular problems are often accompanied by respiratory disorders, blood supply to kidney and liver is reduced; integration and coordination—reaction time slows, greater stimulation is needed for sense receptors to function, decline in bone density, loss of skeletal muscle mass; reproductive system—females undergo menopause, males undergo andropause; **3.** good health habits.

Understanding Key Terms

a. lanugo; **b.** afterbirth; **c.** gerontology; **d.** fertilization; **e.** cleavage

Testing Your Knowledge of the Concepts

12. c; **13.** e; **14.** b; **15.** c; **16.** b; **17.** a; **18.** b; **19.** d; **20.** d; **21.** b; **22.** c; **23.** d; **24.** d; **25.** c; **26.** a; **27.** e; **28.** e; **29.** d; **30.** e; **31.** e; **32.** d; **33.** b; **34. a.** chorion—fetal portion of the placenta, for nutrient, waste, and gas exchange; **b.** amnion—contains fluid to protect and cushion the embryo; **c.** embryo; **d.** allantois—becomes the umbilical vessels; **e.** yolk sac—first site of blood cell formation; **f.** fetal portion of placenta where fetal blood makes exchanges with maternal blood; **g.** maternal portion of placenta; **h.** umbilical cord—connects the embryo to the placenta; **35. a.** arterial duct; **b.** oval opening; **c.** venus duct; **d.** umbilical vein; **e.** umbilical arteries

Thinking Critically About the Concepts

1. It is secreted by the chorion. HCG is filtered from the blood by the nephron during glomerular filtration. It is not reabsorbed during tubular reabsorption, so it remains in the tubule and is a urine component; **2a.** HCG is a hormone distributed by the cardiovascular system; **2b.** It binds to a membrane receptor and activates the second messenger, cAMP; **3a.** Consume calcium-rich foods and participate in bone stressing exercise; **3b.** Osteoblast activity and calcium absorption decline with age, making the building of new bone more difficult; **4a.** FSH; **4b.** decreased; **4c.** FSH stimulates follicle development and estrogen production by the follicle. Without FSH, follicle development doesn't occur and estrogen secretion does not cause the development of the uterine lining that is shed during menstruation.

Chapter 18
Check Your Progress

(18.1) 1. G_1, S, G_2; **2.** G_1:organelles are doubled; S:DNA replicates; G_2: synthesis of proteins needed for cell division. **(18.2) 1.** They are the same; **2.** prophase—chromosomes attach to the spindle, metaphase— chromosomes align at the equator, anaphase—chromatids separate and chromosomes move toward poles, telophase—nuclear envelopes form around chromosomes; **3.** During cytokinesis, a contractile ring pinches the cell in two. **(18.3) 1.** Each daughter cell contains half as many chromosomes as the parent cell; **2.** In prophase I, crossing-over occurs, in metaphase I, independent alignment occurs; **3.** meiosis I—homologous chromosomes pair and then separate, meiosis II— sister chromatids separate, resulting in four cells with a haploid

number of chromosomes. **(18.4) 1.** b; **2.** a; **3.** a; **4a.** a; **4b.** b; **5.** a; **6.** b; **7.** d; **8.** b; **9.** a. **(18.5) 1.** nondisjunction; **2.** trisomy 21; **3.** Turner syndrome, Klinefelter syndrome, poly-X females, Jacobs syndrome; **4.** changes in chromosome structure, including deletions, duplications, inversions, and translocations.

Understanding Key Terms

a. Barr body; **b.** homologous chromosome; **c.** monosomy; **d.** inversion; **e.** haploid

Testing Your Knowledge of the Concepts

12. b; **13. a.** G_1 phase—cells grow, organelles double; **b.** S phase— DNA synthesis; **c.** G_2 phase—cell prepares to divide; **d.** mitosis— nuclear division and cytokinesis; **14.** d; **15.** b; **16.** a; **17.** d; **18.** e; **19.** c; **20.** b; **21.** a; **22.** d; **23.** a; **24.** c; **25.** c; **26.** d; **27.** c; **28.** b; **29.** e; **30.** a; **31.** d; **32.** b; **33.** c; **34.** d; **35.** c; **36.** d; **37.** a; **38.** b

Thinking Critically About the Concepts

1. cell signaling genes that send a cell into the cell cycle or prevent a cell from entering the cell cycle; **2.** It could be that Mechelle's body responds abnormally only to injury, but that normal mitosis functions well. Injury initiates a number of responses besides just mitosis; **3.** Homologous chromosomes separate randomly. Siblings that look similar inherited similar homologous chromosomes; siblings that are dissimilar inherited different combinations of homologous chromosomes; **4a.** They do not undergo the phases of mitosis to form new cells that would replace the damaged cells; **4b.** They are not likely to recover fully.

Chapter 19
Check Your Progress

(19.1) 1. do not undergo apoptosis, have unlimited replicative potential due to telomerase, do not exhibit contact inhibition, have no need for growth factors, undergo angiogenesis and metastasis; **2.** proto-oncogenes, tumor-suppressor genes; **3.** lung cancer, colorectal cancer, prostate cancer, breast cancer. **(19.2) 1.** DNA-linkage studies have revealed breast cancer genes (*BRCA1* and *BRCA2*), a tumor suppressor gene has been associated with retinoblastoma, and an abnormal RET gene predisposes an individual to thyroid cancer; **2.** ionizing radiation, tobacco smoke, pollutants, certain viruses; **3.** good nutrition, and, for example, not smoking. **(19.3) 1.** self-examination, Pap test, mammography, CAT scan, MRI, radioactive scan, ultrasound, biopsy; **2.** tumor marker tests and genetic tests. **(19.4) 1.** surgery, radiation therapy, chemotherapy; **2a.** cancer vaccines and immune cells genetically engineered to bear the tumor's antigen; **2b.** monoclonal antibodies; **3.** raise the level of *p53* in all cells and use an adenovirus to kill cells that lack a *p53* gene; **4.** Antiangiogenic drugs confine and reduce tumors by breaking up the network of new capillaries in the vicinity of a tumor.

Understanding Key Terms

a. telomere; **b.** carcinogen; **c.** proto-oncogene; **d.** angiogenesis; **e.** metastasis

Testing Your Knowledge of the Concepts

10. d; **11.** d; **12.** c; **13.** c; **14.** c; **15.** b; **16.** d; **17.** a; **18.** c; **19.** e; **20.** b; **21.** b; **22.** c; **23.** b; **24.** c; **25.** d; **26.** c; **27.** d; **28.** a; **29.** d; **30.** a; **31.** b

Thinking Critically About the Concepts

1. Chemotherapy and radiation therapy attack all rapidly reproducing cells, including immune cells. The immune response will be diminished; **2.** These cancers are more likely to spread, because the cells are immature. The greater the degree of cell immaturity, the more likely the cancer is to be malignant; **3.** The renal artery, renal vein, and/or any of the arteries within the kidney; **4.** Low-grade fever, abdominal pain, and "fussiness"—irritability in a baby like Cody; **5.** The tumors are more likely to be large and to have spread throughout the body; **6a.** To prevent the virus from ever infecting the girl's body; **6b.** She would develop artificial active immunity; **6c.** Form and remain in a monogamous relationship, and practice safer sex using a condom; **7.** The lymphatic system is often a route used to spread the cancer. Removing lymph nodes will remove cancer cells that may have spread there; **8.** A well-balanced diet, exercise, avoid smoking and excessive alcohol, practice safer sex in a monogamous relationship.

Chapter 20
Check Your Progress

(20.1) 1. Genotype refers to the genes of an individual; phenotype is a characteristic of an individual.; **2.** homozygous dominant *(EE)* and heterozygous *(Ee)* give the dominant phenotype, homozygous recessive *(ee)* gives the recessive phenotype. **(20.2a) 1a.** *W;* **1b.** *WS, Ws;* **1c.** *T, t;* **1d.** *Tg, tg;* **1e.** *AB, Ab, aB, ab;* **2a.** gamete; **2b.** genotype; **2c.** gamete; **2d.** genotype; **3a.** *EeSs;* **3b.** *eeSS.* **(20.2b) 1.** *Ww;* **2.** 75% or 3:1; **3.** *Ee;* **4.** father-*DD,* mother-*dd,* children-*Dd.* **(20.2c) 1.** *EeWw;* **2.** *EeWw;* **3.** $^1/_{16}$; **4.** father-*DdFf,* mother-*ddff,* child-*ddff.* **(20.2d) 1a.** *aa;* **1b.** *A?,* the questions mark means that each could be *AA* or *Aa;* **2.** parents are heterozygous-*Aa,* child is homozygous recessive-*aa;* **3.** 0%; **4.** abnormal hemoglobin stacks up; **5.** Marfan syndrome is dominant. **(20.3) 1.** *AABB* (very dark), *aabb* (very light); **2.** cleft lip, club-foot, congenital hip dislocation, hypertension, diabetes, schizophrenia, allergies, cancer, and behavioral traits; **3.** *aabb* (very light); **4.** familial hyper-cholesterolemia; **5.** codominance—I^A and I^B are fully expressed in the presence of the other; multiple allele inheritance—three alleles for the same gene exist; **6.** child-*ii,* mother-*I^Ai,* father-*I^Ai, I^Bi, ii.* **(20.4) 1.** Color-blindness is a recessive sex-linked genetic disorder, a female must receive two recessive alleles before she receives the condition, the male must only receive the X-linked recessive allele; **2a.** girls: 50% recessive phenotype, 50% dominant phenotype, boys: 50% dominant phenotype, 50% recessive phenotype; **2b.** girls: 100% dominant phenotype, boys: 50% dominant phenotype, 50% recessive phenotype; **3.** mother, mother-X^HX^h, father-X^HY, son-X^hY; **4.** 100%, 0%, 100%; **5.** color blind, husband is not the father.

Figure Questions

Fig. 20.8: The individual does not have the disorder but has a child with the disorder. Therefore, the individual has a recessive allele; **Fig. 20.9:** The individual has the disorder and has a child without the disorder. Therefore, the individual has a recessive allele; **Fig. 20.16:** Type O.

Understanding Key Terms

a. Punnett square; **b.** allele; **c.** dominant allele; **d.** locus; **e.** genotype

Testing Your Knowledge of the Concepts

14. e; **15.** c; **16.** d; **17.** b; **18.** c; **19.** b; **20.** a; **21.** a; **22.** a; **23.** c; **24.** d; **25.** d; **26.** a; **27.** e; **28.** c; **29.** c; **30.** c

Thinking Critically About the Concepts

1. Dad: X^AY Mom: X^AX^a Jeremy: X^aY Jackie: X^A? Dad is normal, so he must have the X^A allele. Jeremy has the disease, so he must have the X^a allele, inherited from his mom, making her genotype X^AX^a. Jackie inherited the X^A allele from her dad, but we cannot tell which allele she inherited from her mother, so we do not know if she is a carrier or not; **2.** Dad: X^+Y Mom: X^+ X^P Jeremy: X^F Y Jackie: X^+? Dad is normal, so he must have the X^+ allele. Jeremy has the disease, so he must have the X^F allele. Mom is normal, so she probably carries one X^+ allele and one X chromosome with the permutation. Jackie inherited the X^+ allele from her father, and because she is normal, she probably did not inherit the X^F allele as Jeremy did. It is impossible to say whether she inherited the X^P or the allele X^+ from her mother. If she did inherit the X^+ allele from her mother, then there is 0% chance of her having a child with fragile X syndrome. If she inherited the X^P allele, then she could pass the X^F allele on to her children; **3.** Full siblings are most likely to have similar alleles to the person needing a transplant, because the same two parents contributed alleles to the siblings.

Chapter 21
Check Your Progress

(21.1) 1. The double-stranded structure of DNA allows each original strand to serve as a template for the formation of a complementary new strand; **2.** Both RNA and DNA are polynucleotides, RNA contains the sugar ribose, the base U instead of T, and is single stranded; **3.** ribosomal RNA, messenger RNA, transfer RNA, small RNAs. **(21.2) 1.** primary, secondary, tertiary, quaternary; **2.** transcription forms an mRNA, translation synthesizes a polypeptide; **3a.** pretranscriptional, transcriptional control, posttranscriptional control, translational control, posttranslational control; **3b.** The use of transcription factors. **(21.3) 1.** genomics—the study of genomes, proteomics—the study of the structure, function, and interaction of cellular proteins; **2.** ex vivo gene therapy and in vivo gene therapy. **(21.4) 1.** recombinant DNA technology or polymerase chain reaction; **2a.** By detecting how many times a short sequence is repeated. **2b.** DNA fingerprinting can be used to identify a virus, a mutated gene, criminals, or remains of bodies; **3.** Can provide resistance to insect predation and herbicides, allow plants to be salt tolerant, improve the food quality of crops, increase productivity, larger animals, production of therapeutic and diagnostic proteins in animal milk, research projects.

Understanding Key Terms

a. polyribosome; **b.** transgenic organism; **c.** template; **d.** proteomics; **e.** vector

Testing Your Knowledge of the Concepts

11. e; **12.** c; **13.** e; **14.** d; **15.** b; **16.** b; **17.** d; **18.** d; **19.** c; **20.** c; **21.** e; **22. a.** ACU'CCU'GAA'UGC'AAA; **b.** UGA'GGA'CUU'ACG'UUU; **c.** thr-pro-glu-cys-lys

Thinking Critically About the Concepts

1. It would have the same amino acid sequence as the natural human product; **2.** The product could be contaminated by the growing process, in other words, with bacteria or bacterial products if grown in bacteria. Also, the organism it is grown in may modify the product in a manner that the human body may not. **3.** The addition of growth hormone may allow the patient to grow to normal stature. However, it treats the symptoms and not the cause. The addition of growth hormone does not correct the reason the patient cannot make growth hormone in the first place. If the growth hormone could be administered via gene therapy, then perhaps the patient could be cured; **4.** recombinant proteins that attack the tumor or enhance the immune system, products that inhibit the growth of the tumor in various ways (such as preventing the tumor from developing a blood supply), vaccines to prevent infection by agents that cause cancer.

Chapter 22
Check Your Progress

(22.1) 1. chemical; **2.** Stanley Miller experiment; **3.** RNA can sometimes be both a substrate and an enzyme during RNA processing. **(22.2) 1a.** change in a population or species over time; **1b.** descent from a common ancestor and adaptation to the environment; **2.** fossil, biogeographical, and anatomical; **3.** natural selecion. **(22.3) 1.** Domain Eukarya, Kingdom Animalia, Phylum Chordata, Class Mammalia, Order Primates, Family Hominidae, Genus *Homo,* Species *Homo sapiens*; **2.** mobile limbs, grasping hands, flattened face, binocular vision, large complex brain, reduced reproductive rate; **3.** human skull is in the midline of the body, human spine is S-shaped, human pelvis is broader, human femur has a longer neck, human toe is not opposable. **(22.4) 1.** bipedal posture, shape of face, and size of brain; **2.** gracile— *A. africanus, A. afarensis*; robust—*A. robustus, A. aethiopicus, A. boisei.* **(22.5a) 1.** *Australopithecus afarensis* (Lucy) **2.** *Homo habilis* because he made tools. **3.** *Homo ergaster.* **(22.5b) 1.** larger brain size, smaller teeth, culture; **2.** larger brain, flatter face, used fire, fashioned advanced tools; **3a.** *H. sapiens* evolved from *H. ergaster* only in Africa, and thereafter *H. sapiens* migrated to Europe and Asia; **3b.** suggests that we all descended from a few individuals and are more genetically similar; **4.** a thoroughly modern appearance, advanced stone tools, possibly first with a language, culture included art.

Understanding Key Terms

a. natural selection; **b.** adaptation; **c.** homologous structure; **d.** chemical evolution; **e.** fossil

Testing Your Knowledge of the Concepts

11. c; **12.** b; **13.** b; **14.** a; **15.** e; **16.** e; **17.** e; **18.** e; **19.** a; **20.** b; **21.** b; **22.** d; **23.** c; **24.** b; **25.** e; **26.** b; **27. a.** chordata; **b.** class; **c.** order; **d.** hominidae; **e.** genus; **f.** *Homo sapiens*; **28.** b; **29.** a; **30.** d; **31.** c

Thinking Critically About the Concepts

1. There are many places to dig on the surface of the Earth where people have inhabited, and without some clues on the surface, it would be difficult to determine where to begin. Human habitations usually are built near water sources, near food sources, on high places (for protection), or near shelter (like a cave). Archaeologists look for unusual dirt structures, such as mounds or evidence of digging/holes. Sometimes aerial surveys and infrared photographs provide evidence of unusual structures that are not visible from the ground; **2.** A survey is an organized way of mapping and then digging for evidence in specific places. It is a sampling technique that allows researchers to narrow their focus to areas that look promising. The probability of finding artifacts depends on the size of the grid and the density and distribution of artifacts under the ground; **3.** type of pottery, decorative items, jewelry, burial artifacts, monuments and memorials, seeds, tools; **4.** brain case size can help determine intellectual capacity, location of the spine's exit from the skull determines whether upright stance, location of the eye sockets determines whether forward vision, size and shape of the jaw determine food source.

Chapter 23
Check Your Progress

(23.1) 1. Because it is a place where organisms interact among themselves and with the physical and chemical environment; **2a.** Autotrophs require only inorganic nutrients and energy to produce food; **2b.** Heterotrophs need a source of organic nutrients and they must consume food; **3.** herbivores, carnivores, omnivores, detritus feeders; **4.** energy flow—occurs because, as nutrients pass from one population to another, all the energy is eventually converted into heat; chemical cycling—inorganic nutrients are returned to the producers from the atmosphere or soil. **(23.2) 1.** food web; **2.** Grazing food web begins with trees and grass, detrital food web begins with detritus, more energy may be found funneling through a detrital food web; **3.** food chain; **4.** ecological pyramid. **(23.3) 1.** reservoir, exchange pool, biotic community; **2a.** carbon, nitrogen; **2b.** phosphorus cycle; **3.** convert nitrogen gas to ammonium; **4.** ground water shortage, global warming, acid deposition, culturual eutrophication.

Understanding Key Terms

a. omnivore; **b.** trophic level; **c.** fossil fuel; **d.** nitrogen fixation; **e.** producer

Testing Your Knowledge of the Concepts

10. d; **11.** a; **12.** b; **13.** c; **14.** c; **15.** b; **16.** c; **17.** b; **18.** b; **19.** e; **20.** e; **21.** c; **22.** b; **23.** b, c; **24.** c, d; **25.** a; **26.** f; **27.** e; **28.** f; **29.** b, c, d; **30.** a, b, c; **31.** d; **32. a.** producers; **b.** consumers; **c.** inorganic nutrient pool; **d.** decomposers

Thinking Critically About the Concepts

1a. Fungicides will kill the fungi and pesticides will kill the insects that are part of the detritus food webs; **1b.** Fungicides kill fungi that decompose dead/discarded tissues and pesticides kill many insects that are detritivores that recycle dead/discarded tissues. The nutrients in the dead/discarded tissues will not be released without the activities of the fungi and insects causing adverse effects to the cycles; **2.** Legumes like clover and soybeans have nitrogen-fixing bacteria living in their roots. These bacteria supply the plant with a form of nitrogen the plant can use. Nitrogen that is not needed by the plants (extra) is added to the soil around the plant, which enriches the soil for other plants. Fertilizer also adds nitrogen to the soil, but it may run off during heavy rainfalls, which contributes to overgrowth in aquatic habitats; **3a.** Answers will vary, but may include recycling newspapers, cans, and so forth or starting a compost pile for food scraps. The compost can be used to enrich the soil used for gardens. **3b.** Answers will vary, but may include turning off the water while brushing your teeth or running the dishwasher/washing machine with full loads only. Others might include minimizing the use of fertilizer and pesticides that can run off into bodies of water; **4.** Organochlorides accumulate in organisms higher up on a food chain, discontinuing the use of pesticides that contain organochlorides would improve the health of all organisms in food chains.

Chapter 24

Check Your Progress

(24.1) 1. LDCs; **2.** Because of its past exponential growth. **(24.2a) 1.** land, water, food, energy, and minerals; **2.** beaches, semiarid lands, tropical rain forest; **3a.** damming rivers and withdrawing water from aquifers; **3b.** land subsidence and saltwater intrusion; **4a.** planting few genetic varieties, heavy use of fertilizers, generous irrigation, excessive fuel consumption; **4b.** A single parasite can cause devastation of a crop, agricultural runoff, water shortage, soil loss, salinization. **(24.2b) 1.** The consumption of nonrenewable energy supplies results in environmental degradation; **2.** Strip mining minerals results in land devoid of vegetation, allowing rain to wash toxic waste deposits into nearby streams and rivers; **3.** heavy metals, synthetic organic compounds, and raw sewage. **(24.3) 1.** habitat loss, alien species, pollution, overexploitation, and disease; **2.** direct value—medicinal value, agricultural value, consumptive use value; indirect value—waste disposal, provision of fresh water, prevention of soil erosion, biogeochemical cycles, regulation of climate, ecotourism. **(24.4) 1.** large portion of land used for human purposes, agriculture uses large amounts of nonrenewable energy and creates pollution, more freshwater is used in agriculture than used in homes, almost half of the agriculture yield goes toward feeding animals, decrease in surface water, use of nonrenewable fossil energy, expansion of the population into all regions of the planet; **2.** To make rural areas sustainable: plant cover crops, plant multiuse crops, use low flow irrigation, use precision farming to reduce habitat destruction, use integrated pest management, plant multipurpose trees, restore wetlands, use renewable forms of energy. To make urban areas sustainable: use energy-efficient modes of transport, use solar or geothermal energy to heat buildings, use green roofs, improve storm-water management, plant native grasses for lawns, create greenbelts, revitalize old sections of cities, use more efficient light fixtures, recycle business equipment, use low-maintenance building materials. **3.** ISEW, GPI, and additional ways ecological economists are in the process of developing.

Understanding Key Terms

a. sustainable; **b.** carrying capacity; **c.** greenhouse gases; **d.** biological magnification; **e.** aquifer

Testing Your Knowledge of the Concepts

9. a; **10.** c; **11.** a; **12.** d; **13.** d; **14.** a; **15.** c; **16.** c; **17.** b; **18.** d; **19.** d; **20.** a; **21.** c, d, e, f; **22.** a, b; **23.** a, b, c; **24.** c, d, e, f; **25.** b; **26.** a; **27.** c; **28.** c; **29.** e; **30.** e; **31.** b; **32.** e; **33. a.** habitat loss; **b.** alien species; **c.** pollution; **d.** overexploitation; **e.** disease

Thinking Critically About the Concepts

1a. MDC; **1b.** There may not be enough young people to support the aging population; **2.** Less energy is needed to grow plant material than animals for human consumption; **3a.** Answers will vary; **3b.** Answers will vary. Example: drive less or drive an energy efficient vehicle, use alternative means of heating/cooling a home, purchase locally grown food; **4.** Most items in a landfill are not exposed to decomposers, so the nutrients do not cycle back; **5a.** Answers will vary; **5b.** Answers will vary. Examples: recycling programs, ride sharing/bus use, water/electricity use, events associated with Earth Day.

Glossary

A

absorption Taking in of substances by cells or membranes. 159

acetylcholine (ACh) (uh-seet-ul-koh-leen) Neurotransmitter active in both the peripheral and central nervous systems. 281

acetylcholinesterase (AChE) (uh-seet-ul-koh-luh-nes-tuh-rays) Enzyme that breaks down acetylcholine bound to postsynaptic receptors within a synapse. 281

acid Molecules tending to raise the hydrogen ion concentration in a solution and to lower its pH numerically. 27

acid deposition The return to Earth in rain or snow of sulfate or nitrate salts of acids produced by commercial and industrial activities. 555

acidosis Excessive accumulation of acids in body fluids. 220

acne Inflammation of sebaceous glands. 79

acquired immunodeficiency syndrome (AIDS) (im-yuh-noh-dih-fish-un-see) Disease caused by HIV and transmitted via body fluids; characterized by failure of the immune system. 381

acromegaly (ak-roh-meg-uh-lee) Condition resulting from an increase in growth hormone production after adult height has been achieved. 334

acrosome (ak-ruh-sohm) Cap at the anterior end of a sperm that partially covers the nucleus and contains enzymes that help the sperm penetrate the egg. 356

actin (ak-tin) One of two major proteins of muscle; makes up thin filaments in myofibrils of muscle fibers. See myosin. 259

actin filament Cytoskeletal filaments of eukaryotic cells composed of the protein actin; also refers to the thin filaments of muscle cells. 54

action potential Electrochemical changes that take place across the axomembrane; the nerve impulse. 278

active immunity Resistance to disease due to the immune system's response to a microorganism or a vaccine. 149

active site Region on the surface of an enzyme where the substrate binds and where the reaction occurs. 56

active transport Use of a plasma membrane carrier protein and energy to move a substance into or out of a cell from lower to higher concentration. 50

acute bronchitis (brahn-ky-tis) Infection of the primary and secondary bronchi. 199

adaptation Organism's modification in structure, function, or behavior suitable to the environment. 5, 518

Addison disease Condition resulting from a deficiency of adrenal cortex hormones; characterized by low blood glucose, weight loss, and weakness. 340

adenine (A) (ad-uh-neen) One of four nitrogen bases in nucleotides composing the structure of DNA and RNA. 38

adhesion junction Junction between cells in which the adjacent plasma membranes do not touch but are held together by intercellular filaments attached to buttonlike thickenings. 74

adipose tissue (ah-duh-pohs) Connective tissue in which fat is stored. 67

ADP (adenosine diphosphate) (ah-den-ah-seen dy-fahs-fayt) Nucleotide with two phosphate groups that can accept another phosphate group and become ATP. 34

adrenal cortex (uh-dree-nul kor-teks) Outer portion of the adrenal gland; secretes mineralocorticoids, such as aldosterone, and glucocorticoids, such as cortisol. 337

adrenal gland (uh-dree-nul) An endocrine gland that lies atop a kidney; consisting of the inner adrenal medulla and the outer adrenal cortex. 337

adrenal medulla (uh-dree-nul muh-dul-uh) Inner portion of the adrenal gland; secretes the hormones epinephrine and norepinephrine. 337

adrenocorticotropic hormone (ACTH) (uh-dree-noh-kawrt-ih-koh-troh-pik) Hormone secreted by the anterior lobe of the pituitary gland that stimulates activity in the adrenal cortex. 332

aerobic Requiring oxygen. 59

afterbirth Placenta and the extraembryonic membranes, which are delivered (expelled) during the third stage of parturition. 410

agglutination (uh-gloot-un-ay-shun) Clumping of red blood cells due to a reaction between antigens on red blood cell plasma membranes and antibodies in the plasma. 126

aging Progressive changes over time, leading to loss of physiologic function and eventual death. 411

agranular leukocyte White blood cell that does not contain distinctive granules. 122

agricultural runoff Water from precipitation and irrigation that flows over fields into bodies of water or aquifers. 570

albumin (al-byoo-mun) Plasma protein of the blood having transport and osmotic functions. 118

aldosterone (al-dahs-tuh-rohn) Hormone secreted by the adrenal cortex that decreases sodium and increases potassium excretion; raises blood volume and pressure. 219, 339

alien species Nonnative species that migrate or are introduced by humans into a new ecosystem; also called exotics. 576

alkalosis Excessive accumulation of bases in body fluids. 220

allantois (uh-lan-toh-is) Extraembryonic membrane that contributes to the formation of umbilical blood vessels in humans. 396

allele (uh-leel) Alternative form of a gene; alleles occur at the same locus on homologous chromosomes. 466

allergen (al-ur-jun) Foreign substance capable of stimulating an allergic response. 152

allergy Immune response to substances that usually are not recognized as foreign. 152

all-or-none law Law that states that muscle fibers either contract maximally or not at all, and that neurons either conduct a nerve impulse completely or not at all. 262

alveolus (pl., alveoli) (al-vee-uh-lus) Air sac of a lung. 190

Alzheimer disease (AD) Brain disorder characterized by a general loss of mental abilities. 289, 295

amino acid Organic molecule having an amino group and an acid group, which covalently bonds to produce peptide molecules. 34

amnion (am-nee-ahn) Extraembryonic membrane that forms an enclosing, fluid-filled sac. 396

ampulla (am-pool-uh, -pul-uh) Base of a semicircular canal in the inner ear. 322

amygdala (uh-mig-duh-luh) Portion of the limbic system that functions to add emotional overtones to memories. 288

amylotrophic lateral sclerosis (ALS) Chronic, progressive motor neuron disease characterized by the gradual degeneration of the nerve cells resulting in death. 267

anabolic steroid (a-nuh-bahl-ik) Synthetic steroid that mimics the effect of testosterone. 345

anaerobic Growing or metabolizing in the absence of oxygen. 58

analogous structure Structure that has a similar function in separate lineages but differs in anatomy and ancestry. 521

anaphase Mitotic phase during which daughter chromosomes move toward the poles of the spindle. 423

anaphylactic shock Severe systemic form of allergic reaction involving bronchiolar contriction, impaired breathing, vasodilation, and a rapid drop in blood pressure with a threat of circulatory failure. 152

androgen (an-druh-jun) Male sex hormone (e.g., testosterone). 344

anemia (uh-nee-mee-uh) Inefficiency in the oxygen-carrying ability of blood due to a shortage of hemoglobin. 121

aneurysm Saclike expansion of a blood vessel wall. 105

angina pectoris (an-jy-nuh pek-tuh-ris) Condition characterized by thoracic pain resulting from occluded coronary arteries; precedes a heart attack. 106

angiogenesis (an-jee-oh-jen-uh-sis) Formation of new blood vessels; one mechanism by which cancer spreads. 448

angioplasty (an-jee-uh-plas-tee) Surgical procedure for treating clogged arteries, in which a plastic tube is threaded through a major blood vessel toward the heart and then a balloon at the end of the tube is inflated, forcing open the vessel. A stent is then placed in the vessel.   108

anorexia nervosa (a-nuh-rek-see-uh nur-voh-suh) Eating disorder characterized by a morbid fear of gaining weight.   179

anterior pituitary (pih-too-ih-tair-ee) Portion of the pituitary gland controlled by the hypothalamus and that produces six types of hormones, some of which control other endocrine glands.   332

anthropoid Group of primates that includes monkeys, apes, and humans.   525

antibiotic resistance A characteristic of pathogens that causes them to survive treatment with chemicals (antibiotics) that normally would kill them.   391

antibody (an-tih-bahd-ee) Protein produced in response to the presence of an antigen; each antibody combines with a specific antigen.   116

antibody-mediated immunity Specific mechanism of defense in which plasma cells derived from B cells produce antibodies that combine with antigens.   145

antibody titer Amount of antibody present in a sample of blood serum.   150

anticodon (an-tih-koh-dahn) Three-base sequence in a tRNA molecule base that pairs with a complementary codon in mRNA.   497

antidiuretic hormone (ADH) (an-tih-dy-uh-ret-ik) Hormone secreted by the posterior pituitary that increases the permeability of the collecting ducts in a kidney.   219, 332

antigen (an-tih-jun) Foreign substance, usually a protein or a polysaccharide, that stimulates the immune system to produce antibodies.   122, 144

antigen-presenting cell (APC) Cell that displays the antigen to the cells of the immune system so they can defend the body against that particular antigen.   147

anus Outlet of the digestive tract.   168

aorta (ay-or-tuh) Major systemic artery that receives blood from the left ventricle.   103

apoptosis (ap-uh-toh-sis, -ahp-) Programmed cell death involving a cascade of specific cellular events leading to death and destruction of the cell.   145, 422, 446

appendicular skeleton (ap-un-dik-yuh-lur) Portion of the skeleton forming the pectoral girdles and upper extremities and the pelvic girdle and lower extremities.   242

appendix In humans, small, tubular appendage that extends outward from the cecum of the large intestine.   159

aquaporin Protein membrane channel through which water can diffuse.   218

aqueous humor (ay-kwee-us, ak-wee-) Clear, watery fluid between the cornea and lens of the eye.   311

aquifer (ahk-wuh-fur) Rock layers that contain water released in appreciable quantities to wells or springs.   549, 568

arteriole (ar-teer-ee-ohl) Vessel that takes blood from an artery to capillaries.   93

arteriovenous shunt A pathway, usually abnormal, that connects an artery directly to a vein.   94

articular cartilage (ar-tik-yuh-lur) Hyaline cartilaginous covering over the articulating surface of the bones of synovial joints.   230

assimilation Action of chemically changing absorbed substances.   555

association area One of several regions of the cerebral cortex related to memory, reasoning, judgment, and emotional feelings.   286

aster Short, radiating fibers about the centrioles at the poles of a spindle.   422

asthma (az-muh, as-) Condition in which bronchioles constrict and cause difficulty in breathing.   200

astigmatism (uh-stig-muh-tiz-um) Blurred vision due to an irregular curvature of the cornea or the lens.   315

atherosclerosis (ath-uh-roh-skluh-roh-sis) Condition in which fatty substances accumulate abnormally beneath the inner linings of the arteries.   105

atom Smallest particle of an element that displays the properties of the element.   2, 20

atomic mass Mass of an atom equal to the number of protons plus the number of neutrons with the nucleus.   20

atomic number Number of protons within the nucleus of an atom.   20

ATP (adenosine triphosphate) (uh-den-uh-seen try-fahs-fayt) Nucleotide with three phosphate groups. The breakdown of ATP into ADP + (P) makes energy available for energy-requiring processes in cells.   39

atrial natriuretic hormone (ANH) (ay-tree-ul nay-tree-yoo-ret-ik) Hormone secreted by the heart that increases sodium excretion and, therefore, lowers blood volume and pressure.   219, 339

atrioventricular (AV) bundle (ay-tree-oh-ven-trik-yuh-lur) Group of specialized fibers that conduct impulses from the atrioventricular node to the ventricles of the heart; also called AV bundle.   98

atrioventricular (AV) valve Valve located between the atrium and the ventricle.   95

atrium (ay-tree-um) One of the upper chambers of the heart, either the left atrium or the right atrium, that receives blood.   95

auditory canal Curved tube extending from the pinna to the tympanic membrane.   317

auditory (Eustachian) tube Extension from the middle ear to the nasopharynx that equalizes air pressure on the eardrum.   187, 317

australopithecine (aw-stray-loh-pith-uh-syn) Any of the first evolved hominids; classified into several species of *Australopithecus*.   528

autoimmune disease Disease that results when the immune system mistakenly attacks the body's tissues.   153

autonomic system (aw-tuh-nahm-ik) Branch of the peripheral nervous system that has control over the internal organs; consists of the sympathetic and parasympathetic systems.   292

autosome (aw-tuh-sohm) Any chromosome other than the sex chromosomes.   481

autotroph Organism that can capture energy and synthesize organic nutrients from inorganic nutrients.   544

AV (atrioventricular) node Small region of neuromuscular tissue that transmits impulses received from the sinoatrial node to the ventricles.   98

axial skeleton (ak-see-ul) Portion of the skeleton that supports and protects the organs of the head, the neck, and the trunk.   237

axon (ak-sahn) Elongated portion of a neuron that conducts nerve impulses typically from the cell body to the synapse.   277

axon terminal Small swelling at the tip of one of many endings of the axon.   280

B

bacillus A rod-shaped bacterial cell.   136

bacteria One of three domains of life; prokaryotic cells other than archaea with unique genetic, biochemical, and physiological characteristics.   136

Barr body Dark-staining body (discovered by M. Barr) in the nuclei of female mammals that contains a condensed, inactive X chromosome.   436

basal nuclei Nerve cells that integrate motor commands to ensure balance and coordination.   287

base Molecules tending to lower the hydrogen ion concentration in a solution and raise the pH numerically.   27

basement membrane Layer of nonliving material that anchors epithelial tissue to underlying connective tissue.   72

basophil (bay-suh-fil) White blood cell with a granular cytoplasm; able to be stained with a basic dye.   122

B cell (B lymphocyte) Lymphocyte that matures in the bone marrow and, when stimulated by the presence of a specific antigen, gives rise to antibody-producing plasma cells.   141

B-cell receptor Molecule on the surface of a B-lymphocyte to which an antigen binds.   145

benign Form of tumor not capable of spreading throughout the body.   447

bicarbonate ion Ion that participates in buffering the blood; the form in which carbon dioxide is transported in the bloodstream.   196

bile Secretion of the liver temporarily stored and concentrated in the gallbladder before being released into the small intestine, where it emulsifies fat.   165, 167

binge-eating disorder Condition characterized by overeating episodes that are not followed by purging.   181

binomial name Two-part scientific name of an organism. The first part designates the genus, the second part the specific epithet. 524

biocultural evolution Passage of culture from one generation to the next. 536

biodiversity Total number of species, the variability of their genes, and the communities in which they live. 7, 575

biogeochemical cycle (by-oh-jee-oh-kem-ih-kul) Circulating pathway of elements such as carbon and nitrogen, involving exchange pools, storage areas, and biotic communities. 548

biogeography Study of the geographical distribution of organisms. 520

bioinformatics Computer technologies used to study the genome. 503

biological evolution Change in life-forms that has taken place in the past and will take place in the future; includes descent from a common ancestor and adaptation to the environment. 518

biological magnification Process by which substances become more concentrated in organisms in the higher trophic levels of a food web. 557, 575

biology Scientific study of life. 2

biomass The number of organisms multiplied by their weight. 548

biosphere (by-oh-sfeer) Zone of air, land, and water at the surface of the Earth in which living organisms are found. 4, 543

biotechnology product Product created by using biotechnology techniques. 506

biotic potential Maximum reproductive rate of an organism, given unlimited resources and ideal environmental conditions. Compare with environmental resistance. 564

bipedal posture Ability to walk upright on two feet. 528

birth control method Prevents either fertilization or implantation of an embryo in the uterine lining. 365

birth control pill Oral contraceptive containing estrogen and progesterone. 366

blastocyst (blas-tuh-sist) Early stage of human embryonic development that consists of a hollow, fluid-filled ball of cells. 397

blind spot Region of the retina lacking rods or cones where the optic nerve leaves the eye. 313

blood Type of connective tissue in which cells are separated by a liquid called plasma. 68

blood doping Practice of boosting the number of red blood cells in the blood to enhance athletic performance. 120

blood pressure Force of blood pushing against the inside wall of a vessel. 100

blood transfusion Introduction of whole blood or a blood component directly into the bloodstream. 126

body mass index (BMI) Calculation used to determine whether a person is overweight or obese. 171

bolus Small lump of food that has been chewed and swallowed. 161

bone marrow transplant A cancer patient's stem cells are harvested and stored before chemotherapy beings. Then, the stored cells are returned to the patient by injection. 459

bone remodeling Ongoing mineral deposits and withdrawals from bone that adjust bone strength and maintain levels of calcium and phophorus in blood. 234

brain Enlarged superior portion of the central nervous system located in the cranial cavity of the skull. 285

brain stem Portion of the brain consisting of the medulla oblongata, pons, and midbrain. 288

Braxton Hicks contraction Strong, late-term uterine contractions prior to cervical dilation; also called false labor. 408

breech birth Birth in which the baby is positioned rump first. 406

Broca's area Region of the frontal lobe that coordinates complex muscular actions of the mouth, tongue, and larynx, making speech possible. 287

bronchiole (brahng-kee-ohl) Smaller air passages in the lungs that begin at the bronchi and terminate in alveoli. 190

bronchus (pl., bronchi) (brahng-kus) One of two major divisions of the trachea leading to the lungs. 190

buffer Substance or group of substances that tend to resist pH changes of a solution, thus stabilizing its relative acidity and basicity. 28, 221

bulbourethral gland (bul-boh-yoo-ree-thrul) Either of two small structures located below the prostate gland in males; each adds secretions to semen. 353

bulimia nervosa (byoo-lee-mee-uh, -lim-ee-, nur-voh-suh) Eating disorder characterized by binge eating followed by purging via self-induced vomiting or use of a laxative. 179

bursa (bur-suh) Saclike, fluid-filled structure, lined with synovial membrane, that occurs near a joint. 255

bursitis (bur-sy-tis) Inflammation of any of the friction-easing sacs called bursae within the knee joint. 267

C

calcitonin (kal-sih-toh-nin) Hormone secreted by the thyroid gland that increases the blood calcium level. 336

calorie Amount of heat energy required to raise the temperature of 1 g of water 1°C. 25

cancer Malignant tumor whose nondifferentiated cells exhibit loss of contact inhibition, uncontrolled growth, and the ability to invade tissue and metastasize. 446

capsule Gelatinous layer surrounding the cells of blue-green algae and certain bacteria. 136

carbaminohemoglobin Hemoglobin carrying carbon dioxide. 196

carbohydrate Class of organic compounds that includes monosaccharides, disaccharides, and polysaccharides. 29

carbonic anhydrase (kar-bahn-ik an-hy-drays, -drayz) Enzyme in red blood cells that speeds the formation of carbonic acid from the reactants water and carbon dioxide. 196

carcinogen (kar-sin-uh-jun) Environmental agent that causes mutations leading to the development of cancer. 451

carcinogenesis (kar-suh-nuh-jen-uh-sis) Development of cancer. 447

carcinoma (kar-suh-noh-muh) Cancer arising in epithelial tissue. 449

cardiac cycle One complete cycle of systole and diastole for all heart chambers. 97

cardiac muscle Striated, involuntary muscle found only in the heart. 70, 254

cardiovascular system (kar-dee-oh-vas-kyuh-lur) Organ system in which blood vessels distribute blood powered by the pumping action of the heart. 80

carnivore (kar-nuh-vor) Consumer in a food chain that eats other animals. 544

carrier Heterozygous individual who has no apparent abnormality but can pass on an allele for a recessively inherited genetic disorder. 475

carrying capacity Maximum number of individuals of any species that can be supported by a particular ecosystem on a long-term basis. 564

cartilage (kar-tul-ij, kart-lij) Connective tissue in which the cells lie within lacunae separated by a flexible proteinaceous matrix. 67, 230

cecum (see-kum) Small pouch that lies below the entrance of the small intestine and is the blind end of the large intestine. 168

cell Smallest unit that displays the properties of life; always contains cytoplasm surrounded by a plasma membrane. 2

cell body Portion of a neuron that contains a nucleus and from which dendrites and an axon extend. 277

cell cycle Repeating sequence of cellular events that consists of interphase, mitosis, and cytokinesis. 421

cell-mediated immunity Specific mechanism of defense in which T cells destroy antigen-bearing cells. 149

cell theory One of the major theories of biology; states that all organisms are made up of cells and cells come only from preexisting cells. 44

cellular respiration Metabolic reactions that use the energy primarily from carbohydrates but also from fatty acid or amino acid breakdown to produce ATP molecules. 56

cellulose (sel-yuh-lohs, -lohz) Polysaccharide that is the major complex carbohydrate in plant cell walls. 30

central nervous system (CNS) Portion of the nervous system consisting of the brain and spinal cord. 276

centriole (sen-tree-ohl) Cellular structure, existing in pairs, that possibly organizes the mitotic spindle for chromosomal movement during mitosis and meiosis. 422

centromere (sen-truh-meer) Constriction where sister chromatids of a chromosome are held together. 421

centrosome Central microtubule organizing center of cells. In animal cells, it contains two centrioles.   54, 422

cerebellum (ser-uh-bel-um) Part of the brain located posterior to the medulla oblongata and pons that coordinates skeletal muscles to produce smooth, graceful motions.   287

cerebral cortex (suh-ree-brul, ser-uh-brul kor-teks) Outer layer of cerebral hemispheres; receives sensory information and controls motor activities.   286

cerebral hemisphere One of the large, paired structures that together constitute the cerebrum of the brain.   285

cerebrospinal fluid (sair-uh-broh-spy-nul, suh-ree-broh-) Fluid found in the ventricles of the brain, in the central canal of the spinal cord, and in association with the meninges.   283

cerebrum (sair-uh-brum, suh-ree-brum) Main part of the brain consisting of two large masses, or cerebral hemispheres; the largest part of the brain in mammals.   285

cervix (sur-viks) Narrow end of the uterus, which projects into the vagina.   358

cesarean section Birth by surgical incision of the abdomen and uterus.   406

chancre Sore that appears on the skin; first sign of syphilis.   373

chemical evolution Increase in the complexity of chemicals over time that could have led to the first cells.   516

chemical signal Molecule that brings about a change in a cell, tissue, organ, or individual when it binds to a specific receptor.   330

chemoreceptor (kee-moh-rih-sep-tur) Sensory receptor sensitive to chemical stimuli— for example, receptors for taste and smell.   194, 304

chlamydia (kluh-mid-ee-uh) Sexually transmitted disease, caused by the bacterium *Chlamydia trachomatis;* can lead to pelvic inflammatory disease.   372

chlorofluorocarbons (CFCs) (klor-oh-floor-oh-kar-buns) Organic compounds containing carbon, chlorine, and fluorine atoms. CFCs, such as Freon, can deplete the ozone shield by releasing chlorine atoms in the upper atmosphere.   558, 574

cholesterol Form of lipid: Structural component of plasma membrane, precursor for steroid hormones.   167

chondrocyte Type of cell found in the lacunae of cartilage.   230

chordae tendineae (kor-dee ten-din-ee-ee) Tough bands of connective tissue that attach the papillary muscles to the atrioventricular valves within the heart.   95

chorion (kor-ee-ahn) Extraembryonic membrane that contributes to placenta formation.   396

chorionic villi (kor-ee-ahn-ik vil-eye) Treelike extensions of the chorion that project into the maternal tissues at the placenta.   400

choroid (kor-oyd) Vascular, pigmented middle layer of the eyeball.   310

chromatin (kroh-muh-tin) Network of fine threads in the nucleus composed of DNA and proteins.   52

chromosome (kroh-muh-som) Chromatin condensed into a compact structure.   52

chronic bronchitis Obstructive pulmonary disorder that tends to recur; marked by inflamed airways filled with mucus and degenerative changes in the bronchi, including loss of cilia.   199

chyme (kym) Thick, semiliquid food material that passes from the stomach to the small intestine.   163

ciliary body (sil-ee-air-ee) Structure associated with the choroid layer that contains ciliary muscle and controls the shape of the lens of the eye.   310

cilium (pl., cilia) (sil-ee-um) Short, hairlike projection from the plasma membrane, occurring usually in large numbers.   55

circadian rhythm (sur-kay-dee-un) Biological rhythm with a 24-hour cycle.   345

circumcision Removal of the prepuce (foreskin) of the penis.   354

cirrhosis (sih-roh-sis) Chronic, irreversible injury to liver tissue; commonly caused by frequent alcohol consumption.   168

citric acid cycle Cycle of reactions in mitochondria that begins with citric acid; it breaks down an acetyl group as CO_2, ATP, NADH, and $FADH_2$ are given off; also called the Krebs cycle.   58

cleavage Cell division without cytoplasmic addition or enlargement; occurs during the first stage of animal development.   396

cleavage furrow Indentation that begins the process of cleavage, by which human cells undergo cytokinesis.   424

clonal selection model Concept that an antigen selects which lymphocyte will undergo clonal expansion and produce more lymphocytes bearing the same type of antigen receptor.   145

cloning Production of identical copies; can be either the production of identical individuals or, in genetic engineering, the production of identical copies of a gene.   504

clotting Process of blood coagulation, usually when injury occurs.   123

coccus Spherical bacterial cell.   136

cochlea (kohk-lee-uh, koh-klee-uh) Portion of the inner ear that resembles a snail's shell and contains the spiral organ, the sense organ for hearing.   317

cochlear nerve Either of two cranial nerves that carry nerve impulses from the spiral organ to the brain; also called the auditory nerve.   320

codominance Inheritance pattern in which both alleles of a gene are equally expressed.   480

codon Three-base sequence in mRNA that causes the insertion of a particular amino acid into a protein or termination of translation.   495

coenzyme (koh-en-zym) Nonprotein organic molecule that aids the action of the enzyme to which it is loosely bound.   56

collagen fiber (kahl-uh-jun) White fiber in the matrix of connective tissue; gives flexibility and strength.   66

collecting duct Duct within the kidney that receives fluid from several nephrons; the reabsorption of water occurs here.   214

colon (koh-lun) The major portion of the large intestine, consisting of the ascending colon, the transverse colon, and the descending colon.   168

colony-stimulating factor (CSF) Protein that stimulates differentiation and maturation of white blood cells.   121

color blindness Deficiency in one or more of the three types of cone cells responsible for color vision.   483

color vision Ability to detect the color of an object; dependent on three types of cone cells.   312

columnar epithelium (kuh-lum-nur ep-uh-thee-lee-um) Type of epithelial tissue with cylindrical cells.   74

community Assemblage of populations interacting with one another within the same environment.   2

compact bone Type of bone that contains osteons consisting of concentric layers of matrix and osteocytes in lacunae.   68, 230

complement system Series of proteins in plasma that form a nonspecific defense mechanism against a microbe invasion; it complements the antigen-antibody reaction.   144

complementary DNA (cDNA) DNA that has been synthesized from mRNA by the action of reverse transcriptase.   505

complementary paired bases Hydrogen bonding between particular bases; in DNA thymine (T) pairs with adenine (A), and guanine (G) pairs with cytosine (C); in RNA, uracil (U) pairs with A, and G pairs with C.   38, 490

compound Substance having two or more different elements united chemically in a fixed ratio.   22

conclusion Statement made following an experiment as to whether the results support the hypothesis.   8

cone cell Photoreceptor in retina of eye that responds to bright light; detects color and provides visual acuity.   312

congenital hypothyroidism Condition resulting from improper development of the thyroid in an infant; characterized by stunted growth and mental retardation.   336

connective tissue Type of tissue that binds structures together, provides support and protection, fills spaces, stores fat, and forms blood cells; adipose tissue, cartilage, bone, and blood are types of connective tissue.   66

constipation (kahn-stuh-pay-shun) Delayed and difficult defecation caused by insufficient water in the feces.   169

consumer Organism that feeds on another organism in a food chain; primary consumers eat plants, and secondary consumers eat animals.   544

contraceptive (kahn-truh-sep-tiv) Medication or device used to reduce the chance of pregnancy.   365

contraceptive implant Birth control method using synthetic progesterone; prevents ovulation by disrupting the ovarian cycle.   367

contraceptive injection Birth control method using progesterone or estrogen and progesterone together; prevents ovulation by disrupting the ovarian cycle.   367

contraceptive vaccine Under development, this birth control method immunizes against the hormone HCG, crucial to maintaining implantation of the embryo.   367

control group Sample that goes through all the steps of an experiment but lacks the factor or is not exposed to the factor being tested; a standard against which results of an experiment are checked.   9

convulsion Sudden attack characterized by a loss of consciousness and severe, sustained, rhythmic contractions of some or all voluntary muscles.   267

cornea (kor-nee-uh) Transparent, anterior portion of the outer layer of the eyeball.   310

coronary artery (kor-uh-nair-ee) Artery that supplies blood to the wall of the heart.   95

coronary bypass operation Therapy for blocked coronary arteries in which part of a blood vessel from another part of the body is grafted around the obstructed artery.   107

corpus callosum Bridge of nerve tracts that connects the two cerebral hemispheres.   285

corpus luteum (kor-pus loot-ee-um) Yellow body that forms in the ovary from a follicle that has discharged its secondary oocyte; it secretes progesterone and some estrogen.   360

cortisol (kor-tuh-sawl) Glucocorticoid secreted by the adrenal cortex that responds to stress on a long-term basis; reduces inflammation and promotes protein and fat metabolism.   338

cough Sudden expulsion of air from the lungs that clears the air passages; a common symptom of upper respiratory infections.   188

covalent bond (coh-vay-lent) Chemical bond in which atoms share one pair of electrons.   24

cramp Muscle contraction that causes pain.   267

cranial nerve Nerve that arises from the brain.   291

creatinine (kree-ah-tuhn-een) Nitrogenous waste; the end product of creatine phosphate metabolism.   209

Cro-Magnon (kroh-mag-nun) Common name for first fossils to be designated *Homo sapiens*.   534

crossing-over Exchange of segments between nonsister chromatids of a tetrad during meiosis.   430

cuboidal epithelium (kyoo-boyd-ul) Type of epithelial tissue with cube-shaped cells.   73

cultural eutrophication Enrichment of water by inorganic nutrients used by phytoplankton. Often, overenrichment caused by human activities leads to excessive bacterial growth and oxygen depletion.   557

culture Total pattern of human behavior; includes technology and the arts, and depends upon the capacity to speak and transmit knowledge.   7, 530

Cushing syndrome (koosh-ing) Condition resulting from hypersecretion of glucocorticoids; characterized by thin arms and legs and a "moon face," and accompanied by high blood glucose and sodium levels.   340

cutaneous receptor Sensory receptors for pressure and touch found in the dermis of the skin.   306

cyclic adenosine monophosphate (cAMP) (sy-klik, sih-klik) ATP-related compound that acts as the second messenger in peptide hormone transduction; it initiates activity of the metabolic machinery.   330

cyclin Protein that regularly increases and decreases in concentration during the cell cycle.   448

cystic fibrosis A generalized, autosomal recessive disorder of infants and children in which there is widespread dysfunction of the exocrine glands.   475

cystitis Inflammation of the urinary bladder.   222

cytokine (sy-tuh-kyn) Type of protein secreted by a T cell that stimulates cells of the immune system to perform their various functions.   143

cytokinesis (sy-tuh-kyn-ee-sus) Division of the cytoplasm following mitosis and meiosis.   422

cytoplasm (sy-tuh-plaz-um) Contents of a cell between the nucleus and the plasma membrane that contains the organelles.   46

cytosine (C) (sy-tuh-seen) One of four nitrogen bases in nucleotides composing the structure of DNA and RNA.   38

cytoskeleton Internal framework of the cell, consisting of microtubules, actin filaments, and intermediate filaments.   54

cytotoxic T cell (sy-tuh-tahk-sik) T cell that attacks and kills antigen-bearing cells.   148

D

data Facts or pieces of information collected through observation and/or experimentation.   8

daughter cell Cell that arises from a parent cell by mitosis or meiosis.   422

dead air space Volume of inspired air that cannot be exchanged with blood.   193

defecation (def-ih-kay-shun) Discharge of feces from the rectum through the anus.   168

deforestation (dee-for-eh-stay-shun) Removal of trees from a forest in a way that continuously reduces the size of the forest.   567

dehydration reaction Chemical reaction resulting in a covalent bond with the accompanying loss of a water molecule.   29

delayed allergic response Allergic response initiated at the site of the allergen by sensitized T cells, involving macrophages and regulated by cytokines.   152

deletion Change in chromosome structure in which the end of a chromosome breaks off or two simultaneous breaks lead to the loss of an internal segment; often causes abnormalities (e.g., cri du chat syndrome).   438

denaturation (dee-nay-chuh-ray-shun) Loss of normal shape by an enzyme so that it no longer functions; caused by a less than optimal pH or temperature.   35

dendrite (den-dryt) Branched ending of a neuron that conducts signals toward the cell body.   277

denitrification Conversion of nitrate or nitrite to nitrogen gas by bacteria in soil.   555

dense fibrous connective tissue Type of connective tissue containing many collagen fibers packed together; found in tendons and ligaments, for example.   67

dental caries (kar-eez) Tooth decay that occurs when bacteria within the mouth metabolize sugar and give off acids that erode teeth; a cavity.   160

deoxyhemoglobin Hemoglobin not carrying oxygen.   119

depolarization When the charge inside the axon changes from positive to negative.   278

dermis (dur-mus) Region of skin that lies beneath the epidermis.   78

desertification Denuding and degrading a once-fertile land, initiating a desert-producing cycle that feeds on itself and causes long-term changes in the soil, climate, and biota of an area.   567

detrital food chain (dih-tryt-ul) Straight-line linking of organisms according to who eats whom, beginning with detritus.   547

detrital food web (dih-tryt-ul) Complex pattern of interlocking and crisscrossing food chains, beginning with detritus.   547

detritus feeder Any organism that obtains most of its nutrients from the detritus in an ecosystem.   544

development Group of stages by which a zygote becomes an organism or by which an organism changes during its life span; includes puberty and aging, for example.   4

diabetes insipidus Condition caused by deficiency of antidiuretic hormone from the pituitary gland, characterized by excessive urination.   221

diabetes mellitus (dy-uh-bee-teez mel-ih-tus, muh-ly-tus) Condition characterized by a high blood glucose level and the appearance of glucose in the urine, due to a deficiency of insulin production and failure of cells to take up glucose.   261, 341

dialysate Material that passes through the membrane in dialysis.   223

diaphragm (dy-uh-fram) Dome-shaped horizontal sheet of muscle and connective tissue that divides the thoracic cavity from the abdominal cavity.   83, 161 Also, a birth control device consisting of a soft rubber or latex cup that fits over the cervix.   367

diarrhea (dy-uh-ree-uh) Excessively frequent bowel movements.   169

diastole (dy-as-tuh-lee) Relaxation period of a heart chamber during the cardiac cycle.   97

diastolic pressure (dy-uh-stahl-ik) Arterial blood pressure during the diastolic phase of the cardiac cycle.   100

diencephalon (dy-en-sef-uh-lahn) Portion of the brain in the region of the third ventricle that includes the thalamus and hypothalamus.   287

differentiation Cell specialization.   396

diffusion (dih-fyoo-zhun) Movement of molecules or ions from a region of higher to lower concentration; it requires no energy and stops when the distribution is equal.   49

digestion Breaking down of large nutrient molecules into smaller molecules that can be obsorbed.   158

digestive system Organ system including the mouth, esophagus, stomach, small intestine, and large intestine (colon) that receives food and digests it into nutrient molecules. Also has associated organs: teeth, tongue, salivary glands, liver, gallbladder, and pancreas.   80

dihybrid Individual that is heterozygous for two traits; shows the phenotype governed by the dominant alleles but carries the recessive alleles.   472

diploid (2n) Cell condition in which two of each type of chromosome are present in the nucleus.   422

disaccharide (dy-sak-uh-ryd) Sugar that contains two units of a monosaccharide (e.g., maltose).   30

distal convoluted tubule Final portion of a nephron that joins with a collecting duct; associated with tubular secretion.   214

diuretic (dy-uh-ret-ik) Drug used to counteract hypertension by causing the excretion of water.   220

diverticulosis A condition in which portions of the digestive tract mucosa have pushed through other layers of the tract forming pouches where food may collect.   159

DNA (deoxyribonucleic acid) Nucleic acid polymer produced from covalent bonding of nucleotide monomers that contain the sugar deoxyribose; the genetic material of nearly all organisms.   37, 490

DNA ligase (ly-gays) Enzyme that links DNA fragments; used during production of rDNA to join foreign DNA to vector DNA.   504

DNA replication Synthesis of a new DNA double helix prior to mitosis and meiosis in eukaryotic cells and during prokaryotic fission in prokaryotic cells.   491

domain The primary taxonomic group above the kingdom level; all living organisms may be placed in one of three domains.   6

dominant allele (uh-leel) Allele that exerts its phenotypic effect in the heterozygote; it masks the expression of the recessive allele.   466

dopamine Neurotransmitter in the central nervous system.   281

dorsal-root ganglion (gang-glee-un) Mass of sensory neuron cell bodies located in the dorsal root of a spinal nerve.   291

double helix Double spiral; describes the three-dimensional shape of DNA.   490

drug abuse Dependence on a drug, which assumes an "essential" biochemical role in the body following habituation and tolerance.   296

Duchenne muscular dystrophy Chronic progressive disease affecting the shoulder and pelvic girdles, commencing in early childhood. Characterized by increasing weakness of the muscles, followed by atrophy and a peculiar swaying gait with the legs kept wide apart. Transmitted as an X-linked trait, and affected individuals, predominantly males, rarely survive to maturity. Death is usually due to respiratory weakness or heart failure.   268, 483

duodenum (doo-uh-dee-num) First part of the small intestine where chyme enters from the stomach.   165

duplication Change in chromosome structure in which a particular segment is present more than once in the same chromosome.   438

E

ecological pyramid Pictorial graph based on the biomass, number of organisms, or energy content of various trophic levels in a food web—from the producer to the final consumer populations.   547

ecosystem (ek-oh-sis-tum, ee-koh-) Biological community together with the associated abiotic environment; characterized by energy flow and chemical cycling.   4, 543

ectopic pregnancy Implantation of the embryo in a location other than the uterus, most often in an oviduct.   398

effacement During the first stage of labor, the uterine contractions of labor occur in such a way that the cervical canal slowly disappears as the lower part of the uterus is pulled upward toward the baby's head.   409

effector Muscle or gland that responds to stimulation.   277

egg Female gamete having the haploid number of chromosomes fertilized by a sperm, the male gamete.   357

elastic cartilage Type of cartilage composed of elastic fibers, allowing greater flexibility.   68

elastic fiber Yellow fiber in the matrix of connective tissue, providing flexibility.   66

electrocardiogram (ECG) (ih-lek-troh-kar-dee-uh-gram) Recording of the electrical activity associated with the heartbeat.   99

electron Negative subatomic particle, moving about in an energy level around the nucleus of an atom.   20

electron transport chain Passage of electrons along a series of membrane-bound carrier molecules from a higher to lower energy level; the energy released is used for the synthesis of ATP.   59

element Substance that cannot be broken down into substances with different properties; composed of only one type of atom.   20

elimination Process of expelling substances from the body.   159

embolus (em-buh-lus) Moving blood clot that is carried through the bloodstream.   106

embryo (em-bree-oh) Immature developmental stage not recognizable as a human being.   398

embryonic development Period of development from the second through eighth weeks.   397

embryonic disk Stage of embryonic development following the blastocyst stage that has two layers; one layer will be endoderm, and the other will be ectoderm.   398

emerging diseases Diseases caused by pathogens that are newly recognized in the last twenty years.   390

emphysema (em-fih-see-muh) Degenerative lung disorder in which the bursting of alveolar walls reduces the total surface area for gas exchange.   199

emulsification (ih-mul-suh-fuh-kay-shun) Breaking up of fat globules into smaller droplets by the action of bile salts or any other emulsifier.   31

endochondral ossification Ossification that begins as hyaline cartilage subsequently replaced by bone tissue.   233

endocrine gland (en-duh-krin) Ductless organ that secretes (a) hormone(s) into the bloodstream.   74, 328

endocrine system Organ system involved in the coordination of body activities; uses hormones as chemical signals secreted into the bloodstream.   81

endomembrane system A collection of membranous structures involved in transport within the cell.   53

endometrium Mucous membrane lining the interior surface of the uterus.   358

endoplasmic reticulum (ER) (en-duh-plaz-mik reh-tik-yuh-lum) System of membranous saccules and channels in the cytoplasm, often with attached ribosomes.   52

eosinophil (ee-oh-sin-oh-fill) White blood cell containing cytoplasmic granules that stain with acidic dye.   122

epidemic More cases of a disease than expected in a certain area for a certain period of time.   381

epidemiology The study of diseases in populations, includes the causes, distribution, and control of these diseases.   380

epidermis (ep-uh-dur-mus) Region of skin that lies above the dermis.   76

epididymis (ep-uh-did-uh-mus) Coiled tubule next to the testes where sperm mature and may be stored for a short time.   353

epiglottis (ep-uh-glaht-us) Structure that covers the glottis during the process of swallowing.   161, 188

epinephrine (ep-uh-nef-rin) Hormone secreted by the adrenal medulla in times of stress; adrenaline.   338

episiotomy (ih-pee-zee-aht-uh-mee) Surgical procedure performed during childbirth in which the opening of the vagina is enlarged to avoid tearing.   409

episodic memory Capacity of brain to store and retrieve information with regard to persons and events.   289

epithelial tissue (ep-uh-thee-lee-ul) Type of tissue that lines hollow organs and covers surfaces; also called epithelium.   72

erectile dysfunction Failure of the penis to achieve or maintain erection.   354

erythropoietin (EPO) (ih-rith-roh-poy-ee-tin) Hormone, produced by the kidneys, that speeds red blood cell formation.   120, 210

esophagus (ih-sahf-uh-gus) Muscular tube for moving swallowed food from the pharynx to the stomach.   161

essential amino acids Amino acids required in the human diet because the body cannot make them.   173

essential fatty acid Fatty acid required in the human diet because the body cannot make them.   174

estrogen (es-truh-jun) Female sex hormone that helps maintain sex organs and secondary sex characteristics.   344, 362

eukaryotic cell Type of cell that has a membrane-bound nucleus and membranous organelles.   46, 518

evaporation Conversion of a liquid or a solid into a gas.   549

evolution Descent of organisms from common ancestors with the development of genetic and phenotypic changes over time that make them more suited to the environment.   6

evolutionary tree Diagram that describes the evolutionary relationship of groups of organisms; a common ancestor is presumed to have been present at points of divergence.   526

excretion Removal of metabolic wastes from the body.   208

exocrine gland Gland that secretes its product to an epithelial surface directly or through ducts.   74, 328

exophthalmic goiter (ek-sahf-thal-mik) Enlarge ment of the thyroid gland accompanied by an abnormal protrusion of the eyes.   336

experiment Artificial situation devised to test a hypothesis.   9

experimental variable Value expected to change as a result of an experiment; represents the factor being tested by the experiment.   9

expiration (ek-spuh-ray-shun) Act of expelling air from the lungs; also called exhalation.   186, 191

expiratory reserve volume (ik-spy-ruh-tor-ee) Volume of air that can be forcibly exhaled after normal exhalation.   193

exponential growth Growth at a constant rate of increase per unit of time; can be expressed as a constant fraction or exponent.   564

external respiration Exchange of oxygen and carbon dioxide between alveoli and blood.   196

exteroceptor Sensory receptor that detects stimuli from outside the body (e.g., taste, smell, vision, hearing, and equilibrium).   304

extinction Total disappearance of a species or higher group.   7

extraembryonic membrane (ek-struh-em-bree-ahn-ik) Membrane that is not a part of the embryo but is necessary to the continued existence and health of the embryo.   396

F

facial tic Involuntary muscle movement of the face.   267

facilitated transport Use of a plasma membrane carrier to move a substance into or out of a cell from higher to lower concentration; no energy required.   50

familial hypercholesterolemia (FH) Inability to remove cholesterol from the bloodstream; predisposes individual to heart attack.   480

farsighted Vision abnormality due to a shortened eyeball from front to back; light rays focus in back of retina when viewing close objects.   315

fat Organic molecule that contains glycerol and fatty acids; found in adipose tissue.   31

fatty acid Molecule that contains a hydrocarbon chain and ends with an acid group.   31

female condom Large polyurethane tube with a flexible ring that fits onto the cervix. Functions as a contraceptive and helps minimize the risk of transmitting infection.   367

fermentation Anaerobic breakdown of glucose that results in a gain of two ATP and end products, such as alcohol and lactate.   59

fertilization Union of a sperm nucleus and an egg nucleus, which creates a zygote.   394, 427

fetal development Period of development from the ninth week through birth.   403

fiber Structure resembling a thread; also, plant material that is nondigestible.   169

fibrin (fy-brun) Insoluble protein threads formed from fibrinogen during blood clotting.   125

fibrinogen (fy-brin-uh-jun) Plasma protein that is converted into fibrin threads during blood clotting.   118

fibroblast (fy-bruh-blast) Cell in connective tissues that produces fibers and other substances.   66

fibrocartilage (fy-broh-kar-tul-ij, -kart-lij) Cartilage with a matrix of strong collagenous fibers.   68

fibromyalgia Chronic, widespread pain in muscles and soft tissues surrounding joints.   268

fibrous connective tissue Tissue composed mainly of closely packed collagenous fibers and found in tendons and ligaments.   232

fimbriae Small bristlelike fiber on the surface of a bacterial cell, which attaches bacteria to a surface.   136, 357

first messenger Chemical signal, such as a peptide hormone, that binds to a plasma membrane receptor protein and alters the metabolism of a cell because a second messenger is activated.   330

flagellum (pl., flagella) (fluh-jel-um) Slender, long extension that propels a cell through a fluid medium.   55, 136

floating kidney Kidney that has been dislodged from its normal position.   208

fluid-mosaic model Model for the plasma membrane based on the changing location and pattern of protein molecules in a fluid phospholipid bilayer.   48

focus Bending of light rays by the cornea, lens, and humors so that they converge and create an image on the retina.   311

follicle (fahl-ih-kul) Structure in the ovary that produces a secondary oocyte and the hormones estrogen and progesterone.   360

follicle-stimulating hormone (FSH) Hormone secreted by the anterior pituitary gland that stimulates the development of an ovarian follicle in a female or the production of sperm in a male.   344, 356

fontanel (fahn-tun-el) Membranous region located between certain cranial bones in the skull of a fetus or infant.   237, 403

food chain Order in which one population feeds on another in an ecosystem, from detritus (detrital food chain) or producer (grazing food chain) to final consumer.   547

food web In ecosystems, complex pattern of interlocking and crisscrossing food chains.   547

foramen magnum (fuh-ray-mun mag-num) Opening in the occipital bone of the vertebrate skull through which the spinal cord passes.   238

formed element Constituent of blood that is either cellular (red blood cells and white blood cells) or at least cellular in origin (platelets).   116

fossil Any past evidence of an organism that has been preserved in the Earth's crust.   518

fossil fuel Fuels, such as oil, coal, and natural gas, that are the result of partial decomposition of plants and animals coupled with exposure to heat and pressure for millions of years.   561, 572

fossil record History of life recorded from remains from the past.   519

fovea centralis Region of the retina consisting of densely packed cones; responsible for the greatest visual acuity.   311

fragile X syndrome Most common inherited form of mental retardation; results from mutation to a single gene and results in deficiency of a protein critical to brain development.   483

functional genomics Study of all the nucleotide sequences, including structural genes, regulatory sequences, and noncoding DNA segments, in the chromosomes of an organism.   502

G

GABA (gamma aminobutyric acid) Major inhibitory neurotransmitter in the CNS.   281

gallbladder Organ attached to the liver that serves to store and concentrate bile.   167

gallstone Crystalline bodies formed by concentration of normal and abnormal bile components within the gallbladder.   167

gamete (ga-meet, guh-meet) Haploid sex cell; the egg or a sperm, which join in fertilization to form a zygote.   370, 427

gamma globulin Large proteins found in the blood plasma and on the surface of immune cells, functioning as antibodies (IgG).   150

ganglia Collections of nerve cell bodies found in the peripheral nervous system.   291

ganglion Collection or bundle of neuron cell bodies usually outside the central nervous system.   291

gap junction Junction between cells formed by the joining of two adjacent plasma membranes; it lends strength and allows ions, sugars, and small molecules to pass between cells.   76

gastric Pertaining to the stomach.   163

gastric gland Gland within the stomach wall that secretes gastric juice.   163

gastrulation Stage of animal development during which germ layers form, at least in part, by invagination.   399

gene Unit of heredity existing as alleles on the chromosomes; in diploid organisms, typically two alleles are inherited—one from each parent.   4

gene cloning Production of one or more copies of the same gene.   504

genetic engineering Alteration of DNA for medical or industrial purposes.   507

genotype (jee-nuh-typ) Genes of an individual for a particular trait or traits; often designated by letters, for example, *BB* or *Aa*.   466

gerontology (jer-un-tahl-uh-jee) Study of aging.   411

gland Epithelial cell or group of epithelial cells specialized to secrete a substance.   74

glaucoma (glow-koh-muh, glaw-koh-muh) Increasing loss of field of vision; caused by blockage of the ducts that drain the aqueous humor, creating pressure buildup and nerve damage.   311

global warming Predicted increase in the Earth's temperature, due to human activities that promote the greenhouse effect.   553

globulin Type of protein in blood plasma. There are alpha, beta, and gamma globulines.   118

glomerular capsule (gluh-mair-yuh-lur) Double-walled cup that surrounds the glomerulus at the beginning of the nephron.   213

glomerular filtrate Filtered portion of blood contained within the glomerular capsule.   216

glomerular filtration Movement of small molecules from the glomerulus into the glomerular capsule due to the action of blood pressure.   215

glomerulus (gluh-mair-uh-lus, gloh-mair-yuh-lus) Cluster; for example, the cluster of capillaries surrounded by the glomerular capsule in a nephron, where glomerular filtration takes place.   213

glottis (glaht-us) Opening for airflow in the larynx.   161, 188

glucagon (gloo-kuh-gahn) Hormone secreted by the pancreas that causes the liver to break down glycogen and raises the blood glucose level.   341

glucocorticoid (gloo-koh-kor-tih-koyd) Type of hormone secreted by the adrenal cortex that influences carbohydrate, fat, and protein metabolism; see cortisol.   338

glucose (gloo-kohs) Six-carbon sugar that organisms degrade as a source of energy during cellular respiration.   29

glutamate Major excitatory CNS neurotransmitter.   281

glycemic index (GI) Blood glucose response of a given food.   173

glycogen (gly-koh-jun) Storage polysaccharide composed of glucose molecules joined in a linear fashion but having numerous branches.   30

glycolysis Anaerobic breakdown of glucose that results in a gain of two ATP molecules.   58

Golgi apparatus (gohl-jee) Organelle, consisting of saccules and vesicles, that processes, packages, and distributes molecules about or from the cell.   53

gonad (goh-nad) Organ that produces gametes; the ovary produces eggs, and the testis produces sperm.   344

gonadotropic hormone (goh-nad-uh-trahp-ic, -troh-pic) Chemical signal secreted by the anterior pituitary that regulates the activity of the ovaries and testes; principally, follicle-stimulating hormone (FSH) and luteinizing hormone (LH).   332

gonadotropin-releasing hormone (GnRH) Hormone secreted by the hypothalamus that stimulates the anterior pituitary to secrete follicle-stimulating hormone (FSH) and luteinizing hormone (LH).   356

gout Joint inflammation caused by accumulation of uric acid.   210

granular leukocyte (gran-yuh-lur loo-kuh-syt) White blood cell with prominent granules in the cytoplasm.   122

gravitational equilibrium Maintenance of balance when the head and body are motionless.   322

gray matter Nonmyelinated axons and cell bodies in the central nervous system.   283

grazing food chain Straight-line linking of organisms according to who eats whom, beginning with a producer.   547

grazing food web Complex pattern of interlocking and crisscrossing food chains that begins with populations of autotrophs serving as producers.   547

greenhouse effect Reradiation of solar heat toward the Earth because gases, such as carbon dioxide, methane, nitrous oxide, and water vapor, allow solar energy to pass through toward the Earth but block the escape of heat back into space.   553

greenhouse gases Gases involved in the greenhouse effect.   553, 572

growth Increase in the number of cells and/or the size of these cells.   396

growth factor Chemical signal that regulates mitosis and differentiation of cells that have receptors for it; important in such processes as fetal development, tissue maintenance and repair, and hematopoiesis; sometimes a contributing factor in cancer.   448

growth hormone (GH) Substance secreted by the anterior pituitary; controls size of individual by promoting cell division, protein synthesis, and bone growth.   334

growth plate Cartilaginous layer within an epiphysis of a long bone that permits growth of bone to occur.   233

growth rate A percentage that reflects the difference between the number of persons in a population who are born and the number who die each year.   564

guanine (G) (gwah-neen) One of four nitrogen-containing bases in nucleotides composing the structure of DNA and RNA; pairs with cytosine.   38

H

hair cell Cell with stereocilia (long microvilli) that is sensitive to mechanical stimulation; mechanoreceptor for hearing and equilibrium in the inner ear.   315

hair follicle Tubelike depression in the skin in which a hair develops.   79

haploid (n) (hap-loyd) The n number of chromosomes—half the diploid number; the number characteristic of gametes, which contain only one set of chromosomes.   427

hard palate (pal-it) Bony, anterior portion of the roof of the mouth.   160

hay fever Seasonal variety of allergic reaction to a specific allergen. Characterized by sudden attacks of sneezing, swelling of nasal mucosa, and often asthmatic symptoms.   152

heart Muscular organ located in the thoracic cavity whose rhythmic contractions maintain blood circulation.   94

heart attack Damage to the myocardium due to blocked circulation in the coronary arteries; also called a myocardial infarction (MI).   106

heartburn Burning pain in the chest that occurs when part of the stomach contents escape into the esophagus.   161

heart failure Syndrome characterized by distinctive symptoms and signs resulting from disturbances in cardiac output or from increased pressure in the veins.   108

helper T cell T cell that secretes cytokines that stimulate all types of immune system cells.   149

hemodialysis (he-moh-dy-al-uh-sus) Cleansing of blood by using an artificial membrane that causes substances to diffuse from blood into a dialysis fluid.   222

hemoglobin (hee-muh-gloh-bun) Iron-containing pigment in red blood cells that combines with and transports oxygen.   34, 118

hemolysis (he-mahl-uh-sus) Rupture of red blood cells accompanied by the release of hemoglobin.   121

hemolytic desease of the newborn Destruction of a fetus's red blood cells by the mother's immune system, caused by differing Rh factors between mother and fetus.   121

hemophilia (he-moh-fil-ee-uh) Genetic disorder in which the affected individual is subject to uncontrollable bleeding.   125, 484

hemorrhoid (hem-uh-royd, hem-royd) Abnormally dilated blood vessels of the rectum.   169

hepatic portal vein Vein leading to the liver and formed by the merging blood vessels leaving the small intestine.   104

hepatic vein Vein that runs between the liver and the inferior vena cava.   104

hepatitis (hep-uh-ty-tis) Inflammation of the liver. Viral hepatitis occurs in several forms.   167

herbivore (hur-buh-vor) Primary consumer in a grazing food chain; a plant eater.   544

heterotroph Organism that cannot synthesize organic molecules from inorganic nutrients, and therefore must take in organic nutrients (food).   517, 544

heterozygous Possessing unlike alleles for a particular trait.   466

hexose Six-carbon sugar.   29

hippocampus (hip-uh-kam-pus) Portion of the limbic system where memories are stored.   288

histamine (his-tuh-meen, -mun) Substance, produced by basophils in blood and mast cells in connective tissue, that causes capillaries to dilate.   143

HLA (human leukocyte antigen) Protein in a plasma membrane that identifies the cell as belonging to a particular individual and acts as an antigen in other organisms.   147

homeostasis (hoh-mee-oh-stay-sis) Maintenance of normal internal conditions in a cell or an organism by means of self-regulating mechanisms.   4, 84

hominid (hahm-uh-nid) Member of the family Hominidae, which contains australopithecines and humans.   528

Homo erectus (hoh-moh ih-rek-tus) Hominid who used fire and migrated out of Africa to Europe and Asia.   531

Homo habilis (hoh-moh hab-uh-lus) Hominid of 2 MYA who is believed to have been the first tool user.   530

homologous chromosome (hoh-mahl-uh-gus, huh-mahl-uh-gus) Member of a pair of chromosomes that are alike and come together in synapsis during prophase of the first meiotic division.   427

homologous structure Structure similar in two or more species because of common ancestry.   521

homologue Member of a homologous pair of chromosomes.   427

Homo sapiens (hoh-moh say-pe-nz) Modern humans.   533

homozygous dominant Possessing two identical alleles, such as *AA*, for a particular trait.   466

homozygous recessive Possessing two identical alleles, such as *aa*, for a particular trait.   466

hormone (hor-mohn) Chemical signal produced by one set of cells that affects a different set of cells.   166, 234, 328

human chorionic gonadotropin (hCG) (kor- ee-ahn-ik, goh-nad-uh-trahp-in, -troh-pin) Hormone produced by the chorion that functions to maintain the uterine lining.   364, 398

human immunodeficiency virus (HIV) Virus responsible for AIDS.   381

Huntington disease Genetic disease marked by progressive deterioration of the nervous system due to deficiency of a neurotransmitter.   477

hyaline cartilage (hy-uh-lin) Cartilage whose cells lie in lacunae separated by a white, translucent matrix containing very fine collagen fibers.   67

hydrogen bond Weak bond that arises between a slightly positive hydrogen atom of one molecule and a slightly negative atom of another, or between parts of the same molecule.   25

hydrolysis reaction (hy-drahl-ih-sis re-ak-shun) Splitting of a compound by the addition of water, with the H$^+$ being incorporated in one fragment and the OH$^-$ in the other.   29

hydrolyze To break a chemical bond between molecules by insertion of a water molecule.   158

hydrophilic (hy-druh-fil-ik) Type of molecule that interacts with water by dissolving in water and/or forming hydrogen bonds with water molecules.   27

hydrophobic (hy-druh-foh-bik) Type of molecule that does not interact with water because it is nonpolar.   27

hypertension Elevated blood pressure, particularly the diastolic pressure.   105

hypothalamic-inhibiting hormone (hy-poh-thuh-lah-mik) One of many hormones produced by the hypothalamus that inhibits the secretion of an anterior pituitary hormone.   332

hypothalamic-releasing hormone One of many hormones produced by the hypothalamus that stimulates the secretion of an anterior pituitary hormone.   332

hypothalamus (hy-poh-thal-uh-mus) Part of the brain located below the thalamus that helps regulate the internal environment of the body and produces releasing factors that control the anterior pituitary.   287, 332

hypothesis (hy-pahth-ih-sis) Supposition that is formulated after making an observation; it can be tested by obtaining more data, often by experimentation.   8

I

immediate allergic response Allergic response that occurs within seconds of contact with an allergen, caused by the attachment of the allergen to IgE antibodies.   152

immune system White blood cells and lymphatic organs that protect the body against foreign organisms and substances and also cancerous cells.   80, 121

immunity Ability of the body to protect itself from foreign substances and cells, including disease-causing agents.   142

immunization (im-yuh-nuh-zay-shun) Use of a vaccine to protect the body against specific disease-causing agents.   150

immunosuppressive Inactivating the immune system to prevent organ rejection, usually via a drug.   153

implantation Attachment and penetration of the embryo into the lining of the uterus (endometrium).   357, 398

incomplete dominance Inheritance pattern in which the offspring has an intermediate phenotype, as when a red-flowered plant and a white-flowered plant produce pink-flowered offspring.   480

incus (ing-kus) The middle of three ossicles of the ear that serve to conduct vibrations from the tympanic membrane to the oval window of the inner ear.   317

infant respiratory distress syndrome Condition in newborns, especially premature ones, in which the lungs collapse because of a lack of surfactant lining the alveoli.   191

infectious diseases Diseases caused by pathogens such as bacteria, viruses, fungi, parasites, protozoans, and prions.   380

infectious mononucleosis Acute, self-limited infectious disease of the lymphatic system caused by the Epstein-Barr virus and characterized by fever, sore throat, lymph node and spleen swelling, and the proliferation of moncytes and abnormal lymphocytes.   123

inferior vena cava (vee-nuh kay-vuh) Large vein that enters the right atrium from below and carries blood from the trunk and lower extremities.   103

infertility Inability to have as many children as desired.   368

inflammatory response Tissue response to injury that is characterized by redness, swelling, pain, and heat.   142

ingestion The taking of food or liquid into the body by way of the mouth.   158

initiation Mutation of a single cell that may lead to cancer development.   447

inner cell mass An aggregation of cells at one pole of the blastocyte, destined to form the embryo proper.   397

inner ear Portion of the ear consisting of a vestibule, semicircular canals, and the cochlea, where equilibrium is maintained and sound is transmitted.   317

insertion End of a muscle attached to a movable bone.   255

inspiration (in-spuh-ray-shun) Act of taking air into the lungs; also called inhalation.   186, 191

inspiratory reserve volume (in-spy-ruh-tohr-ee) Volume of air that can be forcibly inhaled after normal inhalation.   193

insulin (in-suh-lin) Hormone secreted by the pancreas that lowers the blood glucose level by promoting the uptake of glucose by cells, and the conversion of glucose to glycogen by the liver and skeletal muscles.   341

integrase Viral enzyme that enables the integration of viral genetic material into a host cell's DNA.   384

integration Summing up of excitatory and inhibitory signals by a neuron or by some part of the brain.   282, 305

integumentary system (in-teg-yoo-men-tuh-ree, -men-tree) Organ system consisting of skin and various organs, such as hair, found in skin.   76

intercalated disk (in-tur-kuh-lay-tud) Region that holds adjacent cardiac muscle cells together; disks appear as dense bands at right angles to the muscle striations.   70, 254

interferon (in-tur-feer-ahn) Antiviral agent produced by an infected cell that blocks the infection of another cell.   144

interkinesis Period between meiosis I and meiosis II, during which no DNA replication takes place.   427

interleukin (in-tur-loo-kun) Cytokine produced by macrophages and T cells that functions as a metabolic regulator of the immune response.   152

intermediate filament Ropelike assemblies of fibrous polypeptides in the cytoskeleton that provide support and strength to cells; so called because they are intermediate in size between actin filaments and microtubules.   55

internal respiration Exchange of oxygen and carbon dioxide between blood and tissue fluid.   196

interneuron Neuron located within the central nervous system that conveys messages between parts of the central nervous system.   277

interoceptor Sensory receptor that detects stimuli from inside the body (e.g., pressoreceptors, osmoreceptors, and chemoreceptors).   304

interphase Cell cycle stage during which growth and DNA synthesis occur when the nucleus is not actively dividing.   421

interstitial cell (in-tur-stish-ul) Hormone-secreting cell located between the seminiferous tubules of the testes.   356

intervertebral disk (in-tur-vur-tuh-brul) Layer of cartilage located between adjacent vertebrae.   241

intramembranous ossification Ossification that forms from membranelike layers of primitive connective tissue.   233

intrauterine device (IUD) (in-truh-yoo-tur-in) Birth control device consisting of a small piece of molded plastic inserted into the uterus; believed to alter the uterine environment so that fertilization does not occur.   367

invasive Cells, such as tumor cells, that invade normal cells.   576

inversion Change in chromosome structure in which a segment of a chromosome is turned around 180°; this reversed sequence of genes can lead to altered gene activity and abnormalities.   439

ion (eye-un, -ahn) Charged particle that carries a negative or positive charge.   23

ionic bond (eye-ahn-ik) Chemical bond in which ions are attracted to one another by opposite charges.   23

iris (eye-ris) Muscular ring that surrounds the pupil and regulates the passage of light through this opening.   310

isotope (eye-suh-tohp) One of two or more atoms with the same atomic number but a different atomic mass due to the number of neutrons.   21

J

jaundice (jawn-dis) Yellowish tint to the skin caused by an abnormal amount of bilirubin (bile pigment) in the blood, indicating liver malfunction.   167

joint Articulation between two bones of a skeleton.   230

juxtaglomerular apparatus (juk-stuh-gluh-mer-yuh-lur) Structure located in the walls of arterioles near the glomerulus; regulates renal blood flow.   219

K

kidney Organ in the urinary system that produces and excretes urine.   208

kingdom One of the categories used to classify organisms; the category above phylum.   6

kinocilium The largest stereocilium.   322

L

lacteal (lak-tee-ul) Lymphatic vessel in an intestinal villus; it aids in the absorption of lipids.   166

lactose intolerance (lak-tohs) Inability to digest lactose because of an enzyme deficiency.   166

lacuna (pl., lacunae) (luh-koo-nuh, -kyoo-nuh) Small pit or hollow cavity, as in bone or cartilage, where a cell or cells are located.   67

Langerhans cell Specialized epidermal cells that assist the immune system.   77

lanugo (luh-noo-goh) Short, fine hair that is present during the later portion of fetal development.   406

large intestine Last major portion of the digestive tract, extending from the small intestine to the anus and consisting of the cecum, the colon, the rectum, and the anal canal.   168

laryngitis (lar-un-jy-tis) Infection of the larynx with accompanying hoarseness.   199

larynx (lar-ingks) Cartilaginous organ located between the pharynx and the trachea that contains the vocal cords; also called the voice box.   188

learning Relatively permanent change in behavior that results from practice and experience.   289

lens Clear, membranelike structure found in the eye behind the iris; brings objects into focus.   311

leptin Hormone produced by adipose tissue that acts on the hypothalamus to signal satiety.   345

leukemia (loo-kee-mee-uh) Cancer of the blood-forming tissues leading to the overproduction of abnormal white blood cells.   123, 450

ligament (lig-uh-munt) Tough cord or band of dense fibrous connective tissue that joins bone to bone at a joint.   67, 230

limbic system Association of various brain centers, including the amygdala and hippocampus; governs learning and memory and various emotions, such as pleasure, fear, and happiness.   288

lineage Evolutionary line of descent.   528

lipase (ly-pays, ly-payz) Fat-digesting enzyme secreted by the pancreas.   165

lipid (lip-id, ly-pid) Class of organic compounds that tends to be soluble only in nonpolar solvents, such as alcohol; includes fats and oils.   31

liver Large, dark red internal organ that produces urea and bile, detoxifies the blood, stores glycogen, and produces the plasma proteins, among other functions.   167

locus (pl., loci) Particular site where a gene is found on a chromosome. Homologous chromosomes have corresponding gene loci.   466

long-term memory Retention of information that lasts longer than a few minutes.   289

long-term potentiation (LTP) (puh-ten-shee-ay-shun) Enhanced response at synapses within the hippocampus; likely essential to memory storage.   290

loop of the nephron (nef-rahn) Portion of the nephron lying between the proximal convoluted tubule and the distal convoluted tubule that functions in water reabsorption.   214

loose fibrous connective tissue Tissue composed mainly of fibroblasts widely separated by a matrix containing collagen and elastic fibers.   67

lumen (loo-mun) Cavity inside any tubular structure, such as the lumen of the digestive tract.   159

lung cancer Malignant growth that often begins in the bronchi.   200

lungs Paired, cone-shaped organs within the thoracic cavity; function in internal respiration and contain moist surfaces for gas exchange.   190

luteinizing hormone (LH) Hormone that controls the production of testosterone by interstitial cells in males and promotes the development of the corpus luteum in females.   344, 356

lymph (limf) Fluid, derived from tissue fluid, that is carried in lymphatic vessels.   68, 105, 140

lymphatic nodule Mass of lymphatic tissue not surrounded by a capsule.   142

lymphatic organ Organ other than a lymphatic vessel that is part of the lymphatic system; includes lymph nodes, tonsils, spleen, thymus gland, and bone marrow.   141

lymphatic system (lim-fat-ik) Organ system consisting of lymphatic vessels and lymphatic organs that transport lymph and lipids, and aids the immune system.   80, 92

lymph node Mass of lymphatic tissue located along the course of a lymphatic vessel.   141

lymphocyte (lim-fuh-syt) Specialized white blood cell that functions in specific defense; occurs in two forms—T cell and B cell. 122

lymphoma Cancer of lymphatic tissue (reticular connective tissue). 450

lysosome (ly-suh-sohm) Membrane-bound vesicle that contains hydrolytic enzymes for digesting macromolecules. 53

lysozyme Enzyme found in tears, milk, saliva, mucus, and other body fluids that destroys bacteria by digesting their cell walls. 142

M

macromolecule Extremely large biological molecule; refers specifically to proteins, nucleic acids, polysaccharides, lipids, and complexes of these. 29

macrophage (mak-ruh-fayj) Large phagocytic cell derived from a monocyte that ingests microbes and debris. 143

mad cow disease Slowly progressive fatal disease affecting the central nervous system of cattle; transmissible to humans. 138

Major Histocompatibility Complex (MHC) Cluster of genes on chromosome 6 concerned with self-antigen production; matching these is critical to success of organ transplants. The MHC includes the human leukocyte antigen (HLA) genes. 147

male condom Sheath used to cover the penis during sexual intercourse; used as a contraceptive and, if latex, to minimize the risk of transmitting infection. 367

malignant Form of tumor capable of spreading throughout the body. 447

malleus (mal-ee-us) The first of three ossicles of the ear that serve to conduct vibrations from the tympanic membrane to the oval window of the inner ear. 317

Marfan syndrome Congenital disorder of connective tissue characterized by abnormal length of the extremities. 477

mass Sum of the number of protons and neutrons in an atom's nucleus. 21

mass number The number of nucleons (protons and neutrons) in the nucleus of an atom. 21

mast cell Cell to which antibodies, formed in response to allergens, attach, causing it to release histamine, thus producing allergic symptoms. 122, 143

mastoiditis (mas-toyd-eye-tis) Inflammation of the mastoid sinuses of the skull. 237

matter Anything that takes up space and has mass. 20

matrix (may-triks) Unstructured semifluid substance that fills the space between cells in connective tissues or inside organelles. 66

mechanoreceptor (mek-uh-noh-rih-sep-tur) Sensory receptor that responds to mechanical stimuli, such as that from pressure, sound waves, and gravity. 304

medulla oblongata (muh-dul-uh ahb-lawng-gah-tuh) Part of the brain stem that is continuous with the spinal cord; controls heartbeat, blood pressure, breathing, and other vital functions. 288

medullary cavity (muh-dul-uh-ree) Cavity within the diaphysis of a long bone containing marrow. 230

megakaryocyte (meg-uh-kar-ee-oh-syt,-uh-syt) Large cell that gives rise to blood platelets. 123

meiosis (my-oh-sis) Type of nuclear division that occurs as part of sexual reproduction in which the daughter cells receive the haploid number of chromosomes in varied combinations. 427

melanocyte Melanin-producing cell found in skin. 77

melanocyte-stimulating hormone (MSH) Substance that causes melanocytes to secrete melanin in lower vertebrates. 334

melatonin (mel-uh-toh-nun) Hormone, secreted by the pineal gland, that is involved in biorhythms. 345

membrane attack complex Group of complement proteins that form channels in a microbe's surface, thereby destroying it. 144

memory Capacity of the brain to store and retrieve information about past sensations and perceptions; essential to learning. 289

memory T cell T cell that differentiates during an initial infection and responds rapidly during subsequent exposure to the same anitgen. 149

meninges (sing., meninx) (muh-nin-jeez) Protective membranous coverings about the central nervous system. 84, 283

meningitis Condition that refers to inflammation of the brain or spinal cord meninges (membranes). 84

menopause (men-uh-pawz) Termination of the ovarian and uterine cycles in older women. 362

menstruation (men-stroo-ay-shun) Loss of blood and tissue from the uterus at the end of a uterine cycle. 363

messenger RNA (mRNA) Type of RNA formed from a DNA template that bears coded information for the amino acid sequence of a polypeptide. 493

metabolism All of the chemical reactions that occur in a cell. 4, 56

metaphase Mitotic phase during which chromosomes are aligned at the equator of the mitotic spindle. 423

metastasis (muh-tas-tuh-sis) Spread of cancer from the place of origin throughout the body; caused by the ability of cancer cells to migrate and invade tissues. 448

microtubule (my-kro-too-byool) Small cylindrical structure that contains 13 rows of the protein tubulin around an empty central core; present in the cytoplasm, centrioles, cilia, and flagella. 54

micturition Emptying of the bladder; urination. 209

midbrain Part of the brain located below the thalamus and above the pons; contains reflex centers and tracts. 288

middle ear Portion of the ear consisting of the tympanic membrane, the oval and round windows, and the ossicles; where sound is amplified. 317

mineral Naturally occurring inorganic substance containing two or more elements; certain minerals are needed in the diet. 175, 574

mineralocorticoid (min-ur-uh-loh-kor-tih-koyd) Type of hormone secreted by the adrenal cortex that regulates water-salt balance, leading to increases in blood volume and blood pressure. 338

mitochondrion (my-tuh-kahn-dree-un) Membrane-bound organelle in which ATP molecules are produced during the process of cellular respiration. 56

mitosis (my-toh-sis) Type of cell division in which daughter cells receive the exact chromosomal and genetic makeup of the parent cell; occurs during growth and repair. 420

mitotic spindle Microtubule structure that brings about chromosomal movement during nuclear division. 422

mole A unit of scientific measurement for atoms, ions, and molecules. 27

molecular clock Mutational changes that accumulate at a presumed constant rate in regions of DNA not involved in adaptation to the environment. 528

molecule Union of two or more atoms of the same element; also, the smallest part of a compound that retains the properties of the compound. 2, 22

monoclonal antibody One of many antibodies produced by a clone of hybridoma cells that all bind to the same antigen. 151

monocyte (mahn-uh-syt) Type of agranular white blood cell that functions as a phagocyte and an antigen-presenting cell. 122

monohybrid Individual that is heterozygous for one trait; shows the phenotype of the dominant allele but carries the recessive allele. 469

monosaccharide (mahn-uh-sak-uh-ryd) Simple sugar; a carbohydrate that cannot be decomposed by hydrolysis (e.g., glucose). 29

monosomy One less chromosome than usual. 435

morphogenesis Emergence of shape in tissues, organs, or entire embryo during development. 396

morula Spherical mass of cells resulting from cleavage during animal development, prior to the blastula stage. 397

mosaic evolution Concept that human characteristics did not evolve at the same rate; for example, some body parts are more humanlike than others in early hominids. 529

motor neuron Nerve cell that conducts nerve impulses away from the central nervous system and innervates effectors (muscles and glands). 277

motor unit Motor neuron and all the muscle fibers it innervates. 262

movement Motion. 158

MRSA Methicillin-resistant *Staphylococcus aureus*, a type of bacterium that causes "Staph" infections that is no longer susceptible to certain antibiotics including methicillin. 392

mucosa Membrane that lines tubes and body cavities that open to the outside of the body; mucous membrane.   159

mucous membrane (myoo-kus) Membrane lining a cavity or tube that opens to the outside of the body; also called mucosa.   84

multicellular Organism composed of many cells; usually has organized tissues, organs, and organ systems.   2

multifactorial trait Controlled by several allelic pairs; each dominant allele contributes to the phenotype in an additive and like manner.   479

multiple allele (uh-leelz) Inheritance pattern in which there are more than two alleles for a particular trait; each individual has only two of all possible alleles.   480

multiple sclerosis (MS) Disease in which the outer myelin layer of nerve fiber insulation becomes scarred, interfering with normal conduction of nerve impulses.   153

multiregional continuity hypothesis Proposal that modern humans evolved independently in at least three different places: Asia, Africa, and Europe.   533

muscle dysmorphia Mental state where a person thinks his or her body is underdeveloped and becomes preoccupied with body-building and diet; affects more men than women.   181

muscle fiber Muscle cell.   254

muscle tone Continuous, partial contraction of muscle.   263

muscle twitch Contraction of a whole muscle in response to a single stimulus.   262

muscular dystrophy (mus-ku-lar dis-tro-fee) Progressive muscle weakness and atrophy caused by deficient dystrophin protein.   268

muscularis Two layers of muscle in the gastrointestinal tract.   159

muscular system System of muscles that produces movement, both within the body and of its limbs; principal components are skeletal, smooth, and cardiac muscle.   81

muscular (contractile) tissue Type of tissue composed of fibers that can shorten and thicken.   69

mutagen (myoo-tuh-jun) Agent, such as radiation or a chemical, that brings about a mutation.   451

mutation Alteration in chromosome structure or number and also an alteration in a gene due to a change in DNA composition.   492

myalgia Muscular pain.   268

myasthenia gravis (mi-as-thee-ne-ah grav-is) Chronic disease characterized by muscles that are weak and easily fatigued. It results from the immune system's attack on neuromuscular junctions so that stimuli are not transmitted from motor neurons to muscle fibers.   153, 268

myelin sheath (my-uh-lin) White, fatty material, derived from the membrane of Schwann cells, that forms a covering for nerve fibers.   277

myocardium (my-oh-kar-dee-um) Cardiac muscle in the wall of the heart.   94

myofibril (my-uh-fy-brul) Contractile portion of muscle cells that contains a linear arrangement of sarcomeres and shortens to produce muscle contraction.   258

myoglobin Pigmented molecule in muscle tissue that stores oxygen.   264

myosin (my-uh-sin) One of two major proteins of muscle; makes up thick filaments in myofibrils of muscle fibers. See actin.   258

myxedema (mik-sih-dee-muh) Condition resulting from a deficiency of thyroid hormone in an adult.   336

N

NAD (nicotinamide adenine dinucleotide) Coenzyme that functions as a carrier of electrons and hydrogen ions, especially in cellular respiration.   56

nail Protective covering of the distal part of fingers and toes.   79

nasal cavity One of two canals in the nose, separated by a septum.   187

natural selection Mechanism resulting in adaptation to the environment.   522

Neandertal (nee-an-dur-thahl) Hominid with a sturdy build who lived during the last ice age in Europe and the Middle East; hunted large game and has left evidence of being culturally advanced.   534

nearsighted Vision abnormality due to an elongated eyeball from front to back; light rays focus in front of retina when viewing distant objects.   315

negative feedback Mechanism of homeostatic response in which a stimulus initiates reactions that reduce the stimulus.   86

nephron (nef-rahn) Microscopic kidney unit that regulates blood composition by glomerular filtration, tubular reabsorption, and tubular secretion.   212

nerve Bundle of long axons outside the central nervous system.   70

nerve signal Action potential (electrochemical change) traveling along a neuron.   277

nervous system Organ system consisting of the brain, spinal cord, and associated nerves that coordinates the other organ systems of the body.   81

nervous tissue Tissue that contains nerve cells (neurons), which conduct impulses, and neuroglia, which support, protect, and provide nutrients to neurons.   70

neuroglia (noo-rahg-lee-uh, noo-rohg-lee-uh) Nonconducting nerve cells that are intimately associated with neurons and function in a supportive capacity.   71, 276

neuromuscular junction Region where an axon terminal approaches a muscle fiber; the synaptic cleft separates the axon terminal from the sarcolemma of a muscle fibre.   261

neuron (noor-ahn, nyoor-) Nerve cell that characteristically has three parts: dendrites, cell body, and axon.   70, 276

neurotransmitter Chemical stored at the ends of axons that is responsible for transmission across a synapse.   280

neutron (noo-trahn) Neutral subatomic particle, located in the nucleus and having a weight of approximately one atomic mass unit.   20

neutrophil (noo-truh-fil) Granular leukocyte that is the most abundant of the white blood cells; first to respond to infection.   122

niche (nich) Role an organism plays in its community, including its habitat and its interactions with other organisms.   545

nitrification Process by which nitrogen in ammonia and organic molecules is oxidized to nitrites and nitrates by soil bacteria.   554

nitrogen fixation Process whereby free atmospheric nitrogen is converted into compounds, such as ammonium and nitrates, usually by bacteria.   554

node of Ranvier (rahn-vee-ay) Gap in the myelin sheath around a nerve fiber.   277

nondisjunction Failure of homologous chromosomes or daughter chromosomes to separate during meiosis I and meiosis II, respectively.   435

nonrenewable resource Minerals, fossil fuels, and other materials present in essentially fixed amounts (within human time scales) in our environment.   566

norepinephrine (NE) (nor-ep-uh-nef-rin) Neurotransmitter of the postganglionic fibers in the sympathetic division of the autonomic system; also, a hormone produced by the adrenal medulla.   281, 338

nuclear envelope Double membrane that surrounds the nucleus and is connected to the endoplasmic reticulum; has pores that allow substances to pass between the nucleus and the cytoplasm.   52

nuclear pore Opening in the nuclear envelope that permits the passage of proteins into the nucleus and ribosomal subunits out of the nucleus.   52

nucleolus (noo-klee-uh-lus, nyoo-) Dark-staining, spherical body in the cell nucleus that produces ribosomal subunits.   52

nucleoplasm (noo-klee-uh-plaz-um) Semifluid medium of the nucleus, containing chromatin.   52

nucleotide Monomer of DNA and RNA consisting of a 5-carbon sugar bonded to a nitrogen-containing base and a phosphate group.   38

nucleus (noo-klee-us, nyoo-) Membrane-bounded organelle that contains chromosomes and controls the structure and function of the cell.   20, 46, 288

nutrient Chemical substances in foods that are essential to the diet and contribute to good health.   171

O

obesity (oh-bee-sih-tee) Excess adipose tissue; exceeding ideal weight by more than 20%.   170

oil Substance, usually of plant origin and liquid at room temperature, formed when a glycerol molecule reacts with three fatty acid molecules.   31

oil gland Gland of the skin associated with hair follicle; secretes sebum; also called sebaceous gland.   79

olfactory cell (ahl-fak-tuh-ree, -tree, ohl-) Modified neuron that is a sensory receptor for the sense of smell.   309

omnivore (ahm-nuh-vor) Organism in a food chain that feeds on both plants and animals.   544

oncogene (ahng-koh-jeen) Cancer-causing gene.   448

oncologist Physician who specializes in one or more types of cancers.   449

oncology The study of cancer.   449

oogenesis (oh-uh-jen-uh-sis) Production of an egg in females by the process of meiosis and maturation.   360, 434

opportunistic infection Infection that has an opportunity to occur because the immune system has been weakened.   381

optic chiasma X-shaped structure on the underside of the brain formed by a partial crossing-over of optic nerve fibers.   314

optic nerve Either of two cranial nerves that carry nerve impulses from the retina of the eye to the brain, thereby contributing to the sense of sight.   311

optic tract Groups of neurons from the optic nerve that sweep around the hypothalamus. Most fibers synapse with neurons in nuclei within the thalamus.   314

orbitals Pathways in which electrons travel around the nucleus of an atom.   20

organ Combination of two or more different tissues performing a common function.   2, 76

organelle (or-guh-nel) Small membranous structure in the cytoplasm having a specific structure and function.   46

organic Molecule that always contains carbon and hydrogen, and often contains ozygen as well; organic molecules are associated with living things.   29

organic molecule Type of molecule that contains carbon and hydrogen—and often contains oxygen also.   29

organism Individual living thing.   2

organ system Group of related organs working together.   2, 76

origin End of a muscle attached to a relatively immovable bone.   255

osmosis (ahz-moh-sis, ahs-) Diffusion of water through a selectively permeable membrane.   49

osmotic pressure Measure of the tendency of water to move across a selectively permeable membrane; visible as an increase in liquid on the side of the membrane with higher solute concentration.   50, 117

ossicle (ahs-ih-kul) One of the small bones of the middle ear—malleus, incus, and stapes.   317

ossification Formation of bone tissue.   232

osteoblast (ahs-tee-uh-blast) Bone-forming cell.   232

osteoclast (ahs-tee-uh-klast) Cell that causes erosion of bone.   232

osteocyte (ahs-tee-uh-syt) Mature bone cell located within the lacunae of bone.   230, 232

osteoporosis Condition in which bones break easily because calcium is removed from them faster than it is replaced.   176

otitis media (oh-ty-tis mee-dee-uh) Infection of the middle ear, characterized by pain and possibly by a sense of fullness, hearing loss, vertigo, and fever.   198

otolith (oh-tuh-lith) Calcium carbonate granule associated with ciliated cells in the utricle and the saccule.   322

outbreak A disease epidemic that is confined to a local area.   381

outer ear Portion of ear consisting of the pinna and auditory canal.   317

out-of-Africa hypothesis Proposal that modern humans originated only in Africa; then migrated out of Africa and supplanted populations of early *Homo* in Asia and Europe about 100,000 years ago.   533

oval window Membrane-covered opening between the stapes and the inner ear.   317

ovarian cycle (oh-vair-ee-un) Monthly follicle changes occurring in the ovary that control the level of sex hormones in the blood and the uterine cycle.   360

ovary Female gonad that produces eggs and the female sex hormones.   344, 357

oviduct (oh-vuh-dukt) Tube that transports eggs to the uterus; also called uterine tube.   357

ovulation (ahv-yuh-lay-shun, ohv-) Release of a secondary oocyte from the ovary; if fertilization occurs, the secondary oocyte becomes an egg.   360

oxygen debt Amount of oxygen needed to metabolize lactate, a compound that accumulates during vigorous exercise.   264

oxyhemoglobin (ahk-see-hee-muh-gloh-bin) Compound formed when oxygen combines with hemoglobin.   119, 196

oxytocin (ahk-sih-toh-sin) Hormone released by the posterior pituitary that causes contraction of uterus and milk letdown.   332

ozone hole Seasonal thinning of the ozone shield in the lower stratosphere at the North and South Poles.   558

ozone shield Accumulation of O_3, formed from oxygen in the upper atmosphere; a filtering layer that protects the Earth from ultraviolet radiation.   558

P

pacemaker See sinoatrial (SA) node.   98

pain receptor Sensory receptor that is sensitive to chemicals released by damaged tissues or excess stimuli of heat or pressure.   304

pancreas (pang-kree-us, pan-) Internal organ that produces digestive enzymes and the hormones insulin and glucagon.   166, 341

pancreatic amylase (pang-kree-at-ik am-uh-lays, -layz) Enzyme in the pancreas that digests starch to maltose.   166

pancreatic islets (islets of Langerhans) Masses of cells that constitute the endocrine portion of the pancreas.   341

pandemic An increase in the occurrence of a disease within a large and geographically widespread population (often refers to a worldwide epidemic).   380

Pap test Analysis done on cervical cells for detection of cancer.   358

parasympathetic division That part of the autonomic system that is active under normal conditions; uses acetylcholine as a neurotransmitter.   294

parathyroid gland (par-uh-thy-royd) Gland embedded in the posterior surface of the thyroid gland; it produces parathyroid hormone.   337

parathyroid hormone (PTH) Hormone secreted by the four parathyroid glands that increases the blood calcium level and decreases the blood phosphate level.   337

parent cell Cell that divides so as to form daughter cells.   422

Parkinson disease Progressive deterioration of the central nervous system due to a deficiency in the neurotransmitter dopamine.   287

parturition (par-tyoo-rish-un, par-chuh-) Processes that lead to and include birth and the expulsion of the afterbirth.   409

passive immunity Protection against infection acquired by transfer of antibodies to a susceptible individual.   149

pathogen (path-uh-jun) Disease-causing agent.   136, 380

pectoral girdle (pek-tur-ul) Portion of the skeleton that provides support and attachment for an arm; consists of a scapula and a clavicle.   242

pelvic girdle Portion of the skeleton to which the legs are attached; consists of the coxal bones.   243

pelvis Bony ring formed by the sacrum and coxae.   243

penis External organ in males through which the urethra passes; also serves as the organ of sexual intercourse.   354

pentose (pen-tohs, -tohz) Five-carbon sugar. Deoxyribose is the pentose sugar found in DNA; ribose is a pentose sugar found in RNA.   29

pepsin (pep-sin) Enzyme secreted by gastric glands that digests proteins to peptides.   163

peptide bond Type of covalent bond that joins two amino acids.   35

peptide hormone Type of hormone that is a protein, a peptide, or derived from an amino acid.   330

pericardium (pair-ih-kar-dee-um) Protective serous membrane that surrounds the heart.   94

periodontitis Inflammation of the periodontal membrane that lines tooth sockets, causing loss of bone and loosening of teeth.   160

periosteum (pair-ee-ahs-tee-um) Fibrous connective tissue covering the surface of bone.   230

peripheral nervous system (PNS) (puh-rif-ur-ul) Nerves and ganglia that lie outside the central nervous system.   276

peristalsis (pair-ih-stawl-sis) Wavelike contractions that propel substances along a tubular structure, such as the esophagus. 161

peritonitis (pair-ih-tuh-ny-tis) Generalized infection of the lining of the abdominal cavity. 159

peritubular capillary network (pair-ih-too-byuh-lur) Capillary network that surrounds a nephron and functions in reabsorption during urine formation. 213

Peyer's patches Lymphatic organs located in the small intestine. 142

phagocytosis (fag-uh-sy-toh-sis) Process by which amoeboid-type cells engulf large substances, forming an intracellular vacuole. 51

pharynx (far-ingks) Portion of the digestive tract between the mouth and the esophagus that serves as a passageway for food and also for air on its way to the trachea. 161, 187

phenotype (fee-nuh-typ) Visible expression of a genotype—for example, brown eyes or attached earlobes. 466

phenylketonuria (PKU) Result of accumulation of phenylalanine, characterized by mental retardation, light pigmentation, eczema, and neurologic manifestations unless treated by a diet low in phenylalanine. 476

pheromone Chemical signal released by an organism that affects the metabolism or influences the behavior of another individual of the same species. 330

phospholipid (fahs-foh-lip-id) Molecule that forms the bilayer of the cell's membranes; has a polar, hydrophilic head bonded to two nonpolar, hydrophobic tails. 33

photoreceptor Sensory receptor in retina that responds to light stimuli. 304

photovoltaic (solar) cell An energy-conversion device that captures solar energy and directly converts it to electrical current. 572

pH scale Measurement scale for hydrogen ion concentration. 27

pilus Elongated, hollow appendage on bacteria used to transfer DNA from one cell to another. 136

pineal gland (pin-ee-ul, py-nee-ul) Endocrine gland located in the third ventricle of the brain; produces melatonin. 345

pinna Part of the ear that projects on the outside of the head. 317

pituitary dwarfism (pih-too-ih-tair-ee, -tyoo-) Condition in which an affected individual has normal proportions but small stature; caused by inadequate growth hormone. 334

pituitary gland Endocrine gland that lies just inferior to the hypothalamus; consists of the anterior pituitary and posterior pituitary. 332

placebo Treatment that is an inactive substance (pill, liquid, etc.), administered as if it were a therapy in an experiment, but that has no therapeutic value. 10

placenta (pluh-sen-tuh) Structure that forms from the chorion and the uterine wall and allows the embryo, and then the fetus, to acquire nutrients and rid itself of wastes. 364, 396

plaque (plak) Accumulation of soft masses of fatty material, particularly cholesterol, beneath the inner linings of the arteries. 105

plasma (plaz-muh) Liquid portion of blood; contains nutrients, wastes, salts, and proteins. 116

plasma cell Cell derived from a B lymphocyte specialized to mass-produce antibodies. 145

plasma membrane Membrane surrounding the cytoplasm that consists of a phospholipid bilayer with embedded proteins; functions to regulate the entrance and exit of molecules from the cell. 46

plasma protein Protein dissolved in blood plasma. 117

plasmid (plaz-mid) Self-replicating ring of accessory DNA in the cytoplasm of bacteria. 136, 504

platelet (thrombocyte) (playt-lit) Component of blood necessary to blood clotting; also called a thrombocyte. 68, 123

pleura Serous membrane that encloses the lungs. 84, 190

pneumonectomy (noo-muh-nek-tuh-mee, nyoo-) Surgical removal of all or part of a lung. 200

pneumonia (noo-mohn-yuh, nyoo-) Infection of the lungs that causes alveoli to fill with mucus and pus. 198

polar Combination of atoms in which the electrical charge is not distributed symmetrically. 25

polar body In oogenesis, a nonfunctional product; two to three meiotic products are of this type. 435

pollution Any environmental change that adversely affects the lives and health of living things. 566

polygenic trait Trait is controlled by several allelic pairs; each dominant allele contributes to the phenotype in an additive and like manner. 479

polymerase chain reaction (PCR) (pahl-uh-muh-rays, -rayz) Technique that uses the enzyme DNA polymerase to produce millions of copies of a particular piece of DNA. 505

polyp (pahl-ip) Small, abnormal growth that arises from the epithelial lining. 170

polypeptide Polymer of many amino acids linked by peptide bonds. 35

polyribosome (pahl-ih-ry-buh-sohm) String of ribosomes simultaneously translating regions of the same mRNA strand during protein synthesis. 52, 497

polysaccharide (pahl-ee-sak-uh-ryd) Polymer made from sugar monomers; the polysaccharides starch and glycogen are polymers of glucose monomers. 30

pons (pahnz) Portion of the brain stem above the medulla oblongata and below the midbrain; assists the medulla oblongata in regulating the breathing rate. 288

population Organisms of the same species occupying a certain area. 2

positive feedback Mechanism in which the stimulus initiates reactions that lead to an increase in the stimulus. 87, 332

posterior pituitary Portion of the pituitary gland that stores and secretes oxytocin and antidiuretic hormone, which are produced by the hypothalamus. 332

precapillary sphincter Smooth muscle ring that controls blood flow through a capillary bed. 93

precipitation Water deposited on the Earth in the form of rain, snow, sleet, hail, or fog. 549

pre-embryonic development Development of the zygote in the first week, including fertilization, the beginning of cell division, and the appearance of the chorion. 397

prefrontal area Association area in the frontal lobe that receives information from other association areas and uses it to reason and plan actions. 287

primary germ layer Three layers (ectoderm, mesoderm, and endoderm) of embryonic cells that develop into specific tissues and organs. 399

primary motor area Area in the frontal lobe where voluntary commands begin; each section controls a part of the body. 286

primary somatosensory area (soh-mat-uh-sens-ree, -suh-ree) Area dorsal to the central sulcus where sensory information arrives from skin and skeletal muscles. 286

primate (pry-mayt) Animal that belongs to the order Primate; includes prosimians, monkeys, apes, and humans, all of whom have adaptations for living in trees. 525

principle Theory generally accepted by an overwhelming number of scientists; a law. 8

prion An infectious particle that is the cause of diseases, such as scrapie in sheep, mad cow disease, and Creutzfeldt-Jakob disease in humans; it has a protein component, but no nucleic acid has been detected. 138

producer Photosynthetic organism at the start of a grazing food chain that makes its own food (e.g., green plants on land and algae in water). 544

product Substance that forms as a result of a reaction. 56

progesterone (proh-jes-tuh-rohn) Female sex hormone that helps maintain sex organs and secondary sex characteristics. 344, 362

progression In cancer, a second mutation that allows cells to invade surrounding tissues. 447

prokaryotic cell Type of cell that lacks a membrane-bounded nucleus and organelles. 46, 518

prolactin (PRL) (proh-lak-tin) Hormone secreted by the anterior pituitary that stimulates the production of milk from the mammary glands. 334

promotion In cancer, development of a group of cells from a single mutated cell. 447

prophase (proh-fayz) Mitotic phase during which chromatin condenses so that chromosomes appear; chromosomes are scattered. 423

proprioceptor (proh-pree-oh-sep-tur) Sensory receptor in skeletal muscles and joints that assists the brain in knowing the position of the limbs. 306

prosimian Group of primates that includes lemur and tarsiers and may resemble the first primates to have evolved.   525

prostaglandin (prahs-tuh-glan-din) Hormone that has various and powerful local effects.   345

prostate gland (prahs-tayt) Gland located around the male urethra below the urinary bladder; adds secretions to semen.   353

protease Enzyme capable of breaking peptide bonds in a protein.   384

protein Molecule consisting of one or more polypeptides.   34

protein-first hypothesis In chemical evolution, the proposal that protein originated before other macromolecules and allowed the formation of protocells.   517

proteomics The study of all proteins in an organism.   503

prothrombin (proh-thrahm-bin) Plasma protein converted to thrombin during the steps of blood clotting.   123

prothrombin activator Enzyme that catalyzes the transformation of the precursor prothrombin to the active enzyme thrombin.   123

protocell In biological evolution, a possible cell forerunner that became a cell once it could reproduce.   517

proton Positive subatomic particle, located in the nucleus and having a weight of approximately one atomic mass unit.   20

proto-oncogene (proh-toh-ahng-koh-jeen) Normal gene that can become an oncogene through mutation.   448

provirus Latent form of a virus in which the viral DNA is incorporated into the chromosome of the host.   384

proximal convoluted tubule Highly coiled region of a nephron near the glomerular capsule, where tubular reabsorption takes place.   214

pseudostratified columnar epithelium Appearance of layering in some epithelial cells when, actually, each cell touches a baseline and true layers do not exist.   74

pulmonary artery (pool-muh-nair-ee, puul-) Blood vessel that takes blood away from the heart to the lungs.   96

pulmonary circuit Circulatory pathway that consists of the pulmonary trunk, the pulmonary arteries, and the pulmonary veins; takes O_2-poor blood from the heart to the lungs and O_2-rich blood from the lungs to the heart.   102

pulmonary fibrosis (fy-broh-sis) Accumulation of fibrous connective tissue in the lungs; caused by inhaling irritating particles, such as silica, coal dust, or asbestos.   199

pulmonary tuberculosis Tuberculosis of the lungs, caused by the bacillus *Mycobacterium tuberculosis*.   198

pulmonary vein Blood vessel that takes blood from the lungs to the heart.   96

pulse Vibration felt in arterial walls due to expansion of the aorta following ventricle contraction.   99

Punnett square (pun-ut) Gridlike device used to calculate the expected results of simple genetic crosses.   469

pupil (pyoo-pul) Opening in the center of the iris of the eye.   310

Purkinje fibers (pur-kin-jee) Specialized muscle fibers that conduct the cardiac impulse from the AV bundle into the ventricles.   98

pus Thick, yellowish fluid composed of dead phagocytes, dead tissue, and bacteria.   143

pyelonephritis Inflammation of the kidney due to bacterial infection.   222

R

radioisotope Unstable form of an atom that spontaneously emits radiation in the form of radioactive particles or radiant energy.   21

reactant (re-ak-tunt) Substance that participates in a reaction.   56

recessive allele (uh-leel) Allele that exerts its phenotypic effect only in the homozygote; its expression is masked by a dominant allele.   466

recombinant DNA DNA that contains genes from more than one source.   504

rectum (rek-tum) Terminal end of the digestive tube between the sigmoid colon and the anus.   168

red blood cell (erythrocyte) Formed element that contains hemoglobin and carries oxygen from the lungs to the tissues; also called erythrocyte.   68, 118

red bone marrow Blood-cell-forming tissue located in the spaces within spongy bone.   141, 230

reduced hemoglobin (hee-muh-gloh-bun) Hemoglobin carrying hydrogen ions.   196

referred pain Pain perceived as having come from a site other than that of its actual origin.   307

reflex Automatic, involuntary response of an organism to a stimulus.   292

refractory period (rih-frak-tuh-ree) Time following an action potential when a neuron is unable to conduct another nerve impulse.   280

renal artery (ree-nul) Vessel that originates from the aorta and delivers blood to the kidney.   208

renal cortex (ree-nul kor-teks) Outer portion of the kidney that appears granular.   212

renal medulla (ree-nul muh-dul-uh) Inner portion of the kidney that consists of renal pyramids.   212

renal pelvis Hollow chamber in the kidney that lies inside the renal medulla and receives freshly prepared urine from the collecting ducts.   212

renal vein (ree-nul) Vessel that takes blood from the kidney to the inferior vena cava.   208

renewable resource Resources normally replaced or replenished by natural processes; resources not depleted by moderate use. Examples include solar energy, biological resources such as forests and fisheries, biological organisms, and some biogeochemical cycles.   566

renin (ren-in) Enzyme released by kidneys that leads to the secretion of aldosterone and a rise in blood pressure.   219, 339

replacement reproduction Population in which each person is replaced by only one child.   565

repolarization When the charge inside the axon resumes a negative charge.   278

reproduce To produce a new individual of the same type.   4

reproductive cloning Genetically identical to the original individual.   404

reproductive system Organ system that contains male or female organs and specializes in the production of offspring.   81

residual volume Amount of air remaining in the lungs after a forceful expiration.   193

respiratory control center Group of nerve cells in the medulla oblongata that sends out nerve impulses on a rhythmic basis, resulting in involuntary inspiration on an ongoing basis.   194

respiratory pump Mechanism whereby reductions in thoracic pressure during the breathing cycle tend to aid the return of blood to the heart from peripheral veins.   101

respiratory system Organ system consisting of the lungs and tubes that bring oxygen into the lungs and take carbon dioxide out.   80

resting potential Polarity across the plasma membrane of a resting neuron due to an unequal distribution of ions.   278

restriction enzyme Bacterial enzyme that stops viral reproduction by cleaving viral DNA; used to cut DNA at specific points during production of recombinant DNA.   504

reticular fiber (rih-tik-yuh-lur) Very thin collagen fibers in the matrix of connective tissue, highly branched and forming delicate supporting networks.   66

reticular formation (rih-tik-yuh-lur) Complex network of nerve fibers within the central nervous system that arouses the cerebrum.   288

retina (ret-n-uh, ret-nuh) Innermost layer of the eyeball that contains the rod cells and the cone cells.   311

retinal (ret-n-al, -awl) Light-absorbing molecule that is a derivative of vitamin A and a component of rhodopsin.   312

retrovirus RNA virus containing the enzyme reverse transcriptase that carries out RNA to DNA transcription.   384

reverse transcriptase Enzyme that speeds the conversion of viral RNA to viral DNA.   384

rheumatic fever Disease caused by bacterial infection, characterized by fever, swelling and pain in the joints, sore throat, and cardiac involvement.   153

rheumatoid arthritis Persistent inflammation of synovial joints, often causing cartilage destruction, bone erosion, and joint deformities.   153

rhodopsin (roh-dahp-sun) Light-absorbing molecule in rod cells and cone cells that contains a pigment and the protein opsin.   312

ribosomal RNA (rRNA) (ry-buh-soh-mul) Type of RNA found in ribosomes where protein synthesis occurs.   493

ribosome (ry-buh-sohm) RNA and protein in two subunits; site of protein synthesis in the cytoplasm.   52

rigor mortis Contraction of muscles at death due to lack of ATP.   257

RNA (ribonucleic acid) (ry-boh-noo-klee-ik) Nucleic acid produced from covalent bonding of nucleotide monomers that contain the sugar ribose; occurs in three forms: messenger RNA, ribosomal RNA, and transfer RNA.   37, 492

RNA-first hypothesis In chemical evolution, the proposal that RNA originated before other macromolecules and allowed the formation of the first cell(s).   517

RNA polymerase (pahl-uh-muh-rays) During transcription, an enzyme that joins nucleotides complementary to a DNA template.   496

rod cell Photoreceptor in retina of eyes that responds to dim light.   312

rotational equilibrium Maintenance of balance when the head and body are suddenly moved or rotated.   322

rotator cuff Tendons that encircle and help form a socket for the humerus, and also help reinforce the shoulder joint.   242

round window Membrane-covered opening between the inner ear and the middle ear.   317

rugae Deep folds, as in the wall of the stomach.   163

runoff Water, from rain, snowmelt, or other sources, that flows over the land surface, adding to the water cycle.   549

S

SA (sinoatrial) node (sy-noh-ay-tree-ul) Small region of neuromuscular tissue that initiates the heartbeat; also called the pacemaker.   98

saccule (sak-yool) Saclike cavity in the vestibule of the inner ear; contains sensory receptors for gravitational equilibrium.   322

salivary amylase (sal-uh-vair-ee am-uh-lays, -layz) Secreted from the salivary glands; the first enzyme to act on starch.   160

salivary gland Gland associated with the mouth that secretes saliva.   160

saltatory conduction Movement of nerve impulses from one neurofibral node to another along a myelinated axon.   280

saltwater intrusion Movement of salt water into freshwater aquifers in coastal areas where groundwater is withdrawn faster than it is replenished.   569

sarcolemma (sar-kuh-lem-uh) Plasma membrane of a muscle fiber; also forms the tubules of the T system involved in muscular contraction.   258

sarcoma Cancer that arises in muscles and connective tissues.   259, 449

sarcomere (sar-kuh-mir) One of many units, arranged linearly within a myofibril, whose contraction produces muscle contraction.   259

sarcoplasmic reticulum (sar-kuh-plaz-mik rih-tik-yuh-lum) Smooth endoplasmic reticulum of skeletal muscle cells; surrounds the myofibrils and stores calcium ions.   258

saturated fatty acid Fatty-acid molecule that lacks double bonds between the atoms of its carbon chain.   31

Schwann cell Cell that surrounds a fiber of a peripheral nerve and forms the myelin sheath.   277

science Development of concepts about the natural world, often by using the scientific method.   7

scientific method Process of attaining knowledge by making observations, testing hypotheses, and coming to conclusions.   8

scientific theory Concept supported by a broad range of observations, experiments, and conclusions.   8

sclera (skleer-uh) White, fibrous, outer layer of the eyeball.   310

scoliosis Abnormal laterial (side-to-side) curvature of the vertebral column.   240

scrotum (skroh-tum) Pouch of skin that encloses the testes.   353

secondary oocyte In oogenesis, the functional product of meiosis I; becomes the egg.   435

second messenger Chemical signal such as cyclic AMP that causes the cell to respond to the first messenger—a hormone bound to a receptor protein in the plasma membrane.   330

selectively permeable Having degrees of permeability; the cell is impermeable to some substances and allows others to pass through at varying rates.   49

semantic memory Capacity of the brain to store and retrieve information with regard to words or numbers.   289

semen (see-mun) Thick, whitish fluid consisting of sperm and secretions from several glands of the male reproductive tract.   353

semicircular canal (sem-ih-sur-kyuh-lur) One of three tubular structures within the inner ear that contain sensory receptors responsible for the sense of rotational equilibrium.   317, 322

semilunar valve (sem-ee-loo-nur) Valve resembling a half moon located between the ventricles and their attached vessels.   95

seminal vesicle (sem-uh-nul) Convoluted structure attached to the vas deferens near the base of the urinary bladder in males; adds secretions to semen.   353

seminiferous tubule (sem-uh-nif-ur-us) Long, coiled structure contained within chambers of the testis; where sperm are produced.   354

sensation Conscious awareness of a stimulus due to nerve impulses sent to the brain from a sensory receptor by way of sensory neurons.   304

sensory adaptation Phenomenon of a sensation becoming less noticeable once it has been recognized by constant repeated stimulation.   305

sensory neuron Nerve cell that transmits nerve impulses to the central nervous system after a sensory receptor has been stimulated.   277

sensory receptor Structure that receives either external or internal environmental stimuli and is a part of a sensory neuron or transmits signals to a sensory neuron.   277, 304

septum Wall between two cavities; in the human heart, a septum separates the right side from the left side.   95

serosa Membrane that covers internal organs and lines cavities without an opening to the outside of the body.   159

serotonin A neurotransmitter.   281

serous membrane (seer-us) Membrane that covers internal organs and lines cavities without an opening to the outside of the body; also called serosa.   84

Sertoli cell Cell associated with developing germ cells in seminiferous tubule; secretes fluid into seminiferous tubule and mediates hormonal effects on tubule.   355

serum (seer-um) Light yellow liquid left after clotting of blood.   125

severe combined immunodeficiency disease (SCID) Congenital illness in which both antibody- and cell-mediated immunity are lacking or inadequate.   123, 153

sex chromosome Chromosome that determines the sex of an individual; in humans, females have two X chromosomes, and males have an X and Y chromosome.   481

sex-linked Allele that occurs on the sex chromosomes but may control a trait that has nothing to do with the sex characteristics of an individual.   481

short-term memory Retention of information for only a few minutes, such as remembering a telephone number.   289

sickle-cell disease Genetic disorder in which the affected individual has sickle-shaped red blood cells subject to hemolysis.   121, 476

simple goiter (goy-tur) Condition in which an enlarged thyroid produces low levels of thyroxine.   336

sinkhole Large surface crater caused by the collapse of an underground channel or cavern; often triggered by groundwater withdrawal.   569

sinus (sy-nus) Cavity or hollow space in an organ such as the skull.   237

sinusitis (sy-nuh-sy-tis) Infection of the sinuses, caused by blockage of the openings to the sinuses and characterized by postnasal discharge and facial pain.   198

sister chromatids One of two genetically identical chromosomal units that are the result of DNA replication and are attached to each other at the centromere.   420

skeletal muscle Striated, voluntary muscle tissue found in muscles that move the bones.   69, 254

skeletal muscle pump Pumping effect of contracting skeletal muscles on blood flow through underlying vessels.   101

skeletal system System of bones, cartilage, and ligaments that works with the muscular system to protect the body and provide support for locomotion and movement.   81

skill memory Capacity of the brain to store and retrieve information necessary to perform motor activities, such as riding a bike.   289

skin Outer covering of the body; can be called the integumentary system because it contains organs such as sense organs.   76

skull Bony framework of the head, composed of cranial bones and the bones of the face.   237

sliding filament model An explanation for muscle contraction based on the movement of actin filaments in relation to myosin filaments.   260

small intestine Long, tubelike chamber of the digestive tract between the stomach and large intestine.   165

small RNA Short RNA molecules that help to regulate gene expression.   493

smooth (visceral) muscle Nonstriated, involuntary muscle tissue found in the walls of internal organs.   69, 254

sodium-potassium pump Carrier protein in the plasma membrane that moves sodium ions out of and potassium ions into cells; important in nerve and muscle cells.   278

soft palate (pal-it) Entirely muscular posterior portion of the roof of the mouth.   160

somatic system That portion of the peripheral nervous system containing motor neurons that control skeletal muscles.   291

spasm Sudden, involuntary contraction of one or more muscles.   266

species Group of similarly constructed organims capable of interbreeding and producing fetile offspring; organisms that share a common gene pool.   2

sperm Male gamete having a haploid number of chromosomes and the ability to fertilize an egg, the female gamete.   355

spermatogenesis (spur-mat-uh-jen-ih-sis) Production of sperm in males by the process of meiosis and maturation.   354, 434

sphincter (sfingk-tur) Muscle that surrounds a tube and closes or opens the tube by contracting and relaxing.   161

spinal cord Part of the central nervous system; the nerve cord that is continuous with the base of the brain plus the vertebral column that protects the nerve cord.   283

spinal nerve Nerve that arises from the spinal cord.   291

spiral organ Organ in the cochlear duct of the inner ear responsible for hearing; also called the organ of Corti.   318

spirillum A group of bacteria that exhibit a variety of spiral or undulating shapes or are comma-shaped.   136

spleen Large, glandular organ located in the upper left region of the abdomen; stores and purifies blood.   141

spongy bone Porous bone found at the ends of long bones where red bone marrow is sometimes located.   68, 230

sprain Injury to a ligament caused by abnormal force applied to a joint.   267

squamous epithelium (skway-mus, skwah-) Type of epithelial tissue that contains flat cells.   73

standard error Number used in evaluating statistical data to show the range of error in the data.   14

stapes (stay-peez) The last of three ossicles of the ear that serve to conduct vibrations from the tympanic membrane to the oval window of the inner ear.   317

starch Storage polysaccharide found in plants that is composed of glucose molecules joined in a linear fashion with few side chains.   30

stereocilium (pl. stereocilia) Long, flexible microvilli that superficially resemble cilia. Within the inner ear, these signal changes in body position and help to maintain balance and equilibrium.   317

steroid (steer-oyd) Type of lipid molecule having a complex of four carbon rings; examples are cholesterol, progesterone, and testosterone.   33

steroid hormone One of a group of hormones derived from cholesterol.   330

stimulus Change in the internal or external environment that a sensory receptor can detect, leading to nerve impulses in sensory neurons.   278, 304

stomach Muscular sac that mixes food with gastric juices to form chyme, which enters the small intestine.   162

strain Injury to a muscle resulting from overuse or improper use.   267

striae gravidarum Linear, depressed, scarlike lesions occurring on the abdomen, breasts, buttocks, and thighs due to the weakening of the elastic tissues during pregnancy.   408

striated (stry-ayt-ud) Having bands; in cardiac and skeletal muscle, alternating light and dark crossbands produced by the distribution of contractile proteins.   69

stroke Condition resulting when an arteriole in the brain bursts or becomes blocked by an embolism; also called cerebrovascular accident.   106

subcutaneous layer (sub-kyoo-tay-nee-us) Tissue layer that lies just beneath the skin and contains adipose tissue.   76

submucosa Layer of connective tissue underneath a mucous membrane.   159

subsidence Occurs when a portion of the Earth's surface gradually settles downward.   569

substrate Reactant in a reaction controlled by an enzyme.   56

sudden infant death syndrome (SIDS) Any sudden and unexplained death of an apparently healthy infant aged one month to one year.   194

superior vena cava (vee-nuh kay-vuh) Large vein that enters the right atrium from above and carries blood from the head, thorax, and upper limbs to the heart.   103

surfactant Agent that reduces the surface tension of water; in the lungs, a surfactant prevents the alveoli from collapsing.   191

sustainable Ability of a society or ecosystem to maintain itself while also providing services to human beings.   581

suture (soo-chur) Type of immovable joint articulation found between bones of the skull.   245

sweat gland Skin gland that secretes a fluid substance for evaporative cooling; also called sudoriferous gland.   79

sympathetic division The part of the autonomic system that usually promotes activities associated with emergency (fight-or-flight) situations; uses norepinephrine as a neurotransmitter.   294

synapse (sin-aps, si-naps) Junction between neurons consisting of the presynaptic (axon) membrane, the synaptic cleft, and the postsynaptic (usually dendrite) membrane.   280

synapsis (sih-nap-sis) Pairing of homologous chromosomes during prophase I of meiosis I.   427

synaptic cleft (sih-nap-tik) Small gap between presynaptic and postsynaptic membranes of a synapse.   280

syndrome Group of symptoms that appear together and tend to indicate the presence of a particular disorder.   421

synovial joint Freely movable joint having a cavity filled with synovial fluid.   84, 245

synovial membrane Membrane that forms the inner lining of the capsule of a freely movable joint.   84

systemic circuit Blood vessels that transport blood from the left ventricle and back to the right atrium of the heart.   102

systemic lupus erythematosus (SLE) Syndrome involving the connective tissues and various organs, including kidney.   153

systole (sis-tuh-lee) Contraction period of the heart during the cardiac cycle.   97

systolic pressure (sis-tahl-ik) Arterial blood pressure during the systolic phase of the cardiac cycle.   100

T

T cell (T lymphocyte) Lymphocyte that matures in the thymus. Cytotoxic T cells kill antigen-bearing cells outright; helper T cells release cytokines that stimulate other immune system cells.   141

T-cell receptor Molecule on the surface of a T-lymphocyte to which an antigen binds.   147

T (transverse) tubule Membranous channel that extends inward.   258

taste bud Sense organ containing the receptors associated with the sense of taste.   308

Tay-Sachs disease Lethal genetic disease in which the newborn has a faulty lysosomal digestive enzyme.   475

technology The science or study of the practical or industrial arts.   14

tectorial membrane (tek-tor-ee-ul) Membrane that lies above and makes contact with the hair cells in the spiral organ.   318

telomere (tel-uh-meer) Tip of the end of a chromosome.   446

telophase (tel-uh-fayz) Mitotic phase during which daughter chromosomes are located at each pole.   424

template (tem-plit) Pattern or guide used to make copies; parental strand of DNA serves as a guide for the production of daughter DNA strands, and DNA also serves as a guide for the production of messenger RNA.   491

tendinitis (ten-din-eye-tis) An inflammation of muscle tendons and their attachments.   267

tendon (ten-dun) Strap of fibrous connective tissue that connects skeletal muscle to bone.   67, 255

testes (sing., testis) (tes-teez, tes-tus) Male gonads that produce sperm and the male sex hormones.   344, 353

test group Group exposed to the experimental variable in an experiment, rather than the control group.   9

testosterone (tes-tahs-tuh-rohn) Male sex hormone that helps maintain sexual organs and secondary sex characteristics.   344, 356, 407

tetanus (tet-n-us) Sustained muscle contraction without relaxation.   262

tetany (tet-n-ee) Severe twitching caused by involuntary contraction of the skeletal muscles due to a calcium imbalance.   337

thalamus (thal-uh-mus) Part of the brain located in the lateral walls of the third ventricle that serves as the integrating center for sensory input; it plays a role in arousing the cerebral cortex.   287

therapeutic cloning Used to create mature cells of various cell types. Also, used to learn about specialization of cells and provice cells and tissue to treat human illnesses.   404

thermoreceptor Sensory receptor that is sensitive to changes in temperature.   304

threshold Electrical potential level (voltage) at which an action potential or nerve impulse is produced.   278

thrombin (thrahm-bin) Enzyme that converts fibrinogen to fibrin threads during blood clotting.   123

thrombocytopenia Insufficient number of platelets in the blood.   125

thromboembolism (thrahm-boh-em-buh-liz-um) Obstruction of a blood vessel by a thrombus that has dislodged from the site of its formation.   106, 125

thrombus (thrahm-bus) Blood clot that remains in the blood vessel where it formed.   105, 125

thymine (T) (thy-meen) One of four nitrogen-containing bases in nucleotides composing the structure of DNA; pairs with adenine.   38

thymosin Group of peptides secreted by the thymus gland that increases production of certain types of white blood cells.   345

thymus gland Lymphatic organ, located along the trachea behind the sternum, involved in the maturation of T lymphocytes in the thymus gland. Secretes hormones called thymosins, which aid the maturation of T cells and perhaps stimulate immune cells in general.   141, 345

thyroid gland Endocrine gland in the neck that produces several important hormones, including thyroxine, triiodothyronine, and calcitonin.   336

thyroid-stimulating hormone (TSH) Substance produced by the anterior pituitary that causes the thyroid to secrete thyroxine and triiodothyronine.   332

thyroxine (T_4) (thy-rahk-sin) Hormone secreted from the thyroid gland that promotes growth and development; in general, it increases the metabolic rate in cells.   336

tidal volume Amount of air normally moved in the human body during an inspiration or expiration.   192

tight junction Junction between cells when adjacent plasma membrane proteins join to form an impermeable barrier.   74

tissue Group of similar cells that perform a common function.   2, 66

tissue fluid Fluid that surrounds the body's cells; consists of dissolved substances that leave the blood capillaries by filtration and diffusion.   68, 92

tonicity (toh-nis-ih-tee) Osmolarity of a solution compared with that of a cell. If the solution is isotonic to the cell, there is no net movement of water; if the solution is hypotonic, the cell gains water; and if the solution is hypertonic, the cell loses water.   49

tonsillectomy (tahn-suh-lek-tuh-mee) Surgical removal of the tonsils.   198

tonsillitis Infection of the tonsils that causes inflammation and can spread to the middle ears.   198

tonsils Partially encapsulated lymph nodules located in the pharynx.   142, 198

total artificial heart (TAH) A mechanical replacement for the heart, as opposed to a partial replacement.   110

toxin Poisonous substance produced by living cells or organisms. Toxins are nearly always proteins that are capable of causing disease on contact or absorption with body tissues.   137

tracer Substance having an attached radioisotope that allows a researcher to track its whereabouts in a biological system.   21

trachea (tray-kee-uh) Passageway that conveys air from the larynx to the bronchi; also called the windpipe.   189

tracheostomy (tray-kee-ahs-tuh-mee) Creation of an artificial airway by incision of the trachea and insertion of a tube.   190

tract Bundle of myelinated axons in the central nervous system.   283

transcription Process whereby a DNA strand serves as a template for the formation of mRNA.   494

transcription factor In eukaryotes, protein required for the initiation of transcription by RNA polymerase.   501

trans fat Fats, which occur naturally in meat and dairy products of ruminants, that are also industrially created through partial hydrogenation of plant oils and animal fats.   32

transfer RNA (tRNA) Type of RNA that transfers a particular amino acid to a ribosome during protein synthesis; at one end, it binds to the amino acid, and at the other end it has an anticodon that binds to an mRNA codon.   493

transgenic organism Free-living organism in the environment that has a foreign gene in its cells.   506

translation Process whereby ribosomes use the sequence of codons in mRNA to produce a polypeptide with a particular sequence of amino acids.   495

translocation Movement of a chromosomal segment from one chromosome to another nonhomologous chromosome, leading to abnormalities; e.g., Down syndrome.   439

triglyceride (trih-glis-uh-ryd) Neutral fat composed of glycerol and three fatty acids.   31

triplet code Each sequence of three nucleotide bases in the DNA of genes stands for a particular amino acid.   495

trisomy One more chromosome than usual.   435

trophic level Feeding level of one or more populations in a food web.   547

trophic relationship In ecosystems, feeding relationships such as grazing food webs or detrital food webs.   547

tropomyosin (trahp-uh-my-uh-sin, trohp-) Protein that functions with troponin to block muscle contraction until calcium ions are present.   262

troponin (troh-puh-nin) Protein that functions with tropomyosin to block muscle contraction until calcium ions are present.   262

trypsin (trip-sin) Protein-digesting enzyme secreted by the pancreas.   166

tubal ligation Method for preventing pregnancy in which the uterine tubes are cut and sealed.   367

tubular reabsorption Movement of primarily nutrient molecules and water from the contents of the nephron into blood at the proximal convoluted tubule.   216

tubular secretion Movement of certain molecules from blood into the distal convoluted tubule of a nephron so that they are added to urine.   216

tumor (too-mur) Cells derived from a single mutated cell that has repeatedly undergone cell division; benign tumors remain at the site of origin, and malignant tumors metastasize.   446

tumor marker test Blood test for a substance, such as a tumor antigen, that indicates a patient has cancer.   457

tumor-suppressor gene Gene that codes for a protein that ordinarily suppresses cell division; inactivity can lead to a tumor. 448

tympanic membrane (tim-pan-ik) Located between the outer and middle ear where it receives sound waves; also called the eardrum. 317

U

umbilical cord Cord connecting the fetus to the placenta through which blood vessels pass. 400

unsaturated fatty acid Fatty-acid molecule that has one or more double bonds between the atoms of its carbon chain. 31

uracil (U) (yoor-uh-sil) The base in RNA that replaces thymine found in DNA; pairs with adenine. 38, 492

urea (yoo-ree-uh) Primary nitrogenous waste of humans derived from amino acid breakdown. 167, 209

uremia High level of urea nitrogen in the blood. 222

ureter (yoor-uh-tur) One of two tubes that take urine from the kidneys to the urinary bladder. 208

urethra (yoo-ree-thruh) Tubular structure that receives urine from the bladder and carries it to the outside of the body. 209, 353

urethritis Inflammation of the urethra. 222

uric acid (yoor-ik) Waste product of nucleotide metabolism. 209

urinary bladder Organ where urine is stored before being discharged by way of the urethra. 208

urinary system Organ system consisting of the kidneys and urinary bladder; rids the body of nitrogenous wastes and helps regulate the water-salt balance of the blood. 81

uterine cycle (yoo-tur-in, -tuh-ryn) Monthly occurring changes in the characteristics of the uterine lining (endometrium). 362

uterus (yoo-tur-us) Organ located in the female pelvis where the fetus develops; also called the womb. 358

utricle (yoo-trih-kul) Saclike cavity in the vestibule of the inner ear that contains sensory receptors for gravitational equilibrium. 322

V

vaccine Antigens prepared in such a way that they promote active immunity without causing disease. 150

vagina Organ that leads from the uterus to the vestibule and serves as the birth canal and organ of sexual intercourse in females. 358

valve Membranous extension of a vessel of the heart wall that opens and closes, ensuring one-way flow. 94

vas deferens (vas def-ur-unz, -uh-renz) Tube that leads from the epididymis to the urethra in males. 353

vasectomy Method for preventing pregnancy in which the vasa deferentia are cut and sealed. 367

vector (vek-tur) In genetic engineering, a means to transfer foreign genetic material into a cell (e.g., a plasmid). 389, 504

ventilation Process of moving air into and out of the lungs; also called breathing. 186

ventricle (ven-trih-kul) Cavity in an organ, such as a lower chamber of the heart or the ventricles of the brain. 95, 283

venule (ven-yool, veen-) Vessel that takes blood from capillaries to a vein. 94

vermiform appendix Small, tubular appendage that extends outward from the cecum of the large intestine. 168

vernix caseosa (vur-niks kay-see-oh-suh) Cheeselike substance covering the skin of the fetus. 406

vertebral column (vur-tuh-brul) Series of joined vertebrae that extends from the skull to the pelvis. 240

vertebrate (vur-tuh-brit, -brayt) An animal with a vertebral column. 6

vesicle Small, membrane-bounded sac that stores substances within a cell. 53

vestibule (ves-tuh-byool) Space or cavity at the entrance of a canal, such as the cavity that lies between the semicircular canals and the cochlea. 317

vestigial structure Remains of a structure that was functional in some ancestor but is no longer functional in the organism in question. 521

villus (pl., villi) (vil-us) Small, fingerlike projection of the inner small intestinal wall. 165

virus Noncellular, parasitic agent consisting of an outer capside and an inner core of nucleic acid. 137

visual accommodation Ability of the eye to focus at different distances by changing the curvature of the lens. 311

vital capacity Maximum amount of air moved in or out of the human body with each breathing cycle. 193

vitamin Essential requirement in the diet, needed in small amounts. They are often part of coenzymes. 177

vitamin D Required for proper bone growth. 78

vitreous humor (vit-ree-us) Clear, gelatinous material between the lens of the eye and the retina. 311

vocal cord Fold of tissue within the larynx; creates vocal sounds when it vibrates. 188

vulva External genitals of the female that surround the opening of the vagina. 358

W

water (hydrologic) cycle Interdependent and continuous circulation of water from the ocean to the atmosphere, to the land, and back to the ocean. 549

Wernicke's area Brain area involved in language comprehension. 287

white blood cell (leukocyte) Type of blood cell that is transparent without staining and protects the body from invasion by foreign substances and organisms; also called a leukocyte. 68, 121

white matter Myelinated axons in the central nervous system. 283

X

XDR TB Extensively drug-resistant tuberculosis, a type of bacterium that causes tuberculosis (TB) that is no longer susceptible to almost all of the drugs normally used to treat TB. 392

xenotransplantation Use of animal organs, instead of human organs, in human transplant patients. 153, 511

X-linked Allele located on an X chromosome, but may control a trait that has nothing to do with the sex characteristics of an individual. 481

Y

yolk sac Extraembryonic membrane that encloses the yolk of birds; in humans, it is the first site of blood cell formation. 396

Z

zygote (zy-goht) Diploid cell formed by the union of sperm and egg; the product of fertilization. 357, 394, 427

Credits

Photo Credits

Chapter 1

Opener: © The McGraw-Hill Companies, Inc./Susie Ross, photographer; 1.1(mold): © Gary R. Robinson/Visuals Unlimited; 1.1(meerkat): © J & B Photo/Animals Animals/Earth Scenes; 1.1(gloeocapsa): © Michael Abbey/Photo Researchers, Inc.; 1.1 (cotton): © Dale Jackson/Visuals Unlimited; 1.1(giardia): © Dr. Fred Hossler/Visuals Unlimited; 1.1(leech): © St. Bartholomews Hospital/Photo Researchers, Inc.; 1.3a: © John Cancalosi/Peter Arnold; 1.3b: © Vol. 124/Corbis RF; 1.4a(tree): Courtesy Paul Wray, Iowa State University; 1.4a(seedling): © Herman Eisenbeiss/Photo Researchers, Inc.; 1.4b(fetus): © Derek Bromhall/OSF/Animals Animals/Earth Scenes; 1.4b(inset): © Rawlins-CMSP/Getty Images; 1.6(left): © Don and Pat Valenti/DRK Photos; 1.6(right): © William Smithey Jr.; 1.7: © Ryan McVay/Getty RF; 1.8(Dr. Marshall): © Tony McDonough/epa/Corbis; 1.8(bacteria): © Eye of Science/Photo Researchers, Inc.; 1.9a: © blickwinkel/Alamy; 1.9d: © Phanie/Photo Researchers, Inc.; 1A: © Corbis Sygma; 1.11a: © Edgar Bernstein/Peter Arnold; 1.11b: © Luiz C. Marigo/Peter Arnold; 1B: © Bettman/Corbis

Chapter 2

Opener: © Jim West/PhotoEdit; 2.3a: © Biomed Commun./Custom Medical Stock Photo; 2.3b(patient): Courtesy National Institutes of Health (NIH); 2.3b(brain scan): © Mazzlota et al./Photo Researchers, Inc.; 2.4: © Martin Dohrn/Photo Researchers, Inc.; 2.5b(crystals): © Charles M. Falco/Photo Researchers, Inc.; 2.5b(shaker): © The McGraw-Hill Companies, Inc./Evelyn Jo Johnson, photographer; p. 23(Rontgen): © Popperfoto/Getty Images; 2.8a: © The McGraw-Hill Companies, Inc./Jill Braaten, photographer; 2.8b: © Amanda Langford/SuperStock RF; 2.10a: © Ray Pfortner/Peter Arnold; 2.10b: © Frederica Georgia/Photo Researchers, Inc.; 2.12: © Jeremy Burgess/SPL/Photo Researchers, Inc.; 2.13: © Don W. Fawcett/Photo Researchers, Inc.; 2.17b: © Warren Toda/epa/Corbis; 2.17c: © Tony Marsh/Reuters/Corbis; p. 34(student): © Comstock/PunchStock RF; p. 34(blood cells): © P. Motta & S. Correr/Photo Researchers, Inc.; p. 34(runner): © Duomo/Corbis; 2A: © David H. Wells/Corbis

Chapter 3

Opener: © David Madison/Getty Images; 3.1(nerve cells): © Dr. Dennis Kunkel/Visuals Unlimited; 3.1(red blood cells): © Prof. P. Motta, Dept. of Anatomy, Univ. LaSapienza Rome/SPL/Photo Researchers, Inc.; 3.1(hyaline): © Ed Reschke; 3.3a: © David M. Phillips/Visuals Unlimited; 3.3b: © Alfred Pasieka/Photo Researchers, Inc.; 3.3c: © Warren Rosenberg/Biological Photo Service; 3.3d: © Inga Spence/Visuals Unlimited; 3.4: © Alfred Pasieka/Photo Researchers, Inc.; 3.9a(all): © Dennis Kunkel/Phototake; 3.13(nuclear pores): Courtesy E.G. Pollock; 3.13(ER): © R. Bolender & D. Fawcett/Visuals Unlimited; 3.15b: © Y. Nikas/Photo Researchers, Inc.; 3.15c: © David M. Phillips/Photo Researchers, Inc.; 3.16: © Dr. Don W. Fawcett/Visuals Unlimited; 3A: © Biophoto Associates/Photo Researchers, Inc.

Chapter 4

Opener: © JupiterImages/BananaStock/Alamy; 4.2 (all), 4.5a: © Ed Reschke; 4.5b: © The McGraw-Hill Companies, Inc./Dennis Strete, photographer; 4.5c, 4.6, 4.7(all): © Ed Reschke; 4B: © AFP/Getty Images; 4.10a: © John D. Cunningham/Visuals Unlimited; 4.10b: © Ken Greer/Visuals Unlimited; 4.10c: © James Stevenson/SPL/Photo Researchers, Inc.; 4C: © Neil McAllister/Alamy

Chapter 5

Opener: © JGI/Blend Image RF; 5.2(left): © Ed Reschke; 5.2(right): © Biophoto Associates/Photo Researchers, Inc.; 5.3b: © SIU/Visuals Unlimited; 5.4b: © Dr. Don W. Fawcett/Visuals Unlimited; 5.5d: © Biophoto Associates/Photo Researchers, Inc.; 5.6b, c: © Ed Reschke; 5.6d: © Mark Harmel/Getty Images; 5.7: © David Young-Wolff/PhotoEdit; 5.14(plaque): © Biophoto Associates/Photo Researchers, Inc.; 5.14(normal): © Ed Reschke; p. 107(Thomas): Courtesy of The Alan Mason Chesney Medical Archives of The Johns Hopkins Medical Institutions; 5.15a: © Pascal Goethgheluck/SPL/Photo Researchers, Inc.; 5.16(right): Courtesy of SynCardia Systems, Inc.; 5A: © Bill Aron/PhotoEdit

Chapter 6

Opener: © moodboard/Corbis RF; 6.1: © Doug Menuez/Getty; 6.2: © Michael Keller/Corbis; 6.3a: © Andrew Syred/Photo Researchers, Inc.; 6.3c: © Lennart Nilsson, *Behold Man*, Little Brown and Company, Boston; p. 121(sickled cell): © Phototake, Inc./Alamy; 6.5(all): © Dr. Fred Hossler/Visuals Unlimited; 6.7b: © Eye of Science/Photo Researchers, Inc.; 6A: © Gary Conner/PhotoEdit; 6.11: © Corbis RF

Chapter 7

Opener: © Dynamic Graphics/JupiterImages RF; 7.1b: © Dr. David M. Phillips/Visuals Unlimited; 7.1c: © Dr. Dennis Kunkel/Visuals Unlimited; 7.1d: © Dr. Gary D. Gaugler/Phototake; 7A: © Bettman/Corbis; 7.7(thymus, spleen): © Ed Reschke/Peter Arnold; 7.7(marrow): © R. Valentine/Visuals Unlimited; 7.7(lymph): © Fred E. Hossler/Visuals Unlimited; 7.11b: Courtesy Dr. Arthur J. Olson, Scripps Institute; 7.14a: © Michael Newman/PhotoEdit; 7.15a: © John Lund/Drew Kelly/Blend Images RF; 7.15b: © Digital Vision/Getty RF; 7.15c: © Photodisc Collection/Getty RF; 7B(pollen): © David Scharf/SPL/Photo Researchers, Inc.; 7B(girl): © Damien Lovegrove/SPL/Photo Researchers, Inc.; 7.17: © Richard Anderson

Chapter 8

Opener: © Marissa Kaiser/Getty Images; 8A(bike): © Corbis RF; 8A(man): © Stockbyte RF; 8.5c: © Dr. Fred Hossler/Visuals Unlimited; p. 164(band): © ISM/Phototake; p. 164(St. Martin): © Bettman/Corbis; 8.6(villi): © Manfred Kage/Peter Arnold, Inc.; 8.6(microvilli):Reprinted from *Medical Cell Biology*, Charles Flickinger, copyright 1979, with permission from Elsevier; 8B: © Stockdisc/PunchStock RF; p. 172(couple): © BananaStock/age fotostock RF; p. 172(woman): © Photodisc/Getty RF; p. 172(vegetables): © Photolink/Getty RF; 8.12: © Cole Group/Getty RF; 8.13: © Volume 20/Photodisc/Getty RF; p. 176: © The McGraw-Hill Companies, Inc./Evelyn Jo Johnson, photographer; 8.16a: © Tony Freeman/PhotoEdit; 8.16b: © Donna Day/Stone/Getty Images; 8.16c: © Corbis RF

Chapter 9

Opener: © Holloway/Getty Images; 9.4: © CNRI/Phototake; 9.5: © Dr. Kessel & Dr. Kardon/Tissues & Organs/Visuals Unlimited; 9.9: © Burger/Photo Researchers, Inc.; 9.13(both): © Martin Rotker/Martin Rotker Photography

Chapter 10

Opener: © Keith Eng, 2008; 10.3b: © Ralph T. Hutchings/Visuals Unlimited; 10.4(top left): © Prof. P.M. Motta & M. Castellucci/Science Photo Library/Photo Researchers, Inc.; 10.4(top right, lower left):Reprinted from *Journal of Ultrastructure Research*, Vol. 15, A.B. Maunsbach, pages 242–282, copyright 1966, with permission of Elsevier.; 10.5a: © Joseph F. Gennaro Jr./Photo Researchers, Inc.;

10B: © Ian Hooton/Photo Researchers, Inc.; 10.10: © AJPhoto/Photo Researchers, Inc.; 10.12: © James Cavallini/Photo Researchers, Inc.; 10C: © AP Photo/Brian Walker

Chapter 11

Opener: © Tom & Dee Ann McCarthy/Corbis; 11.1(osteocyte): © Biophoto Associates/Photo Researchers, Inc.; 11.1(hyaline, compact bone): © Ed Reschke; 11A: © Jose Luis Pelaez, Inc./Corbis; 11Aa, b: © Michael Klein/Peter Arnold; 11Ac: © Bill Aaron/PhotoEdit; 11.5b: © Tony Freeman/PhotoEdit; 11.8b: © Corbis RF; 11.13a: © Gerard Vandystadt/Photo Researchers, Inc.; p. 247: © Courtesy of the National Library of Medicine (NLM)

Chapter 12

Opener: © Helen King/Corbis; 12.1(all): © Ed Reschke; 12.2c: © David Young-Wolff/PhotoEdit; 12.3: © David R. Frazier Photolibrary, Inc./Alamy; 12.5(gymnast): © Corbis RF; 12.5(myofibril): © Biology Media/Photo Researchers, Inc.; 12.6a: © Victor B. Eichler; 12.11(woman): © Corbis RF; 12.11(man): © Lawrence Manning/Corbis; 12.11(muscle fibers): © G.W. Willis/Visuals Unlimited; 12A(both): © Bettman/Corbis; 12.13: © Julie Lemberger/Corbis; 12B: © AP Photo/Adam Butler

Chapter 13

Opener: © Peter Duddek/Visum/The Image Works; 13.2(cell body): © Manfred Kage/Peter Arnold, Inc.; 13.2(myelin): © M.B. Bunge/Biological Photo Service; 13.5: Courtesy Dr. E.R. Lewis, University of California Berkeley; 13B: © David Becker/Photo Researchers, Inc.; 13.7a: © Karl E. Deckart/Phototake; 13.7d: Posterior aspect of the cervical segment of the spinal cord (right side) from *The Photographic Atlas of the Human Body* by Branislav Vidic and Faustino R. Suarez. The C.V. Mosby Co., St. Louis, 1984.; 13.7a: © Karl E. Deckart/Phototake; 13.8b: © Colin Chumbley/Science Source/Photo Researchers, Inc.; 13.13(all): © Marcus Raichle; 13.14: © Dr. Richard Kessel & Dr. Randy Kardon/Tissues & Organs/Visuals Unlimited; 13.17: © Betts Anderson Loman/PhotoEdit; 13.18: © Vol. 94 PhotoDisc/Getty RF

Chapter 14

Opener: © SIME SRL/eStock Photo; 14.4a: © Coneyl Jay/Photo Researchers, Inc.; 14.4(all tastebuds): © Omikron/SPL/Photo Researchers, Inc.; 14.8a: © Lennart Nilsson, from *The Incredible Machine*; 14.9b: © Biophoto Associates/Photo Researchers, Inc.; 14A: © Pascal Goetgheluck/Photo Researchers, Inc.; 14.13: © P. Motta/SPL/Photo Researchers, Inc.; 14B(both): Robert S. Preston and Joseph E. Hawkins, Kresge Hearing Research Institute, University of Michigan; 14.15: © Myrleen Ferguson Cate/PhotoEdit

Chapter 15

Opener: © Art-Line Productions/Brand X/Cort RF; 15.7a: © AP Photos/Ruth Fremson; 15.7b: © Ewing Galloway, Inc.; 15.8(all): From Clinical Pathological Conference, "Acromegaly, Diabetes, Hypermetabolism, Proteinura and Heart Failure", *American Journal of Medicine* 20 (1956) 133, with permission from Excerpta Medica, Inc.; 15.9a: © Bruce Coleman, Inc./Alamy; 15.9b: © Medical-on-Line/Alamy; 15.9c: © Dr. P. Marazzi/Photo Researchers, Inc.; 15.13a: © Custom Medical Stock Photos; 15.13b: © NMSB/Custom Medical Stock Photos; 15.14(both): Atlas of Pediatric Physical Diagnosis, Second Edition by Zitelli & Davis, 1992. Mosby-Wolfe Europe Limited, London, UK; 15.15: © Peter Arnold, Inc./Alamy; 15A: © Bettman/Corbis; 15.19: © The McGraw-Hill Companies, Inc./Evelyn Jo Johnson, photographer

Chapter 16

Opener: © Markus Moellenberg/zefa/Corbis; 16.3b: © Anatomical Travelogue/Photo Researchers, Inc.; 16.4b: © Ed Reschke; 16.4d: © Dr. Dennis Kunkel/Visuals Unlimited; 16A: © Getty Images; 16.8a: © Ed Reschke/Peter Arnold; 16.13b: © Michael Keller/Corbis; 16.13c: © LADA/Photo Researchers, Inc.; 16.13d: © SIU/Visuals Unlimited; 16.13e: © Keith Brofsky/Getty Images; 16.13g: © Phanie/Photo Researchers, Inc.; 16.13h: © The McGraw-Hill Companies, Inc./Christopher Kerrigan, photographer; 16B: © Bubbles Photolibrary/Alamy; 16.15: © CC Studio/SPL/Photo Researchers; 16.16: © Scott Camazine/Photo Researchers; 16.17(both): Courtesy Centers for Disease Control, Atlanta, GA. (Crooks, R. and Baur, K. *Our Sexuality*, 8/e, Wadsworth 2000, 17.4, pg. 489; 16.18a: © Science VU/CDC/Visuals Unlimited; 16.18b: © Carroll H. Weiss/Camera M.D. Studios; 16.18c: © Centers for Disease Control and Prevention; 16.18d: © Science VU/Visuals Unlimited; 16C(left): © Vol. 161/Corbis RF; 16C(right): © David Raymer/Corbis

Infectious Disease Supplement

Opener: © Frank Gunn/Canadian Press/Phototake; S.5: © Elmer Koneman/Visuals Unlimited; SA: © 2000–2006 Custom Medical Stock Photo; S.6: © ISM/Phototake; S.8: © Shehzad Noorani/Peter Arnold, Inc.; S.9: © China Photo/Reuters/Corbis; S.10: © AP Photo/Xinhua, Huang Benquiang; S.12: © Cordelia Molloy/Photo Researchers, Inc.

Chapter 17

Opener (couple): © Digital Vision/PunchStock RF; Opener (pregnancy test): © Jennifer Salls; 17.1: © David M. Phillips/Visuals Unlimited; 17.6: © Lennart Nilsson, *A Child is Born*, Dell Publishing; 17A: © AP Photo/John Chadwick; 17.8: © James Stevenson/SPL/Photo Researchers, Inc.; 17.10: Courtesy Dr. Howard Jones, Eastern VA Medical School; 17.11: © Rune Hellestad/Corbis; p. 411(Semmelweis): © Getty Images; 17.12(grandma): © David Young-Wolff/PhotoEdit; 17.12(grandpa): © Laura Dwight/PhotoEdit; 17.13: © Ronnie Kaufman/Corbis

Chapter 18

Opener: © Jack Hollingsworth/Getty Images; 18.1: © CNRI/SPL/Photo Researchers; 18.6(early prophase, prophase, metaphase, anaphase, telophase): © Ed Reschke; 18.6(early metaphase): © Michael Abbey/Photo Researchers, Inc.; 18.7a(top): © Scott Camazine/Photo Researchers; 18.7a(bottom): © Edward Kinsman/Photo Researchers, Inc.; 18.7b: © Dr. Ken Greer/Visuals Unlimited, Inc.; 18.15a: Courtesy Chris Burke; 18.15b: © CNRI/SPL/Photo Researchers, Inc.; 18.18: Courtesy The Williams Syndrome Association; 18.19b(both): From N.B. Spinner et al., *American Journal of Human Genetics* 55 (1994):p. 239. The University of Chicago Press; 18A: © Corbis/Sygma

Chapter 19

Opener: © Strauss/Curtis/Corbis; 19.1a: © Biology Media/Photo Researchers, Inc.; 19.1b: © SPL/Photo Researchers, Inc.; 19.1c: © Martin M. Rotker/Photo Researchers, Inc.; 19.8a: © Custom Medical Stock Photo, Inc.; 19.8b: © ISM/Phototake; 19.8c: © Dr. P. Marazzi/Photo Researchers, Inc.; 19.8d: © Custom Medical Stock Photo, Inc.; 19.8e: © Dr. Ken Greer/Visuals Unlimited, Inc.; 19.9: © Kings College Hospital/Photo Researchers, Inc.; 19.10: © BSIP/Phototake; 19.11: © Simon Fraser/Royal Victoria Infirmary/Photo Researchers, Inc.; 19D: © Obstetrics & Gynaecology/Science Photo Library; 19E: © Visuals Unlimited/Corbis

Chapter 20

Opener: © Monique Renee Perrine; 20.2a: © Corbis RF; 20.2b: © Dynamic Graphics/PictureQuest RF; 20.2c-f: © The McGraw-Hill Companies, Inc./Bob Coyle, photographer; 20.2g: © Corbis RF; 20.2h: © Creatas/PunchStock RF; 20.11: © Pat Pendarvis; 20.12(woman): © Steve Uzzell; 20.12(normal and Huntington brain): Courtesy Dr. Hemachandra Reddy, The Neurological Science Institute, Oregon Health & Science University; 20.13: Courtesy University of Connecticut/Peter Morenus, photographer; 20A(both): © Brand X/SuperStock RF; 20.14: © H. Reinhard/Arco Images/Peter Arnold; 20.15: © Mediscan/Medical-On-Line; 20.19(boy): Courtesy Muscular Dystrophy Association; 20.19(tissue: left, right): Courtesy Dr. Rabi Tawil, Director, Neuromuscular Pathology Laboratory, University of Rochester Medical Center; 20B(queen): © Stapleton Collection/Corbis; 20B(prince): © Huton Archive/Getty Images

Chapter 21

Opener: © David Young-Wolff/Alamy; 21A: © Science Source/Photo Researchers, Inc.; 21.12: Courtesy Alexander Rich; 21.14: From M.B. Roth and J.G. Gall, *Cell Biology*, 105:1047-1054, 1987. © Rockefeller University Press; 21.15a(top): © Digital Vision/Getty RF; 21.15a(bottom): © Getty RF; 21.15b(left): © Digital Vision/Getty RF; 21.15b(right): © The McGraw-Hill Companies, Inc./Bob Coyle, photographer; 21.15c(left): © Digital Vision/Getty RF; 21.15c(right): © Getty RF; 21.20a: © Nita Winter; 21.20b: Courtesy Robert H. Devlin, Fisheries and Oceans Canada; 21.20c: © Richard Shade; 21.20d: © Dr. Brad Mogan/Visuals Unlimited, Inc.; 21.20e: © Jerry Mason/Photo Researchers, Inc.; 21.21(both): Courtesy Monsanto; 21Aa: © Larry Lefever/Grant Heilman Photography, Inc.; 21Ab: © James Shaffer/PhotoEdit; 21Ac: Courtesy Carolyn Merchant/UC Berkeley

Chapter 22

Opener: © Bill Aron/PhotoEdit; 22.3(paleontologists): © Richard T. Nowitz/Corbis; 22.4(photo): © Jean-Claude Carton/Bruce Coleman/PhotoShot; 22.4(art): © Joe Tucciarone; 22.5: J.G.M. Thewissen, http://www.neoucom.edu/DEPTS/ANAT/Thewissen/index.html; 22.7(both): © Carolina Biological Supply/Phototake; 22.11(gibbon): © Hans & Judy Beste/Animals Animals/Earth Scenes; 22.11(orangutan, chimps, gorillas): © Creatas/PunchStock RF; 22.14a: © Dan Dreyfus and Associates; 22.14b: © John Reader/Photo Researchers, Inc.; 22.16: © National Museum of Kenya; 22Aa(left): © Mick Tsikas; 22Ab(right): © AP Photo/Richard Lewis; 22.18: © The Field Museum #A102513c; 22.19: Transp. #608 Courtesy Department of Library Services, American Museum of Natural History; 22.20a: © PhotoDisc/Getty RF; 22.20b: © Sylvia S. Mader; 22.20c: © B & C Alexander/Photo Researchers, Inc.; 22B: © Peter Bowater/Photo Researchers, Inc.

Chapter 23

Opener: © Ragnar Schmuck/Getty Images; 23.1(temperate, desert, taiga): © Vol. 63/Corbis RF; 23.1(rain forest, tundra): © Vol. 121/Corbis RF; 23.1(prairie): © Vol. 86/Corbis RF; 23.1(savanna): © Gregory Dimijan/Photo Researchers, Inc.; 23.2a: © Vol. 121/Corbis RF; 23.2b: © The McGraw-Hill Companies, Inc./Carlyn Iverson, photographer; 23.2c: © David Hall/Photo Researchers, Inc.; 23.2d: © Vol. 121/Corbis RF; 23.3a(left): © Ed Reschke/Peter Arnold; 23.3a(right): © Authors Image/PunchStock RF; 23.3b(left): © Digital Vision/Getty RF; 23.3b(right): © Gerald C. Kelley/Photo Researchers, Inc.; 23.3c(left): © Bill Beatty/Visuals Unlimited; 23.3c(right): © Joe McDonald/Visuals Unlimited; 23.3d(left): © SciMAT/Photo Researchers, Inc.; 23.3d(right): © IT Stock/age fotostock RF; 23.5: © George D. Lepp/Photo Researchers, Inc.; 23A(left): © Brand X Pictures/PunchStock RF; 23A(right): © Barry Lewis/Corbis; 23.13a: © Thomas Kitchin/Tom Stack & Assoc.; 23.13b: © John Shaw/Tom Stack & Assoc.; 23.13c: © Sue Baker/Photo Researchers, Inc.; 23B: Courtesy of NASA

Chapter 24

Opener: © Christi Carter/Grant Heilman Photography; 24.2c: © Still Pictures/Peter Arnold; 24.3(land): © Vol. 39/PhotoDisc/Getty RF; 24.3(water): © Evelyn Jo Johnson; 24.3(food): © The McGraw-Hill Companies, Inc./John Thoeming, photographer; 24.3(energy): Photo Link/Getty RF; 24.3(minerals): © T. O'Keefe/PhotoLink/Getty RF; 24.4b: © Melvin Zucker/Visuals Unlimited; 24.5b: © Carlos Dominguez/Photo Researchers, Inc.; 24.6b: Courtesy Carla Montgomery; 24.7b: © Jeremy Samuelson/FoodPix/Getty Images; 24.7a: © Comstock Images/Alamy RF; 24.7c: © Stockbyte/PunchStock RF; 24.8: © AP Photo/Richmond Times—Dispatch; 24.9a: © Jodi Jacobson/Peter Arnold; 24.9b, c: © Peter Essick/Aurora; 24.10a: © Laish Bristol/Visuals Unlimited; 24.10b: © Inga Spence/Visuals Unlimited; 24.10c: Courtesy V. Jane Windsor, Division of Plant Industry, Florida Department of Agriculture & Consumer Services; 24.11: © Jim Richardson/Corbis; 24.12a: © Corbis RF; 24.12b: © Glen Allison/Getty RF; 24.12c: © Argus Fotoarchiv/Peter Arnold, Inc.; 24.12d: Photo Link/Getty RF; 24.13a: © Matt Stiveson/DOE/NREL; 24.13b: Courtesy DaimlerChrysler; 24.13c: © Warren Gretz/DOE/NREL; 24.15b: © Gunter Ziesler/Peter Arnold; 24.16a: © Shane Moore/Animals Animals/Earth Scenes; 24.16b(both): © Peter Auster/University of Connecticut; 24.17(armadillo): © PhotoDisc/Getty RF; 24.17(periwinkle): © Steven P. Lynch; 24.17(fishing): © Herve Donnezan/Photo Researchers, Inc.; 24.17(bat): © Merlin D. Tuttle/Bat Conservation International; 24.17(rubber): © Bryn Campbell/Stone/Getty Images; 24.17(lady bug): © Anthony Mercieca/Photo Researchers, Inc.; 24A(bees): © Lila De Guzman; 24.18(deforestation): © Carlos Dominguez/Photo Researchers, Inc.; 24.18(overfishing): © Shane Moore/Animals Animals/Earth Scenes; 24.18(accumulation): © Michael Gadomski/Animals Animals/Earth Scenes; 24.18(pollution): © Thomas Kitchin/Tom Stack & Assoc.; 24.18(urban sprawl): © AP Photo/West Central Tribune/Bill Zimmer; 24.18(erosion): © USDA/Nature Source/Photo Researchers, Inc.; 24.19(farming): © Inga Spence/Visuals Unlimited; 24.19(delta restoration): © AP Photo/Peter DeJong; 24.19(recycling): © Jeffrey Greenberg/Photo Researchers, Inc.; 24.19(bus): Courtesy DaimlerChrysler; 24.19(garden): © Reuters/Corbis; 24.19(IPM): Courtesy V. Jane Windsor, Division of Plant Industry, Florida Department of Agriculture & Consumer Services; 24.20: © Dan McCoy/Rainbow; p. 586(land): © Vol. 39/PhotoDisc/Getty RF; p. 586(water): © Evelyn Jo Johnson; p. 586(food): © The McGraw-Hill Companies, Inc./John Thoeming, photographer; p. 586(energy): Photo Link/Getty RF; p. 586(minerals): © T. O'Keefe/PhotoLink/Getty RF

EOB Credit Lines

Chapter 5

Fig 5.11: From Moore & Agur- Essential Clinical Anatomy, 2/e, 0-7817-6274-X. Lippincott, Williams, Wilkins. Reprinted with permission. http:lww.com.; Fig 5.15: Source: American Heart Association.; Fig 5.16: Figure from www.texasheart.org AB/OMED Replacement Heart. Reprinted with permission from ABIOMED, Inc.

Chapter 19

Fig 19.6: Porth, C.M. Essentials of Pathophysiology. Lippincott, Williams, Wilkins, 2/e 2005. Reprinted with permission. http://lww.com

Index